Proceedings of the World Congress on Vegetable Protein Utilization in Human Foods and Animal Feedstuffs

Edited by
Thomas H. Applewhite
(retired)

Kraft, Inc.
Research & Development
801 Waukegan Rd.
Glenview, Illinois USA

American Oil Chemists' Society
Champaign, Illinois

Mention of firm names of trade products does not imply endorsement
or recommendation by the editors or contributors over other firms
or similar products not mentioned.

Copyright © 1989 by the American Oil Chemists' Society. All
rights reserved. No part of this book may be reproduced or
transmitted in any form or by any means without written permission
of the publisher.

Library of Congress Cataloging-in-Publication Data

World Congress on Vegetable Protein Utilization in Human Foods and
　Animal Feedstuffs (1988 : Singapore)
　　Proceedings of the World Congress on Vegetable Protein Utilization
in Human Foods and Animal Feedstuffs / edited by Thomas H.
Applewhite.
　　　p.　　cm.
　　Congress held in Singapore in Oct. 1988.
　　ISBN 0-935315-25-X
　　1. Proteins—Congresses.　2. Proteins in animal nutrition-
-Congresses.　I. Applewhite,　Thomas H.　II. Title.
TP453.P7W67　1988
664–dc20　　　　　　　　　　　　　　　　　　　　　89-31527
　　　　　　　　　　　　　　　　　　　　　　　　　　　CIP

Printed in the United States of America

Contents

Preface		xii

Keynote Session

Chairperson:	**Bryce Bell,** *Australian Oilseed Crushers Association*, Sydney, Australia, and **Kenneth Beery,** *Central Soya Co.,* Fort Wayne, Indiana	
Chapter 1	**World Vegetable Protein Supply and Demand** *Thomas Mielke*	1
Chapter 2	**World Trade in Vegetable Proteins in the 1990's** *Lindsay Jolly, Brent Borrell and Peter Perkins*	10
Chapter 3	**Government-Imposed Restrictions on International Trade in Proteins** *Donald E. deKieffer*	17
Chapter 4	**World Vegetable Protein Marketing Perspectives** *David H. Swanson*	25
Chapter 5	**World Vegetable Protein Quality Concerns** *Takaaki Takebe*	28

Oilseeds Extraction—Foods/Feeds

Chairpersons:	**E.W. Lusas,** *Texas A&M University,* College Station, Texas, and **K.C. Rhee,** *Texas A&M University,* College Station, Texas	
Chapter 6	**Handling, Shipping, and Storage of Oilseeds and Meals to Meet World Market Quality Requirements** *Lowell D. Hill and Martin G. Patterson*	31
Chapter 7	**Preparation of Soybeans Prior to Solvent Extraction** *Heinz Schumacher*	37
Chapter 8	**Extrusion-Expansion of Oilseeds for Enhancement of Extraction, Energy Reduction and Improved Oil Quality** *L.R. Watkins, W.H. Johnson and S.C. Doty*	41
Chapter 9	**Oilseeds Extraction and Meal Processing** *George Anderson*	47

Preparation of Vegetable Food Protein Ingredients

Chairpersons:	**Dale W. Johnson,** *Food Ingredients (Minnesota) Inc.,* Golden Valley, Minnesota; and **Kinji Uchida,** *Kikkoman Corporation,* Noda City, Japan	
Chapter 10	**Full-Fat Soya-Products—Manufacturing and Uses in Foodstuffs** *Jens Heiser and Thomas Trentelman*	52
Chapter 11	**The Preparation and Properties of Defatted Soy Flours and Their Products** *Richard W. Fulmer*	55

Chapter 12	**Preparation of Soy Protein Concentrate Products and Their Application in Food Systems** *Kenneth E. Beery*	62
Chapter 13	**Processing for Producing Soy Protein Isolates** *Dale W. Johnson and Saburo Kikuchi*	66
Chapter 14	**Trends in Preparation and Uses of Fermented and Acid-Hydrolyzed Soy Sauce** *Kinji Uchida*	78
Chapter 15	**Manufacturing Soy Protein Isolate by Ultrafiltration** *Fu-Kuang Liu, You-Hua Nie and Bei-Ying Shen*	84

Processing for Feedstuffs I

Chairpersons:	**Don Bushman,** *American Soybean Association,* Beijing, China; and **Colin Kirkegaard,** *Triple F Inc.,* Des Moines, Iowa	
Chapter 16	**Technology for Full-Fat Soya Products and Extracted Soymeal with Different Contents of Water Soluble Protein** *Heinz Schumacher*	91
Chapter 17	**Low Cost Dry Extrusion of Feeds** *Thomas F. Welby*	94
Chapter 18	**Improvement in the Protein Content of Palm Kernel Meal by Solid State Fermentation** *Suan-Choo Cheah, Leslie C.L. Ooi and Augustine S.H. Ong*	96
Chapter 19	**Extrusion of Rice Bran** *Maurice A. Williams*	100
Chapter 20	**Production and Utilization of High Ruminal Bypass Proteins** *Marshal D. Stern and Paul M. Windschitl*	103

Preparation and Uses of Non-Oilseed Vegetable Food Proteins

Chairpersons:	**Mark Uebersax,** *Michigan State University,* East Lansing, Michigan, and **Geoffrey Grace,** *Manildra Starches Pty. Ltd.,* Auburn, New South Wales, Australia.	
Chapter 21	**Preparation of Vital Wheat Gluten** *G. Grace*	112
Chapter 22	**World Food Uses of Vital Wheat Gluten** *J.M. Hesser*	116
Chapter 23	**Utilization of Dry Field Beans, Peas and Lentils** *Mark A. Uebersax and Songyos Ruengsakulrach*	123
Chapter 24	**African Uses of Cowpeas, Pigeon Peas, Local Protein and Oil Seeds** *Bene W. Abbey*	131
Chapter 25	**Development of Processes and Uses of Lupins for Food** *José Miguel Aguilera*	138
Chapter 26	**Progress in Development of Leaf Proteins for Use in Foods** *P. Fantozzi and A. Sensidoni*	143

Processing for Feedstuffs II

Chairpersons:	**Tom Welby,** *Triple F Inc.,* Des Moines, Iowa; and **Choo Su-Hoon,** *Seng Heng Chang Pte. Ltd.,* Singapore	

Chapter 27	**Preparation of Fish and Shrimp Feeds by Extrusion** *Joseph P. Kearns*	152
Chapter 28	**Pelleting for Aquaculture Feeds** *Ronnie K.H. Tan*	162
Chapter 29	**Processing of Wet Shrimp Heads and Squid Viscera with Soy Meal by a Dry Extrusion Process** *Larry A. Carver, Dean M. Akiyama, and Warren G. Dominy*	167
Chapter 30	**Production of Extruded Pet Foods** *Maurice A. Williams*	171
Chapter 31	**Status of Vegetable Food Proteins from Lesser-Used Sources** *E.W. Lusas, K.C. Rhee and S.S. Koseoglu*	175

Vegetable Protein Utilization—Foods/Feeds

Chairpersons: **Daniel E. Shaughnessy**, *Export Processing Industry,* Coalition, Arlington, Virginia; and **Nadir Godrej**, *Godrej Soaps Pvt. Ltd.,* Bombay, India

Chapter 32	**Allergenicity of Soy Proteins** *Hans Elbek Pedersen*	204
Chapter 33	**Protein Sources Made Available by New Technology** *Daniel E. Shaughnessy*	213
Chapter 34	**Nutrition Complementation with Vegetable Protein** *Preeya Leelahagul and Vichai Tanphaichitr*	216

Vegetable Proteins—Nutrition and Regulation

Chairpersons: **Osman M. Galal**, *Nutrition Institute,* Cairo, Egypt; and **John Vanderveen**, *U.S. Food and Drug Administration,* Washington, D.C.

Chapter 35	**Vegetable Proteins in Planning for Child Nutrition** *Osman M. Galal*	220
Chapter 36	**Dietary Soybean Protein and Cholesterol Metabolism** *Anton C. Beynen*	227
Chapter 37	**Concerns in Regulating Vegetable Food Proteins** *R.J. Dawson*	230
Chapter 38	**Preparation and Use of Dry Soy Products and Nutritional Beverages** *Glen Blix*	234
Chapter 39	**Evaluation of Nutritive Value of Local and Soy-Beef Hamburgers** *Abdul Salam Hj. Babji and Selvakumari Letchumanan*	237

Nutritional/Antinutritional Considerations for Diverse Species I

Chairpersons: **Dean Akiyama**, *American Soybean Association,* Singapore; and **Eooi Eng-Teong**, *KFC Holdings Bhd.,* Malaysia.

Chapter 40	**Limiting Amino Acids in Poultry Diets** *M.E. Blair*	243
Chapter 41	**Utilization of Plant Proteins by Warmwater Fish** *Chhorn Lim and Warren Dominy*	245
Chapter 42	**Soybean Meal Utilization by Marine Shrimp** *Dean M. Akiyama*	252

Vegetable Proteins—Nutrition and Related Technology

Chairpersons: Maurice Bennink, *Michigan State University,* East Lansing, Michigan; and Glenn Patterson, *Appropriate Technologies International,* Washington, D.C.

Chapter 43	Digestibility of Dry Legume Starch and Protein *Maurice R. Bennink and Naruemon Srisuma*	266
Chapter 44	Development of Glandless Cottonseed and Nutritional Experience in The Ivory Coast *J.L. Bourely*	273
Chapter 45	Appropriate Technologies for Producing Vegetable Protein *Arleen Richnau and Glenn Patterson*	281
Chapter 46	Nutrition Intervention with Protein Foods *Daniel E. Shaughnessy*	296

Nutritional/Antinutritional Considerations for Diverse Species II

Chairpersons: M.Y. Huang, *Farmers Hybrid,* Des Moines, Iowa; and Goh Kee-Seng, *Agri-Nutritional Asia Pte. Ltd.,* Singapore

Chapter 47	Antinutritional Factors in Vegetable Proteins for Poultry *A.J. Mudd*	299
Chapter 48	Nutritional and Antinutritional Considerations of Soybean Processing for Swine *Colin Kirkegaard*	303
Chapter 49	Utilization of Full-Fat Soybeans for Dairy Cattle *Karl H.G. Sera*	308
Chapter 50	Nutrition of Feedlot Cattle *Larry Dodson*	312
Chapter 51	*In Vitro* Method for Estimating Digestibility of Swine Feeds Using Various Protein Sources *Shi Xue-Shi and Nie Guang-Da*	315

Vegetable Protein Processing for Foods I and II

Chairpersons: Steve Chen, *American Soybean Association,* Taipei, Taiwan; **F.G. Winarno,** *Bogor Agricultural University,* Bogor, Indonesia; **Susani Karta,** *American Soybean Association,* Singapore; **R.W. Fischer,** *Soypro International Inc.,* Cedar Falls, Iowa; **A.H. Chen,** *3I Corporation,* Plano, Texas; and **Paul Gibson,** *A.E. Staley Co.,* Decatur, Illinois

Chapter 52	Determining and Modifying Protein Functionality *Khee Choon Rhee*	323
Chapter 53	Preparation and Uses of Dietary Fiber Food Ingredients *Thomas H. Smouse*	334
Chapter 54	Preparation of Fluid Soymilk *Steve Chen*	341
Chapter 55	Extrusion of Texturized Proteins *Joseph P. Kearns, Galen J. Rokey, and Gordon R. Huber*	353
Chapter 56	Production and Utilization of Tempeh in Indonesian Foods *F.G. Winarno*	363
Chapter 57	Miso Preparation and New Uses *Toshio Hanaoka*	369

Chapter 58	**New Food Proteins, Extrusion Processes and Products in Japan** *Akinori Noguchi and Seiichiro Isobe*	375
Chapter 59	**Traditional Chinese Soyfood** *Susani K. Karta*	382

Nutritional/Antinutritional Factors of Specific Protein Sources I and II

Chairpersons:	**Neoh Soon Bin,** *Soon Soon Oil Mills Sdn. Bhd.,* Butterworth, Province Wellesley, Malaysia, and **David Thomason,** *American Soybean Association,* Singapore	
Chapter 60	**Application for Vegetable Proteins in Processed Pet Foods** *Douglas A. Thompson*	388
Chapter 61	**Efficiency in Feed Resource Utilization and Animal Production** *C. Devendra*	392
Chapter 62	**Mycotoxins in Oilseeds and Risks in Animal Production** *H. Müschen and Karl Frank*	401
Chapter 63	**Utilization of Vegetable Oilseed Cakes and Meals for Livestock Production in China** *Shen Zai-Chun*	405
Chapter 64	**The Use and Outlook of Full-Fat Soybean and Soybean Meal in Japanese Commercial Feeds** *Karl H.G. Sera*	408
Chapter 65	**Lupins as an Energy-Rich Protein Source for Feed and Food** *Rory Coffey*	410
Chapter 66	**Utilization of Sunflower Meal for Swine and Poultry Feeds** *Utahi Kanto*	415

Utilization of Vegetable Protein Food Ingredients

Chairperson:	**Thomas H. Smouse,** *Archer Daniels Midland Co.,* Decatur, Illinois; **Tai-Wan Kwon,** *Korea Food Research Institute,* Seoul, Korea; and **Richard Fulmer,** *Cargill Inc.,* Minneapolis, Minnesota	
Chapter 67	**Uses of Soy Proteins in Bakery and Cereal Products** *Richard W. Fulmer*	424
Chapter 68	**The Utilization of Soy Proteins from Hot Dogs to Haramaki** *Alexander T. Bonkowski*	430
Chapter 69	**Vegetable Protein Foods in Korea** *Seung Ho Kim and Tai-Wan Kwon*	439
Chapter 70	**Dealing with the Shortage of Protein Resources in China** *Zhu Xiang-Yuan*	443
Chapter 71	**Novel Traditional and Manufactured Soy Foods in Japan** *Hitoshi Taniguchi*	444

Nutritional/Antinutritional Factors for Specific Protein Sources III

Chairperson:	**Woo Kok-Kuan,** *Federal Flour Mills Bhd.,* Kuala Lumpur, Malaysia	
Chapter 72	**Canola Meal for Livestock and Poultry** *P.J. McKinnon and D.A. Christensen*	449
Chapter 73	**Palm Kernel Cake as Ruminant Feed** *D. Mohd. Jaafar and A. Yusof Hamali*	463

Chapter 74	Soybean Meal as the Only Supplementary Protein Source for Poultry Feeds in the People's Republic of China (PRC) *Don H. Bushman and C.M. Collado*	468
Chapter 75	Amino Acid Composition of Feedgrade Rice By-products From Several Countries *David Creswell*	474

Biotechnology Applications in Plant and Animal Systems

Chairperson:	**Neils Nielsen**, *Purdue University*, West Lafayette, Indiana; **Frank Orthoefer**, *Riceland Foods*, Stuttgart, Arkansas	
Chapter 76	DNA Sequences Controlling Nutritional and Functional Properties of Cereal Storage Proteins *M. Giband, B. Potier, S. Dukiandjiev, V. Burrows and I. Altosaar*	480
Chapter 77	*In Vitro* Modification and Assembly of Soybean Glycinin *N.C. Nielsen*	487
Chapter 78	The Biochemical Genetics of Lipoxygenases *Rod Casey, Claire Domoney, Paul Ealing and Helen North*	491
Chapter 79	Isolation and Primary Structure for a Novel, Methionine-rich Protein from Sunflowerseeds (*Helianthus annus.* L) *G.G. Lilley, J.B. Caldwell, A.A. Kortt, T.J. Higgins, and D. Spencer*	497
Chapter 80	Enzyme Use in the Food Industry with Potential Applications to Vegetable Protein Utilization in Human Foods *Svend Eriksen*	503
Chapter 81	Soy Protein Fractionation and Applications *Paul W. Gibson and Walter C. Yackel*	507
Chapter 82	Biotechnology in Rice Breeding *Timothy P. Croughan*	510
Chapter 83	Applications of Biotechnology to Soybean Improvement *Suzan S. Croughan*	512
Chapter 84	Biotechnology and Livestock Systems with an Emphasis on Porcine Somatotropin *David Meisinger*	515

Volunteer Presentations

Chairpersons:	**E.W. Lusas**, *Texas A&M University*, College Station, Texas, **K.C. Rhee**, *Texas A&M University*, College Station, Texas, and **Mark Uebersax**, *Michigan State University*, East Lansing, Michigan	
Chapter 85	Hot Dehulling System: "Popping" *W. Fetzer*	518
Chapter 86	A New Generation of Flaking Mills *W. Fetzer*	521
Chapter 87	Studies on the Expansion of Vegetable Proteins in China *Yang Mingduo*	523
Chapter 88	Aqueous, Membrane and Adsorptive Separations of Vegetable Proteins *S.S. Köseoğlu and E.W. Lusas*	528
Chapter 89	Drying of Vegetable Food Proteins *Chan Kwee Chew*	548

Chapter 90	**Preparation and Uses of Vegetable Food Proteins Made by Dry Processes** *Frank W. Sosulski*	553
Chapter 91	**Novel Soybean Food in China** *Shen Zai-Chun, Xu Jing-Kuan and Jiang Xue-Li*	559
Chapter 92	**Production and Markets for Soy Protein Industrial Ingredients in Brazil** *Wilson L. Canto and Jane M. Turatti*	560
Chapter 93	**Sunflower Proteins in Dairy Products** *E. Baldi, L. Lencioni, A.M. Pisanelli, R. Fiorentini and C. Galoppini*	564
Chapter 94	**Leaf Protein in Human Diet: Potential and Perspectives** *C. Galoppini, R. Fiorentini and F. Favati*	567
Chapter 95	**Enzymatic Incorporation of Arginine into Soybean Meal by the Plastein Reaction** *S. Divakaran*	571
Chapter 96	**A New Process for Protein Protection** *J.Å.H. Dahlén, L.A. Lindh and C.G.S. Münter*	573

Preface

"Tuan, Puan Selamat Datang," a Malaysian greeting to you in a Singaporian tongue. As general chairman and a more or less recent resident of Singapore, I had the opportunity to welcome the congress to that island metropolis with its beautiful sights and wonderfully exotic taste in foods. To those of you who were unable to attend this congress, I welcome you to these proceedings and trust that its reading will be as interesting and useful to you as the congress was to us in Singapore. For all of us it should prove to be a valuable reference source.

Three days of technical utilization programming could not possibly cover our broad subjects. However, when complemented by the keynote presentations, which set a tone for the international economic and market situation, and a concluding futuristic, biotechnological outlook, we believe that a well balanced, quality program was achieved.

These utilization discussions were divided into two parts, the concern with human foods and animal feedstuffs. In reality, the fundamental subject was food for human consumption, via the direct or indirect utilization of vegetable protein sources. The specific challenges and opportunities for vegetable protein utilizations were obvious from the many discussions during the congress week. Nevertheless, I do think it appropriate to identify a few, more subtle challenges suggested by those congress activities.

If one has had the opportunity to attend international conferences on world food production and distribution one would have heard from people who decry the wasteful conversion of vegetable proteins into animal proteins. The reality, however, is that that is what people want. Several keynote speakers indicated that the demand for animal protein products is increasing.

In developed countries animal products contribute more than one-third of the daily calories and over a third of the major nutrients in the food supply. In developing countries people will buy, by choice, animal proteins or more animal proteins when they have some extra money to do so. That describes in part the challenge for vegetable oil proteins.

This is hardly the place to have a philosophical discussion on how to feed the world; however, we are a part of that supply picture and we need to appreciate the challenge posed by those growing demands. Unfortunately, this demand is not equalized in terms of purchasing power and so there is a great disparity in this apparent demand. Every twelve days the world's population goes up by the equivalent of Singapore's population. Eighty percent of that growth is in developing countries and 80% of that increase occurs in the lower socioeconomic group which purchased very little yesterday, will buy very little today and cannot be expected to buy more tomorrow. Somehow, I feel that we need to have some appreciation for that observation when we look at the growing demand picture.

The keynote talks focused on the importance of international trade policies and their challenges. These changing policy practices will certainly influence who does what: who provides the *supply*. Liberalization of trade policies, as we are now experiencing in East Asia, means more imported foods and significantly challenges local or indigenous productions. Demand is still there and increasing, and so is the competitiveness of supply. For some countries this is an opportunity to increase production for export, and for others it means increased imports and reduced productions. In the short term the challenge is for increased productivity management.

The final observation I want to make is an imminent challenge. During the final day of this congress we spent considerable time discussing biotechnology and its implicit importance for change. Many of the genetic and biotechnological tools are in place and more are being rapidly developed. These tools can significantly alter the compositional characteristics of vegetable protein crops. The people who will ultimately make or implement these changes need guidance. There have to be intelligent and practical decisions for how much and what to change. The challenge for us is to be a part of those decisions. If wrong or bad decisions are made it will have a negative impact on utilization, especially for major vegetable protein crops. We need to be involved in some responsible way.

"Sayonara"

Lars H. Wiedermann
Congress General Chairman
Tokyo, Japan

World Vegetable Protein Supply and Demand

Thomas Mielke

Oil World, ISTA Mielke GmbH, PO Box 90 08 03, 2100 Hamburg 90, Federal Republic of Germany

Abstract

The effects of adverse climatic conditions in North America in particular but also in China, parts of Europe, Mexico and other countries—coupled with emotions and speculations—have caused price gyrations in the past few months not seen since 83/84. The paper focuses on the 88/89 oilseed supply first, the latest data on yields and crops in major countries and the consequences on the world trade and the regional distribution of crushings resulting from the massive drought damage experienced in the United States this year. There are many bearish and offsetting influences in other parts of the world. Record stocks available in Argentina and Brazil as of Sept. 1 this year as well as prospects of bumper crops in early 1989 should make it possible for both countries to boost their combined oilseed disposals, thus offsetting a major portion of the U.S. shortfall. The prospective world supply and demand of oilmeals are dealt with in the second part of the paper. Growth rates in demand registered in the past four years cannot be sustained, as world production is insufficient. Higher prices have and will continue to do the job of rationing and substitution.

Once again Mother Nature has been interfering heavily with our markets this year. Amplified effects of heat and drought in North America have taken their toll of yields and will cause North American oilseed production to plunge steeply. Additional tightness stems from severe flood and drought damage in most of the major Chinese provinces; characterized as the worst in several years. Mexico, East Europe, the Philippines and North Africa have also suffered from unfavorable weather conditions, cutting oilseed production below initial estimates.

In the European Community the "everlasting" boost in oilseed production registered during the past ten years has been reversed. Less generous state subsidies caused farmers to cultivate less. Unfavorable weather conditions contributed to the decline of 0.8 million metric tons (MMT) in this year's oilseed output of the Community (largely in rapeseed and to a small extent also in soybeans).

88/89 Soybean Output

The September forecast of the USDA confirmed a production shortfall of 11.2 MMT in the U.S. soybean crop alone. The implied average yield (25.9 bu per acre) is around 20% below trend and represents the highest percentage decline recorded in the past 30 years. Although there is the strong likelihood that additional downward revisions will have to occur in the forthcoming official reports, I prefer to work with the current official estimate of 40.1 MMT (against 51.3 MMT last year).

Not only bad weather conditions, but also smaller plantings contributed to the substantial decline in U.S. soybean production during the past three years. The U.S. share of world soybean output has declined sharply from 59% only three years ago to as low as 43% this year (Fig. 1). But, contrary to the recent past, soybean production in the rest of the world cannot rise sufficiently and global soybean output is likely to decline by more than 8 MMT this season.

Combined production gains of 3–4 MMT in next year's South American soybean output can only slightly ease the tightness stemming from smaller crops in the USA, China, the EEC and Canada.

In Brazil the area expansion may at best be 10%, as the new government policy is discouraging soybeans (via less financial assistance) in favor of corn. The rainfall of late has been beneficial, but limited to the southern and central areas, while it is still too dry in the north. Insufficient supplies of good quality seed are another limiting factor.

10 Major Oilseeds

Sizeable declines of 1.3 MMT in world rapeseed and of 0.5 MMT in linseed production will add to the shortage in soybeans. The expansion expected for cottonseed, groundnuts and sunseed cannot prevent world production of 10 major oilseeds from declining by 6.5 MMT this season.

As can be seen from Fig. 2, the situation expected for 88/89 resembles two previous tight seasons, namely 83/84 and 80/81. World production will again fall sizeably below disappearance. Demand can only be satisfied by significantly cutting carryover stocks to or close to minimum pipeline requirements.

Inventories of 10 major oilseeds are therefore likely to plunge by 7.0 MMT in the course of this season (Fig. 3). This is by far the sharpest reduction during the past ten years. Forgotten are the surpluses of 1986, when commodity prices were in the doldrums and at their lowest levels in more than a decade.

Soybean Crush and Exports

The crushing industries of the European Community are likely to lose market shares this season. Crush margins are dismal, to say the least. They were even negative in August and September with the price of soybeans exceeding the combined product value. (Fig. 4 shows the position of the German crusher relative to his competitor in the USA.)

Soybean net imports into the EEC probably will fall by 2.0 MMT or 15%, to the lowest level in many years and soybean crushings by probably 1.6 MMT.

In Argentina, the graduated tax structure favors the exports of products vis-a-vis soybeans—a disadvantage for the crushers in the importing countries.

It will be interesting to see to what extent Argentina and Brazil will replace U.S. supplies on the world market (Fig. 5). I expect that combined Argentine and Brazilian soybean crushings and net exports will increase by almost 9 MMT in Sept/Aug 88/89. This will be made possible by the record-large old-crop carryover stocks available as of

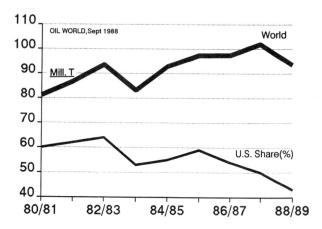

Fig. 1. Soybean output (million metric tons)

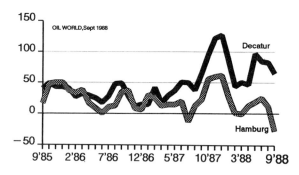

Fig. 4. Soybean crushing margins (cents per bushel)

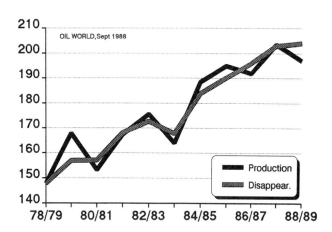

Fig. 2. 10 major oilseeds, world production and disappearance (MMT)

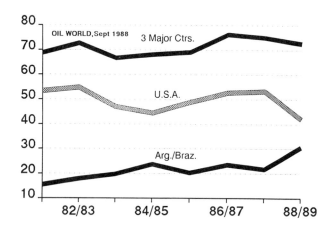

Fig. 5. Soybean crush and exports (disposals of three major countries)

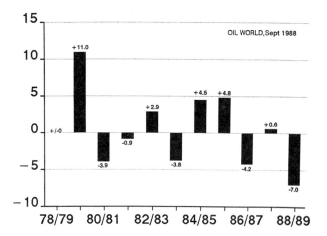

Fig. 3. 10 major oilseeds, excess of deficit of world production (MMT)

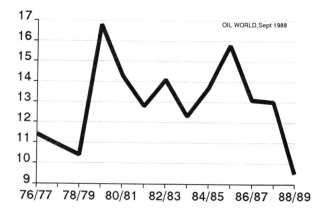

Fig. 6. 10 major oilseeds, ending stocks as percent of usage

Sept. 1, 1988 (11.7 MMT against 7.2 MMT a year ago) as well as a bumper soybean crop early next year, weather permitting.

On the other hand, U.S. soybean disposals are likely to plunge by the unprecedented quantity of 11 MMT or 21%. As shown in Fig. 5, the combined disposals of the three major countries will be declining in the second consecutive season. The moderate slope of the curve for the three major countries does not adequately mirror the forthcoming tightness. It is equivalent to a reduction of no less than 2.4 MMT and requires a lot of demand rationing via high prices.

Oilseed Stocks/Usage Ratio the Lowest in At Least 15 Years

The correlation of ending stocks and usage has always been a good barometer for prices. When we look at Fig. 6 we realize the rapid depletion of inventories since the autumn of 1986. The prospective substantial decline in oilseed inventories to only 19 MMT at the end of this season will probably cut stocks to less than 10% of usage. This is the lowest in at least 15 years and will make our markets extremely price sensitive to any surprises. All buffer stocks have been disposed of and price rallies are inevitable if any crop scare occurs in South America during the growing season.

Most of the depletion in inventories is occurring in soybeans (largely in the USA and Brazil), but significant reductions are also likely in Canadian rapeseed and linseed as well as in the EEC and China.

The decline in stocks is most obvious for U.S. soybeans. Their ending stocks in percent of usage are likely to fall from a peak of 28% only two years ago to a minimum of 8% next autumn (Fig. 7). The situation will thus resemble the tightness registered in 1976/77.

The bullishness in U.S. soybeans is moderated by growing supplies in South America. The prospective aggressive selling by Argentina and Brazil in the second half of this season may exert a bearish price pressure. An effort to dispose of as much as possible can of course be expected in view of the likelihood of an upswing in U.S. oilseed plantings and production next year.

With Argentina and Brazil moderating the U.S. soybean tightness, the soybean stocks/usage ratio of the three countries is currently expected to fall to 18% next autumn, provided current optimistic production forecasts materialize (Fig. 8).

World Oilmeal Supplies

Outstanding is the sharp decline in world supplies of soybean meal by 3.2 MMT this season (Fig. 9). Production is even likely to plunge by more than 4 MMT in the USA and by an estimated 1.3 MMT in the EEC. Record processing in South America cannot prevent the reversal in the downtrend of world soybean meal output. Following good year-to-year expansions during the past four seasons, output is now declining for the first time since 83/84.

Despite good improvements of global supplies of cotton, groundnut and sunflower meals by 0.9, 0.6 and 0.2 MMT, respectively, the substantial shortfall in soymeal cannot be offset.

Worldwide rapeseed crushings and meal output probably will stagnate although India, Canada and the Soviet Union are likely to harvest record crops. But on the other hand, the substantial crop shortfall in China and the EEC by a combined 2.2 MMT is a negative factor to be considered.

Supplies of copra and palm kernel meals are likely to recover only slightly. Copra trees in the Philippines are still suffering from previous droughts and improvements in yields may not be registered before early next year. World export supplies of copra meal will stagnate around the low level registered last season.

In the case of corn meal, gluten meal and corn gluten feed the USA will again be responsible for a prospective shortfall of 0.2 MMT. Following many years of dynamic growth, the combined output will be declining for the first time in more than a decade.

Also fish meal should remain in relatively tight supply this season. Stocks right now are the lowest in at least six years and sizeably below last year in South Africa, the USA and the EEC and also comparatively low in Chile and Peru. Unless the recovery in fish meal output exceeds the moderate expectations in South America and other major producing countries, 88/89 world fish meal supplies will at best stagnate at a comparatively low level.

Expanding aquaculture and fresh water fish raising have so far been the driving force behind the steep increase of fish meal imports in Asia. The unprecedented boost in Chinese and Taiwanese fish meal import demand during the past two seasons contributed to the developing tightness in fish meal and the widening price premiums vis-a-vis soymeal. A further gain in fish meal imports of China, Taiwan and other Asian countries would correspondingly reduce the supplies available for the rest of the world.

The regional distribution of oilmeal output in 88/89 confirms that the bulk of the declines this season is occurring in the USA (off 4.4 MMT) as well as in the EEC and China (for details see Fig. 10). Slight reductions are also taking place in the Soviet Union, Japan and Taiwan on account of smaller soybean imports. On the other hand, the boost in soybean and total oilseed disposals is likely to result in a gain of 3.2 MMT in Argentine and Brazilian output this season.

Indian farmers have been enjoying excellent monsoon rains since June and are looking forward to an increase in oilseed output by at least 2 and probably by 2.5 MMT. As a result, India may achieve the second best increase in meal output this season.

World Oilmeal Demand Relatively Strong in 88/89, Despite High Prices

The job of demand rationing this season is reminiscent of a similar situation five years ago. World usage will have to decline by around 1.3 MMT even if oilseed stocks are reduced to or close to minimum in most countries (Fig. 11). The necessary reduction in demand to be achieved this season follows an average gain of almost 6 MMT during the past four years.

Oilmeal prices in September averaged around 40% higher than a year ago, but I doubt that they are high enough to fulfill the job of rationing.

Similar to the tight season 83/84 all of the adjustment will be occurring in soybean meal alone.

But the 88/89 world supply and demand situation will be different from 83/84 in many respects: (a) Soviet oilmeal demand will continue to increase this season, despite high prices (Fig. 12). Unlike five years ago, Soviet soymeal

Fig. 7. Soybean ending stocks as percent of usage in United States

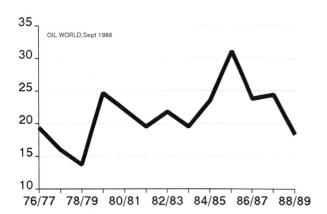

Fig. 8. Soybean ending stocks as percent of usage in U.S., Brazil and Argentina

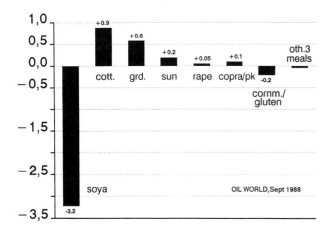

Fig. 9. Changes in supply of 12 meals from Oct/87 to Sept/88 (MMT)

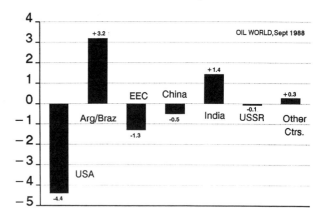

Fig. 10. Changes in output of 12 meals from Oct/87 to Sep/88 (MMT)

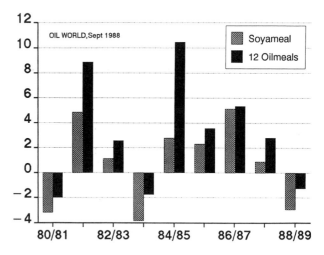

Fig. 11. Annual change in world demand for meal (MMT)

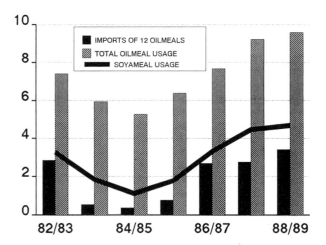

Fig. 12. USSR: Situation in 12 oil meals (MMT)

imports probably will rise to a new record of 3.4 MMT in the new season. This is 0.65 MMT or almost one quarter above last season. The Soviets have aggressively bought South American meal since May and it is considered likely that the soymeal exports of Argentina and Brazil to the USSR will almost double from last year to close to 2 MMT in July/Dec 1988. Although the record imports of soybean meal will probably be somewhat offset by smaller imports and processing of soybeans, total Soviet oilmeal usage is expected to rise by 0.4 MMT this season.

Perestroika has raised new hopes in the minds of Soviet consumers, particularly for larger food supplies. The Soviet Government is thus facing pressure and has to take the necessary steps to improve productivity in the livestock sector and raise total food output. With domestic feed supplies not yet sufficient to improve the quality of feeding and thus raise the production of meat, the Soviets are forced to continue to step up imports of soybean meal this season, despite higher world market prices.

In the previous tight season five years ago, the Soviets contributed to the necessary curbing of demand by reducing their own domestic oilmeal disappearance by 1.5 MMT. But this season they appear to be committed to continue to expand, thus passing the buck to the rest of the world. In fact, oilmeal demand outside the USSR will have to decline by more than five times the quantity necessary five years ago.

(b) China cannot relieve the tightness on the world market as it did back in 83/84 when combined exports of soybeans and meal were boosted by 0.3 MMT (meal basis). For details see Fig. 13.

On the contrary, the significant drought and flood damage has obviously exceeded earlier fears and will make it necessary to reduce exports of soybeans and soymeal more or less sharply, despite the high-ranking desire of earning foreign exchange. Soybean meal exports probably will decline by around one quarter or 0.5 MMT after having peaked at 2.1 MMT last season.

The strong domestic economic performance and the quickly growing disposable income and per capita requirements are also felt in the food industry. Since pork prices were freed from government regulation, pig breeding took off and domestic feed requirements were stepped up accordingly. Also the expanding domestic mixed feed production is absorbing growing quantities of oilmeals.

Total Chinese exports of all oilmeals (Fig. 14) are therefore likely to continue to decline this season at an accelerated pace. In the years ahead they should continue to decline unless the recently growing consumption of meat is reversed (which is unlikely).

(c) Larger exports are likely to come from India (Fig. 15). Unless the weather turns negative from now on, the comfortable domestic supplies should make it statistically possible to sharply boost exports of soybean meal and other oilmeals. The higher world market prices make such a development even more likely. Inferior quality and too high aflatoxin levels in groundnut extraction pellets have, however, been limiting factors recently.

(d) In the USA domestic oilmeal demand does not yet show sufficient rationing, to say the least. This season it will probably decline only by at best half the rate registered in 83/84, causing exports of soybeans and meal to decline even more sharply than five years ago (Fig. 16). Concerning exports, I project steep declines by 3 MMT in soybean meal and almost 7 MMT in soybeans (compared with reductions of 1.6 and 5.2 MMT, respectively, in 83/84).

Total U.S. pig inventories right now are up around 7–8% from last year. The hog/feed ratio has been declining but is probably not yet low enough to trigger liquidation of breeding hogs. The pig cycle is there to maintain soybean meal consumption at the current high level for the next few months.

Also the U.S. poultry and turkey industries are still producing more than a year ago, indicating that obviously most of them are still making profits.

Most of the soybean meal is fed to poultry, turkeys and pigs, and it probably will require a further squeeze of profitability before liquidation begins. If oilmeal prices do not strengthen and/or meat prices decline, actual domestic U.S. usage may finally exceed our estimates and enforce an even sharper decline in exports (Fig. 17). The decline in total U.S. oilmeal exports probably will even approach 4 MMT this season (of which soymeal off 3 MMT and corn germ & corn gluten meal off 0.4 MMT).

(e) Argentina should be in a position to expand exports of all oilmeals by 1.1 MMT this season (Fig. 18). They probably will more than double from 3 MMT in 82/83 to close to 7 MMT this season, with soymeal the driving force. A soybean crop in the vicinity of 11 MMT should also make it possible to boost oilseed exports by 1.6 MMT this season.

A similar development is likely to occur in Brazil (Fig. 19). Following a fluctuation of oilmeal exports between 8–9 MMT in the past six seasons, an unprecedented growth by 2.2 MMT is possible this season with practically all of it occurring in soymeal. Also Brazil may be interested in disposing of as much as possible before the next U.S. soybean crop starts moving in September 1989.

It is doubtful, however, whether the combined expansion of soybean and soymeal exports of more than 3 MMT each in Argentina and Brazil will sufficiently relieve the shortage this season (Fig. 20). Although the growth is exceeding that registered in 83/84, it will still be insufficient to satisfy demand at the current pace.

(f) The European Community will probably have to achieve most of the required demand adjustments this season. The gap between domestic prices of soymeal and of barley and other feed grains in West Germany have widened sizeably since early this year (Fig. 21). Contrary to the USA, government-regulated feed grain prices have not risen. They have even weakened for the new-crop positions since July. Although soymeal prices are still lower than those registered in the tight 83/84 season, the price ratio vis-a-vis grains has deteriorated significantly.

The deterioration is reflected in Fig. 22. On the average of the 86/87 season soymeal prices were below those of feed grains. But they accelerated sizeably vis-a-vis grains last season and, particularly, since early this year. Just look at the figure to see the steep increase of soymeal prices from April to September.

It can thus be expected that the *grain usage* in the EEC will recover again by up to or above 3 MMT. A substantial decline in mixed feed output in the vicinity of 2–3 MMT cannot be precluded, as a growing number of farmers may favor on-farm feeding over commercial mixed feeds.

We consider it likely that total oilmeal usage in the 12 member countries will fall by close to 2 MMT this season (Fig. 23). While the usage of rape and sun meal probably will stagnate, there may be a slight decline in the other 8 meals in addition to the steep reduction in soybean meal. Farmers have generally produced a comparatively high amount of fodder with a good quality this year. Finally,

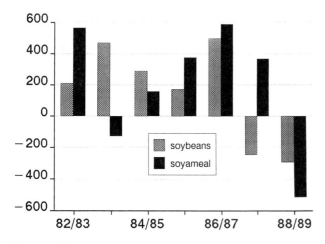

Fig. 13. China: Annual change in exports of soybeans and soybean meal (thousand metric tons)

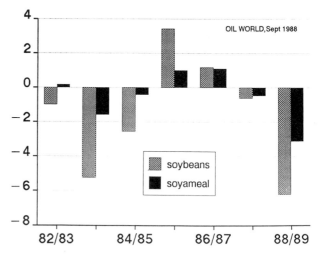

Fig. 16. USA: Annual changes in exports of soybeans and meal (MMT)

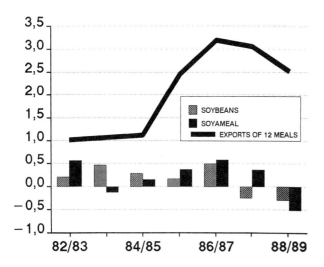

Fig. 14. China: Annual change in exports of soybeans, soybean meal and 12 oil meals (MMT)

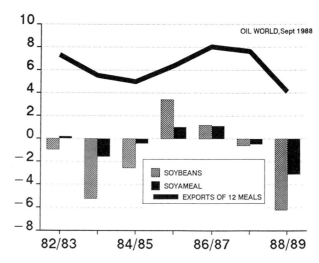

Fig. 17. USA: Annual changes in exports of soybeans, soybean meal and 12 oil meals (MMT)

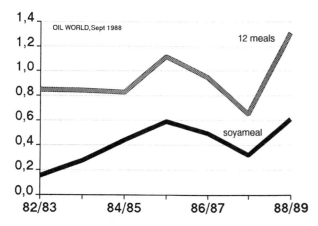

Fig. 15. India: exports of oil meals (MMT)

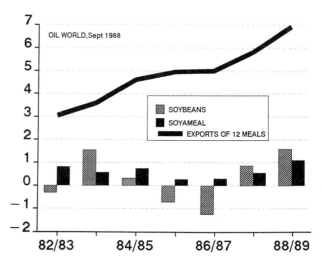

Fig. 18. Argentina: Annual changes in exports of soybeans and meal (MMT)

the boost of around 0.7 MMT in this year's feed bean and pea production is also adding to the negative environment for soybean meal.

The liquidation of pig breeding has already begun in West Germany and the Netherlands and the pig cycle in these two countries is turning down. For the whole EEC, however, total pig numbers are probably off only 2% from last year.

(g) In Central and South America soymeal import demand is likely to continue to expand (Fig. 24). In view of the declining oilseed production in Mexico and expanding domestic demand in most countries, there appears to be no leeway for smaller imports, despite the tightness in world market supplies. We expect most of the growth in imports to occur in Mexico and Venezuela, while the import demand of Cuba may stagnate at a comparatively high level.

(h) In Asia it will be difficult to curb import demand. Expanded livestock production and meat demand have boosted import requirements of oilseeds and, particularly, oilmeals to new record levels especially sharply in Japan, South Korea, Indonesia and Taiwan this season. The slight decline I currently project for 16 major importing countries in Asia may be difficult to achieve without higher prices, though it resembles the performance in 83/84 (Fig. 25).

Conclusion and Price Outlook

In concluding let me summarize the major messages:

(a) The unprecedented decline in U.S. soybean disposals this season cannot be relieved sufficiently by offsetting increases in other exporting countries. The resulting tightness on the world market will cause oilseed stocks—relative to usage—at the end of this season to decline to the lowest level in at least 15 years.

(b) The job of meal demand rationing will be more difficult this season than in 83/84. The unprecedented Soviet soymeal imports and the still too high U.S. domestic oilmeal disappearance will require most of the demand adjustment to take place in the European Community.

The recent additional Soviet tapioca purchase brings the total quantity bought from Thailand so far to 650,000 metric tons (compared with none last year). The Soviets are thus applying the popular EEC soy protein/tapioca blend in their livestock industry. Additional Soviet soymeal tenders in Argentina are reportedly on the agenda within the next few weeks (October 1988) for nearby shipment.

It cannot be precluded that the Soviets will boost soymeal purchases beyond the estimates given here. But the more they buy, the more demand rationing will have to occur elsewhere.

(c) Nevertheless, soymeal remains by far the largest protein source in the world (Fig. 26). Although its share of total oilmeal usage may drop slightly by 2 percentage points to 51% this season it will remain an essential ingredient of livestock feeding almost everywhere. But it cannot be precluded that the recent price rally may force some developing countries to scale down livestock production targets and thus reduce the demand for soybean meal below earlier intentions in the years ahead.

(d) Relief from other protein sources is limited. Noteworthy expansions are taking place only in cotton and groundnut meal supplies. But in the case of fish meal relatively tight supplies in the coming season plus record-high Asian imports could add to the protein shortage on the world market.

(e) The current pace of world meal demand is obviously still too high and it cannot be sustained at this level throughout the season. The low volume of U.S. soybean exports so far (with cumulative shipments since Sept. 1 down 56% from a year ago) is *misleading* and no good indicator of the trend in world market demand this year. Booming South American exports are to a large extent offsetting the U.S. decline. Unless Soviet meal import demand suddenly dries out (which is unlikely) higher prices particularly for meals (but also for oilseeds) will have to enforce the necessary rationing of the limited meal supplies.

(f) Soybean meal has the potential of recovering to US$ 330 on the world market (per ton cif Rotterdam, 44%, U.S. origin) and possibly test or pass the highs registered in June if any lasting weather deterioration and crop damage occurs in South America. The price strength need not necessarily take place in October or November. But the longer it is postponed, the higher may the price peaks be.

In view of the tight stocks/usage ratio, consumers worldwide will focus on the development in South America and our markets will react sensitively to any unfavorable news. Any lasting weather problems would force markets "to jump from the frying pan into the fire" and spark a resurging price explosion to or (in the worst case) beyond the June peaks.

It has always been challenging to analyze the global supply and demand of oilseeds and oilmeals. For the buyers and sellers among you it has even been more challenging to trade and take positions in these interrelated markets. I have painted a largely constructive scenario. Although I may have overestimated demand and underestimated the forthcoming supplies in the southern hemisphere, it is my opinion that the downward risk of prices is relatively limited once most of the U.S. harvest pressure is over whereas the risk of price strength is comparatively greater.

Anyway, I may be wrong and overestimate the upward scope of prices. But I hope you will be right in your decisions in the new, challenging season—at least most of the time.

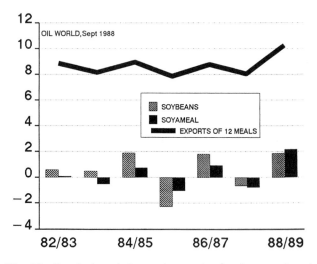

Fig. 19. Brazil: Annual changes in exports of soybeans and meal (MMT)

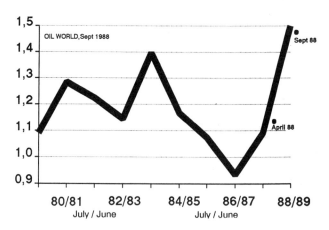

Fig. 22. Soy meal/grain price ratios in four EEC countries

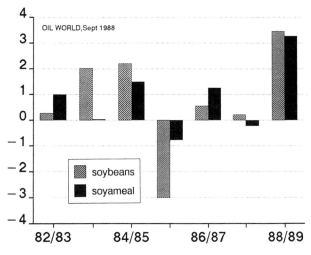

Fig. 20. Combined Argentina and Brazil annual changes in exports of soybeans and meal (MMT)

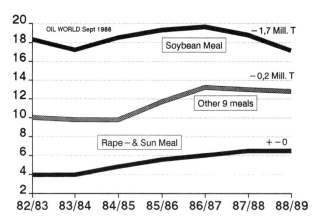

Fig. 23. EEC oil meal usage (MMT)

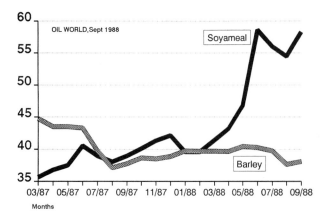

Fig. 21. Price trends in West Germany for soy meal and barley (Deutschmarks per 100 kilograms)

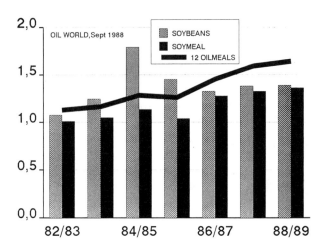

Fig. 24. Central and South American protein imports (MMT)

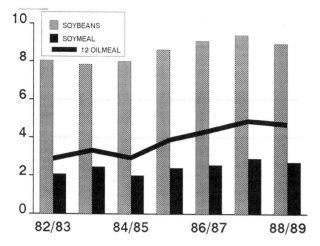

Fig. 25. Protein imports in 16 Asian nations (MMT)

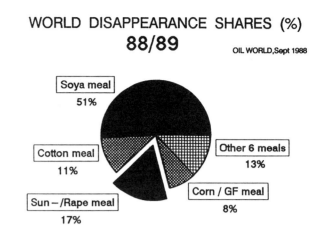

Fig. 26. 1988/89 World disappearance of oil meals

World Trade in Vegetable Proteins in the 1990's

Lindsay Jolly, Brent Borrell and Peter Perkins (speakers)

Australian Bureau of Agricultural and Resource Economics, Canberra ACT, Australia 2602

Abstract

Two factors are most pertinent to future developments in the world market for vegetable proteins. The first is the potential strong growth in demand for livestock products in developing countries and some centrally planned economies. The second is the likelihood of changes to government support for agriculture and to trade policies in developed countries. Such changes, which may be unilateral or multilateral, are likely to substantially affect the volume and direction of trade in vegetable proteins in the 1990's.

A large number of vegetable proteins for animal feeding are traded; oilseed meals are the dominant group of products in the market (Fig. 1). The United States and Brazil are the two most important producers of oilseed meals, particularly soymeal. The European Community (EC) and Japan are the two most important consuming regions. In recent years the European Community, China and India have increased their importance as producers, and the USSR and developing countries of Southeast Asia have emerged as increasingly important consumers.

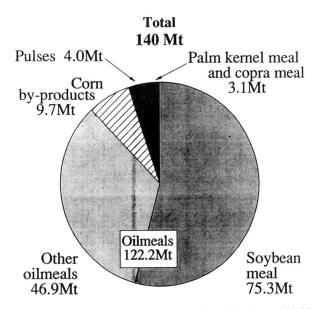

Fig. 1. World vegetable protein production 1984–85 to 1986–87.

World trade in vegetable proteins has grown rapidly since the 1960's relative to that of other agricultural commodities. In the 1970's strong economic growth in major developed countries, advances in the technology of livestock husbandry and feeding, and government assistance to livestock producers in the European Community and Japan stimulated a rapid increase in the demand for and production of vegetable proteins. Because there are few trade restrictions against vegetable proteins, trade grew rapidly, particularly from the United States to the European Community and Japan.

In the 1980's growth in vegetable protein trade has slowed, but is still about double that for food commodities in general. EC demand for imported vegetable proteins has stagnated, but imports by developing countries and some centrally planned economies have increased.

The Vegetable Proteins Market

Demand for Vegetable Proteins

The demand for livestock products is of prime importance to vegetable protein markets. By far the most important use for vegetable proteins is in intensive livestock production. Consumption of soymeal has been shown to be highly responsive to changes in the price of livestock and the number of animals on feed. Changes in the soymeal price and the price of its major substitute or complement, feed grains, also affect demand for soymeal (1).

The demand for meat and dairy products influences prices of livestock products and the number of livestock, as do government policies. Demand for meat and dairy products is directly influenced by changes in the levels of consumer income. Income growth can lead to increases in demand for livestock products, and therefore vegetable proteins (2).

Government assistance policies that support the prices received by livestock producers have a large bearing on the prices of livestock products and the number of livestock. Government assistance is provided mostly by policies which raise domestic livestock prices or livestock product prices, or both. As shown in Fig. 2, in OECD

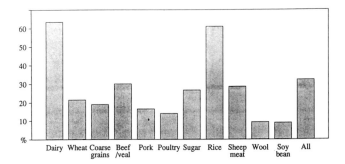

Fig. 2. Percentage producer subsidy equivalents (OECD average 1979–1981). Source: OECD

countries livestock products, especially dairy and beef, receive higher levels of government assistance than do most crops. Soybeans receive low levels of assistance.

Livestock industry assistance policies can have a large bearing on the volume and direction of world trade in vegetable proteins. Without the high assistance to livestock producers in the European Community and Japan, other countries might take over the production of a large amount of the meat and dairy products consumed in Europe and Japan. This would reduce imports of vegetable proteins to EC countries and Japan. One of the factors giving some countries a comparative advantage in beef and dairy products is that they use cheaper feedstuffs containing less vegetable protein.

Assistance to cereal producers in the European Community also has important effects on trade in vegetable proteins. In the European Community high domestic prices for cereals are an integral part of support for cereal producers. Because livestock producers and feed compounders have to pay high prices for cereals, demand for vegetable proteins and cereal substitutes is increased. The European Community is by far the world's largest importer of oilseeds and protein meals because of the high support to both its cereal and livestock sectors, and the low or zero tariffs on imports of vegetable proteins. Trade in oilseeds and oilseed meals would be lower if EC cereals prices were to become more closely aligned with world market prices.

Supply of Vegetable Proteins

Soybeans compete with other crops in production. Movements in the price of soybeans relative to prices for alternative crops have a strong influence on year to year changes in world vegetable proteins supplies.

The ratio of soybean price to corn price generally has been a strong indicator of how crop area in the midwest of the United States is allocated between the two crops (3). Few farmers specialize in the production of soybeans; farmers usually grow them in rotation with other crops, principally corn. However, during periods of low world prices for both corn and soybeans, such as the mid-1980's, the provisions of the U.S. corn program can isolate soybean planting decisions from market conditions (4). The high price and income support offered to growers who cooperate in the corn program encourages corn rather than soybean production. To qualify for the corn program benefits, growers must participate in area reduction programs, whereby agricultural land is withdrawn from production. When farmers set aside land so as to qualify for support under the corn area reduction program, the land that is set aside cannot be sown to other crops covered by the provisions of the U.S. farm bill, including soybeans. Without the support provided by the corn program, farmers would probably use the land for soybean production if not using it for corn. Thus the corn program distorts the pattern of response to corn and soybean prices.

Direct assistance to vegetable protein production is not a major influence on supply in major exporting countries. In the United States, little assistance is provided to oilseed growers (5). Between 1982 and 1986, U.S. soybean producers received no direct income support and relatively light assistance through Commodity Credit Corporation loan activities, credit subsidies and crop insurance. In Argentina and Brazil, soybean producers have been taxed through export controls, export taxes, and sometimes by overvalued currencies.

On the other hand, oilseed production is subsidized in most developing countries which import edible oils, oilmeals and oilseeds. But such countries do not greatly affect the supply of vegetable proteins. However, the subsidization of production in the European Community during the 1980's has resulted in marked increases in EC production of oilseeds and pulses.

The lack of direct government intervention in vegetable protein markets in the major exporting countries means that changes in world prices for protein feeds affect producer prices directly. Producers are not receiving subsidized prices, so they are not isolated from market signals. Intervention in the U.S. corn market can, however, partly determine production of soybeans. In years of low corn and soybean prices, U.S. soybean producers participating in the corn program withdraw land from soybean production, because of their obligation to keep land out of production in order to qualify for corn program benefits. However, on a worldwide basis vegetable protein supply adjusts to market requirements and is probably more flexible than supply from other cropping enterprises that are more highly protected.

Policy interventions are not the only important influences on the supply of vegetable protein. For instance, technical change and land development have contributed significantly to the growth in oilseed production. New agricultural land has been opened in Brazil, and yield increases in both the United States and South America have been important sources of increased production.

Historical Developments

Growth in world trade in vegetable proteins during the 1970's was rapid. World imports of protein meals grew at an average rate of around 11% a year. This compares with growth of around 5% a year for agricultural commodities as a whole. Growth in vegetable protein trade was a response to the growing demand in countries where production of soybeans was not economically efficient, particularly in the European Community and Japan.

In Western Europe strong economic growth stimulated a rapid increase in the demand for meats and other livestock products. Rising meat prices, partly due to EC support to its livestock sector, stimulated the expansion of the EC intensive livestock industry, particularly pig and poultry production. The EC dairy policy encouraged a build-up in cattle numbers. Technical change which increased the efficiency and profitability of intensive livestock feeding also contributed to the expansion in livestock production (6). European demand for vegetable proteins was increased further by EC cereals policy, as discussed earlier.

During the 1970's soybeans dominated oilseed production worldwide, and most soybean production was in the United States. U.S. research and development during the 1950's and 1960's had provided improved soybean cultivars and better processing technology. Brazil and Argentina were emerging as significant producers at this time. In Brazil, most of the growth in soybean production during the 1970's was due to area expansion in response to high soybean prices, the introduction of several cultivars suitable for new areas, indirect effects arising from government assistance to the production of wheat, and government investment in rural transport (7).

Trade in vegetable proteins was relatively free of restriction in the 1970's. Trade was predominantly from North and South America to Western Europe and Japan. As a result of the Dillon Round of the General Agreement on Tariffs and Trade (GATT) in 1961, the European Community had agreed to the free entry of soymeal and soybeans, and low tariffs on imports of other nongrain feed ingredients. Similarly, Japan agreed to the free entry of soybeans and soymeal during the 1973–79 Tokyo Round of GATT negotiations. Growth in world trade in vegetable proteins grew at a faster rate than production, and international prices for vegetable proteins remained high (Fig. 3).

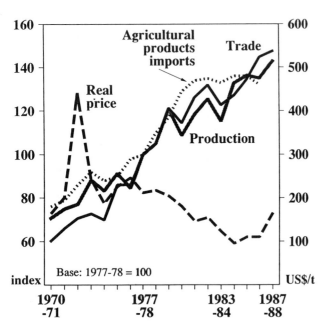

Fig. 3. World vegetable protein production and trade. Excludes pulses and grain proteins. Production is meal equivalent of oilseeds production; trade is oilmeal trade plus meal equivalent of oilseeds trade. Source: ISTA Mielke, USDA

The 1980's have been characterized by much slower growth in import demand for vegetable proteins (Fig. 3). Nevertheless trade has grown at about twice the rate of that for food commodities generally. Import demand for vegetable proteins grew about 4% a year during 1981–87, less than half the rate of growth during the 1970's. In the European Community consumption of protein meals stagnated. Growth in world trade in vegetable proteins was due mainly to increasing imports by developing countries and some centrally planned economies.

The stagnation in protein meal consumption in the European Community resulted from policy changes in the livestock and oilseeds sectors of the European Community, which caused growth in EC net imports of vegetable proteins to stall between 1981–82 and 1987–88 (Fig. 4). EC dairy policy, which had led to surpluses of butter and skim-milk and a build-up in cattle numbers during the 1970's and early 1980's, was changed in April 1984. Subsidies are no longer paid on all milk production, but are limited to a set quota of production. Production surplus to quota attracts penalty levies. Dairy cattle numbers have fallen, so feed demand from that sector has been reduced.

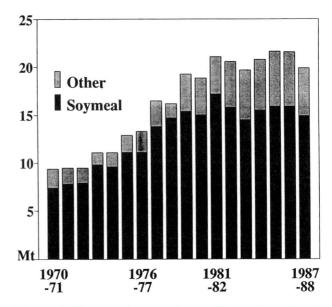

Fig. 4. EEC-10 net imports of vegetable proteins (oilmeals, oilseeds converted to meal basis and corn gluten feed).

The European Community strongly encouraged production of oilseeds and pulses during the 1980's, so as to reduce dependence on imports. EC production of oilseeds, mainly rapeseed and sunflowerseed, increased from 3.2 metric tons (MT) in 1982–83 to an estimated 11.6 MT in 1987–88. During 1981–85 the U.S. dollar was stronger against European currencies, increasing the cost of importing protein meals, making protein meals less attractive in compound feeds, and leading to some substitution of grains for imported protein meals.

In developing countries, consumption of vegetable proteins generally has increased during the 1980's. Imports have increased into countries in East and South Asia, where economic growth has remained generally strong and demand for livestock products has continued to grow. In other developing countries, growth in import demand has been limited by low or negative per capita income growth and by debt servicing problems in the early to mid-1980's. At the same time, many developing countries have been providing incentives to local intensive livestock industries by imposing much greater restrictions on the import of livestock products than on feed ingredients (7). As a consequence, these incentives encourage the importing of feed ingredients, including protein meals, as well as the growth of the domestic livestock industries.

The USSR began importing substantial quantities of oilseeds and oilseeds meal in the early 1980's. The imports have been in response to shortfalls in domestic oilseed production and the introduction of policies to increase growth in livestock protection. In particular, policies aimed at improving the quality and increasing the quantity of compound feeds fed to livestock have been initiated during the mid-1980's.

Import demand for food commodities slumped dramatically in the early 1980's, partly because of a significant downturn in economic activity in the industrial world which spilled over to the less developed world. Demand for imported food by developing countries declined and was further constrained by high real interest rates during the 1980–85 period, which increased the burden of debt servicing for some countries.

The fall in food commodity prices that resulted was exacerbated by support for agricultural production in the United States, the European Community, and Japan, as well as by export subsidies in the United States and the European Community, and restrictions on imports of agricultural commodities in Japan. The high levels of support held prices received by producers in these countries above world prices. The high prices induced further production which, in the United States and the European Community, was added to stocks held under government programs. The high prices also curtailed growth in import demand in Japan. In China and India, government policies to enhance the growth in food production also contributed to the decline in prices of food commodities.

The high level of stocks of other feedstuffs, and their low prices, meant that for vegetable proteins to remain competitive, prices had to fall (Fig. 5). As discussed earlier, in major producing countries government assistance to vegetable proteins production, especially soybeans, is lower than for the related feed grains and livestock sectors, so vegetable protein producers have the incentive to respond directly to world prices. Consequently stocks of oilseeds did not build to the same extent as did other commodities, particularly wheat, coarse grains, sugar and butter.

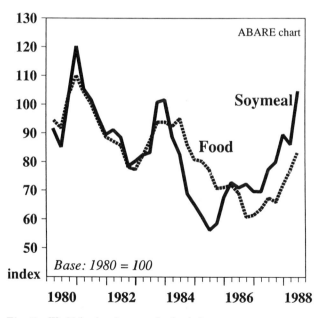

Fig. 5. World food and soymeal price indexes.

Soybean production in the United States has stagnated during the 1980's. Between 1980–81 and 1987–88, production of soybeans increased by only 3 MT. Area sown had fallen to 22.8 million hectares (Mha) in 1987–88, compared with 28.9 Mha in the record year of 1979–80. In Brazil, area sown to soybeans has shown little re-cent growth, reflecting lower world soybean prices and changes in government policy.

Agricultural commodity prices began to recover in 1987. The exceptionally low commodity prices of the mid-1980's initiated a response in terms of supply adjustment and increasing consumption. With declining production and improved demand, stocks of major food commodities had started to decline by late 1986.

Stocks of some commodities, such as oilseeds, sugar, coarse grains and wheat, now have been reduced to the point where the markets for these commodities have become very sensitive to expected change in supplies. Evidence of such sensitivity were the speculative price rises seen during May–June 1988 in response to the drought in the United States and Canada.

Whether agricultural commodity prices continue to be high into the 1990's depends significantly on developments in the world economy, especially the level of economic activity and changes in trade policies. Grain and oilseed prices will be influenced by the speed of the supply expansion to fill the void created by the 1988 North American drought, and by the policy environment in which farmers make such decisions.

The Future for Protein Meals

There appears to be sufficient flexibility in supply of vegetable proteins to ensure production can adjust readily to long term shifts in demand. Major shifts in demand or supply due to technological factors or land development are not expected over the medium term. Changes in technology may not be as important in the 1990's as in the past. Technological improvements such as increased efficiency in livestock feeding and the adoption of improved oilseeds cultivars appear to have slowed. The rate of land development for soybeans in Brazil also appears to have slowed (8). Macroeconomic factors affecting income growth in developing countries and changes in agricultural policies in major developed countries could, however, cause major shifts in demand for vegetable proteins.

Income and Population Growth in Developing Countries

In developed countries demand for livestock products has reached high levels and will be only moderately responsive to future growth in income levels. But there is significant growth potential in developing countries. Consumption of protein meals by developing countries that import vegetable proteins, especially in East and South Asia, and the Middle East, has shown strong growth over the past decade, as seen in the growth in consumption by the "other countries" group (mainly developing countries) shown in Figure 6.

In the developing world the most important factors boosting demand for livestock and poultry products are the upward trend in incomes (2), and changes associated with urbanization. A study of the impact of economic development on global food demand patterns (9) has shown that, as per capita income increases, the percentage of meat in the diet increases until very high income levels are reached and then decreases. With urbanization, diets are no longer confined to local sources of staple foods, and food markets in urban areas are broadened by the presence of food processing and distribution establishments. In addition, as developing

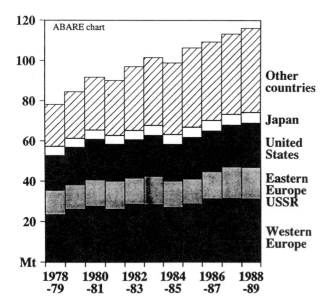

Fig. 6. World apparent consumption of eight major oilseed meals.

countries become more urbanized, more intensive livestock production methods are adopted so demand for vegetable proteins increases.

The potential for growth in livestock production and vegetable protein demand in developing countries during the 1990's can be appreciated by considering expected growth in per capita income and population. The World Bank (10) projects that real per capita gross domestic product in developing countries will grow at an annual rate of 2.2% until 1995. This compares with an average of 3% a year during the 1970's and 1.8% a year in the first half of the 1980's. In particular, economic growth in Asia is projected to be strong.

Population growth in developing countries, projected at around 2.5% during the 1990's, should reinforce increasing demand for livestock products and vegetable proteins. Developing countries should be a major source of growth in the vegetable proteins market during the 1990's, especially while they continue to impose greater restrictions on meat imports than on feed ingredients.

Outside the developing countries, the USSR and some other centrally planned economies show strong potential for growth in vegetable proteins trade. The average protein content of feeds in the USSR is low. The ratio of oilmeals to feed grains usage is only 5%, compared with 26% in Western Europe. However, imports by the USSR have been very erratic. Between 1980 and 1983, soybean meal imports rose from 400 kilotons (KT) to 2.3 MT, but in 1984, fell to 100 KT. Such variations seem to have been the result of transport, storage and distribution problems within the USSR (11). Renewed emphasis on improving the diet of Soviet consumers and the efficiency of its livestock sector is likely to lead to an expanding market for protein meals. Despite the recent upsurge in protein feed imports, more is needed to fully balance Soviet mixed feeds (12). Because the Soviet strategy for livestock growth calls for increased per head productivity rather than larger inventories of livestock, improvements in feed mixes will be a major goal for the Soviet livestock sector. The large Soviet purchases of U.S. soybeans and soybean meal in November 1987 may indicate the commencement of more stable and growing demand for vegetable proteins in the USSR.

Changes in Policy

In recent years growing concern over the consequences of protectionist agricultural policies has created international pressures for trade reform. Agricultural trade reform is the issue under discussion in the Uruguay Round of Multilateral Trade Negotiations currently taking place in the GATT. Countries negotiating in the Uruguay Round have agreed that overcoming the restrictions and distortions that affect agricultural trade would benefit most countries.

Should agricultural trade reforms result from the Uruguay Round negotiations there are likely to be important implications for the world vegetable protein market. Attempts to estimate effects of multilateral reductions in assistance in OECD countries on consumption, production, trade and price of agricultural commodities have been made in recent studies by the OECD (13) and the U.S. Department of Agriculture (14). The results of the two studies generally indicate that large reductions in the trade of soybeans and soymeal, particularly into the European Community and Japan, are likely with multilateral reduction of protectionist agriculture policies.

The OECD study analyzed the effect of a 10% reduction in assistance to all agricultural industries in OECD countries. The U.S. study assessed the consequences of the full removal of trade-distorting agricultural support policies in developed countries. Both studies point to reduced beef and dairy production in Europe and Japan. Exports of soybeans from the United States would fall correspondingly. Production of beef and dairy products would, however, be likely to increase in regions with a comparative advantage in producing livestock products, such as Australia, New Zealand, Canada, the United States and some developing countries. Part of these countries' comparative advantage may involve the use of lower cost feeds. The likely effects on world vegetable proteins prices of multilateral reductions in support to agriculture are, however, estimated to be small. Such changes would be likely to lead to only small reductions in world consumption and production of vegetable proteins for livestock feeding.

Overall, both studies highlight the fact that reductions in support prices for the beef and dairy industries in OECD countries could lead to moderate reductions in the volume of trade in vegetable proteins, and substantial changes in the direction of trade. They are likely to have less impact on production, consumption and prices of vegetable proteins on a worldwide basis.

European Economic Community

The European Community's place as the largest importer of vegetable proteins is due to its GATT bindings negotiated in the Dillon Round, which allow virtually unrestricted entry of protein meals and nongrain feed substitutes, and its high support to both the cereals and livestock sectors. Imported protein meals and other nongrain cereal substitutes displace EC grains in compound feeds, adding to domestic surpluses of grains. Export subsidies are paid to enable disposal of unwanted grains to the world market. This is a major cost to the EC Commission budget.

The use of cheaper imported feed ingredients in compound feeds leads to lower feed costs and increased profitability of livestock production. This, in turn, leads to increased production of livestock products. The disposal of surplus EC meat and dairy products on world markets at low prices also imposes a cost on the EC budget because of the export subsidies involved. Further, vegetable oils extracted from imported oilseeds are used for margarine manufacture. Because margarine displaces butter on the domestic market, additional butter must be exported. This means an additional budgetary cost to the EC Commission.

The EC Commission therefore has an incentive to reduce the supply of relatively cheap imported vegetable proteins and grain substitutes. Assuming no change in EC assistance to the livestock sector, reducing these cheap imports would increase the use of domestic grains in animal feeds and reduce the domestic surplus of high priced grains as well as the surplus of livestock products. It also would reduce the export subsidies paid to assist the disposal of surplus grains and livestock products.

The EC has stated its desire to "harmonize" its agricultural policies, by lowering variable levies on cereal imports but removing its zero tariff bindings on soybean and soybean meal imports. Assistance to EC grain producers would be reduced, tariffs would be placed on vegetable protein imports. One study, assessing the impact of a 10% tariff on vegetable proteins (represented by soybeans and soybean meal) coupled with a 20% decrease in EC support prices for cereals, estimated that world prices of vegetable proteins would fall by around 2% with EC use of protein meals falling by around 8% (15).

Similarly, reducing high EC cereal prices alone could have significant effects on vegetable protein markets. Grains would displace imported nongrain feed substitutes, especially manioc, and would reduce the attractiveness of vegetable proteins in feed mixes. Using an econometric model, Schmidt, Fronberg, and Maxwell (16) estimated that with the gradual alignment of EC grain prices toward world prices, demand for protein feeds in Europe would decline around 25%, as the amount of wheat and coarse grains fed in the European Community would rise by 17% and 24% respectively. Overall, they estimated that world trade in vegetable proteins would fall 8.5%. As a result of declining world import demand, U.S. exports of vegetable proteins would decrease by an estimated 13%, and exports from Brazil would increase only slightly. If the predicted shifts into additional wheat and coarse grain production occurred, the resultant reductions of soybean production in the United States and Brazil would ultimately lead to world prices of vegetable proteins increasing by around 2%.

Japan

Japan is the second largest importer of vegetable proteins, although it imports less than a quarter of that imported by Western Europe. Much of Japan's imported vegetable protein is used in the pig and poultry industries. Recent changes in support arrangements to the Japanese beef sector could have substantial effects on world trade in vegetable proteins.

The Beef Import Agreement made with the United States and Australia in June 1988 provides for the removal of Japan's beef import quotas on April 1, 1991. The quota has been the major policy instrument used to meet the objective of stable wholesale prices for beef and a high level of self-sufficiency in beef. The removal of volume quotas will, however, be set against an increase in the *ad valorem* tariff applied to beef imports, from 25% to 70%. Even with the increase in tariffs, the removal of the beef import quota could reduce the protection to the Japanese beef sector to about half the level in 1986 (17).

The reduction in assistance to Japanese beef producers could result in reduced production of beef in Japan. In response to lower beef consumer prices, Japanese consumers may also substitute imported beef for pork, leading to a reduction in pig production. Reductions in pig and beef production in Japan are likely to lower imports of vegetable proteins. The OECD study (13) indicated that with a 10% reduction in assistance to agriculture, the livestock sector in Japan would contract and net imports of beef would increase. Japanese imports of vegetable proteins would decline by around 5%. Reduced trade in vegetable proteins to Japan could be partly compensated for by increased demand for feedstuffs, including vegetable proteins, in beef exporting countries. Although, as stated earlier, countries with a comparative advantage in producing livestock products may gain their cost advantage from using feed mixes which contain less vegetable proteins.

Discussion

Vegetable protein markets can be seen as entering a new phase of development. In the 1970's strong income growth and protectionist agricultural policies in the European Community and Japan led to strong growth in demand and production of livestock products. The absence of trade restrictions against vegetable proteins provided unique opportunities for rapid growth in trade in vegetable proteins. In the 1980's growth in trade slowed. Policy changes in the European Community resulted in stagnating imports of vegetable proteins. Such growth in world trade as occurred was due to increased imports by developing countries and some centrally planned economies.

In the 1990's, pressures for further policy changes in agriculture, particularly in the European Community and Japan, could lead to declines in import demand from developed countries for vegetable proteins. Income and population growth in developing countries, and improvements in livestock feeding in the USSR, are likely to provide a stimulus for growth in trade.

On balance, trade growth during the 1990's can, at best, be expected to match that of the 1980's. The high level of growth which occurred during the 1970's was due in large part to government policies providing assistance to agriculture and to rapid advances in livestock feeding technology. These factors are unlikely to be repeated in the 1990's.

References

1. Meyers, W., M. Helmar, and S. Devadoss. *FAPRI Trade Model for the Soybean Sector: Specification, Estimation, and Validation*, CARD Staff Report #86-SR2, Center for Agricultural and Rural Development, Iowa State University, Ames, 1985.
2. Centro Internacional do Mejoramiento do Maiz y Trigo, *1984 CIMMYT Maize Facts and Trends, Report Two: An Analysis of changes in Third World food and feed uses of maize*, Mexico, 1984.

3. *Oilcrops: Outlook and Situation Yearbook* OCS-9, Economic Research Service, US Department of Agriculture, Washington, D.C., 1988.
4. Glaser, L.K. *Provisions of the Food Security Act of 1985*, Agricultural Information Bulletin 498, Economic Research Service, US Department of Agriculture, Washington, D.C., 1986.
5. *Estimates of Producer and Consumer Subsidy Equivalents: Government Intervention in Agriculture, 1982–86*, Agriculture and Trade Analysis Division, Economic Research Service, Washington, D.C., 1987.
6. Leuck, D.J. *An Analysis of the Feed-Livestock Sector of the European Community*, ERS Staff Report AGES840529, Economic Research Service, US Department of Agriculture, Washington, D.C., 1985.
7. Williams, G.W., and R.L. Thompson in *World Soybean Research Conference III: Proceedings*, edited by R. Shibles, Westview Press, Boulder, Colorado, 1985, pp. 49–56.
8. Vocke, G. *Research and Development Affects US and Third World Soybean Trade*, World Agriculture: Situation and Outlook Report, WAS-51, Economic Research Service, US Department of Agriculture, Washington, D.C., 1988.
9. Marks, S.M. *The Impact of Economic Development on Global Food Demand Patterns*, World Agriculture: Situation and Outlook Report WAS-51, Economic Research Service, US Department of Agriculture, Washington, D.C., 1988.
10. World Bank, *World Development Report*, Oxford University Press, New York, 1988.
11. Bickerton, T.W. *USSR Oilseed Production, Processing and Trade*, Foreign Agricultural Economic Report 232, Economic Research Service, US Department of Agriculture, Washington, D.C., 1987.
12. Cook, E.C. *The Soviet Livestock Sector: Performance and prospects*, Foreign Agricultural Economic Report 235, Economic Research Service, Agriculture and Trade Analysis Division, US Department of Agriculture, Washington, D.C., 1988.
13. *National Policies and Agricultural Trade*, Organisation for Economic Co-operation and Development, Paris, 1987.
14. Webb, A.J., Roningen, V.O., and Dixit, P. Analysing agricultural trade liberalization for the Pacific Basin. Economic Research Service paper presented at the Livestock and Feedgrains Working Group of the Pacific Economic Cooperation Conference, Napier, New Zealand, Oct. 20–22, 1987.
15. Koester, U. *Disharmonies in EC and US Agricultural Policy Measures*, Commission of the European Community, Brussels, 1987.
16. Schmidt, S.C., Frohberg, K.K., and Maxwell, D. L. *Implications of Grain Trade Liberalisation in the European Community*, Agricultural Economics Research Report 202, University of Illinois, 1987.
17. Teal, F., Dickson, A., Porter, D. and Whiteford, D. *Japanese Beef Policies: Implications for Trade, Prices and Market Shares*, Australian Bureau of Agricultural and Resource Economics Occasional Paper 102, Australian Government Publishing Service, Canberra, 1987.

Government-Imposed Restrictions on International Trade in Proteins

Donald E. deKieffer

Pillsbury, Madison & Sutro, Suite 1100, 1667 K St. NW, Washington, D.C. 20006, USA

Abstract

To thwart the diversification policies of their trading partners, countries have created massive price supports and protectionist devices. The growing degree of government intervention now threatens the operation of the protein markets. For example, Argentina maintains a differential export tax system applying higher rates to raw soybeans than to soybean products in order to block the export of unprocessed commodity. The United States supports the Food for Peace and GSM credit programs that, despite their benevolent purposes, result in government assisted sales, increasing world oversupply. Countries may slow the growth of government intervention somewhat through the inadequate vehicles of cartelization and deregulation. Producers, on the other hand, should agree to trading rules that approximate those for manufactured products in the milieu of the Subsidies Code and under the rubric of the General Agreement. Otherwise, the world faces the perpetuation of anarchy and increased trade wars.

For at least 150 years, economists have postulated the theory of "comparative advantage." This theory suggests it is economically advantageous for all countries to produce those things in which they are most efficient and to import those products which other countries can produce more cheaply or efficiently. Thus, for example, "comparative advantage" suggests that countries such as Thailand, which is endowed with plenty of flat land, a warm climate, adequate water and cheap labor should produce rice while countries endowed with iron ore and coal make steel. Although one can debate the relative merits of this theory, it is astounding that even today economics students are taught the theory of comparative advantage as if it were a "law." Nothing could be further from the truth.

"Comparative advantage" did not operate at the time the theory was propounded and certainly does not today. Even 150 years ago, transportation costs, financial expenses, subjective demands which are not cost-driven and nascent political imperatives "skewed" the comparative advantage "model" into something quite different than the theoretical form economists had envisioned. Today, there are increasingly complicated political and economic factors at work. The world economic system has yet to adjust to the collapse of imperialism and the emergence of the Third World countries as independent political forces. While a small group of European nations is no longer able to dictate where products will be produced and sold to maximize "comparative advantages," their economic interests, and domestic political imperatives, favor maintenance of the old economic order. At the same time, the newly emerging nations have become painfully aware that exclusive reliance on a single product or commodity that it produces "best" is exceedingly dangerous. Countries which have but one "cash crop," be it bananas, bauxite or petroleum are increasingly vulnerable to collapse in prices, political or economic whim. Therefore, most countries, particularly those that have traditionally been "producers" of raw commodities, have attempted to repeal or at least to amend the "law" of comparative advantage by diversifying their economies. To do so is neither painless nor inexpensive.

Developing countries have adopted various plans to encourage the creation of new industrial and agricultural sectors, usually by some form of direct or indirect subsidization. To meet this new competition, and to preserve their traditional sectors, developed countries have responded with massive price supports and a sophisticated array of protectionist devices to thwart the diversification policies of their trading partners.

Developed countries also have adopted social policies having little to do with the economics of production which would otherwise render large segments of their economies noncompetitive. Social policies favoring traditional "family farms" are an example. To offset the artificial disadvantage of these policies, countries in the First World have increasingly turned to even more draconian subsidy and protectionist practices to "preserve" many sectors of their economies. This "having your cake and eating it too" attitude, however, has led not only to huge tax bills at home, but in many instances has prevented countries with more efficient production from entering potentially lucrative markets.

Today, there is hardly an industrial sector in the world which is not affected more by government policy than by the pristine model of "comparative advantage." Although the costs of protectionism and subsidization are not totally calculable, they far exceed the real costs of production and distribution in a "comparatively advantageous" world.

Since World War II, developed countries (and, increasingly, the developing countries as well) have sought to slow the growth of government intervention in the marketplace. These efforts have focused primarily on two contrasting vehicles: cartelization and deregulation.

Cartelization: In an attempt to regulate both price and supply, numerous cartels have been established in the last half century for products as diverse as tin, coffee, sugar and petroleum. While most price/supply cartels have been in commodities, there is a significant trend toward applying these principles to manufactured products such as steel and textiles.

The success of these price/supply cartels has been limited due, in large measure, to the willingness of some members to cheat on their commitments. All cartels rely upon rigorous discipline among all producers as the *sine qua non* for their successful operation.

Further, consumers have rebelled against artificial price/supply restraints either by choosing alternative products or by refusing to purchase altogether. Nevertheless, price/supply cartels have been partially successful in certain products that are geographically concentrated. Their success encourages other industries to adopt cartels as a reasonable alternative to the uncertainties of the free market.

Deregulation: At the other extreme are those who advocate "fair competition" and a "level playing field." The "level playing field" advocates, admittedly, mostly hail from developed countries. But to draw much from this fact would be an error. Most Western democracies believe the free market theory incorporates the "freedom to fail." This attitude is neither widely held nor politically viable in many developing countries that are attempting to diversify their industrial and agricultural sectors and for whom the price of failure for a major development project is intolerable.

This is not to suggest the Western Democracies totally embrace the "freedom to fail" theory. But they are already diversified sufficiently to absorb some losses while those in the developing world are not. Further, governments in many Western countries have a stake in social programs which are so immense that to abandon sections of their economy would mean the destruction of the social systems which they themselves have built. This is politically untenable.

The case of proteins provides a striking example of a global industry at the crossroads. Rapidly growing for the past twenty years, protein production now faces the prospect of increasing regulation either by individual governments or joint action to control prices and production. It is neither a cartel nor an example of free trade. Choices, however, will have to be made within the next few years or the third alternative, economic anarchy, is a certainty. Let us review briefly the problems that have arisen in the protein sector, and the growing degree of government intervention that now threatens the orderly operation of the markets.

The Problem: World Oversupply of Protein Products

A number of factors coalesced in the 1980's which resulted in a world surplus in agricultural commodities in general and in the oversupply of oilseeds and oilseed products in particular. These factors included: Unexpectedly high crop yields as the result of improved plant varieties and improved farm techniques; the ability of farmers to "double crop" oilseeds such as soybeans, and thus to bring additional supply to the market without the need for additional acreage; large increases of acreage in production in many countries including the United States, Brazil, Argentina and Malaysia; and a levelling of demand growth for proteins below predicted levels.

These factors provided substantial supply pressures in the market but would have been ameliorated by natural economic compensations in the form of lower prices, lower profits and concomitant decline in production and processing investment but for unjustified market intervention and subsidy programs developed by a number of governments around the globe, which have continued the upward movement in supply, ignoring natural market signals. The result has been a continued downward spiral in protein product prices. The orderly marketing of protein-based products is now threatened by artificial barriers and destructive subsidies; at this point, I would like to address some of the major impediments to free trade that put us all at risk.

Government Interventions in Oilseeds
The European Community

The EC is a major market for unprocessed oilseeds and for protein meal to supply its substantial pork and poultry industries. However, while the EC is a major net importer of certain edible oils, such as palm and coconut oils, it has also become a major net exporter of other edible oils such as soybean and rapeseed oil. These exports are due in large part to EC processing of imported beans. Whenever possible, however, the EC encourages the use of domestic beans. The current oilseed policy focuses upon domestic production and incentives to encourage domestic crushing as well as incentives to displace oilseed products from non-EC countries. For example, industry sources have indicated that France has provided government assistance to builders or purchasers of oilseed processing plants. Sources claim that France offered a U.S. $6 million grant to restructure an oilseed crushing firm and a U.S. $1 million subsidy for a new sunflower seed crushing facility.

Proposed EC Consumption Tax: In 1987 the EC Commission approved a plan to tax consumption of most edible oils. The tax, if adopted, would both stabilize price differences between various oils and fund the EC's massive subsidization of domestic oil production and EC enlargement, which I mentioned a few minutes ago. Hardest hit by this plan will be soybean oil, with a proposed tax of 89 percent. The tax of about $375 per metric ton on oils would double the price of soybean oil in the EC and raise about $2.3 billion annually. Imported oils will bear the brunt of this tax, because their producers do not enjoy the offsetting benefits of domestic subsidy programs. U.S. soybean exporters pressured the Reagan Administration to collect some $470 million from the EC as GATT-required compensation for the proposed tax. Protectionist programs such as these only exacerbate the problems associated with world agricultural surplus. In the wake of threatened U.S. retaliation, the EC Council rejected passage of the tax, although it is still pending before the Council. Strong opposition within the EC, from the U.K., West Germany, the Netherlands, and Denmark, claiming that the tax would hurt trade relations with the U.S. and other exporters and would increase costs to consumers helped to undermine passage of the tax.

European Import Barriers: Europe continues to limit access to its domestic markets by foreign-sourced vegetable oils through the imposition of a 10% import duty on crude edible oils. This import duty has been cited by Brazil and Argentina as a justification for their own subsidy practices. In 1986 the EC threatened to curb bound duty-free U.S. imports of soybeans, corn gluten and other animal feeds, an action that was strongly resisted by the U.S. government. After receiving the U.S.'s retaliatory threat to impose 200% tariffs on EC wines and cheeses, the EC agreed to guarantee a minimum purchase of non-EC animal feed.

Recently, the American Soybean Association, a U.S. not-for-profit organization of soybean farmers, brought a claim before the United States government against the EC, charging the EC with unfair trade practices that discriminate against the importation and utilization of U.S. soybeans and soybean meal. The EC had granted the United States, under the Dillon Round of Multilateral Trade negotiations, duty-free import bindings on soybeans and soybean meal in exchange for U.S. concessions on certain EC exports. During the 1970's and 1980's the EC created subsidies for oilseeds and protein crops. The EC establishes minimum prices for crops. Processors and compounders must pay the minimum price or growers sell to the EC governments for the minimum price. Moreover, the EC Commission has implemented a processor subsidy to assure EC oilseed processors a higher profit for using EC-grown crops than for using imported U.S. soybeans and soybean meal.

The EC also shifted land out of cereals production and into oilseed and protein production. The EC has accomplished this by raising EC minimum prices for oilseeds and protein crops while reducing cereal supports. For example in 1987, the EC minimum grower price for soybeans was $15/bushel at then current exchange rates while the U.S. soybean loan rate was $4.77/bushel. The EC price was three times the world price.

EC Policy on Rapeseed: The EC currently provides a subsidy that offsets a part of the cost to EC oilseed crushers of purchasing EC-grown rapeseed. European rapeseed production is currently uneconomic, and without this subsidy little rapeseed would be grown. Although this subsidy is apparently intended to provide income support for EC rapeseed farmers, it has the additional impact of creating a substantial supply of rapeseed oil for which there is no EC demand, in part, because of other EC policies to foster their dairy and olive oil industries. As a result, the EC now exports about 500,000 metric tons of subsidized rapeseed oil which competes in world markets with soybean and rapeseed oil from other sources. A similar program subsidizing the purchase of high-priced EC soybeans by European oil processors has received sharp criticism in the U.S. The conflict over soybeans is one of the major issues precipitating the recent agricultural trade war between the U.S. and the EC, which caused a chain-reaction of protective measures around the world.

Spain's Domestic Consumption Quota: Spain and Portugal became members of the EC on January 1, 1986. The transition to EC agricultural policies will last 7–10 years for many commodities. Oilseed prices in both countries have increased over the past few years in expectation of EC enlargement. Spain continues to intervene in the market to protect its olive oil producers. Spain is a substantial importer of unprocessed soybeans, and has developed a large processing industry to meet its large need for protein feed. However, Spain has placed a 90,000 metric ton domestic consumption quota for soybean oil, thus forcing more than 300,000 metric tons of Spanish soybean oil into the export market at low prices.

Spain's Export Subsidies: Prior to its accession to the EC, Spain granted its oilseed processors a subsidy, in the form of a rebate on its turnover tax, equal to 7% of the FOB value of all soybean oil exported. Soybean oil did not, in fact, bear the rebated tax. This was intended to enable Spanish crushers to reduce the prices for soybean oil forced into export by the consumption quota. Although forced to abandon its tax rebate practices under terms of the EC accession agreements, Spain has nonetheless convinced the EC to formulate a regulation which permits a direct subsidy to producers equal to 22.8 ECUs per metric ton of exported soybean oil. Spain's import restrictions, moreover, will be in effect until Dec. 31, 1990 and cannot be challenged under the EC enlargement negotiation process because of the duration for which such controls have been in operation.

Portugal's Domestic Consumption Quota: Prior to its accession to the EC, Portugal limited access to its markets through the requirement of dealing through a state-run trading organization (IAPO). Although Portugal largely dismantled IAPO prior to its EC accession, the EC has now implemented a Portuguese consumption quota for soybean oil similar to that used by Spain.

Portugal limits domestic sales of vegetable oil made with imported oilseeds. During 1986–1987, only 50,000 tons of soybean oil using 285,000 tons of soybean imports could be sold in Portugal.

The U.S. Trade Representative has indicated that the EC is violating a commitment made in 1986 not to curtail U.S. soybean exports to Portugal. Ambassador Yeutter charges that Portugal limits U.S. imports to Portugal by denying requests for licenses needed for entry. Portuguese oil processors, in 1987, asked the government for permission to import 459,156 metric tons but only got licenses to buy 367,201 tons. Thus, U.S. exports were slashed by 20 percent.

Brazil

In the Third World, the types of government interventions that have disrupted the orderly marketing of oilseeds and oilseeds products have involved not so much protection of threatened farming interests as much as the promotion of value-added export industries. The first nation to develop substantial subsidy programs to assist oilseed processing and oilseed product exports was Brazil in the 1970's. The result was an unprecedented boom in both processing capacity and exports of protein meal and vegetable oil through the 1970's and into the early 1980's. For example, despite the fact that the Brazilian farm sector produces roughly one-third the number of soybeans produced in the United States, Brazil has surpassed the United States as the world's leading exporter of soybean meal and oil. In fact, government incentives in Brazil fostering the development of its soybean crushing capacity have been so powerful that Brazil's current crushing capacity of approximately 25 million metric tons per year far exceeds its capacity to grow soybeans which is projected for the 1988–1989 crop year at 19.5 million metric tons.

In the 1987–88 year, however, Brazil's exports of soybean oil declined to 600,000 tons from 950,000 tons in 1986–87. Brazilian officials attributed this decrease to the increase in U.S. exports of soybean oil under the Export Enhancement Program, which they contend violated trade guidelines under the General Agreement on Tariffs and Trade.

This tremendous processing overcapacity has resulted in under-utilization of plant capability in Brazil. In 1987 Brazil utilized only 50% of its crushing capacity. Government incentives have resulted in an uneconomic investment at a time when Brazil is facing a monumental international debt crisis. This has forced Brazilian exporters to sell at low prices to service the debt. At the same time, the cost to the Brazilian government of funding this unwise growth continued to increase throughout the early 1980's. The downward price spiral has intensified as cheap exports of oilseed products from other countries—principally soybean meal and oil from Argentina and palm oil from Malaysia—further added to oversupply. In the past several years Brazil has reduced or eliminated most of its subsidy practices. However, to fully appreciate the degree to which government intervention in Brazil has contributed to the current situation, we need to review the full list of subsidy practices that have been involved.

Differential Export Taxes: Brazil imposed a higher rate of export tax (13%) on unprocessed soybeans than it did on either soybean meal (11.1%) or soybean oil (8.0%). The effects of this was to cause Brazilian-grown soybeans to trade in the domestic market at less than world-market prices, providing an additional crushing margin for Brazilian oilseed processors. The Brazilian practice provided an advantage similar to, although not as great in value, as that provided the crushing industry in Argentina. Although challenged by U.S. soybean processors, the differential export taxation system still favors exports of oilseed products.

Preferential Export Financing: Throughout the 1970's, Brazil formulated a number of preferential export financing programs to assist its soybean processing industry. Many of these programs have been substantially cut back or eliminated in recent years, but their impact in promoting the enormous growth of Brazilian crushing capacity is undeniable.

Under Resolution 674 and its successor programs (Resolutions 882 and 950), Brazil provided below market rate working capital financing of "production to be exported." Soybean crushing firms were eligible for this program based on an established percentage of the F.O.B. value of the previous year's exports. While these resolutions provided tremendous subsidies in the late 1970's and early 1980's, this program now appears to be greatly diminished. By August 1986, Resolution 950 financing was limited to vegetable oil at levels of 15% of the previous year's export FOB value, and, by the end of the year, had been suspended.

Similar financing was available under Resolution 643 to soybean crushers who sold to export trading companies. Resolution 643 is still legal under the Brazilian system of trade, although insufficient financing has forced the government to scale back those subsidies.

Brazil also established a duty drawback system that enabled processors of soybean meal and oil to obtain preferential financing to import raw materials for domestic processing and re-export. Again the rates available were well below commercial levels, and the program was specifically amended to indicate that soybeans were included.

While the preferential financing was eliminated in 1983, recent reports from Brazil indicate that the government may again be considering intervention through the grant of ICM tax credits on imported soybeans. The impact of this tax credit system is currently being studied.

Brazil also granted preferential credits to a major foreign customer under a clearing arrangement with Hungary. This permitted Brazilian exporters to capture the lion's share of the Hungarian import market for soybean meal, approximately 600,000 metric tons per year in the early 1980's. Trade sources report this practice may be expanded to other East European countries as well. The value in 1985 was estimated in the range of $150 million in sales by Brazil to Hungary.

Finally, Brazilian crushers also benefitted from a national system of rural credits (EGF loans) which provided funds at subsidized interest rates for the storage of agricultural commodities. Soybean processors have been the leading recipients of the EGF loans.

In 1986 the Brazilian government initiated a massive government program to create an infrastructure to enhance the competitiveness of the domestic soybean farm sector. The government provided investment capital to develop transportation, irrigation, and storage systems. It is anticipated that by 1989 Brazilian soybean acreage will increase by 700,000 hectares over the average 8.6 million hectares planted during 1981–1985.

Tax Exemptions and Deductions: Brazil also provided special tax treatment to enhance the export sales of soybean products. Export earnings on soybean oil were exempt from Brazil's corporate income tax. However, these exemptions effectively benefitted canned oil only and thus have not been recently challenged. Special deduction rules were developed to allow Brazilian traders to exclude commodity futures hedging net profits from taxable income, while at the same time deducting hedging net losses as a business expense.

Tax rebates for exports have been largely eliminated except for a program that provides companies who commit to reach a certain level of exports with authorization and financing for imports.

Minimum Price System: Brazil maintains a minimum price system in which the government provides direct loans to farmers or indirect loans for crop storage. The minimum price system operates by establishing a soybean price before harvest. The farmer may redeem loans at the support price or forfeit to the government at the minimum price.

The government controls the authorization to export or import through the Bank of Brazil where trade must be registered. The *ad hoc* tightening and loosening of registrations inhibits world prices from correcting the Brazilian market. For example, if the government wants to lower domestic prices and increase supply, it cuts off exports and opens the door for imports. To curb price increases, the government may import commodities but sell them domestically at a loss.

Argentina

Argentina, which had no significant exports of oilseed products prior to 1977, has become a major exporter of soybean and sunflower seed oil and meal in less than a decade. Although Argentina appears to have the soil, weather conditions and available land to make it a significant producer of soybeans and sunflower seeds, its meteoric rise as an exporter of processed protein meal and vegetable oils (rather than the raw soybeans) is linked to government intervention designed to foster a processing industry. The statistics show, for example, that Argentina's share of the world's export market for unprocessed soybeans have remained fairly constant since 1977/78, increasing from 8.7% in that year to a high of 10.8% in the 1985/86 crop year. Soybean exports declined slightly

World Vegetable Protein Marketing Perspectives

David H. Swanson

Central Soya Co. Inc., Po Box 1400, Fort Wayne, Indiana 46801-1400, USA

Abstract

World supply and demand for vegetable proteins will continue to increase and will be driven by a number of demographic, economic, social, and technological factors in the coming decades. This equation for marketing opportunities in the food, feed, and industrial sectors includes such unknowns and variables as macroeconomic factors (exchange and interest rates, budget and trade deficits, monetary and fiscal policy, etc.); LDC development and debt issues: growth in traditional OECD markets; trade barriers and related protectionism issues; and biotechnology and other technological advances. Perhaps the most dynamic of these is the accelerated development of technology and the new market opportunities that may be created by environmental considerations. To position ourselves to take advantage of future opportunities will require heavy investments in research and development, innovative strategies to insure an economic return on these investments, a willingness and talent for managing in an environment of dramatic change, and an international cooperation that recognizes our world market and gives diminished weight to national boundaries.

I'm tempted, when looking at my topic . . . . "World Vegetable Protein Marketing Perspectives" to begin with some sort of disclaimer like Sam Goldwyn's old line . . . . "Never make predictions. Especially about the future." I won't. But, I will offer the opinion that five-year marketing plans, when marketing vegetable proteins in today's environment, can be dangerous things. Those of you charged with marketing vegetable proteins should wake up every morning just a little bit uncertain about your company's marketplace. It is invariably changing, and the forces that bring that change are the same forces that make it imperative that we build into our marketing strategies a large measure of operating flexibility. Take your strategic plans seriously, very seriously, but make sure they can accommodate the unpredictable.

There are some key assumptions upon which we can base our plans. For instance, we know that worldwide demand for food, feed, and industrial proteins will grow. The world's farmers must be prepared to feed 88 million new people each year (that's 242,000 new people each and every day). And I think it is clear that our markets will continue to become more globalized each year. That very slowly, but surely, the strengths of each country can be developed, and directed to the most efficient uses, producing mutual advantages for both producer companies and consumer countries. The laws of comparative advantages will move closer toward development of a healthier equilibrium. It's also quite evident that concerns for the environment will be central in our marketing strategies and will open up new opportunities for vegetable proteins. These long-term trends give us general direction, but a host of other factors in today's marketing equations must be dealt with. Each of these could negate the trends for the company that does not build the mechanisms for change into his marketing plans. Let's look at just a few:

Macroeconomics: The macroeconomics we contend with in today's business world, because it is truly a global environment, are fickle indeed. Exchange rate fluctuations, interest rate swings, trade deficits, budget deficits, and monetary policy are some of the forces that will challenge any marketing manager. While I believe that gradually the trading countries of the world will be able to move toward more internationally coordinated policies, don't expect this to happen next year or the year after. It will happen slowly, and sometimes painfully. In the meantime, the outlook for any given product is going to depend upon things like exchange rates, export programs, and economic growth in a region, among other things.

Trade barriers: Access to markets, or the lack thereof, will continue to determine to a large extent the opportunities to market vegetable proteins. Trade policies in any given country will affect the overall mix, volume, and value of the protein products that cross its borders. Let us all hope that the desire for self-sufficiency can be tempered enough to create opportunities. Trade liberalization is always a slow and arduous process, but the resulting opportunities can be in the best interest of producers and consumers alike, the world over.

LDC debt and development: This is another area in which international cooperation can alleviate hardship and improve growth prospects for the marketer of vegetable proteins. Debtor nations that adopt austerity programs, curtail imports, and export at all cost to earn foreign exchange to meet their foreign debt payments are just one more element in a scenario that dampens marketing opportunities for many. Poorer countries, while they have the potential for large growth, can't import much because of stagnant economies, subsistence incomes, large national debts and unfavorable trade balances. These countries will require substantial long-term assistance, including food aid and debt relief.

Biotechnology: If the theme of my remarks is "plan for drastic change," biotechnology must surely be the wild card that will, more than anything else, give us the means to accommodate that change. In a sense, it has double marketing implications in that technology not only will take us to markets that we've identified, but also create new ones that we never envisioned until a new technology is discovered, sometimes by fortunate accident. Biotechnology also overlays nicely with another dynamic that impacts our opportunities today, and that is concern for our environment. It's because of this concern for environment, along with available technology that improves almost daily, that the matrix of future product demand could be dramatically altered for markets of the 1990's and beyond.

Alfred Sloan's observation many years ago that "there is no resting place for an enterprise in a competitive economy" has never been more true than today. And never has it been more important for a company to keep one foot in the marketplace and one foot in the technology that will be coming down the road in a few years. It takes time and money to hold yourself ready for changes that might never occur; for opportunities that may or may not happen. But the alternative can be deadly. Markets will be created in the blink of an eye, and others will disappear just as quickly. Those caught sleeping may well find themselves with nothing. Only a dozen years ago there were no personal computers, no compact discs, no video cassette recorders, and no genetically engineered vaccines. Some informed people are saying the next dozen years could bring 10 times that much progress.

For decades, chemical companies invested fortunes to make plastics more durable. They succeeded so well that plastic is clogging the world's dumps and killing wildlife. Now they've switched directions and are making biodegradable plastics that rot—egg cartons, disposable diapers, milk jugs, trash bags, and grocery bags, for instance. Concern for the environment, and the accompanying legislation, will ultimately make this a big business, according to many analysts. This, of course, is a promising market for producers of corn starch, a market that didn't exist in anyone's five-year plan just a few short years ago. The biodegradable bags will even cost more. Five years ago, would you have invested in garbage bags that decompose and that cost more, when all the commercials were touting their bags as the heftiest?

In another 10 years, assuming the greenhouse theory is legitimate, (and I believe it is) we may well be contending with atmospheric conditions that make ethanol a much more attractive market for corn. The extra few cents per gallon of fuel may look like money well-spent. I'm not saying there will be palm trees in New York and cotton in Toronto, but chances are there will be considerably more atmospheric pollution in New York, Toronto, and most other areas, and reducing our use of fossil fuels and substituting cleaner alternatives may become an imperative.

And what about soybean-based ink? Who would have thought a few short years ago that by 1988 one-third of all newspapers based in the U.S. would have tested the product. Experts think that by 1995 most American newspapers will be printing colors with soybean ink.

Other new technologies will bring better disease resistance without the environmental problems (ground water, etc.) caused by pesticides and herbicides. Technology will bring increased yields; newly-bred crops designed for specific soils and specific purposes; and, increasingly, proteins for more industrial uses; the as yet unknown "ethanols" of the decades ahead. Billions of dollars per year are being invested by business, by government agencies, and by universities, on biotechnology.

Industrial filters, medical cold packs, absorbants for use in everything from baby powder to diapers to agricultural uses are just a few markets for vegetable products in the 1990's.

For those marketing proteins, the implications translate into a much greater need for flexibility. Technology will bring an astonishing diversity to the marketplace and to its potential customers. New products will bring new market sectors, new niches, and tough problems for companies used to catering just to mass markets.

Marketing in an age of technology, and thus an age of diversity, means more options for goods producers and more choices for customers. It will be the marketer's job to make sure there is perceived differentiation for his or her product. He or she will have to deal with intense competition, without the cost benefits of mass production because markets probably will be narrower. He or she will have to know the customer's business and work with him to tailor the product specifically to the customer's needs. This will mean keeping decision-making very close to the customer. Not staying on top of technology and not knowing your competition will be unforgivable sins. Plan to call on your customers frequently. Chances are, you will find that your competition was there a day or two before you, or has planned a visit for next week. It will be crucial to anticipate evolving trends (like the growing use of canola, or rapeseed products in the United States). And it will be crucial to have a flexible operating plan to capitalize on those trends. The old military maxim, "It's a poor plan that can't be changed," will take on a new meaning.

Customer service, always a necessity even in the good old days, will take on an even more critical importance. The product is no longer just the feed itself, or the oil itself, or the soy protein itself. The product is the research support behind it. It is the service and the knowledge and the assurance that if specifications need to be altered slightly to accomplish a specific task or property, that your organization can do it. If you can't, there will be a dozen others who can.

With so much choice backed by so much service, I'm afraid that world vegetable protein customers will have the luxury to be fickle in this era of technology. They know that they can and should be involved with you in product design and development. This will be no place for rigid, bureaucratic, headquarters' marketing decisions. The rule of thumb is that the more you work with your customers during the early stages of product development, the more value you can add to the products they buy; value specifically designed for them, and therefore likely to keep them in the fold of satisfied users of your proteins. This will be true whether your customer manufactures animal feed, computer tapes, or produces swine or raises fish. Keep a close eye on who your customers are and will be, what they want and will want, and provide that to them better, quicker, and in a cost-efficient way, and you will be successful.

At my own company, we have found that certain marketing oriented research projects can best be handled in very small, extremely responsive and flexible organizations. To this end, we've teamed up in several joint ventures, combining the benefits of our own manufacturing and marketing organizations with their specific technology and research capabilities, and the results have been gratifying. I believe that you will see a continuation of larger corporations teaming up with smaller technology firms, for very rewarding results.

This will continue simply because it's a very effective way to give the customer what he wants and is willing to pay for. Our consumers are no longer a homogenous group, and as we develop marketing strategies for the years to come, we will need to think in terms of individual attention. Mass production will be a source of a

in 1986/87. Its share of the processed meal and oil markets has skyrocketed over the same period, soybean meal increasing from 2.3% to 15.4%, and soybean oil from 2.9% to 23%. Intuitively, one would expect the largest increases to be in the unprocessed product area where Argentina has a natural advantage, rather than in the processing sector where no advantages are apparent. The answer, again, is significant government intervention.

The Reembolso: Prior to 1982, Argentina utilized a tax rebate system for oil and meal exports known as the Reembolso to spur growth in its processing sector.

The Differential Export Tax System: Argentina's most dramatic increases in soybean product export have come since 1982, when Argentina replaced the Reembolso with a differential export tax system which applied significantly higher tax rates to raw soybeans than to soybean products. The differential tax system known as a "retention tax" in Argentina effectively blocks the export of raw soybeans and retains a surplus of beans in the domestic market at prices well below world market levels. The export taxes on soybeans have averaged 25% in recent years compared with 12% on oil and soybean meal. In the third quarter of 1988 the differential between export taxes applied to Argentine soybeans and Argentine soybean oil and meal was 11%. The system, in effect, forces Argentine farmers to accept lower-than-world-market prices for their oilseeds, making those low-priced inputs available to the Argentine crushing industry. A similar differential export tax system is also applied to the sunflower and linseed sectors. The Argentine differential export tax system is similar in structure to the practice earlier adopted by Brazil, although more effective in fostering rapid processing industry growth because the tax rate differentials historically have been much greater in Argentina than in Brazil.

The Argentine differential export system presents perhaps the most divisive single issue among the major oilseed producing and processing countries. This practice was one of several subsidy practices complained of in Section 301 petitions brought by the U.S. soybean and sunflower seed industries in 1983 and 1986.

The U.S. and Argentina held negotiations over the export tax issue and in May 1987, the U.S. government suspended the Section 301 investigation because Argentina agreed to reduce the differential between soybeans and soybean products from 11% to 8%, and promised to eventually reduce it to zero. The U.S. had threatened to retaliate against Argentine exports of corned beef and finished leather. In July 1988, however, the Argentines began giving an export tax rebate, called a "tributary return" of 3% to Argentine processors. Because of this rebate the total differential in the two export taxes is now back to 11%, the figure that originally prompted the U.S. soybean processors to file a section 301 case against Argentina in 1986. At meetings between U.S. and Argentine officials in September 1988, the U.S. officials stated that a failure by the Argentines to adhere to their promise of eliminating the differential between the tax on soybeans and soybean products would likely lead to a reopening of the pending section 301 case against Argentina. The Argentine differential export tax system was also the subject of a recent complaint to the European Commission brought by FEDIOL, the European oilseed crushers association. Recent articles and statements from Brazil have cited the Argentine differential export tax system as a major source of price depression in the world market for soybean meal and oil impacting the value of Brazilian exports.

Price Support System: The government of Argentina guarantees soybean and sunflower seed farmers a certain price. Oilseeds are covered by a reference price which is adjusted as export prices change. Reference prices are based on the average market price of the three previous days in various domestic markets. Generally farmers receive no less than 85% of the export price.

Malaysia

Over the past decade Malaysia has made major strides in the development of its palm oil industry. In fact, Malaysian palm oil is by far the leading edible oil in world trade. Substantial amounts of palm oil are being produced at prices that make palm oil an attractive substitute in many cases for other types of vegetable oil. However, despite its natural advantages in production of palm oil, Malaysia has persisted in assisting its processing industry by developing subsidy practices which significantly lower the costs of processing and refining. For example, the government supports the palm oil industry by creating economic incentives to invest in refining and fractionating facilities as well as by government funded research on applied uses. Malaysia also applies a 13% surcharge on soybean imports, and applies import duties of 30–40% on other processed goods.

Differential Export Duty System: Malaysia, like Argentina and Brazil, has attempted to promote a value added processing industry through the use of a differential export tax system. In the case of Malaysia, the export tax rates are applied inversely to various grades of palm oil, the tax rate decreasing with increased levels of refining. In other words, the highest export tax rate is applied to the unrefined or crude form of palm oil, and the tax rate is progressively lowered as the oil undergoes increased levels of refinement. As with the differential export tax system in Argentina and Brazil, the system in Malaysia serves to insure that raw products are retained in Malaysia and available to Malaysian processors at lower-than-world-market prices. Over 98% of Malaysian palm oil is processed locally. This permits Malaysian exports of processed palm oil to compete much more favorably against exports of alternative vegetable oils from other sources.

Japan

Japan is a major importer of oilseeds and oilseed products. Its imports from the U.S., for example, totaled $860 million in 1987. Imports of U.S. soybeans for crushing alone amounted to $784 million last year. Japan, however, maintains a monopolistic import regime that combines high tariffs and nontariff trade barriers designed to protect Japan's processing industries.

The Japanese government requires that formula feed contain specific amounts of domestic cornmeal and 2% fish meal for on-farm mixing intended for resale. These requirements limit the incorporation of alternative products in the mixture. The U.S., for example, has been able to export soybeans to Japan but not soymeal. If Japanese farmers were able to eliminate expensive fish meal from feed, exporters argue, they could replace it with imported soybean meal.

Japan also maintains restrictions on feed mills. Only licensed mills may import corn duty free, for example, and it is very difficult to obtain a license. These restrictions limit competition in the feed sector because farmers cannot buy corn and mix their own meal. Another trade barrier is the requirement that in order to build a new mill, the geographic area must be designated a "feed deficit" region. The elaborate Japanese system of import barriers protects domestic processing industries from competition from cheaper imported substitutes.

The United States

Agricultural and trade policies in the United States have tended, as in the case of Europe, to be focused primarily on protecting the traditional farm production base. Many of the major crop support and subsidy programs that assist a large segment of U.S. agricultural production have not been extended to oilseeds, which in the United States means principally soybean production. U.S. output of soybeans is falling for the third consecutive year; in September 1988 the U.S. Department of Agriculture forecast a 23% drop in the 1988 crop. The U.S. share of the world soybean trade, declining at a rate of $10 billion a year, is expected to decrease to less than 60% in 1989 from 86% in 1982. The number of U.S. processing plants, crushers and refiners, has fallen 28% in the past 10 years to 88.

U.S. import figures for soybean oil have been confusing. Originally the USDA calculated 1987–1988 soybean oil imports at 152 million pounds and forecast 1988–89 imports at 195 million pounds. But in August 1988 USDA adjusted these figures to 135 million pounds and 175 million pounds, respectively.

The Guaranteed Loan Program: The single exception for soybeans is the guaranteed crop loan program which, during the mid-1980's, had a negligible effect on the oilseed production or price because the loan guarantee price traditionally has been pegged at a level comfortably below the prevailing market. However, during recent years world oversupply of oilseeds has driven market prices below the prevailing loan rate for soybeans, with many farmers ceding their crops to the U.S. government for payment of nonrecourse loans. For example, the U.S. "purchased" 92 million bushels of soybeans from the 85/86 crop. In recent years, acquisitions have dropped due to higher world prices. As of July 1, 1988, the U.S. had 11.5 million bushels in storage and does not expect to acquire any of the commodity under the loan program this year.

The guaranteed loan program does not, of course, have the same type of effect on international trade as does, for example, the EC rapeseed policy or the Spanish consumption quota. For example, the U.S. program for soybeans provided a producer subsidy of up to 9% of the market price in recent years, while comparable EC subsidies amounted to 25% for soybeans and 49% for rapeseed.

In the short term, at least, this policy tends to remove oversupply from the market and to stabilize world prices at a substantial cost to U.S. taxpayers who pay for the program. Moreover, it is often argued that the loan program, when set above market-clearing levels, deters rationalization of U.S. farmland use and serves to make the U.S. soybean farmer uncompetitive with foreign producers. However, there are also longer range problems with the program that interfere with normal market operation and which U.S. policymakers will have to address. Primarily these are the price-depressing effects on world markets of having so large a quantity of a commodity in storage overshadowing the market. A large supply of commodities also creates inevitable pressure to create new concessionary programs which, if not strictly administered, can interfere with normal market behavior.

In the summer of 1988 the Brazilians, alarmed by Brazil's decreasing exports of soybean oil, challenged U.S. soybean oil subsidies as being violations of the GATT, and request consultation with the U.S. under GATT supervision. U.S. officials, while conceding that the subsidies are not "totally consistent" with trade guidelines under the GATT, countered that the Brazilians themselves subsidize their exports of soybean oil by imposing high export taxes on soybeans and lower duties on soybean oil and soybean meal, and point to the pending Section 301 complaint filed by the National Soybean Producers Association against Brazil with the U.S. Trade Representative's Office. U.S. soybean industry sources questioned the motive of Brazil in challenging the U.S. program since the drought makes it highly unlikely that the U.S. industry would be able to carry out a similar program next year. Brazil may be responding to the recent announcement by the White House that it intends to retaliate against Brazil's violations of patents.

The U.S. is scheduled to meet with Geneva in Brazil in October 1988 to take up the Brazilian complaint, which is also directed against subsidies for other types of oil, such as cottonseed and sunflower oil.

PL-480 and GSM Credit Programs: The U.S. practices affecting oilseed and oilseed product trade most severely criticized by other countries have been the use of the PL-480 "Food for Peace" program and the GSM credit programs.

PL-480 was developed in the 1950's as a method of providing food to emerging countries that simply could not participate in the ordinary cash market. Similarly, GSM credit was developed to assist countries that could not arrange reasonable credit sales, although they did not, for other reasons, qualify for PL-480. These programs have been utilized, *inter alia*, to promote U.S. soybean oil sales. Recent Senate proposals would expand the programs, to boost the ability of debt-ridden, least-developed countries to buy U.S. agricultural products. By subsidizing U.S. production earmarked for export to LDC's, the Government hopes to ease both the foreign debt crisis and the domestic farm crisis.

Historically these programs have been a combination of both food assistance and market development. Both U.S. government and industry spokespersons have long noted, in regard to the market development aspects of PL-480, that it has benefitted not only U.S. processors, but exporters from other countries as well. For example, the Indian market for imported vegetable oil was first developed by the United States under PL-480, and is now a major commercial market for Malaysian palm oil and soybean oil from Brazil and Argentina. Despite the food-aid and market development objectives of these programs and their focus on creating market additionality rather than interfering in competitive markets, these programs are often viewed by other exporters as creating their own set of distortions in the markets.

The major arguments against the PL-480 and GSM credit programs is that, in spite of their benevolent purposes, they do in fact achieve government assisted sales which "subsidize" the U.S. soybean farmer and processor to the extent that they provide income, and thus again provide an incentive in the general calculus toward world oversupply. Although these arguments are economically more attenuated than the arguments criticizing, for example, the EC rapeseed policy, the resulting volumes of export create something of a mirror image problem. During the 1987/88 crop year, the U.S. exported about almost a million metric tons of soybean oil under government-assisted programs in an effort to counteract the EC's rapeseed subsidies.

Another major criticism of PL-480 and GSM credits has been that they reduce foreign market prices to the point where it becomes uneconomic for foreign producers to manufacture even in their own countries. It is charged that this represses the development of indigenous agricultural production and undermines ultimate self-sufficiency in agriculture. While these arguments with regard to PL-480 and GSM credit are less direct than with regard to other subsidy practices that are discussed, they are nonetheless part of the perceived problem.

Tropical Oils Bill: Congressman Dan Glickman (D-Kansas) in April, 1987 introduced a bill in the U.S. House of Representatives to require special labeling on packages of food that contain palm oil, palm kernel, or coconut oil, the so-called tropical oils. Packagers would have to identify the ingredients as containing a "saturated fat," and disclose the amount of fat per serving. The alleged purpose of the bill is to inform consumers of their intake of saturated fat which has been linked to increased blood cholesterol levels and thus increases the risk of heart disease. The United States, however, imports almost 90% of its palm oil from Malaysia, and over 90% of coconut oil from the Philippines. Critics of the bill suggest that the labeling requirement would operate as a nontariff trade barrier to discourage demand for tropical oils. The use of the warning label might create an unnecessary deterrent to trade and eventually eliminate those imports.

Sen. Tom Harkin (D-IA) sponsored the Senate version, S. 1109, which is substantively the same as the House bill. On June 22, 1988, there was a hearing before the Energy and Commerce subcommittee on Health and the Environment. According to supporters of the legislation, two-thirds of the subcommittee and a majority of the full Committee favor passage. The Senate Committee on Labor and Human Resources has jurisdiction over the subject matter but has not yet held a hearing. The Reagan Administration, however, as evidenced by a letter from the U.S. Trade Representative, opposed the tropical oils labeling requirements. The Administration said that the bill would be discriminatory against specific oils and would thus operate as an unfair trade barrier.

The Drought Bill: This bill contains a provision that will benefit U.S. soybean producers by allowing them to plant soybeans on so-called program acres in 1989 previously reserved for commodities such as corn, rice and wheat. Under the $3.9 billion drought-relief package enacted in August 1988, U.S. farmers can without penalty shift at least 10% of their land from program crops into soybeans for one year, and perhaps as much as 25%, at the Agriculture Secretary's discretion. The added acreage available for soybean production could have an impact on the world market by moderating prices.

Import Barriers: Other oilseed trading nations, especially Brazil and Argentina, often point to the higher duty barriers in the United States and Europe preventing import of alternatively sourced vegetable oils, as a justification for their own subsidy practices. While the United States permits imports of unprocessed soybeans duty free, and imposes a fairly low duty on soybean meal, its 22.5% *ad valorem* duty on soybean oil is viewed as significant protectionism to exclude foreign competition from a major market.

Canada

The U.S. and Canada have agreed to a bilateral Free Trade Agreement (FTA) which, if approved, will facilitate American-Canadian business. The FTA creates trade benefits to both Canadian and U.S. agricultural producers. The two countries have agreed to eliminate all agricultural tariffs within 10 years. Canada will also eliminate import licenses for U.S. grain and grain products when the support programs of both countries are equal. Subsidies provided by the Canadian Transportation Act for products from Western Canadian ports to U.S. markets will be discontinued. Both countries have agreed to harmonize technical regulations that previously operated as nontariff trade barriers. Of course, ratification of the FTA does not mean that subsidy programs will be prohibited in the short term. As will be noted later regarding government restraints on dairy commodities, Canada has recently imposed additional import quotas on certain products despite its commitment to the Free Trade Agreement.

In the past, Canada has engaged in subsidy practices to assist its rapeseed industry. In 1982, the Province of Alberta provided a one-year processing subsidy of $40 per metric ton to its local crushers, whose estimated annual crush was about 600,000 metric tons. In addition, both the Alberta provincial government and the Canadian federal government have provided financial assistance to underwrite the cost of shipping Canadian rapeseed oil and meal by rail.

The Agricultural Stabilization Act provides for government payments to producers of certain commodities such as soybeans, when the price falls below 90% of the average price for the preceding five years. In 1986, the government announced a $1 billion assistance program to grain and oilseed farmers to cushion the impact of the subsidy war between the U.S. and the EC. The program involves direct cash payments to farmers. These subsidies have angered American producers of edible oils, whose products are being passed over in favor of cheaper imported oils produced from Canadian seeds. Imports of rapeseed and rapeseed oil from Canada in 1987 were valued at almost $29 million compared with $19 million in the preceding year. Similar trade disputes over subsidized Canadian lumber and American grain corn have made headlines and strained friendly relations between the two countries.

Government Interventions in Dairy Trade

European Community

The EC leads the world in dairy exports amounting to 40–50% of total world trade. Within the EC, prices are higher for dairy goods than prices on the world market.

The costs associated with dairy supports are the highest of EC government programs. For example, in 1986–1987 the EC spent the equivalent of U.S. $5.6 billion, and 1987–1988 costs are projected at U.S. $6.8 billion.

The EC maintains a target price for milk. For example, certain products are imported from Italy at predetermined prices. In 1987, the EC instituted a two-year butter disposal program which involves, among other provisions, enhancement of exports to the USSR.

The EC maintains high tariffs on imports to support its expensive agricultural subsidy program. The EC requires import licenses on milk and milk products from third countries. The combined restrictions effectively have eliminated any import market except where specific arrangements have been negotiated with individual countries.

United States

As a result of the Tokyo Round, the U.S. increased controls on imports under Chapter 22 of the Agricultural Adjustment Act. On June 16, 1988, President Reagan placed quantitative restrictions on most of the cheeses that previously had been allowed entry, quota-free. The annual quota for cheese subject to licensing will be 111,000 metric tons, and the unlicensed cheese quota will be 833 metric tons. Although the U.S. was allowed under a 1955 GATT decision to impose quotas on agricultural products, the EC has objected to U.S. restrictions on imports of dairy products. The EC complains that the intention of the 1955 agreement was to help the U.S. solve its surplus problem. Because the U.S. has since increased production and created price support programs, the EC claims that the U.S. trade policy is no longer fair nor consistent with the terms of the GATT.

Canada

The Canadian government operates several dairy support programs that limit the importation of dairy products. Cheese quotas and import licenses are the primary means by which trade doors are closed. These restrictions are largely unaffected by the Free Trade Agreement between Canada and the U.S.

Despite the FTA, Canada recently added certain products to the list of import-restricted goods: skim milk, buttermilk products, casein, whey, ice cream, and yogurt. Until this intervention, the goods would have been subject to the lifting of tariff barriers under the FTA. The dairy industries, however, claimed that they could not survive the flood of imports expected to cross Canadian borders in the post-FTA environment. The U.S. government has strongly protested the Canadian action as inconsistent with the spirit and intent of the FTA, and claims that the Canadians may have threatened Senate ratification.

Discussion

As can be seen, the world protein market is facing a tumultuous decade. With increasing government intervention among the major producing and consuming countries, almost everyone appears to be losing. Taxpayers in both the developed countries and the Third World are spending millions to subsidize otherwise profitable operations; Third World farmers are receiving a fraction of the world market price for their crops; processors and crushers are faced with increasing barriers to new and traditional markets; and in some countries consumers are paying higher prices than are warranted by free market conditions for products. The developing pattern is of complaint by one government about another's subsidy programs, coupled with the threat to retaliate with increased subsidizations or trade barriers of their own.

Sooner or later the pressures created by subsidization and protectionism will result in economic disaster. Otherwise profitable enterprises will flounder, and important capital investment will be lost because fair competition has been prevented. Government budgets will be strained by the increased costs of their intervention programs. International trust in the trading system will be badly damaged.

Rather than bemoan the inevitability of this scenario however, it is not too late for world oilseed producers to ponder the experience of other industries and carefully analyze their options. As noted previously, they can either attempt to: cartelize world production and prices which is as surely more unworkable and as subject to cheating as is petroleum; allow totally "free market" forces to predominate, which, of course, carries with it the hazards of destroying entire market sectors in both developing and developed countries; or it can agree to a set of trading rules which recognize the development needs of the Third World and the political imperatives of consuming countries and moderate the opportunities for retaliatory actions.

The opportunities provided by the last option seem obvious to most observers but require a spirit of compromise and negotiation which has been lacking to date. Nevertheless, the opportunity to consider these matters in the new round of GATT negotiations, and in the free-trade talks between the U.S. and Canada, seem too good to miss.

Both developing and developed countries have too much to lose not to seriously consider the kinds of trading rules that at least approximate those for manufactured products in the milieu of the Subsidies Code and the general rubric of the General Agreement. Should countries not address these questions, the result will be perpetuation of anarchy and trade wars which would be the result of either of the other options.

World Vegetable Protein Marketing Perspectives

David H. Swanson

Central Soya Co. Inc., Po Box 1400, Fort Wayne, Indiana 46801-1400, USA

Abstract

World supply and demand for vegetable proteins will continue to increase and will be driven by a number of demographic, economic, social, and technological factors in the coming decades. This equation for marketing opportunities in the food, feed, and industrial sectors includes such unknowns and variables as macroeconomic factors (exchange and interest rates, budget and trade deficits, monetary and fiscal policy, etc.); LDC development and debt issues: growth in traditional OECD markets; trade barriers and related protectionism issues; and biotechnology and other technological advances. Perhaps the most dynamic of these is the accelerated development of technology and the new market opportunities that may be created by environmental considerations. To position ourselves to take advantage of future opportunities will require heavy investments in research and development, innovative strategies to insure an economic return on these investments, a willingness and talent for managing in an environment of dramatic change, and an international cooperation that recognizes our world market and gives diminished weight to national boundaries.

I'm tempted, when looking at my topic.... "World Vegetable Protein Marketing Perspectives" to begin with some sort of disclaimer like Sam Goldwyn's old line.... "Never make predictions. Especially about the future." I won't. But, I will offer the opinion that five-year marketing plans, when marketing vegetable proteins in today's environment, can be dangerous things. Those of you charged with marketing vegetable proteins should wake up every morning just a little bit uncertain about your company's marketplace. It is invariably changing, and the forces that bring that change are the same forces that make it imperative that we build into our marketing strategies a large measure of operating flexibility. Take your strategic plans seriously, very seriously, but make sure they can accommodate the unpredictable.

There are some key assumptions upon which we can base our plans. For instance, we know that worldwide demand for food, feed, and industrial proteins will grow. The world's farmers must be prepared to feed 88 million new people each year (that's 242,000 new people each and every day). And I think it is clear that our markets will continue to become more globalized each year. That very slowly, but surely, the strengths of each country can be developed, and directed to the most efficient uses, producing mutual advantages for both producer companies and consumer countries. The laws of comparative advantages will move closer toward development of a healthier equilibrium. It's also quite evident that concerns for the environment will be central in our marketing strategies and will open up new opportunities for vegetable proteins. These long-term trends give us general direction, but a host of other factors in today's marketing equations must be dealt with. Each of these could negate the trends for the company that does not build the mechanisms for change into his marketing plans. Let's look at just a few:

Macroeconomics: The macroeconomics we contend with in today's business world, because it is truly a global environment, are fickle indeed. Exchange rate fluctuations, interest rate swings, trade deficits, budget deficits, and monetary policy are some of the forces that will challenge any marketing manager. While I believe that gradually the trading countries of the world will be able to move toward more internationally coordinated policies, don't expect this to happen next year or the year after. It will happen slowly, and sometimes painfully. In the meantime, the outlook for any given product is going to depend upon things like exchange rates, export programs, and economic growth in a region, among other things.

Trade barriers: Access to markets, or the lack thereof, will continue to determine to a large extent the opportunities to market vegetable proteins. Trade policies in any given country will affect the overall mix, volume, and value of the protein products that cross its borders. Let us all hope that the desire for self-sufficiency can be tempered enough to create opportunities. Trade liberalization is always a slow and arduous process, but the resulting opportunities can be in the best interest of producers and consumers alike, the world over.

LDC debt and development: This is another area in which international cooperation can alleviate hardship and improve growth prospects for the marketer of vegetable proteins. Debtor nations that adopt austerity programs, curtail imports, and export at all cost to earn foreign exchange to meet their foreign debt payments are just one more element in a scenario that dampens marketing opportunities for many. Poorer countries, while they have the potential for large growth, can't import much because of stagnant economies, subsistence incomes, large national debts and unfavorable trade balances. These countries will require substantial long-term assistance, including food aid and debt relief.

Biotechnology: If the theme of my remarks is "plan for drastic change," biotechnology must surely be the wild card that will, more than anything else, give us the means to accommodate that change. In a sense, it has double marketing implications in that technology not only will take us to markets that we've identified, but also create new ones that we never envisioned until a new technology is discovered, sometimes by fortunate accident. Biotechnology also overlays nicely with another dynamic that impacts our opportunities today, and that is concern for our environment. It's because of this concern for environment, along with available technology that improves almost daily, that the matrix of future product demand could be dramatically altered for markets of the 1990's and beyond.

Alfred Sloan's observation many years ago that "there is no resting place for an enterprise in a competitive economy" has never been more true than today. And never has it been more important for a company to keep one foot in the marketplace and one foot in the technology that will be coming down the road in a few years. It takes time and money to hold yourself ready for changes that might never occur; for opportunities that may or may not happen. But the alternative can be deadly. Markets will be created in the blink of an eye, and others will disappear just as quickly. Those caught sleeping may well find themselves with nothing. Only a dozen years ago there were no personal computers, no compact discs, no video cassette recorders, and no genetically engineered vaccines. Some informed people are saying the next dozen years could bring 10 times that much progress.

For decades, chemical companies invested fortunes to make plastics more durable. They succeeded so well that plastic is clogging the world's dumps and killing wildlife. Now they've switched directions and are making biodegradable plastics that rot—egg cartons, disposable diapers, milk jugs, trash bags, and grocery bags, for instance. Concern for the environment, and the accompanying legislation, will ultimately make this a big business, according to many analysts. This, of course, is a promising market for producers of corn starch, a market that didn't exist in anyone's five-year plan just a few short years ago. The biodegradable bags will even cost more. Five years ago, would you have invested in garbage bags that decompose and that cost more, when all the commercials were touting their bags as the heftiest?

In another 10 years, assuming the greenhouse theory is legitimate, (and I believe it is) we may well be contending with atmospheric conditions that make ethanol a much more attractive market for corn. The extra few cents per gallon of fuel may look like money well-spent. I'm not saying there will be palm trees in New York and cotton in Toronto, but chances are there will be considerably more atmospheric pollution in New York, Toronto, and most other areas, and reducing our use of fossil fuels and substituting cleaner alternatives may become an imperative.

And what about soybean-based ink? Who would have thought a few short years ago that by 1988 one-third of all newspapers based in the U.S. would have tested the product. Experts think that by 1995 most American newspapers will be printing colors with soybean ink.

Other new technologies will bring better disease resistance without the environmental problems (ground water, etc.) caused by pesticides and herbicides. Technology will bring increased yields; newly-bred crops designed for specific soils and specific purposes; and, increasingly, proteins for more industrial uses; the as yet unknown "ethanols" of the decades ahead. Billions of dollars per year are being invested by business, by government agencies, and by universities, on biotechnology.

Industrial filters, medical cold packs, absorbants for use in everything from baby powder to diapers to agricultural uses are just a few markets for vegetable products in the 1990's.

For those marketing proteins, the implications translate into a much greater need for flexibility. Technology will bring an astonishing diversity to the marketplace and to its potential customers. New products will bring new market sectors, new niches, and tough problems for companies used to catering just to mass markets.

Marketing in an age of technology, and thus an age of diversity, means more options for goods producers and more choices for customers. It will be the marketer's job to make sure there is perceived differentiation for his or her product. He or she will have to deal with intense competition, without the cost benefits of mass production because markets probably will be narrower. He or she will have to know the customer's business and work with him to tailor the product specifically to the customer's needs. This will mean keeping decision-making very close to the customer. Not staying on top of technology and not knowing your competition will be unforgivable sins. Plan to call on your customers frequently. Chances are, you will find that your competition was there a day or two before you, or has planned a visit for next week. It will be crucial to anticipate evolving trends (like the growing use of canola, or rapeseed products in the United States). And it will be crucial to have a flexible operating plan to capitalize on those trends. The old military maxim, "It's a poor plan that can't be changed," will take on a new meaning.

Customer service, always a necessity even in the good old days, will take on an even more critical importance. The product is no longer just the feed itself, or the oil itself, or the soy protein itself. The product is the research support behind it. It is the service and the knowledge and the assurance that if specifications need to be altered slightly to accomplish a specific task or property, that your organization can do it. If you can't, there will be a dozen others who can.

With so much choice backed by so much service, I'm afraid that world vegetable protein customers will have the luxury to be fickle in this era of technology. They know that they can and should be involved with you in product design and development. This will be no place for rigid, bureaucratic, headquarters' marketing decisions. The rule of thumb is that the more you work with your customers during the early stages of product development, the more value you can add to the products they buy; value specifically designed for them, and therefore likely to keep them in the fold of satisfied users of your proteins. This will be true whether your customer manufactures animal feed, computer tapes, or produces swine or raises fish. Keep a close eye on who your customers are and will be, what they want and will want, and provide that to them better, quicker, and in a cost-efficient way, and you will be successful.

At my own company, we have found that certain marketing oriented research projects can best be handled in very small, extremely responsive and flexible organizations. To this end, we've teamed up in several joint ventures, combining the benefits of our own manufacturing and marketing organizations with their specific technology and research capabilities, and the results have been gratifying. I believe that you will see a continuation of larger corporations teaming up with smaller technology firms, for very rewarding results.

This will continue simply because it's a very effective way to give the customer what he wants and is willing to pay for. Our consumers are no longer a homogenous group, and as we develop marketing strategies for the years to come, we will need to think in terms of individual attention. Mass production will be a source of a diminishing piece of the pie. Production of a greater selection of products, to smaller regional markets, or "micromarkets," will be a much larger piece of the pie. Many large feed markets will disappear as feed manufacturers become integrated into the larger food chain much as was seen in the poultry industry over the past decade. You may well see the same thing in the swine industry.

One thing we can be sure of as we move forward into the coming decade. Opportunities will be created for some firms and taken away from others. Making and identifying new markets won't be for the slow or the unimaginative. Those markets are often going to be fragmented, specialized, and complex. But they will be rewarding and profitable for those who have vision, those who have real courage, and for those who know already that today's five-year plan may have been a good exercise to go through, but it ain't necessarily so. Flexibility is perhaps our only defense and offense against unforeseen developments. The pace of change will quicken with every passing year, and victories in the marketplace will go to adaptable teams that are prepared to deal with that rapid and continuous change. For those who do, there will be great opportunities indeed.

World Vegetable Protein Quality Concerns

Takaaki Takebe

Oils and Fats Department, Mitsubishi Corporation, 5-2 Marunouchi 2-chome, Chiyoda-ku, Tokyo 100, Japan

Abstract

Stable supply and high quality for vegetable proteins, especially those extracted from grains, are becoming increasingly important for both human and animal consumption. While modernization of agriculture is progressing throughout the world, there exists an increasing danger of toxic chemical residues such as aflatoxins. There is no over-emphasizing the importance of safety of vegetable proteins to be consumed by humans and animals. To enhance global protein business, the time is ripe to incorporate additional quality specifications that are both practical and feasible. By doing so, all parties involved—producers, processors and consumers—could benefit. Such improvements could be developed by creating new grading specifications in the grade trade or by additional specifications to those now used for No. 1 grade quality products.

As some of the earlier papers have noted, consumption of vegetable protein has been increasing steadily for both feed and food applications. I believe that consumption will continue to grow both in the developed and developing region of the world.

With technological advancement, the ranges of application for feed and food purposes have expanded, and users consume vegetable protein depending upon its functions. It has hitherto been the case, and I believe it still is, that the gross amount of crude protein is the only criterion for determining value as feed. Recently, however, specific kinds of protein are being selected and the protein thus selected is used after further processing.

For food application as well, both the amount of protein content and the type of protein itself have become important. I have been informed that research work to control protein content in grains is being conducted at several laboratories. While ranges of protein applications expand, grains containing protein have greater potential of being used as food at any moment.

I would like to point out the following: Excessively protective agricultural policies are being implemented in the form of domestic price support and export subsidies, without referring you to the deliberation at the GATT Uruguay Round. Because of this, the major interest of producers is to attain the best yield, that is to say, to get the maximum weight out of unit acreage. This causes two concerns to those who are involved in the protein business, namely, the problems of constant deterioration of quality of grains and of agrochemical residue.

As the level of subsidy based only on weight is abnormally high, the producers of grains are going for volume production, paying little attention to the quality of their crop. The central concern of exporters is the amount of export subsidy.

Figure 1 shows the content of protein of soybeans imported into Japan on a year-to-year basis. As you can see, the level of protein has been declining constantly for the past 20 years. Inferior quality necessitates longer processing and pushes up cost. There was a case where double washing was necessary, instead of single washing, to reach a minimum quality level. There also was another case where as much as 6–8% of hulls of soybeans had to be removed, which of course resulted in higher processing costs.

The majority of the producers of vegetable proteins at the moment are crushers or millers with basic infrastructures. They select better quality from among various raw materials to avoid quality problems as much as they can manage.

However, as the protein market grows and the number of protein manufacturers increases, there will be more of those who have to use raw materials as delivered. To them, quality of grains as raw materials becomes a very important economic issue.

I now turn to the situation in Japan. Japan, I believe, consumes soybean protein as food far more than any other country in the world. We not only consume protein as ISP/CSP, or in isolate or concentrate form, but also as tofu, soybean paste, soy sauce, and natto. We constantly are in search of soybeans with high protein content in the United States, China and South America. However, much to our regret, the average quality for the past 20 years obviously has become inferior. The price differential between types of soybean we brought in and imported soybeans of usual grade is widening and we know of no sign that this trend is changing.

We can see in Figure 2 how the world agrochemical market has grown since 1970. Monetary values are used because it was rather difficult to grasp quantitative trends due to numerous varieties of agrochemicals consumed. I may draw one conclusion from this graph, however, the growth of volume of consumption of agrochemicals is outpacing that of production of grains.

Now kindly look at areas of consumption. As you can see in Figure 3, the level of consumption in developed regions is far more than that of the rest of the world. I might dare say that agricultural crops of advanced regions could be more dangerous. Of course adequate technologies have been applied and, as far as I know, no problems have occurred. In order to control the agrochemical residues, each country sets its own allowance to stop the inflow of contaminated grains.

Maximum allowances of, let's say, 0.1 ppm or 0.5 ppm are applied to more than ten different chemicals. However, there are no regulations for accumulated agrochemicals. As consumption of vegetable proteins increases, the use of agrochemicals also will grow in order for production to keep pace with increasing demand that naturally result in higher risks of contamination (Fig. 4).

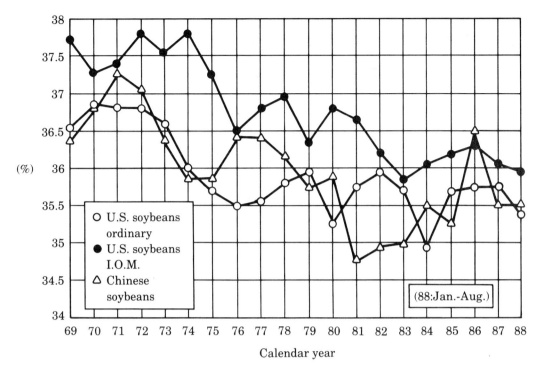

Fig. 1. Tendency of protein content.

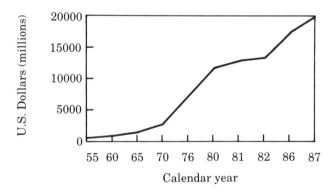

Fig. 2. World consumption of agrochemicals.

It is very difficult to judge how much of agrochemical residues can be removed through processing. It probably would be correct to say that grains would be considerably safer if the hulls are removed. But it is quite possible that the contamination level is higher with the hulls removed and how to dispose of them would become a problem.

As a supporter of the growth of the vegetable protein business, I have pointed out two major concerns as to the quality of grains. The current rules for the international grain trade specify only minimum quality requirements. As was pointed out earlier, there is a lack of incentive for grain producers to improve quality because of agricultural policies of various nations on the one hand, and there is increased usage of agrochemicals to attain effective yield on the other. This is a vicious cycle and we have to stop it somewhere.

I believe that the time has come for all of us involved in the vegetable protein business to give more positive evaluation to the quality of grains. I would like to propose, for instance, to add food grade (that can be consumed directly) or to incorporate protein level in the specifications for the No. 1 Grade, or maybe to introduce bonus/penalty clauses that would be based on protein content. It probably would be necessary to discuss the development of economical instruments for rapid analysis.

I believe that we all have to start making our utmost effort to cut the vicious cycle by introducing incentives into grain trade rules in order for producers to produce better quality raw materials.

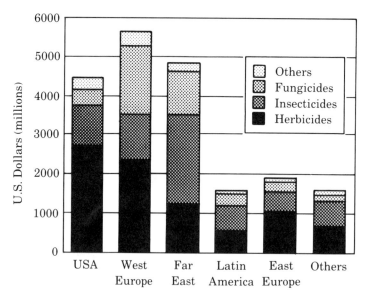

Fig. 3. Agrochemicals sales area/product sector split—1987.

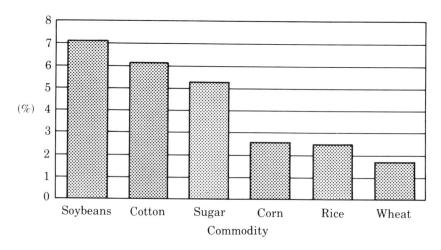

Fig. 4. Agrochemicals share in total income.

Handling, Shipping, and Storage of Oilseeds and Meals to Meet World Market Quality Requirements

Lowell D. Hill and Martin G. Patterson

Department of Agricultural Economics, 305 Mumford Hall, 1301 W. Gregory Dr., Urbana, IL 61801
USA

Abstract

The quantity and quality derived from grains and oilseeds depends upon maintaining the quality of the raw grain during storage and shipment. Losses of grain quality in the marketing channels result from lack of incentives for management practices. Breakage during handling could be reduced by low-temperature drying. Spoilage during transit is related to moisture variability, which is often increased by blending diverse moistures. The grade standards of most countries do not reward efforts to reduce moisture variability or breakage susceptibility. In addition, grades for corn and soybeans in all major producing countries contain no measures of intrinsic value; thus no reward for selecting varieties yielding more protein, oil, or other processed products. Changes in grades are needed to increase the quality and volume of grain and oilseed sources of protein and other feed and food products.

Vegetable proteins are a major component in the human diet. Nearly 240 billion pounds of protein are consumed per year with plant protein making up 64% of the protein consumed (1). The proportion of protein consumption from plant and animal sources differs among the geographical regions of the world and is related to the stage of economic development. The majority of protein for developing countries comes from plant sources—nearly 80% (Fig. 1). Centrally planned economies include more animal protein in their diets but still rely on plant sources for the majority of their protein needs. Animal sources provide the majority of protein for developed nations. A case for further concern is that in many developing economies, the level of total protein consumed each day falls below what many nutritionists consider the minimum daily requirement.

Grain provides 43% of the total protein needs in the world and although these sources do not always provide the proper balance of amino acids, they are nonetheless the primary source for much of the world population (2).

Quality Determines Value

Most vegetable protein is produced as a joint product and most plants provide other important feed and food products (Fig. 2). In some grains and oilseeds, protein is the primary product while carbohydrates and other ingredients are of secondary importance. In other cases, protein is a byproduct that just happens to be produced in the process of deriving other foods or feeds. In soybeans, the relative importance of protein depends upon the relative prices of meal and oil. The market for raw soybeans is sometimes dictated by the demand for protein and sometimes by the demand for oil (Fig. 3). In 1985, the value of the oil in a ton of soybeans was equal to the value of meal. In all cases, the production of protein through grains and oilseeds cannot be treated in isolation from the requirements for profitable processing of the whole grain.

The total value of grains and oilseeds to the processor is influenced by their composition and by the quality of the grain as delivered to the processing plant. An adequate supply of protein of the highest quality at the lowest price to consumers is a common goal in developed as well as developing nations. Quality deterioration of the grain or

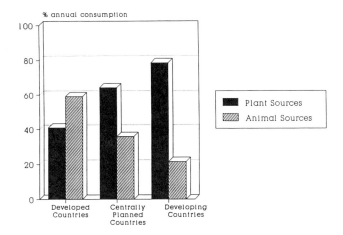

Fig. 1. Protein consumption from plant sources.

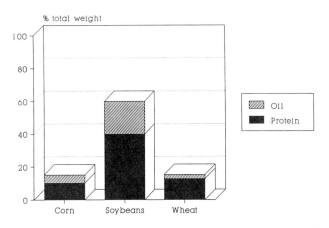

Fig. 2. Average protein and oil composition of corn, soybeans and wheat.

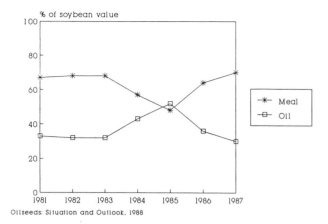

Fig. 3. Relative value of meal and oil from soybeans.

oilseed lowers the quantity and the quality of the products derived during processing and thus lowers their value in the market. Maintaining the quality of the raw grains and oilseeds during drying, storage and transporting in both domestic and international markets is a major concern.

The term quality is used in many ways by many people. For the purposes of this paper, the word quality will encompass three aspects of grains and oilseeds that relate to their ultimate value in the market place, in processing, and in consumption. Those three aspects are: (a) Quality as determined by cleanliness, purity and sanitation considerations. This means freedom from weed seeds, pesticides and deleterious products of any kind. (b) Quality as reflected in the nutritive (chemical) composition of the grains and oilseeds. This is sometimes referred to as intrinsic value or end-use value. It includes many factors such as the oil and protein content of corn and soybeans, the ability to withstand handling and transport without disintegrating and the milling properties when wheat and corn are transformed into food and feed products. (c) Quality in terms of physical characteristics of the grain such as kernel size and shape, kernel damage and kernel density.

Purposes of Grades and Standards

Too often the word quality is associated with numerical grade or grade factors. In fact, only a few of the quality attributes represented by the three categories listed above are explicitly incorporated in the grading standards of any country. Most grades include factors describing impurities and damage, but few contain information on intrinsic value.

Quality defined this broadly can include a long list of attributes. Selection of the appropriate factors for measuring quality is difficult and often arbitrary unless a set of objective and consistent criteria is developed. This requires agreement on the objective of grades and standards.

Most marketing textbooks and government agencies agree that the objective of grades and standards is to provide information needed by buyers and sellers to determine value. In the United States, the 1986 Grain Quality Improvement Act extends this objective into four specific purposes: (a) Define uniform and accepted descriptive terms to facilitate trade; (b) Provide information to aid in determining grain storability; (c) Offer end users the best possible information from which to determine end-product yield and quality; and (d) Create the tools for the market to establish quality improvement incentives.

Quality Attributes Important to Processors

With the emphasis on domestic and foreign users of grain and oilseeds, some method for identifying their preferences is required. Between 1986 and 1988 surveys were conducted in several countries to determine what factors are important in the daily operations of processors. The question was formulated as "What factors are sufficiently important that you would measure them on every delivery of corn or soybeans to your plant?" Corn and soybean processors in Japan and western Europe were surveyed by means of personal interviews and mail surveys. Additional information was received from soybean processors in all major processing regions of the world, through the assistance of country directors of the American Soybean Association. The Office of Technology Assessment, U.S. Congress, also assisted with surveys of domestic and foreign processors.

There were significant differences among industries but relatively similar factors were identified in all countries for any one industry.

Soybean Processors

The majority of soybean processing industry firms indicated that nongrade factors, such as moisture content, oil and protein, were measured at the plant on each delivery (Fig. 4). Over 90% of the respondents regularly tested samples for moisture; over 80% tested for oil content and foreign material content. Over 60% checked for protein content, damaged kernels and broken beans. In contrast, density, as reflected in test weight, was important to less than 30% of the respondents.

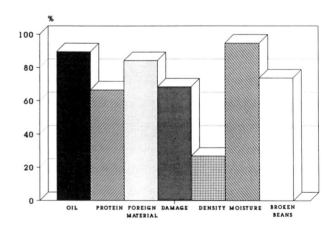

Fig. 4. Percentage of soybean processors measuring characteristics at delivery.

This information was supplemented by a second question asking domestic and foreign processors to assign relative importance to characteristics using values from one to seven, with seven being "most important." Among soybean processors, oil, protein, and heat damage were all ranked above six; test weight and splits below five (Fig. 5).

Handling, Shipping, and Storage of Oilseeds and Meals to Meet World Market Quality Requirements

Lowell D. Hill and Martin G. Patterson

Department of Agricultural Economics, 305 Mumford Hall, 1301 W. Gregory Dr., Urbana, IL 61801
USA

Abstract

The quantity and quality derived from grains and oilseeds depends upon maintaining the quality of the raw grain during storage and shipment. Losses of grain quality in the marketing channels result from lack of incentives for management practices. Breakage during handling could be reduced by low-temperature drying. Spoilage during transit is related to moisture variability, which is often increased by blending diverse moistures. The grade standards of most countries do not reward efforts to reduce moisture variability or breakage susceptibility. In addition, grades for corn and soybeans in all major producing countries contain no measures of intrinsic value; thus no reward for selecting varieties yielding more protein, oil, or other processed products. Changes in grades are needed to increase the quality and volume of grain and oilseed sources of protein and other feed and food products.

Vegetable proteins are a major component in the human diet. Nearly 240 billion pounds of protein are consumed per year with plant protein making up 64% of the protein consumed (1). The proportion of protein consumption from plant and animal sources differs among the geographical regions of the world and is related to the stage of economic development. The majority of protein for developing countries comes from plant sources—nearly 80% (Fig. 1). Centrally planned economies include more animal protein in their diets but still rely on plant sources for the majority of their protein needs. Animal sources provide the majority of protein for developed nations. A case for further concern is that in many developing economies, the level of total protein consumed each day falls below what many nutritionists consider the minimum daily requirement.

Grain provides 43% of the total protein needs in the world and although these sources do not always provide the proper balance of amino acids, they are nonetheless the primary source for much of the world population (2).

Quality Determines Value

Most vegetable protein is produced as a joint product and most plants provide other important feed and food products (Fig. 2). In some grains and oilseeds, protein is the primary product while carbohydrates and other ingredients are of secondary importance. In other cases, protein is a byproduct that just happens to be produced in the process of deriving other foods or feeds. In soybeans, the relative importance of protein depends upon the relative prices of meal and oil. The market for raw soybeans is sometimes dictated by the demand for protein and sometimes by the demand for oil (Fig. 3). In 1985, the value of the oil in a ton of soybeans was equal to the value of meal. In all cases, the production of protein through grains and oilseeds cannot be treated in isolation from the requirements for profitable processing of the whole grain.

The total value of grains and oilseeds to the processor is influenced by their composition and by the quality of the grain as delivered to the processing plant. An adequate supply of protein of the highest quality at the lowest price to consumers is a common goal in developed as well as developing nations. Quality deterioration of the grain or

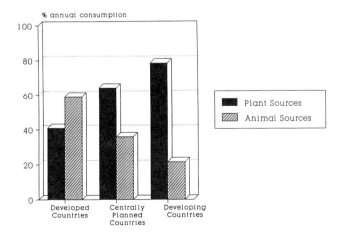

Fig. 1. Protein consumption from plant sources.

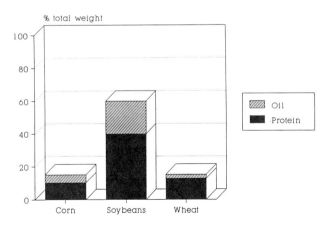

Fig. 2. Average protein and oil composition of corn, soybeans and wheat.

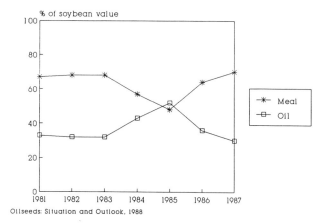

Fig. 3. Relative value of meal and oil from soybeans.

oilseed lowers the quantity and the quality of the products derived during processing and thus lowers their value in the market. Maintaining the quality of the raw grains and oilseeds during drying, storage and transporting in both domestic and international markets is a major concern.

The term quality is used in many ways by many people. For the purposes of this paper, the word quality will encompass three aspects of grains and oilseeds that relate to their ultimate value in the market place, in processing, and in consumption. Those three aspects are: (a) Quality as determined by cleanliness, purity and sanitation considerations. This means freedom from weed seeds, pesticides and deleterious products of any kind. (b) Quality as reflected in the nutritive (chemical) composition of the grains and oilseeds. This is sometimes referred to as intrinsic value or end-use value. It includes many factors such as the oil and protein content of corn and soybeans, the ability to withstand handling and transport without disintegrating and the milling properties when wheat and corn are transformed into food and feed products. (c) Quality in terms of physical characteristics of the grain such as kernel size and shape, kernel damage and kernel density.

Purposes of Grades and Standards

Too often the word quality is associated with numerical grade or grade factors. In fact, only a few of the quality attributes represented by the three categories listed above are explicitly incorporated in the grading standards of any country. Most grades include factors describing impurities and damage, but few contain information on intrinsic value.

Quality defined this broadly can include a long list of attributes. Selection of the appropriate factors for measuring quality is difficult and often arbitrary unless a set of objective and consistent criteria is developed. This requires agreement on the objective of grades and standards.

Most marketing textbooks and government agencies agree that the objective of grades and standards is to provide information needed by buyers and sellers to determine value. In the United States, the 1986 Grain Quality Improvement Act extends this objective into four specific purposes: (a) Define uniform and accepted descriptive terms to facilitate trade; (b) Provide information to aid in determining grain storability; (c) Offer end users the best possible information from which to determine end-product yield and quality; and (d) Create the tools for the market to establish quality improvement incentives.

Quality Attributes Important to Processors

With the emphasis on domestic and foreign users of grain and oilseeds, some method for identifying their preferences is required. Between 1986 and 1988 surveys were conducted in several countries to determine what factors are important in the daily operations of processors. The question was formulated as "What factors are sufficiently important that you would measure them on every delivery of corn or soybeans to your plant?" Corn and soybean processors in Japan and western Europe were surveyed by means of personal interviews and mail surveys. Additional information was received from soybean processors in all major processing regions of the world, through the assistance of country directors of the American Soybean Association. The Office of Technology Assessment, U.S. Congress, also assisted with surveys of domestic and foreign processors.

There were significant differences among industries but relatively similar factors were identified in all countries for any one industry.

Soybean Processors

The majority of soybean processing industry firms indicated that nongrade factors, such as moisture content, oil and protein, were measured at the plant on each delivery (Fig. 4). Over 90% of the respondents regularly tested samples for moisture; over 80% tested for oil content and foreign material content. Over 60% checked for protein content, damaged kernels and broken beans. In contrast, density, as reflected in test weight, was important to less than 30% of the respondents.

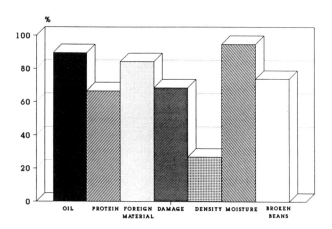

Fig. 4. Percentage of soybean processors measuring characteristics at delivery.

This information was supplemented by a second question asking domestic and foreign processors to assign relative importance to characteristics using values from one to seven, with seven being "most important." Among soybean processors, oil, protein, and heat damage were all ranked above six; test weight and splits below five (Fig. 5).

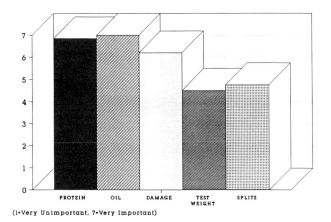

Fig. 5. Relative importance of characteristics to soybean processors.

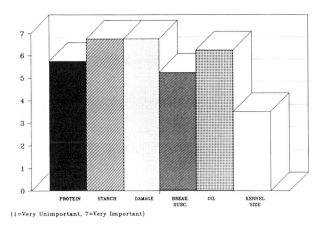

Fig. 7. Relative importance of characteristics to corn wet millers.

Corn Wet Milling

The corn wet milling industry relied more on traditional grade factors in their measurements (Fig. 6). Over 85% of the firms responding indicated that moisture, broken corn, damage and foreign material were measured on every delivery. About 50% of the respondents made regular measurements of protein. Oil, starch and fiber content were all checked by less than 40%.

eign material (80%) (Fig. 8). Protein, oil, fiber content, floaters and stress cracks were less important; only 40% reported making these measurements on every delivery. All respondents said they checked moisture and damage on every load. When asked to rate the characteristics in terms of relative importance, damage was given the highest rating (Fig. 9). Breakage susceptibility, stress

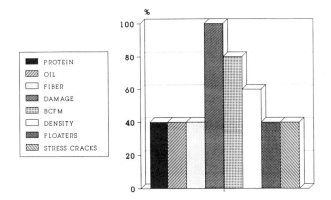

Fig. 8. Percentage of corn dry millers measuring characteristics at delivery.

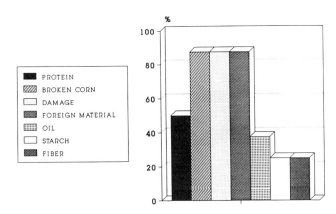

Fig. 6. Percentage of corn wet millers measuring characteristics at delivery.

When asked to evaluate characteristics on a scale from one to seven, domestic and foreign processors listed starch content and damage (mold) as the most important characteristics and kernel size as least important (Fig. 7). Protein for the wet millers was given an average value of 5.75.

Corn Dry Milling

The dry milling industry has a different set of requirements than most other processors. They are trying to construct flaking grits from the corn they process which requires a high proportion of hard endosperm and intact kernels. It is not surprising to find that most dry millers were measuring density (60%) and broken corn and for-

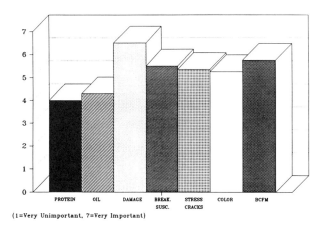

Fig. 9. Relative importance of characteristics to corn dry millers.

cracks, color and foreign material were all above 5. Oil and protein were of minor importance to the dry millers.

Feed Processing

The feed industry, not surprisingly, was the least concerned with testing properties other than the contract grade of the raw grain. None of the quality factors except moisture content (checked by 72% of the respondents) was regularly measured by more than 65% of the respondents. Broken corn and foreign material were listed regularly by *ca.* 60%; protein by 50% of the respondents (Fig. 10). No other chemical properties were mentioned by the respondents frequently enough to justify reporting.

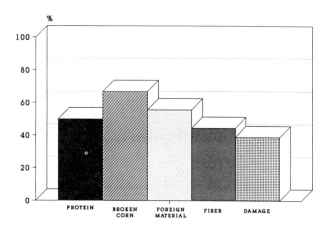

Fig. 10. Percentage of feed industry measuring characteristics at delivery.

When asked to rate the characteristics, domestic and foreign processors of feed put damage at the top, followed closely by protein and oil (Fig. 11).

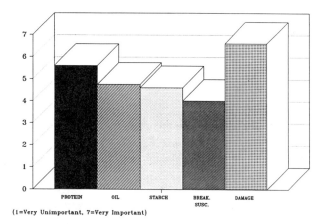

(1=Very Unimportant, 7=Very Important)

Fig. 11. Relative importance of characteristics to the feed industry.

Grade Factors Around the World

Because numerical grades and grading factors generally are associated with price discounts most processors reported measuring grading factors as the grain was received in their plants. It is significant to note that protein was one of the most frequently measured properties outside of those factors contained in numerical grades. From 40 to 67% of the respondents checked protein on every inbound load even though protein is not incorporated in the contracts or grading standards of any major exporting country.

In evaluating the performance of grading standards around the world, the overriding deficiency is the failure to incorporate measures of end use properties (sometimes referred to as intrinsic value) within the grading standards. The grade standards for corn from the United States, Argentina, Canada, and China, all include some obvious physical characteristics such as mold damage, foreign material (sometimes called impurities) and broken kernels (Fig. 12). But none includes a measure of protein, digestible nutrients, or other intrinsic properties related to the yield of processed products. The same is true of soybeans (Fig. 13) with the exception that several foreign buyers now include oil and protein contents as part of their export contracts. Taiwan now uses a contract for soybeans with discounts attached for oil and protein content below a base level. Despite the problems that foreign buyers have encountered with low protein and oil content in soybeans, there has been little progress in introducing measurements of chemical or nutritive properties in the official certificate information or in the contracts between

Country	Nutrient Composition	Test Weight	BC + FM	BC	FM
United States	—	X	X	—	—
Argentina	—	—	—	X	X
Canada	—	X	X	—	—
China	—	—	—	X	X

Fig. 12. Selected grade factors of corn, by country.

Country	Nutrient Composition	Test Weight	Damage	FM	Impurities
United States	—	X	X	X	—
Argentina	—	—	X	X	X
Canada	—	X	X	X	X
Brazil	—	—	X	X	X

Fig. 13. Selected grade factors of soybeans, by country.

buyers and sellers. This is a deficiency that must be corrected.

The lack of measurement of intrinsic value is not the only deficiency related to quality. All exporting countries include in their grades, some measure of cleanliness or absence of impurities or percent of foreign material. Yet the level of foreign material is one of the most frequent complaints received from foreign buyers in their receipts from nearly all exporting countries. This suggests that the current marketing system does not deliver grain to the foreign buyer in the condition that meets buyers' needs and preferences.

Grain Quality Problems

Problems of grain quality center on two basic problems: (a) excess foreign material (measured as broken corn and foreign material combined, in U.S. standards) and (b) development of mold and heated corn as a result of biological activity related to moisture and temperature. The source of these problems has been identified in recent studies in the United States. The results can be transferred directly to the identification and solution of similar problems in other countries, including Argentina and Brazil.

The quality of the grain at destination often looks much different than its appearance when it was loaded at the export point in the United States. The explanation for this quality change between origin and destination requires following the corn through the marketing system.

Quality of grain begins with the selection of the variety to be planted because genetics determines the intrinsic characteristics such as oil and protein content. It also determines many related characteristics such as hardness, kernel size, shape and color. At the time of harvest, corn and soybeans are at their maximum in terms of quality. Whatever is done to the grains following harvest, quality will never increase in the sense of intrinsic value. Grain can be cleaned, dried and sized but the intrinsic characteristics are in general on a downward trend as the marketing year progresses.

Breakage is primarily a response to handling following high temperature drying. High temperature drying at the farm or the elevator creates stress cracks internal to the kernel. Whether we are discussing U.S. grain dryers or Argentina grain dryers, dent corn or flint, the results are the same. From 85 to 100% of the kernels subjected to high temperature drying contain stress cracks; natural air drying produces 5 to 15%. This corn breaks during loading and handling. Studies have shown that broken corn and foreign material (BCFM) can increase by 2 to 7% between origin certificate and the processing plant in the destination country (3). Drying also creates brittle soybeans that break during handling. Broken corn and broken beans increase the probability of mold development and deterioration of oil and protein qualities during storage or delays in transit.

Handling throughout the market channel adds breakage, which in USDA corn grades is called BCFM. In soybean grades any broken bean that will fall through the grading sieve is called foreign material. Once past the farm truck, there is almost no opportunity for additional impurities to enter the grain. The increase in so-called foreign material between country elevators and the foreign buyer's plant is almost always broken kernels.

It is important to identify that in the United States, grain and soybeans are graded at every point where they change ownership. The farmer is discounted for deficiencies such as foreign material or mold damage. He is not penalized or rewarded for any of the quality characteristics not included in the grades. When the country elevator sells the grain to the exporter, the same type of discounts apply.

As the grain moves through the market channel, it is subjected to many impacts, during loading into barges or rail cars, and unloading at the port elevator by marine legs. The export elevators have sophisticated equipment to enable them to blend grain of different qualities to meet the contract specifications or grade requirements. All grain exported from the United States must receive an official grade issued by the United States Department of Agriculture as an objective and impartial reviewer.

Regardless of how carefully the grain is blended, inspected, graded and loaded, its quality at destination always will be different from that of origin. The primary losses in quality as determined by numerous studies are:

a. Increased breakage due to handling, associated with high speed handling devices, high drying temperatures and impacts during unloading.

b. Mold and heating as a result of high temperatures in the hold and a blend of moisture contents above safe storage levels even though average moisture content may be at safe levels (14–14.5% in warm weather). In most commercial shipments the maximum moisture of any individual kernel is generally two percentage points above the average. These high-moisture kernels do not equilibrate with the low-moisture kernels in the blend (4). Exposure to high temperatures invariably will cause breakage during handling. The exposed starch provides the food supply needed to initiate growth of bacteria and mold resulting in losses and deterioration before being received at the final processing plant.

c. Segregation within the vessel results in wide differences in sublots at destination. Normal loading and unloading procedures segregate particle sizes into pockets of fines or spout lines. Most buyers receive only portions of the total lot, and they can never expect the sublots to be exactly identical to the average for the vessel. Even when the average breakage and spoilage for the entire vessel is well within acceptable limits, individual barge loads or truck loads from that vessel may well exceed the limits most buyers consider reasonable (5). This process is inevitable with any handling method other than containerization and identity preservation. Research has demonstrated that this segregation takes place only on the factors related to particle size—broken kernels, splits and impurities. Pockets of high moisture within the hold of a vessel are seldom due to segregation but are due to bacterial action that has been generated by the presence of high moisture kernels in association with high temperatures (6).

These same problems exist for nearly all exporting countries; problems with breakage, splits, segregation and moisture content. The extent of this problem differs among countries. Brazil generally dries all soybeans to 13% moisture as they come from the farm. There is little opportunity for blending high- and low-moisture soybeans in Brazil. The same is true of corn in Argentina with 14.5% as a maximum moisture for legal transactions in the market channel. However, records from some countries of destination indicate a high proportion of damaged kernels in shipments from Argentina as well.

As most Argentine corn is artificially dried, stress cracks and breakage are as prevalent in their shipments as in those of the United States. Flint type corn is not immune to stress cracks and breakage and may be more susceptible because of its hardness. The presence of the smaller amounts of floury endosperm in the flint corn and the relatively short market channel from Argentine farms to the port does provide corn at destination with less dust and fine materials. However, actual percent broken corn through the 12/64-inch sieve is quite similar in shipments from either country (7).

Of more importance to potential buyers is whether there are significant differences in the intrinsic value of grains and oilseeds from the various countries of origin. No systematic scientific data have been collected over a significant period of time to answer that question definitively. However, there are indications that Brazil soybeans contain higher oil and lower protein and lower test weight relative to U.S. soybeans. They also contain significant quantities of red dust. They generally contain lower levels of foreign material than U.S. origin soybeans.

Corn from Argentina generally is drier with less dust and slightly higher protein content. However, genetic shifts from flint varieties to dent varieties in much of the Argentine corn belt is resulting in less difference in protein content on the average compared to the U.S. and a greater variability between shipments in the protein content.

Corn from China is superior in terms of low breakage but it tends to exhibit greater genetic diversity and presumably a wider range of chemical properties. There also have been numerous complaints of Chinese corn arriving with considerable quantities of burlap, as the result of deliveries going directly from burlap bags on rail cars or trucks onto the ocean vessel.

In summary, the competition among countries is based primarily on price and cost of delivery to the point of destination. All exporting countries should continue to search for improved measures of quality that will better reflect the true value of the grain to the user. This requires rapid and accurate measures of chemical composition—currently not present in the grades of any of the major exporting countries. New measures of quality must be accompanied by inspection agencies and a grading system that will assure both buyer and seller of accurate, unbiased measurement and reporting of results. Inspection agencies must be independent of the merchandising firms or agencies.

At the present time about the only option is for buyers to specify within their contracts more of the economically important quality characteristics. This is a more expensive method of identifying value because the information is not readily available and meeting the contract specifications may involve identity preservation even as far back as selecting the variety to be planted. For factors outside the grade standards there is no outside, objective agency at the national level in most countries, ready to provide tests with quality information. The University of Illinois and the Illinois Crop Improvement Association, are now operating an identity-preserved grain laboratory which provides quality assurances of sublots of grain at a low fee to the buyer.

The alternative to contract specification and identity preservation is to change the grading standards of the various countries around the world such that the important characteristics will be inserted in place of characteristics now included but which have relatively little importance. Research has demonstrated that test weight and broken beans are of minor importance in evaluating soybean quality or the quantity and value of products derived from them, i.e., meal and oil. Chemical composition, remaining storage life and drying temperatures are all more important measures of quality than test weight and broken beans. Moisture variability may also be of greater importance than average moisture of the entire lot. All of these measures are within the limits of current technology. What is required is the economic incentives in the form of pressure from buyers and a willing response from sellers to incorporate important quality characteristics within current grades throughout the world as a substitute for individual contracts and the expense associated with it. We can improve the efficiency of the market system in communicating information about value, from buyer to seller to producer to genetic developer. It requires relatively minor changes in the marketing channel but it requires major changes in attitudes in the grain industry and dedication by many people committed to the improvement of the efficiency of the market.

Acknowledgements

This work was supported in part by research grants from the American Soybean Association and the Illinois Corn Marketing Board. Assistance was also provided by the Illinois Crop Improvement Association. The research is a part of Hatch project 371 at the University of Illinois Agricultural Experiment Station.

References

1. Fauconneau, G., "World Protein Supplies: The Role of Plant Protein," Qualitas Plantarum: Plant Foods for Human Nutrition, 1983, Vol. 32, p. 206. (Calculated from Table 1).
2. Ibid, p. 209 (Calculated from Table 3).
3. Hill, L.D., M.R. Paulsen, G.C. Shove, and T.J. Kuhn. "Changes in Quality of Corn Between U.S. and Japan, 1985," AE-4609, Department of Agricultural Economics, Agricultural Experiment Stations, College of Agriculture, University of Illinois at Urbana-Champaign, December 1985.
4. Hill, L.D., M.R. Paulsen, T.J. Kuhn, B.J. Jacobsen, and R.A. Weinzierl. "Corn Quality Changes During Export from the United States to Japan," AE-4636, Department of Agricultural Economics, Agricultural Experiment Station, College of Agriculture, University of Illinois at Urbana-Champaign, January 1988, p. 39.
5. Hill, L.D., M.R. Paulsen, and M. Early. "Corn Quality: Changes During Export," Special Publication 58, Agricultural Experiment Station, College of Agriculture, University of Illinois at Urbana-Champaign, September 1979.
6. Hill, L.D., M.R. Paulsen, G.C. Shove, and T.J. Kuhn. "Changes in Quality of Corn Between U.S. and Japan, 1985," AE-4609, Department of Agricultural Economics, Agricultural Experiment Station, University of Illinois at Urbana-Champaign, December 1985.
7. Hill, L.D., and M.R. Paulsen. "Maize Production and Marketing in Argentina," Bulletin 785, Agricultural Experiment Stations, College of Agriculture, University of Illinois at Urbana-Champaign, July 1987, p. 25.

Preparation of Soybeans Prior to Solvent Extraction

Heinz Schumacher*

Hoeperfeld 26, 1050 Hamburg 80, Federal Republic of Germany

Abstract

The basic system is described and also several systems of hull separation that have been developed in the last 10 years. The special treatment in the preparation in respect to production of degummed crude soybean oil with a low content of phosphorus is considered. The aforementioned processes have been published before, and for that reason they are not described in detail. Late developments not published before are described more in detail. These later developments permit a very flexible production and in addition electric power consumption, and steam consumption can be reduced per ton of seed processed. The reduction in steam consumption is achieved by recycling heat which exists as steam in air from drying operations.

There are many operating plants working according to the process shown in Figure 1; some even with insufficient equipment. Seed cleaning, for instance, takes out only coarse foreign matter bigger than 0.5″ in diameter, and conditioning means warming the seed only to 30 or 35°C. Working in this manner means high wear on equipment as well as higher power consumption, especially for the flakers. Depending on the characteristics desired for the finished products to be made—meal and oil—the preparation equipment changes a little bit.

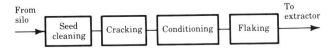

Fig. 1. Basic flow diagram for seed preparation.

The seed should be cleaned of foreign material such as tramp iron, stones, gravels, dust and sticks left from harvesting. When high-protein content meal is to be produced, the hulls must be separated from the meats, and a thorough seed cleaning is a must! Also, when expanders are used prior to extraction a good cleaning of the seed should be done.

There is no doubt that good cleaning of seed improves qualities of meal, oil and lecithin. Furthermore, plant performance improves with less down time.

Seed Cracking

After cleaning, the seed is cracked. This is done on corrugated rolls. The corrugation changed some years ago

*Deceased. Dr. Schumacher's presentation was read by George Anderson of Crown Iron Works Co.

to save power and also to get a more uniform granulation with fewer fines in the cracked seed.

On the left side of Fig. 2 the outdated corrugation is shown, and on the right side the modern one. The rolls

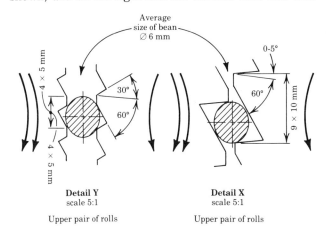

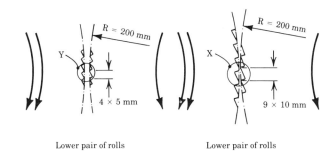

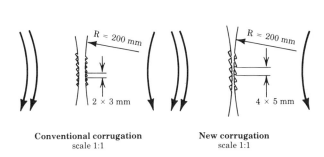

Fig. 2. Comparison of outdated corrugation (left) with modern cut (right). **Fig. 2a** Enlargement of Fig. 2; Detail Y (left) shows conventional corrugation; Detail X (right) shows modern cut. **Fig. 2b.** Enlargement of Fig. 2 for lower pair of rolls; conventional corrugation at left; modern cut at right.

with one arrow have a lower circumferential speed, and the rolls with two arrows have a higher circumferential speed.

In Detail "Y" (Fig. 2a), which is an enlargement 5:1 of conventional corrugation, it can be seen clearly that the seed is not cut into pieces but squeezed into pieces. "Squeezing" means higher power consumption especially when processing seed with a high moisture content; and, on the other hand, many fines are found in cracked seed when seed with low moisture content is processed. The arrows signify that the roll with one arrow has less circumferential speed than the roll with two arrows. By widening the pitch between teeth in Detail "X", a cleaner cut or shearing is achieved that results in less fines when processing seed with lower moisture content and much less power consumption when processing seed with higher moisture content. The modern corrugation has about twice the pitch between teeth, so that the beans can "enter."

In Figure 2b, the enlargement of the pitch from 2.5 mm for the conventional corrugation to 4.5 mm for the improved corrugation can be seen. The top pair of rolls cuts the beans into half pieces, and the second pair of rolls cuts the half beans into quarter beans. It would be ideal to have a third pair of rolls with a corrugation pitch of half the second pair of rolls in order to produce "1/8-beans." The advantage of this corrugation is to produce fewer fines in the cracked beans and also reduce power consumption, especially when processing beans with higher moisture content. With the same power consumption the capacity of the cracking rolls can be increased by about 25% to 35%. In North America the modern corrugation is called "The European Cut" and was introduced by the author and Bauermeister as the manufacturer of cracking machines in many oil mills in Europe more than 10 years ago.

Separation of Hulls

If the plant is to produce an extracted meal with a high protein content, it is necessary to separate the hulls from the meats. This is done after the seed has been cracked into pieces.

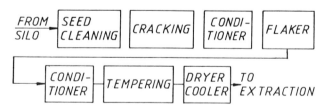

Fig. 3. Flowsheet for conventional front-end separation of hulls and meals (1).

Figure 3 shows the basic scheme of some 25 years ago that is still applied in some factories. First the soybeans are dried and cooled to a moisture content of about 10% and then the beans are kept for two to three days for tempering and maturing. The idea is that the meats will shrink, and after cracking an easier separation of the hulls from meats will be achieved.

Later on Escher-Wyss omitted the bothersome drying-cooling and storage stage by using hot air for the sep-

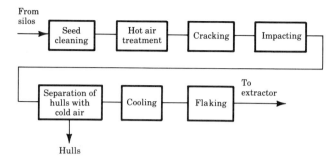

Fig. 4. Flowsheet for hot air dehulling (Escher-Wyss) (2).

aration of hulls from meats. In the latest development, dehulling is combined with preheating the soya prior to flaking (Fig. 4).

Dehulling equipment is also delivered by Bühler of Switzerland. After very thorough cleaning the whole beans enter a fluidization-bed and are treated with very hot air in order to reach a "popping"-treatment (Fig. 5). Afterward the beans are passed into a machine called an "impact dehuller." This machine breaks the beans up into sizes of half beans, and the hulls become loose from the meats. In a second fluidization-bed the hulls are separated from the meats. The plant described produces mainly toasted, fullfat soya. However, the process also can be used as a preparation step before extraction. The dehulling also can be combined with conditioning the seed for flaking.

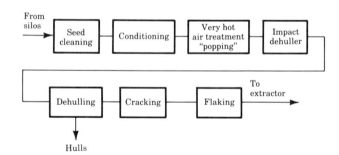

Fig. 5. Flowsheet for "popping"-dehulling process (3).

In each of these optional dehulling processes it is necessary to be sure the beans are heated before flaking. Equipment for heating the cracked beans (if not done during dehulling) is often done by traditional "stack cookers" and such an equipment is still delivered by some machinery suppliers today. A stack cooker is a vertical cylinder with built-in horizontal plates which are heated indirectly by steam. A vertical axis in the center moves and transports the cracked beans, which fall by gravity from deck to deck. Other equipment for heating can consist of horizontal rotating tube-bundles. This type of machinery is costly to repair as the tube bundle often is worn out in less than 10 years.

Preheating has been done also in several plants by countercurrent hot air (4). This type of equipment has been altered recently and shall be described more closely later on.

A good preheating before flaking reduces the power

consumption of the flaker. Flaking cold beans (25° or 30°C) requires a power of about 5 to 6 KWh per ton of seed, whereas flaking at 55° to 60°C requires only 3 to 4 KWh per ton of seed. Furthermore, power consumption of the flakers depends on the average size of beans.

About 8 years ago Lurgi announced the Alcon-process. Since then, five plants are in operation ranging from 250 to 1,000 tons of soya per day. The aim of this process is to produce a degummed crude oil with a low content of phosphorus in order to make physical refining possible, as physical refining is by far more economical than chemical refining with caustic soda and the resulting soapstock-splitting. The Alcon-process increases the yield of lecithin by about 100%. Further advantages are better percolation in the extractor and the resulting possibility of increased capacity. As can be seen from Figure 6, the equipment for the Alcon-process is installed between the flakers and the extractor.

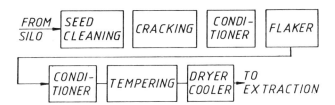

Fig. 6. Flowsheet for Alcon-process (5,6).

Latest Developments

The latest novelty for soya preparation is a new way of saving steam in the conditioner of the cracked seed, in case there is no hot air dehulling system. Stack cookers and rotating tube bundle bean heaters consume steam, of course, and besides that this equipment has a relatively high investment cost per ton of seed. The construction has to be done under code and supervision by the authorities. Transmission of heat is relatively low from steel to beans. From experience in operations with a DTDC, a new economical way was found to save steam by using steam vapors in spent air from the meal drier of a DTDC or from a drier-cooler installed between an expander and extractor. As heat transfer is very quick from steam in the air to the seed, the conditioner can be relatively small. If a preroasting of cracked beans before the extraction is desired, then the volume should be a little bigger. It is possible to construct a bean heater without coding as it can be built without any indirect heating facility. Indirect heating is done only on the air.

Figure 7 shows the heat recovery system. The lowest tray is a cooler. The spent air from the cooler has a temperature of 40° to 45°C and leaves to the atmosphere because it contains very little steam, and heat recovery from this spent air is not economical. From the drier, however, spent air contains about 60 to 70 kg of steam per ton of seed processed, and this steam can be utilized in the top compartment, or even two top compartments in the bean heater, to preheat the cold beans to about 70°C. By turning on the indirect steam on air heater 100°C or a little bit less can be achieved in the preheating section for the beans. In case a real pretoasting is wanted, live steam is injected in the toasting section. Depending on how much toasting is wanted, the toasting section should have a retention time for the beans of somewhere between 10 and 35 minutes.

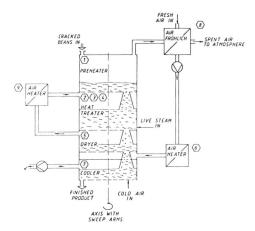

Fig. 7. Toasting system for cracked soybeans.

There is a plant in operation using the above mentioned system, however, without the attached drier and cooler. In other words, the cracked and heated seed enters the flaker at a temperature of 100°C, and the flaker is operating at a low power consumption, about 3 to 4 KWh per ton of processed seed. Naturally a strong aspiration for the flaker is necessary.

From the flaker the flakes enter an expander for increasing the bulk-weight. Afterward the material is dried and cooled. In such a case, the spent drying air containing the steam of evaporated water from the soya can be carried to the conditioner for preheating the cracked beans. Of course, a flow of material as shown in Fig. 8 can be used prior to extraction as well. The cooler could then be omitted.

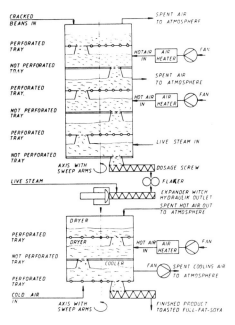

Fig. 8. Toasted fullfat soya plant.

It may also be possible, with appropriate safety measures, to use the moisture-laden air from the drier section of a DTDC for preheating the beans in the bean conditioner. It should then be possible that toasting prior to flaking can be done in some cases with very little additional steam consumption, perhaps 10 to 20 kg per ton of seed processed. The degree of toasting can be regulated by a temperature range of 80° to 100°C and also by the retention time in the toasting section, e.g. regulating the height of meal layer on the toasting trays. At 90°C in the "toasting section" most of the enzymes are inactivated, with the exception of urease and antitrypsin. A heat treatment in a conditioner prior to flaking helps also to improve the quality of degummed extracted oil. If the temperature is kept a little bit lower than 100°C the lecithin recovery will not be raised much in comparison to standard preheating of seed prior to flaking, but it does help to lower the phosphorus content in degummed oil, so that trading rules for a phosphorus content of 200 ppm in degummed oil can be met easily.

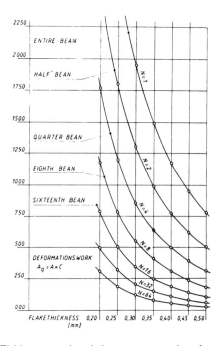

Fig. 9. Flaking power in relation to average size of cracked soybeans.

Flaking is displacement of soymeats from a spheric shape to a flake thickness of various thicknesses. Common practice is to flake the seed to a flake thickness of 0.25 mm minimum and 0.30 mm maximum. The curves in the diagram correspond to the cracking of the seed. N = 1 means entire soybean uncracked, N = 2 means soybean cracked into 2 pieces, N = 4 into 4 pieces, N = 8 into 8 pieces and so on. Soybean broken into 4 or 8 pieces has a very irregular shape, and therefore the volume is recalculated to a spheric shape again. The "center slice" of each spheric piece is considered stationary, whereas the rest of the contents of the spheric piece has to be displaced in order to get a flake of the various sizes mentioned from 0.20 to 0.50 mm.

It is not possible to flake uncracked soybeans, however, rolls of 800 mm diameter of a flaker can flake half beans (a half bean = 1 soybean cracked into 2 pieces). From the diagram it can be seen clearly that beans cracked into 4 pieces consume two and a half times as much power than beans cracked into 16 pieces for flake thickness of 0.30. When processing very dry seed with 8% or 9% water content, especially with conventional corrugation, the seed is easily "overcracked," producing a high percentage of very fine pieces of soya. With these very small pieces of soya there is very little work necessary to displace material in order to make a flake. But with very little displacement of soy-material while making a flake there is not enough rupture of cells to facilitate extraction. The ideal solution for flaking will remain an average size of cracking between 1/8th and 1/16th of beans at a high temperature. Flaking half or quarter beans at temperatures of 30° to 40°C requires power consumption of some 9 to 11 KWh per ton of seed processed, whereas bean sizes of 1/8th to 1/16th at high temperatures near 100°C requires power consumption of 3 to 4 KWh per ton of processed seed.

Modern preparation for soya seed is much different from a preparation many years ago, and the latest development gives a great flexibility in the preparation in order to produce products of desired qualities.

References

1. Moore, N.H., *J. Am. Oil. Chem. Soc. 60*:82 (1983).
2. Florin, G., and H.R. Bartesch, *J. Am. Oil. Chem. Soc. 60*:193 (1983).
3. Fetzer, W., *J. Am. Oil. Chem. Soc. 60*:203 (1983).
4. Schumacher, H., *Fette-Seifen-Anstrichmittel*, Vol. 78, No. 2 (1976).
5. Kock, M., *J. Am. Oil. Chem. Soc. 60*:196 (1983).
6. Penk, G., in *Proceedings of the World Conference on Emerging Technologies in the Fats and Oils Industry*, edited by A.R. Baldwin, 1986, p. 38.

Extrusion-Expansion of Oilseeds for Enhancement of Extraction, Energy Reduction and Improved Oil Quality

L.R. Watkins, W.H. Johnson and S.C. Doty

Food Protein Research and Development Center, Texas A&M University, College Station, TX 77843-2476

Abstract

Preparation of oilseeds prior to extraction by expansion enhances efficiency by releasing oil during the cooking and produces a porous collet with reduced solvent retention. Flake thickness and extraction temperature requirements are improved while bulk density increases of 50% are common. Extraction plant capacity has been increased by 50% with little capital expenditure. Solvent required for extraction is reduced by less solvent holdup and the increased efficiency in wash dilutions. Processes for many oilseeds have been developed utilizing the expander in different modes: extrusion following prepressing; extrusion after flaking/grinding; extrusion preceeding prepressing; and extrusion in lieu of prepressing for high oil content seeds.

Preparation of oil seeds prior to extraction always has been important for good extraction results. Expanders accomplish so much in the preparation stage that better extraction is accomplished while reducing the requirements for cracking and flaking. We start with the microscopic explanation of what is accomplished in the expander. With magnification of only 50 and fat stain of Sudan IV we find cottonseed flakes showing tiny red spheres of oil. (Fig 1). After extrusion or expansion, the oil is found in pools (Fig. 2). At the same time that the oil is being freed, the material is compacted, and then expanded into a dense but porous material.

Conditions of expansion show that some collets extract well. Other collets that lack porosity do not extract well. Referring to Table 1 (1), we find that Farnsworth et al.

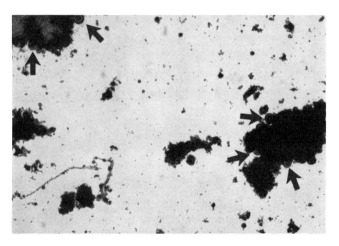

Fig. 1. Cottonseed flaked: fat stain-Sudan IV (Red) magnification 50×.

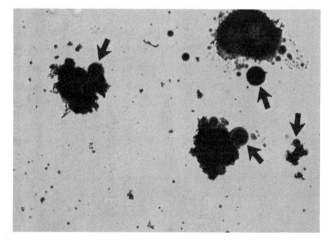

Fig. 2. Cottonseed extruded: fat stain-Sudan IV (Red) magnification 50×.

tested many variables of expansion on cottonseed and found that almost all conditions resulted in lower residual oil than flaked cottonseed (Table 2), and Table 3 shows the results on soybeans. Table 4 illustrates one of the reasons why extraction efficiency is improved so much.

With these results at hand, we tested flake thickness requirements as a preparatory step to expansion, and found that flake thickness, a critical consideration in extraction of raw soybean flakes, was no longer as important (Fig. 3). This reduction in flaking requirements leads us to think about increased capacity with little capital expenditure, or maintaining the same capacity with reduced energy and maintenance costs. In the series of figures to follow, we will attempt to show some of the effects in solvent plant design.

If we allow the flake thickness to be increased from 0.012'' to 0.020'' we can reduce the tempering temperature while allowing a coarser crack on the cracking rolls. All of these changes can usually allow a 50% increase in capacity of the prep-area.

The expander, when initially utilized for soybeans in Brazil, was an electrical energy hog, until control of load by injection of live steam was utilized. Motor load dropped dramatically so expander speeds were increased and capacity jumped from 60 TPD to 300 TPD. Our initiation into the expander world occurred in Brazil where we saw a plant that they said was rated at 600 TPD, but was then extracting at the rate of 1200 TPD. They said: same crackers; same flakers; added expander; added dryer, cooler for collets; same extractor; same D.T. and the meal dryer was no longer required. Needless to say we had a hard time being convinced. With time and our own experiences we became believers.

TABLE 1

Extrusion Conditions for Preparing Cottonseed Meats

Condition	Treatment						
	A	B	C	D	E	F	G
Die exit temperature (°F)	190	190	208	222	232	235	220
Screw speed (rpm)	150	300	300	—	400	300	400
Flake feed rate (lb./hr)	560	375	460	460	560	400	340
Moisture content of feed (%)	12.0	12.0	7.5	7.5	7.5	7.5	24.0
Steam on jacket (turns)							
Free end	0.0	0.0	0.5	0	0	0	0
Die end	0.0	0.0	1.5	0	0.75	0	1.0
Number of dies	3	3	3	3	3	3	1

TABLE 2

Residual Oil Contents of Extracted Cottonseed Meals

	Treatment							
	Flakes	A	B	C	D	E	F	G
Crude free fat (%)								
Initial	29.8	30.7	23.9	28.5	22.8	32.8	28.5	15.0
After 6 stages	2.77	1.44	1.08	0.82	0.70	0.80	0.77	3.16
After 8 stages	1.09	0.71	0.77	0.20	0.64	0.60	0.59	3.10
Acid hydrolyzed fat (%)								
Initial	28.9	28.2	22.6	27.4	22.5	32.0	24.3	17.1
After 6 stages	2.82	2.01	3.53	2.06	2.03	1.59	2.46	4.2
After 8 stages	2.99	1.96	2.84	1.74	2.05	1.55	2.48	3.7

TABLE 3

Residual Oil Contents of Extracted Soybean Meals

	Flakes	Pellets
Crude free fat (%)		
Initial	21.6	21.9
After 6 stages	1.83	0.37
After 8 stages	1.51	0.10
Acid hydrolyzed fat (%)		
Initial	19.8	20.8
After 6 stages	3.06	1.69
After 8 stages	2.78	1.70

TABLE 4

Bulk Densities and Solvent Hold-up of Flakes and Extruded Pellets

Treatment	Bulk density (lb/ft^3)	Solvent hold-up (%)
Cottonseed		
Flakes	26.0	39.5
Pellets A	40.4	27.5
Pellets B	33.6	20.7
Pellets C	44.0	29.1
Pellets D	36.0	23.2
Pellets E	40.7	27.9
Pellets F	N/A	18.1
Pellets G	36.7	18.2
Soybeans		
Flakes	23.5	38.9
Pellets	34.3	20.9

Energy requirements for the expander are approximately 75 HP for 8″ expander operating at 330 RPM and producing 300 TPD. The expander is driven by a 75-100 HP motor. Steam injection into the expander is normally 60-80 lbs/ton of soybeans or 750-1000 lbs/hour, at the 300 TPD rate. The newer 10″ expander has capacity in excess of 500 TPD with a slightly less HP required per ton capacity.

The following curves have been developed for soybean processing, whatever the process design. Similar curves for cottonseed were reported by Lusas (2). No consideration has been shown for heat losses as they are too variable for comparison purposes.

The first curve "Solvent Flow to Extractor" (Fig. 4) is the sum of solvent in miscella plus the solvent in the marc to the DT. Note that solvent to extractor with flakes at typical hold-up of 33% hexane in marc and 30% oil in miscella is approximately 300 gallons/ton and the expanded collets have normal hold-up of 20% and 35% oil in miscella with 200 gallons/ton. This would allow 50%

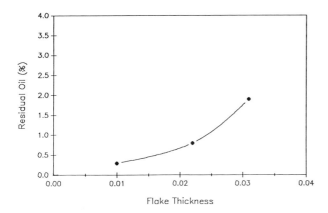

Fig. 3. Comparison of extraction vs. flake thickness.

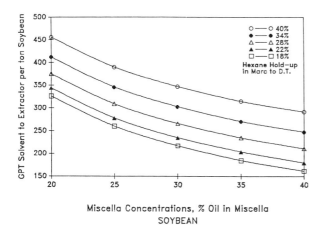

Fig. 4. GPT solvent to extractor.

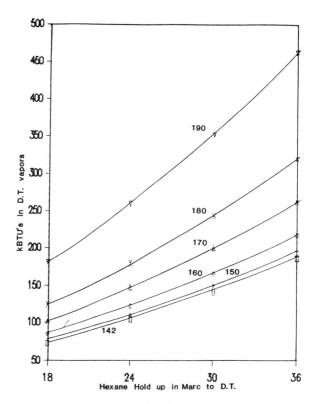

Fig. 5. BTU content of desolventizer vapors at different temperatures.

increase in capacity with no change in hexane flow to the extractor.

Curves illustrating the BTU content of desolventizer vapors (Fig. 5) are derived from hexane hold-up and the vapor composition curve (Fig. 6). Combining BTU content curves with BTUs required to concentrate miscella to 65% in the first effect evaporator (Fig. 7) shows that the difference between availability and requirements for flaked soybean extraction results in a large surplus of DT vapors (240K BTU's-95K BTU's) and extraction of expanded collets is closer balanced (120K BTU's-70K BTU's) but sufficient.

Of the live steam injected in the DT, condensation produces excess moisture to be removed in the meal dryer. The curve showing the energy required to dry the meal is not exact as it does not consider any drying in the DT, but it is useful to illustrate that the hexane hold-up greatly affects the dryer requirements (Fig. 8). Actually, there are many expander extraction systems with a meal dryer not being utilized. Just the steam saved in meal dryer more than compensates for the steam used in the expander.

Finally, hexane hold-up and DT vapor temperature affects greatly the total energy requirements of the solvent plant. There is almost a 50% reduction in steam required. Some plants with overloaded DT's are able to reduce the DT vapor temperature and save even more, i.e., a DT with 50% additional production operated at 25% hexane load reduction.

Historically, we should start with Maurice Williams in 1965 for a discussion on oil quality. He first used the expander for enzymatic inactivation in rice bran and found improved extractability and oil quality. Very few people were interested at that time in rice bran oil, and the potential of the process was not realized (3,4).

Later the Alcon process was found to be effective in inactivating the phospholipase enzymes in soybeans allowing the extraction of an oil almost devoid of nonhydratable phosphotides (5).

From our many years of experience in cottonseed extraction we remember constantly battling for rapid heating through the active zone of the phospholipase. Activation of the enzyme was triggered by rupture, moisture, and temperature above 130°F. Inactivation was accomplished by wet heat about 180°F. The Alcon process accomplishes the rapid inactivation in their conditioner by live steam injection at atmospheric pressure. The expander process accomplishes this inactivation within the expander with steam at 40–60 psig, during the 10–15 seconds that the product is in the expander.

The expander and Alcon processes can accomplish inactivation of the enzymes, but cannot reverse any formation of nonhydratable phosphotides accomplished beforehand. Field-damaged beans seem to be more susceptible to enzymatic action during the drying stage prior to storage. Utilization of higher temperature drying for high moisture beans should be avoided. Conditioning prior to flaking also should be kept as low as can be handled by the flaking rolls. (Tables 5, 6, and 7).

Table 5 shows that when conditioning of cracked beans prior to flaking is accomplished at 180°F, no conditions in the expander operation can prevent the formation of nonhydratable phosphotides.

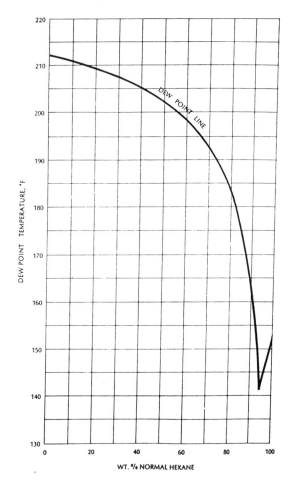

Fig. 6. Vapor-liquid equilibrium for the system of normal hexane-water at one atmosphere.

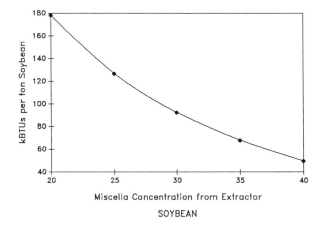

Fig. 7. BTUs required to concentrate miscella to 65% in first effect evaporator.

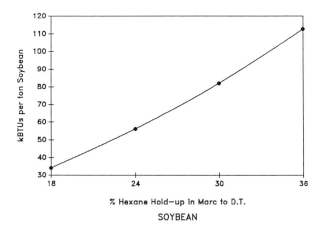

Fig. 8. BTUs required to dry desolventized meal to 13%.

Tables 6 and 7 demonstrate that if the enzyme is kept dormant by low temperature prior to expansion, the cooking within the expander can prevent the formation of the nonhydratable phosphotides. The degummed oils produced can go directly to the deodorizer after light treatment with bleaching earth, thus eliminating caustic refining, water wash, and pollution.

With the HTST treatment within the expander, the urease activity and trypsin inhibitors have not been greatly affected. Even when the samples are maintained at 200°F for periods up to one hour, the urease and trypsin inhibitors are still active. In the Alcon process (4), the inactivation of the urease enzyme and the trypsin inhibitors is accomplished immediately following the conditioning in a stack cooker that they call tempering equipment which maintains the high moisture, high tempera-

TABLE 5

Cracked Beans Heated to 180°F

Sample #	T.I.U.	Urease	N.S.I.	Calcium[a]	Phosphorus[a]	Magnesium[a]
1016-A	73,400	2.09	51.03	13.1	464.1	17.9
1016-B	82,725	2.08	59.61	2.6	68.4	3.8
1016-C	71,545	2.10	54.95	5.0	151.9	7.0
1016-D	61,256	2.13	43.92	1.9	52.8	2.8
1016-E	48,669	2.13	40.10	6.2	181.2	8.6
1016-F	52,050	2.04	43.70	0.9	22.7	1.1
1016-H	54,548	2.08	40.50	0.8	16.5	0.5
1016-I	57,169	2.10	40.46	6.3	233.7	10.4
1016-J	67,960	2.10	60.43	1.7	20.6	1.6

[a] nonhydratable

TABLE 6
Cracked Beans Heated to 150°F.

Sample #	T.I.U.	Urease	N.S.I.	Calcium[a]	Phosphorus[a]	Magnesium[a]
321-NM	34,072	1.70	27.11	2.4	12.2	1.3
321-0 mins	34,387	1.91	37.21	2.2	13.6	1.4
321-5 mins	41,446	1.53	42.6	1.7	11.9	1.4
321-10 mins	34,105	1.78	28.66	1.2	5.5	0.7
321-15 mins	36,990	1.79	31.02	0.3	7.9	0.3
321-30 mins	37,199	1.82	28.91	2.0	8.1	1.1
321-60 mins	32,388	2.11	25.95	1.1	6.7	0.9

[a] nonhydratable

TABLE 7
Cracked Beans Heated to 130°F.

Sample #	T.I.U.	Urease	N.S.I.	Calcium[a]	Phosphorus[a]	Magnesium[a]
320-0 mins	38,821	2.02	36.76	0.2	6.1	0.1
320-5 mins	45,469	2.00	36.76	0.2	6.1	0.1
320-10 mins				1.6	13.2	2.3
320-15 mins	44,812	2.04	44.65	2.4	14.8	1.8
320-60 mins	48,258	1.99	46.3	0.5	8.1	0.4

[a] nonhydratable

ture conditions produced by the steam injection of the conditioner. This exposure completes the inactivation of the urease and trypsin inhibitors before extraction.

Several paths can be followed when the expander process is used. In Brazil, it is common practice to retain the collets in a pellet dryer that has been divided into three sections (Fig. 9). The first section is used as recycle cooker; the second section uses heated fresh air to dry the collets; and the third section uses fresh air to cool the collets for extraction at 140°F. Expander discharge can vary a great deal from slow even discharge. The first section accomplishes some of the reduction of urease activity, as is accomplished in the Alcon process tempering equipment.

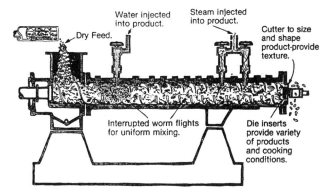

Fig. 9. Sectional view of typical apron-type dryer/cooler.

The simpler soybean process utilizes a hood over the first few feet of the collet conveyor to draw off the flash steam and cool the collets to below 140°F (Fig. 10). The urease activity is reduced in the DT as in the classical

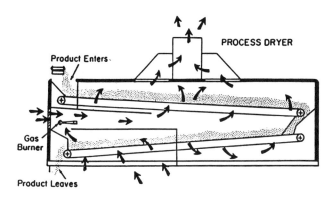

Fig. 10. Cross-section of expander.

extraction of raw flakes. In some cases moisture must be added to the DT in order to reach the moisture level required for inactivation of the urease. When the hexane hold-up in marc drops to 20% hexane (or less), the condensing steam may not be sufficient to raise the moisture level in the DT enough. When this occurs, a spray of water in the vapor scrubber can be used to replace the hexane normally utilized as the scrubbing liquid. Expander discharge can vary as shown in Figures 11a-11d.

Differences found when converting to an expander process: Coarser cracking: allowing more capacity; conditioning: lower temperature flaking allowing more capacity; flaking: .020" instead of .012"; expanding: 70 HP/300 tons/day capacity; cooling: 10-25 HP cooling fan; conveying: with bulk density increased, same conveyor with more HP can carry 150% of rated capacity; extractor: for same extraction time, capacity of extractor is increased to 150% of rated capacity; extraction time required for .020" flakes that have been expanded approximately same as for

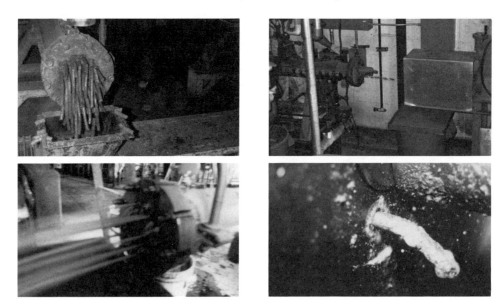

Fig. 11. Expander discharge, A = slow discharge; B = high velocity discharge; C = intermittent discharge; D = foamy discharge.

raw flakes at .012"; conveyor to DT: may need speeding up, but only slightly; DT: reduced HP because material does not go through "doughy" stage; water balls: almost always eliminated (raw protein that caused water balls has been denatured in expander); water-ball breaker: usually by-passed; vapor scrubbers: can be run dry in most cases (hulls incorporated in collets); conveyor from DT: need larger as bulk density is reduced in DT to same as process using flakes; meal dryer: usually not required with expander process unless climatic conditions are very humid reducing the moisture removal during meal cooling; meal cooler: due to the agglomeration of protein in collets, micron size particles of protein have been eliminated and many sock filters are operating without socks; meal conveyor: no difference in bulk density of meal from DT with change in process to expander; if capacity is increased, conveyor may need enlarging; and seed cleaners: need more capacity.

Solvent plant changes are: stage pumps: Collets may improve drainage too much; collets should be mixed with broken collets in approximately 50% fines 50% collets to control drainage; first effect evaporator: normally can handle the 15% increase in load when capacity is increased by 50%; second effect evaporator: load on second effect evaporator is increased by 50% as capacity is increased by 50% and the addition of pre-heater between first effect and second effect may allow this increase; condenser load of second effect also increased but increased load may be condensed in DT condenser; oil stripper: Needs to be enlarged in most cases; increased phosphotides in crude oil increase susceptibility of polimerization of oils onto the disk and donut stripper if higher temperatures are used to increase capacity; oil cooler: cool immediately from stripper to avoid precipitation of gums; cooling tower: reduced load even with 150% rated capacity; mineral oil system: probably needs enlarging as capacity is increased; and boilers: usually no added load unless cold climate causes very high heat losses.

In summary, it is possible to increase capacity 25% to 50% when converting to the expander process. Along with the reduced energy costs, which normally can pay for the conversion in one to two years, the increased capacity can allow use of the oil mill for other oilseeds if not enough soybeans are available. With the use of drainage cages (similar to those found on pre-press expellers), the expanders can be utilized for processing sunflower, canola, and almost all high oil seeds. Our only failure to date has been copra. We will be reporting on those results when our data base builds up.

Capacity of the expander is four or five times the capacity of the prepress machine with the horsepower requirements of one pre-press machine. Maintenance requirements estimated at one-tenth to one-eighth of pre-press maintenance. Residual oil is higher than pre-press but allows better extraction because the collets are porous and extraction is quicker. Hexane hold-up will be higher than pre-press, and load on the evaporation system will increase.

Cooking requirements before the expander are reduced so a pre-press converted to an expander uses less steam and much less electrical energy and maintenance. Just when we have started to learn the details of expander operation, the door has been opened again by Maurice Williams to expander operation for "almost all oil seeds".

Acknowledgement

This work was partially funded by the American Soybean Association.

References

1. Farnsworth, J.T., L.A. Johnson, J.P. Wagner, L.R. Watkins and E.W. Lusas, *Oil Mill Gazeteer* 94:(5)30 (1986).
2. Lusas, E.W., and L.R. Watkins, *J. Am. Oil Chem. Soc.* 65: 1109 (1988).
3. Williams, M., and S. Baer, *J. Am Oil Chem. Soc.* 42:151 (1965).
4. Penk, G., *J. Am. Oil Chem. Soc.* 62:627 (1985).

Oilseeds Extraction and Meal Processing

George Anderson

Crown Iron Works, 1600 Broadway St. NE, PO Box 1364, Minneapolis, Minnesota 55440, USA

Abstract

The solvent extraction and desolventizing of oilseeds will be discussed, with emphasis on soybeans. The paper will note energy consumption and solvent losses in some detail, and will refer to extractors with a shallow bed and more continuous flow, methods of "pre-desolventizing" spent meal, fully counterflow "Schumacher DT" desolventizing machines, and "Schumacher DC" meal dryer coolers. The effects of expanders on the solvent extraction process will be discussed.

This paper is a review of modern solvent extraction techniques to remove the oil from oilseeds. Figure 1 shows the entire process for seed receiving and storage, seed preparation, solvent extraction, meal finishing, and meal storage and shipment. This paper will comment briefly on preparation and focus primarily on solvent extraction.

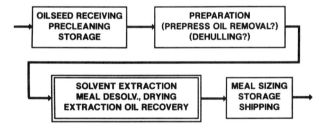

Fig. 1. Typical oilseed extraction process.

Seed Preparation

It is often said that "It all begins in preparation." The seed must be properly prepared to allow for efficient extraction and moisture control. The preparation process may include a dehulling system to remove the outer covering of the raw seed. Preparation for seeds with high oil content usually includes prepresses which mechanically squeeze out half or more of the oil. Most prepared material, regardless of raw oil content, enters the solvent extractor at from 14 to 21% oil content.

Each oilseed has its own characteristics and requires specific operating parameters. For example, soybean flake moisture should be ca. 9.5% to 11% at the extractor. Temperature should be ca. 60 to 68°C at the extractor. Flakes should be appropriate to the extractor type and bed depth. With shallow bed extractors, flakes of 0.25 to 0.4 mm are reasonable. With beds over ca. 1.5 meters flakes of 0.3 to 0.45 mm are reasonable. Thin flakes extract to the desired low oil content much faster, if the extractor can properly handle them. There are various published estimates of the effect of flake thickness (1,2,3). Table 1 shows our estimates of the effect of flake thickness on extraction time.

TABLE 1

Effect of Soybean Flake Thickness and Extraction Time on Residual Oil Content of Extracted Flakes

Average thickness (mm)	Extractor wash-drain time to attain residuals			
	20 min.	30 min.	40 min.	50 min.
0.20	0.39%	0.30%	—	—
0.25	0.68	0.42	0.33%	—
0.30	1.23	0.67	0.46	0.40%
0.36	—	1.09	0.71	0.57
0.41	—	1.69	1.09	0.85
0.46	—	—	1.61	1.25

BASIS: Soybeans conditioned to 71°C (160°F) and 9.5–11% moisture, extracted at 58°C (136°F) in a Crown extractor. Data will be similar for other extractors with shallow beds, efficient miscella staging and good drainage. Expect to use thicker flakes and longer miscella exchange times in deep bed units. Residuals are in spent flakes on a 12% moisture basis.

From the solvent extraction viewpoint, preparation is intended to produce a material with these characteristics: low content of dust or fines; proper content of hulls for the protein level desired; proper heat treatment to make materials more extractable (2,4); proper mechanical treatment to ensure rupture of oil-containing cells; permeable flakes, cake particles or extruded collets; thin flakes or small cake chunks to allow more area for solvent contact; a high bulk density to allow more material in the extractor and DT; correct moisture for extraction and ease of meal drying; proper temperature at the extractor to reduce oil viscosity; and a dry surface for cleanliness and to aid extraction and meal drying (5).

Use of Expanders

Many plants worldwide use expanders to prepare soybeans (and increasingly other seeds) for extraction. The advantages include: higher density in the extractor, better drainage, good permeability to solvent, and lower solvent content to the DT. These translate into more consistent operations, higher capacity, reduced steam use in the DT and meal dryer, and possible improvements in oil quality (9). Disadvantages include: the cost of expanders and cooking additions; power and maintenance of the expander; occasional problems in distillation reported on some seeds due to unusual extracted materials or fines; and uncertainty as to whether the advantages will be sufficient to pay out with a given oilseed or combination of equipment. An expander will usually be of most benefit to a plant with high tonnage running in a small extractor or DT, or an extractor with drainage problems.

The Solvent Extraction System

Figure 2 focuses on part of the total process, the solvent extraction system. The prepared material comes by conveyors to the extraction area. Here, it passes through the solvent extractor, an optional pre-desolventizer, then to a desolventizer toaster, and finally to a meal dryer and cooler. From there the meal—almost oil free, solvent free, dry and cool—moves by conveyor out of the extraction area and to the meal finishing area.

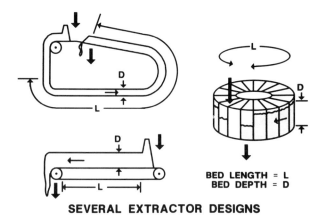

Fig. 3. Typical solvent extractor designs.

been reported that soaking in miscella with 10% oil followed by a pure solvent rinse is just as effective as countercurrent extraction (1,4). However, field data and tests indicate that in commercial practice, with perhaps 25-30% oil in the full miscella, improved countercurrent staging leads to faster extraction.

Shallow Beds versus Deep Beds: Although the shapes of extractors are often quite different, there are several features, measurements, or ratios which have comparable effects in these different machines. The first such measurement is the bed depth. We use two definitions in categorizing extractors as deep or shallow bed. First is the simplest definition: any bed over about 1 meter is considered "deep." At 1 meter some of the characteristics of a "deep bed" are becoming evident, and above 1.5 or 2 meters these characteristics in some cases have become quite significant. Some extractors have beds in excess of 3.7 m. Deep bed machines use somewhat different techniques to get good extraction; they can in fact extract very well. It is clear, however, that each type has certain advantages and disadvantages.

Generally, if the bed is shallow, drainage is facilitated because: There is less distance of material to drain through; more screen area per volume of material; more screen area per volume of fine dust; and less weight of bed and liquid on top of the bed to pack the lower bed down into a less penetrable mass (up to 800 kilograms per square meter on the bottom layer of material even in a 1 m bed). Shallow bed units require less time for drainage and therefore require less time to change miscella at each stage and for final drainage. For this reason they require somewhat less total extraction time. They also usually drain to lower solvent content between stages and at final drainage. The disadvantage is that sometimes on ideal products drainage is excessive and requires more pump volume. It is often reported that the deepest bed units require thicker flakes for good operation (1). This, however, has both good and bad effects: for example, it saves horsepower in flaking, but it increases extraction time and the volume of material in process.

Another definition of a "shallow bed" is mostly of importance in machines with continuous beds, without walls which separate the meal into cells. This definition is based on the ratio of the length of the bed to the depth of the bed. If the "length to depth" ratio is greater than 20 to 1, we call the unit a shallow bed unit. If the ratio is less

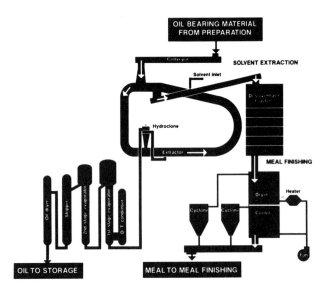

Fig. 2. Typical solvent extraction process.

Extractor

Figure 3 shows several types of extractors. On the left side there is a diagram of the loop type extractor made by our company. There is also a diagram of a single pass, "linear" travel extractor. Some companies have "round" extractors with cells, as shown on the right diagram. There are also variations on these themes, such as two "linear" path extractors mounted one above the other in one shell.

Most modern extractors are similar in their basic principle. There is a bed of material moving through the extractor; as it passes along, solvent or miscella percolates down through the bed and screens. The solvent or miscella dissolves and removes the oil from the bed of material. Finally, there is a drainage area in which no new solvent or miscella is introduced, and the bed is allowed to drain to as low a content of solvent as possible. The meal bed then leaves the extractor.

Most extractors have the flow of miscella approximately countercurrent to the flow of solids. It has long

than 20:1 we call it a deep bed unit. A long and shallow bed is advantageous in machines without cells or walls in providing more length for better separation of stages of miscella. Deep or shallow bed? Existing machines might be categorized as follows: The Crown or "loop" type extractor always has a bed of less than 1 m and typically has a L/D ratio of about 50:1. It is definitely a "shallow bed" machine. The straight-line "linear" units often have a ratio on the order of 15:1, for example a 20 m long bed, 1.2 m deep; in this example the unit is "deep bed" according to both definitions. Other linear units may have less than 1 m beds, cells and/or a high L/D ratio and are "shallow bed" units. The "round" units with cells we class as "deep bed" when the depth is more than 1 m—which is usually true at medium to high capacities.

Another thing we can compare in different kinds of machines is the type of screen. The use of properly designed stationary, self-cleaning screens helps reduce problems with drainage and eliminates problems with the sliding-contact seals required on some machines at the bottom corner of the bed. There seems to be a trend to adapt some form of stationary screens to various types of extractor.

Experience suggests that simplicity of internal mechanisms, good visibility of the machine internals, and ease of access are extremely important in any extractor. There is likely to be greater reliability if there are fewer internal mechanisms, for example gates, hinges, latches, track-wipers, levers, bearings, moving screens, bottom corner seals, and etc. The trend in new extractors appears to be towards simplicity.

A fairly steady flow of solid material from the extractor to the DT is very advantageous. There is less stress on the discharge hoppers and conveyors. The desolventizer can run more steadily, with less variation in its bed depths, and with more constant steam flow, temperatures, and dome vapor heat content. These result in more efficient desolventizing and in steadier operation of meal systems and distillation. The meal usually flows into conveyors more uniformly from a loop or linear extractor. The meal is more uniform in solvent content and generally averages less solvent content with any shallow bed machine. The loop also has some less obvious advantages. For example, the bed slopes up slightly to the discharge, such that it is very unusual and difficult for large volumes of liquid on the bed surface to be trapped and discharge with the meal or to run over into the discharge.

Automatic Level Control: Several extractors, I believe only of the loop and linear type, have automatic level controls at the inlet hopper. The control senses the feed rate to the extractor and adjusts the extractor speed to match it exactly. The extractor always maintains a full bed, and always uses its maximum volume and retention to efficiently process the tonnage supplied from preparation. The solvent in the extractor bed is about half of the solvent in the entire plant! An important side benefit of the automatic speed control and of the constant bed depth in the extractor is that the amount of solvent held up in the flake bed is nearly constant even if capacity changes. This, in turn, means that the flow of solvent through the entire plant is very steady, greatly reducing surges through distillation or extractor stages. The solvent work tank and storage tanks stay at very constant levels. The entire operation is made a little bit steadier.

Flake Bed Redistribution: Several extractors—some double path linear units, some basket extractors and the loop type—share an advantage. They either redistribute the bed at the mid-point of the cycle or they turn the bed over as in the loop type. This increases the chances that all particles will experience good contact with miscella somewhere in the cycle.

The Desolventizer Toaster and Pre-Desolventization

Desolventizing is a key operation in solvent extraction, and the key machine in this operation is usually a Desolventizer Toaster (DT). The DT is an important factor in controlling energy use and solvent loss. The DT often uses about half of the steam used by the entire plant. In addition, much of the solvent loss is from the DT; one rapeseed plant estimated about 80% (6), and we estimate about 35% to 65% for typical soybean plants.

Extracted meal usually enters the DT with about 25 to 35% solvent content depending on the soybean quality, preparation technique, type of extractor and other parameters. This solvent must be recovered for re-use and to avoid an extreme fire hazard in the meal.

Soybeans also require toasting or heat treatment to produce meal with the proper nutritional characteristics. The amount of toasting is usually measured as "urease activity." If toasting is not enough, the DT may need more time, more moisture or higher temperature (usually over 100°C for two-thirds of the time in the DT). Urease activity of from just above zero to about 0.15 pH rise is usually acceptable.

In 1950 the first commercial DT was developed and patented by Central Soya (3). It combined the compact and reliable stacked cooker with the injection of sparge steam into the top tray meal to do most of the desolventizing and to raise the moisture to the level needed for cooking of soy meal. The steam jacket heat of the lower trays served to complete the desolventizing, continue the toasting, and do a little drying as well. It should be noted that the slight drying of meal in the lower trays provided steam from the meal itself which stripped most of the last traces of solvent from the meal.

A properly sized, modern, conventional DT (Fig. 4) will usually do a good job of both desolventizing and toasting, with reasonable steam efficiency. However, its steam use is not optimum, it has a difficult time reducing solvent content in many oilseeds below 500 ppm, and it tends to lose some solvent in the vapor between particles of meal at the discharge (6,7).

Schumacher Counterflow DT: In 1982 Schumacher introduced his new DT. When he first described the idea to me, I thought he was heading for a problem. It was a much different concept than anything that came before, and I was worried that the hollow staybolts his machine required would cause turbulence and dust carryover or perhaps would plug with meal. Nevertheless, in a few short years about 100 units have been built by several licensees of his patents and technology and they have worked beautifully! Units are operating on soybeans, corn germ, rapeseed, and other seeds.

This unit, shown in Figure 5, is somewhat similar in basic construction to the conventional DT. However, all the steam is introduced at the bottom of the lower meal bed; each upper tray has hollow staybolts to allow the

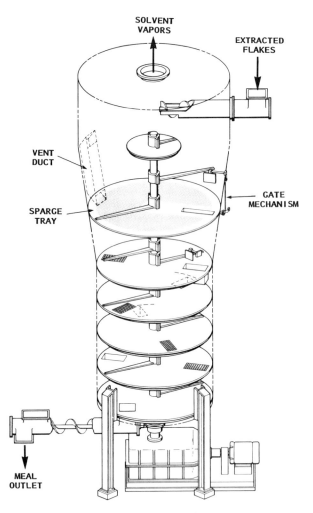

Fig. 4. Modern, conventional DT for desolventizing and detoasting.

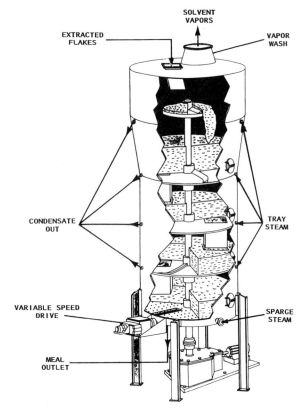

Fig. 5. Improved Schumacher DT.

passage of that steam through each bed in succession, counterflow to the flow of meal. There are several benefits: The counterflow principle works well, and less steam is required. The new DT can often run safely with discharge vapors at 69°C which nearly minimizes the loss of energy in the DT vapors. If an error is made and solvent gets down into the more central or lower beds, the full sparge flow is available to pass through those beds, providing considerable desolventizing power, unlike the conventional DT. The flow of steam through the hollow staybolts increases the heat transfer from the hot surfaces to the meal, resulting in lower moisture. The solvent laden vapors are forced up into higher trays and out the top, rather than some being lost out the bottom. The lower tray meal sees almost pure, superheated steam and tends to desolventize much more completely—often in the 100–250 ppm range and usually well below 500 ppm. The solvent vapors around the meal are swept away just before discharge. And, the meal is also slightly surface dried by the superheated steam. A properly designed Schumacher DT will usually result in noticeable lower solvent loss and steam consumption than a conventional DT in the same facility.

Pre-Desolventizing: Once the extractor has drained off as much solvent as possible before discharge of meal to the DT, there is still perhaps 30% solvent content to remove by evaporation. This evaporation requires heat energy. If the energy is supplied by direct sparge steam condensation, it will raise the moisture of the meal and require more meal drying later. If the energy is supplied in part by contact with steam heated plates, the moisture will not rise, the meal dryer will have less work, and there will be less total steam used. Therefore, during the past 15 years or so, we have seen more emphasis on an added piece of equipment between the extractor and the DT called a Pre-Desolventizer. The device is basically a vessel designed to put the meal in contact with a large surface area of hot steel plates and vent off solvent that evaporates.

Heinz Schumacher and others have used large diameter, steam heated screws in front of the DT, i.e., larger and more reliable versions of the older 'Schnecken' desolventizers. Several manufacturers have used extra flat trays in the top of the DT, before the meal contacts sparge steam. Still others have designed heated tube bundles or discs rotating in the meal. In all cases the intention is the same, to provide just enough dry heat so that just the right amount of sparge will condense and provide balanced toasting and drying with minimal steam use. At this "balance point" we have an optimum DT and meal drying system. Calculations show that a "good, old DT with rotary dryer and cooler" for soybeans may use 199 kg of steam per metric ton of seed processed. A "new Pre-DT and Schumacher DTDC" running correctly may use 117 kg/mt. You will note that this is a 41% reduction is steam per ton of raw seed processed.

Additional Processes

The Dryer Cooler or DC: The Crown/Schumacher DC is an alternative to the commonplace horizontal rotary steam tube units. The DC often costs less, uses less energy, has lower dust emissions, and takes up less space. The flow of air through each stage of the unit is counterflow, resulting in excellent cooling, usually below 38°C or within 10 degrees of ambient. Such units often require about two-thirds as much steam as an older style system.

Miscella Distillation: The miscella from the extractor contains about 22-30% oil and the rest is solvent. The basic process is to heat this mixture under a vacuum and evaporate the solvent, leaving the oil behind. The oil is then steam stripped in a counterflow process and goes to storage as crude oil. The solvent vapors are condensed, separated from water, heated, and sent back to the extractor. In an efficient system about 75% of the energy for this evaporation process is provided by the "waste heat" of the DT discharge vapors.

Mineral Oil Absorber: Air is carried into the system along with the flakes or cake, or finds its way in through leaks. Air and other "non-condensable" vapors must be removed from the system or they will fill the condensers and stop the process. The mineral oil system is used to remove the solvent from this air as it is discharged from the system.

Water Reboiler: The plant process will collect some water from steam or seed. This water is separated from the solvent in a gravity separator. It then goes to a "reboiler" to heat it and remove the last traces of solvent before discharge outside the plant. It is important for solvent recovery and for safety that this reboiler function properly.

Safety

Figure 6 shows a unique feature of the extraction area. It is isolated from the other processes and from public access by fences, by distances of roughly 30 m, by conveyors and seals carefully designed to prevent the escape of solvent into other process areas, and by the systems mentioned above which prevent the escape of solvent into the environment. These features are essential because the typical extraction plant uses flammable hexane as a solvent. Figure 6 is adapted from a publication of the National Fire Protection Association called NFPA #36, Solvent Extraction Plants (8). This booklet is an excellent safety standard for the design and operation of commercial solvent extraction plants. It is required for plants built in the USA and I highly recommend it as a guide for constructing plants in any country.

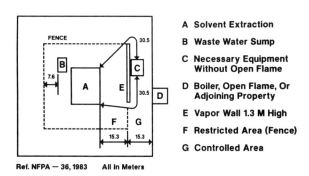

Fig. 6. Distances required for safe design of solvent extraction facilities.

References

1. Myers, N., *J. Am. Oil. Chem. Soc. 54:*491 (1977).
2. Becker, K.W., *J. Am. Oil. Chem. Soc. 55:*754 (1978).
3. Sipos, E.F., and N.H. Witte, *J. Am. Oil. Chem. Soc. 38:*11 (1961).
4. Good, R., *Oil Mill Gazetteer 75:*3:14 (1970).
5. Good, R., "Solvent Extraction Preparation as Practiced Today," presented at Cooperative Oil Mill Superintendents Meetings, Atlanta (March 1971).
6. Dahlen, J.H., and L.A. Lindh, *J. Am. Oil. Chem. Soc. 60:*2009 (1983).
7. Anderson, G.E., *Oil Mill Gazetteer 82:*2.
8. NFPA 36 *Standard for Solvent Extraction Plants*, National Fire Protection Association, Batterymarch Park, Quincy, MA 02269.
9. Lusas, E.W., and L.R. Watkins, *J. Am. Oil. Chem. Soc. 65:*1109 (1988).

Full-Fat Soya-Products—Manufacturing and Uses in Foodstuffs

Jens Heiser and Thomas Trentelman

Lucas Meyer Co., Ausschlaeger Elbdeich 21, 2000 Hamburg 28, Federal Republic of Germany

Abstract

Full-fat soya products are becoming increasingly important in human nutrition today. In its native homeland of Asia, soybeans are as normal an ingredient in many recipes as are flour, milk and starch products elsewhere. The expression "renewable resource" applies to a significant extend to soy. It is especially appropriate to represent the various soya products as nutritious natural products with functional properties. Steadily developing production know-how and intensive advice for applicantions have predestined soya products for such a wide range of applications in the foodstuffs industry, as a food ingredient and as a foodstuff as well.

The soybean is one of the oldest cultivated plants on Earth. Especially in the Far East, the growing and eating of soybeans is a tradition that can be traced back more than 4,700 years. At that time the soybean was mentioned in traditional lore as an important part of the diet in large areas of Asia. The plant was known there for many years and even enjoyed protection as a medicinal remedy. However, it was not until the present century that it emerged to take the Western World by storm as a useful agricultural plant. As recently as 50 years ago the soybean was a relatively insignificant plant for the American farmer.

As a result of the massive increase in worldwide demand for protein, the soybean now has become the world's most important source of vegetable protein and vegetable oil. One-half of total world production of oilseeds is accounted for by soybeans.

Composition of the Soybean

For a long time the use of the soybean as an agricultural crop was confined solely to processing for soybean oil. The residual soybean cake or meal was then processed further before being used for animal feed purposes. Even in the Western world, it was only during the past two decades that the excellent quality of the protein was recognized. Today the status of soybean protein reflects its great nutritional value as a most interesting component of our daily diet.

It is beyond the scope of this paper to detail all the various constituents of the soybean. The major constituents of full-fat soy products with respect to food-processing systems are: protein, fat, lecithin, and tocopherol.

Protein: Among the familiar leguminous plants, the soybean stands out for its high protein content. Until only a few years ago it was held by scientists that animal protein was better suited than vegetable protein to meeting man's protein requirements. The grounds for this were that the amino acid content of animal protein was more in line with human needs. This view was based on animal feeding trials in which the animal protein components in the feed were replaced by a vegetable protein. Since then it has been recognized that, firstly, the results of such feeding trials are not necessarily applicable in their entirety to the protein absorption capacity of human beings, and, secondly, that a combination of suitable vegetable protein sources can indeed produce results that are on a par with the protein value of animal protein.

A striking feature of the amino acid spectrum is the high level of lysine and methionine. Since these proportions are reversed in many varieties of grain, cereals are highly suitable for combining with soybean protein to supplement its amino acid content. The current view of nutritional science is that a desirable diet should consist of about two-thirds vegetable protein and one-third animal protein. The present situation as regards our daily diet is the exact opposite, namely, two-thirds is animal protein and one-third is vegetable protein.

Fat: With a fat content of about 20 percent the soybean is the most important source of edible oils and fats. Fat is also of great importance in human nutrition. The high level of polyunsaturated fatty acids in soybean oil is unmistakable. According to the latest recommendations the human organism should be supplied with about 10 g of polyunsaturated fatty acids every day, which means that this requirement can easily be met with quite small amounts of soybean oil. Soybean oil contains no cholesterol. Especially in view of extremely abundant supply of saturated fats of animal origin in our daily diet, experts believe that 15 to 20 percent of the population of the Federal Republic of Germany have a high plasma cholesterol level. This risk factor of the first order can be countered by changing to a sensible diet and controlling the cholesterol situation as necessary with the recommended quantity of unsaturated fatty acids.

Lecithin: Crude soybean oil contains polyunsaturated fatty acids and soybean lecithin. Thanks to their wide-ranging functional properties, soybean lecithins are used in numerous areas of the food industry. In chocolate manufacture, for example, soybean lecithins have become an indispensible ingredient. And in the fields of pharmaceuticals, cosmetics and various technical areas lecithin plays an important part as a natural emulsifier, an antioxidant, and a synergist. Lecithin is one of the central active substances in the human and animal organism. It supports fat metabolism, combats liver and gall-bladder diseases, and is suggested to prevent the deposition of cholesterol on artery walls.

Tocopherol: Tocopherol, which is also present in soybean oil, is a substance that protects fats. Also known as vitamin E, and tocopherol even passes on its antioxidant properties to those products that contain animal or vegetable fats as part of the recipe.

Manufacture of Full-Fat Soy Products

Whereas in Asian countries soybean products are found mostly on the market as straightforward fresh products or as fermented products, the Western world has increasingly focused its interest more on the product/constituent "protein" in the form of soy flours of various grades. In recent years, the latest manufacturing technologies, comprehensive consumer education based on scientific findings, and a more health-conscious life style have resulted in more and more products becoming established on the market. Products that correspond closely in their composition to ideal notions of a varied and balanced diet. On the basis of numerous studies it has now been shown that only an ideally balanced diet with the right proportions of animal and vegetable protein can help prevent the nutritional disorders that are becoming increasingly common. In Western countries, when we speak of soy we generally mean soy flours, i.e. milled products of varying protien and fat content. Normally, a broad distinction is made between full-fat and defatted soy flours. This discussion will be confined to the production and use of full-fat soy products.

Processes for Full-Fat Soy Products

Since the soybean is the starting material and nothing is added or taken away in the production process the quality of the full-fat soy product will depend almost totally on the quality of the beans. It is obvious that there is a significant variation in the quality of the raw material from cargo to cargo and indeed from season to season. Selection of the beans used to produce full-fat soy products is therefore essential.

As mentioned, the soybean contains ancilliary substances that can considerably impair its use in the untreated state. In addition to the trypsin inhibitor these include enzymes that are responsible for the characteristic smell and taste of soybeans that is so foreign to the European palate. By treating the beans appropriately in a mild process it is possible to inactivate these enzymes and give the subsequent full-fat soy products a pleasantly nutty and slightly sweet smell and taste.

This treatment process, the toasting or debittering process has the effect of inactivating the inhibitors, bringing about a pleasant flavor, killing off undesirable bacteria, and improving the quality of the protein. The soybeans require the application of heat and steam to enable the full release of the nutrients, the elimination of trypsin inhibitors and the improvement of protein digestibility. The degree of heat and steam required as well as the fineness will depend on the particular process employed as well as the desired final product.

Areas of Application of Full-Fat Soy Products

Thanks to the ease with which they can be processed, the functional properties of full-fat soy products, e.g., moisture binding, solubility, emulsification, fat absorption, and gelling and foaming properties, can be exploited to the fullest within a food production process. Particularly in the sugar confectionery, and bakery products sectors, as well as in dietetics and children's foods, toasted full-fat soy products have come to be more important than virtually any other raw material used here. Owing to protein, fat, lecithin, and tocopherol, full-fat soy products form a natural and physiologically highly effective complex whose benefits can be exploited even without complicated processing technology.

The large number of products that are available today in the sugar confectionery and bakery products industry make it almost impossible to generalize about functional properties. I shall therefore select a few areas as examples to show how full-fat soy products work in these foods and where they can be used.

Chocolate and Confectionery Compositions: One of the main uses of toasted full-fat soy products in the confectionery industry is for chocolate and confectionery compositions. Here, special attention has been given to the influence of full-fat soy flour on the final products. Whereas comparable raw materials often can lead to a decline in the quality of the product as well as the desired advantages, this effect is not found with full-fat soy products. Tests have shown that admixtures of between 3–5 percent full-fat soy make it possible to save 1.5 percent cocoa butter and 3 percent whole milk powder depending on the formulation. The addition is possible without any detectable discrepancies occurring during mixing, rolling, and conching.

Viscosity measurements have also shown that compositions with added full-fat soy products display lower viscosities, making it easier to eliminate moisture and undesirable flavoring matter during conching. No differences from conventional compositions as regards gloss, fracture, and contraction capacity are encountered during tempering either.

Spreads: Spreads were originally developed in England after World War II. They were very popular with children in particular and cornered a substantial share of the market. As a rule a conventional spread contains sugar, fat, cocoa powder, and milk powder components. Replacing some of the milk powder with full-fat soy is feasible without any sacrifice in quality, and the addition of full-fat soy flour also has the effect of "softening" the taste of the often very large amounts of sugar used in these products.

Wafer Fillings: In the manufacture of wafer fillings, particularly high standards of stability are required. The incorporation of full-fat soy results in the simultaneous addition of lecithin and hence in improved emulsification of fats, allowing more dry matter, such as milk powder, broken wafer material and nuts, to be used. This also brings out the characteristic flavors of such products especially well. Other benefits are protection of fat components against oxidation, more machinable compositions and improved adhesion of the wafers,

Bakery Products: There are also a wide variety of applications for full-fat soy products in the bakery sector. Just as for sugar confectionery the main reasons for using such products here are also of a functional nature. The use of enzyme-active soy flour as a crumb-lightening component in white-bread doughs, for example, was patented as early as 1934. In conjunction with rye doughs, enzyme-active, full-fat soy flour (our product Soyapan) also makes the doughs more pliable, thus giving them better machining and moulding properties.

Toasted full-fat soy flour (our product Nurupan) is used frequently for flour confectionery in order to improve the moisture retention capacity of the dough in the product and thereby make it stay fresh longer. Since such confectionery products also frequently have a very high fat content, the emulsification of this fat with the other ingredients is favored by the lecithin contained in the full-fat soy flour. The quantities for products such as short pastries, puff pastries, cake, and also sponge batters are usually in the region of 5-8 percent of the fat content of the product. In the case of wafer biscuits, on the other hand, Nurupan in quantities as low as 3 percent of the milled products is enough to produce a marked reduction in wafer breakage and in the wafers' tendency to cling to the baking iron.

Soy breads: With their gritty character, full-fat soy grits are highly suitable for the production of rustic mixed-grain bread in which mild acidification is required. It was the use of full-fat soy grits that first made it possible to produce breads with admixtures of up to 30 percent soy protein without any disadvantages from the point of view of taste or baking problems. Soy breads are becoming increasingly popular in Western Europe, where they are regarded as a welcome addition to the already wide range of breads available.

I could go on with this list of the application for which full-fat soy products are used, but this small selection should be sufficient to demonstrate a few areas of practical application. In this connection it is worth taking a brief look at a new development in the field of full-fat soy products that has resulted from the steadily increasing quality standards demanded by the food industry. The increasingly nutrition-conscious mentality of the consumer, and also the increasing raw material prices for milk and similar products have prompted a development that has led to even better functional properties and even better flavor neutrality in a commercial product.

A completely new production process has made it possible to develop a product in which all the valuable components are subjected to a minimum of strain by the production process, while allowing a product to be produced that is optimal from the point of view of taste and smell. This has added numerous new potential applications to the uses already known for full-fat soy products.

You will see that certain essential steps in this production process are taken from the manufacture of soybean milk. In this way it becomes possible, by making simple modifications to the process, to achieve changes in the properties of the product to suit the required application. These alterations include, for example, reducing or increasing the protein and carbohydrate concentration, adjusting the content of dietary fiber, and modifying dispersibility and emulsifying properties.

With a full-fat soy product capable of being modified in this way, new possibilities are opened up in the field of product development, since among other things it has been possible to virtually eliminate the limiting factor, "taste."

Thus, it is even possible to make products such as ice cream, instant drinks, and also cheese-like products, in which the sole protein source is full-fat soy protein. Unlike the traditional full-fat soy products with fixed specifications and standardized production processes, it is possible here to start from one basic process and produce completely different end products that are still entitled to be called full-fat soy products. In this way it is possible to make, on the basis of soy protein, products whose taste is indistinguishable from the version using their traditional source of protein but which are of considerably greater value from a nutritional point of view.

As you can see, the present state of development in the processing of the versatile raw material known as the soybean opens up many new possibilities to the food industry. Not the least of these possibilities is greater independence in the choice of raw materials. The foundation would seem to have been laid for new and interesting product ideas, so that the protein obtainable from the soybean is no longer the cheap substitute for milk and other other animal products that it was formerly labelled as. Indeed, it must now be regarded as an interesting and highly nutritious food that will find its place in a modern and up-to-date diet.

The Preparation and Properties of Defatted Soy Flours and Their Products

Richard W. Fulmer

Cargill, Inc., Research Department, P. O. Box 9300, Minneapolis, Minnesota 55440

Abstract

This review of defatted soy flours will discuss preparation, composition and uses of various flours. Soybean selection and processing conditions result in products with different functional properties which provide utility in many end uses. These functional properties will be related to applications especially in food and further process uses.

For many years the soybean has been cultivated for production of food and feed materials and for the value derived from separated components of the bean. Diverse societies have learned to use soybeans in numerous food products—as whole beans and in fermented products. Further, soybeans are a source of valuable food materials. Through modern processing plants, the bean can be fractionated into useful products. Thanks to the techniques of analytical chemistry, the bean and its components can be characterized by their chemical constituents.

However, the increased acceptance of soy protein is due to its versatile qualities, good functional properties in food applications, high nutritional value, availability and low cost. Functionality and not protein food value is the major reason for use of soy in formulated food compositions. Protein food value is of more importance in moderate cost world feeding and in animal feeds. Selection of the proper soy protein for a particular use is very important and understanding soybean processing and heat treatment is essential in the selection process (1,2). Today's modern soybean processing plant will process between three and four million pounds of soybeans daily, or roughly 75,000 bushels or 2,500 acres of soybeans each day. Further, it is not uncommon to have in excess of one million bushels of bean storage at the processing plant. The sheer size of the processing plant allows comingling of varieties and beans with different composition, qualities, growing and storage histories, which will minimize fluctuations in the major component quantities by blending.

Processing alternatives enable the manufacture of defatted soy products with varying degrees of heat treatment and with varying granulations. These variables affect the functional and nutritional properties of these products. Untoasted products have maximal functionality. Fully toasted products have optimal nutritional value. By closely controlling the heat treatment and granulation, it is possible to regulate the functional and nutritional properties of the soy flour so they are optimized for each application.

Soy flour and grits are used in food products because of their functionality, nutritional value and low cost. Functional properties are characteristics such as water absorption, fat absorption, fat emulsifying capacity, protein binding capacity and adhesiveness. These properties can be regulated so that they are similar to those of milk, meat or eggs. Therefore, properly processed soy products can functionally supplement and be substituted for animal proteins in certain applications in foods. Because soy flour contains a high percentage of good quality protein, it can make a significant contribution to the nutritional value of foods. The quality of the protein in toasted soy flour is superior to the quality of protein in other grains as judged by the generally accepted procedures and indices used to make such determinations. Because the protein in soy flour is a rich source of lysine, it offers special value as a supplement to all protein sources deficient in this essential amino acid. Both the quality and quantity of the protein are raised in cereal-based foods to which soy flour has been added.

The preparation of beans for processing is important to achieve good hull removal. Beans should be dried carefully to shrink the kernel from the hull. When properly conditioned, the beans are sent through corrugated cracking rolls (0.075 inch clearance). Beans are cracked into six to eight pieces. As cracked beans go over screen shakers, hulls are removed in primary and secondary aspirators. Hulls generally are high fiber with some protein and are toasted and ground for use in animal feeds and as a dietary fiber additive.

Cleaned cracked meats go into a conditioning cooker which raises moisture and temperature to approximately 77°C. Conditioned particles are then pressed between flaking rolls to a thickness of .005 to .010 thousandths of an inch in order to facilitate rapid extraction of the oil with a solvent, generally hexane.

After sufficient contact time with the hexane to reduce the residual oil to a low level, the spent flakes are desolventized to remove residual hexane. Because of the importance of subsequent heat treatment, a description of the two methods of removing solvent from extracted flakes is important.

The standard DT (desolventizer-toaster) uses live steam to drive off hexane. A system called a flash desolventizer or vapor desolventizer (also referred to as a white meal system) produces a less heat-treated product.

In the flash desolventizer, hexane wet flakes go directly from the extractor into super-heated vapor blown at high velocity through a long tube. Hexane, b.p. 160°F, is heated under pressure to 240° to 280°F. As this vapor is blown through the tube, flakes are conveyed in and picked up by the vapor stream. In short seconds, the superheated hexane will flash off the residual liquid hexane leaving desolventized flakes to be collected in the cyclone. A raw or untoasted flake results, which can be further heat treated to any desired degree of cook depending on the final intended application.

Figure 1 outlines procedures for the manufacture of full fat and defatted soy flours from whole soybeans. In preparing full fat soy flour, the solvent extraction step is omitted. The cracking and dehulling will remove soybean hulls. Table 1 shows the comparison of soybeans, soybean meal, and dehulled soybean meal in their proximate composition. The protein content rises from 41.1 in full soybeans to 49.4 in meal due to the removal of the fat component. In the dehulled soybean meal, the protein is further enhanced because of the removal of some of the crude fiber contained in the hull.

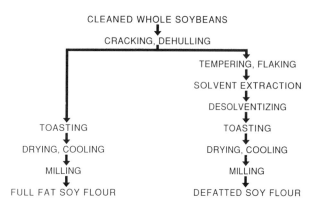

Fig. 1. Manufacture of full fat and defatted soy flour.

TABLE 2
Carbohydrate Content of Dehulled Soybean Meal

Constituent	% of meal
Oligosaccharide content, total	15
Sucrose	6–8
Stachyose	4–5
Raffinose	1–2
Verbascose	Trace
Polysaccharide content, total	15–18
Acidic polysaccharides	8–10
Arabinogalactan	5
Cellulosic material	1–2
Starch	0.5

harvest. The total carbohydrate in the dehulled soybean meal falls between 30% and 38%.

Defatted soybean meal ground to flour consistency (Table 3) is described as soy flour with approximately the same protein content of slightly over 50%, moisture content lower than meal due to moisture loss during the grinding step, low total residual fat, and approximately 33.5% total carbohydrates. Total bacteria count in edible soy flour must be under 10,000 count/gm, and the product must test negative for salmonella.

TABLE 1
Proximate Composition of Soybeans, Soybean Meal, and Dehulled Soybean Meal on a Dry Matter Basis (NRC, 1977)

Component	Soybeans	Soybean Meal	Dehulled Soybean Meal
Protein	41.1	49.4%	53.9
Ether extract	20.0	0.9%	1.1
Crude fiber	6.1	8.2%	4.3
Ash	5.4	5.9%	6.5
Nitrogen-free extract	27.4	35.6%	34.2

The fat content of whole soybeans is about 20% and tends to fluctuate depending on growing season, variety, and other agronomic factors. Although soybean oil contains a number of minor fatty acid components, the primary fat components are palmitic 11.5%, stearic 4.0%, oleic 24.5%, linoleic 53%, and linolenic 7%. Solvent extraction of this oil leaves defatted soy material with less than 1% fat, most of which is tied up as a lipo-protein or lipo-carbohydrate, or in the residual lecithin gums.

Current genetic selection research at several universities has lowered the linolenic content closer to 3.5% to 4% in some varieties, and, in a different direction, plant molecular geneticists are seeking to understand and thereby be able to modify the enzyme system in order to avoid the production of saturated fatty components in the soybean complex.

The carbohydrate content of dehulled soybean meal is given in Table 2 with ranges indicative of the variations in growing conditions, varieties, and plant maturity at

Earlier reference was made to processing steps carried out to inactivate certain enzymes that have a deleterious effect on flavor, and to reduce antinutritional components that tend to interfere with the pancreatic/trypsin digestion of proteins. Figure 2 shows that both lipoxygenase and peroxidase are considered to have a negative influence on flavor development and that trypsin inhibitor activity must be considered in determining the amount of heat treatment given to soybean materials.

Table 4 illustrates the effect of various lengths of heat treatment involving live steam at 100°C on the nitrogen solubility index (NSI), the trypsin inhibitor activity, the protein efficiency ratio, and the ratio of pancreas weight to body weight of fed animals. In each case the longer the heat treatment, the lower the NSI, the lower the trypsin inhibitor activity, and the increase in the protein efficiency ratio. Protein dispersibility index (PDI) is another measure of the extent of heat treatment, with the least heat treated soy at 90+ PDI and the heavily toasted product

TABLE 3
Typical Analysis of Soy Flour

		Test Method
Protein (N × 6.25)	52.5%	AOCS BC 4-49
Moisture	6.0%	AOCS BC 2-49
Fat (ether extract)	0.9%	AOCS BC 3-49
Ash	6.0%	AOCS BC 5-49
Fiber	2.5%	AOCS BC 6-49
Carbohydrates	31.0%	By difference
Total bacterial count	< 10,000/gm	FDA aerobic test method
Salmonella	Negative	FDA standard test

(90, 70 or 20 PDI, AOCS Ba 10-65, 100 or 200 mesh)

CONTROL MEASUREMENT	PROPERTY AFFECTED FLAVOR	NUTRITIVE VALUE
LIPOXYGENASE	+	-
PEROXIDASE	+	-
NSI	-	+
PDI	-	+
UREASE	-	+
TI	-	+
AVAILABLE LYSINE		+

NSI = nitrogen solubility index
PDI = protein dispersibility index
TI = trypsin inhibitor

Fig. 2. Property measurement used to control processing of edible soy flour.

TABLE 4
Processing and Nutritional Parameters of Heat-treated Soy Flours

Heat,[a] min	NSI[b]	TI, TIU/mg[c]	PER[d]	Pancreas wt, g/100 g body wt
0	97.2	96.9	1.13	0.68
1	78.2	74.9	1.35	0.58
3	69.6	45.0	1.75	0.51
6	56.5	28.0	2.07	0.52
9	51.3	20.5	2.19	0.48
20	37.9	10.1	2.08	0.49
30	28.2	8.0	—	—

[a] Live steam at 100°C
[b] NSI = nitrogen solubility index
[c] TI = trypsin inhibitor and TIU = trypsin inhibitor units
[d] Protein efficiency ratio, corrected on a basis of PER = 2.50 for casein

at PDI of around 20. Table 5 relates PDI to use in food applications.

In foods, soy flour can compete with animal protein on the basis of nutritional value and functional characteristics. In these applications, the economic factors are the major incentive to the food processor for using soy. In mixtures with cereals, soy is competitive on an economic basis. Now the incentives are nutrition and functionality. The major factor which limits the use of soy flour and grits in foods is the beany flavor. Untoasted flours have the strongest beany character. Fully toasted products are quite bland and are characterized as having a nutty flavor. Table 5 lists the major applications of defatted soy flour in the food industry.

The principal components of the soybean, namely the fat, protein and carbohydrate components, have been discussed. Figure 3 shows the conversion of defatted soy flour into a variety of further processed materials that go into human and animal feed compositions. One example is the addition of lecithin to soy flour at several different levels.

Samples in Table 6 of 3%, 6% and 15% re-lecithinated flours illustrate the decrease in protein and the increase in fat component as lecithin is added back to soy flour.

TABLE 5
Applications of Defatted Soy Flour in Foods

PDI[a]	Application
90+	White bread-bleaching agent
	Fermentation
	Soy protein isolates, fibers
60–75	Doughnut mix
	Bakery mix
	Pasta
	Baby foods
	Meat products
	Breakfast cereals
	Soy protein concentrates
30–45	Meat products
	Bakery mix
10–25	Baby foods
	Protein beverages
	Meat products
	Hydrolyzed vegetable proteins

[a] Protein Dispersibility Index is a standard AOCS method (Ba 10-65) for measuring the amount of heat treatment used in the processing of soybean meal products

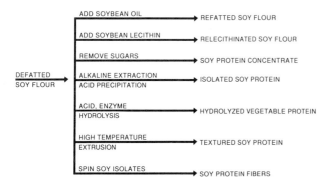

Fig. 3. Use of soy flour as a substrate for further processed food ingredients.

lysine content of soybean meal. The figure shows that the destruction of lysine is much more rapid at either 16% or 12% moisture than at 4%. Moist heat treatment caused protein denaturation at a faster rate (4).

Figure 5 correlates cooking time with the decrease in trypsin inhibitor (TI) activity and again shows that at 0% moisture trypsin inhibitor is quite resistant to protein denaturation, but at 12%, 16% or even 8% H_2O, trypsin inhibitor activity decreased rapidly. This balance of desirable reduction in TI and undesirable reduction of available lysine by heat treatment is a most important nutritional factor.

Figure 6 shows that the cooking time is related to the urease index and also again shows a very decided falloff in urease index at the higher moisture levels with cooking time.

TABLE 6

Typical Analysis of Relecithinated Soy Flour

	Lecithin level			
	3%	6%	15%	Test method
Protein (N × 6.25)	51.0%	50.0%	45.0%	AOCS Bc 4-49
Moisture	5.0%	5.0%	4.2%	AOCS Bc 2-49
Fat (ether extract)	3.0%	6.0%	15.0%	AOCS Bc 3-49
Ash	5.8%	5.6%	4.9%	AOCS Bc 5-49
Fiber	2.4%	2.3%	2.1%	AOCS Bc 6-49
Carbohydrates	32.0%	31.0%	27.5%	By difference
Total bacterial count	< 10,000/gm	< 10,000/gm	< 10,000/gm	FDA aerobic test
Salmonella	Negative	Negative	Negative	FDA standard test

In the same manner, edible oils including soybean oil are frequently added back to soy flour to produce desired fat levels varying from 1% to 15% added fat. These products provide advantages to food and infant animal feed formulators.

Figure 4 refers to the 100°C cooking time in minutes that effects a decrease in the content in the available

The extent of heat treatment will also determine the functionality of various soy protein fractions and will control these functionality properties: water absorption, fat absorption, emulsification capacity, protein binding capacity, adhesiveness or viscosity, and other functional properties. Table 7 illustrates the functional applications of soy flour, concentrate and isolate products in various general food categories (3).

Proteins are made up of amino acids in sequence and

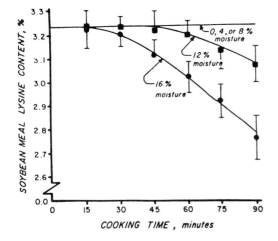

Fig. 4. Effect of moisture and cooking time on total lysine content of soybean meal.

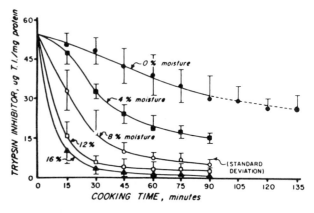

Fig. 5. Effect of moisture and cooking time on soybean meal trypsin inhibitor contents.

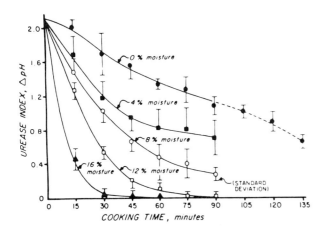

Fig. 6. Effect of moisture and cooking time on soybean meal urease index.

it is the balance of the essential amino acids that makes soybeans a particularly attractive food material.

The total protein component of soybeans can be analyzed for its amino acid composition (Table 8). The crude protein of defatted dehulled soy flour will analyze from approximately 18% of its total protein as glutamic acid down to cystine at less than 1% of the protein total.

Within the protein family there are groups of lower molecular weight proteins constituting the enzymes and the trypsin inhibitor and hemagglutination activity factors. Also scientists have identified and characterized various globulin fractions, such as the 2.3S, 2.8S, 7S, and 11S globulin or clycinin protein fractions, in addition to the lectins and other smaller molecular weight proteins. These protein components vary in molecular weight and amino acid sequence.

Complete chemical analysis requires the hydrolysis of the protein and other components in order to identify the individual amino acid, fatty acid or other entity. All combi-

TABLE 7

Functional Properties of Soy Protein Products in Food

Functional Property	Mode of Action	Food System Used	Product
Solubility	Protein solvation, pH dependent	Beverages	F,C,I,H
Water absorption and binding	Hydrogen-bonding of water, entrapment water (no drip)	Meats, sausages, breads, cakes	F,C
Viscosity	Thickening, water binding	Soups, gravies	F,C,I
Gelation	Protein matrix formation and setting	Meats, curds, cheeses	C,I
Cohesion-adhesion	Protein acts as an adhesive material	Meats, sausages, baked goods, pasta products	F,C,I
Elasticity	Disulfide links in deformable gels	Meats, bakery items	I
Emulsification	Formation and stabilization of fat emulsions	Sausages, bologna, soups, cakes	F,C,I
Fat absorption	Binding of free fat	Meats, sausages, doughnuts	F,C,I
Flavor-binding	Adsorption, entrapment, release	Simulated meats, bakery items	C,I,H
Foaming	Forms film to entrap gas	Whipped toppings, chiffon desserts, angel cakes	I,W,H
Color control	Bleaching (lipoxygenase)	Breads	F

F,C,I,H,W denote Soy flour, concentrate, isolate, hydrolyzate and soy whey, respectively.
Based on ref. (3).

TABLE 8
Typical Composition of Soy Flour

Amino acid	(mg per 100 g)
Lysine*	3300
Methionine*	720
Phenylalanine*	2600
Proline	3040
Serine	2520
Threonine*	2170
Tryptophan*	740
Tyrosine	1700
Valine*	2730
Alanine	2220
Ammonia	1050
Arginine	3790
Aspartic Acid	5690
Cystine	720
Glutamic Acid	9160
Glycine	2100
Histidine	1350
Isoleucine*	2510
Leucine*	4075

* Considered essential Amino Acids

Carbohydrates	
Acid polysaccharide complex	5–7%
Sucrose	8.2%
Stachyose	5.5%
Arabinan	1%
Neutral arabiongalacton	8–10%
Raffinose	1.2%

TABLE 9
Typical Analysis of Soybean for Mineral Content

	Content	
Element	Mg/100 gm	Percent RDA in 20 gm Protein[a]
Macro–		
Sodium Na^+	0.9	—
Potassium K^+	2090.0	—
Calcium Ca^{++}	252	10.0
Magnesium Mg^{++}	302	30.2
Phosphorus PO_4^-	740	29.6
Chloride Cl	0.07	—
Micro–		
Iron Fe^{+++}	9.6	21.0
Cobalt Co^{++}	10^{-7}	—
Copper Cu^{++}	1.6	32.0
Manganese Mn^{++}	3.6	—
Zinc Zn^{++}	5.6	14.9
Aluminum Al^{++}	0.9	—
Iodine I^-	5^{-5}	1.0

[a] RDA of adults and those 4 years and older

TABLE 10
Carbohydrate Constituents of Dehulled Defatted Soybean Meal

Carbohydrate	Source	Percent
Monosaccharides		
Glucose	Cotyledons	0.3
Arabinose	Hull	Trace–0.1
Ribose	Nucleic Acids	Trace–0.1
Oleyosaccharide		
Sucrose	Cotyledons	8.1
Maltose	Cotyledons	0.6
Raffinose	Cotyledons	1.1
Stachyose	Cotyledons	4.9
Verbascose	Cotyledons	Trace
Polysaccharides		
Arabinan	Cotyledons	15.0
Arabinogalactan	Cotyledons	5
Xylan (hemicellulose)	Hulls	3.5
Galactomannans	Hulls	Trace
Cellulose	Hulls	1–2

nations of carbohydrates and proteins, fats and proteins, and fats and carbohydrates exist in the soybean. Further, there is an ash component of about 6% in soybean meal. This mineral content has been analyzed for each of the elements and most of them are present even if in extremely small quantities (Table 9).

Moreover, the soybean contains many minor constituents including vitamins, phytic acid, a few glycosides that have been identified, saponins, phenolic constituents, and phospholipids, Tables 10–12 (5).

TABLE 11

Vitamins of Defatted Soy Flour (5)

Vitamin	Quantity (mg/100 gm)
Thiamin	1.10–1.50
Riboflavin	0.24–0.44
Niacin	4.09–6.70
Vitamin B_6	0.48–1.20
Folic Acid	0.03–0.09
Vitamin B_{12}	0.06–0.20
Biotin	0.17–0.66
Pantothenic Acid	1.3 –5.1
Choline	2.2 –3.8

TABLE 12

Mineral in Defatted Soy Flour (5)

Mineral	Quantity (mg/100 gm)
Aluminum	2.33
Calcium	220.00
Cobalt	0.05
Copper	2.30
Iodine	0.001
Iron	11.00
Magnesium	309.00
Manganese	2.8
Phosphorus	680.0
Potassium	2360.00
Sodium	25.4
Zinc	6.1

References

1. *Soy Protein Products*, published by Soy Protein Council, Washington, D.C. 20037, 1987.
2. Sipos, E.F., "Edible Uses of Soybean Protein," in Proceedings of Soybean Utilization Alternatives Conference, Center for Alternate Crops and Products, University of Minnesota, 1988, pp. 57–95.
3. Kinsella, J.E., *J. Am. Oil Chem. Soc., 56*:242 (1979).
4. McNaughton, J.L., *Ibid., 58*:322 (1981).
5. Kellor, R.L., *Ibid., 51*:771–80A (1974).

Preparation of Soy Protein Concentrate Products and Their Application in Food Systems

Kenneth E. Beery

Central Soya Company, PO Box 1400, Fort Wayne, IN 46801

Abstract

Soy Protein Concentrate (SPC) is one of the growing segments of the soy protein industry. Applications are significant in all sectors of the food industry. This ranges from use in infant foods through pizza toppings and baked foods to meat and dairy products, etc. The widespread application of SPC is enhanced by its high protein level and varied textural or functional types. Thus, SPC adds significant nutrition and/or physical attributes to food products. These physical attributes principally include: structure, texture and moisture and fat holding capacity. Additionally, the functional SPC is being used as a replacement ingredient for the more expensive soy protein isolates and dairy proteins in many food systems.

Soy protein concentrate (SPC) is manufactured by selectively removing the soluble carbohydrates from soy protein flour (54% protein, dry basis) by either aqueous alcohol or isoelectric leaching. The protein content of SPC is typically in the range of 70%, on a dry basis. The textured SPC is manufactured using extrusion type technology. The definition, preparation and application of SPC products will be reviewed in this paper. Several additional reviews on soy protein concentrate (1,2) also should be consulted to gain a more thorough understanding of SPC products, their manufacture, composition, nutrition and application.

Definitions

Soy Protein Concentrates (SPC) contain not less than 65% protein (N × 6.25, dry basis) as defined by the Food and Nutrition Service of the United States Department of Agriculture. This definition was promulgated in January 1983 (3).

Internationally, through the joint FAO/WHO Food Standards Programme, Codex Alimentarius Commission, the Codex Committee on Vegetable Proteins, proposed the following "draft recommendations" for crude protein (N × 6.25) at their 1987 meeting in Havana, Cuba: Soy protein flour (SPF), 50% or more and less than 65% protein; Soy protein concentrate (SPC), 65% or more and less than 90% protein; and Soy protein isolate (SPI), 90% or more protein (4).

In each case above, the protein is calculated on a dry weight basis.

Nutrition

Protein Efficiency Ratio (PER): Nutritionally, the SPC products maintain the same excellent nutritional profile of soy flour. Thus, the PER of these products is 2.0 (5). These values are higher than those reported for soy protein isolates which range from 1.1–1.7.

Dietary Fiber: The SPC products also contain significant quantities of dietary fiber. The total quantity of these complex carbohydrates was reported by Eldridge et al. (6) to be in the range of 15.5% to 20.3% depending on the method of preparation.

Soy Protein Concentrate Products

This paper will review three of the principal types of SPC products. The first type is the traditional SPC product. This product is made by the aqueous alcohol leaching of defatted soy flakes. In this paper, the traditional SPC product will be called SPC. The second type of SPC product is functional SPC. There are many types of functional SPC products. Typically these products are differentiated from SPC by their ability to emulsify large quantities of fat and water. These products will be called FSPC products throughout this paper.

The third type of SPC products are those that are thermoplastically extruded. These textured SPC products will be referred to as TSPC. Again, there are a wide range of TSPC products. Typically, the TSPC products are made from SPC. Several types of extrusion systems are employed to make these TSPC products.

Manufacture of SPC Products

The manufacturing process for SPC is shown in Figure 1. The process starts with clean, dehulled cracked soybeans which are then flaked and extracted with hexane. The defatted soy flakes are then processed either through an aqueous alcohol or isoelectric leaching process to produce the SPC products. The first route is the process for making traditional SPC and then functional SPC from this traditional SPC. The second route is the process for making functional SPC only. As a result of either process, the predominant soluble soy sugars, sucrose, stachyose and raffinose are leached from the defatted flakes. Thus, as the soluble sugars decrease, the protein increases.

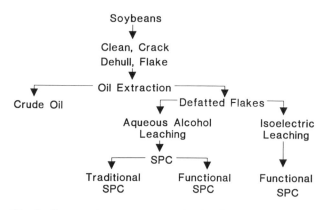

Fig. 1. Soy protein concentrate production flow diagram.

Traditional SPC: In the production process employing aqueous alcohol the proteins remain insoluble while the soy sugars are solubilized and leached away. The composition of the defatted flakes according to Campbell et al. (1) is shown in Table 1. The composition of the aqueous alcohol leached SPC is also shown in Table 1. The protein level of the SPC is approximately 72%. Accordingly, the percentage of sugars has decreased significantly.

TABLE 1

Approximate Composition of Soy Protein Flakes and SPC

	Soy protein flakes*	SPC*
Protein	56	72
Sugars	16	1
Polysaccharides	15	20
Ash	6	5
Other	6	1
Fat	1	0.3

* Percentage on moisture free basis.

Two principal SPC products are available commercially. The first is a finely ground powder of approximately 100 mesh. This product is produced by grinding the aqueous alcohol extracted SPC flakes. The second product is from the partially ground and sieved aqueous alcohol extracted SPC flakes. Typically, this product passes through a 20 mesh screen and is retained on a 60 mesh screen.

Functional SPC: Figure 2 shows the flow diagram for producing the two types of FSPC products. The first type of FSPC is made from SPC following the process as described by Howard et al. (7). This process utilizes a series of steps involving solubilizing the SPC at a slightly alkaline pH, increasing the temperature and employing a shear device. These steps are then followed by neutralizing the mixture and spray drying it. The process essentially converts the low nitrogen solubility index (NSI) SPC into a high NSI FSPC. The resulting FSPC is described as having functional properties similar to milk proteins.

The second method of making FSPC is through the route of isoelectrically leaching the defatted soy protein flakes at pH 4.5. This process, as described by Sair (8), causes the majority of the soy proteins to be in their most insoluble form and allows the soluble soy sugars to be leached from the protein and fiber. Following the extraction phase, the resulting slurry is then neutralized and spray dried.

Textured SPC: The TSPC products are produced on the same equipment and in nearly the same manner as the textured SPF products. Following the thermoplastic extrusion of the SPC, the extruded pieces are either dried as large chunks or post sized with a cutting device to make a granule or flake shaped piece. Additionally, the density of these products can be varied by using extrusion technology.

Through the extrusion and post sizing methods, a wide array of different chunk, granule and flake sizes can be achieved. This process produces "natural colored" TSPC products. By the addition of caramel coloring to the extruder, caramel-colored products of varying intensity can be produced by increasing or decreasing the caramel addition feed rate. Colored pieces can also be made by adding food grade colors to the extruder. The most common colored piece is red. Often this piece is used for ham or bacon bit applications.

Applications

In this section, I will refer to three types of SPC products. The first type is the traditional SPC products. The second type is the functional SPC products, and the third type is the textured SPC products. A summary of these applications is shown in Table 2.

Traditional SPC: These are the aqueous alcohol leached products. These products can be in either a powder (100 mesh) or grit (20/60) form. The powdered form often is used in food systems to increase their protein levels. The grit form is primarily used to add structure to and to limit texture of food systems. Both products help to hold additional water in the food. In meat systems they help to retain the meat juices and fat and assist in improving palatability.

The higher protein content of SPC versus SPF allows it to be used where higher protein or more functionality is required. In addition to providing a higher protein content than SPF, the SPC also has a lower flavor profile and greatly reduced flatulent sugar level. Additionally, because of the reduction of sugars, the browning reactions are less and the color of the SPC is lighter. This allows the SPC to be more useful in poultry and fish systems.

Functional SPC: Although traditional SPC is available in the greatest abundance, the FSPC products are very useful in more demanding applications to replace the higher priced milk proteins and SPI. This is exciting as the FSPC products are used to replace caseinates and SPI on a kilo-for-kilo replacement basis in meat systems.

Comminuted meat systems: Due to FSPC's greater emulsification and water binding capacity than SPC, the FSPC products are very useful in comminuted meat systems like frankfurters, bologna, etc. These applications

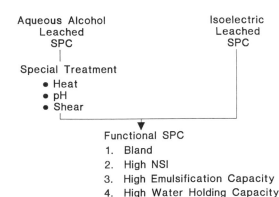

Fig. 2. Functional soy protein concentrate production flow diagram.

TABLE 2

Typical Food Applications for SPC Products

	SPC*	FSPC*	TSPC*
Meat products			
Coarse ground meat systems			
Meat patties			
Beef, pork, poultry, fish	X	X	X
Meat balls	X	X	X
Pizza toppings		X	X
Chili	X	X	X
Comminuted meat systems			
Frankfurters		X	
Bologna		X	
Whole muscle products			
Ham		X	
Corned beef		X	
Baked foods			
Cereals	X	X	
Bread, cakes, doughnuts	X	X	
Dairy products			
Beverage powders		X	
Infant formulas		X	
Coffee whiteners		X	

*Local regulations may determine the type and amount of these SPC products which may be added to these food systems and the name of the finished product.

are shown in Table 2. In these comminuted meat systems, the FSPC mixes with the salt soluble meat proteins to form a stable emulsion. Of importance, the FSPC products are not sensitive to the salt used to solubilize the myosin type meat proteins. Therefore, the FSPC can be added either dry or as a fat emulsion after the extraction of the meat proteins with the salt in the chopper bowl. This allows the application of meat science principles in making these sausage-type comminuted meat products.

Coarse ground meat systems: In addition to the application of FSPC in comminuted meats, FSPC is often used in coarse ground meat systems such as meat patties, pizza toppings, nuggets, etc. to add firmness, higher nutritional value, retention of meat juices and fat and to aid in manufacturing operations. Additionally, the use of the FSPC will increase the yield of cooked meat product. In these applications, the FSPC usually is added dry or with four to five times its weight of water to the meat system during the blending phase of the manufacturing process. Typically, 2–3% of the dry FSPC with or without additional water is added to the meat product.

Ham and other brine-added meat products: Internationally, ham products are increasingly using greater levels of water than ever before. These higher levels of water addition are usually aided by the addition of soy proteins in the pumping, massaging and/or tumbling cure brine. Thus, for these applications, the FSPC is ideal as it is very dispersible and also its hydration and function is not affected by the salt of the brine solution. In these applications, the FSPC is used at a rate of 2–3% in the finished ham product. The total addition rate is usually at 40% to 100% of the original weight of the meat used for these ham products or other water added meat products.

Textured SPC: The TSPC products are unique as they are the second generation of the original and very popular textured soy protein flour (TSPF) products. Nevertheless, they have significantly increased usefulness. Table 3 shows the improved properties of the TSPC versus the TSPF products. Each of these properties helps to allow for the greater versatility and higher percentage replacement applications of the TSPC versus the TSPF.

TABLE 3

Properties of Textured Soy Protein Concentrate Versus Textured Soy Protein Flour

Light color
Bland flavor
Reduced flatulent sugars
Increased hydration levels
Better texture

The most popular applications of the TSPC products are in coarse ground meat systems: products such as meat, poultry and fish patties, pizza toppings, meat balls, nuggets and other products. In these systems, the TSPC is used quite advantageously to replace higher cost meat. Additionally, the blander flavor of TSPC allows the flavor of the meat to be more pronounced. Also, the yield of the meat system is increased by using TSPC. In most applications, TSPC is used at a level of 3–8% (dry) of the finished meat product. The regular hydration rate for TSPC in meat applications is 3:1. For some special applications, the hydration rate is reduced to 2–2.5:1 to allow the TSPC to further hydrate with the meat juices while the meat is thawing and during the cooking process. This further helps to conserve the proteins that would be lost with the drip and to enhance the meaty flavor.

Combination of TSPC and FSPC: For pre-cooked coarse ground meat systems, an improvement in yield can be achieved by the simultaneous application of both a textured and functional SPC. This dual application is important in many countries as pre-cooked meat items are gaining popularity in food service operations. For this application, because the meat processor is selling a partially cooked item, it is important that he achieve a maximum yield. With the addition of 2% of FSPC along with the TSPC, the yield of the pre-cooked patty may increase by as much as 4% over the use of TSPC alone.

Cholesterol content: Because the SPC products do not contain cholesterol, the cholesterol content of the finished meat product is decreased by the same percentage as the meat product is increased by the use of SPC products and water. In some pumped meat products the amount of cholesterol may be reduced to nearly one half of its original level.

In summary, the SPC products are an exciting class of soy protein products. Because of their significant decrease in flatulent soy sugars, the SPC products are considered an excellent choice of soy ingredients to be used in new food products where the nutritional and functional qualities of soy proteins are required. The cholesterol dilution effect, caused by the application of SPC products in meat systems, may also have considerable future value.

Additionally, because of the exceptional emulsification and dispersibility properties of the FSPC products, these products are finding an increasing role in meat systems. The TSPC products are replacing the TSPF products in many improved and upgraded coarse ground meat products. Today, TSPC is the starting material of choice for new product development in the meat industry. Further, the combination of TSPC and FSPC products is adding a new dimension toward improving the yields of processed meat products.

The production capacity for SPC products is significantly increasing as current producers are expanding their facilities and building new plants while new producers are entering the industry. This is an exciting time for soy protein concentrate products and their applications in food systems.

References

1. Campbell, M.F., C.W. Kraut, W.C. Yackel and H.S. Yang, "Soy Protein Concentrate," Chapter IX in *New Protein Foods*, Vol. 5 of Seed Storage Proteins, edited by A.A. Altschul and H.L. Wilcke, Academic Press, Orlando, Florida, 1985.
2. Anonymous, *Soy Protein Products: Characteristics, Nutritional Aspects and Utilization*, Soy Protein Council, Washington, D.C., 1987.
3. Food and Nutrition Service, U.S. Department of Agriculture, *Federal Register*, Jan. 7, 1983.
4. Codex Alimentarius Commission, *Report of the Fourth Session of the Codex Committee on Vegetable Proteins*, Appendix VI, *Proposed Draft Recommended International Standard for Soy Protein Products*, Alinorm 87/30, FAO/WHO, Rome, 1987.
5. Rakosky, J., Jr., and E.F. Sipos in *Food Service Science*, edited by Smith and Minor, Avi Publishing Co., Westport, Connecticut, 1974, p. 383.
6. Eldridge, A.C., L.T. Black and W.J. Wolf, *J. Agri. Food Chem. 27*:799 (1979).
7. Howard, P.A., M.F. Campbell and D.T. Zollinger, U.S. Patent 4,234,620 (1980).
8. Siar, L., U.S. Patent 2,881,076 (1959).

Processing for Producing Soy Protein Isolates

Dale W. Johnson[a] and Saburo Kikuchi[b]

[a] Food Ingredients (Minnesota) Inc., 2121 Toledo Ave. N., Golden Valley, Minnesota 55422 USA, and Department of Chemical Engineering and Material Sciences, University of Minnesota, Minneapolis, Minnesota 55455
[b] The System New Life Inc., 1201 2-19-17 Takanawa, Minato-ku, Tokyo 108, Japan

Abstract

There are many factors which must be considered in order to be able to manufacture isolated soy protein with the physical and functional properties desired in a variety of different food applications. Most isolated soy protein products are produced by slurrying flakes with water, separating the insoluble material and the water soluble protein (referred to as "mother liquor") acidification of the mother liquor by adding acid to about the isoelectric point of the protein (about pH-4.5) washing the curd so formed, neutralizing the curd to a pH of about 7.0 followed by spray drying. The various conditions of processing such as water to flake ratio, pH, temperature, equipment, etc. will be covered. Possible alternative methods of producing isolates will also be discussed, including ultrafiltration and a new swollen gel technology.

There are a number of ways that have been described for producing isolated soy and other vegetable proteins, but the only commercial procedure currently being utilized basically consists of extraction of defatted soyflakes with water, followed by centrifugation of the slurry to separate the insoluble materials (spent flakes) to obtain a dispersion containing soluble protein and some nonprotein solutes (mother liquor). The mother liquor is acidified to precipitate the protein, the curd is washed to remove as much of the nonprotein solubles as possible and neutralized to make a protein isolate that is spray dried. Mention will be made of some of the other technologies which may be considered for producing isolated soy proteins.

The term "isolated soy protein" is a general term referring to a number of different products produced by varying the processing conditions so that each product will have somewhat different chemical, physical and functional characteristics desired for given applications in food systems.

The definition of "isolated soy protein" which has been recommended by the National Soybean Processors Association, and has been accepted generally by the U.S. Food and Drug Administration (FDA) and the United States Department of Agriculture (USDA), states that it is "the major proteinaceous fraction of soybeans prepared from high quality, sound, clean, dehulled soybeans by removing the preponderance of the nonprotein components and shall contain not less than 90% protein (N × 6.25) on a moisture-free basis (1).

The raw materials for producing soy isolates are defatted soy flakes, grits or flour that have been processed to give a desolventized flake with the desired characteristics to produce soy isolates in an economic manner.

Most solvent plants have been built to produce soybean meal for animal feed and oil, i.e., for further processing with minimum considerations for sanitary construction and the sanitation needed to produce edible soy protein products. Most processors, at least in the United States, use a separate solvent plant, for producing the defatted flakes which are further processed to make soy flour and grits, certain types of textured soy flour products, soy protein concentrates and isolated soy proteins. For more detailed information on the reasons behind this, a discussion of solvent plant considerations was given at the Short Course on "Food Uses of Whole Oil and Protein Seeds" held in Hawaii in 1986 (2).

Isolated soy protein products have good nutritional value, particularly when used in combination with protein products low in lysine, but the primary interest in soy isolates has been in utilization of their functional properties. Flavor is of primary concern because without satisfactory flavor any other characteristic would be unimportant. Over the years, improvements have been made in the flavor of isolated soy protein products, but for some uses flavor is still a problem.

An isolated soy protein plant should be designed in a manner similar to plants used for processing cow's milk for distribution as fluid milk or for use in making cheese or other dairy products. In designing the equipment and planning the layout, it should be done with all facets of sanitation in mind, using CIP (cleaning-in-place) equipment with a minimum cleaning by hand required. If there should be an accidental spill during operations, there should be strict operating procedures for immediate cleanup of spills, whether liquid or dry material. It is desirable to have the processing area under a slight positive air pressure. Because the process for producing isolated soy protein is a wet process, it is important that the starting raw material be as low as practical in microbial contamination, including pathogens, thermophilic and mesophilic spores and other organisms. Some people have recommended that the beans used to make edible products should be washed with a chlorine or hydrogen peroxide solution, followed by rinsing and, if necessary, drying before they are cracked, dehulled, flaked and solvent extracted. To the authors' knowledge, none of the major producers of soy protein products are doing this. Hydrogen peroxide would be particularly effective in destroying spores.

Properties of Soy Protein Isolates

The important properties of isolated soy proteins which are of practical interest, depending on their end use, are listed in Table 1. These properties will vary from product to product and can be varied depending on a number of factors, including characteristics of the original soy flakes, extraction conditions, the use of certain chemicals at different stages of the production operation, heat treatment given to dispersions and/or curd at different stages of production and even storage conditions for the dried products.

TABLE 1
Properties of Isolated Soy Protein of Practical Interest

Flavor	Fiber forming characteristics
Odor	Gelation properties with heat
Color	Protein dispersibility
Particle size	Clarity of dispersions
Microbiology	Viscosity of dispersions
Wetability	Gelatin properties with heat
Cohesion	Water holding capacity
Mineral content	Film forming properties
Fat absorption	Foaming ability (whipability)
Emulsifying ability	Foam stability
Density	Fiber forming characteristics
Dough forming ability	Freeze/thaw properties

Typical Analyses of Commercial Protein Isolates

Typical analyses of commercial soy protein isolates are given in Table 2. The analytical differences are minor, but the functional differences may be great.

TABLE 2
Typical Analyses of Commercial Soy Isolate

	Range
Protein (as is)	88.3–91.8
Protein (dry bases)	95.0–96.5
Moisture	4.9–7.0
Fat	0.1–1.0
Fiber	0.01–1.0
Ash	2.4–3.8
Carbohydrates	traces

Comparison of Some Functional Properties of Isolates

Of the functional characteristics listed in Table 1, in comparing any two isolated soy protein products, the functional characteristics of each product may be quite different. Depending on the characteristics desired in the protein for a given application, the processing conditions are varied to obtain specific functional characteristics. To illustrate one of the varying characteristics one may obtain, Table 3 presents some data on the viscosity characteristics of five different commercial soy protein isolates, whereby the dispersions were subjected to certain heat treatment for comparison to the same dispersions without heat treatment. The dispersions were prepared with distilled water at concentrations of 10 and 14 w% and viscosities were determined at room temperature. The same dispersions were heated in flasks for 30 min in boiling water, immediately cooled in cold water, held at 4°C over night, and the viscosities were re-determined. As will be noted, the differences in viscosity between the 10 and 14% dispersions varied considerably for each of the protein products. The same was the case for those which had been heated and cooled.

TABLE 3
Comparison of Viscosity-in Poises-of Dispersions of Commercial Isolated Soy Protein Products in Distilled Water

	Viscosity at room temperature—not heated	
Product	10% Dispersion	14% Dispersion
A	0.6	2300
B	65.0	1900
C	17.0	700
D	14.0	2100
E	.05	.07

Dispersions heated 30 min. in boiling water, cooled and stored at 4°C overnight

	Viscosity at room temperature	
Product	10% Dispersion	14% Dispersion
A	2100	>33,000
B	340	16,000
C	5400	>33,000
D	150	19,000
E	1.5	270

Discussion of Processing Conditions

Figure 1 presents a typical flow sheet for producing isolated soy protein as currently practiced.

Raw Material

The starting raw material is a defatted soy flake which may be used as a flake or ground to a flour. If flour is used, it is usual to grind it to about a 200 US mesh product. If flakes or grits are used, it is desirable to remove the fine material, giving what may be termed a dedusted flake or grit. The use of flakes saves a grinding cost, cuts down on bin storage space and alleviates dust and other handling problems.

In order to obtain satisfactory yields of isolated soy protein, it is desirable to have a flake with as high a PDI (protein dispersibility index) as possible, commensurate with obtaining satisfactory yields of protein. Depending on a number of factors, the PDI of the starting raw material for producing isolates generally is more than 70, with some processors using a material with a PDI of about 90. When the full fat flakes are subjected to hexane extraction to remove the soybean oil in order to have a high PDI flake, it is common to use a flash desolventizing or a vapor desolventize-deodorizer system to obtain a high PDI product of about 90. If a Schnecken system is used, the highest PDI one can obtain is about 70. There are other systems which can be used for desolventizing.

The PDI of the initial starting material can have an effect on the functional characteristics of the isolated soy

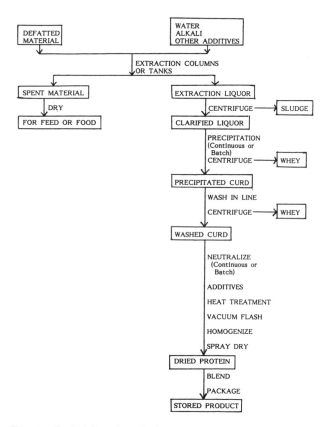

Fig. 1. Typical flow sheet for isolated soy protein production.

protein produced. To obtain the desired properties in the finished isolate, the processing conditions must be adjusted for materials with different PDI values.

Depending on the origin of the bean, the protein in the defatted flakes may run about 49-52% protein, on a 12% moisture basis. If a flour is used, since some moisture is lost during the grinding operation, the moisture content of flour will be slightly lower and the protein content slightly higher from that in the flakes from which the flour was made. Using a flake with a PDI of about 90, with the processes currently being used, the yields of isolate may run in a range of 33-38% based on the original flake weight, although some people claim that higher yields are obtained. Using a starting material with a 70 PDI, the yields will run in the range of 28-32%. The protein content of the original soy beans will vary from year to year in a given growing area and will also vary from area to area in a given year, depending on the variety and the conditions in the area where the beans are grown. Usually beans grown in the northern part of the U.S. and some other parts of the U.S. are lower in protein and oil than those grown in the Middle West. Beans currently grown in India have a higher protein content than those in the U.S.

Water Characteristics

The quality of the water, from the standpoint of flavor and odor, is very important, since considerable quantities of water are used in extraction of the flakes and in washing of the curd. The water hardness preferably should be under 200 ppm, calculated as calcium carbonate, and iron below 0.5 ppm, because iron can affect flavor and color. In some areas it might be necessary to treat the water to have satisfactory water for isolated soy protein processing. Too high a chlorine level may also effect flavor. In the later stages of the process, steam may be used in processing. The steam should be culinary steam, such as is used in the dairy industry and other types of food processing operations. It is wise to check the water supply at intervals for odor and flavor properties.

Extraction Conditions

For the extraction of the flakes, the pH range used in different processes may be from 6.8 to 10, with a value between 8.5 and 9.0 being most common. The use of too high a pH and temperature can have an adverse influence on flavor, nutritional value and functionality of the protein produced. The pH may be adjusted with sodium hydroxide, lime, trisodium phosphate, or combinations, as well as other alkaline agents approved for food use. In some cases sodium sulfite or bisulphite has been used in the initial extraction, partly as a bactericide but also to control viscosity during the processing operations. Its use may have an effect on the functional and flavor characteristics of the dried isolated soy protein. Due to certain allergy problems with some individuals there may be restrictions on its use. The major part of the sulfite is washed out in subsequent operations or largely removed during the vacuum treatment of the neutralized curd and in spray drying. It is reported that the usual methods for determining sulfites in food products may give erroneous results when applied to isolated soy protein.

The alkaline extraction may be carried out in continuous cocurrent or countercurrent extraction systems or in a batch tank system. Depending on the operation, the water to flake ratio is usually in a range from 6:1 to 20:1. The time for the initial extraction may be in the range of 25-35 minutes. Because there is quite a bit of air in the interstices of the flakes, this may be a factor in the rate which the water penetrates the flake and the rate and amount of solubles which move from the flakes into the liquid portion. Consideration might be given to deaerating the flakes prior to wetting, to improve total extractability as well as the rate of extraction. In a countercurrent extraction column, sometimes there is a tendency for the flakes to float to the surface, so deaeration might help this problem. If a number of tanks are used in sequence in a countercurrent batch system, one can move the extraction liquor countercurrent to the movement of the flakes, so that the final wash of the flakes is done with plain water. The fresh flakes are washed with the liquor with the highest solids concentration from previous washes. With the tank system, the water is removed from the cake using centrifuges, and the cake reslurried with extraction water from the previous wash. The spent flakes from the final fresh water wash can be dried for feed or food purposes.

Fiber in Spent Flakes

With the interest in high fiber diets in the U.S., the spent soy material is being promoted as an edible fiber source for use in food. Table 4 gives the proximate analyses of some commercially produced spent soy material being promoted for food use. The dietary fiber consists of cellulosic compounds, lignins, hemi-celluloses, gums, mucilages, etc. (3,4,5). These products are different com-

TABLE 4

Proximate Analyses of Commercial Spent Flakes Products Resulting from the Manufacture of Isolated Soy Protein

	Per cent
Moisture	6–7
Protein (dry basis)	16–36
Ash	4–6
Fat	0.5–1.0
Crude fiber	9–13
Total dietary fiber	45–75

positionally from the soybean hull material, which also has been promoted for dietary fiber use.

Temperature for Processing

Flake extraction is usually carried out in the temperature range of 27-66°C. Temperatures in the range of 35–50°C are more conducive to microbiological growth with its concomitant problems. As low a temperature as possible would be less conducive to microbial growth, but the time required to get the most complete extraction of protein would be increased considerably, thus the equipment and space requirements would increase the plant cost. At about 70°C and higher, the protein becomes progressively less soluble as the temperature and time of exposure is increased, to the point where it is almost completely insoluble above about 88°C (6). If the spent flake material is to be used for food purposes, the drying system must be selected in such a way as to have a sanitary operation with ease of cleaning, as well as drying under conditions so as not to overheat the product, which could result in undesirable color and flavor. If the product is to be used for animal feed, any type of steam tube dryer will do.

Curd Processing

The extraction liquor separated from the spent material may or may not be subjected to further clarification by centrifugation, depending on the desired end use. If the protein is intended for spinning to produce edible protein fibers for food use, it is important to have a clarified extraction liquor to remove as much crude fiber as possible, because, if the crude fiber in the isolate is too high there will be a problem of plugging the spinnerettes used for producing spun protein fiber, and it will cause wear on the spinnerette nozzles.

The extraction liquor, depending on a number of factors, is usually heated to 27-54°C prior to precipitation. Precipitation usually is carried out at a pH around 4.5 using hydrochloric or phosphoric acid, although other acids may be used. The precipitation may be carried out in a continuous process or in a batch system. Following precipitation, the curd slurry is centrifuged to concentrate the cake to a solids level of up to about 30%. With some centrifuges, it is possible to concentrate the curd to 50% solids, but this generally is not recommended because of the difficulty of redispersing the curd and in order to obtain effective washing and neutralization in subsequent processing steps. The curd from the first centrifuging may be washed continuously using a ratio of one part of water to one part of wet curd, followed by a second centrifugation. The curd is usually washed with water at a temperature of 50-55°C. In some cases, the curd from the first centrifugation might be slurried in a tank using a water to wet curd ratio of about 3:1, followed by centrifugation to obtain a curd with about 30% solids. The curd may then be reslurried in a tank or an enclosed pipe system as a continuous neutralizing system using an appropriate alkali, to the point that when the spray dried product is redispersed in water, it has a pH of 6.5 to 7.0. The alkaline agent may be any edible alkali such as sodium hydroxide, potassium hydroxide, trisodium phosphate, lime, etc. If lime is used, the resulting dried protein has a very low PDI, but the other alkalies result in a protein with high dispersibility. If desired, the redispersed material might be centrifuged to remove other insoluble materials. This usually is not done commercially. Depending on the desired characteristics of the final spray dried product, certain other additives, such as fats or lecithin, may be mixed into the dispersion before spray drying. Generally, the lower the pH of the dispersion the higher the bulk density of the dried product, but if the pH is too low, the protein dispersibility of the dried product will be low. The additives may also influence the bulk density. There have been cases where the isoelectric curd has been reslurried and spray dried without neutralization.

Spray Driers

Although spray driers of the spinning disc type are used for drying isolated soy protein concentrates, most companies seem to prefer a nozzle type of drier. The liquid going to the spray drier may run about 12-20% solids, depending on the viscosity of the dispersion. Naturally, the higher the solids in the dispersion going to the drier, the lower the drying cost. Additives to the neutralized dispersion, such as hydrogen peroxide or bisulphites and pH can influence the viscosity characteristics of the dispersions before spray drying and effect certain functional characteristics of the redispersed, spray dried isolates. For example, sulfites can be used to decrease the gelling properties. Hydrogen peroxide can improve gelation properties of isolates, but it can have an adverse effect on flavor of the dried product and on the PER, in that to some extent there is oxidation of methionine and cystine. However, if the isolated soy protein is used in conjunction with other proteins which have adequate quantities of sulfur-containing amino acids, this effect would be relatively unimportant. The important thing is that lysine is not affected, because soy proteins in nutritional products usually are used in conjunction with protein products that are low in lysine but have adequate sulfur-containing amino acids. Using too high a temperature and pH at certain stages can result in the formation of lysinoalanine, which is undesirable.

In order to help minimize the undesirable flavor frequently associated with some isolated soy protein products, before spray drying it is common to subject the protein dispersion to a flash vacuum system, by injecting live steam into the dispersion and subjecting the dispersion to a vacuum treatment, to steam distill off some of the components responsible for the undesirable flavor and odor. The heat treatment is usually done in a continuous manner through a heating tube or other device where the temperature of the dispersion is increased to 104–154°C with a holding time of 10 to 15 seconds, followed by a vacuum flash set up, so the dispersion is flashed to a temperature of 50–55°C. The dispersion is then passed through a homogenizer to the spray drier, using an inlet temperature of 205–218°C and an outlet temperature of about 82–92°C.

Flavor

It appears that some of the flavor problems associated with soy protein products result from oxidation of fat-containing moieties which exist as lipoproteins in these products. If one were to carry out fat analyses using the usual procedure with petroleum ether, the fat content of the isolates will run from about 0.1 to over 1%, depending on the fat content of the original starting raw material. If analyses on isolated soy protein are carried out extracting with hot ethanol followed by petroleum ether, or by petroleum ether extraction of an acid hydrolysate of the protein, one will find that there is about 2 to 3% of fatty material associated with the protein. If that fatty material is recovered by evaporating the solvent, it has a waxy appearance with an unpleasant flavor and odor. It is known that if the soy flakes or the precipitated curd is washed with aqueous ethanol, the resulting flakes or spray dried products from the neutralized alcohol washed curd will result in products with much more acceptable flavor than those prepared without the alcohol treatment (6,7).

In recovery of the alcohol for reuse, it is necessary to have a good rectifying column with a take-off for certain fractions of the ethanol in which the flavor components seem to be concentrated. On reuse of the alcohol, if the fraction is not removed, there is a gradual build up in the recovered alcohol going to storage, and eventually the alcohol will become more concentrated with the flavor agents and become ineffective in removing the undesirable flavor agents from the curd or flakes (7).

Ultrafiltration Studies and Phytate Removal

There has been some concern about the presence of phytates and phytic acid in soy products. The soy protein products produced as protein isolate or isoelectric proteins from acid curd, contain 2 to 3 % of phytates, w/w, based on the protein. For some purposes, the phytic acid or phytates are undesirable because of their ability to sequester certain cations which might result in affecting the availability of trace metals needed for metabolism in the human dietary. This is particularly important for infants.

Phytic acid also has an effect on the isoelectric point of the precipitatable protein. Soy proteins with 2 to 3% phytic acid have an isoelectric point in the range of about pH 4.5, but phytic acid "free" soy proteins have an isoelectric point of about pH 5.0. In certain types of acid food products it may be desirable to have phytic acid free soy proteins so, particularly in certain fruit beverages, if proteins are added, they don't precipitate.

Because of the presence of phytates and the loss of albuminous proteins in the acid precipitation process, in the early 1970's the interest in possible removal of phytates by various procedures, such as ultrafiltration and diafiltration, came under serious consideration (8–27). While the various processes which have been described would appear to be practical methods of preparing improved isolated soy protein products, for reasons unknown to the authors, no commercial operations using these techniques have been installed. Perhaps this may change in the future with improved membranes and technology.

There have been a number of patents issued to researchers at Mead Johnson & Company, a division of Bristol-Meyers Co. (10–14) dealing with phytate removal and nutritionally improved soy products using ultrafiltration and diafiltration by using certain procedures for treating the mother liquor. They carried out water extraction of defatted flakes at a pH range of 7 to 10, followed by centrifugation to remove the insolubles and obtain the mother liquor (10). The mother liquor was adjusted to a pH of 11.0 to 12.0 at a temperature of 25-35°C followed by a second centrifugation to remove other insoluble materials including phytin while maintaining the temperature in the range indicated. The clarified solution was then neutralized to a pH of 6.5 to 7.5 and purified by ultrafiltration. They preferred using a neutralized extract containing about 3.5% protein (w/w) and a carbohydrate content of about 2% and ash and other solubles at 1% or less.

With extracts greater than 12% of protein by weight, they found the solution was quite viscous and inconvenient to handle and inefficiently processed during centrifugation, washing and in the ultrafiltration steps. If the protein is less than 1%, the expense of ultrafiltration would be substantially increased due to the prolonged filtration needed. They used a semipermeable membrane having the capacity of retaining proteins in the range of 10,000 to 50,000 daltons. They generally used a pressure of 25 psig. They state that the reverse osmosis process uses membranes which have much lower porosity and retain much lower molecular weight materials, such as carbohydrate constituents of the soybean, which it is desired to eliminate in their processes. They further state that reverse osmosis is considerably more expensive to operate in that operating pressures are higher and generally lower flux rates are involved.

For the ultrafiltration, they prefer a temperature of about 45°C in order to increase the flux rate and reduce the time required to achieve the desired concentration of protein. They found that straight ultrafiltration left too much carbohydrate in the retentate, so they followed the ultrafiltration with diafiltration (a form of ultrafiltration where the retentate is continually diluted with water in an amount equal to the water being removed). Once the desired protein concentration was established in the retentate by simple ultrafiltration, so that the protein concentration was about 7%, they then used diafiltration, which consists of a washing operation in which soluble low molecular weight constituents are washed from the retentate. Using one part of the original extract and reducing the volume of the retentate by one half of the original volume of the clarified extract, they then subjected this half volume to 2.5 volumes of water to dilute the retentate during ultrafiltration, until the total permeate collected is up to three volumes. Table 5 shows the effect of diafiltration on the carbohydrate content as the material is subjected to ultrafiltration followed by diafiltration (10). The original clarified product contained about 77% protein (dry basis) and by ultrafiltration increased to 87%, and subsequent diafiltration increased the protein content to about 97% protein.

TABLE 5

Effect of Ultrafiltration and Diafiltration on the Carbohydrate Content of the Original Mother Liquor and Retentate at Various Stages of Processing (10)

Volume^a ratio	Per cent protein in retentate (dry basis)
0.0^b	77
0.5^c	87
1.0^d	92
1.5^d	95
3.0^d	97

^{a} Clarified neutralized extract prior to ultrafiltration.
^{b} Ultrafiltration only.
^{c} Diafiltration of material from 2.
^{d} Ratio of total permeate volume to neutralized extract volume.

They found that the pH was an important factor in removal of the phytate. Table 6 shows the phytate content, as per cent of protein, where the pH of the mother liquor was adjusted upward to about pH 11 (10). It will be noted that where the process was carried on at a pH of 8.5 to 10.0, there was very little difference in phytate removal. However, above pH 10 up to about 11, the major part of the phytate (about 95%) was removed.

TABLE 6

Effect of pH and Clarification of the Slurry on the Phytate Content of the Solids in the Mother Liquor (10)

pH	g Phytate/ 100 g solids
8.5	2.18
9.5	2.11
10.0	2.14
10.5	1.45
11.0	0.05

The temperature of the mother liquor also has an effect on phytate removal as shown in Table 7 where the mother liquor was adjusted to a pH of 11 at different temperatures (10). They used a procedure for determining phytate removal as described in the patent.

In the Mead Johnson work, in some cases, they used defatted flakes, but in one study they used full fat flakes. They claim that whether they used defatted or full fat flakes made little difference in the problems of ultrafiltration. With the full fat flakes all of the fat remained in the retentate (12).

Iacobucci et al. (9) used three different procedures to lower the phytic acid content of isolated soy protein prod-

TABLE 7

Effect of Temperature for Adjustment of pH of Mother Liquor at pH 11.0 on Phytate Removal (10)

pH Adjustment temperature	% Phytic acid remaining by volume mother liquor
5°C	69
10°C	46
15°C	15
20°C	10
25°C	5
30°C	0

ucts. Essentially they started with the same type of mother liquor used in the acid precipitation process. In one procedure they adjusted the pH to about 4.5 to 7.0 and added a phytase enzyme which they allowed to act for an appropriate period of time, followed by ultrafiltration. In another procedure they adjusted the pH to 2.0 to 4.5 in the presence of large excesses of calcium and magnesium ions at these low pH values and the liquor so prepared was subjected to ultrafiltration. In another procedure they adjusted the pH to in the range of 7.0 to 11.0 with the addition of chelating agents such as ethylenediaminetetraacetic acid (EDTA), nitrilotriacetic acid (NTA), and some other chelating agents, with ultrafiltration being carried out in the alkaline range. Comparing these results with those of the Mead Johnson workers, it would appear that the Mead Johnson procedures are more effective.

Puski et al. (14) used a slightly different procedure than Goodnight et al., and other workers. For the flake extraction, they preferred not to exceed a pH of 10.0 or temperatures in excess of 60°C for periods of longer than about 30 minutes. They state that the loss of sulfur-containing amino acids, such as cystine, occur during excessive conditions of temperature and alkaline pH. At 20–30°C they found no loss of cystine in the dissolved protein during periods of 6 hours or longer. They state that homogenization of the slurry is not necessary to effect efficient extraction of the protein and is really undesirable, in that the flux rate is reduced during ultrafiltration. They suggest 8 to 16 parts of water per part of defatted material by weight. After separation of the insolubles the mother liquor was adjusted to a pH of 6.5–7.5 prior to membrane filtration. For the ultrafiltration they preferred a temperature of 60–65°C. At temperatures over 75°C certain chemical decomposition and condensation reactions occur, such as formation of lysinoalanine. They prefer a temperature above 60°C to slow down the growth of microorganisms. They state that below 45°C the flux rate diminishes. In their process they prefer a HTST treatment of the extract, because it alleviates microbiological problems and the flux rate is increased. They further suggest that the heat treatment be split into two phases with a relatively mild heat treatment before the ultrafiltration and a more intense treatment of the retentate after removal of a significant part of the carbohydrates. Such a treatment minimizes the browning reaction. They suggest mild heat treatment, using 60°C for 30 minutes or 130°C for one minute followed by cooling to a temperature of

45 to 75°C before ultrafiltration. The retentate from the ultrafiltration may be heated to 110°C for one minute and up to 150°C for one second, followed by cooling.

They define low phytate soy protein isolate as containing at least 88% protein and less than 0.3% phytate by weight of protein. After ultrafiltration, they acidify the retentate to a pH of around 5.0 to 5.5, and the precipitate is recovered and neutralized. At this pH, the phytate stays with the whey. In a number of tests, as shown in Table 8, they found that the isolates contained phytates in the range of 0.02 to 0.11% (w/w) with most of the values in the range of 0.05%, in comparison to most commercial products which contain in the range of 2.0–2.5% phytate. The yield of protein based on 100 parts of soy material was in the range of 30%. The PDI of the original starting material was not given.

TABLE 8

Percent Phytate and Yield of Soy Isolates by the Procedure Described in Several Test Runs (14)

% Phytate[a]	% Yield based on soy material
0.02	31.3
0.03	28.9
0.05	30.5
0.05	30.9
0.11	29.9
0.07	30.4

[a] Commercial soy isolates (acid precipitation process) contain 2–2.5% of phytate.

Table 9 shows the effect of temperature, extraction pH, and pH for precipitation on the phytate content and yield using the procedure they describe (14). With this process they also end up with an isolate with a relatively low aluminum content as shown in Table 10.

TABLE 10

Comparison of Aluminum Content of Soy Isolates Prepared According to the Process (14) to a Commercial Isolate

	Sample 1	Sample 2	Commercial Isolate
Aluminum mcg/g	7.5	6.2	24–40
Phytate %	0.11	0.13	>2

Table 11 shows that the protein product given the double heat treatment to destroy the antitrypsin factor results in a product with a higher PER than the acid precipitated product or the ultrafiltered product which was not heat treated. The whole protein recovered from ultrfiltration is nutritionally superior to acid precipitation of proteins where the albuminous proteins are lost.

Other significant work on ultrafiltration and reverse osmosis for processing vegetable proteins, including soy, has been carried out at Texas A&M University, Food Protein Research and Development Center, College Station, Texas. In much of that work researchers utilized ultrafiltration membranes to concentrate the protein and reverse osmosis systems to treat the retentates, whereby the treated water from the reverse osmosis system might be reused in the initial processes. The process may be used for cotton, soy, peanut (groundnut), sesame, etc. Details of their work are presented in a number of publications (15–27).

Swollen Gel Technology

Our group at the University of Minnesota has been working on a new technology based on the use of certain polymers which have the ability to absorb water and low molecular weight solutes to the exclusion of high molec-

TABLE 9

Results from Treatment Described Showing Effects of Temperature and pH of Extraction and pH for Precipitation on Phytate Content and Yield on Acid Precipitated Protein (14)

Sample No.	Extraction Temperature C	pH	Pptn. pH	% Phytate	% Protein Yield
1	75	9.0	5.3	0.02	31.3
2	65	9.0	5.3	0.46	32.7
3	88	9.0	5.3	0.04	29.9
4	75	8.0	5.3	0.20	28.6
5	75	10.0	5.3	0.08	32.5
6	75	9.0	5.2	0.06	31.4
7	75	9.0	5.5	0.02	29.4
8	40	8.0	5.3	0.41	28.0
9	75	9.0	4.5	0.54	31.3
10	40	8.0	4.5	1.96	32.9

TABLE 11

Nutritional Quality of Soy Protein Isolates Processed by Ultrafiltration with the Mother Liquor Not Heat Treated in One Case and in Another with the Mother Liquor Given Mild Heat Treatment and the Retentate Given a Stronger Heat Treatment in Comparison to Acid Precipitated Protein (10).

Sample	4-week weight gain	Protein Efficiency Ratio	
		g gain per g protein consumed	% of casein
Control casein	89	2.9 +/− 0.4	100
Sample 1[a]	72	2.2 +/− 0.2	76
Sample 2[b]	83	2.6 +/− 0.2	87
Acid precipitated	54	1.8 +/− 0.1	61

[a] U.F. but not heat treated.
[b] U.F. and heat treated.

ular weight components such as protein (32–34). This results in an increased protein content in the nonabsorbed liquid portion, which we refer to as a retentate. The retentate can be removed from the swollen gel, as a more concentrated protein solution, and the gel collapsed to release the absorbed water containing the low molecular weight solutes. While a number of gels had been developed which may be considered for such operations, our work has been primarily with the polymer, polyisopropylacrylamide which has interesting swelling and deswelling characteristics. Figure 2 illustrates the gel properties.

Tube 2 shows the gel which has been hydrated to its fullest extent, with a layer of unabsorbed water above the hydrated gel. The gel swelling takes place at a temperature below about 34°C. Tube 3 represents tube 2 which has been subjected to a temperature above 34°C where the heating at or above the critical temperature causes the gel to collapse, freeing the water that was in the swelled gel. The temperature difference to have a swollen gel and a collapsed gel is less than one degree centigrade. The collapsed gel and water in tube 3, if cooled would then rehydrate to a level as indicated in tube 2. The gel used in these experiments will take up about 30 times its weight in water. A collapsed gel will still have some water associated with it to the extent two to three times its dry weight. It is not necessary to dry the collapsed gel, but it can be used preferably as the wet gel for reswelling. When the gels are synthesized, one ends up with a mass of clear gel which is sized to the desired particulate characteristics to facilitate the swelling and collapsing operation.

We have proven that it is possible to concentrate soy proteins in mother liquor using the gels in the laboratory, but there is still a good deal of work to be done to prove it out as a practical process. The collapsed gel products do not coalesce as one solid gel mass on heating but remain as a mass of material with discrete particulate characteristics. The smaller the collapsed gel particles, the more rapid the hydration and swelling will be in the presence of water at the proper temperatures, and the more rapid will be the collapse as the gel is warmed. Figure 3 shows a schematic layout of how the gel is used to concentrate the protein. In step 1 the small circle represents the collapsed gel (in practice there will be a mass of gel particles) which is mixed at 5°C with the mother liquor containing about 5% protein and 3% nonprotein solids, for a total of 8% solids. At the end of the hydration in step 2, the retentate is separated from the swollen gel. If the gel has taken up half the water in the original mother liquor, the retentate removed from the swollen gel, theoretically, will contain 10% protein and 3.0% nonprotein solubles, or about 77% protein on a dry basis. In step 3, the gel is separated from the retentate. The separation can be done by filtration or centrifugation. In step 4, the swollen gel is subjected to a temperature of above 34°C (in actual practice we use higher temperatures to get the heat penetration in the mass and collapse the gel more rapidly) leaving the collapsed gel and

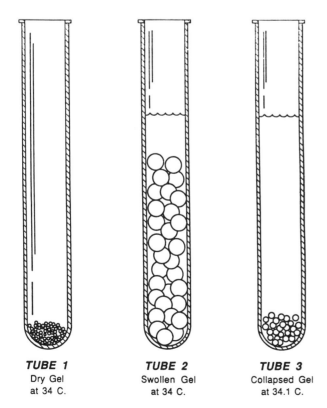

TUBE 1
Dry Gel
at 34 C.

TUBE 2
Swollen Gel
at 34 C.

TUBE 3
Collapsed Gel
at 34.1 C.

Fig. 2. Gel volume changes with temperature in water.

a whey which is then separated. The whey will contain about 3% solids, using a mother liquor with the analysis given above. In step 5, we cool and collapse the gel and remove the whey as a permeate. The cooled, collapsed gel is ready for reuse to complete a new cycle of absorption. At step 6, if the retentate as described above is subjected to a second treatment by absorbing half the water and nonprotein substances, the second retentate will theoretically have 20% protein with 3.0% nonprotein solids or 87% protein on a dry basis. The treatment can be repeated to obtain a protein product containing 90% protein (or higher) on a dry basis. As indicated, there are still a number of problems to work out.

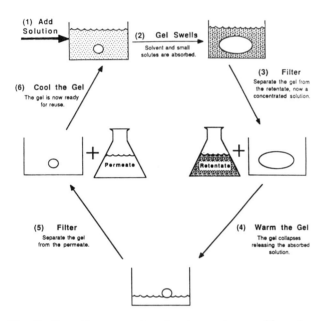

Fig. 3. Separation process based on temperature sensitive gels.

For these studies, the mother liquor was prepared by slurrying 90 PDI flakes in water, usually using a water to flake ratio of 8 to 1, a pH adjustment with sodium hydroxide to about 8.5 and slurrying the mixture at room temperature for about 30 min. The slurry was then centrifuged to separate the undissolved material (spent flakes) to obtain the mother liquor. For the swelling of the gel, we devised a cage which is shown in Figures 4 and 5 (34). To make the cage we used a 40 mesh 304 stainless steel screen made into a cylinder having a diameter of approximately 3 cm and a length of about 30 cm. The basket ends were capped with removable fittings, with a stainless steel rod, equipped with paddles, running down the center of the assembly. It was designed so that the propeller can be set in a position to stir the cage contents or set in a position so the entire cage can be rotated to spin out water from the gel. The cage was fitted loosely into a graduated cylinder containing a feed solution of the mother liquor or retentates.

Gel swelling was performed by placing the graduated cylinder and cage in a temperature bath at 5°C. After swelling, the basket was raised out of the liquid and spun to remove more retentate. Excessive foaming caused by spinning of the basket was controlled using a small amount of silicone antifoam. In some experiments, the swollen gel was washed with cold water to recover protein held

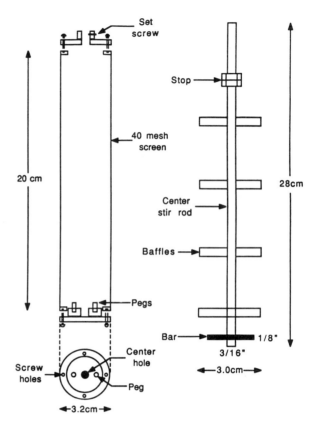

Fig. 4. Gel cage diagram.

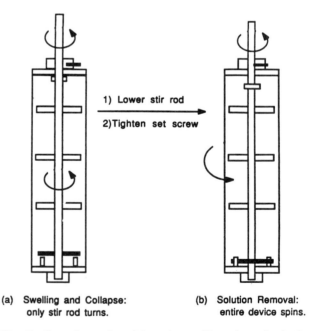

Fig. 5. Operating modes of the gel cage. The gel remains in the cage for swelling, collapse, washing and separation of wheys.

in the interstices of the gel particles or adsorbed on gel surfaces. In these experiments we attempted to have an amount of gel and feed solution so as to absorb about one half of the water in the solution in the swelling cycle.

After the gel swelling and spinning to separate as much of the retentate as possible, the cage containing the swollen gel was placed in another cylinder which was then heated in a water bath at about 40°C to collapse the gel. The collapsed gel was then spun and washed and cooled for subsequent treatment of the retentate for concentrating the protein further or to process more mother liquor. The procedure was repeated on subsequent retentates to concentrate the protein to a desired level and purity.

Separation Efficiency

The efficiency of gel absorption is defined as the concentration increase of protein (or any other material attempting to be concentrated) present in the retentate divided by that expected from the altered solution volume. For example, if the protein concentration in the unabsorbed solution—the retentate—increases by 1.2 times over the concentration in the feed and 50% of the water in the feed was absorbed, the efficiency is 1.2/2 times 100 or 60%. If a solute is completely excluded from the swollen gel, this would be a separation efficiency of 100%. If the solute penetrates the gel completely, so that the concentration of the solute in the retentate is not altered, the separation efficiency would be zero percent. This is roughly a parallel to the percent rejection in ultrafiltration. Table 12 gives the efficiency of the gel obtained with soy protein dispersions.

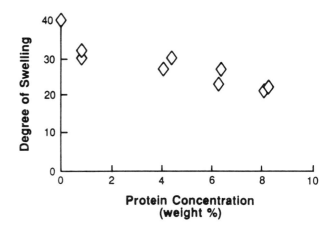

Fig. 6 Gel swelling as a function of protein concentration

on the gel surfaces. As the gel swells, excluded protein will form a more concentrated solution at the surface of the gel particles. If this protein is not removed, it will compromise efficiency in the same manner that concentration polarization can compromise protein ultrafiltration. It was our suspicion that the reduced efficiency shown in Figures 6 & 7 was due to the protein retention on the gel surfaces.

TABLE 12
Gel Extraction From Dilute Solutions

Solute	Feed concentration[a]	Typical retentate concentration[a,b]	Efficiency
Soy proteins			
Precipitate at pH 4.5	4.5	8.91	98%
Soluble at pH 4.5	0.5	.99	98%
Low molecular weight			
Sugars, phytin, salts, etc.	3.0	3.04	1.3%

[a] The concentrations are expressed in weight per cent.
[b] The retentate concentrations shown represent the weight percentage in the solution after half the feed volume has been absorbed by the gel.

As the protein concentrations increased the gel swelling was somewhat reduced, as shown in Figure 6. However, the gel swelling still remained high, being above 10 times the dry gel weight.

As indicated in Figure 7, the protein separation efficiency drops as the protein concentration increases. We feel, however, that with future effort this situation can be improved on or corrected. It appears that the separation of the low molecular weight solutes is essentially unaltered at the high protein concentrations in the feed as shown in Figure 8. The fact that the concentration of the nonprotein solutes in the retentate is about equal to that in the feed indicates that there is almost no concentration of the low molecular weight solutes. The difference between the protein and the small solute separation probably is due to the entrainment of small amounts of retentate between the gel particles and/or adsorption of protein

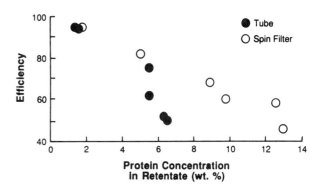

Fig. 7. Separation efficiency as a function of protein concentration.

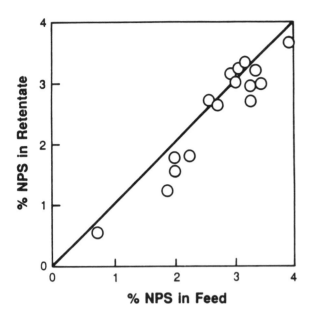

Fig. 8. Nonprotein solids (NPS) in gel-based separations.

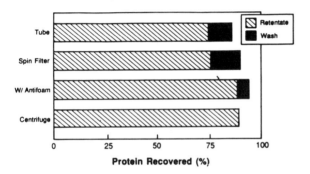

Fig. 9. Protein recovery using different techniques for separation of the retinate from the swollen gels. "Spin filter" refers to the previously described cage.

Figure 9 shows that when we recovered the retentate using a tube whereby we filtered the swollen gel using a screen device about 75% of the total protein in the mother liquor was recovered in the retentate and 12% recovered by washing the swollen gel with cold water, for a total of 87% protein recovery. Using the spin filter, we recovered 75% of the protein in the separated retentate with an additional 15% by washing the swollen gel with cold water, for a total of 90% protein recovery. When we used silicone antifoam, we recovered close to 90% protein in the retentate and an additional 6% in the wash water, for an overall 96% recovery. With centrifugation alone and no washing our recovery was about 89%.

In further work, we carried out a three-step gel extraction. First, we concentrated the mother liquor with gel, washed the gel, and combined the wash and the retentate. Second, we repeated the gel extraction using the combined wash and retentate, washed the gel and discarded the wash, but saved the second retentate. In the third phase we concentrated a second retentate without washing. Table 13 shows that in the first step, when the retentate and the wash are combined, we were purifying the protein. The second step, in which a wash is discarded, seems to both purify and concentrate, but there was a sacrifice in protein yield. The third step is largely one of concentration, producing a retentate containing 17% protein which figured to be 96% protein on a dry basis.

Thus far our work indicates that protein concentration can be carried out using swellable gels, but much more work is needed to show the practicality of the process.

General Comments

In view of the fact that there are variables which may be difficult to control, such as the characteristics of the starting raw materials, holding times at various stages of the operations, processing temperature variations, etc., if one were to take grab samples of product coming from the spray drier at various intervals, one will find that there are some variations in the physical and functional characteristics of the isolated soy protein.

It is recommended that large blenders, perhaps capable of handling as much as 20 metric tons (depending on the size of the operation), be used to blend sizable quantities of product from the spray drier. By so doing, this will even out the differences from bag to bag, as might occur if the product were packed straight from the drier. The product can thus be stored in large lots, requiring a smaller number of analytical tests for the quality control laboratory. Such blending benefits also would apply to producing soy flour and grits and soy protein concentrates.

It is known that during storage the PDI of isolated soy protein products gradually will decrease. The rate of decrease can be slowed by storing the packaged product at low temperatures or under refrigerated conditions.

There are certain quality control procedures which may be carried out in the plant, in order to follow the process and determine whether everything is under control at various stages of the process. Everything possible should be done to insure control of insects, birds and rodents in the processing area, as well as on the property surround-

TABLE 13

Results of Soy Isolate Process Using Gels

	Clarified liquor feed	Retentate (Step 2)	Spent wash	Permeate (Step 2)	Isolate (Step 3)
% Protein	8	13	1	<1	17
% Yield	100	86	7	4	79
% Purity	67	88	47	28	96

ing the plant. Air to the spray drier should be filtered. A CIP system should be installed. The plant should be automated as much as practical, particularly for pH and temperature controls. The processing area should preferably be under a slight pressure. Waste treatment is a major consideration, and how to handle it in the best manner and economically is a subject by itself.

A generally accepted procedure for determining the nutritional value of protein products is by comparing the growth rate of weanling rats with the protein under test with a standard casein control diet to calculate what is known as the PER (protein efficiency ratio). PER values for a properly heat treated soy flour have been reported as being in the range of about 2.2 to 2.8 and unheated soy flour from 1.3 to 1.6 (28,30). Literature figures for PER values may be quite variable depending on many factors. The lower PER values for soy flour that has not been heat treated is primarily due to the presence of antitrypsin factors. When the antitrypsin factors are destroyed by proper heat treatment, higher PER values result. In the case of soy protein isolates produced by acid precipitation, PER values have been reported in the range of 1.4 to 2.0 (31). Some of the reported low PER values for isolated soy protein may be due to the fact that, during their manufacture, the proteins did not receive sufficient heat treatment to destroy the major part of the antitrypsin factors, as well as the loss of albuminous proteins which are lost to the whey in the acid precipitation process. There is some question as to the possible importance of the antitrypsin factors where humans are concerned, with the possible exceptions of infants (29).

PER probably is quite meaningless in so far as the human is concerned because studies are done using rats which have a much higher requirement for sulfur-containing amino acids per unit of body weight in comparison to man, and it is rare that a single protein would serve as the sole source of protein for humans.

In preparing isolated soy protein products using acid precipitation, of the total protein in the mother liquor, roughly 90% of the protein present is a mixture of globular proteins (which are acid precipitatable) and albuminous type proteins (which are acid soluble). Approximately 10% of the total protein present in the mother liquor is lost to the whey and not recovered as part of the protein yield.

Although isolated soy proteins may contain the antitrypsin factor and other antienzymes, during processing it is possible to heat treat the proteins to destroy the antienzymes without appreciably affecting the functional properties of the recovered proteins. However, the heat treatment must be done judiciously to retain functionality in the finished product. Because soy protein isolates are generally used as an ingredient in foods, which are further heat processed, the possible importance of antitrypsin and other antienzymes may not really be of concern.

References

1. *Soybeans–Chemistry and Technology, Vol. 1, Proteins*, edited by S.J. Circle and A.K. Smith, Avi Publishing Co., Westport Connecticut, 1978, p. 443.
2. Johnson, D.W., *Food Use of Whole Oil and Protein Seeds*, edited by E.W. Lusas, American Oil Chemists' Society, Champaign, Illinois, USA (in press).
3. Product Data Sheet FI-Pro-F200-071686, Grain Processing Co., Muscatine, Iowa.
4. General Product Description–Soy Polysaccharide (Dietary Fiber), Ralston Purina Co., St. Louis, Missouri, 1985.
5. Fitrate Data Sheet, Dawson Food Ingredients
6. U.S. Patent 3,926,940 (1975).
7. U.S. Patent 3,897,754 (1975).
8. U.S. Patent 3,728,327 (1973).
9. U.S. Patent 3,736,147 (1973).
10. U.S. Patent 3,995,071 (1976).
11. U.S. Patent 4,072,670 (1978).
12. U.S. Patent 4,088,795 (1978).
13. U.S. Patent 4,091,120 (1978).
14. U.S. Patent 4,697,004 (1987).
15. Lawhon, J.T., S.H.C. Lin, D.W. Hensley, C.M. Cater and K.F. Mattil, *J. Food Process Eng.*, January 1977, p. 15.
16. Hensley, D.W., J.T. Lawhon, C.M. Cater and K.F. Mattil, *J. Food Sci.* 42:812 (1977).
17. Lawhon, J.T., D.W. Hensley, M. Mizukoshi and D. Muslow, *Ibid.* 43:361 (1978).
18. Lawhon, J.T., D.W. Hensley, M. Mizukoshi and D. Muslow, *Ibid.* 44:213 (1979).
19. Hensley, D.W., and J.R. Lawhon, *J. Food Tech.* 33:46 (May 1979).
20. Manak, L.J., J.R. Lawhon and E.W. Lusas, *J. Food Sci.* 45:256 (1980).
21. Manak, L.J., J.R. Lawhon and E.W. Lusas, *Ibid.* 45:197 (1980).
22. Lawhon, J.T., K.C. Rhee and E.W. Lusas, *J. Am. Oil Chem. Soc.* 48:377 (1981).
23. Lawhon, J.T., L.J. Manak, K.C. Rhee and E.W. Lusas, *J. Food Sci.* 46:391 (1981).
24. Lawhon, J.T., L.J. Manak, K.C. Rhee and E.W. Lusas, *Ibid.* 46:912 (1981).
25. Lawhon, J.T., L.J. Manak and E.W. Lusas, "Using Industrial Membrane Systems to Isolate Oilseed Protein Without an Effluent Waste Stream," Texas A&M University, College Station, Texas.
26. Manak, L.J., J.T. Lawhon and E.W. Lusas, *J. Food Sci.* 47:800 (1982).
27. Lawhon, J.T., and E.W. Lusas, *J. Food Tech.* 38:97 (1984).
28. Kellor, R., *J. Am. Oil Chem. Soc.* 51:77A (1974).
29. Liener, I., *Ibid.* 56:121 (1979).
30. Rackis, J., in *Soybeans–Chemistry and Technology*, edited by A. K. Smith and S. J. Circle, Avi Publishing Co., Westport, Connecticut, 1972, pp. 158-202.
31. Bodwell, C., *J. Am. Oil Chem. Soc.* 56:156 (1979).
32. Freitus, R.F.S. and E.L. Cussler, *Chem. Eng. Science* 42:97 (1987).
33. Trank, S.J., D.W. Johnson and E.L. Cussler, "A New Method of Recovery Soy Protein Using Supercritical Gels," minisymposium on Processing and Biotechnology, University of Minnesota, (1987).
34. Trank, S.J., "Design and Application of Temperature-Sensitive Hydrogels," PhD thesis, University of Minnesota 1988.

Trends in Preparation and Uses of Fermented and Acid–Hydrolyzed Soy Sauce

Kinji Uchida

Kikkoman Corporation, 399, Noda, Noda-shi, Chiba-ken 278, Japan

Abstract

Soy sauce is made from soybeans, wheat and salt. Annually, about 200,000 tons of each of these ingredients are used for the production of 1.2 million kl of soy sauce in Japan. However, soy sauce, especially fermented soy sauce, is not a simple hydrolysate but a complexly brewed product fermented with koji-mold, lactic-acid bacteria and yeasts. Recently, a variety of soy sauce-based products have become increasingly popular in the Japanese market. The application of specific microbial activities to fermentation offers an attractive approach for the development of more refined products. Ecological studies on lactic-acid bacteria recently have shown that they are more heterogeneous and metabolically talented than previously realized. The feasibility to improve the chemical quality by adopting certain strains in the fermentation process will be discussed.

In the Orient, especially in east Asia, there are many kinds of indigenous fermented foods made from soybeans (1). In general, the favorite fermented foods were passed down through the ages. In the past, soy sauce was one such traditional food in Japan, together with "Miso," "Natto" etc. But in recent years, it has become increasingly popular throughout the world for its excellent flavor and palatable taste. The prototype presumably was born in China and, after transfer to Japan with Buddhism, it has been extensively improved for hundreds of years to complement Japanese climate and food culture (2,3,4). It has been during the past few decades that chemical or biological details of soy sauce manufacture were studied intensively. By virtue of the scientific understanding of the processes or through engineering innovations, the manufacture of soy sauce has evolved into a highly technological state (5).

The current state of the production and consumption of soy sauces in Japan will be discussed, with the emphasis on the fermented ones along with some topics from a microbiological point of view.

Production and Consumption of Soy Sauce in Japan

Table 1 shows the annual production and consumption of soy sauce over the past ten years. In this decade, the total production and consumption of soy sauce have changed little. The annual consumption per capita was around 10 liters per person annually. In 1987, 1,200,000 kiloliters (kl) of soy sauce was produced. And for this production, about 200,000 tons each of soybeans, wheat and salt were used (6).

TABLE 1

Annual Production of Soy Sauce in Japan 1975–1987 (6,7)

Year	Total production (k-liter)	Per capita/year (liter)
1975	1,120,494	10.0
1980	1,188,970	10.2
1982	1,187,148	10.0
1983	1,193,978	10.0
1984	1,200,063	10.0
1985	1,186,278	9.8
1986	1,200,714	9.9
1987	1,197,100	9.8

In spite of the stable level of the total production, consumption patterns have changed slowly but apparently in this decade. That is, there has been a decrease in home-use and an increase in industrial or commercial use.

Table 2 shows changes in soy sauce purchase from 1975 to 1987. Ten years ago, home-use absorbed nearly 60% of the total, but now, it is below 50%. Nowadays people tend not to use as much soy sauce at home as before, but tend to purchase various pre-cooked foods already seasoned with soy sauce or to dine out more frequently. Along with this trend, a significant change is also observed in the use of package containers of soy sauce (Fig. 1). The use of two-liter glass bottles is decreasing, while the use of one-liter plastic bottles, which are easier to carry or handle, is increasing.

In addition, a variety of soy sauce-based, secondary products have become increasingly popular in the Japanese market. These are several kinds of soup which in Japan are called "Tsuyu," i.e., "men-tsuyu" or noodle soup, sukiyaki-tsuyu, tempura-tsuyu, and many kinds of sauce called "Tare" like "teriyaki"-sauce or steak-sauce and in addition, soy sauce-dressings or soy sauce-vinegar mixtures. The main ingredient of these soups and sauces is fermented soy sauce. Sugars, monosodium glutamate, ribotides, several extracts, spices and various flavoring substances are added. Examples of recipes for typical steak sauce and noodle soup are shown in Tables 3 and 4.

The increase in annual production of these "Tsuyu" and "Tare" in this decade is shown in Figure 2. In 1987, the total production amounted to 140,000 kl and the total sales were nearly 90 billion yen. This corresponds to over one tenth of the total production of soy sauce by volume, and over one third of its total sales.

TABLE 2

Changes in Annual Consumption of Soy Sauce 1975–1987 (6,7)

Year	Total (k-liter)	Rates Home-use %	Rates Commercial use %	Home-use/ person/year (liter)
1975	1,127,300	58.6	41.4	5.1
1980	1,190,500	52.7	47.3	4.5
1982	1,184,400	50.6	49.4	4.2
1983	1,194,700	48.7	51.3	4.1
1984	1,192,000	49.8	50.2	4.2
1985	1,185,000	47.4	52.6	3.9
1986	1,192,000	47.7	52.3	—
1987	1,191,000	47.6	52.4	—

TABLE 3

A Recipe for Typical "Tsuyu" or Noodle-Soup (8)

Ingredients	
Soy sauce	700 l
Salted sweet cooking rice wine	50 l
Corn syrup	60 kg
Sugar	50 kg
Dried bonito	50 kg
Kelp extract	3 kg
Monosodium glutamate	3 kg
5'-Ribonucleotides	100 g

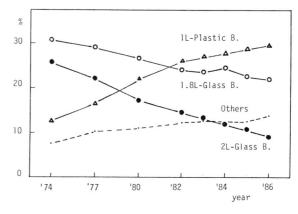

Fig. 1. Changes in the use of package containers of soy sauce 1974-1986 (6,7).

TABLE 4

A Recipe for Typical "Tare" or Steak-Sauce (8)

Ingredients	
Soy sauce	8 l
Salted sweet cooking rice wine	1 l
Sugar	2.5 kg
Salad oil	300 ml
Acetic acid	200 ml
La-oil	150 ml
Ginger	100 g
Garlic	100 g
Onion Powder	100 g
Monosodium glutamate	80 g
Lemon essence	80 ml
Peppermint	1 g

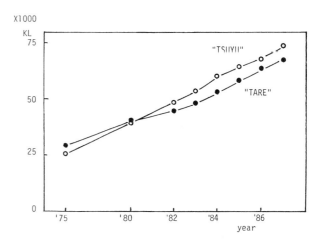

Fig. 2. Increase in the annual production of "Tsuyu" and "Tare," major soy sauce-based products (6,7).

Processes of Manufacturing Soy Sauce (2,4,5)

The Japanese Agricultural Standard (JAS) recognizes five kinds of soy sauce: Koikuchi, Usukuchi, Tamari, Shiro and Saishikomi-types. Among them, the "Koikuchi-type" which means dark in color, is the most popular. Usually it is called simply "shoyu," and it shares more than 85% of the total market. So, I would like to elucidate the manufacturing process of soy sauce, taking the Koikuchi-type as an example.

Fermented soy sauce of Koikuchi-type generally is manufactured through five main processes. They are the treatment of raw materials, the koji-making, the moromi-fermentation and aging, the pressing and the refining (Fig. 3).

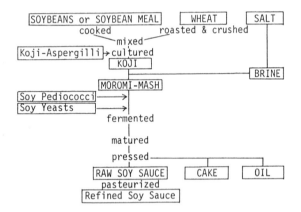

Fig. 3. Flow sheet for manufacture of Koikuchi-type soy sauce.

Treatment of Raw Materials

Whole soybeans or, more commonly, defatted soybean grits are moistened and cooked under steam pressure. This process greatly influences the digestibility of soybean protein, and the cooking conditions have been studied extensively. Wheat, the other half of the raw materials, is roasted at 160–180°C for less than one min, then coarsely crushed into 4 or 5 pieces.

Koji-Making

The above two materials are mixed and inoculated with a pure culture of koji mold, *Aspergillus oryzae* or *A. sojae*. The mixture is spread to a depth of 30–40 cm on a perforated plate in the koji-making plant. The mixture is aerated with temperature- and moisture-controlled air at around 30°C for two to three days. This allows the koji-mold to grow throughout the mass, and produces various hydrolyzing enzymes required for decomposition of the raw materials. The product is called "koji."

Moromi-mash fermentation and aging

The koji is mixed with concentrated (22–23%) saline water of 120–130% volume of the raw materials. The resulted 'moromi' or mash is transferred into deep fermentation tanks and is held for five to eight months with occasional agitation by compressed air.

During the moromi-period, the enzymes from koji-mold hydrolyze substances in the raw materials. The protein is decomposed into amino-acids or small peptides. Starch is converted into sugars which are fermented subsequently by the action of salt-tolerant microorganisms into lactic acid, alcohol and various flavor substances.

At the first stage, a culture of *Pediococcus halophilus* (soy pediococcus) is added. This osmophilic lactic acid coccus propagates and produces lactic acid, which causes a drop of moromi pH. When the pH level reaches around 5.2, a culture of *Saccharomyces rouxii* (soy yeast) is added. The growth and alcoholic fermentation are stimulated by aeration and elevation of the temperature. Another group of salt-tolerant yeasts, known as *Torulopsis* (now classified as species of *Candida*), often occur at the middle or last stage. They produce phenolic compounds such as 4-ethyl guaiacol and add some characteristic aroma to soy sauce. The high salt concentration of moromi, usually 17–18% w/v, effectively excludes the growth of undesirable microorganisms.

Besides all of these biological or enzymatic reactions, some chemical or physico-chemical interactions among the constituents of moromi proceed throughout this stage. So called "browning reactions" between amino acids and sugar components are typical ones and they result in formation of the characteristic reddish-brown color of soy sauce (9). Nearly 300 kinds of flavor components have been identified in soy sauce and the majority of these compounds are thought to be formed during this process (5,10).

Pressing

After five to eight months of fermentation and aging, matured moromi is filtered through cloth under a high hydraulic pressure. The pressure applied is increased in two or three steps, and occasionally, the pressure reaches 100 kg/cm² at the final steps, which makes the moisture content of the pressed cake less than 25%. Automatic loading of moromi into filter cloth or continuous pressing by a diaphragm type machine have been attempted recently for the effective moromi-filtration, instead of the batch type hydraulic press. The filtrate obtained is stored in a tank to separate the sediments at the bottom and the floating oil on the top.

Refining

The clarified raw soy sauce is adjusted to the standard salt and nitrogen concentrations and pasteurized usually at 70–80°C. This heat treatment is essential for the development of the desired color and flavor of the products. After heating, the soy sauce is clarified by sedimentation and the clear supernatant is packed immediately into cans or bottles.

Microbiology of Soy Sauce Manufacture

As stated above, the utilization of a variety of microorganisms, including fungi, yeasts, and lactic acid bacteria, is one of the most notable characteristics of soy sauce manufacture. An application of specific microbial activities to the fermentation offers an attractive approach for the development of more refined products. Two examples of such an approach in microbiology of soy sauce making are presented below.

Breeding of Hybrid Koji-Molds by the Use of Cell-Fusion Technique (11,12)

Degradation of protein, starch and other high molecular substances in the raw materials is the main function of the koji-mold. Their proteolytic activity is particularly the most important. The role of each enzyme of koji during the protein-degradation is shown in Figure 4. Glutaminase is indispensable for effective formation of glutamic acid, because glutamine liberated by peptidases from protein tends to change autonomously into pyroglutamic acid which has no flavor. Therefore, in order to produce flavor-rich soy sauce, it is desirable to adopt a koji-mold providing a sufficient amount of glutaminase as well as the proteolytic enzymes. But in general, a proteinase hyper-producing strain does not provide so much glutaminase, and a glutaminase hyper-producer tends to produce an insufficient amount of proteinases. Although a lot of proteinase-hyperproducing mutants of koji-mold have hitherto been bred through ordinary genetic procedures, most of them produce rather insufficient glutaminase.

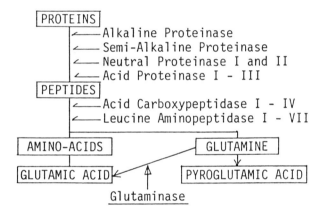

Fig. 4. Role of the proteolytic enzymes of Koji-mold in amino-acids liberation from proteins (3).

In these circumstances, Ushijima of our laboratory attempted to breed a hybrid of proteinase- and glutaminase-hyperproducers by the use of the protoplast-fusion technique (Fig. 5).

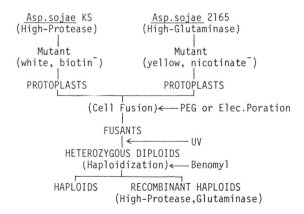

Fig. 5. Breeding of new Koji-molds through cell-fusion.

First, a pair of parental strains (wild), the proteinase-hyperproducer and the glutaminase-hyperproducer were selected from the stock of *Aspergillus sojae*. From each of them, mutants with double genetic marks (in conidial color and in nutritional requirements) were derived. Protoplasts, naked cells devoid of cell wall, carefully were prepared from each mutant, and they were fused under the presence of poly-ethyleneglycol.

After the regenerating process, a number of hybrid-fusants were obtained. They were examined for their proteinase and glutaminase activities. Productivities of both the enzymes varied from strain to strain, but the majority showed properties as intermediates of the parents. In this stage, however, none of the fusants exhibited sufficient production of both enzymes. Ushijima and his coworkers then attempted to haploidize the hybrid-fusants (diploid) with benomyl or FPA. Among a large number of haploids which showed the same properties as either of the parents, some recombinants as to the marked properties also were obtained (Fig. 6). On screening of these recombinants, they succeeded in obtaining a few haploid-recombinants exhibiting high productivities of both proteinase and glutaminase. The cell-fusion technique, especially that followed by the haploidization, was thus shown to be very useful in breeding new koji-molds with more desirable properties.

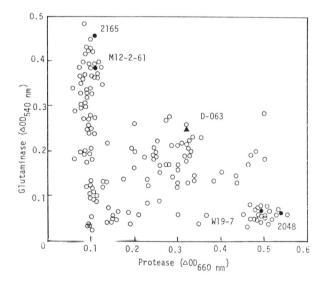

Fig. 6. Distribution of the protease and glutaminase activities of the segregants obtained on haploidization (12).

Isolation and Application of Metabolically Diverse Soy Pediococci in the Fermentation (13)

As stated above, a group of halophilic lactic acid bacteria *Pediococcus halophilus*, propagates in the moromi at an earlier stage and form lactic acid from glucose. Their roles had not been well understood before, except for formation of lactic acid and lowering the moromi-pH as suitable for the following alcoholic fermentation by yeasts. In traditional style soy sauce plants, the lactic acid fermentation generally has been performed by natural flora, and it is difficult to control the fermentation strictly.

In recent years, we have developed a practical and

efficient scheme for discriminating individual strains of soy pediococci based on their differences in sugar fermentation (Table 5). It enables us, for the first time, to analyze the pediococcal flora in a fermenting moromi and to understand more exactly the functions of each strain.

TABLE 5

Heterogeneity of Sugar Fermentation Patterns of Soy Pediococci (13)

Fermentn.[a] A L B S M	Number of Strain	%
– – – – +	223	14.1
– – – – –	221	13.9
– – – + –	169	10.7
+ – – – –	167	10.5
+ – – – +	132	8.3
– – – + +	113	7.1
– – + – –	96	6.1
+ + – – +	74	4.7
– – + + –	67	4.2
– – + – +	60	3.8
Others (19)	262	16.5
Total	1,586	100

[a] A: Arabinose, L: Lactose, B: Melibiose, S: Sorbitol, M: Mannitol

Physiological heterogeneity of the soy pediococci was demonstrated first on their sugar-fermenting ability. Over 1,500 isolates from moromi were examined for their ability to produce acid from five selected sugars: arabinose, lactose, melibiose, sorbitol and mannitol. According to the fermentation pattern, the pediococci were separated into 28 different types. By the use of 10 kinds of sugars for the discrimination, 67 different types of fermentation were recognized. The sugar-fermenting properties were found genetically stable, and being assayed in a high experimental reproducibility.

So we developed a flora-analyzing scheme, in which 50 randomly isolated strains from a specimen were examined for their fermentation on the five sugars. By the use of the flora-analyzing scheme, it was found that a traditionally processed soy sauce-moromi usually harbored at least 10, presumably several tens, of different strains of soy pediococci.

An extensive study on numerous pediococcal isolates from naturally brewed moromi revealed that they were very heterogeneous not only in their sugar-fermenting abilities but also in the other physiological properties such as arginine-metabolism, amino acid-decarboxylation, carboxylic acid-decomposition or environment-reducing ability.

The major flavoring ingredients of soy sauce are amino acids. One hundred representative isolates of soy pediococci were examined for their reactions to amino acids in small scale brewing trials. Some strains split arginine into ornithine and ammonia as known before. The other groups were found to decarboxylate aspartic acid, phenylalanine, tyrosine or histidine.

Generation of two moles of ammonia from arginine, or disappearance of carboxylic moiety from amino acids, results in shift up of the environmental pH, or neutralization of lactic acid formed. Because of the high concentration of amino acids in moromi compared with that of lactic acid to be formed, decomposition of arginine or decarboxylation of aspartic acid affect significantly the lowering of moromi-pH by lactic acid fermentation. It should be very interesting to compare the decarboxylation of aspartic acid in soy sauce fermentation with that of malic acid in wine making, known as MLF, the malo-lactic fermentation (Fig. 7). In the chemical formula, only the OH-moiety of malic acid in the MLF is replaced by NH_2-moiety in the conversion of aspartic acid to alanine. The conversion of a sour amino-acid (Asp) to a sweet amino-acid (Ala) would make the taste of the products milder as observed in MLF.

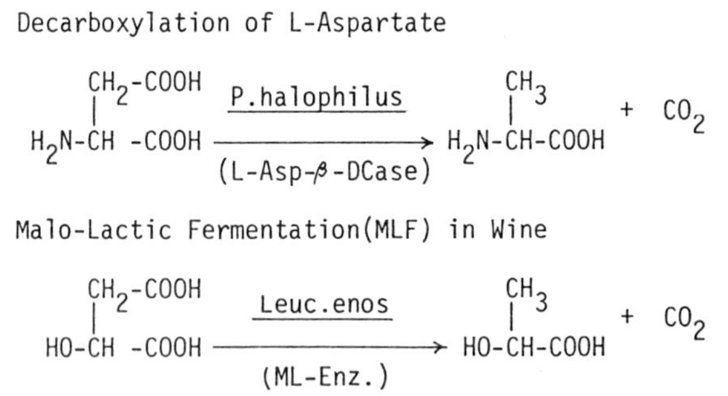

Decarboxylation of L-Aspartate

$$\underset{H_2N-CH-COOH}{\overset{CH_2-COOH}{|}} \xrightarrow[\text{(L-Asp-}\beta\text{-DCase)}]{P.\,halophilus} \underset{H_2N-CH-COOH}{\overset{CH_3}{|}} + CO_2$$

Malo-Lactic Fermentation (MLF) in Wine

$$\underset{HO-CH-COOH}{\overset{CH_2-COOH}{|}} \xrightarrow[\text{(ML-Enz.)}]{Leuc.\,enos} \underset{HO-CH-COOH}{\overset{CH_3}{|}} + CO_2$$

Fig. 7. L-Aspartate decarboxylation by soy pediococci and malo-lactic fermentation (MLF) in wine.

Pediococcus halophilus is a homo-lactic fermentor, and basically produces two moles of lactic acid from one mole of glucose. A certain amount of acetic acid also is produced. Most of the soy pediococci split citric acid into lactic and acetic acids, and convert malic acid to lactic acid. But some strains were found devoid of one or both (15,16).

Soy pediococci are facultative anaerobes and grow faster in an anaerobic condition. Some strains were found having the ability to reduce their environmental substances (Fig. 8). It was observed that, when the reducing strains grew, the redox potential (rH) of the culture decreased below 6.0. Non-reducing strains didn't make the potential of the culture below 8.0. These strongly reducing strains also were found to lighten the brown color of moromi-mash through reduction of the color substances during their growth. So, the utilization of these reducing strains for the fermentation leads to the production of lighter-color soy sauce (17,18).

The soy pediococci thus were found so heterogeneous in their metabolisms that it became possible to improve the chemical composition of the products by the use of strains with special metabolizing properties, within a certain limitation.

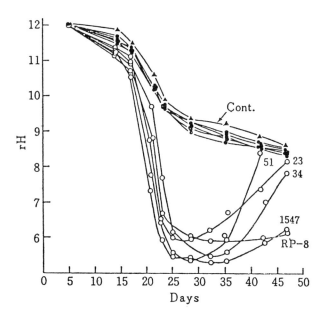

Fig. 8. Oxidation-reduction potentials of soy sauce *Moromi* inoculated with reducing or non-reducing strains of soy pediococci (17).

Soy sauce has long been called an "all-purpose seasoning." In the future, especially in Japan, more specified soy sauce varieties or soy sauce-based products will be needed. Application of microorganisms with more specific abilities or properties will become very useful for their production.

References

1. Beuchat, L.R. in *Biotechnology* Vol. 5, Chapter 11, Edited by G. Reed, Verlag Chemie, 1983, p. 477.
2. Yokotsuka, T. in *Microbiology of Fermented Foods* Vol. 1, Chapter 5, Edited by B.J.B. Wood, Elsevier Applied Science Publishers, 1985, p. 197.
3. Fukushima, D. *Food Review International*, 1:149 (1985).
4. Fukushima, D. in "Indigenous Fermented Food of Non-Western Origin," edited by C.W. Hesseltine, H.L. Wang and J. Cramer, 1986, p. 121.
5. Yokotsuka, T. *Advances in Food Research*, 30:196 (1986).
6. "Shurui Shokuhin Sangyo no Seisan Hanbai 1987," Nikkan-Keizai Tsusin-Sha, 1987.
7. "Shurui Shokuhin Toukei Nenpo 1986," Nikkan-Keizai Tsushin-Sha, 1986.
8. Ohta, S., "Shokuhin Chomi Haigou Reishu," Kougaku-Tosho 1979.
9. Hashiba, H. *Agric. Food Chem.*, 24(1):70 (1976).
10. Nunomura, N. and M. Sasaki, in "Legume-Based Fermented Foods," edited by N.R. Reddy, M.D. Pierson and D.K. Salunkhe, CRC Press, 1986, p. 5.
11. Ushijima, S. and T. Nakadai, *Agric. Biol. Chem.*, 51:1051 (1987).
12. Ushijima, S., T. Nakadai, and K. Uchida, *Agric. Biol. Chem.*, 51:2781 (1987).
13. Uchida, K., *J. Japn. Soy Sauce Res. Inst.*, 13(6):251 (1987).
14. Uchida, K., *J. Gen. Appl. Microbiol.*, 28:215 (1982).
15. Kanbe, C., and K. Uchida, *Agric. Biol. Chem.*, 46:2357 (1982).
16. Kanbe, C., and K. Uchida, *Appl. Envir. Microbiol.*, 53:1257 (1987).
17. Kanbe, C., and K. Uchida, *Nippon Nogeikagaku Kaishi*, 58:487 (1984).
18. Kanbe, C., and K. Uchida, *Agric. Biol. Chem.*, 51:507 (1987).

Manufacturing Soy Protein Isolate by Ultrafiltration

Fu-Kuang Liu, You-Hua Nie and Bei-Ying Shen

Wuxi Light Industry Institute, Wuxi, Jiangsu, People's Republic of China.

Abstract

Ultrafiltration has been used in concentrating extracted protein solutions. We have determined: (a) it is profitable to prepress and presoak the ultrafiltration membrane to promote the efficiency of ultrafiltration; (b) pressure and flux rate in ultrafiltration are the determining effects on the concentration polarization. There is a critical pressure, above which the rate of permeation will not increase no matter how high the pressure is increased. Critical pressure is a function of porostiy of membrane, initial concentration of flux material and flow dynamics; (c) the temperature of the flow material and pH exert a great effect; and (d) permeation of ultrafiltration will be maximum under alkaline conditions, acidic condition ranks next and it would be a minimum at the isoelectric point. Experiments have shown that there is no phase change, less energy is consumed and operation is easy in concentrating soybean protein by ultrafiltration. Functionality and nutritional values are higher than those achieved by conventional methods.

Increasing attention has been focused on oilseed crops as an abundant source of food protein to provide needed nourishment for large segments of the world's increasing population. The world population is now over 5 billion. Of course, soybeans are by far the most important source of vegetable protein and calories. The desirable attributes of soybeans are its protein and fat components, but there also are some undesirable components such as oligosaccharides, phytic acid and trypsin inhibitors, etc. (1). The oligosaccharides and phytic acid share one common characteristic, i.e., they are relatively small in terms of molecular size. By proper choice of membranes and operating parameters, the undesirables can be separated selectively from the protein resulting in a purified and highly functional protein concentrate, and thereby avoid generation of large amounts of whey which result from conventional isoelectric precipitations. This increases recovery of protein with enhanced nitrogen solubility.

Porter and Michaels (2) were the first to suggest ultrafiltration (UF) to fractionate protein extracts. Pompei et al. (3), Okuba et al. (4) and Goodnight et al. (5) produced soy protein isolates using a combined ultrafiltration and continuous diafiltration process. Lawhon et al. (6) used a discontinuous diafiltration or re-UF process to produce a soy product having approximately 90% protein (dry basis) and Olsen (7) produced a soy product having 88% protein (dry basis) by concentrating a defatted soy flour extract from 5.6% to 25% total solids using direct ultrafiltration.

Soy protein isolate can be manufactured either by using the method of ultrafiltration or by using a combination of ultrafiltration and reverse osmosis with good possibilities, but those methods have not been commercialized. Now, the commercial companies producing soya protein isolates do not publish information on how they manufacture specific products (7). This paper reports the processing conditions of ultrafiltration found in manufacturing soybean protein isolate so as to provide some information for the benefit of manufacturers of soya protein isolate on a practical production scale.

Experiment

Raw Materials:

There are three kinds of materials to be used for experiments:
1. Extracts of defatted soy flour with protein content of 1.1–3.2% and dissolved solid material 2.3–4.5%.
2. Whey from soy protein isolates with protein content 0.38–0.52% and dissolved solid material 1.2–2.0%.
3. Whey from manufacturing soy tofu with protein content 0.5–1.8% and dissolved solid material 1.9–2.4%.

Systems Tested:

1. Flat plate–consisted of two series, each contains 9 plates. Available area is $0.27\ M^2$.
2. Tubular: there are 10 tubes with a length of 80 cm, and outside diameter of 2 cm. Five tubes compose a set. Two sets are connected in series with an available area of $0.2\ M^2$.

Membranes Employed:

1. CA–10: Cellulose acetate, molecular cut value 10,000 suitable for pH 3–8 at 45°C.
2. CA–50: Cellulose acetate, molecular cut value 50,000 suitable for pH 3–8 at 45°C.
3. CAT-308: Cellulose acetate, molecular cut value 50,000 suitable for pH 3–8 at 65°C.
4. PSA: Polysulfone acylamide, suitable for pH 1–12, molecular cut value 23,000 at 90°C.

Analytical Methods:

General analytical methods were according to AOAC. Foaming ability, emulsifying property, and relative activity of trypsin inhibitor were determined according to published methods.

Membrane Rejection:

When selecting a membrane for a particular concentration process, it is necessary to know its rejection characteristics. The rejection (R) or retention of any substance is defined as:

$$R_{App.} = \frac{C_F - C_P}{C_F}, \qquad R_{Act.} = \frac{C_I - C_P}{C_I}$$

$R_{App.}$ – Apparent retention of any substance.
$R_{Act.}$ – Actual retention of any substance.
C_F – Concentration of substance in the feed.

C_P – Concentration of substance in the permeate.
C_I – Concentration of substance in the solution at any time.

Results and Discussion

Pretreatment of Membrane Affecting Ultrafiltration

In order to get better results, it is necessary to do some suitable pretreatment on the membrane before operating. Soaking and prepressing were two treatments used in our tests.

Membrane PSA and CAT308 on flat plates were soaked and prepressed; the results are shown on the Figures 1, 2, 3 and 4 respectively.

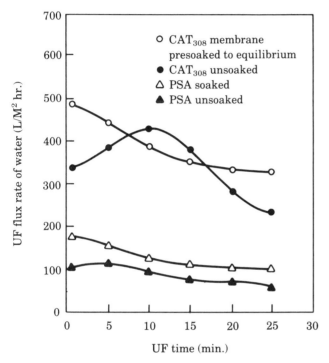

Fig. 1. Effect of soaking on ultrafiltration. UF testing conditions: tap water at 25–30°C, inlet pressure at 3 Kg/cm², outlet pressure 2.6 Kg/cm².

Soaking conditions: The solution was distilled water with formaldehyde (0.5%). Membranes were soaked for 8–10 hours at room temperature.

Prepressing conditions: A prepressing pressure of 20% over that of normal operating ultrafiltration pressure was used for 1.0–1.5 hr.

These experiments showed that after soaking the membrane and reaching equilibrium, the water flux rate of UF was high and stable; with unsoaked membranes the flux rate was low and unstable. Apparently, it is profitable to soak the new membranes to equilibrium before using them in UF operations.

In similar tests, extracts from defatted soy flour using a prepressed membrane were ultrafiltrated and gave higher and more stable flux rates. At the final stage of ultrafiltration, prepressed membranes yielded a flux rate higher

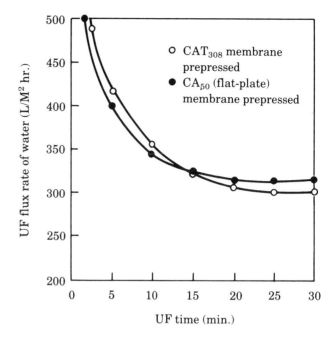

Fig. 2. Relation between flux rate and time of UF for prepressed membrane. Prepressing conditions: distilling water for prepressing, temperature 30–35°C, inlet pressure 3.3 Kg/cm², outlet pressure 3.2 Kg/cm².

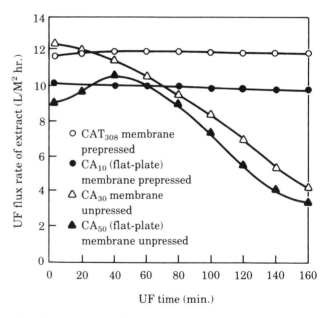

Fig. 3. Comparison of flux rate of pressed and unpressed membrane during UF. Testing conditions: pH 8, temperature 35–40°C, inlet pressure 3.0 Kg/cm², outlet pressure 2.6 Kg/cm², dissolved solid material 3.45%, crude protein 2.2%.

than the flux rate of a nonprepressed membrane. In addition, a prepressed membrane gave higher retention. This has more practical significance in plant operation for production. These experiments proved that pressures 20% higher than that of normal operating pressure applied for 1 hr gave better and more ideal results.

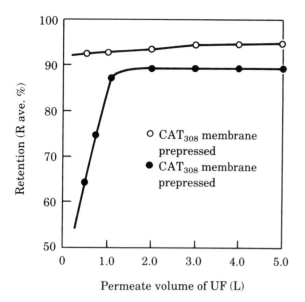

Fig. 4. Effect of pressing on retention. Testing conditions: pH 8, temperature 35–40°C, inlet pressure 3.0 Kg/cm², outlet pressure 2.6 Kg/cm², dissolved solid material 3.45%, crude protein 2.2%.

Effects of Operating Pressure on Ultrafiltration

During ultrafiltration these experiments showed that the pressure was a main factor affecting flux rate. The data on a tubular UF using membranes to retain the protein with a molecular weight of 30,000–50,000 D are shown in Figures 5 and 6.

Experiment conditions: There was 2.9% solid material in extracts with a protein content of 2.07%, viscosity 3.4 cp., pH 8.5, temperature of UF 30–35°C. The permeate was returned to the original storage vessel, the flux rate varied less than 5%.

The old membrane meant that the membrane was operated twice and then was washed and cleaned.

The operating conditions: Temperature of tap water 30°C, the operating conditions were same as in Figure 5.

Comparison of Figures 5 and 6 shows a distinct difference. In Figure 5 the flux rate increases to a certain value and then no longer increases no matter how high the pressure is increased. This pressure is called the critical pressure of UF. This has a more practical significance in production.

On Figure 6, the flux rate of tap water increased as the pressure increased. Figure 5 also shows that critical pressure has a relation with the pores of membrane. Under the same conditions, large pores yield small critical pressure and small pores yield large critical pressure. Therefore, critical pressure is a function of pore size.

Utilizing the same equipment and operating conditions but using an extract with total solid material 4.8% in which the crude protein was 3.65 gave the results shown on Figure 7.

In comparing Figure 5 and Figure 7, it is clear that the higher the initial concentration of extract, the lower the critical pressure will be. This means that it would be impractical to use higher UF pressure at the beginning when the concentration of extract is high. Apparently, the critical pressure of UF is also a function of initial concentration of extract. In addition, the experiments showed the flow dynamics of material would affect the critical pressure. In general the higher the flux rate and the more turbulent, the higher the critical pressure will be. Thus, it is recognized that critical pressure is also a function of flow dynamics of the extract during UF. We, thus, reach the conclusion that critical pressure P is a function of size of membrane pores, initial concentration of extract and the flow dynamics on the surface of the membrane, or $P° = f$ (pore size, initial concentration of extract, and the flow rate of extract).

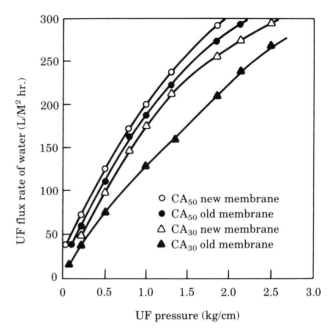

Fig. 6. Effect of pressure on UP flux rate of tap water.

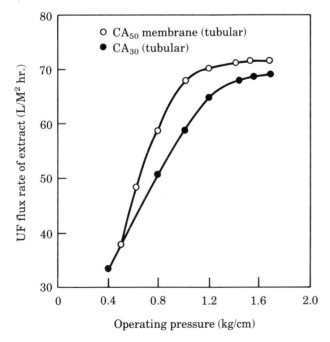

Fig. 5. Effect of pressure on flux rate of extract.

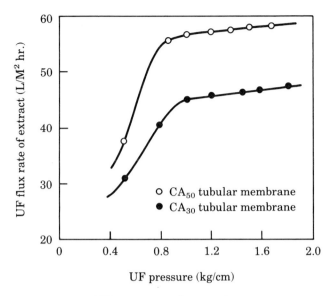

Fig. 7. Effect of UP pressure on flux rate of extract.

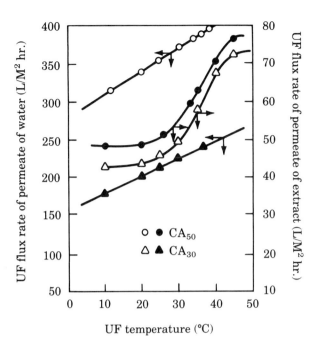

Fig. 8. Effect of temperature on UF flux rate of extract and water for different membranes.

Moreover, it was shown in the experiments that P (critical pressure) had no relation to the temperature and pH value.

Apparently, in order to get high flux rate of UF, it is necessary to carry out the operation under higher pressure leading to a good result with better efficiency.

Relations Among Temperature, pH and Ultrafiltration

In order to study the temperature affecting the action of UF, tubular UF with membranes CA-30 and CA-50 were used for the experiments. The inlet pressure of UF was 2 Kg/cm^2, and the outlet pressure was 1 Kg/cm^2. Under such conditions extracts containing dissolved solid material 2.8% in which the protein content was 2.01% with pH 7.5, was used. The relation of temperature and flux rate was measured and the data are shown in Figure 8.

Experiments indicated that increase of the temperature would increase water flux rate. As for extracts, flux rate would increase 1.2–1.8% for every increase of 1°C. Experiments also indicated when the temperature was above 45°C, an increase of flux rate was not apparent. This may be due to the fact that higher temperature causes the membrane material to swell and the pore size to decrease. Thus for the CA membrane the operating temperature at 40–45°C was more suitable.

Greater viscosity would cause higher resistance to both flow of material and transfer of material.

pH Value Affecting the Flux Rate

In order to find the relation between pH value and flux rate a series of experiments were carried out according to following conditions:

1. For whey of tofu: The whey of tofu was adjusted to pH 4.3, 7.2 and 9.0 using a CA-30 membrane in a tubular UF, the inlet pressure was 2.2 Kg/cm^2, outlet pressure 1.0 Kg/cm^2, and the temperature 35–45°C.

2. For extracts of soy protein: The solid material in an extract of soy protein was 3.15% in which the crude fiber was 1.89%, pH was 5.2, 7.1 and 9.0 using CAT308 flat plate membrane on a flat plate UF to carry the experiments with inlet pressure 4.0 Kg/cm^2, outlet pressure 3.0 Kg/cm^2, and temperature 37–45°C. The data are shown in Figures 9 and 10.

From Figures 9 and 10, these data clearly indicate that under alkaline condition the flux rate reaches the highest, but when approaching the isoelectric point the

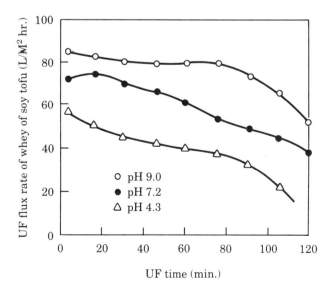

Fig. 9. Whey of soy tofu ultrafiltrated on UF w11-1 model under different pH.

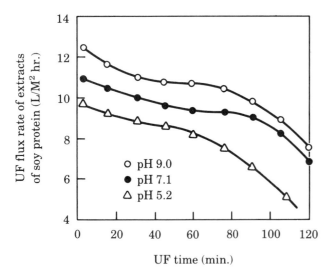

Fig. 10. Extracts of soy protein ultrafiltrated on the flat plate under different pH.

flux rate would be the lowest. As the membrane materials CA and PSA carry negative charges but proteins are amphoteric, protein molecules carrying negative charges in alkaline solution are more efficiently processed by UF. This is due to negative charges on the protein molecules. The negative charges on the membrane repel the protein molecules that move from the surface to the main current of extracts, thus decreasing the degree of polarization and the formation of a jelling layer with a resulting increase in the flux rate of permeate.

When the solution pH = pI, that is, at the isoelectric point, the charges on the protein molecules are equal to zero, thus, the repelling force would be the lowest at which the protein molecules associate easily into larger molecules. Under such conditions the protein retentate would be the highest. Due to the associated protein forming a layer on the membrane surface the solubility would be decreased greatly and a precipitated layer formed there would increase the membrane resistance and the flux rate of permeate would decrease.

Comparison of UF Method and Conventional Acid Precipitation Method

From data listed on Table 1, experiments in the laboratory indicate the yield using UF is 10–13% greater than the yield obtained at the isoelectric point. If the operation was carried out in a plant which causes more losses, the yield would increase to 15%. With the extracts after ultrafiltration, using reverse osmosis (RO) to recover low molecular protein in UF permeate, the yield of nitrogen can reach as high as about 99%.

The composition and nutritive value of the products were compared. Of the four kinds of products obtained in the laboratory, three were obtained by extracting the protein under different pH and then precipitating them at pI, and the fourth one was obtained by UF only. The data analyzed are listed in Table 2.

Analytical data indicate that the protein isolate manufactured by UF contains less oligosaccharides with a higher percentage of protein, less phytic acid and removal of more trypsin inhibitor. Sulfur containing amino acids in this protein isolate are higher than the rest of them. This suggests that it is profitable to promote the quality of protein isolate by using UF.

Chemical, physical and functional properties data deter-

TABLE 1
Comparison of Yields of Protein Manufactured by UF and PI Precipitated Method

		Protein content of extracts (g)	N 6.25 Recovered (g)	N 6.25 Lost (g)	Error (± g)	Protein content in product %	Yield of protein recovered %
UF CAT-308 membrane (flat plate)	UF once	105	99.28	6.12	0.4	88.4	94.55
	Concentrate diluted and UF repeated	123	113.16	10.14	0.3	95.0	92.00
CA-30 membrane (UF tubular)	UF once	347.5	323.1	25.1	0.7	87.8	93.99
	Concentrate diluted and UF repeated	347.5	314.8	33.7	1.03	95.2	91.61
PI precipitated (pH 4.2)	pH 4.2 (not washed by water)	112.77	91.79	21.47	0.5	87.0	81.41
	pH 4.2 (washed by water twice)	15.72	12.64	2.75	−0.33	91.2	80.20
PI-UF combined pH 4.2 precipitated using CAT 308 flat plate UF	pH 4.2 (not washed by water)	205.30	166.11	39.19			
	Whey (UF once)	39.19	32.12	7.07			
	Total	205.3	198.23	7.07			96.56

TABLE 2

Comparison of Four Products Produced in Our Laboratory with Soy Protein Isolate Produced by Purina USA

Components	Extract at pH 7 and precipitated at pH 4.2	Extract at pH 9 and precipitated at pH 4.2	Extract at pH 11 and precipitated at pH 4.2	Extract at pH 9, then UF	Purina 660 USA	Raw material soy flour
Water %	9.23	7.7	6.21	7.07	5.0	12.07
Crude protein %	93.09	90.04	87.47	96.01	96.32	56.07
Ethyl ether extract %	0.078	0.086	0.109	0.077	0.53	2.22
Total sugar %	3.46	4.64	5.50	1.29	—	33.08
Oligosaccharide %	0.75	0.56	0.65	0.48	—	12.23
Crude fiber %	—	0.20	0.24	0.21	0.22	2.89
Ash %	2.76	4.52	5.81	3.13	4.02	6.71
Total phosphorus %	0.88	0.89	0.92	0.80	0.85	0.71
Total %	100.26	100.38	100.04	101.39	101.94	101.46

mined are listed in Table 3. From the analytical data listed on Table 3, it is shown that protein isolate manufactured by UF will give a product of excellent chemical properties, color and normal odor. As there are no phase changes and no excessive overheating, the PDI of protein isolate product is apparently higher than those obtained by other methods. The emulsifiability and stability of emulsifiability are excellent and the product is suitable for thickening agents and emulsifiers. In comparison with egg, the protein isolate manufactured by UF possesses excellent properties with a cheap price and can be used by adding it to replace eggs. Protein isolate manufactured by UF possesses excellent foaming and foam stabilizing properties. These also are advantages of protein isolates manufactured by the UF method.

References

1. Omosaiye, O., and M. Cheryan, *J. Food Sci.* 44:1027–1031 (1979).
2. Porter, M.C., and A.S. Michaels, *Applications of Membrane Ultrafiltration to Food Processing*, presented at the 3rd International Congress of Food Science & Technology. Washington, D.C., Aug. 9–14, (1970).
3. Pompei, C., M. Lucisano and S. Maletoo, *Production*

TABLE 3

Comparison Physico-Chemical and Functional Properties of Products

	Extract at pH 7 and precipitated at pH 4.2	Extract at pH 9 and precipitated at pH 4.2	Extract at pH 11 and precipitated at pH 4.2	Extract at pH 9 and UF to get product	One soy protein concentrate	Egg white
Viscosity (CP)	3.6	3.5	2.3	3.6	3.1	—
Whiteness	58.00	51.30	29.60	55.50	56.50	—
PDI %	25.98	63.74	8.08	72.47	64.68	—
pH of 1:10 dispersion	6.56	6.93	6.79	6.84	7.8	—
Odor	Normal	Normal	Normal	Normal	Unpleasant	—
Foaming ability	66	70	73	95	65	—
Emulsifying ability	58.97	71.61	50.61	80.59	84.71	77.71
Emulsion stability	44.0	70.8	49.1	80.54	83.75	70.87

d'iolates de Soja par Ultrafiltration et Diafiltration, presented at the 4th International Congress of Food Science and Technology, Madrid, Spain, Sept. 23–27, (1974).
4. Okuba, K., A.B. Weldrop, A.G. Iacobucci and D.V. Meyers, *Cereal Chem.* 52:263, (1975).
5. Goodnight, K.C., G.H. Hartman and R.F. Marquarat, Aqueous Purified Soy Protein and Beverage. U.S. Patent 3,995,071.
6. Lawhon, J.T., D.W. Hensley, D. Mulsow and K.F. Mattil, *J. Food Sci.* 43:361 (1978).
7. Olsen, H.J., *Lebensm-Wiss U-Technol.* 11:57 (1978).

Technology for Full-Fat Soya Products and Extracted Soymeal with Different Contents of Water Soluble Protein

Heinz Schumacher*

Schumacher Consulting Engineers, Hoeperfeld 26, 1050 Hamburg 80, Federal Republic of Germany

Abstract

Today's conventional method of producing toasted full fat soya (FFS) products are described. From literature it is known that young chickens and young hogs digest water soluble protein more efficiently than denatured protein. For that reason production methods are described and suggested on how to keep proteins in the natural native state, e.g., water soluble, during processing. Trypsin-inhibitor will be reduced sufficiently during heat treatments while desolventizing or toasting FFS.

For proper feeding of animals one has to consider that ruminants need proteins that are denatured, whereas monogastric animals digest the proteins in the natural state with better results. This is known from many publications and shall not be discussed in this paper. This paper will deal with the processing of soya in respect to feeding for monogastric animals with soya products with a higher content of water-soluble proteins.

As mentioned before, the proteins in soya are not denatured. The proteins are soluble in water and therefore more easily digestible. During the processing of soybeans, e.g., extraction and toasting, the proteins are denatured and coagulated and become insoluble in water. The solubility is measured by the NSI value (nitrogen solubility index), or by the PDI value (protein dispersibility index).

Figure 1 shows the relationship between PDI and NSI. In the soybean itself the protein is highly water-soluble measured as NSI or PDI. Both methods show about 90% to 95% solubility. During processing, especially during desolventization, the proteins are converted from the water-soluble state into an insoluble state. The diagram shows that in the region of 30 to 65 PDI the values of NSI are much lower relatively. The aim of this paper is to show ways of preventing the denaturation of protein.

It is well known from many publications about feeding tests that soy in its natural form cannot be fed to animals without heat treatment. There are many enzymes in soybeans, most of them completely harmless. There is one protease called antitrypsin or trypsin-inhibitor that is the most harmful to animals and also human beings. This enzyme must be inactivated, and this can be done only by heat treatment. There is another enzyme called urease. This enzyme is harmless. However, if urea also is contained in a diet (and compounders sometimes add urea in feed compositions for ruminants) then the urease will react with the urea, and this would be harmful to the animals. All other enzymes are neglectable, for practical purposes.

The antitrypsin is very difficult to analyze, whereas urease activity can be measured in the laboratory more easily. By experience it is known that antitrypsin is well inactivated when the urease is inactivated also. Figure 2 shows that first the urease is activated and, a short time afterward, the antitrypsin also is inactivated. This diagram is valid if the heating is done at atmospheric pressure at 100°C. At higher pressures, higher temperatures can be applied. How this can be effected will be shown later. Also the moisture content while toasting influences the results.

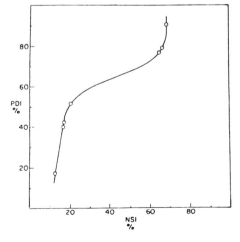

Fig. 1. Relationship between PDI and NSI.

* Deceased.

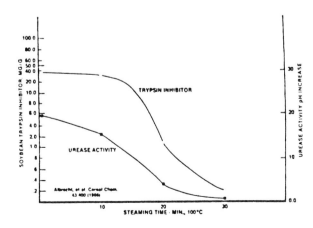

Fig. 2. Reduction of urease-activity and antitrypsin.

In Figures 3 and 4, the influence of moisture content on the process of reducing urease content and antitrypsin during heat treatment at 100°C.

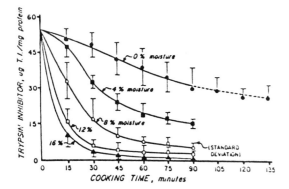

Fig. 3. Influence of moisture content on trypsin inhibitor.

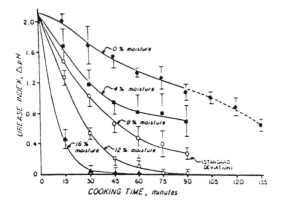

Fig. 4. Influence of moisture content on urease index.

We will review today's system of heat treatment when producing toasted full fat soy meal and later on the suggested process to keep the content of water-soluble protein at a higher level.

The standard procedure to produce toasted FFS is shown in Figure 5. Cleaned seed is cracked and heated in a stack cooler or toaster with direct steam applied. The direct steam condenses on the cracked beans and heats the beans up to 100°C or slightly above 100°C. By condensation of the directly applied steam the moisture content is raised from 12% to about 19%. A retention time of 30 to 40 minutes at 103°C is usual. If the beans are well cracked (maximum size of 1/8 beans), then a retention time of 20 to 25 minutes will do. The urease activity then will be lowered from 2.0 delta pH-rise originally in the untreated beans to about 0.1 to 0.3 delta pH-rise after the heat treatment. The antitrypsin or trypsin inhibitor units (TIU) are lowered by the heat treatment from originally untreated beans with 50 to 55 TIU/mg to about 2 to 8 TIU/mg. A reduction of 80% to 85% is considered sufficient by feed technologists. The water solubility of the proteins is lowered from about 95% PDI to about 15% to 25% PDI. The aim of this paper is to show how to prevent a reduction of PDI that far.

The process as shown in Figure 5 needs about 140 to 180 kg steam per ton of soybeans processed. Only very little of this steam is recovered as most of the steam condenses in the seed and is later dried out of meal and leaves with the air used for drying. Now let's consider a more steam efficient system.

In Figure 6, much of the steam used is recovered from the spent air of the drying section for the preheating of the beans. The analysis of the finished product is about the same as with the equipment shown in Figure 5. The total steam consumption of a plant like this (Figure 6) is only 50 to 60 kg, compared to the former 140 to 180 kg of steam per ton of seed processed, because of reutilization of steam.

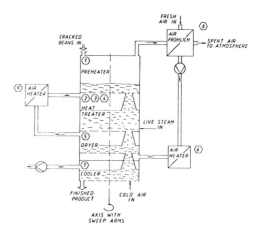

Fig. 6. Conventional full fat soy plant with heat recovery.

It should be noted that the stack cooker or toaster as shown in Figures 6 and 7 does not have any means of indirect heating. The hot air supplied in the drying section which picks up the moisture content from the beans is used to preheat the beans before the direct steam is applied.

In order to reach a high content of water-soluble protein in the finished product, it is necessary to keep the moisture in the product low while heat is applied.

For that reason, the beans to be processed should be dried to a maximum moisture content of 10%. The stack cooker or toaster should have sufficient indirectly heated plates and walls. Preheating the beans by hot air should be done in the top part of the toaster, and the spent air should not be reused as the moisture contained therein would condense on the beans. The beans easily can be heated up to 80°C by hot air, and the moisture content in the beans will be reduced to about 9% simultaneously. The direct steam is supplied to heat them up to 103°C, however the moisture content will not rise higher than 12% to 14%. At this low moisture content denaturing of the protein will be retarded, whereas the antitrypsin will be reduced.

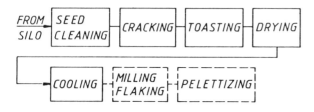

Fig. 5. Generalized flowsheet for producing full fat toasted soy meal in conventional manner.

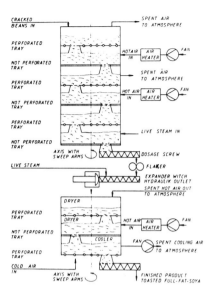

Fig. 7. Full fat soya schematic.

There is a plant of this type in operation and the hot beans are flaked at this moisture content and at a temperature of 103°C. In an expander direct steam is supplied in order to raise the temperature for only 2 or 3 seconds to about 130°C. Right after the outlet of the expander the material flash cools to a little less than 100°C. The finished product has a yellow color and appears granulated with no hulls visible. The product after drying and cooling has no dust, and during storage no separation of hulls from meats occurs. The flowability is good, and there is no tendency to bridging at moisture content of 12% or less. The water solubility of proteins is 50% to 60% PDI and TIU is about 4 to 8 per mg. Urease activity might range from 0.4 to 0.6 delta ph-rise.

To produce an extracted meal with higher content of water-soluble protein, the preparation system should be the same as described in Figure 6 prior to extraction with one exception, e.g. the expander is operated with very little steam, and the material is pelletized in the expander and then expanded. After expanding the material has to be dried and cooled well to 30° to 40°C at a moisture content of 8% to 9%. The material is then ready for solvent extraction.

After extraction the material is desolventized, and this means another heat treatment. To make the material solvent free, a DTDC can be used, but in a modified way. The top part of the desolventization stage should be equipped with "predesolventization" decks where most of the hexane evaporation takes place with indirect heat, in order to keep the water content low while desolventization takes place with direct steam. Instead of using indirectly heated decks inside the desolventizing section, predesolventization could be done also in "Schneckens," e.g. indirectly heated screw conveyors. In this way an extracted meal can be produced with a content of 50% PDI water-soluble protein and higher. In this case the TIU are sufficiently low also, but urease activity is higher than usual. This urease activity higher than usual is something like a barrier to people in the trade. As mentioned earlier, the urease activity is something completely unharmful and unimportant as long as there is no urea added in the feed composition. This however is still hardly understandable to compounders and completely unbelievable to people who have heard anything about trypsin inhibitor. It will take time to reeducate people to new thinking, and it is a big handicap that there is at this time no convenient method of analysis for TIU-content.

Let me make some remarks on expander design. This type of machine has been in use for about 20 years in South America and there are hardly any soy extraction plants in South America where expanders are not in use. However, expanders in South America are essentially the same today as they were some 30 years ago. In South America expanders in all factories need supervision, and power consumption is ca. 6 to 9 KW/ton of seed processed. By technical changes in various aspects this has been reduced ca. 50%, and expanders operating now do their job reliably without any supervision except at start up and shut down. However, there are essentials for trouble free operation, e.g. (a) sufficient cleaning of seed, (b) constant feed supply, and (c) constant steam pressure.

In South America a real expanding of the material is done. If, for instance, the die is bored with conical borings with an outlet diameter of 16 mm, the pellet leaving the die has a diameter of 22 mm. It can be assumed that because of this phenomenon the word expander originated. During continuous trial work in 1988, it was found that pelletizing with expanders is more beneficial than expanding with respect to solvent retention and oil extraction. It can be expected that the use of expanders in oil mills in only very few years from now will be a common practice everywhere.

For the production of toasted full fat soya with a high content of water-soluble protein it has been shown that steam consumption per ton of seed is higher than for the production of ordinary full fat meal with low content of water-soluble protein. The same applied for extracted meal of higher or lower content of water-soluble protein. Today's average steam consumption for a complete extraction plant including all sections is approximately 300 kg of steam per ton of seed processed. This can be reduced by more than one-third to 160 or 180 kg per ton of seed processed when using systems as described previously—reusing steam contained in spent air from drying section for the preheating of seed in the beginning. On the other hand, it is desired to produce a meal with high content of water-soluble protein, then the steam consumption will be about 300 kg per ton of seed processed.

References

1. Horan, F., *J. Am. Oil Chem. Soc., 51:*67A (1974).
2. Wright, K.N., *Ibid., 58:*294 (1981).
3. McNaughton, J.L., *Ibid., 58:*321 (1981).

Low Cost Dry Extrusion of Feeds

Thomas F. Welby

Triple F Inc., P.O. Box 3600, Urbandale Branch, 10104 Dougland Ave., Des Moines, Iowa 50322, USA

Abstract

Use of dry extruders is explained, along with processing and economic advantages in cooking. Dry extruders permit production of new livestock feed products, including full fat soya feeds with stable natural tocopherols combined with ready availability of amino acid content for monogastric and ruminant animals.

The Insta-Pro Dry Extrusion process was developed by Triple F Inc. in the mid-1960's. After numerous attempts of soybean processing using infrared and microwave, the idea of friction as the source of heat was taken into consideration. A prototype machine was built, and this idea proved to be economical and safe. After many improvements, the first extruder was sold in 1969, starting the dry extrusion revolution.

The initial purpose for designing our dry extruder was to develop a method of processing whole soybeans so that the soybean could be fed directly into the final ration. This meant that the temperature, dwell time and pressure had to be such as to allow a decrease or destruction of the growth inhibitors contained naturally in the raw soybean, e.g., trypsin inhibitor, urease and lipase enzymes. The trypsin inhibitor will be decreased to ca. 12 to 15 mg/gm with urease at approximately 0. With this extruder, we found this to be a unique process and received patents on the process and equipment soon thereafter.

Dry Extruders

When we refer to a dry extrusion we mean that no additional steam is introduced and lubrication is provided by water, oil or low-melting point ingredients. The extruder has two types of feeding apparatus: one is a vibratory feeder with a holding hopper above it; the second is a D.C. variable speed motor or A.C. frequency drive variable speed motor controlling the screw feeder. This is the main method of feeding the extruder. The Model 2000R unit is operated on a 75 horsepower electric motor. The product to be extruded is fed into the inlet chamber of the barrel either by gravity or directly into the barrel through a side feeding hopper.

Working Action

The product is moved forward in the inlet chamber with a screw to a restriction point called a steamlock. It continues to move internally causing a build-up of friction and pressure until it develops enough pressure to pass the restriction allowing entry into the second chamber, called a compression chamber. In the second and third compression chambers, the product continues to build up more pressure and temperature. The extrudate is moving in a back-and-forth motion. The size of the restrictions and the configuration of the internal screws can be changed to allow flexibility in the amount of friction, temperature and pressure developed.

After it exits the compression chamber, the extrudate enters the last chamber containing a nose bullet and nose cone. The nose cone can be adjusted while the machine is in motion. This adjustment allows for the final pressure setting to fine tune the desired process temperature. The extruder's main shaft attached to the screws, restrictions and nose bullet revolves at approximately 540 rpm.

Dry Extrusion From a Mechanical Viewpoint

There is low capital investment required as the dry extrusion equipment can be installed in a very simple manner with no boiler or dryer added. Other advantages include: (a) low energy inputs as the extruder can be operated by electric or combustion engine, using portable mounts if desired; (b) controlled process with the equipment designed to be as simple and foolproof as possible, using a standard speed main shaft controlling the dwell time along with simple temperature readings; (c) manual on/off automatic run, designed to be started and properly adjusted to achieve desired product and then let run; the extruder can be operated on an hourly basis or for weeks without stopping (we do recommend that someone be in the general area to check the equipment on a regular basis.); and (d) expandability and flexibility as with the small size of one unit (and normally little auxiliary equipment other than pregrinding, mixing and post cooling), very little space is necessary; any number of units can be operated in combination with multiple units in a large plant providing flexibility to increase or decrease production to match demand; if a mechanical failure happens to one unit in a multi-unit installation, only a portion of the production is lost.

Minimum training is required as all of our new equipment includes start-up and training on-site. This normally requires one or more days depending upon the number of extruders to be started and the number of products to be produced. Post-sales worldwide service is available with trained personnel. During this initial installation and training, maintenance personnel are taught how to maintain the equipment. Since the extruder is designed with simplicity in mind this does not present a problem, nor are special skills needed for routine maintenance. The extruder is designed with the principle of friction to develop the heat and pressure. All parts are designed to minimize wear by using top quality materials to insure durability and safety.

Advantages of Dry Extrusion

The advantages of cooking, sterilization, expansion and dehydration together allow for stabilization.

Cooking: The cooking process allows for reducing inhibitors or cooking raw products. This is accomplished in approximately 25 seconds. This short processing time is important in regard to heat damage to protein, energy and vitamins, along with less "browning" reaction in an oxygen-free atmosphere. The process can be controlled by the variation of the internal extruder parts allowing for temperature control from low to approximately 350°F.

Sterilization: The sterilization effect adds many advantages such as reducing bacterial contamination, and destroying molds, aflatoxins and yeast.

Expansion: The expansion of the extrudate causes positive advantages in the gelatinization of starch, rupturing of oil cells, and allowing for shaping and texturization.

Dehydration: The dehydration caused by the extrudate being cooked to high temperatures under pressure. When the material exits the extruder barrel, into normal atmospheric pressure and temperature, it quickly decreases in temperature to the 212°F boiling point. This quick reduction in temperature causes a flashing off of moisture up to approximately 50% of the preextrusion level.

Stabilization: The stabilization effect is caused by the enzyme destruction, dehydration and sterilization.

Effects of Dry Extrusion

The effects of cooking, sterilization, expansion, dehydration and stabilization lead to a whole new world of possibilities in livestock feeding. For instance, the effects on fat are unique. When an oleaginous product such as whole soybeans is passed through our extruder with a temperature reading at the last compression chamber of approximately 300°F, the fat cells are ruptured allowing for high digestibility of the oil and utilization of the natural tocopherols, lecithin and linoleic acids. This extruded soybean, normally called full energy soybean or full fat soybean, will consist of approximately 38% protein, 18% fat and 6 to 7% moisture content. The natural tocopherols act as antioxidants allowing the product to be stored for long periods of time even in hot conditions as long as the product is kept dry. The cooking effect of the protein allows for the reduction of the natural growth inhibitors yet provides excellent amino acid availability for both monogastric and ruminant animals. The fiber is also affected by increase of the bulk density and the improvement of digestibility.

Suitable materials for processing include: oilseeds (soybean, sunflower, rapeseed, cottonseed and peanut); grains (corn, wheat, barley, rice, oats, beans and peas); high moisture fresh ingredients, e.g., fruits and vegetables; animal, fish and milk proteins; combinations, i.e., corn and soy combinations, rice and bananas, etc.; complete feeds, i.e., pig starters, calf starts, aquatic food, fur foods, horse and pet foods; and raw fats, oils and processing residues.

Products now being produced by dry extrusion include: pet foods; complete feeds; by-pass protein; controlled release urea; meat meal; low trypsin inhibitor products for milk protein substitutes; aquaculture products, i.e., fish and shrimp food.

Types of wet feedstuffs available include packing plant wastes, e.g., pork, beef and poultry. The poultry feather process is one that has become a very valuable process utilizing dry extrusion methods. We have developed a complete process to obtain more than 90% pepsin digestibility of the feathers and this is being fed as a protein source directly back to the poultry. Others are hatchery wastes, spent hens, dead animals, fish waste or scrap fish, dairy by-products, and vegetables and fruits.

Reasons for processing include stability, palatability, sterilization, seasonal availability, shelf life, energy savings on processing, and elimination of environmental problems.

Profit incentives include low spoilage, value added, unique (marketing), nutritional advantages, palatability, low investment for return, and ready-to-use.

Potential markets are animal feed and human foods.

Wet waste can be processed with soy. Many of the wet materials are too high in moisture to process properly without addition of a carrier. Extracted soybean meal makes an excellent carrier to develop high protein ingredients using dry extrusion.

Improvement in the Protein Content of Palm Kernel Meal by Solid State Fermentation

Suan-Choo Cheah, Leslie C.L. Ooi and Augustine S.H. Ong

Palm Oil Research Institute of Malaysia (PORIM), P.O. Box 10620, 50720 Kuala Lumpur, Malaysia

Abstract

Palm kernel meal (PKM) is used mainly in the formulation of compound feed for dairy cattle. The meal is classified as a high energy feed rather than a protein feed. Its oil content ranges from 4.5% to 17% while the crude protein content averages 14%. PKM has been used as a substrate in the solid state culture of *Trichoderma reesei* QM 9414 and other fungal strains for the production of cellulase, xylanase and β-glucosidase. After culture and enzyme extraction, the protein content of the solid substrate increased to over 20% showing that the process is also useful for converting PKM to a protein feed.

The fruits of the oil palm (*Elaeis guineensis*) yield three economically important products, namely, mesocarp oil (termed generically as palm oil), kernel oil and kernel meal. Palm kernel meal (PKM) is the residue left after mechanical or solvent extraction of oil from the kernels (1). Mechanically extracted PKM forms about 85% of the Malaysian PKM production and it is used mainly in the formulation of compound feed for dairy cattle in Europe. Table 1 shows the formula of a typical feed used in The Netherlands (2). The meal can also be incorporated into feed for swine and poultry (3).

TABLE 1

Typical Dairy Cow Compound Feed in The Netherlands

Ingredients	Quantity (%)
Corn gluten pellets	30.64
Citrus pulp	19.63
PKM, mechanically extracted	11.26
Calcium	0.32
German rapeseed	12.00
Molasses	9.00
Soybean hulls	8.00
Filipino coconut	8.71
Feedgrade tallow	0.50

PKM is valued as a high energy feed. The oil content of the mechanically extracted meal ranges from 4.5% to 17.3% while that of the solvent extracted is much lower, ranging from 0.5% to 5% (4). The value of PKM as protein feed is limited, however, by its low protein content, which is about 14% on the average for both the mechanical and solvent extracted meals. The process of solid state fermentation has been shown to be useful for enhancing the nutritive value of animal feeds by increasing the protein content (5,6). We have employed PKM as substrate for the solid state culture of several fungal strains initially for the production of enzymes. However, subsequently it was found that the process is also applicable for increasing the protein content of PKM. This paper describes both aspects of the study.

Experimental

Strains: The *Trichoderma reesei* strains were obtained from NRRL, Peoria, Illinois, USA, except for *T. reesei* Rut C-30 which was a gift from Dr. B.H. Kim of KAIST, Seoul, South Korea. Other fungal strains were obtained from ATCC, Rockville, Maryland, USA.

Culture Conditions: The fungal strains were maintained on potato dextrose agar (PDA) slants. Growth temperatures were as follows: *Sporotrichum thermophile*, 45°C; *Chaetomium cellulolyticum*, 37°C; *Trichoderma koningii* and *T. reesei* strains, 30°C. For the experiments described, the substrate used was PKM mechanically extracted by screw pressing. The protein content of the meal samples ranged from 13.7% to 16.9%. Growth substrates for solid state culture consisted of either PKM + distilled water, PKM + Toyama's minimal salts (TMS) (7), or, PKM + Mandels' minimal medium (CPM) (8). For each experimental flask, 15 ml of liquid supplement was added to 10 g of PKM. The flask was inoculated with a spore suspension of the growth organism and incubated at the appropriate temperature for seven days, unless stated otherwise.

Enzyme Extraction and Recovery of Fermented Biomass: 100 ml of extraction buffer (0.1% Tween 80 in 0.05M sodium citrate, pH 4.8) was added to each flask. The flask was shaken at 150 rpm at ambient temperature (approximately 25°C) for 30 min before the contents were centrifuged at 13,000g at 4°C. The supernatant was decanted and filtered through glass microfibre filter (Whatman GF/A) prior to being used as crude enzyme extract in the assay procedures. The pellet in the centrifuge tube was then resuspended in distilled water and the solids retained on a glass microfibre filter. The fermented biomass was dried at 75°C and ground in a mortar before analysing it for the total nitrogen content.

Enzyme Assays: Filter Paper Activity (FPase) was determined according to the method of Mandels and Sternberg (9). Cottonwool Activity (CWase) was estimated as for FPase except that cottonwool was used as the substrate. Carboxymethyl-cellulase (CMCase or endo-β-D-1, 4-glucanase, EC 3.2.1.4) was measured by incubating 0.5 ml of diluted enzyme extract with 1.0 ml of

3% carboxymethyl-cellulose (Sigma) in 0.05M sodium citrate buffer, pH 4.8 for 10 min. The reaction was carried out at 60°C for *S. thermophile* and at 50°C for the other strains. The amount of reducing sugar produced was estimated by the method of Miller (10) with glucose as the standard. The assay for xylanase activity was modified from that of Khan et al. (11) with 3% xylan from oat spelts (Sigma) as substrate. The reaction temperature for each strain was as for the assay of CMCase. The reducing sugar released was measured with D-xylose as standard. β-glucosidase (β-1,4-glucanase, EC 3.2.1.21) was assayed by the method modified from that of Ishaque and Kluepfel (12) with p-nitrophenol-β-D-glucopyranoside as substrate. The amount of p-nitrophenol produced was measured spectrophotometrically at 400 nm.

Analytical Procedures: Total crude protein was estimated by multiplying total Kjeldahl N by 6.25. The composition of amino acids was determined in acid hydrolysates of the biomass remaining after enzyme extraction. Analysis was carried out using a Technicon TSM Model amino acid analyser.

Results and Discussion

Solid State Culture of Fungal Strains on PKM: The primary objective of our experiments was to produce cellulase, xylanase and β-glucosidase using PKM as substrate in the solid state culture of cellulolytic fungi. When *T. reesei* QM 9414 was grown on PKM supplemented with Toyama's minimal salts, cellulase and xylanase activities were detected after 3 days of culture (Fig. 1). It was found that the protein content of the growth substrate after enzyme extraction increased from the initial level of 15% to over 20% after 7 days of growth.

Several other fungal strains were then tested for their abilities to grow on PKM alone, or on PKM supplemented with minimal salts. Cellulase (as measured by FPase, CWase and CMCase activities), xylanase and β-glucosidase production and protein content of the fermented biomass after enzyme extraction were determined in 7-day old cultures. Table 2 shows that growth of the *T. reesei* strains did not increase the protein content of the substrate unless it was supplemented with TMS, except for *T. reesei* Rut C-30. On the other hand, *T. koningii*, *S. thermophile* and *C. celluloyticum* increased the protein content of PKM to 20% and above even without the addition of minimal salts. However, the *T. reesei* strains, Rut C-30 and MCG 77, were better enzyme producers as shown in Tables 3, 4 and 5. The levels of enzyme activities induced were generally improved when PKM was supplemented with minimal salts.

TABLE 2

Protein Content (% Dry Weight) of Palm Kernel Meal After Solid State Culture

Growth Organism	Medium		
	A	B	C
	% Dry Weight of Protein		
Trichoderma reesei			
QM 9414	11.57	22.82	11.88
Rut C-30	25.00	22.50	11.25
MCG 77	18.13	21.26	17.50
Trichoderma koningii ATCC 26113	20.00	21.88	22.50
Sporotrichum thermophile ATCC 42464	20.63	23.75	20.63
Chaetomium cellulolyticum ATCC 32319	21.88	23.13	21.25

A: PKM
B: PKM supplemented with Toyama's minimal salts.
C: PKM supplemented with Mandels' minimal medium.

Protein Quality: The amino acids lysine and threonine are limiting in PKM (13). It was therefore of interest to determine whether solid state culture was capable of improving the levels of these two amino acids. The amino acid profiles of PKM fermented with *T. reesei* QM 9414 and *C. cellulolyticum* showed that the process increased the level of threonine but did not enrich the substrate with lysine (Table 6).

Growth of T. Reesei QM 9414 on Mixtures of PKM and Palm Oil Meal: In mesocarp oil extraction, the crude oil from screw pressing contains approximately 53% oil, 40.5% water and 6.5% non-oily solids. The crude oil is sometimes fed into decanter centrifuges which separate the solids from the oily liquor. The solids are dried to

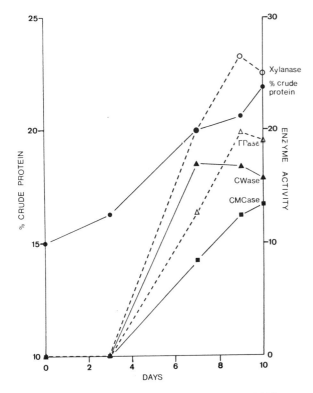

Fig. 1. Time course of solid state culture of *T. reesei* QM 9414 on PKM supplemented with Toyama's minimal salts

TABLE 3

Enzyme Production During Solid State Culture in PKM

Growth Organism	Enzyme Activities				
	1	2	3	4	5
T. reesei					
QM 9414	4.33	0.10	0.0	0.04	9.07
Rut C-30	48.67	1.16	16.34	0.24	46.27
MCG 77	28.89	1.21	20.89	0.61	67.22
T. koningii	4.82	0.22	3.55	nd	13.73
S. thermophile	3.20	0.27	28.39	0.10	20.73
C. cellulolyticum	3.20	0.15	3.86	0.03	7.00

1 CWase in μmoles glucose released per h per ml extract.
2 FPase in Filter Paper Units per ml extract.
3 CMCase in μmoles glucose released per min per ml extract.
4 β-glucosidase activity in μmoles p-nitrophenol released per min per ml extract.
5 Xylanase activity in μmoles xylose released per min per ml extract.
nd = not determined.

TABLE 4

Enzyme Production During Solid State Culture in PKM Supplemented with Mandels' Minimal Medium

Growth Organism	Enzyme Activities				
	1	2	3	4	5
T. reesei					
QM 9414	5.82	0.19	4.51	0.04	21.13
Rut C-30	50.82	1.22	12.27	0.26	41.07
MCG 77	26.43	1.09	20.54	0.48	48.28
T. koningii	5.33	0.36	4.60	nd	16.80
S. thermophile	3.44	0.34	30.57	0.22	22.87
C. cellulolyticum	2.75	0.0	0.0	0.03	7.20

1 CWase in μmoles glucose released per h per ml extract.
2 FPase in Filter Paper Units per ml extract.
3 CMCase in μmoles glucose released per min per ml extract.
4 β-glucosidase activity in μmoles p-nitrophenol released per min per ml extract.
5 Xylanase activity in μmoles xylose released per min per ml extract.
nd = not determined.

TABLE 5

Enzyme Production During Solid State Culture in PKM Supplemented with Toyama's Minimal Salts

Growth Organism	Enzyme Activities				
	1	2	3	4	5
T. reesei					
QM 9414	17.19	0.46	7.70	0.04	20.53
Rut C-30	35.35	0.93	9.81	0.07	26.53
MCG 77	34.14	1.19	23.49	0.49	52.82
T. koningii	7.90	0.38	8.01	nd	20.67
S. thermophile	6.35	0.44	28.79	0.40	21.47
C. cellulolyticum	3.66	0.22	10.23	0.10	9.60

1 CWase in μmoles glucose released per h per ml extract.
2 FPase in Filter Paper Units per ml extract.
3 CMCase in μmoles glucose released per min per ml extract.
4 β-glucosidase activity in μmoles p-nitrophenol released per min per ml extract.
5 Xylanase activity in μmoles xylose released per min per ml extract.
nd = not determined.

TABLE 6

Amino Acid Composition (g per 16 g of N) of Palm Kernel Meal

	PKM		
Amino Acid	A	B	C
	g per 16 g N		
Lysine	3.31	2.28	2.90
Histidine	3.31	1.02	1.25
Arginine	11.00	6.20	10.44
Aspartic acid	6.63	10.58	8.68
Threonine	2.44	3.89	4.22
Serine	3.38	5.73	6.90
Glutamic acid	15.94	21.99	12.55
Proline	3.00	3.45	3.26
Glycine	3.69	3.05	5.21
Alanine	3.56	4.51	6.48
Valine	4.63	2.83	5.58
Methionine	2.13	1.98	—
Isoleucine	3.31	2.22	1.84
Leucine	5.38	3.12	4.06
Tyrosine	2.25	2.71	—
Phenylalanine	3.44	3.22	3.66
Total	77.40	78.78	77.03

A: PKM (13)
B: PKM after growth of *T. reesei* QM 9414
C: PKM after growth of *C. cellulolyticum*

produce a sludge cake known as palm oil meal (POM) (14). The protein content of this meal is lower than that of PKM and is in the range of 11% to 13%. We have used mixtures of PKM and POM for the solid state culture of *T. reesei* QM 9414. Figure 2 shows that growth in mixtures with higher POM content favoured xylanase production while that in higher PKM content produced higher levels of cellulase (as measured by FPase). Moreover, increasing the amount of POM in the growth substrate decreased protein enhancement by microbial culture. Inclusion of POM up to a level of 25% in the medium appeared suitable for the production of a fermented biomass with more than 20% protein content.

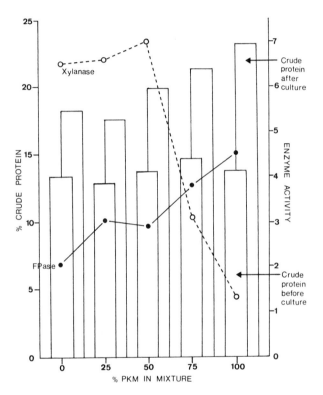

Fig. 2. Growth of *T. reesei* QM 9414 on mixtures of PKM and POM

Acknowledgments

Sharifah Samsiah Rahman and Mohamad Razali M. Noor provided technical assistance. Part of this study was carried out by Teng-Huat Chan, a student trainee from the National University of Malaysia. We thank Mee-ling Lan for total nitrogen analyses.

References

1. Tang, T.S., and P.K. Teoh, *J. Am. Oil Chem. Soc. 62*:254 (1985).
2. Pantzaris, T.P., *Palm Oil Dev. Palm Oil Res. Insti. Malaysia No. 6*, 1987.
3. Yeong, S.W., T.K. Mukherjee and R.I. Hutagalung in "Proceedings of National Workshop on Oil Palm By-product Utilization," Palm Oil Research Institute of Malaysia, Kuala Lumpur, Malaysia, 1981.
4. Siew, W.L., and A.S.H. Ong, *PORIM Report PO(110c) 86 General*, 1986.
5. Bajon, A.M., T.L.S. Tse Hing Yuen, J-C. Li Sui Fong and G.M. Olah, *Biotechnol. Lett. 7*:203 (1985).
6. Smith, R.E., C. Osothsilp, P. Bicho and K.F. Gregory, *Biotechnol. Lett. 8*:31 (1986).
7. Shamala, T.R., and K.R. Sreenkantiah, *Enzyme Microb. Technol. 8*:178 (1986).
8. Sternberg, D., *Biotechnol. Bioengin. Symp. No. 6*, 1976, pp. 35–53.
9. Mandels, M. and D. Sternberg, *J. Ferment. Technol. 54*:267 (1976).
10. Miller, G.L., *Anal. Chem. 31*:426 (1959).
11. Kahn, A.W., D. Tremblay and A. Leduy, *Enzyme Microb. Technol. 8*:373 (1986).
12. Ishaque, M., and D. Kluepfel, *Can. J. Microbiol. 26*:183 (1980).
13. Kim, S.H., and A.N. Mah, *Palm Oil By-products for Food Use*, first annual report submitted to Palm Oil Research Institute of Malaysia, April 1987.
14. Jorgensen, H.K., in "Proceedings of Regional Workshop on Palm Oil Mill Technology and Effluent Treatment," Palm Oil Research Institute of Malaysia, Kuala Lumpur, Malaysia, 1982, pp. 201–212.

Extrusion of Rice Bran

Maurice A. Williams

Anderson International Corp., 6200 Harvard Ave., Cleveland, Ohio 44105 USA

Abstract

Expansion-cooking of rice bran can convert a powdery dusty material with very poor shelf-life into a nondusty shelf-stable product fit for direct utilization as an ingredient into food and feed formulations, or for further processing to obtain a high quality rice oil at lower cost than extracting raw rice bran. There is a disadvantage with this method compared with alternative methods in that the moisture is elevated during processing requiring a subsequent drying step. But there are offsetting advantages in that extremely long shelf life is possible with an expander-cooker for feed applications, and very high throughputs using relatively inexpensive equipment is possible for oil extraction applications.

The use of pretreatment procedures to stabilize rice bran and to prepare it for solvent extraction has become an important issue. Untreated rice bran rapidly undergoes a deterioration process that is initiated immediately upon removal of the bran from the rice kernel and that results in the breakup of the rice oil into free fatty acids due to the action of the enzyme, lipase. This deterioration is a rapid process causing about 3 to 5% of the oil to be converted into free fatty acid each day. If unchecked, the free fatty acid level might reach 50% in about 60 days. In addition, the freshly produced rice bran is in a fine, powder-like form that is difficult to handle in a solvent extractor because of problems with poor percolation and with fines removal from the miscella.

Much work was done in the 1940's and 1950's with heat treatment of rice bran to slow down the activity of the enzyme and to try to crisp and harden the particles somewhat. These methods proved to be helpful, but they didn't go far enough. The enzyme was slowed down rather than destroyed, and the particles were too small to provide sufficient improvement in solvent extraction.

In the early 1960's, the V.D. Anderson Company began experimenting with rice bran in a newly developed extruder-cooker which had successfully been used for the cooking and agglomeration of finely ground animal feed formulations into larger particles that puffed, or expanded, upon discharge from the extruder. They found that processing raw rice bran in an expander-cooker destroyed the activity of lipase and provided a product stable against any further rise in free fatty acid.

The expanded collets had a larger particle size than the raw bran and had porous interiors, both of which provided for much better percolation characteristics, with considerable less problems with fines, than one would experience with raw bran. Anderson filed for a patent in 1962 covering this method of preparation. The patent was issued in 1966 (1).

Expander-cookers were then offered to the industry in a process described as the "Expandolex" process that involved the use of an expander-cooker, a dryer/cooler, and a solvent extractor. The idea was to set up expander-cookers and dryer/coolers in plants that polished rice and ship the shelf-stable extruded collets to a central solvent extraction plant. Data on the process was published in 1965 (2). As far as I know, this was the first article ever published describing the preparation of any oilseed by expansion prior to solvent extraction.

Some expander-cookers were sold for this process on rice bran, but the industry as a whole was not at that time interested in such large-scale processing plants. In the United States, for example, there was no real interest in extracting the bran because there was an adequate market for high-free-fatty-acid rice bran for cattle feed, and no real demand for the rice oil from the bran. In Southeast Asia, because of the wartime conditions in the 1960's, nobody was interested in building solvent extraction plants there, and in the Philippines the rice polishing mills were of very small capacity, too small for even the smallest scale expander-cookers available at that time. Several expander-cookers, however, were placed in plants in Central and South America during the 1960's.

Since that time, various other firms have offered other kinds of extruders for this application, and considerable research was funded by the U.S. Department of Agriculture at their Western Regional Research Center in California (3) and at Colorado State University at Fort Collins, Colorado (4) for developmental work with small-scale rice bran stabilization systems.

In this paper, I would like to describe an expander-cooker and how it performs on rice bran to provide a better overview of the various options in processing rice bran.

Experimental Procedures

The expander-cooker used to procure the data reported in this paper is manufactured by Anderson International Corp. in Cleveland, Ohio. It consists of a rapidly rotating worm shaft having an interrupted flight positioned in a cylindrical barrel having stationary projections intermeshing with the rotating flights. At the end of the barrel is a die plate containing replaceable die inserts of various sizes through which the product extrudes. Live steam and water are injected directly into the rice bran within the barrel.

Figure 1 shows an extruder-cooker in a processing plant preparing flaked soybean for solvent extraction. Figure 2 shows a close up of the collets discharging from the end of the extruder-cooker.

Rice bran would process in a manner similar to soybean, except that the diameter of the collets would be smaller and, in some cases, the collets would be a mixture of coarse meal and broken collets rather than what is seen in Figure 2. Figures 3 and 4 show close up photos of extruded rice bran of large collet size and smaller collet size. The measuring scale in each photo is in inches.

Fig. 1. The Anderson Expander-extruder-cooker on soybean.

Fig. 3 Extruded rice bran.

Fig. 2. Close up of collet formation.

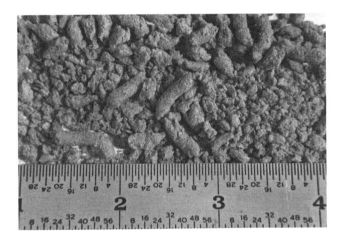

Fig. 4 Extruder rice bran—smaller collets.

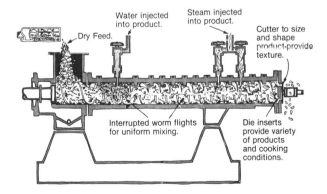

Fig. 5 Sectional diagram of extruder.

An expander-cooker subjects the bran to a high pressure cook utilizing live steam, as well as liquid water, to elevate the moisture and temperature and cause the agglomeration of the fine particles into porous collets. The conditions attained in the expander-cooker are sufficient to inactivate many enzymes including lipase and are usually held between 15 to 25% moisture, 104°C to 132°C (220 to 270°F), 10 to 30 seconds residence time and about 13 atmospheres (200 PSIG) of pressure. Figure 5 shows a sectional view of the expander-cooker.

Results

In the early work with extrusion cooking of rice bran, raw bran at 9% moisture was fed into an expander-cooker along with sufficient live steam and liquid water to elevate the moisture to between 17% to 30% by weight and the temperature to between 103°C to 116°C (218 to 242°F).

The extruder-cooker was operated within the above ranges to provide seven samples extruded under different moisture and temperature conditions. Each sample was divided into two portions, one was dried to less than 10% moisture, the other was not dried. Each of those portions was then divided into two sub-portions, one was stored at 5°C (41°F) in a refrigerator, the other was stored at room temperature. Care was taken to prevent infestation of the samples, but not to prevent them from absorbing moisture if they were hygroscopic. All of the dried samples showed no rise in free fatty acid after 84 days of storage. Most of the undried samples became moldy during the 84-day storage period, but the undried samples showed no signs of rise in free fatty acid until they became moldy; and then the free fatty acid levels quickly rose to 16.7 to 60.5%. One dried sample was continued in long-term storage before the final portions were tested. A test at 387 days and a final one at 681 days showed no rise in free fatty acid.

The raw bran stored "as is" at room temperature showed the most rapid rise in free fatty acid, reaching 30% in 6 days and 55% in 60 days. Thermal drying of the raw bran to 1.4% moisture slowed down the action of the enzyme so that there was a modest rise in free fatty acid for the first 28 days (going from 7% to 9.6%) and a slow rise thereafter, going from 9.6% to 15% during the 28th to the 60th day of storage. Raw, unheated bran stored in the refrigerator also showed a slower rise in free fatty acid, but not as slow as the thermally dried bran. It reached 9.2% on the 2nd day and about 27% on the 60th day of storage.

The samples had considerably higher percolation rates than raw bran, being on the order of 35 to 140 gpm/ft^2 of screen area. Extraction tests showed the collets would reach 0.5% residual oil levels in about 40 minutes extraction time, and the collets, even though larger than raw bran had a higher bulk density, .4 SpG (25#/ft^3) compared to .33 SpG (20#/ft^3) for the raw bran.

Several years later, a sample of stabilized rice bran was received from a processing plant using an expander-cooker. This sample was stored at room temperature in jars with loose lids. There was no rise in free fatty acid for nine months, at which time the sample was depleted.

Some recent work preparing a series of samples at moistures ranging from 24 to 30% and temperatures ranging from 113 to 117°C (235 to 243°F) also showed no rise in free fatty acid for all of the samples for six months, at which time the samples were depleted.

Discussion

It appears from these data that processing through an expander-cooker at elevated moisture levels followed by subsequent drying can effectively destroy the activity of lipase and produce a shelf-stable, full-fat rice bran suitable for subsequent solvent extraction or for use "as is" in foods and feeds formulations. The lipase is inactivated completely; the fat extracted from expanded collets shows no rise in powdery dusty material with very poor shelf-life into a nondusty, shelf-stable product fit for direct utilization as an ingredient into food and feed formulations, or for further processing to obtain a high quality rice oil at lower cost than extracting raw rice bran.

There is a disadvantage with this method compared with alternative methods in that the moisture is elevated during processing requiring a subsequent drying step. But there are off-setting advantages in that extremely long shelf life is possible with an expander-cooker for feed applications, and very high throughputs using relatively inexpensive equipment is possible for oil extraction applications.

References

1. U.S. Patent 3,255,220 (1966).
2. Williams, M.A., and S. Baer, *J. Am. Oil Chem. Soc. 42:*151 (1965).
3. Sayre, R.N., R.M. Saunders, R.V. Enochian, W.G. Schultz and E.C. Beagle, *Cereal Foods World 27:*317 (1982).
4. Harper, J.M., D.A. Cummings, J.D. Kellerby, R.E. Tribelhorn, G.R. Jansen and J.A. Magor, *Evaluation of Low-cost Extrusion Cookers for Use in LDCs*, in L.E.C. Report No. 4, Colorado State University, Fort Collins, Colorado, 1978, p. 144.
5. Williams, M.A., *Oil Mill Gazetteer 91:*24 (1986).
6. Lusas, E.W., and L.R. Watkins, *J. Am. Oil Chem. Soc. 65:*1109 (1988).

Production and Utilization of High Ruminal Bypass Proteins

Marshal D. Stern and Paul M. Windschitl

Department of Animal Science, University of Minnesota, St. Paul, Minnesota, USA

Abstract

In general, research with protected soybean meal and other high ruminal bypass proteins indicates the potential for their use, which includes maintenance of milk production at a lower protein intake; however, results are variable. Reasons for the lack of response in animal performance in some studies may be found in the type of diet fed, level of feeding or productive state of the animals. In many cases much of the dietary protein is protected naturally (e.g., corn), or the level of resistant protein in the control diet already is adequate. Some studies also have used experimental animals in a lower productive state and consequently had a low protein demand, e.g., non-pregnant, mid- to late-lactation or mature ruminants where protein requirements are low or where energy intake is restricted. In order to elicit an animal response to protein or AA, two main criteria must be satisfied. First the animal must be capable of a response; that is, microbial protein should be insufficient to meet the host animal requirement. Secondly, protection must be effective in increasing AA supply to the host animal. At this time, there is insufficient information to allow proper description of requirements for, or production responses to, individual AA, especially in dairy cows. This type of information is necessary before recommendations can be made with confidence for feeding high ruminal bypass proteins or protected AA to ruminants.

Ruminants derive their intestinal protein supply from dietary protein which bypasses ruminal degradation and from microbial protein which is synthesized in the rumen. Although microbial protein alone may be adequate for low producing ruminants, it is inadequate for supporting higher levels of growth, wool or milk production. The use of high ruminal bypass proteins in diets fed to ruminants with high protein requirements improves the amino acid supply to the animal and concurrently decreases excessive ammonia production, thereby reducing stress on liver metabolism. Because soybean meal is a high-quality protein that is extensively degraded in the rumen, various processing methods and treatments have been used to increase the bypass value of the soybean meal protein. Treatments include physical or chemical processes such as heat, aldehydes (formaldehyde, glutaraldehyde and glyoxal), sodium bentonite, tannins, sodium hydroxide, propionic acid, blood, alcohol, xylose and calcium lignosulfonate. Animal responses to these treatments have been inconsistent and in some cases may be due to underprotection and in other cases to overprotection so that the protein becomes indigestible. In addition to the processing or treatment of soybean meal, other sources of high ruminal bypass protein will be discussed.

Absorbed amino acids (AA) from the small intestine of ruminants are supplied by microbial protein synthesized in the rumen, undegraded dietary protein and endogenous protein. Microbial protein usually accounts for the largest proportion of the total amino acid-nitrogen (AA-N) entering the small intestine of ruminants. Microbial protein synthesis in the rumen is dependent mainly on the nitrogen and energy supply to the rumen microbes, which is determined by quantity and ruminal fermentability of protein and carbohydrates. If the energy supplied by the diet is not sufficient, there may be a corresponding decrease in microbial protein synthesis due to less ammonia uptake by the microbes. Therefore, deficiency of energy may lead to a reduced intestinal protein supply and at the same time may precipitate excessive ammonia levels in the rumen, especially when feeding highly degradable proteins. Use of high ruminal bypass or protected proteins in diets fed to ruminants with high protein requirements improves the AA supply to the animal and concurrently decreases the surplus of ammonia, thereby reducing stress on liver metabolism (1).

Even when nutrients are nonlimiting in the rumen, the rumen system may not supply sufficient microbial protein to meet the animal's need for maximum production. Under conditions of high production (fast growth, late pregnancy, or early lactation), the animal depends on an additional exogenous supply to the duodenum, e.g., feeding proteins that because of their physical state escape ruminal fermentation. The fact that protein passes through the rumen undegraded and reaches the small intestine for digestion does not necessarily mean that it is digested efficiently nor, once digested, that the AA profile is such that it provides a better balance of AA for milk production and growth. Protecting protein from microbial degradation in the rumen will be successful only in affecting animal performance if proteins are not denatured to the extent that intestinal absorption of AA is diminished so that the net effect on AA supply is reduced and the animal has the metabolic capacity to respond to an increase of AA supply; that is, requirements for AA have not been met (2).

Amino Acid Release in the Rumen

It is difficult to develop a ruminant protein system based on AA because of an inability to adequately define requirements of the microbial mass and the animal and because it becomes difficult to know what the most appropriate AA combination is for ration formulation (3). To complicate the issue further would be to make the assumption that the AA profile of protein escaping ruminal degradation is more or less identical to the profile in the dietary protein originally ingested. MacGregor et al. (4) suggested on the basis of in vitro studies that alterations in the AA compositions of feed proteins could occur due to rumen microbial activity. Data compiled with 19 feedstuffs showed that the AA pattern of the soluble protein generally deviates sharply from that of the total protein; the soluble protein often containing more lysine. The hypothesis that AA are not degraded equally during ruminal protein degradation is supported by *in vivo* (5,6), *in vitro* (7) and *in situ* work (8,9) using various feedstuffs. In contrast, *in vivo* (10) and *in situ* work (11,12,13,14) using mainly untreated and treated (heat, formaldehyde) soybean meal found little difference in AA composition between original and undegraded protein.

Craig and Broderick (7) used an *in vitro* procedure to study release of 12 AA during degradation of 10 protein sources. Individual AA were not released in proportions equal to amounts present in the protein. For example, they indicated that lysine, arginine, histidine, phenylalanine and alanine were degraded to an extent greater than total degradation while valine, isoleucine, leucine, methionine, threonine, glycine and tyrosine were similar to or lower than total degradation. *In situ* incubation of untreated soybean meal in the rumen resulted in reduction of lysine, arginine, histidine and leucine (9). Stern and Satter (6) measured *in vivo* degradation of essential amino acids from soybean meal, corn gluten meal, brewers dried grains and distillers dried grains. The basic amino acids were shown to be relatively more degradable than the overall degradability. All of the above studies agree that lysine, histidine and arginine were degraded to an extent greater than total degradation.

Nutritive Value of Ruminal Microbial Protein to Ruminants

Under most dietary conditions, microbial protein synthesized in the rumen comprises a substantial part of the protein entering the small intestine (60–85% of total AA-N) where enzymatic digestion releases AA that are absorbed to furnish the animals' needs. Therefore, in general, the AA composition of duodenal digesta usually reflects that of microbial protein except on diets where significant amounts of dietary protein have avoided microbial degradation (15). Storm *et al.* (16) observed that the composition of microbial protein appeared to be constant irrespective of dietary and animal conditions. They also concluded that the mean true digestibility of microbial AA-N is 84.7% in the small intestine and that digestibility of the different AA varies little, with the exception of diaminopimelic acid which is a constituent of bacterial cell walls. Only cystine and histidine showed significantly lower values than the average.

Because microbial protein accounts for such a large proportion of total AA-N supply to ruminants, it would be meaningful to determine any limiting AA in rumen microbes. Due to the complexity of AA and nitrogen metabolism of the ruminant and its microbes, it is difficult to quantify limiting AA. Storm and Ørskov (17) described a method to quantitatively determine the order of limitation of essential AA in rumen microbes or, more precisely, the limiting AA in those absorbed from the small intestine in sheep nourished by infusions of volatile fatty acids and given rumen microbes as the only source of protein. They found that methionine was the most limiting AA in ruminal microbial protein and that lysine was the second limiting AA followed by arginine and histidine.

Based on these observations and other available information, it should be theoretically possible *via* protein or AA supplementation to considerably improve the utilization of microbial protein provided that complementary AA are provided to the animal in such a way that the rumen is bypassed, or that they are protected from ruminal degradation. Because of the findings of Storm and Ørskov (17) and evidence indicating that the basic amino acids are relatively more degradable than other amino acids, further research concerning supplementation of protected methionine, lysine, histidine and arginine needs to be considered.

Treatment of Protein to Increase Ruminal Bypass

Dietary proteins have been protected against microbial degradation in the rumen by introducing linkages by various physical and chemical treatments. These include the use of tannins (18) and more commonly the use of aldehydes (19) and heat (20). Animal responses to these treatments have been inconsistent and in some cases may be due to underprotection and in other cases to overprotection so that protein becomes indigestible. The tanning process involves reactions of mainly hydrogen bonds between hydroxyl groups of tannin and peptide groups of protein. The tannin-protein complex is a reversible reaction, and the complex is hydrolyzable by proteases and in particular by trypsin. Irreversible reactions with quinones, formed by tannin oxidation also may occur (21) and reduce digestibility and availability of some AA. Formaldehyde, a very reactive substance reacts with activated hydrogen-containing molecules and forms hydroxyl methyl derivatives which can react further with formaldehyde or other AA side chains to form cross-links in the form of methylene bridges. The extent to which these linkages can be reversed is not known (19). Heating facilitates the Maillard reaction between sugar aldehyde groups and the free amino groups of protein to yield amino-sugar complexes. These linkages are more resistant than normal peptides to enzymatic hydrolysis. Reversibility of this reaction is dependent upon temperature and time of heat exposure. In addition to heat, formaldehyde and tannins, other treatments have been successful in decreasing degradability of protein in the rumen. These include the use of other aldehydes such as the dialdehydes, glutaraldehyde and glyoxal (19), sodium bentonite (22), sodium hydroxide (9), calcium lignosulfonate (23), propionic acid (24), blood (9), fish hydrolysate (9), alcohol (25) and xylose (26). Most of the above treatments used soybean meal as the test protein source.

Influence of Heating on Protein Utilization

The most commonly used treatment for reducing ruminal microbial protein degradation has been the use of heat processing. Controlled heating of proteins can reduce ruminal degradation of the protein without adversely affecting intestinal protein digestibility (27). Tagari *et al.* (28) were the first researchers to report that heat treatment of soybean meal resulted in decreased protein solubility and increased efficiency of utilization of the soy protein. Animal production responses as a result of heat treatment are due primarily to decreased ruminal degradation of the dietary protein, although heat destruction of inhibitors in some protein sources, such as trypsin inhibitor in soybeans, also can increase animal performance (29).

Factors involved in the heating process are not fully understood. Both temperature and length of heating time are important factors. Tagari *et al.* (28) reported that heating soybean meal at 120°C for .33 h was effective in decreasing the solubility of the soy protein and improving the efficiency of utilization of the soy protein. In contrast, Mir et al. (9) reported that 120°C for .33 h was not effective in reducing *in situ* ruminal degradation of soybean meal protein. Nishimuta et al. (18) and Stern et al. (20) reported that treatment of soy protein at temperatures of 132°C and 149°C resulted in increased quantities of AA available to the small intestine. McMeniman et al. (30) reported that a temperature of 105°C for 24 h did not affect the flow of protein to the duodenum.

When sugars or carbohydrates are present, heating of proteins causes carbonyl groups of the sugars to combine with free amino groups of proteins (31). The AA most affected during the heating process is lysine, presumably due to its free ε-amino group. It has been suggested (32) that even in the absence of sugars or carbohydrates, extensive heating causes unnatural amide bonds to form between the ε-amino group of lysine or other free amino groups and carbonyl groups of proteins.

It has been reported (33,34) that alkaline or heat treatment of proteins can lead to racemization and cross-linking of amino acyl residues. Schwass and Finley (34) suggested that reduced bioavailability of lysine can occur via isopeptide and cross-link formation and racemization of amino acyl residues when proteins are subjected to high temperature or high pH. The dipeptide lysinoalanine is formed in a large variety of food proteins when they are treated with heat or alkali (33). Lysinoalanine, or a metabolite of this compound, was found to be nephrotoxic in rats (34). Chen et al. (35) also suggested that heat treatment of soybean meal can increase lysine-glutamyl cross-linking and hydrophobic interactions within the protein. As a result, the soy protein became less degradable in the rumen.

Heating of cottonseed protein can reduce the bioavailability of the lysine present in the cottonseed (36). Cottonseed contains the compound gossypol, which is a yellow phenol found in the pigment glands of the seed. Gossypol can react with lysine when cottonseed is heated, resulting in decreased bioavailability of the lysine. Schiff bases of gossypol and lysine ε-amino groups have been characterized in heated cottonseed protein (37). However, Craig and Broderick (38) suggested that heat treatment must be more severe than that which normally occurs in commerical cottonseed meal processing to cause substantial, selective loss in lysine availability.

Heat Processing Methods

Expeller processing: Most soybean meal fed to livestock in the U.S. is produced using solvent extraction to remove the oil. An alternative method which generates considerable heat during oil removal is the expeller process. This method involves heating to a maximum of 163°C which results in the Maillard reaction between sugar aldehyde groups and free amino groups. Broderick (39) reported that expeller processed soybean meal provided about 65% more bypass dietary protein than solvent soybean meal and improved milk to feed ratio in lactating dairy cows. He also found that smaller amounts of expeller soybean meal could replace solvent extracted soybean meal without reducing milk production.

Jet-sploding: This process uses a high temperature of 315°C for a short period and utilizes only the moisture within the seed. Deacon et al. (40) found that jet-sploding of whole canola seed reduced protein degradability from 83.5 to 43.2% without dramatically decreasing intestinal digestibility.

Extrusion: The effect of extruding whole soybeans at 132°C and 149°C compared to soybean meal and raw soybeans on ruminal protein degradability measured *in situ* is shown in Figure 1 (20). Raw soybeans were clearly more degradable than the extruded soybeans or soybean meal at all intervals of rumen exposure. At one hour of ruminal exposure time, soybean meal and the two extruded soybean products appeared to be similar in readily available or soluble nitrogen. However, as time in the rumen increased, the extruded soybeans were more resistant to microbial degradation than soybean meal. These same four soybean sources were fed to ruminal and intestinal cannulated lactating Holstein cows to measure protein degradation in the rumen and AA flow and absorption from the small intestine (Table 1). Flow of total AA to the duodenum and subsequent absorption from the small intestine were lowest for the diet containing unprocessed whole soybeans. Extrusion of whole soybeans at 132°C and 149°C increased the flow of AA to the duodenum approximately 10% and caused a 17% higher absorption (g/day) from the small intestine compared with unprocessed soybeans. This effect probably was due to increased resistance of protein in extruded whole soybeans to microbial degradation. Amino acid absorption from the small intestine, expressed as a percentage of AA flow to the duodenum, was higher for the extruded soybean diets, indicating that heat treatment did not overprotect the protein. Lower digestion in the intestine with unprocessed soybeans could possibly be attributed to higher trypsin inhibitor activity compared to the heat processed soybeans. Mielke and Schingoethe (41) determined that trypsin inhibitor activities of extruded soybeans and raw soybeans were 2.7 and 24.0 trypsin inhibitor units/mg, respectively.

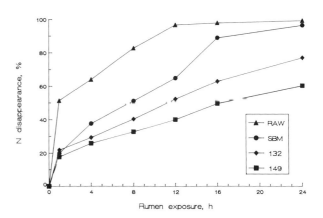

Fig. 1. Nitrogen disappearance from dacron bags suspended in the rumen containing soybean meal (SBM), whole soybeans (RAW) or whole soybeans extruded at 132°C (132) or 149°C (149).

Calcium lignosulfonate: The term lignosulfonate is used to describe any product derived from the spent sulfite liquor that is generated during the sulfite digestion of wood and containing a percentage of lignosulfonic acid or its salt as well as hemicellulose and sugars. Because lignosulfonates can bind and precipitate protein, it was hypothesized that

TABLE 1

Daily Amino Acid Intake, Flow and Digestion in the Digestive Tract of Cows fed Diets Containing Various Soybean Sources

	Diets containing:			
			Whole soybeans extruded at	
Amino acid	Soybean meal	Whole soybeans	132°C	149°C
Intake (g/day)	2081	2064	2085	2097
Degradation in stomach (% of intake)	71.7^a	73.5^a	58.7^b	57.7^b
Flow to duodenum (g/day)				
Total	2265^a	2090^b	2314^a	2361^a
Bacterial	1679	1535	1456	1476
Bypass	586^a	554^b	857^a	885^a
Absorption from small intestine,				
(g/day)	1617^{ab}	1459^b	1749^a	1777^a
(% entering)	71.4^b	69.8^b	75.7^a	75.4^a

a,b Means in the same row not having a common superscript differ (P < .05).

soybean meal treated with lignosulfonates could be rendered less degradable in the rumen. Winowiski and Stern (42) examined various processing factors involved in the lignosulfonate-soybean meal reaction and concluded that heat and the presence of wood sugars in the lignosulfonate preparation were necessary to reduce ruminal protein degradation. In general, calcium lignosulfonate contains a variety of wood sugars, with the main sugar being xylose. Cleale et al. (26) found that treatment of soybean meal with xylose (3 mol xylose/mol lysine) was effective in reducing degradation of soybean protein by rumen microorganisms. It was concluded that controlled nonenzymatic browning improved efficiency of soybean protein utilization by ruminants.

Stern (23) used an artificial rumen system to determine the effects of calcium lignosulfonate (CL) on nitrogen utilization by rumen bacteria. Ammonia-N concentration decreased when soybean meal was pelleted with CL and a modified CL at 4 g/100 g SBM and was further reduced at 8 g CL/100 g SBM (Table 2). This decrease was due to increased resistance of protein to bacterial degradation. In addition, total volatile fatty acid (VFA) production

TABLE 2

Effect of Calcium Lignosulfonate-Treated Soybean Meal on Carbohydrate and Nitrogen Metabolism by Ruminal Bacteria

	Diets^a			
Item	1	2	3	4
NH$_3$-N (mg/100ml)	5.14^b	4.17^{cd}	4.36^c	2.82^d
VFA (mol/d)	$.35^b$	$.32^{bc}$	$.29^c$	$.30^c$
Digestion (%)				
True organic matter	52.3^b	50.6^b	45.5^{bc}	42.4^c
Protein	75.7^b	69.0^{bc}	58.4^c	57.5^c
Nonstructural carbohydrate	91.7	88.9	86.7	84.3
Cellulose	28.5^b	20.5^{bc}	20.6^{bc}	11.0^c
Effluent N (g/d)				
Ammonia	$.129^b$	$.104^{cd}$	$.110^c$	$.097^d$
Bacterial	1.95^b	1.76^b	1.46^c	1.46^c
Bypass	$.72^c$	$.94^{bc}$	1.24^b	1.27^b

^{a} Diet 1, pelleted SBM without any CL addition; Diet 2, CL included in pelleting process at 4 g/100 g SBM; Diet 3, modified CL included at 4 g/100 g SBM; Diet 4, CL included at 8 g/100 g SBM.

b,c,d Means in the same row not having a common superscript differ (P < .05).

and organic matter (OM) digestion were lower with diets containing treated SBM, indicating a possible effect of CL on carbohydrate digestion. Similar observations were made by Faichney (43) who showed less OM digestion and VFA production in the rumen of sheep receiving diets with formaldehyde treated casein. Stern (23) showed no differences for nonstructural carbohydrate digestion, however, cellulose digestion was lower in diet 4 and had a tendency to be lower in diets 2 and 3 when compared to diet 1. Decreases in VFA production, OM digestion, cellulose digestion and bacterial N synthesis with CL treatment may have been due to a deficiency of degradable N from the protected SBM which would be similar to effects with diets deficient in N. Folman et al. (44), and Windschitl and Stern (45) also reported a decrease in the quantity of bacterial protein synthesized in dairy cows when protected (formaldehyde and CL-treated) proteins were fed. It is possible that ammonia-N concentrations may have been insufficient to meet the requirements of the ruminal microbial population. The extent of these effects may have been diminished if a fermentable N source such as urea had been included in the diet. Windschitl and Stern (46) found that ammonia-N concentrations, bacterial protein synthesis, OM digestion and cellulose digestion were higher as the level of degraded N supplied to the bacterial population was increased via increasing levels of urea supplementation.

In a recent study, Stern et al. (47) fed starter diets consisting of ground corn and alfalfa supplemented with the following nitrogen supplements: urea (U), soybean meal (SBM), whole soybeans (RAW), whole soybeans extruded at 149 °C (ES), and extruded whole soybeans plus urea (ES+U) to Holstein steer calves. Results from this study are presented in Table 3. Calves fed starter diets containing extruded soybeans plus urea gained at a faster rate to 181 kg than calves fed all other diets. Calves fed urea and raw soybeans had the lowest ADG, while feed intake was lowest for the calves fed raw soybeans. The advantage gained by calves fed extruded soybeans plus urea up to 181 kg was maintained throughout the experiment compared to the urea and raw soybean diets. It also took approximately three weeks less time for steers fed extruded soybeans to reach 476 kg compared to steers fed raw soybeans. These results indicate that it might be beneficial to feed readily fermentable nitrogen (urea) in conjunction with protected or high ruminal bypass proteins.

Chemical Methods

Formaldehyde treatment: Phillips (47) suggested that the optimum level of formaldehyde treatment was between .3 and .5 g/100 g soybean meal in order to maximize N utilization in ruminants. Formaldehyde levels of .5 to 1 g/100 g soybean meal completely inhibited *in vitro* ammonia production. He further suggested that the formaldehyde-protein reaction may become very stable with time and result in a bond that is not susceptible to enzymatic digestion or cleavage by the acid conditions of the abomasum. The proposed mechanism of action with formaldehyde-treated proteins is that the complex is stable under ruminal conditions but is unstable under acid conditions of the abomasum, allowing the protein and AA to be digested and absorbed from the small intestine. However, Finlayson and Armstrong (48) recently suggested that exposure of formaldehyde-treated casein to low pH values does not completely release bound formaldehyde from casein, resulting in reduced intestinal protein digestion.

The initial step in the formaldehyde-protein reaction, occurring at neutrality or alkaline conditions, is the addition of a methylol group (hydroxymethyl) on the terminal amino group of protein chains and on the ϵ-amino group of lysine to form an amino-methylol derivative (37). These amino-methylol derivatives (hydroxymethyl derivatives) can react further with another molecule of formaldehyde to form dihydroxymethyl derivatives, or with other AA side chains to form cross-links in the form of methylene bridges. Initially, these reactions are readily reversible. Upon longer standing of the reaction mixture, the methylol groups undergo further reactions leading to stable fixation of increasing amounts of the formaldehyde on the amino groups (47). Formaldehyde can react not only with primary amino groups in protein, but also with sulfhydryl groups, imidazole groups and amides (37). Because sulfhydryl groups are better nucleophiles than amino groups, they react more rapidly. One molecule of formaldehyde is capable of reacting with one sulfhydryl. The imidazole group of histidine also can be affected by

TABLE 3

Performance of Growing Holstein Steers Fed Various Nitrogen Supplements

	Nitrogen supplement				
Item	U	SBM	RAW	ES	ES+U
Up to 181 kg,					
ADG (kg/d)	1.04^c	1.12^b	1.01^c	1.11^b	1.18^a
Feed intake (kg/d)	3.51^a	3.47^a	3.20^b	3.42^{ab}	3.61^a
Efficiency (feed/gain)	3.37^a	3.09^b	3.17^{ab}	3.09^b	3.07^b
ADG from:					
181 to 476 kg	1.19^a	1.18^a	1.22^a	1.25^a	1.20^a
weaning to 476 kg	1.14^{bc}	1.17^{ab}	1.12^c	1.18^a	1.20^a
Days on feed from weaning to 476 kg	365.2^a	351.4^{ab}	366.8^a	347.2^b	342.0^b

a,b,c Means in the same row not having a common superscript differ (P < .05).

formaldehyde treatment. Amide groups of glutamine and asparagine also react readily with formaldehyde to form relatively stable derivatives (37).

From a nutritional standpoint, the amino acids lysine, tyrosine and cysteine (cystine) appear to be most affected by formaldehyde treatment. Feeney et al. (37) suggested that formaldehyde treatment caused covalent binding of formaldehyde to tyrosine, as well as to lysine. A number of researchers (19, 48–50) have suggested that lysine and tyrosine may be lost after protein is treated with formaldehyde. Crooker et al. (51) further reported that treating soybean meal with .3 g formaldehyde/100 g may decrease availability of soy protein for use by lactating cows, resulting in no response in milk production. Finlayson and Armstrong (48) showed that apparent intestinal absorption of all AA in casein was reduced due to formaldehyde treatment. The decrease in biological value with formaldehyde-treated casein in their study was attributed to decreased availability of lysine. Reis and Tunks (52) also indicated that in formaldehyde-treated casein, cross-linked lysine is absorbed and appears as N-methyl-lysine in the blood plasma. The extent to which N-methyl-lysine can substitute for lysine in ruminants is not known, however, it is a poor substitute for lysine in mice (19).

Aldehydes other than formaldehyde, such as glutaraldehyde and glyoxal, can react with proteins to afford some protection against ruminal degradation (19). Glutaraldehyde and glyoxal can react with side chains of AA such as lysine to form Schiff bases, whereas formaldehyde forms hydroxymethyl derivatives during the initial reaction with AA side chains. Ashes et al. (19) reported that the proportions of lysine, tyrosine and leucine that were unavailable as a result of aldehyde treatment of casein were higher with glutaraldehyde and glyoxal compared to formaldehyde.

Tannin treatment: Tannins contain phenolic-hydroxy groups capable of binding and precipitating proteins *via* the formation of cross-links between proteins and other molecules (53). Naturally occurring tannins are located in the organelles within the cytoplasm of plants and can react with extracellular proteins upon rupturing of the cell wall. The types of bonds that may be involved in protein-tannin complex formation include covalent bonds, ionic bonds, hydrogen bonds, and hydrophobic interactions. However, hydrogen bonding appears to be the primary bonding mechanism in protein-tannin reaction complexes, although hydrophobic interactions also may play a role (54). This bonding occurs between the tannin phenolic hydroxyl groups and peptide carbonyl groups (53). One tannin molecule can complex with more than one peptide, resulting in the formation of cross-links between protein chains. The protein-tannin complex is stable at pH 3.5 to 7.0, but is unstable and dissociates at pH less than 3.0 and greater than 8.5. Therefore, the tannin-protein complexes would be stable at rumen pH (5.8–6.8) but unstable at abomasal pH (less than 3.0) and possibly at small intestinal pH (7.5 to 8.5), resulting in the cleavage of the tannin-protein complex and allowing the protein to be digested in the small intestine.

Other treatments: Ruminal protein protection resulting from alcohol treatment of proteins is probably due to denaturation of the protein (55). Alcohol treatment causes a change in the three dimensional structure of proteins, resulting in the protein being less soluble in the rumen

(25). Sodium bentonite is known to absorb protein and thus may afford some protection against ruminal protein degradation (22). Sodium hydroxide (alkaline) treatment of proteins was reported to cause recemization and cross-linking in proteins, possibly affording some ruminal protection (33,34).

Physical Methods

The mode of action of blood treatment for protecting proteins may be provided via physical protection (56). Blood protein is considered to be a relatively low degradable protein, therefore, the coating of proteins with blood may provide some physical protection against degradation. In addition, drying of blood-treated protein at relatively high temperatures also may afford some protection. A similar physical protection also may be achieved when proteins are treated with fish hydrolysate, since fish protein also is relatively resistant to ruminal degradation (9).

Evaluation of Various Methods for Protecting Soybean Meal Protein

Waltz (57) used the *in situ* technique and an artificial rumen system (fermenters) to study the effects of protection methods on protein degradation of soybean meal by ruminal bacteria. Treatments included solvent extraction (control), sodium hydroxide, formaldehyde, expeller processing, propionic acid extrusion and lignosulfonate. Results from the *in situ* study (Fig. 2) showed that expeller processing, calcium lignosulfonate treatment and formaldehyde treatment were most effective in reducing ruminal protein degradation. Diets provided to ruminal bacteria in the fermenters contained approximately 17% crude protein, with 50% of the crude protein coming from the respective treated soybean meal. Crude protein degradation of formaldehyde treated, expeller processed, propionic acid treated, extruded and lignosulfonate treated soybean meal diets were lower than the control diet (Table 4). Total bacterial N output was lowest for soybean meal protected by formaldehyde, expeller

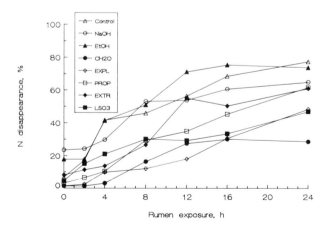

Fig. 2. Nitrogen disappearance from dacron bags suspended in the rumen containing solvent extracted SBM (control), sodium hydroxide treated SBM (NaOH), ethanol treated soy flakes (EtOH), formaldehyde treated SBM (CH2O), expeller processed SBM (EXPL), propionic acid treated SBM (PROP), extruded SBM (EXTR) and lignosulfonate-treated SBM (LSO3).

TABLE 4

Influence of Soybean Meal Treatment on Nitrogen Metabolism by Ruminal Bacteria

Item	Treatment							
	Control	NaOH	EtOH	CH2O	EXPL	PROP	EXTR	LSO3
Ammonia-N (mg 100/ml)	21.7^a	21.3^a	20.2^a	12.9^b	18.9^{ab}	18.8^{ab}	17.2^{ab}	14.5^b
Degradation of dietary crude protein (%)	85.5^a	79.0^{ab}	68.7^{abc}	32.0^d	40.9^{cd}	57.8^{bcd}	65.7^{bc}	52.3^{cd}
Effluent N flow (g/d)								
Total	2.89	2.79	2.89	2.85	2.86	2.82	3.06	2.83
Nonammonia	2.28^a	2.33^a	2.39^{ab}	2.53^c	2.39^{ab}	2.40^{ab}	2.60^c	2.47^{bc}
Bacterial	1.96^a	1.95^a	1.56^{ab}	$.68^c$	$.92^{bc}$	1.27^{ab}	1.67^{ab}	1.10^{bc}
Bypass	$.32^a$	$.38^a$	$.83^{abc}$	1.85^d	1.47^{cd}	1.13^{bcd}	$.93^{abc}$	1.37^{bcd}
Bacterial synthesis (g N/kg OM truly digested)	38.8^a	39.6^a	33.3^{ab}	15.8^c	20.7^{bc}	27.6^b	35.7^{ab}	23.9^{bc}

a,b,c,d Means in the same row having a common superscript differ (P < 0.05).

processing and lignosulfonate treatments. Bypass dietary N in the effluent was highest for soybean meal protected by formaldehyde, expeller processing, propionic acid and lignosulfonate treatments.

In addition, protection by formaldehyde, expeller processing, propionic acid and lignosulfonate treatments increased total AA flow compared to the control. Different protection methods had varying effects on individual AA. The most obvious difference was between formaldehyde and expeller treatments. They both increased arginine and leucine flows but formaldehyde also increased histidine, threonine and phenylalanine whereas expeller processing increased valine and isoleucine. Because the AA composition of the protein escaping degradation could be as important as the total amount of AA in deciding which protection method to use, differences in individual AA flow and availability in the small intestine between methods needs to be investigated further.

Animal Proteins

Animal proteins or byproducts of animal processing such as fish meal, meat and bone meal, blood meal and feather meal have been shown to be high ruminal bypass proteins. Erfle et al. (58) pointed out that although fish meal generally is resistant to microbial degradation in the rumen, there are considerable differences in degradability of various fish meals due to processing. Heating of fish protein can induce formation of S-S cross-linking from -SH oxidation (59). Fish protein heated for 20 min at temperatures ranging from 50°C to 115°C showed a linear decrease in the content of -SH (sulfhydryl) groups and a concomitant increase in the content of S-S (disulfide) bonds. The AA most affected during heating of fish protein is cysteine. Opstvedt et al. (59) determined that heating at 115°C caused a loss in cysteine and cystine. At temperatures of 95°C or greater, protein and AA digestibility of fish protein in rainbow trout was reduced compared to raw fish protein. Moderate heat, as used in the processing of fishmeal, can result in a decrease in the rate of ruminal proteolysis of fish protein (35). Fish meal has shown a positive response in milk yield when compared to soybean meal or other usual protein supplements. Oldham et al. (60) observed an advantage of fish meal compared to urea, soybean meal and formaldehyde treated soybean meal for milk production.

Klopfenstein and Goedeken (61) indicated that protein degradability of blood meal, meat meal and feather meal was 17.6, 36.1 and 30.9% respectively. Bas et al. (62) showed that high ruminal bypass proteins such as lignosulfonate treated soybean meal, blood meal and feather meal can potentially improve AA supply (Fig. 3) to ruminants. Animal byproducts could serve as sources of high ruminal bypass protein but data on their effects are limited. Varying AA composition and palatability are potential difficulties. Craig and Broderick (62) reported no advantage of meat meal over urea or soybean meal in diets of 15% crude protein for cows producing in excess of 30 kg/d. Klopfenstein and Goedeken (61) found that performance of steers in a growth experiment indicated that calves consuming blood meal, feather meal and a combination of the two animal proteins gained faster than steers fed urea. The improved protein efficiency for blood meal

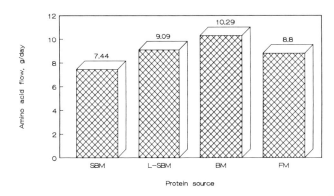

Fig. 3. Total amino acid flow from fermenters provided with soybean meal (SBM), lignosulfonate-treated soybean meal (L-SBM), blood meal (BM) and feather meal (FM) as the major dietary N sources for ruminal bacterial metabolism.

and feather meal compared to either fed alone may be due to sulfur AA supplied by the feather meal and/or other AA supplied by blood meal. Waltz et al. (64) found that feather meal or blood meal alone did not increase the supply of available AA to the small intestine of lactating cows over that supplied by untreated soybean meal. However, the combination of feather meal with an equal amount of blood meal did increase the amount of available AA supplied to the small intestine compared to soybean meal. These observations signify the potential importance of supplying complementary AA to the small intestine from high ruminal bypass proteins.

References

1. Kaufmann, W., and W. Lupping, in "Protein Contribution of Feedstuffs for Ruminants," edited by E.L. Miller, I.H. Pike and A.S.H. Van Es, Butterworths, England, 1982, pp 36–75.
2. Oldham, J.D., and S. Tamminga, *Livest. Prod. Sci. 7:*437 (1980).
3. Sniffen, C.J., and D.E. Hogue, *Proc. Cornell Nutr. Conf.:*152 (1984).
4. MacGregor C.A., C.J. Sniffen and W.H. Hoover, *J. Dairy Sci. 61:*566 (1978).
5. Tamminga,S., *J. Anim. Sci. 49:*1615 (1979).
6. Stern, M.D. and L.D. Satter, in "Protein Requirements for Cattle," edited by F.N. Owens, Stillwater, Oklahoma, 1982, pp 57–71.
7. Craig, W.M., and G.A. Broderick, *J. Anim. Sci. 58:*436 (1984).
8. Crooker, B.A., J.H. Clark, R.D. Shanks and G.C. Fahey Jr., *Can. J. Anim. Sci. 67:*1143 (1987).
9. Mir, Z., G.K. MacLeod, J.G. Buchanan-Smith, D.G. Grieve and W.L. Grovum, *Can. J. Anim. Sci. 64:*853 (1984).
10. Kung, L. Jr., J.T. Huber, W.G. Bergen and D. Peticlerc, *J. Dairy Sci. 67:*2519 (1984).
11. Ganev, G., E.R. Ørskov and R.I. Smart, *J. Agric. Sci. 93:*651 (1979).
12. Demeyer, D., E. Clayes, C.J. VanNevel and R. DeWinne, *Z. Tierphys. 47:*251 (1982).
13. Varvikko, T., J.E. Lindberg, J. Setala and L. Syrjala-Qvist, *J. Agric. Sci. 101:*603 (1983).
14. Weakley, D.C., M.D. Stern and L.D. Satter, *J. Anim. Sci. 56:*493 (1983).
15. Stern, M.D., *Proc. Minn. Nutr. Conf. 42:*23 (1981).
16. Storm, E., D.S. Brown and E.R. Ørskov, *Br. J. Nutr. 50:*479 (1983).
17. Storm, E. and E.R. Ørskov, *Br. J. Nutr. 52:*613 (1984).
18. Nishimuta, J.F., D.G. Ely and J.A. Boling, *J. Anim. Sci. 39:*952 (1974).
19. Ashes, J.R., J.L. Mangan and G.S. Sidhu, *Br. J. Nutr. 52:*239 (1984).
20. Stern, M.D., K.A. Santos and L.D. Satter, *J. Dairy Sci. 68:*45 (1985).
21. Van Sumere, C.F., T. Allrecht, A. Dedonder and H. DePooter, in "The Chemistry and Biochemistry of Plant Proteins," edited by J.B. Harborne and C.F. Van Sumere, Academic Press, London, 1975, pp 211–264.
22. Britton, R.A., D.P. Colling and T.J. Klopfenstein, *J. Anim. Sci. 46:*1738 (1978).
23. Stern, M.D., *Can. J. Anim. Sci. 64(Suppl.):*27 (1984).
24. Waltz, D.M., and S.C. Loerch, *J. Anim. Sci. 63:*879 (1986).
25. Van der Aar, P.J., L.L. Berger and G.C. Fahey Jr., *J. Anim. Sci. 55:*1179 (1982).
26. Cleale IV, R.M., R.A. Britton, T.J. Klopfenstein, M.L. Bauer, D.L. Harmon and L.D. Satterlee, *J. Anim. Sci. 65:*1319 (1987).
27. Chalupa, W., *J. Dairy Sci. 58:*1198 (1975).
28. Tagari, H., I. Ascarelli and A. Bondi, *Br. J. Nutr. 16:*237 (1962).
29. Goering, H.K. and D.R. Waldo, *Proc. Maryland Nutr. Conf.:*52 (1974).
30. McMeniman, P. and D.G. Armstrong, *J. Agric. Sci. 93:*181 (1979).
31. Mauron, J., *Prog. Food Nutr. Sci. 5:*5 (1981).
32. Bjarnason, J. and K.J. Carpenter, *Br. J. Nutr. 23:*859 (1969).
33. Maga, J.A., *J. Agric. Food Chem. 32:*955 (1984).
34. Schwass, D.E. and J.W. Finley, *J. Agric. Food Chem. 32:*1377 (1984).
35. Chen, G., C.J. Sniffen and J.B. Russell, *J. Dairy Sci. 70:*983 (1987).
36. Anderson, P.A., S.M. Sneed, G.R. Skurray and K.J. Carpenter, *J. Agric. Food Chem. 32:*1048 (1984).
37. Feeney, R.E., G. Blankenhorn and H.B.F. Dixon, *Adv. Protein Chem. 29:*135 (1975).
38. Craig, W.M., and G.A. Broderick, *J. Anim. Sci. 52:*292 (1981).
39. Broderick, G.A., *J. Dairy Sci. 69:*2948 (1986).
40. Deacon, M.A., G. DeBoer and J.J. Kennelly, *J. Dairy Sci. 71:*745 (1988).
41. Mielke, C.D., and D.J. Schingoethe, *J. Dairy Sci. 64:*1579 (1981).
42. Winowiski, T.S., and M.D. Stern, *J. Anim. Sci. 65(Suppl. 1):*468 (1987).
43. Faichney, G.J., *Australian J. Agric. Res. 25:*599 (1974).
44. Folman, Y., H. Neumark, M. Kaim and W. Kaufmann, *J. Dairy Sci. 64:*759 (1981).
45. Windschitl, P.M., and M.D. Stern, *J. Dairy Sci.* (in press).
46. Windschitl, P.M., and M.D. Stern, *J. Anim. Sci.* (in press).
47. Phillips, W.A., *J. Anim. Sci. 53:*1616 (1981).
48. Finlayson, H.J., and D.G. Armstrong, *J. Sci. Food Agric. 37:*742 (1986).
49. Crooker, B.A., J.H. Clark, R.D. Shanks and E.E. Hatfield, *J. Dairy Sci. 69:*2648 (1986).
50. Erfle, J.D., F.D. Sauer, S. Mahadevan and R.M. Teather, *Can. J. Anim. Sci. 66:*85 (1986).
51. Crooker, B.A., J. H. Clark and R.D. Shanks, *J. Dairy Sci. 66:*492 (1983).
52. Reis, P.J., and D.A. Tunks, *Australian J. Biol. Sci. 26:*1127 (1973).
53. Kumar, R. and M. Singh, *J. Agric. Food Chem. 32:*447 (1984).
54. Oh, H.I., J.E. Hoff, G.S. Armstrong and L.A. Hoff, *J. Agric. Food Chem.* (1980).
55. Fukushima, D., *Cereal Chem. 46:*156 (1969).
56. Ørskov, E.R., C.R. Mills and J.J. Robinson, *Proc. Nutr. Soc. 39:*60A (1980).

57. Waltz, D.M., PhD thesis, Univ. of Minnesota (1988).
58. Erfle, J.D., S. Mahadevan, R.M. Teather and F.D. Sauer, *Can. J. Anim. Sci. 63:*191 (1983).
59. Opstvedt, J., R. Miller, R.W. Hardy and J. Spinelli, *J. Agric. Food Chem. 32:*929 (1984).
60. Oldham, J.D., R.J. Fulford and D.J. Nepper, *Proc. Nutr. Soc. 40:*30A (1981).
61. Klopfenstein, T. and F. Goedeken, *Proc. National Renderers Assoc. 1:*12 (1986).
62. Bas, F.J., M.D. Stern and N.R. Merchen, *J. Anim. Sci. 66(Suppl.):*345 (1988).
63. Craig, W.M. and G.A. Broderick, *J. Dairy Sci. Suppl. 1, 66:*148 (1983).
64. Waltz, D.M., M.D. Stern and D.J. Illg, *Proc. Rumen Function Conf. 19:*12 (1987).

Preparation of Vital Wheat Gluten

G. Grace

Manildra Starches Pty. Ltd., Auburn, New South Wales, Australia

Abstract

Vital wheat gluten has unique characteristics and functional properties. Food industry recognition of these properties has led to a substantial increase in world production capacity. As a result there is a renewed interest in the development and operation of efficient methods of separation, from a range of raw materials. The design and operation of commercial separation and drying equipment can have a major impact on the functional properties of vital wheat gluten. The high B.O.D. (biological oxidation demand) effluent generated by wheat gluten extraction is also receiving much attention.

Wheat gluten is a complex of proteins, carbohydrates and lipids that forms when flour is hydrated with water. Additional washing with more water removes starch and solubles and leaves an insoluble rubbery mass containing about 70% water which can be dried to a light tan powder. Gluten protein constitutes from 7% to 14% of total wheat, depending on wheat variety and climatic growing conditions. Gluten protein surrounds the starch granules as a matrix in the floury endosperm of the wheat berry and is thought to be uniformly distributed.

The quantity of gluten which can be recovered from different wheats varies widely, for example: American spring wheats can yield up to 17% commercial gluten, whereas soft European wheat varieties may yield 9–10% gluten. Regardless of the yields, all wheat glutens have the property of absorbing approximately twice their dry weight of water to produce a hydrated elastic mass. It is this property which provides the structure and texture in baked bread.

The molecular structure of gluten is quite complex but it is thought that it consists of coiled molecules of linear polypeptides cross-linked with some disulphide bonds. With this structure, large deformations are possible with relatively small forces. The elastic properties of vital wheat gluten are particularly heat sensitive and the operation of commercial drying plant must be carefully controlled to minimize damage to the gluten's functional properties.

World Production

In recent years world production of gluten has risen sharply, with most of the capacity increase occurring within the EEC countries (Fig 1).

World production in 1987 is estimated as follows in Table 1. In addition, there is some production in Eastern European countries and in mainland China, for which statistics are not available. World production in 1980 was only 80,000 metric tons, so there has been a sharp increase in capacity in the past eight years.

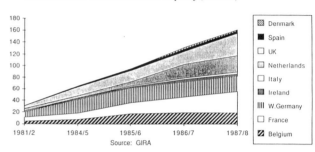

Fig. 1. Growth of EEC gluten production capacity (thousand metric tons).

TABLE 1
World Production of Wheat Gluten (1987)

	(Metric tons)
EEC Countries	140,000
Australia	40,000
United States	30,000
Japan	10,000
Canada	9,000
Argentina	8,000
Finland	5,000
TOTAL	242,000

Methods of Manufacture

In the industrial manufacture of wheat gluten it is usual to start with a conventional dry milling process to separate the skin and the germ fraction from the floury endosperm. Depending on the milling process used some 20% to 25% by weight of wheat is separated and classified as mill feed. It is used principally as animal feed although some is being used in the production of high dietary fiber foods.

This separation is not 100% efficient, because some of the endosperm and, hence, potential gluten yield is lost with the bran fraction. There have been at least two attempts to avoid this loss by processing whole wheat grain. Although both processes were put into commercial production, both have ceased operation.

Farmarco Process

This process is designed to bypass the normal flour dry milling stage and to replace it with a simple pin milling process of wheat tempered to 14–22% moisture. This produces a rather crude flour compared with the roller milling process. The pin milled flour was made into a dough and separated by the Martin process.

Pillsbury Hydromilling Process

Pillsbury started by steeping high protein (15%) wheat in diluted hydrochloric acid. The softened wheat was then put through a wet grinding process which produced an acid suspension of starch and gluten from which the bran was screened out. The pH of the screened slurry was then adjusted to the isoelectric point for gluten (pH 4.5–4.8), and the gluten precipitate was removed by screening. In practice, it proved very difficult to separate the bran from the gluten and the color and functionality of the gluten was poor due mainly to the acid steeping process.

This plant ceased operations in 1979 when it was purchased by Manildra Milling Corporation. It has been converted to a conventional process using wheat flour and is currently operating successfully.

Current Production Methods

Wheat Selection

In general, higher protein wheats are easier to process and produce a higher yield of gluten. However, high protein wheats command a price premium so it is necessary to find a compromise between wheat cost, yields, and ease of processing.

Flour Milling

A straight grade roller milled flour of 75% to 80% extraction is the raw material of choice for most producers. The normal roller milling flour process used to produce bread flours is designed to produce a degree of starch damage as this increases the water absorption of the flour and enhances baking performance. For gluten production damaged starch has a negative effect on yields and some producers use a modified milling process which limits starch damage in the flour. The reduced energy input may assist also in preventing heat damage to the protein in the flour.

In the U.S. particularly, a high-ash, lower-grade milling stream known as 1st and 2nd class is occasionally used for the production of vital wheat gluten. If used alone, clear flour produces a darker colored, high ash gluten of reduced functionality.

Wet Gluten Separation

Gluten is separated from flour by two basic processes either a dough system or by a batter system. There are many variations in the plant and equipment used but each variation can be traced back to its origins either in the dough system or the batter system.

Dough System: The "Martin Process" is the best example of the dough type separation, which is in common use in the industry. It is especially suited to processing low protein (7%–10%), weak flour. Flour is mixed with water and kneaded to form a smooth cohesive elastic dough, containing 55–60% dry substance. It is quite usual to use water at up to 35°C to speed the dough development process. There are many commercial mixers which can be operated on a continuous basis to provide suitably developed dough for the washing process (Fig. 2).

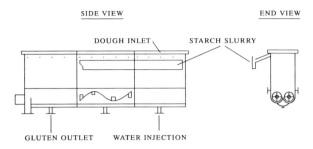

Fig. 2. Martin Process. A, continuous dough mixer; B, rest hopper; C, extractor; D, dewater rolls; E, cooler; F, sieve; G, D. S. M. screen; H, continuous centrifuge; I, basket centrifuge; J, ring drier; K, heater; L, evaporator; M, heat exchanger; N, filter; O, roller drier; and P, flash drier.

Dough Washing

The aim of the washing stage is to release starch from the gluten without dispersing or breaking the gluten into small fragments. Many devices may be used for this process, such as ribbon blenders, twin screw troughs and rotating screens. A popular choice is a modified ribbon blender known as a "Martin" washer (Fig. 3).

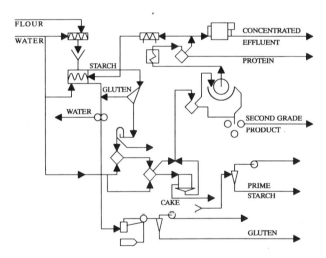

Fig. 3. Martin process.

This is a deep narrow vessel with twin open paddle rotors extending the length of the vessel. The paddles contrarotate at a slight speed differential. Fresh water and/or recycled processed liquor is injected into the bottom of the washer. The starch released by the kneading process overflows into a launder and the wet gluten is discharged from the bottom through a variable speed positive displacement pump which acts as a seal. The rate of dough feed, wash water and paddle speed is controlled so that the gluten leaving the washer is at a minimum protein content of 75% on a dry solids basis.

The major disadvantage of the Martin Process is that it requires a comparatively large volume of water for its operation, particularly when higher protein flour is used.

Eight to twelve tons of water per ton of flour input is common. The large volume of water complicates starch recovery and efficient handling.

Batter Systems

The batter process is a variant of the dough system, and is in its various forms the mainstay of the industry today. Flour and water in roughly equal proportions is mixed to a smooth sloppy dough with a solids content of 48% to 55%. Higher protein flours can handle lower solids without significant gluten fragmentation. The batter is developed by mechanical working to ensure that the gluten is able to withstand the vigourous washing treatment.

When the batter is fully developed, more water is added with additional mixing. The gluten strands coalesce to form a curd-like structure, which can be screened out of the starch liquor. Gyratory screens, sieve bends and rotary screens are all used for this purpose (Fig. 4). The process of water additions and screening may be repeated two or more times until the protein level in the separated gluten reaches a minimum of 75% dry basis.

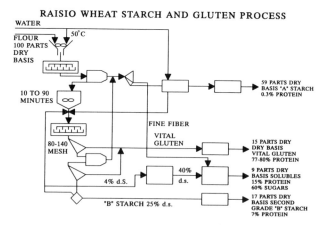

Fig. 5. Raisio wheat starch and gluten process.

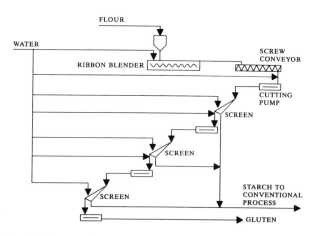

Fig. 4. Improved batter process.

There have been many attempts to refine and improve the batter process for the separation gluten. Among the more successful are:

Raisio/Alfa-Laval Process: In this process flour is dispersed in water at a ratio of 5:6 up to 5:10 depending on the type of flour used. This mixture is processed at high shear in a pin mill to obtain a smooth dispersion without gluten clots. The dispersed mixture is separated in a horizontal decanting centrifuge to give a starch fraction and a gluten fraction. The gluten fraction is allowed to mature to the point where it will coalesce to a form that can be recovered by the usual screening process. Major advantages of this process are high throughout for a low space requirement and low water consumption (Fig. 5).

Gluten Drying

Over the years there have been many methods used to dry gluten. Because of its heat sensitivity, wet gluten is very easily denatured and it is necessary to ensure that the design and operations of a gluten dryer do not apply excessive heat to the product. Product temperatures should be less than 65°C, and preferably in the range 55–60°C.

Spray Dryers: Spray dried gluten has excellent functionality and is particularly suited to some applications. In order to operate the spray dryer, gluten has to be uniformly dispersed at a viscosity appropriate to the dryer system. This is accomplished by the addition of water and either ammonia solution or acetic acid. It is usual to use ammonia solution which readily flashes off from the dryer exhaust. Solids in the feed are 12% to 14% so it is necessary to evaporate nearly 6 metric tons of water for every ton of dry gluten output.

Hydrocyclone Process (K.S.H.): Koninklijke Scholten Honig Lo of The Netherlands has developed a process for producing gluten using a series of hydrocyclones. A well-dispersed batter of flour in water is introduced into a battery of 10 mm hydrocyclones. Starch and fiber are discharged in the underflow and a gluten-rich fraction forms the overflow. The gluten-rich fraction is mixed with water and/or recycled process liquor and is screened to recover gluten (Fig. 6). This can be done external to the drier or actually within the dryer system.

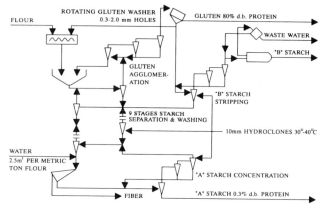

Fig. 6. K.S.H. wheat starch and gluten process.

The most common equipment in use today is the "Barr and Murphy" ring drier (Fig. 7) in which wet gluten in the form of small pellets is injected into recycling dry gluten in the disintegrator mill. Typical specifications for a dryer with a 1000 kgm/hour dry product output are shown in Table 2. This means that the dryer is very large and a high energy input is necessary. The dried product is bulky due to low density. It is only possible to put 8 to 10 metric tons in a standard 20-foot shipping container. The advantages of the spray drying process are: (a) Easy incorporation of other components, e.g., lecithin; (b) very rapid rehydration due to small particle size; and (c) very good retention of gluten vitality.

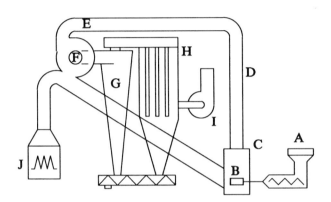

Fig. 7. Ring dryer. A, supply pump; B, extruder; C, disintegrator mill; D, ring duct; E, manifold; F, manifold exit; G, cyclone; H, bag dust collector; I, fan; and J, air heater.

TABLE 2

Operating Conditions for Ring Dryer

Wet gluten feed	3100 kg/hr
Evaporation	2100 kg/hr
Product out	1000 kg/hr
Air flow	55,000 CFM
Main fan	400 HP
Disintegrator drive	800 HP (600 KW)
Steam 150 PSI	12,000 lbs/hr
Air inlet temp	170 C
Product temp	60 C

Flash Drying: More than 90% of world gluten output is flash dried in specially designed equipment. Because wet gluten vitality is lost by prolonged exposure to high temperature, rapid drying at controlled temperature is essential. In order to minimize the effect of high temperature, it is usual to mix incoming wet gluten at 70% moisture, with a quantity of dry gluten at 8% moisture, so that the moisture content of the blend is about 20% to 30%. Wet gluten injected into the disintegrator is mixed with partly dried gluten circulating in the drier ring. It is swept up the drying leg into the classifier. The larger, wetter and heavier particles are separated from the dry product and recycle around the ring again. The dry gluten exits from the classifier into the collection cyclone and bag filter.

In many cases the particle size of the gluten leaving the dryer is too large for most applications, and it is necessary to regrind and screen it to a finer particle size. A screen size between 150 and 250 microns is normal and a pin mill is often used for the regrind operation. Typical analyses are shown in Table 3.

TABLE 3

Typical Analysis for Dry Vital Wheat Gluten

Moisture	8.0%
Protein (%N×5.7)	77.0% (dry basis)
Ash	0.7%
pH	5.8
Water absorption	180.0%
Hydration time	40 seconds
Fiber	0.4%
Solubles	0.5%
Starch	18.0%

Wheat gluten contains 15% to 18% starch, and can, if conditions are right, ignite to produce an explosion in the dryer system. Most driers are constructed with explosion relief panels which will vent the explosion outside the building. In addition most driers also will have an inert gas blanketing system activated by high pressures or temperatures.

Wheat Starch and Effluent

No discussion of the production of wheat gluten is complete without mention of the co-products. Wheat gluten cannot be produced without also producing wheat starch and a liquid effluent. For every ton of wheat gluten produced, one recovers 5.25 tons of wheat starch and generates up to 30 tons of liquid effluent containing 1100 kg of solubles (starch and fiber). BOD of the effluent is usually in the range of 15,000 to 20,000 PPM. The wheat gluten industry has invested considerable effort to overcome the problems posed by the effluent and several systems are in operation as follows: Anaerobic digestion; spray irrigation; fermentation feedstock for ethanol production; evaporation for animal feed; and discharge to city sewer systems (when permitted).

References

1. Knight, J.W., and R.M. Olson, in *Starch Chemistry and Technology* (2nd Ed.), edited by R.L. Whistler, J.N. Bemiller and E.F. Paschall, Academic Press, 1984, Chap. 15.
2. "The Present and Future Market for Vital Wheat Gluten in the EEC and Summary of Key International Markets 1985–1990," in Strategic Studies in Agribusiness series, Gira S.A., Geneva, Switzerland, 1986.
3. Knight, J.W., *The Starch Industry*, Pergamon Press, London, pp. 44–58.

World Food Uses of Vital Wheat Gluten

J. M. Hesser

International Wheat Gluten Association, 4510 W. 89th St., Prairie Village, Kansas 66207 USA

Abstract

Vital wheat gluten is a unique vegetable protein of considerable commercial significance. Its specific property of forming an elastic mass when hydrated and its thermosetting ability account for its unique baking characteristics. Vital wheat gluten is produced industrially from wheat flour by various wet separation methods. These methods produce essentially similar glutens, but raw material selection and manufacturing process control are important in optimizing functionality and yield. Vital wheat gluten is used in a wide variety of food products including breakfast cereals, meat and poultry products, pasta, and pet foods. Increasing interest is also occurring in the areas of cheese, seafood analogues and in aquaculture.

Functional Properties

Composition and General Properties: Wheat gluten is commercially marketed as a cream colored, free flowing powder. The vegetable protein committee of the *Codex Alimentarius*, at a meeting held in Havana, Cuba, in February 1987 adopted the draft standard, proposed by the International Wheat Gluten Association, for wheat gluten and was advanced to Step 8 (of 9) in the Codex process. Those standards are reflected in Table 1 (1).

TABLE 1

Codex Alimentarius International Standard for Vital Wheat Gluten

Protein (N × 6.25) dry basis (d.b.)	80.0% Min
Moisture	10.0% Max
Ash	2.0% Max
Fat (ether extracted)	2.0% Max
Fiber	1.5% Max

Wheat gluten is the water insoluble complex protein fraction separated from wheat flours. It is primarily a mixture of two types of protein from the wheat kernel, glutenin and gliadin (Table 2) (2).

Viscoelasticity: These two major protein components of wheat gluten interact in an aqueous system to produce a unique property known as viscoelasticity. The ability of wheat gluten to form a viscoelastic mass when fully hydrated sets it apart from all other commercially available vegetable proteins (Fig. 1) (2).

The viscoelastic behavior of hydrated wheat gluten (Fig. 2) (2) persists even in the presence of excess water

TABLE 2

Gluten Composition

Gliadin	Glutenin
Highly extensible	Less extensible
Less elastic	Highly elastic
Soluble in alcohols	Insoluble in alcohols
Low molecular weight (less than 100,000)	High molecular weight (greater than 100,000)
Intramolecular bonds	Intra and intermolecular bonds

Fig. 1. Elastic and film forming characteristics of wheat gluten.

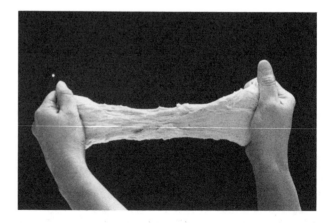

Fig. 2. Fully hydrated wheat gluten.

because of the physico-chemical status of these molecular structures in aqueous systems. The dramatic difference in properties of wheat gluten in comparison with almost all other food proteins is due largely to the low level of polarity of the total amino acid structure. The practical result of this reduced polarity is that excess water is repelled and the wheat gluten molecules associate closely together and resist dispersion. Such behavior is crucial to food systems in that it results in the ability to form adhesive and cohesive masses, films and three-dimensional film networks (2).

Water Absorption: Good quality powdered vital wheat gluten rapidly absorbs about twice its weight of water. Because powdered vital wheat gluten has a protein content of 75–80% (dry basis), the hydrated version will drop to 25–27% protein at full hydration of about 65% moisture. Wheat gluten's capacity for holding water results in increased yield and extended shelf-life in food systems into which it is incorporated. The combination of speed of water absorption and the degree of viscoelasticity produced are evidence of "vitality"; hence, the commonly used term "Vital Wheat Gluten." Deliberate cooking of wet wheat gluten prior to drying results in "devitalized" wheat gluten. This product has undergone irreversible denaturation and does not revitalize or absorb water. Devitalized wheat gluten is a popular substrate for hydrolysis in vegetable protein hydrolysate flavor manufacture. Dispersibility and speed of hydration of wheat gluten can vary with different processes of manufacture and are controllable (2).

pH Effects: Product development and food scientists contemplating the use of wheat gluten in food systems recognize that the low positive electrostatic charge is modified by interaction with ionic ingredients. Relatively minor changes in charge magnitude may have functionally significant effects. Indeed, much of bakery technology is oriented to prevent this from occurring. Functional sensitivity to ionic impact thus becomes both a challenge and an opportunity for the food scientist in controlling and adjusting wheat gluten behavior in a specific product environment.

Flavor: Properly produced and given reasonable care in storage, wheat gluten exhibits a flavor note variously described as "bland" or "slight cereal." Wheat flavors enjoy wide acceptance and wheat gluten merges perfectly into all cereal based products. Blending with meats in various binding, adhesive and extension roles need not result in off-flavor notes, even at high percentage use levels.

Blending of wheat gluten with other food proteins which do possess characteristic flavor notes can result in improved total flavor as, for example, when soy/wheat gluten blends are used for textured vegetable protein manufacture. Low and acceptable flavor levels of wheat gluten are the result of careful selection of flours, good manufacturing procedures and proper storage at normal ambient temperatures (2).

Nutritional Properties: Wheat gluten alone, when measured against the standard casein reference in rat bioassay, rates rather low on the protein efficiency ratio (P.E.R.) scale. Vegetable proteins in general rate low on the P.E.R. scale due to low content of one or more essential amino acids. Blends of different vegetable proteins often result in higher P.E.R. values than the arithmetic average of the components. Such is the case with wheat gluten, low in lysine but high in methionine and cystine, when it is blended with soy flour.

Soy flour, high in lysine but low in methionine, has a P.E.R. value of about 2.0. Table 3 (2) demonstrates the synergistic result of blending soy flour and wheat gluten. At a wheat gluten/soy flour protein ratio of 30:70, a P.E.R. value of 2.4 is achieved, or roughly three times the rating of wheat gluten alone. Similar effects are predictable for other food systems, where excess lysine is available from meat, milk solids or other sources. Wheat gluten in combination with other proteins can have solid nutritional impact.

TABLE 3

Protein Efficiency Ratio (P.E.R.) of Wheat Gluten/Soy Flour

	Wheat gluten protein/soy flour protein blend ratio				
	100/0	55/45	45/55	30/70	0/100
P.E.R.	0.8	2.1	2.3	2.4	2.0

The demand for vegetable proteins with special functional properties is growing rapidly in line with the increasing consumption of fabricated foods. The high protein content, unique structural, thermosetting and adhesive characteristics of wheat gluten, plus its water absorption properties offer opportunities to the food scientist for innovative product formulation.

The major use of wheat gluten traditionally has been and continues to be in bakery products. However, with an increasing awareness of wheat gluten's unique structural and functional properties, an expanding and diverse range of applications have developed. These uses and the functionality of wheat gluten will be discussed.

Applications

Milling and Flour Fortification: Perhaps the most fundamental use of wheat gluten is in the adjustment of flour protein level, by either the baker or the flour miller. Recently, many millers have used gluten to streamline production and supplement lower protein local wheats to meet bakery flour standards. This practice has become increasingly common in parts of Europe, where gluten fortification of weaker flours offers an attractive alternative to blending with expensive, imported strong wheats to satisfy functional performance requirements. With increased emphasis on plant breeding to produce higher yielding wheats, there has been an associated decline in wheat protein content, thereby increasing the use of gluten supplementation. As a natural additive, there are no limits to the amounts that can be used in achieving product quality objectives (3).

The functionality of wheat gluten for this application is for vital wheat gluten to: compensate for the extra quality and performance otherwise provided by the imported flour protein; allow protein control at either the flour mill or user end, producing a high degree of flexibility and increased cost effectiveness; and effect significant savings with respect to duties levied on imported wheats.

Considerable research continues to be devoted to create satisfactory bakery products involving composite or nonwheat flour for application in third world countries. Millet, cassava and maize are some of the grains which can be used along with wheat gluten to produce satisfactory bakery products.

Bakery Products: Perhaps no area of food processing enjoys greater benefits from wheat gluten usage than does the baking industry. Wheat gluten's unique viscoelastic properties improve dough strength, mixing tolerance and handling properties. Its film forming ability provides gas retention and controlled expansion for improved volume, uniformity and texture; its thermosetting properties contribute necessary structural rigidity and bite characteristics; and its water absorption capacity improves baked yield, softness and shelf-life.

Wheat gluten is commonly used in hard rolls and multi-grain, high fiber and other specialty breads at levels ranging from 2% to as high as 10%. If used at approximately 2% in presliced hamburger and hot dog buns, wheat gluten improves strength of the hinge and provides desirable crust characteristics when the buns are stored in a steamer. Yeast raised donuts have greater volume and increased strength during subsequent glazing; and, in virtually all baking applications, wheat gluten's natural flavor enhances consumer acceptability (1).

Breakfast Cereals: Wheat gluten-fortified breakfast cereals have been widely accepted by consumers, because they are very flavorful and nutritious, especially when consumed with milk. Wheat gluten provides not only the protein desired for nutritional claims but also helps bind any vitamin mineral enrichment components to the cereal or grain berry in processing and contributes to the strength of flaked cereal (2).

Meat, Fish and Poultry Products: The unique adhesive, cohesive and film forming characteristics of hydrated vital wheat gluten and its thermosetting properties form the basis of various types of applications in prepared meats, fish and poultry products. Wheat gluten is very effective in binding meat chunks or trimmings together to form reconstructed steaks or chops and may be applied simply by dusting the meat pieces with dry gluten. It is an excellent binder in poultry rolls, canned hams and other nonspecific loaf-type products where it also improves slicing characteristics and minimizes cooking losses during processing and/or preparation. Typical use levels are 2–3.5% by weight (3).

Vital wheat gluten also is useful as an extender in ground meat patties and as a protein binder in sausages and other meat emulsion products. When hydrated, it can be extruded, texturized by expansion, drawn into fibers and spun, or cast into films. Simulated meat products may be produced for a variety of applications, and wheat gluten has been used to produce crab meat analogs and even artificial caviar. Aqueous alcoholic dispersions of wheat gluten have been prepared to form stripable, edible coatings such as sausage casings. Other examples are: textured protein meat extenders; meat analogues; beef, pork, chicken and fish sausage products; and pizza toppings.

Textured protein products fabricated from soy/wheat gluten blends have the triple virtues of P.E.R. values above 2.0, highly acceptable flavor and structural integrity in hot, moist environments such as canning or steamtable serving. A soy/wheat gluten textured meat extender has approval for school lunch programs in the USA.

Pasta: Although durum wheat is preferred for the production of pasta because of its particular rheological properties and color, other more available flours can be used effectively if wheat gluten is added. Gluten addition can reduce cooking loss and stickiness in cooked pasta, provide good cooked firmness, increase resistance to breakage, and improve heat tolerance in canned retorted products. High-protein pasta formulations using vital wheat gluten in combination with soy protein and lactalbumin have been developed; they satisfy the protein level and protein quality (P.E.R.) requirements of the USA Type A School Lunch Program, for example (2).

Cheese Analogs and Pizza: Wheat gluten's viscoelastic properties can be used in preparing simulated cheese products with the characteristic texture and eating quality of natural cheese. Studies sponsored by the International Wheat Gluten Association have shown that vital wheat gluten, alone or in combination with soy protein, can be used to replace approximately 30% of the more expensive sodium caseinate used in imitation American and Mozzarella cheese products.

Wheat gluten also can be used to strengthen pizza crust, making it possible to produce both thin and thick crusts from the same flour. The incorporation of gluten provides crust body and chewiness and reduces moisture transfer from the sauce to the crust. Typical use levels (flour basis) range from 1 to 2% (3).

Nutritional Snacks: The charge of "junk" foods or "empty" calories often leveled at snack foods have prompted fabricators of snacks to examine wheat gluten closely as a worthwhile ingredient. Many international examples of such wheat gluten based snacks are now being marketed, including gluten balls in Europe, vegetarian items in the United States, *Yachi-fu* cakes in Japan, fried gluten items in China and wafers of 30–45% wheat gluten content in Australia. In extruded snacks, wheat gluten provides nutritional value, crispness and desired texture. Use levels are generally 1 to 2% (3).

Breadings, Batter Mixes, Coatings and Flavorings: Adhesion of batter crusts and breadings to fried foods always has been a quality problem, particularly in frozen products. Predusting of such food portions with wheat gluten significantly improves adhesion, reduces cooking loss and enhances appearance. When wheat gluten is added to a batter mix, the film formation reduces liquid loss and also produces crisp, appetizing surfaces.

Vital wheat gluten also has been used in coating dry roasted nuts, providing adhesion for the salt and other seasonings. Acid hydrolysis of wheat gluten has been used to prepare hydrolyzed vegetable proteins as flavor enhancers (3).

Pet Foods: Significant quantities of vital wheat gluten are used by the pet food industry. The usage may be in preparing simulated meat for canned pet food or in both canned and intermediate moisture-type products where the water absorption and fat binding properties of wheat gluten improves yields and quality. In dry dog biscuits, incorporating gluten into the dough before baking can improve resistance to breakage during packaging and shipping. Because it is a relatively inexpensive source of protein, wheat gluten can significantly contribute to nutritional labeling claims (2).

Aquaculture Feeds: The farming of aquatic organisms, including fish and crustacea, is a rapidly expanding industry—worldwide. Modern aquaculture increasingly depends on feeding to maximize production, and wheat gluten offers desirable properties for such usage. The adhesive properties of wheat gluten provide the binding needed for the pellet or granule forms of feed commonly used; its water insolubility reduces pellet breakdown; its viscoelastic properties can provide a chewy texture preferable to an extremely hard pellet; it lends itself to extrusion and air incorporation, depending on whether surface or bottom feeding is desired; and it provides nutritional value.

New Applications

Modern industrial processes with high technological requirements for chemically modified wheat gluten are not only setting new standards out of the generally recognized responsibility for maintaining a natural and safe environment for life, but are also demonstrating the growing need for improvements in the tapping of sources for raw materials which heretofore have received little attention. These processes indicate the value of gluten in the following applications: for use as a co-binder in paper coating mixtures; a filler in urea-formaldehyde resin adhesives; a component in synthetic, edible sausage skins; and as a possible constituent of plastics.

There is certainly not the slightest doubt that the successes that have been achieved justify further intensive endeavors to develop and optimize the chemical modification of wheat gluten so as to open up new and theretofore unexploited applications in the chemical and industrial field. Continual research is essential, not only to broaden theoretical knowledge but also to investigate how knowledge gained to the present, primarily in the laboratory, can be applied on a pilot plant scale and finally in the field.

Recent results obtained in research laboratories in Europe justify investment because they have shown that it is possible to exploit agriculturally derived products such as wheat gluten successfully. The chemical and industrial use of gluten, which, after starch, is the largest component of wheat, as an additive raw material cannot noticeably reduce the immense overproduction in agriculture and/or perceptibly counteract the increasing shortage of fossil resources; however, it can make a reasonable contribution to solving these problems (4).

Wheat gluten also has been used as a chewing gum base (5), in cosmetic products, such as mascara; and in pharmaceutical tableting (1).

Chemical or enzymatic modifications of wheat gluten have shown potential for beverage fortification, manufacture of biodegradable surfactants, use in wallpaper paste, and in producing pressure-sensitive adhesive tapes and other adhesives.

Additionally, a new generation of gluten products have been developed for the food industry. These products are comprised of gluten, which has been improved by the use of certain biotechnology. Some of these products have been modified by the use of protease enzymes after the gluten washing steps. Other products have been modified through the complexing of enzyme modified phospholipids with the gluten prior to drying. These products, when proven commercially, may offer the wheat gluten user an expanded product line from which to choose for his application (6).

It is hoped that work by the commercial, private, academic, governmental and industry associations will continue to improve this unique and natural protein through new and improved manufacturing techniques, plant breeding and increased knowledge of cereal chemistry.

World Production, Consumption and Capacity

As a vegetable food protein, wheat gluten ranks second to soy based protein in terms of volume of product produced.

In 1980, world production of wheat gluten was approximately 90,000 tons; by 1985 it had grown to 191,000 tons. Production in 1987 was 245,000 tons (7) and output is expected to increase to 260,000 tons by the end of 1988 (Table 4).

TABLE 4

Comparison of Wheat Gluten Production by Major Geographical Producing Areas—1980–1987

Country	Year/Amount (tons)	
	1980	1987
Australia	24,000	45,000
North America (Can/USA/Mex)	30,000	54,000
Europe	29,500	132,300
(EEC)	(25,000)	(118,300)
(Non-EEC)	(4,500)	(14,000)
Japan	3,000	6,700
South America	2,000	7,000
Totals	88,500	245,000

This increase is more apparent in Western Europe than any other geographical location, where 1987 capacity reached 159,000 tons (Table 5); over six times the 25,000 tons reported in 1980. Current industry and trade resources estimate the total productivity of wheat gluten manufactures in Western Europe to be at 118,300 tons (Table 6).

TABLE 5

1987 EEC Wheat Gluten Capacity

Country	Amount (tons)
Belgium	20,000
France	24,000
Federal Republic of Germany	30,000
Holland	29,000
United Kingdom	37,000
Others	19,000
Total	159,000

TABLE 6
1987 EEC Wheat Gluten Production

Country	Amount (tons)
Belgium	18,000
France	20,000
Federal Republic of Germany	22,000
Holland	24,000
United Kingdom	22,500
Others	11,800
Total	118,300

Examining data for other geographical areas, Australia production in 1987 was 45,000 tons up from 15,000 tons in 1980. North America, i.e., Canada, the United States and Mexico aggregated 54,000 tons in 1987; up 80% from 30,000 tons in 1980. Europe, including both the EEC and non-EEC producers, increased from 29,500 tons in 1980 to over 132,000 tons in 1987. Japan and South America also have increased production significantly during this timeframe (8). Changes in end-use patterns for wheat gluten between 1980 and 1987 are reflected in Table 7.

TABLE 7
Comparison of Worldwide End-use History for Wheat Gluten by Percentage

Category	Year/Per cent 1980	1987
Baking	77	63
Milling (flour fortification)	4	14
Meats	0	5
Pet foods	10	8
Cereals	3	2
Aquaculture feeds	0	1
Pasta	0	1
Cheese analogs	0	1
Seafood analogs	0	1
Other animal feeds	4	1
Devitalized	1	1
Others	1	2
	100%	100%

World Market

Most of the European gluten production has been taken up in fortification of the EEC's low protein soft wheats which otherwise would be unsuitable for many commercial baking needs. However, it is unlikely that internal-Community requirements will consume the enormous increases in production that have occurred over the past 2–3 years. Thus, it is inevitable that the EEC will seek export markets for its excess production, of which the largest target is the USA. Some U.S. manufacturers of wheat gluten have proposed that a quota of 1 million lbs (approximately 450 tons) should be imposed on wheat gluten exports from the EEC and that this would be an "appropriate response" to EEC restrictions on imports of U.S. agricultural products since the accession of Spain and Portugal (despite the negotiated "truce" giving temporary special accession).

However, Canada is clearly in a privileged position, particularly with regard to the new unilateral trade agreement proposed by the United States and Canada. Australia sees both the Pacific rim and the USA as natural markets (Table 8) (9).

TABLE 8
Overview of 1987 Supply and Trade by Major Geographical Regions

Country	Amount (tons)		
	Production	Net imports	Consumption
Australia	45,000	−26,900	18,100
North America (Can/USA/Mex)	54,000	35,600	89,600
South America	7,000	−2,500	4,500
Japan	6,700	3,300	10,000
Europe (ECC)	118,300	−20,000[a]	98,300

[a] Exports outside EEC only.

Australia, Canada and the USA: Western countries and Australia, with diets and culture similar to that of Europe, also have a similar pattern of gluten consumption. The largest sector of gluten demand remains the baking industry. In North America and Australia gluten additions to flours for baked goods mainly is accomplished at the bakery and is not via flour fortification at the mill or bakery improvers, as is usual now in Europe.

There are two main reasons for this difference: the high protein content and baking performance of North American and Australian wheats eliminates the need for gluten in standard white bread flours; and the manufacture of a wide variety of baked products (breads, morning goods, biscuits, cakes, crackers, etc.) requires flexibility in the bakery operations.

It is believed by some experts that protein levels in North American wheats will decline over the next 5–10 years. A trend toward lower wheat protein levels, if realized, would undoubtedly stimulate bakery demand for vital wheat gluten and provide a market for some of the large surplus forecast for the future.

It is forecast that the U.S. market will grow about 5% per year and that 2% would be accessible to the EEC. This would give rise to an additional demand of some 5,000 tons by 1990 (9).

The pattern of consumption of wheat gluten between Australia, North America and Europe are contrasted in Tables 9 and 10. The leadership role of Australia in both the areas of technology and applications are apparent in these tables. The wheat oriented starch industry of Australia results in a single-mindedness for technological and marketing innovation for wheat gluten promoting the product and creating a demand for this unique vegetable protein. This can be further demonstrated by the

TABLE 9

Australian, European and North American Wheat Gluten Consumption by End-use Amount

	Wheat gluten consumption (000's tons)		
End-use	AUS[a]	NA[b]	EUR[c]
Baking	15.1	60.5	44.6
Milling (flour fortification)	0	9.8	40.7
Meats	0.9	0.3	0.1
Pet foods	1.0	11.9	7.4
Cereals	0.3	1.9	1.0
Aquaculture feeds	0.2	0.6	0.4
Pasta	0.1	0.5	0
Cheese analogs	0.1	0.7	0
Seafood analogs	0.1	0.9	0
Other animal feeds	0	0	0.8
Devitalized	0.1	0.9	1.0
Others	0.2	1.6	2.3
Totals	18.1	89.6	98.3

[a] AUS—Australia.
[b] NA—North America (Canada/USA/Mexico).
[c] EUR—Europe (EEC only).

TABLE 10

Australian, European and North American Wheat Gluten Consumption by End-use Percent

	%Total		
End-Use	AUS[a]	NA[b]	EUR[c]
Baking	83.7	67.5	45.4
Milling (flour fortification)	0	11.0	41.4
Meats	4.8	0.3	0.2
Pet foods	5.9	13.3	7.5
Cereals	1.9	2.1	1.0
Aquaculture feeds	1.2	0.7	0.4
Pasta	0.1	0.5	0
Cheese analogs	0.5	0.8	0
Seafood analogs	0.5	1.0	0
Other animal feeds	0	0	0.8
Devitalized	0.4	1.0	1.0
Others	1.0	1.8	2.3
Totals	100	100	100

[a] AUS—Australia.
[b] NA—North America (Canada/USA/Mexico).
[c] EUR—Europe (EEC only).

more than one kilo (2.4 pounds) per capita consumption of wheat gluten by the Australians (10).

Argentina: Argentina is somewhat of an enigma in that meat products for domestic and export markets comprise the major outlet for wheat gluten in Argentina; but it is unlikely that overall consumption will show significant growth in the near future. Table 11 shows the Argentine demand for wheat gluten by application. Its comparison with the previous two tables comparing the end-uses for wheat gluten in Australia, North America and Europe is most interesting (8).

TABLE 11

Argentine Wheat Gluten Demand by End-use

Application	% Total	Wheat gluten consumption (tons)
Meats	60	1,800
Baking	15	450
Milling (flour fortification)	15	450
Pasta	7	210
Others	3	90
Total	100	3,000

Japan: Japan has a highly developed technology-based economy with a sophisticated standard of living and a diet that is a mixture of eastern and western influences. As food producers in these countries adopt newer industrial technology the functional/structural benefits of gluten will become apparent and increased consumption may be realized.

As a result of the shortage of animal protein, wheat gluten has important applications in the area of meat extension and substitution. Although wheat gluten consumption in Japan, China and other Asian countries is currently quite small, this level will undoubtedly increase. In simple terms, these regions have a combined population of approximately two billion, and government policies are tending to support the addition of protein to improve the quality of food products.

Current indications are, however, that gluten from Europe would be unlikely to gain more than 5,000 tons of sales in the forecast by 1990, starting from a virtual total state of ignorance today (8).

Table 12 indicates the relative end uses and consumption of wheat gluten in Japan. These consumption patterns

TABLE 12

Japanese Wheat Gluten Demand by End-use

Application	% Total	Wheat gluten consumption (tons)
Baking	30	3,000
Imitation meats/fish	25	2,500
Processed foods (goyza)	20	2,000
Sausages	12	1,200
Noodles	10	1,000
Others	3	300
Total	100	10,000

parallel the mixture of Eastern and Western influence present there today.

Product Quality

Many factors determine the quality of vital wheat gluten and its suitability for applications. These factors include raw materials and, probably most important, drying. One of the primary needs of this industry is the standardization of quality assurance and development of new methods of analysis. This need is being addressed by many groups at research centers around the world. Today, however, the most dependable method of evaluating vital wheat gluten is still the bake test, particularly, considering the significant use of wheat gluten in baked goods.

The Farinograph method has been suggested and is a fairly good measure of absorption requirement and, to a lesser extent, relative mixing requirement, but it is no measure of baking quality of wheat gluten.

Another test which is frequently used is the Nitrogen Solubility Test developed at the U.S.D.A. Northern Regional Research Center.

The Alveograph has been used extensively in Europe for evaluation of gluten quality, and a recent article published in a French journal stated, "Gluten powders vary in characteristics, and choice of appropriate quality gluten for different bakery applications is considered with reference to four commercial glutens of widely differing Alveograph characteristics" (11).

Other instruments available include: AB Falling Number's Glutomatic Gluten Washer; Labor MIM's Gluten Washer; and C.W. Brabender's Gluto-Graph. Other instruments and methods are in various stages of development in the laboratories of several producers and users of wheat gluten, as well as instrument manufacturers around the world.

Acknowledgment

The author thanks the International Wheat Gluten Association (IWGA) for assistance in preparing this paper.

References

1. Hesser, J.M., in *Proceedings of the 3rd International Workshop on Gluten Protein*, edited by R. Lasztity and F. Bekes, World Scientific, Singapore, 1987, pp. 441–445.
2. *Wheat Gluten: A Natural Protein for the Future—Today.* International Wheat Gluten Association, Kansas City, MO, 2nd Ed., 1986.
3. Magnuson, K.M., *Cereal Foods World. 30:*179 (1985).
4. Kempf, W., B. Pelech and W. Bergthaller, in *Proceedings of the Wheat Industry Utilization Conference*, edited by Y. Pomeranz (in press) 1988.
5. Lutz, H.J., U.S. Patent 2,586,675,1952.
6. Pfefer, D.N., in *Proceedings of the 63rd Annual Meeting of the ASBE* edited by J.A. Aaron, American Society of Bakery Engineers, Chicago, IL, 1987, pp. 81–89.
7. Sosland, N. N., *World Grain. 4:*11 (1986).
8. Booth, M., 8th International Cereals and Bread Congress, Lausanne, 1988.
9. *Vital Wheat Gluten Update 1986–1987–1990: EEC & World Perspectives*, Gira, S.A., Geneva, Switzerland, 2nd Ed., 1987.
10. Sosland, N.N., *Milling & Baking News*, Jan. 12, 1982, p. 65.
11. Hesser, J.M., in *Proceedings of the International Association of Cereal Chemists Symposium on Amino-Acid Composition and Biological Value of Cereal Proteins*, edited by R. Lasztity and M. Hidvegi, Hungarian Academy of Sciences, Budapest, 1983, pp. 529–542.

Utilization of Dry Field Beans, Peas and Lentils

Mark A. Uebersax and Songyos Ruengsakulrach

Department of Food Science and Human Nutrition, Michigan State University, East Lansing, Michigan 48824

Abstract

Dry field beans (*Phaseolus vulgaris*), peas (*Pisum sativum*) and lentils (*Lens culinaris*) encompass numerous cultivars which are diversely processed and utilized in many parts of the world. The common methods of preparation and processing include water cooking, cooking followed by frying, roasting, puffing, baking, germination and sprouting, fermentation, and canning in brine or sauce. Food legumes possess a dense source of protein, complex carbohydrate, vitamins and minerals. Nutritional quality of legumes is influenced by losses which occur during preparation and processing, the activity level of antinutritional components, and complex factors associated with low digestibility. Treatments that alleviate these limitations will enhance legume utilization. This paper provides an assessment of processing factors influencing product performance, nutritional quality and food utilization.

Dry beans (*Phaseolus vulgaris*), peas (*Pisum sativum*) and lentils (*Lens culinaris*) are the three grain legumes in the family Leguminosae. Dry beans originated from Mexico and Guatemala and peas and lentils are from Southeast Asia and Mediterranean regions. Eleven different classes of dry beans have been identified and are similar botanically; however, their color, size, shape and flavor characteristics vary widely between classes. World production of major grain legumes exceeds 150 million mT per annum (Fig. 1). The United States production distribution of major commercial classes of dry edible beans is illustrated in Fig. 2. In addition, seven different classes of dry beans grown in Michigan account for nearly one-third of U.S. annual production. These are navy beans, cranberry beans, dark red kidney beans, light red kidney beans, pinto beans, black turtle soup beans and yellow eye beans. Approximately 40% of all dry beans exported from the U.S. are shipped from Michigan during an average marketing year. California produces about half of the kidney beans grown in the U.S., and pinto beans are grown extensively in the western states: North Dakota, Idaho, Colorado, and Wyoming (1). Nearly 95% of peas and lentils are commercially produced in the Pacific Northwest states. The major commercial classes include whole green peas, green split peas, whole yellow peas, yellow split peas, Austrian winter peas, Chilean lentils and decorticated red chief lentils (2). Approximately 75% of U.S. dry pea and lentil annual production is exported to the United Kingdom, and to numerous South American and Far Eastern countries. Dry beans, peas and lentils are priced ranging from $10–30/cwt, $10–13/cwt and $17–36/cwt, respectively (3,4).

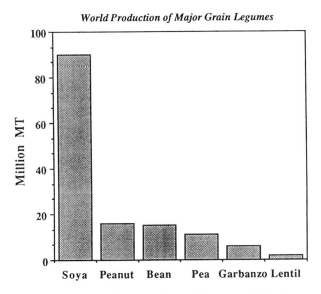

Fig. 1. World production of major grain legumes. (U.S. Department of Agriculture)

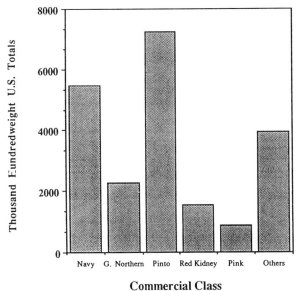

Fig. 2. U.S. Commercial dry bean production. (U.S. Department of Agriculture)

Legumes as a Food Resource

Food legumes contribute a good source of several important nutrients. They provide variety to the human diet and, more importantly, an economical source of supplementary protein for many populations lacking animal proteins. Variability in the chemical composition of the legumes occurs among cultivars, geographic locations, and growing conditions (5,6,7). The proximate composition of dry beans, peas and lentils is shown in Table 1. In general, the dry field beans, peas and lentils contain 20–30% protein which is rich in lysine but limiting in methionine content, thereby complementing the amino acid pattern found in cereal. Furthermore, legumes are also recognized as a major source of complex carbohydrates including dietary fiber.

TABLE 1

Proximate Analyses (%db) of Dry Beans, Peas and Lentils[a]

Legumes	Protein	Fat	Carbohydrates	Ash
Dry beans				
(Navy)	25.4	1.5	69.3	3.8
Peas	24.1	13.0	60.3	2.6
Lentils	25.7	1.1	70.3	2.9

[a] Ruengsakulrach and Uebersax, 1988, unpublished data.

Several factors are found to limit legume utilization. These include the low levels of the sulfur amino acids (8,9), the low digestibility of unheated proteins (10), the presence of antinutrients (11,12), the high level of phytic acid (13–18), various flatulence factors (19,20) and the hard-to-cook phenomenon which may develop during bean storage (21,22). Although legumes are a crucial source of amino acids, especially lysine, in the lesser-developed countries their production is far below that of the cereal grains. The most important reason for this is that cereals are much more productive than legumes. Hulse (23) states that on average yields of corn, rice and wheat are 2.8, 2.2 and 1.7 tons/hectare compared to a yield of 0.5 ton/hectare for legumes (excluding soybeans). Because of the difference in yields, legumes are less likely to provide a satisfactory economic return to farmers than the higher yielding cereal crops. This is especially evident in Asia where the production ratio of cereals to legumes had increased to 9:1 by 1977. Because the dietary legume to cereal ratio for optimum protein quality should be approximately 1:2 (24) it is readily apparent that legume production should be stimulated. The most effective means to stimulate increased production is to develop higher yielding legume varieties that are more economically attractive to small farmers. Raw legumes are poorly digested, but adequate heat treatment improves the digestibility significantly (25,26,27). However, in many parts of the world the thermal treatment that can be provided for bean preparation in the home setting is not sufficient to inactivate toxic lectins (28) and is often just sufficient to heat and hydrate the beans. Gomez Brenes et al. (25) reported that peak digestibility and Protein Efficiency Ratios (PER) of dry P. vulgaris were obtained after soaking them for 8 or 16 hours and cooking them at 121°C for 10 to 30 minutes. Heating these legumes for longer than this resulted in lowered protein quality and decreased available lysine.

Beans contain a number of antinutrients and potentially toxic substances. Bressani (8) categorized the toxic substances present in legumes into seven groups: trypsin inhibitors, hemagglutinins or lectins, goitrogenic factors, cyanogenic glucosides, lathyric factors, compounds that cause favism, and other nonassociated factors. Of these factors, trypsin inhibitors and hemagglutinins are considered primarily responsible for causing the growth retardation observed in laboratory animals fed raw beans (29–35).

Trypsin inhibitors are found in several legumes including *Phaseolus*. As the name implies, these are protein fractions that strongly inhibit the enzymatic activity of trypsin in the intestine (11) thereby reducing the digestion of proteins and hence, the subsequent absorption of their constituent amino acids. Besides the inhibition of the enzyme trypsin, the poor digestibility of bean proteins may result from the nature of native bean protein structure (35). Soaking (12,32) or germination (36) plus heating (37–40) have been found to improve the digestibility of bean protein in two ways: by denaturation which makes the proteins more susceptible to the actions of enzymes, and by destroying trypsin inhibitors. Liener (30,35) also reported that the protein of unheated beans resists proteolysis in the intestine; however, after heating, the true digestibility increases and trypsin inhibitory activity decreases. Further, Gatfield (41) reported that trypsin inhibitors were reduced by water soaking of beans prior to cooking them.

Hemagglutinins also are present in beans and are destroyed by heat (42). These compounds cause agglutination of red cells and impair absorption (35). Kakade and Evans (32) found that merely soaking the beans did not result in any loss of hemagglutinating activity. Aguilera et al. (43,44) proposed that special preheating and grinding control during processing may aid inactivation and reduction of both hemaglutinin and trypsin activity. In addition, legumes contain relatively large amounts of phytic acid, up to 5% by weight (45). The phytic acid (myoinositol hexaphosphate), the major phosphorus-bearing compound is recognized to chelate divalent cations and thereby restrict bioavailability of essential elements (13–18).

The ability of legume seeds to stimulate intestinal gas formation (flatulence production) limits their consumption by humans. It has frequently been suggested that the galactose-containing oligosaccharides including raffinose, stachyose and verbascose are the components in the legume seeds responsible for flatulence (19,20,46). These sugars have been considered colon fermentable because the α-galactosidase enzyme, although not produced by the human digestive system, is secreted by the indigenous gut microflora and hydrolyzes these sugars yielding abdominal gas production (47,48). Fleming (49) studied the flatus potential of seven types of legume seeds and reported that hydrogen production was significantly and positively correlated to oligosaccharides and acid hydrolyzable pentosan content, but negatively correlated to the starch and lignin content.

Processing Technologies

Post-harvest Handling and Preparation: Numerous factors influence the quality of the food legume product. These include variety, seed source, agronomic conditions, handling and storage of the dry product, and processing procedures. Dry beans, peas and lentils are harvested in the mature, dry stage and stored until processed (Fig. 3).

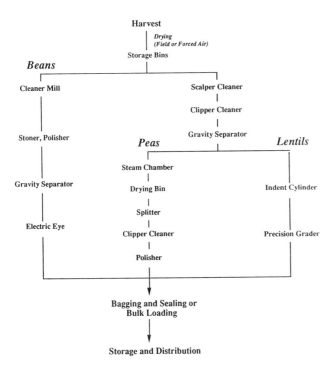

Fig. 3. Dry processing of beans, peas and lentils.

These legumes for processing typically require dry cleaning, sorting and handling to assure a minimum degree of mechanical damage. Moisture and storage temperature play a major role in controlling the final product quality. Moisture control must be maintained within optimal limits to assure a stable product. The moisture status is influenced by the physiological seed characteristics which control water imbibition. High moisture associated with high temperature or heating during storage ("bin burn") will result in brown, discolored and off-flavored beans. Further, beans subjected to adverse storage conditions may undergo a hardening or an inability to hydrate resulting in a bean which is termed "hard-to-cook." Various research reports and the general experience of storage facility operators and processors demonstrate that adverse storage conditions dramatically affect the color, flavor and texture of dry beans. Various investigators have reported the effects of bean moisture content on quality loss. Muneta (50) reported that "Michelite" beans stored at 13.4% moisture for more than one year required twice the cooking period of beans stored at lower moisture contents. Molina et al. (51) reported the interrelationships between storage and processing conditions on the nutritive value of Guatamalan black beans. Although nutrient composition of beans has been shown to be affected by the cultural and storage environment, limited results have been reported for processing quality characteristics comparing bean genotypes grown and stored under varying conditions. Nutritional value encompassing protein quality, vitamins and minerals of processed canned beans is affected by storage time and temperature, cooking time, soaking time and canning medium (51,52).

Dry beans, peas and lentils traditionally have been prepared for dry-pack retail sales intended for home preparation and use. The majority of legumes in the lesser-developed nations are grown for home consumption.

Typically, dry beans are soaked for a few hours, cooked in an open container and consumed whole or as a mashed bean paste with cereal grains or tubers. A preliminary soaking process is carried out to moisten and soften the seed so as to shorten the cooking time and to reduce the toxin content. Legume seeds, prepared by soaking them to initiate the germinating or sprouting process followed by seed coat removal, can be roasted or ground for use in soups or side dishes. Lentil sprouts are mixed in vegetable salads. Germination also has profound effect on chemical-physical composition changes in dry beans. Snauwaert and Markakis (53) reported that both stachyose and raffinose decreased in concentration during the germination process. This procedure has potential for developing diverse bean foods. Bressani and Elias (54) discussed many factors involved with the cookability of beans. Cooking whole legumes in boiling water or steam is the most common method used to obtain palatable and flavorable products in addition to improving their nutritional value. In Latin America, beans are usually boiled as whole seeds, optionally dehulled, and then mashed or pureed. Cooked dry beans, split peas and whole dry peas are used in various recipes such as salads combined with other vegetables. No soaking step before cooking is necessary for lentil preparation. In many countries, dry beans, peas and lentils are commercially canned in brine, tomato sauce, or in sauce mixed with other vegetables. Novel processing operations may allow improved utilization of dry beans, peas and lentils. In recent years, these legumes have gained much interest for their potential uses as high protein ingredients in formulated foods by offering distinctive functional properties such as color, flavor, texture, solubility, viscosity, cohesion, gelation, foamability, water or oil absorption, emulsification and improved nutritional quality. In addition, in Asian cultures, soybeans have been manipulated to produce a variety of nutritious food products. Some important soybean based foods are *tofu, natto, miso, tempeh* and *tamari*. If other legumes could be prepared by methods such as protein curd production or fermentation, the quality and digestibility might be improved. Swanson and Raysid (55) described the changes in protein quality in tempeh made from red beans (*P. vulgaris* L.) and corn indicating that there was a significant increase in digestibility and PER in these fermented products.

Cooking: Processing methods for canned beans, peas and lentils have been studied extensively. Beans require some means of cooking before they are eaten. This thermal process provides tenderization of the cotyledon which increases palatability and inactivates endogenous toxic factors that would markedly limit the final nutritional value. The preparation of dry beans generally involves hydration in water to a moisture content ranging between 53–57% which is followed by extended cooking. The initial soaking treatment ensures uniform expansion of the seed coat and cotyledon matrix and cellular hydration to aid in heat transfer and subsequent tenderization. The rate and extent of hydration has a marked effect on cooked product yield and on the dietary satiety provided by a given quantity of dry beans. Traditionally, beans have been soaked overnight (8–12 hours) in cold water; however recently, high temperature soaking has been used to accelerate hydration. Cold water soaking is time-consuming and may result in dramatic increases in potential hazardous microflora; however, cold soak procedures require relatively low levels of energy input and provide low levels of nutrient leaching. Hot water treatments are rapid, sanitary, yet result in substantial nutrient losses due to leaching. Alternate methods designed to accelerate hydration and tenderization include vacuum, ultrasonic sound, the use of soak water additives and γ-irradiation. The degree and rate of hydration of the starch/protein matrix influences the cooking rate and final texture of cooked beans. Processing studies have been aimed at decreasing soaking and cooking time, reducing nutrient loss and improving appearance and textural quality attributes. The mechanisms that govern the rate and amount of water uptake during soaking are not clearly understood. Pectic substances, hemicelluloses and proteins are primary functional components affecting absorption and water holding capacity in plant tissue. Studies have implicated pectin and divalent metal ions as agents which reduce water uptake and increase bean firmness through the formation of tough pectic/metal ion complexes (56,57). Seed coat permeability and starch and protein characteristics have been shown to affect the soaking and tenderization processes. Smith and Nash (58) concluded that the seed coat is the principle factor controlling the rate of water uptake in soybeans. Powrie et al. (59) reported the results of chemical and anatomical studies of dry beans and demonstrated that water migration occurred over the cotyledon after water entry by imbibition occurred. Acidity of the soaking medium affects bean hydration characteristics (60). At low pH, starch and protein swelling was inhibited causing a reduction in hydroscopicity. Addition of citric acid to canning brine resulted in denaturation of some proteins and suppression of the hydration of the starch/protein matrix resulting in increased firmness of canned beans. Choline esterase activity, phosphatase and phytin containing salts have been implicated as influencing water uptake. Polyphosphates increase the rate and extent of imbibition by chelating divalent cations ($Ca++$, $Mg++$) and by solubilizing and dissociating metal complexes of proteins. Ethylenediaminetetraacetic acid (EDTA) and a mixture of alkaline carbonates increases water uptake apparently by softening the seed coat. Wetting or surfactant agents fail to influence water uptake.

Hoff and Nelson (61) studied physical and chemical methods of accelerating water uptake of dry beans. Physical methods of releasing surface gas improved the rate of water uptake. Chemical additives improved water uptake and increased tenderization. Powers et al. (62) subjectively and objectively evaluated the gelatinization characteristics of canned beans. Rockland and his group (63) have conducted extensive studies on processing techniques designed to reduce cooking time of legumes. During the course of this work much fundamental information has been developed concerning gelatinization properties of bean starch and solubilities of bean seed protein. The cooking rate of dry legumes is temperature dependent (64).

Dry beans contain various metabolic inhibitors. Increases in nutritional quality have been demonstrated by heating beans to destroy the heat labile trypsin inhibiting compounds. Increases in moisture content of beans during cooking increases the thermal conductivity and the inhibitor is rapidly inactivated. Soaking in media other than water may improve nutritive value. It is commonly known that lye-treated corn is nutritionally superior to untreated corn. The high pH soak increases the availability of niacin and lysine, which is the first limiting amino acid in corn. Soaking beans in water is recommended as it reduces the time necessary for adequate cooking (51). Recent research by Coffey, (65) and Coffey et al. (28) has demonstrated that alkali cooking of dry beans may be a possible approach for reducing hemagglutinating activity and improving the cooking characteristics of beans.

Sufficient cooking is necessary to soften tissue; excessive cooking results in a loss of protein biological value. The excessive heating of beans can cause a loss of nutritive value because of the highly reactive nature of several of the amino acids. Lysine, which accounts for about 90% of the free amino groups in the form of ϵ-amino lysl residues, and methionine are reactive amino acids. The specific loss of ϵ-amino groups of lysine occurs by a condensation reaction with reducing sugars. Maillard browning reactions may occur in beans stored under adverse environmental conditions. Browning will result in decreased nutritional value and reduced acceptability.

Canning: In developed nations, beans are generally prepared by commercial food processing operations and consumed as canned beans in sauce. Beans may be processed in water or brine, tomato sauce or molasses. Beans to be processed should contain a moisture level of about 12% to 16%, be of uniform size, fully mature and free from foreign materials and seed coat defects. Traditionally, dry beans have been soaked for 8 to 16 hours at room temperature or soaked using high temperature (180°F to 212°F) for 20 to 40 minutes. Either procedure is designed to provide beans with a final moisture range of 53% to 57%. The high temperature short time soaking is generally preferred because it reduces labor costs, reduces floor space requirements, provides more readily controlled schedules, and reduces potential bacteriological problems, which may occur during a long soaking period. Hydrated beans may be processed in a standard rotary water blancher for 3 to 8 minutes. After blanching, beans are continuously washed with cold water sprays utilizing rod washers or vibrating sieves to facilitate removal of broken beans and skins. A final inspection of cooled, blanched beans is provided prior to filling. Soaked beans will further hydrate during the thermal process and equilibrate with the sauce or brine. This equilibration will require two to four weeks and final moistures of the bean solids will be in the range of 65% to 70% moisture. Recent research conducted at Michigan State University evaluated the effects of several soaking/blanching methods and calcium soak/blanch water treatments on the final processing quality of navy beans. A three-level factorial experiment was designed to include: four soak methods, two soak mediums and two brine mediums, each at 25 or 125 ppm calcium ion. The four soak or blanch methods used were as follows: Method I—overnight soak at ambient temperature for 14 hours followed by a sub-boiling blanch for 3 minutes; Method II—77°C blanch for 60 minutes; Method III—99°C blanch for 30 minutes; Method IV—two stage process with 21°C soak for 30 minutes, followed by an 88°C blanch for 30 minutes. The navy beans (Seafarer cv) were processed using the procedure described by Uebersax (66). After processing, the cans were held at 21°C for 14 days prior to quality assessment. Increased calcium ion concentration produced decreased drained weights, increased total bean solids, increased bean texture, and decreased splitting of the bean seed coat (Figs. 4 and 5). Addition of calcium in the soak water had a greater effect than in the

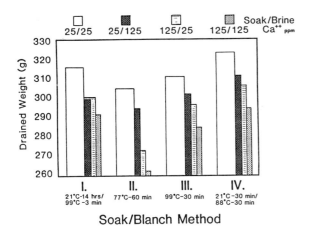

Fig. 4. Relationship of drained weight to process method.

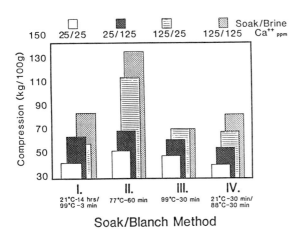

Fig. 5. Relationship of bean texture to process method.

brine. The soak/blanch process and calcium interaction which demonstrated the greatest calcium firming of beans was Method II (77°C for 60 min.). The drained weight or processor's yield and compression force or bean firmness showed an inverse relationship (Fig. 6).

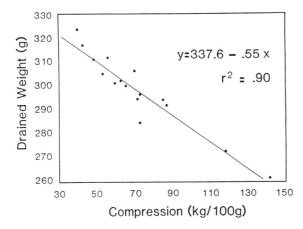

Fig. 6. Relationship of drained weight to bean texture.

Extrusion: Extrusion recently has gained increased technical focus toward development of novel food products. This process utilizing continuous high pressure, high temperature cooking of starch or flour from cereals and legumes has provided numerous diverse product applications. The understanding of physico-chemical properties of various ingredients related to their rheological behaviors during processing is the key factor in designing food products with improved physical and nutritional quality characteristics and possessing diverse marketability.

The use of this technology and proper ingredient formulation may provide for pre-cooked or quick cooking legume products. Starch gelatinization occurs primarily during the later stages of cooking of the beans. The amount of water available is critical to the gelatinization process. When the cooking and extrusion process are combined, the increased pressure differential will facilitate expansion of the final product. This formed product of open porosity may require additional air drying to assure shelf-stability. The dry formed bean will readily hydrate and soften. Die configuration can be designed to produce a variety of intricate product shapes which can be further optimized to produce simulated beans. Appropriate binders and coating agents can be added to improve appearance of pre-cooked or quick cooking bean products.

Many puffed snack foods and breakfast cereals currently are made from corn or wheat bases. The potential for starch products to swell and form an expanded cellular structure during extrusion is an important characteristic in the manufacture of puffed snack products. High starch bean flours from navy, pinto and black beans yield good expansion (67). In this process, bean starches were extruded with various levels (0–20% total dry weight) of ground and unground hulls. After exiting the die, the puffs were oven-dried to remove excess moisture. Flavoring was accomplished by applying a fine spray of vegetable oil (10% of puff weight), followed by coating them with salt (2.5% by weight) and one of the following flavor powders: onion, hickory, barbecue (all 7.5%), and cheese (15%). The unflavored puffs were bland and flavorless, thus making them compatible with a variety of flavored products. The extruded bean puffs may be improved nutritionally by incorporating high protein bean flour as well as hulls. Product quality can be optimized by further studies with bean fraction combinations. Properly balanced levels of bean starch, bean hulls, and bean protein flours could yield an acceptable, nutritious snack food or breakfast cereal.

Food Ingredient Preparation: A shelf-stable flour has been proposed as a food ingredient to provide increased versatility and to improve bean utilization. Initial efforts to produce bean ingredients were aimed at keeping cell rupture at a minimum in order to retain the same texture, appearance and taste, simulating conventionally prepared beans. Instant precooked bean powders have been prepared by soaking, cooking, slurrying, and drum or spray drying beans (68). Furthermore, dry beans, peas and lentils possess a good source of plant protein which are currently receiving much research interest for their potential uses in high protein food ingredient production. High protein food ingredient products are, for example, flour processed by pin milling and air classification and concentrates and isolates processed by alkali, salt and acid extraction together with isoelectric precipitation or ultrafiltration. Chang and Satterlee (69) produced bean protein concentrates containing 72% and 81% protein by wet processing using water extraction techniques. Molina and Bressani (70) prepared protein isolates containing about 90% protein and starch products with almost 50% starch. However, several disadvantages exist in using wet processing for fractionation and concentration: (a) significant energy is used in drying the final product; (b) waste by-products are produced which contain significant amounts of organic matter; and (c) yields are reduced by the losses in wet by-product streams. Much research has been conducted to investigate the functionality of these high protein products to broaden their use in various foods. Recent product development has included spray dried powders from peas (71), imitation milk replacers from peas and lentils (72), protein curd similar to *tofu* from peas (73), pasta (noodles and spaghetti) ingredients from dry beans, peas and lentils (74) and meals and concentrates mixed in hamburger preparations (75,76). In addition, the high fiber or high protein bean flour fractions were substituted (20%) for wheat flour to produce acceptable quality in cookies, donuts, quick breads and leavened doughs (77).

Dry milling of dry beans, peas and lentils into whole flour and air-classifying them into high starch and high protein fractions have received increased interest as efficient processing methods (78–81). The compartmentalization of protein (friable protein bodies) and starch (denser and less brittle granules) enables size reductions and fractionation of dry beans into a coarse fraction which contains high starch and a fine fraction which contains high protein. High protein flour fractions of dry beans, peas and lentils generally contain two times the original whole seed protein content. Air-classified high protein flour contains a residual starch which offers some particularly desirable functional applications. Air-classification utilizes the principle of differential terminal velocities inherent among different particle sizes. A simple fractionation technique, sieving, has been proposed; however, it becomes progressively more difficult as the size separation point is reduced. Very fine material has the tendency to agglomerate, especially if oil is present. Such powders tend to build-up on the screen causing blinding of the sieves and a loss of separation efficiency.

It has been observed that, when raw legumes are ground without pretreatments, they develop undesirable odors and flavors. Lipoxidase has been held responsible for the appearance of off-flavors by catalyzing formation of peroxides from unsaturated fatty acids (82). However, treatment with dry heat for six to eight minutes at 104°C to 105°C completely inactivates this enzyme (83). Aguilera *et al.* (44) found that heat treatment also destroys trypsin inhibitor and hemagglutinin activities. Therefore, it is advantageous to apply a rapid heat treatment to beans prior to bean flour production. Conventional air heating is attractive due to ease of handling and lack of product contamination. However air does not provide a high rate of heat transfer. Carvalho *et al.* (84) produced instant navy bean powders by roasting dry beans in a bed of salt, followed by grinding them, resulting in a reduction of trypsin inhibitory activity. Peterson and Harper (85) developed a granular bed roaster. Based on this previous work, Aguilera *et al.* (44) developed a particle-to-particle heat exchanger for roasting navy beans and observed the functional characteristics of several roasted bean products. These roasted products showed reduced water-soluble nitrogen content, good gel forming capacity, low trypsin inhibitor and hemagglutinin activities, increased water holding capacity and cold paste viscosities, and no changes in available lysine and only a limited degree of starch damage. Air classification of roasted beans produced high fiber, starch and protein fractions each being suitable as food ingredients in a variety of food products (86–90).

References

1. Conklin, N.C. and B. Peacock, *Bean Commission J.* 9:8 (1985).
2. Anonymous, in *Growing into the Future: USA Dry Peas and Lentils*, USA Dry Pea & Lentil Industry, Moscow, Idaho (1985).
3. Conklin, N.C. and B. Peacock, *Michigan Dry Bean Digest* 10:2 (1985).
4. Pederson, L.E. and K.L. Casavant, Washington State University College of Agriculture Research Center Circular 0626, Pullman, WA (1980).
5. Hosfield, G.L. and M.A. Uebersax, *J. Am. Soc. Hort. Sci.* 105:246 (1980).
6. Hosfield, G.L., A. Ghaderi and M.A. Uebersax, *Can. J. Plant Sci.* 64:285 (1984).
7. Salunkhe, D.K., S.S. Kadam and J.K. Chavan, in *Postharvest Biotechnology of Food Legumes*, CRC Press Inc., Boca Raton, Fl., 1985, p. 29.
8. Bressani, R., in *Nutritional Improvement of Food Legumes by Breeding*, edited by M. Milner, John Wiley-Interscience Publication, New York, 1975, p. 381.
9. Molina, M.R., G.D.L. Fuente and R. Bressani, *J. Food Sci.* 40:587 (1975).
10. Chang, K.C., and L.D. Satterlee, *Ibid.* 46:1368 (1981).
11. Gomes, J.C., U. Kock, and J.R. Brunner, *Cereal Chem.* 56:525 (1979).
12. Tyler, R.T., C.G. Youngs, and F.W. Sosulski, *Ibid.* 58:144 (1981).
13. Sathe, S.K., and K. Krishnamurthy, *Ind J. Med. Res.* 41:453 (1953).
14. O'Dell, B.L., and J.E. Savage, *Proc. Soc. Exp. Biol. Med.* 103:304 (1980).
15. Roberts, A.H., and J. Yudkin, *Nature* 185:823 (1960).
16. O'Dell, B.L., C.E. Burpo, and J.E. Savage, *J. Nutr.* 102:653 (1972).
17. Davies, N.T., and R. Nightingale, *Br. J. Nutr.* 34:243 (1975).
18. Maga, J.A., *J. Agric. Food Chem.* 30:1 (1982).
19. Fleming, S.E., *J. Food Sci.* 46:794 (1981).
20. Fleming, S.E., *Ibid.* 47:12 (1981).
21. Stanley, D.W., and J.M. Aguilera, *J. Food Biochem.* 9:277 (1985).
22. Srisuma, N., R. Hammerschmidt, M.A. Uebersax, S. Ruengsakulrach, M.R. Bennink and G.L. Hosfield, *J. Food Sci.* In Press (1988).
23. Hulse, J.H., K.O. Rachie, and L.W. Billingsley, in *Nutritional Standards and Methods of Evaluation for Food Legume Breeders*, International Development Research Center, Ottawa, Canada, 1977.
24. Bressani, R., and L.G. Elias, in *Nutritional Standards and Methods of Evaluation for Food Legume Breeders*, International Development Research Center, Ottawa, Canada, 1977, p. 51.

25. Gomez-Brenes, R., L.G. Elias, M.R. Molina, G. de la Fuente, and R. Bressani, *Arch. Latinoam. Nitr., Proceedings of Annual Meeting, 1973*:93–108 (1975).
26. Wolzak, A., R. Bressani, and R. Gomez-Brenes, K.W. King, and R. Bressani, *J. Food Sci. 41*:661 (1976).
27. Wolzak, A., L.G. Elias, and R. Bressani, *J. Agric. Food Chem. 29*:1063 (1981).
28. Coffey, D.G., M.A. Uebersax, G.L. Hosfield, and J.R. Brunner, *J. Food Sci. 50*:78 (1985).
29. Honovar, P.M., C. Shih, and I.E. Liener, *J. Nutr. 77*:109 (1962).
30. Liener, I.E., *Am. J. Clinc. Nutr. 11*:281 (1962).
31. Kakade, M.L., and R.J. Evans, *Brit. J. Nutr. 19*:269 (1965).
32. Kakade, M.L., and R.J. Evans, *J. Food Sci. 31*:781 (1966).
33. Jaffe, W.G., in *Toxic Constituents of Plant Foodstuffs*, edited by I.E. Liener, Academic Press, New York, NY, 1969.
34. Liener, I.E., and M.L. Kakade, in *Toxic Constituents of Plant Foodstuffs*, edited by I.E. Liener, Academic Press, New York, NY, 1969.
35. Liener, I.E. in *Nutritional Improvement of Food Legumes by Breeding*, edited by M. Milner, John Wiley & Sons, New York, NY, 1975.
36. Gupta, K., and D.S. Wagle, *J. Food Sci. 45*:394 (1980).
37. Ellenrieder, G., H. Geronazzo, and A.B. de Bojaski, *Cereal Chem. 57*:25 (1980).
38. Ellenrieder, G., S. Blanco, and A. Bondoni, *Ibid. 58*:291 (1981).
39. Johnson, L.A., C.W. Deyoe, W.J. Hoover, and J.R. Schwenke, *Ibid., 57*:376 (1980).
40. Chang, C.R., and C.C. Tsen, *Ibid. 58*:211 (1981).
41. Gatfield, I.L., *Lebensm.-Wiss. U.-Tech. 13*:46 (1980).
42. Thompson, L.U., R.L. Rea, and D.J.A. Jenkins, *J. Food Sci. 48*:235 (1983).
43. Aguilera, J.M., E.W. Lusas, M.A. Uebersax, and M.E. Zabik, in *Process Development, Characterization and Utilization of Dry Heated Navy Bean Products*, Research Agreements No. 59-2481-0-2-001-0, United States Department of Agriculture Science and Education Administration, Arlington, VA, 1981.
44. Aguilera, J.M., E.W. Lusas, M.A. Uebersax, and M.E. Zabik, *J. Food Sci. 47*:1151 (1982).
45. De Boland, A.R., G.B. Garner, and B.L. O'Dell, *J. Agric. Food Chem. 23*:1186 (1975).
46. Sathe, S.K., S.S. Deshpande, and D.K. Salunkhe, *CRC Crit. Rev. Food Sci. Nutr. 20*:1 (1983).
47. Calloway, D.H., C.A. Hickey, and E.L. Murphy, *J. Food Sci. 36*:251 (1971).
48. Wagner, J.R., R. Becker, M.R. Gubmann, and A.C. Olson, *J. Nutr. 106*:466 (1976).
49. Fleming, E.E., *J. Food Sci. 47*:12 (1982).
50. Muneta, P., *Food Technol. 18*:1240 (1964).
51. Molina, M.R., M.A. Baten, R.A. Gomez-Brenes, K.W. King and R. Bressani, *J. Food Sci. 41*:661 (1976).
52. Nordstrum, C.L., and Sistrunk, W.A., *Ibid. 42*(3):795 (1977).
53. Snauwaert, F., and P. Markakis, *Lebensm.-Wiss. U.-Tech. 9*:93 (1976).
54. Bressani, R. and L.G. Elias, in *Food Nutrition Bull.* UNU 1 (4), 1979, pp. 23–34.
55. Swanson, B.G., and A. Raysid, 44th Annual Meeting of the I.F.T. Abstract #310 (1984).
56. Matz, S.A. 1962, in *Food Texture*, The AVI Publishing Co., Westport, Connecticut, pp. 109–113.
57. Morris, H.J., and R.M. Seifert, Constituents and treatments affecting cooking of dry beans. Fifth Annual Dry Bean Res. Conf. USDA, 1961.
58. Smith, A.K., and A.M. Nash, *J. Am. Oil Chem. Soc. 38*:120–123 (1961).
59. Powrie, J.J., D.E. Pratt and J.B. Joiner, *Agron. J. 52*:163 (1960).
60. Snyder, E., in *Some Factors Affecting the Cooking Quality of Pea and Great Northern Types of Dry beans*. Nebraska University Agr. Expt. Station, Research Bulletin 85., 1936.
61. Hoff, J.E. and P.E. Nelson in *An Investigation of Accelerated Water-uptake in Dry Pea Beans*. Indiana Agric. Expt. Sta. Res. Progress Rept. 211, 1965.
62. Powers, J.J., E.G. Pratt, J.B. Joiner, *Food Technol. 15*:41 (1961).
63. Rockland, L.B., and F.T. Jones, *J. Food Sci. 39*:342 (1974).
64. Quast, D.C., and S.D. daSilva, *Ibid. 42*:370 (1977).
65. Coffey, D.G., in *Studies of Phytohemagglutinin, The Lectin of Phaseolus vulgaris*. PhD Dissertation, Michigan State Univ., 1985.
66. Uebersax, M.A., in *Proceedings: Technical Conference on Dry Bean Research*, The Food Processing Institute, San Francisco, 1985.
67. Zabik, M.E., M.A. Uebersax, J. Benzinger and G. Agbo, *Michigan Dry Bean Digest. 8*:12 (1983).
68. Bakker, F.W., R.J. Patterson, and C.L. Bedford, in *Nutritional Aspects of Common Beans and Other Legume Seeds as Animal and Human Foods*, edited by W. Jaffee, Archivos Latinoamericanos de Nutricion, Venezuela, 1973.
69. Chang, K.C., and L.D. Satterlee, *J. Food Sci. 44*:1589 (1979).
70. Molina, M.R., and R.Bressani, in *Nutritional Aspects of Common Beans and Other Legume Seeds as Animal and Human Foods*, edited by W. Jaffe, Archivos Latinoamericanos de Nutricion, Venezuela, 1973.
71. Patel, P.R., C.G. Youngs and D.R. Grant. *Cereal Chem. 58*:249 (1981).
72. Sosulski, F.W., P. Chakraborty and E.S. Humbert. *Can. Inst. Food Sci. Technol. J. 11*:117 (1978).
73. Gebre-Egziabher, A., and A.K. Sumner. *J. Food Sci. 48*:375 (1983).
74. Nielson, M.A., A.K. Sumner and L.L. Whalley. *Cereal Chem. 57*:208 (1980).
75. McWatters, K.H., and E.K. Heaton, *J. Am. Oil Chem. Soc. 56*:864 (1979).
76. Vaisey, M., L. Tasses and B.E. McDonald, *Can. Inst. Food Sci. Technol. J. 8*:74 (1975).
77. Zabik, M.E. and M.A. Uebersax, in *Plant Proteins: Application, Biological Effect, and Chemistry*, edited by R.L. Ory, ACS Symposium Series, American Chemical Society, Washington, DC, 1986, pp: 190–205.

78. Gueguen, J., *Qual. Plant. Plant Foods Hum. Nutr. 32*:267 (1983).
79. Kon, S., *J. Food Sci. 44*:1329 (1979).
80. Ekpenyong, T.E. and R.L. Borchers, *Ibid. 45*:1559 (1980).
81. Reddy, N.R. and D.K. Salunkhe, *Cereal Chem. 57*:356 (1980).
82. Kon, S., *J. Food Sci. 44*:1329 (1979).
83. Smith, A.K., and S.J. Circle, in *Soybeans: Chemistry and Technology, Vol. 1, Proteins*, AVI Publishing Co. Westport, CT, 1972.
84. Carvalho, C.C.C., G.R. Jansen and J.M. Harper, *J. Food Sci. 42*:553 (1977).
85. Peterson, D.G., and J.M. Harper, U.S. Patent 4,094,633 (1978).
86. Defouw, C.L., M.E. Zabik, M.A. Uebersax, J.M. Aguilera and E.W. Lusas, *Cereal Chem. 59*:245 (1982).
87. Defouw, C.L., M.E. Zabik, M.A. Uebersax, J.M. Aguilera and E.W. Lusas, *Ibid. 59*:229 (1982).
88. Dryer, S.B., S.G. Phillips, T.S. Powell, M.A. Uebersax and M.E. Zabik, *Ibid. 59*:319 (1982).
89. Zabik, M.E., M.A. Uebersax, J.P. Lee, J.M. Aguilera and E.W. Lusas, *J. Am. Oil Chem. Soc., 60*(7):1303 (1983).
90. Lee, J.P., M.A. Uebersax, M.E. Zabik, G.L. Hosfield and E.W. Lusas, *J. Food Sci. 48*(6):1860 (1983).

African Uses of Cowpeas, Pigeon Peas, Local Protein and Oil Seeds

Bene W. Abbey

Department of Biochemistry, University of Port Harcourt, Rivers State, Nigeria

Abstract

Legumes and oilseeds are cheap sources of protein and energy in the African diet. The cowpea is by far the most popular of all the legumes cultivated in West Africa. Apart from the traditional methods of preparing sumptuous dishes from them, technologies have been developed to produce prototype products that have similar nutritional quality to that of the traditional products. The pigeon pea can replace cowpea in most dishes and has been reported to possess some medicinal properties. The authenticity of these claims is being investigated. Groundnut, and the oil palm are cultivated widely and are processed commercially for food and export. Other oil seeds (water melon, African locust bean, African oil bean, castor oil and conophor nut) still are cultivated and processed by traditional family art which are time and energy intensive. Generally, oil seeds are utilized as fermented products and serve as condiments in foods or salad. Because they have a high economic potential, technologies for improving their cultivation and utilization should be investigated.

TABLE 1

Current Retail Prices of Selected Foodstuffs in Some Parts of Nigeria

Product	Quantity (Kg)	Average Prices (in Naira)	(US $)
Rice (imported)	50	285	60.6
Rice (local)	50	220	46.8
Beans (red)	50	300	63.8
Beans (white)	50	260	59.6
Garri (white) (fermented cassava)	50	180	38.3
Garri (colored)	50	190	40.4
Yam	50	190	40.4
Plantain	50	200	45.6

Legumes such as cowpeas, pigeon peas, groundnuts and melon seeds are important food crops in Africa. They long have been recognized as subsidiary crops to be relied on during the "hungry season" (1). In most rural areas they are planted interspersed with other economic crops such as yams, maize, cassava and sorghum. Prices are relatively high in these places when compared to urban centers. For instance in Nigeria, prices of these legumes have soared so high when compared to other food crops that they can no longer be regarded as "cheap" sources of protein (Table 1). Legume seeds are important primarily for their supply of protein in diets in many developing countries of Africa. They are important sources of energy for they contain both oil and carbohydrate. They contain reasonable levels of nicotinic acid, thiamin, calcium and iron.

However, legume grains contain a large variety of antimetabolites (2). Notable among them are phytate which can lead to hard-to-cook phenomenon and mineral unavailability. Others include tannins and saponins that lead to digestive and physiological disturbances, respectively, lectins that interfere with the protective barrier of the intestinal membranes and enzyme inhibitors that disrupt the digestive process and lead to other undesirable physiological reactions (3). Despite all these antimetabolic effects, legumes such as cowpeas, pigeon peas, groundnuts and melon seeds still remain formidable crops to be reckoned with in Africa. This paper reviews how these crops and some oilseeds are used in Africa.

Cowpeas *(Vigna unguiculata Walp)*

Cowpeas also are called blackeyed peas or southern peas. The cowpea is indigenous to Africa and is widespread in the tropics and subtropics. Cowpeas are grown in 16 African countries. The major producers are Nigeria, Uganda, Niger, Senegal, Upper Volta and Tanzania; Nigeria being the highest producer (850,000 tons in 1981). It is a staple crop for many developing countries. It is high in protein, energy and other essential nutrients. Due to the combined effort of several research groups, e.g. the International Institute of Tropical Agriculture (IITA) which has a mandate for cowpea research and the Bean/-Cowpea Collaborative Research Support Program of the United States Land Grant University System which provides grants for cowpea research and scientists from various countries of the world, cowpea production has increased tremendously within recent decades, and it is estimated that by the turn of the century about 2 million tons will be produced annually (4).

The crop adapts well to stressful conditions. On account of its ability to fix nitrogen efficiently (240kg N/ha), the cowpea provides a high proportion of its own nitrogen requirements as well as leaving a N deposit of 60–70kg/ha for the succeeding crop (4). It is planted interspersed with other crops. At present new varieties being sold in the market mature in 60–65 days and they are pest resistant and drought tolerant.

Storage of Cowpeas: The composition of cowpea seed is given in Table 2. It shows that cowpea seeds are high in protein and soluble carbohydrate but low in crude fibre and oil. The seeds contain a fair amount of vitamins. However, when the grains are harvested and stored they become infested by bruchids or weevils (*Callosobruchus maculata*) resulting in severe losses (5). In West Africa the price

TABLE 2

Average Chemical Composition of Cowpeas, Pigeon Peas and other Local Oil and Protein Seeds (g/100 g dry matter)

	Protein	Ether extract	Crude fiber	Ash	Nitrogen free extract
Cowpea	24.5	2.1	5.6	3.8	64.0
Pigeon pea	23.0	1.1	6.7	5.8	63.4
Groundnut	26.9	50.4	9.4	2.3	11.0
Palm kernel	21.4	8.6	8.9	6.1	55.0
Water melon	29.7	58.5	3.1	3.9	4.8
African locust bean	28.8	19.6	8.5	6.2	36.9
African oil bean	29.8	21.5	5.7	5.9	37.1

of beans doubles during the rainy season due to scarcity, and those available are usually infested with bruchids.

The effect of pest infestation of cowpea was investigated by feeding experiments on rats (6). The results showed that the infested cowpeas had poorer biological scores than the uninfested. It is advisable to remove the insects before cooking as the insect could introduce unpleasant tastes to the cooked beans.

Storage is done mostly in households by using enamel dishes with sealed lids, in airtight biscuit tins and clay pots or in earthen ware jars with sealed lids (sealed with cereal paste, beeswax or cowdung). Some households leave the grains in the pods and cover them with wood ash in granaries (7). In some households the grains are dried in the sun and inert material such as dried pepper, rice grains, etc. are mixed with the grain and stored in enamel or plastic storage vessels. This restricts the movement of the weevils through stored grains and limits ovipositing. Others have used lemon oil or palm oil to control the weevil (7, 8). For commercial purposes phostoxin is used as a preservative.

Prolonged storage of cowpeas under high humidity and high temperatures hardens the seed and impairs cookability because the affected seeds do not become tender during cooking even after prolonged cooking periods (9,10). However, when cowpeas were stored at 2°C, 65% relative humidity in high density polyethylene bags flushed with 100% CO_2 no major adverse qualities occurred (11). The benefits of the CO_2 packaging were reported to include low cost, simplicity, effectiveness, low energy requirements and adaptability for small or large operations. Such a packaging will be suitable to the developing countries where technology is at its infancy.

Utilization of cowpeas: In tropical Africa cowpeas are primarily cultivated as a pulse for human consumption. For this purpose, cowpea is prepared in a variety of dishes (7). In many areas of both West and East Africa the tender green leaves are cooked like spinach or as a relish. Cowpeas also are cultivated to a lesser extent as fodder for livestock, green manure and ground cover. The husks are mixed with maize and guinea corn and cut for silage or hay (12).

Processing of Cowpeas: There are many varieties of cowpeas, however, preference in West Africa is for brown, white or cream seeds with a small eye and wrinkled or rough seed coat (7). The seeds are utilized whole or processed into flour.

Using Whole Cowpea Seeds: The seeds are cleaned by placing them on a flat surface made of enamel, plastic, wood, tin or basket and the stones are picked out by hand. The seeds are then washed so that any stray stones can be removed. The washed beans are placed in enamel, aluminium or iron pots and cooked over an open fire, gas or electric cooker until tender. The cowpea varieties such as Adzuki, Ife brown and Prima become tender in 30–40 min. The tender beans are eaten with rice, fried plantain or yam and stew.

Cowpea also can be prepared into porridge. In the preparation of cowpea porridge the seeds are cooked until tender. Pepper, onions, shrimp, fish, salt and palm oil or groundnut oil are added and the mixture is simmered for 10 min. The ingredients are mixed into porridge and served alone or with fried plantain yam, rice or 'gari' (fermented and fried cassava).

Using Cowpea Flour and Paste: The traditional method of processing cowpea into flour is characterized by waste and drudgery (7,13,14,15). The laborious time consuming nature of traditional seed coat removal and grinding has been emphasized as one of the constraints on increased consumption of beans (7). Removal of the seed coat and hilium is necessary to obtain a light coloured flour (7,16). The seed coats contain pigments such as tannins and if not removed, these impart a bitter taste and leave black or brown specks in the flour and paste. Removal of the seed coat upgrades the quality of the grain, improves its appearance, quality, texture, palatability, cooking properties as well as digestion and absorption of nutrients after it is eaten (17).

Decortication of the seed coat normally is done by dry or wet methods (7,18,19). Traditionally, the dry method involves breaking up the seeds into smaller pieces, using a mortar and pestle or grinding stone. The split seeds are then placed in a calabash, tray or any flat surface and winnowed to remove the husks. At the cottage industry level removal of the husk commonly is done in small machines including both hand and power operated under-run disc shellers or blunt blade mills. In many cases hulling is accompanied by splitting. The husks are removed by aspiration while unhulled grains are separated easily from the split cotyledons by sieving (17). Recently, a collaborative research project being conducted by the University of Georgia and the University of Nigeria under the auspices of the Bean/Cowpea Collaborative Research Support Program (U.S. Agency for International Development) is developing technologies to improve the ease and efficiency of removing the cowpea seed coat (15). The main focus is to produce meal or flour that consumers can use simply by adding water (20).

The wet method of removing the seed coat involves soaking the seeds in water for about 30 mins. or longer depending on the variety of the bean. The seeds with hard shiny seed coats are soaked for longer periods. The seeds then are rubbed manually to loosen the seed coat. The seed coats float on the water and are drained off. The process of manual rubbing and addition of water is repeated until all the seed coats are removed. Combination of both wet and dry methods are used sometimes to facilitate the removal of seed coats.

Grinding Cowpeas to Paste or Flour: In a number of African countries whole legumes or decorticated split seeds are ground dry or wet into flour or paste. Traditionally, grinding is done in mortar or stone grinders with the addition of water to obtain pastes of varying consistencies. When larger quantities are handled mechanized mortars are available. At the cottage industry level, hammer mills are used. Abrasive, rather than attrition grinding of cowpeas has been emphasized because most African varieties of cowpeas have tightly adhering seed coats which are not readily released in the absence of water (19). Roller mills are also used but losses are high if the grains are not graded properly by size (17). The flour or paste thus produced is used in preparing a number of sweet and savoury dishes (7). Some of the popular dishes will be discussed.

Preparation of Akara (Fried Cowpea Paste): Akara is prepared from cowpea paste or hydrated flour by stirring or whipping the paste in a mortar or blender to incorporate air. Seasonings such as pepper (fresh or dried), salt, onions and at times tomatoes and crayfish are added. The paste is then scooped up with a spoon and fried in deep hot vegetable oil until brown. The fried product called by various names, akara, kosai, akla or accara in West Africa is consumed mainly as breakfast food. Apart from home preparation akara is sold along market places by street vendors.

Akara prepared from commercial Nigerian flour has not been well-received by consumers, because of its poor water absorption capacity; its products are heavy, lack crispness, sponginess and possess flavour different from akara prepared from fresh paste (7). A major difference found between traditional paste and commercial flour obtained in Nigeria was particle size distribution (14). The commercial flour was more finely milled than traditional paste with 48% of the flour particles riding 400 mesh screen compared to 16% at 400-mesh size for the traditional paste. The greatest concentration of paste particles (64%) was in the 50–100 mesh range whereas most of the flour particles (68%) were concentrated in the 200–400 mesh range. Furthermore, traditional paste made from soaked beans contains about 61% water and has a viscosity value after whipping of about 302 poise. Studies have shown that akara made from 1 mm screen flour hydrated to 60% moisture content before cooking were acceptable when compared to traditional akara (15). However, akara made from such hydrated flour still has a drier texture and mouth feel than traditional akara. McWatter (21) has attributed this difference to fat content of the akara. On a dry weight basis, traditional akara contains about 38% fat whereas akara made from hydrated meal (60% moisture content) contains 29% fat.

Furthermore, when cowpeas were conditioned for mechanical abrasive decortication by adjusting the moisture content to 25%, holding for 30 mins with occasional stirring and drying on a rotary air dryer to a moisture content of approximately 10%, the 50°, 70° and 90°C treatments produced akara that compared favourably with untreated cowpeas (16).

Preparation of Moimoi (A steamed cowpea paste): Hydrated cowpea flour or paste is sometimes seasoned with shrimp, pepper, salt, onions, tomatoes and vegetable oil. Small quantities are wrapped in leaves (*Thaumatococcus* sp. or *Sarcophrynium* spp.), tin foil or placed in tin cans covered with a lid and are steamed in an earthen, iron or aluminum pot lined with sticks or leaves to prevent direct contact with the heat and the paste. Traditional *moimoi* preparation has been described in detail (7,18).

Steamed cowpea paste known as *moimoi, ole-le alele* or *tabani* in various parts of West Africa is served as a breakfast food or served with rice at home for lunch/dinner and in ceremonial occasions. To preserve the shelf life of the product, canned *moimoi* with excellent organoleptic characteristics when compared to traditional *moimoi* has been produced (22). This product is not yet sold in the markets.

Novel food products from Cowpea: In recent times, attempts have been made to prepare ready-to-eat foods from cowpeas by various dehydration techniques (23). The great versatility of cowpeas as a base material for many food products such as '*moimoi*' and *akara,* soups, gravies and stews calls for cowpea processing into instant, or semi-instant forms that will be easier to process into the various cowpea based foods (24,25).

In order to expand cowpea use, to avoid storage losses and the laborious processes used for converting the whole grains to popular food items, novel foods are being produced from cowpea by extrusion cooking (26,27,28,29). These products are, however, still undergoing investigations.

Cowpeas are incorporated into infant weaning food (30,31,32) to increase the protein content of such foods. Cowpeas are used also in preparing cake-type doughnuts (33) and in replacement of milk protein in baking powder biscuits (34). Natto is traditionally made from soybeans (*Glycine max*) using either *Bacillus natto* or *Aspergillus oryzae*. Natto-like products have been made from cowpeas though the products need further investigation to fully characterize the organoleptic qualities (35).

The production of novel foods is a logical extension of the traditional methods of preparing these foods. Such products may be made more convenient, more stable and more nutritious than traditional dishes, thus expanding the use of cowpeas and increasing the total market for the raw commodity rather than competing with traditional uses (27).

Pigeon Peas (Cajanus cajan)

The pigeon pea is also known as Congo pea, red grain, noneye pea. It is widely grown in the tropics and subtropics. It has a protein content of 18–20% (Table 2). Pigeon pea is important nutritionally in the Caribbean but not in Africa where it virtually grows wild in some places (36).

The plant is drought resistant, tolerating drought areas with less than 65 cm annual rainfall. In Nigeria, it is found mostly in the middle belt region. It is an erect plant, a short-lived perennial shrub with compound leaves and yellow and red flowers. The plants produce seeds profusely and the crop matures early and pest damage is low. Depending on location and time of sowing, flowering occurs in 100–430 days. Threshing is done by trampling on the floors. The grains are cleaned, winnowed and sun dried. It is a good soil improver in addition to its fertilizer value as a legume because its prolific growth smothers weeds and checks soil erosion (12).

Utilization of Pigeon Peas

Pigeon peas, though less popular than cowpeas, can replace cowpeas weight for weight in all the cowpea dishes discussed by Dovlo (7). The pigeon pea seeds are smaller in size and so would result in higher milling losses when milled by machines. The seeds are nutritious and wholesome; the green seeds serve as vegetable and are excellent substitutes for English peas when used for human consumption (12). Ripe seeds are used as a source of flour in preparing *moimoi* and *akara* as discussed for cowpeas. Whole seeds take a minimum of 2 hours to cook unless potash ($K_2CO_3KHCO_3$) is added to facilitate softening. The pods and leaves of pigeon peas are used as excellent fodder for cattle (12). The pigeon pea meal can also be substituted into poultry diets (36).

Pigeon pea may not be as popular as cowpeas in human consumption but it has more traditional medicinal value than cowpeas. Some trado-medicalists claim that powdered leaves of pigeon peas expel bladder stones and salted leaf juice cures jaundice. The flowers are used to cure bronchitis while the seeds alleviate liver and kidney ailments.

Recently, in our laboratories, Ekeke et al. (37) have reported antisickling and membrane stabilizing properties of methanolic extract of the crushed beans *in vitro*. *In vivo* experiment with 40 confirmed sicklers (HBSS) using aqueous extract of cooked beans showed significant reduction in the percentage of sickle cells in venous blood and there was drastic drop in the frequency of sickle cell crisis.

It will be recalled that 20% of Africans are sickle cell disease sufferers. Sickle cell anaemia is an inherited disease that is caused by a single mutation in the β-subunits of oxygen carrying protein haemoglobin (Hb). This one hydrophilic Glutamate-6β to hydrophobic valine 6β alteration causes the deoxy form of Hb to polymerize and gelatinizes the erythrocyte, distorting its normal elliptical shape to a sickle shape. Upon oxygenation in the lungs, most of the sickled cells return to their normal shape. Those cells that remain in an abnormal shape regardless of the state of oxygenation of HbS are termed irreversibly sickled cells. Polymer formation is the primary factor underlying the manifestation of this disease (38).

It thus means that the edible pigeon peas possess some bioactive principle whose antisickling and other possible properties may offer great potential to sickle cell sufferers.

Oilseeds

The oilseeds commonly found in Africa are groundnuts or peanuts (*Arachis hypogea*) water melon (*Citrullus vulgaris*), African locust bean (*Parkia filicoides, Parkia biglobosa*), African oil bean (*Pentaclethra macrophylla*), castor oil bean (*Ricinus communis*) and the oil palm (*Elaeis guineensis*).

Groundnut

Groundnut is known also as peanut, earth nut, monkey nut, and in various Nigerian languages *Epa (Yoruba), Gya' da (Hausa), Okpa (Ibo)* and *Apapa (Kalabari)*.

Groundnut is a crop plant of South American origin and possibly was brought to West Africa by the Portuguese (39). Until recently, peanut has been a major crop for export in Nigeria. Though the quantity has dropped sharply, efforts are being made to recrop large areas in Nigeria (40).

Investigation on the nature and biological value of the plant has received a great deal of attention (12,18,39). Groundnut contains 27% protein (Table 2) with a digestibility coefficient of about 90% (12). The carbohydrate content is low (10%). It is an excellent source of vitamin B.

Utilization: Traditionally, the whole seeds are roasted, cooked or steamed and eaten as snacks. Sometimes the seeds are ground and used in preparing soups and stews. The oil is expressed out of the seeds manually after grinding them in a mortar or on a stone. The cakes so produced are fried and eaten and the oil is used in preparing other food items such as rice, yam, plantain, stews and soup. At the cottage industry level, the groundnut is crushed for its oil and residual cake using several industrial processing methods, i.e., hydraulic pressing, continuous horizontal screw pressing and prepress solvent extraction or a combination of these methods. The residual cake is richer than whole kernel and forms valuable livestock cakes for commerce. Groundnut flour is used as food in soup and stew and in enriching infant foods.

A variety of new products have been developed from groundnuts. Such products include snack foods (cake type doughnuts and cookies) (33,41), extenders for ground meal patties (McWatters 1977) and as substrates in preparing natto, a fermented product (35).

Water Melon (Citrullus vulgaris)

The water melon is a widely cultivated plant in Nigeria. The seeds are early-maturing, short season plants. In developed countries the water melon is principally cultivated as fruit while in Nigeria, the fruit is cultivated mostly for its seeds (12,18). The seeds contain 29.7% protein (Table 2) and provide an important source of protein to the diet of many people in southern Nigeria.

The seeds are shelled by hand, ground and used as condiment in soups and stews. The oil is expressed after grinding the seeds and used in preparing soups and stews and the cake is fried and eaten as a snack. In the southern parts of Nigeria, the melon seeds are fermented into a condiment known as '*ogiri*.' The production of *ogiri* is still a traditional family art done in homes in a crude manner. The quality is variable and shelf life is short. The fermentation process has been described by Abbey *et al.* (18) and Odunfa (44).

In some African countries such as Cameroon, the seeds are shelled and ground to paste and steamed after adding pepper, salt, onion and fish, wrapped in plantain or banana leaves or foil. This pudding can be served with boiled plantain, yam or eaten with fermented cassava (Miondo).

African Oil Bean Seed (Pentaclethra macrophylla)

The oil bean belongs to the legume family. The tree is either cultivated or wild. The seeds are oval but flat in shape, dark in color and have tough seed coats (43). The flat seeds are roasted, decorticated and eaten. It is popularly consumed as a fermented product. The fermented product is eaten alone as hors d'oeuvres and added as condiment in soup and stews to improve the nutritional quality. The method of preparation has been described by Abbey et al. (18), Achinewhu (43) and Odunfa (44). The fermented product is used in preparing salads (*Ugba*) or mixed with other food ingredients.

Studies have shown that fermentation increased all the vitamins, particularly riboflavin and biotin of the seed (44). The increase in riboflavin was attributed to the fermenting organism, *Bacillus subtilis*, which is unique in producing high levels of riboflavin synthetase (45). This product has been exploited to produce riboflavin commercially. Riboflavin has been reported to be the most limiting nutrient in the diet of many West Africans (46). The quantity of *ugba* normally used as soup condiments is enough to meet the recommended daily allowance for riboflavin (44).

African Locust Bean (Parkia filicoidea weku)

African locust bean tree is distributed fairly widely in Africa particularly in the Sahel and Savanna regions (12,47). They are perennial plants with large pods. The pod contains a yellow dry powdery pulp inside of which are embedded a number of dark brown or black seeds (12). The yellow pulp is rich in sweet carbohydrate and can be mixed with cereal, meat, stew or soup. The seeds are hard when raw but can be fermented into a palatable product called '*dawa dawa*' or '*iru*'. This product is rich in fat, protein and lysine (47,48,49). The production procedure for *iru* or *dawa dawa* has been described (18,50). The *iru* or *dawa dawa* is used normally as a soup condiment and serves as a low cost meat substitute in poor families (50).

Castor Bean (Ricinus communis)

Castor seed contains 44–46% oil and 20% protein. Castor seed is used in producing fermented products. The traditional method of preparing *ogiri* has been described (44). Briefly, the castor seeds are decorticated and wrapped in small packets with banana leaves with holes to allow penetration of water. The packets are placed in a pot and boiled. To facilitate cooking, '*kaun*' (containing K_2CO_3 and $KHCO_3$ is added. The packets are removed from the boiling water after 6 hours and kept in a warm place at about 32°C to ferment for 4–5 days. The fermented seeds are removed from the leaves and ground in a mortar or grinding stone to a fine paste called '*ogiri*'. Salt (about 5%) is added as preservative. The *ogiri* is used mostly as a condiment for soups, stew etc.

The Oil Palm (Elaeis guineensis Jacq)

The oil palm is grown extensively in West Africa as a food or cash crop for domestic and export trade. Nigeria was one of the most important sources of world supplies of palm kernels and palm kernel oil accounting for 80% of Nigeria's foreign exchange earnings up until 1960. However, presently, palm products account for less than 0.1% of Nigeria's foreign exchange earnings. Most of the oil palm groves in West Africa are the wild forms, although large and small estates exist owned by government and the private sector and financed by foreign agents in some cases. In Nigeria, the wild palm groves cover 2–4 million hectares. Organized private and public small holdings and private/public nucleus estates have cultivated 96,783 and 71,841 hectares respectively (51).

Utilization of the Oil Palm: The major products obtained from the oil palm are palm oil, palm kernel and palm kernel oil. These products can be obtained by traditional methods of processing which vary from household to household. Mechanized processes are used for commercial purposes. Flow charts of traditional and industrial methods of processing commonly used in Nigeria are shown in Figs. 1 and 2. These processing methods result in the production of palm oil of a special grade since they contain free fatty acid of less than 5% and less than 0.5% moisture. The impurities present in oils processed as described fail to meet the 0.01–0.05% standard required by many industries utilizing palm oil in food formulations (52).

In addition to the major primary products, there are other important byproducts of the palm oil industry (Table 3).

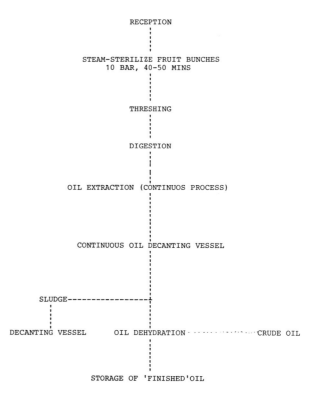

Fig. 1. Flow chart for mechanically processed palm oil (Denenu and Nze, 1983).

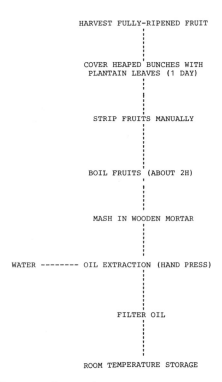

Fig. 2. Flow chart for traditional process of palm oil refining (Denenu and Nze, 1983).

TABLE 3

Traditional Uses of Various Parts of the Oil Palm Tree (Akpokoje, 1988)

Part or product	Uses
Palm sap of the male influorescence	Palm wine
Leaflets	Thatch for roofing
Leaf rachises	For fencing, reinforcing buildings and basket making
Mid-ribs of leaflets	Broom making
Bunch refuse	Rich source of potassium and locally used for soap making
Fiber residue after extraction of oil from fruit and oil palm shell from cracked palm nut (oil palm)	Used as fuel and as aggregates for concrete
Palm trunk	Sawn into timber and used for roofing houses, constructing fences and reinforcing building

Palm oil is an important vegetable oil in the diets of many West Africans. It is obtained from the mesocarp of the fruit and consists of mixture of palmitic acid, which constitutes the yellowish/orange solid fraction and unsaturated fatty acids (oleic and linoleic) that constitute the reddish liquid fraction. The separation and further processing of these fractions results in various grades of oils that meet either specific industrial requirements or a wide range of edible end use products (51).

Palm kernel oil is obtained by traditional methods (18) or mechanized extraction to give palm kernel meal, cake and pellet. The oil has about 47% of lauric acid while palm oil has only 0.4%. The high content of saturated fatty acids gives this oil its good creaming properties, chemical stability and bland flavor. The various ways in which the oil palm is used both traditionally and industrially in Africa are summarized in Figure 3.

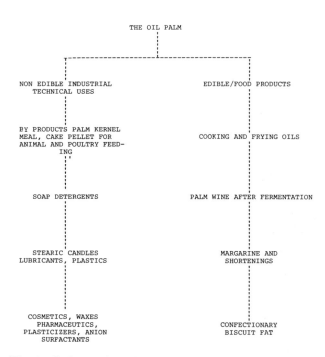

Fig. 3. End uses of the oil palm.

There are other lesser known oilseeds in West Africa that are gaining prominence. Among them are *Treculia africana* 'ukwa' and conophor (*Teracarpidium conophorum*) nut. It is not cultivated widely. The plant is very prolific and the technology for shelling the nut has been developed. When conophor flour was incorporated in wheat composite flour for baking, bread loaves of good rheological properties were produced (40).

References

1. Aykroyd, W.K., and J. Doughty, "Legumes in Human Nutrition," Food and Agriculture Organization of the United Nations, Rome, Italy, 1964.
2. Liener, I.E., "Toxic Constituents of Plant Foodstuffs," Academic Press, New York and London, 1969.
3. Boulter, D., *Proc. Nutr. Soc. 41*:1 (1982).
4. Rachie, K.O., in "Cowpea Research, Production and Utilization," edited by S.R. Singh and K.O. Rachie, John Wiley and Sons, New York, 1985.

5. Singh, S.R., and L.E.N. Jackai, in "Cowpea Research, Production and Utilization," edited by S.R. Singh and K.O. Rachie, John Wiley and Sons, New York, 1985.
6. Ecokakpan, O.U., O.U. Eka and E.J. Ifon, *Nigerian J. Nutritional Sci. 7:*(2)67 (1986).
7. Dovlo, F.E., C.E. Williams and L. Zoaka, "Cowpeas: Home Preparation and Use in West Africa," IDRC publication 055e, International Development Center, Ottawa, Canada, 1984.
8. Su, H.C.F., R.D. Speirs, and P.G. Mahavy, *J. of Econ. Entomol. 65:*(5)1433 (1972).
9. Sefa-Dedeh, S.D., D.W. Stanly and P.W. Voisey, *J. Food Sci. 44:*790 (1979).
10. Swanson, B.G., J.S. Hughes, and H.P. Rasmussen, *Food Microstructure 4:*115 (1985).
11. McWatters, K.H., M.S. Chinnan, R.E. Worthington and L.R. Beuchat, *J. Food Process. Preserv. 11:*63 (1987).
12. Oyenuga, V.A., editor, "Nigeria's Food and Feeding Stuffs," Ibadan University Press, 1968.
13. Williams, C.E., *Niger. J. Nutritional Sci. 5:*(2)129 (1984).
14. McWatters, K.H., *Cereal Chem 60:*(5)333 (1983).
15. McWatters, K.H., and M.S. Chinnan, *J. Food Sci. 50:*(2) 444 (1985).
16. McWatters, K.H., M.S. Chinnan, Y.S. Hung and A.L. Branch, *Cereal Chem. 65:*(1)23 (1988).
17. Anonymous, "Post-harvest Food Losses in Developing Countries," Board on Science and Technology for International Development Commission on International Relations, National Research Council, National Academy of Sciences, Washington, D.C., 1978.
18. Abbey, B.W., R.D. Phillips and K.H. McWatters, "Proceedings of AOCS Short Course on Food Uses of Whole Oil and Protein Seeds," (in press).
19. Phillips, R.D., *J. Am. Oil Chem. Soc. 59:*351 (1982).
20. McWatters, K.H., in "Cowpea Research, Production and Utilization," edited by S.R. Singh and K.O. Rachie, John Wiley and Sons, New York, 1985, pp. 361–366.
21. McWatters, K.H., in "Plant Proteins: Applications, Biological Effects, and Chemistry," edited by R.L. Ory, American Chemical Society, Washington, D.C., 1986, pp. 8–18.
22. Adeniji, A.E., and N.N. Potter, *J. Food Sci. 45:*1359 (1980).
23. Onayemi, O., and N.N. Potter, *J. Food Sci. 41:*48 (1976).
24. Ukhun, M.E., *Food Chem. 14:*35 (1984).
25. Akpapunam, M.A., and P. Markakis, *J. Food Sci. 46:*972 (1981).
26. Phillips, R.D., M.S. Chinnan and M.B. Kennedy, *J. Food Sci. 49:*(3)916 (1984).
27. Phillips, R.D., M.B. Kennedy, E.A. Baker, M.S. Chinnan, V.N.M. Rao, in "Cowpea Research, Production and Utilization," edited by S.R. Singh and K.O. Rachie, John Wiley and Sons, New York, 1985, pp. 367–373.
28. Kennedy, M.B., R.D. Phillips, V.N.M. Rao and M.S. Chinnan, *J. Food Process Eng. 8:*193 (1986).
29. Phillips, R.D., and E.A. Baker, *J. Food Sci. 52:*(3)696 (1987).
30. Oyeleke, O. A., I.D. Morton and A.E. Bender, *Br. J. Nutr. 54:*343 (1985).
31. Abbey, B.W., and U.B. Nkanga, *Nutr. Rep. Int.* (in press).
32. Abbey, B.W., and T. Mark-Balm, *Nutr. Rep. Int.* (in press).
33. McWatters, K.H., *Peanut Sci. 9:*46 (1982).
34. McWatters, K.H., *Cereal Chem. 57:*(3)223 (1980).
35. Beuchat, L.R., T. Nakayama, R.D. Phillips and R.E. Worthington, *J. Ferment. Technol. 64:*(4)319 (1985).
36. Nwokolo, E., *Niger. J. Nutritional Sci. 7:*(2)75 (1986).
37. Ekeke, G.I., F.O. Shode and O. Oguranti, at International Symposium on Sickle Cell, 1985.
38. Abraham, D.J., D.M. Gazze, P.E. Kennedy and M. Mokotof, *J. Med. Chem. 27:*1549.
39. Anonymous, in "West African Crops," edited by F.R. Irvine, Oxford University Press, London, 1969.
40. Ogusua, A.O., *Niger. Food J. 4:*(1)45 (1986).
41. McWatters, K.H., *J. Food Sci. 42:*(6)1492.
42. Odunfa, S.A., *Die Nahrung 25:*(9)811 (1981).
43. Achinewhu, S.C., *Food Chem. 19:*105 (1986).
44. Odunfa, S.A., *Ibid. 19:*129 (1986).
45. Bacher, A., R. Baur, V. Eggers, H. Harders, N.K. Otto and H. Schapple, *J. Biol. Chem. 255:*632 (1980).
46. Whitby, P., "A Review of Information Concerning Food Consumption in Ghana," Food Research and Development Unit and Food Research Institute, Accra, Ghana, 1968.
47. Ikenebomeh, M.J., R. Kok, and J.M. Ingram, *J. Sci. Food Agric. 37:*27.
48. Schery, R.W., "Plants for Man, 2nd ed." Prentice-Hall, Englewood Cliffs, New Jersey, 1972, p. 461.
49. Platt, B.S., *Food Technol. 18:*68 (1964).
50. Odunfa, S.A., and E.Y. Adewuyi, *Chem. Microbiol. Technol. Lebensm. 9:*6 (1985).
51. Akpokodje, E.G., *Raw Materials 1:*(1)18 (1988).
52. Denenu, E.O., and J.N. Eze, *Niger. Food J. 1:*(1)123 (1983).

Development of Processes and Uses of Lupins for Food

José Miguel Aguilera

Department of Chemical Engineering, Universidad Católica de Chile, P.O. Box 6177, Santiago, Chile

Abstract

Lupins as a protein-rich legume have been used for human food in some Mediterranean and South American countries. In the past decade lupin has been utilized as a refined vegetable protein source and several processes have been proposed to adequately fractionate its components. This presentation will update information regarding conventional uses of sweet and bitter lupins, as well as extraction methods for protein, oil and alkaloids. Functional and nutritional characteristics of lupine proteins also will be reviewed.

Lupins, in particular, *L. albus* in the Mediterranean region of the Old World and *L. mutabilis* in the Central Andean region of South America, have been important crops on poor acid soils for many centuries. *L. luteus* and *L. angustifolius* are now also finding a role as green manure, soil improvers or as feed grain in coastal areas of Northwest Europe, New Zealand, Australia and South Africa. Their development has taken place in spite of the presence of toxic concentrations of bitter alkaloids in the grain. However, in the past 20 years plant breeders around the world have given systematic attention to the development of "sweet" lupin varieties for food and feed uses.

Utilization of lupins has expanded in the EEC countries as substitute for imports of soybeans, in Western Australia as an export crop and, to a limited extent, in Latin America for direct human consumption. Table 1 presents the regional distribution and average yields of different lupin species around the world. The total area of lupin cultivation probably exceeds 1.5 million hectars now and it is divided in a 60:40 ratio between grain production and green forage. The most important producing countries and species under cultivation are Australia (*L. angustifolius*), the U.S.S.R. (*L. luteus*), and Poland (*L. luteus*). The development of the narrow-leafed lupin (*L. angustifolius*) in Australia is a success case among the recent crop introductions in the world and represents a heartening example of the potential of lupins as a high-protein crop. Since its start 20 years ago, Australia's lupine production has grown to account now for over 60% of the world production of lupins, representing approximately 14% of the total worldwide exports of pulses (1).

The increasing importance of lupins as a promising new crop prompted the establishment of the International Lupin Association (ILA). Five International Conferences have been held in different parts of the world since April 1980 and proceedings from these meetings are available.

Chemical and Structural Aspects of Lupins

Seeds of lupin varieties of commercial interest have shapes that range from almost round and flat to nearly spherical. Low-alkaloid varieties give an overall color impression of white to cream; some bitter varieties are mottled with dark gray to dark brown background; cotyledons of mature seeds are yellow.

Lupins have differences in composition peculiar to each species and variety (Table 2). High-protein lupins such as *L. mutabilis* and *L. luteus* have protein contents above 40%, higher than all commercial grain legumes and similar to soybeans. *L. mutabilis* combines this desirable characteristic with a respectable oil content, in most cases similar to or even higher than soybeans. *L. luteus* on the other hand has the lowest reported oil content of all grown species. *L. angustifolius* exhibits a reduced oil content (5.5 to 6.7%) and about 30% protein. Intermediate values

TABLE 1
Regional Distribution and Average Yield of Lupin Production in 1985

Region	Area (1000/ha)	Average yield (MT/ha)
USSR	280	1.43
Poland	115	0.98
Other European countries	39	0.87
South Africa	20[a]	0.35
Other African	7	1.12
Latin America	12[b]	1.64
Australia	606[c]	0.97
World total	1,073	10.8

[a] In the early 1980's was about 200,000 ha.
[b] Chile accounts for ca. 8,000 ha.
[c] Estimates for 1988 are ca. 0.8 million ha.

TABLE 2
Average Chemical Composition (% dry weight) of Cultivated Lupin Species

Species	Protein	Fat	Fiber	Ash
L. albus	36.7	11.5	9.8	3.4
L. angustifolius	31.1	6.0	14.7	3.5
L. luteus	41.8	5.4	15.8	4.1
L. mutabilis	42.6	18.7	7.3	3.7

for protein and oil characterize *L. albus* varieties with an average protein content of 37% and about 11% oil (2).

A significant proportion of the seed weight in lupins is hull. It ranges from about 12% in *L. mutabilis* up to 20–25% in some commercial varieties of *L. albus*. Because the hull content is high, the protein and oil content of dehulled seeds is enhanced significantly, so a full-fat flour from *L. albus* may contain over 45% protein. However, the yield of the oil plus protein fraction from whole seeds is much lower than in soybeans, which have only 8–9% hulls. Cell walls in lupins are also thicker than in soy (Fig. 1) and represent between 18–23% of the weight of the seed in *L. albus* and *L. luteus* but only 7–11% in *L. mutabilis* (3).

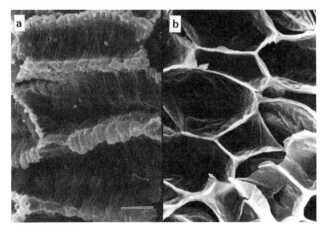

Fig. 1. Scanning electron photomicrograph of cells from lupin (a) and soybean (b). Marker: 10 μm.

Compared to soy, lupin proteins have been studied in any detail only during the past decade. Lupin proteins can be classified broadly as globulins (about 87% of the protein fraction) and albumins (approximately 5–13%), with their relative proportion varying depending on the species (Table 3). Globulins correspond to the storage proteins, are extracted at high ionic strength and can be grouped in three categories: vicilin-like proteins (44%), which correspond to globulins 4, 5, 6 and 7; legumin-like proteins (33%) and other globulins (23%). Most legumins and vicilins are glycoproteins. Albumins are extracted in water at 4°C and natural pH (5.0–5.5) and contain most of the enzymes present in the seeds (4).

The different protein fractions have distinct amino acid patterns which influence directly the nutritional quality. Although lupins are limited in S-containing amino acids, some sulphur-rich protein fractions, such as conglutin-δ, have been found in *L. angustifolius* (5). Vicilins are poor nutritionally, lacking both S-containing amino acids and tryptophan, but legumins are better balanced showing only minor deficiencies in lysine, S-containing amino acids and valine. Albumins display a well-balanced amino acid pattern with a high content of lysine and methionine which improves the overall profile (6).

Other components in lupin also influence technological processes and applications. Among the most important are the oil content of the seed, presence and level of alkaloids, types and levels of antinutritional factors, if any, and the amount of flatus-inducing carbohydrates.

The oil content of lupin seeds varies widely depending on the species from a low value of less than 5% in *L. luteus* to a maximum of around 25% in *L. mutabilis* (Table 4). A distinctive fatty acid pattern also exists for each species with a common high unsaturated fatty acid fraction. *L. mutabilis* and *L. albus* are high in oleic acid (over 50%) and linoleic acid is predominant in *L. luteus* and *L. angustifolius*. Significant amounts of erucic acid have been found only in some varieties of *L. albus*, *L. luteus* and *L. angustifolius* (7).

TABLE 4
Average Content of Some Fatty Acids in Cultivated Lupin Species

Species	16:0	18:0	18:1	18:2	18:3
L. albus	8.3	2.5	55.0	17.7	9.1
L. angustifolius	10.2	5.5	32.9	43.5	5.0
L. luteus	6.3	2.4	25.4	49.4	8.3
L. mutabilis	10.7	7.3	50.9	28.1	2.4
Soybeans	10.5	3.2	22.3	54.5	8.3

Alkaloids present in all *Lupinus* species are quinolizidine derivatives of varying complexity which constitute a resistance factor against herbivores and microorganisms. They occur as tertiary bases or N-oxides in concentrations in the seeds of up to 4%, and are water soluble. The most important alkaloids are lupanine, lupinine, angustifoline, 13 hydroxilupanine and sparteine. In toxic doses, lupin alkaloids induce a range of effects including respiratory depression and failure, a general hypotensive action and inhibition of muscular transmission, and fibrillation in mammals (8). Lethal dose estimates for children varied from 11–25 mg/kg body weight, although adults have survived poisoning at levels of 30–40 mg/kg body weight (9).

The alkaloid content is governed by polygene control (at least three genes). Stresses affect the alkaloid content in the seed, in particular, dry seasons increasing it and

TABLE 3
Types of Proteins in *Lupinus albus*

Albumins (13%)			MW 6–17 kD	
Globulins (87%)		Sub units	% total globulins	MW kD
		Conglutin		
		1 (γ)	6.0	199
	Vicilins	4 (β)	10.0	270
		6 (β)	30.2	335
	Legumins	8 (α)	21.2	240–470
		9a (α)	12.0	240–470
		9b (δ)	12.5	44

reducing yield. Breeding programs have been successful in removing most alkaloids from commercial varieties of *L. albus*, *L. angustifolius* and *L. luteus*. Of particular importance is a new experimental variety of *L. mutabilis* with an alkaloid content of 0.0075%, a high protein content (52% d.m.) and interesting oil content (15%)(10). A lupin seed is considered to be sweet if the alkaloid content is lower than 0.05%.

Low molecular weight oligosaccharides of the α-galactoside family such as raffinose, stachyose and verbascose are known to cause flatulence activity in monogastric animals. They are present in mature lupin seeds, almost exclusively in the cotyledons, in levels ranging from 8–14%. *L. mutabilis* appears to have a remarkably high α-galactoside content while *L. consentinii* has only about 7% or half the quantity of *L. mutabilis*. This wide difference in α-galactoside content suggests that it should be possible to reduce its content in the seed genetically. Flatulence-producing compounds can reduce drastically by leaching the seed with water and by germination. In this latter case, it has been shown that after a one day germination of *L. albus* var Multolupa, the α-galactoside content decreased from 13.7 to 3.4% (11).

Several reports continue to demonstrate that most commercial lupin varieties are relatively free of haemagglutinins, trypsin inhibitors and saponins, at least at levels that are toxic to humans. However, there is increasing evidence that *L. albus* concentrates manganese up to unphysiological levels (12).

Nutritional and Functional Aspects of Proteins

Published work and actual applications continue to confirm that once the alkaloids are removed from the seed (either by breeding or processing) the protein in lupins is highly digestible and when supplemented with methionine, similar to that in soybeans.

The protein efficiency ratio (PER) of *L. albus* and *L. angustifolius* were 24% and 29% of the value of casein, quite low compared to that of soy (48%). However, lupins were more responsive than soy to supplementation with 2% methionine, increasing their PER to 73 and 88%, respectively, and supplemented soy had a PER of 65% (13). Table 5 shows data comparing the biological value of different lupin species and soy (14). Cooking and leaching significantly improved the biological value of soy and *L. mutabilis* while addition of 0.5% methionine brought all values to 80% or higher, except for *L. consentinii*.

Functional properties of proteins connote those physicochemical properties which affect the behavior of proteins in food systems during preparation, processing

TABLE 5
Biological Value of Lupin Species and Soy

	Raw	Cooked leached (C & L)	C & L 0.5% met
Soybeans	—	73.3	85.6
L. albus	70.0	61.2	79.7
L. angustifolius	53.5	52.0	84.7
L. consentinii	43.4	57.9	58.5
L. mutabilis	—	79.7	87.9

storage and consumption (15). Only scarce information is available on the functional properties of lupin proteins and their relation with the basic protein structure. Emulsifying and water binding capacity of lupin concentrates and isolates vary depending on the preparation procedure and between lupin varieties, possibly because of different proportions of the globulin fractions (16). Lupin protein isolates prepared at laboratory level from *L. albus* var. Multolupa showed higher water solubility than soy isolates and similar emulsification capacity while swelling and gelation were found to be inferior (17). These results coincide with others reporting excellent emulsifying and solubility properties of isolates prepared from *L. albus* (18). Concentrates prepared from *L. mutabilis* had good water and oil absorption characteristics and gelation properties. Solubility increased rapidly in the pH range 4 to 6, a fact that seems to be peculiar to lupin products and should be exploited since this is an important pH range of foods. Foaming and emulsion capacity were both concentration and pH dependent (19).

Debittering and Other Processes

It has been pointed out that alkaloids in their salt form are water soluble. Traditionally, bitter lupins have been debittered in the highlands of South America and in Mediterranean countries by soaking, cooking and leaching out of the alkaloids with running water. The debittered grain is then consumed either wet as a snack or as a flour after sun drying. High losses in dry matter, particularly protein, make this method unattractive for industrial applications.

Technologies used for the production of soy isolates and concentrates based on solubilization or immobilization of the protein as a function of pH, respectively, have been amply applied to lupins (17,20,21,22). Resulting products have protein contents over 90% and 55%, respectively, and low alkaloid contents. The main problem is economical since the starting raw materials are low-heat defatted flakes which are by-products of the oil extraction operation. In the case of lupins it has been demonstrated that unless the oil content in the seed is at least 18%, solvent extraction of oil is uneconomical. This leaves out all species except *L. mutabilis*.

Simultaneous aqueous extraction of oil and protein was evaluated due to its simplicity, use of nonorganic solvents and ease of adaptation to existing milk processing plants. Basically, the process consists of dispersing finely comminuted seed in water followed by centrifugal separation of the dispersion into oil, solid (residue) and aqueous phases. Although the process has been proven feasible for soybeans after many years of pilot plant development work, straight adaptation to lupins yielded satisfactory results only in the case of bitter lupins, where about 55% percent of the extracted protein was recovered as a concentrate practically free of alkaloids (23). Further research is needed on extraction techniques and demulsification procedures to improve recovery of oil and protein. Since over 80% of the protein ends up in the extract, membrane processing may be used simultaneously to concentrate the protein, a technique that has proved economical in the case of soy (24).

Alcohols, in particular ethanol and isopropanol, and alcohol/water mixtures present several interesting characteristics as solvents for whole lupin meal extraction. Solubility of oil increases as water content decreases (becomes less polar) and temperature increases. Conversely, the ability to extract phosphatides, pigments, sugars and alkaloids, decreases. So, selective control of the extraction process can be achieved. Several researchers have used pure alcohols (methanol, ethanol, isopropanol), alcohol/water mixtures with or without addition of acid or base, and in combination with hexane (25,26,27). Lupin concentrates containing over 70% protein and 0.1–0.2% alkaloids were obtained at laboratory level by staged extractions of hexane-defatted flakes of *L. mutabilis* using aqueous methanol and ethanol. It is questionable whether such processes would have a positive economic evaluation unless the alkaloids obtained from the debittering process are proved valuable as potential natural pesticides (28).

Very interesting data were generated during the pilot plant removal of oil with boiling azeotropic aqueous isopropyl alcohol. A high protein defatted flour containing 64% protein, 1.3% residual oil and low alkaloids was produced from *L. mutabilis*. The oil fraction contained 96.3% of the original oil in the flakes and after alakali refining a yield of 92.4% was obtained (29).

Uses of Lupins as Foods

Consumption of high-alkaloid *L. mutabilis* or "*tarwi*" by the ancient Incas in pre-columbian Peru after removal of the bitterness and elimination of the toxic alkaloids is well-reported. This was achieved by a short boiling process which coagulated the protein, followed by leaching of it in running water or diluted lime and sun drying. A small debittering plant with a capacity of 250 kg/day has been setup in Peru with assistance from the GTZ (Deutsche Gesellschaft für Technische Zusammenarbeit) (30). The Institute for Nutrition in Peru has been promoting the use of debittered lupin products for many years.

In the Mediterranean region, and particularly in Spain, lupins have been consumed cooked and wet as snacks. A more recent use of lupin flour has been as an adjunct in cereal products. Addition of lupin flour at levels below 10% appears to increase water absorption, loaf stability, and freshness without a reduction in loaf volume (31). Marked increases in PER of enriched bread have also been reported. Higher levels of incorporation (20–25%) can be achieved in biscuits, cakes and pasta products. In Australia, addition of up to 10% lupin seed flour to bread and cereal products is permitted, provided the alkaloid content does not exceed 0.02% (32). Recently, the allowance has been extended for unlimited food consumption (33).

Australian sweet lupin, *L. angustifolius*, has been evaluated in Korea as substitute for soybeans. The large proportion of seed coat in this variety and the thick cell walls appeared to hinder microbial growth during fermentation but use in some typical products such as *Meju* and lactic fermented beverages were found promising (34).

In Chile products from sweet lupins (*L. albus* var Multolupa) have also been cleared for use as food supplements in milk substitutes for nutrition intervention programs. An industrial plant has been producing up to 50% of the protein requirements for milk substitutes. A typical formula includes 68–73% wheat flour, 12% *L. albus* meal, 12% skim milk and 5–6% soybean oil. A more recent commercial use of lupin grain has been as a coffee substitute after roasting. Other uses are as extrusion-precooked lupin flours, and as additives in sausages and jams.

Lupins definitively represent a viable alternative crop to soy in poor soils of temperate zones. The main drawbacks for full technological utilization of lupins as food are clearly identified and refer to economical (e.g.: high hull and low oil content), nutritional (e.g.: presence of toxic alkaloids, deficiency of S-containing amino acids) and some functionality aspects. Positive aspects relate to their nitrogen fixing ability, wide genotype variation that may allow improvements by breeding, the high protein content and responsiveness to amino acid fortification.

References

1. William, W., *Proceedings 4th International Lupin Conference,* Geraldton, Western Australia, p. 1, 1986.
2. Aguilera, J.M. and A. Trier, *Food Technol.* 32(8):70 (1978).
3. Brillouet, J.-M. and D. Riochet, *J. Sci. Food Agric.* 34:861 (1983).
4. Cerletti, P., "Lupine seed proteins," in *Developments in Food Proteins- 2*, edited by B.J.F. Hudson, Applied Science Publishers, London, 1982, pp. 133–171.
5. Lilley, G.G., *J. Sci. Food Agric.* 37:20 (1986).
6. Hill, G.D., *Nutrition Abstracts and Reviews,* B47:511 (1977).
7. Cerletti, P. and M. Duranti, *J. Am. Oil Chem. Soc.* 54:460 (1979).
8. Culvenor, C.C.J. and D.S. Petterson, *Proceedings 4th International Lupin Conference,* Geraldton, Western Australia, p. 188, 1986.
9. Schmidlin-Meszaros, J. *Mitteilungen aus dem gebiete dei Lebensmitteluntersuchung und Hygiene* 64:194 (1973).
10. von Baer, E., R. Ibañez, and D. von Baer. *Proceedings 4th International Lupin Conference,* Geraldton, Western Australia, p. 283, 1986.
11. Trugo, L.C., D.C.F. de Almeida, and R. Gross, *Lupin Newsletter* No. 11, p. 20, February 1988.
12. Gross, R., J. Auslitz, P. Schramel, and H.D. Payer, *Proceedings of the 3rd International Lupin Conference,* La Rochelle, France, p. 386, 1984.
13. Hove, E.L., S. King, and G.D. Hill, *N.Z.J. Agric. Res.* 21:457 (1978).
14. Savage, G.P., J.M. Young, and G.D. Hill, *Proceedings of the 3rd International Lupin Conference,* La Rochelle, France, p. 629, 1984.
15. Kinsella, J.E. and D. Srinivasan, in *Criteria of Food Acceptance,* Forster Verlag AG, Switzerland, 1981.
16. Manrique, J. and M.A. Thomas, *J. Food Technol.* 11:409 (1976).
17. King, J., C. Aguirre, and S. de Pablo, *J. Food Sci.* 50:82 (1985).
18. Malgarini, G. and B.J.F. Hudson, *Riv. Ital. Sost. Grasse* 57:378 (1980).
19. Sathe, S.K., S.S. Deshpande, and D.K. Salunkhe, *J. Food Sci.* 47:491 (1982).
20. Pompei, C. and M. Lucisano, *Lebensm. Wiss. Technol.* 9:338 (1976).
21. Ruiz, L.P. and E.L. Hove, *J. Sci. Fd. Agric.* 27:667 (1976).
22. Bouthelier, V., J. Cabanyes, and M. Muzquiz, *Qualitas Plantarum* 33(2), 145 (1983).
23. Aguilera, J.M., M.F. Gerngross, and E.W. Lusas, *J. Food Technol.* 18:327 (1983).
24. Lawhon, J.T., K.C. Rhee, and E.W. Lusas, *J. Am. Oil Chem. Soc.* 58:377 (1981).

25. Hatzold, T., J. Gonzales, M. Bocanegra, R. Gross, and I. Elmafda, in *Agricultural and Nutritional Aspects of Lupines*, edited by R. Gross and E.S. Bunting. GTZ, Germany, 1980, p. 333.
26. Blaicher, F.M., R. Nolte, and K.D. Mukherjee, *J. Am. Oil Chem. Soc. 58*:761 (1981).
27. Pompei, C., *Proceedings of the 3rd International Lupin Conference*, La Rochelle, France, p. 398, 1984.
28. Johnson, L.A., *Alcoholic extraction of lupin*. Food Protein R&D Center, Texas A&M University, College Station, Texas, 1983.
29. Wink, M., *Proceedings of the 3rd International Lupin Conference*, La Rochelle, France, p. 325, 1984.
30. Mohr, U., Abstracts *5th International Lupin Conference*, Poznan, Poland, p. C-19, 1988.
31. El Dash, A.A., V.C. Sgarbieri, and J.E. Campos, *Cereal Chemistry 57*:9 (1982).
32. Pristley, B., *Proceedings 4th International Lupin Conference*, Geraldton, Western Australia, p. 240, 1986.
33. Gladstones, J.S., *5th International Lupin Conference*, Poznan, Poland, 1988.
34. Lee, C.H., *Proceedings 4th International Lupin Conference*, Geraldton, Western Australia, p. 64, 1986.

Acknowledgments

Parts of this work have been financed in part by the Dirección de Investigación, Universidad Católica (DIUC).

Progress in Development of Leaf Proteins for Use in Foods

P. Fantozzi[a] and A. Sensidoni[b]

[a]Istituto di Industrie Agrarie, Universitá di Perugia, San Constanzo, 06100-Perugia, Italy, and
[b]Istituto di Technologie Alimentari, Universitá di Udine, via Marangoni 97, 33100-Udine, Italy

Abstract

Although at present the major sources for vegetable protein for food utilization are seeds, in the near future leaf protein could play an important role because of its particular amino acid composition, functional properties and, in some cases, purity. In particular, the chloroplastic enzyme, ribulose 1-5 diphosphate carboxylase, is easily crystallized from some plants, qualitatively surpasses all other storage seed proteins and can be proposed for different food and specific pharmaceutical uses. Studies on the introduction of this enzyme into special diets for some patients open up a wide, interesting and highly remunerative market, which is precluded for seed proteins because of their poor compositional and functional properties. The development and evolution of existing processes and the actual utilization of different leaf proteins is discussed.

The ability of plants to transform CO_2 into organic substances using solar energy is the basis of the trophic chain. Agricultural production usually makes use of this process, increasing by various means the production potential. But apart from these observations of a general nature, in the more industrialized countries it is the market demand that influences production. The Western countries have developed an agricultural system using petroleum derivatives that without doubt places them among the major consumers of this energy source. For example, in France, two-thirds of the energy resources are destined for animal feeding. From the 300×10^{12} kcal set aside for animal feedstuffs, products are obtained corresponding to 25×10^{12} kcal, giving a ratio of about 1 to 12.

Thus, given the scarce conversion index for animal utilization of plant feedstuffs, there is great interest in a system for the fractionation of forages, with a view to separating the components and a consequent diversification of their use. It also must be remembered that in a modern agricultural system dietetics play an important role and interest is increasing in concentrates with a plant-protein base.

During the past 10 years, studies on the extraction and use of vegetable protein have intensified greatly. One of the main reasons is the considerable nutritional interest involved in amino acid composition, and, furthermore, that these proteins are a nonconventional and easily acquired source of food. The use of vegetable protein from seeds and legumes has been common for a long time in certain geographic regions (beans in South America, soya in Asia), but leaf proteins are a very recent novelty in human nutrition. Herbivores are able to utilize plant foods because of the abundant quantities ingested and the particular physiological and morphological structure of their digestive system (e.g., ruminants) that enable them to digest forage efficiently.

Wet-fractionation of leaf-protein constituents is used for two purposes: to obtain fibers that are still suitable for animal nutrition (polygastrics) (19,20) and to obtain structural lipoproteins for the feeding of monogastric animals, as well as purified proteins for human consumption.

The main difference between seed and leaf proteins is related to their composition, particularly that of the amino acid components (Table 1) (16).

TABLE 1

The Amino Acid Composition of Ribulose 1-5 Diphosphate Carboxylase Obtained from Tobacco in Comparison to Other Proteins

	Tobacco	Soy	Egg	FAO/WHO reference pattern
Isoleucine[a]	4.1	4.9	5.8	4.0
Leucine[a]	8.4	7.7	9.0	7.0
Lysine[a]	5.7	6.1	6.7	5.5
Phenylalanine[a]	6.1	5.4	5.3	—
Tyrosine[a]	3.3	3.7	4.3	—
Phenylalanine + Tyrosine	9.4	9.1	9.6	6.0
Methionine[a]	1.8	1.1	3.0	—
Cysteine	1.6	1.2	2.1	—
Methionine + Cysteine	3.4	2.3	5.1	3.5
Threonine[a]	3.7	3.7	5.3	4.0
Tryptophan[a]	1.1	1.4	1.8	1.0
Valine[a]	5.8	4.8	7.2	5.0
Histidine	1.7	2.5	2.6	—
Aspartic acid	7.9	11.9	10.7	—
Serine	3.6	5.5	7.7	—
Glutamic acid	11.0	20.5	3.8	—
Proline	3.0	5.3	7.7	—
Glycine	4.4	4.0	12.3	—
Alanine	4.5	3.9	—	—
Arginine	4.6	7.8	6.4	—

[a] Essential amino acid.

In the higher plants (Phanerogams), photosynthesis is carried out by the leaves, with tissue cells containing chloroplasts with all the enzymes needed to perform this function, and with cytoplasm containing other functional proteins. The principal function of seed proteins is that of storage and thus their extraction is hindered by the presence of structures related to the composition of the seed tissues themselves—for example, the lignified integuments, starches, lipids and polyphenols.

In addition, protein separation processes generally are not carried out on whole seeds, but on material that already has undergone a previous process (e.g., oil cake, in the case of soya, maize and sunflower), thus constituting a further source of income. The economic and qualitative evaluation must relate the protein concentrates to the whole dried cakes. The intended use for humans or animals is a reason to consider cost. In the case of leaf proteins, the question is quite different. Up until a few years ago, one spoke of forage concentrates only in terms of animal feed, but now international research is aimed at the preparation of compounds intended exclusively for human use. The main objectives and the practical results obtained in recent years are reported here.

Products for Developing Countries

In many developing countries the lack of protein constituents in human nutrition is still strongly felt. Thus, the ability to make concentrates from leaves would be very interesting. Leaf proteins were considered from a nutritional point of view by some authors (12,13,58). It was shown that it is possible to use vegetable protein concentrates for human nutrition as ingredients in traditional foods (12,13,24,28,33,35,37,44,58). Pirie is, without doubt, an authority in this field (45–48). However, the high cost of the machinery cannot be neglected. This has been overcome in part by using more labor and by using solar energy in the drying process.

Market Products for Industrialized Countries

Leaf Proteins for Alimentary Use: Modern foodstuff formulations provide for the preparation of products with a high biological value and commercial image—characteristics linked to the typical consumption of the more advanced countries. Thus, leaf protein products must compete not only with proteins of animal origin (usually from milk and eggs), but also with proteins from soybeans that already have a valid and accredited use both in the meat-derivative industry (frankfurters, hamburgers, snacks, food pastes, etc.) and in the confectionery and baking industries. Furthermore, much progress has been made during the past 20 years in developing biotechnology for algal mass culture of *Spiruline*, as a source of food and feed due to its high content of protein and vitamins (38,39,62–64). Finally, the Italian group headed by Galoppini set up an original application of polyelectrolytes to the juice extracted from vegetable by wet fractionation. This system was called Poly-Protein and these authors obtained interesting results (27).

The protein concentrate precipitates rapidly and can be recovered by decantation-filtration or by continuous centrifugation. This shows interesting compositional characteristics, making it useful not only as an additive for animal feed, but also as an edible leaf protein concentrate (LPC). Their nutritional value, functional properties and food uses are interesting (27). The latter can be achieved only by finding raw materials of plant origin so as to obtain different products for feedstuffs and alimentary use. In this connection, plants other than alfalfa have been investigated, representing one of the most recent developments in this field (Table 2). For example, there has been research on the extraction of leaf protein from sugar beets, optimizing the best harvest period. Additionally, leaf proteins have been extracted from Jerusalem artichoke.

Leaf Protein for Pharmaceutical Use: When some American researchers (68) reported that it was possible to obtain crystallized leaf proteins on an industrial scale, research in this field proliferated and the nutritional and pharmaceutical uses of these products were hypothesized. The phenomenon of spontaneous crystallization that occurs in juices extracted from tobacco plants is dealt with later. The extreme purity that can be had with these products suggests that they could have a pharmaceutical use, which would be an interesting prospect for the application of leaf proteins. The crystals consist of aggregates of ribulose 1,5-diphosphate (or biphosphate) carboxylase/oxygenase molecules, an enzymatic protein that plays a part in the conversion of CO_2 into organic matter during the photosynthetic cycle and, paradoxically, its inherent oxygenase activity takes part in the photorespiration. RDC, RuBisCO and Fraction 1 protein (F_1P), are synonymous, the names are used interchangeably (1). Its amino acid composition and functional properties definitely support its use in the field of medicine and pharmaceutics. Some promising market research has been carried out in this direction.

The functional properties of protein condition their utilization as these properties affect their behavior in the alimentary system. Acquaintance with these properties makes it possible to single out the feasible market possibilities of a product. For example, the following functional properties were evaluations on leaf proteins (RDC) obtained in two different processing lines (Fig. 1) situated in Italy and France, respectively (21): solubility (Fig. 2), minimum at pH 5–6 and almost total at pH < 4 and > 9; gel-forming capacity (Fig. 3), similar to soy protein; emulsifying capacity (Fig. 4) with maximum emulsion at pH 5–6; and foaming capacity (Fig. 5) maximum from pH 6 to 8, almost absent with acid pH.

From examination of the values obtained we believe that tobacco protein can be used as an additive for foods with a high fatty content, in preparations in which the formation of gel or foam plays an important role, or in the pharmaceutical field, e.g., in the composition of oily medicines, pills, etc.

Leaf Proteins for Animal Use: In the United States, Europe and other regions, processing plants have been constructed for the wet fractionation of vegetable material. Among these are the Pro-Xan (Protein-Xanthophyll) process and Pro-Xan II. These are widely known systems relying on the technology of the thermal coagulation of the protein components (35). One of the most important properties for the qualitative evaluation of proteins is solubility, as it is related to purity and to the level of denaturation of the extracted fraction (Table 3). Because the products obtained were lacking in these characteristics, other systems were studied to optimize the temperature and pH parameters of coagulation, in order to improve the solubility of the protein concentrates.

In France, the Society "France Luzerne" devised pro-

TABLE 2

Positive Utilizations of Leaf Proteins for Human Use

A. In human food

Vegetable source	Food	Country	Reference
Alfalfa	Sweets	India	(12,40)
	Various cakes	India	(33)
	Various	Pakistan	(58)
	Various	Italy	(27)
	Snacks	Mexico	(34)
	Cakes	Mexico	(25)
	Pasta	Japan	(67)
Sugarbeet	Flour	India	(37)
Cabbage	Snacks/bread	India	(40)
Radish	Bread/fried food	India	(40,50)
Mulberry	Various	India	(60)
Nettle	Various	Bulgaria	(9)
Amaranthus	Isolate	France	(4)
Jerusalem artichoke	Protein concentrate	Italy	(18,22)
		USA	(51)
Tobacco	Pasta	Italy	(57)
Cassava	Biscuits	Philippines	(41)
Various	Meat/flour extender	India	(55)
	Flour	India	(42)
	Various	Ghana	(24,50)
	''	Sri Lanka	(24,50)
	''	Nigeria	(43)
	''	Bolivia	(48)
	''	Mexico	(24,28)
	''	Egypt	(48)
	''	Kenya	(48)
	''	Zambia	(48)
Berseem	''	Pakistan	(58)
Banana tree	Protein concentrate	Italy	(18)
	''	Venezuela	(52)

B. For dietetic/medical purposes

Vegetable source	Suggested action	Country	Reference
Tobacco	Liquid diets	USA	(36)
	Hyperproteic liquid diets	Italy	(5,22)
	Action of nicotine on Central Nervous System	Italy	(5,22)
Alfalfa	Hypocholesterolemic effect	Japan	(31)
	Purified enzyme or Hydrolisates	France	(15)
	Purified products	Australia	(66)
Berseem	Nutrition utilization	Pakistan	(58)
Various	not known	USA	(53)

cesses for the fractionation of forage using the P.X.1. system (essentially a modification of the American Pro-Xan), an interesting approach that produces concentrates of high nutritional value for monogastric animals. The high pigment content, (e.g., xanthophyll) distinguishes it from the alimentary products based on soybeans. Further research was carried out to make a tasteless formulation intended for human consumption, using membrane techniques (ultrafiltration and dialysis).

A large amount of research has been carried out both in Great Britain and in Ireland, with Pirie an authority in this field. Previously, research was aimed mainly at obtaining protein concentrates for animal feed; lately, encouraging results have shifted interest to their use for humans (45,46).

In addition, the Vepex System (from Vegetable Protein Extract) was perfected by Hungarian researchers (29,30,59). The system differs from Pro-Xan mainly in

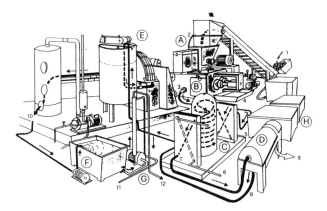

Fig. 1. Basic technological protocol of the pilot plant for wet-extraction of leaf proteins from tobacco plants and for its actual assembly. A = grinding; B = pressing; C = heat exchanger; D = centrifugation with decanter; E = vacuum filtration with rotating filter; F = resting period of the brown juice with successive crystallization; G = ultrafiltration; H = collection of serum and F_2P precipitate. Products: 1 = tobacco; 2 = antioxidant solution; 3 = green juice; 4 = fiber; 5 = heating fluid; 6 = cooling fluid; 7 = green juice after heat treatment; 8 = green juice after centrifugation; 9 = cytoplasmic protein concentrate; 10 = brown juice; 11 = permeate following ultrafiltration; 12 = concentrate after ultrafiltration. Arrows indicate direction and course of product flow.

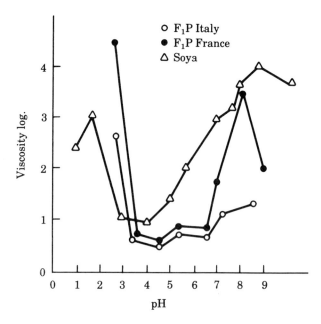

Fig. 3. Gel foaming capacity related to the pH of different nonconventional proteins.

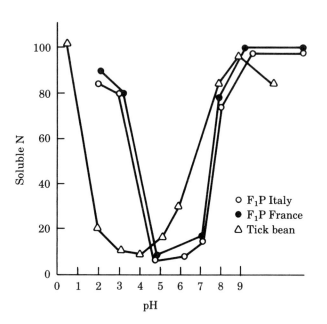

Fig. 2. Solubility related to the pH of different nonconventional proteins.

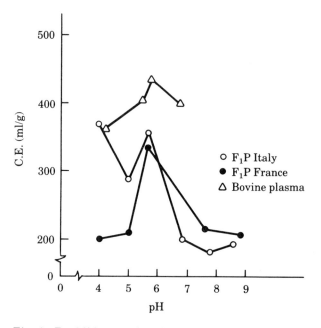

Fig. 4. Emulsifying capacity related to the pH of different nonconventional proteins.

that the yield of extracted juice is better, as more pressure is applied and the deproteinated serum is then fermented to produce a reusable biomass. In Italy, Galoppini carried out the separation of leaf proteins from forage to use then as animal feeds (26,27).

Economy in these systems can be reached by diversifying the end products in relation to commercial competition from animal and seed proteins (fish and soybean meals) and by ensuring that the costly plants work the whole year round.

Tobacco Leaf Proteins

During the late 1970's, several authors mentioned tobacco plants as a possible unconventional source of leaf proteins (3,10,11,65,68). The aim of the proposed technology was to obtain, after deproteinization, cured tobacco for smoke with a low toxicity (homogenized leaf curing process).

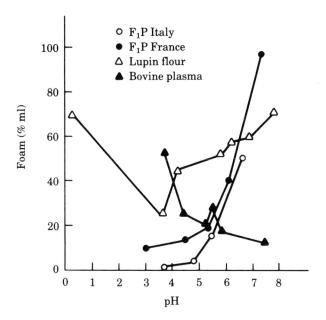

Fig. 5. Foaming capacity related to the pH of different nonconventional proteins.

TABLE 3
PRO-XAN and PRO-XAN II Technological Processes[a]

	Dry matter %	Protides %
Fresh grass	100	100
Fibers	68	43
Residual serum	17	15
LPC green (PRO-XAN)	15	42
LPC green (PRO-XAN II)	11	26
LPC white (PRO-XAN II)	3.3	16
Washing solution	0.7	—

[a] Values for dry matter and proteins present in the products obtained using the two processes expressed as % of the dry matter and of the proteins present in the initial product (35).

Once it was established that the technological procedure would fully utilize the particular ability of the RDC to crystallize, technological protocols were proposed and studied in some countries. Although the first works referred to the theoretical protein yields per hectare based on extrapolation of results obtained in the laboratory, in France and Italy pilot plants were set up so the actual protein yield could be measured (Table 4). It was found that the quantities that could be obtained, although lower than the theoretical values, were nonetheless interesting. But elaboration on an industrial scale would increase the losses.

Wet Fractionation

In wet fractionation, the vegetable tissues are triturated to yield a juice containing the different protein fractions (56). Fresh tobacco was transported rapidly from the field to the plant, loaded onto an elevator-conveyor belt and sprinkled with an antioxidant solution (sodium metasulphate in water, 6 g/l). Next, it dropped through a hopper into a hammer-mill triturator and from there to a screw-torque press where the green juice was separated from the fibrous material. Using a "mono"-type pump, the juice immediately was driven into a plate-type heat exchanger where it was brought to 50°C in a few seconds. It then passed through an insulated tube, remaining at this temperature for exactly 5 minutes before being cooled to 20°C in a second exchanger. Following this stage, it was vacuum filtered through a Dicalite plate which separated the green coagulated fraction resulting from the heat treatment, from the brown juice filtrate. This latter was collected in 300-l stainless steel containers. It became cloudy after a few hours; and 12 hours later a crystalline precipitate F_1P was collected and passed through a vertical centrifugal separator, producing as waste filtrate a juice, named serum, and retaining the protein fraction in the form of a compact paste. This paste was frozen and then lyophilized. Concentrated HCl was added to the serum to bring it to pH 3.5, at which point the soluble proteins rapidly precipitated and could be collected, frozen and lyophilized (F_2P) crystals under the scanning electron microscope (SEM). Because the yield of F_1P in g/l was quite low (0.67 as an average), it was hypothesized that part of Fraction 1 did not separate out and later was collected together with Fraction 2. In fact, this was confirmed by SEM pictures.

First, the structural proteins are precipitated and subsequently separated by filtration. A brown juice is thus obtained which contains the soluble proteins. Of these, Fraction 1 (F_1P) consists of the enzyme ribulose diphosphate carboxylase (RDC), which is found in the chloroplast stroma and gives rise to the primary process in the photosynthetic cycle (Calvin cycle). This protein crystal-

TABLE 4
Values of Leaf Protein Production per Hectare (Maximum)

Author	Year	Ref. No.	No. Plants/ha	Structural proteins kg/ha	kg/ha F_1P	kg/ha F_2P
Wildman	1977	(68)	400.000	1650	325	1325
Staron	1980	(61)	400.000	600	700	200
Gopalam	1982	(28)	80.000	120	25	100
Chouteau	1983	(7)	250.000	1000	137	74
Fantozzi	1985	(17)	200.000	971	122	80

lizes from the aqueous medium. A second fraction (F_2P) can then be recovered from the residual serum by means of acidification and is precipitated through denaturation. Many researchers have examined the feasibility of obtaining RDC from different vegetable sources. To date, tobacco is the only plant that has yielded large quantities of this enzyme through a spontaneous crystallization process.

Technological Results

Tables 5 and 6 show an increase in the extraction yields of tobacco leaf protein over a three-year period. Note that the quality has been improved considerably through reduction of pollutants such as nicotine and polyphenols. The addition of SO_2 to the brown juice induces further positive effects and leads to an increase in the F_1P crystallization and purity, as well as a greater homogeneity and size of the crystal (56). In addition, trials were carried out to improve yield and to increase the purity of the proteins obtained. Thus, the effect of bubbling carbon dioxide through the brown juice was tested, as was the use of activated carbon and/or polyvinylpolypyrrolidone (PVPP) to reduce the polyphenol residues in the protein fraction, but the results were disappointing.

Ultrafiltration Tests

Barbeau and Kinsella (2) confirmed how a strong bond of a covalent nature forms between protein and polyphenolic substances (mainly chlorogenic acid and rutin in the case of tobacco). However, no work has been carried out on the type of bond that forms between protein and nicotine. An electrostatic type of bond could be hypothesized between the alkaloid nitrogen and the acid group of the amino acids of the protein chain. In fact, analyzing the nicotine content of the brown juice and of the crystallized fraction, it was found that a greater concentration was present in the fraction. Thus, as reported (14), trials were carried out with the ultrafiltration process (the UFSI pilot plant of the Alfa Laval Company, Sweden) with a hollow-fiber filtering cartridge (Tomicon, cut-off 10.000) in line immediately after the vacuum filtration.

In that trial, the effects of the following treatments were verified: UF, as such; UF with diafiltration; UF with the prior addition of SO_2 (3 g/l); UF with the prior addition of PEG (1 g/l); UF after alkalization (pH = 10); and, redissolving the F_1P in an alkaline solution (pH = 10), followed by UF. Unsatisfactory results were obtained (Table 7) even if the ultrafiltration procedure undoubtedly improved protein purity. Treatment with diafiltration, however, had a positive effect on the F_1P, although this procedure is

TABLE 6

Evaluation of the Effect of Washings on Proteins, PFT (Total Polyphenols) and Nicotine Content

	PFT (mg/g)	Nicotine (μg/g)
Sample A[a]	4.97	28.11
Sample B[b]	0.79	15.16
Sample C[c]	0.39	10.90

[a] = 4 washings with H_2O
[b] = 2 washings with H_2O + 2 washings with EtOH
[c] = washing with EtOH

TABLE 5

Total N_2 content in Tobacco Pulp and F_1P and F_2P Production (Years 1984, 85, 86)

	Min.	Average	Max.
Total nitrogen in the pulp (% D.M.)			
Year 1984	1.6	2.0	2.2
1985	1.1	2.1	3.3
1986	1.4	2.5	3.5
F_1P in the juice (g/l)			
Year 1984	0.3	0.7	1.2
1985	0.4	1.1	2.1
1986	0.8	1.3	2.2
F_1P: total yield (kg/ha)			
Year 1984	67.2	78.5	100.1
1985	27.4	76.7	149.4
1986	59.6	93.9	156.4
F_2P in the juice (g/l)			
Year 1984	0.3	0.4	0.5
1985	0.3	0.4	0.8
1986	0.3	0.5	0.8
F_2P: total yield (kg/ha)			
Year 1984	—	42.7	—
1985	21.8	28.9	56.6
1986	22.0	36.9	52.9

TABLE 7

Evaluation of the Effect of Ultrafiltration

U.F. trial	Protein[a] yield		Nicotine[b]		Polyphenols[c]	
	F_1P	F_2P	F_1P	F_2P	F_1P	F_2P
A	95	61	19	62	47	90
B	86	227	21	215	39	82
C	158	297	193	50	178	92
D	120	95	614	94	243	101
E	114	95	510	105	325	65

The data are expressed as percentages based on the weight of the nonultrafiltered sample = 100.
[a] Taking = 100 the g/l of protein in the control.
[b] Taking = 100 the content in $\mu g/g$ of protein of the nicotine present in the control.
[c] Taking = 100 the content in mg/g of protein of the polyphenols present in the control.
A) UF as such;
B) UF with diafiltration;
C) UF with the prior addition of SO_2 (3 g/l);
D) UF with the prior addition of PEG (1 g/l);
E) UF after alkalization (pH = 10);
F) redissolving the F_1P in an alkaline solution (pH = 10) followed by UF.

impractical due to the time and high cost involved, without considering the large quantity of liquid to treat. The last experiment consisted of redissolving the F_1P, collection by centrifugation in an alkaline solution (pH = 10 with NaOH) and a successive ultrafiltration (Table 8).

TABLE 8

Evaluation of the Ultrafiltration Treatment on the F_1P Redissolved in an Alkaline Solution (Di Lucente F., et al., 1987)

Samples	Polyphenols		Nicotine	
	mg/g	Variation %	$\mu g/g$	Variation %
T. Q.	8.11	100	129.21	100
U. F. (A)[a]	3.99	49.2	0	0
U. F. (F)[b]	1.63	20.1	0	0

[a] Test A (average of 4 trials).
[b] Redissolved, ultrafiltered and recrystallized protein.

Results and Agronomic Yields

The agronomic practices which offer considerable advantages in increasing tobacco protein yield are the following (16,17): nitrogen fertilization with doses of 200–300 kg/ha giving the highest protein yield; planting density with up to 200,000 plant/ha shows no competition problems between the tobacco plants; harvesting period is important because young plants give scant yields until the floral bud phase protein concentration in the leaves increases; multiple harvests through the year exploit the plant's capacity to regenerate its epigeal organs after cutting; and, genetic selection to exploit varieties selected for their greater RD synthesizing characteristics.

Evaluation of the Possible Mutagenic Capacity of Tobacco Proteins

A recent experiment using the Ames test demonstrated that F_1P (RDC) and F_2P have no mutagenic effect nor cytotoxic activity, because all the *in vivo* tests carried out gave negative results.

Nutritional Value and Amino Acid Composition of the Proteins

The fundamental difference between seed proteins and leaf proteins is that the former have a storage function in the plant, whereas the latter have enzymatic functions. Thus, if the biological qualities of some of the most common proteins are compared, it is apparent that both the F_1P and the F_2P have a high nutritional value (8,16). The balanced amino acid composition is undoubtedly the main reason for this characteristic. Furthermore, the F_1P fraction obtained through crystallization is extremely pure which prompts its use in the medico-pharmaceutical field (i.e., the administration of this protein in liquid suspension form to selected patients) (Table 9).

TABLE 9

Possible nutritional and dietary uses of the proteins extracted from the tobacco leaf (5,16,17)

A) Nutritional uses
— Additives in nonalcoholic beverages to increase their nutritional value
— Milk substitutes for those who cannot metabolize lactose
— Integration into some cereal products to improve the amino acid profile
B) Therapeutic uses
— Nutrition for people with renal insufficiency: the absence of Na^+ and K^+ in these proteins can severely limit the frequency of hemodialysis; possibility of using proteins rather than mixtures of simple amino acids in diets of equal nutritional value, and therefore reducing the volume of the diets
— Nutrition for people with gastro-intestinal problems (GI or esophageal tumors, pylorospasma, etc.): the purity of the protein allows for the use of complete, wholesome liquid diets, with low residue
— Nutrition for people who must severely restrict Na^+ and K^+ intake
— Hypercalorie and hyperproteinaceous diets for neurosurgical patients in the state of stress (trauma, fever, coma, infections, etc.).

Economic Evaluation

To date, merely a rough economic estimate can be made because there is still no real market value for the products. However, today there is a definitely commercial interest in both F_1P and F_2P, and it is possible to make a comparison between the costs and proceeds of one hectare of tobacco cultivation for the production of protein concentrate (Table 10).

TABLE 10

Cost-Return statement (thousand of Lire/ha)

	Traditional cultivation smoke	Cultivation for protein extraction	
		by seedling transplant	by direct seeding
1. Cost			
A. Production	11.600	10.500	3.500
B. Processing	—	2.500	2.500
C. Total	11.600	13.000	6.000
2. Returns			
A. without EEC support	4.400	4.300	4.300
B. with EEC support	13.000	(13.000)	(13.000)
3. Difference of cost-return			
A.	−6.800	−8.700	−1.700
B.	+1.400	(nul)	(+7.000)

[a] Numbers in parentheses are hypothetical.

Thus, we can stress the great interest on the part of tobacco producers and processors regarding the proposed technology. Many and important possibilities of utilization are suggested and they can become a reality if the pharmaceutical and medical industries will understand the real future impact of the novelty of these products on the market (Table 2).

References

1. Akazawa, T., in *Proceeding of the International Grassland Congress XV*, Kyoto, Japan, 1985, pp. 67–68.
2. Barbeau, W., and J.E. Kinsella, *J. Agric. Food Chem. 31*:993 (1983).
3. Bourque, D.P., in *Nonconventional Proteins and Foods*, NSF-RA-770278, 1977, pp. 159–164.
4. Burghoffer, C., for *Proceedings of the Third International Conference on Leaf Protein Research*, Pisa Perguai, Viterbo, Italy, (in press).
5. Casotto, A., C. Castrioto and P. Fantozzi, Somministrazione di dieta ipercalorica iperproteica derivata da piante verdi in pazienti neurochirurgici in condizione di stress, EEC/Agrimed Meeting, Perugia, Italy, 1988.
6. Chiesara, E., M. Gervasoni and R. Rizzi, Proteini estratte dalle foglia di tabacco: studio sulla eventuale mutagenicità nel test di Ames, EEC/Agrimed Meeting, Perugia, Italy, 1988.
7. Chouteau, J., *Final Report of the Institut Experimentalu Tabac de Bergerac*, Project EEC/IETB (Agrimed), 1983.
8. Corcos-Bendetti, P., V. Gentili and P. Fantozzi, Valutazione della qualità nutrizionale di proteine estratte da foglia di tabacco, at 17th General Meeting of Italian Society for Human Nutrition, Abano Terme, Italy, Sept. 27–29, 1984.
9. Daley, P., for *Proceedings of the Third International Conference on Leaf Protein Research*, Pisa, Perugia, Viterbo, Italy, (in press).
10. Dejone, D.W., and J.J. Lam, *Tob. Res. 5*:1 (1979).
11. Dejong, D.W., J. Lamb, R. Lowe, E. Yoder and T.C. Tso, *Beitr. Tabakforsch. 8*:93 (1975).
12. Devdas, R.P., G. Kamlanthan and P. Vijayalakshmi, *Ind. J. Nutr. Dietets 18*:427 (1981).
13. Devdas, R.P., and P. Vijayalakshmi, in *Proceedings of the International Grassland Congress XV*, Kyoto, Japan, 1985, pp. 152–153.
14. De Lucente, F., M. Bacinelli and P. Fantozzi, "Purificazione di preparati proteici da tabacco mediante ultrafiltrazione," *Annali della Facoltà di Agrarig, Università di Perugia*, Vol. 41 (in press).
15. *Proceedings of the Third International Conference on Leaf Protein Research*, Pisa, Perugia, Viterbo, Italy (in press).
16. Fantozzi, P., and A. Sensidoni, *Qual. Plant. Food Hum. Nutr. 32*:351 (1983).
17. Fantozzi, P., in *Proceedings of Round Table on Tobacco Protein Utilization Perspectives*, held Oct. 28–31, 1985, at Salsomaggiore Terme, Parma, Italy.
18. Fantozzi, P., and A.S. Bermudez, "Edible proteins from the leaves of *Helitanthus tuberosis* and *Musa paradisiaca*," in the Proceedings of 7th World Congress of Food Science and Technology, Sept. 28–Oct. 2, 1987, Singapore.
19. Fantozzi, P., "Leaf protein technologies: various utilizations of the tobacco plant" in *Proceedings of 4th AAAP Animal Science Congress*, Hamilton, New Zealand, Feb. 1–6, 1987.
20. Fantozzi, P., P. Pollidori and F. Constantini, "Utilization of tobacco plant residues as feed for mono and polygastrics" in *Proceedings of 4th AAAP Animal Science Congress*, Hamilton, New Zealand, Feb. 1–6, 1987.
21. Fantozzi, P., "Functional properties of tobacco proteins" in *Proceedings of the 2nd World Congress of Food Technology*, Barcelona, Spain, March 3–6, 1987.
22. Fantozzi, P., in *Proceedings of the Third International Conference on Leaf Protein Research*, Pisa, Perugia, Viterbo, Italy (in press).
23. *FAO/WHO Energy and Protein Requirements*, World

Health Organization Tech Rep. Series No. 52, FAO/WHO, Geneva, Switzerland (1973).
24. Fellows, P.J., M.N.G. Davys and W. Bray, in *Proceedings of the International Grassland Congress XV*, Kyoto, Japan, 1985, pp. 127–131.
25. Find Your Feet Newsletter report, London.
26. Galoppini, C., and R. Fiorentini, in *Proceedings of Congress Prospettive delle proteaginose in Italia: aspetti nutrizionali agronomici, tecnologici ed economici della utilizzazione delle leguminose da granella*, Perugia, Italy (1979).
27. Galoppini, C., and R. Fiorentini, in *Proceedings of the International Grassland Congress XV*, Kyoto, Japan, 1985, pp. 50–57.
28. Gopalam, A., *Curr. Trends in Life Sci. 11*:41 (1982).
29. Hollo, J., and L. Koch, in *Leaf Protein: Its Agronomy, Preparation, Quality and Use*, edited by N.W. Pirie, Blakewell Science Publishers, Oxford, England, 1971, p. 63.
30. Hollo, J., and L. Koch, *Process Biochem. 5*:37 (1970).
32. Jones, A., in *World Protein Resources*, edited by A. Jones, MTP Publishers, Lancaster and London, England, 1974, p. 26.
33. Joshi, R.N., V.A. Savangikar and B.W. Patunkar, *Indian Bot. Reptr. 3*(2):136 (1984).
34. Kennedy, D., in *Proceedings of the Third International Conference on Leaf Protein Research*, Pisa, Perugia, Viterbo, Italy (in press).
35. Kohler, G.O., and B.E. Knuckless, *Food Techn. 31*:191 (1977).
36. Leaf Proteins International, Raleigh, N.C., USA, 1982.
37. Matai, S., S.T.A.T. Calcutta (India) Report No. 051/8/DST/87 (1987).
38. Materassi, R., "Prospettive della coltura di *Spirulina* in Italia," in *Ricerca di nuove fonti proteiche e di nuove formulazioni alimentari*, edited by R. Materassi, Consiglio Nazionale delle Ricerche (CNR), Rome, Italy (1980).
39. Materassi, R., M.R. Tredici and W. Balloni, *Appl. Microbiol. Biotechnol. 19*:384–386 (1984).
40. Mathur, B., in *Proceedings of the Third International Conference on Leaf Protein Research*, Pisa, Perugia, Viterbo, Italy (in press).
41. Meimban, E.S., *Philippine J. Nutr. 35*:82 (1982).
42. Mohan, M., in *Proceedings of the Third International Conference on Leaf Protein Research*, Pisa, Perugia, Viterbo, Italy (in press).
43. Oke, O.L., *Nutrition (Pois) 20*:18 (1986).
44. Ostrowski-Meissner, T.M., and H.T. Ostrowski-Messner, in *Proceedings of The International Grassland Congress XV*, Kyoto, Japan (1985), pp. 63–66.
45. Pirie, N.W., *Leaf Protein and Other Aspects of Fodder Fractionation*, Cambridge University Press, USA, 1978.
46. Pirie, N.W., *Qual. Plants: Plant Fds. Hum. Nutr. 34*:229 (1984).
47. Pirie, N.W., "Leaf Protein in the Human Diet," in *Proceedings of The International Grassland Congress XV*, Kyoto, Japan (1985).
48. Pirie, N.W., in *Proceedings of the Third International Conference on Leaf Protein Research*, Pisa, Perugia, Viterbo, Italy (in press).
49. Pompei, C., and M. Lucisano, *Riv. Ital. Sost. Grasse 51*:149–154 (1974).
50. Ramappa, B.S., in Proceedings of The International Grassland Congress XV, Kyoto, Japan (1985), pp. 116–126.
51. Tawate, P.K., in *Proceedings of The International Grassland Congress XV*, Kyoto, Japan (1985), pp. 24–28.
52. Rosas Romer, A.J., and A. Christina Diaz, *Progress in Leaf Protein Research 11*:41–47 (1982).
53. Rubisco Corp., Berthoud, Colorado, USA.
54. Saccomandik, V., and A. Cordella, "Estrazione delle proteine dal tabacco: una analisi economica," EEC/Agrimed Meeting, Perugia, Italy, July 22, 1988.
55. Savangikar, V.A., in *Proceedings of the Third International Conference on Leaf Protein Research*, Pisa, Perugia, Viterbo, Italy (in press).
56. Sensidoni, A., M. Bacinelli and P. Fantozzi, *Industrie Alimentari 26*(9):781–786 (1987).
57. Sgrulletta, D., and E. De Stefanis, "Le proteine vegetali nella pasta alimentare: primi risultati," EEC/Agrimed Meeting, Perugia, Italy, July 22, 1988.
58. Shah, F.H., *Progress in Leaf Protein Research 11*:317–322 (1982).
59. Skole, R., *Chem. Eng.*:68 (Dec. 10, 1973).
60. Srivastava in *Proceedings of the Third International Conference on Leaf Protein Research*, Pisa, Perugia, Viterbo, Italy (in press).
61. Staron, T., *Med. et. Nutr. 16*:337–352 (1980).
62. Tomaselli, L., *Biotechnologies for the Production of Spirulina*, IPRA Consiglio Nazionale delle Ricerche, Monograph No. 17, Rome, Italy (1987).
63. Torzillo, G., B. Pushparay and G. Florenzano, *Ann. Microb. 35*:165–173 (1985).
64. Torzillo, G., B. Pushparay, F. Bocci, W. Ballono, R. Materassi and G. Florenzano, *Biomass 11*:61–74 (1986).
65. Tso, T.C., R. Lowe and D.W. Dejong, *Beitr. Tabakforsch. 8*:44 (1975).
66. TTD International, PO Box 425, Seven Hills, N.W., Australia.
67. *Proceedings of the 2nd International Conference on Leaf Proteins*, Nagoya and Kyoto, Japan, 1985.
68. Wildman, S.G., P. Kwanyuen and B.H. Ershoff, in *Proceedings of Nonconventional Proteins and Foods Conference*, Madison, Wisconsin, USA (1977).

Preparation of Fish and Shrimp Feeds by Extrusion

Joseph P. Kearns

Wenger International, Inc., Kansas City, Missouri

Abstract

Preparation of fish and shrimp feeds by extrusion has become very popular in the aquatic feed industry for the many advantages this production method has to offer. Extrusion cooking can produce floating, sinking, slow sinking and semi-moist aquatic feeds as well as other products useful to feed mills, such as full fat soy, piglet feeds, cooked grains and pet foods. Aquatic product diameters vary between 1.5 and 10.0 mm, while formulations can include high levels of soybean meal, high levels of fish meal or combinations of both. Liquid ingredients, such as aquatic or animal wastes, can be included, levels dependent on moisture content, as well as fish oils and animal fats up to 12% in the raw mixture. Subsequent drying, external coating and cooling have an impact on finished product quality and as a result are important steps in the extrusion process.

Extrusion cooking of aquatic feeds is a very broad topic when considering the number of aquatic species being raised for profit in the world today and the variety of feed formulations and product specifications required to fulfill those needs. Expertise in all aspects of the aquaculture industry has improved, from nursery designs through handling of the final aquatic products at harvest. As a result of advances in feeding techniques and better understanding of the nutrient requirements of various aquatic species, feed producers are challenged to make higher quality, cheaper, nutritious feeds. Feed millers that utilize equipment with the proper design criteria are able to control and influence the phenomena that occur inside the extruder barrel. Controllability of the extrusion cooking equipment is what allows extruders to produce a wide range of products such as breakfast cereals, breadings, textured soy protein, snacks, instant rice, instant pasta, modified starches, animal feeds such as piglet feed, cooked corn, full fat soy, dog food, cat food and aquatic feeds.

Aquatic Feed Production

Production of aquatic feeds falls into one of four categories: floating feeds, slow sinking feeds, sinking feeds and soft, moist feeds. Extruders are also capable of producing ingredients that can be used in aquatic feeds. The production process utilizing an extrusion cooker for all of these categories is very similiar.

The flow diagram (Fig. 1) can be separated into processing areas: raw material processing area, extrusion processing area, post extrusion processing, drying and cooling processing, post drying and cooling processes. A closer look at each processing area follows.

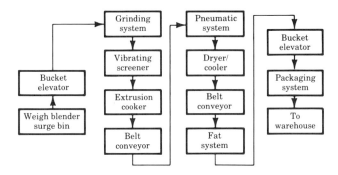

Fig. 1. Flow diagram for aquatic feeds.

Raw Material Area and Processes

The function of this area is to store and prepare the formulations for processing. There are a number of methods and strategies for preparing raw materials. This area can be completely automated or labor extensive. In either case, the raw materials are stored in bags or bulk. The raw material may need cleaning and pregrinding at this stage depending on its condition and size when received. The ingredients are weighed and mixed and generally passed through a final grinder and sieved prior to the extrusion cooker to ensure that particles that could plug the extruder die are not present.

The particle size specification prior to the extrusion cooker for final products with sizes smaller than 4 mm is that the raw material must be ground so that a minimum of 99% passes through a screen with 840 micron openings (U.S. Standard 20 Mesh). If this specification is met, then the geometric mean diameter of the particles will be approximately 400 microns (approximately 40 U.S. Standard Mesh).

In producing semi-moist aquatic feeds, liquid ingredients need to be prepared which are generally used to increase the moisture content as well as control the water activity. There are cases when liquified raw materials may be used for producing dry feeds. In this case, these are prepared and mixed so that they may be pumped into the extruder system. The vitamin solutions and marine oil top coatings are prepared to be applied to the surface of the final product after extrusion and drying.

There is a very wide range of formulations that can be utilized for aquatic feeds. Examples are:

Extruded eel feed. Fish meal, 69.10%; wheat flour, 28.62%; vitamins and minerals, 2.28%; approximate total protein, 52%.

Dough type eel feed. Fish meal, 70%; Alpha starch, 20%; fish oil, 5%; vitamins and minerals, 5%; approximate total protein, 46%.

Low aquatic meal shrimp feeds. Soybean meal, 40%; wheat products, 23%; aquatic meals, 20%; vitamins, minerals and fillers, 11%; marine oils and fish solubles, 6%; approximate total protein, 35%.

High aquatic meal shrimp feeds. Aquatic meals, 57%; wheat mids, 21%; corn, 14%; vitamins and minerals, 4%; fish oil and solubles, 4%; approximate total protein, 45%.

Carp feeds. Aquatic meal, 15%; soybean meal, 17%; wheat, 10%; sorghum, 56%; vitamins and minerals, 2%; approximate total protein, 25%.

Tilapia feeds. Aquatic meal, 5%; soybean meal, 45%; corn, 30%; rice bran, 18%; vitamins and minerals, 2%; approximate total protein, 28%.

Extrusion Processing Area

This area begins with the finished raw material being placed in the holding bin of the extrusion cooker. Figure 2 shows a cutaway view of a high-capacity extrusion cooker.

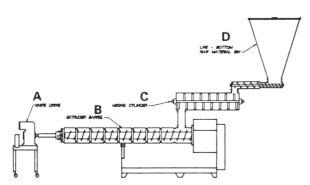

Fig. 2. Cut-away view of a high capacity extrusion cooker. A, knife drive; B, extruder barrel; C, mixing cylinder; D, live-bottom raw material bin.

The live bottom raw material bin keeps the raw materials free flowing so they are easily fed into the feeder screw. The feeder screw is the device that controls the final output of the extrusion cooker. It utilizes a varispeed motor for rate control into the mixing cylinder.

Mixing cylinders are high speed devices that are designed to incorporate steam, water and other liquid additives or slurries into the dry feed mix. It is desirable to have a homogeneous, evenly-moistened and preheated mix prior to the extrusion cooking barrel. This is achieved by controlling the moisture level, steam injection level and retention time. The thermal conductivity of the feed at the proper moisture level allows easier heat penetration from the steam injection. Atmospheric pressures in conditioning cylinders (Fig. 3) are preferred and beneficial, as certain anti-flavor and anti-nutritional factors are allowed to volatilize and escape from the substrate as the preconditioned material passes from the conditioning cylinder through the vented transition into the extruder barrel.

Conventional conditioning cylinders as shown in Figure 3 have retention times of approximately 40 to 60 seconds when producing animal or aquatic feeds at relatively high capacities of 4 to 5 tons/hr. The normal level of pre-

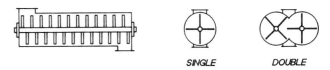

Fig. 3. Conventional conditioning cylinders.

conditioning in a conventional mixing cylinder achieved by steam injection and water addition will elevate the moisture level of the raw material by 7 to 8%. Starting with a dry feed moisture of 12% yields moisture contents out of the conditioning cylinder of 19 to 20%. The degree of cook achieved by these conventional cylinders alone will be no higher than 15%.

A new differential diameter conditioning (DDC) cylinder (Fig. 4) has greater retention times and more intense mixing of the water and steam added to the dry feed, allowing the moisture level of the material to be maintained at an optimum during the preconditioning process. Numerous tests in the Wenger Technical Center have shown that optimum preconditioning times of formulas with fat levels of 7% range from 120 to 180 seconds. Preconditioning times up to 290 seconds are required to optimize formulas containing 22% internal fat. The differential diameter and differential speed conditioning cylinder are required for adequate retention time and moisture blending of this magnitude. Further testing has shown that the degree of cook when utilizing the new DDC design varies nearly in a linear fashion for most materials in moisture ranges of 20 to 30%. Preconditioning at a moisture content of 25% for 150 seconds results in a degree of cook of 33%. Increasing the moisture level to 30% results in a degree of cook of 47%. These are typical moisture and associated degrees of cook levels achieved in the DDC cylinder. This equipment is designed to achieve the maximum retention time, moisture and heat penetration without reducing the nutritional value of the final product. As a warning, over-preconditioning in atmospheric or pressurized systems can lead to destruction of certain amino acids and also contributes to excess fines in the final product as well as to reduced durability of aquatic feeds.

Fig. 4. Differential diameter and differential speed conditioning cylinder (U.S. Patent 4,752,139).

Other benefits seen with proper preconditioning at optimum moisture levels are reduction in extruder barrel wear and reduction of extruder shaft horsepower per ton of product produced. This is a result of complete penetration of the moisture through each and every particle of the raw material, allowing an increase in rate of heat transfer through friction into each particle and also reducing abrasiveness of the substrate. The homogeneous mixture then enters into the extrusion cooking barrel. Up to this point the processes for producing floating aquatic feeds, sinking aquatic feeds and soft, moist feeds are relatively the same. The ability to understand the extruder screw and head geometry has allowed the production of a wide range of products with exacting product specifications. A wide range of extruder screws, heads and shearlocks exists to perform specific functions in the extruder barrel. Correct selection of these heads and screws and placement of these parts in the extruder barrel configuration results in optimization of the process for the final product being produced. The idea is to maximize the throughput while maintaining product quality to ensure the maximum amount of profitability based on investment, utilities and manpower involved.

Additional water and steam may be injected into the extruder barrel. Extruder heads also are equipped with water jackets on their external diameter. This allows the flow of water to be circulated through the jackets cooling the heads. The amount of cooling can be controlled by reducing or increasing the water flow to each extruder head.

The extruder die plate, the last device the product passes through in the extruder barrel, governs the final shape, density, texture, appearance and capacity of the extruder. As a general overview, the extrusion process can be compared to the bread-baking process as shown in Figure 5.

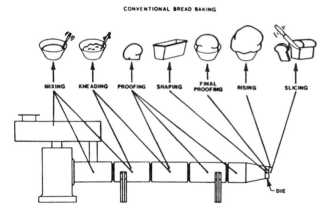

Fig. 5. Extrusion/baking (1).

Manufacturers also produce twin screw extruders with the same preconditioning strategy that is utilized in single screw extruders. Comparing the extruder barrel of a single screw extrusion cooker with that of a twin screw will show that approximately the same transformation occurs as the product passes through the extruder barrel. It is noted that the twin screw extrusion cooker does have advantages for specific products due to the positive pumping action of the intermeshing screw flights. "The most apparent gain in the final extruded product from this more uniform flow and die pressure is a more uniform product shape and length. This advantage is not important when producing a chunk type final product" (2).

High fat formulas also benefit from twin screw extrusion cookers. Twin screw extruders can handle fat levels approaching 25%, while single screw extrusion cookers designed specifically for high fat formulas can handle up to 20% fat. The cost of twin screw extrusion cookers being 60 to 100% more than singles and energy costs between 20 and 50% higher permits justification of their use for products that allow value added to the raw material substrate in sums sufficient to cover the added initial equipment investment and operating costs.

The use of moisture in the form of steam and/or water in the extrusion process is technology applied to all extrusion cookers. It results in reduction of processing costs, denaturing of protein, increased extruder capacity, improvement of cook and reduction of extruder wear. Moisture injected into the preconditioning area of the extruder and into the extruder barrel in the form of steam carries additional energy for cooking. This energy increases production capacity and reduces the size of the main drive motor. The presence of moisture aids in the gelatinization of the starches and denatures the proteins. Preconditioning softens and moistens the raw material and thus reduces the abrasiveness of the material as it passes through the extrusion cooker reducing wear and thus operating costs.

Extrusion cooking at reduced moisture levels results in reduced acceptance of the feed, loss of vitamins and destruction of amino acids resulting from increased shear in the extruder barrel. Figure 6 shows that the optimum moisture content is approximately 27%. The extruder operating cost increases exponentially below 27% and levels off at and above 27% moisture. When extruding at approximately 27% moisture the product must be reduced in moisture content for safe storage. This drying phase is paid for quickly when you realize the exponential increase of electrical cost and wear cost when reducing moisture from 27% to 15% in the extruder barrel while the water, steam and drying costs decrease in a linear fashion (Fig. 6). These inverse cost relationships permit rapid capital recovery of the investment cost of the drying equipment when operating in the moisture processing range of 22 to 29%.

Floating Aquatic Feeds. These feeds typically are in the density range of 320 to 400 g/l. They are expanded pellets varying in diameter from 1.5 to 10 mm. Fish farmers have proven that floating feeds result in better feed conversion because feed consumption can be monitored and adjusted so that no feed is wasted. Many species that consume floating feeds are fed based on eating all the food in a certain time frame. For example, if all the feed is eaten in less than 10 min, then more feed must be given to the fish. The opposite is true; if feed is still on the pond after 10 min then the feeding amount should be reduced. Typically, floating feeds are extrusion cooked at approximately 24 to 27% moisture and expand upon exiting the die to approximately 125 to 150% of the original hole size.

Sinking Aquatic Feeds. These feeds are typically in the density range of 400 to 600 g/l. These products are generally 1.5 to 4 mm in diameter and are used predominantly for shrimp feeds. Sinking aquatic feeds are designed to feed slow-eating, bottom-feeding species. As a result, the major product specification is for the product to hold together in water for two to four hr so that it can be consumed.

A brief history of how sinking aquatic feeds were produced on extruders will give a better understanding of the improvements that have been made. Sinking feeds

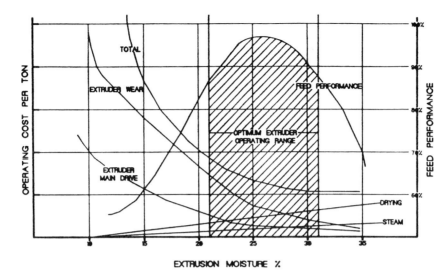

Fig. 6. Relationship of operating costs, extrusion moisture and feed performance.

were produced with the same basic extruder set-up as used for floating feeds except that the die open area was increased and the extruder was run at higher moisture levels, approximately 30-32%. Throughput or capacity of the extruder was also reduced in order to produce a sinking, stable product. For example, a four-metric-ton-per-hour floating product extruder arrangement would have a capacity when producing sinking feeds of 1.2 to 1.8 metric tons of product per hour.

As technology advanced and a better understanding of the requirements in the sinking feed area became apparent, we modified the extruder barrel design in order to improve the profitability of using our equipment on these products. These improvements have allowed an increase in production rates up to levels of 3.3 metric tons (MT)/hr (on our four-ton-per-hour extruder) and higher, depending on the formulas. We also have reduced the moisture content necessary to produce these feeds down to the 22 to 25% range. This has saved costs in drying as the product needs to have only approximately 10 to 12% moisture removed as opposed to 20 to 22%.

Another benefit has been the ability of this design to handle very difficult formulas when compared to what can be produced on a pellet mill. Sinking, water-stable feeds have been produced with as little as 5 to 10% starch-bearing ingredients present in the formula.

Slow-Sinking Aquatic Feeds. These feeds are typically in the density range of 390 to 410 g/l. This is the range at which products begin to sink in salt water. The product size will have an effect in the density measurement and thus exact densities for each formula must be developed to produce slow-sinking pellets. These types of feeds are being used in the salmon-raising industry, especially in the floating ponds in the ocean where net bottoms are used. The slow-sinking feeds thus have a better chance to be consumed prior to their exiting the net bottom.

Soft, Moist Feeds. These products are being produced in small scale operations. They have the attention of researchers, as some aquatic species have shown that they will accept soft feeds but will not consume hard pellets. The technology involved in producing these feeds is very similar to producing floating aquatic feeds on the extruder. The only major difference would be the addition of liquids in the mixing cylinder that increase the moisture content up to approximately 30 to 32%. These liquids contain ingredients such as propylene glycol, potassium sorbate or acids which act as preservatives and control the water activity of the product controlling molds and spoilage.

Feed Ingredient Production

There are applications for extrusion cooking systems that greatly benefit single ingredients or waste products that would normally not be utilized in their natural state.

Animal or Marine Waste

The discussions above regarding preconditioning mentioned the addition of steam, water, other liquids or slurries into the mixing cylinder area of the extrusion cooker. Liquids or slurries can be ingredients that are added on a percentage basis into finished feeds such as liquified animal waste, ground fish slurries, chicken offal, ground shrimp heads, squid viscera and others. In the case of processing single ingredients or combinations of ingredients, the above liquids can be combined with various substrates to produce a stable value-added ingredient. For example, extrusion of mechanically extracted soybean meal with scrap waste fish that has been ground and emulsified into a slurry would yield an ingredient that would be beneficial especially in areas where imported fish meal is expensive. Data in Table 1 show what theoretically happens to moisture, protein and three amino acids, glycine, lysine and methionine, when soybean meal and scrap waste fish are combined to form a stable feed ingredient. The waste fish in this table is 80% moisture in the live state. However, it is assumed that these fish are caught in nets offshore and stored on deck where they dry out. The valuable fish are iced down and kept fresh for sale. The major concern with the above process is the amount of moisture in the liquid portion, as this dictates the percentage allow-

TABLE 1
Combination of Soybean Meal and Waste Fish

	% Moisture	% Protein	% Glycine	% Lysine	% Methionine
Soybean meal	10	43	2.38	2.79	0.65
Scrap waste fish	50	34.1	2.42	2.48	0.92
Combination 50/50	30	38.5	2.40	2.63	0.78
Combination 50/50 dried to 10% H_2O	10	49.5	3.08	3.38	1.00

able in the extrusion cooker and also affects the cost of drying.

Alpha Starch

Alpha starch is an ingredient used in making aquatic feeds for eels. The older, more traditional method of producing eel feeds is to produce a large dough ball by combining various ingredients such as fish meal, fish oil, vitamins and minerals and alpha starch to form a dough ball. The alpha starch gives a firm dough ball that holds its shape in water, allowing the eels to consume it in bites without having great losses due to their voracious feeding habits. Alpha starch is traditionally produced on drum drying equipment utilizing raw, cold weather potato starch as the starch source. The uniqueness of drum dryers in aquatic feeds limits their use to production of alpha starch and flaked fish foods.

Recent work has shown that extrusion cooking equipment can produce alpha starch from raw potato starch on equipment designed for producing floating aquatic feeds. An extruder die change is required.

Figure 7 shows an amylograph of drum dried alpha starch as marketed in Mainland China for eel feeds. Figure 8 shows an amylograph of extrusion cooked alpha starch.

These amylographs utilized 12% solutions, 45 g of starch and 375 ml of water. The heating cycle increased for 45 min by 1°C/min followed by a 10-min holding period and concluding with a 45-min cool down period. The amylographs show that the starches had approximately the same degree of shear and cook even though the processes used to make these products are totally different. Another test used to show that the alpha starch has been processed correctly is the stretch test where 50% water and 50% alpha starch are mixed and formed into a ball. This

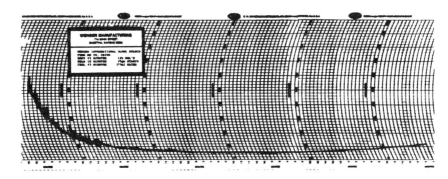

Fig. 7. Amylograph of drum-dried alpha starch.

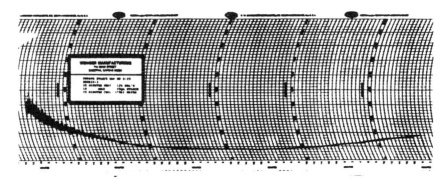

Fig. 8. Amylograph of extrusion-cooked alpha starch.

ball then should be able to be stretched approximately one ft in length, or .3 m.

Post Extrusion Processes

Post extrusion processes start after the product is cut at the extruder die and end before the dryer. With regard to aquatic feeds, the only process used in this area has been a flaking roll to produce flaked fish food which is typically for tropical fish.

Drying and Cooling Processes

Discussions pertaining to extrusion cooking of aquatic feeds generally center on the extrusion equipment itself. The moisture reduction of aquatic feeds is as important as extrusion cooking when overall finished product quality is concerned. The goal in any drying process is to remove moisture so that the product can be bagged and stored safely.

There are many dryer/cooler designs, but for the purpose of discussion we will review the tray-type dryer/cooler (Fig. 9), as it is the most popular style for aquatic feed production. Tray-type dryer/coolers can be built with various lengths, widths, number of passes, materials, conveyor styles, air flows, sanitary considerations and heating systems. The ability to construct dryer/coolers of various lengths, widths and number of passes allows proper sizing for the extruder output as well as offering flexibility in equipment placement in existing or new facilities. The two-pass design not only saves floor space but allows the product to be turned over in the dryer, improving evenness of drying. Using varispeed motors on the dryer passes allows for changes in the product bed depth and, thus, retention time of the product in the dryer.

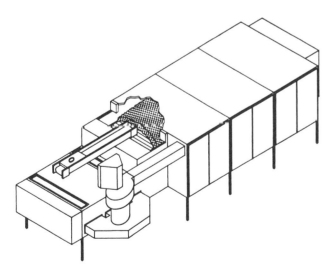

Fig. 9. Dryer/cooler.

Dryer/cooler construction materials include mild steel, stainless steel and a combination of both. In aquatic feed production the dryer/coolers normally are equipped with mild steel construction. However, stainless steel is used where the product comes in contact with the dryer if all stainless steel construction is not available. Product-carrying conveyors are perforated mild steel which can be galvanized or stainless steel trays that can be coated with Teflon or Silverstone. Standard trays have 9/64'' (3.6 mm) as standard and 3/32'' (2.4 mm) as optional perforations. Also available are wire mesh screens which are carried on special supports. The wire mesh design has a greater open area for even air flow and allows for smaller particles to be dried and retained on the conveyor, such as 1.5 mm shrimp feeds.

Air flow is generally up through the product with the heating elements mounted in a single side plenum chamber above air recirculation fans. These recirculation fans create the static pressure below the dryer trays, resulting in evenness of drying across the product bed. A percentage of air is exhausted to control the humidity in the dryer while the balance is recirculated, which permits maximum efficiency in energy consumption. The heating elements mounted in the side plenum are generally steam coils or gas burners. This usually is based on the location of the plant and the cost of the fuel to be used. Steam coils are generally the most popular. Full insulation on all sides of the dryer coupled with the heating systems allows for safe operating temperatures up to 350°F (175°C).

Sanitary considerations include doors wherever possible for cleaning access and maintenance. Fines recovery systems remove fines from the bottom of the dryer/cooler and deposit them in a cross auger which removes them to the side of the dryer/cooler.

Dryers are commonly equipped with integral coolers or extensions of the botton dryer pass. Internal airlocks are used to reduce the amount of air leakage between these two zones. Air flow is generally down through the product, and a percentage of this exhausted air is used as make-up air for the dryer, yielding a more efficient dryer and reducing energy consumption.

In smaller production systems we utilize a combination dryer/cooler where the product is dried and cooled in the same machine. Larger systems operate most efficiently on aquatic feeds when the drying and cooling steps are separated. The reasons for this are ease in controlling the drying and cooling independently and ability to apply vitamins and marine oils to the external surface of the products while they are still warm, allowing good penetration of these liquids into the product.

Post Drying/Cooling Processes

Post drying/cooling processes with regard to aquatic feeds include crumbling, sizing, external coating and packaging. Crumbling is a process by which the feed pellets are reduced in size by passing the product through a set of crumbling rolls. The size of crumbles produced will depend on the gap of the rolls. This process is used in order to produce a variety of very small feeds for young fish and shrimp. The sizing or sifting of the different fractions is required for crumbled feeds. This process simply separates the product according to size. Manufacturers sometimes also sift the final product to remove fines of the larger grow out diets.

External coating systems can vary greatly depending on the application. For aquatic feeds two tanks and an application reel are used. One tank is designed to handle the vitamin and water solution which is sprayed via air atomizing nozzles onto the product at the inlet end of the application reel. The other tank handles the aquatic oils or fat that will be sprayed via air atomizing nozzles in the second half of the application reel, over-coating the vitamins for additional protection against loss of vitamins in the water and as an energy source and attractant for some aquatic feeds.

Equipment Selection and Capacities

Companies and individuals interested in setting up a facility to produce aquatic feeds have two basic strategies, i.e., use extrusion cooking equipment or use pelleting equipment. What are some of the advantages of using extrusion systems to produce shrimp feeds and other aquatic feeds? Advantages of extrusion cooking include versatility of extrusion equipment, increased conversion rate, water stability, density control (floating/sinking), mechanical durability and higher production yields.

Versatility of Extrusion Cooking

Versatility of an extrusion system has great advantages for a feed miller. Aquatic feeds, cooked, ground grain, full fat soy, dog and cat foods, creep and piglet feeds as well as other specialty feeds can all be produced on the same basic equipment with only slight modifications. The versatility of extruders allows many formulas to be extruded into acceptable feeds, resulting in a cost advantage to the feed formulator over pelleting systems.

The pelleting process for shrimp feeds requires that approximately 30–35% of the formula contains ingredients designed to hold the balance of the formulation together. This fraction is referred to as starch-bearing ingredients and binders. As a result of this requirement and the fact that most shrimp feeds require 40–45% protein for excellent growth performance, the feed miller selects expensive raw materials with a high protein content to achieve the 40–45% protein level. These ingredients tend to be fish meals and other aquatic meals. Extrusion cookers are capable of producing shrimp feeds with approximately 10% starch-bearing ingredients. This gives the feed miller the benefit of filling the remainder of the formula with lower cost ingredients that have been proven acceptable to shrimp and still maintain a 40–45% protein content in the final feed.

The data in Table 2 show the approximate percentages that are required for pelleted and extruded shrimp-type diets. As shown there, a pelleted ration needs to utilize ingredients that average 60% protein in order to reach 40–45% protein content diets. The extruded products can use ingredients that average 50% in protein content.

What does the ingredient cost difference do to the final cost of a feed? The formulations in table 3 can provide ideas. Note that the prices used are based on U.S. prices August 87.

As can be seen these two formulas, which are designed within the abilities of the processing method used, yield a lower cost formula by $93.90 US or less expensive by 26.6% for the extruded product per ton. Cost of raw materials can vary, so calculations of local costs are recommended. A savings of this magnitude over the course of one year with a 1,500 kg/hr extruder operated 270 days per year and averaging 8 hours a day (3240 MT/Y) would be over $300,000.00 US. This is a savings to the feed miller in terms of formulation costs which can be passed on to feed users. The savings would be even greater in areas where aquatic meals are more expensive than in the U.S.

Increased Conversion Rates

Conversion rates of extrusion-cooked feeds are traditionally higher due to the cooking process which causes an increase in digestibility of the raw materials, particularly the starch fraction, and resistance to breakdown in water (3).

Aquatic feeds are normally about 90% cooked based on the modified glucoamylase method of testing. What does this mean to the shrimp feed producers and shrimp farmers? Equipment users producing this type of feeds report that the extrusion cooker reduces the amount of feed fed to the amount of weight gained by the animal, the conversion ratio, by .1 to .2. For example, for a feed conversion of 2.5 to 1 then one year's production of feed, 3,240 tons based on above example, would yield 1,296 tons of shrimp. If the conversion ratio is reduced to 2.4 or 2.3, then 1,350 to 1,408 tons of shrimp can be produced from the same 3,240 tons of feed. This means an additional 54 to 112 tons of shrimp. Assuming a world market price of $6.00 US per kilo of shrimp converts this into an additional $324,000.00 to $672,000.00 US in the farmer's pocket.

The above example is true in theory. However, it assumes many things such as increased stocking density in the same surface area of water, proper feeding technique, good water quality and other factors that cannot be controlled by the feed producer. More accurately, a shrimp farmer with a defined pond area has the capability to raise a certain amount of fish or shrimp. If, over the course of a growing season, the farmer can raise five MT of shrimp with a feed conversion ratio (FCR) of 2.5, then he will require 12.5 metric tons of feed. If the farmer switched to a feed that reduces the FCR to 2.3 then he only requires 11.5 tons of feed for the same production level. This means the farmer reduced his feed requirement by 8%. Compound this with lower cost feeds as discussed above, and the shrimp farmers will be able to produce shrimp for a much lower cost.

TABLE 2

Ingredient and Process Comparison

Ingredient sources	Process comparison			
	Pelleted	% Protein	Extruded	% Protein
Nutrients and protein	68% at 60% P.	40.8%	88% at 50% P.	44.0%
Binder and starches	30% at 12% P.	3.6%	10% at 12% P.	1.2%
Vitamins and minerals	2% at 0% P.	0.0%	2% at 0% P.	0.0%
Total % protein		44.4%		45.2%

TABLE 3
Formula Cost Comparison Chart

Ingredient	% P	US$/ Ton	Extruded			Pelleted		
			% In form.	% P	Cost	% In form.	% P	Cost
Soybean meal	44	175.00	35	15.40	61.25	10	4.40	61.25
Cottonseed meal	41	170.00	5	2.05	8.50	—	—	—
Fish meal (men.)	60	385.00	18	10.80	69.30	44	26.40	169.40
Meat and bone meal	50	225.00	10	5.00	22.50	—	—	—
Shrimp meal	40	390.00	10	4.00	39.00	7	2.80	27.30
Squid meal	75	570.00	5	3.75	28.50	5	3.75	28.50
Wheat mids	15.5	60.00	10	1.55	6.00	—	—	—
Corn gluten meal	60	240.00	—	—	—	5	3.00	12.00
Wheat flour	12	200.00	—	—	—	25	3.00	50.00
Brewers yeast	45	500.00	2	0.90	15.00	—	—	—
Fish oil	—	450.00	2	—	9.00	1	—	4.50
Vitamins and minerals	—	—[a]	3	—	—[a]	3	—	—[a]
		Totals	100	43.45	259.05	100	43.35	352.95

[a] Vitamin and mineral costs omitted due to variations in cost. All other costs in this column are based on feedstuffs prices, August 1987.

Water Stability

Water stability or resistance to breakdown in water is very important to the shrimp farmer as shrimp are slow, scavenger-type, bottom feeders and therefore the longer the feed holds together the greater the chance it will be consumed. The average accepted stability time is approximately two to four hr; however, extruded formulas have shown stability of 12 hr without binders and 24 hr and above with the use of binders.

A stability study has been conducted recently in which three formulas were extrusion cooked with various binders to study the water stability and to determine which binder is most appropriate for extrusion cooking. The binders were none, guar gum, carrageenan, sodium alginate, carboxymethyl-cellulose, lignin sulfonate, blood plasma, whole blood, beef stock, basfin and gelatin. The formulations used were low-aquatic, high-soybean meal, high-aquatic meal and a proprietary test diet from the Ocean Institute (4). We also included a commercial Japanese shrimp feed in the stability test for comparison to feed in the industry. The bar charts in Figures 10–12 show the results.

Other areas important to the feed miller are bacteria and water quality. An extrusion cooker eliminates bacteria to a great degree when a formulation is passed through it. We performed a test run on May 6, 1987, where we

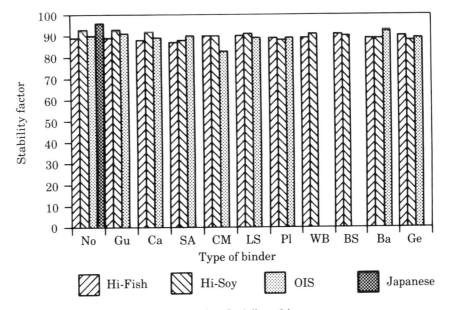

Fig. 10. Pellet stability of extruded shrimp food diets, 2 hr.

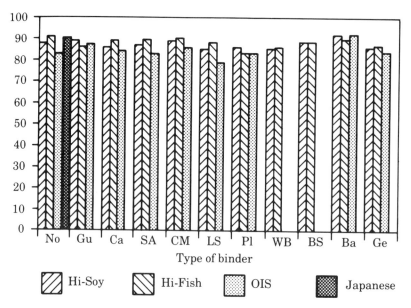

Fig. 11. Pellet stability of extruded shrimp food diets, 8 hr.

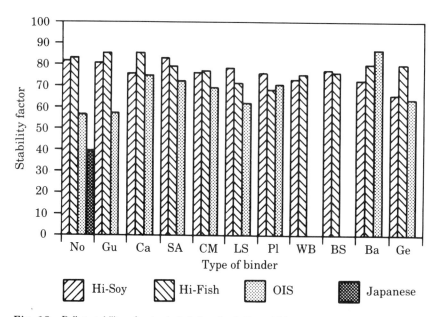

Fig. 12. Pellet stability of extruded shrimp food diets, 16 hr.

extrusion-cooked a shrimp diet. We performed a total bacteria plate count on the raw formula that had 1,200,000 microbes/gram. We also performed a plate count on the extruded version of this same diet, and the total plate count yielded 1,000 microbes/gram. A general guideline is that feed mill products with a plate count 10,000 microbes/gram are considered pasteurized. The animal and fish meals used in the formulations are generally the ingredients that bring a great percentage of microbes into the feed. As a result, formulations that are high in fish meal and meat and bone meal benefit greatly from extrusion cooking due to the reduction in the total bacterial count. This helps maintain water quality and reduces the stress on the shrimp or fish while also reducing mortality.

Other Results

Density Control. Control of density of the finished extruded product allows the feed miller to produce products that float or sink so that the optimum feed can be produced for the species intended. This is controlled by changing the expansion rate and pressure of the extrudate.

Mechanical Durability. Mechanical durability is also important as any breakage or degradation of the product before use results in fines in the pond which affects water quality as well as reduced weight gain. Extruded products have an internal matrix system which tends to increase resistance to mechanical handling in screw conveyors, automatic feeders or simply bag handling.

Pelleted feeds normally generate 5–8% fines in the handling of the feeds in bulk or in bags. Extrusion-cooked feeds produce approximately 1–2% fines, reducing by 75% the amount of material that enters the water and decays on the bottom of the pond. Fines also increase the conversion ratio due to the fact that these particles are rarely eaten in a production pond. Decreasing the fines to 1–2% instead of 6–8% will increase the productivity and efficiency of the feed. On a yearly basis the farmers would save about 194 MT of feed (6% of 3,240 MT).

Extrusion-cooked feeds also have a prolonged storage time. The reason is that extrusion cooking tends to stabilize the raw materials by reducing bacteria and oxidation. Oils do not go rancid and vitamins in extruded feeds are more stable than in pelleted rations. Overdosing of vitamins in an extruded diet is required due to heat liability of vitamins and pelleted rations also required overdosing to get acceptable shelf lives.

Higher Production Yields. It is reported that extruders produce a greater percentage of marketable products. This is called an increased production ratio, the ratio of raw material entering a plant to the amount of finished product produced in the plant not counting the reworked material.

References

1. *Technical and Practical Processing Conditions with Single Screw Cooking Extruders*, Johnston, Gary L., Cooking Extruding Techniques, ZDS, Solingen-Grafrath, Germany, November 1978.
2. Hauck, B.W., *Petfood Industry*, March 1988.
3. Wood, J.F., *Trop. Sci. 22*:(4):351 (1980).
4. *Properties of Extrusion Cooked Shrimp Feeds Containing Various Commercial Binders*, Kearns, J.P., Gordon R. Huber, Warren G. Dominy, Donald W. Freeman, presented at the World Aquaculture Society 19th Conference, January, 1988.

Pelleting for Aquaculture Feeds

Ronnie K. H. Tan

Gold Coin Singapore Pte Ltd, 14 Jalan Tepong, Singapore 2261

Abstract

Apart from raw material quality and formulation, quality shrimp feeds depend on processing which determines the physical characteristics. The farmer requires a feed that must sink, have a good water stability, have an attractive smell, have an even raw material grind, have raw materials finely ground and be dust free. Shrimp feeds may be produced from a specially designed plant or a converted commercial (poultry/pig) feedmill. However, production must be on a dedicated plant to prevent raw material contamination. A schematic flow in a shrimp feed plant is shown and respective processing steps are explained. The important steps of grinding and conditioning lead to good water stability. Grinding should achieve a particle size below 400 microns. Problems faced will be choking or blocking of the screen thus reducing productivity. Two stage grinding is discussed to enhance productivity. Conditioning prepares the mash for better pelletability and aids in binding through starch gelatinisation. Gelatinisation requires heat and water which can be derived from and determined by conditioning time, steam pressure and temperature. In shrimp feeds where accurate and discerning smaller sizes are required, the sifter is required to sieve more than three products.

Quality feed is important to the success and development of aquaculture. Feed can comprise up to 60% of the production cost in shrimp and fish culture. It is thus important that feed used should be as efficient as possible. Shrimp are nibblers and slow feeders. The feed must remain intact for a period of time. An optimum stability should be aimed for. Extreme water stability is not required as long immersion times will reduce the attractiveness and palatability of the feed due to the leaching of nutrients. Fish can be trained to take compounded feed but the texture is important. A hard pellet would result in regurgitation and loss of feed intake. Apart from raw material quality and formulation, quality aquaculture feeds depend on processing which determines the physical characteristics. Dry feeds can be manufactured by the pelleting process. This paper will describe the normal pelleting process and modifications required in the processing of shrimp and fish feeds.

Feed Processing by Pelleting

Pelleting feed has a number of distinct advantages for the animal. Pelleting reduces toxic and microbial content and degradation of feed and ensures that raw materials are well mixed such that the animal has no choice of selection. Pelleting also improves ingestion of the feed and allows for better digestibility leading to improved utilization of nutrients. A standard pelleting system involves the main processes, namely conditioning, pelleting and cooling. Fig. 1 shows a standard pelleting system.

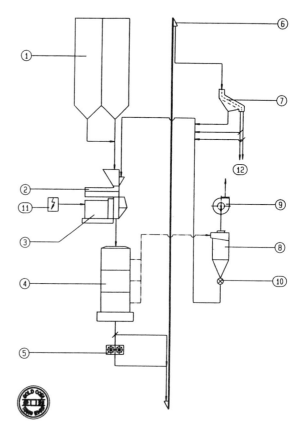

Fig. 1. Pelleting system (standard diagram). 1, meal bins; 2, screw feeder and conditioner; 3, pellet mill; 4, cooler; 5, crumbler with feeder; 6, elevator; 7, fines screen; 8, separator; 9, cyclone with fan; 10, rotary air seal; and 11, control system.

The formulated mash is delivered through a feeder at a variable feed rate to the conditioner. In conditioning, steam (heat and moisture) is injected into the formulated mash to facilitate a workable meal. This allows lower energy consumption during pelleting and increases pellet durability. The finished pellets then undergo cooling and possibly crumbling.

Conditioning is the process of preparation of the mash with heat and moisture from steam before pelleting (Fig. 2). Water, molasses and fats can be added at this point.

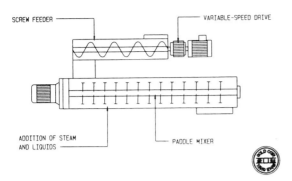

Fig. 2. Short-term conditioning.

Conditioning influences the pelleting and pelletability of the formulated mash and the durability of the pellets and quality of feed. Conditioning the mash is important in order to balance the the differences in moisture content of various raw materials in the formulation; improve the resilience of the particles, reduce friction during pelleting thereby decreasing energy consumption, improve adhesiveness through moisture bridges and partially gelatinize the starches leading to adhesion and increased utilization of the nutrients. Conditioning requires heat, moisture and time. Heat and moisture are derived from steam. The rate of steam addition should vary with the formulated meal to achieve the required physical pellet quality. Moisture levels of more than 16% of the mash before the die increase the chances of choking.

There are different types of steam resulting from the conditioning process. Wet steam is condensate in the form of water droplets and is unsuitable for normal pelleting due to high moisture content. Saturated steam occurs when all the moisture is vapourized. This is very good for conditioning as there is a fixed relationship between temperature increase and moisture increase of the feed. Superheated steam is saturated steam that is further heated. The temperature is higher than the boiling temperature at that particular pressure.

In pelleting, it is necessary to increase the temperature of the feed while only slightly increasing the moisture. A 10°C temperature relates to a moisture increase of approximately 0.6%. To ascertain moisture of the mash, the feed temperature must be determined pre- and post-conditioning. In the pelleting process, condensation of steam produces a thin film of liquid on the surface of the particles. This surface moisture equalizes irregularities and produces adhesive mechanisms (moisture bridges) during the compression of the meal in the die holes. The mash fed uniformly on the die ahead of the rollers forms a bed of material. The rollers in contact with the die compress this bed of material when they roll over it. The material moves into the individual die holes. What is left over is a thin precompressed layer. Each time a roller passes over this layer, a new platelet of material is forced into the die hole. The pressure increases until the plug of material moves through the die hole (Fig. 3). Knives on the outside of the die cut or break off plugs of material forced through die perforations so that pellets are formed. The pellet length is normally 1.5–2.0 times the pellet diameter.

To make pellets storable and transportable, the heat generated by the pelleting process must be dissipated

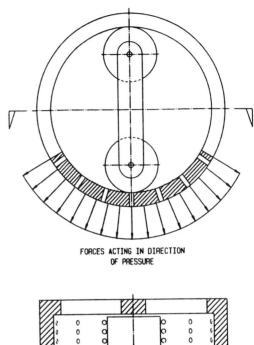

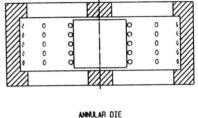

Fig. 3. Forces acting in a pellet mill, with annular dies.

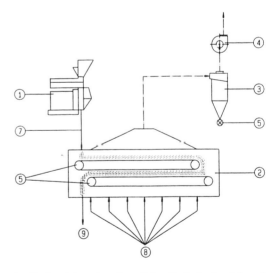

Fig. 4. Horizontal cooler. 1, pellet mill; 2, horizontal cooler; 3, separator; 4, fan; 5, rotary air seal; 6, motors with variable-speed gearings; 7, hot pellets; 8, cooling air; and 9, cooled pellets.

and moisture extracted. Cooling is done with ambient air flowing through the bed of pellets (Fig. 4). At the end of the cooling systems, pellets should have a temperature of 5–8° above that of cooling air. It is important to maintain the migration of moisture from the inside to the outside of the pellets. Otherwise cracks will occur on the pellet surface resulting in higher abrasion.

Processing Shrimp Feeds

Processing determines the physical characteristics of shrimp feeds. The physical characteristics required by a shrimp farmer are that the feed must sink, have good water stability, have an attractive smell, have evenness of the raw material grind, have raw materials finely ground and be dust free. There are two options in setting up a shrimp feed mill. The first is to build a new plant specially designed for aquafeeds. The second is to convert an existing feedmill which currently produces terrestrial animal feeds into one that produces aquafeed. It should be stressed that shrimp feeds should be produced on a dedicated plant to prevent contamination of raw materials. The raw materials used for aquafeeds and poultry feeds are different not so much in the type but the standard of quality.

Fig. 5 shows a schematic flow in a prawn feed plant. The plant is of a post grinding system which includes grinding to extremely fine granulations as required for the pelleting process to achieve good water stability. The plant has a designed mixing capacity of 0.5 MT/hr with two shift operations. Monthly production of about 200 MT can be achieved. The complete plant could be installed inside an 800m² steel structure.

Raw materials: In order to aid grinding capacity, it is advisable to purchase raw materials of a finer grind (particle size distribution).

Blending: Raw materials are manually weighed on platform blending scales. A smaller scale is required for the microingredients in order to achieve weighing accuracy.

Grinding: To achieve 400 μm particle size of the final grind, two options are available: for new feedmills, a pulverizer will do the job; and for those converting an established feed plant to an aquafeed plant, two-stage grinding is possible with a hammermill.

As mentioned, raw material selection should be toward those of less than 2.5 mm particle size. First stage grinding will reduce the particle size using a 1.0 mm screen on a hammermill. In normal animal feedmilling, a 1.0 mm screen is considered small. The chances of choking or blocking the holes of the screen is high. It is made worse when heat emitted from grinding causes the reduction in the viscosity of the oil from materials such as fishmeal and full fat soya. This tends to clump particles together and clog screen holes. Therefore air assistance with a jet filter to create a suction behind the screen is a necessity.

The raw materials used would have varying proximate analyses, some easier to grind than others. It is always advisable to premix ingredients manually in the dumping pit before the grinding. This would smooth the grinding process and prevent choking of the hammermill screen.

In the second stage grinding, the hammermill screen can be changed to 400 μm. The same principles apply as for the first stage grinding.

Sieve analysis using a sieve shaker should be carried out periodically to check particle size distribution of the grind (Fig. 6). Vitamins and minerals which are sensitive should not undergo grinding but are added manually into the mixer.

Mixing: The ready ground raw materials together with the required premix and microingredients are discharged into the mixer. Due to the fineness of ground materials, the homogenity of this mixer has to be 1:100,000. The option of liquid addition such as oils should be provided for at the mixer stage.

Conveying system: The conveying system after fine grinding should be *via* suction pneumatic due to the fineness of particles to be transported. It is interesting to note that there could be demixing as the finer lighter particles are transported faster than the heavier larger particles. This results in the lighter particles (possibly higher percentage of certain raw materials) at the lower layers of the storage bin.

Conditioning: Apart from preparing the mash for better pelletability, conditioning has a very important objective in the manufacture of aquatic feeds, i.e., the gelatinization of starch to increase binding and produce a water stable pellet. Gelatinization of starch requires heat and water. Heat is derived from high temperatures over a certain period of time. Water can be derived from steam or actual addition of water.

In order to achieve the above conditions, three important requirements need to be met. First, a long condition-

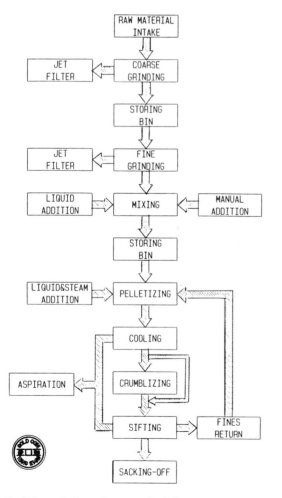

Fig. 5. Schematic flow of a prawn feed plant.

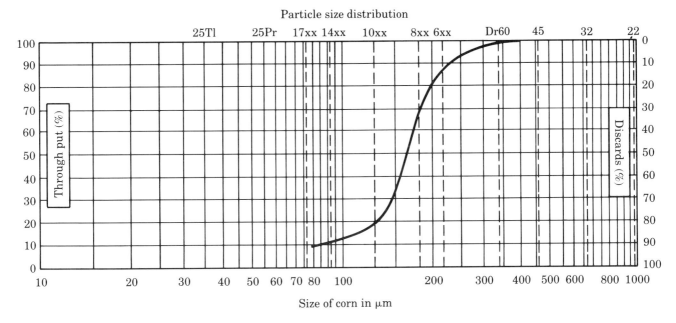

Fig. 6. Particle size distribution chart.

ing time of more than 30 seconds. Second, saturated steam at 1.5 kg/cm³. It should be noted that dry steam is achieved at high steam pressure while wet steam is achieved at lower steam pressures. However, condensate in the form of water droplets should be avoided as the distribution of water droplets is uneven resulting in clumping. This increases the risk of choking the die and uneven water stability if it were able to be pelleted. Third, temperatures of the conditioned mash before the die should be more than 90°C. The moisture content of the conditioned mash before the die should be 15–16%. The moisture of the mash before the conditioner is normally 9–10% which indicates that 6% moisture is added during the conditioning stage. In a situation where moisture from steam is insufficient, water addition may be considered. However, this should only be attempted if there is atomizing equipment producing a mist to ensure even addition of water.

Pelleting: In pelleting, maximum compaction should be achieved in the die which would contribute to increased density and therefore sinking rate, water stability and pellet durability. In choosing the correct die, the criteria for selection are that the hole diameter should produce a pellet suitable for the prawn. Sizes of 2.2–2.5 mm are therefore recommended. Moreover, the die thickness should be optimum to keep the mash in the die hole as long as possible to increase compaction. For a size 420 m die, a thickness of 40 mm is recommended. The cutter should be adjusted so that the length of the pellet is 5 mm for the grower and 10 mm for the finisher.

Drying/cooling: Shrimp feeds are usually smaller diameter products which pack into an area with less air space between each piece. In the dryer the static pressure required to pass air through the product bed depth is greater. As it is a dense product, other considerations have to be taken into account. Moisture migrates much slower through a dense compacted pellet. Thus longer drying times are usually required.

Sifting: After drying, the product is conveyed into a shaker/screen where fines and overs are removed. Overs can be broken up and put back in the system. Fines can be reground and used as a percentage of the formula. This can be done up to approximately 5% of the raw mix.

Crumbling & sifting: In shrimp feeds where accurate and discerning smaller sizes are required, the product has to be pelleted and then crumbled. The crumbling apparatus would be normal but the sifter is required to sieve probably more than three products. The requirement by the farmer may be as shown in Table 1.

TABLE 1
Particle Size of Shrimp Feed

	Post larva 1	Post larva 2	Starter
Shape	Fine crumble	Fine crumble	Coarse crumble
Size	0.6–1.0 mm	1.0–1.5 mm	1.5–2.5 mm

The three products may be of the same formulation so what is required is a sifter with an efficient capability in sifting small volumes to produce more than one product. Such a sifter is a rotational sifter (Fig. 7).

Assuming the requirement in sales is as follows: starter = 60%; post larval 2 = 30%; post larval 1 = 10%, the crumbling flutings can be adjusted in order to raise the percentage of the product in the starter region i.e. 1.5–2.5 mm.

The rotational sifter vibrates in three planes. Vibration in the two horizontal planes causes a rotational movement. This together with vertical vibration ensures efficiency.

Bagging off: The bagging off system will follow that of the normal animal feed system. However bag sizes will be different. Starter, Grower and Finisher feeds should be

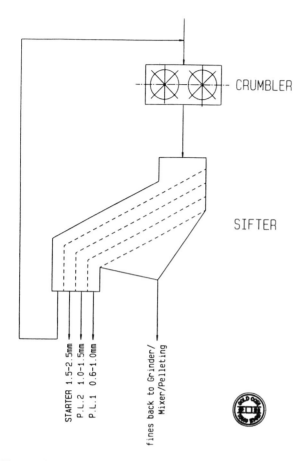

Fig. 7. A rotational sifter.

in 20–25 kg bags as these are easier to transport manually around the farm. Furthermore, if the farm is small and feed usage is low, the time a bag is left open is shorter. Post larval and starter diets should be in 2–10 kg bags. When feed usage is low, the shorter the time a bag is left open, the better the freshness of the feed.

Processing of Pelleted Fish Feeds

The production of pelleted fish feeds is similar to that of processing shrimp feeds. However, fish feeds do not need to achieve such high water stability, e.g., 10 minutes is sufficient. Nor do raw materials in fish feeds require extremely fine grinding as compared to shrimp feeds. And unlike shrimp feed, the texture of the pellet is an important consideration in fish feeds. Certain fish require water stable and durable pellets but spit the feed out if it is too hard. The same schematic flow chart as that for shrimp feed processing can be used.

Raw materials: Raw materials do not have to be of a finer grind but should pass a 3.5 mm sieve to aid grinding capacity.

Blending: Due to the larger volumes of fish feed production, the normal batching system in animal feedmilling is recommended. A small scale for microingredients should be used.

Grinding: One stage grinding is sufficient as the particle size of the raw materials should be below 1 mm. The hammermill will be fitted with a 1 mm screen and air assistance is employed. Again, premixing manually before grinding will smooth the process and prevent choking of the screen. Monitoring of the grind can be done by sieve analysis with a particle size distribution chart.

Mixing: Microingredients that are already finely ground are added at this stage together with liquids. More liquids are used in fish feeds than in shrimp feeds for the following major reasons: fish require higher fat levels. Depending on the raw material proximate analysis, up to 3% fat can be added into the mash; and some fish require softer texture pellets. The addition of molasses is helpful.

Conveying system: The choice for the conveying system is much wider. Chain conveyers can be used as the main advantages are dust free, sanitary operation, gentle handling of the material, high conveying rates, minimized maintenance and self supporting.

Conditioning: Again, gelatinization of starch is required for binding and producing a water stable diet. The requirements are a long conditioning time of more than 20 seconds, saturated steam at 1.5 kg/cm^3, and temperatures of conditioned mash before the die should be more than 90°C.

Pelleting: Maximum compaction does not need to be achieved for fish feeds. In fact, with feeds where a slow sinking rate is required, bulky raw materials such as copra cake are used. These would expand after pelleting decreasing density. Where bulky materials cannot be used, lower compaction is considered for slow sinking rates.

In die selection, fish feed sizes can vary between 2.5–10 mm for a pellet. The length of the pellet should be of a 2:1 ratio in relation with the pellet diameter.

Drying: Process is similar to that of prawn feed.

Crumbling & sifting: Discerning smaller sizes is required for the initial part of fish farming. A rotational sifter identical to that used in shrimp feeds is required for the small volume of crumbles.

Bagging off: The system will follow that of the normal animal feed system. 50 kg bags can be used for the pellets due to the high consumption of fish feed. Furthermore, fish can be trained to feed in certain areas; thus manual transportation of heavy bags around the farm is reduced. For the crumbles, 25 kg bags can be used if consumption of the feed is low. This is to maintain freshness and avoid a bag being opened for long periods.

Processing of Wet Shrimp Heads and Squid Viscera with Soy Meal by a Dry Extrusion Process

Larry A. Carver,[a] Dean M. Akiyama,[b] and Warren G. Dominy[c]

[a]Medipharm USA, 10215 Dennis Drive, Des Moines, Iowa 50322, [b]American Soybean Association, 541 Orchard Road #11-03, Liat Towers, Republic of Singapore 0923; and [c]The Oceanic Institute, Makapuu Point, P.O. Box 25280, Honolulu, Hawaii 96825.

Abstract

Wet marine by-products can be used to enhance soy as a feed ingredient for aquatic animals. Raw and ensiled shrimp heads and raw and ensiled squid viscera were coextruded with 47% solvent-extracted soybean meal to produce dry soy/squid and soy/shrimp feed ingredients. Four products with a ratio of 25% (by weight) wet marine by-products to 75% dry soy meal were produced. These were: raw shrimp head + soy, ensiled shrimp head + soy, raw squid viscera + soy and ensiled squid viscera + soy. The 25%/75% mixtures were processed with a single pass through an Insta-Pro 2000R cooking extruder. An additional amount of wet waste was mixed with the 25/75 air dried product. A second pass through the extruder yielded four more products with a ratio of 46% wet marine by-products to 54% soy meal. The moisture content of the products after overnight air-drying was approximately 9%. Analysis indicates the coextrusion process creates a valuable high quality and inexpensive protein ingredient.

Given the decreasing supply and increasing cost of both animal and plant protein, alternate sources of nutritious but inexpensive proteins for animal feeds need to be developed. These may be provided by by-products of commercial seafood processing operations (1–5). Utilizing high moisture waste presents difficulties, however, because of the high cost of transportation, the extremely liable nature of the product by microbial and enzymatic degradation and the absence of or limited access to drying or refrigeration facilities near the processing plant. These problems currently preclude the economical use of seafood processing wastes.

This might be resolved by "adding value" to feed ingredients that are already produced. The application of a coextrusion process to solvent-extracted soybeans mixed with a raw or an ensiled seafood processing by-product may eliminate a costly waste disposal problem and create an added value feed ingredient (6).

The objectives of this study were to: coextrude solvent-extracted soybean meal with a wet marine by-product (raw and ensiled shrimp heads and squid viscera by a dry extrusion process); determine the processing parameters and the nutrient profiles of the coextruded products produced; and estimate the cost of production.

Materials and Methods

Frozen shrimp heads and squid viscera were obtained from Texas A&M University's shrimp mariculture project and Boston's Calamari Fisheries, respectively. These were air shipped to Des Moines for processing. Frozen wet waste by-products were reduced through a meat grinder with a 10 mm hole die plate and then mixed with soy (Fig. 1). The wet waste was either used raw or put through a lactic acid-ensiling process. The raw or ensiled waste was then coextruded with an Insta Pro 2000R extruder to produce a 25/75 product after the first pass and a 46/54 product after the second pass.

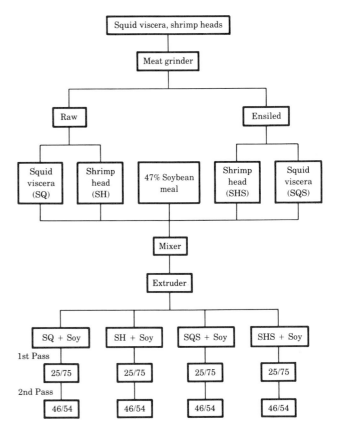

Fig. 1. Extrusion of wet by-products.

The ensiling process (Fig. 2) of the raw by-products utilized lactic acid bacterial fermentation: a fermentable substrate (molasses) was mixed with the raw ground wet waste. Molasses was added at 7.5% by weight to the ground squid viscera and 10% by weight to the ground shrimp heads. Stabisil, a commercial freeze dried lactic

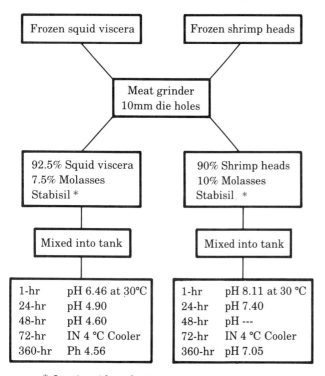

Fig. 2. Procedure used to ensile squid viscera and shrimp heads.

* Lactic acid producing organisms fermented for 24 hours at room temperature. 1 ounce per ton.

acid bacteria starter culture, was fermented for 24 hrs and added at a level of approximately 28 gm/liter of water per ton of wet by-product waste. The pH of the ensiled products was taken at 0, 24, 48 and 360 hrs. Because of the limited amount of frozen squid viscera and shrimp heads, the ensiled products were placed into a 4°C cooler after 72 hrs in order to have an acceptable ensiled material to coextrude 12 days later. The ground raw squid and raw shrimp heads were kept in a −12°C freezer until they were mixed and coextruded.

The Insta Pro 2000R extruder was set up with a steam lock configuration of 11-6-6, double flight screws, and a 9-mm nose cone. The extruder was heated with whole soybeans to obtain a starting temperature of 120–130°C prior to running the coextrusion mixes. Whole soybeans also were used to flush the extruder between coextrusion runs. All the wet waste by-products mixed with soy were calculated to contain a pre-extrusion moisture content of about 22%. After extrusion, the hot extruded products were then spread on a concrete floor and left to cool and dry overnight. For the second pass, the air-dried material of the 25/75 mix was again mixed with the wet by-product waste to a moisture content of 22%, then extruded and air-dried to create the 46/54 mix.

Analyses of dry matter, ash, crude protein, and crude fat of ingredients and coextruded products were by AOAC methods (7). Mineral analysis was made by inductive coupled atomic spectrophotometry using a Perkin-Elmer 650. Lipids for fatty acid determination were extracted from the sample using petroleum ether. Fatty acid methyl esters (FAMES) were prepared following standard AOAC procedures (8). The samples were then injected into a Hewlett-Packard GC with a 30 m × 0.32 mm WCOT capillary column with helium carrier gas at 50 ml/min. Runs were 90 min. with oven temperatures increasing from 100 to 190°C at the rate of 1°C per minute. Fatty acids were identified from retention time obtained by Supelco (Bellefonte, PA) oil standards. Free fatty acid analysis followed AFAD procedures (9).

Results and Discussion

The pH of the lactic acid-ensiled squid viscera decreased from 6.46 to 4.6 after 48 hrs, with a final level of 4.5. Ensiled shrimp heads had an initial pH of 8.11, decreasing to 7.4 after 24 hrs. The buffering action of carbonate in the shrimp shells kept the pH of the ensiled shrimp heads too elevated for long-term storage (10,11). Additional research will be needed to resolve this problem.

The first extruder run had temperatures ranging from 146–181°C (Table 1). This flashed off approximately 50% of the water in the mix. Moisture content of the 25/75 mix the next morning ranged from 8–10.2% after overnight air drying. Extrusion temperatures on the second pass ranged from 163–177°C. The coextruded products from the second pass also were spread on the floor and air-

TABLE 1

Extrusion Processing Run Data

First pass	SQ +SOY	SH +SOY	SQS +SOY	SHS +SOY
Ratio	25/75	25/75	25/75	25/75
Mix moisture (%)	23	20	23	21
Extrusion temp. (°C)	166–177	146–152	166–181	146–152
Air dry overnight (%)	10.2	9.2	8.5	8.0
Second pass	SQ + SOY	SH + SOY	SQS + SOY	SHS + SOY
Ratio	46/54	46/54	46/54	46/54
Mix moisture (%)	22	23	23	23
Extrusion temp. (°C)	163–168	166–171	166–177	166–171
Air dry overnight (%)	9.5	9.0	9.5	9.5

SQ = Squid; SH = Shrimp heads; SQS = Squid viscera silage; SHS = Shrimp head silage

dried overnight. Moisture content of the coextruded 46/54 products ranged from 9–9.5%.

Nutrient profiles of the raw and ensiled by-products as well as the coextruded products are reported on a dry matter basis for a better comparison between basal ingredients and coextruded products. The solvent-extracted soybean meal (Table 2) is an excellent protein source, and has two of the four fatty acids essential for marine crustaceans and fish (12). Squid viscera and shrimp heads are excellent protein components. They also contain several other nutrients that are quite valuable and also necessary for shrimp nutrition.

The 18:2n6, 18:3n3, 20:5n3 and 22:6n3 fatty acids, a high level of cholesterol (13), the available phosphorus, and attractants in both by-products make them excellent candidates for coextrusion with soy. Because soybean meal lacks attractants and some of the nutrients essential to aquatic animals, the coextrusion of a wet marine waste with soy enhances the use of soy in cultured marine species. Analysis of the coextruded products shows improved essential fatty acids and phosphorus content of the soy meal (Tables 2, 3 and 4).

Table 5 presents estimated cost of production of the coextruded products. The cost of the by-product waste will vary according to the supply and demand of the product. In some areas, processing plants will pay to have it hauled away. If the ensiling process is used, the cost of the lactic acid bacteria and the molasses is additional. If the raw product is used, the estimated cost of production less depreciation is about $13.35/ton (not including refrigeration costs) versus $24.65/ton for the ensiled material.

TABLE 2

Analysis of Ingredients (Dry Matter Basis)

	Soy meal (%)	Squid viscera (%)	Squid viscera silage (%)	Shrimp head (%)	Shrimp head silage (%)
Dry matter	90.0	17.8	19.9	19.4	21.0
Crude protein	55.1	75.3	60.8	54.8	45.6
Crude fat	1.0	7.4	7.3	2.7	4.8
Ash	6.5	6.8	9.5	27.4	24.3
Calcium	0.3	0.1	0.4	8.1	6.0
Phosphorus	0.7	1.1	1.0	2.0	1.6
FFA	—	1.9	4.3	0.3	1.25
EFA[a]					
18:2n6	55.6	3.0	NA	1.2	NA
18:3n3	8.4	—	NA	—	NA
20:5n3	—	15.2	NA	10.6	NA
22:6n3	—	25.2	NA	8.9	NA

[a] EFA % extracted fat

TABLE 3

Analysis of Coextruded Shrimp Head Products (Dry Matter Basis)

	SH+SOY 25/75 (%)	SH+SOY 46/54 (%)	SHS+SOY 25/75 (%)	SHS+SOY 46/54 (%)
Dry matter	89.1	89.4	92.0	91.0
Crude protein	53.3	54.7	53.1	54.5
Crude fat	0.9	0.7	0.6	0.7
Ash	8.2	8.9	7.9	9.1
Calcium	1.0	1.2	0.9	1.2
Phosphorus	0.9	0.9	0.8	0.9
FFA	0.07	0.08	0.06	0.29
EFA[a]				
18:2n6	52.5	40.1	46.7	43.5
18:3n3	7.1	5.5	6.8	6.2
20:5n3	1.7	2.3	1.3	2.2
22:6n3	1.7	2.4	1.5	2.2

[a] EFA % extracted fat.

TABLE 4

Analysis of Coextruded Squid Viscera Products (Dry Matter Basis)

	SQ +SOY 25/75 (%)	SQ +SOY 46/54 (%)	SQS +SOY 25/75 (%)	SQS +SOY 46/54 (%)
Dry matter	88.0	88.0	89.9	88.4
Crude protein	55.4	56.2	53.3	53.7
Crude fat	0.6	1.1	1.0	1.7
Ash	7.0	7.4	7.1	7.3
Calcium	NA	0.6	0.9	0.6
Phosphorus	NA	0.9	0.6	0.9
FFA	0.06	0.3	0.14	0.3
EFA[a]				
18:2n6	44.2	37.8	42.4	36.5
18:3n3	6.2	5.2	6.0	4.9
20:5n3	3.0	4.6	2.2	3.7
22:6n3	5.5	8.1	4.1	6.7

[a] EFA % extracted fat.

TABLE 5

Estimated Cost of Products Produced

Estimated ingredient cost	Cost/ton ($)
Cost of by-product waste	—
10% Molasses @ $73.00/ton	7.30
1 oz Stabisil @ $4.00/oz	4.00
Estimated production cost	
Energy @ $.107/kw hour (extruder, cutter, feeder, etc.)	8.65
Maintenance and repair	1.50
Depreciation (building and equipment)	—
Labor cost @ $12.00/hr (mix, grind, extrude)	3.20
	24.65 + ensiled material or 13.35 + raw material

References

1. Meyers, S.P., and J.E. Rutledge, *Feedstuffs* 43:49, 31 (1971).
2. Meyers, S.P., S.C. Sonu and J.E. Rutledge, *Ibid.* 45:47,34 (1973).
3. Meyers, S.P., Utilization of Shrimp Processing Wastes in Diets for Fish and Crustacea in Proceedings of Seafood Waste Management in the 1980's, Orlando, Florida, Sept. 23–25, 1980 (1981).
4. Meyers, S.P., *Infofish Marketing Digest*, No. 4, 1986.
5. Barratt, A., and R. Montano, *Ibid.*, No. 4, 1986.
6. Robinson, E.H., J. Miller and V.M. Vergara, *The Progressive Fish-Culturist, 45(2)*:102–109 (1985).
7. Horwitz, W. (ed.), *Official Methods of Analysis of the Association of Official Analytical Chemists*, 13th edition, AOAC, Washington, D.C., p. 1018 (1980).
8. AOAC, *Official Methods of Analysis of the Association of Official Analytical Chemists*, 14th edition, Arlington, VA., p. 1141 (1984).
9. Anonymous (AFAD publication), in "Test Methods for Fatty Acids," The Association of Fatty Acid Distillers, Liverpool, England, 1968, p. 22.
10. Meyers, S.P., and G. Benjamin, *Feedstuffs*, March 30, 1987.
11. Tacon, A.G.J., and A.J. Jackson, Utilization of Conventional and Unconventional protein sources in practical fish feeds, in *Nutrition and Feeding in Fish*, edited by C.B. Cowey, A.M. Mackie and J.B. Bell, Academic Press, New York, NY (1985), pp. 118–145.
12. Kanazawa, A., Penaeid nutrition, in *Proceedings of the Second International Conference on Aquaculture Nutrition: Biochemical and Physiological Approaches to Shellfish Nutrition*, edited by G.D. Pruder, C. Langdon and D. Conklin, Delaware, U.S.A. (1981), pp. 87–105.
13. Sidwell, V.D., "Chemical and Nutritional Composition of Finfishes, Whales, Crutaceans, Mollusks and Their Products," NOAA Technical Memorandum, NMFS F/SEC-11 (1981).

Production of Extruded Pet Foods

Maurice A. Williams

Anderson International Corp., 6200 Harvard Ave., Cleveland, Ohio 44105 USA

Abstract

Single screw, high temperature, high moisture, self-emptying extruders have found wide acceptance in the production of pet foods since their inception in 1955. They were first used in the production of "kibbled" corn and of fully formulated dry dog foods. Through the years, they also were applied in the production of cat foods, semi-moist and intermediate-moist dog and cat foods, and various cereal grains or mixtures of cereal grains for use as additives into pet foods. This type of extruder also is used in the production of fully formulated fish feeds, both floating and sinking varieties, and feeds for shrimp and for livestock animals including full fat soybean for use as an ingredient in piglet feeds and chicken feeds. This type of extruder is described, and the means discussed by which it can convert ground raw formulations into cooked particles having specific size, shape, texture, and specific degrees of cook to insure adequate starch gelatination, protein denaturation, enzyme inactivation and flavor enhancement without undue damage to vitamins and amino acids.

Fig. 1. The Anderson Expander-extruder-cooker

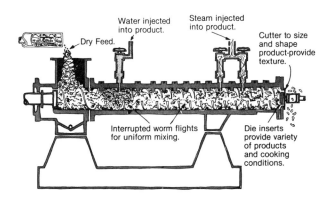

Fig. 2. Sectional view of extruder

Today there is a large variety of extruders offered to the animal feed industry for the manufacture of pet foods and livestock feeds. They range from relatively simple machines requiring very little in preparation ahead of the extruder or in post-processing after the extruder to very sophisticated machines requiring a more extensive processing line. There is a considerable variation in the cost of these extruders and a considerable variation in the range of capability of each kind of extruder.

A single screw, high moisture, high temperature cooker-extruder of a design that permits it to self empty once the die plate is removed offers a high range of capability in producing many kinds of pet foods and animal feeds including floating and sinking fish feeds and shrimp feeds at moderate cost compared to more sophisticated machines (1).

The single screw cooker-extruder used to prepare the animal feeds that are described in this paper was made by Anderson International Corp. of Cleveland, Ohio. A photo of it is shown in Fig. 1. Fig. 2 is a sectional drawing showing the internal mechanism of the extruder.

The extruder internal mechanism consists of a rapidly rotating worm shaft having individual worms with an interrupted flight positioned inside a smooth walled barrel equipped with removable stationary pins protruding from the barrel and intermeshing with the interruptions of the flights. The purpose of the intermingling of the rotating worms with stationary pins is to provide for a high shear, turbulent mixing action, which kneads the solid formulation with the injected water and steam to provide for a rapid and uniform absorption of the injected moisture into all of the solid matter. As the steam is absorbed, it releases its heat of vaporization which helps to elevate the temperature of the mixture. Frictional heat also is generated by the rapid motion of the flights, further elevating the temperature as they compact and work the mixture subjecting it to higher and higher pressure as it is forced through the length of the barrel. By the time the mixture reaches the end of the barrel it is thoroughly cooked and placed under high pressure, high enough that all of the water contained is in the liquid state even though the temperature may be 120° to 150°C (248° to 302°F).

The degree of cook can be varied by varying the final moisture and final temperature achieved within the barrel just before discharge. Typical moisture levels after injection of water and steam range from about 20% by weight to 30% or more. Moisture levels are influenced by the amount of water and steam injected, and the temperature is influenced by the portion of the total moisture that is injected as steam and also by the amount of horsepower consumed for any given capacity, which is influenced by the number and size of the discharge openings selected for the tonnage being extruded. Attractive features of this kind of extruder design are that premoistening or preheating of the formulation are not required and that the extruder, because of the relatively deep channel (distance between the hub of the shaft and the wall of the barrel) is self emptying once the die plate is removed.

The only preparation that is required for this type of extruder is to pregrind the material, preferably so that all of it passes through a 20 wire-per-inch screen (U.S. Series 20 mesh screen), and the only post-extrusion activity required is drying and cooling of the product.

A wide variety of dog and cat foods are made this way as well as floating and sinking fish feeds (2). Figs. 3, 4 and 5 show a typical pet food, floating fish food, and shrimp food made on this kind of extruder. Cross sectional dimensions of the collets are controlled by the configuration of the die openings, and length is controlled by the speed of the flyknife. The cleanness of the cut is controlled by the sharpness of the blades and the clearance between the blades and the discharge dies. Fig. 6 shows a close up of the die plate and cutter assembly. Fig. 7 shows a close up of a typical die for producing a square shaped collet with a rectangular shaped opening in the center. This would be representative of the many different kinds of dies used in this kind of extruder.

The bulk density and texture of the product is influenced by formulation and by the moisture level during extrusion. Formulations high in starch puff to lower bulk densities, and have more friable textures. Formulations high in fat puff to higher bulk densities. Judicious selec-

Fig. 5. Extruded sinking shrimp feed

Fig. 3. Extruded cat food

Fig. 4. Extruded floating fish feed

Fig. 6. Die plate and cutter assembly

tion of starch level and fat level provides an effective way to predetermine what the product's bulk density and texture will be. This when augmented by moisture level of extrusion provides for a wide range of bulk density and texture capabilities. Animal feeds prepared in a single screw, cooker-extruder can vary from a specific gravity of .28 to .48 (bulk density of 18 to 39 lbs/ft^3) and vary from a dense, hard texture to a texture that is easily crumbled in the animal's mouth. The product always puffs or expands to some extent as it exits the extruder (provided it reached cooking temperature). If the moisture level is low, it loses some moisture due to flash as it puffs and it becomes rigid when it reaches maximum puff. If the moisture level is high, the product doesn't

Fig. 7. Closeup of die

lose sufficient moisture to become rigid right away and will collapse somewhat before it finally does become rigid. Taking advantage of this phenomenon provides for quite a bit of latitude in bulk density and texture even from identical formulations. Usually, during operation, the modulation of a single steam injection valve is sufficient to make corrective adjustments in product bulk density. This type of extruder is also used to make some forms of textured vegetable protein from materials like defatted soy flour, particularly a small granule collet used as a substitute for ground meat.

In addition to cooking and puffing animal feed formulations, this type of extruder is used for precooking of starchy cereal grains and the starchy products separated from cereal grains to attain specific levels of starch gelatinization or dextrinization. High moisture, low temperature cooks provide mild gelatinization while low moisture, high temperature cooks provide for a high degree of gelatinization and also some dextrinization of the starch. This is important when extrusion is used to cook starch for use as binders and adhesives. When producing binders, one can obtain a very mild cook when operating at a temperature of 88°C (190°F) and a moisture of 30% and a very strong cook where the temperature reaches 204°C (400°F) or higher and the moisture is around 15%.

Proteins are denatured easily without doing excessive damage to the essential amino acids. Lysine is a very important amino acid in animal nutrition and it is heat sensitive. It is typical of animal feeds produced on this type of extruder that all of the lysine is supplied through animal and vegetable protein sources, all of which pass through the extruder, and the product still contains nutritionally available lysine at the levels required by the animal. To insure that this is so, animal feeds are usually processed at temperatures of 115° to 140°C (240° to 284°F) and moisture levels of 23% to 27%. These conditions are sufficient to inactivate enzymes, destroy bitter, beany, flavors and inactivate certain anti-nutritional factors, and, yet, not do significant damage to protein nutritional quality.

There is a very short residence time (20 to 30 sec) in this type of extruder. If there are variations in the rate of formulation flow, steam flow, or water flow, the equilibria within the extruder particularly the moisture level will fluctuate causing a drift in product quality. This is prevented by assuring uniform formulation flow using a variable speed screw feeder ahead of the extruder and by providing regulators to hold the pressure of the injection water and steam uniform because the flow of water and steam is directly proportional to the pressure. With proper control on the three input flows this type of extruder easily levels out into steady-state operation. It is easy to stabilize the operation of this type of extruder because the formulation can be handled in a dry, free flowing condition as it enters the extrusion barrel.

Some animal feed products are coated with fat after extrusion and some are dusted with a powdered gravy mix after the fat is applied. The gravy mix turns into a gelatinous gravy-like consistency as soon as the pet owner pours water onto the product. The gravy mix usually contains flavor and aroma additives and vitamin and mineral supplements.

Products extruded as an all-in-one formulation are supplemented with vitamins prior to extrusion. Some vitamins, particularly A and C, are damaged by the cooking conditions but not to such an extent that it is impractical to add them ahead of extrusion. An excess of heat-sensitive vitamins are added to allow the product to contain the desired level of usable vitamins (3).

Flavoring additives can be used to complement the natural flavors of the formulation or to mask its off flavors. The additives usually are not added to the raw feed going to the extruder because the high-heat, high-moisture conditions present during extrusion cooking can break down the chemicals changing the flavor of the product or can distill the aroma out of the product when the steam vaporizes from the product during discharge through the die. The introduction of microencapsulated flavoring agents offers a greater heat stable aroma additive that may in some cases be added to the feed entering the extruder.

Dry pet foods, semi-moist (marketed at 20% to 35% moisture) and intermediate moist (marketed at 12% to 18% moisture) are produced on this type of cooker extruder. In addition, full fat soybean for use as an ingredient in poultry and swine feeds also is produced. Grinding is the only required pretreatment, preferably to 20 mesh, although acceptable results are attained with coarser grinds. Sufficient steam is injected to bring the temperature to 121° to 135°C (250° to 275°F) and the product discharges in the form of a meal or crumb without requiring a cutter (Fig. 8). It is then thermally dried to about 10% moisture. Both the enzyme urease and the trypsin inhibitor are thoroughly inactivated at these temperatures if the internal moisture level is around 18% at the higher temperatures or around 25% at the lower temperature. Typical results are 0.0 to 0.09 pH rise (urease activity) and 2 to 5 trypsin inhibitor units per milligram. Some feed manufacturers prefer to reduce cook to allow 0.3 pH rise in order to avoid any possibility of heat damage to the protein while the enzyme and inhibitor are

Fig. 8. Extrusion of full fat soybean

inactivated. This is done by reducing moisture level at any given temperature or temperature at any given moisture level.

References

1. Horn, R.E., *Cereal Foods World 24*:140, 144 (1979).
2. Williams, M.A., *Infofish Marketing Digest*, No. 4, pp. 43–44, 1986.
3. William, M.A., R.E. Horn and P.P. Rugala, "Extrusion—An In-Depth Look at a Versatile Process," *Food Engineering 49*:(10)99, (11)87 (1977).

Status of Vegetable Food Proteins from Lesser-Used Sources

E.W. Lusas, K.C. Rhee, and S.S. Koseoglu

Food Protein Research and Development Center, Texas A&M University System, FM-183, College Station, Texas 77843-2476, USA

Abstract

Development of vegetable food proteins from sources other than soybeans during the past 10 years is described. Citations are provided on investigations of 140 species of oilseeds, cereals, dry legumes and pulses, leaf protein sources, and nontraditional protein sources like rubberseed, buffalo gourd and others.

The objectives of this article are to review the research and development status of potential edible vegetable protein sources and to provide a bibliography through which the reader can access information on food proteins from sources other than soybeans.

As used here, "vegetable food protein" (VFP) means a processed high-protein content flour, concentrate, isolate or derivative intended for use as an ingredient in factory- or home-made foods. "Flours" are the defatted, dehulled, ground solids of oilseeds, or the dehulled and ground solids of low-fat seeds like legumes ("pulses") or cereals; "concentrates" are low-fat content preparations from which hulls and all or substantial amounts of the natural sugars and starch have been removed; and "isolates" are further refined products from which hulls, fat, carbohydrates, and insoluble fibers have been removed. Soybean industry standards require that defatted flours contain 50% or more protein on a dry matter basis (DMB); concentrates over 70% protein, and isolates over 90% protein. VFPs from other oilseeds generally approach these protein levels, but pulse and cereal products are considerably lower in protein content even though the designations "flour," "concentrate" or "isolate" are used.

The Proceedings of the AOCS World Conference on Vegetable Food Proteins, held in Amsterdam in 1978 (1) mention preparation and utilization of food proteins from 32 plant species, including 7 oilseeds, 5 cereals, 16 legumes (dry beans, peas and lentils), and 5 other sources, including coconut, grapeseed, alfalfa leaf, and potato. In preparing for this review, over 2,000 citations were found in *Food Science and Technology Abstracts* on vegetable food protein research in species other than soybeans during the past decade. References are given in this article for 140 species, including 8 oilseeds, 7 cereals, 46 legumes, 69 leaf sources, and 10 others. While neither the Amsterdam Conference, the 1981 European Congress on Vegetable Proteins for Human Foods in Nantes, France (2), or this review attempt to exhaustively cover all plant species which man has investigated as protein sources, it is obvious that the major new areas of food protein research during the past decade have been in dry legumes and leaf proteins.

Current World Protein Resources

Approximately 200 million metric tons of oilseeds are produced annually in the world (3). Based on tonnage, the major species are: soybean, 50%; cottonseed, 14%; peanut (groundnut), 11%; rapeseed/canola, 10%; and sunflowerseed, 10%. The leading producers of the major oilseeds are: (a) soybean: United States, 54%; Brazil, 18%; China, 12%; and Argentina, 7%; (b) cottonseed: China, 22%; Soviet Union, 18%; United States, 13%; India, 12%; and Pakistan, 10%; (c) peanuts: India, 30%; China, 29%; and United States, 8%; (d) rapeseed/canola: China, 30%, Canada, 19%; European Economic Community 12, 19%; India, 14%; and East Europe, 12%; (e) sunflowerseed: Soviet Union, 27%; European Economic Community 12, 17%; Argentina, 13%; China, 8%; and United States, 6%. Approximately 110 million metric tons of protein meal are produced annually, consisting primarily of soybean, 61%; rapeseed, 10%; cottonseed, 9%; sunflowerseed, 7%; and peanut, 4%. The rank order of nations producing meal is not the same as for seed production because of differences in ratios of whole seed and "value added" products shipped, and substantial in-transit processing industries in nations like The Netherlands and Portugal.

Approximately 1.6 billion metric tons of cereal and coarse grains are produced annually (wheat, 31%; [rough] rice, 28%; corn [maize], 27%; barley, 10%; and oats, 3%), and 45 million metric tons of legumes and pulses. In theory, the world's current production of cereals, oilseeds and pulses alone could provide the protein requirements for about 14 billion people annually at a 45 g protein/day basis, and 10 billion people annually at a 65 g protein/day basis if extracted and consumed directly. In practice, the majority of oilseed meals, corn and oats are used in feeding animals. In addition to these crops, man's diet includes proteins from products of domesticated animals, vegetables, tubers, fruits, fish and game. The average per capita protein intake is about 99 g/day in affluent countries, 58 g/day in developing countries, and 84 g/day world-wide (4).

By far, soybeans are the major source of commercial VFPs, with an estimated annual world production of 1.78 million metric tons (1.5 million tons of soy flour, 90,000 metric tons of soy protein isolate, and 70,000 tons of soy protein isolate (5). The second major commercial VFP is wheat gluten, with an annual estimated world production of 275,000 metric tons. In addition, significant quantities of dry bean and pea flours (for which statistics are not available), plus smaller quantities of dry pea concentrates, and bean, cottonseed and peanut flours are consumed.

Changes in the state-of-the-art of VFP production and utilization during the past decade may be characterized as: continued growth in usage of soy food proteins and wheat gluten; massive efforts in characterizing the world's bean and pulse crops and fortifying regional diets; limited marketing of cottonseed, peanut, corn (zein), and microbial single-cell protein products; continued development of sunflower, rapeseed and leaf proteins; increasing interest in newly available crops and protein sources like lupin, corn germ and extracted rice bran; and declining interests in safflower, grapeseed and potato protein products.

Objectives of Vegetable Food Protein Usage

Utilization of VFPs varies among localities, but is primarily driven by economic considerations. The industrial nations have been the first to develop VFPs, primarily as a means of obtaining lower-cost functional ingredients for centrally processed convenience foods. Properties like heat-setting, water and fat binding, emulsification capacity, aeration, viscosity induction, imparting of texture, light color, bland flavor, and compatibility with other ingredients are of prime interest. Nutritional characteristics have been less important in the industrial societies which already exceed recommended minimum daily requirements of proteins, even though demonstration of ingredient wholesomeness before authorization for use and nutritional labeling of the final product are often required.

In contrast, the developing countries have not had the capital to develop modern VFP industries. Instead, emphasis has been on improving usage of local crops, primarily through complementation of local cereal or tuber calorie and protein sources with legume and oilseed products to achieve improved essential amino acid profiles, especially lysine and the sulfur-bearing amino acids. Special attention has been given to development of low-cost and effective ways of removing or inactivating antinutritional and toxic compounds inherent in pulses and some oilseeds, including trypsin inhibitors, hemagglutinins and specific toxic components; and instructing the public in home, village-level and factory food preparation techniques for utilizing the new protein sources. Although functionality has not been as important a consideration in developing countries, proposed new VFP ingredients still have to be compatible with local native foods and preparation techniques.

Usage of VFPs is expected to increase in the next decade, but the specific applications cannot be predicted. Production of partially or fully preprocessed foods utilizing varieties of ingredients is increasing in all countries. High animal protein diets are coming under closer medical scrutiny in the affluent nations, whose populations are sufficiently wealthy and informed to choose between food sources. The long-established tradition of increasing animal protein consumption as standards of living improve remains in the developing nations, but probably receives less encouragement from nutritionists than two decades ago. Good nutritional performance using traditional plant protein foods has been documented, and a seemingly inexhaustible supply of high quality leaf proteins also is potentially achievable if the world learns to use it. The balancing of import-export trade payments is becoming a major priority in all countries, and often has resulted in subsidizing or protection of local agricultural production and local processing industries. Through this mechanism, the European nations now have significant pools of locally-grown soybean, canola and sunflowerseed proteins to exploit for food uses.

Matz (6) has reviewed U.S. patents issued for protein food supplements for the period 1974-1981. Processes for preparing VFPs typically are generic and adaptable to other oilseeds with minor adjustments, for example, the procedures developed by Lawhon and coworkers on industrial membrane isolation of proteins and aqueous processing for soy (7) have been applied experimentally to glandless cottonseed, peanuts, sunflowerseed, sesame and lupine at the Food Protein Research and Development Center and implemented in selected industries. A process for producing bland, light-colored oilseed proteins, in which a filterable solution of proteins is passed through ultrafiltration membranes with molecular weight cutoffs of 70,000 or larger, has been patented by Lawhon (8). Most of the research on VFP functionality has been conducted on soy, but also is applicable to proteins of other species. Comprehensive reviews on functionality have been prepared by Kinsella (9), Hermansson (10), Kinsella et al. (11), and Pomeranz (12), and on chemical and enzymatic modification of functionality by Feeney and Whitaker (13).

Typical compositions of experimentally prepared soybean, peanut, glandless cottonseed, sunflowerseed and sesame defatted flours, protein concentrates and isolates, and air-classified high protein fractions of navy and pinto beans are shown in Table 1. Uses of VFPs in traditional foods have been reviewed by Lusas and Rhee (14).

Reviews have been prepared on nutrient contents of oilseeds, legumes, pulses and oriental food products by Haytowitz, et al. (15); nutrition and complementation of vegetable and legume food proteins by Steinke and Hopkins (16); nutrition of oilseed proteins by Bodwell and Hopkins (17); trypsin inhibitors by Mounts and Rackis (18); nutritional and enzyme inhibitors by Friedman (19); and phenolic compounds in oilseed flours by Dabrowski and Sosulski (20). Amino acid compositions of proteins of the major oilseeds are shown in Table 2. A computer-based graphical method for evaluating protein quality of food blends relative to cost has been developed by Traver et al. (21).

Multiple Crop and Comparative Research

Considerable comparative research has been done simultaneously on multiple crop materials, and is presented here instead of by individual species.

Hron (22) reported development of a differential settling test to assess protein-classification efficiencies of cottonseed, corn germ and peanut samples utilizing commercial-size liquid cyclones. El-Sherbiny et al. (23) reported on production and evaluation of protein isolates from selected legumes (broad beans, soybeans, peas, white beans, and peanuts). Ibrahiem et al. (24) found that protein isolates made from infested wheat, corn, lentil, and horse beans were unsuitable for human consumption.

Saunders and Kohler were granted a U.S. patent for a general process for extraction of protein concentrates from oilseed, cereal and legume flours by mixing them with whey, adjusting the pH to 9-10 with alkali, separating a liquid fraction containing soluble proteins, adjusting the pH 3-4, adding sodium hexa-metaphosphate to precipitate the protein, separating the protein concentrate solids, adjusting to pH 7, and spray drying (25). In later publications, Kozlowska and Borowska (26) reported on preparation of protein concentrates from field pea and faba

TABLE 1
Per Cent Protein Content of Various Fractions of Several Oilseeds and Legumes (14)

Fractions	Soybeans	Peanuts	Glandless cottonseed	Sunflower seed	Sesame	Navy beans	Pinto beans
Whole seed	34	27	39	20	18	26	26
Dehulled seed (kernel)	43	30	43	24	26	30	29
Classical process							
Defatted flour	55	52	63	48	50	—	—
Protein concentrate	72	71	71	70	77	52[c]	53[c]
Protein isolate	93	92	91	90	90	—	—
Aqueous extraction process							
Protein concentrate	68	70	67	68	71	—	—
Protein isolate	86	92	91	89	87	—	—
Membrane process							
Protein concentrate	68	71	71	—	—	—	—
Protein isolate	87	92	92[a] 80[b]	—	—	—	—

[a] Storage protein; [b] Nonstorage protein; [c] Air-classified high-protein fraction.

TABLE 2
Essential Amino Acid Profiles (g/16g N) and Protein Efficiency Ratios of Various Protein Food Ingredients (14)

	Essential Amino Acid								PER
	Lys	Leu	Val	Ileu	Thr	Phe + Tyr	Met + Cys	Try	
FAO/WHO Reference protein	5.5	7.0	5.0	4.0	4.0	6.0	3.5	1.0	—
Soybean									
Defatted flour	6.9	7.7	5.4	5.1	4.3	8.9	3.2	1.3	2.2
Concentrate	6.3	7.8	4.9	4.8	4.2	9.1	3.0	1.5	1.8
Isolate	6.1	7.7	4.8	4.9	3.7	9.1	2.1	1.4	1.6
Peanut									
Defatted flour	3.0	6.4	5.3	3.2	2.6	8.4	1.9	1.0	1.8
Concentrate	3.0	6.7	4.5	4.3	2.5	10.0	2.4	1.1	1.6
Isolate	3.0	6.6	4.4	3.6	2.5	9.9	2.4	1.0	1.4
Glandless cottonseed									
Defatted flour	4.0	6.0	4.5	3.1	3.2	8.5	3.8	1.5	2.2
Concentrate	4.0	6.2	4.9	3.2	3.1	8.8	3.5	1.5	2.0
Isolate, classical	4.0	6.1	4.7	3.2	3.2	8.7	3.7	1.5	1.8
Nonstorage protein	6.2	6.4	4.6	3.4	3.4	7.4	5.0	1.6	2.4
Storage protein	2.9	5.6	4.6	3.0	2.6	9.2	2.4	1.0	1.6
Sunflowerseed									
Defatted flour	4.2	7.2	5.7	4.5	3.9	8.7	3.6	1.1	2.1
Concentrate	4.2	6.9	5.7	4.6	3.7	8.7	3.6	1.0	2.0
Isolate	4.1	6.4	5.5	4.3	3.7	8.6	3.4	1.0	1.9
Sesame									
Defatted flour	3.5	7.4	4.6	4.7	3.9	10.6	5.6	1.9	1.7
Concentrate	3.0	7.1	4.5	4.2	3.6	8.4	4.9	1.9	1.6
Isolate	2.1	6.6	4.6	3.6	3.3	7.9	3.7	1.8	1.4
Whole navy bean flour	7.2	7.6	4.6	4.2	4.0	7.7	1.9	1.0	—
Whole pinto bean flour	6.5	7.5	5.4	5.0	3.8	6.7	1.8	0.9	—

beans using sweet whey or effluents of potato processing, and (27) preparation of faba bean and soybean protein isolates with reduced phytic acid and trypsin inhibitor contents using potato processing effluents.

A U.S. patent was awarded to Campbell et al. (28) for production of vegetable protein concentrates by extracting defatted flakes with aqueous alcohol solution, and desolventizing them in a humid gas atmosphere to produce a vegetable protein concentrate that is very bland and light colored. The process is applicable to soybeans, rapeseed, sesame seed, cottonseed, sunflowerseed, peanuts, corn, yellow field peas, and faba beans. Bradford et al. (29) were granted a European patent on treatment of proteinaceous materials from sunflower seed, rapeseed, corn, soybeans and other legumes with aluminum at pH 2-4 to render the phytate-containing proteins water soluble. The process also reduces trypsin inhibition effects, enhances digestibility, and alters the chemical, nutritional and functional properties of proteins.

Last (30) in Australia evaluated the suitability of producing texturized protein products or expanded cereal-based foods from soybean, lupin, vital wheat gluten, peanut, white mustard, brown mustard, subterranean clover, chick pea, split pea and broad bean flours and flour blends. Most of the flours needed to be blended with defatted soy flour to facilitate their texturization. Amiot et al., (31) evaluated the nutritive value of sodium caseinate, wheat gluten, and soy isolate proteins textured with alginate by extrusion into calcium chloride solutions. Low-cost extrusion studies have included preparation of blended cereals, oilseed flours and dried milk to prepare low-cost products in Sri Lanka (32), and production of precooked foods in developing countries (33).

McWaters and Cherry (34) compared emulsion and foam capacities, emulsion viscosity, protein solubility and gel electrophoretic properties of distilled water suspensions of defatted soybean, peanut, field pea and pecan flours. McWaters (35) conducted an extensive comparison of cookie baking properties of defatted peanut, soybean and field pea flours. Fleming and Sosulski (36) conducted microscopic evaluations of breads fortified by replacing 2% of the flour by vital wheat gluten, or by 15% soy, sunflower, faba bean and field pea concentrates, and reported that supplemented breads showed decreased loaf volumes, more compact or coarse crumb grains, and textures more resistant to compression.

Simultaneous comparisons of several protein sources, alone or as mixtures, for nutritional value include: evaluation of 10 protein-enriched formulas for infants and pre-school children utilizing combinations of lentils, chickpeas, skim milk, yeast, sesame, soy flour, and fenugreek (37), soy, rapeseed and wheat proteins (38), and melon seed, cowpea and soybean preparations as sole sources of protein in rat feeding trials to evaluate infant weaning foods (39). Kozlowska et al. (40) characterized and quantified phenolic acids in oilseed flours, including soybeans, cottonseed, peanuts, rapeseed, white mustard, flax and sesame seed.

Problems of crop identification became apparent in preparing this review, with identical common names often used for different dry legumes and leafy plants. Thus, the Latin botanical names reported by the respective researchers have been included to minimize confusion, although in some cases there may be reason to question their correctness.

Oilseed Proteins

Soybean (Glycine max L. Merr.)

More effort has been invested in developing the technology of soy food proteins than perhaps for any other species. Soybeans are a relatively inexpensive protein source, with less than 2% of the world's crop utilized in production of VFPs. Consequently, they are the cost and performance standards of the VFP industry, and contemplated alternative plant protein ingredients must offer significant benefits to be chosen in the marketplace. The structure of soy proteins has been reviewed by Nielsen (41). Soy protein products, processing functionalities and applications have been reviewed recently by Visser and Thomas (42) and also in a Conference on Soybean Utilization Alternatives (43).

Cottonseed (Gossypium hirsutum L., G. barbadense L., G. arboreum L., and G. herbaceum L.)

Cottonseed is the world's second major oilseed in tonnage produced, and has potential for providing the annual protein needs for approximately 350 million people at the 45 g/day rate or 240 million at 65 g/day. However, cottonseed contains gossypol, a green-brown compound toxic to man and other monogastric species but tolerated in limited amounts by ruminants. Typical domestic cultivars contain 0.7–1.2% gossypol. The FAO/WHO has set 0.060% (600 ppm) of free gossypol and 12,000 ppm of total gossypol as the upper limits allowable in food products; the United States Food and Drug Administration has established an upper limit of 450 ppm of free gossypol. Options for reducing free gossypol content in cottonseed include: "binding" gossypol by moist heating of seeds flaked to rupture gossypol "glands" before solvent extraction; extraction of gossypol using selective solvent systems; physical removal of intact gossypol glands; and growing of genetic varieties of "glandless" cottonseed that do not contain gossypol. A fifth option, treatment of cottonseed meal with iron salts to selectively bind gossypol, has been used in animal feeds. References for these techniques are given in two review articles by Lusas and Jividen (44, 45) on the first 25 years of glandless cottonseed utilization research.

Cottonseed flour deactivated by moist heat, reportedly, is produced in Israel, Mexico, India, China and some African countries. Two Liquid Cyclone Process (LCP) plants, for removing intact gossypol glands by fine grinding dried dehulled cottonseed, suspending in hexane, and separating the denser gossypol glands continuously by liquid cyclones (46), were installed in the mid-1970s in Texas and India, but later dismantled. An air-classification process, consisting of thick flaking dry dehulled cottonseed, extraction of oil by hexane, grinding the flakes by pin mill, and dry separation of intact gossypol glands by air classifier (47), was investigated but not implemented commercially.

Processes have been developed for making glandless cottonseed flour, concentrates and isolates, and roasted kernels ("nuts"). However, although cottonseed accounts for approximately 60% of the weight of seed cotton, it only provides about 10–15% of the gross returns to producers. Growers have been concerned that the benefits from changing to glandless varieties may be more than offset by reduced yields or fiber quality, or increased insect problems. Although it has been repeatedly documented in several countries that these problems do not exist with the latest varieties of glandless cottonseed, farmers have been reluctant to convert to growing glandless cotton. A glandless cottonseed kernel plant established in Texas in the late 1970s has been closed, but production and processing of glandless cottonseed reportedly is being investigated in Cameroon, China, Egypt, India, Israel, Ivory Coast, Mali, and the Soviet Union as well as the United States.

Research has continued on preparation of low-gossypol content cottonseed protein products by solvent extraction (48), preparation and nutritive value of isolate by alkali peptization (49), solvent differential sedimentation of gossypol glands from Egyptian cottonseed (50), and gossypol deactivation by high moisture cooking of a mixture of up to 50% cottonseed (or defatted cottonseed flour) and soybeans using a low-cost extruder (51, 52). India has allowed use of low-gossypol cottonseed flour in Paushtik wheat atta, a ground whole wheat product nutritionally fortified with oilseed proteins, vitamins and minerals (53). Enrichment of local wheat flour with cottonseed flour for making bread has been evaluated in Saudi Arabia (54).

Glandless cottonseed research has included amino acid fortification of glandless cottonseed (55), evaluations of corn-glandless cottonseed blends in feeding Haitian children (56), preparation of soft-textured high protein curds (57), and structure and functionality studies of protein isolates at various levels of succinylation (58).

Peanut, Groundnut (Arachis hypogae L.)

Peanut protein technology and utilization has been reviewed by Rhee (59), and preparation of peanut vegetable food proteins by Natarajjan (60). As with cottonseed and corn, peanut food products share the potential of aflatoxin contamination unless selection of raw materials is closely monitored.

A patent for making a bland slurry of ground peanuts and other components by steam injection, followed by vacuum cooling, was granted to Harris (61); this product can be dried and used in composited foods. Pominski, et al. (62) developed a direct extraction process for production of white defatted food-grade peanut flour. Maw was issued a United States patent for a process for making bland peanut concentrates (63). Ihekoronye in Nigeria, studied the functional properties of red skin peanut meals (64), and prepared an acid-precipitated protein concentrate (65). Lawhon and Lusas (66), using an ultrafiltration membrane process, extracted soy and peanut protein isolates using cheese whey.

Using twin-screw extrusion, Hagan et al. (67) compared texturization of coprecipitated soybean and peanut proteins with dry-mixed proteins, and found that coprecipitated concentrates showed increased rehydration capacity, decreased peak force and work, and modified extrusion characteristics to achieve moderate expansion and a unique ultra-structure. Aboagye and Stanley (68) studied thermoplastic twin-screw extrusion of peanut flour over the range of 120–180°C, and reported that product bulk density, Warner-Bratzler shear, water absorption capacity and microstructures were influenced mostly by process temperature and feed moisture. High temperatures, high feed moisture and low screw speed enhanced product expansion and increased textural integrity. Products with many different characteristics could be produced by manipulating these three variables. Toft and Openshaw have applied for a United Kingdom patent for preparation of snack food half-products using peanut flour (69).

McWatters evaluated substitution of peanut and cowpea meals for wheat flour in cake-type doughnuts (70), and also reported that the oil content of fried doughnuts was increased above that of the control (71). This work was later extended to other fried and baked foods (72). Yogurt has been prepared from buffalo milk and peanut proteins in India (73); and Chinese-type noodles supplemented with peanut flour have been evaluated in Thailand (74).

Studies in Pakistan have included the chemical composition and nutritive value of wheat bread supplemented with peanut flour (75), nutritional and organoleptic evaluations of wheat bread supplemented with peanut flour (76), and wheat and corn breads supplemented with a mixture of peanut and chickpea flours (77). Research has been conducted in the United States on fortification of pearl millet gruels and steamed cakes with peanut flour for infant and preschool children feeding in Cameroon (78), and on protein enrichment of saltine crackers (khara biscuits) with peanut or soy flours (79). Supplementation of cake mixes with peanut, soybean and sweet potato flours has been studied in Egypt (80, 81).

Rapeseed/Canola (Brassica campestris L., B. napus L.)

The chemistry and technology of rapeseed proteins has been reviewed by Ohlson (82). Cultivation of "canola" ("double-low") types of rapeseed is increasing throughout the world. These varieties were developed in Canada, and contain less than 5% erucic acid in the oil and less than 3 mg glucosinolate expressed as 3-butenyl isothiocyanate per gram of oil-and moisture-free meal. However, food utilization research also is being conducted on local and improved rapeseed varieties in many localities. Rapeseed protein has an excellent essential amino acids balance, and generally has displayed Protein Efficiency Ratios (PERs) higher than soybeans and at least equivalent to whole milk protein.

Problems in rapeseed processing and utilization have included separation of hulls from the small kernels; genetic reduction of erucic acid and glucosinolate contents; deactivation of myrosinase which hydrolyzes glucosinolates into glucose, sulfate and isothiocyanate compounds (that then can be transformed into cyanide or vinyloxazolidinethion (OZT); and studies of antinutritional effects of phytic acid (myo-inositol hexaphosphate); and reduction of phenolic compounds (present at 10 times the level in soybeans) which have been shown deleterious to pregnant rats.

Sosulski (83) has fractionated rapeseed meal into flour and hull components using liquid cyclones. Kozlowska and coworkers have studied removal of undesirable antinutritional and flavor compounds from genetically-improved (low erucic acid and glucosinolate) varieties of rapeseed by diffusion-extraction (84), developed HPLC procedures for determination of phenolic acids in rapeseed (85), characterized and quantitated phenolic acids in rapeseed and mustard (86), characterized myrosinase in Polish varieties of rapeseed (87), and published on production of rapeseed vegetable food protein ingredients (88). Schwenke et al. (89) have studied the precipitation of 12S globulin and low-molecular weight albumins with phosphate-containing polyanions. A U.S. patent has been awarded to Cameron and Myers for preparation of protein isolates from rapeseed, sesame, cottonseed and other oilseeds, and peas, by a modified water extraction technique (90).

Government subsidies to oilseeds producers in European Economic Community nations during the past decade have led to major growing of rapeseed and oil-type sunflowerseeds. The ready local availability of both of these oilseeds has resulted in intensified studies on their food uses, and a new wave of utilization publications is expected during the next several years.

Sunflowerseeds (Helianthus annuus L. var. marcocarpus DC.)

The chemistry and utilization of sunflower proteins has been reviewed by Lusas (91). Among the oilseeds considered as sources of vegetable food proteins by man, sunflowerseed is unique for its relative freedom from toxicity problems (Gassmann, (92)). However, sunflowerseed contains chlorogenic acid, a colorless phenolic compound which changes to an irreversible olive-green color when protein isolates are made by traditional alkali extraction-acid precipitation processes. Methods to avoid formation of the green color have included diffusion extraction of chlorogenic acid and use of compounds often characterized as antioxidants (87). Lawhon et al. (93) developed the OEEP (oxygen expulsion and exclusion process), wherein white-colored protein isolate is made by alkali extraction of defatted sunflower flour in an oxygen-free atmosphere using deoxygenated water.

Rossi and Italian coworkers have reported on production of food-grade flour from defatted sunflower meal (94), and found that sunflower protein has greater whippability capacity than soy protein (95). Their studies on effects of extrusion cooking on structural and functional characteristics of sunflower proteins showed that good texture can be obtained at natural pH 7.0, and that noncovalent and disulfide bonding are responsible for protein solubility, texture formation and stabilization (96).

Researchers in the German Democratic Republic have reported on preparation and application of sunflower and other vegetable proteins for human consumption (97), functional properties of sunflower and other plant proteins (98), functional properties of succinylated sunflower proteins (99), and functional properties of sunflowerseed and other oilseeds protein products in meats (100, 101).

Phytic acid has been studied in sunflower seeds, pressed cake and protein concentrates in South Africa (102); functional properties of enzyme-modified concentrates evaluated in Canada (103); and preparation, properties and food uses of sunflower flour and protein isolate (104), and enrichment of biscuits with sunflower seed protein (105) in Australia.

Sunflowerseed flour has been offered occasionally in many markets. Protein concentrates have been made commercially in Italy and evaluated in various products including pastas. When peanut production was decreased in the United States by drought in the early 1980s, a sunflower "butter" was marketed briefly. Sunflower milks and tofus also have been sold.

Sesame (Sesamum indicum L.)

Sesame utilization has been reviewed by Johnson et al. (106), and physical characteristics and functional properties by Kinsella and Mohite (107). Functional and utilization characteristics of sesame flour and proteins have been described in Venezuela (108); and Mexican researchers have reported on effects of sodium stearoyl-2-lactylate on rheological and baking properties of breads containing defatted soy and sesame meals (109).

Studies on fortification of corn-based foods with sesame, chickpeas and soybeans have been conducted in Mexico (110); and U.S. workers have reported that a mixture of equal parts sesame and soy flour has a PER of 2.45 when used in Venezuelan-type corn *arepas* (111).

Safflower (Carthamus tinctoria L.)

Betschart (112) described preparation of safflower protein isolate, whose functional properties were later evaluated in simple systems and breads (113). PER of safflower protein isolate was 2.13, but was increased to 2.26 by supplementation with lysine. Also in the same laboratory, lipid-protein complexes, containing 18% oil and 17% protein, were prepared from safflower expeller cake. These products were free of bitterness and cathartic activities. Bread, in which 10% of the wheat flour was replaced by the lipid-complexes, had acceptable properties and contained 25–36% more protein than the control (114). Production and composition of safflower protein isolate has also been described by Mexican workers (115).

Flaxseed, Linseed (Linum usitatissimum L.)

Food utilization research of flaxseed has been led by Federal Republic of Germany workers, who prepared linseed flour from cold pressed meal by hexane extraction, followed by removal of hulls by grinding and sieving to obtain a flour containing 50% protein (DMB) (116). Flaxseed protein isolate was prepared by alkali extraction at pH 10 and precipitation at pH 4.1. Composition, functionality and microstructural characteristics were determined for both flaxseed flour and protein isolate (117, 118).

Cereal Proteins

Approximately 70% of the proteins consumed as food in the world are from plant sources, and 71% of these are cereal proteins. Protein contents of cereal grains and selected cereal foods have been reported by Betschart (119), and amino acid compositions and biological value of cereal proteins by Lasztity and Hidvegi (120).

Wheat (Triticum aestivum L.)

Utilization of wheat, and triticale, a cross of *Triticum aestivum* and rye (*Secale cereale*) has been described in many books and journals. Generally, the essential amino acid profiles of foods from these cereals are low in lysine but high in sulfur amino acids, and are beneficially complemented by mixing them with oilseed proteins which are high in lysine but low in methionine. The germ fraction of wheat closely resembles oilseeds in content of essential amino acids, and steam treatment of wheat germ has resulted in a product with a PER of 2.84 (121). Also, a United States patent has been issued for a membrane process for removing off-flavors from vital wheat gluten using alcohol-alkali extraction (122).

Corn, Maize (Zea mays L.)

The proteins of corn have been reviewed by Watson (123) and Wilson (124), and account for about 10% of the weight of the corn kernel (DMB), with about 74% occurring in the endosperm, 26% in the germ, and minor amounts in the bran and tip cap. The endosperm proteins consist of approximately 3% nonprotein nitrogen, 3% (water-soluble) albumins, 3% (salt-soluble) globulins, 60% (isopropanol-soluble) prolamin ("zein") and 26% (alkali-soluble) glutelin. The germ proteins consist of approximately 20% nonprotein nitrogen, 35% albumin, 18% globulin, 5% prolamin, and 18% glutelin. The endosperm fraction contains approximately 15% of the fat in the seed, and the germ about 83%. As a result, corn endosperm protein is generally similar to wheat and sorghum in having relatively low protein efficiency ratios, but lacks the elasticity of wheat gluten. In contrast, the corn germ fraction is similar to oilseeds, containing large amounts of fat and having a protein efficiency ratio equivalent to soybeans.

The germ is separated from the endosperm during dry milling of corn flours and grits, and from ground steeped corn in wet milling of starch and production of corn sweeteners and alcohol. The germ stays with the endosperm in direct fermentation processes, and remains part of the dried distiller's grains. Corn gluten is the protein residue remaining after separation of starch by the wet milling process. Zein has unique film-forming properties as food and industrial moisture barriers, and is purified from corn gluten by solvent processes that start with hot alcohol extraction (122); however industrial uses of zein have fluctuated considerably during the past 20 to 30 years.

Corn gluten feed is increasingly in demand as a rumen-escape protein source in feeding dairy cattle. The availability of corn germ, as the result of rapid growth in use of corn sweeteners in the United States and production of alcohol from wet milling products, has led to corn oil becoming the second major domestic oil after soybean. Defatted corn germ flours contain approximately 17–18% protein, 65% total carbohydrates, 1–2% lipids and 8–10% crude fiber. Considerable research on corn germ protein utilization has been done in United States and in Italy (125, 126).

Fractionation of defatted wheat and corn-germ flours by air classification was reported by Stringfellow et al. (127). Approximately 40% protein was obtained in wheat fractions, and 28% in corn, in cuts that gave 33% yield. Blessin et al. (128) prepared high-protein flours from commercially dry-milled yellow, white and high-lysine corn germs by aspiration, flaking, hexane extraction, drying, grinding, and silk bolting. The amino acids composition of protein in the flours was similar regardless of corn type. Lysine accounted for approximately 5% of the amino acids recovered from the flours.

Nielsen et al. (129) showed that protein isolated from solvent-extracted dry-milled corn germ still contains 7–11% bound lipid, which results in stale and rancid flavors. Re-extraction of the defatted corn germ with 80% ethanol before isolation, or of the wet precipitated isolate with 80% ethanol, were effective for removing the bound lipids. In a later publication, Nielsen et al. (130) described preparation of flour containing 30% protein and 5.9% lysine from wet-mill corn germ by drying the germ, removal of hulls by aspiration, flaking, and hexane extraction followed by a secondary extraction of 82:18 hexane/ethanol azeotrope. A U.S. patent has been issued to Lawhon for a process using industrial membranes to recover proteins from corn and other agricultural commodities prior to fermentation into alcohol (131).

Inglett and Blessin (132) reviewed food applications of corn germ products, including defatted corn germ flour, dry-milled corn fractions, flour from wet-milled corn germ, corn germ isolate and zein. Messinger et al. (133) found that trypsin digestion improved water absorption and foaming properties of corn germ protein isolate, but reduced emulsifying capacity and nitrogen solubility. Digestion with pepsin decreased oil absorption, nitrogen solubility, and emulsifying capacity, but improved foaming properties. Treatment with succinic anhydride improved water and oil absorption, nitrogen solubility and foaming capacity, but decreased emulsifying capacity. Jao et al. (134) reported that aqueous or ethyl acetate extracted corn germ protein concentrates were more suitable for extrusion than those extracted with hexane. A blend containing corn protein concentrate and soy flour at a ratio of 15:85 produced good extruded chunks and granules in the pH range of 7.0–7.3. A blend of 21% corn protein concentrate, 20% soy isolate and 59% soy concentrate was suitable for a dry spinning process which required a dough with high cohesiveness for fiber formation.

Barbieri and coworkers in Italy (135, 136) were unable to obtain significant protein enrichment of corn germ flour by milling and sieving, and recommended the use of extrusion cooking to prepare corn germ for food use. Peri et al. (137) reported on twin-screw extrusion of corn germ meal that was obtained by grinding and screening. Barbieri and Casiraghi (138) reported on development of a defatted corn germ flour by milling and screening a commercial defatted corn germ meal. The flour contained about 20% protein of good nutritional quality, and more than 60% starch. In a later paper, Peri et al. (139) described making expanded nutrition-fortified snacks by extrusion cooking of defatted corn germ flour. The addition of 5% milk protein improved the organoleptic characteristics of the extrudates.

Dondero and Meneses (140) prepared a corn germ protein concentrate from defatted corn by alkali extraction at pH 11.5, followed by precipitation at its isoelectric point of pH 4.7, washing and vacuum drying. The resulting product contained 62.8% protein. Restani et al. (141) reported high water and fat absorption and good emulsifying properties for proteins extracted from defatted corn germ by alkali. Peri and Barbieri studied the processing and biological value of corn germ meal (142).

Lucisano et al. (143) evaluated macaroni produced from commercial durum semolina blended with 10, 20 and 30% defatted corn germ meal. The addition of corn germ meal resulted in a longer mixing time of the dough, higher farinograph water absorption, higher protein content of the products and improvement of the essential amino acid patterns. The optimum cooking time and per cent increase in weight of cooked macaroni decreased with increasing levels of wheat substitution. Enrichment with germ flour had no effect on flavor and texture of conventional macaroni as evaluated by a taste panel, but greatly affected physical measurements of cooked product textures, including increased firmness.

Kulakova et al. (144) obtained 24.3% protein in corn germ flour made from press cake in the Soviet Union.

Approximately 79.9% of the protein was alkali-soluble, 7.3% was water-soluble, 4.6% was alcohol-soluble, and less than 1% was alcohol-insoluble. The biological value, as determined by fermentative hydrolysis, was similar to egg albumin and casein. The alkali-soluble fraction showed high foaming capacity and long stability.

Christianson et al. (145,146) reported on supercritical carbon dioxide extraction of dry-milled corn germ. The resulting flour contained approximately 20% protein, with good amino acid balance meeting FAO requirements for food protein supplements. Super critical-CO_2 extraction denatured the proteins, including the oxidative enzymes, and peroxidase activity was decreased approximately tenfold when compared with hexane-extracted flours. Flour prepared from $SC-CO_2$ extracted corn germ contained approximately 0.7% triglycerides and 2% bound lipids, and had longer shelf-life, lighter color, and milder flavor than that from hexane-extracted germ. A U.S. patent on production of food-grade corn germ product was later issued to Christianson and Friedrich (147).

Mesallam et al. (148) reported that the zein fraction of corn weakened the dough mixing quality of bread, and the glutelin fraction improved it. Buck et al. (149) reported that incorporation of corn gluten meal in sugar cookies, white pan bread, pasta and extruded snacks resulted in less desirable flavors in all products except cookies, and poorer textures in all products except pasta. Tonella et al. (150) reported on physical, nutritional and sensory properties of fortified corn-base food products, using sesame, chickpea and soybean products. Ndupuh and Akobundu (151) partially replaced meat in beef patties with mixtures of corn flour, corn protein concentrate, and defatted peanut flour. Patties containing vegetable mixtures at up to 20% were comparable to the all-beef control, and were preferred by the sensory panel; patties containing corn protein concentrate and defatted peanut flour were judged superior to those containing corn flour and defatted peanut flour.

The nutritional quality of corn protein is generally poor, and various studies have been reported on fortification/complementation of corn flour products with other protein sources including sesame, chickpea, soybeans, powdered egg, nonfat dry milk, lysine and methionine (152). Lime treatment has been shown to improve the nutritional quality of corn by increasing availability of lysine in the glutelin fraction of the protein (153).

Considerable work was done at the U.S. Department of Agriculture's Northern Regional Research Center on uses of high-protein byproducts of alcohol grain fermentation in blended food products for the Food for Peace donation program. Materials investigated included corn distillers dried grains, corn distillers dried grains with solubles, and corn protein concentrates obtained from fermenting degermed and dehulled dry-milled corn (154, 155). Although these ingredients fit in well in computer-driven formulations, major problems were experienced in acceptance of products by taste panels.

Rice (Oryza sativa L.)

The proteins of rice have been reviewed by Juliano (156). Conkerton and Ory (157) studied chemical and functional properties of peanut and rice bran flour blends. Increasing the rice bran content increased the methionine, lysine and available lysine contents, and hydration capacity and fat absorption of the mixtures. A rice gluten, obtained from wet-milling of rice starch, has recently been introduced in India. The product is generally inert and has functional properties resembling corn gluten.

Barley (Hordeum vulgare L.)

Matthews and Douglass (158) reported on the nutrient content of barley, oats and rye. Vose and Youngs (159) developed procedures for fraction of barley and malted barley flours by air classification. Brewers malted barley was dehulled and roller milled, and field barley was dehulled and pin milled; the resulting flours were partitioned by air classification into dense, low-fiber starch fractions (84% yield) and less dense, low-fiber, high protein flours (16% yield). The α-amylase activity was partially concentrated in the dense starch fraction, and the malt aroma and flavor in the less dense protein fraction.

Wu et al. (160) prepared a protein concentrate from a high-protein, high-lysine variety of barley by extracting a ground barley:solvent ratio of 1:10 at pH 11.1, with centrifugation to remove the insolubles and starch fraction, and adjustment of pH to 5.3–5.4 to precipitate the protein. Up to 72% of the protein was recovered, with the dried precipitates containing 72–84% protein. The protein concentrate contained 2.9–5.0 g lysine and 2.1–4.2 g sulfur amino acids/16 g N.

Oats (Avena sativa L.)

The proteins of oats have been reviewed by Peterson and Brinegar (161) in a comprehensive book on oats chemistry and technology edited by Webster (162). Oat protein has a PER of approximately 1.9, the highest PER among whole cereal grains with the exception of high-lysine corn.

Cluskey et al. (163) developed an alkali extraction process for making defatted and nondefatted oat protein concentrates which are soluble in neutral or acidic beverages. The gum and water-soluble protein fractions could be removed by a preliminary water extraction if desired. In a later study (164), this group reviewed preparation of oat proteins by wet milling-alkali extraction and air classification.

D'appolonia and Youngs (165) found that oat bran decreased the loaf volume of bread more than wheat bran. Oat concentrate increased absorption and decreased loaf volume, and defatting the concentrate further increased these detrimental effects. Bread containing 10 or 20% oat bran was better accepted by panels than that made with wheat bran.

Ma (166) investigated protein concentrates made from oat groats by alkali extraction. The concentrates contained 60–70% protein, consisting of 50% globulins. Two major subunits with molecular weights of about 37,000 and 22,000 were observed. In later work, Ma (167) reported that modification of oat protein concentrate by potassium linoleate treatment after partial hydrolysis with trypsin improved protein solubilities in the pH range of 4–7, emulsification capacity and stability, water hydration capacity and foaming properties. Fat binding capacity was increased by linoleate treatment, but decreased by enzyme hydrolysis.

Ma (168) prepared isolates containing over 90% protein from oats by either isoelectric precipitation of alkali extracts, or dialysis of salt extracts. The recovery of N was greater (60%) in the alkaline extracts than in the salt isolates (25%). The resulting isolates had high fat binding capacities and good foaming properties. Ma and Khanzada (169) reported that deamidation of oat protein isolates by mild acid hydrolysis resulted in marked improvement in solubility, emulsifying properties and water and fat binding capacities. The pH of heat-induced gelation was lowered, and a firm, elastic gel was produced by mixing egg white with deamidated oat protein isolates.

Goulet et al. (170) reported PERs of 1.97 for oat protein concentrates obtained by alkali extraction at pH 9.5 and isoelectric precipitation. In later work the same researchers (171) reported that treatment with acetic and succinic anhydrides, to produce oat protein concentrates with approximately 36% and 76% acylation of the ε-amino groups respectively, significantly reduced PER values but increased digestibility coefficients.

Sorghum (Sorghum vulagare Pers.)

Grain sorghum protein is low in lysine content, but responds well to complementation by soy, cowpeas and peanut butter (172).

Grain Amaranth (Amaranthus caudatus L.)

During the past decade, considerable interest has been shown in grain amaranth, a crop used by ancient civilizations. An extensive review has been prepared by Saunders and Becker (173); and Teutonico and Knorr (174) have reported on uses of *A. caudatus*, *A. cruentus*, and *A. hypochondriacus* as seeds for food, starch manufacture and baking purposes, and on uses of other varieties as vegetables and leaf protein concentrates.

Becker et al. (175) published a study on grain amaranth composition, and Betschart et al. (176) reported on milling characteristics, composition, and nutritional quality of *A. cruentus* L. In research on the nutritional value of grain amaranths, Pant (177) found that *A. cruentus* and *A. hypochondriacus* contained 14.5–17.9% protein and 4.7–5.8 g lysine/16 g N, and also contained more calcium than wheat. Pedersen et al. (178) reported that amino acid scores and rat feeding performances of wheat, corn or low-tannin sorghum flours were improved by addition of 10, 15 and 50% popped amaranth flour, with improvement especially appreciable for sorghum. Lysine remained the first limiting amino acid in the Chemical Score, but addition of amaranth alleviated the tryptophan deficiency of corn protein. Cooking only slightly impaired the nutritional value of the blends.

Composite Flours

The objective of the international composite flour program has been to extend the functional properties of limited supplies of wheat, while at the same time using local ingredients and improving the nutritional quality of foods. Crabtree and James (179) have summarized the experience of the (English) Tropical Products Institute in developing composite flours from non-wheat materials, including other cereals, roots, tubers and legumes. Fraser Nochera (180) has conducted nutritional evaluations of products prepared from composite flours containing breadfruit flour and soy, peanut or whey.

Legume and Pulse Proteins

Legumes and pulses generally include dry peas, beans, lentils, and seeds of related botanical crops which primarily store their energy reserves in the form of starch rather than oil. Technically, soybeans and peanuts also are legumes, but more commonly are discussed with the oilseeds. During the past decade, under the guidance of International Agriculture Research Centers and assistance programs to developing countries by the industrial nations, great strides have been made in cataloging, improving, and optimizing utilization of the world's legume and pulse crops. Whereas the oilseed processing industry has continued to become centralized in fewer extraction plants, legume production, processing and utilization has often remained dispersed at village, appropriate technology and cottage industry levels.

Because one species may have many common names throughout the world, legumes and leaf proteins are arranged alphabetically by botanical names in order to minimize confusion. Different legumes serve as inexpensive protein sources in different localities, with small white (and navy) beans generally consumed in the cold weather industrial nations of the Northern Hemisphere, dry field peas in the extremely northern nations and the Soviet Union, pinto and black (turtle) beans in Central America, faba beans in Central and Eastern Europe and Northern Africa, garbanzo beans (chickpeas) along the South and East borders of the Mediterranean Sea and eastward to India and Thailand, blackeye peas (cow peas) in Central Africa and Southeast United States, and mung beans in Eastern Asia and the Philippines. Protein and fat contents of these seeds generally are in the ranges of 20–25% and 1–5%, respectively.

A line of field pea products, including pea protein isolate, pea starch, and pea fiber has been introduced recently in Denmark. Typical compositions claimed by the manufacturer are presented in Table 3. The Danish Ministry of Agriculture approved use of these ingredients in preparation of meat products in 1987 (181). A line of split and whole green and yellow pea flours; pregelatinized pea, lentil, white and pinto bean powders, and natural and bleached pea fiber concentrates has been available for several years in the United States from a Minnesota manufacturer; protein contents of split legume seed products are approximately 28%, while those of whole seed products are lower by about 3.5% because of fiber contributed by the hulls. Pea flour contains approximately 25% dietary fiber, and also has been allowed for use as a binder in processed meats in the United States. Up to 2% faba bean flour is allowed in breads in France.

Generic Processes and Characteristics

Papers published during the past decade, which are concerned with several species of legumes, include work in France on preparation and composition of protein isolates from kidney beans, peas and lentils, their uses in bakery products and nutrition (182), and production and evaluation of protein isolates from 5 legumes in Egypt (183).

United States scientists have published on: chemical, nutritional and microbiological qualities of protein concentrate prepared from dehulled dry beans by wet processing (184); effects of processing on sulfur amino acid stabilities (185); and preparation and antinutritional characteristics of protein concentrates of small white, Great Northern, pink, cranberry, and light red kidney beans (186). A United States patent has been issued on a membrane process for removal of flatulence-causing sugars from bean products (187).

TABLE 3

Typical composition of Commercial Field Pea Food Protein Ingredients Made in Denmark[a]

Characteristic	Pea protein isolate	Pea starch	Pea fiber
Composition ("as is"):			
Moisture, %	4.0	12.5	10.0
Protein, %	86.0	0.5	7.0
Carbohydrates, %	2.0	98.0	—
Fat, %	2.0	0.5	0.0
Ash, %	3.0	0.5	2.0
Dietary fiber	—	—	45.0
Starch	—	—	36.0
Nitrogen Solubility Index, %	61.0	—	—
pH (10% solution)	6.8	6.8	6.8
Water binding capacity	—	—	10–12 ×
Minerals:			
Sodium, %	0.04	0.02	0.05
Potassium, %	0.4	0.04	0.7
Phosphorus, %	0.4	0.01	0.15
Calcium, %	0.06	0.01	0.08
Composition, (dry matter basis):			
Protein, %	90.0	—	—
Carbohydrate, %	—	98.0	—
Dietary fiber, %	—	—	50.0

[a] Commercial literature, DeDanske Sukkerfabrikker, Copenhagen K, Denmark, 1988.

Processes have been developed in the United Kingdom for preparation of soybean, chickpea and other protein isolates using food grade salt solutions (188); and in Egypt for preparation of faba bean, chickpea and fenugreek proteins by alkali extraction-isoelectric pH precipitation, extraction with salt and precipitation by reduction of ionic strength, or extraction in aqueous or saline solutions in the presence of pepsin or pancreatin (189). Scientists in Argentina have reported on the effect of heat-moisture treatment on water uptake of field bean flour and protein isolate (190).

Air-classification is a process in which legume seeds are (optionally) roasted, dehulled, ground by an attrition grinder, and air-classified to take advantage of differences in specific gravities between the intact starch granules and the shattered protein and carbohydrate matrix. It is generally possible to more than double protein content in the high-protein fraction over that of the original seed; a low-protein, high-starch content, fraction also is produced. This process is attractive because it is performed dry, and the use of alkalis and acids is avoided.

Canadian researchers have taken the lead in applications of air-classification to legume seed flours, and have published on: factors affecting impact milling of grain legumes (191); impact milling and air-classification (192); effects of seed maturity on air classification of field peas (193); effects of seed moisture on air classification of field peas and faba beans (194); effects of cut-size on air classification of legumes (195); separation efficiency, yield and composition of air-classified legume starch and protein fractions (196); comparison of three air-classifiers for separation of protein and starch in pin-milled legume flours of seven legume species (197); oligosaccharides in 11 legumes, and their air-classified protein and starch fractions (198); fate of antinutritional factors (199); and thermodynamic parameters of eight legume air-classified starch and protein flours (200). In one study, air-classification of five field bean varieties yielded from 37% protein in black bean protein fractions to 61% protein for navy beans, and from 38 to 66 units of α-amylase inhibitors. No α-amylase inhibitors were found in lentil, lima bean, field pea, chickpea, faba bean and mung bean high-protein fractions (201). United Kingdom researchers have published on air-classification of cow pea, faba bean, and pigeon pea flours (202); and French scientists on air-classification of field pea and bean flours (203). United States scientists have reported that minerals, phytic acid, and flatulence sugars in air classification of dry-roasted navy, pinto, and black bean flours generally partition with the protein fraction (204).

Reports on studies in Canada of legume compositions include: electrophoretic patterns and amino acid compositions of eight edible legume species (205); lipoxygenase activities in 14 legumes (206); composition and nutritional qualities of field pea, faba bean and lentil meals, protein concentrates and isolates (207); and gel exclusion HPLC fractionation of phaseolus beans proteins (208). Among 20 food plants checked in Australia, saponin contents were highest in chickpeas, soybeans, alfalfa (lucerne) sprouts, and *Phaseolus vulgaris* varieties. Saponins were not destroyed by processing, baking, cooking, or canning, and the saponin content of fermented soybean tempe was only half that of whole soybeans (209). Chilean researchers reported compositions of raw and precooked legume flours obtained from four local manufacturers (210). Protein qualities of flours from six native legumes in the Philippines (211), and chemical and functional properties of selected legume flours in Egypt (212), have been reported. Also, Brazilian researchers have compared dietary fiber constituents of selected pulses (213).

Legume food proteins are chosen for use in prepared foods primarily on the basis of their functionalities. Canadian researchers reported that 2% vital wheat gluten (flour basis) and 1% dough conditioner (glycolipids, sucrose monolaureate, or polyoxyethylene-8-stearate), were needed to restore bread quality in high protein breads containing 12% sunflower concentrate, or 15% faba bean or field pea concentrate or soy flour (214). In a later study, soy, faba bean and field pea fortified breads showed PERs of 1.7–1.9, compared to 1.3 for sunflower protein-fortified wheat bread and 1.1 for control bread (215). Effects of germination on electrophoretic, functional and bread-baking properties of yellow pea, lentil and faba bean protein isolates have been reported by U.S. researchers (216).

Physicochemical and composition characteristics, and rheological, processing and quality evaluations of spaghettis fortified with navy bean, pinto bean, and lentil flours and concentrates have been reported in the United States (217, 218). Also, soy proteins; sunflower, mustard, rapeseed, peanut, cottonseed, sesame, pea and chickpea flours; and textured navy bean concentrate have been evaluated as extenders in meat emulsions (219).

In evaluating beefburgers containing defatted soybean, faba bean, chickpea, and white rice flours, Egyptian researchers found that soybean flour provided the most protein and absorbed the most free water and fat, but broken rice and faba bean flours were more economical (220). Substitution of defatted glandless cottonseed flour, soybean flour, or faba bean flour for non-fat milk solids has also been evaluated in ice cream in Egypt (221).

Studies in the German Democratic Republic have included preparation of spun protein fibers from faba bean proteins (222); functionality of faba bean and casein fibers (223); nutritional studies of fibers (224); and functional properties of native and denatured proteins isolated from faba beans, soybeans and sunflowerseed (225). French researchers have obtained patents on spun proteins made from casein, soy, sunflower or faba bean proteins (226) and have applied for a patent on food products made from spun fibers (227).

The presence of antinutritional factors in dry seeds of field legumes has been described frequently. During the past decade, U.S. scientists also have published on effects of cooking and food preparation on nutritional values of legumes (228), and effects of germination on functional and nutritional properties of dry peas, lentils and faba beans (229). Also, Hungarian researchers have described trypsin and chymotrypsin inhibitor activities in plant foods from Vietnam and Hungary (230).

Tetrahymena have been used in the United States to determine effects of processing and composition on relative nutritive values of green and yellow peas, lentils and white pea beans (231); and to evaluate effects of sample age and amino acid supplementation on relative nutritive values of lentils, green and yellow split peas and processed flours (232). Japanese researchers have reported on amino acid compositions of protein fractions isolated from adzuki beans, kafae and red kafae beans, lima beans, red lima beans, and faba beans (233).

Malts of eight pulses used to prepare cereal-pulse malt biscuits with high protein content were evaluated by sensory methods in India (234). The nutritional value of patti beans was improved by supplementation with faba beans, soybeans and chickpeas in Egypt (235). The nutritional value of staple Nigerian legumes (brown beans, white beans, and bambara ground nut) in combination with rice was evaluated in rat feeding trials (236). Supplementation of fermented sorghum kisra bread with alkali-extracted protein isolates of bonavist bean (*Dolichos lablab* L.) and white bean (*Phaseolus vulgaris* L.) was reported from the Sudan (237). Studies in Sweden evaluated the nutritional quality of home-prepared weaning foods supplemented with legumes for consumption in Bangladesh (238). Specifications for processed cereal-based weaning foods were developed in Tanzania (239).

Research on food utilization of the following legume and pulse species was reported in the literature during the past 10 years.

Cajanus cajan (Pigeon Pea, Red Gram)

Egypt—phytochemical and nutritional studies on pigeon peas and kidney bean beans (240); chemical studies on pigeon pea protein (22% protein present in pigeon pea meal) (241).

India—protein contents of whole grains and dahls (242); physicochemical properties and amino acids composition (243).

Brazil—production, characterization and functional properties of pigeon pea protein isolates (244); chemical composition of pigeon peas of Kaki cultivar (245).

Canada—nutritional evaluation of pigeon pea meal (246); United States—nutritive and organoleptic evaluations of wheat breads supplemented with pigeon flour (247).

Canavalia ensiformis DC. (Jack Bean, Chickasaw Lima Bean)

Brazil—composition and hemagglutinating activity (248); India—PERs of extrusion-cooked blends of jack bean flour and semolina (249).

Canavalia rosea DC. (Brown Bean)

Nigeria—functional properties of raw and roasted brown bean flour (250).

Cajanus flavus (Pigeon Pea)

Brazil—chemical, nutritional and organoleptic evaluations, and trypsin inhibitor inactivation (251).

Certonia siliqua L. (Carob Seed)

Greece—nutritional evaluations of carob seed germ meal and its protein isolates with rats showed that isolates had high protein content but were not well-balanced in amino acid composition (252).

Cicer arietinum L. (Chickpea, Garbanzo Bean, Bengal Gram)

India—sunflowerseed meal, maize, roasted chickpea flour and sesame meal, in the ratio 65:15:10:10, had a PER of 2.21 and produced significantly greater increases in height and weight when fed to school children (253); nutritional values of trypsin inhibitors (present at 5.03% of the extractable seed protein) (254); formulation of fruit toffee with Bengal gram flour (255).

Pakistan—report on supplementation of *roti* and *nan*, respectively unleavened and leavened pancake-like staple foods made from wheat flour with 20% chickpea flour (256); nutritional evaluation of wheat and corn breads supplemented with a mixture of peanut-chickpea flour (257); nutritional and organoleptic evaluations of wheat bread supplemented with chickpea flour (258). Bangladesh—process for making infant/children food from *dahls* (parched cotyledons) and nutritive values (259, 260).

Egypt—up to 15% flour from raw and parboiled chickpeas, and 6% soy flour found usable in bread supplementation (261); effects of cooking on tryptophan and amino acids contents, protein solubility and vitamin retention (262). Sudan—composition of 24 varieties of Kabuli-type chickpeas (263). Iraq—process for extraction of protein isolate from chick pea flour (264).

Italy—compositions of chickpea flours and protein preparations (265); studies of composition, nutritional value, and oligosaccharide contents of 10 cultivars of Italian yellow-coated chick peas and 10 cultivars of black chick peas from India, showed that black chick peas had higher oligosaccharides contents, and higher *in vitro* digestibilities (266); preparation, composition, nutritional properties of chickpea cake (*Torta di ceci*) of Liguria and Tuscany (267).

Mexico—chemical composition and nutritional values of seven cultivars of local chickpeas showed average compositions of raw chickpeas at 6.9% moisture, 19.5% protein, 5.0% fat, 3.7% fiber, 3.1% ash, and 61.8% N-free extract (268); nutritional supplementation of cookies with chickpea flour (269); supplementation of corn masa and tortillas with chickpea flour to improve protein quality (270).

United Kingdom—use of chickpea flour in sausages (271); nutritional evaluation (272). Panama—effect of processing conditions on chemical and nutritive properties of chickpeas for infant foods (273). Argentina—effect of thermal treatments on moisture absorptions of flour, starch and protein isolate (274).

Cyamopsis tetragonoloba (Guar)

India: functional properties (fat absorption, emulsification capacity, foaming capacity and foam stability of guar meal and protein isolate) (275); functional properties of guar meal protein isolates (276); trypsin inhibition and protein digestibility of raw and detoxified guar meal (277); depression of serum cholesterol, total lipids and uric acid levels, and significant rises in total serum proteins from feeding guar flour to 50 healthy men (278).

Dolichos biflorus L. (Horsegram, Twin Flower Bean)

India—protein range of 16.5–26.9% determined for 27 varieties (279).

Lathyrus sativus L. (Kesari Dahl, Chicking Vetch)

India—a legume grown during droughts, which uniquely causes nervous paralysis of the lower limbs of young adult men. Over 90% of the toxin can be removed by cooking seeds in excess water which is then discarded. Soaked, steamed, roasted or sun-dried seeds can be ground into flour to make chapatis, an unleavened Indian bread (280). Indian standards for besan (chickpea flour) do not permit contamination with *L. sativus*.

Lens culinaris L. (Lentil)

Canada—chemical composition and characteristics of lentils (281). Iraq—process for preparation of protein isolates from lentils (282). Pakistan—characterization of lentil protein fractions (283). Egypt—effects of cooking on chemical composition of lentils, rice and their blend *Koshary* (284).

Lens esculenta L. (Lentil)

Argentina—preparation, composition and nutritional values of navy bean and lentil meals and protein isolates were compared. Biological values of lentil proteins were lower than for navy beans due to lower tryptophan levels (285).

Leucaena leucocephala

Mexico—toxic and antinutritional factors (286).

Lupinus spp.

Four types of lupins are cultivated for food use. Three of these, *Lupinus albus*, *L. luteus* and *L. angustifolius*, have white to cream-colored blossoms and low alkaloids content in the seed and can be used directly for foods. Considerable genetic improvement has been done on these "sweet" species, with *L. albus* and *L. luteus* developed primarily in Europe and *L. angustifolius* (narrow-leaf lupin) in Australia. *L. mutabilis* is a blue-flowering species that has long been grown in Northern Europe, often as a green manure crop. Its seeds contain 2–3 times the oil content of the white flowering species. However, *L. mutabilis* also contains large quantities of "bitter" toxic alkaloids, which must be removed, typically by soaking, before the seed can be used as food. Never the less, *L. mutabilis* is extremely well-suited to some climates, and research on using its seed for food has continued.

Lupin seed proteins have been reviewed extensively by Cerletti (287). Comparative compositions have been reported for the four major lupin species grown in South Africa, *L. albus* L., *L. luteus* L., *L. angustifolius* and *L. mutabilis* L., respectively, on a dry matter basis were: crude protein—36.7, 41.8, 31.1, 42.6%; ether extract—11.5, 5.4, 6.0, 18.7%; crude fiber—9.8, 15.8, 14.7, 7.3%; ash—3.4, 4.1, 3.5, 3.7%; and N-free extract—37.8, 35.0, 43.1, 27.3%. Lupin seed proteins generally are deficient in the sulfur amino acids (288). Although the following lupin papers have been arranged on the basis of the species emphasized, they often contain mention of the other lupin types.

Lupinus albus L. (Sweet Lupin, White Lupin)

Italy—amino acid composition (289); subunit composition of seed globulins (290); heterogeneity of subunit composition in lupin globulins (291); changes in globulins during germination (292); emulsifying and gelling properties of lupin proteins (293); effect of bound carbohydrates on action of lupin seed glycoproteins (294); composition and nutritional performance (295); baking properties of lupin flour (296); use of lupin flour in pasta manufacture (297); use of lupin endopeptidase for proteolytic processing of pea and Jack bean storage proteins (298).

Chile—baking and nutritional studies of bread supplemented with full-fat lupin flour (299); successful multigeneration study of lupin flour on rats (300); wheat and oat flour fortification with sweet lupin flour for nutritional evaluation (301); composition and nutritional quality of sugar cookies with full-fat sweet lupin flour (302); functional properties of lupin protein isolates (303).

United States—preparation of lupin (*L. albus* and *L. mutabilis*) protein concentrates by aqueous extraction (304); cooking quality of macaroni was increased by 6% substitution of high-protein lupin flour, but decreased with soy flour (305). France—separation and characterization of protein and cell walls from dehulled white lupin and *L. mutabilis* (sweet) meals (306).

Brazil—dough properties, bread quality, and amino acid composition of lupin flour in bread making (307). Egypt—nutritious beverages from light-colored lupin, soy and sunflower protein isolates (308). Japan—patent on use of lupin flour in conjunction with wheat flour in preparation of filament-type food products (309). Korea—lactic acid fermentation of lupin seed milk (310).

Lupinus angustifolius (Narrow-Leaf Lupin)

Australia—comparisons of aqueous extracts of seed proteins of *L. angustifolius* and *L. albus* L. (311); isolation of conglutin DELTA, a sulfur-rich protein (312); manganese content and distribution in lupin seed, flour and protein products (313).

Lupinus mutabilis L. (Blue Lupin, "Bitter" Lupin)

Federal Republic of Germany—preparation of lupin protein concentrates using aqueous alcohols (314); production of alkaloid-debittered protein concentrate (315); production of edible oil and protein concentrate (316); rat feeding study of water-debittered Andean *L. mutabilis* (317).

Peru—composition and protein quality (318); composition of native tarhui (319). United States—functional properties of flour and protein concentrates (320). Italy—removal of 60% of alkaloids by 4 stages of hexane extraction (321).

Lupinus termis

India—effect of debittering treatment on composition of flour (322).

Mucuna pruriens DC. (Utilis Bean)

Ghana—composition of raw beans: 6.4% moisture, 31.6% protein, 4.2% fat, 7.9% fiber and 3.0% ash. Content of lysine is relatively high, and methionine is the first limiting amino acid (323).

Mucuna utilis (Velvet Bean)

India—chemical composition and antinutritional factors, 26% protein and 14% crude fat content (324).

Parkia filicoidea Welw. (Locust Bean)

Nigeria—composition and nutritive value of lipid-protein concentrate (325); effects of fermentation on nutritional qualities (326); chemical changes during fermentation of African locust bean seeds for preparation of *daddawa* (327).

Pentaclethra macrophylla Benth. (African Oil Bean, Owala Oil Tree Bean)

Nigeria—composition and food potential of African oil bean and velvet bean (*Mucuna uriens*) (328).

Phaseolus acutifolius Gray. var. *latifolius* (Escumite Bean, Tepary Bean)

Mexico—studies on toxicity factor of Escumite bean, which is highly toxic in the raw state (329). United States—composition: 41% starch, 26% protein, physicochemical properties of starch indicate potential usefulness in puddings (330).

Phaseolus aureus (Mung Bean, Green Gram)

Philippines—preparation of protein isolate, composition and functional properties (331); functional properties of raw, blanched, sprouted, and air-classified high-protein mung bean fractions and flours, containing 23–54% protein (332); preparation of mung bean beverage from washings of Sotanghon bean noodles (333).

India—physical and hydrodynamic properties of salt-soluble proteins (334); viscosities of precooked, dried, green gram, Bengal gram and peanut weaning food ingredients (335); best results in improving the protein quality of local rice diets obtained with soy and mung bean flours (336); use of mung bean and black gram flours in biscuits (337).

Sri Lanka—preparation of "cheeses" from protein extracts of soybeans, mung beans and tur dhals (a type of lentil) (338). Thailand—production, composition and utilization of protein isolate (339). Iraq—cellulase enzyme treatment solubilized 40% of mung bean flour nitrogen, compared to 85% extracted with NaOH (340). Iraq—preparation of protein isolates and chemical, nutritional and solubility characteristics (341). Egypt—use of mung bean protein concentrate and flour in wheat bread (342).

Japan—extraction of protein and starch from mung beans and sprouts (343); studies of physicochemical properties of starches from mung bean, faba bean and commercial harusame noodles showed that mung bean starch prepared by fermentation was 4 times higher in protein content, more opaque when gelatinized, and formed a weaker gel (344). United States (Hawaii)—mung bean and soybean flours improved extrudability and nutritional qualities of taro products (345).

Phaseolus coccineus L. (Runner Bean)

Mexico—nutritional value of runner beans, with and without methionine supplementation (346).

Phaseolus lunatus L. (Lima Bean)

Nigeria—composition differences among varieties (347); hemagglutinin activities in raw, sprouting and processed lima beans (348). Mexico—nutritional quality and toxin contents of wild and cultivated lima beans (349).

Phaseolus mungo L. (Black Gram)

India—role of arabinogalactan in producing high viscosity and stabilizing foams in fermented or leavened foods and steamed puddings (350); surface activity of proteins or viscosity of polysaccharides was not affected by autofermentation of black gram (or dhal) flour for 20 hr (351); Chemical Scores of essential amino acids (352). United States—functional properties of black gram protein concentrate (353).

Phaseolus vulgaris L. (Field Bean)

This species includes varieties of many colors and sizes, such as black, cranberry, Great Northern, kidney, navy, pink, pinto, and small and large white beans. Brazilian scientists have reported on antinutritional factors and toxicity in 12 Brazilian raw dry bean cultivars (354). United States researchers have reported on use of roasted air-classified, navy, pinto and black bean protein flours in cake doughnuts (355).

Black (Turtle Soup) Beans: Guatemala—process for making precooked flour from black beans (356); contamination of black and other beans by aflatoxin in dry seasons (357).

Mexico—preparation of tempe-like product using the mold *Rhizopus oligosporus* (358). Chile—hard-to-cook defect in black beans (359).

Great Northern Beans: United States—solubilization and electrophoretic patterns of Great Northern Bean protein (360); trypsin and chymotrypsin inhibition, hemagglutinating activity, and sugars content (361); isolation and characterization of major protein fraction (362); isolation and characterization of phytic acid-rich particles (363); removal of phytate from flour by dialysis (364); functional properties (emulsification, foaming, viscosity and gelation) of food protein fractions (365, 366); functional properties of proteins, amino acid composition, *in vitro* digestion, and use in cookies (367); effects of bean flour and protein concentrates on dough properties and baking quality of bread (368).

Kidney Beans: A review on kidney beans has been published by Deshpande et al. (369).

Chile—digestibility improvement, and trypsin inhibitor inactivation by heat treatment of kidney beans (370); extruded kidney bean products for use in infant foods (371); starch characteristics of three *P. vulgaris* beans (cvs. Assoz, Tortola, and red kidney) (372); protein quality and hemagglutinin content of fresh and dry Coscorron kidney beans (373).

United Kingdom—protein quality and heat-stable antinutritional factors (374); functional properties of protein isolate (375); freeze drying and water-protein interactions of protein isolates (376). United States—poor utilization by rats of protein methionine and cystine in autoclaved kidney beans (377). Canada—lipase activity and fatty acid composition in stored full-fat French kidney bean flour (378).

German Democratic Republic—isolation and characterization of pectinase inhibitor from kidney beans (379). Greece—isolation of methionine-rich protein (380); nutritional significance of trypsin inhibitors from kidney beans (381). Argentina—effects of heat treatment on sorption isotherms and solubilities of flour and protein isolates (382). Mexico—unavailable carbohydrates in cotyledons of Canario group kidney beans (383). Brazil—effects of storage on nutritional value of whole kidney beans (384); physicochemical and nutritional properties of kidney bean cv. Rosinha G2 (385).

Kintoki Bean: Japan—characteristics of lethal lectin from kintoki beans (386); isolation and characterization of lethal protein (387); heat inactivation of trypsin inhibitor (388).

Navy Bean: United States—development of food ingredients from navy beans by roasting, pin milling and air classification (389); preparation of protein isolates (wet alkali extraction process) from navy and pinto beans and their use in macaroni products (390); effects of extraction conditions on yield of isolated proteins, with malic acid solution found the best performer (391); physicochemical characteristics of dry-roasted navy bean flour (392); characteristics and utilization of dry-roasted, air-classified, navy bean protein fraction (393); amino acid and mineral profiles of air-classified navy bean fractions (394); optimizing nutritive value of navy bean protein by complementation with cereal proteins (395); methionine supplementation of navy beans with Brazil nuts (396); isolation of trypsin inhibitor from navy beans (397); role of heat-stable inhibitors *in vitro* digestibility of dry bean proteins (398).

United States—incorporation of dry roasted navy bean flour in a quick bread (399); navy bean flour substitution in master mix for muffins and cookies (400); preparation of sugar cookies with wheat-navy bean flour blends (401); supplementation of short bread cookies with roasted whole navy bean flour and high-protein content flour (402); microbiological evaluation of navy bean protein concentrate and its blend with retail ground beef (403); air classification and extrusion of navy bean fractions (404); texturized navy bean protein concentrate as a meat extender in frankfurters (405).

Pink Beans: United States—preparation of canned refried beans from quick-cooking pink beans (406).

Pinto Beans: United States—evaluation of protein quality of corn and wheat flour tortillas supplemented with pinto beans and soybeans (407); extraction and characterization of soluble and insoluble fiber fractions from pinto beans (408).

White Beans: United States—process for making bean chip snacks from California small white beans (409). Canada—composition and functional characteristics of air-classified white bean protein and starch fractions (410).

Pisum sativum L. (Field Pea, Garden Pea)

Canada—storage stability studies of pea flour, protein concentrate and starch (411); wet method for production of pea protein isolate (412); production and functionality of wet-processed field pea and horse bean starches and proteins (413); preparation of high protein content curd (414); preparation of 33–60% protein concentrates by air classification of peas (415); effects of heating on emulsifying properties of field pea isolate and selected food proteins (416); isolation and characterization of cell wall material from wet-processed dehulled pea flours and concentrates (417); functionality of field pea and faba bean flours, protein fractions and isolates (418).

Denmark—process for preparing field pea and faba bean protein concentrates and isolates (419); European Patent Application for improved process for making pea protein isolate employing wet milling, ultrafiltration, diafiltration, and protein coagulation by enzyme (rennet), excellent emulsification properties claimed in preparation of meat sausages (420).

United States—functional properties of pea globulin fractions (421); use of field pea flours as protein supplements in foods (422). France—production of meat analogues from field pea flour (423). United Kingdom—binding of diacetyl by pea proteins (424). Hungary—composition and dough characteristics of yellow pea flour bakery mixes (425). Poland—inhibition of lard oxidation in dry soup by use of yellow pea flour (426).

Prosopis juliflora DC. (Mesquite Bean)

Mexico—kernels found to contain 38% protein in composition and nutrition studies (427); development of procedure for making protein concentrate (428).

Psophocarpus tetragonolobus DC. (Winged Bean, Asparagus Bean, Goa Bean)

The winged bean was first popularized in developing countries as a completely usable crop, being relatively protein-rich in the leaves, stems, green pods and roots. However, attention turned to processing whole seeds as crops matured. Winged bean seeds contain 33–37% protein and 16–17% fat, with the hulls constituting about 17% of dry seed weight. Winged beans handle somewhat like soybeans, and have been evaluated for the functional and nutritional properties of their proteins in making bread (429).

United States—biochemical composition of mature winged bean seeds (430,431); chemical composition of dry seeds and fresh leaves grown in Puerto Rico (432); functional properties of proteins (433); rheological characteristics of composite flours containing winged bean meal (434), baking and organoleptic characteristics (435) and nutritional qualities of winged bean meal, composite breads (436); presence of a nonprotein trypsin inhibitor in soy and winged beans (437); nutrient and antinutrient contents, and solubility profiles of protein, phytic acid and minerals in flour (438); review and recent advances in winged beans as source of food protein (439).

Japan—chemical composition of four winged bean varieties (440); comparison of winged bean and soy proteins (441); comparison of winged bean and other legume proteins (442); effects of presoaking and cooking on production of tofu from winged beans (443).

United Kingdom—preparation and properties of protein isolates (444); effects of blanching and soaking on winged beans (445). Australia—amino acids of winged bean seed meal are similar to soybean, but the storage globulins are quite different (446); isolation of acidic and basic lectins from winged bean seed (447).

India—food value of root tubers (448); purification of lectin from winged bean tubers (449). Sri Lanka—improvement of nutritional value of wheat flour by fortification with full-fat winged bean flour (450); comparative studies of chemical and nutritional properties, and biological evaluation of soybean and winged beans grown in Sri Lanka (451). Peru—composition of winged beans (452).

Stizolobium cinerium (Velvet Bean)

Mexico—nutritional evaluation of velvet bean alone and with supplementation (453).

Trigonella foenum graecum L. (Fenugreek)

India—composition (24.5% protein, 7.9% fat), and nutritional evaluation of protein quality (454).

Vicia faba L. (Faba Bean, Broad Bean, Horsebean, Field Bean)

After soybeans, the most intensive research in seed food proteins during the past decade probably has been on the faba bean. Many of the publications are from the German Democratic Republic, and the technology of faba bean protein functionality modification and utilization is highly advanced. The faba bean has been reviewed in detail in the book by Hebblethwaite (455).

German Democratic Republic—three processes for preparing faba bean meals (456); effects of electrolyte content and pH on solubility of globulins from sunflowerseed and faba bean (457); properties of faba bean starch (458); functional properties of native and denatured isolates of faba beans, soybeans and sunflowerseed (459); influence of denaturing processing conditions on functional properties of protein isolates (460); effects of isolation conditions on foaming properties of faba bean protein isolates (461); foaming properties of modified faba bean proteins (462); determination of emulsification characteristics of protein preparations (463); emulsification properties of enzymatically-modified faba bean isolate (464); effect of degree of acylation of faba bean protein (465); characteristics of acylation of faba bean protein isolate on functional properties (466); relationships between concentration and degree of acylation of field bean protein isolates and their viscosity (467); interfacial and emulsifying behavior of acetylated faba bean protein isolate (468); effects of succinylation on physicochemical properties of some foods (casein, faba bean isolate, sunflowerseed) (469). GDR patent on preparation of oil-in-water emulsions using acetylated proteins (470).

GDR—effects of easily-soluble and sparingly-soluble protein isolate fractions on functional properties of fibers (471); effects of broad bean protein on functional properties of noodles (472); GDR process patent for obtaining a transparent gel-forming protein fraction from faba bean (473); effects of extruded faba bean proteins on properties of sausages (474), fried hamburgers (475), heated sausage emulsions (476); extruded faba bean isolates in sausage meat (477,478,479); formation of plastein gels from pepsin-hydrolyzed faba bean protein isolate (480); effects of combinations of enzyme and mechanical modification on functional properties of proteins (481); GDR patent on enzyme treatment of proteins from faba beans (482); incorporation of methionine into peptic partial-hydrolyzate from faba bean protein isolate by plastein reaction with thermitase (483); influence of enzyme treatment on lysinoalanine formation in alkali-treated plasteins of faba bean protein (484); GDR patent on formation of cold-gelling (succinylated) vegetable protein (485); GDR patent on preparation of solid transparent gel products (486); in vitro digestibilities of original and enzyme-modified protein isolates (487); gas chromatography determination of sugar in native and enzyme-modified faba bean protein isolates (488); testing systems for extenders and additives for frankfurter-type sausages (489); use of rocket immunochemical method to detect or identify faba bean protein isolate (490).

Poland—characteristics of carbohydrates in faba beans, protein concentrates and waste products (491); chemical characteristics and properties of faba bean starch byproduct from protein extraction (492); faba bean and soybean protein isolates with lowered phytic acid and trypsin activity (493).

Canada—purification and properties of vicin and covicin from faba bean protein concentrate (494); fate of antinutritional factors during formation of faba bean protein isolate using salt extraction-micellization technique (495); effect of moisture content on thermal stability of faba bean protein (496); effect of salt on thermal stability of faba bean storage proteins (497); effects of heat and storage on flavor of faba bean flours (498).

Egypt—bread making properties of composite flours of wheat and faba bean protein preparations (499); effects of faba bean flour addition on rheological properties of dough and baking quality of *Balady* bread (500); improvement of nutritional and flavor properties of faba beans by blending them with other protein sources (an 80% faba bean, 20% sesame butter combination was found best) (501); characteristics of bread made with wheat flour fortified with faba bean and lentil protein concentrates (502); solubility and gel filtration chromatography of water-extractable faba bean proteins (503).

France—description of turbo-separation of faba bean proteins (504); effects of processing on faba bean products (505); solubilities of faba bean and pea (*P. sativum*) proteins (506); elimination of antinutritional factors in preparation of faba bean and pea protein isolates (507); preparation of faba bean protein isolates by ultrafiltration (508).

Denmark—continuous pilot plant production of faba bean proteins by extraction, centrifugation, ultrafiltration and spray drying (509); vicine and covicine contents in faba beans and isolated faba bean proteins (510).

U.S.—faba bean protein as source of aroma and flavor in making Moroccan-type bread (511); uses of faba bean flour and air-classified protein concentrate in baked goods and pasta products (512). U.K.—protease inhibitor activity of faba beans (513). Italy—biological and chemical assays for hemolytic and toxic factors in faba beans (514). Ethiopia—replacement of chick peas in *faffa* (a commercially-produced infant formula) with faba beans and kidney beans (515). India—characteristics of enzymatic hydrolyzate of faba bean protein (516).

Vigna aconitifolia (Jacq) Marcehal (Moth Bean)

India—solubilization and functional properties of moth bean and horsegram (*Macrotyloma uniflorum*) proteins (517).

Vigna angularis (Adzuki Bean)

U.S.—preparation and characterization of adzuki bean protein and starch (518).

Vigna radiata (Mungo Bean, Mung Bean, Green Gram)

Philippines—evaluation of nutritional quality of flour from mung bean sprouts (519); preparation of mung bean protein meat analog by combining protein isolate, wheat flour, agar, salt, dry yeast, sugar, soy sauce and vegetable oil, steaming in a shallow pan, chopping mechanically, and oven-drying (520); condensed tannins of mung bean shown to decrease *in vitro* protein digestibility (521); changes in polyphenolic compounds of 10 cultivars of mungbeans during processing: 24–50% during soaking, 73% by boiling for 30 min, 17% by roasting; and 36% by 24 hr sprouting (522). India—chemistry and technology of mung beans (523).

Vigna sinensis Endl. (Cow pea, Cream Pea, Blackeye Pea)

U.S.—preparation of a variety of appealing foods for children (524); effects of mild alkali treatment during dry roasting on protein quality of flour (525).

Vigna umbellata (Rice Bean)

India—compositions of 6 strains, with 18.7–25.2% protein (526).

Vigna unguiculata (Cow pea, Blackeye Pea, Southern Pea, Texas Cream Pea, Chinese Red Pea)

Nigeria—mean chemical compositions of 10 varieties: crude protein—25.2%, ether extract—2.0%, crude fiber—3.2%, ash—4.2%, N-free extract—65.1% (527); carbohydrate compositions of 20 cow pea varieties, and protein ranges of 24–33% (528); quality changes of cow pea flour in storage (529); effects of cooking, germination and fermentation on chemical composition of cow peas (530); use of cooked cow pea powder for improving Nigerian weaning foods (531).

U.S.—replacement of milk protein in baking powder biscuits with air-classified cow pea and field pea flours (532); preparation of *akara*, a deep-fried cow pea paste popular in Africa (533); preparation and composition of dry-milled cow pea flour (534); composition, physical and sensory characteristics of *akara* (deep-fat fried balls of cow pea paste) (535); development of weaning foods from locally grown seeds in Nigeria (536); preparation and composition of dry milled cow pea flour (537); development of high protein low-cost Nigerian foods including *akara* (538); effects of temperature and moisture on kinetics of trypsin inhibitor activity, protein *in vitro* digestibility and nitrogen solubility of cow pea flour (539); protein supplementation of southern pea *moin-moin* with melon seed or pollock fish (540,541); nutritional and organoleptic evaluation of corn-cow pea mixture infant food (542).

Canada—development of surface response methodology for extracting cow pea protein (543); microscopic, nutritional and functional properties of flours and concentrates during storage (544); changes in microbiological and lipids characteristics (545). Philippines—reduction of protein digestibility by condensed tannins in cooking water from seed coats of darker cow peas, as determined by Tetrahymena (546). Saudi Arabia—use of cow pea flour and protein isolate in bakery products (547). Thailand—development of local home-processed foods from cow peas (548).

Voandzeia subterranea Thon. (Bambara Groundnut)

U.K.—preparation of milk-like beverages and curds (549). Nigeria—rat feeding comparisons of bambara groundnuts and other staple local legumes supplemented with rice flour (550).

Wild Legumes

India—chemical and biological evaluation of *Mucana prurita, M. capitata, Cassia tora* L. and *C. occidentalis* L. proteins (551). U.S.—antinutritional factors in 10 edible legumes of the Sonoran Desert (552). Chile—comparative studies of food qualities of lentils (*L. culinarias*) and vetch (*Vicia sativa* L. supsp. *aborvata*) containing 25.9 and 35.6% protein, and 0.9 and 0.8% fat (553). Nigeria—chemical composition of 16 unexploited native leguminous seeds (554).

Leaf Proteins

Leaf proteins have the potential for yielding up to several metric tons of purified high quality protein per hectare. Research on their extraction and utilization has continued from the 1970's, and has been broadened to survey the potential usefulness of many plant species. Variations exist among processes, but procedures generally include maceration of leaves and shoots, pressing the juice, centrifuging it to remove the green chloroplasts (sometimes after coagulation by heating), precipitation by acid, heat or sodium sulfate, and centrifugation to recover a white leaf protein concentrate. In contrast to oilseed proteins, which contain 10–20% albumins, with the remainder being primarily globulins, the ratio of albumins to globulins in leaf proteins often is about 6:1. Also, unlike oilseed proteins, high-purity leaf proteins are commonly obtained in crystalline form. Earlier developments of this technology have been summarized in a book edited by Telek and Graham (555). An *in vitro* method to assess nutritive value of leaf protein concentrates has been described by Maliwal (556). The most intensive research on leaf food proteins has been conducted with alfalfa and tobacco. A small-scale commercial cooperative is reported to be producing fractionating food proteins from local leaves in Ghana (557).

Leaves of the following species have been studied for composition, processing and/or utilization as sources of food proteins: *Amaranthus tricolor, A. cruentus, A. caudatus* L., *A. gracilis, A. graecizans,* and *A. spinosus* L. (Vegetable Amaranth)—(147); *Atriplex numularia*—Argentina (558); *A. numularia* and *A. repanda*—Chile (559); *Arachis hypogae* L. (Peanut, Groundnut)—U.S. (560); *Brassica oleracea* L. (Cabbage)—Poland (561); *Brassica oleracea* L. var. *botrytis* (Cauliflower)—India (562,563,564); *Celosia argentea* L. (Vegetable Green)—Nigeria (565); *Centilla asiatica* L. (Gotukola, Vegetable Green)—Sri Lanka (566); *Cynon dactylon* L. (Bermudagrass)—U.S. (567); *Dolichos biflorus* L. (Horsebean) and *D. lablab* L. (Field Bean)—India (568); *Dystaenia takeshimana* Nakai—South Korea (569); *Eichhornia crassipes* (Water Hyacinth)—Egypt (570), Guatemala (571), India (572); *Ipomoea aquatica* Forsk. (Water Spinach)—U.S. (573); *Ipomoea batatas* Poir (Sweet Potato)—U.S. (574); *Lactuca sativa* L. (Lettuce)—Egypt (575); *Manihot utilissima* Pohl. (Cassava)—Nigeria (15 varieties, (576)), Brazil (577); *Manihot esculenta* Cranz. (cassava)—Philippines (578); and *Marantha arundinacea* L. (Arrowroot)—South Korea (579).

Medicago sativa L. (Alfalfa, Lucerne).

Research reported in the U.S. on development of alfalfa leaf proteins for food uses includes: functional properties of leaf protein (580); succinylation of the ϵ-amino groups of lysine, which has increased solubility of isolated protein over ten-fold, enhanced emulsifying activity by 32%, and foaming capacity three-fold, and acetylation of leaf protein which has improved solubility and foaming capacity to a lesser extent (581); leaf protein fractionation by ultrafiltration (582); use of recycled dilute alfalfa solubles to increase yield of recovered leaf protein concentrate (583); isolation and partial characterization of the major (Fraction I protein) component (584); functional properties of edible leaf protein concentrates (585); heat and alkali damage to leaf protein including racemization and lysinoalanine formation (586); and formation of a food ingredient gel from a heated emulsion of alfalfa leaf protein and peanut oil (587).

Researchers in India have reported on: development of a screw press for obtaining alfalfa leaf juices (588); modified processes for preparing leaf protein concentrate (589); fractionation of leaf proteins with organic solvents (590); and nutritional evaluation (591). Spanish researchers have reported on solvent extraction of pigments, polyphenols, and unsaturated fatty acid degradation products from alfalfa leaves (592); and have employed microbial coagulation to recover protein curd from alfalfa green juice (593). Italian scientists have reported on relationships between enzyme levels and extractable proteins in alfalfa (594), and on pilot plant production of an edible alfalfa protein concentrate (595).

Additional research has been reported from: France on improvements in making white protein (596); New Zealand on egg yolk pigmenting properties of alfalfa leaf protein concentrate and paprika (597); Sweden on the nutritive value of white leaf protein (598); and Yugoslavia on production of hydrolyzed vegetable proteins from soy grits and dehydrated alfalfa flour (599).

Morus albus L. (Mulberry Bush)

India (600).

Nicotiana ssp. (Tobacco)

Most of the publications on development of tobacco leaf proteins for food uses have been from U.S. scientists, and include: extraction of protein fractions from fresh tobacco leaves (601); characterization of Fraction I protein and other products from tobacco (602); functional properties of crystalline Fraction I protein (603), which include the ability to produce heat-setting foams similar to egg white; characterization of Fraction I protein degradation by chemical and enzymatic treatments (604); and characterization of the water-soluble glycoproteins of tobacco leaves (605). Staron (606) in France has also reported on preparation and potential food uses of tobacco leaf protein.

Pennisetum purpureum Schum., cv. napier (Napier grass, elephant grass, tropical C 4 grass)

Brazil (607, 608).

Pereskia aculeata Mill. (Gooseberry, Lemon Vine)

Brazil (609, 610).

Pisium sativum var. arvense (Alaska Peas)

Venezuela (611).

Psophocarpus tetragonolobus DC. (Winged Bean, Asparagus Bean, Goa Bean)

India (612), Sri Lanka (613).

Saccharum officinale L. (Sugar Cane)

Leaves and shoots (614), Mauritius (615).

Trifolium repens L. (White Clover)
Japan (616).

Trifolium resupinatum L. (Persian Clover)
Pakistan (617).

Urtica dioica L. (Common Nettle)
U.K. (618).

Multiple Studies
Many researchers have included several species in their publications on characterization, processing and utilization of leaf proteins. Egyptian scientists have studied bean, cabbage, tomato and sugar cane leaves (619). Ghana researchers have published on coco yam, amaranth, adansonia, and hibiscus leaves (620).

Indian scientists have reported on: tree leaves of *Cassia fistula, Sesbania grandiflora, Gliricidia maculata, Morus alba, Moringa oleifera,* and *Leucaena leucocephala* (621); the water weeds *Eichhornia crassipes, Limnanthemum cristatum,* and *Ipomoea reptens* (622); leaves of *Madhuca indica* (Gmd), *Millettia ovalifolia,* and *Sesbania sesban* (Linn Merr) (623); and leaves of 9 miscellaneous plants (624).

Japanese researchers have studied the green juices of alfalfa, white clover and Italian rye grass (625), and anaerobically fermented leaf extracts to prepare protein concentrates of *Dolichos lablab,* maize, and sorghum (626). An aqueous extract of comfrey, perilla and field mint leaves, wheat or soy protein, and ginseng saponin has been developed as a health food, and is claimed to have potential use in reducing the harmful effects of smoking (627). Korean scientists have reported on fatty acid and sterol compositions of leaf proteins of Italian ryegrass, red clover, oats and alfalfa (628). Taiwanese researchers have characterized the leaf proteins of 7 low-cost vegetables, wastes of 18 vegetables and other crops, and 19 grasses and weeds (629).

Nigerian researchers have published on tannins and nutritive value of leaf protein concentrates from the tropical plants *Manihot esculenta, Leucaena* spp., *Desmodium distortum, Cassia tora, Phaseolus calcaratus, P. sathywide, Psophocarpus tetragonolubus* (winged bean) and *Brassica napus* (rapeseed) (630). Also, ten leafy vegetables have been studied; *Abelmoschus esculentus, Corchorus olitorius, Talinum triangulare, Amaranthus hybridus, Piper guineense, Ocimum bascilicum, Curcubita pepo, Vernonia amygdalina, Telfairia occidentalis* and *Marsdenia latifolia* (631). Tanzanian scientists have surveyed leaf proteins of 22 wild and cultivated green leafy vegetables (632).

In the U.S., Florida scientists have studied leaf proteins of *Potamogeton illinoensis, Eichhornia crassipes, Pistia stratioties, Hydrilla verticillata,* and *Typha* ssp. aquatic weeds (633), and determined lipids distributions in green leaf protein concentrates from four tropical leaves, chaya, a sorghum-sudan cross, cassava and sauropus (634). Thirty-four species of tropical plant leaves have been screened in Puerto Rico, with *Manihot esculenta, Sauropus androgynus, Cnidoscolus chyamansa, Canavalia ensiformis, Lablab niger,* and *Vigna unguiculata* selected for further studies (635). Solubility properties of Fraction I proteins of corn (maize), cotton, tobacco and spinach leaves also have been described (636).

Other Potential Food Protein Sources

Researchers are looking continuously for additional potential sources of vegetable food proteins. Nutritional values of watermelon (*Citrullus vulgaris*), locust bean (*Parkia filicoides*) and walnut (*Conophorum tetracarpidium* Welw.) protein isolates were determined in Nigeria by amino acid analysis, enzyme digestion and animal feeding trials. All isolates were highly (74–97%) digestible with pepsin followed by pancreatin. Methionine was the first limiting amino acid in walnut and locust bean, and lysine in watermelon. Watermelon had the highest PER (2.40) compared to the casein standard of 2.50, followed by locust bean (1.81). Functionality assessments showed that locust bean protein isolate was insoluble between pH 4–10, watermelon isolate between pH 5–8, and walnut isolate between pH 4–7. When used as protein and energy supplements in the local weaning food *ogi*, these isolates double the protein content at 10% supplementation and triple it at 20% (637). Other studies showed that defatted watermelon flour contained 52% protein compared to 51% for defatted sesame flour and 28% for dehulled and autoclaved chickpea flour. Cow pea flour was limiting in S-amino acids, while the oilseed flours were limiting in lysine (638).

Korean scientists have been working on preparation of toxin- and allergen-free protein isolates of castor bean (*Ricinus communis* L.) (639). Studies have been conducted in the Philippines on the processing stability of coconut (*Cocus nucifera* L.) milk using added mungo protein isolate and nonfat dry milk (640), and on protein enrichment of coco spread with mung beans, soybean, cow peas, and rice bean (641).

Research has been conducted in Nigeria on potential food uses of rubber tree (*Hevea brasiliensis,* Muell.) seed. Rubberseed protein is lower in lysine content than soybean protein isolate (3.5 vs 6.4 g/100 g protein), but higher in cystine and methionine (3.7 vs 1.1 and 2.5 vs 1.2, respectively) (642). Saponin contents of Nigerian oilseeds *Pentaclethra macrophylla, Mucuna uriens,* and *Hevea brasiliensis* were found to be similar to those of soybeans (643). Rubberseed meal has been fed successfully to layers in Indonesia (644), and to pigs in Malaysia (645).

Protein isolates of buffalo gourd (*Curcubita foetidissma*) and Hubbard squash (*C. maxima*) have been prepared by alkali extraction. Their essential amino acid patterns were comparable to sunflower, safflower and flax isolates, but inferior to soybean and rapeseed, with lysine and threonine being the limiting amino acids. Biological evaluation showed that curcubit protein was minimally effective as a sole source of protein to supplement sorghum and millet diets (646). Further descriptions of composition and physicochemical properties of buffalo gourd proteins, oil and starch have been reported by Scheerens and Berry (647).

The white potato (*Solanum tuberosum* L.) contains only about 10% protein on a dry weight basis, but a U.S. patent has been issued for making flours from potatoes and other crops for persons suffering from allergies (648). An "appropriate technology" has been developed in Peru for making low-cost dried potato products in developing countries (649). Effects of processing on functionality of protein concentrates made from potato processing wastes have been studied in the U.S. (650), and a process developed in Poland for recovery of protein concentrate from potato juice by ultrafiltration (651). It is common practice in West European potato starch plants to recover high molecular weight proteins in starch washings by heat coagulation and centrifugation, and it is estimated that about 10,000 metric tons of potato protein product (containing 80% protein, N × 6.25) is recovered in the potato starch plants in the Federal Republic of Germany (652). Most of these products currently are used for feeding in animals.

Interest also exists in development of food proteins from the sweet potato (*Ipomoea batatas* Poir). Although relatively content on a dry matter basis, the protein in sweet potato flour has a PER of 2.2–1.3 depending upon the specific cultivar and processing (653). Sweet potato and cassava flours have been substituted for wheat in making soy sauce in the Philippines (654), and flours made from 26 varieties of sweet potato have been evaluated for food uses in Puerto Rico (655).

References

1. Proceedings: World Conference on Vegetable Food Proteins, A.R. Baldwin, Ed. *J. Am. Oil Chem. Soc. 56*:99 (1979).
2. *Qual. Plant.-Plant Foods Hum. Nutr. 33*:121 (1983).
3. *World Oilseed Situation and Market Highlights*, U.S. Department of Agriculture FAS Circular FOP 6-88. June 1988.
4. Food Balance Sheets. FAO, Rome, 1984.
5. Sipos, E.F., in *Soybean Utilization Alternatives*, Laura McCann, ed., Center for Alternative Crops and Products, U. Minnesota, St. Paul, MN, 1988, p. 57.
6. Matz, M.A., ed. *Protein Food Supplements—Recent Advances*, Noyes Data Corporation, Park Ridge, NJ, 1981.
7. Lawhon, J.T., K.C. Rhee and E.W. Lusas, *J. Am. Oil Chem. Soc. 58*:377 (1981).
8. Lawhon, J.T., U.S. Patent 4,420,425 (1983).
9. Kinsella, J.E., *J. Am. Oil Chem. Soc. 56*:242 (1979).
10. Hermansson, A.-M., *Ibid. 56*:272 (1979).
11. Kinsella, J.E., S. Damodaran and B. German, in *New Food Proteins, Vol. 5, Seed Storage Proteins*, A.M. Altschul and H.L. Wilcke, eds., Academic Press Inc., New York, 1985, p. 108.
12. Pomeranz, Y., *Functional Properties of Food Components*, Academic Press, Inc., New York, 1985.
13. Feeney, R.E., and J.R. Whitaker, in *New Protein Foods, Vol. 5, Seed Storage Proteins*, A.A. Altschul and H.E. Wilcke, eds., Academic Press, Inc., New York, 1985, p. 181.
14. Lusas, E.W., and K.C. Rhee, *ACS Symp. Ser. 312*:32 (1986).
15. Haytowitz, D.B., A.C. Marsh and R.H. Matthews, *Food Technol 35*(3):73 (1981).
16. Steinke, F.H., and D.T. Hopkins, *Cereal Foods World 28*:338 (1983).
17. Bodwell, C.E., and D.T. Hopkins, in *New Protein Foods, Vol. 5, Seed Storage Proteins*, A.A. Altschul and H.E. Wilcke, eds., Academic Press, Inc., New York, 1985, p. 221.
18. Mounts, T.L., and J.J. Rackis, eds., "Trypsin Inhibitor," *Qual. Plant.-Plant Foods Hum. Nutr. 35*:183 (1985).
19. Friedman, M., ed., *Nutritional and Toxicological Significance of Enzyme Inhibitors in Foods*, Plenum Press, New York, 1986.
20. Dabrowski, K.J., and F.W. Sosulski, *J. Agric. Food Chem. 32*:128 (1984).
21. Traver, L.E., G.N. Bookwalter and W.F. Kwolski, *Food Technol. 35*(6):72 (1981).
22. Hron, R.J., *Cereal Chem. 58*:334 (1981).
23. El-Sherbiny, G.A., S.S. Rizk, and M.A. El-Shaity, *Egypt. J. Food Sci. 14*:373 (1986).
24. Ibrahiem, N.A., A.H. Fahmy, M. Khairy, S. Morsi and S.A. Hassan, *Ibid. 14*:421 (1986).
25. Saunders, R.M. and G.O. Kohler, U.S. Patent 4,204,008 (1980).
26. Kozlowska, H. and J. Borowska, *Nahrung 28*:151 (1984).
27. Borowska, J. and H. Kozlowska, *Ibid. 30*:11 (1986).
28. Campbell, M.F., R.J. Fiala, J.D. Wideman and J.F. Rasche, U.S. Patent 4,265,925 (1981).
29. Bradford, M.M., K.N. Wright and F.T. Orthoefer, European Patent EP 0 083 175 B1 (1987).
30. Last, J., *CSIRO Food Res. Quarterly 39*(2):25 (1979).
31. Amiot, J., G.T. Brisson, F. Castaigne, G. Goulet and M. Boulet, *Can. Inst. Food Sci. Technol. J. 12*(1):23 (1979).
32. Jansen, G.R., and J.M. Harper, *Food Nutr. 6*(2):15 (1980).
33. Harper, J.M., and G.R. Jansen, *Food Rev. Intern. 1*:27 (1985).
34. McWaters, K.H., and J.P. Cherry, *J. Food Sci. 42*:1444 (1977).
35. McWaters, K.H., *Cereal Chem. 55*:853 (1978).
36. Fleming, S.E., and F.W. Sosulski, *Ibid. 55*:373 (1978).
37. Morcos, S.R., Z. El-Hawary and G.N. Gabriel, *Z. Ernaehrungswiss. 20*:275 (1981).
38. Deslisle, J., J. Amiot, C. Goulet, C. Simard, G.J. Brisson and J.D. Jones, *Qual. Plant.-Plant Foods Hum. Nutr. 34*:243 (1984).
39. Fashakin, J.B., M.B. Awoyefa and P. Fuerst, *Z. Ernaehrungswiss. 25*:220 (1986).
40. Kozlowska, H., R. Zadernowski and F.W. Sosulski, *Nahrung 27*:449 (1983).
41. Nielson, N., in *New Protein Foods, Vol. 5, Seed Storage Proteins*, A.A. Altschul and H.E. Wilcke, eds., Academic Press, Inc., New York, 1985, p. 27.
42. Visser, A., and A. Thomas, *Food Rev. Intern. 3*(1/2):1 (1987).
43. McCann, L., ed., *Soybean Utilization Alternatives*, Univ. Minnesota, St. Paul, MN, 1988.
44. Lusas, E.W., and G.M. Jividen, *J. Am. Oil Chem. Soc. 64*:839 (1987).
45. Lusas, E.W., and G.M. Jividen, *Ibid. 64*:973 (1987).
46. Gardner, H.K., R.J. Hron and H.L.E. Vix, *Cereal Chem. 53*:549 (1976).
47. Decossas, K.M., R.S. Kadan, J.J. Spadaro, G.M. Ziegler Jr. and D.W. Freeman, *J. Am. Oil Chem. Soc. 59*:488 (1982).
48. Hanumantha Rao, K., H.N. Chandrasekhara and G. Ramanatham, *J. Food Sci. Technol., India 24*(4):190 (1987).
49. Hanumantha Rao, K., H.N. Chandrasekhara and G. Ramanatham, *Ibid. 24*:190 (1987).

50. Osman, H.O.A., M. Khalil, A.R. El-Mahdy and S. El-Iraki, *Food Chem. 24*(2):109 (1987).
51. Valle, F.R. del, M. Escobedo, P. Ramos, S. de Santiago, R. Becker, H. Bourges, K.C. Rhee, Y.R. Choi, M. Vega and J. Ponce, *J. Food Sci. 51:*1242 (1986).
52. Valle, F.R. del, M. Escobedo, P. Ramos, S. de Santiago, H. Bourges, K.C. Rhee, Y.R. Choi, M. Vega and J. Ponce, *J. Food Processing Preservation 9*(1):35 (1985).
53. Indian Standard, IS:10901 (1984).
54. El-Shaarawy, M.T. and A.S. Mesallam, *Z. Ernaehrungswiss., 26*(2):100 (1987).
55. Yoo, J.H., and A.M. Hsueh, *Nutr. Rep. Int. 31:*147 (1985).
56. Hayes, R.E., C.P. Hannay, J.I. Wadsworth and J.J. Spadaro, *ACS Symp. Ser. 312:*138 (1986).
57. Choi, Y.R., and K.C. Rhee, in *Annual Progress Report*, Food Protein Research and Development Center, Texas A&M Univ., College Station, TX. 1984, p. 263.
58. Choi, Y.R., and E.W. Lusas and K.C. Rhee, *J. Food Sci. 48:*1275 (1983).
59. Rhee, K.C., in *New Protein Foods, Vol. 5, Seed Storage Proteins*, A.A. Altschul and H.E. Wilcke, eds., Academic Press, Inc., New York, 1985, p. 359.
60. Natarajan, K.R., *Adv. Food Res. 26:*215 (1980).
61. Harris, H., U.S. Patent 4,362,759 (1982).
62. Pominski, J., J.J. Spadaro and H.M. Pearce, U.S. Patent 4,355,951. (1982).
63. Maw, W.A., U.S. Patent 4,650,857 (1987).
64. Ihekoronye, A.I., *J. Sci. Food Agric. 37:*1035 (1986).
65. Ihekoronye, A.I., *Ibid. 38:*49 (1987).
66. Lawhon, J.T., and E.W. Lusas, *Food Technol. 38*(12):97 (1984).
67. Hagan, R.C., Dahl, S.R. and R. Villota, *J. Food Sci. 51:*367 (1986).
68. Aboagye, Y. and D.W. Stanley, *Can. Inst. Food Sci. Technol. J. 20*(3):148 (1987).
69. Toft, G. and D.W. Openshaw, U.K. Patent Application GB 2,166,638 A (1986).
70. McWaters, K.H., *Peanut Sci. 9*(1):46 (1982).
71. McWaters, K.H., *Ibid. 9:*46 (1982).
72. McWaters, K.H., *ACS Symp. Ser. 312:*8 (1986).
73. Venkateshaiah, R.V., A.M. Natarajan and K. Atmaram, *Cheiron 11*(6):294 (1982).
74. Chompreeda, P., A.V.A. Resurreccion, Y.C. Hung and L.R. Beuchat, *J. Food Sci. 52:*1740 (1987).
75. Khalil, J.K., and M.I.D. Chushtai, *Pak. J. Sci. Res. 43*(3/4):149 (1982).
76. Khalil, J.K., I. Ahmad, P. Iebal and S. Mufti, *Pak. J. Sci. Ind. Res. 26*(2):87 (1983).
77. Khalil, J.K. and M.I.D. Chushtai, *Qual. Plant.-Plant Foods Hum. Nutr. 34:*285 (1984).
78. Som, J.N., *Diss. Abstr. Int. B. 42*(2):572 (1981).
79. Sathe, S.K., D.V. Tamhane and D.K. Salunkhe, *Cereal Foods World 26:*407 (1981).
80. El-Samahy, S.K., M.M. Morad, H. Seleha and M.M. Abdel-Baki, *Baker's Dig. 54*(5):32 (1980).
81. Morad, M.M., M.M. Abdel-Baki, H. Seleha and S.K. El-Samehy, *Ibid. 54*(5):34 (1980).
82. Ohlson, R., in *New Protein Foods, Vol. 5, Seed Storage Proteins*, A.A. Altschul and H.E. Wilcke, eds., Academic Press, Inc., New York, 1985, p. 339.
83. Sosulski, R., *J. Am. Oil Chem. Soc. 58:*96 (1981).
84. Kozlowska, H., R. Zadernowski and B. Lossow, *Nahrung 26:*857 (1982).
85. Zadernowski, R., H. Kozlowska and K. Lysakowski, Proceedings 6th International Rapeseed Congress, Paris. 1983. p. 1351.
86. Kozlowska, H., and D.A. Rotkiewicz, R. Zadernowski, and F.W. Sosulski, *J. Am. Oil Chem. Soc. 60:*1119 (1983).
87. Kozlowska, H., H. Nowak and J. Nowak, *J. Sci. Food Agric. 34:*1171 (1983).
88. Kozlowska, H., and R. Zadernowski, Proceedings of 6th International Rapeseed Conference, Paris, p. 1412 (1983).
89. Schwenke, K.D. and R. Mothes, K. Marziger, J. Borowska and H. Kozlowska, *Nahrung 31:*1001 (1987).
90. Cameron, J.J., and C.D. Myers. U.S. Patent 4,418,013.
91. Lusas, E.W., in *New Protein Foods, Vol. 5, Seed Storage Proteins*, A.A. Altschul and H.E. Wilcke, eds., Academic Press, Inc., New York, 1985, p. 394.
92. Gassmann, B., *Nahrung 27:*351 (1983).
93. Lawhon, J.T., R.W. Glass, L.J. Maniak and E.W. Lusas, *Food Technol. 36*(9):76 (1982).
94. Rossi, M. and I. Germondari, *Lebensm.-Wiss. Technol. 15:*309 (1982).
95. Rossi, M. and I. Germondari, *Ibid. 15:*313 (1982).
96. Rossi, M., and C. Peri, in *Food Engineering and Process Applications, Vol. 2, Unit Operations*, M. Lemaguen and P. Jelen, eds., Elsevier Applied Science Publications, London, 1986, p. 197.
97. Gassmann, B., *Nahrung 27:*351 (1983).
98. Brueckner, J., G. Mieth and G. Muschiolik, *Ibid. 30:*428 (1986).
99. Schwenke, K.D. and E.J. Rauschal, *Ibid. 27:*1015 (1983).
100. Brueckner, J., G. Mieth, K. Dabrowski, S. Gwiazda and A. Rutkowski, *Ibid. 26:*457 (1982).
101. Dudonis, W. and R. Lasztity, *Ibid. 30:*434 (1986).
102. Miller, N., H.E. Pretorius, L.J. du Toit, *Food Chem. 21*(3):205 (1986).
103. Jones, L.J. and M.A. Tung, *Can. Inst. Food Sci. Technol. J. 16*(1):57 (1983).
104. Kabirullah, M., *Diss. Abstr. Int. B. 44*(2):446 (1983).
105. Willis, R.B.H., Kabirullah, M. and T.C. Heyhoe, *Lebensm.-Wiss. Technol. 17*(4):205 (1984).
106. Johnson, L.A., T.M. Sulieman and E.W. Lusas, *J. Am. Oil Chem. Soc. 56:*463 (1979).
107. Kinsella, J.E. and R.R. Mohite, in *New Protein Foods, Vol. 5, Seed Storage Proteins*, A.A. Altschul and H.E. Wilcke, eds., Academic Press, Inc., New York, 1985, p. 435.
108. Rivero de Pauda, M., *J. Food Sci. 48:*1145 (1983).
109. Serna-Saldivar, S.O., G. Lopez-Ahumada, R. Ortega-Ramirez and R. Abril Dominguez, *Ibid. 53:*211 (1988).
110. Tonella, M.L., M. Sanchez and M.G. Salazar, *Ibid. 48:*1637 (1983).
111. Brito, O.J., *Diss. Abstr. Int. B. 41*(10):3726 (1981).
112. Betschart, A.A., *J. Am. Oil Chem. Soc. 56:*454 (1979).

113. Betschart, A.A., R.Y. Fong and M. Hanamoto, *J. Food Sci. 44:*1022 (1979).
114. Lwon, C.K., G.O. Kohler and M.M. Hanamoto, *J. Am. Oil Chem. Soc. 59:*119 (1982).
115. Paredes-Lopez, O. and C. Odorica-Falomir, *J. Sci. Food Agric. 37:*1097 (1986).
116. Dev, D.K., E. Quensel and R. Hansen, *Ibid. 37:*199 (1986).
117. Dev, D.K., T. Sienkiewicz, E. Quensel and R. Hansen, *Nahrung 30:*391 (1986).
118. Dev, D.K. and E. Quensel, *Lebensm.-Wiss. Technol. 19:*331 (1986).
119. Betschart, A.A., *Cereal Foods World 27:*396 (1982).
120. Lasztity, R. and M. Hidvegi, eds., *Amino Acid Composition and Biological Value of Cereal Proteins*, D. Reidel Publishing Co., Dordrecht, The Netherlands (1985).
121. Kumar, G.V., T. Emilia Vershese, P. Hardias Rao and S.R. Shurpalekar, *J. Food Sci. Technol., India 17*(6):256 (1980).
122. Lawhon, J.T., U.S. Patent 4,645,831 (1987).
123. Watson, S.A. in *Corn Chemistry and Technology*, S.A. Watson and P.E. Ramstad, eds., American Association of Cereal Chemists. St. Paul, MN, 1987, p. 53.
124. Wilson, C.M., in *Corn: Chemistry and Technology*, S.A. Watson and P.E. Ramstad, eds., American Association of Cereal Chemists, St. Paul, MN, 1987, p. 273.
125. Cerletti, P., and P. Restani, in "Proceeding International Society of Chemistry Symposium on Amino Acid Composition and Biological Value of Proteins, Budapest, Hungary," R. Lasztity and M. Hidvegi, eds., 1983, p. 467.
126. Riccardi, A., and P. Cerletti, *Qual. Plant.-Plant Foods Hum. Nutr. 33:*209 (1983).
127. Stringfellow, A.C., O.L. Brekke, V.F. Pfeifer, L.H. Burbridge, and E.L. Griffin, *Cereal Chem. 54:*415 (1977).
128. Blessin, C.W., W.L. Deatherage, J.F. Cavins, W.J. Garcia and G.E. Inglett, *Ibid. 56:*105 (1979).
129. Nielsen, H.C., J.S. Wall, J.K. Mueller, K. Warner, and G.E. Inglett, *Ibid. 54:*503 (1977).
130. Nielsen, H.C., J.S. Wall and G.E. Inglett, *Ibid. 56:*144 (1979).
131. Lawhon, J.T., U.S. Patent 4,624,805 (1986).
132. Inglett, G.E. and C.W. Blessin, *J. Am. Oil Chem. Soc. 56:*479 (1979).
133. Messinger, J.K., J.H. Rupnow, J.G. Zeece and R.L. Anderson, *J. Food Sci. 542:*1620 (1987).
134. Jao, Y.C., A.H. Chen and W.E. Goldstein, *Ibid. 50:*1257 (1985).
135. Peri, C., and R. Barbieri, *Riv. Ital. Sostanze Grasse 57:*185 (1980).
136. Peri, C., R. Barbieri, C. Brambilla and G. Tomassone. *Riv. Ital. Sostanze Grasse 57:*240 (1980).
137. Peri, C., R. Barbieri, C. Brambilla and G. Tommasone, *Ibid. 17:*240 (1980).
138. Barbieri, C., and E.M. Casiraghi. *J. Food Technol. 18:*35 (1983).
139. Peri, C., R. Barbieri and E.M. Casiraghi, *Ibid. 18:*43 (1981).
140. Dondero, C.M., and R.E. Meneses, *Alimentos 6*(3):19 (1981).
141. Restani, P., A. Riccardi and P. Cerletti, *Annales de Technologie Agricole 29:*409 (1980).
142. Peri, C., and R. Barbieri, *Riv. Ital. Sostanze Grasse 57:*185 (1980).
143. Lucisano, M., E.M. Casiraghi and R. Barbieri, *J. Food Sci. 49:*482 (1984).
144. Kulakova, E.V., E.S. Vainerman and S.V. Rogozhin, *Nahrung 27:*721 (1983).
145. Christianson, D.D., J.P. Friedrich, E.B. Bagley and G.E. Inglett, in *Maize: Recent Progress in Chemistry and Technology*, G.E. Inglett, ed., Academic Press, Inc, New York, 1982, p. 231.
146. Christianson, D.D., J.P. Friedrich, G.R. List, K. Warner, E.B. Bagley, A.C. Stringfellow and G.E. Inglett, *J. Food Sci. 49:*229 (1984).
147. Christianson, D.D. and J.P. Friedrich, U.S. Patent 4,495,207 (1985).
148. Mesallam, A.S.F., A.E. Salem, Z. Mohasseb and M.E. Zoueil, *Alex. J. Agric. Res. 27*(2):387 (1979).
149. Buck, J.S., C.E. Walker and K.S. Watson, *Cereal Chem. 64:*264 (1987).
150. Tonella, M.L., M. Sanchez and M.G. Salazar, *J. Food Sci. 48:*1637 (1983).
151. Ndupuh, E.C. and E.N.T. Akobundu, *J. Food Sci. Technol., India 21*(2):108 (1984).
152. Tonella, M.L., M. Sanchez and M.G. Salazar, *J. Food Sci. 48:*1637 (1983).
153. Trejo-Gonzales, A., A. Feria-Morales and C. Wild-Altamirano, *Advances in Chemistry Series 198:*245 (1982).
154. Wall, J.S., Y.V. Wu, W.F. Kwolek, G.N. Bookwalter, K. Warner and M.R. Gumbmann, *Cereal Chem. 61:*504 (1984).
155. Bookwalter, G.N., K. Warner, J.S. Wall, Y.V. Wu and W.F. Kwolek, *Ibid. 61:*509 (1984).
156. Juliano, B.O., in *Rice: Chemistry and Technology*, American Association of Cereal Chemists, St. Paul, MN, 1985, p. 59.
157. Conkerton, E.J., and R.L. Ory, *Peanut Sci. 10*(2):56 (1983).
158. Matthews, R.H., and J.S. Douglas, *Cereal Foods World 23:*606 (1978).
159. Vose, J.R., and C.G. Younga, *Cereal Chem. 55:*280 (1978).
160. Wu, Y.V., K.R. Sexson and J.E. Sanderson, *J. Food Sci. 44:*1580 (1979).
161. Peterson, D.M. and A.C. Brinegar, in *Oats: Chemistry and Technology*, F.H. Webster, ed., American Association of Cereal Chemists, St. Paul, MN, 1986, p. 153.
162. Webster, F.H., *Oats: Chemistry and Technology*, American Association of Cereal Chemists, St. Paul, MN, 1986.
163. Cluskey, J.E., Y.V. Wu, G.E. Inglett and J.S. Wall, *J. Food Sci. 41:*799 (1976).
164. Cluskey, J.E., Y.V. Wu, J.S. Wall and G.E. Inglett, *J. Am. Oil Chem. Soc. 56:*481 (1979).
165. D'Appolonia, B.L., and V.L. Youngs, *Cereal Chem. 55:*736 (1978).
166. Ma, C.Y., *Ibid. 60:*36 (1983).
167. Ma, C.Y. *Can. Inst. Fd. Sci. Technol. J. 18*(1):79 (1985).
168. Ma, C.Y., *Ibid. J. 16*(3):291 (1983).

169. Ma, C.Y., and G. Khanzada, *J. Food Sci. 52:*1583 (1987).
170. Goulet, G., J. Amiot, D. Lavergne, V.D. Burrows and G. J. Brisson, *J. Food Sci. 51:*241 (1986).
171. Goulet, G., R. Ponnampalam, J. Amiot, A. Roy and G. J. Brisson, *J. Agri. Food. Chem. 35:*589 (1987).
172. Okeiyi, F.C., and M.F. Futrell, *Nutr. Rep. Int. 28:*451 (1983).
173. Saunders, R.M., and R. Becker, *Advances in Cereal Technol. 6:*357 (1984).
174. Teutonico, R.A., and D. Knoor, *Food Technol. 39*(4):49 (1985).
175. R. Becker, E.L. Wheeler, K. Lorenz, A.E. Stafford, O. K. Grosjean, A.A. Betschart and R.M. Saunders, *J. Food Sci. 46:*1175 (1981).
176. Betschart, A.A., D.W. Irving, A.D. Shephers and R. M. Saunders, *Ibid. 46:*1181 (1981).
177. Pant, K.C., *Nutr. Rep. Int. 28:*1445 (1983).
178. Pedersen, B., L. Hallgren, I. Hansen and B.O. Eggum, *Qual. Plant.-Plant Foods Hum. Nutr. 36:*325 (1987).
179. Crabtree, J., and A.W. James, *Trop. Sci. 24*(2):77 (1982).
180. Fraser Nochera, C.L., *Diss. Abstr. Int. B. 47*(2):578 (1986).
181. Commercial literature. De Danske Sukkerfabrikker, Copenhagen K, Denmark, 1988.
182. Staron, T., *Medicine Nutrition 18*(1):21 (1982).
183. El-Sherbiny, G.A., S.S. Rizk and M.A. El-shiaty, *Egypt. J. Food Sci. 14:*373 (1986).
184. Chang, K.C., and L.D. Satterlee, *J. Food Sci. 44:*1589 (1979).
185. Marshall, H.F., K.C. Chang, K.S. Miller and L. D. Satterlee, *Ibid. 47:*1170 (1982).
186. Deshpande, S.S., and Cheryan, M., *Qual. Plant.-Plant Foods Hum. Nutr. 34:*185 (1984).
187. Lawhon, J.T., and E.W. Lusas, U.S. Patent 4,645,677 (1987).
188. Murray, E.D., T.J. Maurice, L.D. Barker and C. D. Meyers, U.K. Patent Application 2,077,739A (1981).
189. El-Sayed Abdek-Aal, M., A.A. Shetata, A.R. El-Mahdy and M.M. Youssef, *J. Sci. Food Agric. 37:*553 (1986).
190. Pilosof, A.M.R., R. Boquet and G.S. Bartholomai. *Cereal Chem. 63:*456 (1986).
191. Tyler, R.T., *J. Food Sci. 49:*925 (1984).
192. Tyler, R.T., "Impact Milling and Air Classification of Grain Legumes," *Diss. Abstr. Int. B. 45*(1):119 (1984).
193. Tyler, R.T., and B.D. Panchuk, *Cereal Chem. 61:*192 (1984).
194. Tyler, R.T., and B.D. Panchuk, *Ibid. 59:*31 (1982).
195. Tyler, R.T., C.G. Youngs, and F.W. Sosulski, *Can. Inst. Food Sci. Technol. J. 17*(2):71 (1984).
196. Tyler, R.T., C.G. Youngs, and F.W. Sosulski, *Cereal Chem. 58:*144 (1981).
197. Sosulski, F.W., A.F. Walker, P. Fedec and R.T. Tyler, *Lebensm.-Wiss. Technol. 20:*221 (1987).
198. Sosulski, F.W., K. Elkowicz and R.D. Reichert, *J. Food Sci. 47:*498 (1982).
199. Elkowicz, K. and F.W. Sosulski, *Ibid. 47:*1301 (1982).
200. Sosulski, F.W., R. Hoover, R.T. Tyler, E.D. Murray and S.D. Arntfield, *Starch/Staerke 37:*257 (1985).
201. Hoover, R., and F. Sosulski, *Ibid. 36:*246 (1984).
202. Cloutt, P., A.F. Walker, and D.J. Pike, *J. Sci. Food Agric. 38:*177 (1987).
203. Lallemant, J., *Ind. Cereales 27:*23 (1984).
204. Tecklenburg, E., M.E. Zabik, M.A. Uebersax, J.C. Dietz and E.W. Lusas, *J. Food Sci. 49:*569 (1984).
205. Bhatty, R.S., *J. Agric. Food Chem. 30:*620 (1982).
206. Chang, P.R.Q., and A.R. McCurdy, *Can. Inst. Food Sci. Technol. J. 18:*94 (1984).
207. Bhatty, R.S., and G.I. Christison, *Qual. Plant.-Plant Foods Hum. Nutr. 34:*41 (1984).
208. Musakhanian, J., and I. Alli, *Food Chem. 23:*223 (1987).
209. Fenwick, D.E., and D. Oakenfull, *J. Sci. Food Agric. 34:*186 (1983).
210. Romeo, M., B. Escobar, L. Masson and M.A. Mella, *Alimentos 8*(1):3 (1983).
211. Mabesa, L.B., E.O. Atutubo and M.E. Castro-Sandoval, *Philippine Agriculturalist 65:*245 (1982).
212. Abdel-Aal, E.M., M.M. Youssef, A.A. Shetata and A.R. El-Mahdy, *Egyptian J. Food Sci. 13:*201 (1985).
213. Mendez, M.H.M., and M.A. Pourchet Campos, *Cienc. Technol. Alimentos 4*(1):95 (1984).
214. Fleming, S.E., and F.W. Sosulski, *Cereal Chem. 54:*1124 (1977).
215. Fleming, S.E., and F.W. Sosulski, *Ibid. 54:*1238 (1977).
216. Hsu, D.L., H.K. Leung, M.M. Morad, P.L. Finney and C.T. Leung, *Ibid. 59:*344 (1982).
217. Bahnassey, Y., K. Khan and R. Harrold, *Ibid. 63:*210 (1986).
218. Bahnassey, Y., and K. Khan, *Ibid. 63:*216 (1986).
219. Mittal, G.S., and W.R. Usborne, *Food Technol. 39*(4):121 (1985).
220. Moharram, Y.G., M.A. Hamza, B. Aman and O. El-Akary, *Food Chem. 26:*189 (1987).
221. El-Deeb, S.A., and A.E. Salam, *Alex. Sci. Exch. 5*(1):87 (1984).
222. Schmandke, H., and G. Schmidt, *Nahrung 21:*205 (1977).
223. Schmandke, H., and G. Schmidt, *Ibid. 21:*781 (1977).
224. Proll, J., J. Uhlig, H. Schmandke and R. Noack, *Ibid. 20:*743 (1976).
225. Schwenke, K.D., L. Prahl, E. Rauschal, S. Gwiazda, K. Dabrowski and A. Rutkowski, *Ibid. 25*(1):59 (1981).
226. Fabre, A., British Patent 1,577,994 (1980).
227. Magnat, R., and H. Bertrand, U.K. Patent Application 2,066,644A (1981).
228. Rockland, L.B., and T.M. Radke, *Food Technol. 35*(3):79 (1981).
229. Hsu, D.L., "Effect of Germination on Functional and Nutritional Properties of Dry Peas, Lentils, and Faba Beans," *Diss. Abstr. Int. B. 42*(5):1807 (1981).
230. Le Tien Winh and E. Dworschak, *Nahrung 30:*53 (1986).
231. Davis, K.R., *Cereal Chem. 58:*454 (1981).
232. Davis, K.R., M.J. Costello, V. Mattern and C. Schroeder, *Ibid. 61:*311 (1984).
233. Kanamori, M., T. Ikeuchi, F. Ibuki, M. Kotaru and K. K. Kan, *J. Food Sci. 47:*1991 (1982).
234. Vaidehi, M.P., M.S. Kumar and N. Joshi, *Cereal Foods World 30:*283 (1985).
235. El-Sherbiny, G.A., Rizk, S.S. and M.A. El-Shaity, *Egypt. J. Food Sci. 14*(1):111 (1986).
236. Obizoba, I.C., *Ecology Food Nutr. 14*(1):71 (1984).

237. El-Tinay, A.H., Z.M. El-Mahdi and A. El-Soubki, Sudan, *J. Food Technol. 20:*679 (1985).
238. Ahmed, R.V., M. Loewgfren, N. Velarde, L. Abrahamsson and L. Hambraeus, *Ecology Food Nutr. 11:*93 (1981).
239. Tanzanian Standard TZS 180:1983 (1983).
240. Habib, F.G.K., G.H. Mahran, S.H. Hilal, G.N. Gabrial and S.R. Marcos, *Z. Ernaehrungswiss. 15:*224 (1976).
241. Taha, F.S., *Grasas Aceites 38:*169 (1987).
242. Singh, U. and R. Jambunathan, *J. Sci. Food Agric. 32:*705 (1981).
243. Gopala Krishna, T., R.K. Mitra and C.R. Bhattia, *Qual. Plant.-Plant Foods Hum. Nutr. 27:*313 (1977).
244. Sant'anna Filho, R., E.R. Vilela and J.C. Gomez, *Cienc. Techo. Alimentos 5*(2):94 (1985).
245. Teixeira, J.P.F., D.S. Spoladore, N.R. Braga and E.A. Bulisani, *Bragantia 44*(1):457 (1985).
246. Nwokolo, E., *Qual. Plant.-Plant Foods Hum. Nutr. 37:*283 (1987).
247. Gayle, P.E., E.M. Knight and J.S. Adkins, *Federation Proc. 43*(3):475 (1984).
248. Clemente, P.R., and M.D.V. Penteado C., *Cienc. Technol. Alimentos 2:*84 (1982).
249. Vaidehi, M.P., and H.B. Shivaleela, *Nutr. Rep. Int. 30:*1173 (1984).
250. Abbey, B.W., and G.O. Ibeh, *J. Food Sci. 52:*406 (1987).
251. Sales, A.M., N.R. Braga, I. dos S. Draetta, E.M. Mori, M.M.E. Travaglini and A. Pizzinatto, *Boletim do Instituto de Alimentos, Brazil 17*(2):181 (1980).
252. Drouliscos, N.J., and V. Malefaki, *Br. J. Nutr. 43:*115 (1980).
253. Devadas, R.P., U. Chandrasekhar, G. Vasanthamani and V. Gaya Thri, *Indian J. Nutr. Dietetics 14:*291 (1977).
254. Shamanthaka Sastry, M.C., and D.R. Murray, *J. Sci. Food Agric. 40*(3):253 (1987).
255. Shastri, P.N., V.J. Daru and B.Y. Rao, *Indian Food Packer 33*(5):15 (1979).
256. Ikramul Hag, M.Y., and M.S. Chandry, *Pak. J. Sci. Ind. Res. 19:*66 (1976).
257. Khalil, J.K., and M.I.D. Chughtai, *Qual. Plant.-Plant Foods Hum. Nutr. 34:*285 (1984).
258. Akbar, S., M. Siddiq and P. Iqbal, *Pak. J. Sci. Ind. Res. 29*(2):126 (1986).
259. Kabirullah, M., R. Ahmed and O. Faruque, *Bangladesh J. Sci. Ind. Res. 11*(1/4):14 (1977).
260. Khaleque, A., A. Rahman and A. Rashid, *Bangladesh J. Sci. Ind. Res. 11:*103 (1977).
261. Yousseff, S.A.M., A. Salem and A.H.Y. Abdel-Rahman, *J. Food Technol. 11:*599 (1976).
262. Abdel-Rahman, A.H.Y., *Food Chem. 11*(2):139 (1983).
263. El-Hardallou, S.B., and F.A. Salih, *Legume Res. 4*(1):14 (1981).
264. Shehata, A.A.Y., A.M. Thanoun and H.N. Eldin, *Mesopotamia J. Agric. 12*(2):59 (1977).
265. Gattuso, A.M., G. Fazio, G. Arcoleo and V. Cilluffo, *Riv. Soc. Ital. Sci. Alimentazione 15:*451 (1986).
266. Rossi, M., I. Germondari and P. Casini, *J. Agric. Food Chem. 32:*811 (1984).
267. Fiorentini, R., M. Carcea and C. Galoppini, *Ind. Aliment. 20:*605 (1981).
268. Sotelo, A., F. Flores and M. Hernandez, *Qual. Plant.-Plant Foods Hum. Nutr. 37:*299 (1987).
269. Hernandez, M., and A. Sotelo, *Nutr. Rep. Int. 29:*845 (1984).
270. Hernandez, M., *Ibid. 36:*213 (1987).
271. Verma, M.M., D.A. Ledward and R.A. Lawrie, *Meat Sci. 11*(2):109 (1984).
272. Verma, M.M., R.J. Neale and D.A. Ledward, *Ibid. 15*(1):31 (1985).
273. Khaleque, A., L.G. Elias, J.E. Braham and R. Bressani, *Arch. Latinoamer. Nutr. 35:*496 (1985).
274. Gerschenson, L.N., R. Boquet and G.B. Bartholomai, *Lebensm.-Wiss. Technol. 16:*43 (1983).
275. Nath, J.P., and M.S. Narasinga Rao, *J. Food Sci. 46:*1255 (1981).
276. Tasneem, R., and N. Subramanian, *J. Agric. Food Chem. 34:*850 (1986).
277. Kochar, G.K., *Indian J. Anim. Res. 19:*45 (1985).
278. Khalsa, N., and P.K. Sharma, *Indian J. Agric. Sci. 17*(8):297 (1980).
279. Pore, M.S., *Ibid. 49:*532 (1979).
280. Liener, I., *J. Amer. Oil. Chem. Soc. 56:*121 (1979).
281. Bhatty, R.S., A.E. Slinkard and F.W. Sosulski, *Can. J. Plant Sci. 56:*787 (1976).
282. Shehata, A.A.Y., H. Nour El-Din and Abd-El-Mottaleb, *Alex. J. Agric. Res. 26:*369 (1978).
283. Siddiqi, S.H., *Pak. J. Sci. Ind. Res. 25*(5):167 (1982).
284. Shekib, L.A.E., M.E. Zouil, M.M. Youssef and M.S. Mohammed, *Food Chem. 18*(3):163 (1985).
285. Kaba, H., and J.C. Sanahuja, *Arch. Latinoamer. Nutr. Arch. 28:*169 (1978).
286. Perez-Gil, R.F., M.L. Arellano, R.H. Bourges and B. A. Llorente, *Tecnol. Alimentos 22*(4):3 (1987).
287. Cerletti, P., in *Developments in Food Proteins—2*, B.J.F. Hudson, ed., Applied Science Publishers, London, 1983, p. 133.
288. Gross, R., *S. African Food Rev. 9* (2 suppl.):73 (1982).
289. Duranti, M., and P. Cerletti, *J. Agric. Food Chem. 27:*977 (1979).
290. Restani, P., M. Duranti, P. Cerletti and P. Simonetti, *Phytochemistry 20:*2077 (1981).
291. Casero, P., M. Duranti and P. Cerletti, *J. Sci. Food Agric. 43:*1113 (1983).
292. Duranti, J., E. Cucchetti and P. Cerletti, *J. Agric. Food Chem. 32:*490 (1984).
293. Riccardi, A., P. Cerletti and G. Bertolotti, in *Perspectives for Peas and Lupins as Protein Crops*, Martinus Nijhoff Publishers, The Hague, 1983, p. 328.
294. Semino, G.A., P. Restani and P. Cerletti, *J. Agric. Food Chem. 33:*196 (1985).
295. Cerletti, P., M. Duranti and P. Restani. *Qual. Plant.-Plant Foods Hum. Nutr. 32:*145 (1983).
296. Lucisano, M., and C. Pompei, *Lebensm.-Wiss. Technol. 14:*323 (1981).
297. Pompei, C., M. Lucisano and N. Ballini, *Sci. Aliments 5:*665 (1985).
298. Duranti, M., J.A. Gatehouse, D. Boulter and P. Cerletti, *Phytochemistry 26:*627 (1987).
299. Baluster, D., I. Zacarias, E. Garcia and E. Yanez, *J. Food Sci. 49:*14 (1984).

300. Baluster, D.R., O. Brunser, M.T. Saitua, J.I. Egana, E.O. Yanez and D.F. Owen, *Food Chem. Toxicol. 22:*45 (1984).
301. Yanez, E., D. Vallester and D. Ivanoic, *Nutr. Rep. Int. 31:*493 (1985).
302. Baluster, D., P. Carreno, X. Urrutia and E. Yanez, *J. Food Sci. 51:*645 (1986).
303. King, J., C. Aguirre and S. de Pablo, *Ibid. 50:*82 (1985).
304. Aguilera, J.M., M.F. Gerngross and E.W. Lusas, *J. Food Technol. 18:*327 (1983).
305. Morad, M.M., S.B. El-Magoloi, and S.A. Afifi, *J. Food Sci. 45:*404 (1980).
306. Davin, A., and J.M. Brillouet, *Sci. Aliments* 6(1):61 (1986).
307. Campos, J.E., and A.A. El-Dash, *Bol. Tec. Cent. Tecnol. Agric. Aliment. 13:*40 (1978).
308. Taha, F.S., S.S. Mohamed and A.S. El-Nockrashy, *Grasas Aceites* 37(1):8 (1986).
309. Japanese Patent 5,227,221 (1977).
310. Han, O., W.T. Tae, Y.W. Kim, J.K. Lee, and C.H. Lee, *Korean J. Appl. Microbiol. Bioengr.* 13(3):191 (1985).
311. Yu, R.S.T., W.S.A. Kyle, T.V. Hung and R. Zeckler, *J. Sci. Food Agric.* 41(3):205 (1987).
312. Lilley, G.G., *Ibid. 37:*20 (1986).
313. Hung, T.V., P.D. Handson, V.C. Amenta, W.S.A. Kyle and R.S.T. Yu, *J. Sci. Food Agric.* 41(2):131 (1987).
314. Blaicher, F.M., R. Nolte and K.D. Mukherjee, *J. Am. Oil Chem. Soc. 58:*761 (1981).
315. Hatzold, T., R. Gross and I. Elmadfa, *Fette Seifen Anstrichm.* 84(2):59 (1982).
316. Hatzold, T., I. Elmadfa and R. Gross, *Qual. Plant.-Plant Foods Hum. Nutr. 32:*125 (1983).
317. Schoenberger, H., S. Moron and R. Gross, *Ibid. 37:*169 (1987).
318. Schonenberger, H., R. Gross, H.D. Cremer and I. Elmadfra, *J. Nutr. 112:*70 (1982).
319. Blasco Lamenca, M., F.D.R. Horqque and J. de Cabanyes, *Turrialba* 31(3):258 (1981).
320. Sathe, S.K., S.S. Deshpande and D.K. Salunkhe, *J. Food Sci. 47:*491 (1982).
321. Lucisano, M., C. Pompei and M. Rossi, *Lebensm.-Wiss. Technol. 17:*324 (1984).
322. Rahma, E.H., and M.S.N. Rao, *J. Agric. Food Chem. 32:*1026 (1984).
323. Dako, D.Y., and D.C. Hill, *Nutr. Rep. Int. 15:*239 (1977).
324. Janardhanan, K., and K.K. Lakshmanan, *J. Food Sci. Technol., India 22:*369 (1985).
325. Balogun, O.O., and A.A. Odutuga, *Nutr. Rep. Int. 25:*867 (1982).
326. Eka, O.U., *Food Chem. 5:*303 (1980).
327. Ibrahim, M.H., and S.P. Antai, *Qual. Plant.-Plant Foods Hum. Nutr. 36:*179 (1986).
328. Achinewhu, S.C., *J. Food Sci. 47:*1736 (1982).
329. Gonzales-Garza, M.T., V. Sousa and A. Sotelo, *Qual. Plant.-Plant Foods Hum. Nutr. 31:*319 (1982).
330. Abbas, I.R., "Physicochemical Properties of Tepary Bean Starch," PhD Thesis, Univ. Arizona, Tucson, AZ, (1987).
331. Coffmann, C.W., and V.V. Garcia, *J. Food Technol. 12:*473 (1977).
332. Rosario, R.R. del, and D.M. Flores, *J. Agric. Food Chem. 32:*175 (1981).
333. Rosario, R.R. del, and O.M. Maldo, *Philipp. Agric.* 65(1):103 (1982).
334. Narang, A.S., Bains, G.S. and I.S. Bhatia, *Cereal Chem. 58:*92 (1981).
335. Chandrasekhara, H.N., and C. Ramanathan, *J. Food Sci. Technol., India* 20(3):126 (1983).
336. Chavan, J.K., and S.K. Suggal, *J. Sci. Food Agric. 29:*230 (1978).
337. Diwan, M.A., P.N. Sastri and B.Y. Rao, *Indian Food Packer* 36(6):71 (1982).
338. Pulle, M.W., *J. National Agric. Soc. Ceylon* 11/12:45 (1974–1975).
339. Bhumiratana, A., *Food* 9(4):51 (1977).
340. Shedata, A.A.Y., and M. Thannoun, *Z. Lebensm.-Unters.-Forsch. 172:*206 (1981).
341. Adel, A., Y. Shehata and A. Mohammed Thannoun, *J. Plant Foods 3:*265 (1981).
342. Mesallam, A.S., and M.A. Hamza, *Qual. Plant.-Plant Foods Hum. Nutr. 37:*17 (1987).
343. K. Takahanshi, *J. Japanese Soc. Food Sci. Technol. 26:*396 (1979).
344. Takashi, S., R. Kobayashi, K. Kainuma and M. Nakamura, *Ibid. 32:*18 (1985).
345. Moy, J.H., W.K. Nip, A.O. Lai, W.Y.J. Tsai and T.O.M. Naka Yama, *J. Food Sci. 45:*652 (1980).
346. Infante, M.H., and A. Sotelo-Lopez, *Arch. Latinoamer. Nutr. 3346.0(1):*99 (1980).
347. Ologhobo, A.D., and B.L. Fetuga, *Food Chem. 10:*297 (1983).
348. Ologhobo, A.D., and B.L. Fetuga, *J. Agric. Sci., U. K. 102:*241 (1984).
349. Vegam, A. de la, and A. Sotelo, *Qual. Plant.-Plant Foods Hum. Nutr. 36:*75 (1986).
350. Susheelamma, N.S., and M.V.L. Rao, *J. Food Sci. 44:*1309 (1979).
351. Susheelamma, N.S., and M.V.L. Rao, *J. Food Technol. 14:*463 (1979).
352. Vasi, I.G., and R.S. Chintalapudi, *Comp. Physiol. Ecol. 6:*277 (1981).
353. Sathe, S.K., S.S. Deshpande and D.K. Salunke, *Lebensm.-Wiss. Technol. 16:*69 (1983).
354. Durigan, J.F., V.C. Sgarbieri and L.D. Almeida, *J. Food Biochem.* 11(3):185 (1987).
355. Spink, P.S., M.E. Zabik and M.A. Uebersax, *Cereal Chem. 61:*251 (1984).
356. Bressani, R., L.G. Elias, M.T. Huezo and J.E. Braham, *Arch. Latinoamer. Nutr. 27:*247 (1977).
357. Campos, M. de, and A.E. Olszyna-Marzys, *Bull. Environ. Contamination Tox. 22:*350 (1979).
358. Paredes-Lopez, O., G.I. Harry and R. Montes-Rivera, *Biotech. Letters* 9(5):333 (1987).
359. Hohlberg, A.I., and D.W. Stanley, *J. Agric. Food Chem. 35:*571 (1987).
360. Sathe, S.K., and D.K. Salunke, *J. Food Sci. 46:*82 (1981).
361. Sathe, S.K., and D.K. Salunke, *Ibid. 46:*626 (1981).

362. Chang, K., and L.D. Satterlee, *Ibid. 46:*1368 (1981).
363. Reddy, N.R., and M.D. Pierson, *Ibid. 52:*109 (1987).
364. Reddy, N.R., S.K. Sathke and M.D. Pierson, *J. Food Sci. 53:*107 (1988).
365. Sathe, S.K., and Salunkhe, D.K., *Ibid. 46:*71 (1981).
366. Sathe, S.K., and D.K. Salunke, *Ibid. 46:*1910 (1981).
367. Sathke, S.K., V. Iyer and D.K. Salunkhe, D.K., *Ibid. 47:*8 (1982).
368. Sathke, S.K., J.G. Ponte Jr., P.D. Rangnekar and D. K. Salunkhe, *Cereal Chem. 58:*97 (1981).
369. Deshpande, S.S., S.K. Sathe and D.W. Salunkhe, *CRC Crit. Rev. Food Sci. Nutr. 21*(2):137 (1984).
370. Pak, N., A. Mateluna and Araya, H., *Arch. Latinoamer. Nutr. 28:*184 (1978).
371. Pak, N., and Araya, H., *Ibid. 31:*371 (1981).
372. Figuerola R., F., A.M. Estevez A., E. Sepulveda E., and P. Villarroel S., *Alimentos 9*(4):5 (1984).
373. Pak, N., H. Araya and C. Cafati, *Arch. Latinoamer. Nutr. 27*(1):91 (1977).
374. Phillips, D.E., M.D. Eyre, A. Thompson and D. Boulter, *J. Sci. Food Agric. 32:*423 (1981).
375. Romo, C.R., and G.B. Bartholomai, *Lebensm.-Wiss. Technol. 11:*35 (1978).
376. Romo, C.R., and G.B. Bartholomai, *Ibid. 10:*279 (1977).
377. Evans, R.J., and D.H. Bauer, *J. Agric. Food Chem. 26:*779 (1978).
378. Kermasha, S., F. R. van de Voort and M. Metche, *Can. Inst. Food Sci. Technol. J. 19*(3):92 (1986).
379. Goeel, H., and W. Bock, *Nahrung 22:*809 (1978).
380. Apostolatos, G., *J. Food Technol. 19:*233 (1984).
381. Apostolatos, G., *Ibid. 19:*561 (1984).
382. Pilosof, A.M.R., G.B. Bartholomai, J. Chirife and R. Boquet, *J. Food Sci. 47:*1288 (1982).
383. Lena-Valdiva, C.B., and M.L. Ortega-Delgado, *Qual. Plant.-Plant Foods Hum. Nutr. 34:*87 (1984).
384. Amaya, J., *ABIA (Assoc. Brazil Indust Aliment.) 51:*20 (1980).
385. Sgarbieri, V.C., P.L. Antunes and R.G. Junqueria, *Ciencia e Techologia de Alimentos 2*(1):1 (1982).
386. Miyoshi, M., J. Nakabayashi, T. Hara, T. Yawata, I. Tsukamoto and Y. Hamaguchi, *J. Nutr. Sci. Vitaminol. 28:*255 (1982).
387. Hamaguchi, Y., N. Yagi, A. Nishino, T. Mochizuki, T. Mizukami and M. Miyoshi, *Ibid. 23:*525 (1977).
388. Tsukamoto, I., M. Miyoshi and Y. Hamaguchi, *Cereal Chem. 60:*194 (1983).
389. Aguilera, J.M., E.W. Lusas, M.A. Uebersax and M.E. Zabik, *J. Food Sci. 47:*1151 (1982).
390. Seyam, A.A., O.J. Banasik, and M.D. Breen, *J. Agric. Food Chem. 31:*499 (1983).
391. Alli, I., and B.E. Baker, *J. Sci. Food Agric. 32:*503 (1981).
392. Lee, J.P., M.A. Uebersax, M.E. Zabik, G.L. Hosfield and E.W. Lusas, *J. Food Sci. 48:*1860 (1983).
393. Zabik, M.E., M.A. Uebersax, J.P. Lee, J.M. Aguilera and E.W. Lusas, *J. Am. Oil Chem. Soc. 60:*1303 (1983).
394. Patel, K.M., C.L. Bedford and C.W. Youngs, *Cereal Chem. 57:*123 (1980).
395. Yadav, N.R., and I.E. Liener, *Legume Res. 1*(1):17 (1977).
396. Antunes, A.J., and P. Markakis, *J. Agric. Food Chem. 25:*1096 (1977).
397. Gomes, J. C., U. Koch and J.R. Brunner, *Cereal Chem. 56:*525 (1979).
398. Deshpande, S.S., and S.S. Nielsen, *J. Food Sci. 52:*1330 (1987).
399. Dryer, S.B., S.G. Phillips, T.S. Powell, M.A. Uebersax and M.E. Zabik, *Cereal Chem. 59:*319 (1982).
400. Cady, N.D., A.E. Carter, B.E. Kayne, M.E. Zabik and M.A. Uebersax, *Ibid. 64:*193 (1987).
401. Hoojjat, P., and M.E. Zabik, *Ibid. 61:*41 (1984).
402. Dreher, M.L., and J.W. Patek, *J. Food Sci. 49:*922 (1984).
403. Duszkiewicz-Reinhard, W., K. Kahn and B. Funke, *Ibid. 53:*88 (1988).
404. Aguilera, J.M., E.B. Crisafulli, E.W. Lusas, M.A. Uebersax and M.E. Zabik, *Ibid. 49:*543 (1984).
405. Patel, K.M., R.A., Merkel, A.E. Reynolds and C.G. Youngs, *Can. Inst. Food Sci. Technol. J. 13*(1):5 (1980).
406. Zaragosa, E.M., L.B. Rockland and D.G. Guadagni, *J. Food Sci. 42:*921 (1977).
407. Valencia, M.E., M.G. Vavich, C.W. Weber and B.L. Reid, *Nutr. Rep. Int. 19:*195 (1979).
408. Monte, W.C., and J.A. Magna, *J. Agric. Food Chem. 28:*1169 (1980).
409. Kon, S., and J.R. Wagner, *Food Prod. Development 13*(7):49 (1979).
410. Sahasrabudhe, M.R., J.R. Quinn, D. Paton, C.G. Youngs and B.J. Sukra, *J. Food Sci. 46:*1079 (1981).
411. Sumner, A.K., L.L. Whalley, G. Blankenagel and C.G. Youngs, *Can. Inst. Food Sci. Technol. J. 12*(2):51 (1979).
412. Sumner, A.K., M.A. Nielsen and C.G. Youngs, *J. Food Sci. 46:*364 (1981).
413. Vose, J.R., *Cereal Chem. 57:*406 (1980).
414. Gebre-Egziabher, A., and A.K. Sumner, *J. Food Sci. 48:*375 (1983).
415. Reichert, R.D., *Ibid. 47:*1263 (1982).
416. Voutsinas, L.P., E. Cheung and S. Naki, *Ibid. 48:*26 (1983).
417. Reichert, R.D., *Cereal Chem. 58:*266 (1981).
418. Sosulski, F.W. and A.R. McCurdy, *J. Food Sci. 52:*1010 (1987)
419. Bramsnaes, F., and H.S. Olsen, *J. Am. Oil Chem. Soc. 56:*450 (1979).
420. Madsen, R.F. and E. Buchbjerg, European Patent Application EP O 238 946 A2 (1987).
421. Koyoro, H. and J.R. Powers, *Cereal Chem. 64:*97 (1987).
422. Klein, B.P. and M.A. Raidl, in "AOCS Symposium Series" 312:19 (1986).
423. Gueriviere, J.F. de La, *Rev. Fr. Corps Gras 23*(2):67 (1976).
424. Dumont, J.P. and D.G. Land, *J. Agric. Food Chem. 34:*1041 (1986).
425. Saeed, B.M., E. Gelencser and J. Petres, *Suetoeipar 29*(1):3 (1982).
426. Grzeskowiak, B., Z. Pazola and M. Gogolewski, *Acta Alimentaria Polonica 13*(2):147 (1987).
427. Valle, F.R. del, M. Escobedo, M.J. Munos, R. Ortega and H. Bourges, *J. Food Sci. 48:*914 (1983).

428. Valle, F.R. del, E. Marco, R. Becker and R.M. Saunders, *J. Food Processing Preservation 11:*237 (1987).
429. Kailasapathy, K., "Utilization Studies on Winged Bean (Psophocarpus tetragonolobus (L.) DC) with Special Emphasis on Functional and Nutritional Properties When Used in Making Bread," *Diss. Abstr. Int. B. 43*(10):3180 (1983).
430. Garcia, V.V., "Biochemical Composition of Mature Winged Beans (Psophocarpus Tetragonolobus (L.) DC)." *Diss. Abstr. Int. B. 40*(5):2102 (1979).
431. Garcia, V. V. and J.K. Palmer, *J. Food Technol. 15:*469 (1980).
432. Onuma Okezie, B., and F.W. Martin, *J. Food Sci. 45:*1045 (1980).
433. Sathe, S.K., S.S. Deshpande and D.K. Salunkhe, *Ibid. 47:*503 (1982).
434. Okezie, B.O., and S.B. Dobo, *Baker's Dig. 54*(1):35 (1980).
435. Blaise, D.S., and B.O. Okezie, *Ibid. 54*(6):22 (1980).
436. Nmorka, G.D., and B.O. Okezie, *Cereal Chem. 60:*198 (1983).
437. Hazes, Y.S., and A.I. Mohammed, *J. Food Sci. 48:*75 (1983).
438. Kantha, S.S., N.S. Hettiarachchy and J.W. Erdman, Jr., *Cereal Chem. 63:*9 (1986).
439. Sri Kantha, S., and J.W. Erdman, Jr. AOCS Symposium Series 312:206 (1986).
440. Ibuki, F., M. Kotaru, K.K. Kan, T. Ikeuchi and M. Kanamori, *J. Nutr. Sci. Vitaminol. 29:*621 (1983).
441. Yangi, S.O., *Agric. Biol. Chem. 47:*2273 (1983).
442. Yanagi, S.O., *Japan Agric. Res. Quart. 18*(1):53 (1984).
443. Omachi, M., E. Ishak, S. Homma and M. Fujimaki, *J. Japan. Soc. Food Sci. Technol. 30:*216 (1983).
444. Dench, J.E., *J. Sci. Food Agric. 33:*173 (1982).
445. King, R.D., and P. Puwastien, *Ibid. 35:*441 (1984).
446. Gillespie, J.M., and R.J. Blagrove, *Austral. J. Plant Physiol. 5:*357 (1978).
447. Kortt, A.A., and J.B. Caldwell, *J. Sci. Food Agric. 36:*863 (1985).
448. Vaidehi, M.P., M.L. Annapurna, M.R. Gururaja and C. N. Uma, *J. Food Sci. Technol., India 19*(4):136 (1982).
449. Shet, M.S., and M. Madaiah, *J. Sci. Food Agric. 41:*287 (1987).
450. Kailaspathy, K., P.A.J. Perera and J.H. MacNeil, *J. Food Sci. 50:*1693 (1985).
451. Kailaspathy, K., *Sri Lanka J. Agric. Sci. 23*(1):105 (1986).
452. Gross, R. *Qual. Plant.-Plant Foods Hum. Nutr. 32:*117 (1983).
453. Vega, A. de al., G. Giral and A. Sotelo, *Nutr. Rep. Int. 24:*817 (1981).
454. Udayasekhara Rao, P., and R.D. Sharma, *Food Chem. 24*(1):1 (1987).
455. Hebblethwaite, P.D., ed., *The Faba Bean*, Butterworths, London, UK, (1983).
456. Schneider, C., M. Schultz, H. Schmandke and J. Borowska, *Nahrung 31:*863 (1987).
457. Schwenke, K.D., K.D. Robowsky and D. Augustat, *Ibid. 22:*425 (1978).
458. Schierbaum, F., S. Radosta, W. Vorwerg and B. Kettlitz, *Ibid. 29:*867 (1985).
459. Schwenke, K.D., L. Prahl, E. Rauschal, S. Gwiazda, K. Dabrowski and A. Rutkowski, *Ibid. 25:*59 (1981).
460. Schwenke, K.D., L. Prahl, B. Raab, K.S. Robowski, K. Dabrowski, J. Kocon, and A. Rutkowski, *Ibid. 27:*79 (1983).
461. Muschiolik, G., Hoerske, C. Schneider, M. Schultz and H. Schmandke, *Ibid. 30:*431 (1986).
462. Schwenke, K.D., E.J. Rauschal and K.D. Robowski, *Ibid. 27:*335 (1983).
463. Muschiolik, G., K. Ackermann and H. Hahnemann, *Ibid. 30:*101 (1986).
464. Ludwig, I., and E. Ludwig, *Ibid. 29:*949 (1985).
465. Schmandke, H., T.M. Bikov, E.M. Belavtseva, L.G. Radschnko, R. Maune, M. Schultz, V. Ya Grinberg and V.B. Tolstoguzov, *Ibid. 25:*263 (1981).
466. Schmandke, H., R. Maune, S. Schumann and M. Schultz, *Ibid. 25:*99 (1981).
467. Schmidt, G., and H. Schmandke, *Ibid. 31:*809 (1987).
468. Muschiolik, G., E. Dickinson, B.S. Murray and G. Stainsby, *Food Hydrocolloids 1:*191 (1987).
469. Schwenke, K.D., and E.J. Rauschal, *Nahrung 24:*593 (1980).
470. Muschiolik, G., K. Ackermann and C. Schneider, German Democratic Republic Patent DD 232 191 (1986).
471. Schmandke, H., G. Schmidt, M. Schultz, G. Muschiolik, S. Schumann, Yu. A. Antonov and V.B. Tolstoguzov, *Nahrung 25:*379 (1981).
472. Schmandke, H., G. Karpati, G. Muschiolik, R. Maune and Z. Szabo, *Ibid. 22:*907 (1978).
473. Schmandke, H., T.I. Bikbow, R. Maune, W.J. Gringberg and W.B. Tolstogusov, German Democratic Republic Patent DD 150 981 (1981).
474. Muschiolik, G., and H. Schmandke, *Nahrung 26:*65 (1982).
475. Muschiolik, G., H. Schmandke, J. Petres and G. Karpati, *Ibid. 26:*177 (1982).
476. Muschiolik, G., C. Schneider, M. Schultz and H. Schmandke, *Ibid. 28:*141 (1984).
477. Muschiolik, G., Schmandke, J. Petres and G. Karpati, *Ibid. 26:*169 (1982).
478. Muschiolik, G., H. Schmandke and J. Petres, *Ibid. 26:*157 (1982).
479. Muschiolik, G., H. Schmandke, A.K. Zadziennicke, D. Rotkiewicz and J. Borowska, *Ibid. 28:*133 (1984).
480. Ludwig, E. and H. Winkler, *Ibid. 27:*319 (1983).
481. Kroll, J., B. Gassmann, J. Proll and B. Gruetter, *Ibid. 28:*389 (1984).
482. Behnke, U., M. Schultz, H. Ruttloff and H. Schmandke, German Democratic Republic Patent DD 201 247 (1983).
483. Winkler, H., H. Noetzold and E. Ludwig, *Nahrung 28:*1029 (1984).
484. Noetzold, H., H. Winkler, B. Wiedemann and E. Ludwig, *Ibid. 28:*299 (1984).
485. Schwenke, K.D. and L. Prahl, German Democratic Republic Patent DD 206 607 (1984).
486. Muschiolik, G., L. Kraut, K. Ackermann, B. Hartmann and K. Valdeig, German Democratic Republic Patent DD 223 630 (1985).

487. Krause, W., E. Ludwig and H. Schneider, *Nahrung* 28:889 (1984).
488. Ludwig, I. and E. Ludwig, *Ibid.* 28:581 (1984).
489. Gassmann, B., K. Hoppe and J. Kroll, *Fleisch* 40(5):95 (1986).
490. Tebling, F., *Nahrung* 29:658 (1985).
491. Fornal, L., M. Sormal-Smietana and H. Kozlowska, *Przem. Spozyw.* 34(3):106 (1980).
492. Fornal, L., M. Soral-Smietana and J. Fornal, *Nahrung* 29:793 (1985).
493. Borowska, J., and H. Kozlowska, *Ibid.* 30:11 (1986).
494. Marquardt, R.R., D.S. Muduuli, and A.A. Frohlich, *J. Agric. Food Chem.* 31:839 (1983).
495. Arntfield, S.D., M.A.H. Ismond and E.D. Murray, *Can. Inst. Food Sci. Technol. J.* 18(2):137 (1985).
496. Arntfield, S.D., E.D. Murray and M.A.H. Ismond, *Ibid.* 18(3):226 (1985).
497. Arntfield, S.D., E.D. Murray and M.A.H. Ismond, *J. Food Sci.* 51:371 (1986).
498. Hinchcliffe, C., M. McDaniel, M. Vaisey and N.A.M. Eskin, *Can. Inst. Food Sci. Technol. J.* 10(3):181 (1977).
499. Youssef, M.M., and W. Bushuk, *Cereal Chem.* 63:357 (1986).
500. Abdel-Hamid, N.A., M.O. Osman and A.A. El-Farra, *Egyptian J. Food Sci.* 14:429 (1986).
501. El-Sherbiny, G.A., S.S. Rizk and H.A. Heikal, *Ibid.* 8(1/2):61 (1980).
502. Khairy, M., S. Morsi, A.A. El-Farra, N.A. Ibrahim and S.A. Hassan, *Egyptian J. Food Sci.* 14:435 (1986).
503. Elmorsi, E.S., *Annals Agric. Sci. Ain Shams Univ.* 27(1/2):23 (1982).
504. Vallery-Masson, D., *Ind. Aliment. Agric.* 97:583 (1980).
505. Bau, H.M., P. Caudy, M. Gay and G. Debry, *Can. Inst. Food Sci. Technol. J.* 12(4):194 (1979).
506. Gueguen, J., *Lebensm.-Wiss. Technol.* 13:156 (1980).
507. Geuguen, J., B. Quemener and P. Valdebouze, *Ibid.* 13:72 (1980).
508. Berot, S., J. Gueguen and C. Berthaud, *Ibid.* 20:143 (1987).
509. Olsen, H.S., *Lebensm.-Wiss. Technol.* 11:57 (1978).
510. Olsen, H.S., and Anderson, J.H., *J. Sci. Food Agric.* 29:323 (1978).
511. Patel, K.M., J.F. Caul and J.A. Johnson, *Cereal Chem.* 54:379 (1977).
512. Lorenz, K., Dilsaver, W. and M. Wolt, *Baker's Dig.* 53(3):39 (1979).
513. Griffiths, D.W., *J. Sci. Food Agric.* 30:458 (1979).
514. D'Aquino, M., G. Zaza, E. Carnovale, S. Gaetani and M.A. Spadoni, *Nutr. Rep. Int.* 24:1297 (1981).
515. Besrat, A., *J. Food Biochem.* 5:233 (1981).
516. Chakraborty, P., F. Bramsnaes and A.N. Bose, *J. Food Sci. Technol., India* 16(4):137 (1979).
517. Borhade, V.P., S.S. Kadam and D.K. Salunkhe, *J. Food Biochem.* 8:229 (1984).
518. Tjahjadi, C., "Isolation and Characterization of Adzuki Bean (*Vigna angularis*) Protein and Starch," *Diss. Abstr. Int. B.* 44(9):2739 (1984).
519. Prudente, V.R., and L.B. Mabesa, *Philippine J. Nutr.* 34(4):199 (1981).
520. Payumo, E.M., E.F. Fabian, B.S. Corpuz and L.S. Manguiat, *ASEAN Food J.* 1(2):94 (1985).
521. Barroga, C.F., A.C. Lauresn and E.M.T. Mendoza, *J. Agric. Food Chem.* 33:1157 (1985).
522. Barroga, C.F., A.C. Loaurena, and E.M.T. Mendoza, *Ibid.* 33:1006 (1985).
523. Adsule, R.N., S.S. Kadam and D.K. Salunkhe, *CRC Crit. Rev. Food Sci. Nutr.* 25(1):73 (1986).
524. Sales, M.G., "Characteristics of Processed Foods from Whole Cowpeas (Vigna sinensis)," *Diss. Abstr. Int. B.* 41(10):3727 (1981).
525. Sales, M.G., C.W. Weber, R.R. Taylor and J.W. Stull, *Nutr. Rep. Int.* 29:243 (1984).
526. Singh, S.P., and B.K. Mirsa, *J. Food Sci. Technol. India* 17(5):238 (1980), 526.
527. Ologhobo, A.D., and B.L. Fetuga, *Nutr. Rep. Int.* 25:913 (1982).
528. Longe, O.G., *Food Chem.* 6(2):153 (1980).
529. Ukhun, M.E., *Nutr. Rep. Int.* 30:933 (1984).
530. Akpapunam, M.A., and S.C. Achinewhu, *Qual. Plant.-Plant Foods Hum. Nutr.* 35:353 (1985).
531. Oyeleke, O.A., I.D. Morton and A.E. Bender, *Br. J. Nutr.* 54:343 (1985).
532. McWatters, K.H., *Cereal Chem.* 57:223 (1980).
533. McWatters, K.H., and F. Flora, *Food Technol.* 43(11):71 (1980).
534. Phillips, R.D., *J. Am. Oil Chem. Soc.* 598:351 (1982).
535. McWatters, K.H., *Cereal Chem.* 60:333 (1983).
536. Odum, P.K., L.A. Adamson, L. Morange and E.H. Edwards, *Nutr. Rep. Int.* 23:1005 (1981).
537. Phillips, R.D., *J. Am. Oil Chem. Soc.* 59:351 (1982).
538. Reber, E.F., L. Eboh, A. Aladeselu, W.A. Brown and D.D. Marshall, *J. Food Sci.* 48:217 (1983).
539. Phillips, R.D., M.S. Chinnan and L.G. Mendoza, *J. Food Sci.* 48:1863 (1983).
540. Lasekan, J.B., M.L. Harden and S.P. Yang, *Federation Proc.* 43(3):477 (1984).
541. Lasekan, J.B., M.L. Harden and S.P. Yang, *Lebensm.-Wiss. Technol.* 20:115 (1987).
542. Akobundu, E.N.T., and F.H. Hoskins, *J. Food Agric.* 1:111 (1987).
543. Sefa-Dedeh, S., and D. Stanley, *J. Agric. Food Chem.* 27:1238 (1979).
544. Sosulski, F.W., E.N. Kasirye-Alemu and A.K. Sumner, *J. Food Sci.* 52:700 (1987).
545. Sosulski, F.W., E.N. Kasirye-Alemu and A.K. Sumner, *Ibid,* 52:707 (1987).
546. Laurena, A.C., T.V. Den and E.M.T. Mendoza, *J. Agric. Food Chem.*
547. Mustafa, A.I., M.S. Al-Wessali, O.M. Al-Basha and R.H. Al-Amir, *Cereal Foods World* 31:756 (1986).
548. Ngarmsak, T., M.D. Earle and A.M. Anderson, *Internatl. Dev. Res. Center Monographs* IDRC-195e:36 (1982).
549. Poulter, N.H., and J.C. Caygill, *J. Sci. Food Agric.* 31:1158 (1980).
550. Obizoba, I.C., *Ecology Food Nutr.* 14:71 (1984).
551. Ninrahan, G.S., and S.K. Katiyar, *J. Indian Chem. Soc.* 58:70 (1981).

552. Thorn, K.A., A.M. Tinsley, C.W. Weber and J. W. Berry, *Ecology Food Nutr. 13:*251 (1983).
553. Ciudad, B.C., K.C. Cafati and A. Moyano, *Agric. Tec. 43*(3):185 (1983).
554. Balogun, A.M., and B.L. Fetuga, *J. Agric. Food Chem. 34:*189 (1986).
555. Telek, L. and H.D. Graham (eds.), *Leaf Protein Concentrates*, Avi Publishing Co., Westport, CT. 1983.
556. Maliwal, B.P., *J. Agric. Food Chem. 31:*315 (1983).
557. Fellows, P., *Trop. Sci. 27*(2):77 (1987).
558. Mucciarelli, S.I.L. de, J.A. Cid, M.A.L. de Arellano, S. Fernandez, N.G. de Luquez and M.A. Chirino, *Arch. Latinoamer. Nutr. 35:*458 (1985).
559. Silva, S.E., and C.C. Pereira, *Ciencia e Investigacion Agraria, 3*(4):169 (1976).
560. Neucere, N.J., and M.A. Godshall, *J. Agric. Food Chem. 27:*1138 (1979).
561. Hanczakowski, P., *Rocz. Nauk. Zootechniki 4*(1):227 (1977).
562. Kalra, C.L., O.P. Beerh and J.S. Pruthi, *Indian Food Packer 36*(6):42 (1984).
563. Goel, U., B.L. Kawatra and S. Bajaj, *Ibid. 32*(6):19 (1978).
564. Goel, U., B.L. Kawatra and S. Bajaj, *J. Sci. Food Agric. 28:*786 (1977).
565. Oke, O.L., B.A. Osuntogun and I.B. Umoh, *Nutr. Rep. Int. 25:*887 (1982).
566. Padma Kumarsinghe, L. Abrahamsson, *Indian J. Nutr. Dietetics 18:*442 (1981).
567. Yorks, T.P., "Table Grass: The Extraction of Protein from Bermudagrass and Its Implications," *Diss. Abstr. Int. B. 37*(8):3852 (1977).
568. Singh, A.K., *Vegetable Sci. 9*(1):42 (1982).
569. Cho, Y.S. and J.K. Kim, *J. Korean Soc. Food Nutr. 12:*251 (1983).
570. Abo-Bakr, T.M., N.M. El-Shemi and A.S. Mesallam, *Qual. Plant.-Plant Foods Hum. Nutr. 34:*67 (1984).
571. Lareo, L. and R. Bressani, *Food Nutr. Bull. 4*(4):60 (1982).
572. Francis, H.J., S. Subbiah and C.R. Lakshminarasimhan, *Madras Agric. J. 65*(6):412 (1978).
573. Bruemmer, J. H. and B. Roe, *Proc. Florida State Hort. Soc. 92:*140 (1979).
574. Walter, W.M. Jr., A.E. Purcell and G.K. McCollum, *J. Agric. Food Chem. 26:*1222 (1978).
575. Moharram, Y.G., M. Hamza and M.H. Abd-El-Aal, *Nahrung 31:*941 (1987).
576. Fafunso, M.A., and O.L. Oke, *Nutr. Rep. Int. 14:*629 (1976).
577. Luiza, M., V.C. Tupynamba and E.C. Viera, *Nutr. Rep. Int. 19*(2):249 (1979).
578. Meimban, E.J., J.G. Bautista III, and M.R. Soriano, *Philipp. J. Nutr. 35*(2):82 (1982).
579. Lee, K.S., K.Y. Yim, W.Y. Choi and M.J. Oh, *J. Korean Soc. Food Nutr. 14:*345 (1985).
580. Wang, J.C., and J.E. Kinsella, *J. Food Sci. 41:*286 (1976).
581. Franzen, K.L., and J.E. Kinsella, *J. Agric. Food Chem. 24:*914 (1976).
582. Eakin, D.E., P.R. Singh, G.O. Kohler and B. Knuckles, *J. Food Sci. 43:*544 (1978).
583. Edwards, R.H., D. de Fremery and G.O. Kohler, *J. Agric. Food Chem. 26:*738 (1978).
584. Hood, L.L., S.G. Cheng, U. Koch and J.R. Brunner, *J. Food Sci. 46:*1843 (1981).
585. Knuckles, B.E., and G.O. Kohler, *J. Agric. Food Chem. 30:*748 (1982).
586. Schwass, D.E., and J.W. Finley, *Ibid. 32:*1377 (1984).
587. Barbeau, W.E., and J.E. Kinsella, *J. Food Sci. 52:*1030 (1987).
588. Ahmed, S.Y., R. Harendranth and Narendra Singh, *J. Food Sci. Technol., India 21*(2):97 (1984).
589. Maliwal, B.P., *J. Food Biochem. 7*(2):93 (1983).
590. Singh, G., and N. Singh, *J. Food Sci. Technol., India 22*(1):69 (1985).
591. Singh, G., and N. Singh, *Indian J. Nutr. Dietetics 22:*308 (1985).
592. Hernandez, A., G. Gonzales and C. Martinez, *J. Sci. Food Agric. 43*(1):67 (1988).
593. Godessart, N., R. Pares and A. Juarez, *Appl. Environ. Microbiol. 53:*2206 (1987).
594. Camici, M., E. Balesteri, M.G. Tozzi, D. Bacciola, I. Saracchi, R. Felicioli, and P.L. Ipata, *J. Agric. Food Chem. 28:*500 (1980).
595. Fiorentini, R., and C. Galoppini, *J. Food Sci. 46:*1514 (1981).
596. Burghoffer, C., C. Costes, J.C. Rambourg, I. Gastineau and O. de Mathan, *Sci. Aliments 7*(1):111 (1987).
597. Johns, D.C., *New Zealand J. Agric. Res. 29:*677 (1986).
598. Carlsson, R., and P. Hanczakowski, *J. Sci. Food Agric. 36:*946 (1985).
599. Dzanic, H., I. Mujic and V. Sudarski-Hack, *J. Agric. Food Chem. 33:*683 (1985).
600. Srivastava, G.P. and M. Mohan, *Proc. 6th International Congress Food Sci. Technol. 3:*104 (1983).
601. Knuckles, B.E., G.O. Kohler and D. de Fremery, *J. Agric. Food Chem. 27:*414 (1979).
602. Wildman, S.G., and P. Kwanyuen, in *Photosynthetic Carbon Assimilation*, H.W. Siegelman and G. Hind, eds., Plenum Press, New York, NY, 1978, p. 1.
603. Sheen, S.J., and V.L. Sheen, *J. Agric. Food Chem. 33:*79 (1985).
604. Sheen, S.J., and V.L. Sheen, *Ibid. 35:*948 (1987).
605. Cheng, A.L.S., and H.A. Skoog, *Ibid. 31:*362 (1983).
606. Staron, T., *Medicine Nutri. 16*(5):337 (1980).
607. Carlsson, R., L. Jokl and C. Amorim, *Nutr. Rep. Int. 30:*323 (1984).
608. Jokl, L., R. Carlsson and C.F. Barbosa, *Ibid. 30:*331 (1984).
609. Souza Dayrell, M. de, and E. Vieira, *Ibid. 15:*529 (1977).
610. Sousa Dayrell, M. de, and E. Cardillo Vieira, *Ibid. 15:*539 (1977).
611. Romero, A.R., and J. Vanbalen, *Acta Cientifica Venezolana 31:*425 (1980).
612. Banerjee, A., A. Chandra, B.C. Sasmal and D.K. Bagchi, *J. Sci. Food Agric. 37:*783 (1986).
613. Kailasapathy, K., and C. Sandrasegaram, *Sri Lanka J. Agric. Sci. 23*(1):129 (1986).
614. Revuelta, J.R., *ATAC 46*(2):44 (1985).
615. Li Sui Fong, J.C., *Int. Sugar J. 84*(997):5 (1982).

616. Satake, I., S. Makino, R. Kayama and S. Koga, *J. Japan. Soc. Food Sci. Technol. 32:*705 (1985).
617. Nazir, M., and F.H. Shan, *Qual. Plant.-Plant Foods Hum. Nutr. 37:*3 (1987).
618. Hughes, R.E., P. Ellery, T. Harry, V. Jenkins and E. Jones, *J. Sci. Food Agric. 31:*1279 (1980).
619. Abo-Bakr, T.S., M.S. Mohamed, E.K. Moustafa, *Food Chem. 10*(1):15 (1983).
620. Dako, D.Y., *Nutr. Rep. Int. 23:*181 (1981).
621. Mohan, M., and G.P. Srivastiva, *J. Food Sci. Technol., India 18*(2):48 (1981).
622. Chakrabarti, S., D.K. Bagchi and S. Matai, *Indian J. Nutr. Dietetics 19*(4):125 (1982).
623. Singh, A.K., *Sci. Culture 47:*326 (1981).
624. Maliwal, B.P., *J. Agric. Food Chem. 31:*315 (1983).
625. Gwiazda, S., A. Noguchi and K. Saio, *J. Japan. Soc. Food Sci. Technol. 33:*213 (1986).
626. Satake, I., M. Sugiura, T. Kubota, T. Kobayashi, T. Sasaki, S. Makino and Y. Kayama, *Ibid. 31*(3):161 (1984).
627. Harima, S., European Patent Application EP 0 242 976 A1 (1987).
628. Kim, J.K., *J. Korean Soc. Food Nutr. 12:*259 (1983).
629. Sun, C.T., S.Y. Tseng, J.J. Lu and W.H. Chang, *J. Chinese Agric. Chem. Soc. 15*(1/2):78 (1979).
630. Osuntogun, B.A., S.R.A. Adewusi, L. Telek and O.L. Oke, *Human Nutr. Food Sci. Nutr. 41*F(1):41 (1987).
631. Ifon, E.T., and O. Bassir, *Food Chem. 5:*230 (1980).
632. Seeramulu, N., *J. Plant Foods 3:*265 (1981).
633. Roe, B., and J.H. Bruemmer, *Proc. Florida State Hort. Soc. 93:*338 (1980).
634. Nagy, S., H.E. Nordby and L. Telek, *J. Agric. Food Chem. 26:*701 (1978).
635. Martin, F.W., L. Telek and R. Ruberte, *J. Agric. Univ. Puerto Rico, 61*(1):32 (1977).
636. Bahr, J.T., D.P. Bourque and H.J. Smith, *J. Agric. Food Chem. 25:*783 (1977).
637. Stafford, W.L., I.B. Umoh, E.O. Ayalogu and O. L. Oke, *Nutr. Rep. Int. 18:*69 (1978).
638. Akaapunam, M.A., *Diss. Abstr. Int. B. 41*(9):3373 (1981).
639. Yoon, J.O., *Korean J. Food Sci. Technol. 12:*263 (1980).
640. Escueta, E.E., *Philippine J. Coconut Studies 5*(1):63 (1980).
641. Mabesa, L.B., *Philippine J. Coconut Studies 4*(4):45 (1979).
642. Achinewhu, S.C., *Food Chem. 21*(1):17 (1986).
643. Achinewhu, S.C., *Qual. Plant.-Plant Foods Hum. Nutr. 33:*3 (1983).
644. Karossi, A.T., in "Proceedings of the 3rd AAAP Animal Science Congress, Vol. 2," National Institute for Chemistry, Bandung, Indonesia, p. 958 (1985).
645. Ong, H.K., and J. Radem, *MARDI Res. Bull. 9*(1):78 (1981).
646. Henderson, C.W., "Preparation and Isolation of a Protein Isolate from *Curcubita foetidissima* Seed," *Diss. Abstr. Int. B. 45*(8):2502 (1985).
647. Scheerens, J.C., and J.W. Berry, *Cereal Foods World 31:*183 (1986).
648. Slimak, K.M., PCT International Patent Application. WO 87/04599 A1 (1987).
649. Keane, P.J., R.H. Booth, N. Beltran, "Appropriate Technology for Development and Manufacture of Low-Cost Potato-Based Food Products in Developing Countries," International Potato Center, Lima, Peru, 1986.
650. Knoor, D., *J. Food Proc. Engr. 5*(4):215 (1982).
651. Wojnowska, I., S. Poznanski and W. Bednarski, *J. Food Sci. 47:*167 (1982).
652. Meuser, F., (Private Communication) Technical University of Berlin, Berlin, Federal Republic of Germany, 1988.
653. Walter, W.M. Jr., G.L. Catignani, L.L. Yow and D.H. Porter, *J. Agric. Food Chem. 31:*947 (1983).
654. Data, F.S., J.C. Diamante and E.E. Forio, *Anals. Trop. Res. 8*(1):42 (1986).
655. Martin, F.W., *J. Agric. Univ. Puerto Rico 68*(4):423 (1984).

Allergenicity of Soy Proteins

Hans Elbek Pedersen

Aarhus Oliefabrik A/S, P. O. Box 50 DK-8100 Aarhus C, Denmark

Abstract

Saline extracts of soybean have been reported to contain up to 34 different antigenic proteins which can stimulate the rabbit systemic immune system after injection. Soy proteins have also shown an ability to orally sensitize guinea pigs, calves, pigs and humans. In hypersensitive individuals the presence of allergenic soy proteins in the diet will cause severe adverse reactions in the gastrointestinal tract. The extension of soy allergy and the underlying mechanisms seem to vary with the species. Although there are several reported cases of human allergy to soy-containing foods, the problem is often regarded as minor since it affects only a limited number of people. Soy-based products, such as baby formulas, burgers, curd, soy milk, soy sauce and lecithin, have all shown a high level of soy antigens, but the processing conditions for preparing these products may influence the antigenic activity; only a minor reduction in antigenicity is achieved during drying or toasting, but the use of a high temperature treatment, organic solvents or proteolytic enzymes can reduce the antigenicity almost completely.

Allergens are common constituents of food, and in the case of soy proteins, are used increasingly in the human diet. In contrast to antinutritional factors and toxins associated with foodstuffs, allergens display their effects only in those individuals who have an altered reactivity (allergy) to otherwise innocuous substances. This can result in unpleasant symptoms such as flushing of the face, skin disorders, respiratory problems and gastrointestinal disturbances. The intensity of the reaction depends on the degree of hypersensitivity of the individual consuming the soy protein rather than the quantity consumed (1). Food constitutes the most concentrated antigenic stimulus that an organism normally encounters, but the basis for the immune response to food proteins is poorly understood (2). Soy and other vegetable proteins are among the most common causes of hypersensitivity reactions (3,4). Peanut is named as the vegetable protein with the greatest allergenic potential (1), and allergy to peanut has been reported to cause fatal anaphylaxis (5). Furthermore sensitization to one leguminous plant protein may sometimes cause sensitization toward another also of legume origin (6). The increased use of soy protein products in prepared foods and feeds has raised concern about the possible immunological consequences of a high intake of soy protein in man (7,8,9,10,11,12,13) and in animals (14,15).

Soy Proteins As Antigens

Von Pirquet (16) introduced the term "allergie" in 1906 to describe the altered capacity of a human to react to a second injection of horse serum. Since then, all types of hypersensitivity in humans have been called "allergies" to identify the body's response to any allergen-antibody reaction that causes release of the chemical mediators of hypersensitivity. In the context of this review the definitions that follow will be used for immunological reagents and reactions.

Antibody or immunoglubolin: A binding protein that is synthesized by the immune system of the organism in response to either the invasion of a foreign organism or the injection of an antigen; may be divided into a number of different classes: IgG (the principal one used in immunoassays), IgE (involved in certain types of allergy), IgA and IgM.

Antigen: A substance that will stimulate antibody production through the immune system, and will react with the specific antibodies.

Epitope: The immunological active site(s) of an antigen. Catsimpoolas (17) used the reaction between different soy proteins and their antibodies to identify and characterize the different soy proteins; glycinin, α-, β-, and γ- conglycinin. Glycinin corresponds to the 11S fraction from ultracentrifugation studies, β and γ-conglycinin to the 7S fraction, whereas α-conglycinin is part of the 2S fraction. Immunological methods can be used not only for the detection, quantitation and evaluation of conformational changes of proteins, but also for their physiochemical characterization. Soy specific antibodies have been used with success to develop highly sensitive assays for the detection of soy proteins in processed foods (18,19) and for the detection of Kunitz trypsin inhibitor (20).

Allergen: An antigen with biological activity to provoke an allergic response via its reaction with specific antibodies. Allergens are generally large molecular weight compounds, nondialyzable, and are most often identified with the protein moiety of the material (21). It is important to note that lectins in processed soy and other foodstuffs may behave as "pseudo-allergens." They can have a direct pharmacological activity or alter mucosal permeability to food proteins, and thereby cause allergy-like symptoms without the involvement of the immune system. However, the activity of lectins of soy is eliminated by a normal toasting procedure (22).

Allergy or hypersensitivity: An immunological mediated adverse reaction giving symptoms like skin reactions or gastrointestinal disorders.

Intolerance: Failure to tolerate certain dietary constituents resulting in adverse reactions which could be either immunological or nonimmunological mediated.

In Vitro Studies of Soy Antigenicity

A range of studies has been made concerning the different antigenic proteins in soy. Firstly because the use of immunochemical methods has provided information about alterations in the protein molecule consequent to industrial or biological processing, and secondly there is much interest in the possibility that properties of such immunologically active proteins may cause harmful hypersensitivity reactions. There are several immunological methods such as crossed immunoelectrophoresis (CIE) available to study the antigenic fractions in soy products. Some of the results reported are listed in Table 1. Using rabbit anti-

TABLE 1

Antigenic Fractions of Soy Products Measured by Immunoelectrophoresis with Rabbit Anti-soy Antibodies

Soy product	Number of antigenic fractions	Reference
Soy flour	at least 10	(27)
	at least 12	(17)
	20–34	(23)
	at least 19	(89)
	25	(25)
	30	(24)
	multiple	(86)
Concentrate	1–3	(23)
	4–8	(24)
Isolate	6–15	(23)
	26	(24)

TABLE 2

Antigenic Glycinin Activity of Different Soy Products Measured by HIA (28) or ELISA (29). Values are Expressed as Titervalues Defined as the log2 to the Sample Dilution Giving End-point Reaction with Rabbit Anti-glycinin Antibodies

Product	Glycinin		Reference
	HIA	ELISA	
Non-toasted soy flour	13		(42)
	13	14	(44)
Toasted soy flour	12		(42)
	10		(42)
	12	13	(44)
Water extracted concentrate	8		(42)
	15	15	(44)
	9	12	
Alcohol extracted concentrate	12		(14)
	<1		(42)
	1		(42)
	<1	1	(44)
	<1	5	
Acid precipitated isolate	9		(42)
	7	10	(44)
	<1	7	
	9	12	

bodies to an extract from nontoasted soy flour made it possible to show that soy flour contains up to 34 different antigenic protein fractions. Ethanol-extracted soy concentrate contains 1–8 antigens, whereas isolates contain 6–26 antigenic structures possessing distinct activity (23,24). Brandon et al. (6) demonstrated that the Kunitz trypsin inhibitor has at least two distinct epitopes, one of which is retained under denaturing conditions. The epitopes of glycinin and β-conglycinin have not yet been identified. Soy antigens crossreact with antibodies to other leguminous proteins like peanut, lentil, kidney bean and pea (25,26), and anti-soy antibodies have been shown to crossreact with other leguminous protein (25,27).

The activity level of the different antigenic fractions can be quantified by methods like haemagglutination inhibition assay (HIA) (28) and enzyme linked immunosorbant assay (ELISA) (29). These measurements mainly have been carried out to quantify the activity of the major globular fractions glycinin and β-conglycinin. The level of antigenic activity of different types of soy products can be seen from Table 2. It is obvious that there is a great difference in the antigenicity of different soy products.

According to the definition, the study of the allergenic effects of proteins can be done only with in vivo experiments. However, it is possible to estimate the "allergen potential" by testing the reaction between different soy products or protein fractions and serum from individuals who already have developed a soy allergy. The multiplicity of potential allergens in various foods can be assessed in vitro by several immunoassays including crossed radio-immunoelectrophoresis (CRIE), radio-allergo sorbent test (RAST) and sodium dodecyl sulfate polyacrylamide gel electrophoresis (SDS-PAGE).

Soy Intolerance in the Pre-Ruminant Calf

Soy protein has a great potential for replacement of milk since it is readily available and has a reasonably balanced amino acid composition. However, it is well established that feeding calves on liquid diets containing large amounts of crudely processed soybean products leads to diarrhea, loss of appetite, high mortality and poor growth amongst surviving calves (30,31,32,33). Early investigations of this problem examined the possibility that certain antinutritional factors, such as protease inhibitors and lectins, may have been responsible for the unsatisfactory performance, but heating of soybean products to destroy these substances failed to improve the nutritive value of the soy protein for the calf (34,35,36,37). Calves cannulated in different regions of the digestive tract and given a succession of liquid feeds containing heated soy flour often developed abnormalities in digesta movements and nutrient absorption (38,39). These observations led to speculation that calves receiving protein as crudely processed soy flour may develop a gastrointestinal allergy to some factor present in the product. Support for this idea was given by the detection of serum IgG antibodies specific for soybean globulins, mainly glycinin and β-conglycinin, in calves fed soybean protein (13,38,40,41,42,43,44).

The digestive disturbances appear to correlate with immunological sensitization of the calf to glycinin and β-conglycinin (14,37,43,44). The development of a systemic antibody response to the soy protein was observed prior to the changes in digestive physiology. Calves acquired soy-specific antibodies from their dam via colostrum, but this passive immunity did not result in the calf showing tolerance of antigenic soy products (ASP). Calves sensitized to ASP showed significantly higher anti-soy antibody titers, disturbed motor function of the antrum, villous atrophy and crypt elongation of the small intestine, increased number of mast cells in the lamina propria, decreased small intestinal transit time, increased ileal digesta flow, electromyographical changes in the motor pattern of the small intestine, and decreased net nitrogen absorption compared with nonsensitized calves. The contraction of villi in response to antigenic soy proteins

(Fig. 1) would be expected to reduce the area available for absorption and may limit the extent of uptake of nutrient from the intestine. Recordings of intestinal motility in calves given antigenic soy protein showed that allergic reactions can cause gut muscles to contract in a spasmodic manner. These disturbances were quite different from other abnormal, but more gentle motions seen during an "osmotic" scour caused by the indigestible oligosaccharides. These motor disorders were linked with the production of antibodies to antigens of soy protein and to damage of gut tissue (45).

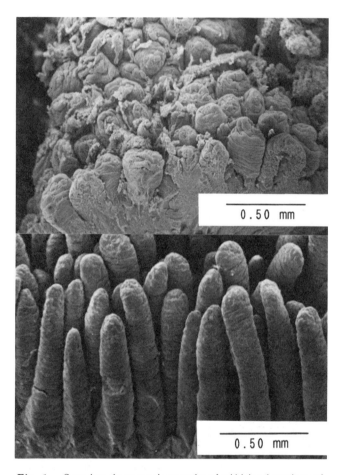

Fig. 1. Scanning electron micrographs of mid-jejunal sections of the small intestine of two soy-sensitive calves challenged with either a milk replacer containing soy flour (top) or an ethanol extracted soy concentrate (bottom) (44).

All calves fed ASP developed systemic antibodies specifically against the soy antigens Kunitz trypsin inhibitor, 2S fraction, glycinin and β-conglycinin, which were found to survive proteolysis *in vitro* and *in vivo*. Penetration of the gut wall by dietary soy antigens has not been demonstrated directly, but increased absorption of β-lactoglobulin has been shown to occur in soy sensitive calves (46). The fact that there was a strong stimulation of the humoral immune system following feeds of ASP indicates a failure of the mucosal membrane to act as a barrier to these macromolecules. It is not possible from the experiments to conclude if the increased antibody level was a result of more protein penetrating the gut wall due to the disorders in mucosal structure, or if these mucosal changes were the result of reactions mediated by the immunological response. Histamine is often involved as a chemical mediator in allergic reactions (47). But measurements of urinary excretion of histamine from soy fed calves suggest that histamine is not involved in the hypersensitivity reactions seen in soy sensitive calves (44). The response shown in the calves seems to be a failure of the exclusion-tolerance system. Recent passive transfer-studies suggest that IgG antibody in the absence of oral priming of the immune system is not directly responsible for gastrointestinal disorders in soy fed calves, and an IgE mediated reaction could be involved (Heppell, personal communication).

The abnormalities in the digestive processes of the gut were not observed with an "antigen-free" soy concentrate prepared by extracting soy flour with hot aqueous ethanol to remove oligosaccharides and inactivate antigenic proteins (14,39,44,48). Feeds containing "antigen-free" soy flour had effects on digestive processes which were intermediate between effects known by ASP and ethanol extracted soy concentrate. Similar observations that immunologically inactive soy proteins, when fed to calves, did not provoke circulatory anti-soy antibody production have been reported by other workers (43,49,50). Levels of immunological active glycinin and β-conglycinin in different soy products measured by *in vitro*-assay were positively related to the severity of gastrointestinal disorders by feeding the products to calves (14). For example antigenic soy protein gave significantly lower protein digestibility compared with "antigen-free" soy products (49). Thus immunochemical assay provides an important "tool," when developing and evaluating new protein products for calf milk replacers.

Post-Weaning Diarrhea in the Piglet

Rearing piglets through weaning without loss of body weight or diarrhea remains a challenging problem for pig producers. Persistent and severe scours in early weaned piglets undoubtedly involve pathogenic micro-organisms such as *E. coli* on many farms. The frequency of *E. coli* has been shown to be higher in pigs fed soy-based diets (51). During recent years a number of reports imply that an allergic reaction to dietary soy protein may partly be responsible for the post-weaning diarrhea problem in piglets, though there is more debate on the particular mechanisms involved.

Pigs weaned with antigenic soy diet at 3 weeks of age showed significant soy sensitivity through skin testing with soy and changes in absorption of xylose (15). This is associated with a period of malabsorption, diarrhea, crypt cell hyperplasia, villous atrophy and an increased susceptibility to *E. coli* enteritis. The results concurred with the hypothesis that following weaning, pigs experience a transient immune hypersensitivity to dietary soy antigens. Giesting (52) demonstrated that sows, like the cows, can show high levels of soy specific antibodies in their colostrum. These antibodies are consequently transferred passively to the suckling piglet resulting in titers against soy protein in most pigs at 1 d of age. Serum from unsuckled pigs showed no antibodies against soy. Pre-weaning regimen with or without soy protein did not affect anti-soy antibody titers measured at 2 and 3 weeks of age. Weaning diets of either soy or fish proteins did not affect the soy antibody titer measured at 4 and 5 weeks of age. The skin swelling response to intradermal injection of sterile saline solutions, and different proteins was evaluated as an indicator of cell mediated immunity. The skinfold response to soy protein was significantly greater for pigs exposed to soy flour than controls, 3 h post injection. Pre-weaning diet did not affect post-weaning swelling responses, and post-weaning treatment had no consistent effect on post-weaning skin responses. In addition to the minimal effect that the pre-weaning diet had on systemic immune responses, the performance of pigs was not affected to a large extent of the pre- or post-weaning diet (52). The findings do not provide support for the hypothesis that the post-weaning lag often observed in pigs fed soy-based diets is a result of a systemic immune response to soy protein (53). However, early weaned piglets may experience a localized immune response at the level of the intestinal mucosa following ingestion of soy protein, as seen in calves.

Soy products have been reported to cause morphological changes in the gut of the piglet around five days after weaning (53,54,55). Results of a recent physiological study in piglets have indicated a possible link between disorders in gut motility, measured electro-myographically, and the ingestion of antigenic soy protein in the weaner diet (56). Preliminary findings of a histological study of piglet gut tissue suggest that IgE antibodies could be involved in an inflammatory reaction to antigenic soy protein at weaning (Heppell, personal communication). Piglets suffering from diarrhea had large numbers of eosinophils and mast cells in connective tissue and there was some evidence of degranulation. Apart from dietary antigens, the piglet may have problems in coping with the unusual oligosaccharides found in soy flour. The accumulation of these components in the intestine not only provides a substrate for the growth of undesirable bacteria, it could also lead to an "osmotic" diarrhea which occurs when body fluid is drawn into the lumen of the gut to counter a raised osmotic pressure. Undigested protein in the gut also would add to the osmotic problem (57). Diarrhea could arise, therefore, from "osmotic" and "allergic" reactions occurring at the same time. Unfortunately it is not possible to distinguish between the two simply by looking at the feces.

Soy Protein Allergy in Humans

Duke (58) first reported cases of soy allergy in workers at a soybean processing plant in USA. In all five cases the patients suffered from coughing and asthma when working in the dust-filled factory, and on the basis of this it was concluded that the reason was soy-allergy with the lungs as effect organ. After Duke's publication there have occasionally been reports of soydust-allergy cases in the literature (5,59,60,61,62).

For a long time soy products were considered as being hypoallergenic (63,64). Soy-based infant formulas are often recommended for babies who have developed cow's-milk-protein intolerance, but some of these infants then develop hypersensitivity reactions to the soy protein (7,8,65,66,67,68,69,70,71,72,73,74,75), with observations of morphological changes in the small intestine similar to those seen in soy fed calves (Fig. 1). Several researchers have reported infants showing adverse reactions to soy products on the first time of challenge following the transfer to antigens from the mother to the infant via colostrum (13,76,77). Although cases of soybean allergy in man have been well documented, there is much controversy in the literature about the allergenicity of soybean proteins. It has been suggested that feeding soy rather than cow's milk from birth results in a lower incidence of allergic disease (78), but this has been disputed by a number of authors (79,80). There is some evidence that soy products can be less antigenic than cow's milk protein (79,81). However, the situation is not clear since from studies of the serum antibody response of babies given soy or cow's milk based formulas, Eastham et al. (82), concluded that soy protein is as least as antigenic as cow's milk protein.

There are very few investigations of the size of the soy allergy problem, incidence and prevalence, the mechanism of the allergy reactions, and the factors responsible. Soybeans have been placed 11th in order of their allergenicity in a list of foods (83). Foods of animal origin were high on the list. The first studies of the allergenic properties of soybean products in man were published in two articles by Ratner (63,64). It was concluded (64) that soy products heated to 127°C for 60 minutes gave positive skin reactions only in a few cases, while soy products given less heat treatment (e.g. spraydrying) gave positive skin reactions in all of the persons participating in the experiment. In an epidemiologic study (79) 1,753 children were examined regularly up to the age of 7 years to investigate the development of allergy to different foods. Of these, 632 children received diets based partly or totally on soy based baby formulas during the first months of life, and among these only 3 (0.5%) reacted with allergic symptoms to soy. No in vitro analyses were done.

Most reported cases show a great variation in sensitivity between individuals, but there is most often a relation between the clinical diagnosis of soy allergy and the soy specific systemic response of immunoglobulin IgE and IgG (8,23,24,83,84,85,86). However, the serum antibodies to soy protein in children with or without adverse reactions to soy products were not definitive in a study by May et al. (81). A group of 38 healthy adult persons had only low activities of antibody to soy in amounts that probably reflected no more than harmless exposure (87). However, Goulding et al. (10) found that 9% of a group receiving at least four main meals per week based on textured soy protein did show a marked increase in soy specific IgG antibody levels. In a 12-week metabolic study, Beer et al. (88) reported a slight increase in the IgG soy titer to ethanol extracted soy protein concentrates when fed to adult men as the only protein source. There were no symptoms or clinical signs of any allergic reactions, and there was no development in the IgE titer values. In contrast, Goodwin (9) tested another soy product and reported a significant increase in IgE titers in some subjects during a four-week period with the soy containing diet. Fries (5) reported that 27 out of 30 allergic children gave positive skin reactions to soybean and to one to four other legumes.

Different soy products were tested on guinea pigs by injection of protein extracts (63). Only in few cases was it possible to sensitize the animals, and it was concluded that soy protein was only a weak antigen. Preliminary results have indicated, that feeding antigenic soy flour stimulates the immune system of rats and consequently induces the development of diabetes mellitus, whereas "antigen-free" soy concentrates do not provoke these disorders (Brogren, personal communication). In a recent study (12) four soy-based baby formulas were tested for antigenicity and allergenic potential using animal models. The four formulas showed considerable differences in antigenicity. *In vivo* studies using guinea pigs, rabbits and calves were in good agreement and broadly correlated with the immunochemical assessment of antigenicity. It was suggested that these variations in antigenicity of different commercial products prepared from soy isolates may be important, when interpreting the results from studies of the development of allergy in infants given soy based formulas. From studies of the serum antibody response of bottle fed infants, Eastham et al. (82) concluded that soy protein is at least as antigenic as cow's milk protein. The soy based formula used in this earlier study was tested by Heppell et al. (12) who found it to be just as antigenic as nontoasted soy flour and having the highest allergenic activity in animal models. It is possible that the conclusions of Eastham et al. (82) might have been different if the formula had been based on a low-antigenic soy product.

Different *in vitro* immunoassays have been used in an attempt to identify the soy protein fractions responsible for the allergic reactions in man. Table 3 summarizes this work. It is basically the same proteins as involved in the adverse reactions seen in calves: the two globular proteins glycinin and β-conglycinin and their subunits as well as proteins from the 2S fractions like the Kunitz trypsin inhibitor (8,23,24,25,86,89). The latter protein has also been shown as an allergen from *in vivo* studies (90).

TABLE 4

Antigenic Glycinin Activity of Different Commercial Soy Related Foods or Food Ingredients Using a Competitive Inhibition ELISA. Titer Values are Expressed as the log2 to the Sample Dilution which Gives 50%Inhibition with Rabbit Antisoy Antibodies (44)

Product	Glycinin titer
Soy milk 1.	12
Soy milk 2.	4
Baby formula liquid concentrate	10
Soy bean curd in gravy	11
Soy tofu	< 1
Soy sauce (Chinese style)	12
"Burger" with 6% soy flour	10
"Burger" with 6% soy concentrate	< 1
Vegetable paté	< 1
Soy lecithin	6

TABLE 3

In vitro Determination of Potential Allergenic Fractions of Soy Products Using Sera from Soy-hypersensitive Humans

Soy product	Methods	Number of fractions	Type of fractions	Reference
Flour	RAST	3	11S +7S +2S	(8)
	CRIE	7–14	11S +7S +2S	(23)
	CRIE	1–25	11S +7S	(25)
(non-toasted)	SDS-PAGE	15	11S +7S, subunits	(24)
(toasted)	SDS-PAGE	7	11S +7S, subunits	(24)
	ELISA	> 2	11S +7S	(86)
Concentrate	CRIE	3	?	(23)
	SDS-PAGE	0–5	11S, subunits	(24)
Isolate	SDS-PAGE	10	11S +7S, subunits	(24)

Antigen Activity of Soy-Containing Foods

Although there have been several reported cases of allergic reactions to soy products as a food ingredient (76,91,92,93,94), only very few cases report reactions to soy containing foods: baby formulas (7,12,71), soy milk (5) and tuna fish salad extender (96). However, in early studies of Duke (58), it was reported that soy proteins were present in 250 or more different food products. As can be seen from Table 4, it is possible to detect high levels of soy antigen activity in many common foods. The RAST test has been used to evaluate the allergen potential of different commercial soy products (Pedersen, unpublished). As seen from Table 5 there is a significant difference between different soy products confirming the data from Ratner (64) and Edslev (89), and also the data from calf studies by Sissons et al. (14,96). The two soy protein concentrates showed a very low reaction with sera from soy allergic patients. These concentrates have been manufactured by the extraction of oligosaccharides from soy flour using hot aqueous ethanol.

It is unclear whether soy oil can provoke allergic response in soy sensitive individuals. Cooking food in soy oil has been reported to give severe anaphylactic reactions (90). Porras et al. (96) detected antigenic soy protein in some soybean oil samples and the active protein content varied between 0 and 3.3 mg per g of sample. In another study (98), the protein content of crude soybean oil and processed soybean oil was found to be 1.9 and 0.7 mg per kg of sample respectively. However, there are studies (64,85) where soy allergic patients did not react to com-

TABLE 5

In vitro Determination of the Potential Allergen Activity of 5 Commercial Soy Protein Products. The RAST Test was Performed to Show IgE Antibodies Against the Different Soy Products in Sera from 13 Soy Allergic Adults with Positive Skin Reactions and Oral Challenge to Non-toasted Soy Flour. Values are Given as Mean Values from the 13 Double-determinations Expressed as Phadebas RAST Units per ml

Soy product	Phadebas RAST value (PRU/ml)	
	Mean	+/- SD
Flour	1.91	0.42 a
Textured concentrate	0.25	0.03 c
Ethanol extracted concentrate	0.38	0.09 bc
Isolate	0.72	0.26 b
Isolate	0.65	0.15 bc

Values with different letters (a, b, c) are significantly different (P < 0.05).

mercially available soybean oils or olive oil. Sera from two soy allergic children have been used to demonstrate antigenic soy protein in soy lecithin and some samples of soy oil and margarine using an inhibition ELISA (97). The content of antigenic soy protein in soy lecithin was reported to be between 1 and 27 mg per g of sample. Other data (Pedersen, unpublished) have confirmed a content of protein (N×6.25) in soy lecithin of 2–5% giving ELISA glycinin titer values of 5–7. Overall the data demonstrate the presence of active soy antigens in a number of foods. It cannot be excluded that this exposure can contribute to sensitization and untoward reactions.

Effect of Processing

There are often conflicting reports about the intolerance or tolerance to soy products in animal and human nutrition as can be seen from the reported cases. The data presented in Tables 1, 2 and 4 clearly demonstrate that different soy products show a wide difference in both antigenic fractions and in levels of antigenic activity. These differences might explain the different results observed in clinical trials with humans or animals using different soy products. Unfortunately, it is only very seldom that the reported cases include any immunological evaluation of the soy products tested. Processing conditions can influence the concentration of antigens and allergens by altering the immunochemical structure of proteins and by influencing the digestibility of antigenic proteins (43,44,48,49,64,99). The epitopes of soy antigens may be affected in a variable manner according to the conditions encountered during food processing. For example, heat treatment at 80°C, 100°C or 120°C for 30 min reduced the RAST reactivity of the 7S and 11S fractions of soy proteins by up to 75%, whereas the 2S fraction was enhanced slightly at 80°C (8). In a similar study complete elimination of the allergen activity in an extract from nontoasted defatted soy flour required a heat treatment of 180°C for 30 min. (1). This is more heat than that required to inactivate other antinutritional factors (e.g. lectins, trypsin inhibitors) in nontoasted soy flours and may be sufficient to lower nutritional quality. Tests for allergic responses to foods should therefore take into account the changes in antigenicity that accompany processing food proteins (6,87,88).

Laboratory experiments (44,96) have shown from inactivation curves that it is possible to eliminate the antigenic activities of both glycinin and β-conglycinin by physiochemical denaturation of the proteins. Wetting the protein product with an alcohol-water mixture led to a lower activity remaining after the heat-treatment. A concentration of ethanol between 50 and 80% (w/w) appeared to be optimal for penetrating the hydrophobic parts of the soy globulins (Fig. 2). Thellersen (100) studied the use of proteolytic enzymes of sodium hydrogen sulfite to eliminate the antigenicity of soy proteins. The results showed only limited effects with trypsin and sodium hydrogen sulfite, whereas pepsin and Alcalase(R) were effective in reducing the antigenic activity of a soy extract. It has been speculated, that industrial processing and digestion within the gastrointestinal tract, while destroying some native antigenic determinants, may lead to the creation of new antigenic epitopes. Studies (101,102,103) have shown that hydrolysis of cow's milk protein with pepsin and trypsin can generate peptides with new antigenic determinants. However, the importance of such new antigenic fragments in oral sensitization is unclear. Serum from soy-allergic individuals was shown to give positive skin reaction in passive transfer tests using native soybean as the challenge antigen (64). More recently, McLaughlan *et al.* (104) and Heppell *et al.* (105) showed that heat treatment of cow's milk protein reduced its oral sensitizing capacity in parallel with a reduction in the level of native antigenic deter-

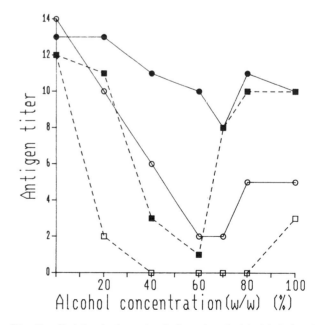

Fig. 2. Variation in the antigenically active glycinin (circles) and β-conglycinin (square) in a pre-wetted soy flour after heating for 15 min at either 40°C (solid dots) or 70°C (open dots). Antigen titers are measured by HIA (28).

minants. Heppell *et al.* (12) also were unable to demonstrate any new antigenic activities resulting from processing of soy-based infant formulas which were capable of evoking an oral immune response. Thus, there is no clear evidence of any secondary antigenic determinants, which after industrial processing or gastrointestinal digestion, can have importance in stimulating an immune response by the oral route.

The foregoing discussion indicates that soybean proteins do have immunological activity and can act as allergens in humans, calves, guinea pigs, rabbits, and maybe in piglets and rats. However, as pointed out by Kilshaw and Sissons (42), there is no evidence that soy proteins possess any special allergenic property compared with other proteins. Great variations in the sensitivity toward antigenic soy proteins are observed in all the named species. The reactions of piglets are generally transient, whilst those of calves are persistent and severe. Results suggest that the development of an immunological control mechanism (oral tolerance) in piglets and the apparent lack of such a mechanism in calves, explain the difference in severity of gastro-intestinal hypersensitivity observed in the two species. The mechanisms in human soy allergy are still unclear, but it has been shown that the allergic response is different to the different soy products. The use of immunoassays like ELISA can predict the sensitizing capacity of soy proteins *in vivo*, and can be used to ensure the lowest level possible in the commercial soy products. The level of immunological activity has been shown to be another quality parameter which has to be taken into account when evaluating soy products for human and animal consumption.

Acknowledgements

This work was supported in part by research grants from The Royal Agricultural University of Denmark. Dr. John Sissons, AGAP, England, provided valuable advice.

References

1. Perlman, F., in "Toxic Constituents of Plant Foodstuffs," edited by I.E. Liener, Academic Press, N.Y. (1980).
2. Buckley, R.H., *J. Am. Med. Assoc.* 248:2627 (1982).
3. Kamke, W., and B. Frosch, *Z. Ernahrungswiss.* 22:65 (1983).
4. Barnett, D., B.A. Baldo, and M.E.H. Howden, *J. Allergy Clin. Immunol.* 72:61 (1983).
5. Fries, J.H., *Annals of Allergy* 29:1 (1971).
6. Brandon, D.L., S. Haque, and M. Friedman, in "Nutritional and Toxicological Significance of Enzyme Inhibitors in Foods," edited by M. Friedman, Plenum Press, N.Y. (1986).
7. Halpin, T.C., W.J. Byrne, and M.E. Ament, *J. Pediatr.* 91:404 (1977).
8. Shibasaki, M., S. Suzuki, S. Tajima, H. Nemoto, and T. Kuroume, *Int. Archs. Allergy Appl. Immun.* 61:441 (1980).
9. Goodwin, B.F.J., *Clin. Allergy* 12:55 (1982).
10. Goulding, N.J., M.J. Gibney, P.H. Gallagher, J.B. Morgan, D.B. Jones, and T.G. Taylor, *Qual. Plant Food Hum. Nutr.* 32:19 (1983).
11. Taudorf, E., A. Bundgaard, S. Hancke, L.V. Hansen, P. Prahl, and B. Weeke, *Allergy* 39:203 (1984).
12. Heppell, L.M.J., J.W. Sissons, and H.E. Pedersen, *Br. J. Nutr.* 58:393 (1987).
13. Pitts, M.J., "The Immunological Consequences of the Oral Ingestion of Soybean Proteins," PhD thesis, University of Southampton, UK, 1983.
14. Sissons, J.W., R.H. Smith, D. Hewitt, and A. Nyrup, *Br. J. Nutr.* 47:311 (1982).
15. Miller, B.G., T.J. Newby, C.R. Stokes, D.J. Hampson, P.J. Brown, and F.J. Bourne, *Am J. Vet. Res.* 45:1730 (1984).
16. Von Pirquet, C., *Muench. Med. Wochenschr.,* 53:1457 (1906).
17. Catsimpoolas, N., E. Leuther and E.W. Meyer, *Arch. Biochem. Biophys.* 127:338 (1968).
18. Hitchcock, C.H.S., F.J. Bailey, A.A. Crimes, D.A.G. Dean, and P.J. Davis, *J. Sci. Food Agric.* 32:157 (1981).
19. Rittenburg, J.H., A. Adams, J. Palmer, and J.C. Allen, *J. Assoc. Off. Anal. Chem.* 70:582 (1987).
20. Brandon, D.L., A.H. Bates, and M. Friedman, *J. Food Sci.* 53:102 (1988).
21. Ory, R.L., and A.A. Sekul, in "Seed Proteins," edited by J. Daussant, J. Mossé, and J. Vaughan, Academic Press, N.Y. (1983).
22. Jaffé, W.G., in "Toxic Constituents of Plant Foodstuffs," edited by I.E. Liener, Academic Press, N. Y. (1980).
23. Nyrup, A., "Soyprotein Products. II. Analysis and Development based on Immunochemical Analytical Methods," PhD thesis, Technical University of Denmark (1980).
24. Sørensen, K.W., F. Edslev, L.K. Poulsen, and B. Weeke, XXVI Nordic Congress of Allergology, Tromsö, Norway, 1987.
25. Rasmussen, K., "Antigenic and Allergenic Crossreactions between Five Leguminous Plants," MSc thesis, University of Aarhus, Denmark (1987).
26. Barnett, D., B. Bonham, and M.E.H. Howden, *J. Allergy Clin. Immunol.* 79:433 (1987).
27. Crawford, L.V., J. Roane, F. Triplett, and A.S. Hanissian, *Annals of Allergy* 23:303 (1965).
28. Herbert, W.J., in "Handbook of Experimental Immunology, Vol. I, Immunochemistry," edited by D.M. Weir, Blackwell Scientific Publications Ltd., Oxford (1978).
29. Voller, A., D.E. Bidwell, and A. Bartlett, *Bull. World Health Org.* 53:55 (1976).
30. Shoptaw, L., *J. Dairy Sci.* 19:95 (1936).
31. Williams, J.B., and C.B. Knodt, *J. Dairy Sci.* 33:809 (1950).
32. Stein, J.F., C.B. Knodt, and E.B. Ross, *J. Dairy Sci.* 37:373 (1954).
33. Lassiter, C.A., G.F. Fries, C.F. Huffmann, and C.W. Duncan, *J. Dairy Sci.* 42:666 (1959).
34. Gorrill, A.D.L. and J.W. Thomas, *J. Nutrition* 92:215 (1967).
35. Nitsan, Z., R. Volcani, A. Hasdai, and S. Cordin, *J. Dairy Sci.* 55:811 (1972).
36. Roy, J.H.B., I.J.F. Stobo, S.M. Shotton, P. Ganderton, and C.M. Gillies, *Br. J. Nutr.* 38:167 (1977).
37. Seegraber, F.J., and J.L. Morrill, *J. Dairy Sci.* 65:1962 (1982).
38. Smith, R.H., and J.W. Sissons, *Br. J. Nutr.* 33:329 (1975).

39. Sissons, J.W., and R.H. Smith, *Br. J. Nutr. 36:*421 (1976).
40. Van Adrichem, P.W.M., and A.M. Frens, *Tijdschr. Diergeneesk. 90:*525 (1965).
41. Barratt, M.E.J., P.J. Strachan, and P. Porter, *Clin. Exp. Immunol. 31:*305 (1978).
42. Kilshaw, P.J., and J.W. Sissons, *Res. Vet. Sci. 27:*361 (1979).
43. Srihara, P., "Processing To Reduce the Antigenicity of Soybean Products for Pre-ruminant Calf Diets," PhD thesis, University of Guelph, Canada (1984).
44. Pedersen, H.C.E., "Studies of Soyabean Protein Intolerance in the Pre-ruminant Calf," PhD thesis, University of Reading, UK (1986).
45. Sissons, J.W., H.E. Pedersen, C. Duvaux, S.M. Thurston, S. Starkey, and J.A.H. Wass, in "Food Allergy," edited by R.K. Chandra, Nutrition Research Education Foundation Publishers, St. John's, Newfoundland (1986).
46. Kilshaw, P.J., and H. Slade, *Clin. Exp. Immunol. 41:*572 (1980).
47. Roit, I.M., "Essential Immunology," Blackwell Scientific Publications Ltd., Oxford (1984).
48. Sissons, J.W., R.H. Smith, and D. Hewitt, *Br. J. Nutr. 42:*477 (1979).
49. Stobo, I.J.F., P. Ganderton, and H. Conners, *Anim. Prod. 36:*512 (1983).
50. Guilloteau, P., T. Corring, J.A. Chayvialle, C. Bernard, J.W. Sissons, and R. Toullec, *Reprod. Nutr. Develop. 26:*717 (1986).
51. Armstrong, W.D., and T.R. Cline, *J. Anim. Sci. 42:*1042 (1977).
52. Giesting, D.W., "Utilization of Soy Protein by the Young Pig," PhD thesis, University of Illinois, USA (1987).
53. Miller, B.G., A.D. Phillips, T.J. Newby, C.R. Stokes, and F.J. Bourne, *Proc. Nutr. Soc. 43:*116A (1984).
54. Kenworthy, R., *Res. Vet. Sci. 21:*69 (1976).
55. Gay, D.D., I.K. Baker, and P. Moore, *Proc. Int. Vet. Soc. V:*11 (1976).
56. Heppell, L.M.J., J.W. Sissons, C. Duvaux, and S.M. Banks, Annual Report 1986 from Institute for Grassland and Animal Production 1:80 (1988).
57. Anonymous, *Agricultural and Research Council Science News 2:*2 (1988).
58. Duke, W.W., *J. Allergy 5:*300 (1934).
59. Wightman, H.B., *J. Allergy 9:*601 (1938).
60. Peters, G.A., *Annals of Allergy 23:*270 (1965).
61. Bush, R.K, and M. Cohen, *Clin. Allergy 7:*369 (1977).
62. Davidson, A.G., M.G. Britton, J.A. Forrester, J.R. Davies, and D.T.D. Hughes, *Clin. Allergy 13:*553 (1983).
63. Ratner, B., and L.V. Crawford, *Ann. Allergy 13:*289 (1955).
64. Ratner, B., S. Untracht, L.V. Crawford, H.J. Malone, and M. Retsina, *Am. J. Dis. Child. 89:*187 (1955).
65. Wergeland, H., *Acta Paediatr. 35:*321 (1948).
66. Cook, C.D., *New Eng. J. Med. 263:*1076 (1960).
67. Mortimer, E.Z., *J. Pediatr. 58:*90 (1961).
68. Mendoza, J., J. Meyers, and R. Snyder, *Pediatr. 46:*774 (1970).
69. Whittington, P.F., and R. Gibson, *Ibid. 59:*730 (1977).
70. Powell, G.K., *J. Pediatr. 93:*553 (1978).
71. Perkkiö, M., E. Savilahti, and P. Kuitunen, *Ibid. 137:*63 (1981).
72. Ament, M.E., and C.E. Rubin, *Gastroent. 62:*227 (1972).
73. Baudon, J.J., J. Boulesteix, B. Lagardere, and J.L. Fontaine, *Arch. Francais. Pediatr. 33:*153 (1976).
74. Goel, K., F. Lifshitz, E. Kahn, and S. Teichberg, *J. Pediatr. 93:*617 (1982).
75. Beyreiss, K., W. Hoepffner, G. Scheerschmidt, and G. Kuchta, *Deutsche Gesundheitswesen, 34:*2337 (1979).
76. Kuroume, T., M. Oguri, T. Matsumura, I. Iwasaki, Y. Kanbe, T. Yamada, S. Kawabe, and K. Negishi, *Ann. Allergy 37:*41 (1976).
77. Cant, A.J., J.A. Bailes, and R.A. Marsden, *Acta Paediatr. Scand. 74:*467 (1985).
78. Glaser, J., and D.E. Johnstone, *J. Am. Mec. Assoc. 153:*620 (1953).
79. Halpern, R.S., W.A. Sellars, R.B. Johnson, D.W. Anderson, S. Saperstein, and J.S. Reisch, *J. Allergy Clin. Immunol. 51:*139 (1973).
80. Kjellman, N.I., and S.G.O. Johansson, *Clinical Allergy 9:*347 (1979).
81. May, C.D., S.J. Fomon, and L. Remigio, *Acta Paediatr. Scand. 71:*43 (1982).
82. Eastham, E.J., T. Lichauco, M.I. Grady, and W.A. Walker, *J. Pediatr. 93:*561 (1978).
83. Wraith, D.G. and G.V.W. Young, in "The Mast Cell: Its Role in Health and Disease," edited by J. Pepys and A.M. Edwards, Piman Medical, London (1979).
84. Hattevig, G., B. Kjellman, S.G.O. Johansson, and B. Björkstén, *Clin. Allergy 14:*551 (1984).
85. Bush, R.K., S.L. Taylor, J.A. Nordlee, and W.W. Busse, *J. Allergy Clin. Immunol., 76:*242 (1985).
86. Burks, A.W., J.R. Brooks, and H.A. Sampson, *Ann. Allergy, 60:*148 (1988).
87. Heaney, M.R., B.J.F. Goodwin, M.E.J. Barratt, N. Mike, and P. Asquith, *J. Clin. Pathol. 35:*319 (1982).
88. Beer, W.H., W. Murray, S.H. Oh, H.E. Pedersen, R. Wolfe, and V.R. Young, *Am. J. Clin. Nutr.* (in press).
89. Edslev, F., I. Søndergaard, and B. Weeke, International Congress of Allergology and Clinical Immunology, London, UK (1982).
90. Moroz, L.A., and W.H. Yang, *New Eng. J. Med. 302:*1126 (1980).
91. Dahl, R., *Allergy 33:*120 (1978).
92. Dahl, R., and O. Zetterström, *Clin. Allergy 8:*419 (1978).
93. Mortimer, E.Z., *J. Pediatr. 58:*90 (1961).
94. Powell, G.K., *Ibid. 93:*553 (1978).
95. Gunn, R.A., P.R. Taylor, and E.J. Gangarosa, *J. Food Protection 43:*525 (1980).
96. Sissons, J.W., A. Nyrup, P.J. Kilshaw, and R.H. Smith, *J. Sci. Food Agric. 33:*706 (1982), *J. Food Prod. 43:*525 (1980).
97. Porras, O., B. Carlsson, S.P. Fällström, and L.Å. Hanson, *Int. Archs. Allergy Appl. Immunol. 78:*30 (1985).
98. Klurfeld, D.M. and D. Kritchervsky, *Lipids 22:*9 (1987).
99. Barratt, M.E.J., P.J. Strachan, and P. Porter, *Proc. Nutr. Soc. 38:*143 (1979).

100. Thellersen, M., "*In Vitro* Modification of Soyprotein to reduce the Antigenicity," MSc thesis, Technical University of Denmark, (1988).
101. Spies, J.R., M.E. Stevan, W.J. Stein, and E.J. Coulson, *J. Allergy 45:*208 (1970).
102. Wright, R.N., and R.M. Rothberg, *J. Immunol. 107:*1410 (1971).
103. Haddad, Z.H., S. Verma, and V. Kalra, *J. Allergy Clin. Immunol., 63:*198 (1979).
104. McLanghlan, P., K.J. Anderson, E.M. Widdowson, and R.R.A. Coombs, *Arch. Dis. Child. 56:*165 (1981).
105. Heppell, L.M.J., A.J. Cant, and P.J. Kilshaw, *Br. J. Nutr. 51:*29 (1984).

Protein Sources Made Available by New Technology

Daniel E. Shaughnessy

Export Processing Industry Coalition, 1008 N. Randolph St., Suite 2006, Arlington, Virginia 22201, USA

Abstract

Technological developments in recent years that have focused on such methodologies as high temperature extrusion have resulted in striking improvements in protein content levels in food and feed. Major commodities, including corn, rice, wheat, cotton, and sunflower have all experienced, to varying degrees, the effects of improved technology resulting in higher pure levels of protein in the processed products. The development of feed products such as corn gluten feed resulting from alcohol processing of corn as well as corn sweeteners for human consumption are notable examples. Of particular importance is the "value-added" effect of these processing technologies.

It is fair to say that the past several years have witnessed an array of technological developments that have resulted in striking improvements in protein content levels in both food and feed. Practically every major commodity including rice, corn, soybeans, wheat, cotton, and sunflower have all experienced, to varying degrees, the effects of improved technology, which in turn, has resulted in higher pure levels of protein in the final processed product.

Research and development in this area is considerable and continues on a broad front; for example, work with soy protein in the form of concentrates and isolates to make them more functional and more acceptable for incorporation into a broader base of food systems, is one area of concentration. I would like to focus upon the impact these developments have had upon the food and food delivery chain and the impact such technology has upon the food system generally. I also would like to focus on the "value-added" impact that these technologies and new protein sources have had upon the entire agricultural complex.

Current Examples

Allow me to provide a few illustrative examples of products that have emerged from this changing technology: In corn processing and applications, technology related to corn starch production is used to make ethanol and fructose. Protein by-products of these processes such as corn gluten feed and corn gluten meal, remain in the food cycle as protein animal feeds. These are interesting products, and, in the case of corn gluten feed, their availability and use have wide implications for international trade, and in the feed industry generally. Corn sweeteners, also, have had a profound influence on the food and beverage market with new applications becoming evident each year.

Corn gluten feed (CGF) is a feed ingredient with a medium protein level and is palatable to all classes of livestock and poultry. In spite of its name, CGF does not contain any gluten. It commonly contains a minimum of 21% crude protein and approximately 15% starch. Although the energy content of CGF varies with the species and application, it can be used in large quantities in beef cattle fattening rations because of its high-energy dietary fiber content, high rate of digestibility, and semibulkiness.

Corn gluten meal (CGM) is the dehydrated protein stream resulting from starch separation. It has a high nutrient density and usually is sold containing a minimum of 60% total protein. Its crude protein is highly digestible and the amino acid pattern of soybean meal complements that of CGM very well.

In soybean processing technology the increasing and diversified uses of lecithin are nothing short of amazing. Lecithin is familiar to many people in the forms of capsules and granules which are sold by health food stores and by supermarkets. Found naturally in eggs, butter and soybeans, lecithin is added to a variety of commercially prepared foods as an emulsifier. Lecithin keeps the fat and the cocoa from separating in chocolate, for instance, and it keeps margarine from splattering when hot. It's also the active ingredient in nonstick, noncaloric vegetable spray coatings. Interestingly, lecithin is used equally as an emulsifier in ice cream products or as a dispersing and wetting agent for the pigments in paints and magnetic media. In the paint application, the lecithin both wets and disperses the pigments and keeps them in solution for a longer period of time. The magnetic media area uses the higher quality, more refined de-oiled lecithin as a wetting and dispersing agent for the metal oxides (pigments) used in video, audio, and computer tapes and discs. Although the use of lecithin in magnetic media is not new, its application is becoming much greater because of the smaller pigments being used in the higher-grade tapes.

Soy processing technology continues to provide a wide array of traditional high protein food and feed products, ranging from soybean meal and flour to highly specialized soy-fortified foods used in worldwide feeding programs. From lecithin to animal feed, the applications of soybean processing technology are highly diverse, to say the least.

Biotechnology represents a new frontier for agriculture. While still in its infancy, prospects are bright for the development of new and better protein systems through this growing science.

Recent efforts to find new uses for corn starch have resulted in a modified corn starch that can be used by the plastics industry to make biodegradable plastics. The implications of this technology are enormous in terms of its environmental impact. Here again, protein feed will be one of the by-products of this technology.

The growing availability of these new and traditional protein sources or by-products from protein processing, has had an enormous effect on the application of these products through new and traditional delivery systems, in both the food and feed sectors. In commercial sales and deliveries, in various government supported sales and export credit arrangements, and in food aid applications the use of protein products has expanded considerably. At the same time, advances in processing technology and growing interest on the part of many countries and regions throughout the world in making better use of their basic grain and oilseed products in processed form also continues to grow. There is, in addition, growing realization that expanded market growth in the world market for basic grains will slow and that the real potential for increased market growth lies in the area of processed and higher value and value-added commodities. It is obvious that in many instances high protein food and feed products fall within this category.

Commercial sales of such products as corn gluten feed, soybean meal and other high protein feed ingredients and components have grown considerably over the past decade. Traditional reliance on basic grains for animal feed is being replaced with a growing realization that the use of high protein feeds such as corn gluten feed or soybean meal, in combination with locally available commodities, is both an economical and efficient assist to livestock, poultry and swine production. Remember, that when a concentrate or higher protein feed is produced, it compacts the protein or nutrient density of the product. This permits delivery of the protein nutrient in compact form. This results in a transfer of nutrient value at a much lower cost than would otherwise occur. Thus, if the demand is for protein, it does not make sense to deliver protein in the form of corn or wheat. Why pay for the transport and handling of 10% protein in a system where 90% of the product is carbohydrate and fat? Compacting the needed nutrient and delivering it in that form, is highly cost effective and increasingly attractive to many users.

In recent years, government supported export programs, both from North America and Europe, have begun to emphasize greater support for protein products. In the United States, export credit programs such as Title I of Public Law 480, CCC loan guarantee programs, and the United States Export Enhancement Program, have gradually moved toward greater concentration on the export of processed, high protein products. Recently for example, the United States concluded a P.L. 480 Title I arrangement with the government of El Salvador under which funds were loaned on long-term credits for the purchase and use of soybean meal. In the United States Export Enhancement Program, animal and poultry feeds are a substantial portion of exports to the Middle East with a heavy concentration of poultry feed being made available to Yemen. Short-term and intermediate credit programs have also begun to focus on protein feed with an emphasis on meals and other products.

In food aid programs, the use of processed, fortified, high protein products for direct human consumption has been a mainstay of such programs for many years. However, it is very clear that the growing availability of new sources of protein both for fortification in grain products or in direct supplemental programs is an attractive feature to many food aid program managers. In addition, there is considerable attention being devoted to the application of various specialty products for particular requirements.

The "value-added" nature of high protein products and their importance to an exporting country's economy and farm sector is considerable. In North America, today's farms are larger, more capitalized, and provide raw materials to a wide range of industries which process, distribute, and market them. The result is that commodity production concerns are the concern of the agri-processing and distribution sector, the supporting industries, and indirectly, all economic sectors. Those factors which affect the health of the farm sector in general, and its productivity base in particular, directly and indirectly affect the health of a variety of businesses due to the linkages of the farm sector with other economic sectors. The continued growth of protein product utilization, therefore, has a direct relationship to the farm sector.

Further, the food and agricultural sector is becoming highly specialized. There are numerous stages of production involved in delivering a product to the ultimate consumer. Numerous industries sell their output to other industries rather than to final markets. For example, interindustry transactions account for 50% of the total U.S. economic activity. Associated with that activity are gross economic output, corporate output, jobs, personal income, and tax revenues.

When it comes to processed protein products, both the farming and processing sectors are often dependent upon exports for their health and viability. When raw commodities are processed and then exported, the economic benefits which accrue to the economy are maximized. Raw commodity exports involve exporting in bulk with relatively little labor input and involvement of only a few sectors of the economy outside of agriculture. When those commodities are processed further and then exported, the economic benefits associated with the processing are much greater and more broadly distributed throughout an economy. This increasing interest in processed products comes at a time when the prospects for market expansion in the raw commodity sector, the traditional outlet for most farm production, are not bright. In the 1980's, the value of bulk exports has fallen dramatically. For example in 1981, U.S. bulk exports alone were valued at $30 billion. In 1987, that value had fallen to $16 billion.

Also, as mentioned earlier, world demand for cereal grains is projected to grow at only 2% per year for the remainder of the century. This is hardly enough to absorb world production capacity. In addition, North American exports will command a shrinking share of the total low-value commodity trade due to growing competition from other producers.

Thus, the U.S. and other major producers face a potential situation of declining volume and value for farm production if there is a continued concentration of marketing activities in the bulk market. An alternative to this scenario is an expansion of market share in processed protein and other products. In other words, as more farm production is traded in the form of value-added products, greater economic benefits ensue.

This is why trade in processed products such as protein feeds provides an excellent outlet for farmers. For example, in the United States, when 48 pounds of soybean meal, or 11 pounds of soybean oil are exported, that means an export of 60 pounds of soybeans. When 14.5 pounds of corn gluten feed are sold overseas, that amounts to an export of 56 pounds of U.S. corn. Consequently, far greater economic activity accompanies the export of a pound of soybean meal/oil, or corn gluten, than the export of a pound of the raw commodity. There are more jobs created, more business profits, and more revenues for government treasuries. For other countries involved in processed and high protein product production, the effects are the same.

This is what the economists call the "multiplier effect." A USDA study published in 1981, reported that "sizeable economic and employment advantages" derive from having a greater share of processed products in agricultural exports. That study estimated, for example, that increases in economic output can be considerable when raw products are exported in processed form. When USDA looked at that economic output in U.S. dollars, the results were astounding! For example, $1.00 (US) worth of raw soybean exports were calculated to be worth $2.79 to the U.S. economy when those beans were exported in processed form. A dollar's worth of exported corn was calculated to be equivalent to $44.90 in our economy when exported in the form of dressed poultry! The study also looked at processed product exports and their effects on jobs. For example, every million dollars worth of processed soybean exports meant 183 additional jobs; one million dollars worth of corn exports in the form of dressed poultry means 1100 additional jobs! No wonder then that producing countries are paying more attention to high protein and other processed products! Not only are the products themselves attaining more widespread use, but also, they are becoming major factors in international trade relationships.

Negotiations of great significance for international agricultural trade and the national and rural economies that depend on it, are in process under GATT, the General Agreement on Tariffs and Trade. In the Uruguay Round over the past year, the United States has proposed phasing out all agricultural subsidies over a 10-year period and eliminating all import barriers and restrictions against agriculture. This is, frankly, an overambitious proposal. However, the world's major traders had an initial deadline of December 1988 when they had to agree on a framework for agricultural reform. We will see what this portends for agricultural trade in the future, but in any case we can be sure that trade in high protein foods and feed will be affected.

In developing countries, increased emphasis on the use of protein products is evident in both food and feed applications. Local and government supported feeding and nutrition programs have long recognized the value of adding protein supplements to local foods made available in direct feeding programs. Fortification projects such as the development of balahar, a weaning and child feeding product developed in India in the late 1960's, focused on the blending of indigenous foods with external protein sources. At the same time, experience gained through the use of the blending and fortification technology enhanced awareness of the importance of processing generally, and contributed to increased knowledge in protein technology.

In other countries, increasing awareness of the effectiveness of higher protein feed supplements is contributing to better livestock and poultry production as well as improved techniques in animal husbandry and subsequent product processing. Even major grain producers such as the Soviet Union recently have purchased external protein feeds, which have contributed to an improved animal production sector.

Further, government policies throughout the world are beginning to reflect an awareness of quality considerations in agriculture and the importance of processed products such as protein feeds and food to their own national development. For example the government of India has recently established, at the ministerial level, a government agency responsible for national policies and guidance in food processing. Other governments, including the United States, also have laws and policies reflecting attention to the value of processed products, and their place in national development.

And this brings us back to our immediate interest. At the center of many of these activities is increasing protein availability resulting from new technologies. Some of the product examples which I cited earlier are but a few of the immense range of new and developing products which are available. Their importance, both to producer and receiver, continues to grow and the "value-added" characteristic of their production and use certainly enhances that importance.

Thus, while technological and scientific progress continues almost daily on protein availabilities, the effects of that progress go far beyond the laboratory, the production plant or even the end user. Protein product usage is now an element of international trade and government relationships; it is an essential ingredient in agriculture development generally; it affects consumers in a manner previously not considered and the use of protein sources has become an element in everyday life, ranging from food to feed and thus to jobs and personal income.

Nutrition Complementation with Vegetable Protein

Preeya Leelahagul and Vichai Tanphaichitr

Division of Nutrition and Biochemical of Medicine, Department of Medicine and Research Center, Faculty of Medicine, Ramathibodi Hospital, Mahidol University, Rama 6 Road, Bangkok 10400, Thailand

Abstract

Protein-energy malnutrition (PEM) is a public health problem of developing countries and usually is caused by the consumption of cereal-based diets. Because legumes are good sources of protein and energy with low cost, several supplementary food mixtures including soy-cereal formulas have been developed and shown to be beneficial for improving protein status in children. Our study in adolescents has shown the benefits of supplementing their daily rice-based diet with soybean milk and soybean residue. Their serum total protein and albumin levels at wk 4 and 8 and nitrogen balance at wk 8 after soybean supplementation were significantly higher than the initial values. We also have shown that soybean can be used as the major dietary protein source in adults. Seven healthy men were fed isocaloric and isonitrogenous diets containing soybean and rice (90:10) or soybean, rice, and egg (72:10:18) as the protein sources for 10 days and had no significant difference in nitrogen balances (37.1 vs 36.7 mg/kg/day) and serum transport protein levels. There was also no significant difference in nitrogen balances in healthy men fed isocaloric and isonitrogenous enteric formulas containing milk protein or soy protein isolate as the protein sources for 8 days (68 vs 54 mg/kg/day). Patients with carcinoma of the larynx undergoing laryngectomy also exhibited positive nitrogen balance (94 mg/kg/day) when they received the enteric formula containing soy protein isolate through nasogastric tubes.

Protein-energy malnutrition (PEM) is one of the major problems affecting the life and well-being of various populations in developing countries. Their total protein intake is usually lower than those in developed countries. Besides, in the former countries cereals are their main protein sources whereas those in the latter are meat and dairy products (1). The prevalence of PEM is alarming, ranging from 20% to 80% of preschool children (2). Supplementation with protein-rich foods has been recognized as the important mode to alleviate the problem of PEM. It is the purpose of this paper to review and present our studies on nutrition complementation with vegetable protein.

Functions and Requirements of Protein

In the body, protein is used for the structural formation of cells and tissues, the production of various essential compounds such as enzymes, antibodies, hormones, and protein mediators, and for regulating fluid and electrolyte balances as well as blood neutrality. It also can be utilized as an energy source; 1 g of protein provides 4 kcal.

Dietary protein is used for depositing new protein in tissues of pregnant women, infants and children; for the protein secretion in milk of lactating women; and for the maintenance of body protein synthesis in adults. Thus, inadequate protein intake causes diminished protein content in cells and organs and deterioration in the cellular capacity to perform their normal functions. This leads to increased morbidity and mortality. On the other hand, protein intake in excess of physiologic need also is disadvantageous (1). Therefore, an adequate diet must contain an appropriate level of protein for the assurance of long-term health.

Safe levels of protein intake according to sex and age groups have been proposed recently by FAO/WHO/UNU (3). The greatest need of protein in proportion to body size is for infants (1.85–1.50 g/kg/day for 3–12 mo) and young children (1.20–1.10 g/kg/day for 1–5 yr). As individuals proceed through childhood the requirement diminishes gradually (1–0.95 g/kg/day for 5–14 yr; 0.94–0.80 g/kg/day for 14–18 yr) until the static requirement of adulthood is reached (0.75 g/kg/day). Additional protein intake should be provided during pregnancy (+6 g/day) and lactation (+17.5–13 g/day). However, these numbers apply to highly digestible proteins such as those of eggs, milk, meat or fish.

Usually it is accepted that the relative concentration of essential amino acids is the major factor determining the nutritional value of food protein. Most animal proteins have a satisfactory essential amino acid pattern in relation to the amino acid requirements. Therefore animal proteins are of high quality. On the other hand vegetable proteins may be of lower nutritional value because they tend to be limiting in one or more of the essential amino acids.

Lysine is the first limiting amino acid in rice, maize, and wheat, whereas that in soybean is methionine (4). The limiting amino acid may lead to poor utilization of amino acids by humans and relatively more of the protein is required to meet the minimum nutritional requirement. Such deficiencies can be corrected either by adding the limiting amino acids or by complementation with other protein sources. However, this paper will concentrate on the benefit of nutrition complementation with vegetable protein for supporting the needs of body protein and nitrogen metabolism because this is an economical way to provide high quality protein in the diets for populations in the developing countries.

Legume Supplementation in Children

Legumes and seeds such as soybean, mungbean, groundnut, sesame, and cottonseed have higher protein contents than cereals (4). They have been employed to improve the nutritional value of cereal-based diets that can lead to PEM.

The pioneer work of Scrimshaw et al. (5) has shown that Incaparina, a vegetable mixture consisting of 29% whole ground corn, 29% whole ground sorghum, 38% cottonseed flour, 3% Torula yeast, 1% calcium carbonate, and 4,500 IU of vitamin A per 100 g is an effective protein source for the treatment of kwashiorkor. When compared with isonitrogenous amounts of milk in alternate 5-day nitrogen balance periods in young children at adequate levels of intake, the amount of nitrogen retained as percent of intake was not significantly different. Subsequent reports by various investigators who studied the nitrogen retention in children receiving soy-cereal diets expressed in relation to nitrogen retention with milk have shown the nutritional value of soy in combination with rice or corn flour or with mixed sources such as milk solids and corn meal closely approaches that of milk (6).

To alleviate the problem of PEM in Thai preschool children, the Division of Nutrition, Ministry of Public Health has implemented the production and dissemination of supplementary food mixtures in more than 10,000 rural villagers by community participation. Among these supplementary food mixtures are rice:soybean:groundnut (70:15:15), rice:soybean:sesame (70:15:15), rice:mungbean:groundnut (60:15:20) and rice:mungbean:sesame (60:20:15). The energy contents in these mixtures are 437, 448, 443, and 451 kcal/100 g, respectively, whereas the respective protein contents are 16.5, 14.8, 14.5, and 13.2 g/100 g. The formulation of these mixtures is based on the Thai regulation for infant foods stating that the protein content should not be less than 2.5 g/100 available kcal, the protein quality expressed in term of amino-acid score should not be less than 70% of the FAO/WHO reference pattern, the fat content should not be less than 2 g and should not be more than 6 g/100 available kcal, and linoleate content should not be less than 300 mg/100 available kcal. As a consequence of this implementation, there is a decline in the prevalence of PEM by weight for age in these preschool children from 51% in 1982 to 24% in 1986 (7).

Soybean Supplementation in Adolescents

In 1983, there was an outbreak of beriberi in institutionalized adolescents at the Observation Protection Center (OPC) of Rayong Province, eastern Thailand (8). Dietary factors causing inadequate thiamine intake in these institutionalized adolescents include high consumption of milled rice, loss of thiamine due to methods of cooking, and lack of consumption of rich sources of thiamine. Dietary assessment revealed that 10, 11, and 79% of their total energy intake before supplementation (wk 0) was derived from protein, fat, and carbohydrate, respectively, and 65% of their protein intake came from vegetable sources, mainly rice (Table 1).

Due to the constraint of food budget at OPC we had introduced the soybean supplementation program in these adolescents to improve their nutritional status. Soybean was chosen because it is locally available, relatively inexpensive, and highly nutritious. The soybean supplementation included 164 males and 19 females, aged 10–19 yr. Each subject took 200 ml of soybean milk which was prepared fresh each morning, except Saturday and Sunday. The soybean residue left after the preparation of soybean milk was used to mix with animal protein to be served at lunch and dinner. Each 100 g of soybean milk provided 48 kcal, 2 g protein, 0.7 g fat, and 8 g carbohydrate whereas each 100 g of soybean residue provided 75.8 kcal, 6 g protein, 2.8 g fat, and 7 g carbohydrate.

Dietary assessment at wk 4 and 8 after soybean supplementation revealed an increase in total energy and protein intakes in these adolescents (Table 1). Soybean milk and residue constituted 3–4% of their energy intake and 5–7% of their protein intake during the supplementation. Table 2 shows that their serum total protein and albumin

TABLE 1

Mean Energy Intake and Its Distribution in Institutionalized Adolescents

Nutrient	Wk 0		Wk 4		Wk 8	
	Amount	%	Amount	%	Amount	%
Energy (kcal)	2,627	100.0	3,048	100.0	3,116	100.0
Protein (g)	66	10.0	74	10.0	84	11.0
Animal (g)	23	3.5	19	2.6	28	3.3
Vegetable (g)	43	6.5	55	7.4	56	6.7
Fat (g)	31	11.0	51	15.0	40	12.0
Carbohydrate (g)	510	79.0	562	75.0	587	77.6

TABLE 2

Effect of Soybean Supplementation on Serum Transport Protein and Nitrogen Balance Levels in Institutionalized Adolescents

Parameter	Mean ± SEM		
	Wk 0	Wk 4	Wk 8
Serum total protein (g/l)	71.3 ± 0.5	78.3 ± 0.4[a]	78.0 ± 0.3[a]
Serum albumin (g/l)	39.1 ± 0.3	43.9 ± 0.2[a]	44.5 ± 0.2[a]
Nitrogen balance (g/day)	5.51 ± 0.19	5.47 ± 0.16	7.05 ± 0.14[a,b]

[a] Wk 0 VS Wk 8 : P < 0.001.
[b] Wk 4 VS Wk 8 : P < 0.001.

levels at wk 4 and 8 and nitrogen balance at wk 8 were significantly higher than those at wk 0. These data indicate the improvement of protein status in these adolescents by soybean supplementation.

Soybean as the Major Dietary Protein Source in Adults

Inadequate protein-energy status is also common in rural Thai adults (9). Their protein and energy needs are supplied mainly from a rice-based diet. The bulkiness of this diet makes it difficult to meet their energy requirement (10,11). The nutritional quality of their protein intake is also low because rice is the major dietary source of protein evidenced by low serum albumin, low serum and urinary carnitine levels suggesting lysine deficiency (10). In addition to carnitine deficiency, essential fatty acid depletion usually coexists in this free-living population (11). Thus, we were interested in investigating the effect of soybean as the major dietary protein source on protein-energy status in adults.

Seven healthy males, aged 30–35 yr, were fed two types of diets. The vegetarian diet (VD) contained 90% soybean and 10% rice as the protein sources. The ovovegetarian diet (OV) contained 72% soybean, 10% rice, and 18% egg as the protein sources. Energy distributions of VD and OV were 15, 30, and 55% derived from protein, fat, and carbohydrate, respectively. Each diet was fed for 10 days. Energy intakes for each subject during receipt of these two diets were held constant according to his energy intake during the baseline period of 14 days. The protein intakes of each subject between the two diets were isonitrogenous (12.1 ± 0.5 g/day). Their nitrogen balances derived from the latter five days of each dietary period were positive and not significantly different. This finding is consistent with the nonsignificant differences in arm muscle circumferences, serum total protein, albumin, transferrin, and retinol-binding protein (RBP) levels between the two dietary periods (Table 3). Subjects accepted and tolerated various menus of VD and OV. Our data indicated that soybean can be used as the major dietary protein source in adults (12).

TABLE 3

Nitrogen Balance, UAMC, and Serum Transport Protein Levels in Seven Healthy Men on Vegetarian or Ovovegetarian Diet

Parameter	Mean ± SEM	
	Vegetarian	Ovovegetarian
Nitrogen balance (g/day)	2.3 ± 0.4	2.2 ± 0.6
UAMC (cm)	24.7 ± 0.3	24.8 ± 0.3
Total protein (g/l)	74.3 ± 2.2	72.9 ± 1.3
Albumin (g/l)	55.0 ± 1.6	55.5 ± 1.0
Transferrin (g/l)	1.9 ± 0.1	2.1 ± 0.2
RBP (mg/l)	37.6 ± 2.0	37.7 ± 3.2

Soy Protein Isolate as the Sole Protein Source in Health and Disease

It is now evident that rat bioassay of protein appears to greatly underestimate the nutritional value of some food proteins such as well processed soy protein (1,6). Young and his colleagues (13–15) have shown that soy protein isolate is highly qualitative in adults. This was also observed in children by Torun (16). Along this line of interest we had evaluated the nutritional value of a soybean-based formula (SBF) compared with a milk-based formula (MBF) in 8 healthy men aged 22–31 yr.

The energy distribution of SBF is 17.6, 21.6, and 60.8% derived from protein, fat, and carbohydrate, respectively, whereas the respective figures of MBF are 18, 15.3, and 66.7%. The protein source of SBF is soy protein isolate whereas that of MBF is nonfat milk solids. Lactose in MBF is enzymatically hydrolyzed to contain less than 6 g of lactose per 100 g. The subjects were fed isocaloric and isonitrogenous intakes of SBF and MBF. The feeding period of each formula was eight days. The results show that the subjects had positive nitrogen balances while consuming SBF and MBF and the data were not significantly different. In addition there were no significant differences in their serum total protein, albumin, transferrin, and RBP levels between the two formulas (Table 4). However, we had observed that SBF was more accepted and tolerated than MBF by the subjects. Some developed lactose intolerance during the first 3 days of milk-based period (17).

TABLE 4

Nitrogen Balance and Serum Transport Protein Levels in Eight Healthy Men on Soybean-Based or Milk-Based Formula

Parameter	Mean ± SEM	
	Soybean	Milk
Nitrogen balance (g/day)	3.2 ± 0.9	4.0 ± 0.6
Total protein (g/l)	75.4 ± 1.8	74.6 ± 1.5
Albumin (g/l)	51.0 ± 1.1	48.7 ± 0.8
Transferrin (g/l)	2.7 ± 0.1	2.7 ± 0.2
RBP (mg/l)	52.1 ± 2.4	48.4 ± 2.0

PEM is common in hospitalized adult patients. The prevalences based on weight for height in medical and surgical patients at Ramathibodi Hospital were 73 and 79%, respectively (18). Thus the prescription of proper feeding formulas and selection of the correct feeding method are necessary to prevent or treat the inadequate protein-energy status in these patients (19,20).

In order to evaluate the nutritional value of the aforementioned SBF in hospitalized patients, we had conducted a study in patients with carcinoma of the larynx who underwent laryngectomy. The study shows that when six patients postlaryngectomy with the mean (±SEM) age of 65 ±3 yr were tube-fed with SBF providing the mean energy intake of 38 kcal/kg/day and protein intake of 1.68 g/kg/day during days 8–12 postoperatively they were in positive nitrogen balances with the mean value of 4.4 g/day. In addition, the patients could tolerate this enteric formula.

Acknowledgments

This work was supported by a grant from American Soybean Association, Bristol-Myers Company International Division, and Archavadis-Sirirat Fund.

References

1. Young, V.R., and P.L. Pellet, *Am. J. Clin. Nutr. 45*:1328 (1987).
2. Valyasevi, A., S. Wichaidit, and A. Stuckey, in *Human Nutrition Better Nutrition Better Life,* edited by V. Tanphaichitr, W. Dahlan, V. Suphakarn, and A. Valyasevi, Aksornsmai Press, Bangkok, 1984, pp. 335–338.
3. FAO/WHO/UNU, *Energy and Protein Requirements,* World Health Organization. Tech. Repser. No. 724, Geneva, 1985.
4. FAO *Amino-Acid Content of Foods and Biological Data of Proteins,* 3rd Edition, Food and Agriculture Organization of the United Nations, Rome, 1981.
5. Scrimshaw, N.S., M. Behar, D. Wilson, R. De Leon and R. Bressani, in *Progress in Meeting Protein 843,* National Academy of Sciences-National Research Council, Washington, D.C., 1961, pp. 57–67.
6. Young, V.R., and N.S. Scrimshaw, *J. Am. Oil Chem. Soc. 56*:110 (1979).
7. Tontisirin, K., and U. Valaiphatchara, in *Proceedings of Fifth Asian Congress of Nutrition,* edited by K. Yasumoto, Y. Itokawa, H. Koishi and Y. Sanno, Center for Academic Publications Japan, Tokyo, 1987, pp. 347–350.
8. Tanphaichitr, V., in *Sixth Nutricia-Cow & Gate Symposium: Child Health in the Tropic,* edited by R.E. Eeckels, O. Ransome-Kuti, and C.C. Kroonenberg, Martinus Nijhoff Publishers, Boston, 1985, pp. 157–166.
9. Tanphaichitr, V., in *Recent Advances in Clinical Nutrition 2,* edited by M.L. Wahlqvist, and A.S. Truswell, John Libbey, London, 1986, pp. 79–84.
10. Tanphaichitr, V., N. Lerdvuthisopon, S. Dhanamitta and H.P. Broquist, *Am. J. Clin. Nutr. 33*:876 (1980).
11. Tanphaichitr, V., and W. Chaiyaratana, *Prog. Lipid. Res. 25*:225 (1986).
12. Tanphaichitr, V., P. Leelahagul, R. Summasut, S. Kulapongse and S. Songchitsomboon, in *International Soyfoods Symposium,* edited by F.G. Winarno, Food Technology Development Center, Bogor Agricultural University, Jogyakarta, 1986, pp. 323–332.
13. Young, V.R., M. Puig, E. Quieroz, N.S. Scrimshaw and W.M. Rand, *Am. J. Clin. Nutr. 39*:16 (1984).
14. Young, V.R., A. Wayler and C. Garza, *Ibid. 39*:8 (1984).
15. Istfan, N., E. Murray, M. Janghorbani, W.J. Evans and V.R. Young, *Ibid. 113*:2524 (1983).
16. Torun, B., in *Soy Protein and Human Nutrition,* edited by H.L. Wilcke, D.T. Hopkins, and D.H. Waggle, NY: Academic Press, New York, 1979, pp. 101–119.
17. Leelahagul, P., and V. Tanphaichitr, in *Abstracts of Original Communications: XIII International Congress of Nutrition,* Brighton, 1985, pp. 90.
18. Tanphaichitr, V., and S. Kulapongse, in *Clinical Biochemistry: Principles and Practice,* edited by A.S. Eng and P. Garcia, Second Asian & Pacific Congress of Clinical Biochemistry, Singapore, 1983, pp. 101–110.
19. Tanphaichitr, V., P. Leelahagul, O. Puchaiwatananon and S. Kulapongse, in *Recent Advances in Clinical Nutrition 2,* John Libbey, London, 1986, pp. 215–219.
20. Tanphaichitr, V., P. Leelahagul, V. Tantibul and A. Kanjanapiak, in *Proceedings of Fifth Asian Congress of Nutrition,* edited by K. Yasumoto, Y. Itokawa, H. Koishi, and Y. Sanno, Center for Academic Publications Japan, Tokyo, 1987, pp. 248–251.

Vegetable Proteins in Planning for Child Nutrition

Osman M. Galal

Nutrition Institute, Cairo, Egypt, and Department of Family and Community Medicine, University of Arizona, Tucson, Arizona, USA

Abstract

Child nutrition problems in developing countries, including chronic undernutrition, iron-deficiency anemia, and vitamin A deficiency, can be counteracted at least in part by effectively targeted supplemental foods of appropriate composition. By developing such foods from locally available cereal and legume bases, the need for special child-targeted food(s) can be met in an economically sustainable manner. Local agricultural and other industries can be supported. The further advantage to this approach is that such products are likely to be used as supplements rather than replacements for breast milk. Milk-based formulas and other products that replace breastfeeding endanger the health of children and mothers under the circumstances prevalent in most of the developing world. Vegetable-based products, however, need not have these adverse impacts as they can be designed to complement rather than replace traditional patterns of infant feeding.

Children from developing countries usually are smaller than children from developed countries. While small size alone is not necessarily a handicap, the significance of the size differential lies in increased rates of illness and death and in functional deficits for which there is abundant evidence. The main contributing factors leading to this situation are (a) inadequate availability of energy, protein and other nutrients; (b) the excessive burden that children bear due to repeated episodes of illness. The relative inadequacy of energy and protein hinders the natural development and growth of these children, and contributes to the high infant and child mortality rates we witness in developing countries.

Most children in developing countries depend on breast milk as their main source of energy and protein for a relatively long time in their life. In some countries, children are exclusively breast fed up to the age of two years. The process of weaning, i.e., the change from breast milk to household food, always is met with great difficulty and commonly is accompanied by gastrointestinal disturbances. This is partly due to the mothers' lack of awareness of how to handle this transition period. Absence of appropriate inexpensive weaning foods to be used as supplements is also a contributing factor.

The aim of this presentation is to illustrate the importance and the role of plant-based protein and energy in providing appropriate foods for children and preventing their nutritional deterioration. We will discuss the role of supplementation with plant proteins as a complement to breastfeeding to enhance the child's diet and to protect the mother's health.

Lastly, a question of great importance is "Should health planners in developing countries, be at all concerned with the role that plant proteins can play in either promoting a child's diet and/or accelerating growth rates of the chronically malnourished children to catch-up?" That role, if planned properly, can then contribute to the factor for child survival strategies.

Body Size and Nutrition

Body size as a reflection of the state of nutrition often has been used as an indicator of general health of children (1). Information on body size is considered to be a useful tool to assess health intervention programs. Heights and weights are the basic anthropometric data used to identify body size in most surveys of nutritional status and surveillance systems for monitoring the nutritional situation (2). The accumulation of data from many parts of the world using these anthropometric indices has made it clear that chronic malnutrition is prevalent in most developing countries. One estimate puts the number of children aged 0–6 years with chronic energy-protein deficits at 500 million (3). Growth faltering of children in developing countries starts very early in life, as early as three months of age (4). Figure 1 shows that the growth velocity of Egyptian infants starts to drop as early as the second to the third month of age.

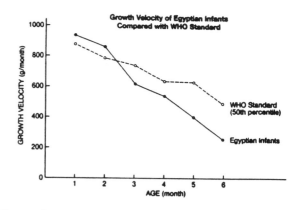

Fig. 1. Growth velocity of Egyptian infants compared with WHO standards

In addressing questions about the significance of growth faltering, one must distinguish between wasting and stunting (5). Wasting results from reduced mass relative to length and reflects acute episodes of stress and their duration. Stunting, on the other hand, refers to retardation in linear growth as measured by total body length or height and reflects slow, chronic stress. Both are caused basically by deficient food intake. Stunting and wasting start during the first two years of life when growth rate is rapid (6). Infectious disease has been shown to be one of the key factors that limits normal growth rates; diarrheal disease is on the top of the list of infectious diseases (7,8). The two crucial years that involve the difficult transition period from breast milk to household food are a time of high susceptibility to diarrheal diseases and other infections, thus increasing the risk of undernutrition.

Figures 2 to 4 show the body size of children from birth up to 12 years of age in three developing countries, Egypt, Kenya and Mexico (9). These figures portray the weights, heights and weight for height expressed as Z-scores in relation to the U.S. reference population. The children in all three populations have, on average, birth weights which are only slightly lower than those of U.S. children (Z-scores close to zero), but between

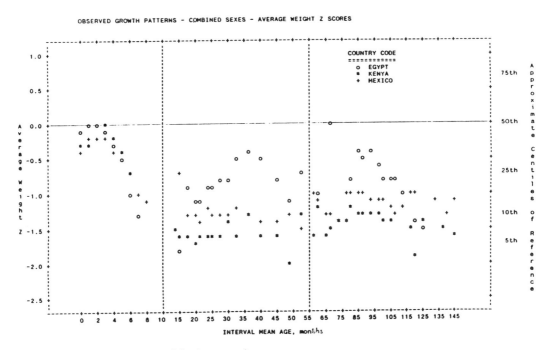

Fig. 2. Z scores for average weight, by age and country

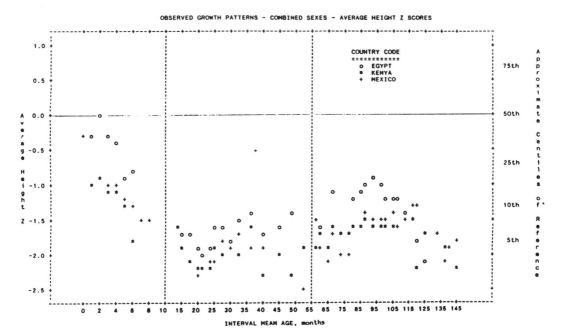

Fig. 3. Z scores for average height, by age and country

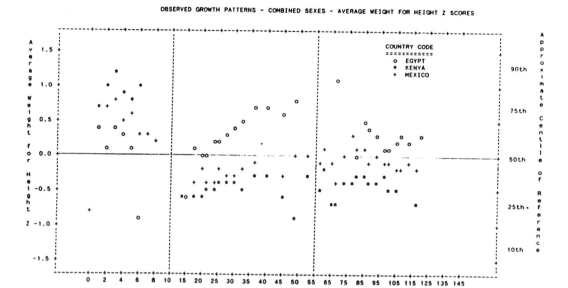

Fig. 4. Z scores for average weight, by height, by age and country

about 3 to 4 months and 12 months of age, the average Z-scores for weight fall dramatically. They are then consistent thereafter. In Egypt weights continue to fall after the first year and then tend to rebound. At 8 years of age children show a deficit that ranges from 2 to 3.5 kg in the three countries. The picture for height is generally similar to that of weight. By the age of 8 years the loss of height represents 5 to 8 cm in the three countries. When weight and height are considered together, a different pattern emerges. Children in all three countries have relatively normal weights for their reduced heights.

These patterns of growth denote clearly that there is pressure on growth-promoting factors during the whole of childhood, which starts early in life and is not released. This pressure is distributed equally on linear growth and weight, resulting in normal weight for the reduced height. The main causes are deficits of energy and protein intake which leads to chronic malnutrition or marginal malnutrition.

Fortunately, children are capable of growth spurts (catch-up) characterized by utilizing more than usual amounts of energy and protein to grow for a period of time at an accelerated rate. This period of catch-up is accompanied by increased efficiency of energy and protein utilization (10), but cannot occur without abundant food in terms of energy and protein at the right time (11). This catch-up growth is very limited once the child reaches 3 to 5 years of age. If the child already is stunted at these ages, in all probability he will remain small throughout the growing years, eventually becoming a small adult. The role that plant products could play in reducing the gap during this period should always be emphasized to and remembered by health care planners.

The morbidity burden that young children face during the first two years in developing countries is high. Many studies have shown that lower growth velocity occurs with peaks of prevalence of diarrhea. In an Egyptian village it has been demonstrated that growth velocity is decreased in summer when diarrhea is more prevalent (12). Areas with very high incidences of infection show greater prevalence of wasting and decelerated growth rates (13).

To improve nutritional status during the first two years of life it is necessary to make energy and protein available to those children. Besides energy and protein that children can utilize from breast milk, plant products should be thought of as additive sources of food, particularly important during those catch-up phases during which relatively large amounts of energy and protein are needed. Well-designed plant-based food products in the presence of breast milk can potentially increase nutritional intake of children in developing countries, leading to a general improvement of health.

Feeding Practices and Supplementation

The practice in most parts of the world is to encourage breastfeeding followed by a gradual weaning period after which a child can continue feeding on a totally nonbreast-milk diet by the age of 2 to 3 years. In Islamic countries, religion supports breastfeeding the child up to the age of 2 years. This strong support is a factor in determining the feeding practices that takes place in these countries.

An important policy issue in the weaning process is to determine the optimal age at which supplementation should start. This decision must take into consideration not only the nutritional aspects but also the safety of food as contaminated food can be the cause of infections when feeds other than breastmilk are introduced. The younger the age of the child is when he is introduced to nonbreast-milk foods the higher the potential risk for infection. This is always true in countries where hygienic conditions are unfavorable. The assumption in most developing countries is that children are exclusively breastfed up to the age of 6 months, but it is not unusual for these children to be given small amounts of complementary foods, usually fluids. These feeds contribute negligible amounts to the infant's total energy intake (14) and at the same time are potential sources of infection through contamination.

Earlier weaning and shorter breastfeeding therefore increase infant mortality rates especially when there is no concomitant improvement in the standard of living. In some developing countries when the standard of living has improved substantially, earlier weaning has not contributed much to the infant mortality rates (15). Examples of such countries are Indonesia and Malaysia.

In Egypt, baby cereals are used by a minority of mothers, almost all of whom live in the urban areas. Some of these cereals are fortified with dry milk. Older children more often are fed foods characteristic of the family diet, primarily vegetable and cereal products rather than foods of animal origin (16).

Our own data from a village in lower Egypt (9) show most of the infants were found to receive water and sugar and some of them had rice water as a supplementary food. In addition to water and sugar, children received other supplementary feeds. Over the six-month period, 28% of the infants did not receive any supplements other than water and sugar, sugary concoctions, and rice water. The infants who were supplemented tended to fall in somewhat lower weight percentiles at 6 months of age and to show greater losses in percentiles for weight than unsupplemented infants (Table 1). Infants supplemented with water and sugar, sugary concoctions and rice water were excluded from the analysis. It is also important to note that infants who were supplemented early, i.e., from 0-3 months of age, tended to be from households of higher socioeconomic status (Table 2). Morbidity (per cent of days with illness) also was greater in children who were supplemented than in the nonsupplemented group.

When children are ill, mothers may withhold feeding and resort to fluid diets with negligible amounts of protein. This is most common in cases of diarrhea (17). In Egypt, foods are categorized by mothers into "heavy" and "light" foods (18). Table 3 shows the common foods given to toddlers by their designation. Children are given "light" foods which are all low in calories and protein.

These findings demonstrate the importance of the presence of food supplements. The unavailability of high-protein, low-price foods for the young increases the risk of malnutrition, hence increases morbidity and leaves the child in a state of malnutrition. The ideal supplement should be a source of protein and micronutrients as well as calories. This food should not be substituted for breast milk but should be used as a supplement.

A further important aspect that should be considered during the weaning process is maternal health. Breastfeeding is known to many mothers to be a natural birth control device in the absence of contraceptives and contributes to prolonging birth intervals (child spacing). Early weaning will encourage repeated pregnancies and shortens the birth interval. In turn, the mother's nutrient stores are depleted and the nutritional status of the fetus affected. Low birth weight in developing countries ranges between 12% in a country like Egypt to 50% in a country like Bangladesh (19). Lactation can be sustained even under very harsh dietary conditions and there is a strong drive by mothers toward milk synthesis. One of the energy-sparing adaptive mechanisms that operates during lactation when calories are scarce is a decrease in maternal basal metabolic rate causing an overall energy saving. The other area for potential energy savings is a reduction of physical activity. The reduction in basal metabolic rate negatively affects the health of mothers which in turn reduces her physical activity. Reduction in physical activity will reduce the time allocated for child care, perhaps leading to increased periods of illnesses of the child. Time

TABLE 1

Comparison of the Morbidity of Breastfed Infants Who Were Supplemented and Not Supplemented at 1, 3 and 6 Months of Age

	Age of infant					
	1 month		3 months		6 months	
Illness[b]	Supplement[a] n = 5	No supplement n = 62	Supplement n = 19	No supplement n = 47	Supplement n = 27	No supplement n = 39
	Number of days ill					
All except skin problems	1.0 ± 2.24[c]	1.6 ± 4.32	2.6 ± 3.42	2.2 ± 3.78	7.6 ± 6.45	5.0 ± 5.88 $p < 0.10$
Diarrheal	0.0 ± 0.00	0.4 ± 2.52	0.8 ± 1.90	0.8 ± 2.15	2.1 ± 4.06	2.5 ± 4.60
Respiratory	1.0 ± 2.23	1.2 ± 3.96	1.3 ± 2.83	1.2 ± 2.25	3.5 ± 5.50	1.6 ± 3.68
	% of time ill					
All except skin problems	3.7 ± 8.3	7.3 ± 18.2	12.0 ± 17.0	8.5 ± 13.82	23.2 ± 19.6	17.2 ± 20.5
Diarrheal	0.0 ± 0.0	1.7 ± 9.0	3.0 ± 7.6	3.1 ± 8.5	5.9 ± 10.3	7.5 ± 12.9
Respiratory	3.7 ± 8.3	5.6 ± 17.2	6.3 ± 14.3	4.7 ± 8.6	11.8 ± 19.7	5.7 ± 12.3

[a] Sugar water, sugary concoctions, and rice water were excluded.
[b] Morbidity data included only the month during which supplementary food intake was assessed.
[c] Mean ± SD.

TABLE 2

Relationship of Household Socioeconomic Status to the Total Number and Types of Supplements Introduced to Infants from Birth to 6 Months of Age

Age of infant, months	Household socioeconomic status			
	High n = 11	Upper intermediate n = 8	Lower intermediate n = 33	Low n = 16
	Total number of supplements[a]			
0–3	1.3 ± 1.2[b]X 0–3[c]	1.0 ± 1.4X** 0–4	0.4 ± 0.8** 0–3	0.7 ± 1.2X** 0–4
3–6	3.6 ± 3.0 0–9	3.1 ± 3.0 0–9	2.0 ± 3.3 0–14	3.8 ± 5.3 0–22
0–6	4.9 ± 3.5 0–11	4.1 ± 4.2 0–13	2.4 ± 3.7 0–17	4.4 ± 5.3 0–22
	Number of milk and weaning foods			
3–6	1.6 ± 1.6 0–4	2.0 ± 2.0 0–5	1.0 ± 1.7 0–5	1.1 ± 1.8 0–6
	Number of legumes and vegetables			
3–6	0.5 ± 1.0 0–3	0.3 ± 0.7 0–2	0.3 ± 0.7 0–2	0.7 ± 2.0 0–8

[a] Excluding sugar water, sugary concoctions, and rice water
[b] Mean ± SD
[c] Range
X** Means with different superscripts differ significantly (p < .05)

for food preparation will be reduced further, causing the child to be more prone to malnutrition.

Earlier Supplements: Dried Milk

It is important to mention that with the advancement of food technologies during the 1960's and the increased productivity of dairy products in developed countries there were attempts from scientists to supplement or replace breastmilk. The abundance of milk and the production of powdered milk made it easy for the dairy industry to provide developing countries with inexpensive powdered milk to combat malnutrition in these countries.

During the following two decades it was quite clear to pediatricians and nutritionists that the imposed trends on feeding practices were not beneficial. Nutritional surveys in developing countries did not show any improvement in the nutritional status of children. Utilization of powdered milk was low in many countries, due to cultural inappropriateness and perhaps to biological intolerance.

A further economic constraint was evidenced by the unsustainability of providing developing countries with inexpensive powdered milk. These factors and others encouraged scientists to re-evaluate the situation and turn to locally sustainable products for nutritional intervention.

Role of Supplementation

Most surveys show that 25% of the child population in the world is moderately or severely undernourished. There is a need to prevent these children from deteriorating nutritionally and keep them from falling into the high-risk category. What is needed is for children who are in the high risk category will recover quickly. Both of these levels of prevention need the use of food from plant origin, in terms of protein and fat (oils). Children who are wasted can grow more rapidly than normal children. It is estimated that the supplement should have a caloric density of 120 to 135 calories per 100 ml. The food should also have a high level of high-quality plant protein and it should have special amounts of selected vitamins and minerals. The supplement should be easy for the mother to prepare. And it also will be advantageous for the food to have palatability characteristics that will make it attractive to the young child.

Some have recommended that limitations should be put on the amount of carbohydrate used, as simple carbohydrate ingested in too great amounts can overload the absorptive capacity of the intestine. The main source of energy in an appropriate supplemental food would be fat from plant origin. The child can tolerate up to 50% of energy from fat sources. The supplement would seek to keep osmolality down by using starch as its principal carbohydrate source. Limiting the amounts of solute is an important issue. These are primarily nonmetabolizable dietary components ingested in excess of body needs and metabolic end products resulting from digestion of protein. The level of protein should be adjusted to be sufficient for growth, but also low enough to limit the solute load.

TABLE 3

Per Cent of Infants Fed Various Supplementary Foods at Different Ages During the First 6 Months

Supplementary food	Age of infant months					
	1 (99)[a]	2 (95)	3 (91)	4 (90)	5 (87)	6 (91)
	% of infants					
Water and sugar	100	100	100	100	100	100
Sugary concoction	35	5	7	3	11	7
Rice water	37	43	36	30	29	22
Water	—	—	—	—	—	1
Soft drinks	1	—	3	4	11	9
Tea	—	—	—	2	1	7
Orange/tomato juice	—	—	—	1	—	—
Cow milk (fresh)	—	—	—	—	—	1
Skim milk	3	6	8	9	13	9
Powdered milk	—	—	1	3	—	—
Full cream cheese	2	9	12	8	7	9
Processed cheese	—	—	—	—	2	1
Yogurt	—	—	—	—	—	1
Mehallabia[b]	1	6	5	10	13	9
Supramine[c]	—	—	—	2	8	9
Biscuits	—	—	—	—	—	1
Bread	—	—	—	—	7	12
Rice	—	—	—	—	—	4
Sprouted beans	—	—	—	1	3	4
Stewed beans	—	—	—	3	3	15
Other legumes	—	—	—	—	2	1
Cooked vegetables	1	—	—	—	—	—
Fresh vegetables	—	—	—	1	5	5
Eggs	—	—	1	2	—	1
Rabbit	—	—	—	—	1	—
Sesame cake[c]	—	—	—	1	7	11
Other cakes and sweets	—	1	2	1	—	—
Fats/oils	—	—	—	1	—	4
Not supplemented[d]	92	79	73	63	60	54

[a] Number of infants; [b] Mixture of approximately 20g cornstarch, 10g sugar and 200 ml water heated to a thick consistency and cooled; [c] The Nile Company, Cairo, Egypt (composition by wt %: chick peas, 31.5; wheat flour, 23.1; powdered milk, 20.0; lentils, 15.4; sucrose 8.0; and minerals and vitamins, 2.0%; [d] Most of the oil is extracted from sesame seeds leaving a residue to which sugar is added; [e] Water and sugar, sugary concoction, and rice water excluded; [f] 1–6 month period.

The cost of the supplement will depend on the ingredients and processing procedures. It is essential the foods should utilize local ingredients and simple, nonexpensive technologies. In addition, packaging in small bags might be very important in helping to create a strong image for the food.

Plant protein based supplemental feedings for children should be considered as a major component in health care systems. They act in a dual way by improving the nutritional status of children and reducing their morbidity and at the same time affecting the health of the mother.

Supplemental foods should be introduced in programs which have fairly good targeting and nutrition education. They should never be used as substitutes for breast milk. Mothers would be told that the food should be given to the child to help him and supplement her breast feeding. The supplement also would seek to support a broader strategy of education. It is recommended that a weight chart approach be used which will intensify child care in general. A successful supplemental food should have positive effects on mothers' health as well. The first is by delaying subsequent pregnancies by encouraging continuation of breastfeeding. Decreased morbidity state of the child will give more time to the mother to use her resources in a more effective and positive way including more time allocated for food preparation or even for educating the young children.

References

1. Martorell, R., H.L. Delgado, V. Valverde and R.E. Klein, *Hum. Biol. 53*:303 (1981).
2. Mason, J.B., J.P. Habicht, H. Tabatabai and V. Valverde, *Nutritional Surveillance*, World Health Organization, Geneva (1984).
3. Lathan, M.C., *International Nutrition Problems and Policies*, in World Food Issues, 2nd Ed., Program in International Agriculture, Ithaca, NY (1984).
4. Serdulo, M.R., J. Sewerd, J.S. Marks, N. Stachling, O.M. Galal and E. Trowbridge, *Am. J. Clin. Nutr. 44*:405 (1986).
5. Waterlow, J.C., *British Med. 3*:566 (1972).
6. Martorell, R., and J.P. Habicht, in *Human Growth*, 2nd Ed., edited by F. Falkner and J.M. Tanner, Plenum Press, New York, 1987.
7. *Diarrhea and Malnutrition: Interactions, Mechanisms and Interventions*, edited by L.C. Chen and N. Scrimshaw, Plenum Press, New York, NY, 1983.
8. Martorell, R., in *Social and Biological Predictors of Nutritional Status, Growth and Neurological Development*, edited by L.S. Green and F.S. Johnson, Academic Press, New York, (1980) p. 81.
9. *Food Intake and Human Function: A Cross-Project Perspective*, The Collaborative Research Support Program (CRSP), University of California, Berkeley, CA, 1988.
10. Thomson, E.F., H. Bickel and A. Schurch, *J. Agric. Sci. 98*:183 (1982).
11. Solomons, N.W., *J. Amer. Dietetic Assoc. 85*:28 (1985).
12. Afifi, Z.E., *Human Biology 57*:649 (1985).
13. Brown, K.H., R.E. Black and S. Becker, *Am. J. Clin. Nutr. 36*:303 (1982).
14. Hagazy, M.I., O.M. Galal, M.T. El-Mongy, S. Wallace-Cabin and G.G. Harrison, *Ecology of Food and Nutrition 19*: 247 (1987).
15. Huffman, S.L., and B.B. Lamphere, *Breastfeeding Performance and Child Survival*, 1984.
16. Harrison, G.G. and O.M. Galal, *Assessment of Child Feeding Practices: Relationship to the Design of Appropriate Interventions*, MEAwards Workshop on Assessment of Health Intervention, Aswan, Egypt, Oct. 13–17, 1987.
17. Jelliffe, P., D.B. Jelliffe, K. Feldon and N. Ngokwey, *World Rev. Nutr. Diet 53*:218 (1987).
18. Sukkary-Stolba, S., *Soc. Sci. Med. 25:*401 (1987).
19. *The State of the World's Children, A Statistical Picture*, UNICEF, 1987.

Dietary Soybean Protein and Cholesterol Metabolism

Anton C. Beynen

Department of Laboratory Animal Science, State University, P. O. Box 80.166, 3508 TD
Utrecht and Department of Human Nutrition, Agricultural University, P. O. Box 8129, 6700 EV
Wageningen, The Netherlands

Abstract

In hypercholesterolemic patients lowering of serum cholesterol occurs after the transfer from a mixed protein diet to a diet containing predominantly soy protein. Thus, a diet enriched with soybean protein has cholesterol-lowering activity in humans. The rabbit is the most popular model for studying the underlying mechanisms. Using diets with the protein source as the only variable, soy protein prevents hypercholesterolemia and atherosclerosis, whereas the animal protein casein induces these disorders. When compared with casein, soy protein decreases the absorption of intestinal cholesterol and reduces the re-absorption of bile acids. Evidence is presented that these effects are the key to the hypocholesterolemic action of soy protein.

Dietary Proteins and Serum Cholesterol in Rabbits

The nature of the protein source in the diet of rabbits is an important determinant of the concentration of cholesterol in the blood serum (1–3). Certain protein sources, such as extracted whole egg, casein and beef protein concentrate produce rather high concentrations of serum cholesterol, whereas other sources of protein, such as wheat gluten, peanut meal and soybean protein are able to maintain low levels of serum cholesterol (Table 1).

It cannot be decided from the data presented in Table 1 whether the differing effects of dietary proteins on serum cholesterol in rabbits are due to the proteins themselves or to other constituents of the protein preparations. Most of the preparations used in the reported experiments contained no more than 65% of protein. Even the purest preparations of casein and isolated soybean protein contained up to 20% of nonprotein material. To see whether the protein determines the level of serum cholesterol, rabbits were fed semipurified diets containing an amino acid mixture resembling the composition of either casein or soybean protein (Table 2). The amino acid mixture equivalent in composition to casein produced concentrations of serum cholesterol similar to those obtained with casein, whereas the mixture imitating soy protein induced higher levels of serum cholesterol than the intact protein, but the levels were still lower than those seen with casein (Table 2). Thus, at least part of the differential effect of casein and soy protein on serum cholesterol appears to be related to differences in amino acid composition of these proteins.

The induction of hypercholesterolemia in rabbits fed diets containing casein is associated with the development of arterial lesions. In contrast, dietary soybean protein not only produces low levels of serum cholesterol but it also prevents atherogenesis (1).

TABLE 1

Concentration of Serum Cholesterol in Rabbits Fed Semipurified Diets Containing Various Animal and Plant Proteins

Dietary protein	Serum cholesterol (mmol/L)
Extracted whole egg	6.1
Skim milk	6.0
Lactalbumin	5.6
Casein	5.2
Beef protein concentrate	4.2
Pork protein concentrate	2.9
Raw egg white	2.7
Wheat gluten	2.1
Peanut protein concentrate	2.1
Peanut meal	1.9
Soybean protein concentrate	0.7
Soybean protein isolate	0.4

Results are presented as means for 4 to 6 animals per dietary group. The rabbits, aged 8 to 12 weeks at the beginning of the experiment, were fed isonitrogenous (4.23%, w/w, of N from the proteins) for 28 days. Serum cholesterol:1 mmol/L = 38.7 mg/100 ml. After Carroll and Hamilton (4) and Hamilton and Carroll (5).

TABLE 2

Effect of Intact Proteins and Amino Acid Mixtures on Serum Cholesterol Levels in Rabbits

Dietary nitrogen source	Serum cholesterol (mmol/L)		
	Day 0	Day 14	Day 28
Intact casein	1.0	4.3	5.5
Amino acids imitating casein	1.4	4.1	5.5
Intact soy protein	1.6	1.3	1.8
Amino acids imitating soy	1.5	3.3	3.2

Results are means for 6 to 10 animals per dietary group. The diets provided nitrogen equivalent to 25% (w/w) protein. Serum cholesterol:1 mmol/L = 38.7 mg/100 ml. After Huff *et al.* (6).

Dietary Soybean Protein and Serum Cholesterol in Humans

In 1977 Sirtori *et al.* (7) reported the hypocholesterolemic activity of the soybean protein diet in type II hypercholesterolemic patients. The soybean protein diet contained

13% of energy as soybean protein, 6% as other vegetable protein and 1.5% as animal protein. When compared to a diet containing 8% of energy as vegetable protein and 13% as animal protein, the soybean protein diet reduced serum total cholesterol by 10 to 20% after three weeks (Table 3). Further studies from the same laboratory (8) and from other laboratories (9–12) have confirmed the cholesterol-lowering activity of high soy protein diets. Diets rich in protein from other beans such as *Vicia faba* also exert cholesterol lowering activity in type II hypercholesterolemic patients (13). The soybean protein diet specifically lowers cholesterol in the low density lipoproteins (LDL) whereas that in the high density lipoproteins (HDL) is left unchanged or somewhat increased.

TABLE 3

Effect of Diets Rich in Either Soybean Protein or Mixed Animal Proteins on Serum Cholesterol Concentrations in Type II Hypercholesterolemic Patients

Dietary protein	Serum cholesterol (mmol/L)	
	Before cross-over	After cross-over
Animal protein → Soy	8.7	6.7
Soy → Animal protein	6.6	7.2

Means for 10 patients per group. The patients consumed one type of diet for three weeks followed by the other diet for another three weeks. After Sirtori et al. (7).

The soybean protein diet is only marginally effective in normocholesterolemic subjects (7,14). Moreover, it has been shown that purified soybean protein preparations exert negligible antihypercholesterolemic activity (15–17). However, the diet containing soybean protein isolate produced a shift in cholesterol from LDL to HDL, resulting in a 7% (16) to 14% (15) increase in the HDL:LDL-cholesterol ratio. This effect of soybean protein versus casein may be favorable with regard to the risk for atherosclerotic diseases as LDL cholesterol is considered to be atherogenic and HDL-cholesterol to be antiatherogenic.

Metabolic Basis for the Cholesterol-Lowering Activity of Soybean Protein

The rabbit is the most popular model for studying the differential cholesterolemic effects of dietary proteins. In most studies soybean protein has been compared with the prototypical animal protein, casein. For the rabbit model the mechanism underlying the cholesterol lowering activity of soybean protein, when compared to casein, can be described as follows (18,19). Soybean protein decreases the absorption of intestinal cholesterol (20), which is of endogenous and/or exogenous origin, and probably also reduces the re-absorption of bile acids. This results in the observed increase in fecal excretion of neutral steroids and bile acids (20,21). Thus, soybean protein causes a diminished feed-back inhibition of the hepatic conversion of cholesterol into bile acids, and more cholesterol will be channeled into the bile acid synthetic pathway. This in turn tends to deplete liver cholesterol pools. Liver cholesterol concentrations have been shown to be lower in rabbits fed a soybean protein diet than in their counterparts fed casein (20,22).

Thus dietary soybean protein reduces liver cholesterol in rabbits. The liver responds by an increase in the number of LDL receptors and by enhancing *de novo* cholesterol synthesis. Indeed, Sirtori et al. (23) have shown that the binding of apoprotein B containing β-VLDL particles to liver membranes of rats is increased when the donor animals had been fed a cholesterol-rich diet containing soybean protein, compared with casein. Stimulation of hepatic cholesterol synthesis in soybean protein fed animals has been demonstrated both directly and indirectly. Liver microsomal HMG-CoA reductase activity has been found to be increased in rats fed soybean protein when compared with casein (23,24). Cholesterol turnover is much faster in rabbits fed soybean protein when compared with casein (20,21).

The increased number of LDL receptors induced by soybean protein is responsible for the fall in serum cholesterol. The extra LDL cholesterol taken up by the liver can be used for the production of bile acids which may be lost in feces: the absorption of bile acids is depressed. However, in order to prevent the body from depletion of cholesterol, *de novo* synthesis has to be activated. A new steady state will be reached, at which hepatic cholesterol synthesis is increased and fecal excretion of bile acids is also increased. Thus, cholesterol turnover is enhanced. At this new steady state serum cholesterol is low and the number of LDL receptors high. It can be concluded, from the rabbit studies, that the cholesterol lowering activity of soybean protein may well reside in its ability to interrupt the enterohepatic cycle of cholesterol and bile acids. How this effect is brought about is not yet known.

References

1. Kritchevsky, D., *J. Am. Oil Chem. Soc.* 56:135(1979).
2. Carroll, K.K., *Ibid.* 58:416(1981).
3. Van der Meer, R., and A.C. Beynen, *Ibid.* 64:1172(1987).
4. Carroll, K.K., and R.M.G. Hamilton, *J. Food Sci.* 40:18(1975).
5. Hamilton, R.M.G., and K.K. Carroll, *Atherosclerosis* 24:47(1976).
6. Huff, M.W., R.M.G. Hamilton and K.K. Carroll, *Ibid.* 28:187(1977).
7. Sirtori, C.R., E. Agradi, F. Conti, O. Mantero and E. Gatti, *Lancet* 1:275(1977).
8. Sirtori, C.R., E. Gatti, O. Mantero, F. Conti, E. Agradi, E. Tremoli, M. Sirtori, L. Fraterrigo, L. Tavazzi and D. Kritchevsky, *Am. J. Clin. Nutr.* 32:1645(1979).
9. Descovich, G.C., C. Ceredi, A. Gaddi, M.S. Benassi, G. Mannino, L. Colombo, L. Cattin, G. Fontana, U. Senin, E. Mannarino, C. Caruzzo, E. Bertelli, C. Fragiacomo, G. Noseda, M. Sirtori and C.R. Sirtori, *Lancet* 2:709(1980).
10. Wolfe, B.M., P.M. Giovanetti, D.C.H. Cheng, D.C.K. Roberts and K.K. Carroll, *Nutr. Rep. Int.* 24:1187(1981).
11. Vessby, B., B. Karlström, H. Lithell, I.-B. Gustafsson and I. Werner, *Hum. Nutr. Appl. Nutr.* 36A:179(1982).
12. Widhalm, K., in *Nutritional Effects on Cholesterol Metabolism*, edited by A.C. Beynen, Transmondial, Voorthuizen, 1986, p. 133.

13. Weck, M., M. Hanefeld, W. Leonhardt, H. Haller, K.-D. Robowsky, R. Noack and H. Schmandke, *Die Nahrung* 27:327(1983).
14. Carroll, K.K., P.M. Giovanetti, M.W. Huff, O. Moase, D.C.K. Roberts and B.M. Wolfe, *Am. J. Clin. Nutr.* 31:1312(1978).
15. Van Raaij, J.M.A., M.B. Katan, J.G.A.J. Hautvast and R.J.J. Hermus, *Ibid. 34:*1261(1981).
16. Van Raaij, J.M.A., M.B. Katan, C.E. West and J.G.A.J. Hautvast, *Ibid. 35:*925(1982).
17. Grundy, S.M., and J.J. Abrams, *Ibid. 38:*245(1983).
18. Beynen, A.C., R. Van der Meer and C.E. West, *Atherosclerosis 60:*291(1986).
19. Beynen, A.C., R. Van der Meer, C.E. West, M. Sugano and D. Kritchevsky, in *Nutritional Effects on Cholesterol Metabolism,* edited by A.C. Beynen, Transmondial, Voorthuizen, 1986, p. 29.
20. Huff, M.W., and K.K. Carroll, *J. Lipid Res. 21:*546(1980).
21. Kuyvenhoven, M.W., C.E. West, R. Van der Meer and A.C. Beynen, *J. Nutr. 116:*1395(1986).
22. Beynen, A.C., G. Den Engelsman, K.E. Scholz and C.E. West, *Ann. Nutr. Metab. 27:*117(1983).
23. Sirtori, C.R., G. Galli, M.R. Lovati, P. Carrara, E. Bosisio and M. Galli Kienle, *J. Nutr. 114:*1493(1984).
24. Nagata, Y., N. Ishiwaki and M. Sugano, *Ibid. 112:*1614 (1982).

Concerns in Regulating Vegetable Food Proteins

R. J. Dawson

Food and Agriculture Organization of the United Nations, Via delle Terme di Caracalla, 00100 Rome, Italy

Abstract

Vegetable food proteins, being nutrients and not additives in most countries are subject to general food legislation, which is based on quality protection, safety, nutritive value and organoleptic criteria. In addition they are subject to regulation for use as food ingredients. There are many countries in the world, e.g., Belgium, Canada, Denmark, Finland, France, Federal Republic of Germany, India, Ireland, Italy, Japan, Luxembourg, Netherlands, New Zealand, Peru, Sweden, U.K., U.S.A. and EEC, which have established national and group regulations for the use of vegetable proteins in food. Some of these are quite extensive and include: (a) nutritional requirements; (b) labelling requirements; (c) compositional requirements; (d) regulations for the use of vegetable proteins in food and (e) regulations for use of vegetable proteins in meat products.

Assessment of the Present Situation

A review of the current national regulations on vegetable proteins leads to a number of observations, which underline the rather complex situation in which the national regulatory agencies appreciate guidance. There appears to be a clear cut distinction of approach in the use of vegetable proteins between developed economies and developing countries. The regulatory provisions in developed economies concern primarily the use of vegetable proteins in processed meat, seafood, poultry and dairy products. In developing countries the use of vegetable proteins in the form of defatted flours is practically confined to increasing the protein content and accordingly to improving the nutritional quality of staple foods. The regulatory provisions in this case are simple referring essentially to the composition and safety of the defatted flours with no restrictions on the levels of use or other complications which obviously prevail in the regulatory provisions for vegetable proteins used in processed products.

The areas where there appears to be a substantial divergence of opinion and approaches are: the definitions of the basic types of vegetable protein products; the levels of use of vegetable protein products as binders, as substitutes or replacers for meat or other animal products; the use in the baking, the dietary food, the baby food and in other food processing industries for nutritional and/or technological purposes; the nutritional quality in terms of vitamins and minerals; the biological quality; the nutritional equivalency; the labelling provisions, which vary from a simple declaration of the source of vegetable proteins to a very detailed descriptive name and ingredient listing; and, the wholesomeness and safety in use of the protein source, an area which appears to be neglected in most of the national regulatory provisions.

The preceding limited analysis of national regulatory provisions and practices on the use of vegetable proteins in food and the evidence that there is a wide spectrum of diverging practices and approaches to the problem suggest, *prima facie,* the need for coordination and international agreement. National regulatory services appear to be reluctant to come to definitive regulations mainly because of conflicting views and interests in introducing and promoting innovation in the food supply system. At the same time, some governments came to grips with the problem either collectively as in the case of the Commission of the European Communities or individually as in the case of the United States and the United Kingdom. For years the problem has been under study and review and still no conclusive results have been obtained. This situation also suggests that guidance for national regulations and international harmonization is more than desirable. The Codex Alimentarius Commission certainly represents the unique international organization to undertake the task.

The Codex Alimentarius Commission

The Codex Alimentarius Commission is an intergovernmental body established in 1962 by FAO and WHO with the purpose of ensuring fair practices in international trade and protecting the health of the consumer. It has in the intervening period established itself as the international forum for elaborating standards for foods moving in international trade and for providing guidance to countries wishing to create their own national food laws and regulations. The Commission, after 25 years, has elaborated more than 200 individual commodity standards and 35 codes of hygienic and technological practice. The impact of this work on the quality and safety of foods has helped to upgrade food manufacturing and processing standards all over the world. It has improved prospects for the facilitation of trade, and encouraged governments and industry alike to take cognizance of consumers' expectations. Currently there are 134 member countries in the Codex.

Codex standards for raw and processed food commodities moving in international trade cover aspects such as: uniform labelling requirements; use of food additives; the presence of contaminants or residues of pesticides or veterinary drugs; sanitary (or hygiene) requirements; composition and analysis, etc. These internationally agreed upon food standards assure the consumer of the safety and quality of food since they take into consideration all aspects needed to protect the health of the consumer.

Protecting the Consumers' Health: One of the principal purposes of food quality control and standards activities of the Codex Alimentarius Commission is protecting the consumer against health risks and commercial fraud. Safe and adequate food supplies are essential for proper nutrition. Not only must foods have appropriate nutritional content and be available in sufficient variety, but they also must not endanger consumer health through chemical or microbial contamination and be presented honestly. Food safety and quality starts at the farm and continues throughout the processing and distribution chain to storage and final preparation by the consumer. Good agricultural, processing, distribution, and marketing practices are essential to ensure consumer protection. The commission contributes to protecting consumer health by obtaining international agreement on food standards and codes of hygienic practice which are concerned with essential composition, chemical and microbial contaminants, nutritional quality and consumer information.

The work of the Codex Alimentarius Commission is consistent with the 1985 UN General Assembly resolution on consumer protection. Moreover, the objectives of food safety and improved quality appeal to the general public as well as to specialized groups. In fact, work in the area of food quality and standards can represent a complete congruence of public and commercial interests.

Creation of a Codex Committee on Vegetable Proteins (CCVP)

The Codex Alimentarius Commission at its 12th session (1978) recognized that: vegetable proteins intended for human food, whether in developed or developing economies had to meet definite nutritional requirements and be safe in use; vegetable proteins, to be used as food or food ingredient, had to offer economic incentives to both producer and consumer; the use of vegetable proteins in improving the diets of populations at nutritional risk was of particular economic and social interest when protective foods such as milk, meat, fish, were in short supply or were beyond the economic reach of such groups; and in most developing countries edible fats and oils generally were in short availability and supply and encouragement in expanding the production of oilseeds in such countries offered the additional advantage that the proteins of the press cake or the extracted oilseed meal could add to the food supply of the country. To achieve this goal it was necessary that regulatory provisions for the safe use of the vegetable proteins be prepared and promulgated.

Therefore it created a Codex Committee on Vegetable Proteins, which has been hosted by Canada, with the following term of reference: to elaborate definitions and worldwide standards for vegetable protein products deriving from any member of the plant kingdom as they come into use for human consumption, and to elaborate guidelines on utilization of such vegetable protein products in the food supply system, on nutritional requirements and safety, on labelling and on other aspects as may seem appropriate.

The committee has held five sessions to date. The fifth session was held in Ottawa, Feb. 6–10, 1989.

Activities of the Codex Committee on Vegetable Proteins

Elaboration of International Standards for Vegetable Protein Foods: Since its inception, the committee was interested in elaborating international standards for vegetable protein foods and regulate their use in other foods. As referred to earlier, Codex standards are quality standards which assure safety to the consumer since they take into consideration all aspects needed to protect the health of the consumer.

The committee undertook to elaborate standards for wheat gluten, soy protein products and for vegetable protein products. So far it has finalized elaboration of the standard for wheat gluten, which shortly will be sent out to governments for acceptance. Other standards are presently at different stages of elaboration. It should be noted that the Codex does not undertake elaboration of standards unless the Codex criteria as given below are fulfilled.

Codex Criteria Applicable to Commodities

Criteria applicable to all commodities include: Consumer protection from the point of view of health and fraudulent practices; volume of production and consumption in individual countries and volume and pattern of trade between countries; diversification of national legislations and apparent resultant impediments to international trade; international or regional market potential; amenability of the commodity to standardization; number of commodities which would need separate standards indicating whether raw, semi-processed or processed; work already undertaken by other international organizations in this field, and the type of subsidiary body envisaged to undertake the work.

Regulation of the Use of Vegetable Proteins in Food: The Codex Committee on Vegetable Proteins is elaborating guidelines for the use of vegetable proteins in food, which are presently in their final stage. The committee undertook this work since a review of existing regulations had shown that harmonization was required with regard to the use, nutritional value and labelling of vegetable proteins and that practical guidelines on these points were required.

Vegetable protein products (VPP), in their powdered, hydrated or structured forms rarely are consumed directly as human food. As a rule, they are used as components in food products for their specific functional properties, to supplement the protein content and/or improve the nutritive value of staple foods and for their ability to extend, substitute for, or simulate other food products.

Use of VPP for Their Functional Properties: Vegetable protein products are used for their functional properties in that their physio-chemical characteristics influence the technological behavior of the protein in food systems during the preparation of foods as well as during their processing, storage and consumption. The increased utilization of VPP is attributed to their properties in water, oil, absorption, retention and emulsification; solubility, viscosity and foaming: in gel forming, cohesion, adhesion, elasticity, and film forming. Usually the addition of VPP for functional purposes in the processing of food products ranges between 0.5–3.5%.

The CCVP recommends the following guidelines to be observed when VPP are used in food for functional and optional purposes: "When VPP are used at low relative levels for functional purposes, or as optional ingredients, their use should not result in any replacement of prinicpal protein and associated nutrients in the food to which they are added."

For the purpose of defining VPP as a functional or optional ingredient in Codex standards the level of VPP should be calculated on a dry weight basis in the final product. The actual level of use will vary according to the nature of the protein and of the product concerned. The use of VPP as a functional or optional ingredient should be regulated in the same way as other functional or optional ingredients with no required change in the name of the product. However, a declaration of the presence of VPP should be given in connection with the name of the product if its omission would mislead the consumer.

Use of VPP To Increase Content of Utilizable Protein: The use of VPP to increase content of utilizable protein depends on nutritional aims and socioeconomic conditions of target populations and acceptability within the pattern of traditional food habits.

The CCVP recommends the following guidelines to be observed when VPP is used to increase the content of utilizable protein: VPP may be used to improve the protein intake of populations by increasing the content of utilizable protein in the diet. This can be done by increasing the protein content of the diet or increasing the protein quality of the proteins in the diet, or a combination of both. It should be noted that increasing the protein quantity and/or quality of a diet will be ineffective if energy requirements are not met.

In general, the minimum aim of supplementation and/or complementation should be to increase utilizable protein by 20%. Addition of amino acids should be considered only when protein complementation and supplementation have proved impracticable for economic and technological reasons. Only L forms of amino acids should be used. Because a variety of VPP are available for use for this purpose, the choice of VPP should favor products which have been processed in such ways and to such extents as to optimize both the nutritional contributions and economic considerations. The addition of vitamins and minerals should be in accordance with the Codex general principles for the addition of essential nutrients to foods. The need for such addition should be considered on a case-by-case basis.

The need for fortification of VPP with vitamins and minerals should be considered in the following instances: when the VPP is a suitable vehicle for fortification in regions where there is a demonstrated need for increasing the intake of one or more vitamin(s) or mineral(s) in one or more population groups; and, when the VPP contains anti-nutritional factors (e.g., phytate) which may interfere with the bioavailability or utilization of nutrients.

The need for nutritional equivalence of the VPP should be considered in those instances in which the VPP replaces staple ingredients which are higher in vitamins and minerals than the VPP. When VPP is used in a food to increase the content of utilizable protein, its presence need not be indicated in the name of the food unless its omission would mislead the consumer.

The protein content of a food in which VPP has been added to increase the content of utilizable protein should be declared in accordance with the Codex guidelines on nutrition labelling. Where claims are made with respect to the protein quality, the protein nutritional value should be assessed according to the established methods for protein quality measurement.

Use of VPP in Partial or Complete Substitution of the Animal Protein in Foods: The addition of VPP in food products aims at substituting, partly or wholly for the original protein component of a food or at extending its protein content to desired levels. The problems connected with the use of VPP as partial or complete substitute for animal protein in food are evidenced clearly in the regulatory provisions of those countries permitting the use of VPP. These problems are connected with levels of use, definition of ratio of substitution, and the types of VPP used.

The CCVP recommends the following guidelines to be observed when VPP is used as partial or complete substitute of animal protein in food. "The use of VPP to partially or completely substitute animal protein in foods should be permitted, provided that the final partially or completely substituted product is nutritionally adequate and provided that the presence of VPP is clearly indicated on the label."

The nutritional adequacy of a product can be defined in terms of protein quality and quantity and content of minerals and vitamins.

Such a product should be considered nutritionally adequate if: its protein quality is not less than that of the original product or is equivalent to that of casein; and it contains the equivalent quantity of protein / ($N \times 6.25$) / and those vitamins and minerals which are present in significant amounts in the original animal products.

The nutritional adequacy of a partially substituted animal product can be achieved by any of the following three methods: by using a VPP which meets the nutritional adequacy requirements of protein quantity and quality and levels of vitamins and minerals; by using a VPP which is nutritionally adequate with respect to levels of vitamins and minerals, but placing the requirements for protein quantity and quality on the final product; or, by the addition of the required nutrients to the partially substituted product (i.e., by placing all the nutritional requirements on the partially substituted product).

The second approach is considered the most satisfactory because the first method does not make allowance for the complementary effect of animal-VPP mixtures on protein quality. For example, according to its amino acid score, wheat gluten (which would require the addition of several amino acids before it could meet the protein quality requirement for partial substitution) could be used to substitute meat protein up to 30% without any significant deleterious effect on adequacy of the final product in protein quality. The third method would require that the vitamin and mineral content of the animal portion of the partially substituted product be known and accounted for in each instance. Moreover, the expertise and control facilities for ensuring proper addition of nutrients and stability of vitamins may not exist in places where VPP would be utilized in animal products such as retail outlets and meat packing plants.

In the case of completely substituted (simulated) animal products, all the nutritional adequacy requirements (i.e., protein quantity and quality as well as vitamins and minerals) should be placed on the final product.

When VPP partially substitutes for the protein of an animal product, the following nomenclature criteria should apply: the presence of the VPP should be indicated in the name of the food; the name of the substituted product should describe the true nature of the product; it should not mislead the consumer; and it should enable the substituted product to be distinguished from products with which it could be confused; in cases where the substitution results in an amount of the animal protein product lower than that required by a Codex or national standard, the name of the standardized animal food should not be used as part of the name of the substituted product unless properly qualified; and, the provisions of a Codex standard or a national compositional standard should be taken into full account when determining the name of a food.

In the case of a simulated animal product in which 100% of the protein is from VPP, the established or common name of the food should be the name of the VPP with appropriate flavor designation or other descriptive phrasing.

As mentioned earlier the guidelines for the use of vegetable protein in foods being elaborated by the CCVP are in their final stages and may be adopted by the Codex Alimentarius Commission at its 18th session to be held in Geneva, July 3–12, 1989. It should be noted that the Codex Committee on Processed Meat and Poultry Products (CCPMPP) is elaborating separate guidelines for the use of vegetable proteins in meat and products. These guidelines are consistent with the general guidelines being developed by CCVP.

Use of VPP as Sole Protein Source in Products with New Identities: There is an expanding group of foods made with VPP that are not intended to supplement utilizable protein or to replace traditional protein foods. Each of these foods will develop an identity of its own and will have its own nutrient composition. There need not be specific nutrient requirements for these foods. As with any other foods, these VPP foods should be safe, should be produced in accordance with good manufacturing practices and should be labelled in accordance with the Codex standard for the labelling of prepackaged foods (Codex Stan 1-1985).

Quantitative Methods for the Differentiation of Vegetable and Animal Protein: For control purposes, there may be a need for an internationally accepted quantitative method for the differentiation of vegetable and animal proteins. The CCVP is presently working on standardization of such methodology. The most promising approach in methodology appears to be by techniques based on immunochemistry or liquid chromatography.

These are the efforts being made by the Codex towards harmonization of national regulations for the use of vegetable proteins in foods.

Preparation and Use of Dry Soy Products and Nutritional Beverages

Glen Blix

School of Public Health, Loma Linda University, Loma Linda, California, USA

Abstract

The concern of providing adequate protein for the world's ever-growing population and the relatively inefficient utilization of agricultural space by livestock make the soybean one of the logical alternatives. Soy beverages have established themselves as an important component of the food system. Properly used and fortified they continue to provide a safe and nutritious alternative to their mammalian counterparts. As research progresses it is likely that the soybean will continue to establish its importance in the maintenance of human health and well-being.

Beverages were a part of the repertoire of human food elements long before recorded history. Although those with an alcoholic base may be more interesting and glamorous, at least in their short term effect, the more mundane aqueous variety will provide the theme for this presentation. This category of foodstuffs, while not possessing the notorious image of their alcoholic counterparts, are nevertheless an integral part of our diet. The origins of our fondness for these liquids are lost in the dim recesses of time but it is certain that mankind has at one time or another attempted to make watery blends of every edible substance and, for that matter, some that are not so edible.

As human beings, it is likely that the first exposure to a beverage was the white liquid expressed from a mother's breast or those of her mammalian counterparts. It was the existence of this nursing liquid that undoubtedly led to experimentation with other beverages and the supplementation of the infant's diet. History records many attempts at replacement of mother's milk in the case of her inability to nurse the baby due to illness, physiological problems or untimely demise.

Nutritional composition was, understandably, not a consideration in those early concoctions. Those which met with organoleptic approval were retained, those that did not were refined or discarded.

The original brews were undoubtedly the result of the modification of the broths of soups and stews through intentional and/or unintentional dilution. The resultant beverages most often were the product of the manipulation of an already cooked and known food and probably were not initiated in a dry state.

The various brews indigenous to each culture were largely dependent on the immediate availability of the required raw materials. This effectively limited certain beverages to specific territorial boundaries. One can only speculate on the events which supplied the impetus for the transfer of these beverages from one culture to another. It is likely that as cultures became more sophisticated and interactions among them more common, food products and beverages from distant climes came to be known by the local societies. Man's fascination with the exotic and foreign were no different then and caused in him a longing for a way to provide these gustatory delights to his own culture.

Transportation always has been a major factor in the dissemination of goods and knowledge and carrying gourds or other containers of liquid across rough terrain was not the most desirable or efficient activity particularly when spoilage and spillage were considered. A means of transporting these beverages without the additional encumbrance provided by the water became more and more desirable.

The use of powdered vegetation to make a beverage obviously is not new and may represent the second phase in the development of edible brews. The discovery of the pleasant organoleptic qualities of the dilute solutions of various cooked products undoubtedly led to further experimentation with various powders. Observations regarding spoilage also may have increased the interest in developing a way to provide drink bases which would not change appreciably with storage. The logical approach would be to see if a dried and powdered form of the same ingredients would yield a similar beverage. Many sources of vegetable and animal products have been ground to a powder and brewed, dissolved or suspended in water and thus become an integral part of the food system.

However, it is a relatively recent phenomena to take this brew and remove the water so as to produce a powdered material capable of reconstitution. The essence of all methods is the same; evaporate the water and that which is left becomes a reconstitutable powder with the added advantage of a prolonged shelflife, a discovery made no doubt when a pot of soup was left too long over the camp fire. Early attempts at duplicating the occurrence must have been frustrating. The application of heat would remove the water but unless the container was removed from the fire at precisely the correct moment the residue would burn. The use of less heat as in sun drying would often be so lengthy that spoilage would occur prior to the formation of the dried product.

In modern times it has been the dairy industry that has pioneered the techniques of beverage drying. First with drum driers and more recently with spray drying technology. It was not long before this knowledge was applied to other foods.

It was here that the soybean came to play a significant role. The production of a milk-like liquid extract from the soybean is a standard precursor to the manufacture of tofu. H. Miller, one of the early westerners in China, often recounted his frustration in being unable to adequately feed the orphaned and unwanted infants he often found on his doorstep. His observation of the milk-like constancy of the solution of ground soybeans and water was met with the disclaimer that before it became tofu it was not digestible, a fact which his experimentation soon verified. His interest did not end and through trial and error he discovered that the application of heat transformed the soy liquid into a digestible product.

In a short time, his soy dairy was delivering soy milk to the population of Shanghai. World War II necessitated his return to the United States but he brought with him his knowledge of the soybean and was soon manufacturing his soy milk in Ohio. It was obvious that the soy milk could be transported more economically and that it would keep for a much longer period of time if it were dried to a powder. Spray drying equipment was added to the plant in the late 1940's. This produced a product definitely superior to those other experimenters had obtained using soy flour. In the United States, both liquid and powdered soy milks still are being produced using methods which differ little from those originated by Miller.

In the West, soy milk made its debut disguised as an infant formula for use by infants who could not tolerate the more conventional preparations based on cow's milk. Even today it is this market segment that is the largest consumer of soy beverages. In the United States approximately 25% of the 1.1 billion dollar infant formula market is comprised of soy-based products. Of this amount over 17% is in the powdered form. These numbers indicate that soy milk is being utilized by many infants who do not require it to combat allergy since only 7% to 10% of all infants fall into this category. This prevalence in utilization has had some interesting effects. Infants raised on soy milk formulas often find it difficult to make a later transition to cow milk and thus either refuse to use milk (placing them at nutritional risk) or demand an adult soy beverage substitute. This fact, coupled with a multitude of other factors, has contributed to a proliferation of soy beverages in recent years. Many of these are packaged in ready to drink liquid form in aseptic fiber boxes, but there also is a significant demand for the powdered version.

There is, however, some cause for concern from a nutritional perspective. While many of the products are fortified so as to provide the nutritional equivalent of cow milk others are little more than "bean juice." Unless the consumer is sufficiently aware of the differences and their importance there is danger of malnutrition.

The market for an adult soy beverage in the United States is difficult to determine but best estimates indicate that it approached 8 million liters in 1987. The bulk of this amount was made up of aseptically packaged ready to drink milk (7.2 million liters) with the balance comprised of reconstitutable powder. The preparation of a beverage from the powdered product requires little skill. The most appropriate dilution results in a drink that is about 14% solids. For the best taste the beverage should be reconstituted and allowed to stand in a refrigerator for 8 to 12 hours prior to use. This will allow the entrained air to dissipate and provides a more milk-like mouthfeel. It also may allow for the dissipation of some undesirable volatiles thus producing a better tasting product.

Because powdered product often is exposed to extended periods of storage under a variety of conditions it is appropriate to ask what effect this may have on the nutritional profile. The presence of a relatively high fat content in the infant formula makes it a ready candidate for oxidation and powdered product that is exposed to air does oxidize quite rapidly. This also has implications for some of the anti-oxidative properties of certain of the vitamins which are either naturally present or which have been added as fortification.

Obviously vitamins C and E would be on that list but interestingly enough it is vitamin A that is most susceptible to degradation. Experimental investigation has revealed that in product packaged in hermetically sealed containers, using ambient air in the head space, the level of vitamin A dropped more than 70% in 12 months of storage. Replacing the ambient air in the head space with nitrogen (O_2 at less than 2%) resulted in a stabilization of the product with only a 12% reduction in vitamin A in product from the same lot. When the product was not packaged immediately after drying (within 24 hours) it still experienced substantial loss (56%) even when nitrogen-packed. The exact mechanism involved in this degradation phenomenon has yet to be completely delineated but it would appear that it is related to free radicals generated during the drying process because liquid product from the same lot experiences no such degradation.

The physical parameters experienced during drying have a marked effect on the characteristics of the powder. In general when the product is exposed to lower drying temperatures the solubility is increased. This also is the case when the concentration of the feed liqueur is raised. Because of the viscous nature of the soy milk at relatively low concentrations, feed liquors above 30% solids are not practical. This means that dryer capacities of roughly twice that of those designated for dairy products are required for the same output.

The bulk density of the product is also affected by the drying parameters. For example, bulk density in soy milk dried via indirect heat is considerably lower than that produced in a direct fired dryer.

Solubility is a consideration which must be addressed. A product which will not dissolve is not likely to capture the imagination of the consumer and may indeed effectively destroy a product's marketability. Technically speaking, spray dried soy protein is soluble but has dispersability difficulties. The addition of lecithin to the feed liquor assists in the dispersability but fairly high concentrations are required for acceptable results. This has a confounding effect of making the product less palatable. Some experimentation has been done in adding amorphous silica to the product at a level of less than one per cent with good results. Fluid-bed agglomeration also has been used to achieve more soluble products, but the effect with soy has not approached that achieved with milk.

The most familiar soy powder product is soy isolate. The total volume of soy isolate now produced is difficult to ascertain since the major players in this arena do not divulge their volume figures. A projection by the American Soybean Association in 1983 estimated that 80 million lbs. of isolate were produced in the United States in that year and that the 1987 usage would reach 110 million lbs. Industry officials contacted by Soya Newsletter indicated that the original projections may have been a little too optimistic and that the growth while steady did not live up to anticipation.

In the late 1970's and early 1980's only a handful of U. S. food processors were involved in the manufacture and sale of "tofu." That has changed drastically today and one can now find tofu in virtually any supermarket in the United States. The popularity which tofu enjoys is such that other food processors have begun to see it as a desirable ingredient to incorporate into their products. Tofu in its traditional form does not lend itself to the ideal image of a food ingredient—it is simply too difficult to store and handle. Recently, using a combination of soy and dairy technology, several companies have developed a spray-dried tofu. The demand for this product is expected to reach more than 10 million pounds this year, the equivalent of 25 to 50 million pounds of fresh tofu. The dried tofu finds its primary use in meat and cheese substitutes, salad dressings and dips, soups and baked goods and nondairy deserts including the frozen variety. Dried tofu is also finding a place in nutritional supplements and cosmetics.

A recent survey by Ralston Purina indicates that consumers have done an "about face" in their attitude toward soy protein. The majority of American consumers now perceive soy as enhancing the product to which it is added, a marked contrast to their earlier negative perception. Consumers reported that they believed that products which contained soy were "healthier."

This perception has had some unfortunate side effects. Soy protein often has been utilized as a component in the recently popular liquid diets, both those used as protein supplements and those marketed as weight reducing aids. This usage is aggravated by the still widely held, but erroneous, belief that a high protein diet enhances athletic performance. Nowhere in the world is this less necessary. The average protein consumption in America amounts to about 115 gram/day, which is double the recommended allowance.

The American Dietetic Association has issued statements decrying the unsupervised use of these products in weight reduction diets. In cases of gross obesity where their use may be indicated they should be administered only with medical supervision. It is estimated that between two and four million individuals have utilized this type of diet in an unsupervised fashion. Reported metabolic disturbances with accompanying cardiac arrhythmias and unexplained sudden death in a number of young women who were using these products have resulted in a reduction of use of the liquid protein diet but there has been a subsequent rise in the use of the powdered equivalent. There is no reason to believe that those using the powdered product are any less susceptible to the metabolic disorders. It is obvious that much care should be exercised in this use of such products.

In spite of years of demonstrated efficacy there still remains some concern that as the known nutritional adequacy of milk and meat are replaced by soy and other vegetable products that somehow the consumer will be compromised nutritionally. These fears are totally unnecessary as long as basic nutritional principles are used. Soy protein products, particularly those intended for consumption by growing children should be appropriately supplemented with those nutrients which are required. Foremost on the list would be such nutrients as the essential amino acid methionine, calcium, and vitamin B_{12}.

Failure to heat treat the product properly prior to drying results in the incomplete destruction of trypsin inhibitors. It is therefore important that the feed liquor be exposed to sufficient heat treatment to effectively denature this protein. Careful process control is required to reduce these protease inhibitors to sufficiently low levels without adversely affecting the protein quality.

Fortunately there is more positive than negative to be said about soy protein. As research progresses soy protein products continue to be recognized for their contribution to health and wellness.

The following are a few examples: The hypo allergenic effect of soy based infant formulas have not only made life more pleasant and satisfying for the parents of the afflicted child but in many cases have been the only thing that made it possible for the infant's life to continue at all.

Some animal studies indicate that soy isolate has an hypocholesteremic effect through the regulation of the liver during cholesterol metabolism.

Work presently under way at the University of Alabama is promising in regard to the possibility that soy isofavones may be protective in breast cancer. It is hypothesized that the isoflavones compete with estrogen for binding sites, thus blocking the carcinogenic promotion occasioned by the hormone.

Evaluation of Nutritive Value of Local and Soy-Beef Hamburgers

Abdul Salam Hj. Babji and Selvakumari Letchumanan

Department of Food Science and Nutrition, Universiti Kebangsaan Malaysia Bangi, 43600, Malaysia

Abstract

Protein nutritive value of local market hamburgers together with two reference formulations were evaluated. The two references were pure beef and soy-beef hamburgers. Protein content, protein quality and protein digestibilities were determined. Procedures for evaluation included Protein Efficiency Ratio (PER) using the rat bioassay, *in vivo* apparent digestibilities and *in vitro* digestibilities. Protein content of the local hamburgers studied was relatively lower than the two reference hamburgers. All local hamburgers showed PER values that were significantly lower ($p < 0.05$) compared to the pure beef and soy-beef reference hamburgers. There was no significant reduction in PER value with 30% hydrated, textured soy protein added to beef. Local hamburgers also showed a significantly lower ($p < 0.05$) apparent digestibility, whereas with pure beef and soy-beef references the values were higher. Similar results with a positive correlation ($r = 0.86$) were obtained with the *in vitro* enzymatic procedure. The overall results of this study indicate that the protein nutritive value of the local hamburgers studied are lower than the reference hamburgers formulated.

The fast food industry has made a significant entry into the food sector in Malaysia. One of the biggest meat items seen in the market lately is the hamburger. These include giant franchises such as McDonald's, Wendy's and the A&W chain of restaurants. There also has been a rapid growth of local production of hamburgers. Today there are about ten brand names of hamburgers commonly found in supermarkets, retail outlets and many fast food stalls. However there are major differences between imported hamburgers and those produced locally. Differences include organoleptic properties, chemical composition, formulations, nutritional composition, and overall acceptance of these two classes of hamburgers. Babji et al. (1) used 5–25% soy protein in processed meats such as hamburgers. The amount of soy protein added varies, depending on the technology and know-how of the manufacturers. Babji et al. (2) also showed locally produced hamburgers to contain between 3.1–19.0% soy flour, 8.9–28.3% of carbohydrate, food colors and various spices. In another study on the use of food additives in processed meats, Babji et al. (3) reported the incorporation of many food chemicals, colors and flavors in products such as hamburgers and hot dogs. These include salt, polyphosphates, nitrates, nitrites, ascorbates, benzoic acid and sorbic acid. Babji et al. (4) demonstrated soy protein addition at 30% in hamburgers to be acceptable among Malaysian consumers. Sensory properties showed that meat substitution with textured soy protein increased the intensity of beany flavor and taste, but had no specific effects on quality attributes such as appearance, texture, saltiness and juiciness of the product. Babji et al. (5) further investigated the meat and nonmeat components of locally produced hamburger in Malaysia. They reported hamburgers with 23.5–71.1% meat content, soy protein ranging from 15.5–36.5% and cereal added at 0.9–27.4%. In addition to these three major components in local hamburgers, other substances such as egg, milk, sugar, MSG, spices, onion, ox-fat and mechanically deboned meat also are formulated in hamburgers.

However, there are no data currently available with regard to the nutritional status of locally produced hamburgers. Plant protein additions in meat products will affect the nutritional composition, as is well known. Thus the question arises as to the nutritive value of local hamburgers. This study was initiated with the objective of comparing the nutritive values of mixed beef-soy cereal proteins commonly manufactured locally with that of pure hamburger (100% meat). The methods used are the bioassay of protein efficiency ratio (rat-PER) and the *in vitro* enzymatic assay for digestibility.

Methods and Materials

Four locally processed hamburgers were selected for protein quality determination. In addition, two formulations, i.e., pure burger (100% beef) and beef-soy mixture (70% beef : 30% hydrated textured protein) also were prepared in the laboratory for protein quality evaluation. All samples were kept at $-18°C$ until ready for further testing.

Proximate Analysis

Protein, fat, moisture and ash were determined using the A.O.A.C. method for meat (6).

Rat Diet Preparation

Frozen meat samples were ground using a meat grinder with an 8 mm grinder plate. Diet formulation was followed using the procedure for PER as outlined by A.O.A.C. (6), with casein (USBC) as the reference protein. Other components included in the diet are ash (AIN 76™), vitamin mix (AIN™), cellulose-celufil, D$^+$ sucrose, corn starch and corn oil (Mazola). Calculation of percent of ingredient in diet formulation was based on the proximate analysis of the test protein (Table 1) and calculated as in Figure 1. The diet composition (g/kg) for each test protein is illustrated in Table 2. The total rat diet prepared for each protein source for PER assay is as follows.

Diet requirement = X gm/day × number of days
  × number of rats per treatment
 = 15 g/day × 28 days × 8 rats
 = 3360 gram.

TABLE 1

Proximate Analyses of Local, Formulated Hamburgers and Reference Casein

Source of protein	% Nitrogen	% Protein	% Fat	% Air	% Ash
Pure	11.62 ± 0.14	72.63 ± 0.88	1.02 ± 0.24	18.69 ± 0.55	3.52 ± 0.09
70:30	8.40 ± 0.00	52.50 ± 0.44	1.11 ± 0.03	26.08 ± 3.00	3.42 ± 0.03
Angus	6.69 ± 0.04	41.78 ± 0.22	19.44 ± 1.21	12.91 ± 0.37	7.07 ± 0.05
Fika	6.85 ± 0.02	42.77 ± 0.11	9.55 ± 0.09	18.70 ± 0.02	7.49 ± 0.02
Ramly	6.13 ± 0.23	38.28 ± 1.80	13.63 ± 0.15	15.25 ± 0.08	4.71 ± 0.05
Thrifty	3.99 ± 0.06	24.92 ± 0.33	15.73 ± 1.90	12.07 ± 0.06	5.31 ± 0.02
Casein	13.37 ± 0.14	83.56 ± 0.88	0.23 ± 0.01	8.07 ± 0.08	0.87 ± 0.18

```
Material                        %

Test Protein (A)    X
                    X = 1.60 x 100
                        -----------
                        % N in test protein/reference

Mazola Corn Oil     8 - X x % Fat A
                        -----------
                        100

Ash Mixture         5 - X x % ash A
   TM                   -----------
AIN 76                  100

Moisture            5 - X x % moisture A
                        ----------------
                        100

Vitamin Mixture
   TM
AIN 76              1

Cellulose-Celufil   1

Sucrose mixture     Make up 100%
and Corn starch
(1:1)
```

Fig. 1. Rat diet formulation (6).

After diet preparation for each type of hamburger and the reference protein (casein) another proximate analysis was carried out to ensure the diet formulation was done correctly following the recommendation of A.O.A.C. (6). This is shown in Table 3. Each type of hamburger and casein was fed to eight male weanling rats (Sprague-Dawley Strain) obtained from the Animal Research Laboratory at UKM, Bangi.

The 28-day-old albino rats were placed in individual cages and randomly assigned by treatment to individual cages. The weight ranges between individuals was less than 10 gram and mean weight between treatments was less than 5 gram. The range in weight for rats used is between 46–56 gram with a mean of 50±3 gram. Prior to feeding the experimental diets, the rats were placed on an adaptation diet for a 3-day period.

PER-assay

Food and water were supplied *ad libitum*. Body weight was recorded for 0 day and every two days for 28 days. Total weight of feed intake was recorded at 9, 18 and 28 days. For determination of feed intake, feed wasted and left over were collected daily, dried in oven (100°C) for an hour, then analyzed for moisture content before weighing. From these data, the PER is calculated using the formula:

$$PER = \frac{\text{Increase in body weight (gram)}}{\text{Weight of protein consumed (gram)}}$$

TABLE 2

Rat Diet Composition with 10% Reference/Test Protein (g/kg feed)

Rat diet	Source of protein	Test protein	Mazola corn oil	Ash mixture AIN 76	Moisture	Vitamin mixture AIN 76	Cellulose CELUFIL	Corn starch	D (+) sucrose
1.	Pure	137.69	78.60	45.15	24.27	10.00	10.00	347.15	347.15
2.	70:30	190.48	77.89	43.89	0.32	10.00	10.00	333.91	333.91
3.	Angus	239.16	33.51	33.09	5.28	10.00	10.00	334.48	334.48
4.	Fika	233.58	57.69	32.50	6.32	10.00	10.00	324.96	324.96
5.	Ramly	261.01	44.42	37.71	10.20	10.00	10.00	313.33	313.33
6.	Thrifty	401.00	16.92	28.71	1.60	10.00	10.00	265.89	265.89
7.	Casein	119.67	79.73	48.96	40.40	10.00	10.00	345.62	345.62

TABLE 3
Proximate Analyses of Formulated Rat Diets

Rat diets	Burgers/reference	% Protein	% Fat	% Moisture	% Ash
1.	Burger	10.07 ± 0.44	8.69 ± 0.10	10.05 ± 0.04	3.69 ± 0.11
2.	70:30	10.06 ± 0.00	8.54 ± 0.05	10.53 ± 0.26	3.82 ± 0.1
3.	Angus	10.08 ± 0.23	8.27 ± 0.06	10.98 ± 1.43	4.31 ± 0.01
4.	Fika	10.50 ± 0.44	8.97 ± 0.27	8.52 ± 0.12	4.23 ± 0.23
5.	Ramly	10.46 ± 0.26	8.77 ± 0.03	9.00 ± 0.22	4.10 ± 0.07
6.	Thrifty	10.17 ± 0.11	8.02 ± 0.19	9.66 ± 0.12	4.04 ± 0.20
7.	Casein	10.50 ± 0.00	8.47 ± 0.05	9.09 ± 0.34	3.73 ± 0.12

In Vivo Apparent Protein Digestibility

Food consumption and fecal output data were recorded daily for eight days (day 10–18) of the 28-day study to determine the *in vivo* apparent protein digestibility. It was calculated as follows:

In Vivo Apparent Protein Digestibility =

$$\frac{\text{N in diet (g)} - \text{N in feces (g)}}{\text{N in diet (g)}} \times 100$$

In Vitro Protein Digestibility

The *in vitro* protein digestibilities of various hamburgers and casein were measured using the A.O.A.C. method (6). Casein^{AIN} was used as a reference. The four enzyme systems used were α-chymotrypsin from bovine pancreas (type II), peptidase from porcine mucosa, trypsin from porcine pancreas (type IX) and protease from *Streptomyces grieus* (Type XXI), purchased from Sigma Chemical Company. Solution A was made up of 227,040 units of BAFE trypsin, 1860 units of α-chymotrypsin, and 0.520 units of β-napthalamide L-Leucine peptidase in 10 ml water. Solution B was prepared by dissolving 65 units of protease casein in 10 ml water. Both solutions were kept in ice until ready for use.

Measurement of pH Reduction by the 4 Enzymes System

A reference solution containing 6.25 mg/ml protein casein was prepared. Sample weight (depending on protein content) was placed in a test tube (25 × 95mm) containing 10 ml distilled water. This solution was kept in ice for protein hydration for 1 hour. Using the apparatus shown in Figure 2, the protein solution was adjusted to pH 8±0.05 at 37°C. Solutions A and B were also adjusted to reach pH 8±0.05 at 37°C, then returned to the ice bath. Beginning with the casein standard solution, 1 ml of solution A was added to the reaction vessel. After exactly 10 mins of addition of solution A, 1 ml of solution B was added and the tube transferred to the 55°C water bath. At 19 mins, the tube was transferred immediately to the reaction vessel and the pH was measured at exactly 20 mins. The pH of standard casein should read 6.42±0.05. The same procedure was repeated for the protein test samples and the pH value (X) read at exactly 20 minutes.

In vitro apparent digestibility was calculated as follows:

In vitro apparent digestibility = 234.84 × 22.56(X)

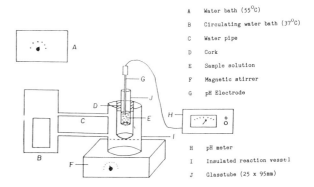

Fig. 2. Schematic presentation of apparatus used for *in vitro* digestibility assay.

Result and Discussion

A pure beef burger and a 70:30 beef:soy protein burger were formulated to serve as references when making comparison with locally processed hamburgers. The 30% soy protein was chosen because that is the level reported to be acceptable by consumers and it is also the maximum level recommended by USDA; Babji et al. (2) and Forbes, (7). Woolford (8) reported problems with texture when meat was substituted with 40–75% soy protein.

The proximate analyses of various hamburgers are shown in Table 4. Pure beefburger contained 19.44% protein followed by the 70:30 burger at 18.17%. These levels were higher than those reported for local burgers such as Angus, Fika, Ramly and Thrifty. They are, however, similar in protein content to those of A&W (19.43%) Wendy's (17.17%) and McDonald's (18.58%) as reported earlier by Babji et al. (2). A study by Cichon et al. (8) showed a beef sirloin cut to contain 20.60% protein and 2.90% fat. Ng (9) in his study reported fat content of 1.40% in local beef cut (lean). This value is close to the fat content of pure beef (1.92%) reported in this study.

The lower protein contents in local burgers indicated the use of binders other than soy protein and meat. The higher fat content in local burgers also means that fat is added in the formulation of local burgers. Babji et al. (2) reported the use of nonmeat components such as potatoes, starch, bread crumbs, eggs and food conditioners in local burger formulation.

Growth and PER data for rats fed local and formulated hamburgers are shown in Table 5. Results from this study

TABLE 4
Proximate Analyses of Raw Hamburgers (% wet weight)

Burger	% Protein (N × 6.25)	% Fat	% Moisture	% Ash	% (CHO + Crude fiber)
Pure	19.44 ± 0.22	1.92 ± 0.03	78.90 ± 0.02	1.19 ± 0.02	—
70:30	18.17 ± 0.22	0.55 ± 0.02	77.12 ± 0.06	5.41 ± 0.02	—
Angus	13.78 ± 0.22	24.67 ± 0.20	54.14 ± 0.22	2.28 ± 0.02	5.13 ± 0.62
Fika	14.73 ± 0.15	25.35 ± 0.03	52.27 ± 4.62	2.35 ± 0.24	5.31 ± 5.03
Ramly	13.54 ± 0.21	27.08 ± 0.11	51.16 ± 0.79	1.81 ± 0.06	6.41 ± 1.16
Thrifty	11.32 ± 0.07	13.29 ± 0.12	56.52 ± 0.74	2.13 ± 0.17	17.13 ± 0.96

TABLE 5
PER Values of Local, Formulated Hamburgers and Casein Reference

Source of protein	Increase in weight g ± S.D.	Total feed intake (g/rat/28 days) g ± S.D.	% Protein in feed (N × 6.25)	Protein consumed (g/rat/28 days) g ± S.D.	PER[1,2] X ± S.D.	Adj. PER = PER × $\frac{2.5}{2.3}$
Pure	99 ± 11	328.1 ± 23.7[a]	10.07	33.04 ± 2.39	2.98 ± 0.23[ab]	3.24
Mixture (70:30)	95 ± 12	320.7 ± 22.1[a]	10.06	32.26 ± 2.23	2.94 ± 0.25[ab]	3.20
Angus	61 ± 10	268.1 ± 19.4[b]	10.08	27.02 ± 1.96	2.26 ± 0.42[d]	2.46
Fika	69 ± 14	273.0 ± 28.9[b]	10.50	28.66 ± 3.03	2.38 ± 0.33[cd]	2.59
Ramly	74 ± 14	285.2 ± 30.8[b]	10.46	29.82 ± 3.22	2.45 ± 0.27[cd]	2.66
Thrifty	80 ± 14	293.2 ± 27.5[b]	10.17	29.82 ± 2.79	2.67 ± 0.30[abc]	2.90
Casein	60 ± 12	288.3 ± 26.6[b]	10.50	28.97 ± 2.67	2.30 ± 0.26[d]	

[1]Mean and standard deviation from 8 rats.
[2]Means with different superscripts are significantly different (P < 0.05).

showed PER values of 2.98 and 2.94 for pure and 70:30 burgers respectively. The values for local burgers were 2.26, 2.38, 2.45 and 2.67 respectively, much lower than the formulated burgers, but close in value to the reference casein. The casein reference used in this study was high nitrogen casein supplied by United State Bio-Chemical Corporation, Cleveland, Ohio, whereas the recommended casein [AOAC (6)] is ANRC (Animal Nutrition Research Council). Pallert and Young (10) reported low sulfur containing amino acid, especially methionine, and suggested enrichment with 1.0% DL-Methionine in rat diet for bioassay. Cichon et al. (8) also reported lower PER values with ANRC casein that is not enriched with methionine, ranging from 2.48–2.63. This study followed the AOAC (6) procedure which makes no mention of enrichment of casein with methionine. This may explain the low PER value of casein in this study.

Reference casein is included to reduce variation in PER value obtained between research laboratories. PER values for tested proteins are corrected using reference casein value of 2.5. Since the actual PER of casein in this study was 2.3, discussion will be based on the actual PER values and not the adjusted PER values as normally discussed.

The PER values for pure burgers and 70:30 beef:soy burgers were found to be 2.98 and 2.94 respectively. These values are similar to those reported by Cichon et al. (8) for sirloin cut, ranging from 2.98–3.05. The beef:soy burger had PER of 2.94, which indicated no significant reduction in protein nutritional quality when soy protein was added up to 30% in the burger formulation. Steinke et al. (11) reported a PER value of 2.90 for 70:30 beef:soy; whereas with all beef the PER was 3.22. Wolford (8) reported PER values of 2.76 and 2.68 for ground beef and ground beef:soy protein (75:25), respectively. The PER values for local hamburgers were significantly lower (P < 0.05) ranging from 2.26 to 2.67, compared to pure burger and beef:soy (70:30) burger formulations.

Lower PER values of locally produced hamburgers could be due to addition of nonmeat components in the formulation, which could alter the protein quality of the products. Thrifty, for example, used egg in its formulation (Table 6), thus its PER value of 2.67 was the highest among local hamburgers. Egg protein is a high quality protein that is particularly rich in sulfur containing amino acid (12,13).

Table 7 shows the *in vivo* apparent digestibility of local, formulated and casein protein. Similar results were reported by Jewel et al. (14), Babji et al. (15) and Hendricks et al. (16). Pure burger (100% beef) had the highest value of 90.04%. Similar studies by Jewell et al. (14) and Happich et al. (17) yielded 92.0% and 93.0% for lean beef, respectively. The digestibility for 70:30 burger was 87.91% which was significantly (P < 0.05) lower than all beef. This indicated addition of 30% soy protein lowered protein digestibility. Bodwell et al. (18) however reported digestibility value of 91.70% for soy isolate. Hsu et al. (19) and Satterlee et al. (20) reported digestibility of 90.0% and 88.0% for soy isolate and concentrated soy protein respectively.

TABLE 6

Ingredient as Stated on the Labels of Hamburgers Studied

Brandnames	Manufacturer	Ingredient
Angus	Cold storage (M) Bhd.	Beef, soy protein, salt, spices, food conditioner, permitted food flavor and color.
Fika	Fika Foods Company	Beef, spices, soy protein, sugar, salt.
Ramly	Mokni Sdn. Bhd.	Beef, beef fat, soy protein, onion, salt, sugar, spices, food conditioner, food flavor and MSG.
Thrifty	Thrifty Supermarket Sdn. Bhd.	Beef, potato, onion, starch, bread crumbs, egg, spices and color.

TABLE 7

% In vivo Apparent Digestibility of Local, Formulated Burgers and Casein Reference

Source of protein	Weight of feed consumed g/rat 8 days g ± S.D.	% Nitrogen in feed	Total nitrogen consumed g/rat 8 days g ± S.D.	Feces Dry Wt. g/rat 8 days g ± S.D.	% Nitrogen in dried feces	Total in nitrogen dried feces g/rat/8 days g ± S.D.	% Apparent digestibility ± S.D.
Pure	106.8 ± 5.5	1.61	1.72 ± 0.09	4.8 ± 0.5	3.56	0.17 ± 0.02	90.04 ± 0.62^a
70:30	104.8 ± 9.0	1.61	1.69 ± 0.15	5.1 ± 0.6	4.01	0.20 ± 0.02	87.91 ± 1.66^b
Angus	95.2 ± 11.4	1.68	1.54 ± 0.19	5.4 ± 0.6	4.14	0.22 ± 0.03	85.50 ± 1.26^c
Fika	91.84 ± 10.3	1.68	1.54 ± 0.17	4.6 ± 0.7	4.65	0.21 ± 0.03	86.18 ± 1.41^{bc}
Ramly	92.2 ± 13.1	1.68	1.55 ± 0.22	5.4 ± 1.1	4.14	0.23 ± 0.04	85.50 ± 1.96^c
Thrifty	97.7 ± 12.8	1.68	1.64 ± 0.22	4.8 ± 2.6	4.84	0.23 ± 0.04	85.91 ± .81^{bc}
Casein	96.4 ± 10.0	1.63	1.57 ± 0.16	4.9 ± 0.9	3.05	0.15 ± 0.03	90.58 ± 1.80^a

Means with different superscripts are significantly different (P < 0.05).

All local hamburgers had lower digestibility value compared to the pure as well as the 70:30 mixture. This could be due to the presence of soy protein in local hamburgers which had been shown to reduce digestibility. According to Rackis et al. (21) 80% of the trypsin inhibitor activity is destroyed via thermal processing, enabling higher digestibility value for most soy products.

Table 8 shows the *in vitro* protein digestibility of local, formulated and casein reference. Again, as *in vivo* protein digestibility, casein had the highest value at 87.45%. Clutterbuck et al. (22) using the same method reported protein digestibility of 88.2% for ANRC casein. Among the burgers studied, the pure burger had the highest value at 85.57% followed by the 70:30 mixture at 84.82%. All local burgers had lower *in vitro* digestibility compared to the pure beef. These results were similar to that reported for *in vivo* digestibility earlier. Figure 3 shows the PER values, *in vivo* and *in vitro* protein digestibilities as a result of this study. Statistical analyses showed a positive correlation between the two methods (r = 0.86). The lower values reported for *in vitro* digestibility is similar to those reported earlier by Bodwell et al. (18).

This study indicated that locally produced hamburgers are lower in protein quality compared to the two references. Protein content was lower because of substitution of carbohydrate components for meat. Locally processed hamburgers had lower PER values than the pure beef and beef:soy (70:30) burgers. This could be due to the addition of other nonmeat proteins in the local hamburger formulation.

The addition of 30% soy protein did not decrease PER

TABLE 8

In vitro Protein Digestibility of Local, Formulated and Casein Reference

Protein source	% In vitro digestibility^a
Pure	85.57 ± 3.58
70:30	84.82 ± 4.26
Angus	82.94 ± 4.15
Fika	83.31 ± 5.55
Ramly	82.56 ± 4.22
Thrifty	83.69 ± 4.61
Casein	87.45 ± 2.96

^{a} Means and standard deviation from three samples.

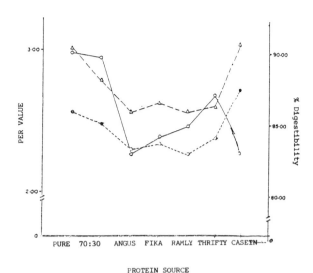

Fig. 3. Relationship between PER (open circles connected by solid line), *in vivo* (triangles) and *in vitro* (circles connected by dotted line) digestibilities of protein.

value significantly, but it decreased the *in vivo* and *in vitro* protein digestibilities. Locally processed hamburgers also had lower *in vivo* and *in vitro* digestibilities as compared to the two references.

References

1. Babji, A.S., A. Aminah and A. Adnan, *PERTANIKA* 7:(2) (1984).
2. Babji, A.S., A. Sayuwa and A. Aminah, "The Need for Standards and Specification for Processed Meats in Malaysia," presented at seminar on Food Technology and Nutrition, July 8–10, 1985, Yogjakarta, Indonesia.
3. Babji, A.S., A. Aminah and D. Muchtadi, "The Use of Food Additive in Local Meat Products in ASEAN Region," presented at Adv. Food Research in Malaysia UPM, May 6–7, 1986, Serdang, Malaysia.
4. Babji, A.S., A. Aminah and Y. Gatimah, *PERTANIKA* 9:(2) (1986).
5. Babji, A.S., A. Aminah and P.H. Ooi, "Determination of Meat and Non-Meat Components in Local Beef Burger," presented at Adv. Food Research Conference II in Malaysia UPM, Feb. 3–4, 1988, Serdang, Malaysia.
6. AOAC Official Methods of Analysis (14th Edition), Association of Official Analytical Chemists, Washington, D.C., 1984, pp. 871–881.
7. Forbes, A.L., personal communication.
8. Woolford, K.M., *J. Am. Oil Chem. Soc.* 51:131A (1974).
9. Babji, A.S., A. Aminah and K.S. Ng, "Evaluation of Soy Protein, Mineral and Phytate Contents of Beef Burgers in Malaysia," presented at ASEAN Food Conference Oct. 24–26, 1988, Bangkok, Thailand.
10. Pallert, P.L., and V.R. Young, Food and Nutrition Bulletin Supplement 4, UNU, Tokyo, Japan, 1980.
11. Steinke, F.H., E.E. Preschr and D.T. Hopkins, *J. Food Sci.*, 45:323 (1980).
12. Amino Acid and Contents of Foods and Biological Data on Proteins, United Nations Food and Agricultural Organization, Rome, Italy, 1970.
13. Energy and Protein Requirements: Report of a joint FAO/WHO ad hoc expert committee, World Health Organization Tech. Rep. Ser. 522, WHO, Geneva, Switzerland, 1973.
14. Jewel, D.K., J.G. Kendrik and L.D. Satterlee, *Nutr. Ref. Int.* 21:25 (1980).
15. Babji, A.S., G.W. Froning and L.D. Satterlee, *J. Food Sci.* 45:441 (1980).
16. Hendriks, J.G., A.W. Mahoney and T.A. Gillert, *J. Food Sci.*, 42:186 (1977).
17. Happich, M.L., R.A. Whitmore, S. Feairheller, N.M. Taylor, C.E. Swift, J. Nagshki, A.N. Booth and R.H. Aismeyer, *J. Food Sci.* 40:35 (1975).
18. Bodwell, C.E., L.D. Satterlee and L.R. Hackler, *Amer. J. Clin. Nutr.* 33:677 (1980).
19. Hsu, H.W., N.E. Sutton, M.O. Banjo, L.D. Satterlee and J.G. Kendrik, *Food Technol.* 32:69 (1978).
20. Satterlee, L.D., J.G. Kendrik and G.A. Miller, *Food Technol.* 31:31 (1977).
21. Rackis, J.J., J.E. MacGhee and A.N. Booth, *Cereal Chem.* 52:85 (1975).
22. Clatterbuck, K.L., N.L. Kehrberg and N.L. Marable, *J. Food Sci.* 45:931 (1980).

Limiting Amino Acids in Poultry Diets

M. E. Blair

Central Soya Company, Decatur, Indiana 46733

Abstract

The appearance of feed grade synthetic amino acids in the marketplace has initiated interest in formulating poultry diets with lower crude protein levels. Supplementing low protein corn-soy diets for broiler chicks with essential amino acids has resulted in inferior performance relative to standard protein control diets. Conversely, feeding turkey poults low protein corn-soy diets supplemented with amino acids has resulted in similar performance relative to poults fed diets with standard protein levels. For both chicks and poults, the common limiting amino acids have been determined to be methionine, lysine, arginine, valine, threonine and sometimes isoleucine. However, their order of limitation varies according to the absence or presence of glutamic acid in the diet and experimental technique (addition versus deletion). Therefore, care must be taken in evaluating studies determining the order of limiting amino acids in poultry diets.

During the early 1940's, Almquist and associates at the University of California conducted experiments to show the essentiality of eleven amino acids for the chick. Researchers in the 1950's established the sulfur amino acids (SAA), methionine and cysteine, to be first-limiting for chicks and poults (1–3). As a result of this research, the fact that the chick or poult required amino acids rather than crude protein *per se* was realized. Methionine and cysteine are deficient relative to poultry requirements in soybean meal, the primary protein source in poultry feeds. The development of synthetic methionine in the 1960's allowed the replacement of some soybean meal by methionine, lowering crude protein levels in poultry diets. Synthetic lysine use increased dramatically in poultry feeds in the 1970's. The advent of the synthetic forms of methionine and lysine allowed the nutritionist to lower the crude protein level of the diet, as no longer were natural proteins needed to totally meet the SAA and lysine requirements of poultry.

The successful feeding of lower protein diets for poultry is dependent upon the identification of the limiting amino acids and their order of limitation. Several studies have been conducted to determine the requirements of the essential amino acids for poultry, yet only a few studies have concentrated on the feasibility of lowering the protein diet using both addition and deletion methods. The addition trials showed Met, Arg and Lys to be first-, second-, and third-limiting, respectively. The deletion trials showed the order of limitation to vary with Glu supplementation. In the absence of supplemental Glu, Arg was first-limiting with Met and Lys equally second-limiting and Val and Thr equally third-limiting. When the 16% protein diet was supplemented with 3% Glu, Met and Lys were equally first-limiting followed by Arg with Val and Thr being equally third-limiting. This study verified that Met, Lys, Arg, Val and Thr are limiting in low protein corn-soybean meal diets for chicks. Also demonstrated was how methodology can affect order of limitation. An antagonism between Lys and Arg has been well established (6,7). Thus, in the deletion method, removal of either amino acid may have exacerbated the deficiency, causing Lys or Arg to become first-limiting. In the addition trial, Val was calculated to be adequate. The lack of a response to Thr in the addition of trial may be due to Val being limiting (indicated in deletion trials), thus causing the calculated order to differ from the actual order of limitation.

In addition to investigating methodology, this study also determined the effects of low protein and low protein plus amino acid supplemented diets on body fat deposition. Based upon regression from body water contents, body fat was predicted to be 11.3, 15.6 and 11.0% for those chicks fed 24, 16 and 16% AA+Glu protein diets. Lowering of the crude protein level of the diet resulted in increased body fat deposition.

Broiler Finisher Diets

The feasibility of feeding low protein finisher diets to broilers supplemented with amino acids also has been investigated (8,9,10). In a recent study (10), optimal performance was obtained with 22% and 18% finisher diets for male and female broilers, respectively. Lowering crude protein below these levels and supplementing the diets with amino acids, while at times supporting maximal growth, did not result in maximal feed efficiency. Additionally, feeding low protein diets supplemented with amino acids resulted in more abdominal fat. When nonessential nitrogen (Glu) was added to the amino acid supplemented diet, abdominal fat deposition was reduced. This finding agrees with those found with broiler starter diets (5).

Turkey Starter Diets

Investigations also have been conducted with the turkey poult, which has the highest amino acid requirements of all species of poultry (11,12,13). Thus, identification of limiting amino acids and their order is of great economic importance. An additional study (12) using high protein (28%), low protein (20%) and low protein (20%) corn-soybean meal diets supplemented with Lys, Val or Thr in all combinations or in an amino acid mixture showed Lys to be second-limiting. An earlier study with poults (11) using a deletion technique showed Lys to be third-limiting after Val and Thr, and that the reductions in performance seen with the Val, Thr and Lys deletions did not equal the reduction seen with feeding a low protein (22%) diet, indicating that nitrogen may be limiting. These results seen with turkey poults are in agreement with those results seen with chicks (5), in that methodology does affect order of limitation.

The determination of the limiting amino acids in a particular feedstuff is also of interest. With soybean meal being the primary protein source in poultry diets, a study was conducted with turkey poults to determine the limiting amino acids in dehulled soybean meal (13). The experiment used a deletion method where high protein (30%), low protein (22%), and low protein plus IAA supplemented diets were compared. Additional test diets were made by deleting individual amino acids from the IAA mixture. Deletion of Val, Thr, Lys and Ile resulted in body weight gains being depressed 19, 16, 11 and 6%, respectively. Those poults fed diets where Val, Lys and Thr were removed had significantly lower feed consumptions. An effect on feed intake also has been observed in deletion trials for chicks (5), indicating that supplementing low protein diets with IAA and removing individual amino acids create imbalances and/or antagonisms which thereby reduce growth rate *via* reduced feed intake.

Although much remains to be elucidated concerning limiting amino acids in low protein diets for poultry, the studies conducted to this date have yielded several important findings: (a) Met, Lys, Arg, Val and Thr have been shown to be limiting for both chicks and poults; (b) with Arg, Thr and Val being limiting, economical synthetic sources of these amino acids will need to be developed for the use of low protein diets commercially; (c) the order of limitation is dependent upon the methodology used. The use of addition trials is more relevant, as deletion trials are associated with imbalances which exacerbate deficiencies and affect feed intake which ultimately affect performance; (d) some amino acids while calculated to be sufficient, were found to be limiting based on performance; and (e) the fact that feeding low protein diets supplemented with amino acids results in greater body fat deposition may be undesirable as the poultry industry moves toward marketing leaner products. Further research is needed to further define limiting amino acids in not only low protein corn-soybean meal diets, but in diets containing a variety of energy and protein sources.

References

1. Almquist, H.J., *Poultry Sci. 31:*966 (1952).
2. Donovan, G.A., E.L. Johnson, S.L. Balloun and R.E. Phillips, *Ibid. 34:*251 (1955).
3. Waibel, P.E., *Ibid. 38:*712 (1959).
4. Schwartz, R.W., and D.J. Bray, *Ibid. 54:*1814 (1975).
5. Edmonds, M.S., C.M. Parsons and D.H. Baker, *Ibid. 64:*1519 (1985).
6. D'Mello, J.P.F., and D. Lewis, *Br. Poultry Sci. 11:*299 (1975).
7. Allen, N.K., D.H. Baker, H.M. Scott and H.W. Norton, *J. Nutr. 102:*171 (1972).
8. Lipstein, B., and S. Bornstein, *Br. Poult. Sci. 16:*189 (1975).
9. Uzu, G., A.E.C. Document No. 252 (1983).
10. Fancher, B., Proc. Georgia Nutr. Conf. 116 (1987).
11. Stas, R.J., L.M. Potter and B.A. Bliss, *Poultry Sci. 61:*933 (1982).
12. Blair, M.E., L.M. Potter and B.A. Bliss, *Ibid. 65:*154 (1986).
13. Blair, M.E., and L.M. Potter, *Ibid. 66:*1813 (1987).

Utilization of Plant Proteins by Warmwater Fish

Chhorn Lim[a] and Warren Dominy[b]

[a]USDA-ARS, Tropical Aquaculture Research Unit. The Oceanic Institute, Makapuu Point, P.O. Box 25280, Honolulu, HI 96825. [b]The Oceanic Institute, Makapuu Point, P.O. Box 25280, Honolulu, HI 96825, USA

Abstract

Unlike domesticated farm animals, fish species currently cultured have a high dietary protein requirement. Consequently, only high protein content feedstuffs are included in fish feeds. Fish meal traditionally has been used as a major protein source because of its high nutritive value and palatability. Recently the high cost and short supply of fish meal has made it necessary to substitute with cheaper plant proteins such as soybean meal, peanut meal, cottonseed meal, sunflower seed meal, rapeseed meal and *Leucaena* leaf meal in order to sustain economical fish production. Soybean meal appeared to be better utilized by most fish species as compared to other plant protein sources. However, high levels of plant proteins in fish diets generally have resulted in reduced growth and poor feed efficiency. This may be the result of an improper balance of essential nutrients such as amino acids and minerals, presence of toxic substances or antinutritional factors, or a decrease of palatability and pellet stability in water.

Traditional fish farming used extensive methods where the animals obtained all their nutrition from the aquatic environments in which they were cultured. By early in the twentieth century, the culture techniques had progressed to stocking fish at densities higher than could be supported by the natural productivity. The nutrients input into the culture systems were limited to fertilizers and crude agriculture and animal products or by-products. Recently, in response to the increased cost of land and labor, as well as increased demand for fish in the world market, fish husbandry has changed its form from extensive to semi-intensive or intensive systems. These latter employ modern facilities, equipment and technology to obtain significantly higher yield per unit area. High production levels depend on good nutrition as well as proper culture management. The quality and quantity of feed used are the major factors in determining profitability because feed represents the largest single expenditure in semi-intensive or intensive culture operations. Thus, economical production depends on availability of least-cost, nutritionally-balanced diets.

Fish meal customarily is used as a major protein source in fish feeds because of its high nutritive value and palatability. To date, fish meal still constitutes a substantial part of the feed formula for aquacultured species. However, the rising cost and uncertain availability of fish meal have forced aquaculture nutritionists and feed manufacturers to use less expensive, readily available plant protein as a substitute for fish meal.

Protein and Amino Acid Requirements

Fish, like shrimp and other animals, do not have an absolute requirement for protein per se, but require a well-balanced mixture of essential and nonessential amino acids. The most effective, economical source of these amino acids is a proper combination of high quality natural proteins in feedstuffs. The minimum amount of dietary protein needed for optimum growth of some warm water species has been determined using practical or purified diets (Table 1). Fish have a much higher dietary protein requirement than domestic animals, but require lower dietary energy (Table 2). The levels of dietary crude protein required by warm water fish range from 30% to 56%. These variations probably reflect differences in species, size of fish, water temperature, culture management (e.g. stocking density and water exchange), daily feed allowance, amount of nonprotein energy, quality of the dietary protein and availability of natural food.

All warmwater fish species studied require the same 10 essential amino acids as do cold water fish, shrimp, and terrestrial animals. These are arginine, histidine, isoleucine, leucine, lysine, methionine, phenylalanine, threonine, tryptophan and valine. The quantitative requirements of the essential amino acids for some warmwater species are presented in Table 3. In some cases, the presence of nonessential amino acids in the diet represents a sparing effect which reduces the need for fish to synthesize them. Two specific examples of sparing action are the conversion of methionine to cystine and phenylalanine to tyrosine. In both cases the nonessential amino acids can be synthesized only from the essential amino acid precursors. In the case of methionine, the fish actually have a requirement for total sulfur-containing amino acids which can be met by either methionine alone or a proper mixture of methionine and cystine. Cystine was found to replace or spare 60% of dietary methionine on a per mole sulfur basis in channel catfish (10). For *Sarotherodon mossambicus*, dietary cystine can substitute up to 50% of the total sulfur amino acid requirement (7). Growth studies with channel catfish fingerlings indicated that tyrosine can replace or spare 50% of the total phenylalanine requirement (11). However, since most protein sources used in practical fish feeds contain adequate levels of phenylalanine and tyrosine, the sum of these two amino acids normally exceeds dietary needs (12).

Nutritional Value of Proteins

The nutritive value of proteins, commonly referred to as protein quality, is based on the amount of amino acids present in the protein source, particularly essential amino acid content and biological availability. A protein with an essential amino acid composition which closely matches the essential amino acid requirements of the fish is described as being of high nutritive value. A protein that is deficient in one or more essential amino acids is of low biological value. The amino acids present at a level below that required by the fish are called limiting amino acids.

TABLE 1
Estimated Dietary Protein Requirements of Warm Water Fish

Species	Protein source	Protein requirement (%)	Reference
Channel catfish (*Ictalurus punctatus*)	whole egg	32–36	(1)
Common carp (*Cyprinus carpio*)	casein	31–38	(2,3)
Japanese eel (*Anquilla japonicus*)	casein, arginine and cystine	44.5	(4)
Milkfish (*Chanos chanos*)	casein	40	(5)
Tilapia			
Sarotherodon aureus	casein and egg albumin	34–56	(6)
S. mossambicus	fish meal	30–50	(7)
S. niloticus	casein and gelatin	36	(8)

TABLE 2
Comparison of Dietary Protein and Digestible Energy Requirements of Some Warm Water Fish and Domestic Animals

Animal	Protein requirement (%)	D.E/Protein kcal/g	Reference
Channel catfish	30	8.8	(9)
Common carp	32	8.3	(3)
S. aureus	34–56	8.2–9.4	(6)
S. niloticus	36	8.3	(8)
Broiler chicken	18	14.4	(9)
Swine	16	20.6	(9)
Beef cattle	11	23.6	(9)

TABLE 3
Essential Amino Acid Requirements of Some Warm Water Fish Species (Percentage of Protein)

Amino acid	Japanese eel[a]	Common carp[b]	Channel catfish[b]	*Tilapia nilotica*[a]
Arginine	4.2	4.2	4.3	3.5–4.4
Histidine	2.1	2.1	1.5	1.3–1.9
Isoleucine	4.1	2.3	2.6	3.1
Leucine	5.4	3.4	3.5	2.8–3.6
Lysine	5.3	5.7	5.1	4.6–5.6
Methionine	3.2	—	—	—
(+ Cystine)	5.0	3.1	2.3	3.2
Phenylalanine	5.6	—	—	—
(+ Tyrosine)	8.4	6.5	5.0	5.0–6.1
Threonine	4.1	3.9	2.0	3.6
Tryptophan	1.0	0.8	0.5	0.7–1.3
Valine	4.1	3.6	3.0	2.3–3.0

[a] Data from reference (9).
[b] Data from reference (12).

Table 4 presents the essential amino acid content of some protein sources commonly used in experimental and commercial fish diets. Casein contains adequate amounts of all essential amino acids required by fish except arginine. In contrast, gelatin is deficient in all essential amino acids except arginine. Thus, the proper combination of these two ingredients meets the amino acid requirements of fish and is often used in experimental diets. Menhaden fish meal is high in lysine and methionine, the two most limiting amino acids in oilseed meals. Soybean protein probably is the most complete of any plant seed proteins for meeting the essential amino acid needs of channel catfish. However, soybean meal is deficient in methionine plus cystine and threonine for eel and methionine plus cystine for *Tilapia nilotica*. The first limiting amino acid of cottonseed meal and sunflower seed meal is lysine but with peanut meal, the sulfur containing amino acids (methionine + cystine) are the most deficient followed by lysine. Rapeseed meal has an amino acid composition similar to that of soybean meal but is slightly deficient in sulfur-containing amino acids.

Although the essential amino acid content can be used as an index for determining the nutritive value of protein sources, it has been found that the individual and average values of the apparent amino acid availability (AAAA) and true amino acid availability (TAAA) vary within and among the various protein sources (13). For example, the average AAAA value of a menhaden fish meal sample was 69.2% while that of another sample was 82.9%. The apparent digestibility value of lysine in cottonseed meal was 27.1% lower than that in soybean meal. Therefore, if amino acid requirement data are to be used in fish feed formulation, individual amino acid availabilities should be used for more accurate and economical diet formulations. The apparent protein digestibility (ADP) values and averages of AAAA and TAAA values of some protein sources for

TABLE 4

Essential Amino Acid Content of Some Protein Sources Commonly Used in Experimental and Commercial Diets of Fish (12)

	Amino acid content (% protein) in:							
	Casein	Gelatin	Menhaden fish meal	Soybean meal	Peanut meal	Cottonseed meal	Rapeseed meal	Sunflowerseed meal
International feed #	5-01-162	5-14-503	5-02-009	5-04-612	5-03-650	5-01-621	5-03-871	5-04-739
Arginine	4.2	8.0	6.1	7.4	9.5	10.2	5.6	9.6
Histidine	3.1	0.9	2.4	2.5	2.0	2.7	2.7	2.7
Isoleucine	6.8	1.6	4.7	5.0	3.7	3.7	3.7	4.9
Leucine	10.5	3.3	7.3	7.5	5.6	5.7	6.8	8.3
Lysine	8.5	4.1	7.7	6.4	3.7	4.1	5.4	4.2
Methionine	3.3	0.8	2.9	1.4	0.9	1.4	1.9	2.5
(+ cystine)	3.7	1.0	3.8	3.1	2.4	3.3	2.7	4.1
Phenylalanine	5.7	2.0	4.0	4.9	4.2	5.9	3.8	5.1
(+ tyrosine)	11.6	2.6	7.2	8.3	7.4	7.9	6.0	8.1
Threonine	4.7	2.0	4.1	3.9	2.4	3.4	4.2	4.2
Tryptophan	1.3	0.1	1.1	1.4	1.0	1.4	1.2	1.3
Valine	8.0	2.4	5.3	5.1	3.9	4.6	4.8	5.6

channel catfish are presented in Table 5. There is a reasonable agreement between the protein digestibility values and average AAAA values.

Utilization of Plant Proteins in Diets of Warmwater Fish

Many plant protein feedstuffs have been used in domestic animal feeds but relatively few are used in fish feeds because fish require high dietary levels of protein. Commercial aquaculture feeds for growout contain 25% to 45% crude protein. Consequently, only high protein content plant feedstuffs such as oilseed residues are used in fish feed. The extent of plant protein utilization is also influenced by its availability, cost, acceptability by fish, ease of processing and nutritive value.

Soybean Meal: Considerable work has been done to evaluate the nutritive value of soybean meal as a substitute for fish meal. Most feeding studies showed that substitution of fish meal with soybean meal in fish diets resulted in reduced growth and feed efficiency. Lovell et al. (14) reported that channel catfish fed all-plant diets grew significantly less than fish fed diets containing fish meal. Andrews and Page (15) found that growth and feed efficiency of channel catfish were substantially reduced when soybean meal was substituted on an isonitrogenous basis for menhaden fish meal. In another study, soybean meal was partially replaced on an isonitrogenous basis with 0, 5, 10, 20, and 40% menhaden fish meal. Weight, protein and fat gains of channel catfish significantly increased as fish meal in the diet increased up to 20% (19). He speculated that the improved fish growth by adding fish meal could be attributed to increased amounts of limiting essential amino acids, namely the sulfur amino acids and lysine. Earlier reports also indicated that the lower nutritive value of soybean meal for channel catfish as compared to fish meal is because of its relatively low sulfur containing amino acids in soybean (20).

Andrews and Page (15) tried to improve the quality of

TABLE 5

Percentages of Apparent Protein Digestibility (APD) and Averages of Apparent Amino Acid Availability (AAAA) and True Amino Acid Availability (TAAA) Values of Some Protein Sources for Channel Catfish

Protein source	International feed number	APD			Av. value of all amino acids (13)	
		Hastings (16)	Cruz (17)	Wilson & Poe (18)	AAAA	TAAA
Menhaden fish meal	5-02-009	74	87	85 ± 3	82.0	86.1
Meat and bone meal	5-00-388	—	75	61 ± 6	74.3	78.4
Cottonseed meal	5-01-621	76	81	83 ± 1	75.1	78.2
Peanut meal	5-03-650	—	—	74 ± 1	88.4	92.4
Soybean meal	5-04-612	72	84	97 ± 3	80.1	84.2

soybean protein by adding methionine, cystine, or lysine. They found that adding crystalline methionine, cystine or lysine did not improve the growth of channel catfish. Murai et al. (21) reported that supplementing soybean meal diets with either coated or uncoated methionine significantly improved growth and feed efficiency of fingerling common carp and channel catfish. In common carp, addition of methionine coated with aldehyde-treated casein significantly enhanced growth over uncoated methionine, while in channel catfish, coated methionine did not significantly improve utilization of the methionine supplement. The discrepancy between these two studies may reflect differences in the levels of total sulfur amino acids in the basal diets, because the diet of Andrews and Page (15) contained at least 20% more methionine and cystine than those of Murai et al. (21). Andrews and Page (15) attributed the superiority of fish meal relative to soybean meal to the following: (a) unidentified growth factors are not proteins or amino acids but are other nonlipid or nonmineral compounds; (b) differences between soy and fish meal protein in availability of various amino acids; or (c) a delicate amino acid balance in fish meal but imbalance in soybean meal. Wilson and Poe (22) indicated that soybean meal contains additional antinutritional factors other than trypsin inhibitor activity which apparently result in reduced utilization by fingerling catfish.

With tilapia (Oreochromis niloticus X O. aureus), Shiau et al. (23) demonstrated that fish meal can be partially replaced by soybean meal when the dietary protein level is below optimum level for growth (24%). At the optimum level of dietary protein (32%), replacement of 30% fish meal with soybean meal significantly depressed growth and feed efficiency. However, these were restored by addition of methionine to the level of the control diet. Tacon et al. (24) reported that supplementation of 0.8% D,L-methionine to a diet in which 75% of brown fish meal was replaced by soybean meal improved the growth performance of O. niloticus to a level comparable to that of a fish meal diet. Viola and Arieli (25) reported that soybean meal can be used to replace up to half of the fish meal in tilapia feeds having 25% crude protein content without requiring any supplementation. Complete substitution of fish meal by soybean meal resulted in significant reduction of weight gain and feed efficiency which were not overcome by supplementation with oil, lysine, methionine and vitamins. With a 30% crude protein diet, Viola et al. (26) found that isonitrogenous substitution of fish meal with 24% soybean meal reduced the growth of tilapia (O. aureus X O. niloticus). When 2-3% dicalcium phosphate was added, the growth rate of fish was comparable to that of the all-fish meal diet. Jackson et al. (27) also observed a growth reduction of Sarotherodon mossambicus fed a diet in which 50% or more fish meal was replaced by soybean meal. They attributed this to the low level of sulfur amino acids and the presence of other factors such as trypsin inhibitor of haemagglutinins. The growth performance and feed efficiency of common carp (Cyprinus carpio) also were reduced when dietary fish meal was replaced by soybean meal. Viola et al. (28) concluded that soybean meal is deficient in available energy and lysine, as well as methionine, for carp. However, these deficiencies can be remedied by supplementation with proper amounts of oil and amino acids. With these supplements, soybean-based diets are nutritionally equivalent to fish meal-based diets, as measured by growth rate, protein utilization and energy retention. The quantity of oil and amino acids required varies with the level of soybean used. Substituting 40% fish meal with soybean meal in pond diets required only supplementation with methionine and 5% oil. When most or all of the fish meal was replaced by soybean meal, supplementation with methionine, 0.4–0.5% lysine and 10% oil was necessary to achieve gains, protein efficiency ratio, net protein utilization, protein retention and energy retention equal to those of the control fish meal ration. In another study (29), common carp were fed diets in which 75% of the white fish meal was isonitrogenously replaced by methanol-treated or untreated soy flour and supplemented with essential amino acids to the levels of a fish meal-torula yeast control diet. Fish on test diets attained approximately 90% of the weight gain obtained with the control. Feed efficiency and percentage of protein deposition of the test diets however, were significantly lower than those of the control. Kim and Oh (30) attributed the poor performance of common carp fed a diet containing 40% soybean meal to lack of phosphorus rather than sulphur amino acids, because the addition of 2% sodium phosphate (dibasic) to the soybean meal diet improved its performance to a level obtained with the best commercial feed. Viola et al. (31) observed that when 18% soybean meal was substituted for fish meal in a 25% crude protein diet, the total phosphorus content in the feed reduced from 0.8% to 0.7% but the growth rate of carp was not affected.

Roasted Full-Fat Soybean Meal: The use of full-fat soybean meal in fish feed has been limited to an experimental scale. Full-fat soybeans normally are processed by dry heating to a temperature of 205°C or by steam cooking under pressure of 0.7 kg/cm^2 for 10 minutes to destroy the antinutritional factors and improve nutritional value. Saad (32) fed channel catfish in ponds with diets in which heat-treated soybean meal replaced 50 and 100% of the soybean meal in a control diet containing 64% commercial defatted soybean meal and 10% fish meal. The amount of protein gain was essentially the same for the three diets, but much more fat was observed in fish fed the full-fat soybean diets. Abel et al. (33) observed that mirror carp fed diets containing 50% heat-treated soybean meals as replacements of half of the fish meal attained only 60–65% of the weight gain obtained with the fish meal control diet. Fat deposition was higher in fish fed the full-fat soybean diets than in fish fed the control diet. The better performance of the control diet is thought to be due to more available and/or better balance of amino acid in fish meal. In a study with Oreochromus niloticus, addition of 0.5% D,L-methionine to a diet containing 50% puffed full-fat soybean meal as a replacement of 75% brown fish meal provided the same growth and feed conversion as those of the fish meal diet (24). However, the body fat percentage was greater in fish fed the full-fat soybean diet.

The apparent benefits of properly heated full-fat soybean meal over defatted commercial soybean meal are that higher heat treatment destroys more of the antinutritional factors and possibly increases nutrient availability, especially amino acids. Full-fat soybean meal also contains much higher energy than commercial soybean meal. Full-fat soybean meal contains approximately 20% fat on a dry basis as compared to only 1% for defatted soybean meal. Lovell (34) indicated that the additional fat in the full-fat soybean meal is beneficial only if it improves the nutritional quality of the diet. Too much fat can cause an imbalance of dietary protein to energy ratio which can reduce nutrient intake and produce fatty fish. It also can have an adverse effect on pellet processing.

Cottonseed Meal: Cottonseed meal is an important protein source for domestic animals, but its use in commercial aquaculture feeds is very limited because of the presence of gossypol and the low available lysine content. The level of cottonseed meal that can be incorporated in fish diets depends mainly on the gossypol content of the meal used. Dorsa et al. (35) reported that growth and feed conversion of channel catfish were significantly depressed when fed diets containing more than 17.4% solvent-extracted, glanded cottonseed meal of free gossypol content higher than 990 ppm. In another study (36), growth and feed conversion of catfish fed diets which contained 26.5% solvent-extracted, glanded cottonseed meal were not significantly different from fish fed the basal diet, although the growth of fish fed the glanded meal was lower. The discrepancy between these two studies may be caused by different levels of free gossypol content. The glanded meal used by Robinson et al. (36) contained approximately one-tenth the free gossypol found in the glanded meal used by Dorsa et al. (35). The performance of fish fed the glanded meal was significantly depressed when compared to fish fed defatted, glandless cottonseed meal as the sole plant protein source (36). This growth depression may not be related to free gossypol level as the glanded meal diet contained only 120 ppm free gossypol. Levels of less than 900 ppm free gossypol have been reported to have no detrimental effect on growth of catfish (35). The higher nutritive value of the glandless cottonseed protein could be due to increased lysine availability. Wilson et al. (13) reported that gossypol bound with the amino groups of lysine, reducing the amount of available lysine in glanded cottonseed meal.

Robinson et al. (36) concluded that defatted, glandless cottonseed meal is an acceptable substitute for soybean meal and peanut meal in channel catfish diets when supplemented with lysine. The safe level of glanded cottonseed meal used depends on the amount of free gossypol. It was recommended that the levels of the glanded cottonseed meal used in channel catfish feed should not exceed 12–15% of the diet (37).

Low gossypol cottonseed meal (0.03%) was reported to be a good protein source for *S. mossambicus*. The growth of tilapia was improved when 50% of fish meal was replaced by cottonseed meal. The growth rate was essentially the same as the control even at 100% substitution level (27). In contrast, Ofojekwu and Ejike (38) reported a much lower weight gain and feed efficiency of *O. niloticus* fed cottonseed cake diet as compared to the tilapia fed fish meal control diet.

Leucaena leucocephala *Leaf Meal: Leucaena leucocephala* is a fast-growing and drought-resistant tropical leguminous tree which offers probably the widest assortment of uses as compared to other legumes. These include reforestation of eroded areas, provision of cover for plantations like coffee or cacao, source of rich organic fertilizer, fodder for livestock, and use as firewood and timber (39). Leaf meal of *Leucaena leucocephala*, which contains approximately 29% crude protein on a dry matter basis has been used in the tropics as a protein source in ruminant and poultry feeds. However, its use has been limited because of the presence of the toxic, nonprotein amino acid, mimosine.

A few studies have been conducted to evaluate the nutritive value of *Leucaena leucocephala* leaf meal as a protein source in fish feeds but the data obtained are conflicting. Pantastico and Baldia (40,41) reported improved growth responses of *Tilapia mossambica* and *T. nilotica* fed diets containing 100% *Leucaena* leaf meal. However Jackson et al. (27) obtained very poor growth of *S. mossambicus* fed diets in which 25% of the fish meal was replaced by *Leucaena* leaf meal. This growth reduction was attributed to the toxic effects of mimosine. A trend of reduced growth performance and feed efficiency with increased dietary levels of *Leucaena* leaf meal was also reported by Wee and Wang (42) for nile tilapia. The reproductive performance and growth of *O. niloticus* broodstock were also affected by the presence of high levels of *Leucaena* leaf meal in the diets. Santiago et al. (43) recommended that *Leucaena* leaf meal should not be incorporated to more than 40% of the diet of nile tilapia broodstock.

The nutritive value of the *Leucaena* leaf meal can be improved if most of the mimosine is removed or degraded to a relatively less toxic compound. About 90% of mimosine can be extracted by soaking the leaves in freshwater for 36 hours (44). Wee and Wang (42) found that soaked *Leucaena* leaf meal gave a significantly better growth of nile tilapia than sundried or commercial leaf meal. However, they indicated that the nutritive quality of soaked leaf meal appears to be limited by other nutritional factors, such as the lack of certain amino acids.

Other Plant Proteins: Peanut meal (groundnut) and sunflowerseed meals have been used to replace part of the soybean meal in channel catfish feeds. At present their use is limited because of low lysine and methionine contents and inconsistent supply (34,35). Peanut meal is highly palatable and has better binding properties for pelleting than soybean meal (34). Peanut meal can replace 25% of fish meal in the diet of *S. mossambicus* without affecting growth performance. At higher levels, growth rate decreased rapidly, probably because of its low methionine content (27). Sunflowerseed meal has been reported to contain a variety of endogenous antinutritional factors, such as a protease inhibitor, an arginase inhibitor and the polyphenolic tannin, chlorogenic acid (46). It has relatively high crude fiber content which can reduce the pelleting quality of the feed if included at high levels. Despite these drawbacks, sunflowerseed meal has been reported to be a good protein source for *S. mossambicus* even at 69.6% of the diet (27).

Rapeseed, the dominant oilseed crop in Canada, is used as a source of edible oil for humans and protein-rich meal for livestock, poultry and to a limited extent in fish feeds. Rapeseed meal is comparable in protein quality to soybean meal but has higher crude fiber content. Jackson et al. (27) reported a good growth performance of *S. mossambicus* fed diets in which 50% of fish meal was substituted by low glucosinolate rapeseed meal. Decreased growth rate was observed at higher levels of substitution.

Rapeseed meal contains anti-nutritional factors such as tannin, phytic acid, glucosinolates and the enzyme, myrosinase. Glucosinolates can be hydrolyzed by myrosinase or intestinal microorganisms to produce a variety of substances which impair thyroid function or cause histopathological changes in liver and kidney (47). However, great improvements have been made in the processing techniques to inactivate or remove the myrosinase and glucosinolates. Also, plant breeders have developed new varieties of rapeseed which contain low levels of glucosinates and erucic acid. The meal from these new varieties is called canola meal (48).

Discussion

Replacing fish meal either partially or totally with less expensive plant proteins in practical diets of various warm water fish species has had varying degrees of success. It is generally observed that plant proteins have a lower nutritive value than fish meal and high levels of inclusion of plant proteins usually results in reduced growth and feed efficiency. The ability of fish to utilize plant proteins also differs among species. Even within the same species, results obtained by different researchers are sometimes contradictory. This discrepancy could be attributed to a number of factors such as genetic variation and age or size of fish, composition and nutrient content of the diet, source or quality of the test ingredients, feeding management, culture system, experimental conditions, and water management. The reasons for poor utilization of plant proteins as compared to fish meal have not been thoroughly investigated, but several hypotheses have been suggested: (a) presence of antinutritional factors or toxic substances; (b) improper balance of essential nutrients such as amino acids, energy and minerals; (c) presence of high amounts of fiber and carbohydrates; (d) decrease of palatability of the feed; and (e) reduction of pellet quality especially their water stability.

Despite these problems, practically all the plant proteins discussed here are being used to some extent in commercial warm water fish feeds. The amount used depends on the species, availability, cost, acceptability by fish, nutrient content and availability, and presence of toxins or antinutritional factors. Among all plant proteins, soybean meal is by far the most commonly used and often constitutes approximately 30 to 40% of the feed for warm water fish.

References

1. Garling, D.L. Jr., and R.P. Wilson, *J. Nutr. 107*:2031 (1976).
2. Ogini, C., and K. Saito, *Bull. Jpn. Soc. Sci. Fish. 36*:250 (1970).
3. Takeuchi, T., T. Watanabe and C. Ogino, *Ibid. 45*:983 (1979).
4. Nose, T., and S. Arai, *Bull. Fresh W. Fish. Res. Lab. Tokyo, 22*:145 (1972).
5. Lim, C., S. Sukhawongs and F.P. Pascual, *Aquaculture 17*:195 (1979).
6. Winfree, R.A., and R.R. Stickney, *J. Nutr. 111*(6):1001 (1981).
7. Jauncey, K., and B. Ross, *A Guide to Tilapia Feed and Feeding*, Institute of Aquaculture, University of Sterling, Scotland (1982).
8. Kubaryk, J.M., Ph.D. thesis, Auburn University, Auburn, Alabama, 1980.
9. Lovell, R.T., in *Fish Nutrition and Feeding*, edited by R.T. Lovell, Van Nostrand Reinhold Co., New York, NY, (in press).
10. Harding, D.E., O.W. Allen Jr. and R.P. Wilson, *J. Nutr. 107*:2031 (1977).
11. Robinson, E.H., R.P. Wilson and W.E. Poe, *Ibid. 110*:1805 (1980).
12. *Nutrient Requirements of Warm Water Fish*, National Research Council, National Academy of Sciences, Washington, D.C., 1983.
13. Wilson, R.P., E.H. Robinson and W.E. Poe, *J. Nutr. 111*:923 (1981).
14. Lovell, R.T., E.E. Prather, J. Tres-Dick and C. Lim, *Proc. Ann. Conf. S.E. Assoc. Game Fish Comm. 28*:222 (1974).
15. Andrews, J.W., and J.W. Page, *J. Nutr. 104*:1091 (1974).
16. Hastings, W.H., *Progress in Sport Fisheries Research: Feed Formulation, Physical Quality of Pelleted Feed, Digestibility*, U.S. Bureau of Sports Fisheries and Wildlife, Publication No. 39, Washington, D.C., 1966.
17. Cruz, E.M., Ph.D. thesis, Auburn University, Auburn, Alabama, 1975.
18. Wilson, R.P., and W.E. Poe, *Prog. Fish-Cult. 47*(3):154 (1985).
19. Mohsen, A.A., M.S. thesis, Auburn University, Auburn, Alabama, 1988.
20. *Nutrient Requirements of Warm Water Fish*, National Research Council, National Academy of Sciences, Washington, D.C., 1977.
21. Murai, T., H. Ogata and T. Nose, *Bull. Jap. Soc. Sci. Fish. 48*(1):85 (1982).
22. Wilson, R.P., and W.E. Poe, *Aquaculture 46*:19 (1985).
23. Shiau, S.Y., J.L. Chuang and C.L. Sun, *Ibid. 65*:251 (1987).
24. Tacon, A.G.J., K. Jauncey, A. Falaye, M. Pantha, I. MacGowan and E. A. Strafford, *Proc. Int. Symp. on Tilagpia in Aquaculture*, Nazareth, Israel, 1983, pp. 356–365.
25. Viola, S., and Y. Arieli, *Bamidgeh 35*(1):8 (1983).
26. Viola, S., G. Zohar and Y. Arieli, *Ibid. 38*(1):3 (1986).
27. Jackson, A.J., B.S. Capper and A.J. Matty, *Aquaculture 27*:97 (1982).
28. Viola, S., S. Mokady, U. Rappaport and Y. Arieli, *Ibid. 26*:223 (1982).
29. Murai, T., H. Ogata, P. Kosutabak and S. Arai, *Ibid. 56*:197 (1986).
30. Kim, I.B., and J.K. Oh, *Bull. Korean Fish. Soc. 18*(5):491 (1985).
31. Viola, S., G. Zoha and Y. Arieli, *Bamidgeh, 38*(2):44 (1986).
32. Saad, C.R.B., M.S. thesis, Auburn University, Auburn, Alabama, 1979.
33. Abel, H.J., K. Becker, C.H.R. Meske and W. Friedrich, *Aquaculture 42*:97 (1984).
34. Lovell, R.T., in *Fish Nutrition and Feeding*, edited by R.T. Lovell, Van Nostrand Reinhold Co., New York, NY, in press.
35. Dorsa, W.J., H.R. Robinette, E.H. Robinson and W.E. Poe, *Trans. Amer. Fish. Soc. 3*:651 (1982).
36. Robinson, E.H., S.D. Rawless and R.R. Stickney, *Prog. Fish-Cult. 46*:92 (1984).
37. Robinson E.H., and W.H. Daniels, *J. World Aquacul. Soc. 18*(2):101 (1987).
38. Ofojekwu, P.C., and C. Ejike, *Aquaculture 42*:27 (1984).
39. Vogt, G., E.T. Quinito and F.P. Pascual, *Ibid. 59*:209 (1986).

40. Pantastico, J.B., and J.P. Baldia, in *Finfish Nutrition and Fishfeed Technology*, edited by J.E. Halver and K. Tiews, Heenenmann, Berlin, Germany, 1979, Vol. I, pp. 587–593.
41. Pantistico, J.R., and J.P. Baldia, *Fish Res. J. Philipp.* 5(2):63 (1980).
42. Wee, K.L., and S.S. Wang, *Aquaculture 62*:97 (1987).
43. Santiago, C.B., M.B. Aldaba, M.A. Laron and O.S. Reyes, *Ibid. 70*(1–2):53 (1988).
44. Pascual, F.P., and V. Penaflorida, *Aquaculture Department, SEAFDEC Q. Res. Rep. 3*(3):4 (1979).
45. Robinson, E.H., and R.P. Wilson, in *Channel Catfish Culture*, edited by C.S. Tucker, Elsevier, New York, 1985, pp. 323–404.
46. Tacon, A.G.J., J.L. Webster and C.A. Martinez, *Aquaculture 43*:381 (1984).
47. Higgs, D.A., J.R. McBride, J.R. Market, B.S. Dosanjh, M.D. Plotnikoff and W.C. Clarke, *Ibid. 29*:1 (1982).
48. *Canola Meal for Livestock and Poultry*, Canola Council of Canada, Winnipeg, Canada, 1981.

Soybean Meal Utilization by Marine Shrimp

Dean M. Akiyama

American Soybean Association, 541 Orchard Road #11-03, Liat Towers, Singapore 0923

Abstract

Fish meal and other marine animal meals traditionally have been components of marine shrimp feeds. However, plant protein meals which are considerably cheaper than marine animal meals can be a cost effective ingredient in marine shrimp feeds. Of these plant protein meals, soybean meal offers the most potential for use, given its nutritional quality and availability. Essential amino acids, energy, fatty acids, minerals, and other nutrients must be considered when utilizing high levels of soybean meal in shrimp feeds. Soybean meal is not nutritionally better than the marine animal meals; it is more cost effective. The apparent protein digestibility and apparent amino acid digestibility are discussed for various protein feedstuffs commonly used in shrimp feeds. There were no differences in digestibility due to animal or plant feedstuff origin. However, protein quality is important for digestibility by marine shrimp. Several growth studies in which fish meal was replaced by soybean meal are reviewed. Several studies have successfully replaced a considerable amount of fish meal with soybean meal with no decrease in animal performance. There are species differences and size differences in the ability of marine shrimp to nutritionally utilize soybean meal. There is little doubt that a considerable amount of soybean meal can be utilized in marine shrimp feeds.

The use of feeds in aquaculture systems has increased production and profits considerably. However, shrimp feeds are expensive and can amount to over two-thirds of the variable cost of a shrimp culture operation. Shrimp feeds usually contain 30% to 55% protein. To obtain these protein levels in shrimp feeds, protein meals are utilized extensively and comprise 60% to 75% of the feed formula. Shrimp feeds were developed on the concept of "marine animals eat marine animals," therefore marine animal meals (fish meal, shrimp meal, squid meal) are a major and traditional component of shrimp feeds. Limited world supplies and the high price of these marine animal meals have forced aquatic nutritionists to consider alternative sources of protein. Plant protein meals are generally cheaper than animal protein meals. Of these plant protein meals, soybean meal is increasingly being utilized in shrimp feeds due to its nutritional quality, lower cost, and consistent availability.

The basic nutritional principles are similar for all animal species, terrestrial and aquatic. It has long been realized that heating of raw soybeans increases its nutritional value and palatability (1). Raw soybeans contain antinutritional substances which decrease animal growth and performance. The predominant antinutritional substance is trypsin inhibitor (TI) (2,3) which may also cause pancreatic hypertrophy with excessive endogenous protein losses (4,5). Several less-studied antinutritional substances have also been identified (5,6). It has been well documented that heat treatment of raw soybeans improves its utilization. Heat treatment primarily inactivates the heat labile TI (5,7) and denatures the soy proteins making them more digestible (8). However, excessive over-heating may be undesirable because of the Maillard reaction; the formation of unavailable sugar-amine complexes (9) and the destruction of heat sensitive amino acids, i.e., lysine and cysteine (10). Due to the quality of commercial processing of soybeans, soybean meal is the most utilized protein source comprising two-thirds of all protein supplements fed to livestock in the United States (11).

The use of soybean meal in shrimp feeds has been limited partly due to the limited knowledge available on marine shrimp nutrition and the infancy of the shrimp culture industry. Soybean meal has been successfully used to replace fish meal and shrimp meal in *Macrobrachium rosenbergii* diets (12,13). The replacement of 50% of the fish meal and shrimp meal by soybean meal produced higher growth rates and better feed conversion ratios in *Penaeus californiensis* (14). Fenucci et al. (15) replaced 50% of squid meal with a purified soy protein and obtained better growth, survival, and feed conversion ratios in *P. setiferus* and *P. stylirostris*. Sick and Andrews (16) reported soybean meal was a superior protein source to fish meal and shrimp meal for *P. duorarum*. Whereas, Forster and Beard (17) reported a reduction in growth when fish meal was replaced completely by soybean meal in *Palaemon serratus*. High levels of a purified soy protein (58%) has been attributed to a reduction of feed intake in *P. aztecus* (18). Because of the limited knowledge in shrimp nutrition these observed differences may be due to dietary factors (i.e., lipids, minerals) other than soybean meal protein or to species of shrimp.

The purpose of this paper is to review the use of soybean meal in marine shrimp feeds.

Digestibility Studies of Soybean Meal and Other Feedstuffs

Akiyama et al. (19) determined the apparent dry matter digestibility (ADMD), apparent protein digestibility (APD), and apparent amino acid digestibility (AAAD) of various feedstuffs for the marine shrimp, *Penaeus vannamei*. The feedstuffs evaluated were casein, corn starch, gelatin, soy protein, wheat gluten, fish meal, rice bran, shrimp meal, soybean meal, and squid meal which comprised 88% of the experimental diets.

The ADMD of the diets containing the purified feedstuffs high in protein: casein, gelatin, soy protein, and wheat gluten were greater than the ADMD value of the high carbohydrate diet containing corn starch (Table 1). This suggests that proteins are more efficiently digested than carbohydrates. The lower ADMD of soybean meal as compared to squid meal and fish meal is probably due to its lower protein content and higher carbohydrate content.

The APD and AAAD are presented in Tables 1–2. The results of this study indicated that APD was not influenced by animal or plant feedstuff origin. For the experimental diets containing practical feedstuffs, soybean meal had a

TABLE 1

Apparent Dry Matter Digestibility and Apparent Protein Digestibility of Experimental Diets by the Marine Shrimp, *Penaeus vannamei*[a]

Major feedstuff in diet	Apparent dry matter digestibility	Apparent protein digestibility
Purified feedstuffs		
Casein	91.4 ± 0.1[c]	99.1 ± 0.1[c]
Wheat gluten	85.4 ± 0.4[d]	98.0 ± 0.4[c]
Soy protein	84.1 ± 0.8[d]	96.4 ± 0.4[c]
Gelatin	85.2 ± 1.2[d]	97.3 ± 0.5[c]
Corn starch	68.3 ± 1.6[e]	81.1 ± 1.1[b,e]
Practical feedstuffs		
Squid meal	68.9 ± 1.0[e]	79.7 ± 1.7[e,f]
Fish meal	64.3 ± 1.4[f]	80.7 ± 1.7[e]
Shrimp meal	56.8 ± 2.0[g]	74.6 ± 1.6[g]
Soybean meal	55.9 ± 1.4[g]	89.9 ± 0.9[d]
Rice bran	40.0 ± 1.5[h]	76.4 ± 0.8[f,g]

[a] Values are means for three replicates (24 shrimp/replicated) ± standard deviations.
[b] Based solely on protein from squid attractant.
[c,d,e,f,g,h] Means in the same column with different superscripts differ (P < 0.05).

higher APD than fish meal, squid meal, rice bran and shrimp meal. There were no differences between squid meal and rice bran or between rice bran and shrimp meal. For the experimental diets containing purified feedstuffs, there were no differences between casein, wheat gluten, gelatin, and soy protein. This digestibility similarity has been reported previously (20–22). Whereas, Nose (23), Forster and Gabbott (24), and Fenucci et al. (25) working with *Penaeus japonicus*, *Palaemon serratus* and *Pandalus platyceros*, and *Penaeus stylirostris*, respectively, reported that protein from animal origins are better digested than protein from plant origins. These conflicting observations concerning APD are possibly related to species examined, ingredient quality and diet composition. The AAAD values further support the contention that feedstuff origin has no effect on protein digestibility since similar trends, as discussed with APD, were observed in the AAAD (Table 2).

The ADMD of the purified feedstuffs were observed as being digested more efficiently compared to the practical feedstuffs. This difference also was observed in the APD and AAAD. This indicates that proteins are more readily digested in the purified form. This is further demonstrated by the comparison of the purified and practical forms of soybean protein. The soy protein had a higher APD than soybean meal. This higher digestibility also was observed for all amino acids measured. This suggests that protein quality is important for the digestibility of this nutrient by marine shrimp.

Another study compared the apparent digestibility of soybean meal by three species of marine shrimp, *P. japonicus*, *P. monodon*, and *P. vannamei* (Akiyama, unpublished data). The experimental diets evaluated contained 94% soybean meal which ranged in trypsin inhibitor from 1.6 to 3.7 mg/gram of diet.

There were only slight species differences in the apparent digestibility of the experimental diets (Table 3). The apparent dry matter digestibility, apparent protein digestibility, apparent lipid digestibility, and apparent carbohydrate digestibility for all species ranged from 61.2% to 63.6%, 90.1% to 91.6%, 70.4% to 75.5%, and 63.4% to 75.8%, respectively. The data suggest that these three species of marine shrimps are able to digest soy protein very efficiently.

An apparent digestibility difference was observed between *P. monodon* and *P. vannamei*. *P. monodon* digested soy lipids better than *P. vannamei* whereas *P. vannamei* digested soy carbohydrates better than *P. monodon*. This suggests there are individual species preferences for certain nutrients as an energy source.

In summary, soybean meal was palatable to marine shrimp. Soybean meal protein digestibility was higher than the marine animal meals, i.e., fish meal, shrimp meal, and squid meal. There seems to be little species differences in soybean meal digestibility by *P. japonicus*, *P. monodon*, and *P. vannamei*. Of concern is the digestibility of the carbohydrate fraction in soybean meal by shrimp. This is reflected in the low apparent dry matter digestibility of soybean meal.

Basic Nutritional Research on Soybean Meal

In 1984, Texas A&M University initiated a comprehensive research program to evaluate the nutritional response of marine shrimp to soybean meal. In previous studies, the replacement of fish meal and other marine animal meals by soybean meal had limited success. These meals contained protein, energy, fatty acids, and minerals of different concentrations as compared to soybean meal. The early replacement studies were primarily concerned with balancing the diets isonitrogenously. The limited success in replacing these meals with soybean meal was due to the lack or imbalance of the other nutrients and not necessarily protein.

Penaeus Shrimp Feeding Trial at Texas A&M University, U.S.A.

In this feed trial, two series of experimental diets containing 25% and 35% protein were evaluated (26) (Tables 4–5). The use of a lower protein feed was designed to limit the protein levels in order to better detect any nutritional differences. The soybean meal level in the test feeds ranged from 15% to 75% and primarily replaced menhaden fish meal and shrimp head meal. The diets were balanced isonitrogenously and cellulose, corn starch, fish oil, and a mineral premix were added or deleted in appropriate amounts to keep the other nutrients balanced.

Three sizes (0.04 gm, 0.5 gm, 5.0 gm) and six species of marine shrimp (*P. aztecus*, *P. duorarum*, *P. setiferus*, *P. schmitti*, *P. vannamei*, *P. stylirostris*) were tested in a laboratory tank system. The results of these experiments are presented in Figures 1–12. The data indicated that there are species differences and size differences in the ability of marine shrimp to nutritionally utilize soybean meal. *Penaeus vannamei* did not require any marine animal meals (other than attractant, fish solubles) and had excellent performance on diets containing 75% soybean meal, whereas *P. duorarum* had a lower tolerance level to soybean meal. In general, smaller shrimps were more

TABLE 2

Apparent Amino Acid Digestibility of the Experimental Diets by the Marine Shrimp, *Penaeus Vannamei*.*

Major feedstuff in diet	Essential amino acids							
	ARG	LYS	LEU	ILE	THR	VAL	HIS	PHE
Casein	99.2^a	99.5^a	99.5^a	99.4^a	99.1^a	99.4^a	99.3^a	99.4^a
	(0.2)	(0.1)	(0.1)	(0.1)	(0.2)	(0.1)	(0.2)	(0.2)
Wheat gluten	98.1^a	96.7^a	$98.5^{a,b}$	$98.3^{a,b}$	$97.2^{a,b}$	$98.1^{a,b}$	98.1^a	$98.7^{a,b}$
	(0.4)	(0.6)	(0.4)	(0.5)	(0.2)	(0.4)	(0.3)	(0.2)
Soy protein	97.5^a	97.5^a	$96.7^{a,b}$	$96.8^{b,c}$	95.3^b	$96.4^{a,b}$	$96.7^{a,b}$	$96.6^{a,b}$
	(0.7)	(0.6)	(0.4)	(0.8)	(0.7)	(0.7)	(0.5)	(0.6)
Gelatin	98.4^a	96.9^a	96.2^b	95.8^c	94.5^b	96.1^b	93.6^b	96.3^b
	(0.9)	(0.7)	(0.8)	(1.0)	(0.4)	(0.7)	(1.7)	(0.9)
Soybean meal	91.4^b	91.5^b	88.4^c	90.2^d	89.3^c	87.9^c	86.3^c	89.6^c
	(2.9)	(2.2)	(1.9)	(2.0)	(1.5)	(2.4)	(4.8)	(1.8)
Fish meal	81.0^d	$83.1^{c,d}$	80.7^d	80.4^e	$80.6^{d,e}$	79.4^d	79.0^e	79.1^d
	(1.3)	(2.6)	(2.6)	(1.8)	(2.7)	(1.4)	(0.8)	(0.8)
Shrimp meal	81.8^d	85.7^c	82.1^d	81.6^e	83.7^d	$79.0^{d,e}$	75.4^f	75.6^e
	(2.9)	(2.8)	(2.7)	(2.9)	(3.8)	(3.2)	(3.2)	(3.1)
Squid meal	79.4^d	78.6^e	79.4^d	77.2^f	79.7^e	79.3^d	73.6^f	74.1^e
	(0.6)	(1.1)	(1.4)	(1.6)	(1.9)	(2.6)	(0.9)	(0.9)
Rice bran	85.1^c	$81.0^{d,e}$	74.9^e	73.4^g	73.2^f	75.9^e	82.6^d	74.9^e
	(2.2)	(5.2)	(3.5)	(1.3)	(4.1)	(3.6)	(2.5)	(3.8)

Major feedstuff in diet	Nonessential amino acids						
	GLU	ASP	GLY	PRO	SER	TYR	ALA
Casein	99.5^a	98.9^a	98.4^a	99.3^a	99.2^a	99.5^a	97.9^a
	(0.1)	(0.2)	(0.1)	(0.1)	(0.1)	(0.1)	(0.1)
Wheat gluten	$99.2^{a,b}$	96.0^a	$97.3^{a,b}$	99.1^a	$98.0^{a,b}$	98.3^a	94.1^a
	(0.2)	(0.7)	(0.5)	(0.3)	(0.2)	(0.3)	(2.1)
Soy protein	$97.7^{b,c}$	97.2^a	95.8^b	97.2^a	96.4^b	97.1^a	94.1^b
	(0.4)	(0.4)	(0.4)	(0.5)	(0.5)	(0.6)	(0.5)
Gelatin	97.0^c	$95.9^{a,b}$	$98.1^{a,b}$	98.4^a	96.2^b	92.2^b	97.0^a
	(0.4)	(0.1)	(0.4)	(0.3)	(0.4)	(3.4)	(0.5)
Soybean meal	91.9^d	92.2^b	87.0^c	89.1^b	88.5^c	91.1^b	85.9^c
	(1.3)	(1.3)	(1.6)	(1.4)	(2.8)	(1.6)	(1.7)
Fish meal	82.4^e	$80.6^{c,d}$	82.2^d	84.1^c	81.6^d	78.4^c	81.4^d
	(1.8)	(2.2)	(0.9)	(0.7)	(1.6)	(1.1)	(0.9)
Shrimp meal	82.0^e	$78.6^{d,e}$	80.3^d	78.8^d	78.0^e	76.7^c	55.4^g
	(1.5)	(4.2)	(3.7)	(3.4)	(2.9)	(3.1)	(0.8)
Squid meal	82.2^e	83.2^c	80.4^d	78.5^d	77.2^e	73.5^d	77.0^e
	(1.0)	(4.5)	(0.7)	(0.5)	(1.4)	(0.1)	(1.3)
Rice bran	79.5^f	75.5^e	75.9^e	68.7^e	72.7^f	$75.8^{c,d}$	71.0^f
	(2.0)	(2.9)	(2.5)	(6.5)	(3.0)	(1.5)	(2.0)

* Values are averages for three replicates (24 shrimp/replicate) ± standard deviation in ( ).
^{a,b,c,d,e,f,g} Means in the same column with different superscripts differ ($P < 0.05$).

TABLE 3

The Apparent Digestibility of Experimental Diets Containing 94% Soybean Meal by Marine Shrimp[a]

Species	Apparent dry matter digestibility	Apparent protein digestibility	Apparent lipid digestibility	Apparent carbohydrate digestibility
P. japonicus	63.6[b]	90.1[c]	—	75.8[b]
P. monodon	60.1[c]	90.4[b,c]	75.5[b]	63.4[c]
P. vannamei	61.2[b,c]	91.6[b]	70.4[c]	73.5[b]

[a] Values are means for three replicates by four diets.
[b,c] Means in the same column with different superscripts differ ($P < 0.05$).

TABLE 4

Ingredient Composition and Chemical Composition of the Experimental Diets at Texas A&M University[a]

Ingredient	SBM content : 25% Protein				SBM content : 35% Protein			
	15	30	45	53	30	45	60	75
Soybean meal	15.0	30.0	45.0	52.7	30.0	45.0	60.0	74.6
Shrimp meal	15.8	9.6	3.3	—	18.7	12.4	6.1	—
Menhaden fish meal	15.8	9.6	3.3	—	18.7	12.4	6.1	—
Corn starch	38.2	33.2	28.7	25.9	22.2	17.1	12.0	6.9
Cellulose	1.4	1.5	1.5	1.6	0.1	0.4	0.7	1.0
Capelin fish oil	1.7	2.5	3.3	3.8	0.4	1.2	2.0	2.9
Mineral premix	3.6	5.1	6.7	7.5	1.4	3.0	4.6	6.1
Basal mixture	8.5	8.5	8.5	8.5	8.5	8.5	8.5	8.5
Protein	25.0	25.0	25.0	25.0	35.0	35.0	35.0	35.0
Fat	8.0	8.0	8.0	8.0	8.0	8.0	8.0	8.0
Fiber	5.0	5.0	5.0	5.0	5.0	5.0	5.0	5.0
Ash	13.0	13.0	13.0	13.0	13.0	13.0	13.0	13.0
Carbohydrate	39.0	39.0	39.0	39.0	29.0	29.0	29.0	29.0
Moisture	10.0	10.0	10.0	10.0	10.0	10.0	10.0	10.0

[a] Per cent as fed basis.

TABLE 5

Basal Mixture Composition of the Experimental Diets at Texas A&M University[a]

Ingredient	Per cent of total feed
Soy lecithin	1.0
Cholesterol	0.5
Vitamin premix	2.0
Fish solubles	2.0
Sodium alginate	2.0
Sodium hexametaphosphate	1.0
TOTAL	8.5

[a] Per cent as fed basis.

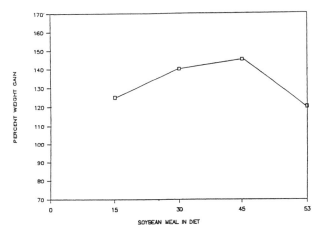

Fig. 1. Per cent weight gain of *Penaeus schmitti* (initial mean weight, 0.47 g) fed 25% protein diet containing 15%, 30%, 45% and 53% soybean meal.

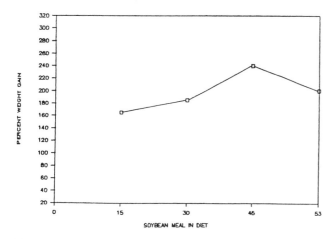

Fig. 2. Per cent weight gain of *Penaeus setiferus* (initial mean weight, 0.50 g) fed 25% protein diets containing 15%, 30%, 45% and 53% soybean meal.

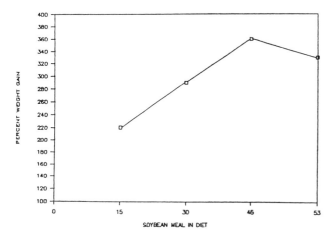

Fig. 3. Per cent weight gain of *Penaeus vannamei* (initial mean weight, 0.58 g) fed 25% protein diets containing 15%, 30%, 45% and 53% soybean meal.

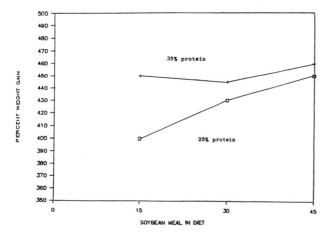

Fig. 4. Per cent weight gain of *Penaeus vannamei* (initial mean weight, 0.40 g) fed 25% and 35% protein diets containing 15%, 30% and 45% soybean meal.

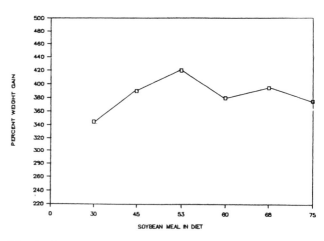

Fig. 5. Per cent weight gain of *Penaeus vannamei* (initial mean weight, 0.54 g) fed 35% protein diets containing 30%, 45%, 53%, 60%, 67.5% and 75% soybean meal.

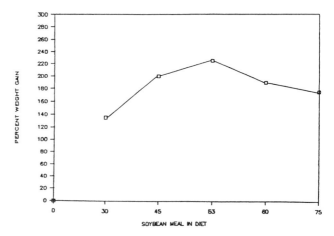

Fig. 6. Per cent weight gain of *Penaeus setiferus* (initial mean weight 0.74 g) fed 35% protein diets containing 30%, 45%, 53%, 60% and 75% soybean meal.

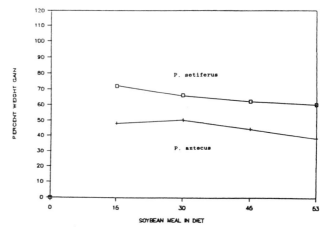

Fig. 7. Per cent weight gain of *Penaeus setiferus* (initial mean weight, 7.4 g) and *Penaeus aztecus* (initial mean weight, 6.4 g) fed 25% protein diets containing 15%, 30%, 45% and 53% soybean meal.

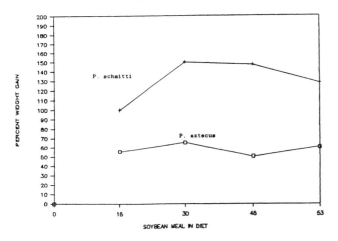

Fig. 8. Per cent weight gain of *Penaeus aztecus* (initial mean weight 7.2 g) and *Penaeus schmitti* (initial mean weight 4.5 g) fed 25% protein diets containing 15%, 30%, 45% and 53% soybean meal.

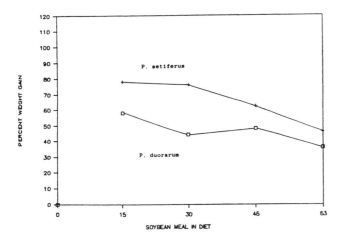

Fig. 9. Per cent weight gain of *Penaeus duorarum* (initial mean weight, 8.0 g) and *Penaeus setiferus* (initial mean weight, 7.4 g) fed 25% protein diets containing 15%, 30%, 45% and 53% soybean meal.

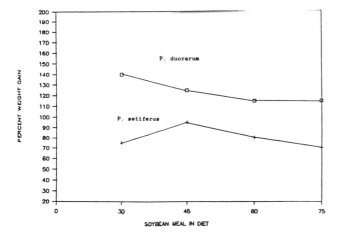

Fig. 10. Per cent weight gain of *Penaeus duorarum* (initial mean weight, 4.4 g) and *Penaeus setiferus* (initial mean weight, 6.5 g) fed 35% protein diets containing 30%, 45%, 60% and 75% soybean meal.

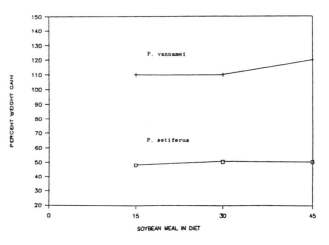

Fig. 11. Per cent weight gain of *Penaeus setiferus* (initial mean weight, 6.6 g) and *Penaeus vannamei* (initial mean weight, 4.9 g) fed 35% protein diets containing 15%, 30% and 45% soybean meal.

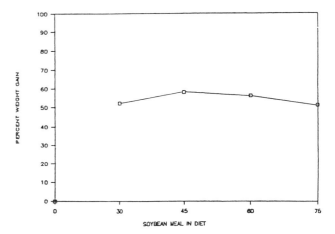

Fig. 12. Per cent weight gain of *Penaeus duorarum* (initial mean weight, 7.2 g) fed 35% protein diets containing 30%, 45%, 60% and 75% soybean meal.

sensitive to the level of soybean meal in the diets as compared to larger shrimp. With the higher protein diets more soybean meal was tolerated by shrimp. This is probably due to the less severe protein stress on the shrimp.

Each species of shrimp and size of shrimp need to be evaluated further to determine soybean meal tolerance levels, but there is little doubt that high levels of soybean meal can be utilized.

Penaeus merquiensis Feeding Trial at National Inland Fisheries Institute, Thailand

This feeding trial examined the replacement of fish meal with soybean meal in post-larval and juvenile *P. merquiensis* (27). The diets were formulated similar to those at Texas A&M in that all diets were constant in protein, fat, fiber, ash, nitrogen free extract, calcium, and phosphorous (Tables 6–7). The soybean meal levels tested were 25%, 35%, 45%, and 55% with concurrent fish meal levels of 23.4%, 15.6%, 7.8%, and 0%, respectively. All diets were laboratory prepared. The

TABLE 6

Ingredient Composition and Chemical Composition of the Feeds at National Inland Fisheries Institute, Thailand[a]

Ingredient	25% SBM	35% SBM	45% SBM	55% SBM
Soybean meal	25.00	35.00	45.00	55.00
Fish meal	23.40	15.60	7.80	—
Starch	9.51	6.34	3.17	—
Cellulose	1.99	1.33	0.67	—
Fish oil	4.08	4.73	5.38	6.04
Soybean oil	0.18	0.12	0.06	—
Crushed shell	—	0.04	0.08	0.12
Di-Ca-P	3.00	4.00	5.00	6.00
Basal mixture	32.84	32.84	32.84	32.84
Protein	39.0	40.0	39.2	38.6
Fat	9.6	10.3	10.9	10.4
Fiber	7.6	8.6	8.7	9.6
Ash	14.1	13.5	13.2	12.8
Nitrogen free extract	21.9	19.6	20.2	20.9
Moisture	7.8	8.0	7.8	8.0
Ca	2.6	2.8	2.3	2.2
P_{Total}	2.0	1.9	1.7	1.8

[a] Per cent as fed basis.

TABLE 7

Basal Mixture of the Feeds at the National Inland Fisheries Institute, Thailand[a]

Ingredient	Per cent of total feed
Shrimp meal	10.00
Squid meal	6.00
Rice bran	3.34
Wheat gluten	10.00
Soy lecithin	0.50
Vitamin/Mineral premix	3.00
TOTAL	32.84

[a] Per cent as fed basis.

study was conducted in 200 l aquaria which were part of a closed recirculating system. The experimental period was 56 days.

Post-larval P. merquiensis had an initial weight ranging from 35 mg to 43 mg and were stocked at 30 shrimp per aquaria (Table 8). There was no significant difference in the final weight, survival rate, and feed conversion ratio between all experimental feeds. The final weight, survival rate, and feed conversion ratio ranged from 900 mg to 1407 mg, 74.2% to 88.9%, and 1.66 to 1.81, respectively. Though there were no statistical differences, there appears to be a decrease in animal performance from the 45% to 55% soybean meal diets.

Juvenile P. merquiensis had an initial weight of approximately 3.0 gm and were stocked at 15 shrimp per aquaria (Table 9). There was no significant difference in the final weight, survival rate, and feed conversion ratio between all experimental feeds. The final weight, survival rate,

TABLE 8

Results of the Post-larval *Penaeus merquiensis* Feeding Trial Conducted in Aquaria at National Inland Fisheries Institute, Thailand[a]

Diets	25% SBM	35% SBM	45% SBM	55% SBM
Initial weight (mg)	43	37	35	36
Experimental time (days)	56	56	56	56
Final weight (mg)	1283[b]	1115[b]	1407[b]	900[b]
Weight gain (%)	2884	2914	3920	2400
Survival rate (%)	88.9[b]	76.7[b]	86.7[b]	74.2[b]
Feed conversion ratio	1.66[b]	1.75[b]	1.70[b]	1.81[b]

[a] Values are means for four replicates.
[b] Means in the same row with different superscripts differ ($P < 0.05$).

TABLE 9

Results of the Juvenile *Penaeus merquiensis* Feeding Trial Conducted in Aquaria at National Inland Fisheries Institute, Thailand[a]

Diets	25% SBM	35% SBM	45% SBM	55% SBM
Initial weight (gm)	3.1	3.0	3.0	3.0
Experimental time (days)	56	56	56	56
Final weight (gm)	5.2[b]	4.7[b]	4.8[b]	5.3[b]
Weight gain (%)	68	56	60	77
Survival rate (%)	81.2[b]	80.5[b]	79.8[b]	79.2[b]
Feed conversion ratio	2.8[b]	3.0[b]	3.1[b]	3.4[b]

[a] Values are means for four replicates.
[b] Means in the same row with different superscripts differ (P < 0.05).

and feed conversion ratio ranged from 4.7 gm to 5.3 gm, 79.2% to 81.2%, and 2.8 to 3.4, respectively. The growth rate of the shrimp was very limited and this may question the reliability of the data.

Applied Nutritional Research on Soybean Meal

The experimental diets produced at Texas A&M University and the National Inland Fisheries Institute were produced with laboratory equipment and contained ingredients which would not be economically feasible on a commercial scale. The following studies were conducted to apply the nutritional principles developed at these research facilities on a commercial basis.

Penaeus vannamei *Feeding Trial at Texas A&M University, U.S.A*

Juvenile *P. vannamei* were stocked in eight 0.1 ha earthen ponds at a density of $13.1/m^2$ (26). The shrimp averaged 0.6 gm at stocking. Two commercially produced shrimp feeds were tested, one containing 10% soybean meal and the other containing 40% soybean meal. The additional soybean meal primarily replaced marine animal meals. The chemical composition of these feeds was similar (Table 10). The production period was 82 days.

TABLE 10

Chemical Composition of the Commercially Produced Feeds for Texas, USA[a]

Ingredient	Feed I	Feed II
Soybean meal	10.0	40.0
Protein	36.0	37.7
Fat	9.3	9.0

[a] Per cent as fed basis.

The results are presented in Table 11. The production time was shortened due to the presence of a red tide organism (*Ptychodiscus brevis*) in the incoming seawater. The low survival rate of shrimp is attributed to this

TABLE 11

Water Quality and Production Data for *Penaeus vannamei* Fed a Commercial Feed Containing Either 10% or 40% Soybean Meal in Earthen Ponds in Texas, USA[a]

Diet	10% Soybean feed	40% Soybean feed
Initial size (gm)	0.6	0.6
Stocking density (#/m²)	13.1	13.1
Production period (days)	82.0	82.0
Harvest size (gm)	12.2	12.6
Survival rate (%)	48.1	49.1

[a] Values are means for four replicates.

organism. However, the harvest size of shrimp and survival rate were similar for both feeds. A commercial growth rate for *P. vannamei* of approximately 1 gm/wk was obtained on both commercially produced feeds.

Penaeus monodon *Feeding Trial at Southeast Asian Fisheries Development Center, Philippines*

This feeding trial compared several feeds in experimental cages ($1 \times 1 \times 1$ m) installed in earthen ponds (28). The feed formulas and chemical compositions are presented in Tables 12–13. The feeds contained soybean meal levels of 15%, 25%, 35%, 45%, and 55% with concurrent fish meal levels of 30%, 23%, 16%, 9%, and 0%, respectively. Shrimp averaging 2.8 gm were stocked at two densities, $10/m^2$ and $20/m^2$. The duration of the feeding trial was 4 months. The shrimp stocked at $10/m^2$ and $20/m^2$ had final weights ranging from 27.5 gm to 28.6 gm and 23.0 gm to 24.2 gm, respectively. There were no differences in growth of shrimp due to the feeds. However, the shrimp stocked at the lower density were larger than the shrimp stocked at the higher density indicating a density effect on growth. The survival rate ranged from 68% to 93% (Table 14).

The production value estimates both the final weight and survival rate. The average production at $10/m^2$ and $20/m^2$ was $221 gm/m^2$ and $405 gm/m^2$, respectively (Fig. 13). These values correspond to 2.21 MT/ha/crop and 4.05 MT/ha/crop, respectively.

TABLE 12

Ingredient Composition and Chemical Composition of the Feeds at Southeast Asian Fisheries Development Center, Philippines[a]

Ingredient	Soybean meal content				
	15	25	35	45	55
Soybean meal	15.0	25.0	35.0	45.0	55.0
Fish meal	30.0	23.0	16.0	9.0	—
Wheat flour	10.8	7.8	4.8	1.8	0.8
Basal mixture	44.2	44.2	44.2	44.2	44.2
Protein	41.7	38.9	40.1	39.4	38.5
Fat	9.3	9.2	8.8	8.6	8.4
Fiber	4.5	5.4	5.9	6.3	7.2
Ash	13.7	12.9	12.1	11.2	9.9
Nitrogen free extract	30.8	33.6	33.1	34.5	36.0

[a] Per cent dry matter basis.

TABLE 13

Basal Mixture Composition of the Southeast Asian Fisheries Development Center, Philippines[a]

Ingredient	Per cent of total feed
Shrimp meal	15.0
Rice bran	15.0
Cod liver oil	5.0
Potato starch	6.0
Vitamin/mineral premix	3.0
BHT	0.2
TOTAL	44.2

[a] Per cent dry matter basis.

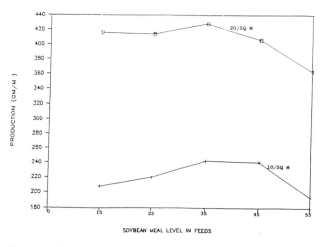

Fig. 13. Production results (final weight × survival) of the *P. monodon* feeding trial conducted in cages at Southeast Asian Development Center, Philippines.

Chemical analysis of the tail muscle of the shrimp fed the various feeds were similar in protein, fat, fiber, ash, and nitrogen free extract at both stocking densities.

Penaeus monodon *Feeding Trial in Indonesia*

This feeding trial compared several feeds in experimental cages installed in earthen ponds and commercially-managed 0.7 ha earthen ponds (29). The feed formulas and chemical composition are presented in Tables 15–16. All feeds were commercially processed. Feeds I and II compared the use of 35% and 45% soybean meal, respectively. Feeds II and III had similar compositions, however, these feeds were processed differently. Feed II was pelleted and Feed III was extruded. Feeds IV and V were commercially marketed shrimp feeds from Indonesia and Taiwan, R.O.C., respectively.

In the cage study, 1 × 1 × 1 m cages were installed

TABLE 14

Results of the *Penaeus monodon* Feeding Trial Conducted in Cages at Southeast Asian Fisheries Development Center, Philippines

Initial stocking density	Soybean meal level (%)	Initial weight (gm)	Final weight (gm)[a]	Suvival (%)[a]
	15	2.8	27.5[b]	75
	25	2.8	27.5[b]	80
10/m²	35	2.8	27.5[b]	88
	45	2.8	27.7[b]	87
	55	2.8	28.6[b]	68
	15	2.7	24.2[c]	86
	25	2.8	24.1[c]	86
20/m²	35	2.9	23.0[c]	93
	45	2.8	23.1[c]	88
	55	2.8	25.3[c]	78

[a] Values are means for four replicates.
[b,c] Means in same column with different superscripts differ (P < 0.05).

TABLE 15

Ingredient Composition and Chemical Composition of the Commercially Processed Feeds in Indonesia[a]

Ingredient	I	II[b]	III[b]	IV[c]	V[d]
Soybean meal	35.0	45.0	45.0	—	—
Fish meal	12.0	6.0	6.0	—	—
Wheat middlings	13.5	9.3	9.3	—	—
Fish oil	1.1	1.8	1.8	—	—
Di-Ca-P	4.0	5.0	5.0	—	—
Oyster shell	0.7	1.4	1.4	—	—
Wheat flour	4.4	2.2	2.2	—	—
Basal mixture	29.3	29.3	29.3	—	—
Protein	41.2	41.8	43.7	42.5	39.6
Fat	3.8	4.0	3.8	5.9	4.8
Fiber	3.0	3.7	3.6	1.8	3.0
Ash	15.8	16.0	15.4	10.5	17.8
Moisture	8.4	8.3	6.0	6.4	10.4
Ca	3.8	3.5	3.5	2.2	—
P_{Total}	2.0	2.2	2.2	1.7	—

[a] Per cent as fed basis.
[b] Feed formulas similar. Diet II was pelleted whereas diet III was extruded.
[c] Indonesia commercial feed. Formula proprietary.
[d] Taiwan, R.O.C. commercial feed. Formula proprietary.

TABLE 16

Basal Mixture Composition of the Commercially Processed Feeds in Indonesia[a]

Ingredient	Per cent of total feed
Shrimp meal	10.0
Squid meal	6.0
Wheat gluten	10.0
Vitamin/Mineral premix	3.3
TOTAL	29.3

[a] Per cent as fed basis.

TABLE 17

Results of the *Penaeus monodon* Feeding Trial Conducted in Cages in Indonesia[a]

Diets	\multicolumn{5}{c}{Diets}				

Diets	I	II	III	IV	V
Initial weight	1.0	1.0	1.0	1.0	1.0
Stocking density (#/m^2)	15.0	15.0	15.0	15.0	15.0
Experimental time (days)	45	45	45	45	45
Final weight (gm)	7.2	5.9	4.8	4.5	5.9
Survival rate (%)	63.3	58.3	45.0	45.0	50.0

[a] Values are means for four replicates.

TABLE 18

Results of the *Penaeus monodon* Feeding Trial Conducted in Earthen Ponds in Indonesia[a]

Times in weeks	Growth of shrimps in grams			
	I	II	III	IV
0	0.15	0.15	0.25	0.25
1	—	—	1.40	1.50
2	2.0	1.8	2.0	2.2
3	2.9	2.9	2.6	3.3
4	3.5	4.0	4.0	4.5
5	4.6	5.0	5.6	6.4
6	6.2	6.3	7.1	8.2
7	10.2	10.2	9.4	9.1
8	13.1	12.9	11.8	10.9
9	14.5	13.2	12.5	11.5
10	15.6	14.7	14.4	12.4
11	16.3	15.4	15.1	13.2
12	17.3	16.1	17.0	14.5
13	18.2	17.0	18.5	14.9
14	20.6	18.6	22.6	15.4
15	23.0	20.3	23.7	16.1
16	26.1	22.1	24.9	16.6
17	28.9	26.3	25.0	17.1
18	29.0	25.0	27.0	16.6
Survival (%)	33.6	32.5	24.9	47.0
Production (gm/m^2)	77.9	65.0	53.7	62.4
Feed conversion ratio	3.1	3.7	4.3	3.3

[a] Values represent one 0.7 ha earthen pond. All ponds were stocked at 8/m^2.

in 0.5 ha ponds. The cages were stocked with 1.0 gm *P. monodon* at a stocking density of 15/m^2 (Table 17). The experimental period was 45 days. The data were not analyzed statistically. However, Feed I had the largest final weight of 7.2 gm, followed by Feeds II, V, III, and IV, respectively. The survival rate was low and ranged from 45% to 63%. This was probably due to inexperienced management techniques.

In the earthen ponds, one 0.7 ha pond was used to test each feed. The ponds were stocked with 0.15 gm and 0.25 gm *P. monodon* at a stocking density of 8/m^2 (Table 18). The production period was 18 weeks (126 days). The growth rates on Feeds I, II, and III were commercially acceptable and similar. Feed IV had a lower growth rate. The survival rate was low and ranged from 24.9% to 47.0%. However, these survival rates were inaccurate due to the significant incidence of poaching of shrimp.

The low survival rate reflects on the production and FCR which ranged from 53.7 gm/m^2 to 77.9 gm/m^2 and 3.1 to 4.3, respectively. Moreover, poor culture conditions such as low pH of the substrate, high organic content of the substrate, and disease seem to have had a greater effect on shrimp performance than the quality of the feed.

The reliability of the data is further questioned for both studies because of the low survival rates.

Penaeus monodon Feeding Trial in Taiwan, R.O.C.

This study was conducted in Taiwan to examine the utilization of soybean meal under intensive culture conditions (Akiyama, unpublished data). The study was conducted over a period of two years (Phase I and II). For Phase I, the ingredient composition and chemical composition of the feeds are presented in Tables 19–20. The feeds compared the replacement of white fish meal with soybean meal. The soybean meal level in the feeds increased from 20% to 50% of the feed with concurrent decreasing levels of white fish meal from 18% to 0%. All test feeds were similar in chemical composition. A commercial Taiwanese feed was used as a control. All feeds were commercially processed by a low temperature extrusion method.

P. monodon of approximately 4 gm were stocked in outdoor 9 m^2 concrete tanks at a stocking density of 30/m^2 (Table 21). The experimental period was 42 days. The final weight of shrimp ranged from 11.8 gm to 14.7 gm for a growth rate of 1.3 gm/wk to 1.8 gm/wk, respectively. There was no statistical difference in per cent weight gain comparing all feeds. There were no statistical differences in survival rate and feed conversion ratio which ranged from 84.8% to 92.8% and 1.1 to 1.5, respectively. Though there was no statistical difference between the test feeds, the growth rate decreased and the FCR increased from the 40% to 50% soybean meal feeds.

In Phase II of this study, the 40% soybean meal feed was further examined. The feed formula and chemical composition of the feeds are presented in Tables 22–23. Feeds A, B, and C had similar formulations, the difference being the extent of processing of the soybean meal. Feed A used commercially available soybean meal similar to Phase I of this study. Feeds B and C used commercial soybean meal which was further processed by extruding the soybean meal at 140°C and 170°C, respectively. Feed D had 86.8% of the feed similar to Feed A with 13.2% proprietary feed formulation information. Feed D was an attempt to incorporate Taiwanese "unknown growth factors" to a 40% soybean meal feed. The same commercial Taiwanese feed was used as a control. All feeds were processed similar to Phase I.

The experimental system and methodology for Phase II was identical to Phase I except that the initial weight of shrimp was approximately 2.3 gm (Table 24). There was no statistical difference in final weight, per cent weight gain, and survival rate which ranged from 7.5 gm to 8.3 gm, 219% to 259%, and 98.3% to 99.3%, respectively. The FCR (1.2) of the Taiwanese commercial feed was similar to the 140°C extruded soybean meal feed

TABLE 20
Basal Mixture Composition of the Commercially Processed Feeds in Taiwan, R.O.C. (Phase I)[a]

Ingredient	Per cent of total feed
Soy lecithin	0.5
Wheat gluten	5.0
Shrimp meal	7.0
Squid meal	7.0
Vitamin/mineral premix	3.0
Yeast	2.0
Sodium phosphate	2.0
Zeolite	2.0
Dehydrated fish solubles	1.0
Squid oil	1.0
TOTAL	30.5

[a] Per cent as fed basis.

TABLE 19
Ingredient Composition and Chemical Composition of the Commercially Processed Feeds in Taiwan, R.O.C. (Phase I)[a]

Ingredient	20% SBM	30% SBM	40% SBM	50% SBM	Control
Soybean	20.0	30.0	40.0	50.0	—
White fish meal	18.0	12.0	6.0	—	—
Wheat bran	12.5	8.3	4.1	—	—
Limestone	—	0.3	0.7	1.0	—
Di-Ca-P	1.0	2.0	3.0	4.0	—
Fish oil	—	0.4	0.7	1.0	—
Wheat flour	16.8	15.7	14.6	13.5	—
Potassium bicarbonate	1.2	0.8	0.4	—	—
Basal mixture	30.5	30.5	30.5	30.5	—
Protein	43.6	42.8	42.2	43.2	40.4
Fat	5.8	5.7	5.6	5.7	8.1
Fiber	3.0	3.1	3.3	2.8	1.2
Ash	15.9	15.3	15.0	15.2	8.0
Ca	2.9	2.7	2.6	2.8	2.4
P_{Total}	2.4	2.0	2.0	2.1	2.0

[a] Ingredient composition: per cent as fed basis. Chemical composition: per cent dry matter basis.

TABLE 21

Results of the *Penaeus monodon* Feeding Trial Conducted in Outdoor Concrete Tanks Results of the *in Taiwan, R.O.C. (Phase I)*[a]

Diets	20% SBM	30% SBM	40% SBM	50% SBM	Control
Initial weight (gm)	3.8	3.8	3.6	4.0	4.0
Stocking density (#/m^2)	30.0	30.0	30.0	30.0	30.0
Experimental time (days)	42	42	42	42	42
Final weight (gm)	12.3	12.3	12.8	11.8	14.7
Weight gain (%)	224[b]	224[b]	256[b]	195[b]	268[b]
Survival rate (%)	92.8[b]	92.5[b]	91.3[b]	85.4[b]	84.8[b]
Final biomass (gm/m^2)	343.5	340.5	351.1	303.2	374.8
Feed conversion ratio	1.2[b]	1.2[b]	1.3[b]	1.5[b]	1.1[b]

[a] Values are means for four replicates.
[b] Means in the same row with different superscripts differ ($P < 0.05$).

TABLE 22

Ingredient Composition of the Commercially Processed Feeds in Taiwan, R.O.C. (Phase II)[a]

Ingredient	A	B	C	D	E[b]
Fish solubles	1.0	1.0	1.0	1.0	—
Lecithin	1.0	1.0	1.0	1.0	—
Wheat gluten	5.0	5.0	5.0	5.0	—
Shrimp meal	7.0	7.0	7.0	7.0	—
Squid meal	7.0	7.0	7.0	7.0	—
Yeast	2.0	2.0	2.0	2.0	—
Sodium phosphate	2.0	2.0	2.0	—	—
Wheat bran	4.1	4.1	4.1	—	—
Di-Ca-P	3.0	3.0	3.0	—	—
White fish meal	6.0	6.0	6.0	6.0	—
Wheat flour	14.6	14.6	14.6	14.6	—
Potassium bicarbonate	0.4	0.4	0.4	—	—
Limestone	0.7	0.7	0.7	—	—
Fish oil	0.7	0.7	0.7	0.7	—
Squid oil	1.0	1.0	1.0	1.0	—
Zeolite	1.5	1.5	1.5	1.5	—
Soybean meal	40.0	40.0[c]	40.0[d]	40.0	—
Vitamin/Mineral premix	3.0	3.0	3.0	—	—
Proprietary information	—	—	—	13.2	100.0

[a] Per cent as fed basis.
[b] Commercial Taiwanese feed.
[c] Commercial soybean meal extruded at 140°C.
[d] Commercial soybean meal extruded at 170°C.

(FCR = 1.4) and better than the other feeds (FCR = 1.5). There may be some advantages in shrimp performance by extruding soybean meal at 140°C.

Discussion

Soybean meal appears to be highly digestible by marine shrimp especially the protein and amino acids. The origin of protein, plant or animal, does not affect digestibility but the quality of protein is important. There appears to be little difference in soybean meal digestibility by three species of marine shrimp, *P. vannamei*, *P. monodon*, and *P. japonicus*. The carbohydrate fraction of soybean meal appears to lower the total dry matter digestibility.

Several studies successfully have substituted fish meal and shrimp meal with soybean meal. When replacing these marine animal meals with soybean meal, one needs to consider not only protein but energy, fatty acids, minerals, and other nutrients which are present in the marine animal meals and not in soybean meal. There are species differences and size/age differences in the ability of marine shrimp to nutritionally utilize soybean meal. Further processing of soybean meal may increase its nutritional value to marine shrimp.

TABLE 23

Chemical Composition of the Commercially Processed Feeds in Taiwan, R.O.C. (Phase II)[a]

Diets	A	B	C	D	E
Protein	45.5	44.1	43.7	46.4	44.7
Fat	4.0	4.7	4.1	4.1	7.7
Fiber	3.9	3.7	3.5	4.1	2.1
Ash	15.3	15.1	14.7	13.1	14.8
Ca	2.4	2.4	2.2	1.6	2.5
P_{Total}	2.0	2.2	1.7	1.4	2.0

[a] Per cent dry matter basis.

There is little doubt that soybean meal can replace a considerable amount of the marine animal meals while maintaining similar shrimp production performance, given that all the required nutrients are available. The expanded use of soybean meal in commercial shrimp feeds would lower the cost of these feeds and make shrimp farming more cost effective. Soybean meal is not nutritionally better than these marine animal meals, but it is more cost effective.

References

1. Osborne, T.B., and L.B. Mendel, *J. Biol. Chem. 32*:369 (1917).
2. Kunits, M., *Science 101:*668 (1945).
3. Birk, Y., A. Gertler and S. Khalef, *Biochem. J. 87*:281 (1963).
4. Booth, A.N., D.J. Robbins, W.E. Ribelin and F. DeEds, *Proc. Soc. Exp. Biol. Med. 104*:681 (1960).
5. Rackis, J.J., *Biologically active components*, in *Soybeans: Chemical and Technology*, A.K. Smith and S.J. Circle (editors), Avi Publishing Co., Westport, CT (1972).
6. Liener, I. E., *Effects of antinutritional and toxic factors on the quality and utilization of legume proteins*. in *Protein Nutritional Quality of Foods and Feeds*, M. Friedman, (editor), Marcel Deckker, New York (1975).
7. Liener, I.E., *J. Am. Oil Chem. Soc. 58*:406 (1981).
8. Kakade, M.L., D.E. Hoffa and I.E. Liener, *J. Nutr. 103*:1772 (1973).
9. Maynard, L.A., J.K. Loosli, H. F. Hintz and R.G. Warner, *The Proteins and Their Metabolism*. in *Animal Nutrition*. McGraw Hill, New York (1979).
10. Smith, A.K., and S.J. Circle, *Soybeans: Chemistry and Technology*. Avi Publishing Co. Westport, CT (1972) pp 470.
11. Hansen, B.C., *Effects of different heat treatments during processing of the soybean meal in a commercial solvent-extraction plant on the performance of pigs weaned at four weeks of age and growing swine*. M.S. Thesis. (1984) Texas A&M University, College Station, TX.
12. Balazs, G.H., E. Ross, and C.C. Brooks, *Aquaculture 2*:369 (1973).
13. Balazs, G.H., and E. Ross, *Ibid 7*:299 (1976).
14. Colvin, L.B., and C.W. Brand, *J. World Maricult. Soc. 8*:821 (1977).
15. Fenucci, J.L., Z.P. Zein-Eldin, and A.L. Lawrence, *Ibid. 11*:403 (1980).
16. Sick, L.V., and J.W. Andrews, *Ibid. 4*:263 (1973).
17. Forster, J. R.M., and T.W. Beard, *Growth experiments with the prawn Palaemon serratus Pennant fed fresh and compounded foods*. Fisheries Investigations, Series II, Vol. 27, No. 7, Ministry of Agriculture, Fisheries and Food, London (1973).
18. Fenucci, J.L., and Z.P. Zein-Eldin, *Evaluation of squid mantle meal as a protein source in penaeid nutrition*. FAO Technical Conference on Aquaculture, Kyoto, Japan (1976).
19. Akiyama, D.M., S.R. Coelho, A.L. Lawrence, and E.H. Robinson, *Bull. Japan. Soc. Sci. Fish.* (in press) (1988).
20. Lee, D.L., *China Fish. Monthly 208*:2 (1970).
21. Condrey, R.E., J.G. Gosselink, and H.J. Bennett, *Fish. Bull. 70*(4):1281 (1972).
22. Ting, Y.Y., *Bull. Taiwan. Fish. Res. Inst.* No. 16:125 (1970).
23. Nose, T., *Bull. Freshwater Fish. Res. Lab.* (Tokyo) *14*:23 (1964).
24. Forster, J. R.M., and P.A. Gabbott, *J. Mar. Biol. Ass., U.K. 51*:943 (1971).

TABLE 24

Results of the *Penaeus monodon* Feeding Trial Conducted in Outdoor Concrete Taks in Taiwan, R.O.C. (Phase II)[a]

Diets	A	B	C	D	E
Initial weight (gm)	2.6	2.3	2.3	2.3	2.2
Stocking density (#/m^2)	30.0	30.0	30.0	30.0	30.0
Experimental time (days)	42	42	42	42	42
Final weight (gm)	8.3^b	7.7^b	7.7^b	7.5^b	7.9^b
Weight gain (%)	219^b	235^b	235^b	226^b	259^b
Survival rate (%)	99.1^b	99.3^b	98.4^b	98.3^b	98.9^b
Final biomass (gm/m^2)	246	229	226	221	234
Feed conversion ratio	1.5^c	1.4^{b,c}	1.5^c	1.5^c	1.2^b

[a] Values are means from three replicates.
[b,c] Means in same row with different superscripts differ ($P < 0.05$).

25. Fenucci, J.L., A.C. Fenucci, A.L. Lawrence, and Z.P. Zein-Eldin, *J. World Maricult. Soc. 13*:134 (1982).
26. Lawrence, A.L., F.L. Castille Jr., L.N. Sturmer, and D.M. Akiyama, *Nutritional response of marine shrimp to different levels of soybean meal in feeds*, USA-ROC and ROC-USA Economic Councils' Tenth Anniversary Joint Business Conference, Taipei, Taiwan, R.O.C., December, 1986.
27. Boonyaratpalin, M., W. Phromkunthong, and K. Supamataya, *J. Sci. Tech. 10*(1) 45 (1988).
28. Pascual, F., E.M. Cruz, and A. Sumalangcay Jr., *Proc. First Asian Fish. Forum*, Manila, Philippines, 26–31 May, 1986 (1986).
29. Akiyama, D.M., J. Ditchon, and J.M. Fox, *The use of soybean meal to replace fish meal in commercial* Penaeus monodon *feeds in Indonesia*, World Aquaculture Society Conference, Honolulu, Hawaii, Jan. 1988.

Digestibility of Dry Legume Starch and Protein

Maurice R. Bennink and Naruemon Srisuma

Food Science and Human Nutrition, Michigan State University, East Lansing, Michigan 48824-1224

Abstract

The digestion of starch and protein in dry legumes is lower than in cereal grains and many other foodstuffs. Factors contributing to decreased digestibility include: increased dietary fiber intake; intact cell walls which hinder the digestive enzymes from gaining access to the starch and protein; residual lectin activity; limited proteolysis of certain protein subfractions; incomplete hydration of the starch granule; and amylose retrogradation. Incomplete protein digestion represents a loss of essential nutrients and research efforts should be devoted toward improving legume protein digestibility. Slow digestion of starch is beneficial as it reduces hyperglycemia and hypertriglyceridemia which are risk factors in the development of cardiovascular disease. On the other hand, it is likely that slow digestion of starch and retrograded amylose are major contributors to the flatulence problem frequently associated with legume consumption. The challenge to the food processor and plant geneticist is to produce legumes that continue to provide a low glycemic index and a good source of fiber with improved protein digestibility and decreased flatulence. However, producing a low glycemic, fiber rich food with minimal flatulence may be mutually exclusive.

Dry legumes are dietary staples in many parts of the world and are important dietary sources of protein, complex carbohydrates (starch and dietary fiber), other essential nutrients, and energy. However, legume protein is generally less digestible (1,2,3,4) than protein in grains and animal products (Table 1) and legume starch probably is less digestible than other starches. The relatively poor digestion of legume protein limits the biological value of the protein and reduces the overall contribution of legumes toward providing a nutritionally well-balanced diet. Although it is clear that starch, the main component of dry legumes, is digested more slowly than starch from grains and root crops, the true digestibility of legume starch is not known. This article will examine some potential benefits and unwanted consequences resulting from slower and poorer digestion of legumes. In addition, possible causes of reduced digestibility will be explored. There are several inherent characteristics of legumes which reduce or slow digestion of both protein and starch. These factors will be discussed first and then factors which specifically limit protein or starch digestibility will be addressed. Even though most of the discussion will be in reference to the common bean, *Phaseolus vulgaris*, the data will in general be applicable for all nonoilseed, dry legumes.

Factors affecting digestion of protein and starch: One factor affecting digestion of both protein and starch is the amount of dietary fiber in the diet. As a person increases dietary fiber intake, nutrient digestibility tends to be decreased (5). This effect is shown in Table 2 (6). When dietary fiber was increased from 16.4 g/day to 37.4 g/day, protein digestion decreased 6%, fat digestion decreased 2% and total energy digestibility decreased by 1.8%. Decreased digestibility of protein and starch presumably is due to

TABLE 1

Average Protein Digestibility of Various Legumes, Grains and Animal Products[a]

Legume	Digestibility	Legume	Digestibility
Bean (*Phaseolus*)	73	Lima bean	78
Broad bean	87	Pea	88
Chick pea	86	Pigeon pea	78
Cowpea	79	Peanut	87
Lentil	85	Soybean	90
Animal products		Grains	
milk	97	wheat	91
beef	99	rice	98
veal	99		

[a] FAO/UN, 1970(1).

TABLE 2

Effect of High (37.4g/day) or Low (16.4g/day) Fiber Diet on Protein, Fat and Energy Digestibility[a]

	% Digestible	
	High fiber	Low fiber
Protein	84 [c]	90
Fat	96 [c]	98
Energy[b]	93.5	95.3

[a] Adapted from Miles et al. (6).
[b] Calculated to exclude energy from fiber.
[c] Significantly different from low fiber group (P < 0.001).

physical entrapment and/or premature release of food particles from the stomach, although direct inhibition of digestive enzymes by fiber (7) cannot be ruled out. Dietary fiber can encase nutrients in a fibrous matrix which impedes digestive enzymes from gaining access to protein and starch, and fiber can reduce absorption of digested protein and starch by slowing or preventing diffusion of the digestive products to the mucosal cells (8,9,10,11,12). Food is normally held in the stomach until it is reduced to particles less than 1 mm (13,14); but fiber can cause the premature release of large particles from the stomach (15). Large particles represent another type of physical entrapment which limits digestion (16,17). The many possible effects of fiber on nutrient bioavailability have been thoroughly reviewed (18,19,20). We find that *Phaseolus vulgaris* beans typically contain 17.5–20% dietary fiber and Table 3 shows that legumes are an excellent source of dietary fiber (21,22,23). Because legumes

TABLE 3
Dietary Fiber Content of Various Foods

Food	Dietary fiber %, dry basis	Reference
Wheat bran	42.3	21
Whole wheat flour	12.9	21
White wheat flour	3.1	21
Oats	12.5	21
Rye bread	5.9	21
Rice	3.7	21
Soya flour	19.5	22
Garbanzo beans, canned	10.2	23
Green beans, canned	34.0	23
Kidney beans, canned	20.9	23
Lima beans, canned	14.4	23
Lentils, dried, cooked	15.7	23
Black-eyed peas, canned	11.1	23
Green peas, canned	21.3	23
Phaseolus vulgaris		
var. seafarer	20.0	*
var. fleetwood	19.5	*
var. C-20	18.7	*
var. 84004	17.5	*

* Srisuma and Bennink, 1988, unpublished data.

provide dietary fiber, we should recognize that protein and starch digestibilities are going to be decreased slightly as a consequence of that fiber.

Another factor that affects digestion of legume protein and starch is the physical form of the protein and starch when it enters the stomach and small intestine. Grains and root vegetables are usually ground and/or cooked prior to consumption. Grinding and cooking serve to break up the cell structures and to release the protein and starch so that proteolysis and amylolysis can occur. Legumes are generally consumed as whole seeds that have been cooked. Figure 1 shows that the cell wall structures in cooked beans are intact and surround the protein bodies and starch granules. Chewing macerates some of the cells, but most of the cells remain whole and enter the digestive tract intact. Thus, the cell structures render legume protein and starch less available to the digestive enzymes (24, 25) except for the small amounts of starch that leach from the cells (Fig. 1).

A potential factor that could limit protein and starch digestion is the presence of naturally occurring protease and amylase inhibitors. During processing and cooking, most of these inhibitors are inactivated; however, residual levels of active inhibitors may remain, particularly in improperly cooked beans. If one is accustomed to consuming beans with residual levels of inhibitors, the pancreas enlarges and secretes more digestive enzymes to compensate for the enzymes which have been inactivated by the inhibitors (26,27). As a result of pancreatic hyperplasia and hypertrophy, adequate amounts of digestive enzymes are secreted and digestion is not limited. The best evidence to support this conclusion is based on the experience with exogenous amylase inhibitors (often referred to as "starch blockers"). Several groups demonstrated that chronic consumption of amylase inhibitors was ineffective in reducing starch digestion and caloric utilization (28,29,30,31,32). Essentially the same results are found with continued consumption of protease inhibitors (27).

Legumes contain lectins which potentially can reduce nutrient digestion and absorption. Cooking beans will inactivate much of the lectin activity in raw beans; however, fully cooked beans can still contain a significant amount of the original lectin activity (33,34). The deleterious effects of raw beans or comparable levels of isolated lectins are well documented (reviewed by Sgarbieri and Whitaker [35] and Liener [36]). But, it is difficult to measure the effect of residual lectins on digestion and absorption. Data provided by Donatucci et al. (37) strongly suggest that small quantities of lectins are sufficient to reduce absorption of glucose and presumably amino acids and small peptides. We conclude that there is sufficient residual lectin activity in cooked beans to cause at least a small decrease in protein and starch digestibility. Data to substantiate or refute this conclusion are not yet available.

Legume protein digestion: There have been two prevalent concepts to explain poor digestion of legume protein: (a) legumes contain protease inhibitors and lectins which reduce legume digestion (38,39,40) and (b) the inherent structure of legume protein is such that it resists proteolysis (41,42,43,44). Point (a) was addressed previously and it appears that protease inhibitors in cooked beans are not a major concern in legume research today. Much of the data to support contention (b) resulted from experiments conducted with raw or native bean protein (43–45). Because beans are not eaten raw, the applicability of that research is questionable (46,47). Deshpande and Nielsen (46) provide evidence that heated legume protein is not resistant to trypsin proteolysis *in vitro*; but, their research does not preclude the possibility that a portion of the trypsin digestion products are resistant to other proteolytic enzymes which are necessary to hydrolyze the protein to absorbable end products (free amino acids, di- and tripeptides). Table 4 shows the digestibility of protein fractions isolated from *Phaseolus vulgaris* and heated before testing them for digestibility. It should be noted that phaseolin, globulin GI and globulin are all the same protein fraction which we will refer to as GI. Both *in vivo* and *in vitro* digestion of GI showed that GI is very

Fig. 1. Scanning electron micrograph of a cross section from a canned *Phaseolus* bean. S denotes starch granule leached from a cell.

TABLE 4

Digestibility of Heated Protein Fractions from *Phaseolus vulgaris* Beans

	% Digestibility		
Fraction	Bean	Control[a]	Reference
In vivo			
Phaseolin	92	95	48
Globulin	95	98	84
In vitro			
Phaseolin	91	100	50
Trypsin inhibitor	38	100	50
GI	100	100	49
GII	60	100	49
Glutelin	40	100	49
Albumin	53	100	49

[a] Control = Casein (48, 49).
Control = Ovalalbumin (50, 84).

TABLE 5

Predicted Protein Digestibility for *Phaseolus vulgaris*

Protein fraction	% of total protein	Digestion coefficient	% digestion
Albumin and globulin GI	60.7	.98	57.0
Globulin GII	10.0	.60	6.0
Glutelins	18.0	.40	7.2
Protease inhibitors	0.3	.38	0.1
Nonextractable proteins	11.0	.50	5.5
		Total	75.8

digestible. In fact, it is as digestible as other plant proteins and nearly as or equally as digestible as animal proteins. Because globulin GI is the major protein fraction of *Phaseolus* beans (50% to 75% of the total bean protein (35,48), some other fraction of the bean protein must be much less digestible to produce a true digestibility of 65–75% (Table 1, (1–4) and our unpublished data). Marquez and Lajolo (49) found that the globulin GII fraction (also called phytohemagglutinin or lectin) was 60% digestible, whereas the glutelin fraction was only 40% digestible. Trypsin inhibitor is also poorly digested (38% digestible [50]). Heated albumin was only 53% digestible, but unheated albumin was 100% digestible. The authors noted that the heated albumin formed large aggregates which probably were inaccessible to the proteolytic enzymes (49). If the albumins were in the bean rather than isolated, it is likely that its digestibility would be approximately 100%. A protein fraction that does not appear in Table 4 is the nonextractable proteins which probably is located in the cell walls and hulls. The nonextractable proteins comprise about 10–12% of the total protein. We are not aware of any reports concerning the digestibility of the nonextractable proteins in *Phaseolus* beans; but, based on the digestibility of proteins in brans and hulls of other plant products, one would expect relatively poor digestion of the nonextractable proteins. If you take into consideration the proportion of total protein associated with each protein fraction and the digestibility coefficient for each fraction, the digestibility coefficient for the total bean protein is approximately 76% (Table 5) which is similar to the higher values reported for the true digestibility for *Phaseolus vulgaris*. Additional factors such as tannins (51,52,53), dietary fiber, residual lectin activity, and protein encasement within cell walls (discussed above) will decrease total protein digestibility to what is report for *Phaseolus vulgaris*. The information presented in Table 5 is extrapolated from data from several research groups and should be regarded as "predicted values" until confirmed or refuted by subsequent research. Marquez and Lajolo (49) found good agreement for digestion of *Phaseolus* bean protein by a rat assay and *in vitro* digestion calculated as for Table 5.

There are several hypotheses to explain decreased protein digestibility for the various fractions. Steric hindrance of proteolysis by the carbohydrate moieties of the glycoproteins (globulin GII and protease inhibitor fractions) is one possibility for these two fractions (41,43,44,54). Some (42,43) have hypothesized that poor legume protein digestibility is due to the compact, dense nature of bean protein which prevents proteolytic enzymes from reaching the internal catalytic sites. Dissociation of protein fractions by urea and/or sulfhydryl agents resulted in a significant improvement in protein digestibility in soy protein (55) and sorghum protein (56). Similar research should be done with the globulin GII, protease inhibitors, and glutelin fractions of *Phaseolus* beans. The presence of carbohydrate-protein bonds and protein-protein bonds have been proposed to reduce protein digestibility in legumes. Carbohydrate-protein and protein-protein bonds are not susceptible to mammalian proteases and reduce overall protein digestibilities. Formation of carbohydrate-protein bonds is thought to occur via oxidative coupling of two activated benzene rings, one from cell wall phenolic compounds and the other from tyrosine (57). Similarly, protein crosslinking is thought to occur via peroxidase-catalyzed coupling of two tyrosine residues. If the nonextractable protein is in cell walls and hulls and if it is poorly digested, at least part of the poor digestibility would likely be due to crosslinking of protein with carbohydrate or other protein.

Digestion of legume starch: We know that legume starch is digested more slowly than starch from most other sources. The rate of *in vivo* starch digestion is illustrated by the rise in blood glucose and insulin following the consumption of starch. The "glycemic index" has been developed to aid dietary treatment of diabetics (58) and is defined as:

$$\frac{\text{area under the glucose response curve for a food}}{\text{area under the glucose response curve for the equivalent amount of glucose}} \times 100$$

Table 6 shows the average glycemic index for several groups of food (58). Canned beans have a low glycemic index compared to other food groups. This means that bean starch is digested more slowly than starch from most other foods. The importance of the glycemic index is that elevated blood glucose promotes triglyceride and cholesterol synthesis by the liver which results in more very low density lipoproteins (triglycerides and cholesterol) being secreted into the blood (59,60,61). Also, elevated blood glucose promotes glycosylation of proteins in the vascular system (62). The net effect of elevated blood lipids

TABLE 6

Average Glycemic Indices by Food Group[a]

Food group	Average glycemic index	Range
Grains and grain products	61	42–80
Root vegetables	72	48–97
Fruit	50	39–64
Canned legumes	31	14–47

[a] From (58).

and unwanted protein glycosylation is an increased risk of cardiovascular disease. Thus, consumption of legume starch reduces the risk of premature cardiovascular disease compared to consuming starch with a high glycemic index.

Although slow starch digestion (a low glycemic index) is desirable, it is undesirable for the starch to be digested too slowly because there will be incomplete digestion in the small intestine. Undigested starch will pass to the colon where rapid fermentation will take place and flatulence will result. Technical difficulties prevent precise determination of starch digestibility. Typical balance trials cannot be used to determine true starch digestion as starch digestion and absorption in the small intestine cannot be differentiated from starch fermentation by microflora in the large intestine. Several approaches have been used to estimate true starch digestion: (a) presence of starch in the cecum (63), (b) monitoring exhaled hydrogen and methane (64,65) which reflects fermentation by microflora, (c) aspiration of intestinal fluid at the ileocecal sphincter and quantitation of starch in the intestinal fluid (66), (d) quantitation of starch in ileostomy effluent (67) and (e) estimates based on *in vitro* digestion (reviewed in (68) and (69)). All of the above approaches have limitations and do not allow true estimates of starch bioavailability. Two methods (70,71) have been used to estimate bioavailability of carbohydrates. However, these procedures were used to estimate nitrogen-free extract (total carbohydrate by difference) digestion rather than starch bioavailability. These approaches could be used to estimate bean starch digestibility if the results are corrected for fermentation of dietary fiber. Estimation of true digestibility for bean starch warrants further study.

Two major factors affect the rate and perhaps the extent of starch digestion. The first factor is gastric emptying time and the second factor is accessibility of the glycosidic bonds in starch to pancreatic amylase and other digestive enzymes. If the rate of starch digestion is constant, starch which is released from the stomach slowly and over an extended time period will be digested and absorbed more slowly than when gastric emptying is rapid and over a short time period. There is a strong negative correlation between postprandial changes in blood glucose and gastric emptying (72). Likewise, if gastric emptying is constant, starch digestion rate determines how high the blood glucose rises in response to starch ingestion. The digestion rate of starch strongly influences the glycemic response to different foods (61,73,74). The glycemic index obviously reflects both gastric emptying and starch digestion rate.

Legume starch digestion rate is affected by parameters which slow or prevent digestive enzymes from gaining access to the glycosidic bonds of starch. As discussed above, much of the legume starch reaches the stomach and small intestine within intact cell walls. Thus, the fiber matrix of cell walls is the first parameter to hinder starch digestion because amylase must penetrate the cell wall before amylolysis can proceed. Another parameter affecting digestion rate is the hydration state of the starch; starch must be hydrated to be digested. Raw potato starch is virtually indigestible (68) because it exists in a highly crystalline, β-pattern with a compact structure that does not allow amylase to penetrate the granule. Cooking the potato destroys the crystalline structure, hydrates the starch and makes the starch digestible.

Most starch granules contain a mixture of amylose and amylopectin. In the nonhydrated state, the chains of amylose and amylopectin exist in tightly packed α-helix or a double α-helix type of configuration. The helix is held together by hydrogen bonds between glucose moieties of the same chain or between glucose moieties of adjacent chains in the case of double helixes. The tightly packed, helical structure prevents amylase from reaching the catalytic sites on the glucose polymer. Starch will imbibe water if sufficient energy is present to break the hydrogen bonds which hold the helixes together. Heating water provides the energy necessary to break the hydrogen bonds. As the bonds within and between the starch chains are broken, new hydrogen bonds are formed between water and the hydroxyl groups on the polysaccharide. Breakage of the hydrogen bonds within and between the starch chains allows the helixes to relax and the granule to swell. With continued swelling and hydration, the granule eventually loses its crystallinity. When the amylose and amylopectin chains are hydrated, amylase can reach the catalytic sites and digestion proceeds.

Many types of starch, including legume starch, must be gelatinized to hydrate the starch and the extent of gelatinization often determines the rate of starch digestion (75,76). When beans are cooked, the starch is not fully hydrated (25) and there are data which suggest that some of the starch still retains its bifringence (77,78,79). It appears that more energy is required to break the bonds between starch chains in legume starch than in most other types of starch (80,81). Figures 2a-c and 3a-d illustrate the differences in energy required to solubilize 10% starch suspensions from potato, tapioca, corn, and bean. Figure 2a shows the starches 5 minutes after vortexing them in 25°C water. When the starch suspensions were heated for 1 hour at 70°C with gentle shaking, the potato starch was solubilized almost completely (Fig. 2b). Much of the tapioca starch was also solubilized during this heat treatment; incompletely dissolved starch formed a translucent layer at the bottom of the tube. The white opaque layers at the bottom of the tubes with corn starch and bean starch show that these starches require more energy for complete hydration and solubilization. Scanning electron micrographs (SEM) of the particulate material from the bottom of the tubes in Fig. 2b are shown in Fig. 3a-d. The SEM for the potato (Fig. 3a) and tapioca (Fig. 3b) starches show that the amylose has been solubilized and the granule structures disintegrated as the amylopectin became solubilized. The SEM of the particulate matter at the bottom of the test tube containing corn starch shows that the starch granules have collapsed as the amylose became soluble and leached out of the granule (Fig. 3c). The remaining amylopectin appears to be melting and in the process of being solubilized. The bean starch granules

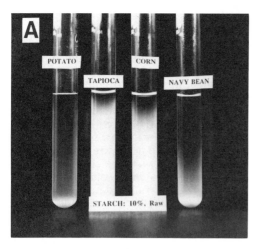

Fig. 2a. Unheated starch suspension, 5 mins. after mixing. Compare with Figs. 2b and 2c to see differential solubility of potato, tapioca, corn and beans starches.

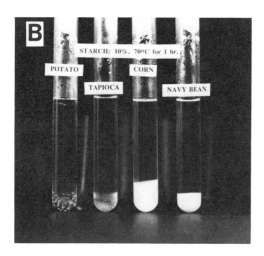

Fig. 2b. Starch suspensions after gentle shaking for 1 hr at 70°C. Compare with Figs. 2a and 2c to see differential solubility of potato, tapioca, corn and beans starches.

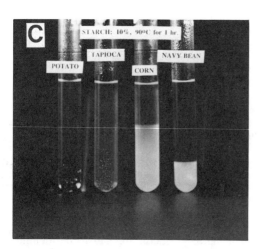

Fig. 2c. Starch suspension after gentle shaking for 1 hr at 90°C. Compare with Figs. 2a and 2b to see differential solubility of potato, tapioca, corn and beans starches.

Fig. 3a. Scanning electron micrograph (200×) of undissolved potato starch material at bottom of tube shown in Fig. 2b.

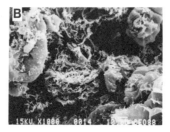

Fig. 3b. Scanning electron micrograph (1000×) of undissolved tapioca starch material at bottom of tube shown in Fig. 2b.

Fig. 3c. Scanning electron micrograph (1000×) of undissolved corn starch material at bottom of tube shown in Fig. 2b.

Fig. 3d. Scanning electron micrograph (1000×) of undissolved bean starch material at bottom of tube shown in Fig. 2b.

remain intact with only a few strands of amylose being leached from the granules. Figure 2c shows the starches after heating them for one hour at 90°C. Potato starch is completely solubilized and the tapioca solution is slightly translucent, indicating the starch is almost completely solubilized. The additional energy applied to the corn starch (90°C vs 70°C) caused more starch to go into solution as shown by the translucent layer at the top of the tube and the greatly swollen opaque material occupying the bottom two-thirds of the tube. The additional energy caused the bean starch to swell compared to the 70°C treatment, but there was no visible indication that the bean starch was being solubilized. These figures along with the results of others (80,81) all demonstrate that considerably more energy is required to break the bonds in bean starch. The net result is that bean starch is digested slower than starch from most other foods (58). If some of the bean starch retains its crystalline structure after cooking as suggested (77–79), it is likely that the crystalline starch will remain indigestible as is the case for raw potato starch.

Another factor which limits starch digestibility is the tendency for some starches to retrograde. When gelatinized starch cools slowly, some of the water between the starch chains is squeezed out and hydrogen bonds within or between starch chains reform. Amyloses with degree of polymerization (DP, glucose units) between 200 and 1200, have a high tendency to retrograde to crystalline "beta sheet" type configurations. The average DP for legume amylose is DP 1000–1400 (82), which is highly conducive to formation of retrograded crystalline starch structures. Retrograded amylose is indigestible within the small intestine (83). Legume starch has a higher percentage of amylose than most starches (Table 7) which increases the potential for formation of indigestible retrograded starch. During cooking some of the amylose leaches from the starch granule. Because the cell structure in cooked beans remains intact, much of the leached amylose remains within the cell and fills spaces between protein bodies and starch granules. If amylose retrogradation occurs, another indigestible matrix (similar to fiber in cell walls) is formed in and around the protein and starch granules. This retrograded amylose matrix is another factor that will slow protein and starch digestion.

TABLE 7

Amylose Content in Various Starches[a]

Starch	% Amylose
Adzuki bean	34.9
Kidney bean	34.1–35.0
Navy bean	32.1–36.0
Black bean	37.3–38.1
Pinto bean	30.2
Potato	20.1–21.0
Corn	22.6–28.0
Waxy corn	0
Tapioca	17.7

[a] From (77, 81, 85, 86).

One often cited reason for limited legume consumption is the problem of flatulence and much research has been directed toward the development of a flatulence-free legume. In this regard, it is important to determine if the slow digestion rate of legume starch and retrogradation of amylose are causing starch to pass to the colon and produce gas. It is likely that indigestible starch is a significant part of the flatulence problem. However, caution must be exercised that in our efforts to eliminate flatulence producing factors that we do not produce a product that has a high glycemic index with little fiber. Thus, the challenge is to minimize flatulence without sacrificing the attributes of being a low glycemic food that is a good source of dietary fiber.

References

1. FAO/UN, *Amino-acid content of foods and biological data on proteins*, Food and Agriculture Organization of the United Nations, Rome, (1970).
2. Tobin, G., and K.J. Carpenter, *Nutr. Abstr. and Rev. 48*:920 (1978).
3. Wolzak, A., L.G. Elias and R. Bressani, *J. Agric. Food Chem. 29*:1063 (1981).
4. Wolzak, A., R. Bressani and R. Gomez Brenes, *Qual. Plant-Plant Food Hum. Nutr. 31*:31 (1981).
5. Kies, C., *J. Agric. Food Chem. 29*:435 (1981).
6. Miles, C.W., J.L. Kelsay and N.P. Wong, *J. Nutr. 118*:1075 (1988).
7. Schneeman, B.O., in *Dietary Fiber in Health and Disease*, edited by G.V. Vahouny and D. Kritchevsky, Plenum Press, N.Y. (1982) p. 73.
8. Southgate, D.A.T., *Proc. Nutr. Soc. 32*:131 (1973).
9. Johnson, I.T., and J.M. Gee, *Gut 22*:398 (1981).
10. Flourie, B., N. Vidon, C.H. Florent and J.J. Bernier, *Ibid. 25*:936 (1984).
11. Elsenhaus, B., U. Sufke, R. Blume and W.F. Caspary, *Digestion 21*:98 (1981).
12. Vahouny, G.V., R. Tombes, M.M. Cassidy et al., *Proc. Soc. Exp. Biol. Med. 166*:12 (1981).
13. Meyer, J.H., J.B. Thomson, M.B. Cohen, A. Schadchehr and S.A. Mandiola, *Gastroenterology 76*:804 (1979).
14. Meyer, J.H., H. Ohashi, D. Jehn and J.B. Thomson, *Ibid. 80*:1489 (1981).
15. Meyer, J.H., and J.E. Doty, *Am. J. Clin. Nutr. 48*:267 (1988).
16. Williams, N.S., J.H. Meyer, D. Jehn, J. Miller and A.S. Fink, *Gastroenterology 86*:1451 (1984).
17. Doty, J.E., and J.H. Meyer, *Ibid. 94*:50 (1988).
18. Vahouny, G.V. and D. Kritchevsky, editors, *Dietary Fiber in Health and Disease*, Plenum Press, New York (1982).
19. Trowell, H., D. Burkitt and K. Heaton, editors, *Dietary Fibre, Fibre-Depleted Foods and Disease*, Academic Press Inc. (London) Ltd. (1985).
20. Vahouny, G. and D. Kritchevsky, editors, *Dietary Fibers, Basic and Clinical Aspects*. Plenum Press, New York (1986).
21. Prosky, L., N. Asp, I. Furda, J.W. DeVries, T.F. Schweizer and B.F. Harland, *J. Assoc. Off. Anal. Chem. 68*:677 (1985).
22. Cummings, J.H., H.N. Englyst and R. Woods, *J. Assoc. Publ. Analysts 23*:1 (1985).
23. Anderson, J.W., and S.R. Bridges, *Am. J. Clin. Nutr. 47*:440 (1988).
24. Wursch, P., S. Del Vedovo and B. Koellreutter, *Ibid. 43*:25 (1986).
25. Golay, A., A.M. Coulston, C.B. Hollenbeck, L.L. Kaiser, P. Wursch and G.M. Reaven, *Diabetes Care 9*: (1979).
26. Schneeman, B.O., and R.L. Lyman, *Proc. Soc. Exp. Biol. Med. 148*:897 (1975).

27. Madar, Z., Y. Tencer, A. Gertler and Y. Birk, *Nutr. Metab. 20*:234 (1976).
28. Savaiano, D.A., J.R. Powere, J.J. Costello, J.R. Whitaker and A.J. Clifford, *Nutr. Rep. Intl. 15*:443 (1977).
29. Granum, P.E., and B. Eskeland, *Ibid. 23*:155 (1981).
30. Bo-Linn, G.W., C.A. Santa Ana, S.G. Morawski and J.S. Fordtran, *N. Engl. J. Med. 307*:1413 (1982).
31. Carlson, G.L., B.U.K. Li, P. Bass, and W.A. Olsen, *Science 219*:393 (1983).
32. Garrow, J.S., P.F. Scott, S. Heels, K.S. Nair and D. Halliday, *Lancet* 60 (1983).
33. Thompson, L.U., R. Rea and D.J.A. Jenkins, *J. Food Sci. 48*:235 (1983).
34. Rea, R., L.U. Thompson and D.J.A. Jenkins, *Nutr. Res. 5*:919 (1985).
35. Sgarbieri, V.C., and J.C. Whitaker, *Adv. Food Res. 28*:93 (1982).
36. Liener, I.E., *The Lectins. Properties, Functions, and Applications in Biology and Medicine*, edited by I.E. Liener, N. Sharon and I.J. Goldstein, Academic, New York, (1986) p. 527.
37. Donatucci, D.A., I.E. Liener and C.J. Gross, *J. Nutr. 117*:2154 (1987).
38. Jaffe, W.G., and C.L. Vega Lette, *Ibid. 94*:203 (1968).
39. Palmer, R., A. McIntosh and A. Pusztai, *J. Sci. Food Agric. 24*:937 (1973).
40. Liener, I.E., *Plant Proteins*, edited by G. Norton, Butterworth's, London, (1978) p. 117.
41. Pazur, J.H., and J.H. Aronson, *Adv. Carbohydrate Chem. Biochem. 27*:301 (1972).
42. Romero, J., and D.S. Ryan, *J. Agric. Food Chem. 26*:784 (1978).
43. Chang, K.C., and L.D. Satterlee, *J. Food Sci. 46*:1368 (1981).
44. Semino, G.A., P. Restani and P. Carletti, *J. Agric. Food Chem. 33*:196 (1985).
45. Seidl, D., J. Jaffe and W.G. Jaffe, *Ibid. 17*:1318 (1969).
46. Deshpande, A.S. and S.S. Nielsen, *J. Food Sci. 52*:1326 (1987).
47. Deshpande, A.S., and S.S. Nielsen, *Ibid. 52*:1330 (1987).
48. Liener, I.E., and R.M. Thompson, *Qual Plant-Plant Foods Hum. Nutr. 30*:13 (1980).
49. Marquez, U.M.L., and F.M. Lajolo, *J. Agric. Food Chem. 29*:1068 (1981).
50. Bradbear, N., and D. Boulter, *Qual. Plant-Plant Foods Hum. Nutr. 34*:3 (1984).
51. Elias, L.G., D.G. DeFernandez and R. Bressani, *J. Food Sci. 44*:524 (1979).
52. Bressani, R., L.G. Elias, A. Wolzak, A.E. Hagerman and L.G. Butler, *Ibid. 48*:1000 (1983).
53. Aw, T-L., and B.G. Swanson, *Ibid. 50*:67 (1985).
54. Marsh, J.W., J. Denis and J.C. Writon, *J. Biol. Chem. 252*:7678 (1977).
55. Rothenbuhler, E., and J.E. Kinsella, *J. Food Sci. 51*:1479 (1986).
56. Hamaker, B.R., A.W. Kirleis, L.G. Butler, J.D. Axtell and E.T. Mertz, *Proc. Natl. Acad. Sci. (USA) 84*:626 (1987).
57. Fry, S., *Ann. Rev. Plant Physiol. 37*:765 (1986).
58. Jenkins, D.J.A., R.H. Taylor and T.M.S. Wolever, *Diabetologia 23*:477 (1982).
59. Anderson, J.W., L. Story, B. Sieling, W.L. Chen, M.S. Petro and J. Story, *Am. J. Clin. Nutr. 40*:1146 (1984).
60. Jenkins, D.J.A., T.M.S. Wolever, J. Kalmusky, *Ibid. 42*:604 (1985).
61. Thompson, L.U., *Food Tech. 42*:123 (1988).
62. Jenkins, D.A.J., T.M.S. Wolever, G. Buckley, et al., *Am. J. Clin. Nutr. 48*:248 (1988).
63. Fleming, S.E., and J.R. Vose, *J. Nutr. 109*:2067 (1979).
64. Anderson, I.H., A.S. Levine and M.D. Levitt, *N. Engl. J. Med. 304*:891 (1981).
65. Fleming, S.E., M.D. Fitch and D.W. Stanley, *J. Food Sci. 53*:777 (1988).
66. Stephen, A.M., A.C. Haddad and S.F. Phillips, *Gastroenterol. 85*:589 (1983).
67. Chapman, R.W., J.K. Sillery, M.M. Graham and D.R. Saunders, *Am. J. Clin Nutr. 41*:1244 (1985).
68. Dreher, M.L., J.W. Berry and C.J. Dreher, *CRC Critical Rev. Food Sci. Nutr. 20*:47 (1983).
69. Reddy, N.R., and M.D. Pierson, *Food Chem. 13*:25 (1984).
70. Lodhi, G.N., R. Renner and D.R. Clandinin, *J. Nutr. 99*:413 (1969).
71. Karimzadegan, E., A.J. Clifford and F.W. Hill, *Ibid. 109*:2247 (1979).
72. Mourot, J., P. Thouvenot, C. Couet, et al., *Am. J. Clin. Nutr. 48*:1035 (1988).
73. Jenkins, D.J.A., M.J. Thorne, K. Camelon et al., *Ibid. 36*:1093 (1982).
74. Jenkins, D.J.A., H. Ghafari, T.M.S. Wolever et al., *Diabetologia 2*:450 (1982).
75. Holm, J., I. Lundquist, I. Bjorck, A.C. Eliasson and N.G. Asp, *Am. J. Clin. Nutr. 47*:1010 (1988).
76. Daniel, J.R. and R.L. Whistler, *Chemical Changes in Food during Processing*, edited by T. Richardson and J.W. Finley, Avi Publishing Co., Inc., Westport, 1985, p. 305.
77. Lai, C.C., and E. Varriano-Marston, *J. Food Sci. 44*:528 (1979).
78. Hahn, D.M., F.T. Jones, I. Akhavan and L.B. Rockland, *Ibid. 42*:1208 (1977).
79. Varriano-Marston, E., and E. DeOmana, *Ibid. 44*:531 (1979).
80. Biliaderis, C.G., D.R. Grant and J.R. Vose, *Cereal Chem. 58*:506 (1981).
81. Hoover, R., and F. Sosulski, *Starke 37*:181 (1985).
82. Biliaderis, C.G., D.R. Grant and J.R. Vose, *Cereal Chem. 58*:496 (1981).
83. Englyst, H.N., and J.H. Cummings, *A.J. Clin. Nutr. 45*:423 (1987).
84. Phillips, D.E., M.D. Eyre, A. Thompson and D. Boulter, *J. Sci. Fd. Agric. 32*:423 (1981).
85. Biliaderis, C.G., T.J. Maurice and J.R. Vose, *J. Food Sci. 45*:1669 (1980).
86. Swinkels, J.J.M., *Starke 37*:1 (1985).

Development of Glandless Cottonseed and Nutritional Experience in The Ivory Coast

J.L. Bourely

Laboratoire de Chimie des Plantes textiles, Institut de Recherches du Coton et des Textiles Exotiques, C.I.R.A.D., B.P. 5035, 34032 Montpellier CEDEX France

Abstract

The Institut de Reserches du Coton et des Textiles Exotiques (IRCT) has developed research programs on glandless cotton in several African countries and mainly in the Ivory Coast where 23,700 hectares (59,000 acres) were cultivated in 1984. This experience was successful from the agronomic aspect. Nevertheless, the technological characteristics of the glandless varieties being lower than those of glanded cottons (especially ginning percentage), glandless cotton were not cultivated during the following years. Nevertheless, the Ivory Coast experience of 1984 demonstrates that glandless cottonseed flours can be produced economically as a source of oil, cakes and flours of high nutritional value for human food and animal feedings. From 1984 to the end of 1987, a research program financed by the European Economic Communities was conducted on glandless cottonseeds with the Institut des Savanes, (IDS) in Bouaké, the Institut de Santé Publique, (INSP) in Abidjan and different French laboratories. New glandless varieties which have agronomic and technological characteristics at least comparable to those of the best cultivated glanded varieties have been created in the Ivory Coast. Nutritional rehabilitation of young children stricken by kwashiorkor marasmus was successfully undertaken on glandless cotton flour from IRCT. This paper describes the main results of all these experiences and develops some technological and economic data related to the glandless cottonseed production and processing. Additionally, the future prospects for glandless cotton in the main French-speaking African countries are examined.

Interest now exists in glandless cotton as a food, oleoproteaginous crop, as well as a cash, fiber crop. The main work on this crop began in 1959 in the United States. Lusas and Jividen recently reviewed the glandless cottonseed processing and utilization research conducted in the world (1).
In Africa, the IRCT-CIRAD has developed research programs on glandless cotton with several African countries, in connection with the corresponding national institutions. Several agronomic and nutritional experiments have been undertaken by IRCT since the first glandless seeds were introduced in Chad by Roux in 1958.
The first glandless varieties were cultivated by Fournier and Roux in 1972, in Chad and in Mali on 200 hectares (2) and then 2,500 hectares were cultivated in Chad in 1974 and 2,000 in 1976.
In the Ivory Coast, research on glandless cotton is carried out at the Institut des Savannes (IDS) in Bouaké by agronomists, geneticists, entomologists and technologists, in connection with the extension service, the Compagnie Ivoirienne des Fibres Textiles, CIDT, the local cotton oil mill Trituraf, and plant protection, textile and agrofood industries.

IRCT's Work on Cotton Development in Africa

Since 1946, IRCT has coordinated a research network linking eleven francophone African countries in the fields of cotton agronomy, genetic, technology and plant protection. In African savanna and sahelian countries cotton is the main cash crop and the principle source of currency. On account of its agronomic technical requirements, mechanization, pesticides and fertilizer use, it obliges the farmer to invest, to modernize and to diversify his farm.
Cotton prices are decided by the government and, before sowing, the peasant knows he is sure to sell all his crop. Consequently, by the security and guarantees it offers to farmers, cotton has a motor action on the agriculture as a whole and has a direct impact on the economic development of the whole country (3).
The cottonseed oil mill of the Ivory Coast was built in 1974 and today the cottonseed oil consumption is regularly increasing there as in many of the francophone countries. This country is already self sufficient in oil but nevertheless continues to import soya cake (about 5,000 tons a year), whereas it exports 95% of its cottonseed cake to Europe for cattle feeding.
In 1985 a colloquium was held in Abidjan, on "glandless cotton, a new source of nutrition." It verified the technical and economic advantages of producing glandless instead of glanded cotton. The collaboration between the various partners of the cotton industry: researchers, development and extension organizations, crushers, feedingstuff's manufacturers and potential local users, gives a good illustration of the Ivory Coast and the agrofood industry intentions to develop local sources of proteins in order to obtain food self sufficiency.

Agronomic behavior of glandless cotton in the Ivory Coast

As seen in Figure 1, in 1986–87, the seed-cotton yield of one hectare under cotton cultivation was 1,335 kg in the Ivory Coast and refined oil represents 1,083 French francs benefit, cakes, 149 francs; and fibers 5,421 francs (4). When glandless cotton was processed instead of glanded cotton, the cultivation cost is the same, whereas the price of cake can be doubled and oil benefit is increased for the crushers.
In 1984, 23,736 hectares (59,000 acres) of glandless cotton were cultivated in the Northern Ivory Coast (5–8). This experience has been successful from the agronomic point of view (Table 1) and demonstrates that glandless varieties raise no particular agronomic problems when they are compared to ordinary cottons. Phytopathologists feared that the glandless plants would be more sensitive to pests than ordinary varieties. This happens when glandless cottons are cultivated among ordinary cottons but

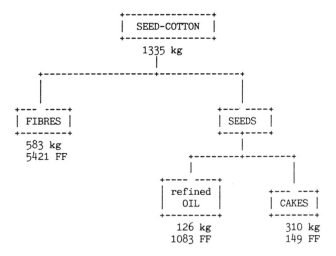

Fig. 1. Ivory Coast yields of cotton cultivation in 1986–87 (per hectare) [from Hau (4)].

TABLE 1

Results of Glandless Cultivation in Ivory Coast in 1984 (ISA BC 2) (from Hau, 6)

Areas	=	23 736 ha
Yields:		
Seed-cotton	=	30 775 tons
Fibers	=	12 433 "
Refined oil	=	3 250 "
Cakes	=	6 600 "
Seed-cotton yield	=	1 297 kg/ha
Ginning percentage	=	40.4%

not when they are cultivated alone on extensive areas (6,7). Nevertheless, since the technological characteristics of the glandless varieties cultivated in 1984 are lower than those of glanded cottons (especially ginning percentage), they were not cultivated during the following years. The genetic research carried out during the last three years has led to the creation of new glandless varieties which have better agronomic and technological characteristics than the more recently cultivated glanded varieties. Consequently, the decision to grow 30,000 ha's of glandless cotton in 1989 is now imminent in the Ivory Coast.

From 1984 to the end of 1987, a research program financed by the EEC has been conducted under the scientific control of IRCT on "glandless cotton seeds: a source of high nutritional quality proteins for human and monogastric animal consumption." The main partners of this research were the Institut des Savanes, IDS, in Bouaké, INSP in Abidjan and French laboratories. This program included technological research on seeds, kernels and flours, biochemical and toxicological controls and the functional properties' analysis of the flours, concentrates and isolates produced both at laboratory and pilot scale. Parallel to this, agronomic, genetic and nutritional experiments were carried out in the Ivory Coast.

Technology as a flour quality factor

The conventional oil mill prepress-solvent process (Fig. 2) is not suitable for the production of high quality glandless cottonseed flours fit for human consumption (9). Pressing requires the addition of hulls to the flakes and is responsible for the high crude fiber content of the cake. During pressing, high temperature leads to molecules linkage in Maillard reactions which develop brown coloration, bitterness and hardening of the cake and make much of the lysine unavailable. Consequently, the nutritional value of the residual cake is decreased. Traditional mills produce cakes for animal feeding with 40–45% proteins and 10–13% "crude cellulose" contents.

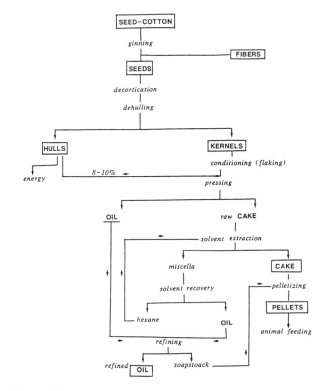

Fig. 2. Traditional prepress-solvent process.

Glandless cottonseeds must be treated by a direct hexane oil extraction without pressing (Fig. 3). The high oil content of the kernels requires particular conditioning and defatting procedures in order to facilitate the oil percolation at a relatively low temperature and to maintain highest protein quality, especially available lysine, and to avoid chemical bonds between proteins and saccharides that could damage color, taste and other organoleptic properties.

The work carried out by IRCT demonstrates that the use of a cold-operated expander running at a high speed is a promising conditioning kernel treatment before oil extraction. After oil extraction by hexane at 60°C of the cold pressed "extrudates" and vacuum desolventation, the cake is dried at 90°C. Being processed at a low temperature, the ground flour is white and has a nutty taste, and contains high available lysine and soluble protein con-

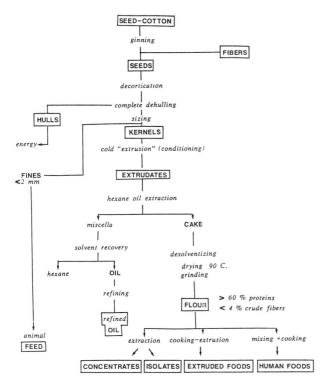

Fig. 3. Proposed technology for glandless cottonseed oil extraction.

tent (0.04 N NaOH soluble proteins), more than 60% total proteins (N × 6.25), and less than 4% of "crude fibers". No protein damage nor chemical bonds occur in the flour which is suitable to prepare human foods, cooked extruded foods, or concentrates and isolates, after adequate extraction procedures (10).

Functional properties of glandless cottonseed derivatives

Concentrates (water or alcohol extracted) and isolates have been processed at the Ateliers Pilotes of INRA in Nantes (10).

The Laboratoire de Biochimie et de Technologie Alimentaire of Montpellier USTL University has analyzed the protein constituents and the functional properties of defatted glandless flour, protein concentrates and isolates (12).

The nitrogen solubility index at pH 6.8 does not exceed 20% for the defatted flour and 10% for the concentrates and the isolate. It increases up to 80% for the defatted flour when ionic strength is raised to 0.5 M by adding sodium chloride. The highest solubility index was obtained, for all the samples, at a pH > 10. Nitrogen solubility of the flour does not reach 50% at acidic pH but reaches 80% for concentrates and about 90% for isolate at pH 2. Fast liquid chromatography and electrophoresis indicate the protein constituents which are soluble at different pH or ionic strengths and the molecular weights of the subunits which constitute the storage globulins, alcalins A and B. Functional properties of glandless cottonseed flour, protein concentrates and isolate were inferior to soya equivalents. Water absorption capacity, emulsifying, and thickening properties increased with ionic strength. At pH 6.8, defatted flour dispersed on water gave no gel but heterogeneous and granular compounds. Adding sodium chloride, even in large quantities, solubilizes most of the proteins but does not give any gel after heating. The highest available lysine content was obtained for glandless cottonseed flour, and decreased in the following order: concentrates, isolates.

Use of glandless cotton cakes in animal feeding

Several nutritional tests have been performed on monogastric animals with glandless cottonseed cakes. In 1966, in Dakar, (Senegal), six different diets were tested on rats. The best nutritional results were given by millet plus "*néou*" (Cayor nut, *Parinari macrophylla*) and a mixture of millet and glandless cottonseed flour.

Experiments run on broilers by Tacher et al. in Chad have shown that the incorporation of up to 30–50% of glandless cottonseed cake in the feedstuff is favorable to the broilers' growth (13). Nevertheless, more than 75% of glandless cottonseed cake in the feed requires lysine enrichment of the diet.

In Abidjan, the Laboratoire Central de Nutrition Animale (LACENA), has tested glandless cottonseed cakes on poultry (14). Two trials were conducted to compare the performances of broilers fed with a starter and a grower finishing diet in which soybean meal was replaced by glandless cotton cake. (Table 2). The data indicated that a direct substitution of soybean meal by glandless cottonseed meal without any supplementation resulted in equal performances in both trials (Table 3). Moreover, the glandless cottonseed diet significantly reduced abdominal fat pad at market age (14). Feed cost efficiency respectively amounts to 64 and 73 for the two trials for the cotton diets compared to 67 and 77 for the corresponding soya diets.

At the laboratoire de Physiologie de la Nutrition, Montpellier U.S.T.L. University (France), Besancon et al. have compared the nutritional efficiency of glandless cotton cake and soya cake produced on the same manner at the Trituraf oil mill in Bouaké (15,16). For three weeks, young rats were fed with a well balanced diet ensuring all the growing needs. In all the diets, the protein content was 10%. Young rats fed with the diet in which cotton cake was the only source of protein presented a better growth curve than those receiving soya cake or standard diets (Fig. 4). The weight curve slopes given by soya cake was improved and that of glandless cotton decreased after steaming the cakes at 112°C for 15 minutes because in the first case the antinutritional components of soya were destroyed, whereas in the second case the cotton cake proteins were damaged by heating. The calculated P.E.R. were respectively 3.27 for raw cotton cake, 2.76 for soya cake and 3.64 for the standard casein diet. Nitrogen digestive utilization coefficient was a little less with the cotton diet than with other diets. This could be due to the cottonseed diet's higher fiber content. The decreased biological value of the cottonseed diet results from its lower lysine content. Steaming reduces the available lysine content of the cotton diet and, consequently, its biological value and the digestive utilization of both nitrogen and dry matter.

At the Laboratoire de Physiologie de la Nutrition, Mirande University, Dijon, (France), Bertrand and Belle-

TABLE 2

Nutritional Tests on Broilers Composition of the Starter and the Grower-finisher Diets (Kouakou Brou, Lacena, (14))

	Starter		Grower-finisher	
Ingredients (g % diet)	Soya	Cotton	Soya	Cotton
Soya cake (44 % proteins)	10		8	
Glandless cottonseed cake (41% proteins)	6	16	6	14
Corn	60	60	64	64
Wheat middling	7.4	7.4	8.0	8.0
Fish meal	14	14	11	11
Dicalcium phosphate	0.4	0.4		
Calcium carbonate			0.8	0.8
Sodium chloride	0.2	0.2	0.2	0.2
Vitamin and mineral mix	2	2	2	2
Calculated % nutrient composition:				
ME (Mcal/kg)	2.92	2.94	2.95	2.96
Crude protein	22.3	22.1	19.9	19.8
Crude fibres	3.6	4.2	3.5	4.0
Calcium	1.0	1.0	1.1	1.1
Total phosphorus	0.9	0.9	1.1	1.1
Lysine	1.28	1.16	0.84	0.86
Methionine + cystine	1.00	0.79	1.08	0.99
			0.72	0.71

TABLE 3

Comparison of Performance of Broilers 7 Weeks Old Fed with Soya Cake and Glandless Cottonseed Cake (Kouakou Brou, Lacena, (14))

	Trial 1		Trial 2	
Traits	Soya	Cotton	Soya	Cotton
Feed intake (g)	3209[a]	3099[a]	4473[a]	4372[a]
Average body gain (g)	27.9[a]	27.3[a]	34.0[a]	33.6[a]
7-Weeks body weight (g)	1402[a]	1370[a]	1705[a]	1687[a]
Feed conversion (Feed/gain)	2.35[a]	2.32[a]	2.69[a]	2.66[a]
Feed cost efficiency (cost/kg gain) ([b]) in French francs	67	64	77	73

[a] Means that rows with same superscript are not different (P 0.05)
[b] for soya cake price at 1.50 FF and glandless cotton seed cake at 0.90 FF

ville have tested several diets on young rats previously submitted to severe protein malnutrition for 30 days with only 2% casein and 0.3% methionine (17,18). The four diets tested contained corn-cotton-milk, rice-cotton-milk, corn-soya-milk (respectively 60/35/5 : W/W/W) and the control diet: 20% casein and 0.3% methionine. It must be underlined that after nutritional rehabilitation, the weight of the rats previously suffering from malnutrition never reached the weight of the rats which were constantly well fed (Fig. 5).

After ten days of rehabilitation, the behavior of all the rats was practically the same. All the weight curve slopes were comparable to that of the well fed rats, but later the weight curve slope was better for the animals fed with corn-soya-milk and rice-cotton-milk diets than the other standard and corn-cotton-milk diets. Apparent nitrogen digestive utilization coefficients were lower with corn-soya-milk than with rice-cotton-milk diet. Amino acids provided by the corn-soya-milk diet caused higher nitrogen utilization and those provided by the rice-cotton-milk diet resulted in the best digestibility.

Cotton flour allowed for higher nitrogen and lipid digestibility coefficients but led to a lower net nitrogen utilization and protein efficiency ratio as compared to soya flour. Net nitrogen utilization and nitrogen digestive utilization coefficients of the rice-cotton-milk diet were higher than those of the corn-cotton-milk-diet. Rice had a better nutritional value than corn. The beneficial asso-

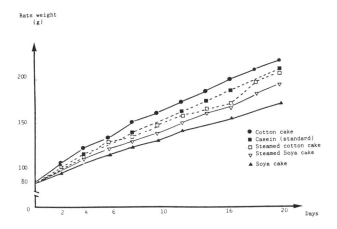

Fig. 4 Nutritional tests on rats [from Besancon et al. (15)].

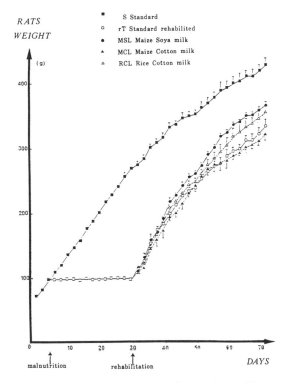

Fig. 5. Nutritional tests on rats [from Bertrand (17,18)].

ciation of rice with cotton is to be noted. These results demonstrate that glandless cotton flour mixed with corn can replace soya flour associated with corn in a nutritional rehabilitation diet (17,18).

Human food uses of glandless cottonseed flour

Previous nutritional experience: Traditionally, several Sahelian groups in Chad, Mali, Cameroon and Burkina-Faso are in the habit of eating cottonseed derivatives, in sauce or compound flours, either regularly or only in critical periods such as severe drought or food shortage. The first nutritional studies carried out on glandless flour by IRCT began in Senegal in ORANA (Organisme de Recherches sur l'Alimentation et la Nutrition Africaine) and at Le Dantec hospital, in Dakar in the 1960's.

Forty-nine malnourished children were fed with a diet in which glandless cottonseed flour was associated with sugar, milk and cereals. The nutritional value of this diet was found to be comparable to the commercial Nesmida standard diet. The formula including 45% millet flour, 25% glandless cotton flour, 10% semi-skimmed milk and 20% sugar was used successfully both as a therapeutic diet and as a weaning food. Kwashiorkor edemas decreased in 4–7 days, as rapidly with the cotton diet as with the standard diet, followed by a weight gain varying from 42 to 200 g per day.

In Mali, in 1967, a test was carried out near Bamako in medical centers on 6-month to 4-year-old children fed with a diet containing 30% glandless cotton flour and sorghum flour. Again in Mali, an experiment was undertaken by the Malian health Ministry from June to October 1970 on a group of 1- to 3-year-olds. Twenty-seven children from a Bambara village had glandless flour added to their diet for four months on the basis of 200 g food a day (36 g protein/day) in a diet containing 80% millet flour and 20% glandless cottonseed flour. Eighteen of the 27 children tested showed weight gains varying from 100 to 1000 g above those of the children of a neighboring village that received no glandless cottonseed flour in their diet.

In Chad, four French nutritionists carrying out an investigation on 2,507 people showed that 53% of the tested subjects had already eaten cotton seeds at least once in their life, sometimes in drought periods. Nutritional tests were carried out in 1975, on 5- to 6-year-old children fed for six months with a diet containing 70% of glandless cottonseed flour, as a supplement to the home diet. The children presented a better weight increase than those only received a millet, sugared isocaloric pap.

Successful tests were performed on volunteer adolescents who consumed a daily average of 40 g of glandless cottonseed flour in sauces and paps.

Another test carried out on 14 Chadian farmer families, on a 6-week period on the consumption of a diet containing glandless cottonseed flour had the same success.

In Chad, 4.5 tons of glandless kernels were promptly sold, at the same price as sorghum, and eaten unprocessed, a proof of Chadian consumers' keen interest.

Recent nutritional experiments: In 1985, the Institut National de Santé Publique in Abidjan undertook treatment of young children stricken by kwashiorkor marasmus. Acceptability tests have shown that glandless cottonseed flour is well appreciated by mother and child when rice flour is added to the diet. They appreciate the taste and the smooth and creamy aspect (19).

Table 4 presents the different diets used at the Institute for the edema melting period and for the rehabilitation period. The standard diet used for melting edema contained casein as hyperprotidine, milk powder and sugar, whereas rice flour replaced casein for the supplementation period. In the glandless fortified diets, glandless cottonseed flour replaced casein in edema melting paps and milk powder in supplementation. Rice flour was added to all the paps whereas five grams of oil were added to the supplementation paps. The nutritional rehabilitation of young children was obtained as rapidly with the cotton fortified diet as with the standard diet usually used in the Institute (Table 5).

TABLE 4

Nutritional Rehabilitation of Children Stricken by Kwashiorkor Marasmus; Composition of the Experienced Diets (in gram, for one pap) (Institut National de Santé Publique) (19,20,21)

Diet components (g %)	For oedema melting (6 paps a day)	Supplementation (3 paps a day)
Standard diet		
hyperprotidine (casein)	8	
Milk powder	20	30
Sugar	16	12
Rice flour		20
Glandless cottonseed flour diet		
Cotton flour	15	20
Milk powder	15	
Sugar	12	16
Rice flour	5	20
Oil		5

TABLE 5

Nutritional Rehabilitation of Children, Main Results of the Biological Tests (Institut National de Santé publique)

Trials	Standard diet	Glandless cottonseed flour diet
Time for oedema melting (days)	6.8	6.4
Protid (g/l)	67.7	68.9
Albumin (g/l)	35.8	36.8
Prealbumin (mg/l)	135.7	127.2
Pseudocholinesterase (IU/ml)	6.90	6.47
Hemoglobin (g/100 ml)	10.1	9.2
Weight/age	97.7	97.2
Month body weight (kg)	1.03	0.96
Height/age	92.13	92.25

Nevertheless, the standard diet seems to be more effective than the glandless cotton fortified diet for the rehabilitation of several biological parameters, such as albumin and pseudocholinesterase. This improved efficiency could perhaps be due to the higher essential amino acids content and their better equilibrium in the standard diet. Nevertheless, no statistical differences separate the main results of these experiences (19–21).

Food preparations with glandless cottonseed flour

Owing to its high protein content, and to the absence of antinutritional factors, glandless cottonseed flour constitutes a source of protein of high nutritional value. It can be used as a protein supplement in traditional dishes that are rich in glucidic compounds, in weaning and children's foods, breakfast, cereals, breads, pastries, dietary biscuits, hyperprotein bars like CSM, traditional dishes, and into minced mixed meat etc...

In bakery, the use of glandless cottonseed flour is limited since the flour has a lower dough raising capacity and because the dough turns an unpleasant greenish color when its cotton flour content is over 18%. This is due to its flavinoid content.

Up to 20% of glandless cottonseed flour can be added to wheat flour to prepare fancy biscuits. Pastries harden and keep well. In our laboratory, we have prepared biscuits such as cookies and short breads, which contained up to 50% cotton flour mixed with various other flours such as wheat, rice, corn, millet and sorghum. No differences in taste and flavor have been noticed by the consumers, both African or European, between the various samples regardless of whether or not they contain wheat. Nevertheless, pastries containing sorghum mixed with cotton flours appear slightly rough. Adding eggs makes for an easier mixing of cereals with cotton flour and improves the aspect and the volume of the product. We have never noticed the bitterness which has been stressed by African consumers in Mali, even when we have added hulls to the dough before cooking. Bitterness does not appear to be due to the residual hulls but possibly to cooking conditions.

Extrusion tests performed at the Laboratoire de Biochimie et Technologie Alimentaire of Montpellier University have made it possible to produce crusty biscuits containing 18% of glandless cottonseed flours (11).

Economic and commercial problems

The present market of glandless cotton cakes in Côte d'Ivoire: Glandless cottonseed cakes can be efficiently used for poultry feeding (22). Although this market might be considered as not very profitable, poultry feeding would be the first commercial utilization to develop glandless cotton in the Ivory Coast. The market for glandless cottonseed cake now available for animal feeding can be estimated at about 7,000 tons which can be used without any reserve for local consumers. Consequently, this local market might make it possible to initiate the cultivation of glandless cotton in the Ivory Coast before developing new markets for human consumption. Today, taking into account the existing oil mill equipment, the locally produced cakes can only be used for animal consumption.

Profitability of glandless cotton growing: The Ivory Coast experience in 1984 has shown that glandless cotton can be produced economically as a source of oil, cakes and flours of high nutritional value. It must be stressed that cotton alone can give this area the possibility of becoming self sufficient in vegetable proteins. It would represent a currency saving estimated at 500 millions CFA francs per year (6). Oil and cake represent about 20% of the total crop benefits.

Owing to a higher oil yield, to easier seed crushing and to a lower oil refining cost due to the absence of gossypol and a lower pigment content, but also to a better oil quality and the recovery of valued subproducts such as lecithin, glandless cottonseeds will provide the crushers with more profits than glanded cotton (Table 6).

TABLE 6

Oil Mill Crushing Yields of Glandless and Glanded Cottonseed (from Hau and Richard, (7))

Yields	Glanded	Glandless
Kernels % seeds	52.19	54.50
Cake % ''	44.24	40.26
Oil % ''	19.41	20.64
Oil acidity (as oleic acid, g %)	2.29	1.36
Neutralization yield	93.70	96.30
Refining oil yield	18.01	19.58

Under present market conditions, the glandless Ivorian varieties are still competitive in providing proteins which are cheaper than the imported soya proteins (22,23,6). The different milling experiments carried out in the Ivory Coast have shown that glandless cottonseed cake can be sold for 1.5 FF a kilogram. The glanded cottonseed cake, either sold on the local market or exported (Abidjan FOB price) is sold at 0.80 FF a kilogram. Consequently, the profit is 0.70 FF a kg, i.e. 0.28 FF per kg of seeds for a cake yield equal to 40% of the seed weight (5).

Future Prospects

Today, only a few countries in the world are working on glandless cotton in order to use its seeds as a food source. In fact, even now, in most of the African countries, glandless cotton varieties are not cultivated since they are not as efficient as ordinary cottons for fiber production.

The development of glandless cotton is possible under three conditions: The first and primary condition is to breed and cultivate glandless varieties with agronomic and technological characteristics at least comparable to those of commercial glanded varieties, in order to avoid any fiber deficit. The second condition is to improve the oil mill process in order to produce flours suitable for human consumption. This requires a financial investment for new mill equipments and a sanitary control of the production. Modifying the present pre-press solvent oil extraction process by a direct hexane extraction is foreseen by the national cottonseed oil mill, but will be justified only when glandless cotton production and the human food market for glandless cotton flour are important enough. Moreover, presumably, during the first years of cultivation it will not be possible to obtain seeds which are totally devoid of gossypol. The problems of seed impurity will disappear when a steady production is set up on large areas (more than 25,000 hectares) and the seed-cotton and seeds are processed in specialized ginning and oil mills. The third condition will be to develop local national and international markets for the utilization of glandless cakes and flours first for monogastric animals and then for human consumption. These conditions imply a permanent technical and commercial collaboration between the different partners of the cotton industry (24,25).

Consequently, the first priority of the research on glandless cotton is now to provide cotton growers and extension services with glandless varieties of high agronomic and technologic characteristics. Good glandless cotton varieties have been created recently in the Ivory Coast. This aim is about to be achieved in several francophone African countries such as Cameroon, Chad and Burkina-Faso.

Taking into account the numerous scientific results that already have been obtained, mainly in the United States and in the Ivory Coast, research now in progress in Africa aims to encourage the different partners of the cotton production industry: government authorities, extension services, oil mills and money lenders to grow glandless cotton on large areas; to build factories specialized in the treatment of glandless seeds in order to produce glandless cottonseed flours of a high nutritional value. When glandless cottons are cultivated exclusively instead of glanded varieties, oil mills will be able to produce flours and cakes available on national and international markets both for human and all animals consumption. These products will contribute to insure an improved food supply and a better value for the fiber subproducts of many cotton producing countries and also to fight against world malnutrition. If successful, the experiment which starts in the Ivory Coast will be an example for all the cotton producing countries who have to face malnutrition.

Acknowledgments

This research was carried out under research contract N° TSD-078-F (MR) 1984 of the Commission of European Communities.

E.W. Lusas, of Food Protein Research and Development Center, Texas A&M University presented this paper. Mrs. Chichester helped us in the English translation of this paper.

References

1. Lusas, E.W., and C.M. Jividen, in *Proceedings of the World Conference on Emerging Technologies in the Fats and Oils Industry*, American Oil Chemists' Society, Champaign, IL, 1985, pp. 221–231.
2. Fournier and Roux, *Cot. Fib. Trop. 27*:251 (1972).
3. Desso and Bisson, Le Coton moteur du développement des zones de savanes. Le cotonnier sans gossypol: L'expérience de la Côte d'Ivoire. C.R. Colloque IDESSA-CIDI-TRITURAF: Le cotonnier sans gossypol: une nouvelle ressource alimentaire; Abidjan, 20-34, 1985.
4. Hau, B., Le cotonnier sans gossypol, nouvelle ressource alimentaire bientôt disponible. 4èmes Rencontres Scientifiques Internationales, Agropolis. SICAD. 1988 (in press).
5. Hau, B., *Cot. Fib. Trop. 39*:fasc. 3, 83–89 (1984).
6. Hau, B. and G. Richard, Le cotonnier sans gossypol: L'expérience de la Côte d'Ivoire. C.R. Colloque IDESSA-CIDI-TRITURAF: Le cotonnier sans gossypol: une nouvelle ressource alimentaire; Abidjan, 35–57, 1985.
7. Hau, B. and G. Richard, *Cot. Fib. Trop. 41*:fasc. 2 97–101 (1986).
8. Hau, B., "Development of Glandless Cotton in Africa" in *Proceedings of Beltwide Cotton Production Research Conference*, Dallas, Texas, 1987.
9. Bourely, J., in *Proceedings of the World Conference on Emerging Technologies in the Fats and Oils Industry*, American Oil Chemists' Society, Champaign, IL, 1985, pp. 406–408.
10. Berot, S., and J. Gueguen, Préparation de concentrats et d'isolat protéiques de coton glandless, C.R. Colloque IDESSA-CIDI-TRITURAF: Le cotonnier sans gossypol: une nouvelle ressource alimentaire; Abidjan, 186–212, 1985.
11. Marquie, C., Utilisation alimentaire des dérivés des cotonniers glandless, C.R. Colloque IDESSA-CIDI-TRITURAF: Le cotonnier sans gossypol: une nouvelle ressource alimentaire; Abidjan, 166–186, 1985.
12. Dumay, E., F. Condet and J.C. Cheftel, *Science des Aliments, 6*:623 (1986).
13. Tacher, G., R. Riviere and C. Landry, Utilisation du tourteau de coton dans l'alimentation de volailles de type chair. C.R. Colloque IDESSA-CIDI-TRITURAF: Le cotonnier sans gossypol: une nouvelle ressource alimentaire; Abidjan, 97–112, 1985.
14. Kouakou, B., Etude comparative du tourteau de soja et du tourteau de coton sans gossypol sur les performances des poulets de chair. Mémoire de fin d'études, Ecole Nationale Supérieure Agronomique d'Abidjan, 1987.
15. Besancon, P., O. Henri and J.M. Rouanet, Valeur nutritionnelle comparée de farines délipidées de coton glandless et de soja. C.R. Colloque IDESSA-CIDI-TRITURAF: Le cotonnier sans gossypol: une nouvelle ressource alimentaire; Abidjan, 63–79, 1985.
16. Bourely, J. and P. Besancon, La graine du cotonnier: une source de protéines de haute valeur pour l'alimentation humaine. Deuxièmes Journées Scientifiques du Groupement d'Etudes et de Recherches sur la Malnutrition (G.E.R.M.) I.N.S.E.R.M., Vol. 136, pp. 561–568, 1985.
17. Bertrand, V. and J. Belleville, *C.R. Soc. Biol. 182*:94 (1988).
18. Bertrand, V., Ph.D. theses, 1988.
19. Sess, Schoepfer, Adou, Roy, Boualou, Darracq, Raffier and Coulibaly, Expérience d'utilisation de la farine de coton glandless dans l'alimentation infantile. C.R. Colloque IDESSA-CIDI-TRITURAF: Le cotonnier sans gossypol: une nouvelle ressource alimentaire; Abidjan, 212–223, 1985.
20. I.N.S.P. Utilisation de la farine de coton glandless dans la réhabilitation nutritionelle d'enfants souffrants de malnutrition protéino-énergétique grave. Rapport du Contrat I.R.C.T.-C.E.E. 1984–1987, Institut National de Santé Publique, Abidjan, 1986.
21. Adoubl, L., Thèse, Faculté de Médecine d'Abidjan, Ivory Coast (1986).
22. Coulibaly, M. and D.C. Odi, Substitution de tourteau de soja par le tourteau de coton sans gossypol en fonction des prix pratiqués. CIDT Service Recherche-Développement Colloque d'Abidjan, 1985.
23. Defromont, C., L'intérêt du coton glandless dans l'industrie agroalimentaire. C.R. Colloque IDESSA-CIDI-TRITURAF: Le cotonnier sans gossypol: une nouvelle ressource alimentaire; Abidjan, 135–166, 1985.
24. Bourley, J., Etat actuel des recherches sur les cotonniers san glande. Troisièmes Journées Scientifiques du Groupement d'Etude et de Recherches sur la Malnutrition (G.E.R.M.), Dakar, 1987.
25. Bourely, J., Les cotonniers sans glande à gossypol, source de protéines pour l'alimentation animale et humaine, Colloque I.E.M.V.T. N'gaounderé, Cameroun, 1987.

Appropriate Technologies for Producing Vegetable Protein

Arleen Richnau and Glenn Patterson

Appropriate Technology International, 1331 H Street, N.W., Washington, D.C. 20005

Abstract

ATI International assists organizations in developing countries with development and commercialization of small scale agriculture processing, mineral resources and farm implement oriented technologies. To optimize technology selection, ATI considers cost effectiveness, compatibility with skills of the people, ecology, suitability to the supply, and marketing infrastructures. In the Cameroon, palm nut oil extraction and in Tanzania, sunflower seed oil extraction upgrades traditional technologies. Fresh coconut processing, now undergoing commercial viability testing in the Philippines, focuses on production of a high quality cooking oil, oil for soapmaking, and other protein by-products having functional properties suitable for production of a wide range of food stuffs. Improvements in traditional processing of shea nuts in Mali has increased extraction rates and reduced labor. A "koji" style two stage (filamentous fungus, *Rhizopus oryzae* and a yeast *Saccharomyces cerevisiae*) solid state fermentation, in Thailand, produced an 8–12% protein cassava feed ingredient for poultry and swine.

Several different types of technologies can be used to process vegetable protein for human food and animal feedstuffs. Some of these technologies and levels of these technologies may be well-suited to industrialized nations; others may be more appropriate to developing countries. The appropriateness of a technology depends on how it is used, what it is used to produce, and who gains and loses as a result of the production and consumption decisions. Compared to conventional technologies, appropriate technologies typically are less capital-intensive, more labor intensive, less dependent on imported goods, and easier to operate, maintain, and repair. Technologies introduced into developing countries should effectively use locally available resources, be affordable, be able to be managed by local people, and generate reasonable returns and benefits.

Many times appropriate technologies are not selected because the availability of these technologies generally is not known or is known only to a small group. Other technologies may be appropriate, but they have not been tested in the marketplace; they may need further R&D before they can be disseminated on a large scale.

Appropriate Technology International, a private, not-for-profit development assistance organization based in Washington, D.C., funded largely by the U.S. Agency for International Development, helps bridge the gap between research and development and viability of appropriate technologies in the fields of minerals utilization, agricultural products processing, and small farm equipment. ATI's projects focus on appropriate technologies that add value to local resources and increase the productivity and income of poor people in rural areas and market towns in developing countries. This paper discusses five ATI projects that introduced appropriate technologies to process wet coconuts, karite nuts for the production of shea butter, palm oil, sunflower seed oil, and protein-enriched cassava.

Palm Oil Processing

Palm oil is the principal cooking oil in Cameroon and an important source of vitamin A and calories. It also is the basis for local soap manufacture. Most of the palm oil production (89%) is for domestic consumption (1). Seasonal variations in palm oil prices are significant and mostly reflect changes in supply. The poor distribution system in Cameroon exacerbates this situation (2).

Traditional methods of processing palm are labor intensive and have low productivity. Typically, fresh fruit bunches (FFB) are harvested, washed, placed in a pit and covered with leaves for up to five days. After this initial fermentation, the fruits are stripped from the stalks and boiled for several hours to loosen the pulp. The hot fruits then are either pounded with a wooden mortar and pestle or stomped by foot. The oil is skimmed off and either cooked to halt the formation of free fatty acids (FFA) or left to ferment for 3–7 more days. The resulting oil is clarified by boiling and decanting and then used or sold (1). Roughly 52 man-hours are required to process 16 FFB of palm fruit by the traditional process (3).

Until 1972, the Colin Company, a French firm, exported a small-scale, hand-powered expeller that used two counter rotating screws in a perforated metal cage to expel oil. The Colin expellers use a relatively sophisticated system of gears to convert the rotary motion of the hand wheels into the rotary motion of the expeller screws. The old Colin expellers are extensively used in parts of Cameroon for commercial and artisanal production. However, the expellers are no longer being produced by the manufacturer. Though the expellers are very robust, replacing broken and worn parts has become a problem.

In order to substantially increase the productivity and income of small-scale (farming less than 10 hectares [ha]) palm oil producers in rural Cameroon, ATI decided to work with a local organization to develop, manufacture and market an integrated processing technology.

Project Description

Originally, the project planned to purchase the broken Colin expellers, recondition them, and then resell to local producers. Despite maintenance difficulties and disrepair, most of these expellers are still being used and their owners did not want to sell them. Only 12 could be purchased at a reasonable price. Imported parts were difficult to obtain because the expellers were no longer being manufactured; thus repair costs were very high.

As a result, the project shifted focus to manufacturing a new, lower cost horizontal axis expeller designed by Carl Bielenberg and the Altech Co. (named the Caltech press). The new press is hand operated and, if desired, can be motorized at a later date. ATI's project partner, the Association Pour la Promotion des Initiatives Communautaires Africaines (APICA), established a workshop,

Outils Pour les Communautes (OPC), to manufacture the expeller, an oil clarifier, and a bunch stripper. Further design innovations produced an even lower cost vertical axis expeller, which can better meet the needs of small-scale oil palm producers.

Palm Fruit

Oil palm varieties vary significantly in their oil content and fruit yields per hectare. The Dura and Tenera varieties have the largest commercial significance. Dura fruit has a thick shell and a medium amount of oil-containing pulp. Typically, the oil content of Dura is from 8% to 11% of the weight of the fresh fruit bunches containing the palm fruit as well as the stalks. Although yields can vary significantly, Dura yields range from 1–3 tons of FFB per hectare, depending on weather and degree of cultivation (2).

Tenera is a dwarf hybrid of the Dura and Pisifera varieties. Tenera palms have a smaller kernel and a thinner shell. The oil yields for Tenera average around 22.5% of the FFB weight. Tenera palms yield an average of 9.0 tons of FFB per hectare in plantations in Cameroon and more than 15 tons of FFB per hectare in Latin America and Asia (2). In general, extraction inefficiencies present in artisanal and traditional processing produce an actual yield far below the potential. In addition to yielding more oil, Tenera palms are much easier to harvest than Dura. Dura palms grow up to 10 meters high, with the bunches located at the crown. It is necessary to climb the trees to harvest the bunches with a knife or machete, which is dangerous and unpopular. Because Tenera palms average between 3 and 5 meters in height, a harvester with a sickle hook can free the bunches while he is standing on the ground.

Technologies

Bunch stripper: A manually operated bunch stripper removes the cooked (sterilized) fruits from the palm stalks. The technology is based on a design by Hartley (4). It consists of a hexagonally shaped barrel with wooden slats for sides (Fig. 1). The drum is filled with three or four cooked FFB and closed. The handle is turned first one way and then back to create a shaking or agitating motion. The bunch stripper reduces the time required to strip palm fruit by 67–75% (5).

Fig. 1. Hand-Operated Bunch Stripper (5).

Caltech Expeller: The horizontal Caltechs were designed to replace the Colin expellers (Fig. 2) that are no longer available in Cameroon. Each consists of a single expeller screw inside a metal cage. The stripped and cooked fruits are fed into the hopper, and a feed screw forces the fruits through depulping knives. The expeller screw picks up the depulped fruit and presses it against the walls of the cage and against the back pressure nut at the end of the cage. Oil is expelled through the cage and into the pan below for later clarifying.

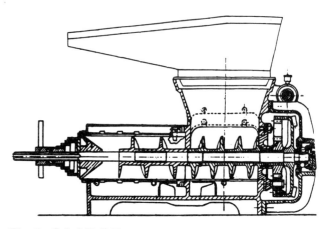

Fig. 2. Colin Mill (3,5).

In the manually operated, horizontal Caltech, the expeller screw is turned by a pair of hand cranks operating through a worm-gear reducer (Fig. 3). In the motorized version the hand cranks are replaced by a motor and system of step-down pulleys and belts (Fig. 4).

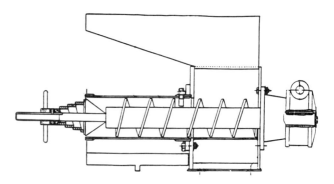

Fig. 3. Horizontal manual Caltech expeller (3).

The vertical Caltech reorients the expeller axis and eliminates the expensive worm-gear reducer (Fig. 5). The reducer, the only imported component of the horizontal expeller, significantly adds to the cost. The expeller screw and depulper on the vertical press are essentially the same as on the horizontal models. The screw is turned by a capstan operated by two people and the oil is deposited in a collection tray at the bottom.

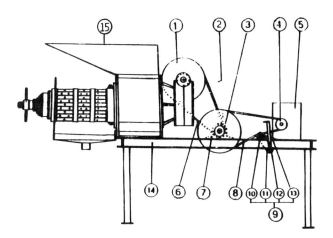

Fig. 4. Horizontal motorized Caltech expeller. 1, gear reducer; 2, intermediate pulley; 3, intermediate pulley (small); 4, motor pulley; 5, 2 HP engine; 6, double belt transmission; 7, intermediate shaft bearing; 8, single transmission belt; 9, clutch system; 10, clutch bearing; 11, clutch handle; 12, clutch lever; 13, clutch spring; 14, chassis; and 15, intake hopper (3).

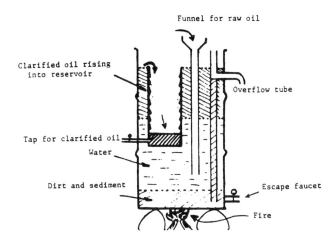

Fig. 6. Drum clarifier (5).

the top while the sediment is deposited on the bottom. Clarified oil spills into the reservoir which is tapped off into individual containers.

Operating Size/Capacity

Large processing facilities are defined as those capable of processing more than 10 tons of FFB per hour. Minimills typically can process between 1 and 10 tons of FFB per hour and artisanal/traditional processing utilizes less than 1 ton of FFB per hour. ATI has focused on artisanal scale production in Cameroon.

Within the artisanal range, a broad mix of technologies is available, each with differing capacities and costs. Small, imported motorized hydraulic presses can process up to 1 ton of FFB per hour, but are expensive to buy and maintain. Smaller screw-operated batch presses can process around 90 kg of FFB per hour and are very inexpensive (6). These compare favorably with the traditional process which only processes 1–2 kg of FFB per hour.

The Caltech expellers are intended for small service pressing organizations, local cooperatives, and small farmers. The motorized, horizontal Caltech expeller can continuously process about 350 kg of FFB per hour. Its hand-powered equivalent can process between 180 and 200 kg of FFB per hour, compared to 120–140 kg with the lower-cost vertical axis version of the Caltech expeller. Oil extraction rates achieved from the Caltech family of presses average from 16% of the FFB weight for Tenera to 7% Dura.

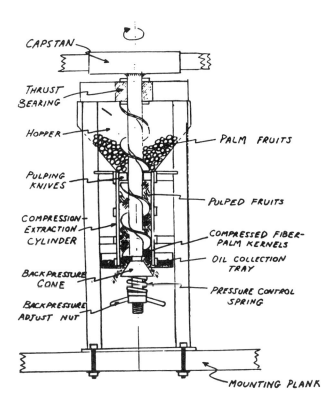

Fig. 5. Vertical Caltech expeller.

Clarifier: The clarifier removes the water, fiber, and sediment contained in the raw processed palm oil. Traditionally, the unclarified oil was boiled to remove water and the lighter oil was skimmed off of the top. The sludge was discarded or used as an animal feed. The drum clarifier shown in Fig. 6, separates the clear oil from the water and sediment. Raw oil is poured into the funnel and heated by the warm water. The warmed oil rises to

Products and Byproducts

The primary product is palm oil which is an important part of the Cameroonian diet. Due to the oil's high FFA content and high production costs, it only serves the domestic market. Expelling of palm fruit also produces dry pulp residues. The residues are currently used only as fuel for the clarification and cooking steps. No other potential use (e.g. animal feed) has been explored or documented.

The palm kernels, which remain after the depulping and pressing, can provide a supplementary source of a different type oil, largely for cosmetic and medicinal uses. Although the kernels are difficult to crack and the kernel itself has a tendency to cream during expression, oil from the palm kernel is a potential byproduct of palm oil processing. Because Dura kernels are larger in absolute size and higher in oil content than Tenera kernels, palm kernel oil processing is only done with Dura. Because intermediate technologies are not yet widely available for processing palm kernels, traditional methods are used by rural processors. Typically, a mortar and pestle are used to crack the kernel shells. The paste made from the kernels is then heated to separate the oil, water, and sludge phases. The kernel oil is stored in containers and the sludge and water are discarded. Generally, however, most artisanal producers do not bother with palm kernel oil production. Instead, producers use the kernels for fuel for cooking palm fruits and for clarifying palm oil.

Problems Encountered, Solutions, and Future Plans

By early 1988, 50 expellers of all three types had been manufactured, sold, and installed. Some smallholders use the expellers on their own harvest, and in addition do service processing of neighbors' palm fruits. Clarified oil is provided in exchange for 20% of the finished product, which increases the income for both the farmer and the processor. Compared to the more mechanized processes, the Caltech is labor intensive. Each expeller can employ 3-4 people part-time. With the Caltech, more palm oil can be processed and the oil can be stored longer than with traditional processes.

Several issues require further analysis as this technology becomes more widespread.

Increased oil processing may have an impact on the distribution of labor and income within a family and a village. Traditionally, palm oil production was largely women's work; the men tended and harvested the palm. The increased processing capability may well change these roles.

The form of labor for the vertical Caltech may be a problem in some localities where walking around in circles while turning a capstan is perceived as animal's work not suited to humans. Some users have indicated that the manual expellers are difficult to operate. Some modifications to increase the mechanical advantage of the hand cranks and capstan have already been incorporated to decrease the required effort.

Due to their desire to be perceived as progressive farmers, some farmers want a motorized expeller, even when it is not economically justifiable at their scale of production. This factor may limit the market for the manual versions in certain areas. The motorized version does not add significantly to the throughput of the process because dwell time and pressures in the expeller cage are more important for good oil output. Increasing power requirements and excessive heat buildup in the pressed material also limit the speed of production.

Most users have elected not to purchase the special clarifier or bunch stripper, and there is some question about the financial viability of this auxiliary equipment as a separate investment.

Future plans call for replication of all three presses in other countries and regions. A replication project in Zaire is being planned using local manufacturing facilities and credit sources. A recent market study exploring the potential for replication or export of the Caltech expeller showed favorable prospects in three of the four countries studied: Benin, Togo, and Guinea Conakry, but not in the Ivory Coast.

Sunflower Seed Oil Processing

People in rural communities in Tanzania traditionally use sunflowerseed oil for cooking and as a source of calories. Unfortunately, traditional processes for sunflower oil extraction are very time-consuming and inefficient. Typically, the undecorticated seeds are sun-dried and crushed with a large wooden mortar. The resulting paste then is heated in a vessel with water. The lighter oil floats to the surface and is skimmed off and bottled (Rukuni, 1988). The process, which is mostly done by women, is arduous, and the remaining residues still contain a large amount of oil. Finally, unlike groundnuts and soybeans, sunflower seeds are not consumed in any other form (i.e., flour, paste), so utilization of sunflower seeds is very low.

The alternatives to traditionally processed sunflowerseed oil are limited. Though sunflower, groundnut and soybean oils are processed in the urban areas, transportation costs to move the seed to the city and the oil back to the rural areas are prohibitive. Rural incomes are low due to the low value of unprocessed agricultural products, and prices for processed oil are high. The rural markets for oil are "thin;" i.e. the availability and prices for oil vary considerably depending on the agricultural season. A project that could increase the availability of reasonably priced cooking oil and raise agricultural incomes would significantly improve the quality of life in rural Tanzania. ATI would capitalize a private factory in Tanzania to manufacture affordable, hand-operated, sunflowerseed oil presses for use in Tanzanian villages.

Project Description

The project began to change almost from the moment of its inception. Originally, ATI was to provide technical assistance to our partners in Tanzania to locally manufacture a screw-operated, scissors-jack batch press. Difficulties arose concerning the mechanical life and strength of several of the components as well as the cost of each press. The shop that was manufacturing the presses earned most of its revenues from other product lines and focused on those products.

An ATI engineer recommended a relatively radical departure from the existing design. A prototype was designed, manufactured, and tested. Both designs were developed in tandem for testing later in the project.

Technologies

The village oil project was to develop a complete processing technology for sunflower seeds, including a decorticator and an oil press. Additional equipment might also include a winnowing machine, though ideally this would be a part of the decorticator.

Decorticator: The decorticator evolved from a Japanese design that used a rotating disk with three fins to hurl the seeds against a hard rubber stop. The disk was hand cranked by two people, and pulleys were used to increase the rotational speed of the disk. Seeds were fed into the center of disk, accelerated to roughly 100 feet per second and ejected onto the stop. The hand crank also powered a small fan which separated the nut meats from the shells. A bicycle powered version that could be operated by one person also was developed.

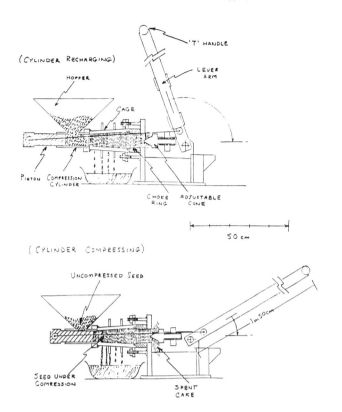

Fig. 7. Bielenberg ram press.

Ram Press: In the Bielenberg ram press (Fig. 7), sun warmed seeds are fed into the charging cylinder from a hopper. The ram is pushed backward to charge the pressing cylinder with fresh seed, then it is pulled forward to compress the seed. The expressed oil squeezes out through a cage assembly, and the spent seed moves through a choke ring and out the end of the cage. The ram operates with a lever arm designed to maximize the mechanical advantage, trading a long stroke for high ram forces. By continuously rocking the level arm back and forth, fresh seed is moved into the cage, compressed, and ejected.

Clarifier. The sunflower oil process is complete when the oil is clarified or filtered to remove suspended matter. Normally, the dark coating of the hulls imparts a fine black sediment to the oil during pressing. A filter screen (using cotton fiber) or sedimentation tank removes these impurities. Any foreign matter that entered during the pressing operation is screened out at the same time.

Operating Size/Capacity

The press is the limiting element in the decorticator-press-filter technology. Capacity of the Bielenberg ram press varies according to a number of factors. The type of seed that is used, hybrid or local variety, will be discussed under inputs. Capacity may also depend on whether the seeds have been decorticated (shelled). Shells increase the volume of material in the press cage; increased effort is required to operate the press; and the amount of oil per ram charge is decreased. There is also concern that the shells absorb a portion of the expressed oil rather than allowing all of the oil to escape, but this has not been tested. One person can operate a well-made press for as many as five hours at a time.

Using undecorticated, hybrid seeds, the ram press can press a 55 kilogram sack in 3 to 4 hours. The sack of seed will yield between 14 and 19 liters of clarified oil. Extraction efficiencies of 60%–75% are typical. Using undecorticated local varieties of seed, it would take more than 6 hours to press a 55 kilogram sack, and the yield would drop to between 8 and 11 liters of oil per sack. Assuming a six-month pressing season and five-day work week, the press could process more than 3000 l of oil per season (8).

Press Inputs

The ram press technically can process undecorticated seed, but decortication is preferred. Pressing undecorticated seed increases both the wear on the cage and bearing parts and the effort required for operation. Local and hybrid varieties of sunflower seed are both available in Tanzania. The local varieties tend to have thicker, harder shells with a smaller meat weight per volume than the hybrids. Consequently, the potential and realized oil yields of local sunflower seeds by volume of undecorticated seed also are smaller. Typically, hybrid seeds have 0.42–0.58 liters of oil per kilogram of seed and the local varieties contain 0.29–0.40 liters per kilogram (8,9).

Products and Byproducts

Oil is the primary economic and nutritional product resulting from sunflower seed pressing. In this regard, the oil produced compares very favorably with the commercially available oil in southern Africa. The lack of further refining leaves the oil with a darker yellow color and stronger flavor that appeals to traditional tastes in the region (8).

The residue remaining after pressing also has substantial value. The pressed mass is often sold as a poultry feed extender. In poultry feed, the residues often are mixed according to a ratio of one part residues and four parts local grasses and feed grains. The relatively high oil content of the residue (typically 25% to 50% of available oil) makes lower ratios impossible (8).

Problems Encountered, Solutions, and Future Plans

Many diverse development agencies in Africa have purchased or are beginning to manufacture presses based on the ram concept. The press design has been copied by a number of other manufacturers in Tanzania (neither ATI nor its project partners own a patent on the press), and they are beginning to manufacture credible ram-actuated presses. Sufficient field testing has been conducted to justify ATI's in-depth engineering analysis of the press. The results of the analysis led to substantial modifications that improve its performance (10).

The technology has remained a cost effective alternative for small-scale, rural pressing operations. From a manufacturing standpoint, the press requires only standard machine tools such as lathes, milling machines, welding equipment, and drill presses and can be made in most rural shops. Relatively little additional capital equipment is needed for its manufacture. The simplicity of the design has kept both manufacturing and maintenance costs low.

Finally, the oil yields have been substantially higher than yields achieved by traditional methods, and the costs of production are lower than for commercially available oils.

The major problems do not pertain to the technology—but to the "software." Because the Bielenberg press is a new technology, the initial phases of product development required substantial technical support and management. The project created a great deal of interest before ATI could provide required technical and managerial assistance. Future projects should try to anticipate or monitor market reactions and provide appropriate levels of technical and managerial support.

Second, quality control and manufacturing issues have hampered the dissemination of the technology. Workers may omit or change important elements if they have not been properly trained and do not thoroughly understand the design concepts. Then the technology becomes discredited by failure and poor performance.

Finally, the press works best on hybrid varieties of sunflowerseed. Wear and tear on the machine is reduced, oil yields are higher, and hence, higher earnings are generated. Using hybrid seeds also reduces the physical effort required to operate the press. Unfortunately, the hybrid varieties of seed are more difficult to obtain and they are more susceptible to bird predation than local varieties. If the technology is to achieve its full potential, access to and protection of good quality seed is essential.

ATI plans to expand the use of the technology into other regions and countries. ATI is replicating the sunflower seed press in Zimbabwe, and other organizations are proposing similar projects in both Rwanda and Malawi.

Protein Enrichment of Cassava Root by Fungal-Yeast Fermentation

In Thailand, some types of animal feed ingredients—broken rice, rice bran, cassava, and fish meal—are produced locally. However, important protein sources such as corn and soy press cake, are imported and therefore are expensive. Beginning in 1978, the Government of Thailand launched a campaign to help small-scale swine farmers compete with highly capitalized, large-scale piggeries. Province-wide cooperatives were set up to organize marketing channels, reduce feed costs by establishing small-scale feed mills, and to reorganize swine producers into specialty groups (e.g., breeders, growers).

The Thailand feed industry was highly concentrated; in 1984 one sixth of the mills produced about one million tons or 72% of all formulated feeds. The feeds were very expensive or otherwise not available for small farmers. The 600,000 tons produced at farm level however did not, in many cases, contain adequate protein for swine and poultry. It therefore appeared possible that a market for protein-enriched cassava existed not only for small farmers but also for small feed mills.

Since 1959 experiments have been conducted on various ways to ferment cassava using various microorganisms to detoxify or otherwise render cassava more fit for food use (11). Cassava however contains only 0.8–1.0% protein on a 65% moisture basis, 0.2–0.5% fat, and 0.8% fiber; cassava is primarily starch and soluble sugars. Therefore, it was necessary to find ways to improve the protein content of this crop which is widely grown in Thailand.

Literature searches by Kasetsart University showed that most of the fermentation processes research that had been carried out in submerged substrates required substantial investments, prohibitive for small-scale commercial and artisanal methods. Koji-type cassava fermentation methodologies were chosen as an alternate approach. This technology could be carried out using mostly locally available equipment and raw materials, would require minimal imported or expensive inputs, and could be managed by persons with limited technical and financial resources.

Objectives and Project Description

The project's objective was to adapt a serial, solid state, two-step fermentation process utilizing dried cassava chips from pilot scale to commercial size that would be manageable by small farmers. It was to commercially test the more promising Koji-type fermenting processes and/or equipment tested at laboratory and pilot level at Kasetsart University. Several serial, solid state, fermentation procedures were tested. These either were too capital intensive, too technical to be managed easily, required extensive imported inputs or produced unacceptably low levels of protein. A brief description of each follows:

Column Fermenter: Wood frame column 25 × 25 × 80 cm incorporated acrylic plastic trays, a water chamber below a stainless steel screen and an air compressor and vents for aeration; fermented 25 kg of cassava per batch.

Ribbon Mixer Reactor: Open top cylindrical tank was fitted with a helical mixing blade and aerator; fermented 50 kg of cassava per batch.

Polyethylene Lined Baskets: Baskets measuring 35 × 40 × 35 cm were lined and covered with 4 mil polyethylene sheets and loaded to 4 cm deep with inoculated cassava.

Bamboo Baskets: Round woven bamboo-strip baskets 60 cm in diameter and 60 cm high, covered with damp jute sack material, utilized 50 kg of cassava per basket.

Fermentation on Concrete Floor: Inoculated cassava chips were spread to 4 cm deep on concrete floor and covered with moistened jute sacks.

Fermentation on Concrete Floor with Continuous Mixing: A crawling mixing machine mixed a bed of inoculated chips 10 m long, 2 m wide and 20 cm deep, 6 times per day for 7 days.

Rotating Drum Fermenter: A 200 L stainless steel drum was mounted to rotate manually on its long axis, with 2.5 cm square wooden cleats bolted to the inner surface. Humidified air was blown through the drum. 50 kg cassava were used per batch.

Motorized Drum Reactor: An 80 cm diameter by 2.4 m long cylinder was fitted with mixing baffles and humidity controlled air, driven by 1/4 HP motor rotating at 15 rpm.

Cabinet Fermenter: A cabinet 70 × 40 × 35 cm was assembled to contain 3 shelves of wire baskets each containing 40 kg of cassava.

In 1984, ATI was asked to support a project to determine the commercial feasibility of enriching cassava with protein through a Koji-type fermentation method to provide a swine feed for small farmers in Thailand (J. Kumnuanta, personal communication). Unpublished experiments carried out at Kasetsart University showed a Koji-type process could increase cassava protein up to 25%. Preliminary feeding trials showed a 15% protein product to be acceptable as a complete grower swine ration, at 70% of a finisher swine ration, and at 50% of a breeder sow ration. The project was designated to scale up the laboratory/pilot scale procedure to a commercial cottage-scale process manageable by swine farmers using dried and chipped cassava. Major inputs were to be the fungal and yeast inoculants, certain chemicals for buffering, and a nitrogen supply for microorganisms.

The staff at Kasetsart University had catalogued a large collection of molds and yeasts, and selected those most efficient for cassava fermentation. A submerged culture is known to be more efficient but capital and energy costs of sterilizing a very large autoclave containing cassava, water, and chemicals plus a process to remove large quantities of water made it less suitable for small rurally located operations. The proposed process, unlike a submerged culture, would not add more than 40–45% moisture or require autoclaving, but was expected to produce, on a commercial scale, a stable product containing 12–15% moisture and 14–20% protein. The process was designed so that 1 cm cube-size, dried cassava plus 5% rice bran (to provide growth factors and minerals for microbial growth), 5–7% ammonium sulfate, and water added to 40–45% moisture would provide optimal fermentation conditions. The prepared substrate was inoculated with mold spores and fermented for three days. The yeast was added and fermented for an additional two days. The fermented product was ready to use wet or dried, or could be added as a feed supplement. Anticipated difficulties were contamination with wild organisms, aeration and carbon dioxide removal, and temperature control.

Two molds (*Aspergillus niger* and *Mucor* sp.), both heat tolerant, were selected for testing in stage 1 of the fermentation of cassava chips. *Saccharomyces cerevisae* and *Candida* sp., which grow on partially hydrolyzed starch, and tolerate temperatures to 42°C were the selected yeasts for the second stage. Lactic acid bacteria, which proliferated at pH 3.5, were present and regarded as beneficial in delaying overgrowth by wild contaminants.

Technical and Inputs

Procedures for preparation of mold and yeast cultures followed those of Kumnuanta (personal communication) and Kumnuanta and Chettanachittara (personal communication). Commercial testing of the Koji fermentation process using these inoculants is shown in Figure 8.

A concrete tank measuring 1.2 × 4.75 × 0.8 m was constructed with a perforated plate fixed 40 cm above the floor of the tank. A one-quarter hp squirrel cage fan exhausted air out of the bottom. The 500 kg cassava substrate was prepared for fermentation by blending in a mixture of 2 L sulfuric acid in 350 L water and 50 kg ammonium sulfate, and then letting the blend equilibrate for two hours. The charge was inoculated with 0.5 kg of mold spores carried on 25 gm of rice bran; then the charge was covered with polyethylene sheets. After 20 hours aeration was started. The observation of mycelial growth starting at about 72 hours indicated the time to add the yeast starter. After 48 hours the charge was unloaded and sun dried to a moisture content of 12–15%.

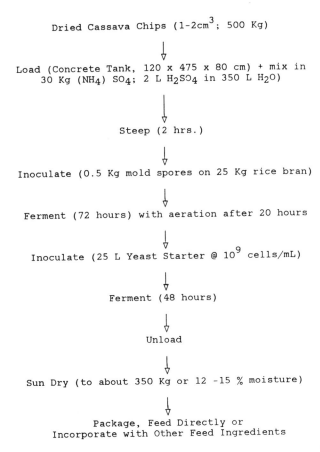

Fig. 8. Koji fermentation process.

Results, shown in Table 1 from 6 trials showed an average mass loss of 32% and protein of 11.5 and 10.3% using *Aspergillus niger* and *Aspergillus niger-Mucor* respectively for the first stage and *Saccharomyces cerevisae* for the second stage.

TABLE 1

Protein Content of PEC in the Koji-Type Reactor

		% Protein by mold species	
Trial No.	Mass loss (%)	A. niger	A. niger & Mucor
1	35	9.8	11.2
2	33	11.8	10.0
3	33	12.3	10.2
4	30	11.6	10.4
5	31	10.9	9.6
6	31	12.7	10.8
x̄	32	11.5	10.3

Temperature control was not entirely satisfactory; 40°C was reached, which cannot be tolerated by *A. niger*. Four strains of *A. niger* namely 50M, PM41, PM50, and PM74 and one *Mucor* W252, grew well at 45°C and demonstrated high amylolytic activity. Strain W252 showed high α-amylase and glucoamylase activity, including raw starch digesting activity, and is capable of activity at 38–50°C. This was used in a mixed culture with *A. niger* in anticipation that *A. niger* would grow first at lower temperatures, and the *Mucor* sp. would grow well when the temperature exceeded 30°C.

Problems, Solutions, and Future Plans

In Thailand, commercially available dried cassava chips vary greatly in quality partly due to the absence of quality standards as evidenced by inclusion of sticks, stones, solid stems, woody portions of roots and other foreign materials. Therefore fermentable carbohydrate varies substantially (40–70%), subsequently affecting protein production. Season of supply affects quality through variations in available moisture and degree of contamination by competing organisms.

Chip size variation influences rates of fermentation since large pieces (3–4cm^3) reduce penetration of microbial enzymes into the interior portions reducing both starch hydrolysis and amylase activity, as evidenced by the residual ammonium sulfate in the interior of large chips.

Optimum moisture after steeping was found to be 40% for inoculation with *niger* alone, or 40–45% with *A. niger* and *Mucor*. More than 45% water caused water logging and dense, sodden areas which encouraged anaerobic or facultative anaerobic growth, evidenced by ammoniacal and putrid odors (12).

Aeration maintained the temperature in the appropriate ranges and encouraged aerobic conditions. A one v/v/minute aeration gave suitable results.

Sporulation is evidently inhibited by the presence of wild yeasts and bacteria (13). These contaminants are thought to metabolize the sugars as quickly as they are produced. Reducing sugars were found in the 2–4% range after 72 hours, although the percentage should be higher (14). After 72 hours *Saccharomyces cerevisae* is inoculated to prevent sporulation and to take advantage of higher sugar levels released by mold enzyme action. Addition of agricultural fertilizer (ammonium sulfate:urea-3:2) as a nitrogen source proved a suitable procedure (15).

Cassava residues from large scale processing, such as pelletizing for feed stuffs, also could be used for producing protein enriched cassava (PEC). Such materials have little economic value, but may be heavily contaminated; the extra care required in processing could be disproportionate to their value. The texture of the fermented material from these residues posed little problem during feedings (no packing or sogginess) and yielded 12% protein.

Examination of all PEC samples for aflatoxins and for hydrocyanic acid was very encouraging. The range of hydrocyanic acid for dried cassava was 13–60 ppm. PEC samples showed a reduction to 4.87–5.98, with a mean of 5.4 ppm; well below any limit of toxicity. Animal feeding trials showed no acute or chronic effects in chicks, mice, or piglets. Buffering the substrate at pH 3.5 and fermenting at high temperature are thought to contribute to the volatilization of hydrocyanic acid.

Aflatoxin analyses of PEC showed a range of 0.15–0.19 ppm with a mean of 0.17 ppm; again well below any limit of toxicity. No toxicity was seen in experimental animals. These results suggest mycotoxin producing fungi may be inhibited by the two molds chosen for starch hydrolysis (16,17).

Many economic and commercial assumptions made to justify the project and to estimate the social and economic benefits at the beginning were not entirely correct. The competing feed components were greatly underestimated. Corn production in Thailand improved dramatically and broken rice quantities and prices were very competitive with PEC. Both competitors contained protein at comparable or competitive percentages. The lower protein yields shown in Table 2 put PEC out of the competition.

TABLE 2

Protein Content and Feed Ingredient Prices, Thailand, 1984

Ingredient	Per cent protein	Price ($U.S./Kg)^a of protein
PEC	11	0.246
Corn	8.9	.089
Fish meal	57.0	.432
Soy press cake	44.0	.272
Broken rice	8.0	.090
Cassava chips	2.0	.092

^{a} Imputed protein and estimated cost of production.

PEC could serve as a feasible protein feed ingredient in other cassava growing areas of the world where corn is not used for feed, where pork or poultry are important foods, and where major protein contributing feed ingredients are imported.

Production of PEC by some Koji methods described here meet the needs and resources of low-income farmers. The larger Koji-type fermenter (due to its higher capital cost) may be more appropriate for a cooperative or feed mill. Future work should be directed toward determining whether raw, fresh, cassava chips can be fermented as easily. In all, the project proved to be a technical success but needs some refinements and more commercial testing if it is to be marketed elsewhere.

Wet or Aqueous Coconut Processing

Coconut farmers and coconut product users are influenced greatly by world export coconut oil and copra prices. The Philippines, Indonesia, Sri Lanka, Papua New Guinea, Pacific Islands, Malaysia, and Singapore depend on coconut-derived products, such as oil and copra for a significant portion of their export earnings. Oil prices did rise in 1974, 1979 and 1984, reached low levels in 1972, and 1982 but generally have fallen over the last 15 years. This is due mainly to increased production from increased yield per tree and an expansion in the planted area. The export demand also depends on world industrial activity and the supply of other vegetable oils, which increased from 37.9 million tons in 1975 to 59.3 million tons in 1985. Improved processing technologies for soybean oil and increased production of palm oil also have reduced the competitiveness of coconut oil (18). In addition, taxes, tariffs and quality standards set by importing countries have reduced market opportunities for many coconut derived products (19,20). All of these factors have resulted in reduced returns to farmers.

Most coconut oil is processed from copra using a dry process. The byproduct, copra meal, is routed to animal feed if the oil content is not too high, contaminated with aflatoxins, or greatly denatured. Copra processing and oil refining using conventional equipment require large investments beyond the means of many village or small-scale investors.

The traditional household wet process extracts oil from fresh mature coconuts, rather than copra. Its main advantage is that it can yield a variety of edible protein, fibrous and water-soluble byproducts. However, without a reduction in edible oil production, only about 50% of the available oil is recovered if only one pressing is done. Because the milk is cooked to separate the oil, the oil sometimes takes on a dark color and a burnt taste and odor. The process is also very laborious and consumes a lot of energy.

Since 1921, improvements have been made to develop a commercial-scale wet process (21,22,23) (J. Banzon, personal communication). Two technical problems have been the center of this research and development: to increase the amount of oil released from the coconut meat in the form of milk above the 80% limit thus far obtained; and, to free the oil from the stable milk, oil in water emulsion, by breaking either the emulsion or by forming an emulsion with a higher oil content. (Banzon and Velasco, personal communication).

To date, depending on the technology used, only about 72% to 80%, of the 32 to 34% oil in the fresh meat has been recovered. The higher oil yields, obtainable from dry copra processing (91–93% using expellers and 97% to 99% using solvent extraction) have inhibited wider commercial use of the wet process. Most of the wet processes developed have not been commercially tested. Those wet processes tested have required larger capital investments than medium to large scale copra mills due to the requisite concentration of sophisticated food processing equipment for other food byproducts. The wet process, which is more labor intensive than the dry process, has been relatively successful in India and in the Philippines (20).

Objectives

The primary objective was to commercially test a simple "wet coconut processing" technology, an improved scaled-up version of the traditional household procedure, which could produce a high quality edible oil and other edible byproducts. A secondary objective was to locate and/or create markets and develop products which would allow more value to be added to byproducts to increase the venture's overall commercial viability.

Project Description

The project was to direct coconut processing away from copra toward a modified version of the more labor-intensive wet process developed by J. Banzon and other Philippine researchers. Edible oil for local consumption was still considered to be the main product. Market needs and processes to produce products to meet those needs were explored to permit production of higher value added products from such byproducts as pressed coconut meat, skim milk, protein curd ("latik"), water, shell, and husk.

A village-scale coconut wet processing operation handling 2,000 coconuts per day was established in Tagbanon in the province of Negros Occidental, Philippines in 1986. Appropriate Technology International supplied the venture capital, the Filipinas Foundation supplied the equity financing, and a private local entrepreneur supplied the land, building, and support facilities. Local coconut farmers would be allowed to purchase shares in the venture at a later date. This operation was an adaptation of an operation in Tulong, Negros Oriental, Philippine; however, more emphasis was placed on better byproduct utilization.

Size (Capacity): The operation had a rated capacity of 2,000 coconuts/day using one shift with possible expansion to 5,000 coconuts per day using three shifts. The 2,000 nut input was expected to provide 720 kg fresh coconut meat, 490 kg husk, 470 kg coconut water, 320 kg shell, and 150 kg reject nuts. The fresh coconut meat was expected to produce 231 kg oil, 185 kg of skim milk and 304 kg of wet-pressed meat.

Raw Material and Utilities

Mature whole coconuts purchased from local farmers were the primary input. A reject rate of 6% occurred, but most of these rejected coconuts were converted to copra and sold to copra buyers. Other needed inputs were water for the pressing operation, soap and steam production, and general cleaning, and sanitation; electricity for operating the graters; and chemicals for soap making. The plant employed 17 persons.

Capital investments included 1.25 hectare of land, a 450 square meter building, and equipment and support facilities (Table 3). Equipment and support were valued at $6382 U.S.; total cost was $15,182 U.S. at mid-1986 values.

Technology Selection

The wet process was chosen over a scaled-down version of the dry or copra process because: the edible oil does not require extensive refining as it has a lower free fatty acid (FFA) level, lighter color and greater stability than semirefined dry process copra oil; after the milk (oil) is removed, the pressed meat ("sapal") is still edible and useful in a variety of food products. If dried, because of low aflatoxin levels, it may be sold on the export market; other edible byproducts such as the water and skim milk could be converted to value added products; and capital costs are closer to investment capabilities of small towns and rural investors.

The process used, illustrated in Figure 9, is as follows: (a) After dehusking and cracking open mature nuts, the fresh meat is removed from the shell with a motorized spindle type grater. Grated particles are 0.15 to 0.4 mm in size. (b) Grated meat is pressed in a hydraulic press at 17.5 kg per square centimeter, to yield milk which is 56 to 58% of meat weight. A second pressing of the meat after mixing it with an equal weight of water or coconut water yields an additional 8 to 9% milk. (c) The milk is mixed with water at 25% of milk volume and allowed to set 2–3 hours to facilitate separation of emulsion cream from the skim milk. (d) The cream is decanted or skimmed off and heated to 100–103°C in an open kettle to break the emulsion. The residual water boils off leaving raw oil and latik; the latik constitutes about 5% of the fresh meat weight. (e) Raw oil is transferred to the steam deodorization and neutralization tank for refining, filtering and packaging in 5 gallon containers. Edible oil accounts for approximately 51 to 56% of the oil contained in the fresh meat.

TABLE 3

Equipment Costs of the Tagbanon Village-Scale, Wet Process Plant (mid-1986 values)

Item	Cost (US)	Expected lifetime (years)
Preparation		
Dehuskers (2)	45.00	10
Shelling knives (3)	22.50	5
Coconut water collectors (3)	35.00	5
Spindle grater (6 stations)	188.30	5
Electric motor for grater (2 hp)	118.30	5
Coconut meat collectors (8)	70.00	5
Oil production		
Hydraulic press	1,351.20	5
Coconut milk collectors (5)	43.75	5
Cream separators (7)	42.00	5
Cooling pans (2)	130.00	10
Heating control	250.00	5
Steaming kettle	300.00	10
Steam boiler	796.25	5
Latik containers (5)	50.00	5
Filter pans (4)	225.00	5
Clay stoves (3)	125.00	5
Plastic drums (7)	94.00	5
Settling tanks (35)	192.50	5
Deodorizer tank	199.35	5
Pasteurizing tank	199.35	5
Soap		
Mixer	436.75	5
Soap molds (2)	200.00	5
Cutting tables (2)	70.20	10
Stamper	50.00	5
Miscellaneous		
Small weighing scales (2)	24.50	2
Large weighing scales (2)	250.00	5
Thermometer	250.00	2
Glassware	200.00	2
Clock	11.25	4
Push carts (2)	12.50	2
Drying beds (4)	35.25	10
Oil container racks (2)	47.25	2
Chimney	216.50	5
Total[a]	6,282.15	

[a] Excludes equipment for making vinegar due to the low marketability of the product.

Sources: (24); M. Cuano (personal communication)

Byproduct Processing

Husks are burned to fuel the cream cooker and steam generator.

Shells are carbonized in a below-ground kiln to yield 28 to 31% charcoal.

Pressed meat or sapal is sun dried for one day and repressed to remove oil for soap. The oil makes up about 12 to 20% of available oil depending on the efficiency of the initial two pressings of fresh meat. The deoiled pressings comprising about 23% of the original fresh meat are sold for ruminant feed.

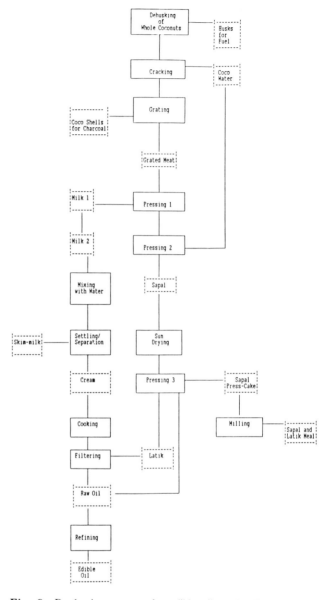

Fig. 9. Production process for edible oil at the Tagbanon wet coconut plant.

Latik is also pressed to recover oil for soap. This oil constitutes 6 to 7% of the available oil. As the protein from the latik is denatured, it is not edible and is discarded.

Coconut water and skim milk could be converted to vinegar, but heavy market competition and the low demand for coconut vinegar made this option unprofitable.

All of the equipment was fabricated domestically and was simple and easy to operate and maintain. Stainless steel was used on those equipment pieces producing edible products. A midterm technical evaluation in 1987 pointed out several changes in procedures and equipment that would increase oil yields and overall productivity.

Problems Encountered, Solutions and Future Plans

The major obstacles reducing commercial viability are listed below:

Raw Materials. The purchase price for mature coconuts was higher than the prevailing equivalent prices for copra. This was done to encourage the farmers to sell mature coconuts to the operation for start-up purposes. The size of the purchased coconuts was also smaller than average. Setting quality and size standards and providing a steady reliable alternative market outlet for farmers is planned once the operation has stabilized.

Processing/Technology. Low oil recovery rate of about 71% (51% edible and 20% inedible) of available oil compared to a theoretical yield of 82% to 85%, was caused by a number of factors, including: (a) Failure to adequately reduce the particle size of fresh meat before pressing reduced recoverable milk and subsequent oil; (b) Poor design and operation of the hydraulic press prevented higher expression pressures from being attained; (c) Poor design and operation (temperature control and agitation) during cream heating to break the emulsion caused FFA levels to increase thus reducing edible oil yields; (d) Adding excess quantities of sodium hydroxide during neutralization caused unnecessary saponification, thus reducing edible oil yields; (e) Open steam deodorization at atmospheric pressure did not deodorize the oil but caused additional FFA formation and hydrolysis; (f) Sun drying rather than roasting reduced the amount of oil recovered from the pressed meat or sapal; and (g) Quality control equipment to monitor oil, moisture and FFA content was lacking. Equipment will be installed and staff trained to conduct the necessary procedures (20). Equipment and procedures modification and staff training are planned to reduce or eliminate these problems.

Management and Labor Skills: Staff members were not sufficiently knowledgeable about the technologies and the reasons why certain procedures were to be used to insure better efficiency and productivity. Technical assistance, training and monitoring will be provided by the University of the Philippines at Los Banos.

Markets: Neither sufficient markets nor products had been developed to allow better utilization of byproducts such as skim milk, coconut water, sapal and latik. Additional market research and product development is now taking place to improve earnings.

Degree of Success: This venture as described proved unprofitable because of low recovery of oil and failure to produce marketable and profitable products from the byproducts. In spite of these problems, ATI feels the operation could be commercially viable if established in a location where raw material supplies and markets are more favorable, with sufficient development of byproducts.

Future Plans. At present, market research is being planned to identify alternative products which will better use products that may result from the wet process. Once market needs are more defined and products developed to meet those needs, appropriate equipment and scale of operations will be assembled.

Shea Butter Production in Mali

Shea butter is a vegetable fat derived from the kernels of the shea nut tree (*Butyrosperum parkii*). It is a major cooking fat in rural areas of the Sahel where the climate is too dry to support oil palm. Shea trees grow wild in parts of West Africa where rainfall averages between 800 and 1,400 mm per year. Most of the production is for domestic consumption and is processed by artisans using traditional methods. Only 20–25% of production is by large-scale firms for export as an ingredient in cosmetics and processed foods. About 250,000 women are involved in shea nut collection and processing in Mali. In a typical year, a household collects approximately 240 kg of dried kernels. There can be a 50% difference between the harvest in an average year and a good or bad one, which affects the economics of processing technologies. Annual shea butter consumption averages 5.5–9.1 kg per Malian, which is 10–15% of the kilocalories in the diet. Per capita consumption of shea butter is higher for the poor than for the middle class; partly because shea butter fat is less expensive than the alternatives of cottonseed oil or groundnut oil.

The traditional method for extracting shea butter is slow and laborious requiring 2.5 to 3.4 person-hours per kg of smoked kernels processed. In addition, nut collection and preprocessing takes 1.5 hours per kg of smoked kernels processed.

During the 1950s and 1960s, the Institut de Recherches pour les Huiles et Oleagineux (IRHO), based in France, introduced semi-industrial equipment to extract shea butter in Mali. Because the imported equipment was expensive, a commercial market did not materialize.

Objectives and Project Description

In 1985, ATI began supporting development of alternative equipment by CEPAZE, the Centre d'Echanges et Promotion des Artisans en Zones a Equiper, a not-for-profit organization in France. The primary objectives were to develop less labor-intensive equipment to increase production and extraction rates, and test the technical and commercial viability of ownership and use of the equipment by village groups. The CEPAZE system covers the whole range of shea kernel pretreatment and butter extraction. Like the traditional method, this is a wet process; however, extraction is based on a density difference rather than breaking of an emulsion.

Raw Material Inputs

Shea nut collection and storage, and butter preparation and sales are traditionally women's tasks in most of Mali. Figure 10 diagrams the traditional wet process for shea butter production in Daban, Mali. The process differs somewhat elsewhere in Mali and in other countries. The pretreatment steps are done soon after collection, but processing into butter may take place months later as time permits.

In areas of good production, shea nuts commonly are collected daily within a 1–3 km radius over a 3–4 month harvest period (25,26). The fruits are buried underground in pits for 12 days or more so that the pulp ferments and falls off of the nuts. After the nuts are separated, they are dried in the sun for two days to reduce the moisture content from 40–50% to 6–7% for satisfactory storage. Then, the kernels are smoked over a fire for 3–4 full days. Decortication (the separation of the shells from the nut) may be done by pounding the nuts in a mortar with a pestle, or on the ground with a stone. Further drying is done on a mudstove for 24 hours to decrease the moisture content of the kernels to 10%. If properly prepared, dried nuts can be stored 3–9 months without significant deterioration.

The Traditional Process for Producing Shea Butter. The traditional wet process of shea butter production relies on human energy for grinding smoked kernels into a paste and extracting oil by emulsification in water. In Mali, the process begins with heating of the kernels overnight. The next day, while they are still warm, the kernels are pounded in a mortar with a pestle into a coarse brown paste. The pounding should be done in the sun which makes the task even more strenuous. Then, the crushed kernels are ground into a fine paste by hand on top of a large stone, either with a small, palm-shaped stone or a wooden roller (27,28).

At Daban in Mali, the fine paste is mixed with hot water for 2–3 hours until it is covered with a white foamy emulsion of fat. Mixing introduces air into the paste and has to be quite vigorous. Mixing may be done with a stick in a cauldron, but is more commonly done by rubbing the paste between the hands in a large gourd "Until you sweat," as one woman described it. The oil that floats to the surface of the emulsion or mousse is skimmed off using a gourd or ladle and then poured into a container of warm water and decanted. A white film of liquid shea butter forms on the surface. This washing is repeated five or more times (25).

Next, the butter is boiled for 1–2 hours to evaporate the water and allow the impurities to settle. The cleaner oil at the top of the cooking pot is poured into a container to solidify into butter and the sludge at the bottom is used for making soap. After cooling, the butter is formed into balls and wrapped in leaves and left overnight to solidify. The butter will keep for six months without a significant change in flavor even though the traditional process is unhygienic and results in a relatively high level of free fatty acids (25).

Estimates of the yield from the traditional process vary: 15–25% (27); 22–25% or 37.5%. Some of this variation is due to local differences in user skills, processing, and the quality and age of the shea nuts.

Other than labor, the cash costs of the traditional process are nil. Although a wooden mortar and pestle is needed, nearly all rural households already have one to pound grain. The fuelwood and water required are collected rather than purchased.

The Upgraded Technology

In 1979, the Agricultural Machinery Division of the Malian Ministry of Agriculture asked CEPAZE to study equipment for upgrading shea butter production. Between 1980 and 1985, equipment was designed and lab tested in France by the Ecole Nationale Superieure des Industries Agricole et Alimentaries due to the limited indigenous capacity for this testing in Mali. In 1985, ATI provided CEPAZE with a grant for further adaptation, field testing, and dissemination of this technology.

Total expenditures for development of the animal-driven and motorized systems amounted to U.S. $367,400 through mid-1988. ATI provided 45% of the cost, the European Economic Community supplied 33% and the remaining 22% was given by Comite Catholique Contre la Faim et Pour le Developpment, Fondation de France, Band-Aid, and CEPAZE itself. The total cost and length

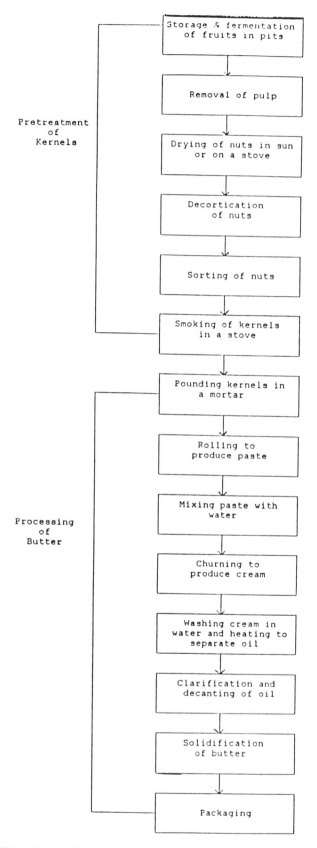

Fig. 10. Traditional Daban, Mali, wet process for shea butter production.

of time to carry out the project were much higher than anticipated due to the experimental nature of the technology and the implementing organizations' lack of a field presence in Mali.

The animal-driven unit was installed at N'Djinina in 1985 and the first prototype of the motorized unit was placed at Dara in 1985–86. The animal-driven unit was intended for villages where 150–300 women are involved in shea production while the motorized unit can serve 300–500 users.

Both systems include a solar dryer, manual decorticator/winnower, metal drums, a piston-type coffee filter, and a weighing scale. The motorized unit has a diesel-powered grinder, water reheater, a centrifuge powered by the same engine, and an electric mixer (Fig. 11). The animal traction unit has a smaller gear-operated grinder and uses a mechanical kneader instead of the centrifuge and electric mixer.

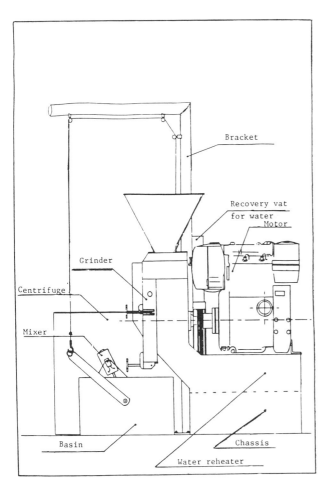

Fig. 11. Side view of CEPAZE equipment (29).

Shea butter produced by the CEPAZE system is similar in taste, appearance, and composition to the product obtained from the traditional process. When properly packaged to reduce the rate of oxidative rancidity, it can be stored for one year.

Technologies

Solar Dryers: The solar dryers have a base composed of cemented stones; sides and ends of compressed earth bricks, polyester fiberglass top and straw or soot for insulation inside. Each dryer module with dimension of 1 × 1.5 cm can hold 20 kg of shea nuts or 12 kg of the kernels. The dryers are easy to construct using village labor. The construction cost is $85 per module (29).

The solar dryers may save 80% of the fuelwood required per kg of butter produced; the 6 kg of wood used for drying the nuts and smoking the kernels, and the 2 kg of wood needed for heating the smoked kernels before grinding them. It takes 2 days to dry the nuts and 5–7 days to smoke the kernels in the solar dryer. The smoked kernels can be reheated in the morning for processing that same afternoon.

The solar dryers were not popular mainly because they required a large amount of maintenance. They also need to be rebuilt every year since the mud bricks deteriorate rapidly, especially in the rainy season. Another important factor is the inconvenience of monitoring the drying and guarding against theft at the communal site of the solar dryers. With the solar dryers, the grinder can only be used in the afternoon once the kernels have heated up because they cool off and reabsorb moisture at night. To match the capacity of the motorized system, the dryers would occupy a considerable amount of land.

Decorticator/Winnower: The decorticator/winnower removes the kernels from the shea nuts. The decorticator also has a blower to separate the kernels from the shells and fragments. The manual decorticator has a capacity of 250 kg/hour and costs $800 (29). One was installed at Dara for centralized use. Its throughput is too slow to match the extraction equipment and the machine is fragile and likely to break at a high capacity use rate. Also, women prefer to transport dehulled kernels to the grinder rather than whole nuts, which are heavier.

Motorized Grinder. The motorized grinder was specially designed for shea processing because a conventional hammermill will not work well with shea kernels. The grinder converts the hot kernels to a fine paste. As it takes eight minutes to grind a 25 kg batch, it can handle 188 kg/hour at full capacity. However, if used with centrifuge and operated by the same motor, only 75 kg/hour can be processed (29).

As a result of the field testing, a cheaper and more effective grinder was developed in late 1987. With production in series, the grinder, base, motor, electric mixer, electric generator, and light would cost $5,400 (29). The expected life of the grinder is 12 years if used for shea 300 hours/year (26). Because of the close clearance between the blades and screens, it can be damaged easily; for example, if a coin drops inside.

After grinding, hot water is added to the paste in a ratio of 15 liters to 12 kg. That is followed by three minutes of agitation with an electric mixer. A water reheater taps some of the waste heat from the motor. Then, the paste is transferred to the centrifuge for extraction of the oil.

Centrifugation. After three minutes of centrifugation, a water feed tube is inserted into the bottom of the centrifuge bowl. As oil is lighter than water, it floats to the top of the bowl. As the volume of material in the bowl increases, the oil overflows into a storage chamber.

When water begins to flow out of the bowl, the centrifuge should be stopped; this usually takes less than six minutes. Approximately 15 liters of water are removed from the centrifuge for reuse. The press cake is detached and stored for fuel use (26).

The first model of the centrifuge, a horizontal axis design, proved unsatisfactory. The throughput and extraction rates were too low relative to the traditional process and the method of discharging the presscake was impractical. In addition, the machine was delicate and the means of holding the bowl inadequate.

Consequently, a vertical axis centrifuge design was developed with an improved system of sedimentation and overflow, loading and unloading, and remixing of the paste after the first extraction. One with a capacity of 52.5 kg/hour was installed in late 1987. A new version with a capacity of 75 kg of paste per hour has been constructed (26). Its extraction rate is higher than the previous model due to an increase in the speed of rotation of the centrifuge bowl and the temperature of mixing (29). It is also safer and easier to use. With full production, the centrifuge should cost $2,200 (29).

A limited number of tests showed an average extraction rate of 38% with the horizontal axis centrifuge, which is comparable to that of the traditional process. However, a subsequent test in June of 1988, also conducted by CEPAZE, was unable to verify that result. The women at Dara perceive that the centrifuge has a lower extraction rate than the traditional process. Some of this perception may be due to their bad experience with the earlier prototype. CEPAZE believes that the motorized technology can only surpass the traditional technology when the quality of shea kernels is good (26). Insufficient operator efficiency and effort may also be a problem. Another source of the variability may be due to the fineness of the paste before centrifugation.

Unlike the grinder, the centrifuge can be operated only with a full batch because oil separation depends on volume displacement. This factor may have discouraged use of the centrifuge since some women might not have enough paste to process at one time. The women resist mixing their paste with someone else's because of differences in the oil content of the shea nuts and the care in pretreatment and storage. Moreover, women may be reluctant to use the centrifuge because they don't want to be considered lazy for spending extra money to avoid work that their peers do by hand, especially when the shea harvest is small.

After centrifugation, the next steps are to decant the sediment from the oil, evaporate the residual water in the oil through heating, and filter the oil. The filtered oil is left to cool before it is manually churned into shea butter. Finally, the butter is weighed and wrapped in leaves.

Problems, Solutions and Future Plans

Some components of the motorized CEPAZE system have been more successful than others and the package need not be adopted as a whole. During the 1986 and the 1987 seasons, the motorized grinder processed more than 23 t of kernels (26). Yet, none of the women at Dara are now using the centrifuge. After they use the grinder, they resume the traditional process. Nevertheless, partial mechanization of the process does eliminate much of the time-consuming and difficult part of the labor, although it may limit use to women living close enough to transport the paste home for further processing. Further work is being done to improve the operation of the centrifuge and the other parts of the package.

Because of its cost, the CEPAZE motorized system is most suitable for service processing in a large village operation or for small-scale commercial production. The equipment is durable and has a sufficient capacity for cost-effective operation. The main advantage is the labor savings. An economic analysis of the technology is contained in Hyman (30). Before this project, the women at Dara, unlike most villagers in Mali, were already accustomed to paying a fee for grinding shea kernels in a motorized plate mill. However, they did not have access to a mechanized service for oil extraction. At the current price of shea butter, the grinding cost amounts to 23% of the income to the producer. For both grinding and centrifuging, the women would have to pay 32% of their proceeds.

Initial reports indicate that the grinder also could be used for milling grain; however, it is not well suited for extracting oil from groundnuts. The ability to use the grinder for both shea and grain would greatly increase its market demand and profitability of the equipment. Further testing of the grinder's production rate and quality of millet or maize flour is needed under various conditions of use—moisture content of the grain, motor speed, speed of passage of the grains, and grill sizes. Some design modifications may be needed to optimize the machine for both uses and to increase its durability or ease of repair when run more intensively year-round.

The centrifuge is not expensive compared to the grinder. If the centrifuge worked well and was accepted by potential users, the marginal cost of including it in the package might be economically warranted. However, some problems with the design still need to be resolved. To be well accepted, the centrifuge's extraction rate would probably have to consistently be at least 10–20% greater than that of the traditional process or its price must be reduced. Further modifications of the equipment are underway and will be field tested during the next shea season due to the generally poor harvest in 1988.

Because the animal-driven unit was less convenient to use than the motorized one and the villagers at N'djinina were aware of its existence, they were dissatisfied with the less convenient animal traction unit. They were dissatisfied with its low throughput (18 kg/hour), relative to its capital costs of $2,700, excluding the decorticator. The manual centrifuge was physically tiring for the animals and extracted too little oil. Obtaining animals for village operation proved difficult. As a result, ATI decided not to provide additional financial support for development of the animal traction version. CEPAZE believes the technology still has potential and has continued limited testing of it.

References

1. Moll, H.A.J., *The Economic of Oil Palm*, Center for Agricultural Publishing and Documentation, Wageningen, The Netherlands, 1987.
2. Hyman, E., *The Range of Choice in Palm Oil Processing Technologies in Cameroon*, Appropriate Technology International, Washington, D.C., 1988.
3. Anonymous, *Etude Compare des Couts et Revenues de Deux Processus d'Extraction d'Huile de Palme: le Processus Traditionnel et le Processus Ameliore avec Unite Villageoise d'Extraction d'Huile de Palme*, Association Pour la

Promotion des Initiatives Communataires Africaines (APICA), Douala, Cameroon (1982).
4. Harley, C., *The Oil Palm*, Longman Press, London, 1977.
5. Anonymous, *Guide d'Utilisation de l'Unite d'Extraction d'Huile de Palme*, Outils Pour les Communautes (OPC), Douala, Cameroon, 1985.
6. Donkor, P., *Appropriate Technology* 5(4):8 (1979).
7. Rukuni, T., *How Can Oilseed Production and Processing Enhance Food Security?*, presented to South African Development Coordinating Conference, 1988.
8. Fernandes, S., et al, *Small Scale Cooking Oil Expression in Rural Zimbabwe Using a Manually Operated Sunflower Seed Press: A Research Report* (draft copy), World University Service, Canada, 1988.
9. Lusas, E.W. et al, *Presentations and Speeches Related to Oilseed Proteins*, Food Protein Research and Development Center, Texas A&M University, College Station, Texas, 1979.
10. Fisher, M., *An Engineering Analysis of Three Manually Operated Sunflower Seed Presses Manufactured in Tanzania*, Appropriate Technology International, Washington, D.C., 1988.
11. Collard, P., and S. Levi, *Nature*(London) *183*(4661)620 (1959).
12. Reader, A.E., and K.F. Gregory, *Appl. Microbiol. 30*:897 (1975).
13. Azoulay, E. et al, *Appl. Environ. Microbiol. 39*:41 (1980).
14. Omstedt, P., and A. von de Decken, in *Single Cell Protein II*, edited by S.R. Tannenbaum and D. I. C. Wangs, MIT Press, Cambridge, Massachusetts, 1975, pp. 533–563.
15. Rimbault, M., F. Deschamps, F. Meyer and J. C. Senex, in *Proceedings GIAM-V*, Bangkok, Thailand, 1979, pp. 425–432.
16. Charlotte, B., and G.W. Rambo, *Phytopath. 62*:749 (1972).
17. Charlotte, B., G.W. Rambo and G.A. Bean, *Ibid. 62*:668 (1972).
18. *Concept Paper for a Coconut Industry Development Strategy*, United Coconut Association of The Philippines, Manila, 1986.
19. Anonymous, *Coconut Statistics Annual*, United Coconut Association of The Philippines, 1986, pp. 117–153.
20. Hyman, E.L., *Technology, Scale and the Coconut Industry in The Philippines*, Appropriate Technology International, Washington, D.C., 1985.
21. Edmonds, M., D. Edwards and P. Mars, *An Economic Evaluation of the Wet coconut Process Developed at the Tropical Products Institute*, Tropical Products Institute, London, 1973.
22. Woodroff, J., *Coconuts: Production, Processing, Products*, Avi Publishing Co., Westport, Connecticut, 1979.
23. Hagenmaier, R., *Coconut Aqueous Processing*, University of San Carlos Publications, Cebu City, The Philippines, 1980.
24. *Village Coconut Processing Project: Investment Plan*, Filipinas Foundation, Manila, 1986.
25. *Point Sur L'Extraction Villageoise du Beurre de Karite*, Division du Machinisme Agricole, Government of Mali, Bamako, 1984.
26. Delabre, G., P. Paris and B. Clamagirand, *Etude de Marche et commercialisation des Equipments Villageois de Fabrication du Beurre de Karite au Mali*, Centre d'Echanges et Promotion des Artisans a Equiper, Paris, 1987.
27. Godin, V., and P. Spensley, in *Crops and Products Digest No. 1*, Tropical Products Institute, London, 1971.
28. Fleury, J-M., *IDRC Reports 10*(2) (1981).
29. *Description du Projet CEPAZE*, Division du Machinisme Agricole, Government of Mali, Bamako, 1988.
30. Hyman, E., *Labor Saving Technologies for Processing Shea Nuts: An Important Edible Tree Product in West Africa*, Appropriate Technology International, Washington, D.C., 1988.

Nutrition Intervention with Protein Foods

Daniel E. Shaughnessy

Export Processing Industry Coalition, 1008 N. Randolph St., Suite 2006, Arlington, Virginia 20201, USA

Abstract

A considerable body of experience has developed over the past 20 years with the use of protein foods in national and international feeding programs. High protein, fortified food products have been developed from all major grain, oilseed and dairy sources, with significant results in a variety of settings, ranging from emergency feeding activities to highly structured nutrition applications. Recent and continuing research underscores the importance of these applications and their continued use in the future. International program examples, case studies and field use of high protein commodities have special implications for future programs of nutrition intervention and direct feeding.

Malnutrition and hunger are two of the principal detriments to health and economic welfare in the world today. Solutions to the malnutrition problem are multifaceted and complex. They include higher levels of production, higher per capita income, better marketing and distribution strategies, better education, greater levels of employment, more sophisticated processing and storage methods, more responsive financial institutions, and many other social, economic, and political factors which provide a stable environment for social and economic development.

Humanitarian food assistance to alleviate hunger has been undertaken throughout man's recorded existence, but formalized food assistance programs on a global basis are a 20th century phenomenon. Major international food relief programs include the World Food Program (WFP), the U.S. P.L. 480, Food for Peace Program, and bilateral and multilateral programs, including CIDA, EEC, UNICEF, and others.

Supplemental feeding is now viewed as a primary means by which national health goals may be achieved. Traditionally, supplemental foods used to be associated with the disposal of agricultural surpluses, meeting emergency and disaster relief requirements, and welfare programs. In that context, attention was diverted from assessing the broader potential of supplemental feeding as a part of national health and economic development.

In 1974, the World Food Conference estimated that 10 million tons of food aid was required annually. That requirement has not changed; indeed, the need may now be greater. At the same time greater attention began to focus on the quality as well as the quantity of food aid.

Supplemental Feeding

Supplemental feeding is well-known in emergency and disaster relief situations. In these instances, the food supplement functioned in a curative capacity. The objective was to correct a crisis situation, overcome the immediate problem and prevent a further decline in nutritional status. The results are often dramatic, but they may also be short term.

However, with the growth of the concept that the prevention of malnutrition may be a prerequisite to national development, supplementary feeding has been expanded to encompass a broader base. It now includes the subclinical malnourished infant and young child. It encompasses as a part of continuing programs the pregnant and nursing mother, both for her nutritional status, as well as the health of the unborn, newborn, and young child.

For these "at risk" groups, food supplementation functions in a "preventive" rather than a "curative" capacity. It provides a form of "immunization" against mental retardation, inadequate growth, and infant mortality. Expanded into this broader long-term role, food supplementation has become a primary instrument in national health and development.

In food aid, a "food supplement" is one that, when combined with other food or foods, will provide the additional nutrients needed for full physical and mental development. Where the diet is lacking in essential nutrients, food supplementation can contribute significantly to better health and national development.

In much of the developing world, the diet is deficient in protein of significant quantity and quality to maintain good health. In other areas, the diet may be deficient in calories in the form of carbohydrates and fats, and both protein and calories may be lacking. Diets may be lacking in essential vitamins or minerals. For example, it is estimated that 200 million children now suffer from Vitamin A deficiency that leads to blindness. Preventable malnutrition takes a heavy toll in human life and suffering, in human physical and mental productivity. It severely retards national growth and development.

Those groups which are most "at risk," are pregnant and nursing mothers, infants, preschool and school age children. In addition, drought and other disaster victims and underemployed and undernourished adults often require supplemental foods that contain high quality protein, calories, vitamins, and minerals to provide sufficient nutrients when taken alone or in conjunction with local foods to maintain good health.

To meet these needs, ideal food supplements should be low in cost; capable of being produced in volume; easily transported and stored; capable of simple preparation; and acceptable to taste. Protein cereal grain/oilseed food blends are ideal food supplements for various types of food aid and nutrition intervention programs. They are rich in quality protein, carbohydrates, and fats. They are also natural carriers of vitamin and mineral premixes.

Nutrient needs depend upon age, stage in the life cycle, and the specific environment in which one lives. High protein blended foods have been engineered to meet the various needs of specific individuals or groups and over the past twenty years, four basic categories of supplemental feeding programs have evolved. These programs and their use of supplemental commodities include:

Maternal-Child Health Feeding Programs (MCH):

These programs offer food supplements to pregnant and nursing mothers and preschool children from low-income families. The programs are typically implemented at feeding or health centers. Feeding rations are distributed to mothers and infants and rudimentary lessons in health or nutrition are provided. In the better run centers, a growth surveillance system is used to ensure adequate levels of intake and to monitor impact.

MCH programs are most effective when attention is given to targeting the program at malnourished children and the nutritionally at risk mothers; to the appropriate quantity and type of ration; to integration of the feeding program with health services; to the inclusion of a nutrition component; and to beneficiary participation.

The pregnant mother needs additional quality protein in her daily diet. This additional protein is required for increases in maternal tissue and for proper fetal development. Inadequate protein intake can result in maternal anemia, poor uteral muscle tone, insufficient lactation and possible fetal abortion. It can also result in reduced infant birth weight, length and general condition of the newborn infant.

In many areas, the diets of young children, and pregnant and nursing women are largely based upon one or two staples such as corn, wheat, and starchy roots. While these foods do contain calories and may contain a reasonable protein content, they are not nutritionally complete foods. Serious problems occur when the staple is a tuber, plantain or breadfruit, since these staples are bulky and contain very low protein levels.

With these diet limitations, it is practically impossible to consume sufficient protein to protect the pregnant mother, the unborn child or the young child from protein malnutrition. Protein supplementation in these instances is vital.

Protein cereal grain food blends can be ideal supplements to the diet of pregnant and nursing women. They have been scientifically engineered to provide quality protein in addition to fats and carbohydrates. Blended foods such as WSB (Wheat Soy Blend), CSM (Corn Soy Milk), and a blend of wheat protein concentrate and soy (WPC-Soy) also contain vitamin and mineral premixes. Although primarily developed to meet the high levels of nutrients required for children, these blends also can materially upgrade the diet of the pregnant and nursing mother. Because the blends contain no animal protein, they can be extremely effective where the maternal diet is subject to cultural restrictions of such items as eggs, fish, poultry, or meat.

The preschool period is a time of rapid growth and development and the nutritional demands are great. It is also the weaning period when the child moves from dependency on the mother to independence for its food supply. It is also a period of great susceptibility to childhood diseases, infection and parasitic infestation.

It is therefore a period of high risk for the various types of malnutrition which adversely affect physical growth, learning ability, and social behavior. It is the period of high mortality. The single biggest contributor to this mortality rate is protein calorie malnutrition.

For the pre-school child, it has long been recognized that a need exists for a low cost food to supplement his or her diet. Such a product needs to contain a high caloric density to meet the energy demands of the growing child, and enough quality protein to provide for maximum growth and mental development. In addition, a vitamin and mineral premix is required to guard against micro-nutrient diseases. WPC-Soy, mentioned earlier, is an example of a commodity designed to meet these needs and to bridge a nutritional gap in the diet of the young infant and preschool child.

The contribution of MCH to development is primarily through improved human health conditions resulting from nutritional awareness and better practices, and an increase in access to health services. Given the nutritional emphasis of these programs, recent evaluations have focused on their nutritional impact. Though methodological difficulties hamper nutritional evaluation, positive nutritional contributions have been found to have occurred in MCH programs in a number of countries.

School Feeding Programs:

The school environment is an excellent structure within which to deliver nutrition and health care programs. In a sense, it is a captive situation. It brings together children at a particular place and time for administration of the programs.

In many areas of the world, school children either do not get enough food or the right kind of food to meet their nutritional needs. Through intense competition for the available family food supply, ignorance, or neglect, the child comes to school in a physical condition ill suited for learning. Supplemental feeding at the school can do much to correct this problem. School lunch programs in the United States and Japan, for example, have done much to guarantee the nutritional health of the school children.

For years, protein cereal grain blended foods have been a dietary mainstay in U.S.-supported school feeding programs throughout the world. Hundreds of thousands of tons of high protein blended foods have been acquired by the U.S. government for donation through the Food for Peace Program. These programs provide meals to school age children who physically are present at educational institutions. In general, school feeding programs are intended to contribute to development through improved health conditions and increased productivity. They also are designed to improve the nutritional status of school-age children, and to enhance school enrollment and attendance.

Food for Work Projects:

One of the greatest assets of a developing nation is its abundance of manpower. One of the greatest liabilities, however, to national development is the inefficient use of that manpower. This may be due either to poor health, unemployment, or underemployment. The rate of national development is, in part, directly proportional to employment and productivity levels. Maximum productivity can be attained only when workers are properly fed.

Locally and internationally sponsored Food for Work Programs deal with employment and worker productivity simultaneously. Under such programs, a worker receives employment to build roads, buildings, irrigation and sanitation systems and other projects designed to stimulate national and regional development. Part of the wages paid are in the form of food. In many cases that food is a high protein product such as bulgur wheat, or soy-fortified bulgur wheat.

Emergency Programs:

In our lifetime, millions of men, women, and children have suffered because of famine, drought, earthquakes, floods, and other natural disasters. The manmade disaster of war has been equally devastating. In such emergency situations, total food supplies for large population groups are provided immediately. These foods must contain the necessary nutrients to sustain life for all levels of the affected population.

High protein, cereal grain food blends also function well in this role. They can be produced quickly in volume; they are adaptable to various transport and storage conditions and can be prepared for immediate consumption even under the most trying of circumstances. Products like WSB, CSM and WPC-Soy also contain vitamins and minerals in addition to the carbohydrates, fats, and protein necessary to sustain life.

Blended and Protein Food Development

The widespread use of protein foods in feeding programs is primarily a phenomenon of the past 20 years. In U.S. feeding programs, a particular point in time was 1966, when the U.S. food aid law (P.L. 480) was amended to place greater emphasis on nutrition. Authority also was provided for enrichment and fortification of commodities to improve their nutritional values. What emerged in subsequent years was a range of processed, fortified and blended products, many of which carry an amino acid profile and utilizable protein content comparable to that of meat and milk and yet are priced at only a fraction of what the animal protein would cost in most developing countries. Field program administrators throughout the world now recognize the nutritional contribution that products such as corn-soy-milk, wheat-soy-milk, bulgur, soy-fortified bulgur, cornmeal, soy-fortified sorghum grits and others make to the overall dietary improvement of recipients. These products, which were developed by U.N., USDA and food industry nutritionists, are used alone or in combination with other locally produced or imported foods.

The introduction of processed, blended and fortified grain products brought a new dimension to supplementary feeding programs. The concept was based on the fact that if the prevention of malnutrition is to become a prerequisite in national development, then supplementary feeding should be expanded to cover a broader base.

The first of the engineered foods, first made available in 1966, was a corn-soya-milk blend (CSM), and consisted of 64% gelatinized cornmeal, 24% soya flour, 5% nonfat dry milk, 5% refined soy oil and 1% enrichment with all known essential vitamins and minerals. CSM contains 20% protein, and has a protein efficiency ratio (PER) of about 2.4, comparable to casein of milk.

In 1968 another blended food, a wheat-soya-blend (WSB), with properties similar to CSM was made available for worldwide distribution under the U.S. P.L. 480 Title II donation program.

The range of processed, blended and fortified foods also includes:

Instant CSM (ICSM): A fully precooked modification of CSM. ICSM is designed primarily as a weaning food and for use in preschool feeding programs, although it may have other uses such as emergency refugee feeding. The nutritive value of the product is essentially the same as CSM. A minimum protein content of 19% is specified for both. Like CSM the PER is about 2.4 and is comparable to casein in milk. Product cost is roughly 1.75 to 2 cents more per pound than regular CSM.

Soy-Fortified Bulgur Wheat: This fortified product is 85% regular bulgur and 15% toasted defatted soy grits. With this fortification, the protein content of bulgur is nearly doubled from 9.3% to 17.3%. The nutritional quality of the soy-fortified bulgur as measured by the PER exceeds 2.0 compared to 1.1-1.3 for regular bulgur.

Soy-Fortified Cornmeal: When regular cornmeal is fortified with soy flour at the 15% level, the protein content of the cornmeal is raised from about 8% to 14.5%. The nutritive value of the product is increased even further with the PER increased from 0.3 to approximately 2.0. The additional cost of the soy flour amounts to about 0.5 to 0.75 cents per pound of fortified product. Soy fortified cornmeal is also enriched with vitamins A, B1, B2, niacin, iron and calcium.

Soy-Fortified Rolled Oats: The minimum protein content of rolled oats is raised from 15% to 20% when fortified at the 15% level with toasted, defatted soy flakes and PER is increased from 1.8 to 2.2.

Soy-Fortified Wheat Flour: Regular bread flour, enriched with vitamins A, B1, B2, niacin, iron and calcium, may be further fortified at either the 6% or 12% rate with soy flour. Fortification at the 6% level increases the protein content by 30% (from 11 to 14%) and nearly doubles the nutritive value of the flour. The PER is increased from 0.7 to 1.3. Soy fortification at the 12% level increases the protein content of the flour from 11 to 17% and the PER of the bread from 0.7 to 1.95. The quality of these fortified processed foods is specified through formulation, analytical, and predetermined performance requirements.

As the years have gone by, extensive field applications and experience has been developed and these foods are now standard features in worldwide supplementary feeding programs.

The levels, in terms of total volume, of these food aid programs, have reached the 1974 World Food Conference target of 10 million tons. However, much of this tonnage still tends to be in basic grains, although a growing percentage of the volume is in the form of high protein, blended and fortified foods.

WFP provides more than 1.5 million tons in a variety of programs throughout the world in all types of feeding activities and a wide variety of products.

The United States government in 1988 has provided nearly eight million tons under the following programs: (a) Title I, 4.5 million tons; (b) Title II, 2.0 million tons; (c) Section 416, 1.5 million tons.

Projected future needs for food aid include requirements for both direct, targeted feeding as well as larger resource transfers of basic grains and edible oils. In the former category, increased emphasis on high protein, fortified foods will be necessary.

Antinutritional Factors in Vegetable Proteins for Poultry

A. J. Mudd

Cyanamid International, One Cyanamid Plaza, Wayne, New Jersey 07470, USA

Abstract

Antinutritional factors associated with vegetable proteins can be divided into two groups: (a) intrinsic factors and (b) extrinsic factors. Examples of intrinsic factors are those components which for genetic or other reasons are a fundamental component of the vegetable protein. Thus a wide range of proteins used for poultry feeds may contain cyanogenetic glycosides. Other examples include the ricin component of castor oil bean and the theobromine content of cacao products. Such factors have been reported to cause retarded growth or mortality in chicks and should be considered as antinutritionals. Extrinsic factors are those components which are either contaminants of vegetable protein, pre- or post-harvest, or those substances which occur as a result of the processing of the vegetable protein. Examples of such extrinsic factors are the mycotoxins frequently found associated with spoilage or poor storage conditions and resulting fungal contamination. This paper will focus on some aspects of detail concerning antinutritional factors and suggest ways to overcome some of the problems that may occasionally arise as a result of using certain types of vegetable protein.

The nutritional parameters associated with the use of vegetable protein have been well researched and there is a comprehensive literature which assists in the definition of nutrient value of a wide range of vegetable proteins. Less well understood, however, are the antinutritional factors present in certain vegetables, and the literature, although extensive, is far from comprehensive and when linear programming for least cost diets, it is necessary to bear in mind a wide range of factors which may limit the extent to which a particular vegetable protein can be used. For the purposes of this review these factors can be divided into intrinsic factors and extrinsic factors.

Intrinsic Factors

These are components on metabolites of plant material which cause depressed performance in poultry or may be even more toxic. Examples of such intrinsic factors are a range of cyanogenetic glycosides which are found in a wide range of plant materials, for example, rapeseed meal, sorghum, java bean. Other types of intrinsic factors are the straight cyanides found in cassava and the toxin ricin found in the castor oil plant.

Extrinsic Factors

These are indirect components which cause toxicity due to factors which are not related to the structural composition of the vegetable protein itself, but are a series of complicating factors which because of the nature of the protein or its method of preparation has resulted in a series of problems.

Many vegetable proteins are prepared in such a way that contaminants of one kind or another can occur. Examples of such contaminants include insects, molds and fungi and their associated mycotoxins. In other cases, the crop itself can be contaminated with a weed which causes problems in the livestock rearing enterprise. Coffee Weed seed is an example.

Early methods of vegetable protein preparation usually involved the use of a particular solvent for oil extraction that were eventually recognized as hazardous. Although not related to a specific poultry problem, the use of trichlorethylene-extracted soybean meal caused the formation of an unidentified toxic factor which produced aplastic anemia and leucopaenia in livestock. For this reason, this method of preparation was withdrawn.

Other indirect effects are associated with vegetable proteins having a particularly high level of electrolyte, which when included in a ration containing an energy source also with a high electrolyte content causes an imbalance in poultry receiving such diets.

In Europe it is well documented that soybean meal from certain places of origin has an abnormally high level of potassium content. When incorporated in rations containing tapioca (cassava) as an energy source also with a high K content the resulting feed causes an electrolyte imbalance and broilers receiving such rations may develop a wet-litter syndrome.

This review focuses on these intrinsic and extrinsic factors and confirms that high quality poultry rations can be formulated providing a certain amount of care is taken to minimize the risk factors.

Cyanogenetic Glycosides

Rapeseed Meal

Many areas of Europe and Canada do not have appropriate climate or soil conditions for soybean production. In the past 20 years, however, there has been a considerable incentive for oil crushers to use vegetable oil based on locally produced seed and, for this reason, rapeseed was grown on an extensive scale. Because of the intense economic factors, there was considerable incentive to overcome the problems associated with rapeseed. An extensive review was published in the U.K. by Fenwick and Curtis, (1980) and further data have been published since that date, e.g., Peckham (1984), and Wight et al. (1987).

Rapeseed meals began to be used extensively during the 1960's and the problems of glucosinolate content were recognized. These are potent goitrogen factors which have an antithyroid effect. From the outset it was recognized that the European strains of *Brassica napus* had a higher content of glucosinolates than the *B. campestris* strains of Canadian origin and the Canadian authorities placed considerable emphasis on the reduction of glucosinolates in *B. campestris* strains. This low content strain resulted in a branded rapeseed meal known as canola. Mitaru et al, (1983) studied the effect of the hull tannins from canola on broiler performance and concluded these had no impact.

In Europe, and particularly the U.K., the consumer has a misguided impression that brown eggs are more nutritious than white eggs. Therefore, there is a high proportion of brown eggs sold in the U.K. supermarket. In the early 1970's in the Vitamealo protein supplements group of Beecham, we were formulating layers rations using increasing levels of rapeseed meal. In the autumn of 1972, we were able to confirm that the use of rapeseed meal in protein supplements for brown egg layers was associated with taint or off-flavor in the egg. Other workers in the U.K. confirmed that this taint was due to the presence of trimethylamine. Canadian work confirmed these findings in brown strains, particularly Rhode Island Red. Subsequent work showed that the taint was associated with the presence of a choline ester, sinapine and, because it was not possible to remove this in processing, rapeseed meal for layers has been included at low or nil levels.

There are other reasons for this as a haemorrhagic liver syndrome has been described in layers. This has not been related directly to the glucosinolate content, but some experiments have demonstrated a 13% mortality in layers with a diet comprising 50% rapeseed meal. When a ration containing 10% rapeseed meal was fed to broilers, there was no evidence of a problem, but 10% in layers caused the hemorrhagic liver condition. Other experiments in Ontario produced liver hemorrhage in broilers with 10% rapeseed meal in the ration. As before, *B. napus* is considered more hazardous, but because there is no direct link between level of glucosinolate and incidence of hemorrhage the only conclusion at this stage is to ensure that rapeseed meal is used at low levels in layer rations.

There have been some suggestions of a carcass taint in broilers associated with rapeseed meal, but this is not proven. Leg abnormalities were seen in broilers receiving a ration with 30% rapeseed meal of German origin. This has been related also to levels of glucosinolates and can be minimized by using meals of low glucosinolate content.

Although rapeseed meal has been researched extensively, many other vegetable proteins contain cyanogenetic glycosides which have caused toxicity at high levels of inclusion. Lancaster and Biely, (1965) list sorghum, linseed cake and Java or Burma bean among these.

Cassava (Manioc, Tapioca)

Gomez et al. (1988) recently have reviewed the levels of cyanogenetic glycoside in cassava. Cassava is used extensively in many European broiler rations particularly in those countries where cereals are limited, such as The Netherlands and Spain. They conclude that broilers are tolerant to high levels of cyanogenetic glycosides.

In recent years, cassava leaf meal has been used as a substitute for other proteins. Ravindran et al. (1986) replaced coconut oil meat in broiler diets with cassava leaf meal. Ten per cent cassava leaf meal improved performance while 20 to 30% depressed performance. This was thought to be due to a methionine deficiency and the presence of antinutritional factors from cassava leaf meal.

Gossypol

Cottonseed Meal

Toxic gossypol is a recognized component of cottonseed cake. According to Garner (1961), gossypol causes damage to the myocardium and liver resulting in cardiac oedema, dyspnoea, weakness and anorexia. When fed to layers, eggs from such hens showed a pink discoloration of the egg on storage. However, Reid et al. (1984) investigated glandless cottonseed meal versus regular cottonseed meal. Fifteen per cent of the regular meal caused discoloration in more than 70% of all eggs, whereas glandless meal did not. The discoloration is thought to be related directly to gossypol.

Nzekwe and Olomu (1984) compared cottonseed meal with groundnut for layers and turkeys. They reported no adverse effects using well-processed cottonseed meal replacing groundnut meal at 0 to 20% levels of inclusion.

Trypsin Inhibitors

A series of vegetable proteins are known to contain trypsin inhibitors if they are improperly processed. Since this was recognized at an early stage, such proteins can be easily rendered safe and the problem controlled.

Soyabean Meal

Ward et al. (1986) reported on the use of raw unprocessed soyabean meal and confirmed the requirement for processing. Antinutritional factors were observed. A certain level of heat treatment is required to deactivate the protease (trypsin) inhibitor and to denature the native soyabean protein.

Faba Beans

Ilian et al. (1985) described the use of faba beans in broilers. Some depressed performance occurred at high levels. Trypsin inhibitor was about one third that of commercial soyabean meal. At 30 or 45% inclusion in the ratio, blood hemoglobin decreased due to haemolytic factors. It is unclear, therefore, that the depressed performance is directly associated with the presence of trypsin inhibitor.

Other Toxic Factors

Lathyrism

This neuromuscular disease has been recognized for many years and is associated with the use of Indian peas *(Lathyrus sativus)*. The particular factors causing the disease are ill-defined.

In man it has been suggested recently that vegetable toxins from an uncultivated strain of tapioca palm can cause motor neurone disease (Lou Gehrig's Disease). It is possible, therefore, that such plant toxins also could cause disease in poultry although the long incubation period for neuromuscular defect to become apparent would mean that the birds already are used for meat. The question remains as to whether many of these toxins of plant origin are transmitted to man via the animal intermediary. Comprehensive data exist for the toxicology of new molecules developed for human or animal pharmaceutical purposes, but there is still a relative paucity of information on detailed toxicology and permissible tissue residues of natural plant toxins.

Ricin (Castor Oil Plant)

Ricin is a highly toxic component of the castor oil plant *(Ricinus Communis)*. On a London main street a few years ago an individual was injected with a minute amount of ricin using a modified umbrella with a hypodermic in the tip. The man collapsed and died within a short time and was a dramatic demonstration of the potency of this toxin. Castor cake contains a small amount of ricin, but the toxin is coagulated by heat and therefore destroyed. However, according to Lancaster and Biely (1965), heat-treated material was found to retard chick growth. Young pullets showed a dullness with drooping wings, ruffled feathers and greyish colored wattles and comb. Adult birds appeared to be less severely affected.

Theobromine

Cacao Products These are derived from the plant *Theobroma cacao* and the toxicity is derived from theobromine. This toxin is used therapeutically, but in uncontrolled amounts acts as a direct cardiac stimulant. Small quantities of meals or cakes can be fed safely to poultry although there are numerous cases on record whereby minuscule amounts in a horse compound feed have shown up in horse doping tests with subsequent penalties to the trainer.

Ten to 30% undecorticated cocoa cake fed in meal form to pullets caused deaths from the fifth day onward. Lower levels can cause nervous excitability in poultry.

Extrinsic Factors

Mycotoxins

In the early 1960's a condition was described in the U.K. which became known as turkey X disease. Detailed work at the Central Veterinary Laboratory at Weybridge showed that this fatal disease was related to the use of groundnut cake on meal and initially it was thought that a toxic factor was present in groundnut.

Subsequently, however, the toxic factor was isolated and found to be a mycotoxin. By 1965, Lancaster and Biely were able to describe the causal pathway which resulted from inadequate storage conditions of groundnut and contamination with the fungus *Aspergillus flavus* which produced the mycotoxin, aflatoxin. Feeding contaminated groundnut to poultry, turkey and ducks that are particularly susceptible results in hemorrhages in liver and kidneys. Because of the widespread occurrence of aflatoxin and its impact for man, many authorities have now set a limit on the amount of aflatoxin permitted in imported feeds.

Wherever vegetable proteins are stored under less than satisfactory conditions of high temperatures in a moist and humid environment, there is a potential hazard due to aflatoxin.

A more recent review by Dalvi (1986) describes the causal factor as Aflatoxin B_1. Between 250 and 500 ppb is known to cause liver damage and results in a reduction of immunological competence. More recent information has concentrated on the potent carcinogenicity of aflatoxin. Dairy cows consuming aflatoxin may be able to withstand the toxin moderately well, but this is passed out in the milk with all the implications as a potent carcinogen.

Giambrone (1985) studied aflatoxin in turkeys and broilers and found that 200 ppb significantly comprised cell-mediated immunity. Below this level was considered relatively safe and for this reason the FDA set a limit of 100 ppb for interstate shipment of feeds.

Huff et al. (1986) have studied the pathology of combinations of aflatoxin and dioxynivalenol (DON) otherwise known as vomitoxin, another mycotoxin. The combined effect was found to be addictive and DON was reported as being more toxic in broilers than in previous reports.

Ergot

Ergot is another fungal contaminant associated with *Claviceps purpurae* of rye. This mycotoxin causes constriction of the arterioles and in extreme cases will restrict the circulation to the extremities of livestock and cause widespread sloughing of tissues.

According to Lancaster and Biely (1965) poultry suffer from a vesicular dermatitis when exposed to ergot.

Crop Contaminants

Other extrinsic factors could include certain pesticides or herbicides sprayed onto crops prior to harvesting or materials used to assist oilseed storage and preservation. These always should be considered as part of an investigation where vegetable protein is suspected of causing poor performance or disease in poultry.

A particular poultry contaminant of corn and sorge crops is *Cassia abtusefolia*, coffee weed seed. The presence of this contaminant in the crop which subsequently goes through the processing cycle has been associated with a reduced feed intake resulting in a reduced liveweight gain and depressed feed efficiency. Egg production in layers is lower.

The potential for contamination of crops is widespread, for example, oilseed rape could be contaminated with mustard seed and there is scope for a wide range of problems associated with this factor.

Intrinsic and extrinsic factors of vegetable proteins have been shown to give reduced performance in all classes of poultry. As industry has developed a wide range of antibacterials, growth promoters, anticoccidials and other feed supplements, these have undergone progressive scrutiny and detailed review by independent scientific panels of experts to ensure all aspects of quality, safety and efficacy have been reviewed. It is only comparatively recently, however, that the same degree of scrutiny has been paid to vegetable proteins and it is likely in the future that consumer demand will insist that all factors which could potentially cause problems in livestock and man are carefully eliminated prior to use. Vegetable proteins have served man and his livestock well for thousands of years. Modern, intensive production systems demand the highest possible quality of feed ingredients and industry will continue to meet this challenge.

References

Dalvi, R.R., *Vet. Res. Commun. 10:*429 (1986).

Fenwick, G.R., and R.F. Curtis, *Animal Feed Sci. Technol. 5:*255 (1980).

Giambrone, J.J., U.L. Diener, N.D. Davis, V.S. Panangala and F.J. Hoerr, *Poult. Sci. 64:*1678 (1985).

Gomez, G., M.A. Aparicio and C. C. Willhite, *Nutr. Rep. Int. 37:*63 (1988).

Huff, W.E., L.F. Kubena, R.B. Harvey, W.M. Hagler Jr.,

S.F. Swanson, T. Philips, and C.R. Creger, *Poult. Sci.* 65:1291 (1986).

Ilian, M.A., M.D. Husseini, A. Al-Awadi and A.J. Salman, *Nutr. Rep. Int.* 31:477 (1985).

Lancaster, J.E., and J. Biely, "Poultry Diseases" (1965).

Mitaru, B.N., R. Blair, J.M. Bell and R. Reichert, *Can. J. Anim. Sci.* 63:655 (1983).

Nzekwe, N.M., and J.M. Olomu, *J. Anim. Prod. Res.* 4:57 (1984).

Peckham, M.C., in "Diseases of Poultry (8th Ed.)," edited by M.S. Hofstad, Iowa State University Press, Ames, Iowa, 1984.

Ravindran, V., E.T. Kornegay, A.S.B. Rajaguru, L.M. Potter and J.A. Cherry, *Poult. Sci.* 65:1720 (1986).

Reid, B.L., S. Galariz-Moreno and P.M. Maiorino, *Poult. Sci.* 63:1803 (1984).

Ward, N.E., J.E. Jones and D.V. Maurice, *Poult. Sci.* 65:106 (1986).

Wight, P.A.L., R.K. Scougall, D.W.F. Shannon, J.W. Wells and R. Mawson, *Res. Vet. Sci.* 43:(3)313 (1987).

Nutritional and Antinutritional Considerations of Soybean Processing for Swine

Colin Kirkegaard

Triple F Inc., PO Box 3600, Urbandale Branch, 10104 Dougland Ave., Des Moines, Iowa 50322, USA

Abstract

The antinutritional factors found in soybeans include protease inhibitors, hemagglutinins, goitrogens, antivitamins, phytates, saponins, estrogens, flatulence factors, lysinoalanine and allergens. While most of these factors have been studied in detail in the rat, chick and man, only trypsin inhibitor (a protease inhibitor) has been researched to any great extent in the pig. The pig's pancreas makes up less than 0.3% of the total body weight and does not enlarge in the presence of raw soybeans or trypsin inhibitor. Recent interest in feeding higher energy diets to sows and growing/finishing pigs has stimulated researchers to look at the feeding of raw soybeans. These studies have resulted in suboptimal animal performance thereby pointing out the importance of properly processed soya protein before using it as a supplement in swine diets. The measurement of urease activity (pH rise, Soy-Chek) is a good indicator of the adequacy of heat treatment of soybeans in regard to their trypsin inhibitor content.

Soybeans presently serve as an excellent source of protein in the U.S. for all species of livestock including swine. In the midwestern section of the U.S. where most of the swine are grown, the "corn-soy" diet is the standard against which all other potential feed ingredients are measured. This is due to the excellent job soya protein performs in balancing the amino acid requirements of the pig when corn is used as the primary energy source as shown in Table 1 (1). This feeding practice is further established by the yield enhancement farmers experience when following the agronomic practice of a corn-soybean rotation.

The full benefit of the soybean's nutritional profile can be obtained only after a certain amount of heat has been applied during processing to deactivate the antinutritional factors contained in the raw product.

Antinutritional Factors

The antinutritional factors inherent to the soybean have been summarized very well by Liener (2) in Table 2 as they affect the human diet. Of these factors, only trypsin

TABLE 2
Antinutritional factors in soybeans

Heat-labile	Heat-stable
Trypsin inhibitors	Saponins
Hemagglutinins	Estrogens
Goitrogens	Flatulence factors
Antivitamins	Lysinoalanine
Phytates	Allergens

TABLE 1
Essential Amino Acids Required and Supplied by Corn and Soybean Meal (SBM) Either Alone or in Combination for Growing and Finishing Swine Weighing from 50 to 110 kg

	Requirements %	Supplied by corn %	Supplied by SBM %	Supplied by 13% crude protein corn-SBM %
Arginine	0.10	0.43	3.20	.78
Histidine	0.18	0.27	1.12	.38
Isoleucine	0.38	0.35*	2.00	.56
Leucine	0.50	1.19	3.37	1.47
Lysine	0.60	0.25*	2.90	.59
Methionine & cystine	0.34	0.40	1.18	.50
Phenylalanine & tyrosine	0.55	0.84	3.60	1.19
Threonine	0.40	0.36*	1.70	.53
Tryptophan	0.10	0.09*	0.64	.16
Valine	0.40	0.48	2.02	.68

* Denotes amino acids found deficient in swine fed a corn only diet.

inhibitors have been studied in any detail in the swine species. Trypsin inhibitors belong to a class of protease inhibitors which cause pancreatic hypertrophy, stimulate pancreatic juice secretion, and inhibit growth in rats (3), chickens (4) and swine (5). It should be noted that the pig's pancreas, which is responsible for trypsin enzyme production, appears to be nonresponsive to the hypertrophic effect of trypsin inhibitors seen in other species as demonstrated in Table 3 (6). Thus the growth inhibition in the pig appears to be due to the inhibition of intestinal proteolysis whereas there may be secondary losses in the rat and the chick as a result of excessive enzyme secretion from pancreatic hypertrophy (8). This makes the extrapolation of research results from such commonly used experimental models as the rat and the chick to the swine species very difficult.

TABLE 3

Relationship Between Size of Pancreas of Various Species of Animals and Response of Pancreas to Raw Soybeans or Trypsin Inhibitor

Species	Size of pancreas (% of body weight)	Pancreatic hypertrophy
Mouse	0.6–0.8	+
Rat	0.5–0.6	+
Chick	0.4–0.6	+
Guinea pig	0.29	±[a]
Dog	0.21–0.24	−
Pig	0.10–0.12	−
Human being	0.09–0.12	−[b]
Calf	0.06–0.08	−

[a] Observed in young guinea pigs but not in adults.
[b] Predicted response.

Vandergrift (7) reported on the relationship between increasing the heat treatment of soy protein with *in vitro* assays and subsequent performance in 25–45 kg pigs fitted with T-cannula near the terminal ileum (Table 4). This study suggests that processing soy to a point where approximately 90% of the trypsin inhibitor has been inactivated (as in Treatment 4, Table 4) may give the best results in growing-finishing swine under full-fed feedlot conditions.

There has been increasing interest in the feeding of heated, full fat soybeans as both a protein and fat source in various phases of swine production. This interest has stimulated researchers to look at the feeding of raw soybeans to determine if at some point increasing animal maturity can compensate for some of the antigrowth factors found in the raw product. We will review some of the more recent research in these areas.

Growing and Finishing Swine

Crenshaw and Danielson (9) evaluated raw soybean diets against a balanced corn-solvent extracted soybean meal basal diet for average daily feed intake, average daily gain and feed conversion. Three groups of pigs were started at 50, 100 and 150 pounds body weight and the trials were terminated when the basal corn-soybean meal pigs reached approximately 210 pound body weight (Table 5). Irrespective of their starting weights, pigs fed raw soybeans consumed less feed, grew slower and converted less efficiently than the pigs fed the soybean meal basal diet. Older pigs (starting weights 100 and 150 pounds) did not appear to utilize raw soybeans any more efficiently than the younger pigs (starting weight 50 pounds). No measurements were made of the trypsin-inhibitors content of the diets used in this study.

Goodband, *et al.* (10) recently reported on the use of raw soybeans and soybean oil as energy sources for finishing pigs while monitoring the increase in trypsin-inhibitor content of the experimental diets. The composition of the diets studied and their effect on pig performance are depicted in Tables 6 and 7. Trial 1 utilized 120 pigs with an average starting weight of 121.7 pounds and Trial 2 utilized 150 pigs with an average starting weight 128.5 pounds. Both trials were 63 days in duration. These data demonstrate that the addition of raw soybeans as just an energy source can have detrimental effects on the performance of finishing pigs. Combs, *et al.* (11) demonstrated the inability of protein and fat supplementation to overcome the growth depression seen in young growing swine fed raw soybeans. These studies suggest that some of the other antinu-

TABLE 4

Performance of Pigs Fed Diets Containing Soyflakes Heated for Different Periods

Treatment	1	2	3	4	5	6
Minutes heated	0	25	35	45	65	105
Urease, pH rise	2.02	0.48	0.19	0.08	0.08	0.05
Trypsin inhibitor mg/g (as fed)	31.1	6.2	4.3	3.2	2.0	1.4
Average daily gain kg/day	0.04	0.36	0.33	0.42	0.31	0.31
Average gain/feed	0.06	0.35	0.31	0.39	0.30	0.29
Apparent digestibilities of nitrogen at end of small intestine (%)	35.1	78.9	76.0	82.0	78.6	81.5
Nitrogen retained g/day	8.26	13.61	13.61	13.77	13.78	13.21

TABLE 5

Performance of Growing and Finishing Pigs Fed Diets Supplemented with Raw Soybeans

Criteria	Diet treatment[a]		
	A	B	C
Group I—50 lb, 98 day termination			
Average daily feed intake, lb.	5.27	3.65	3.83
Average daily gain, lb.	1.63	.81	.82
Feed conversion ratio	3.25	4.49	4.73
Average final weight, lb.	213	134	134
Group II—100 lb, 56 day termination			
Average daily feed intake, lb.	6.90	5.67	5.85
Average daily gain, lb.	1.93	1.15	1.26
Feed conversion ratio	3.58	4.95	4.64
Average final weight, lb.	210	166	172
Group III—150 lb, 28 day termination			
Average daily feed intake, lb.	7.42	6.66	6.48
Average daily gain, lb.	1.83	1.32	1.24
Feed conversion ratio	4.15	5.23	5.32
Average final weight	204	190	188

[a] A—Balanced corn-soybean meal diet
B—Raw soybeans replaced soybean meal in Diet A on equal weight basis
C—Raw soybeans replaced soybean meal on an isonitrogenous basis

TABLE 6

Composition of Diets

	Trial 1					Trial 2				
		Soy oil		Raw soybeans			Soy oil		Raw soybeans	
Criteria	Control	2%	3%	2%	3%	Control	2%	3%	2%	3%
Ingredient (%)										
Milo, ground	79.35	77.35	76.35	68.15	62.55	89.85	87.85	86.85	78.65	73.05
Soybean meal	18.25	18.25	18.25	18.25	18.25	7.50	7.50	7.50	7.50	7.50
Soy oil	—	2.0	3.0	—	—	—	2.0	3.0	—	—
Raw soybeans	—	—	—	11.2	16.8	—	—	—	11.2	16.8
Minerals, vitamins	2.40	2.40	2.40	2.40	2.40	2.65	2.65	2.65	2.65	2.65
Calculated Analysis										
Crude protein (%)	15.1	15.0	14.9	18.3	19.9	11.4	11.2	11.1	14.5	16.1
Lysine (%)	.69	.69	.69	.94	1.06	.61	.60	.60	.85	.97
Ca (%)	.60	.60	.60	.63	.64	.55	.55	.55	.58	.59
P (%)	.56	.55	.55	.59	.61	.52	.52	.52	.56	.57
ME Kcal/lb	1421	1463	1483	1437	1445	1426	1466	1487	1441	1449
Trypsin inhibitor units/mg	17.0	11.0	18.5	50.5	75.0	10.5	15.5	15.0	54.5	67.5

TABLE 7

Effects of Soybean Oil and Raw Soybean Additions to Finishing Diets

	Trial 1					Trial 2				
		Soy oil		Raw soybeans			Soy oil		Raw soybeans	
	Control	2%	3%	2%	3%	Control	2%	3%	2%	3%
Item										
Average daily gain, lbs	1.83	1.77	1.92	1.74	1.74	1.73	1.64	1.66	1.64	1.65
Average daily feed intake, lbs	7.05	6.54	7.01	7.20	6.96	6.51	6.01	6.12	6.28	6.11
Feed efficiency	3.86	3.71	3.65	4.14	4.22	3.77	3.71	3.68	3.83	3.71
Plasma urea, mg/dl	22.8	20.2	17.6	24.7	26.7	13.5	14.0	14.0	18.5	19.8

tritional factors found in raw soybeans, other than the trypsin inhibitors, may have an antigrowth effect in the pig.

Gestating and Lactating Sows

There has been increasing interest in the feeding of higher energy diets to gestating and lactating sows to both decrease the interval from weaning to first estrus and increase the survivability of the piglets and their subsequent weaning weights. Fat sources, which contain anywhere from 2.25 to 2.50 times the caloric content of an equivalent amount of protein or carbohydrate, are an excellent means of increasing the energy density of these diets. Soybeans with their naturally high fat content of 18% represent one method of increasing the caloric content of a sow diet without encountering the handling problems associated with liquid fat sources.

Researchers (12,13,14) have looked at the addition of raw soybeans to gestation and lactation diets to determine if adult swine (sows) can overcome the antinutritional factors in raw soybeans and effectively use them as both a supplemental protein and fat source.

The gestating sow is basically in a maintenance mode for much of her 114-day gestation period and has a relatively low crude protein requirement of 228 grams per day (1). Of all the members of the swine species, it was postulated that the mature sow would stand the best chance of tolerating the antinutritional factors found in raw soybeans.

Jensen *et al.* (13) concluded that second and third parity sows did not efficiently utilize the energy in raw soybeans as the sows receiving the raw soybeans in their diets gained less weight during gestation than the sow fed either cooked, whole soybeans or soybean meal. The sows fed cooked soybeans during gestation also had, on average, about one-half the subsequent lactation weight loss when compared to the sows receiving either raw soybeans or soybean meal.

Crenshaw and Danielson (12,14) recently compared the feeding of raw soybeans on an isonitrogenous basis with 44% soybean meal in limit (four pounds/day), individual stall-fed, gestation sows studied for three consecutive parities. The rations fed and their analyses are depicted in Table 8. The sows were fed the treatment diets through the 110th day of gestation at which point they were placed in the farrowing house where they were switched to the same lactation diet containing 44% soybean meal as the supplemental protein source. The sows were weaned at 21 days, bred on the first estrus cycle and returned to their respective gestation treatments for second and third parity evaluations. There were no significant differences in sow weights, either gestation weight gain or lactation weight losses despite the increased consumption of energy during the gestation period. Raw soybean-fed sows were calculated to receive 6,012 Kcal of M.E. per day (1503 Kcal/lb × 4 lbs/day) while soybean meal fed sows received 5,768 Kcal of M.E. per day (1442 Kcal/lb × 4 lbs/day). There was a treatment × parity interaction for survival percentage because the survival rate for the piglets in the raw soybean fed sows declined by parity while the survival rate for soybean meal fed sows remained relatively constant.

It has been shown in poultry (15,16) that raw soya protein depresses fat digestion and absorption. In fact, as little as 5% protein from raw soybean meal in a diet containing 25% total crude protein caused nearly total depression of fat absorption (15). This fat digestibility depression seen in poultry fed raw soybean meal could be overcome by the addition of trypsin enzyme concentrate

TABLE 8

Composition of Gestation Diets

	Diet	
Ingredients	Soybean meal	Raw soybeans
Corn	77.03	68.01
Soybean meal (44%)	16.42	—
Raw soybeans	—	25.38
Alfalfa hay	2.50	2.50
Mineral-vitamin premix	4.05	4.11
	100.00	100.00
Analysis		
Protein (%)	14.00	14.00
Fat (%)	3.07	7.31
Calcium (%)	.90	.90
Phosphorus (%)	.80	.80
Lysine (%)	.69	.79
ME (Kcal/lb)	1442.00	1503.00

to the diets (16). It is not clear whether the inability of both growing and finishing swine as well as mature sows to efficiently utilize the extra gross energy supplied in their diets by raw soybeans is entirely due to the presence of trypsin inhibitors. The hemagglutinins found in soybeans have been shown to decrease the digestibility of nitrogen-free extract in chickens by interfering with the normal absorption of pancreatic amylase to the intestinal epithelium, thereby allowing the enzyme to be quickly excreted in the feces (17). This may account for the lower metabolic energy found in raw soybeans as compared to adequately heated soybeans.

Weanling Pigs

Simplified corn-soybean meal diets do not maximize weanling pig performance. For many years, it was believed that a proteolytic enzyme insufficiency was responsible for the inability of the pig to adequately digest plant protein (18) when in fact residual trypsin inhibitor may be complexing with the limited trypsin enzyme production by the young pig's pancreas. The inability of the pig's pancreas to respond to the presence of trypsin inhibitor has been noted earlier.

It has been suggested that the presence of certain soluble carbohydrates in soya such as raffinose and stachyose adversely affect the performance of weanling pigs. Soy protein concentrates (SPC), which have a high percentage of the soluble carbohydrate removed, have been substituted for soybean meal with generally favorable results (19). However, the demand for SPC and dried skim milk powders for human food may not allow their consideration in pig starter formulas.

More research is needed in the area of processing soya to improve its digestibility in the young pig.

Quality Control

Quality control is a function normally performed by the soybean processor. It is an extremely important function if maximum digestibility of the soy protein is to be obtained. The feed industry has long used pH rise to test for urease activity as a measure of the proper degree of cooking. Urease is an enzyme which converts urea to ammonia and has no importance in swine except as an indicator of soy processing. Urease activity can be dangerous to ruminants receiving raw soybeans as cases of ammonia toxicity have been documented from soybean overloads and from diets containing urea.

Two simple tests are available for testing soy products in the field at production sites. They are the qualitative Dupont jar test and the quantitative Soy-Chek test. The Dupont test consists of reacting a cup of processed soy and a teaspoon of urea in a closed vessel (jar) with warm water (the water should cover the mixture) for twenty minutes. When the lid is removed, the degree of ammonia detected would indicate the degree of cooking. The Soy-Chek test is a commercially available more sophisticated version where soy products can be tested for both under-cooking and over-processing by means of a colorimetric test. This test takes five to ten minutes to run.

Although urease activity is a sensitive test for normal levels of processing, the actual measurement of trypsin inhibitors is desirable when working with animals that are trypsin inhibitor sensitive such as the young pig.

Lipase enzyme activity normally is not measured since usually it is destroyed in the heating process. Lipase activity can present problems in soybeans which are ground, rolled or flaked without heating.

The role the other antinutritional factors found in soy play in swine nutrition is uncertain and as such, their activity is not usually monitored.

References

1. Anonymous, "Nutrient Requirements of Swine (9th edition)," National Academy Press, Washington, D.C. (1988).
2. Liener, I.E., *J. Am. Oil Chem. Soc. 58:*406 (1981).
3. Rackis, J.J., A.K. Smight, A.M. Nash, D.J. Robbins and A.N. Booth, *Cereal Chem. 40:*531 (1963).
4. Bray, D.J., *Poult. Sci. 43:*382 (1964).
5. Yen, J.T., T. Hymowitz and A.H. Jensen, *J. Anim. Sci. 38:*304 (1974).
6. Liener, I.E., and M.L. Kakade, "Nutritional Significance of Protease Inhibitors" in *Toxic Constituents of Plant Foodstuffs,* edited by I.E. Liener, Academic Press, New York, 1980.
7. Vandergrift, W.L., D.A. Knabe, T.D. Tanksley Jr. and S.A. Anderson, *J. Anim. Sci. 57:*1215 (1983).
8. Yen, J.T., A.H. Jensen and J. Simon, *J. Nutri. 107:*156 (1977).
9. Crewnshaw, M., and M. Danielson, *Raw Soybeans for Growing Finishing Pigs,* Nebraska Swine Report, Nebraska Cooperative Extension Service, Lincoln, Nebraska, 1984.
10. Goodband, R.D., R.H. Hines and R.C. Thaler, *A Comparison of Raw Soybeans and Soybean Oil as Energy Source for Finishing Pigs,* Kansas State University Swine Day Report of Progress 528, 1987.
11. Combs, G.E., R.G. Conness, T.H. Berry and H. D. Wallace, *J. Anim. Sci. 26:*1067 (1967).
12. Crenshaw, M.A., and D.M. Danielsen, *J. Anim. Sci. 60:*163 (1985).
13. Jensen, A.H., D.H. Baker, B.G. Harmon, D.M. Woods and G.R. Carlisle, *Cooked Soybeans and Raw Soybeans as Supplemental Protein Sources in Diets for Gestating Swine,* University of Illinois Cooperative Extension Service AS-658b, Champaign, Illinois (1971).
14. Crewnshaw, M.A., and D.M. Danielson, *Raw Soybean Value for Gestating Swine,* Nebraska Swine Report, University of Nebraska Cooperative Extension Service, 1984.
15. Nesheim, M.C., J.D. Garlich and D.T. Hopkins, *J. Nutri. 78:*89 (1962).
16. Brambila, S., M.C. Nesheim and F.W. Hill, *Ibid. 75:*13 (1961).
17. Scott, M.L., M. Sandholm and H.W. Hockstetler, "Effects of Anitrypsius and Hemagglutinius in Soybeans and Other Feedstuffs Upon Feed Digestion in Chickens," proceedings of Cornell Nutrition Conference, 1976.
18. Catron, D.V., "Factors Influencing the Nutritional Requirements of Swine" in *Bridging the Gap in Nutrition,* edited by R.H. Thayer, Midwest Feed Manufacturers Association, Kansas City, Missouri, 1963.
19. Stahly, T.S., G.L. Cromwell and H.J. Monegue, *Feedstuffs 56:*(4)14 (1984).

Utilization of Full-Fat Soybeans for Dairy Cattle

Karl H. G. Sera

American Soybean Association, Akasaka Tokyu Bldg., 11th Floor, 2-14-3 Nagata-cho, Chiyoda-ku, Tokyo 100, Japan

Abstract

Genetic potential of modern dairy cattle for milk production has improved to the point where maximum feeding levels of grain concentrates between 15 to 18 kilograms per day with a minimum level of forage dry matter intake at a 60:40 ratio respectively is insufficient to meet nutrient requirements. Such is the case today for many cows in early lactation or up to four months after calving. In recent years, interest on feeding heat processed full-fat soybean for dairy cattle has increased in many developed countries and much research is being conducted. A practice of feeding one to two kilograms of heat processed full-fat soybeans per cow per day as an high nutrient density supplement for protein (rumen non-degradable) and energy (non-starch) is becoming popular among top dairy farms in the midwest, north-central, and mountain states of the United States. This concept and feeding practice also is becoming popular in Japan. As the importance of protein utilization, both in the rumen and in the lower gastrointestinal tract, and the limitation on the level of feeding non-structural carbohydrates becomes more apparent, the feeding of heat-processed full-fat soybeans to dairy cattle should increase in dairy cattle feeding practices.

The challenge to the dairy industry in any country is to improve continually on the genetic capability of both sire and dam, and simultaneously to improve the nutritional management of dairy cattle to obtain optimum and economical milk production. Dairy cows in the past survived on pasture, and high level feeding of concentrates was not needed because the genetic capability for milk production was low as was the demand for milk from the dairy industry. Cows supplied colostrum and milk for their calves and sometimes supplied a small additional quantity of milk for consumption by members of the family who raised the cows. Milk production of 300 to 800 kilograms per cow in her lactation was not unusual in European countries in the 19th century. In the 1800's Japanese emperors and their family members consumed a very small quantity of cows' milk only as a source of medication to maintain good health. Here again, high milk production was not pursued.

As the popularity of consumption of milk and dairy products began to pick up among people, particularly in the Western nations, pressure was applied for the industry to improve the milk production capability of the cow. In the United States in the early 1900's, most cows grazed on pasture and were fed hay, and very little concentrate feeding was practiced. Around 1930, some interest in feeding concentrates as part of the feeding program began to develop. For milk production of 2,000 kilograms and above, 10% to 15% of total dry matter intake was fed as concentrates. A book entitled "Cow Philosophy" by Mark Keeney and based on his experiences at Overbrook Dairy Farm pointed out the need for protein and grain feeding to obtain high milk production from cows having good genetic backgrounds. His book and lectures inspired many dairymen in the United States to feed more concentrates to their cows. In the 1950's, 25% to 30% of dry matter intake by cows producing 2,500 kilograms and above was concentrates. In the 1970's, 35% to 40% of dry matter intake was concentrates for cows producing 5,000 kilograms and above. In the 1980's, the concentrate feeding went up to 45-50% of the total dry matter intake for a milk production of 6,000 kilograms and above.

Reid estimated that forage at 60% to 65% digestibility is adequate to support milk production of near 5,000 kilograms per year by Holstein cows. Reid further estimated that forage at 70% digestibility and consumed at 3.3% of body weight (635 kilograms Holstein low) would sustain a production of 34 kilograms of milk per day or a potential yield of 8,156 kilograms per year. It may take a while, but this production is a feasible goal and demonstrates the importance of forage quality and quantity in dairy production.

In the 1990's, the annual average milk production per cow in the United States and other developed nations including Japan is projected to be above 7,000 kilograms. Grain or concentrate feeding will go as high as 60% of total dry matter intake, but that will probably be the highest level it will go.

Limitations on Concentrate Feeding

The rumen of a cow is suited to utilize a large quantity of cellulose in forage. Through microbial degradation, cellulose is converted to volatile fatty acids. Cows use these volatile fatty acids to obtain energy for production and maintenance. For any dairy cow, acetic acid is most important because it is utilized in butter fat synthesis in milk. As indicated (Figure 1), the acetic acid production in the rumen decreases rapidly as the concentrate to forage ratio goes beyond 40:60 in total dry matter intake. Beyond a 60:40 ratio, butterfat begins to drop. If finely chopped forage or silage is fed along with pulverized concentrates, the butterfat content of milk could begin to drop at about 50:50 ratio of concentrate to forage feeding. Buffers, such as sodium bicarbonate and magnesium oxide, are commonly fed by dairymen at 200 grams and 100 grams, respectively, per cow per day during the period of high level concentrate feeding for high milk production and also during the hot summer months when the forage intake drops resulting in high concentrate intake in total dry matter. Even with such practices the feeding concentrates beyond 60% of total dry matter intake is considered unwise.

Recommended energy content of diets on dry matter for dairy cattle in the new 1988 edition of NRC-Dairy

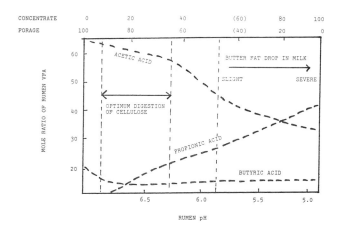

Fig. 1. Concentrate-to-forage ratio (% of DMI).

shows 1.72 Mcal/kg of net energy for lactation or total digestible nutrients (TDN) 75% for a 600 kilograms cow producing 50 kilograms of milk per day. For the same cow in early lactation or up to three weeks post-calving, 1.67 Mcal/kg of net energy for lactation or TDN 73% is recommended. Also, for the same 600 kilograms cow producing 50 kilograms of milk per day, minimum levels of crude fiber (CF), acid-detergent fiber (ADF), and neutral-detergent fiber (NDF) of 15, 19, and 25% respectively are recommended. For early lactation or post-calving period up to three weeks, 17, 21, and 28% are recommended for (CF), (ADF), and (NDF) respectively. Approximately 70~80% of neutral detergent fiber needs to come from forage. Maintaining these minimum fiber levels during the period of high energy needs places a limit on the amount of concentrate feeding.

Recent research indicates also that nonstructural (soluble) carbohydrate level in total dry matter intake of a cow producing above 35 kilograms of milk per day be around 30 to 40%. These levels are still preliminary. But these guideline figures also place a limit on the high level feeding of concentrates. Non-structural carbohydrate of an ingredient is calculated as follows: 100 − (CP % + EE % + NDF % + Ash %).

Protein Feeding

Recommended protein content of diets on dry matter for dairy cattle in the new 1988 edition of NRC-Dairy shows 18% of CP with 6.2% of undergraded intake (crude) protein (UIP) and 10.4% of degraded (crude) protein (DIP) in the rumen for a 600 kilograms cow producing 50 kilograms of milk per day. For the same cow in early lactation or up to three weeks post-calving, 19, 7.0, and 9.7% are recommended for (CP), (UIP), and (DIP) respectively.

Guidelines put out by Cornell University on protein feeding suggest the following: Out of total protein (100% basis) in feed dry matter, 40% should be rumen non-degradable protein and 60% should be rumen degradable protein. Of the 60% rumen degradable protein, one-half or 30% unit is to be soluble protein.

Body Condition Scores

Cornell University also has guidelines for body condition of dry cows and cows in the all three stages of lactation on a one to five scoring. Score one is assigned to cows with severe under-conditioning (very thin) and score five is assigned to over-conditioned, very fat or obese cows. The areas of particular interest when body condition scoring is done by palpation and not by visual appraisal include the loin, backbone, hooks, pins, tail head and the area between the hooks and pins.

Recommended scores for dry cows are between 3+ to 4-; cows in early lactation 3- to 3; cows in mid-lactation 3; and cows in late lactation 3 to 3+ and 3+ to 4- at the time of drying off.

These guidelines are practical and are used by many dairymen in the United States and in Japan to determine whether or not the nutrition is at an optimum level for a given stage of lactation.

Maintaining a 3- to 3 score for early lactation is difficult for cows producing above 45 to 50 kilograms of milk a day during the period. Cows in early lactation are in severe negative energy and protein balances and body fat tissues are mobilized quickly for milk production. There is, however, a limit to concentrate or grain feeding to supply energy lost from high milk production.

Nutritional Management Factors To Consider

Dry matter intake with a balance of nutrition is the key factor for cows producing 6,000 kilograms of milk (Figure 2). As the production increases to 7,500 kilograms, energy supply must be watched. Importance of rumen nondegradable protein in cows' diets begins starting from approximately 7,300 kilograms of milk production. With 9,000 kilograms milk yield, dry matter intake, energy supply, and quantity and type of protein or rumen degradable and rumen nondegradable proteins in diets need to be checked. Importance of rumen by-pass fat in cows' diets begins starting from approximately 9,000 kilograms of milk production depending on the quality of forage, concentrates, and grain by-products fed to cows. For cows producing above 10,000 kilograms, superior genetics have to be considered on top of all of the factors aforementioned.

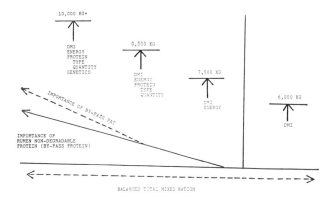

Fig. 2. Management factors to consider in milk production.

Ideal Period For Feeding Full-Fat Soybean

Cows normally reach peak milk production per day at 4-6 weeks after calving (Figure 3). Peak for dry matter intake lags behind peak milk production. It normally appears at 8-12 weeks after calving. Also, for each additional kilogram

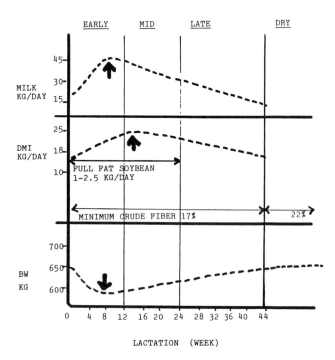

Fig. 3. Ideal period for feeding full fat soybean.

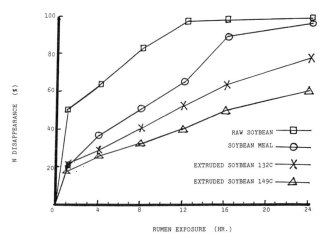

Fig. 4. N disappearance from dacron bags containing soybean meal and extruded soybean.

of milk attained at peak is equivalent to an increase in total lactation yield in 305 days of 200-220 kilograms.

Ideal period for feeding full-fat soybean may be up to 24 weeks post-calving. Up to 12 weeks post-calving, milk production is higher than nutritional dry matter intake for the amount of milk cow is producing. Body fat tissue, and in some cases body lean tissue are mobilized during this period. This will mean that the nutritional density per kilogram of diet dry matter must be increased while maintaining the fiber and soluble carbohydrate levels close to the recommendations and guidelines mentioned earlier. From 12 to 24 weeks post calving, body weight recovery begins to take place while maintaining high milk production after peak milk yield.

Protein and energy contents of full fat soybean are higher than many other ingredients fed to cows. On dry matter basis, crude protein, TDN, digestible energy and net energy for lactation will show 42.2%, 94%, and 4.14 and 2.18 Mcal/kg, respectively. High energy comes from 20% of the ether extractables (lipids) in soybean. Crude fiber and acid detergent fiber show 11.0% and 5.6%, respectively. Lysine, the first limiting amino acids for milk production is high in soybeans. The lysine content of soybean is 2.67% on dry matter basis.

Stern experimented with extruded soybean to find levels of protein degradation (Figure 4). Extruded soybeans were more resistant to degradation than soybean meal or raw soybeans at all intervals of rumen exposure. Flow of dietary amino acids to the duodenum was highest (885 grams) for soybeans extruded at 149°C. Flow of dietary amino acids to the duodenum for soybean meal and raw soybean, was 586 and 554 grams, respectively. Amino acid absorption in the small intestine, expressed as percent of amino acid flow to the duodenum, was higher for the extruded soybean diets. This indicates that heat treatment did not overprotect the protein.

Factors To Consider In Feeding Rumen Non-Degradable Protein

1. Feed adequate amount of rumen degradable protein for microbial protein synthesis in the rumen. Up to crude protein level of approximately 13% in the diet or up to protein requirement for milk production equivalent to body weight $\times$ 4.5% needs to be fed in the form of rumen-degradable protein. Soybean meal, cottonseed meal, and urea are examples of easily degraded protein ingredients to supply ammonia for rumen microbes.

2. After the condition stated above is fulfilled, rumen nondegradable protein (by-pass protein) needs to be fed. Approximately crude protein levels of 13% to 20% in the diet should be fed in the form of rumen nondegradable protein. Heat processed full-fat soybean, corn gluten meal, fish meal, and blood meal are examples of rumen nondegradable protein ingredients.

3. When full-fat soybeans are chosen as an ingredient, the degree of heat processing may be checked using the following: (a) Urease activity at 0.2 to 0.05 pH rise. For improved nondegradation of protein in the rumen, a pH rise of 0.05 is better. Raw soybeans will have a pH rise of 2.0; (b) Nitrogen solubility index of 10. Raw soybeans have an index of 95; and (c) Trypsin inhibitor unit of 5 UG/mg of crude protein. Raw soybeans may give a value close to 50 UG/mg of CP.

Summary On Utilizing Full Fat Soybean

1. Continued improvement to attain higher milk production will come from: (a) Genetic improvement; and (b) Future use of somatotropin (BST).

2. The above will be pushed due to economic reasons: (a) Maintenance cost of feed (nutrition) is fixed for a given weight of cow; and (b) The higher the milk production, the lower the cost to produce a kilogram of milk (generally).

3. Limitation exists in feeding grain concentrates: (a) Maximum grain to forage ratio is at 60:40; (b) Maximum soluble carbohydrate in total diet dry matter at no more than 30-40%; and (c) Minimum fiber content in total diet dry matter at 15-17% for crude fiber, 19-21% for acid-detergent fiber, and 25-28% for neutral detergent fiber.

4. Maintaining body condition: (a) Dry cow: 3+ to 4-; and (b) Early lactation: 3- to 4.

5. Forty per cent of protein intake needs to be rumen nondegradable, by-pass, protein.

With some degree of analysis and ration balancing, full-fat soybean is incorporated into: (a) total mixed ration (TMR), (b) supplement, and (c) fed as top dress. It is fed at one to 2.5 kilograms per day per cow. Full-fat soybean is best suited for feeding generally up to six months post calving. Calcium should be fed at higher levels, 0.8% to 1.0% full fat soybean in total dry matter. More than 2.5 kilograms, and up to 4 kilograms per cow per day can be fed. But, the risk of lowering butterfat in milk becomes much greater at present.

The conclusion is that at present full fat soybeans are highly palatable and an inclusion of up to 2.5 kilograms in dairy ration is a very good way to raise the nutrient density of dairy feed and raise milk production per cow.

Selected References

Seventy Five Years Of Dedication And Progress, J. Am. Dairy Science Assn. (1981, 6).

Illinois-Iowa Dairy Handbook, Univ. of Ill. and Iowa State Univ. (1988).

Nutrient Requirements Of Dairy Cattle, NRC (1988).

Body Condition Scoring, Cornell Univ. (1985).

Full Fat Soybean Use In Ruminants, Stern, M.D., Soybean Utilization Alternatives, Univ. of Minn. (1988, 2).

Feeding The High Quality Producing Dairy Herd, Chase, L., Japan Seminar (1988).

Feed Ingredient And Management To Maximize Use Of Fat In Dairy Rations, Palmquist, D.L., Japan Seminar (1988).

Fullfat Soya, A Regional Conference, Am. Soy. Assn. Italy (1987).

Effect Of Dietary Forage: Grain Ratio On Performance Of Lactating Dairy Cows, Tessmann, N. J., J. Am. Dairy Science Assn. (1988, 6).

Effect Of Diet Fiber Level And Forage Source On Intake And Milk Production Cows In Early Or Late Lactation, Allen, M.S., J. Am. Dairy Science Assn. (1988, 6).

Effects Of Buffer Supplementation On Feed Intake, Milk Production And Milk Composition For Cows Fed A Milk Fat Depressing Diet, Solorzano, L.C., J. Am. Dairy Science Assn. (1988, 6).

Protein And Energy As An Integrated System. Relationship Of Ruminal Protein And Carbohydrate Availability To Microbial Synthesis And Milk Production, Nocek, J.E., J. Am. Dairy Science (1988, 8).

Effect Of Soybean Meal And Fish Meal On Protein Digesta Flow In Holstein Cows During Early And Mid-Lactation, Zerbini, E., J. Am. Dairy Science Assn. (1988, 5).

California Dairy Cattle Day, Univ. Of California (1987, 4).

Lactational Response To Soybean Meal, Heated Soybean Meal, And Extruded Soybeans With Ruminally Protected Methionine, Schingoethe, D.J., J. Am. Dairy Science Assn. (1988, 1).

Methods Of Processing Full-Fat Soybeans, Influence On Nutritive Value, Kohlmeier, R.H., Nutrition Conference For Feed Manufacturers, Univ. Of Guelph (1988, 4).

Nutrition of Feedlot Cattle

Larry Dodson

Dodson Consultancy, Eugowra Road, Canowindra, New South Wales 2804, Australia

Abstract

Composition of feedlot rations is important not only from the theoretical nutritional values, but also in terms of what the cattle like, and don't like, to eat. Environmental conditions can affect cattle's feeding habits, and the feedlot manager must be aware of how to provide nutritional rations palatable to the cattle.

There are volumes of papers written on this subject; accurate, inaccurate and some inconclusive. Space does not permit me to properly cover all the technical aspects of this complex field. I will however try to pass onto you on a layman's level some of the professional management aspects of a feedlot ration that are not normally acknowledged.

I tend to stay pretty close to nature on decisions in feeding cattle. We are virtually altering nature to finish cattle to the degree of carcass acceptability. The cause and effect of this process is a study in itself and definitely alters the end result of good or bad nutrition. Cattle are creatures of habit with personalities, peck order, stress tolerance geared and triggered by nature. Around this scenario a properly formulated ration nutritonally can be an economic success or an absolute disaster depending on the management of that ration and the environment encountered by the animals.

Basic Ration

The basic ration includes (a) grain supplying starch to convert to glucose; (b) energy from molasses, animal fat or other energy producing products other than starch but not necessary to the ration; (c) roughage, the physical fibre that cultures the population of the micro flora (bug in the rumen) that are essential to the digestibility of the ration and the efficiency of feed conversion; (d) micro ingredients such as vitamins, trace minerals, salt, limestone, potassium chloride, dicalcium phosphate plus the supplementary protein, usually processed into a pellet for conveyance to the ration; (e) supplementary protein, either vegetable or nonprotein nitrogen (urea); (f) ionophores as monocin (rumensin), lasalocid (Bovetec) which are rumen modifiers or stimulants that increase the activity level in the rumen and appears to have a nonbloat effect. Trials have proven ionophores are an economic essential to the nutrition. They are introduced to the ration through the above mentioned protein pellet; and (g) antibiotics at a rate of 70mg low level of antibiotic per head per day for the control of liver abscesses. It is also introduced through the protein pellet. That level also contributes some positive rumen stimulant effects that enhance its use. I, however, at this point in time do not recommend the continuous low level use because of the political public view of antibiotic use in foodstuffs. If a problem arises I would recommend a three day level of top dressing the feed with a 5 gram per head basis and only at the proven withdrawal timing before slaughter. Liver abscesses, footrot, and pneumonia complexes may arise to treat the pen rather than individually, thus the inclusion into the individual pens via top dressing on the feed.

Grain

Corn or maize is pretty well accepted as the best grain because of their qualities of being easily processed dry or wet rolled. It turns fat white in 60 days and has exceptional starch utilization and fat marbling characteristics.

Barley is probably the second preference, preferably two-row or malting brands. While lower in protein than six-row, two-row has a better starch availability for fattening. It can be effectively dry or wet rolled. It turns fat white in 60 days and has good marbling characteristics.

Wheat has high energy and exceptional starch release ability above barley, but loses its preference to barley and corn because of inducing "over eating" enterotoxemia or clostridium deaths. Wheat should be blended at no more than 30% of the grain in the ration to reduce the incidence of enterotoxemia. Feeding wheat demands top feeding management.

Sorghum (milo) is the least desirable as it must be either steam flaked or moisture induced to be effective in conversion of ration to live weight. It varies greatly in accepting moisture and while most charts rate it with barley in energy it has consistently been negative in starch utilization to barley, takes longer on feed to whiten the fat and finish the animal to maximum yield. This lack of utilized starch release increases the feeding time to marble the prime cuts. Sorghum can be complemented well by blending it with wheat.

Oats is not a good grain source as it lacks the starch energy and is too high in fibre to be economical to feed in a feedlot. Grain as a protein source contributes as follows: corn 55% to 65% of protein requirements, barley 70% to 80%, sorghum 58% to 68%, and wheat 75% to 85%.

Grain Processing

The maximum efficiency results from processing grains have been steam flaking in the past. This entails cooking the grain in a steam chest or bin at 210° F. for at least 30 to 40 minutes before pressure rolling. The pressure rolling gelatinizes the starch as the moisture in the grain is approximately 20–22%. The percentage of gelatinization can be laboratory tested and with the known bushel weight of the rolled berry, a correlation can be made on site as to the probable degree of gelatinization. If the gelatinization gets above 50%, death loss can intermittently occur from the clostridium type C and D bacteria that thrive in a high starch base. The bacteria metabolize the dietary starch and synthesize α and β toxins. Death usually results from respiratory or circulation complications. By applying less roll pressure in rolling the grain, this reduces the degree of gelatinization in the grain and the death loss is reduced or stopped. The economical issue rides a fine line of the best efficiency, yet reduced death loss at 40–45% gelatinization is a safe area with good management. There are two forms of overeating; the above and the second is lactic acidosis. This normally occurs when cattle are first put on feed or during a change of ration to a higher grain ration. They simply overeat and a greater per cent of digestion takes place in the rumen. The wall of the rumen becomes thickened due to the low pH in the rumen due to a build up of lactic acid. This occurs as the production of lactic acid is greater than it can be converted to propionic acid. Rumen motility occurs and rumen abscesses develop along with liver abscesses.

Poor doers develop if they live. Death can result from shock and respiratory failure. This can be the result of a good ration gone bad through poor management. New cattle coming into the feedlot must be introduced to the starting ration slowly to allow for rumen adjustment in fermentation of the ration. Cattle must not get hungry before feeding as the same thing will happen especially when changing energy levels in moving up on a higher grain ration. The draw back to steam flaking is the cost of firing a boiler and energy costs have forced operators in the U.S. to look to other processes.

Mel Karr, a nutritionist from Texas, has developed an alternative system. A mechanical grain moisturizer scratches the surface of the grain allowing a conditioner additive to be mixed with cold water which is then induced and mixed with the grain. In 30 minutes the moisture can be achieved at 22% in the middle of the berry. The grain is then rolled with the desired effects achieved. The cost is approximately 20% of steam flaking. Corn and barley can be effectively dry rolled and fed but would lose at least 10% efficiency from the wet grain systems. Sorghum would lose 15% or more efficiency.

Hammermilling is ill advisable as cattle do not like fines in the ration.

Roughage

The physical scratching in the rumen is the first priority of good roughage for the micro flora "bug" development. Palatability is another important factor that a roughage may add or detract from a ration and should be judged on that merit in degree of acceptance. Its other factor is its contribution to the protein in the ration. Lucerne (alfalfa) can contribute 10% to 18% of its inclusion in the ration whereas cotton hulls, grain straws and stubbles would contribute nothing to the protein requirements.

Non-Protein Nitrogen

Urea should not be used in excess of 25% of the added protein to balance the ration. The other 75% of supplemental protein should come from vegetable protein sources. High humidity can reduce the protein available in urea through being highly soluble. It should be added to the ration in the form of a pellet with the micro ingredients, salt, vitamins, minerals and the supplementary protein to ensure a thorough mix in the ration and lessen the chance of urea poisoning, through poor feed mixing. Urea restricts somewhat the physical production of a pellet so it is not as desirable if vegetable protein alternatives are available at a least cost basis.

Vegetable Protein

Most of the supplemental protein for cattle comes from the remains of seeds that have been processed for their oil content. We evaluate their merit in a ration based on availability, digestibility, cost per unit of protein, energy contribution, palatability, handling storage qualities and the ability to form in a pellet with the other ingredients mentioned before.

Soybean meal and cottonseed meal are the two most popular sources of protein for feedlot cattle. They are available the year round and meet the foregoing requirements except at times they may not be the lowest cost per unit of protein. Other sources of protein available are peanut meal which can be substituted for 100% of the supplemental protein mix, brewers grain which can be used at 100%, sunflower meal, 25% to 75%, and guar meal, 20% to 50%.

The protein requirements of a ration ideally should be from 12% to 14% on a dry matter basis. Younger cattle need the upper 14% level which at an AS FED basis, not dry matter, equals about 11.25% in the ration. This will vary some as all grains vary measurably from field to field and season to season. A practical monitoring system is advisable. More protein than the 11.25% is of no advantage and in fact if much higher is detrimental to the animal as it must be sluffed off as a byproduct from the system. So cattle do have a maximum effective protein utilization level.

A constant energy ration is essential to keep the consumption on a slow continual rise. A rule of thumb is cattle will eat from 2.5% to 3% of their body weight depending on age and the condition of fat content on them at that time. If an animal is eating 9 kilos of ration per day and he then eats 11 kilos he may the next day reduce to 6 kilos and take several days to be back up on the consumption of 9 kilos. Over flaking the grain more then normal will produce more gelatinization of starch. That abrupt increase will in most instances reduce dramatically the consumption of feed the next day.

The dramatic drop in the barometer is a trigger of nature to induce over eating. If allowed this will back cattle off feed. Good bunk management controls these conditions. If feed is days old and stale, cattle will leave it even though it looks all right to us. You assume they are satisfied but in reality if it were fresh they would eat it. I have a simple rule on that area. If a feeder feeds over yesterday's feed he is fired. It's that important. High roughage rations will not be as palatable as the finishing ration if rained on. Cattle will eat a high grain ration when rained on up to 24 hours then it goes off and must be cleaned out.

It is essential that the ration is mixed thoroughly to ensure every animal gets the total requirements. That is why a pellet is produced to ensure that the animal gets the daily requirements. The pellet mixes well and because it is of a large size it is totally consumed. Cattle do not like fines and if all the micro ingredients etc. listed above were mixed separately, most would end up as fines in the bottom of the bunk left to be thrown out. This is the most important item that needs the daily intake. Protein can vary a little as an optium is built up in the animal. The pellet ingredients do not.

When feeding in a colder season as in the upper U.S. a percent or two of roughage is added to the ration and a like amount of grain reduced. Roughage creates heat in the body thus helping nature to accommodate the animal in the colder weather.

Nutrition many times gets the blame for other adverse effects on performance. High humidity, proper quality and quantity of water, wind and wild cowboys all can effect the consumption. I make a thorough search before I make a ration change and I only make a change when I have a problem. Cattle are creatures of habit and do not respond to starting and stopping in their ration or eating habits.

The expansion of the feedlot business in Australia is surging. The interest in top quality meat to serve the U.S. and Japanese tourists in Asia is immense. This will be a very interesting and viable business throughout Asia. Vegetable protein is a key element in that development and I am sure that out of these proceedings there will emerge some bright and innovative protein sources not yet tapped to complement nutrition in the cattle feedlot business.

In Vitro Method for Estimating Digestibility of Swine Feeds Using Various Protein Sources

Shi Xue-Shi and Nie Guang-Da

Institute of Animal and Veterinary Science, Shanghai Academy of Agricultural Sciences, Shanghai, PRC

Abstract

The modification of the *in vitro* rapid method using intestinal fluid for predicting digestibility of diets and single feedstuffs for swine is discussed. Mixed intestinal fluid of several pigs fitted with cannulae in the jejunum about 150–200 cm from the pylorus was used in the *in vitro* experiments. Depending on DM^1 weight, feeds in question should be added to digestion fluid in the proportion of 0.1 g:2 ml. As each batch of unknown sample was estimated, two standard samples known precisely in *in vivo* digestibilities were estimated simultaneously so that the digestibility of each batch of feed in question could be corrected. Results of experiments on the digestibilities of nine mixed diets showed that a high and significant correlation existed between the *in vitro* method and *in vivo* procedure (r = 0.97 for dry matter; r = 0.96 for digestible energy). The *in vitro* values by the correction of standard feeds are more correct, reasonable and repeatable in comparison with the *in vivo* digestibilities of 16 single feeds which were fed directly or indirectly. The coincidence of the *in vitro* additive calculation values of 6 ingredients according to their percentage in rations with the direct *in vitro* digestibilities of four diets composed of these ingredients further indicated that the *in vitro* procedure is an important and rapid approach to the solution of assessment of single feed for pigs.

The conventional methods of evaluating pig feeds is well known to have a lot of disadvantages; especially that the determination of the digestibility of feeds that can not be fed alone is more complicated, and the different ways of calculating digestibility by difference will obtain different and unreasonable results for the same feed. It is impossible to determine the digestibility of the majority of feeds by the direct feeding procedure. The establishment of the *in vitro* precise method for estimating metabolic energy of ruminant feeds (1) provides an enlightenment for us on similar studies with single stomach animals. With the rapid development of the formula feed industry and the set up of pig feeding standards in China, it becomes an urgent task to find new methods to evaluate the digestible values of pig feeds, especially single feedstuffs.

In 1979, Japanese scientists (2–5) reported for the first time an *in vitro* digestion method in pigs. Using the intestinal fluid obtained from the upper jejunum about 50 cm from the pylorus, they employed a two-stage digestion procedure and obtained the same results as the *in vivo* values. Although, to date, no Chinese papers have been published, studies on the digestion of pigs (6) as well as the duodenal digestive enzymes of the goat pointed out that the *in vitro* method reported in literature may not be exactly correct. The procedure must be proved, modified and perfected.

In this paper, we have compared the effect of intestinal fluid from our cannula site and the site described in the literature on *in vitro* digestibility, and designed some experiments to find an available way to correctly estimate the digestibility coefficients of single feeds which could be fed both directly and indirectly to swine. By the modified approach, digestible energy (DE) of over one hundred kinds of soybean, soybean cake and meal from throughout China were for the first time correctly assessed in a short period in order to set up national standards for plant feedstuffs.

Materials and Methods

Fitting of the Intestinal Cannula

Eight castrated Russian White/Mei-Shan male swine maintained in metabolism cages and of about 36 kg average weight were used in this experiment. After anesthetization with barbital by intravenous injection, they were fitted on the right lateral abdomen wall with two new intestinal cannulae designed by our group or with T-shaped stainless steel cannula. Cannula I was fitted in the upper jejunum about 70 cm from the pylorus according to Furuya's method (2); and cannula II was about 150–200 cm beyond the pylorus. The surgical technique was similar to those of Komarek (7), Han Zheng-Kang (8) and Sauer (9). The cannulated animals were put into the experiments after returning to the pre-surgery diets.

Preparation and Storage of Intestinal Fluid

Cannulated animals were fed twice daily at 8 a.m. and 4 p.m. The proportion of diets to water was 1:3, with relatively constant consumption of feed and water. Intestinal contents (300–400 ml) were collected through the cannula at 9:30 a.m. at two-day intervals and centrifuged immediately after collection for 10 min at 4,000 r.p.m. The midlayer fluid was used for experiment immediately or was stored at −30°C for future use.

The Feeds in Question

The *in vivo* digestibilities of diets I–V were known except for V. Their ingredients are shown in Table 1.

Nine mixed diets and eight single feeds were used in *in vitro* and in *in vivo* comparative experiments. All samples used in *in vitro* assessment were ground to 100 Mesh and kept in bottles for use. Their ingredients and compositions are shown in Tables 2 and 3.

In Vivo Method

Nine Russian White, Mei-Shan pigs equally divided into three groups were used for measuring the *in vivo* digestibilities of diets A–I and ground barley, ground maize and three feed powders. Average animal weight was approximately 25 kg at the beginning of the experiment. Every trial consisted of a 5-day pretest, a 4-day fixed-quantity

TABLE 1
Ingredients of Diets I–V[a] (%)

Ingredients	Diets				
	I	II	III	IV	V
Ground maize	37.0	40.0	36.0	40.0	36.0
Ground barley	28.5	30.0	27.5	28.5	35.0
Wheat meal	7.0	5.5	5.0	8.0	17.0[b]
4 Feed powder	11.0	10.0	11.0	12.0	—
Fish meal	7.0	5.0	11.0	3.0	2.0
Cottonseed meal	6.0	6.0	4.0	5.0	6.5
Soybean meal	2.0	2.0	4.0	2.0	2.0
Mineral powder	1.0	1.0	1.0	1.0	1.0
Salt	0.5	0.5	0.5	0.5	0.5
Total	100	100	100	100	100

[a] Providing (100 kg diet): Cupric sulfate, 100 g; Zinc sulfate, 25 g; Ferrous sulfate, 100 g; Multivitamin for poultry, 5 g.
[b] Wheat bran.

TABLE 2
Ingredients and Composition of Nine Mixed Feeds (%)

Ingredients composition	Feeds								
	A	B	C	D	E	F	G	H	I
Barley meal	39.5	35.6	30.3	28.7	35.6	35.6	29.5	29.5	68.5
Ground maize	38.0	34.0	29.2	27.3	34.0	34.0	29.5	29.5	30.0
Fish meal	5.0	4.5	3.8	3.5	14.5	4.5	—	—	—
Soybean meal	5.0	4.5	3.8	3.5	4.5	14.5	—	—	—
Wheat bran	11.0	9.9	8.5	7.5	9.9	9.9	—	39.5	—
White rice bran	—	—	—	—	—	—	39.5	—	—
Sludge powder	—	10.0	23.0	28.0	—	—	—	—	—
Mineral powder	1.0	1.0	1.0	1.0	1.0	1.0	1.0	1.0	1.0
Salt	.5	.5	.5	.5	.5	.5	.5	.5	.5
Moisture	13.9	11.1	13.1	9.6	9.5	9.9	11.8	12.4	11.6
Crude protein	16.5	17.7	17.7	20.6	24.9	21.8	13.3	13.5	12.3
Energy (Cal/g)	4587	4456	4351	4247	4631	4637	4793	4581	4526
Ether extract	2.4	2.4	1.8	2.1	2.6	2.1	8.1	3.1	2.2
Crude fiber	3.8	3.4	3.0	3.0	3.6	4.6	6.3	6.2	3.9
Ash	4.6	8.5	11.5	14.7	6.5	4.9	6.9	4.9	4.1

TABLE 3
Compositions of Eight Single Feedstuffs (%)

Single feed	Composition					
	Moisture	Crude protein	Energy (Cal/gDM)	Ether extract	Crude fiber	Ash
Ground barley	13.9	12.0	4490	1.3	4.8	4.3
Ground maize	11.6	9.6	4597	3.7	1.6	2.5
3 Feed powder	13.7	13.9	4517	2.2	4.7	4.1
Fish meal	9.4	68.2	4836	8.4	1.1	21.3
Soybean meal	9.1	51.2	4827	4.4	5.2	6.5
Wheat bran	11.8	14.8	4622	4.9	10.6	5.4
White rice bran	11.3	14.6	5217	11.7	3.9	8.7
Sludge powder	8.4	24.3	3683	1.6	0.1	39.3

feeding period and a 6-day faeces collection period. All faeces were analyzed routinely.

The feeds which could not be fed alone were determined by using an indirect procedure of feeding in two trials. The digestibility of a basal ration in which 10–15% of the feed in question was added was determined first and then the proportion of feed under consideration was increased by 20–30% for a second trial. The digestibility was deduced indirectly by difference based on the data from two trials.

The feeding regime was based on the method introduced by Burch et al. (10), i.e. wet-feeding twice daily at 8 a.m. and 4 p.m.

Results

Observations on the Analog Effects of In Vitro Digestion

Based on Furuya's method, 0.2% pepsin chlorohydrate solution and intestinal fluid of cannula I were used to estimate the *in vitro* digestibility of dry matter (DM) and crude protein (CP) of diets I and II.

As shown in Fig. 1, when pepsin solution was used alone the digestibility of CP could reach about 77%; while that of DM was as low as 28%. On the contrary, when the small intestinal fluid was added alone, the DM value was increased to 70%. This shows that most of carbohydrate in diet was digested with the enzymes in the small intestinal fluid. Experiments indicated that the process of *in vitro* digestion was similar to the natural digestion procedure.

Differences in In Vitro Digestibility When the Intestinal Fluids from Different Regions of Small Intestine were Used

Studies on the physiology of pig digestion suggested that there may be great differences in the concentration and activity of enzymes in various regions of the small

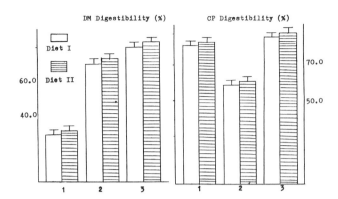

Fig. 1 Tendency of DM and CP digestibility of Diets I and II using the *in vitro* method. 1 = incubation for 8 hrs with 0.2% pepsin solution; 2 = incubation for 8 hrs using intestinal fluid of cannula I; 3 = result of two-stage method (cannula I).

intestine. Therefore, it was assumed that there also were differences in *in vitro* digestibility. The comparison of the activity of intestinal fluids from the regions between those reported in the literature (2) and those selected by our group has been made by diets I–IV with values known from *in vivo* digestibilities.

Table 4 shows that the *in vitro* values of DM and general energy (GE) of four diets estimated by using intestinal fluid (about 150–200 cm from the pylorus) were generally in accordance with the *in vivo* values. The means of *in vivo* vs. *in vitro* were 83.2 vs. 83.6% for DM, 83.7 vs. 82.4% for GE ($P > 0.2$). The results were similar to those in the literature. However, when cannula I was used, the *in vitro* values were much lower than those *in vivo*. The average values are 74.2 and 73.2% for DM and GE, and the relative errors are 108 and 124%, respectively. In view of these facts, it is advisable that the site of cannula for the *in vitro* trials should be about 200 cm beyond the pylorus.

TABLE 4

The Comparison of *in vitro* Values of Intestinal Fluid from Cannula I and II with *in vivo* Digestibilities

Diets	Cannu. no.	DM (%) in vitro	DM (%) in vivo	GE (%) in vitro	GE (%) in vivo	CP (%) in vitro	CP (%) in vivo
I	1.	74.0	82.5	73.4	84.8	83.8	82.5
	2.	83.3		81.8		89.0	
II	1.	74.7	82.3	72.3	82.6	86.8	81.3
	2.	83.6		82.9		89.6	
III	1.	72.5	83.7	74.0	83.4	83.5	83.9
	2.	83.8		82.0		90.8	
IV	1.	75.4	84.3	73.1	84.1	87.0	79.9
	2.	83.8		83.0		90.7	
Mean	1.	74.2	83.2	73.2	83.7	85.3	81.9
	2.	83.6		82.4		90.0	

[a] DM = dry matter; GE = general energy; CP = crude protein

Effects of Sample Size on In Vitro Digestibility

It was reported that different sizes of the same sample in *in vitro* trials may cause an error as high as 10% (2). In order to diminish the error, 5 levels of diet I and II of 0.2, 0.4, 0.6, 0.8, and 1.0 g were used. Every extra 0.1 g dry matter was added into 2 ml digestion fluid.

It can be seen in Table 5 that under the assay procedure, although there was a tendency for the value of digestibility to increase as the sample size decreased, the differences were not significant between samples of different sizes. The results for 0.4–0.8g were essentially the same. So, the 0.5 g ($\pm$ 0.0001 g) size suggested by Furuya was adopted in our experiments. However, we found that using the weight of DM to the constant proportion of digestion solution could avoid the error caused by different moistures in air-dried samples.

TABLE 5

Effect of Sample Size on *in vitro* Digestibility

Digestibility Sample size (g/DM)	Diet I DM(%)	Diet I CP(%)	Diet II DM(%)	Diet II CP(%)
0.20	84.95	94.96	85.15	94.55
0.40	83.65	92.37	84.89	93.68
0.60	83.44	91.55	83.87	92.53
0.80	83.81	91.60	84.16	93.30
1.00	82.40	91.12	83.47	91.88
Mean	83.65	92.32	84.31	93.19

The Differences of Digestibility Among the Individual Animals

The digestibility of diet V was predicted using the intestinal fluids of 6 slaughtered pigs. Table 6 shows that there were differences in the *in vitro* values of intestinal fluid from different individuals. The range of DM and CP was between 80.3–83.4% and 90.5–91.7%.

It is the routine of the *in vivo* experiments that the amount of pigs in each group should be at least three. *In vitro* results also indicated that the intestinal fluid for *in vitro* estimations should be mixed with more than three individuals so that representative results can be obtained.

TABLE 6

Effect of the Individual on *in vitro* Digestibility of the Same Diet (%)

Item	Pig No. 7	14	15	20	22	29	Mean
DM	83.4	81.5	82.4	82.0	80.3	82.9	82.1
CP	91.3	90.5	91.2	91.7	90.6	91.5	91.1

The Proposal of the Correction of Standard Diets

Based on the method of the *in vitro* digestion trials of ruminant animals, a correction of standard feeds was proposed for the *in vitro* procedure for swine. In this experiment, two standard diets with precisely known digestion coefficients were examined simultaneously when the feeds in question were estimated. It was found that different values may be obtained from standard diets in each batch because of some man-made and experimental factors. If the measured values of feeds in question are multiplied by the *in vitro* variation coefficient of standard diets, the error in each batch may be greatly diminished and a relatively correct value can be obtained.

The results in Table 7 showed that the CP *in vitro* corrected digestibilities of nine mixed diets were obviously more similar to those *in vivo* than those *in vitro* values estimated directly. Their mean is 79.0, 78.1 and 86.3% for *in vivo*, *in vitro* (corrected) and *in vitro* (direct), respectively.

The Modified Standardized Approach for Determination of In Vitro Digestibility

A procedure called the "two-stage method" is as follows: First stage: Feeds in question (prepared as four parallel samples) and two standard diets of precisely known *in vivo* values with each sample being 0.5 g ($\pm$ 0.0001g) were put into 100 ml Ehrlenmeyer flasks, then 0.1% pepsin HCl solution was added in the proportion of 0.1g DM to 2 ml fluid. The flask was incubated in a 37°C water bath for 4 hours, shaken at 80 r.p.m.

Second stage: The acid content of the flask was neutralized with 0.2 N NaOH solution, and then mixed intestinal fluid of at least three pigs was added in the same volume as the pepsin solution. The conditions were the same as for the first stage. After 4 hours, the contents were placed in plastic centrifuge tubes and diluted with distilled water, and the tubes were refrigerated at 4°C overnight. The next day, they were centrifuged for 10 min. at 4,000 r.p.m. The supernatant was discarded and this procedure was repeated for several times so that digestion fluid could be removed completely. Then, the residue was rinsed with distilled water onto a piece of quantitative filter paper of known weight and energy value per gram. Two pieces of filter paper with residues were dried in an oven for 2 hours at 65°C and the nitrogen content was determined by the Kjeldahl procedure. The others were dried for 2 hours at 105°C. DM, DE and DCP digestibilities were determined and calculated based on the formula below:

$$\text{DM, DE and DCP digestibility (\%)} = \left(\frac{A_{1,2,3} - B_{1,2,3}}{A_{1,2,3}}\right)$$
$$+/- \left(\frac{A_{1,2,3} - B_{1,2,3}}{A_{1,2,3}}\right) \times C_{1,2,3} \times 100$$

$A_{1,2,3}$ = Sample weight, Energy value of sample and Sample $N \times 6.25$

$B_{1,2,3}$ = Residue weight, Energy of residue and residue $N \times 6.25$

$C_{1,2,3}$ = Standard feed coefficients for DM, DE and DCP

TABLE 7

Development of a Correction Process for CP Digestibility by Two Standard Diets Known Precisely by *in vivo* Digestibilities (%)

Feeds	Items				
	in vivo	*in vitro* (correct)	error	*in vitro* (direct)	error
A	82.5	81.0	−1.5	88.0	5.5
B	75.1	76.4	1.3	83.0	7.9
C	62.2	66.4	4.2	72.2	10.0
D	75.4	73.0	−2.4	76.1	0.7
E	84.6	82.2	−2.4	85.5	0.9
F	85.5	83.5	−2.0	89.7	4.2
G	77.3	76.0	−1.3	82.6	5.3
H	76.2	82.9	6.7	90.1	13.9
I	80.6	81.5	0.9	88.6	8.0
Mean	79.0 ± 6.1	78.1 ± 5.7	0.4	86.3 ± 4.9	6.8
Correlation coefficient with *in vivo* values		r = 0.89		r = 0.80	

Reliability Testing of the Modified Procedure

Data in Table 8 show no significant differences ($P > 0.5$) between the two approaches, although the ingredients and percentages in every diet were different to some extent. The mean is 76.3 vs. 76.3 for DM and 77.9 vs. 77.6 for DE, respectively. The maximum absolute errors are about 4.5%. The relationships between the *in vivo* and *in vitro* methods are $r = 0.97$ and $r = 0.96$ for DM and DE.

The Estimation of 16 Single Feedstuffs In Vivo and In Vitro

Usually, single feeds which cannot be evaluated directly are fed with a base feed and their digestibilities indirectly determined by difference (11). In our trials, the *in vivo* values of single feeds were also deduced by difference for DM, DE and CP, but barley, corn and three feed powders were measured by direct feeding. The *in vitro* digesti-

TABLE 8

DM and DE Digestibility Comparison Between the *in vivo* and *in vitro* (%)

Feeds	Item				
	DM			DE	
	in vivo–in vitro	error		*in vivo–in vitro*	error
A	81.7–82.6	0.9		81.9–83.3	1.4
B	74.8–75.1	0.3		76.5–76.2	−0.3
C	65.4–65.7	0.3		68.4–68.0	−0.4
D	64.8–65.4	0.6		70.5–69.8	−0.7
E	82.1–82.5	0.4		83.0–82.9	−0.1
F	81.5–83.5	2.0		81.8–84.1	2.3
G	77.0–72.6	−4.4		79.7–75.5	−4.2
H	76.3–76.3	0		75.9–77.5	1.6
I	83.3–82.7	−0.6		83.2–80.7	−2.5
Mean	76.3±7.0, 76.3±7.2, −0.1			77.9±5.5, 77.6±5.8, −0.3	
t Test	$P > 0.5$			$P > 0.5$	
r value	r = 0.97			r = 0.96	

bilities of all feeds were estimated and corrected by standard diets. The results are shown in Table 9.

It can be seen that the DM and DE *in vitro* digestibilities of three high energy feeds were close to the *in vivo* values obtained from direct feeding. It seems that *in vitro* values were more reasonable, reliable and accurate as compared with the *in vivo* values deduced by difference for seven feeds. So far as, for example, the comparisons of compositions and theoretical digestible coefficients of corn and white corn are concerned, there are no great differences between them. The *in vivo* values from direct feeding as well as *in vitro* values of two kinds of corn were 90.3, 90.2% for DM and DE; 90.5, 91.7% and 87.5, 87.3% of DM and DE of corns *in vitro*. Obviously, the 77.4 and 78.1% for DM and DE of white corn, deduced by difference, may be too low.

In our work, we found that digestibility of the by-products of oilseeds tends to decrease in the order below: soybean meal, soybean cake, rapeseed meal, rapeseed cake, cotton meal, and cotton cake. The digestibility of same plant seed meals was always consistently higher than that of the cake. Fish meal falls between soy cake and rapeseed meal.

The Additivity Testing of Single Feeds in Mixed Diets (12)

This approach also assumes additivity. Four mixed diets composed of corn, barley, wheat bran, fish meal, soybean meal and hay powder which were of known *in vitro* digestible energy were designed to examine the additivity of DE for each ingredient in compound diets of which DE (Mcal/kg) were exactly predicted *in vitro*. According to the *in vitro* DE of individual ingredients and their percentage in rations, the accumulative values of every diet were calculated.

The DE (Mcal/kg) obtained with the *in vitro* assessment compared favorably with the values obtained in the additive study, namely 3.79 vs. 3.88, 3.76 vs. 3.85, 3.79 vs. 3.86 and 3.86 vs. 3.91 for samples 1–4, respectively (Table 10). From these results, the DE of a mixed diet may be estimated from a knowledge of the *in vitro* digestibility of an ingredient and its exact percentage.

The Application of the in vitro Method in the Set up of Swine Feedstuffs Standards

Soybean and its cake as well as meal are important plant protein sources not only in human food, but also in animal feeds in the world. In China, there were no complete data about the digestible coefficient of nutrients until last year when the *in vitro* method was proved to be highly effective and was carried into practice.

The chemical composition analysis for over a hundred samples from the main districts throughout the country showed that the contents of CP, EE and GE on the average were 40, 42 and 43.5%; 20, 6.5 and 2.0%; and 5.7, 5.0 and 4.5 (Mcal/kg DM) in SB, SBC and SBM. The average digestible coefficient of general energy was gradually increased from 75.0, 77.0 to 78.3% with the decline of EE and GE content, but with the rise of protein (Tables 11 and 12). The correlation coefficients were r = 0.999 between CP and DE means, r = −0.991 and r = −0.987 between EE, GE and DE means. The results indicated that nutrient value of SBM is always better than SBC and SB in a sense. It is a pity that so far, DCP of

TABLE 9

Results of the *in vivo* and *in vitro* Evaluation of 16 Kinds of Single Feedstuffs

	Item					
	DM (%)		DE (%)		CP (%)	
Feeds	*in vivo-vitro*	error	*in vivo-vitro*	error	*in vivo-vitro*	error
Barley	80.5–82.1	1.6	79.5–83.6	4.1	76.2–78.2	2.0
Corn	90.3–90.5	0.2	90.2–91.7	1.5	84.5–80.9	−3.6
3 feed powders	80.9–80.7	−0.2	80.3–80.7	0.4	81.5–85.5	4.0
White corn*	77.4–87.5	10.1	78.1–87.3	9.2	—	—
Wheat*	81.7–91.0	9.3	82.8–92.5	9.7	—	—
White rice bran*	78.7–68.7	−10.0	78.0–68.3	−9.7	76.6–70.2	−6.4
RBGR*	77.6–78.7	1.1	79.4–76.5	−2.9	—	—
Ground rice*	82.1–97.8	15.7	82.6–97.4	14.8	—	—
Sludge powder*	24.0–22.0	−2.0	38.1–29.0	−9.1	79.0–48.0	−31.0
Rapeseed meal*	71.1–63.5	−7.6	72.7–67.7	−5.0	—	—
Wheat bran	63.0		63.7		78.0	
Fish meal	81.3		79.6		82.8	
Soybean meal	80.4		82.9		87.6	
Rapeseed cake			63.9			
Cottonseed cake			59.0			
Cottonseed meal			61.0			

* The *in vivo* digestibilities were indirectly deduced by difference with a base feeding in Jiangsu Academy of Agricultural Sciences except the sludge powder.

TABLE 10

Comparison Between the *in vitro* DE (%) (Mcal/kg) and *in vitro* Additive DE for 4 Diets[a]

Diet No.	Corn	Barley	Wheat bran	Fish meal	Soybean meal	Hay powder	*in vitro* value	Addi. value	Error
1	60.9	25.6	3.1	5.3	2.0	3.1	3.79	3.88	2.4
2	50.4	30.7	5.1	5.6	5.1	3.1	3.76	3.85	2.4
3	48.2	30.7	5.1	6.8	6.1	3.1	3.79	3.86	1.8
4	60.2	27.5	3.1	—	6.1	3.1	3.86	3.91	1.3
GE	4.56	4.64	4.58	5.13	4.96	4.61	r = 0.99		
DE	3.99	4.09	2.32	3.86	4.11	1.58	P > 0.5		
DE (%)	87.5	88.1	50.7	75.2	82.9	35.0			

[a] Unpublished data from X.T. Li. et al., 1986.

TABLE 11

Means and Ranges of Composition and DE for Swine of 30 Soybean Samples from 13 Provinces and Cities[a] (14)

	Soybean (%, Mcal/kg DM)						
Item	DM	GE	CP	EE	CF	DE (%)	DM (%)
$\overline{X}$	91.7	5.7	40.3	19.8	5.0	75.0	73.8
SX	1.2	0.1	2.2	1.7	0.7	3.8	3.7
n	30	30	30	30	30	29	29
C.V%	1.4	1.6	5.4	8.4	14.1	1.3	5.1
Range	88–89	5.5–5.8	34–43	16–23	3.5–6.6	68–81	64–79

[a] Data from S. N. Wang, Thesis for M.S. in Kirin (1986).

TABLE 12

Means and Ranges of Composition and DE for Swine of 39 Soybean Cake and 35 Meal Samples from Throughout China[a]

	Soybean cake (%, Mj/kg)					
Item	DM	GE	CP	EE	CF	DE (%)
$\overline{X}$	89.4	18.4	42.1	5.9	4.9	77.0
SX	2.6	0.6	2.1	1.5	0.9	3.4
n	39	39	39	38	39	37
C.V%	2.9	3.0	4.9	25.6	18.2	4.4
Range	83–96	17.5–19.5	38–45	4.2–11.8	3.5–7.0	67–83
	Soybean meal (%, Mj/kg)					
$\overline{X}$	87.5	17.0	43.3	1.9	4.8	78.3
SX	2.2	0.6	1.8	1.4	0.7	2.7
n	35	35	35	35	35	35
C.V %	2.3	3.4	4.1	70.4	13.4	3.4
Range	82–93	15.4–18.4	41–48	0.1–5.8	3.5–6.0	71–83

[a] Data from S. Q. Zhu, et al., 1987, in Hei Longjiang.

oilseeds and their by-products for swine have not been estimated by the *in vitro* method.

Discussion

Most of the nutrients ingested are digested and absorbed mainly in the small intestine with digestion enzymes from the secretion of intestinal glands and the pancreas, from broken intestine cells which are sloughed off and from the brush border of intestinal mucosa. There are mainly pancreatic enzymes near the duodenal lumen where the biological environment is greatly changeable. This is the reason for the low and poorly repeatable *in vitro* trials when intestinal fluid from near the duodenum was used. Kidder (6) and Z.K. Han (13) reported that nearer the middle of the small intestine, the concentration of digestion enzymes was higher. Our experiments firmly indicated that the key to success in *in vitro* estimation of feeds for swine depends on the exact selection of the cannula site in the small intestine.

The similarity of *in vitro* values to *in vivo* values for compound diets and single feeds indicated the representation of *in vitro* values. We believe that *in vitro* estimation for DM and DE could be applied directly to practical production with standard feed corrections instead of regression equations.

The estimation methods for single feedstuffs which have been reported include determinations by difference, simultaneous regression equations and extrapolation (10). However, each approach has its deficiencies because results obtained are quite inconsistent. A series of experiments in this paper have shown that the *in vitro* assessment is rapid, correct, reliable and highly repeatable as well as independent of various interferences encountered in *in vivo* experiments (15). Thus, it is appropriate for the routine evaluation of a large number of samples, especially single feedstuffs.

In some literature, the *in vitro* digestibility of crude protein was much higher than that of *in vivo*. We obtained similar results during preliminary trials. However, after using the corrections from standard feeds, the *in vitro* digestibility of crude protein was obviously nearer the *in vivo* value. This technique warrants further investigation.

It is definitely impractical to employ complex biochemical methods to determine the activity of various enzymes in intestinal fluids. The simpler method, namely "correction of standard feeds" proved to be easier than other correction methods. However, the representation and precision of the *in vivo* values of standard feeds are fairly important as it would affect the value of feeds in question.

Unlike the experiment for ruminants where the fresh rumen fluid must be used immediately after collection (16), the unique advantage of the *in vitro* method with swine is that the intestinal fluid could be stored at low temperatures for a long time; especially, it could be freeze-dried and used commercially.

It seems that the utilizable value of plant oilseeds and their by-products for swine may be arranged in the following order according to our experiment results: soybean meal > soybean cake > soybean > rapeseed meal > rapeseed cake > cottonseed meal > cottonseed cake.

It follows that nutrient value of meals is always better than that of cakes. DE and DCP coefficients of soybean meal are higher than those of fish meal.

A major single chemical fraction affecting the digestibility of a feed for swine is its fiber content (12). The swine does not possess the enzymes required for the digestion of cellulose, but it may have an active bacterial flora in the large intestine which will do this. The application of this approach may be affected by the high content of crude fiber in feeds. A further modification might be required for fibrous feeds, as the method only assesses stomach and small intestine digestion.

Acknowledgments

In this paper, the *in vivo* digestibility values of six single feeds as well as the data about additivity tests were offered by collaborators: W.J. Chau working in the Jiangsu Academy of Agricultural Sciences, and X.T. Li in the Agricultural College of Hu Nan. The data about soybean cake and meal were generously provided by S.Q. Zhu, et al., working in the Farming Institute of Heilong Jiang. We would like to thank J.S. Li and J.Y. Sen for their correct and skilled analysis of energy and protein.

References

1. Menke, K.H., Animal nutrition teaching material at Peking Agri. University (1982).
2. Furuya, S., and S. Takahashi, *Br. J. Nutr. 41*:511 (1979).
3. Furuya, S., and S. Takahashi, *Jap. Res. Soc. Swine Sci* (3):187 (1980).
4. Furuya, S., and S. Takahashi, *Jap. Res. Soc. Swine Sci.* (2):114 (1980).
5. Furuya, S., and S. Takahashi, *Jap. J. Zootech. Sci.* 52(3):198 (1981).
6. Kidder, D.E., *Digestion in The Pig*, (1980).
7. Komarek, R.J., *J. Anim. Sci.* 53(1):796 (1981).
8. Han, Z.K., *The Experiment Guidance of Farm Animal Physiology*, Peking (1962).
9. Sauer, W.C., H. Jorgensen and R. Berzins, *Can. J. Anim. Sci.* 63(1):233 (1983).
10. Burch, H., Schneider, and W.P. Flatt, *The Evaluation of Feeds Through Digestibility Experiments*, The University of Georgia Press (1975).
11. Yang, S.X., *The Evaluation Methods of Feed Nutrients*, Guansu (1982).
12. Morgam, C.A., and C.T. Whittemore, *Animal Feed Science and Technology*, No. 7, pp. 387–400 (1982).
13. Han, Z.K., and X.Z. Mao, *The Digestive Physiology of Pigs*, Peking (1977).
14. Wang, S.N., *In vitro Method for Estimating Soybean Energy Digestibility for Swine and Poultry*, M.S. Thesis, Kirin (1986).
15. Nordfeldt, S., *Digestibility Experiments with Pigs*, Kungl, Lanttruksshogsk, Ann. 21:1–29 (1954).
16. Osbourn, A., *Proc. Nutr. Soc. 36*:219 (1977).

Determining and Modifying Protein Functionality

Khee Choon Rhee

Food Protein Research and Development Center, Texas A & M University, College Station, Texas 77843-2476, USA

Abstract

Vegetable proteins often are used in foods for functional rather than nutritional attributes. This review will primarily deal with selected chemical modifications of cottonseed proteins that alter physical and biochemical properties of importance in various food and industrial applications. Cottonseed protein products, usually heterogenous protein mixtures lacking in certain desired functional properties, have been modified by various chemicals to alter their structure and physiochemical characteristics. Results indicate that protein functionalities can be altered significantly to make them more desirable for various food and industrial uses. However, all chemically modified proteins should be subjected to rigorous safety tests before they can be considered for food uses.

Proteins are modified intentionally for structure-function relationship studies or for development of new and improved products from underutilized or useless resources. Among the numerous techniques that have been developed for these purposes, acylation has been of interest because of its potential for nutritional and functional improvement of food proteins (1,2). Acylation is the chemical reaction which attaches acyl groups to unprotonated amino groups of protein to form amide derivatives, depending upon the reagents that are used (3,4).

Acylation of proteins can be used to control deteriorative reactions during processing or transportation, to improve physical or functional properties, to eliminate undesirable substances such as pigments, to improve nutritional properties and to increase the acceptibility of unconventional protein resources (2). Also, it may be helpful in the physical separation of proteins from crude animal, plant or microbial materials, in the inactivation of certain antibiological substances such as allergens or enzyme inhibitors, and in the development of new raw materials from natural resources for the chemical and pharmaceutical industries, as well as for human foods and animal feeds. Because of such advantages, acylation of proteins has been investigated to understand the mechanisms, but attempts to apply it to food protein systems have been hampered by refusal of Food and Drug Administration to approve the use of chemically modified proteins in foods.

For successful use, proteins should possess certain critical functional properties required (5). These properties govern the suitability of the proteins as food supplements and as ingredients for new food products. Cottonseed has been considered as a potential source of protein for human consumption because of the increasing cost and limited supply of animal products (6). However, it has been poorly utilized, due to the lack of desirable functional properties. If the appropriate technology for improving functional properties could be developed, cottonseed protein could become a significant source of food protein in the near future. The objective of this research was to develop procedures that will facilitate the practical exploitation of cottonseed protein by the food and chemical industries.

Our attention has focused particularly on development of practical techniques for obtaining protein isolates possessing desirable properties from defatted cottonseed meal. Therefore, practical application of acylation has been studied with defatted cottonseed meal for improving yields and functional properties of protein isolates.

Application of Acylation Techniques to Cottonseed Protein

Cottonseed Protein and Problems

The most important factor limiting the use of glanded cottonseed meal or protein for foods or feeds is the presence of the pigment gossypol. This substance, a polyphenol, is toxic to monogastric animals, particularly swine (7). Glanded cottonseed protein isolate prepared by conventional procedures is deficient in the amino acid lysine because of bonding between lysine and gossypol. This makes the lysine nutritionally unavailable. In glandless cottonseed, yellow flavonoid pigments are also responsible for decreasing lysine availability.

The water-soluble proteins of cottonseed are the predominant functional proteins of the cytoplasm, and they represent 25–30% of the total nitrogen of solvent-defatted flour. The water-insoluble proteins are essentially the storage proteins located in the protein bodies, and these represent 60–65% of the total nitrogen (8). Therefore, unlike soy proteins, cottonseed proteins are not readily dispersible in water; alkali is needed to solubilize them. The nitrogen solubility index for cottonseed, proposed by Lyman (9), is relatively insensitive in the pH range of major interest for food use. Many of the important functional properties of food proteins, such as solubility, viscosity, and gelation, are related to water-protein interactions. Protein solubility in water is an important factor in food application of proteins. Low solubility and emulsion capacity are some of the limiting factors in the application of cottonseed proteins as food ingredients. Choi et al. (10) determined various functional properties of cottonseed protein isolates prepared under pilot scale conditions by various preparation procedures and reported that structural modification is essential to improve functional properties for wider utilization as food ingredients.

Extraction and Recovery of Protein from Acylated Cottonseed Flour

Generally, acylation of proteins is done at pH 7-9 and near 0°C to prevent side reactions of groups other than the amino groups, and the solution is dialyzed against distilled water or buffer solution after the pH has been stabilized (11–13). Some investigators acylated at room temperature and at slightly alkaline pH for mass treatment of protein (14). The excess of acylating agent is then eliminated by dialyzing the solution.

Modified acylation procedures were developed by omitting the dialysis step to mass produce protein isolate (15,16). In these methods, acid anhydrides were added continuously to the cottonseed flour suspension at room temperature, while the pH was maintained at 8.5 with 3N NaOH. Upon completion of acylation, the protein extract was separated from the residue by centrifugation. The pH of the supernatant was then decreased to 4.5 with 3N HCl, in order to precipitate the protein. The protein modifications thus obtained vary depending upon the acylating agent used (Fig. 1).

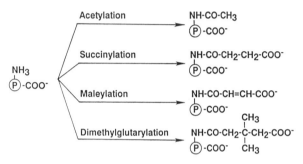

Fig. 1. Major changes on protein molecules by various chemical modifications.

When cottonseed flour was modified with succinic anhydride, protein extractability from the flour was increased remarkably at concentrations greater than 5% succinic anhydride (% of total flour protein) (Fig. 2). At very low anhydride concentrations, the extractability was similar to that of water extraction alone. Maximum extraction occurred at a concentration of 20% succinic anhydride. At this concentration, the extracted protein was approximately 64% succinylated. However, the succinylation ratio of the protein increased further at higher concentrations of succinic anhydride, to a maximum of 85% when using 50% succinic anhydride.

Such increases in extractability are considered to be due to solubilization of storage proteins, and to rupture of the membranes of the protein bodies by the high ionic strength and osmotic pressure that occur during succinylation. Martinez et al. (17) reported that high ionic strengths were necessary to rupture the membranes of protein bodies and to solubilize enclosed proteins. Shetty and Kinsella (18) also reported that the protein extractability from disrupted yeast cells increased with increasing amount of succinic anhydride, up to a ratio of 0.4 (succinic anhydride to yeast, g/g). At this concentration, approximately 90% of the yeast cell protein was extracted.

The amounts of free amino groups in the resulting cottonseed isolates decreased logarithmically with increasing concentration of succinic anhydride during succinylation, showing a linearity with the amount of anhydride added on a semi-log scale (Fig. 3). This indicates that attachment of succinyl groups follows first order kinetics with respect to the amount of anhydride when the reaction is conducted in a constant period of time. A first order kinetic equation can be used to calculate the amount of succinic anhydride required for succinylating protein isolates to levels desired for specific purposes. The reaction constant, as calculated from the slope in Figure 3, was K = 0.03258 per % succinic anhydride added.

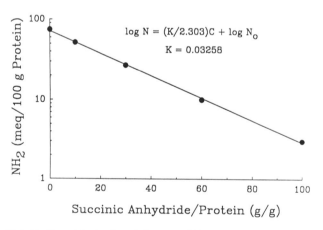

Fig. 3. Logarithmic plot of changes of free amino group concentration of protein isolates with concentration of succinic anhydride in conttonseed flour suspension. Mean of 3 experiments.

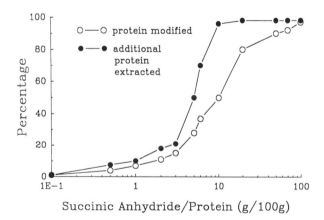

Fig. 2. Effect of succinic anhydride concentration during succinylation of cottonseed flour on degree of succinylation and extractability of resulting protein. Average of 3 experiments with 2 assays each; standard deviations less than 2%.

Succinylation also decreased free sulfhydryl groups of the proteins, thus indicating their modification. However, the concentration of disulfide groups was not significantly affected by succinylation (Table 1). This indicates that disulfide bonds between peptide chains of proteins, or between subunits in quaternary structure, might not be cleaved by succinylation. The reaction of sulfhydryl groups with succinic anhydride is known to be spontaneously reversible, depending upon reaction conditions,

TABLE 1
Changes in Sulfhydryl and Disulfide Groups of Protein Isolates by Succinylation of Cottonseed Flour

Succinic anhydride added as % of total protein	Resulting SH groups µg/g protein[a]	SH groups from free SH and reduction of S-S bonds µg/g protein[b]	SH groups from reduction of S-S bonds µg/g protein
0	12.4 +/− 0.8	106.3 +/− 4.5	93.9
10	11.9 +/− 0.7	116.9 +/− 3.0	105.1
60	8.4 +/− 0.5	117.9 +/− 1.5	109.5
100	6.7 +/− 0.6	108.3 +/− 2.1	101.4

Values are given as means +/− standard deviation
[a] Significantly different at p = 0.01.
[b] Significantly different at p = 0.05.

but disulfide groups show negative reactivity to the acidic anhydrides generally used for acylation (4). Both groups are closely associated with functional properties of proteins, especially gelation (19).

When defatted cottonseed flour was modified at 10% concentrations (% of flour protein) with various acidic anhydrides such as succinic anhydride, dimethylglutaric anhydride, maleic anhydride, acetic anhydride or sodium sulfite, acylation (with the exception of acetylation), and sodium sulfite treatment, caused the extraction of approximately 20% more protein from the modified flour than from untreated flour (Fig. 4).

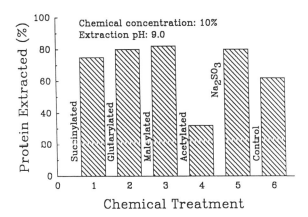

Fig. 4. Extractability of proteins from chemically modified cottonseed flours.

The extractability from the untreated flour suspension was about 60% of the total flour protein at pH 9.0, but acetylation of the flour decreased the extractability to 32%. Acetylation adds neutral hydrophobic groups to ε-amino groups and thereby decreases the water affinity of the proteins. However, succinylation, maleylation and dimethylglutarylation increase the negatively charged, hydrophilic acyl groups (Fig. 1). Sodium sulfite is known to cause cleavage of disulfide bonds (20).

Acylation also influenced protein recovery from the extract by classical isoelectric precipitation. Approximately 30% of the extracted protein was precipitated at pH 4.5 when cottonseed flour was extracted with water alone (Fig. 5). However, more protein was precipitated at the same pH from the succinylated extract. The precipitability of the extracted protein was increased to about 80% by increasing the concentration of succinic anhydride to 100% of the flour protein during succinylation. However, the precipitability remained the same at pH's lower than 4.5. Similar results were obtained with succinylated peanut flour (14). On the acid side of the isoelectric point, nitrogen solubility decreased progressively in peanut flours treated with increasing levels of succinic anhydride. Thus, extensive succinylation would in effect shift the apparent isoelectric point to a more acidic pH.

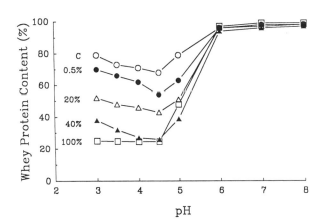

Fig. 5. Recoveries of protein at various pH levels from extracts at various concentrations of succinic anhydride. Numbers represent percentage succinic anhydride concentration; C = extract from untreated flour. Average of 3 experiments with 3 assays each. Standard deviations less than 3%.

The portion of whey protein (nonprecipitable at pH 4.5) to total extracted protein was almost the same for suc-

TABLE 2
Effects of Succinylation of Defatted Cottonseed Flour on Extractability and Precipitability of Protein

Succinic anhydride concentration (%)	Protein extractable (%)	HCL-precipitable protein (pH 4.5) (as % of extractable protein)	HCL-nonprecipitable protein (pH 4.5) (as % of extractable protein)
Control (0%)	28 ± 2.0	35.9 ± 1.5	17
10%	55 ± 4.5	63 ± 3.3	20.1
30%	65 ± 3.7	69 ± 5.0	20

cinylated and untreated samples (Table 2). Evidently, this portion was not altered by succinylation, and the increase in extractability might have been due to conversion of water-insoluble storage proteins into soluble forms. This conversion might increase the portion of protein precipitable at pH 4.5.

When sequential, three-step extractions were used to determine changes in proportions of the water-, 4% NaCl (pH 6.5)-, and 0.2% NaOH (pH 11.5)-soluble protein fractions, a large portion of the protein (43.6% of the total) was recovered as salt-soluble isolate from the untreated flour. Recoveries of water- and NaOH-soluble isolates were only 10.7 and 12.5%, respectively (Table 3). However, amounts of proteins recovered from salt and alkali extracts were markedly decreased by succinylation (30% succinic anhydride), thus indicating their conversion to water-soluble forms (representing 48% of total protein). Yields of salt- and alkali-soluble isolates from succinylated flour were 16.6 and 7.8%, respectively. The converted water-soluble protein isolate contained about 86% protein on a dry basis.

In isoelectric precipitation of protein from extracted solutions, the amounts of precipitated protein and precipitation profiles at various pH's were different, depending upon the acylating agents used. Maximum precipitation was obtained at around pH 4.0–4.5 for the untreated sample, but a pH lower than 4.0 was required for modified samples (Fig. 6). This observation indicates that some change in the isoelectric behavior of the proteins had occurred due to acylation. A rather large change was observed in acetylated samples, whose maximum precipitation was obtained at pH's below 3.0.

Protein extracts from acylated flour or sodium sulfite treated flour contained more soluble protein at neutral pH than did the extract from untreated flour. More than 90% of the extracted protein was in a soluble form in the pH range of 6–7 for modified samples, while only about 60% was soluble for the untreated sample. The results suggest that storage proteins of cottonseed, which have minimum solubility at pH 7.0, were modified into a soluble form at this pH. Maleylated and succinylated extracts contained especially high concentrations of soluble proteins in the pH range of 5–7.

As for the protein recovery, about 80% of the extracted protein in the acylated and sodium sulfite-treated flour was precipitated at around pH 4.0; corresponding values for untreated and acetylated flours were, however, only 60% and 40%, respectively. These results suggest that acetylation of cottonseed flour before extraction of protein should be avoided because of the low protein recovery at pH 4.0, as well as the low protein extraction. To obtain a better yield of acetylated protein isolate, proteins should be extracted first from the flour by the conventional method, and then acetylated instead of during

TABLE 3
Composition and Yield of Protein Isolates

Treatment	Isolate	Composition of isolate[a]			Yield[b] of dry matter (%)	Protein recovery[b] (%)	Free amino group[c] (%)
		Protein (%)	Sugar (%)	Phosphorus (%)			
Succinylated[d]	Water-soluble	86.1	9.4	0.3	31.9	48.1	48
	Salt-soluble	89.2	6.7	0.23	10.6	16.6	43
	Alkali-soluble	90.2	4.3	0.2	4.9	7.8	11
Untreated	Water-soluble	81.5	7.0	1.7	7.5	10.7	100
	Salt-soluble	91.3	2.4	0.3	27.3	43.6	50
	Alkali-soluble	84.7	5.3	0.8	8.0	12.5	60

[a] Data represent means of 2 experiments with 3 determinations. Protein content of defatted cottonseed flour was 57% on a dry wt basis. Flour (100 g) was used.
[b] Data are the average of 2 preparations. Coefficients of variation ranged from 1.0 to 2.6%.
[c] Percentage of untreated water-soluble isolate.
[d] Succinylated at 30 succinic anhydride on protein basis.

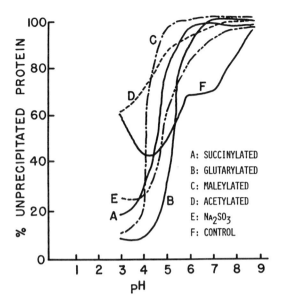

Fig. 6. Protein precipitability profiles of chemically modified proteins at various pH's.

extraction of protein. However, it is better to add dibasic acid anhydrides to the extraction medium of the flour for a high yield of modified protein isolate with negatively charged acyl groups.

Changes of Functional Properties by Acylation

Protein Solubility in Water: Protein isolates prepared from succinylated (30% succinic anhydride) and untreated cottonseed flour were almost 100% soluble at pH 10 (Fig. 7). However, solubility of the untreated isolate decreased rapidly as pH was lowered, whereas succinylated isolate remained 100% soluble until it was acidified to pH 6.0. Both isolates showed minimum solubility around pH 4.0. The succinylated isolate also was highly soluble at lower than pH 3.0.

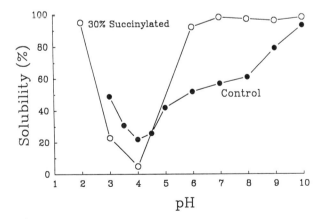

Fig. 7. Solubility profile of succinylated and untreated protein isolates at various pH levels. Average of 3 assays. Standard deviations less than 0.8%.

The solubility of the isolates in water increased logarithmically with respect to concentration of succinic anhydride added during succinylation of the flour (Fig. 8). The unmodified protein isolate (prepared by the conventional method (10)) showed approximately 20% solubility at room temperature when a 2% protein suspension was prepared. The highly succinylated sample was completely solubilized at the same protein concentration.

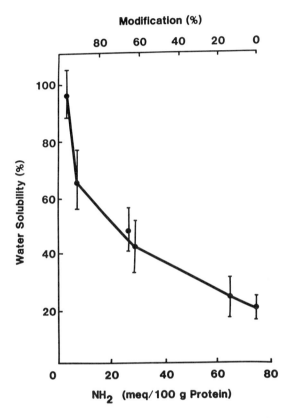

Fig. 8. Changes in water solubility of cottonseed protein isolates by succinylation. Solubility was determined in 2% protein suspension (w/v) at room temperature.

On plotting protein solubility vs. degree of succinylation of the isolates on a semi-log scale, two straight lines, having two distinct slopes and a breaking point at approximately 90% succinylation levels, were obtained (Fig. 9). This observation can be accounted for by two independent exponential components, each having a different constant for solubility. In other words, protein solubility exponentially increases with degree of succinylation of protein up to 90%. However, at higher than 90% succinylation, additional secondary reactions of protein molecules occur, yielding a second straight line with a different constant.

The solubility of the isolates varied, depending upon the acylating agent used. Protein isolates from succinylated, maleylated, and dimethylglutarylated cottonseed flours were highly water-soluble at room temperature. There was no significant difference in solubility among these protein isolates (Fig. 10). Unexpectedly, acetyla-

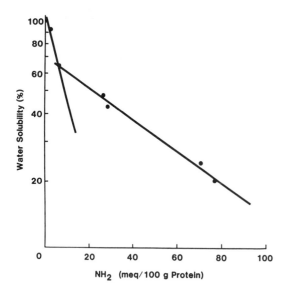

Fig. 9. Semi-logarithmic plot of protein solubility vs. free amino groups of protein isolates. Mean of 3 experiments.

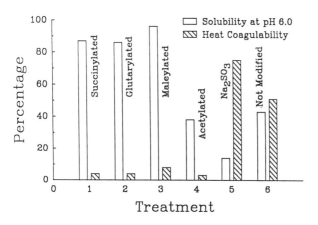

Fig. 10. Water-solubility and heat-coagulability of modified and unmodified protein isolates.

tion of the flour did not significantly affect the solubility of the resultant isolate, but treatment with Na_2SO_3 resulted in a protein isolate which was less soluble in water than the control isolate. These results indicate that the water solubility of proteins is correlated with the amounts of negative charges, but not with neutral groups of protein molecules or cleavage of disulfide bonds.

Succinylation is the most frequently used laboratory method to improve functional properties of various proteins. It increases solubility of soybean (11), sunflowerseed (12), cottonseed (13), peanut (14) and wheat proteins (21). However, it was reported that acetylation did not affect the solubility of soybean (11) and sunflowerseed (22) proteins appreciably. However, Barman et al. (23) reported that acylation decreases water binding and increased solubility of soy protein isolate in the pH 4.5–7.0 range.

Heat Coagulability: Approximately 55% of the protein coagulated to form a turbid solution when a 2% suspension of unmodified protein isolate was boiled in a water bath for 30 min (Fig. 11). However, the succinylated isolate (higher than 80% succinylation) was almost completely soluble under the same conditions.

Heat coagulability of protein isolates decreased with increasing levels of succinylation (Fig. 11). Plotting of coagulability vs. degree of succinylation results in a straight line with a slope of 0.95% coagulability per 1 meq. of NH_2, indicating a decrease of 0.95% coagulability when one milliequivalent of free amino groups per 100g protein is succinylated. This plot also shows that coagulation was inversely proportional to the amount of negative charge on the protein molecules. Haurowitz (24) reported that bi-functional crosslinking between peptide chains of protein could occur through the ϵ-amino groups of lysine by succinylation. Such crosslinking increased the resistance of protein to thermal denaturation involving unfolding of compact molecules.

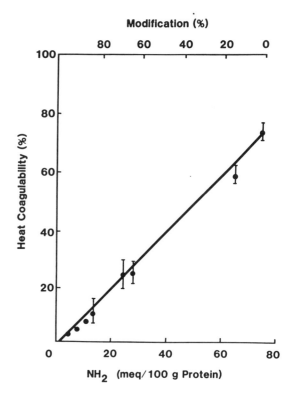

Fig. 11. Changes in heat coagulability of succinylated cottonseed protein isolates. The coagulability was determined in 2% protein suspension by heating in a boiling water bath for 30 min.

Protein isolates acylated with dibasic acid anhydrides, such as succinic, maleic and dimethylglutaric anhydrides, resulted in heat resistant proteins. The proteins were not coagulated by heating their water suspensions at 95–100°C for 30 min. However, the isolate obtained from untreated flour was moderately heat-coagulable (50%). Acylation of the flour also decreased heat coagulability of the resulting isolate to less than 5%. Treatment of the flour with sodium sulfite produced a protein isolate which was highly sensitive to heat. No significant difference in heat coagulability was observed among protein isolates prepared with different acylating agents (Fig. 10). Similar results were reported by Groninger (12), who found that a succinylated fish myofibrillar protein suspension was not coagulated or precipitated by heating at 100°C.

Oil Absorption: The ability of a protein to bind oil is a very important property in many food applications, such as meat extenders/replacers (1). Oil absorption was not changed significantly at low degrees of succinylation, but it increased considerably with succinylation at levels higher than 60% (Fig. 12). The mechanism of improvement is attributed mostly to structural alteration as well as the increased negative ions of succinylated proteins. Swelling and unfolding, which increase bulk density, may accompany enhanced oil absorption.

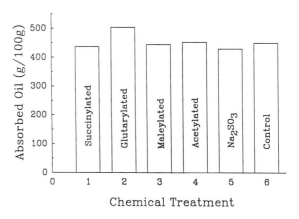

Fig. 13. Oil absorption capacity of modified and unmodified protein isolates.

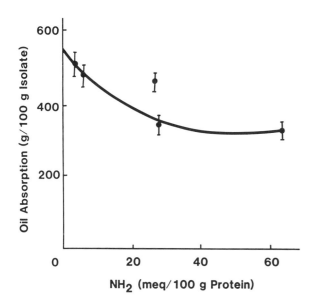

Fig. 12. Changes in oil absorption of succinylated cottonseed protein isolates.

Oil absorption capacity was not significantly changed by maleylation, succinylation, acetylation or sodium sulfite treatment, but the isolate from dimethylglutarylated flour showed slightly higher oil absorption (Fig. 13). This indicates that the long, branched hydrocarbon chain of dimethylglutaric anhydride may have caused the increase of oil absorption capacity. Therefore, oil absorption probably is correlated to the type and number of hydrophobic groups of protein molecules rather than to the distribution of charges and disulfide bonds.

Emulsion Capacity: Emulsifying properties are related to the capacity of a protein to lower interfacial tension between the hydrophobic and hydrophilic components of foods. Low level succinylation (less than 60%) did not significantly alter emulsion capacity of the resulting protein isolates (Fig. 14). However, at higher succinylation levels the emulsion capacity increased markedly. The increase might be related to enhanced unfolding of protein molecules by high level succinylation, thereby increasing the exposure of hydrophobic groups and the effective surface area for interfacial membrane formation with oil. Succinyl ions attached to protein molecules do not seem to be stimulators for improving the emulsion capacity, but secondary changes of protein molecules during high succinylation may be major factors in increasing this capacity. It was reported that emulsion capacity and stability were improved in soy protein (11), sunflower protein (22), and fish protein (25), by succinylation. Presumably, this improvement is due to secondary changes of protein molecules, since, in general, such proteins are succinylated at high concentration of succinic anhydride.

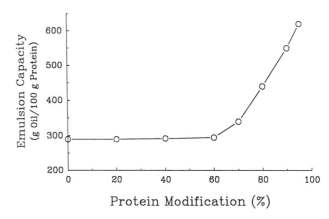

Fig. 14. Changes in emulsion capacity of succinylated cottonseed protein isolates.

Structural Changes in Protein Isolates by Acylation

Protein isolates obtained from acylated cottonseed flours showed slightly lower protein contents (81–87%) and higher ash contents (8–10%) than those of storage and nonstorage protein isolates (93 and 94% protein; and 3 and 5% ash, respectively) (Table 4). These differences are considered to result from adsorption of salt formed from the acid anhydride and sodium hydroxide during modification.

Acylation causes structural changes, such as cross-

TABLE 4

Compositions of Protein Isolates

Isolate	Composition (%)		Free NH$_2$ group (meq/100g protein)
	Protein	Ash	
Succinylated	83	10.4	6.4
Dimethylglutarylated	83	8.1	4.8
maleylated	87	10.1	2.8
acetylated	81	9.5	3.5
Na$_2$SO$_3$	92	8.0	33.0
NSP control[a] (pH 4.5)	93	5.0	46.0
SP control[b] (pH 7.0)	94	3.2	26.0

[a] Nonstorage protein.
[b] Storage protein.

linking or dissociation, depending upon degree of modification. Gel filtration chromatograms of cottonseed protein isolates prepared by sequential extraction with water, 5% salt and 0.2% NaOH, and by isoelectric precipitation at pH 4.5 showed their typical separation patterns (Fig. 15). The water-soluble isolate separated into three peaks, including a major peak at 350 ml on Sephadex G-100 columns. The major peaks of the salt- and alkali-soluble isolates had of 300 ml and 200 ml, respectively, indicating that the major proteins of the salt- and alkali-soluble isolates were larger in molecular size than that of the water-soluble isolate. Isolate from the flour succinylated with succinic anhydride at 30% of the total flour protein was eluted as a large peak at 200 ml, which coincides with the void volume. This elution behavior indicates that proteins isolated from succinylated flour are larger in molecular size than those from untreated flour. It can be hypothesized that increased negative charges on protein molecules by attachment of succinic ions result in increasing intermolecular electrostatic repulsive forces among peptide chains of the proteins, and that these forces may result in increased molecular size of proteins by swelling and unfolding. Similar phenomena have been noted by Grant (21). Succinylated derivatives of wheat protein eluted much earlier in gel filtration chromatography than did untreated protein. However, addition of excess succinic anhydride to wheat flour suspensions enhanced dissociation of protein subunits.

In the Sephacryl S-200 gel chromatogram of protein isolates from cottonseed flours succinylated at various concentrations of succinic anhydride, the isolate from highly succinylated flour (100% succinic anhydride) was eluted as one protein peak (Fig. 16). However, the other isolates from flours succinylated at 0, 30 and 60% concentrations were resolved into three major protein peaks. This suggests that the highly succinylated protein isolates consisted of one homogenous protein agglomerate of uniform molecular size, while the other isolates consisted of at least three different sizes of protein agglomerates. These results were confirmed by polyacrylamide disc gel electrophoresis, which shows one clearly separated protein band for the 100% sample and more than three bonds for the other samples. The faster migration of protein bands in suc-

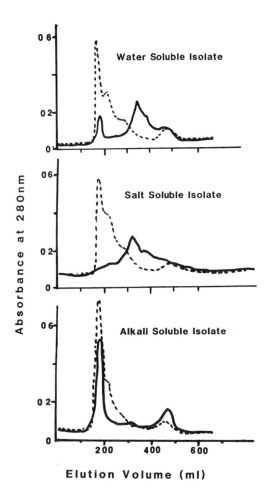

Fig. 15. Gel filtration chromatograms of cottonseed protein isolates extracted sequentially with water, salt (4% NaCl) and alkali (0.2% NaOH) from succinylated and untreated cottonseed flours. The solid line represents a chromatogram of isolates from untreated flour. The dotted line represents isolates from the flour succinylated with 30% succinic anhydride. Sample was eluted on Sephadex G-100 column with 0.2% NaOH containing 4% NaCl.

tein agglomerates due to repulsive forces (26,27,28). However, in cottonseed protein isolates, dissociation of the subunits was observed neither on the gel permeation chromatogram nor on the disc-gel electrophretogram (Fig. 17) of the highly succinylated samples. Possibly, succinylation might cause some type of polymerization or rearrangement of subunits of cottonseed proteins by crosslinking.

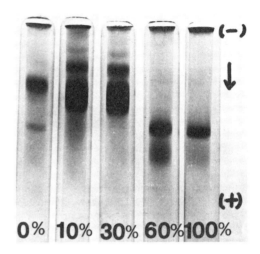

Fig. 17. Polyacrylamide disc gel electrophoretograms of untreated and succinylated cottonseed protein isolates. "Per cent" indicates amount of succinic anhydride added basis protein content of cottonseed flour.

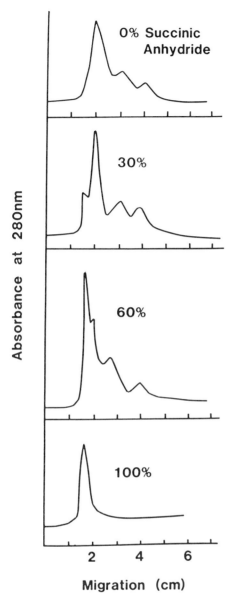

Fig. 16. Sephacryl S-200 column chromatograms of untreated and succinylated cottonseed protein isolates.

cinylated samples than in untreated samples resulted from the increase in negatively charged succinyl groups on the protein molecules, which move faster to the anode during electophoresis.

It is interesting that the highly succinlyated isolate showed one single peak on the gel chromatogram, and that the elution volume of the peak was less than that of the other samples, indicating larger molecular size. It is unknown whether this change is due to the swelling or unfolding of compact protein molecules to larger sizes by repulsive forces between negatively charged succinyl groups of the peptide chains, or to a polymerization or rearrangement of protein subunits by crosslinking through the amino groups of lysine. Some investigators have reported that succinylation frequently causes dissociation of subunits from the quaternary structure of pro-

Deacylation of Acylated Protein Isolate

About 70% of the maleylated cottonseed protein isolate was deacylated by incubating the isolates at pH 2.0 and at 55°C for 1 hour, while 15–25% deacylation was observed for the acetylated, succinylated and dimethylglutarylated isolates (Fig. 18). Deacylation of succinylated protein is very slow under mild conditions (29). However, deacylation of derivatives formed by maleic, citraconic and 2,3-dimethylmaleic anhydrides containing cis-double bonds occurs at a much faster rate under mild conditions (30). The most important structural factor in the deacylation process is the cis-double bond, which maintains the terminal carboxyl group in the spatial orientation that makes a nucleophilic attack on the amide carbon much more probable (29). In succinylated proteins, the freely rotating single bond between carbons 2 and 3 allows the terminal carboxyl group to assume many more orientations, and, as a result, this drastically reduces the probability that it will stay in the proper conformation long enough for deacylation to occur.

Nutritional Aspect of Acylated Proteins

Acylation of protein has been noted to considerably lower the extent of the Maillard reaction (31), the formation of lysinoalanine by alkaline treatment (32) and the polymerization of proteins with oxidizing lipids (33). Also, it improves functional properties of proteins in their utilization as food ingredients. However, acylation of food

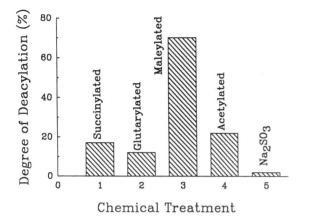

Fig. 18. Deacylation of modified protein isolates incubated at pH 2.0 for 1 hr at 50°C.

proteins has not occurred because of lack of sufficient information on the nutritional quality of such modified proteins.

Conflicting evidence on the nutritional value of acylated protein exists in the literature. However, most investigators agree that succinylation or acetylation of protein reduces overall nutritional value because of unavailability of lysine in animal feeding tests. The PER's (Protein Efficiency Ratios) of acylated proteins have been known to be lower than that of untreated proteins (12,34). However, Melnychyn and Stapley (35) reported no significant change in the nutritive properties of succinylated vegetable proteins. Groninger (12) noted that proteins containing low levels of lysine may not show serious reduction in nutritive value after acylation. Most nutritional studies have been carried out by single feeding with acylated proteins. Nutritional evaluation of acylated proteins as supplemental protein sources has not yet been reported. Furthermore, there are no reports on the nutritional value of proteins acylated with reversible modification agents such as maleic or citraconic anhydrides. Reversible modification of proteins may increase availability of lysine by deacylation of the proteins at the low pH of the stomach. Thus far, only a few nutritional studies on acylated soybean proteins (35), casein (36), fish (37), egg white (38) and beef heart myofibrillar protein (39) have been reported.

There have been reports that the acetylated lysine residues of proteins are utilized nutritionally to some extent (37), but that succinylated lysine residues are not utilized (12). Matoba et al. (36) reported that lysine bonds in acetylated and succinylated proteins were not hydrolyzed by proteases from the digestive glands, and that the acylysine residues might remain as larger peptide fractions. These peptides might be hydrolyzed into amino acids in the intestinal mucosa and/or enter into the intracellular tissues. The acylysine residue might be altered to free lysine, and thus be available nutritionally.

Groninger and Miller (37) investigated the metabolism of acylated proteins by feeding ^{14}C-succinylated and acetylated fish protein to rats, and they noted that the majority of the radioactivity was recovered in the carbon dioxide expelled from the lungs and in the urine. Their finding showed that acylated lysine residues were absorbed into the intestine and metabolized without any physiological problems. However, more nutritional evaluations are needed in order to assess whether acylated proteins are suitable for use as food.

References

1. Kinsella, J.E., *CRC Crit. Rev. Food Sci. Nutr.* 7:219 (1976).
2. Feeney, R.E., R.B. Yamasaki and K.F. Geoghegan, *Adv. Chem. Ser., 198:*1 (1982).
3. Glazer, A.N., *Ann. Rev. Biochem. 39:*101 (1970).
4. Means, G.E., and R.E. Feeney, "Chemical Modification of Protein," Holden-Day, San Francisco, 1971.
5. Adler-Nissen, J., *J. Agric. Food Chem. 24:*1090 (1976).
6. Cater, C.M., K.F. Mattil, W.W. Meinke, M.W. Taranto and J.T. Lawhon, *J. Am. Oil Chem. Soc. 54:*90A (1977).
7. Kornegay, E.T., A.J. Clawson, F.H. Smith, and E.R. Barrick, *J. Anim. Sci. 20:*597 (1961).
8. Berardi, L.C., W.H. Martinez and C.H. Fernandez, *Food Technol., 23:*75 (1960).
9. Lyman, C.M., W.Y. Chang and J.R. Couch, *J. Nutr. 49:*679 (1953).
10. Choi, Y.R., E.W. Lusas and K.C. Rhee, *J. Food Sci. 46(3):*954 (1981).
11. Franzen, K.L., and J.E. Kinsella, *J. Agric. Food Chem. 24:*788 (1976).
12. Groninger, H.S. Jr., *J. Agric. Food Chem. 21:*978 (1973).
13. Childs, E.A., and K.K. Park, *J. Food Sci. 41:*713 (1976).
14. Beuchat, L.R., *J. Agric. Food Chem. 25:*258 (1977).
15. Choi, Y.R., E.W. Lusas and K.C. Rhee, *J. Am. Oil Chem. Soc. 58:*1044 (1981).
16. Choi, Y.R., E.W. Lusas and K.C. Rhee, *J. Food Sci. 47:*1713 (1982).
17. Martinez, W.H., L.C. Berardi and L.A. Goldblatt, Third International Congress of Food Science and Technology, SOS/70 (248), Washington, D.C. Aug. 1970.
18. Shetty, K.J., and J.E. Kinsella, *J. Food Sci. 44:*634 (1979).
19. Saio, K., M. Kajikawa, and T. Watanabe, *Agric. Biol. Chem.* (Japan) *35:*890 (1971).
20. Cole, R.D., *Meth. Enzymol. 11:*206 (1961).
21. Grant, D.R., *Cereal Chem. 50:*417 (1973).
22. Kabirullah, M., and R.B.H. Wills, *J. Food Technol. 17:*135 (1982).
23. Barman, B.C., J.R. Hansen and A.R. Mossey, *J. Agric. Food Chem. 25:*638 (1971).
24. Haurowitz, F., The Chemistry and Function of Proteins, 2nd ed., Academic Press, New York, N.Y., 1963.
25. Chen, L.F., T. Richardson and C.H. Amundson, *J. Milk Food Technol. 38:*89 (1975).
26. Klotz, I.M., and S. Kerosztes-Nagy, *Biochem. 2:*445 (1963).
27. Lenard, J., and S. Singer, *Nature 210:*536 (1966).
28. Sia, C.L., and B.L. Horecker, *Biochem. Biophys. Res. Commun. 31:*731 (1968).
29. Shetty, J.K., and J.E. Kinsella, *Adv. Chem. Seri. 198:*169 (1982).
30. Butler, P.J.G., J.I. Harris, B.S. Hartley and R. Leberman, *Biochem. J. 107:*789 (1976); *112:*679 (1969).
31. Finot, P.A., F. Mottu, E. Bujard and J. Mauron, in *Nutritional Improvement of Food and Feed Proteins,* edited by M. Friedman, Plenum Press, New York, 1978, p. 547.

32. Friedman, M. *Ibid.*, 1978, p. 613.
33. Shimada, K., and S. Matsushita, *Agric. Biol. Chem.* 42:781 (1978).
34. Creamer, K.L., J. Roeper and E.H. Lohrey, *J. Dairy Sci. Technol.* 6:107 (1971).
35. Melnychyn, P., and R. Stapley, U.S. Patent 3,764,711, 1973.
36. Matoba, T., E. Dvi and D. Yonezawa, *Agric. Biol. Chem. (Japan)* 44:2323 (1980).
37. Groninger, H.S., and R. Miller, *J. Agric. Food Chem.* 27:949 (1979).
38. King, A.J., H.R. Ball and J.D. Garlich, *J. Food Sci.* 46:1107 (1981).
39. Eisele, T.A., C.J. Brekke and S.M. McCurdy, *J. Food Sci.* 47:43 (1982).

Preparation and Uses of Dietary Fiber Food Ingredients

Thomas H. Smouse

Archer Daniels Midland Co., 1001 Brush College Rd., Decatur, Illinois 62525, USA

Abstract

Dietary fiber has been an ingredient of foods since Adam ate the forbidden apple. However, in recent years it has become recognized as an important nutritional ingredient in the human diet even though the metabolizable energy (ME) is considered negligible. Nutritionally, dietary fiber has been associated with weight control, diverticulitis, deodenal ulcers, colon cancer, cholesterol reduction, and diabetes mellitus. Therefore, it is an important component food manufacturers must consider in providing balanced diets. This presentation will cover what a dietary fiber is, the methods used to measure the amount of dietary fiber in a fiber source, what sources are there for various types of fiber, how various fibers compare in selected applications, and what are some of the dietary implications and nutritional labeling aspects of edible fiber.

Although dietary fiber (DF) has been an important ingredient in human nutrition since life began, recently it has become the subject of many studies and much research. Only about ten published articles per year appeared in the early 1970's, growing to over 400/year in the early 1980's. This growth of interest in fiber will make it one of the most important nutritional ingredients of the medical profession. Peter Cleave certainly addresses the importance of dietary fiber in his 1974 book entitled, "The Saccharine Disease." Peter Cleave has received international fame as the father of the dietary fiber hypothesis. He felt the human body was maladapted to the artificial foods of civilization, such as refined sugar and white wheat flour. He reasoned that if a person would avoid unnatural foods, then he would avoid unnatural diseases, or those not observed in wild animals or primitive societies. He was one of the first advocates of raw bran in the medical area in 1941. While serving on the battleship "King George V," he would have bags of bran brought aboard that he could use to treat constipation that the sailors would obtain from the lack of fresh fruits and fresh vegetables. Bran was such an integral part of his treatment that he became known in the British Navy as "the bran man." A recent book (1) published in 1986 by Plenum Press entitled, "Dietary Fiber, Basic and Clinical Aspects," was dedicated to Peter Cleave, who passed away in 1983 at the age of 77.

What exactly is dietary fiber? In 1972, Trowell (2) defined dietary fiber as "the skeletal remains of plant cells that are resistant to digestion by enzymes of man." Later in 1976, the definition was modified by Trowell et al. (3), to state that dietary fiber is "the plant polysaccharides and lignin that are resistant to hydrolysis by the digestive enzymes of man." The only plant polysaccharide known to be hydrolyzed by man's digestive enzymes is starch.

Therefore, all nonstarch polysaccharides, or NSP as they are referred to, could be considered as dietary fiber sources. However, some starches are not completely hydrolyzed with α-amylase *in vitro* (4,5,6) and starch that has been retrograded during food processing (7) is not hydrolyzed completely by α-amylase unless extreme hydrolysis is conducted (8). Also, α-amylase inhibitors in various foods can prevent the hydrolysis of starch (9) or starch complexes with fat (10), or protein (11), so under specific conditions starch itself could be included in Trowell's 1976 definition of dietary fiber. After much controversy as to what should or should not be included in dietary fiber, Trowell in 1985 modified his 1976 definition to say, "dietary fiber is the sum of the polysaccharides and lignin which are not digested by the endogenous secretions of the human gastrointestinal tract." (12) Generally, dietary fiber is composed of the components listed in Table 1.

TABLE 1
Dietary Fiber Components

Cellulose
Hemicellulose
Pectins
Lignins
Gums
Cutins
Minerals
Manmade additives
   (polydextrose, sucrose esters, etc.)

The first five components are universally accepted as dietary fiber components. However, due to the vast number of indigestible food components that are available today, components such as the manmade fat substitutes or sucrose esters (13), modified starches that are less digested than unmodified starch (14,15); chitin from seafoods (16), and the manmade fiber polydextrose (17) are not digested by the endogenous secretions of the human gastrointestinal tract and thus are considered by some researchers as dietary fiber. Therefore, one can classify dietary fiber into its conventional and unconventional components. These have been listed by Dreher in his recent book entitled, "Handbook of Dietary Fiber." (18) The conventional components are given in Table 2.

The definition and explanation of these components are thoroughly covered by Dreher (18). Briefly, cellulose is the most common cell wall component of higher plants, and is a linear polysaccharide composed of β-1,4- glucose units having a degree of polymerization (DP) ranging from 300–15,000. The β linkage is not attacked by digestive

TABLE 2

Conventional Dietary Fiber Components

Cellulose
Cellulose derivative
    Microcrystalline cellulose (MCC)
    Carboxymethylcellulose (CMC)
    Hydroxypropylcellulose
    Methyl cellulose
Hemicellulose
Amyloids
Pectic material
beta-Glucans
Carrageenans
Alginates
Lignin
Mucilages
    Guar gum
    Locust bean gum
Plant erudates

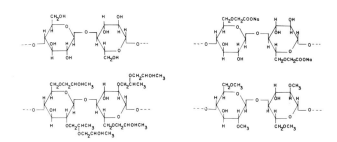

Fig. 1. Structure of cellulose, CMC, hydroxypropyl cellulose and methyl cellulose.

enzymes, therefore cellulose is not hydrolyzed in the human digestive system, which makes cellulose and its derivatives one of the most concentrated dietary fibers available. However, due to the publicity that cellulose comes from trees, its use has been limited in favor of some of the other natural fiber sources such as various brans.

Microcrystalline cellulose (MCC) is a derivative prepared from cellulose by treating α-cellulose with mineral acids, causing depolymerization. It has the following advantages over cellulose: (a) nonfibrous and (b) higher water-binding capacity. Carboxymethylcellulose (CMC) is produced by treating cellulose with sodium hydroxide and sodium monochloroacetate. By thorough washing with aqueous alcohol to remove the excess salts, a food grade CMC is obtained. The DP can be varied from 400–3200 units, or molecular weights of 90,000–700,000.

Hydroxypropylcellulose is made by reacting alkali with cellulose to depolymerize the polysaccharide and then reacting it with propylene oxide which forms an ether linkage at two or more of the primary hydroxyl groups on each glucose unit. This gives this derivative some unique properties over cellulose, such as insolubility in water over 40°C, solubility in polar organic solvents and stability over a wide pH range of 3–10.

Methyl cellulose (MC) is the methyl ether of cellulose, and is made by reacting cellulose with alkali and methyl chloride, which forms the methyl ether and hydrochloric acid. By selecting the degree of substitution with methyl chloride, one can form MC soluble only in alkali, others that are soluble in water, and still others that are soluble in organic solvents. Methyl cellulose, like its cousin hydroxypropyl-cellulose, also shows excellent stability over a wide pH range of 3–11.

Figure 1 shows the chemical structure of cellulose, sodium carboxylmethylcellulose, hydroxypropylcellulose and methylcellulose. As mentioned earlier, the cellulose products are an excellent source of dietary fiber. However, because they are derived from wood, the public initially was not ready to accept them. As they are flavorless and have a very white color, they are very useful fiber sources in baked items. With the recent interest in dietary fiber, the cellulosic products are being received more favorably than in the past.

The hemicellulose components of plant material are the next large category of dietary fiber. This material also has the 1,4-β-linked pyranoside that is similar to cellulose except sugars other than glucose are present—sugars such as galactose, arabinose and uronic acid. Purified hemicelluloses are not typically produced commercially for food use. Nevertheless they are an important dietary fiber in the daily diet, being consumed as arabinoxylans in cereal foods, or glucomannans and galactoglucomannans from plant cell walls.

Everyone has heard "an apple a day keeps the doctor away." It is thought by some researchers that the pectin in apples is the basis for this saying. Pectins are a group of polysaccharides consisting of α-1,4-glycosidic-linked D-galacturonic acid as its major constituent. Pectins also have side chains consisting of galactose, arabinose, xylose, rhamnose and glucose. Foodgrade pectins are classified by their degree of esterification (DE) or methoxy content. Low methoxy pectins (LM) are those with less than 50% methyl esters and high methoxy pectins (HM) are those with more than 50% methyl esters. LM's and HM's have different gelation requirements, which are important attributes in the production of jams and jellies.

The lignins are another group of conventional dietary fiber components. A lignin is a high molecular weight aromatic polymer composed of phenylpropane residues from the condensation of phenolic alcohols such as coniteryl, sinapyl, and p-coumaryl. Molecular weights vary from 1000–4000 and fiber sources vary widely in their lignin content. An entire book on lignins (19) is available for those desiring details on this DF source. Lignin appears to be closely associated with the cellulose and hemicellulose constituents of cell walls of mature plants.

Gums make up the last class of conventional DF's. The common ones used in the food industry are guar, locust bean, gum arabic, gum ghatti, gum karaya, lack gum and gum tragacanth. All of these are a mixed group of complex polysaccharides and associated with the endosperm rather than the cell wall.

Table 3 lists some of the unconventional dietary fiber components. These materials are not as prevalent in our diets as the conventional DF's but they are present, they are not digested, and they are included in a total dietary fiber (TDF) analysis.

The tannins and other polyhydroxy-phenolic compounds are found throughout the plant kingdom. Examples of such materials are the chlorogenic acid found in sunflower seeds, the ferulic acid found in cereals, or the gen-

TABLE 3

Unconventional Dietary Fiber Components

Tannins and phenolic materials
Cutin and suberin
Extensin
Minerals
Phytic acid
Maillard compounds
Resistant starch
Chitin
Manmade dietary fiber

---

istic acid found in soybeans. These materials condense easily and are believed to be precursors to lignin.

Cutin and suberin are complex polymeric mixtures of nonpolar lipids and are normally grouped with the waxes. These materials are generally found on the outer epidermal cells of leaves, stems, fruits and seeds, and serve as a protective covering. They are extremely resistant to digestion and are included in a TDF analysis.

Extensin is a glycoprotein found in all walls and is similar to collagen of the animal kingdom. It is covalently linked with such things as hemicellulose and pectin, and it may function as a bond to hold cellulose fibrils together. The soybean seed coat can contain as much as 10% extensin during germination.

Minerals such as silicon and silica are found in rice and other dry cereals. Phytic acid, which is common in cereal grains, oilseeds and legumes, normally is complexed with metal ions and binds cations such as calcium, magnesium, iron and zinc, the brown pigments that result from the Maillard reaction of a free amino group and a reducing sugar and are commonly called melanoidins. Starch that is modified, retrograded starch, or starch complexes with lipids, are not digested and are referred to as resistant starch. The resistant starch of some foods is shown in Table 4.

TABLE 4

Resistant Starch in Various Foods

Food	Resistant starch (% d.d.)
White bread	1.2
Wholewheat bread	1.1
Whole rye flour	0.2
Cornflakes	3.1
Beans	0.5
Boiled potatoes	2.1
White rice	—

None of these food items is considered an excellent source of DF, but the resistant starch that is present will add to the TDF during a TDF analysis.

Chitin is to fungi and crustaceans what cellulose is to the plant kingdom. It is the principle skeletal matrix and can be obtained from the waste material of shrimp and crab. Although it can be considered as a dietary fiber source, it is not used or processed as such in the United States, and is not officially approved by the F.D.A.

The manmade dietary fibers are also a minor source. These materials were made for other purposes, such as thickeners, noncaloric fat substitutes, and emulsifiers. However, compounds such as the polydextroses, sucrose esters, lactulose, and the glycosylsorbitols have been reported to be nondigestible, and thus dietary fiber sources (17,13,20,21).

The American Chemical Society recently had a complete symposium addressing "Unconventional Sources of Dietary Fiber." A monograph on this subject was edited by Furda (22), and contains chapters on such items as leguminous seed fibers, psyllium seeds, citrus pectic polysaccharides, etc. The data given in Table 5 show the total dietary fiber of some unconventional fiber sources as compared to standard AACC wheat bran. Using wheat bran as a reference material at 44.1% TDF, you can see psyllium seeds, citrus pectins, tobacco leaves, leguminous seeds, and sugar beet fiber are all higher in TDF, while sugar beet pulp and apple pulp are considerably less. The data used for leguminous seeds was from several products—Arsoy or Fibrim, which is not a soybean hull or a soy bran but a product made from the spent soybean cell wall material after oil, protein, and soluble sugars have been removed, and is produced by ADM and Protein Technologies International, respectively. The pea fiber is available from Good Earth Agri Products, Inc., and is being used in bakery items. The major suppliers of sugar beet fiber are a Swedish company called Sockerbolaget Food Ingredients, with a product called Fibrex, and the American Crystal Sugar Company, with a product called Duo-Fiber. Psyllium seed products are available in numerous commerical preparations, such as Mucilose Flakes, Effersyllium, Metamucil, Konsyl, Syllact and Syllamalt. Such products have been fed to humans for generations as a pharmaceutical to increase stool bulk and maintain gastrointestinal tract motility.

TABLE 5

Total Dietary Fiber in Some Unconventional Fiber Materials

Source	TDF
Tobacco leaves	54.7
AACC standard wheat bran	44.1
Citrus pulp	24.8
Sugar beet pulp	13.5
Apple pulp	12.7
Citrus pectin	92.5
Psyllium seeds	87.7
Sugar beet fiber	80.0
Leguminous seeds	
Peanut hulls	80.0
Fibrim	75.0
Pea fiber	34.0

There are a large number of nondigestible materials that when analyzed contribute to the total dietary fiber. However, the structural polysaccharides that make up cell walls of wheat, oat, soy and corn brans are the major sources of dietary fiber consumed in our daily diets.

The data given in Table 6 are on the more conventional sources of dietary fiber. These are the ones that are presently being used in various food products such as bread, cookies, and cereals, and are the major sources of dietary fiber being consumed.

The various types of cellulose: carboxymethylcellulose (CMC), microcrystalline cellulose (MCC), hydroxypropylcellulose, and methylcellulose were discussed earlier. Until recently, these materials were considered wood products and the negative relationship to trees limited their use as a fiber source. Recently, with the extreme interest in dietary fiber and the fact that the cellulose products are pure white and flavorless, these products are being used in more products as fiber sources.

Oats is the fiber source of interest, since data have appeared on its ability to lower blood cholesterol (23). Oats has three fiber sources: the hull, which is mostly cellulose; the bran, which has a large amount of soluble fiber and has been shown to reduce cholesterol in normal-borderline cholesterol patients by a 10–15% reduction over and above the reduction caused by a low fat diet. It is believed the β-glucans or the soluble fiber that are present in this fraction are responsible for this reduction; and rolled oats. Processing steps include physical separation of the hulls followed by grinding. The bran is separated from oats by a dry milling-air classification separation; and the rolled oats product is made by steaming it to inactive enzymes with rolling or flaking and drying (24). These products are major fiber sources for cereals, muffins, cookies and breads.

Corn bran from a milling operation contains large amounts of impurities, such as starch, protein, oil, and lignified tip cap. There are numerous patents that cover various treatments to improve corn bran as a fiber source (25–27). It is produced from the outermost structure of the seed which is a thin, almost invisible membrane termed the seed coat. All tissue exterior to the seed coat is named the pericarp, or hull, and makes up 5–6% of the kernel dry weight. The outermost layer is covered by a waxy cutin layer that provides protection for the seed. Most corn is processed by wet milling, which is a steeping step where the corn is soaked at 52°C for 22–50 hours in large tanks, in the presence of SO_2 and lactic acid. This step is to soften the kernel and facilitate separating corn into its components, such as germ, hulls and starch. After the germ is removed, the corn slurry is screened to separate the fiber which is mostly pericarp from the starch and gluten. One process uses α-amylase to digest residual starch before the corn hulls are dried and ground into corn bran. In another, called the Kraft process, an acid prehydrolysis is used prior to an alkaline delignification step which facilitates the removal of the hemicellulose and the noncarbohydrate fractions. Processing details on wet milling and dry milling can be obtained from a recent AACC monograph (28). Corn bran is composed mostly of hemicellulose (70%) and cellulose (23%). Because the majority of corn is processed by wet milling, any soluble fiber is removed during steeping and washing, and is not found in corn bran.

Although Brewer's dried grain is listed in Table 6 and contains 70.0% TDF, very little has found its way into food products. Nevertheless, large breweries have active research programs on using the spent grain as a fiber source after it has been fermented and the soluble sugars depleted. A product available from Miller Brewing Company called Barley's Best (29) is being used for its dietary effects. The spent barley grain is removed from the wort prior to adding the hops. In doing so the bitter flavors that hops contribute are eliminated. Limited nutritional data are available presently but within two years its effects upon cholesterol, gallstones, and diabetes will be released.

TABLE 6

Total Dietary Fiber (TDF) in Some Conventional Fiber Materials

Source	TDF
Cellulose	100.0
Oat hulls	90.0+
Corn bran	82.3
Soybean bran	65.0
Brewers dried grain	47.9
Wheat bran	44.1
Rice bran	39.7
Oat bran	27.8
Whole grains	10.0–12.0
Dried fruits	7.0–17.0

Soybean bran has always been used as a fiber source in animal and human foods. The bran is mostly cellulose and obtained from the dehulling operation of soybeans. A newer fiber source produced from the internal soybean polysaccharides is called Arsoy or Fibrim. This material has approximately 70% TDF, of which 7% is soluble fiber, and research by Lo has shown its effect upon plasma cholesterol in rabbits (30), rats (31), and humans (32). Several of these products are made from the spent soy residues after dehulling, oil extraction, and the soluble sugars and proteins have been removed. By proprietary processing techniques, the water adsorption/binding can be controlled to make products useful in products like beverages, puddings, retorted soups etc., or in food systems with less water like baked bread, muffins, crackers, cookies or breakfast bars.

Wheat bran (Table 7) has been used for years as a fiber source. Much like casein is used as a standard reference material in protein comparisons, wheat bran is used in dietary fiber comparisons. An excellent review of wheat, its processing and utilization, and its nutritional implications in health and disease appeared in a dozen publications in 1985 (33). A kernel of wheat is composed of the "bran," "endosperm," and the "germ." Millers bran, as wheat bran is often referred to, consists of the pericarp, the seed coat and the aleurone layer. Wheat milling is a dry process and its processing was recently covered by Nelson (34). The milling of wheat starts with tempering of the kernels to moisten the bran, which facilitates separation in the break rolls. By using multiple break rolls, the endosperm is reduced to smaller and smaller pieces, thus making a cleaner bran upon a sifting separation. The dry milling of wheat will produce a 15–16% bran, of which most is used for animal feeds. It is an excellent fiber source and very compatible with bakery products. In fact, whole grains by themselves such as whole wheat, rolled

TABLE 7

Selected Fiber Comparisons

Source	TDF	IF	SF	CF	Supplier	Cost ($/#)
Wheat bran	42.2	38.9	3.3	9.1	Lauhoff Grain	0.15
Oat bran	22.2	11.7	10.5	2.7	Ntl. Oats Co.	0.37
Soy bran	70.0	65.0	5.0	36.0	ADM	0.33
Corn bran	82.3	82.3	n.d.	12.0	A.E. Staley	0.50
Rice bran	38.0	36.0	2.0	11.5	Riviana Foods	0.15
Rolled oats	9.6	4.8	4.8	1.2	Quaker Oats	0.30
Soy fiber	75.0	70.0	5.0	11.0	Protein Tech. Int.	0.75
Barley bran	70.0	67.0	3.0	20.0	Miller Brew.	0.35
Dried fruit	12.0	6.0	6.0	3.0	—	—

oats, or whole rye are used in bakery products and cereals as excellent fiber materials.

Full-fat rice bran always has been an unstable product because of active lipases and hydrolysis with storage. At 100°F storage, rice bran will increase its free fatty acid content by eight times (35). Work done at the Western Regional Research Laboratory of USDA has led to stabilized full-fat rice bran and stabilized deffated rice bran (36). Basically, a commercial extruder was modified that can economically inactivate the lipase enzymes in rice bran and stop the rapid free fatty acid formation. This treatment also reduces the microbial content. Full-fat rice bran has been used in multigrain breads, pancake and waffle mixes and health foods. This material can also be extracted to produce a high quality rice oil and a stabilized defatted rice bran. This type of rice bran is approximately 5% higher in TDF than the full-fat type. It has been found to have better foaming capacity and better foam stability than wheat bran (37), making it an excellent choice in meringues, whipped toppings and bakery items. A recent publication of rice covers production and utilization of rice, as well as rice bran (38). Rice is one of the two major food cereals, and the principle food cereal in Asia, where 90% of the world's crop is grown and consumed. Rice bran has not been a major fiber source in North America to this date.

The last source of conventional dietary fiber is dried fruits. The data shown in Table 6 are an average of dried apples, apricots, figs, raisins and prunes. Dried fruit is an excellent source for fiber and most of it is of the soluble type. Data published in Food Technology (39) lists the dietary fiber content of 23 fruits; however, not all of those listed are adaptable to cereals, breakfast bars, or bakery products where products like raisins, dates, apples, etc. play an important role as dietary fiber sources.

Table 7 shows some comparisons for a few selected sources of dietary fiber. The abbreviations are for TDF-total dietary fiber, IF-insoluble fiber, SF-soluble fiber, and CF-crude fiber. Up until this point I have only referred to the total dietary fiber. However, there are several methods used to measure fiber content and, if one is not careful, it can be confusing when you compare data from several sources. The method being used by most research groups to establish dietary fiber is the AOAC/Prosky Method (39–43). This method is based on enzymatic digestion with α-amylase for gelatinization and protease and amyloglucosidase for solubilizing protein and starch and gavimetric measurement of the remaining residue. Although it is fairly rapid, work done at Beltsville (44) has developed a simplified method that uses a single enzyme, only requires one buffer, filtration time is shorter and the data collected on 13 foods correlate to the AOAC procedure with good agreement ($R = 0.998$). Other methods have been developed and complete chapters have been written (1,18,41) addressing the pros and cons of each method. It is now known that the crude fiber analysis should not be used to evaluate fiber sources. It does not represent the enzymatic digestive process and is only useful for ruminant feeds. Table 7 only shows a few suppliers of several fiber sources. Complete listings are available that give fiber sources, TDF values, suppliers, and costs (45–47). These lists contain over 50 fiber sources and 60 producers. Because the selection of a fiber source for specific application can lead to a major new product development program, companies such as Canadian Harvest and the Specialty Fiber Division of Con Agra-DuPont Joint Venture have specialists in fiber selection and application knowledge that can help select the optimum blends. Consumer demand for high fiber products is increasing rapidly as health and nutrition studies continue to show that dietary fiber appear to reduce or prevent some of the diseases or ailments shown in Table 8. The health implications of dietary fiber are widespread. Numerous books and handbooks of scientific publications exist. The American Institute of Baking (48) also has listed health benefits from the insoluble and soluble dietary fibers. Present thinking is that the sol-

TABLE 8

Health Implications of Fiber

Coronary heart disease (CHD)
Hypertension
Colon cancer
Diabetes mellitus
Hyperlipidemia
Gallstones
Weight control
Duodenal ulcers
Diverticulitis
Varicose veins and vein thrombosis

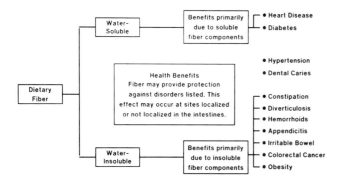

Fig. 2. Fiber classification and health benefits.

uble fibers are associated with decreased risk of coronary heart disease and the insoluble fibers are concerned with cancer, weight maintenance, diverticulosis, etc. (Fig. 2).

Some high fiber products are being marketed in the United States. Dietary fiber is not one entity, but consists of several sources that differ in their chemical and physical characteristics. Although the FDA has not taken a position on specific regulations concerning requirements for labeling, the guidelines being followed in U.S.A. are, any foods containing less than two grams of fiber per serving should not have any fiber claim. Higher levels of fiber, around three grams or more, is needed to justify a high fiber label. The National Cancer Institute has recommended a level of 25–35 grams per day for an American diet. However, the FDA has not supported this RDA value.

References

1. Dietary Fiber, Basic and Clinical Aspects, edited by G.V. Vahouny & D. Kritchevsky, published by Plenum Press, 1986.
2. Trowell, H., *Atherosclerosis, 16:*138, (1972).
3. Trowell, H., D.A.T. Southgate, T.M.S. Wolever, A.R. Leeds, M.A. Gassull, and D.J.A. Jenkins, *Lancet, 1:*967, (1976).
4. Fuwa, H., T. Takaya, and Y. Sugimoto, in "Mechanisms of Saccharide Polymerization and Depolymerization," edited by J.J. Marshall and published by Academic Press, p. 73, (1980).
5. Fleming, S.E., and J.R. Vose, *J. Nutr., 109:*2067, (1979).
6. Palmer, G.H., *J. Inst. Brew., 78:*326, (1972).
7. Englyst, H., H.S. Wiggins, and J.H. Cummings, *Analyst, 107:*307, (1982).
8. Englyst, H.N., and J.H. Cummings, *Analyst, 109:*937, (1984).
9. Shainkin, R., and Y. Brik, *Biochim. Biophys. Acta, 221:*502, (1970).
10. Holm, J., I. Bjorck, S. Ostrowska, A.C. Eliasson, N.G. Asp, K. Larson, and I. Lundquist, *J. Clin. Nutr., 34:*2061, (1983).
11. Anderson, I.H., A.S. Levine, and M.D. Levitt, *Engl. N., J. Med., 304:*891, (1981).
12. Trowell, H., Chapter on dietary fiber in *Dietary Fibre, Fibre-Depleted Foods and Disease*, edited by H. Trowell, D. Burkitt, and K. Heaton, and published by Academic Press, pp. 1-20, (1985).
13. Mattson, F.H., and G.A. Nolen, *J. Nutr., 102:*1172, (1972).
14. Liegwater, D.C., and J.M. Luten, *Die Starke, 23:*420, (1971).
15. Hood, L.F., and V.G. Arneson, *Cereal Chem., 53:*282, (1976).
16. Gordon, D.T., Non-traditional Fiber Sources. Fiber-rich Foods: Their Formulation, Marketing and Nutritional Advantages. Symposium at the University of Delaware, March 12–13, (1985).
17. Torres, A., and R.D. Thomas, *Food Tech., 35:*44, (1981).
18. Dreher, M.L., *Handbook of Dietary Fiber, an Applied Approach*, published by Marcel Dekker Inc., 1987.
19. Sarkanen, K.V., and C.H. Ludwig, *Lignins*, published by Wiley-Interscience, pp. 1–150, (1971).
20. Donner, L.W., and K.B. Hicks, *Lactose and the Sugars of Honey and Maple: Reactions, Properties & Analysis, Food Carbohydrates*, edited by D.R. Lineback & G.E. Inglett, Avi Publishing Co., pp. 83–91, (1982).
21. Layton, R.M., and J.C. Vlazny, *Food Prod. Dev., 12:*53, (1978).
22. Furda, I., Unconventional Sources of Dietary Fiber, Physiological and *in vitro* Functional Properties, ACS Symposium Series 24, pp. 1–305, (1983).
23. Anderson, J.W., and W-J.L. Chen, Cholesterol-Lowering Properties of Oat Products, Chapter 11, in *Oats: Chemistry and Technology*, edited by F.H. Webster and published by AACC, (1986), pp. 309–334.
24. Deane, D., and E. Commers, Oat Cleaning and Processing, Chapter 13 in *Oats: Chemistry and Technology*,

edited by F.H. Webster and published by AACC, (1986), pp. 371–412.
25. U.S. Patent 4,038,481, Method for Treatment of Corn Hulls, by R.L. Antrim and D.W. Harris, (1977).
26. U.S. Patent 4,104,463, Method for Treatment of Corn Hulls, by R.L. Antrim and D.W. Harris, (1978).
27. U.S. Patent 4,239,906, Method for Obtaining a Purified Cellulose Product from Corn Hulls, by R.L. Antrim, Y-C Chan, J.R. Crary and D.W. Harris, (1980).
28. *Corn: Chemistry and Technology*, edited by S.A. Watson and P.E. Ramstad, and published by AACC, pp. 351–398, (1987).
29. Weber, F.E., and V.K. Chaudhary, *Cereal Foods World, 32:*548, (1987).
30. Lo, G.S., R.H. Evans, K.S. Phillips, R.R. Dahlgren, and F.H. Steinke, *Atherosclerosis, 64:*47, (1987).
31. Lo, G.S., S.L. Settle, F.H. Steinke, and D.T. Hopkins, Effects of Soy Polysaccharide Fiber on Lipid Metabolism in Rats, *Fed. Proc., 39:*784, (1980).
32. Lo, G.S., A.P. Goldberg, A. Lim, J.J. Grundhauser, C. Anderson, and G. Schonfeld, *Atherosclerosis, 62:*239, (1986).
33. *The Amer. J. of Clinical Nutr., 41:*1069–1172, (1985).
34. Nelson, J.H., *Am. J. Clinical Nutr., 41:*1070, (1985).
35. Babcock, D., *Cereal Foods World, 32:*538, (1987).
36. Randall, J.M., R.N. Sayre, W.G. Schultz, R.Y. Fong, A.P. Mossman, R.E. Tribelhorn, and R.M. Saunders, *Food Sci., 50:*361, (1985).
37. James, C., and S. Sloan, *J. Food Sci., 49:*311, (1984).
38. *Rice: Chemistry & Technology*, 2nd Edition, edited by B. Juliano and published by AACC, pp. 1–774, (1986).
39. Prosky, L., N-G. Asp, I. Furda, J.W. Devries, T.F. Schweizer, and B.F. Harland, *JAOAC 67:*1044, (1984).
40. Total Dietary Fiber in Foods by Enzymatic Gravimetric Method, First Action, *JAOAC 68:*399, (1985).
41. Prosky, L., and B. Harland, Dietary Fibre Methodology in *Dietary Fibre, Fibre-Depleted Foods and Disease*, edited by H. Trowell, D. Burkitt, and K. Heaton, and published by Academic Press, (1985), pp 57–75.
42. Prosky, L., et al., *JAOAC, 69:*259, (1986).
43. Prosky, L., et al., *Ibid, 69:*297, (1986).
44. Li, B.W., and K.W. Andrews, Simplified Method for the Determination of Total Dietary Fiber in Foods, Accepted for publication on 2-1-88, by *JAOAC*, (1988).
45. *Cereal Food World, 32:*556, (1987).
46. Technical Bulletin, publ. by American Institute of Baking, Fiber Ingredients, Vol. X, Issue 5, (1988).
47. Dietary Fiber Supplements in North America: Challenges and Opportunities for an Emerging Industry, by J.M. Hesser and R.W. Reardon, HRA, Inc., Prairie Village, Kansas pp. 1–150, (1988).

Preparation of Fluid Soymilk

Steve Chen

American Soybean Association, P.O. Box 3512, Taipei, Taiwan ROC 10099

Abstract

Soymilk is a water extract of soybeans, a white colored emulsion resembling cow's milk. Soymilk can be prepared by a simple method of soaking soybeans overnight, followed by wet grinding, filtering, boiling. Then it is ready to serve. Soymilk has been widely consumed in East Asia and it is steadily spreading to the western world. Because of this, sophisticated, modern technology has been used for large scale production of a nonbeany, flavored soymilk. The continuous production systems can control critical parameters, such as taste and nutrition, and achieve a high quality soymilk. The basic steps, various methods and some problem areas of soymilk production are herein described. The nutritional parameters, standards and utilization of soymilk also are presented.

Development of Soymilk

Soybeans are often called the miracle crop and are the world's foremost supplier of vegetable protein and oil. For many years, the soybean has been cultivated for the production of soy foods in the East and animal feeds in the West.

Soymilk has been known in China for centuries. When prepared by a simple, traditional method, it contains a characteristic beany flavor which is acceptable to Chinese of all ages as an enjoyable, nutritious drink. When soymilk is produced by a modern, high technology process, the so-called beany flavor is removed, and it is this nonbeany flavored soymilk which is now more acceptable by Westerners as a natural, healthy drink. In tropical regions, soymilk is used as an inexpensive, refreshing drink, known for its nutritious value as well as being a thirst quencher.

Soymilk is basically a water extract of soybeans obtained by first soaking the soybeans, followed by a wet grinding and filtering process.

Tofu (bean curd) was developed in China by Liu An, King of Huai-Nan of Han Dynasty, in 164 B.C. It is made by the precipitation of soy protein in soymilk by a coagulant, such as calcium sulfate. Because soymilk is the starting product for making tofu, the history of soymilk dates as far back as tofu itself. Soymilk has been used by Chinese for centuries and was produced in thousands of small shops throughout the country. Large-scale soymilk production started only about a half a century ago.

The history of soymilk has been described in detail by Shurtleff and Aoyagi (1). The first real soymilk pioneer in the West was Yu-Ying Li, a Chinese living in Paris. He started the world's first soy "dairy" and was granted the first patent (British) for soymilk production in 1910. An American missionary medical doctor, H.W. Miller, started his first soymilk plant in Shanghai, China in 1936.

The most successful commercial soymilk production, "Vitasoy," was started by K.S. Lo in Hong Kong in 1940. "Vitasoy" soymilk is now produced by Hong Kong Soya Bean Product Co. Ltd. and is marketed not as a milk substitute, but as the world's first soymilk soft drink. By 1974, the sale of "Vitasoy" soymilk reached half a million bottles and surpassed Coca Cola to become the best selling soft drink in Hong Kong.

The development of soymilk continued with the Yeo Hiap Sheng Co. (Vitabean) in Singapore in 1954. In 1958, The Green Spot Co. in Bangkok introduced soymilk into Thailand, whose population did not traditionally consume soymilk.

With assistance from H.W. Miller, the first large-scale soymilk production was started in Taiwan near Taipei in 1962, for the Seventh-Day Adventist Hospital in Taipei. The next large-scale production facility in Taiwan was established by seven members of the Taiwan Farmers' Association producing over three million liters of soymilk in 12 million 260-ml glass bottles. The first modern large-scale commercial soymilk plant in Taiwan has been operated by the President Enterprises Corp. since 1977. By 1982, President Enterprises was producing 300,000 Tetra Briks (250 ml) of soymilk a day during the peak summer months, and in 1983 it sold 75 million packs of soymilk. President Enterprises produces plain soymilk and flavored soymilk, such as egg, milk, strawberry, peanut, chocolate, coffee, orange, apple . . . etc. Among these, egg, strawberry and milk flavored soymilk are the most popular.

Japan and South Korea provide typical examples of the phenomenal growth of soymilk consumption in countries which traditionally did not consume it.

Soymilk has never been regarded as being as important as tofu in the Japanese diet, because of its beany flavor. The undesirable beany flavor of soymilk now can be eliminated through modern food processing technology so that an acceptable soymilk can be produced. In recent years, soymilk has been well accepted by Japanese consumers as a natural, cholesterol-free, health food. Soymilk production in Japan increased from 5,300 MT in 1978 to 137,000 MT in 1984, with an average annual growth rate of 88%. The value of soymilk sold increased from a nominal US$5 million in 1978 to US$146 million in 1984, an average increase of 91% per year. The top five market share holders in 1983 were Kibun Foods (39.1%), Marusan (12.2%), Meiji Milk (12.2%), Mitsubishi Chemical Foods (7.86%) and Yakult (7.0%). Those five soymilk manufacturers held 78.3% of the total soymilk market share in Japan. Currently, the market shares for three types of soymilk in Japan are soymilk, 1%; formulated soymilk, 61%; and soy drink, 38%. The trend of soy beverage consumption in formulated soymilk and soy drink is increasing and in soymilk is decreasing.

As reported by the American Soybean Association in

Seoul, South Korea produced 65,000 MT of soymilk in 1983, a seven-fold increase over the previous four years, with a further increase to 75,000 MT in 1984. In 1987 the soymilk production further increased to 92,000 MT after a setback to 66,000 MT and 67,000 MT in 1985 and 1986, respectively. The major soymilk makers in Korea are Dr. Chung's Food Co., Lotte Chilsung and Dong A Food Co.

Three factors are considered to have contributed to the great momentum of the soymilk boom in the last decade. These are:

Technological Improvements: The major problem in expanding the use of soymilk in the West and in many countries is the elimination of soymilk's so-called beany flavor. Modern food technology produces a quality soymilk more like cow's milk without the beany flavor.

Nutrition: Soymilk contains two essential nutrients, protein and oil, and is highly digestible. Soymilk is the most simple form of soy food and is a good alternative source of nutrition for countries where cow's milk production is insufficient or too expensive.

Health Consciousness: Soymilk contains more protein and less calories than that of cow's milk. In addition, soymilk provides balanced essential fatty acids, lecithin, and is free of cholesterol, lactose and most allergenic factors. Recently, many affluent people have become interested in soymilk because they view it as a natural, healthy food.

Problems of Soymilk Production

There are three major problems to overcome to produce a good quality soymilk (2): elimination of beany flavor, inactivation of antinutritional factors such as trypsin inhibitors, and removal of flatulence-causing oligosaccharides.

Beany flavor

In general, the Chinese tolerate the natural characteristic of a mild beany flavor in soymilk, but Westerners, Japanese, Indians and others do not have such tolerance. Westerners have a strong cultural bias because most of them instinctively compare the flavor of soymilk with that of fresh cow's milk. The natural flavors of soymilk thus mean different things to different cultures depending on their dietary habits and patterns.

The major problem to overcome in expanding the use of soymilk in the West, and in many other countries is the reduction of the so-called beany flavor to develop a bland product more resembling cow's milk. It has been said that soymilk will become a viable commercial product when its flavor and nutritional value are as similar to cow's milk, as margarine is similar to butter.

Nevertheless, unflavored soymilk never will taste like cow's milk, since soymilk is a plant product and its success therefore must be on its own merits. Acceptable soymilk should have no beany/bitter taste and should have a pleasant cereal-like taste.

Wilkens et al. (3) found that the lipoxygenase (or lipoxydase) enzyme present in the soybean is the cause of the beany flavor in soymilk. Soybean lipoxygenase catalyzes the oxidation of *cis, cis* 1,4 pentadiene - containing fatty acids in soybeans, to form 1,3, *cis, trans* hydroperoxides. These hydroperoxides then decompose to form beany or "painty" flavors, such as ethyl vinyl ketone, n-hexanal and pentanal and 1-octene-3-ol (3 & 4). As a result, the lipoxygenase activity is responsible for the undesirable flavors and odors developed upon soybean breakage or grinding.

Lipoxygenase activity can be inactivated by adjusting the pH (below pH 3 or above pH 10) or by heating the beans to temperatures of above 80°C. Lipoxygenase is often inactivated by hot water blanching prior to cracking or hot water grinding of the soybeans to minimize off-flavor development during soymilk preparation. Many methods have been developed to eliminate the off-flavor of soymilk and some of them are described as follows (5):

Hot Grinding (6): A Cornell University group found that simply grinding whole soaked soybeans with boiling water (or steam) to yield a soy slurry having a temperature of 80°C or above, followed by holding the slurry at this temperature for 10 minutes, inactivates the lipoxygenase, thereby preventing the formation of the undesirable beany flavor and producing a fairly bland soymilk.

Blanching (7,8,9): An Illinois university research group discovered that they could blanch soaked soybeans in boiling water for 10 minutes or place dry soybeans directly into boiling water so that bean hydration and lipoxygenase inactivation occurred simultaneously after 20 minutes. In some cases, 0.25 to 0.5% sodium bicarbonate was used in the soaking and/or blanching water. The soybeans were then drained and ground with water to produce a soymilk with little beany flavor.

Rakosky (10) advocated a simple method to remove the beany flavor. This is commonly practiced in Mexico. The process uses three steps of boiling, with the soybeans first boiled for 5 minutes and the water poured off and fresh water added. Then secondly, the beans are soaked for 6 to 8 hours and boiled for a further 5 minutes. The resulting soymilk is simmered for an additional 10 minutes.

Use of defatted soy flour (2): Some research workers have suggested that removing the oil from soybeans will prevent any reaction with the lipoxygenase, thus producing a truly bland soymilk. Steinkraus (11) patented a method for producing defatted soy flour that was suitable for the production of good flavored soymilk. Soymilk made with the resulting soy flour was formulated by adding 2% sucrose and 2.5% refined soy oil and was rated by a taste panel to be as acceptable as cow's milk.

Vacuum deodorization (2): It is possible for soymilk to be passed through a vacuum pan at a high temperature such as 0.526 mm/Hg, 70°C to 115°C (12) in order to remove most of the volatile off-flavor components.

A new deodorizing of beany flavor of soymilk was described in a Japanese patent (13) as follows: Spray saturated steam to whole soybeans for 1–10 minutes, then dry with 100-160°C hot air followed by cooling. Then grind dehulled soybeans with hot water and separation. Under vacuum, squirt soymilk with 120-160°C steam for 1–10 seconds, then deaerate under vacuum.

The use of soy protein isolates or soy protein concentrates: Re-suspending a soy protein isolate (90% minimum protein) or a soy protein concentrate (70% minimum protein), can result in a bland soymilk. The taste of soymilk prepared from spray-dried soy isolate or soy concentrate may not be the same as the one prepared from whole soybeans because a cooked, cereal-like flavor is a characteristic of soymilk prepared from soy isolate and concentrates. An acceptable soymilk can be prepared simply from soy isolates and concentrate, by adding water, emulsifier, refined vegetable oil, sweeteners, flavors, vitamins and minerals. The soymilk thus prepared contains no flatulence-causing oligosaccharides.

Flavor formulation: The beany flavor of soymilk can be masked and the flavor improved by the addition of flavorings and sweeteners, such as vanilla, milk, egg, chocolate, honey, etc.

Alkaline soaking: Nelson et al. (8) reported that soaking and/or blanching the soybeans in an alkaline solution (0.5% $NaHCO_3$) helps to improve the soymilk flavor, remove oligosaccharides and decreases the cooking time. Badenhop (14) found that one of the beany flavor compounds (1-octen-3-ol) may be reduced by alkaline soaking.

Dehulling (2): Many commercial soymilk producers are dehulling their soybeans in an attempt to improve the flavor of soymilk, but, others find that whole soybeans yield the best-flavored soymilk.

Rapid Hydration, Hydrothermal cooking and UHT method: Further improvements in the flavor of soymilk can be achieved by slurrying soy flour in hot water or by inline dispersement of soy flour which minimizes its time in contact with water, and then by hydrothermal cooking at 154°C for 30–40 sec. (15,16). The ultra high temperature (UHT) sterilization of soymilk by steam injected into soymilk (or soymilk injected into the steam) at 140°C for 4 seconds also was reported to improve soymilk flavor.

Trypsin Inhibitors (17)

Trypsin inhibitors present in raw soybeans must be inactivated to facilitate the digestion of proteins in the human body.

From a practical standpoint, trypsin inhibitors do not appear to be a serious problem in foods and feeds since they are largely inactivated by moist heat. Conditions of heating-time, temperature, moisture content and particle size which influence the rate and extent of inactivation in trypsin inhibitors are presented in Table 1. Van Buren et al. (18) indicated that trypsin inhibitors in the soymilk could be inactivated through heating at 100°C for 14 to 30 min, or at 110°C for 8 to 22 min.

Flatulence factor—Oligosaccharides

Whole soybeans contain 5% sucrose and 5.1% of other sugars (arabinose, glucose, verbascose etc). In addition, soybeans also contain two oligosaccharides, raffinose (1.1%) and stachyose (3.8%). Soybeans as well as many other legumes cause flatus when ingested. Studies (17) indicate that flatus is caused by the fermentation of low molecular weight sugars—raffinose (galactose-glucose-fructose) and stachyose (Gal-Gal-G-F)—which are not digested because humans do not have α-galactosidase activity in their digestive tract. When soymilk with low, medium and high levels of raffinose and stachyose was fed to rats, flatus production corresponded to the level of the oligosaccharides present (20).

Oligosaccharides can be removed by alkaline soaking of soybeans (8), heat treatment, removal of soy fiber residue and enzyme treatment (21). Sugimoto and Van Buren (21) used an enzyme system of α-galactosidase and invertase to remove oligosaccharides.

Both the Alfa-Laval and Soya Technology System describe using ultrafiltration to remove oligosaccharides. The STS also advocated that low molecular solutes, such as antitrypsin, lipoxidase, and phytic acid also can be removed by the ultrafiltration process.

Basic Steps of Soymilk Production

Preparation of soymilk by small-scale cottage industries requires a minimum of five steps (2): cleaning of the soybeans, soaking, grinding, filtering soy residue and cooking the soymilk.

The modern, high technology, large-scale soymilk industries require much more sophisticated methods and many more steps to prepare soymilk. Generally, it includes cleaning soybeans, dehulling, soaking, blanching, grinding, filtering, cooking, deodorizing, formulating, homogenization, sterilization and packaging.

Basic steps involved in soymilk production are discussed as follows:

Soybean Varieties

There are many varieties of soybeans grown in different areas. Even when the same variety of soybeans is grown

TABLE 1

Inactivation of Trypsin Inhibitors (TI) (16,19)

Soy product	Inactivation conditions	TI Inactivated	Protein efficiency
Raw defatted soybean flakes	100°C, 15 min	95%	
Soybean flakes (19% moisture)	100°C, 15 min	97%	2.04
Soybean flakes (5% moisture)	100°C, 15 min	97%	1.87
Whole soybeans (20% moisture)	100°C, 15 min	Most of TI	
Whole soybeans (60% moisture)	100°C, 5 min	Most of TI	
soybean slurry	100°C, 30 min	90% TI	
soybean slurry	110°C, 22 min	90% TI	

* Protein efficiency for casein is 2.15.

in the same location, the quality of soybeans often may differ due to various weather, soil and cultivation conditions. The raw material of soybeans affects the quality of soymilk produced. It is essential therefore, for soymilk producers to select available soybeans that will produce soymilk with good flavor, color and the best yield (recovery of protein and solids from soybeans in the soymilk). Studies conducted on soybean varieties from Canada, the Philippines and the United States (2) showed that soymilk yield varied widely from one soybean to another. Typically, the protein recovered in soymilk is in the range of 70 to 80% and the average solids recovery is 55 to 65%. Fresh, white hilum, and high protein variety soybeans will produce soymilks with a high yield. Currently, the majority of soymilk manufacturers are using cleaned U.S. Grade No. 2 yellow soybeans (22) for the production of soymilk, with satisfactory results. In addition to using whole soybeans as raw material, dehulled soybeans, low temperature defatted soybeans, soy flour and soy protein isolate can also be used as raw materials for soymilk production.

Cleaning

Commercial soybeans contain dirt, dust and microorganisms on the surface of the beans; therefore, soybeans should be thoroughly washed. Foreign materials, such as straw, stones, metals and weeds should also be removed. Just as important is the elimination of damaged soybeans. This is necessary because the enzyme lipoxygenase, which is present in the soybean, will act on the unsaturated fatty acids in the damaged soybean cell tissue, producing compounds with the characteristic beany flavor.

Dehulling

Dehulling refers to the removal of the outer seed coat or hull from soybeans which constitutes about 9% of whole soybean by weight. Soybeans are dehulled dry with a burr mill (Bauer Bros.) or a stone mill. A low technology soybean dehuller is described by Fiering (23) and Shyeh et al. (24).

The soybean dehulling process consists of three steps. First, a heat treatment (93°C for 15 min) is required to break the bond between hulls and cotyledons for the most effective and efficient dehulling. The second step is the cracking of the soybean and in the third step the hulls are separated from cotyledons usually by air aspiration with a final dehulling efficiency of up to 88%.

A unique cracking device and dehulling procedure was developed by Nelson et al., and it is covered by U.S. patent 3,981,234 owned by the University of Illinois Foundation. The license for the production of the equipment is owned by Dublin Corporation, 3973 Grove Ave., Gurnee, IL 60031 USA.

Most dehulling is done dry, but wet dehulling is possible also. No preheating is required and the hulls may be floated off in running water, after proper soaking.

Good quality soymilk can be made without dehulling. In fact, the hulls can aid in soymilk filtration by preventing clogging of the processing sack. Dehulling requires an additional operation and equipment, but the advantages of dehulling include: improved protein recovery (yield); improved soymilk flavor by eliminating a beany flavor and subtly bitter component; improved digestibility; slightly whiter soymilk; reduced oligosaccharides, soaking time, grinding energy and heating time; reduced bacteria count coming from the hulls, resulting from a soymilk with better flavor and longer shelf-life; and reduced soy residue (okara) to a minimum.

Soaking

Soaking can be done in cold or hot water, but cold water soaking is preferable as there is less total solids loss. Hot water soaking may inactivate enzymes, but longer hot soaking may promote browning reaction, greater loss of solids (about 2%) and carbohydrates, and can cause water pollution problems (12). Soymilk can be made from either soaked or dry soybeans. Soybeans are soaked in about three times their weight of water with a soaking time that is dependent upon the temperature of the water. The colder the water the longer the required soaking time. Typically, the soaking time is 8 to 10 hours at 20°C in the summer and 14 to 20 hours at 10°C in the winter. Soybeans can also be soaked in circulating, overflowing water continually to rinse them throughout the soaking period. After proper soaking, soybeans will expand about 2.0 to 2.3 times their original weight.

Most commercial soymilk production calls for the soaking of soybeans since soaking reduces the energy input required for grinding, enables better dispersion and suspension of the solids during extraction, increases yield and decreases cooking time.

Various researchers have studied the effect of soaking and/or blanching soybeans in an alkaline solution containing sodium bicarbonate, sodium citrate and other bases. The alkaline solution containing sodium ions is reported to decrease but not eliminate the beany soymilk flavor, and to tenderize the soybeans which in turn shortens cooking time and improves homogenization. Oligosaccharides also are reduced and the inactivation of the soybean trypsin inhibitor is accelerated.

When the soaking temperature is above 45°C, there is a large decrease in total solids (from 60% down to 50%) and carbohydrates (15% to 10%), and a small decrease in the recovery of proteins and fats. The longer the soaking, the greater the loss of soybean components (25). Soaking dehulled soybeans for 24 hours results in a 5% loss of solids (26) as compared with 1.5% loss for whole soybeans (27). The bulk of solids lost are water-soluble carbohydrates (73.2% of total carbohydrates, including 5% of flatulence-causing oligosaccharides).

Dehulled soybeans reach full hydration much faster than whole soybeans by requiring only 2 to 3 hrs at 30°C or 1 hr at 50°C as compared to whole soybeans which require 7 hrs at 30°C or 3–4 hrs at 50°C.

Blanching

Blanching or steaming of soybeans in a solution of sodium bicarbonate at high temperatures inactivates the beany flavor caused by the lipoxygenase enzymes. Blanching also washes out water-soluble oligosaccharides (flatulence-causing) and inactivates trypsin inhibitors (which reduce protein digestibility).

Grinding

Soaked soybeans may be ground with hot or cold water by using a stainless steel horizontal or vertical grinder or disintegrator (Rietz), hammermill (Fitzpatrick), pin mill or large blender (Waring). For soymilk production, a vertical grinder is preferable (12). When the rpm is higher than 1450, it is possible to make a 150 mesh colloid solution. When grinding in a hot water solution above 80°C, the lipoxygenase will be inactivated to prevent the formation of an undesirable beany flavor.

Filtering

The insoluble soy residues (okara) are removed from the soy slurry by a decanter centrifuge to improve flavor, mouth feel and to dispose of oligosaccharides.

A roller with screen and press is used in the separation of soymilk for tofu production. This rough separation causes precipitation of solid matter in the soymilk and poor mouth feel and texture. Therefore, a screen decanter and a clarifier is recommended for quality soymilk separation (12).

Cooking or Heating

The purpose of cooking soymilk is to destroy microorganisms that cause spoilage, improve flavor, and improve nutritive quality by inactivating trypsin inhibitors (TI). Trypsin, a proteolytic enzyme produced by the pancreas, plays a key role in digesting protein in the human body. Trypsin inhibitors, which are present in raw soybeans, inhibit trypsin activity and the growth of animals and can cause enlargement of the pancreas (hypertrophy). Rackis (28) however, reported that no pancreatic hypertrophy occurred in rats fed with soy flour in which 55% to 69% of the trypsin inhibitors had been inactivated. Hackler (29) also studied the effects on rats given soymilk which has been heated to 93°C and 121°C for various lengths of time. He did this by determining the difference in the protein efficiency ratio (PER) as related to the percentage of TI in soymilk. The results showed that the maximum nutritional value (PER) for soymilk was obtained when 80% of the trypsin inhibitors were inactivated. Van Buren et al. (18) indicated that 80% to 90% of trypsin inhibitors were inactivated by heating the soymilk to 100°C for 14 to 30 min, or to 110°C for 8 to 22 min. Miller (30) recommended heating soy slurry or soymilk for about 30 min at 100°C to obtain optimum nutritional value and flavor.

Heating is used not only to inactivate the antinutritional factors present in soymilk, but also to lower its viscosity which increases extraction and gives higher yields of protein and solids.

Deodorization

Soymilk can be passed through a vacuum pan at high temperature (0.526 mm/Hg, 115-70°C) to remove most of the volatile off-flavor components.

Formulation

One key to the increased acceptance of soymilk is that sweetening and flavoring agents have been used to suit local tastes. Both natural and artificial flavorings have been used successfully. These added flavors include vanilla, milk, egg, strawberry, apple, peanut, chocolate, coffee, almond, orange (Taiwan), malt (Hong Kong), Pandan (Singapore, Malaysia), fruit, vegetable, beef, yakult, lactic acid, honey, sesame (Japan) etc. The addition of soy oil, lecithin or an emulsifier to soymilk improves its richness, creaminess and mouth feel. Additional calories are provided when soymilk is enriched with oil, and this factor may be important in many developing countries. Generally, a homogenizer is required when oil is added to soymilk.

Many commercial soymilk producers are mixing soymilk and nonfat dry milk powder with soy oil or adding 20% cow's milk to improve the flavor of soymilk. Actually, soymilk can be used as a cow's milk extender to lower milk production costs or to increase its supply. Soymilk can be nutritionally fortified with vitamins and minerals. The most widely used fortifiers are vitamin B_{12} (cyanocobalamine) and calcium. The mineral most widely used for enrichment is calcium (such as calcium carbonate, mixtures of calcium citrate and tricalcium phosphate), because soymilk contains only 18.5% as much calcium as cow's milk and 53% as much as mother's milk (2).

If the batch system of the blending formulation is used, it may cause fat rancidity—and so it is more desirable to use the continuous system with a line mixer in the pipe line (12).

Homogenization

Homogenization is a process which breaks down fat globules into very fine particles which remain dispersed in the soymilk without separation. Homogenization also disperses any solids which might tend to settle to the bottom of the container. Homogenization thus will make soymilk creamier or more uniform in consistency. For most soymilks, one pass at 2000 to 3500 psi at 90°C in a dairy type homogenization is sufficient to give a good product.

Some soymilks exhibit a chalky taste resulting from a coating of the mouth and throat by fine, grainy particles. The main cause of such chalkiness is the relatively large particles (>150 mesh) contained in the soymilk. Chalkiness often occurs in soymilk manufactured by blanching, whole-bean or hot grinding methods. Chalkiness can be reduced (31) by increasing the alkalinity of the blanching solution, by adjusting the final pH of soymilk to about 7.5, by reducing the soy solid content to make a more dilute soymilk, or by homogenizing at a high temperature and pressure.

Sterilization and Pasteurization

Soymilk is an ideal medium for microbial growth, and therefore, it can spoil easily due to the survival of spoilage microorganisms if inadequate heat treatment is used in processing.

There are three basic types of heat treatments to prolong the shelf life of soymilk. These are namely, pasteurization, sterilization and ultra high temperature treatment (UHT).

Pasteurization at 75°C for 15 sec (typical) is designed to destroy pathogenic microorganisms and also greatly reduces the total microbial count which improves the keeping quality of soymilk. Pasteurized soymilk in Pure Pak or plastic bottle containers requires refrigeration of 4°C to maintain a typical one-week shelf life.

Sterilization in a retort or autoclave is a high temperature heating process (121°C for 15 to 20 mins) is designed to kill all heat resistant microorganisms and spores in soymilk. Such sterilized soymilk in tin cans or glass bottles requires no refrigeration and can maintain a shelf life of six months or more.

Ultra high temperature (UHT) sterilization is done by steam injection of soymilk at 140°C for 4 sec followed by aseptic packaging. The shelf life of UHT sterilized and aseptically packaged soymilk may be three months or more at room temperature. The UHT treatment/aseptic packaging system requires expensive and sophisticated equipment, but it provides a soymilk with an adequate shelf life, and good flavor, while retaining its nutritional value.

There are basically two types of UHT treatment based on either indirect or direct heating. Indirect heating is carried out without contact of soymilk with water or ultra high-temperature steam, such as ultramatic/APV (U.K.), Steritherm/Alfa Laval (Sweden), Steridea/Stork (Holland). Direct heating systems are divided into two types as follows: steam is injected into the soymilk Uperizer/APV and VTIS/Alfa Laval; or soymilk is injected into the steam Flow Sterilizer/Laguilharre (France).

The main purpose of the high temperature heating of soymilk is to deodorize off-flavors, to remove inhibitor, improve digestibility and prolong shelf life. But heating for a longer time may cause a browning reaction and cooking flavor of the soymilk.

Packaging (32)

Selecting proper types of packaging is one of the most important decisions for the successful commercial production of soymilk. In the most simple case, soymilk can be produced inexpensively on a small scale and consumed directly without any packaging or provided in bulk containers for institutional use. On the other hand, soymilk also can be produced on a large scale by high technology, sophisticated and expensive operations such as UHT and aseptic packaging.

Soymilk can be distributed in bulk or in individual containers of various types with or without refrigeration. These are listed in Table 2. In Japan in 1982, 72% of the soymilk was packaged in Tetra Briks; 13% in tin cans; 7% in Pure Paks and 8% in other containers such as glass and plastic bottles.

In Taiwan, soymilk is mostly consumed without packaging. Commercial soymilk production accounted for only about 30% of all the soymilk produced in 1987, with 75% in aseptic Tetra Briks and the other 25% in glass or plastic bottles and tin cans.

In Asia, the most widely used packaging for commercial soymilk production is aseptic Tetra Briks. In the northern part of Asia (Japan and South Korea), 200-ml Tetra Brik packages are mostly used, but in Southern Asia (subtropical and tropical regions) like Hong Kong, Taiwan, Thailand, Singapore and Malaysia, 250-ml Tetra Briks are used.

Methods of Soymilk Production

The principles of soymilk production (33) are aimed at producing a delicious, nutritious, good quality soymilk with high yield at a low cost. Soymilk processors can use one or a combination of the following methods to achieve their objectives.

A summary of five different soymilk processes is presented in Figure 1 for comparison. Because there are many different methods for the production of soymilk, several of these methods are shown in Figures 2 through 9.

Process	Traditional	Cornell	Illinois	RHHC	Alfa-Laval
	(2)	(3,6)	(7,8,9)	(15)	(34)
	SB ↓ SOAK ↓ COLD GRIND ↓ COOK FILTER (100°C/ 20 MINS) ↓ FILTER COOK	SB ↓ SOAK ↓ HOT GRIND (>80 C) ↓ COOK (100°C/ 10 MINS) ↓ FILTER	SB or DSB ↓ ALKALINE SOAK ↓ BLANCH (100°C/10-20 MINS) ↓ GRIND ↓ COOK (to 82°C) ↓ HOMO	SB ↓ GRIND TO FLOUR ↓ SLURRY in H₂O ↓ COOK (154°C/ 30 SECS) ↓ COOL ↓ FILTER	SB or DSB ↓ HOT GRIND ↓ FILTER ↓ COOK ↓ DEODO-RIZE
BEANY FLAVOR	STRONG	IMPROVED	NONE	NONE	NONE
SOLIDS YIELD	61% (55-65%)	65%	89%	86%	-
PROTEIN YIELD	73% (70-80%)	83%	95%	90%	70%

Fig. 1. Summary of soymilk process. SB = cleaned soybeans; DSB = dehulled, cleaned soybeans; RHHC = rapid hydration hydrothermal cooking.

TABLE 2
Soymilk Packages (32)

Container	Size (ml)	Treatment	Refrigeration	One-way Container
Tetra Brik	200, 250	UHT/aseptic	No	Yes
Tin can	250, 350	Sterilized	No	Yes
Glass bottle	260	Sterilized	No	No
Pure pak	250, 500, 1000	Pasteurized	Yes	Yes
PE, PP bottle	350	Pasteurized	Yes	Yes

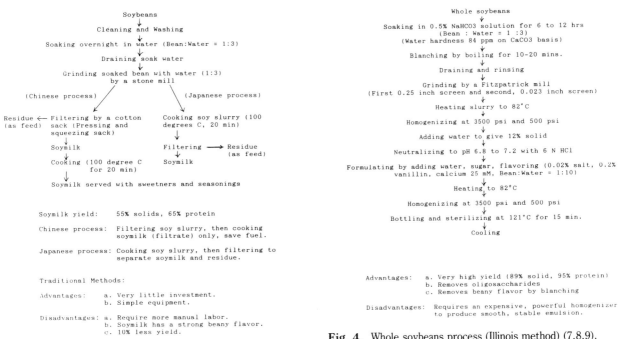

Fig. 2. Traditional Oriental process.

Fig. 4. Whole soybeans process (Illinois method) (7,8,9).

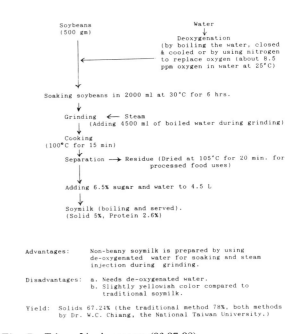

Fig. 3. The hotel water grind process (Cornell method) (3,6).

Fig. 5. Taiwan Liaw's process (36,37,38).

Yields of Soymilk

In typical soymilk production, 70 to 80% of the protein and 55 to 65% of the solids from the soybeans are recovered in the soymilk.

The yields of soymilk may be expressed in several ways. Losses of total solids and of individual components such as carbohydrates and proteins occur in soymilk processing and thus have an impact on the final quality and yield of the soymilk.

The first loss normally occurs during soaking where there is a loss in total solids, carbohydrates and protein, depending upon the method and time of soaking.

Grinding of the soaked soybeans normally does not cause any decrease in any of the above components and basically they are the constituents of the soaked soybean and are in the resulting slurry, preventing any mechanical loss such as spillage. Cooking of the slurry, where employed, also does not contribute to any loss.

The step of filtration or decantation on the other hand,

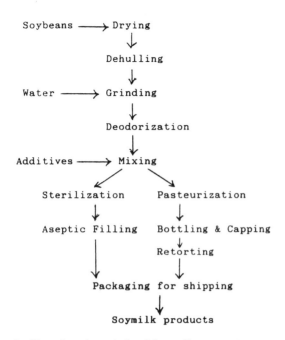

Fig. 6. Flow sheet for an industrial soymilk process (35).

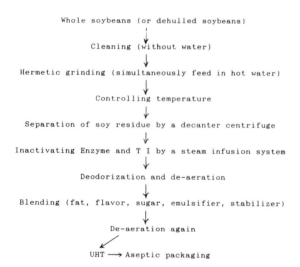

Fig. 7. Alfa-Laval Soyal process. Details available from Alfa-Laval Food, Soy Application Group, P.O. Box 64, Lund, Sweden.

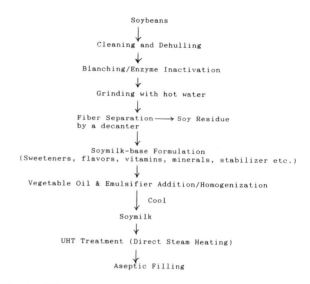

Fig. 8. STS Modern soymilk process. Details available from Danish Turnkey Darries Ltd., Soya Technology Division, PO Box 146, Aarhus, Denmark.

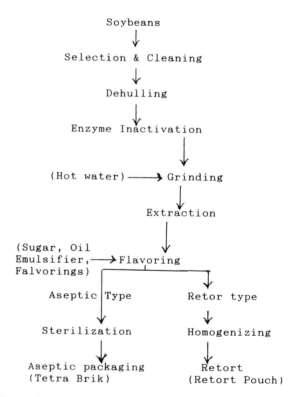

Fig. 9. Takai modern soymilk production. Details available from Takai, Tofu & Soymilk Equipment Co., 1-1 Inai, Nonoichi-machi, Iskikawa-ken 921, Japan.

does contribute to loss of constituents with a consequent negative impact on the soymilk yield. The losses are attributable to the insoluble matter retained on the filter media or removal by decantation. These residues contain some soluble materials depending upon the obtained degree of dryness of the filtered residue and upon the efficiency of filtration.

Once the filtrate is obtained, subsequent losses are again only those due to spills, leaks or other mechanical losses.

The loss profile also will be affected by whether or not the soybeans are dehulled and at what step in the process this is done. For the processes mentioned above, the "dehulling" in effect occurs in the filtration step and this may or may not be more efficient than dry dehulling, depending on the efficiency achieved in either filtration or dry dehulling.

As a final consideration the total solids and/or total protein content of the final soymilk are affected greatly by the amount of these constituents desired in the final soymilk and are controlled mainly by the amount of water used in the process, in relation to the initial amount of soybeans started with.

Normally, the weight of soymilk obtained by the traditional processes is about 6 to 10 times that of soybeans processed. More modern and innovative processes are now available which will allow better yields whatever the target values may be for the final soymilk composition.

In studying yield claims for proposed processes, it is advised that the considerations given be kept in mind and that the basis for calculation of yields be clearly delineated for thorough understanding.

Soymilk Standards and Quality Control

A key factor in determining the quality and cost of soymilk is its richness or thickness, which depends largely on the weight of water added per weight of dry soybeans used (see section on soymilk yields).

In general, higher solids (rich) soymilk is considered to have the best flavor and its cost is proportional to its higher solids or richness (2).

The most common way of expressing the amount of water used in soymilk preparation is by the water:beans ratio, which is the total weight of water added during soymilk preparation (including soaking water) divided by the weight of dry soybeans used.

Soymilk used as a beverage can be divided into three basic consistencies, as follows: (a) higher solids or rich soymilk, (b) dairy-like soymilk and (c) lower solids or economy soymilk as shown in Table 3. It was reported by Shurtleff and Aoyagi (2) that rich soymilk can be further processed into silken tofu or yuba in Japan and dairylike soymilk is best suited to American tastes. Economy soymilk is excellent for use as a soymilk soft drink.

For making the best flavored soymilk, Shurtleff suggests using well or spring water, but did not mention water hardness (2). It may not be practical for commercial soymilk production to use well or spring water. The Illinois soymilk process (7) uses fresh tap water with 84 ppm $CaCO_3$ hardness.

Soymilk Standards

In order to ensure that consumers enjoy quality soymilk and also to discourage the production of a very diluted soymilk, it is important for each country to establish soymilk quality standards. Soymilk standards for Taiwan, Japan, Thailand and Singapore are presented in Table 4 for reference. "Soymilk" was legalized for use in Canada Nov. 10, 1984. Soymilk can be labelled according to Japan Agricultural Standards (JAS), as set up on Nov. 16, 1981, and in Taiwan based on the China National Standards (CNS) established on Nov. 20, 1984. The original JAS standards dated Nov. 16, 1981, for soymilk, formulated soymilk and soy drinks, specified that they be made from whole soybeans. The JAS standard was revised on Oct. 5, 1985, to include soy beverages made from soy proteins, with the provision that they contain at least 1.8% protein from the soy proteins.

TABLE 3

Three Basic Consistencies of Soymilk

Soymilk	Water:beans ratio	Cups/Kg beans	Solids	Protein	Fat
Rich	5:1 or 6:1	15–19	10–11.5%	4.5–5.2%	2.8–3.2%
Dairylike	8:1 or 8.5:1	28–31	7.4–8%	3.3–3.6%	2.1–2.3%
Economy	10:1	37–40	6%	2.7–3.3%	1.2–1.6%

TABLE 4

Soymilk Standards for Japan, Taiwan, Singapore and Thailand (Minimum Requirement)

Country	Product	Protein	Fat	Soybean solids
Japan	Soymilk	3.8%		8%
	Formulated soymilk	3.0%		6%
	Soy drink	1.8%		4%
	Soy protein beverage	1.8% (0.9%, if added more than 5% fruit juice)		
Taiwan	Soymilk	3.4%	2.0%	
	Formulated soymilk	2.0%	1.0%	
	Soy drink	1.4%	1.0%	
Singapore	Soymilk	2%		
	Soy drink	Less than 2%		
Thailand	Soymilk	2%	1% (from Soybeans)	

Quality Control

Quality control procedures for soymilk are not yet well established. In the United States, L.S. Wei of the University of Illinois has suggested that the specifications and quality control procedures for soymilk should follow the model used in the U.S. dairy industry. Taiwan followed exactly the same approach by adopting the CNS analytical inspection methods of dairy products for Taiwan soymilk industry.

In general, seven items are checked in the soymilk plant for their routine quality control purposes: crude protein by Kjeldahl method, crude fat by Gerber's method, bacteria test by plate count method, total solids by oven method, acidity by titration or pH meter method, specific gravity by Quevenne Lactometer or hydrometer and moisture content by Drying method. In addition, water quality for soymilk preparation is maintained by using ion exchange resins to soften hard water.

Cost and Price of Soymilk Production

The material balance of the soymilk plant (2000 L/hr, 3%–3.5% protein content) is presented in Table 5 (39). When using 335 Kg of U.S. No. 2 yellow soybeans as raw material for soymilk production, after cleaning and dehulling, the cleaned, dehulled soybeans ready for processing soymilk are 245 Kg or 73% (by weight) of the original raw material. The final 2000 L soymilk prepared, will give the clean beans to water ratio of 1:8.2 and a protein content of 3% to 3.5%.

TABLE 5

Material Balance of the Soymilk Plant (39)

Item	Kg/hr
Soybean (U.S. No. 2 yellow)	335
Waste raw beans (from cleaning and dehulling)	−90
Cleaned, dehulled soybeans ready for processing	245
Absorbed and added water (during soaking and grinding)	1625
Soy slurry produced[a]	1870
Separation of soy residue	−370
Soymilk-base produced	1500
Additive for standard formulation	500
Soymilk produced[b]	2000

[a] Clean beans:Water = 1:8.2.
[b] Protein: 3–3.5%.

Soymilk production cost in Taiwan, including raw materials, packaging, other costs (manpower, utilities, interest, advertising and depreciation) are presented in Table 6. It can be seen that the raw materials cost is US $0.14/L or 29% of the total soymilk production cost of US $0.48/L. Other costs are about US $0.12/L or 25% of the total cost. The major cost is for packaging, which costs US $0.22/L in Tetra Brik (250 ml) or 46% of the total soymilk production costs. Obviously, the key to lowering the cost of soymilk production is to find a more economical way to package the perishable soymilk.

TABLE 6

Taiwan: Soymilk Production Cost[a]

Item	US Cent/L
Soybeans & other ingredients	14.0 (29%)
Packaging cost (Tetra Brik)	22.0 (46%)
Other costs (manpower, utilities & maintenance)	12.0 (25%)
Total soymilk product cost	48.0 (100%)
or U.S. Cent/250 ml	12

[a] ASA estimate.

In Taiwan, the raw materials for soymilk production is estimated at US Cent 14/L which is only 21% of the cost of fresh cow's milk (US Cent 66/Kg) paid by the local milk processing plants to dairy farmers. The retail price of soymilk or flavored soymilk in Tetra Brik in Taipei is US Cent 35/pak (250 ml). Because of this, the gross profit for soymilk operation is about 66% in Taiwan.

Most large soymilk manufacturers reported that their total cost of soymilk production, pasteurization and homogenization (excluding filling and containers) is about one-third to one-half the farm price of cow's milk. (5)

The equipment cost for soymilk production for a simple, small-scale (500 L/hr) operation is about US $42,000 (FOB/Taiwan) while a small-scale soymilk beverage production plant (500 L/hr with 280 CC/bag of plastic pouch filter and sealer) by Takai Tofu & Soymilk Equipment Co. (1-1, Inari, Nonoichi-machi, Ishikawa-ken 921, Japan, is about Japanese Yen 8.2 to 10.7 million (or US $63,077 to $82,308). The Takai modern soymilk production plant with filter and sealer for a capacity of 2000 L/hr (soy protein 3.5%) will cost for cartons, Japanese Yen 313 million (or US $2.41 million); plastic pouches, Japanese Yen 267 million (or US $2.05 million); retort standing pouches, Japanese Yen 389 million (or US $3 million) and Tetra Briks, Japanese Yen 436.5 million (or US $3.36 million) at an exchange rate of US $1/Japanese Yen 130 in June, 1988. When the cost of land (about 10,000 to 20,000 M square or 2.5 to 5.0 acres), building (about 3,000 M square or 33,000 ft square), working capital, inventory and operational cost are included, the total investment can be as high as US $6–7 million or more.

Soymilk Utilization

Fluid soymilks can be divided into plain, and flavored varieties.

When a company is planning to produce soymilk, it should first decide what types of soymilk and packaging are best suited for its market: Plain soymilk (Contains only soybeans and water.) Flavored soymilks are mixed with sugar and flavors to suit local tastes. The flavored soymilks include vanilla, chocolate, coffee, cocoa, strawberry, malt, fruit and vegetables, sesame, peanut, corn, beef, apple, milk, and egg. Soymilk can also be made by adding a little salt (0.1%), vegetable oil (2–3% soy oil), and fortifying with L-methionine, or adding calcium etc. In most food items which use dry milk powder or fresh milk, these may be replaced by soymilk powder or soymilk. Therefore, there is a great market potential for low cost soymilk or powdered soymilk for bakeries, ice cream makers, and confectioneries.

Other soymilk based products In addition to the plain and flavored soymilk, soymilk can be used to make a second generation of soy foods, such as soy ice cream, soy shakes, soy yogurt, soy pudding, soy dipping, soy mayonnaise, soy cheese, soy salad dressing, soy whipped cream, soymilk-based infant formulas, and other dairy-like products.

Nutrition of Soymilk

It is generally conceded that our world faces a growing food crisis which is a shortage of both calories (energy) and protein. As a protein food that makes fullest use of soybeans, soymilk offers a revolutionary, yet simple, approach to help alleviate the growing world's food shortage (39). Soybeans are a miracle crop which provides two of the most important nutrients—protein and oil (Table 7). Soy proteins are high in quantity as well as in quality and soy oil is the most concentrated energy source. Soybeans rightfully can claim to be the answer to the world food shortage as the right product at the right time with the right price. Unfortunately, only 12% of world soybeans production currently is used directly as soy foods; the remaining 88% of soybeans being used as livestock and poultry feed (40).

The usual benefits attributable to soymilk are as follows: no lactose, cholesterol free, highly nutritious, easily digestible, nonallergenic, low cost, simple technology, versatile applications, and no religious taboos.

The production of a nutritious, organoleptically-enjoyable, pleasant-tasting, good-quality soymilk for human consumption is thus considerably more complicated than just grinding the soybeans with water. It is considered an art as well as a science to produce good-quality soymilk (39). Certain factors have to be controlled during the process, such as: inactivation of lipoxygenase enzyme which causes beany off-flavor development; elimination of flatulence-causing oligosaccharides; maintenance of high protein efficiency ratio (PER, a measure of protein quality); and, removal of other volatile off-flavors.

The composition of soymilk, cow's milk and mother's milk is presented in Table 8. Comparing these three protein beverages in Japan, soymilk is found to contain more protein, iron, unsaturated fatty acids, and less calories,

TABLE 7

Approximate Composition of Soybeans (Dry weight basis)

Component	Per cent
Protein	40
Lipid	20
Cellulose and hemi-cellulose	17
Sugars	7
Crude fiber	5
Ash	6

TABLE 8

Composition of Soymilk, Cow's Milk and Mother's Milk

Item/100 g	Soymilk	Cow's milk	Mother's milk
Calorie	44	59	62
Water (g)	90.8	88.6	88.2
Protein	3.6	2.9	1.4
Fat	2.0	3.3	3.1
Carbohydrates	2.9	4.5	7.1
Ash	0.5	0.7	0.2
Minerals (mg)			
Calcium	15	100	35
Phosphorus	49	90	25
Sodium	2	36	15
Iron	1.2	0.1	0.2
Vitamins (mg)			
Thiamine (B1)	0.03	0.04	0.02
Riboflavin (B2)	0.02	0.15	0.03
Niacin	0.50	0.20	0.20
Saturated fatty acids (%)	40–48	60–70	55.3
Unsaturated fatty acid (%)	52–60	30–40	44.7
Cholesterol (mg)	0	9.24–9.9	9.3–18.6

Source: Standard Tables of Food Composition (Japan)

fat, carbohydrates, and calcium than cow's milk and mother's milk. The great merit of soymilk is that it is highly digestible and contains no cholesterol, no lactose, and is rich in polyunsaturated essential fatty acids and lecithin.

References

1. Shurtleff, W., and A. Aoyagi, "Soymilk Industry & Market," The Soyfood Center (1984), P.O. Box 234, Lafayette, CA 94549, USA.
2. Shurtleff, W., and A. Aoyagi, "Tofu and Soymilk Production - The Book of Tofu," Vol. II, Chapters 5 & 11 (1979) by The Soyfood Center.
3. Wilkens, W.F., L.R. Mattick and D.B. Hand. *Food Tech.*, 21:1630 (1967).
4. Johnson, L.A., "Soy Protein: Chemistry, Processing and Food Applications" presented at 70th annual meeting of Am. Assoc. Cereal Chemists, Sept. 1985.
5. Eriksen, S., *J. Food Sci.* 48:445 (1983).
6. Hand, D.B., K.H. Steinkrans, J.P. Van Buren, L.R. Hackler, I. El Rawi and W.R. Pallesen, *Food Tech.* 23:1963 (1964).
7. U.S. Patent 3,901,978 (1975).
8. Nelson, A.I., M.P. Steinberg, and L.S. Wei, *Food Sci.* 41:57 (1976).
9. U.S. Patent 4,041,187 (1977).
10. Anonymous, *J. Am. Oil Chem. Soc. 61*:1784 (1984).
11. U.S. Patent 3,721,569 (1973).
12. Tsuchiya, K., "Soymilk - The Renaissance of Traditional Food," Tokyo (1980).
13. Japanese Patent 58-5392 (1987).
14. Badenhop, A.F., and W.F. Wilkens, *J. Am. Oil Chem. Soc. 46*:179 (1969).
15. Johnson, L.A., Ph.D., Dissertation, Kansas State Univ. (1978).
16. Johnson, L.A., C.W. Deyoe and W.J. Hoover, *J. Food Sci. 46*:239–243 (1981).
17. Wolf, W.J., and J.C. Cowan, "Soybeans as a Food Source," P. 60 (1971).
18. Van Buren, J.P., K.H. Steinkraus, L.R. Hackler, I. El Rawi and D.B. Hand, *J. Agr. Food Chem. 12*:524 (1964).
19. Hooks, R.D., V.W. Hays, V.C. Speer and J.T. McCall, *J. Animal Sci. 24*:894 (1965).
20. Cristofaro, E., F. Mottu and J.J. Wuhrmann, 3rd Int. Congress, Food Sci & Tech. Washington, Aug. 9–14 (1970).
21. Sugimoto H., and J.P. Van Buren, *J. Food Sci. 35*:655 (1970).
22. Official United States Standards for Soybeans, FGIS/USDA, Revised, Sept. 9, 1985.
23. Fiering S., *Soyfoods*, No. 4, p. 52 (Winter, 1981).
24. Shyeh, B.J., E.D. Rodda and A.I. Nelson, *Transactions of the Am. Soc. of Agr. Eng. 23*:523 (1980).
25. Wilkens, W.F., and L.R. Hukler, *Cereal Chem. 46*:391 (1969).
26. Lo, W.Y., K.H. Steinkraus, D.B. Hand, L.R. Hackler and W.F. Wilkens, *Food Tech. 22*:1188 (1968).
27. Hackler, L.R., D.B. Hand, K.H. Steinkraus and J.P. Van Buren, *J. of Nutr. 80*:205 (1963).
28. Rackis, J.J., First Latin American Soy Protein Conference, Mexico City, Nov. 9–12, 1975.
29. Hackler, L.R., J.P. Van Buren, K.H. Steinkraus, I. El Rawi and D.B. Hand, *J. Food Sci. 30*:723 (1965).
30. U.S. Patent No. 2,078,962 (1937).
31. Kuntz, D.A., A.I. Nelson, M.P. Steinberg and L.S. Wei, *J. Food Sci. 43*:1279 (1978).
32. Chen, S.C., "Soymilk - A drink from the great earth" by American Soybean Association ASA (1983).
33. Chen, S.C., "Principles of Soymilk Production" AOCS Short Course on Food Uses of Whole Oil and Protein Seeds, Honolulu, Hawaii, May 12–14, 1986.
34. Wilson, J., Alfa-Laval Soymilk Processing Presentation, November 1987.
35. Snyder, H.E., and T.W. Kon, "Soybean Utilization," Avi Publishing Co., 1987, p. 226.
36. Japanese Patents 53-69856, 53-69857, 53-69858 (1978).
37. Taiwan Patents 68-11054, 12443 (1980).
38. Japanese Patent 1264121 (1980).
39. Soya Technology Systems Ltd. "Soymilk," 3rd Ed. September 1984.
40. Chen, S.C., "Soyfoods in the Far East and USA," presented at the First European Soyfoods Workshop, Sept. 27-28, 1984.
41. Mustakas, G.C., W.J. Albrecht, G.N. Bookwalter, V.E. Sohns and E.L. Griffins Jr., *Food Tech. 25*:80 (1971).

Extrusion of Texturized Proteins

Joseph P. Kearns

Wenger International, Inc., Kansas City, Missouri, U.S.A.

Galen J. Rokey and Gordon R. Huber

Wenger Manufacturing, Sabetha, Kansas, U.S.A.

Abstract

Extrusion cooking is a continuous process by which many foods are produced on an industrial basis. Texturized proteins, a unique product made by extrusion, can be produced from a wide range of raw ingredient specifications, while controlling the functional properties such as density, rate and time of rehydration, shape, product appearance and mouthfeel. Raw material specifications for extrusion of texturized proteins have increased and now include: PDI ranges from 20 to 80; fat levels from 0.5 to 6.5%; fiber levels up to 7%; and particle size up to 8 mesh (2360 micron). Additional benefits of extrusion cooking are denaturing of the proteins, deactivation of heat liable growth inhibitors, control of bitter flavors and the homogeneous bonding of ingredients that may include colors, chemicals and other additives which can have an effect on appearance or textural quality.

Extrusion of texturized proteins is one of many successful applications of this unique cooking process. There are other methods utilized to produce texturized proteins including spun soy protein isolates and formed meat analogs. Spun soy protein isolates involve redissolving precipitated vegetable proteins and passing them through a spinneret into a precipitating bath. The bundles of fibers resulting from this are compacted, shaped, flavored, cooked and/or dried and packaged. Formed meat analogs are blends of various protein sources such as isolates, glutens, albumin, extrusion-cooked vegetable proteins and others which are blended with oils, flavors and binders before forming them into sheets, patties, strips or disks.

Extrusion-cooked texturized proteins include meat extenders in the form of chunks or small granular pieces which are wet milled or produced directly off the extruder. Extruders also are able to produce a meat analog that has a remarkable similarity in appearance, texture and mouthfeel to meats. The utilization of extrusion cooking throughout the food industry has shown that a variety of products can be made on extrusion equipment. Some of these products include breakfast cereals, breadings, snacks, instant rice, instant pasta, starch modifications, animal and aquatic feeds.

This paper, however, will center on the production of texturized proteins produced by extrusion cooking.

Raw Materials

Traditionally the most popular raw material for production of textured vegetable proteins in an extrusion system has been slightly toasted defatted soy flour. This defatted soy flour usually meets the following characteristics: 50% protein minimum, 3.5% fiber maximum, 1.5% fat maximum and PDI of 60 to 70.

Soy flour with these specifications allows controllable production of textured proteins in chunk and extended form on single screw extrusion cookers. Other vegetable protein sources also have been used as raw materials for texturizing, and these include: glandless cottonseed flour, rape seed or canola concentrates, defatted peanut flour, defatted sesame flour, as well as soybean grits, flakes, meal, concentrates and isolates.

The use and development of twin screw extrusion cookers in the field of texturized proteins has increased the raw material specification range to include raw materials that include: PDI ranges from 20 to 70, fat levels from 0.5 to 6.5%, fiber levels up to 7% and particle sizes up to 8 mesh.

Protein dispersibility index. The Protein Dispersibility Index (PDI) is the percentage of total protein that is dispersible in water under controlled conditions of extraction. The PDI is now the preferred measurement with regard to specification of raw ingredients, as it is a more reliable figure when compared with NSI. Textured soy products have been produced with raw materials ranging from 20 to 70 PDI.

Fat level. Raw materials have been texturized containing 0.5 to 6.5% fat levels. This higher range of fat (6.5%) allows mechanically extracted soybean meal to be texturized into meat extenders and meat analogs. When extruding material with higher fat levels, generally it can be said that increased shear energy input and higher temperatures are required to maintain a desired product integrity.

Fiber level. The presence of fiber in extruded soy proteins inhibits or blocks the interaction or cross-linking of protein molecules necessary for good textural integrity. Changing the extruder configuration to impart more shearing action into soy proteins containing higher levels of fiber may achieve a final product with textural properties similar to soy protein with lower levels of fiber.

Particle size. With regard to successful production using single screw extrusion cookers, the exact limitations of particle size requirements of defatted soy flour have never been determined and the range is very wide. The only limitations encountered are as follows: Very fine flour, below 400 mesh (38 micron), should be limited because of problems in wetting this very fine powder without lumping; also, very coarse product, over 80 mesh (180 micron), requires complicated premoistening, and sometimes, whole granules are seen in the finished product. We, therefore, recommend a product grind of 95% through 100 mesh (150 micron), with a maximum of 50% through 325 mesh (45 micron) for defatted soy flour.

Twin screw extruders can, sometimes, use raw material with a particle size range up to 8 mesh (2360 micron) without affecting the textural properties of the final product.

It is also believed that proper preconditioning has a great effect on the ability of the extrusion cookers to utilize the larger particle size raw materials for textured protein production. Preconditioning is a time, temperature and moisture level relationship and controlling variables allows for the soy flour or grits to be evenly premoistened and pretempered.

Protein levels. The percentage of protein is normally inversely related to the remaining constituents of a raw material such as fat and fiber. For example, soybean protein content goes up as the oil and hulls are removed. Therefore, as a protein level of a raw material decreases, the textural integrity and water holding capacity decrease and the bulk density of the final products increases.

Adjustments in pH. Increasing the pH of vegetable proteins before or during the extrusion process will aid in texturization of the protein. Extreme increases in pH will increase the solubility and decrease the textural integrity of the final product (1). Modifying pH above 8.0 also may result in the production of harmful lysinoalanines (2). Lowering the pH has the opposite effect and will decrease protein solubility, making the protein more difficult to process. Undesirable sour flavors in the texturized vegetable protein products may be evident, if the pH is adjusted below 5.0.

Modifying the pH to the alkaline side will increase the water absorption. This is generally done by using calcium hydroxide or the more widely used sodium hydroxide at about .1% or as required.

Calcium chloride. Calcium chloride ($CaCl_2$) is very effective in increasing the textural integrity of extruded vegetable proteins and also aids in smoothing its surface. Dosing levels for $CaCl_2$ range between 0.5 and 2.0%. With the addition of $CaCl_2$ and small amounts of sulfur, soybean meal containing 7.0% fiber may be texturized, retorted for one hour at 110°C and still maintain a strong meat-like texture. Sodium chloride (NaCl) does not appear to add any benefit to the texture of extruded vegetable proteins. In fact, it tends to weaken the textural strength.

The addition of sodium alginate will increase chewiness, water-holding capacity and density of extruded protein products. Sugar will also disrupt the textural development of soy proteins (3).

Soy lecithin. When added to formulations of vegetable proteins at levels up to 0.4%, lecithin tends to assist smooth laminar flow in the extruder barrel and die configuration, which allows the production of increased density soy products. The ability to make dense vegetable protein products is related to the higher degree of cross-linking that occurs during the extrusion process.

Sulfur and sulfur containing ingredients. Sulfur is known for its ability to aid in the cleavage of disulfide bonding, which assists the unraveling of long twisted protein molecules. This reaction with the protein molecules causes increased expansion, smooth product surface and adds stability to the extrusion process. These benefits, however, are not without some undesirable side effects that include off-flavors and aroma.

Sodium metabisulfite, sodium bisulfite, as well as cystine, can be used with effects similar to those from using sulfur.

The normal dosing levels for sulfur or sulfur derivatives are in the .01 to .02% range. Cystine is used at approximately .5 to 1% level.

Color enhancers. When supplementing light colored meats with meat extenders made from textured vegetable proteins, it is desirable to bleach or lighten the color of the extruded meat extender. Bleaching agents such as hydrogen peroxide are often used for this purpose. Dosing levels for the hydrogen peroxide range from 0.25 to 0.5%. Pigments such as titanium dioxide are also used at levels between 0.5 and 0.75% to lighten color, but at increased levels will weaken the textural properties of extruded vegetable proteins.

Extrusion Process

The above raw materials or combinations are generally mixed prior to the extrusion cooker, except in the cases where small liquid additions can be added directly to the base raw materials in the extrusion cooker itself.

Definition and functions of extrusion of textured proteins. Extrusion cooking has been defined as "the process by which moistened, expansile, starchy and/or proteinaceous materials are plasticized in a tube by a combination of moisture, pressure, heat and mechanical shear. This results in elevated product temperatures within the tube, gelatinization of starchy components, denaturization of proteins, the stretching or restructuring of tractile components and the exothermic expansion of the extruder" (4). Extrusion is widely used to accomplish this restructuring of protein-based foodstuffs to manufacture a variety of textured convenience foods. When mechanical and thermal energy are applied during the extrusion process, the macromolecules in the proteinaceous ingredients lose their native, organized structure and form a continuous, visco-elastic mass. The extruder barrel, screws and die align the molecules in the direction of flow. This realignment "exposes bonding sites that lead to cross-linking and a reformed, expandable structure" that creates the chewy texture in fabricated foods (5).

In addition to retexturing and restructuring vegetable food proteins, the extrusion cooking system performs several other important functions: It denatures proteins. Proteins are effectively denatured during the moist, thermal process of extrusion. Denaturation of protein "lowers solubility, renders it digestible and destroys the biological activity of enzymes and of toxic proteins" (6). It causes deactivation of residual heat-labile growth inhibitors native to many vegetable proteins in a raw or partially processed state. Growth inhibitors exert a deleterious physiological effect on man or animals, as revealed by growth or metabolism studies. It controls raw or bitter flavors commonly associated with many vegetable food protein sources. Many of these undesirable flavors are volatile in nature and are eliminated through the extrusion and decompression of the protein at the extruder die. The use of preconditioning and atmospheric-venting devices in the design of the extrusion system also assists in volatilization and removal of off-flavors. It provides a homogeneous, irreversible, bonded dispersion of all microingredients throughout a protein matrix. This not only insures uniformity of all ingredients such as dyes throughout the product, but provides a means whereby minor ingredients can be intimately associated with potential reaction sites promoting cross-linking or other desirable chemical and physical modifications. And, it controls the shape and size of the final extruded protein in convenient and transportable portions for packaging in the retail or institutional marketplace.

Methodology of extruders relating to texturized proteins. It is convenient to divide the basic components of an extrusion system barrel configuration, and the die and knife assembly. The design of each of these components is engineered to accomplish a specific function in the process of texturizing vegetable food proteins. Within the design features, the operating conditions are adjusted to vary the texture of the finished product. The effects of each processing step on the product will be addressed specifically.

A typical arrangement of the components of an extrusion system is shown in Figure 1. This arrangement includes a live bin/feeder, preconditioner and extrusion cooker. The live bin/feeder provides a means of uniformly metering the raw materials (granular or floury in nature) into a preconditioner or mixing cylinder and subsequently into the extruder itself. This flow of raw materials must be uninterrupted and rate-controllable. This component controls the product rate or throughput of the entire system. The feeder portion also can be used to initiate the preconditioning of raw materials through the injection of steam to control raw material temperature and moisture. Not all extruded products require a preconditioning step. It is particularly useful where the particle size of the raw vegetable protein is large (a grit or flake). This initial steam preconditioning promotes moisture and heat penetration of the individual particles, resulting in uniform moisture application and elevated raw material temperature. The raw material then passes through the atmospheric preconditioner or mixing cylinder where moisture can be uniformly applied in the form of water and/or steam to achieve a moisture content of 10 to 25%. If steam is applied, it is carefully metered into the raw material to precondition it at modest temperatures of 65 to 100°C. The mixing cylinder is vented to avoid excess steam and undesirable volatile flavor components found in the raw vegetable protein. Flavors, coloring agents and other additives may also be introduced at this phase of the process to insure thorough and continuous mixing of all foodstuffs entering the extruder barrel. As mentioned earlier, the preconditioning step is also an effective means of initiating control of growth inhibitors found in many raw vegetable proteins.

Most preconditioners (Fig. 2) contain one or two mixing/conveying elements which consist of rotating shafts with radially attached pitched paddles. Atmospheric or pressurized chambers may be utilized in this preconditioning step. Pressurized preconditioners can achieve higher discharge temperatures, but have the disadvantages of potential nutrient destruction and higher operating costs. Without the use of preconditioning, it is difficult to make good laminar-structured, textured soy protein. Unpreconditioned vegetable proteins have a strong tendency to expand rather than laminate due to nonuniform moisture penetration which does not allow uniform alignment of protein molecules. Uniform moisture penetration of raw ingredients significantly improves the stability of the extruder and final product quality. This moisture-temperature-time history allows the extrusion of raw materials having larger particle sizes without sacrificing final product quality.

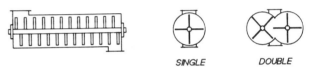

Fig. 2 Typical preconditioners.

The preconditioner discharges directly into the extruder assembly which consists of the barrel and screw configuration. Here, the major transformation of the raw or preconditioned vegetable proteins occurs, which ultimately affects the final product texture. Extruders are popularly classified as either being a single or twin screw design. In both designs, the impact on final product texture is affected by screw and barrel profile, screw speed, processing conditions such as temperature, moisture, pressure, raw material characteristics and die selection.

The initial section of the extruder barrel is designed to act as a feeding or metering zone and simply conveys the raw or preconditioned vegetable protein away from the inlet portion of the barrel and into the extruder. The product then enters a processing zone where the amorphous, free-flowing vegetable protein is worked into a colloidal dough. The compression ratio of the screw profile is increased in this stage to assist in blending water or steam with the raw material. The temperature of this moist, proteinaceous dough is rapidly elevated in the last 2–5 seconds of dwell time within the extruder barrel. Most of this heat is from mechanical energy dissipated through the rotating screw and may be assisted from the direct injection of steam or from external thermal energy sources. The screw profile can be altered by the pitch, flight height and angle, and steamlock diameter, which affects the conveying of this plasticized food material down the screw channel. The net flow patterns of the product within the screw are quite complicated and difficult to understand and describe. Retention times of 5–15 seconds, temperatures of 100 to 200°C and moisture levels of 15 to 30% all influence the protein dough quality just behind the die and the final product expansion. The temperature at this point actually melts the protein into a visco-elastic, plasticized mass having very high viscosities. This plasticized mate-

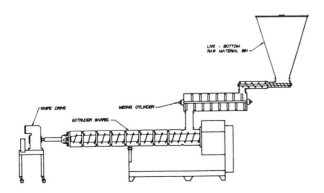

Fig. 1. Cutaway view of a high capacity extrusion cooker.

rial is extruded through the die openings, and expansion occurs as the product is released to ambient pressures. Final product density has been shown to correlate with temperatures and moistures of extrusion.

Extruded meat analogs. It is possible to produce an extruded meat analog which has a remarkable similarity to meat in appearance, texture and mouthfeel (7). Although the process to extrude meat analogs is similar to that used for the production of meat extenders, it often uses two extruders or a long-barreled, single extruder. Through this configuration, the dwell time in the extruder is extended, the moisture content increased to 30% and the pressure maintained below 150 PSIG. The vegetable proteins exposed to this process are stretched, cooled and formed into uniform layers of parallel orientation and cut into bite-sized portions. The lower pressures used inboard of the die reduces product expansion and disruptions to the laminar structure. Due to the dense nature of these analogs, they are able to retain their integrity and meat-like texture even after retorting them for 2–3 hours.

The extruder barrel. The above discussion is valid for processing texturized protein on both single and twin screw extruders, however, there are differences between these two extruder barrel designs. Single screw extruders are the machines that have produced and still produce the greatest tonnage of textured soy protein products in the world. Over the years of extrusion development, the process has become more of a science, which has resulted in the ability to design screws, heads and shearlocks for single screw extruders for specific functions. This variety or selection of parts includes straight and spiral-ribbed jacketed heads which allow the control of friction or product shear, internally, while the flow of cooling water or heating fluids in the external jacket has allowed control of temperatures of the product as it passes through the various extruder sections. Coupling these heads with the wide range of extruder screws that are unrestricted in their number of flights, the flight pitches, cut flights, the screw volumes and the shearlocks between these screws allow for a very versatile machine that has been successful in producing a wide range of products. In the case of texturized proteins, the use of a narrow range of raw materials has been the limiting factor in the use of single screw extruders. Their use requires good uniform raw materials and a maintenance program to keep the extruder barrel parts within tolerances.

The twin screw extruder barrel generally used in the food industry consists of two corotating intermeshing self-wiping sets of screws. Because the screws intermesh they do have restrictions in screw design. They are limited by the flight profile and root diameter. The twin screw extruder has a segmented smooth bore head that is jacketed. The screws are also segmented so that their arrangement can be modified. The twin screw barrel design does have the advantage of a more positive feeding characteristic over the single screw barrel design. The twin screw design, as a result, has been more lenient in raw material specifications.

In either case, single or twin, the combination of preconditioning and the extruder barrel arrangement results in modification to the raw materials and the result is a texturized protein. Hydration and heating cause unraveling of the long twisted protein molecules in vegetable proteins. In the extrusion process, these molecules align themselves along the streamline flows of the screws and dies. The increase in shear temperature and retention time causes cross-linking to occur, yielding a textured product that is layered, and the resulting denaturation or cross-linking can be considered an irreversible endothermic chemical reaction. The extent of cross-linking seems to be a function of the time, temperature and moisture history and can be related to changes in apparent viscosity of the extrudate. The proper exposure to shearing action, as the protein molecules align themselves for cross-linking during the extrusion process, is important.

The rheology of soy proteins or soy protein doughs will not be discussed in this paper. Several papers published in the past describe the effects of shear rate, shear stress, adequate time and temperature history (5,8,9). When a proper screw profile for the extruder is derived, one can meet final product quality requirements with a much broader range of raw materials.

Die. Dies for texturized vegetable proteins should have smooth streamlined flows that do not disrupt or cause shearing effects to the already laminated and cross-linked protein molecules (8). Harper describes in detail the effects of various land lengths and an open area of dies for extruding texturized vegetable protein.

On-Going Research

Various studies have been performed to confirm and demonstrate the use of a broader range of raw materials. These studies include the examination of textural properties, and the work was performed with a Wenger TX-80 twin screw extruder at the Oak B. Smith Extrusion Research Center in Sabetha, Kansas.

Protein dispersibility index range tests. Commercially available ingredients were used in the tests. Soy flour (200 W) from Central Soya with a PDI of 70 and soy grits from A.D.M. (Nutrisoy) with a PDI of 20 and particle size of 20 mesh were mixed together to obtain five mixes with various PDI levels ranging from 20 to 70. These are shown in Table 1.

TABLE 1

Formulas and PDI's for PDI Range Tests

Test #	Formula	Calculated PDI	Tested PDI
1	100% soy flour	70	79
2	80% soy flour, 20% soy grits	60	68
3	60% soy flour, 40% soy grits	50	58
4	40% soy flour, 60% soy grits	40	48
5	100% soy grits	20	27

All the formulations were run under approximately the same extruder conditions and with the same final product densities. Samples from the five formulations were tested for water holding capacity (WHC) and textural integrity (TI). Table 2 shows that for the full range of PDI levels, there is no significant difference in water holding capacity or textural integrity, with the exception of Test #2. The increased WHC and textural integrity values in

Test #2 can be attributed to the processing conditions of the extruder that yielded a lighter density product.

TABLE 2
Effects of Raw Material PDI on Final Product Quality

Test #	PDI of raw materials	Density g/l	WHC	Textural integrity
1	(79)100% soy flour	280	2.03	1.19
2	(68)80% soy flour, 20% soy grits	216	3.12	1.31
3	(58)60% soy flour, 40% soy grits	288	1.93	1.03
4	(48)40% soy flour, 60% soy grits	288	2.01	1.22
5	(27)100% soy grits	280	2.29	1.27

During these energy-conscious times we must point out that using raw materials with decreasing PDI levels also requires additional specific mechanical energy input to maintain the same quality of texture in the extrudate (Table 3). This is due to the reduction in available reactive sites for cross-linking in the lower PDI soy protein, thus requiring additional shearing action to expose those reactive sites.

Fat level tests. It is an advantage in certain situations to have the ability to texturize vegetable proteins having high levels of residual fat. For example, it may be more economically feasible to use mechanically extracted soy meals which have fat levels between 3 and 6.5% to produce meat extenders and meat analogues. It also has been documented that texturization of whole soybeans via extrusion processing improves the feeding quality of soy protein products (Table 4).

Table 5 shows the materials used, and Table 6 shows the processing conditions and product characteristics of extruded raw materials having various fat levels. In general, increased fat levels in vegetable proteins require increased shear energy input and higher temperatures to maintain a desired product integrity.

It can also be noted that the split and whole beans were texturized and they contained 18 to 20% fat. However, these products did not have desirable characteristics after rehydration. The defatted soy flour with 4.5% fat added into the raw material did produce a good laminar product with good integrity.

Fiber levels. Table 7 presents the materials used in these tests. As discussed earlier, fiber inhibits the cross-linking of protein molecules necessary for good textural integrity.

TABLE 5
Fat and Fiber Content of Raw Materials

Raw material	Total % fat	% Fiber
1) Split soybean @ 16% MCWB	18–20	6–7
2) Whole soybeans @ 14% MCWB	18–20	6–7
3) Defatted soy flour [w/4.5% add fat]	6	3.0–3.5

TABLE 3
Energy Requirements for Extruding Soy Proteins at Various PDI Levels

Test #	Raw material actual PDI	Extruder RPM	Required energy KWH/kg	WHC	Density g/l	Textural integrity
1	79	425	.112	2.03	280	1.19
3	58	425	.147	1.93	288	1.03
4	48	450	.181	2.01	288	1.22

TABLE 4
Effect of Processing on Nutrient Performance

Raw materials	Ave TME Kj/gr[a] GE Kj/gr	% Fat digestibility[b]	Available protein ratio
Texturized whole soybeans	.756	98.2	.992
Conventionally processed full fat soy	.754	95.3	.981
Raw soybeans (ground through 1/16" screen)	.632	81.0	.970

[a] Average total metabolizable energy (Ave TME) and gross energy (G.E.) studies performed by McNab (10).
[b] % Fat digestion performed by McNab (10).

TABLE 6

Effect of Raw Materials and Processing Conditions on Rehydrated Product Characteristics

Ingredient	% Fat	Extruder	Mechanical energy input KWH/kg	Temperature	Extruder pressure PSIG	Characteristics after rehydration
1) Split beans	18–20	SX-80	.135	193°C	1500	Soft or mushy
2) Whole beans	18–20	SX-110	.191	127°C	1400	Soft or mushy
3) Flour w/fat	6	TX-80	.181	154°C	400	Firm with good integrity and laminated structure.

TABLE 7

Fiber Content of Commercially Available Soy Proteins

Ingredient	Supplier	PDI	Particle size	% Fiber
Soy grits	A.D.M. Nutrisoy	20 PDI	20 mesh	3.5 max
47% Soybean meal	Lortscher Agri Service	30 PDI	20 mesh	3.5 max
44% Soybean meal	Lortscher Agri Service	30 PDI	20 mesh	7.0 max

TABLE 8

Effect of Fiber Content on Final Product Quality

Formula	Density g/l	WHC	Textural integrity	% Fiber
Soy grits	280	2.29	1.27	3.5 max
40% (44% soybean meal) 60% (47% soybean meal)	288	2.03	1.16	5.0 max

This is overcome by imparting more shear as the product passes through the extruder barrel. Results are in Table 8.

Particle size tests. In time past, particle size variations have been important in achieving the proper textural properties in a textured soy product. Table 9 shows particle size and PDI's of three raw ingredients that were extruded yielding the same bulk densities. It appears from Table 10 that the particle size had no significant effect on the product's textural properties. Preconditioning of the raw ingredients may be the highest contributor to the insensitivity to particle size variation for these test runs.

TABLE 9

Ingredients of Various Particle Size

Soy flour	Central Soya 200-W	70 PDI	flour
Soy grits	A.D.M. Nutrisoy	20 PDI	20 mesh
Soy grits	A.D.M. Nutrisoy	20 PDI	8 mesh

TABLE 10

Effect of Raw Material Particle Size on Final Product Quality

Formulation	Density g/l	WHC	Textural integrity
Soy flour	280	2.03	1.19
Soy grits (20 mesh)	280	2.29	1.27
Soy grits (8 mesh)	280	1.96	1.22

Effects of screw profile and retention time. Table 11 shows the effect of screw profile on final product characteristics. The table has a section on continuous flighting of the extruder screws and a section regarding screw flighting that is interrupted or has breaks in the screw profile.

Three types of products were made on both continuous and interrupted flighting. These are dense chunks, light chunks and blown product (hamburger style TVP).

TABLE 11
Effect of Screw Profile on Final Product Characteristics

Processing conditions	w/Continuous flighting			w/Interrupted flighting		
	Dense chunk	Light chunk	Blown	Dense chunk	Light chunk	Blown
Preconditioning temperature (°C)	82.0	82.0	82.0	82.0	82.0	82.0
Extruder RPM	275.0	350.0	300.0	275.0	350.0	300.0
Extrusion temperature (°C)	105.5	108.0	113.0	105.5	108.0	113.0
Moisture added (Kg/min.)	2.1	1.7	1.4	2.1	1.7	1.4
S.M. Energy input[a] (KWH/kg)	.122	.114	.131	.130	.120	.131
Product characteristics						
Density (g/l)	328.0	216.0	192.0	320.0	216.0	192.0
WHC[b]	1.01	2.24	3.15	1.00	1.57	1.64

[a] Specific Mechanical Energy Input.
[b] Water Holding Capacity by Wenger Mfg.

The processing conditions, preconditioning temperature, RPM, extrusion temperature, moisture added and energy input were kept as close as possible to the same for each case.

Product density was maintained; however, the water holding capacity went down, especially on the light chunk and blown product, as interrupted flighting was used. The test data demonstrated that although adequate shear rates are necessary to enhance the cross-linking effects of protein molecules, over-shearing, once the cross-linking has begun to occur, may disrupt the layered structure of the protein molecules, resulting in decreased water holding capacities.

The effect of moisture on the rheology of soy protein doughs is shown in Table 11. Increased moisture levels increase mobility of the chemically reactive protein molecules, increasing the potential for reactive sights to come in close proximity, which facilitates cross-linking, resulting in increased densities of the extrudate.

Table 12 shows a 36% increase of retention time in the extruder barrel. This allows an approximately equal increase in the maximum capacity of the extruder, while maintaining an equivalent final product quality. At the same time, the required specific mechanical energy input per kilogram of extrudate is reduced.

Product enhancement concepts. Due to consumer preferences, there is a growing demand for marketing gimmicks. Therefore, product shape and design have become an important issue in extrusion processing. Recent developments allow increased flexibility in color development of extruded products.

Dyes or color pigments may be injected with high pressure pumps into the final sections of the extruder barrel to achieve special effects such as marbling, striping and multiple colored products. The conical designed final screw and head section that divides the product flow into separate chambers allows tremendous flexibility of product enhancement, not only with color schemes, but with the addition of heat-sensitive chemicals for textural development or reactions.

Post Extrusion Processes

These are processes that are performed on the extrudate prior to the drying and cooling system. The most predom-

TABLE 12
Effect of Retention Time on Maximum Capacity of Extruder

Extruder barrel length : dia. ratio	Max. capacity to maintain same quality final product	Specific mechanical energy input
16½ : 1	500 kg/hr.	.130 KWH/kg
22½ : 1	800 kg/hr.	.090 KWH/kg

inate process would be particle size reduction. This is normally referred to as "wet milling," as the moisture content is high at this point. There are various devices that can be used for particle size reduction of textured protein and these can include hammermills, rotating cutting heads and shredders. Wet milling tests were performed on light and medium density chunks and the results are shown in Table 13.

TABLE 13

Comparison of Wet Milling Devices for Textured Soy Protein

SAMPLE	Mesh size, %						Bulk density (G/L)
	4	6	8	10	16	PAN	
Run 1, 050583 (light density)							
Raw chunk	—	—	—	—	—	—	144
Fitz, round 1/4″[a]	0.15	11.09	20.83	11.90	22.81	33.22	227
Fitz, square 1/4″	.43	20.77	24.67	10.00	22.33	21.8	219
VGR, 5/32″ square[b]	1.88	9.09	15.16	12.98	30.44	30.44	156
Run 2, 050583 (medium density)							
Raw chunk	—	—	—	—	—	—	192
Fitz, round 1/4″	0.06	3.99	11.83	8.56	22.00	53.55	284
Fitz, square 1/4″	0.39	7.33	14.06	8.96	20.74	48.79	258
VGR, 5/32″ square	1.07	2.73	8.86	9.65	32.73	44.96	211
Urschel, #120 screen[c]	0.91	20.18	28.14	13.38	22.74	14.64	201

[a] Fitzmill Hammermill.
[b] Wenger Shredder.
[c] Urschel Comitrol.

Grinding devices used must be based on desired final product sizes and intended uses. Generally speaking the hammermill style device will yield a higher final bulk density as a result of fines generated. The rotating cutting head and shredder designs will generate less fines and maintain the approximate bulk density of the original sample. There are adjustments that can be made in each of these grinding devices to modify the particle size generated.

Drying and Cooling

In the extrusion processing of textured proteins the moisture content is elevated, and as a result, the cooked product must have the moisture reduced for safe storage prior to its use. Proper design and size selection of the dryer/cooler is important for making good quality products, as well as overall cost of production, because this production step is generally the most costly.

There are many dryer/cooler designs, but for the purpose of discussion, we will review the tray-type dryer/cooler (Fig. 3), because it is the most popular style for textured soy protein production. Tray-type dryer/coolers can be built with various lengths, widths, number of passes, materials, conveyor styles, air flows, sanitary considerations and heating systems.

The ability to construct dryer/coolers with various lengths, widths and number of passes allows proper sizing for the extruder output, as well as offering flexibility in equipment placement in existing or new facilities. The two pass design not only saves floor space, but allows the product to be turned over in the dryer, improving evenness of drying. Using varispeed motors on the dryer passes allows for changes in the product bed depth, and thus retention time of the product in the dryer.

Dryer/cooler construction materials include mild steel,

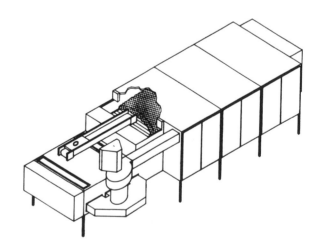

Fig. 3. Dryer/cooler.

stainless steel and combination of both. In human food production the dryer/coolers are normally equipped with stainless steel, at least where the product comes in contact with the dryer, or are all of stainless steel construction. Product-carrying conveyors are perforated mild or stainless steel trays that can be coated with teflon or silverstone. Also available are wire mesh screens that are carried on special supports. The wire mesh design has greater open area for even air flow and allows for smaller particles to be dried and retained on the conveyor.

Air flow is generally up through the product with the heating elements mounted in a single side plenum chamber above air recirculation fans. These recirculation fans create the static pressure below the dryer trays resulting in evenness of drying across the product bed. A percentage of air is exhausted to control the humidity in the dryer, while the balance is recirculated, which permits maximum efficiency in energy consumption. The heating elements mounted in the side plenum are generally steam coils or gas burners. This is usually based on the location of the plant and the cost of the fuel to be used. Steam coils are generally the most popular. Full insulation on all sides of the dryer coupled with the heating systems allow for safe operating temperatures up to 350°F (175°C).

Sanitary considerations include doors wherever possible for cleaning access and maintenance. Fines recovery systems remove fines from the bottom of the dryer/cooler and deposit them in a cross auger that removes them to the side of the dryer/ cooler. Dryers are commonly equipped with integral coolers or extensions of the bottom dryer pass. Internal airlocks are used to reduce the amount of air leakage between these two zones. Air flow is generally down through the product and a percentage of this exhausted air is used as make-up air for the dryer, yielding a more efficient dryer, reducing energy consumption.

Final Products

The products discussed above include texturized proteins in the meat extender form, as well as chunk style products. The meat extenders, when ground to a particle size simulating hamburger, are used at a 25 to 30% level in dishes such as pizza, hamburgers, meatloaf, chili, tacos, enchiladas, sloppy joes and meat sauces. Chunk style products are effectively used in soups, stews, meat pies, dry soup mixes and oriental dishes.

Recently, a new line of natural flavors has come to our attention, and flavor response in meat extender products is remarkable. We have tested beef, ham, chicken, lobster, clam and shrimp flavors and have been quite impressed with their utility, flavor survival through the extrusion cooking process and lack of residual soya flavor in the end product. When blended with a system for producing soups, chowders or gravies, for example, their similarity to meat has proven to be excellent.

Particularly interesting is the idea of using natural flavors as opposed to synthetic flavors. Because the alcohol, fat or water bases of most synthetic flavors are quite volatile, most of the synthetic flavor's effectiveness is lost at the high temperatures necessary for structuring and moisture removal. Additionally, synthetic flavors are often leached out of the product and are often dispersed throughout the liquid phase during the rehydration of textured soy protein in the canning operation or in the making of soups or stews. Conversely, the natural flavor agents we have utilized seem to be enhanced by cooking, drying and rehydration steps and react in much the same manner as natural meats, in that thermal processing is required to bring out their flavor. When included in a flavor system such as soup stock or meat gravy, the product may be frozen or canned to provide stability. Additionally, systems may be designed for convenience type products which could include the textured vegetable protein extender in a gravy or soup stock to be finished by the end user with fresh vegetables, noodles or other constituents to develop nutritious, inexpensive and convenient complete meals, entrees or soups.

Discussions above on textural integrity and water-holding capacity help food scientists determine how texturized proteins will function in the food systems utilizing these products. The following test procedures are the ones used in the studies.

Procedure for determining water holding capacity. (Equipment Needed: Beaker, sieve screen and scales) Weigh out small amount of sample and record weight. Soak in room temperature water for 20 minutes. Let drain on screen for 5 minutes. Reweigh sample. (Rehydrated wt. − Original wt.)/Original wt. = Water absorption.

Test for measuring integrity of textured vegetable proteins. (Equipment needed: U.S. 20 mesh screen; gram scale; 600 ml. beaker; aluminum foil; autoclave, capable of holding 15 psi; food grinder with 1/8 in. holes in plate; and water bath. Ingredients needed: At least 50 g of sample and tap water.) Begin test by soaking at least 50 g of sample in a 600 ml beaker with sufficient water to fully cover all of sample and allow for absorption. Let hydrate one hour, stirring occasionally. After hydration, cover beaker with aluminum foil and place in a preheated retort. Pressure cook for 30 min under pressure at 15 psi (250°F). Cool sample to 70–80°F by putting beaker in running tap-water bath. Do not pour off excess water. Grind hydrated sample in food grinder fitted with plate with 1/8 inch diam. holes. If water comes out of grinder, keep with sample. Weigh out 50 g of the hydrated, cooked and ground sample and spread evenly on a U.S. 20 mesh screen. Spray rinse the sample on the screen for one minute with tap water at ambient temperature under 12 psi. Spray nozzle should be approximately 9 in. from the screen and slowly moved back and forth and rotated so that the spray water contacts the entire screen surface during the rinsing operation. Flow of water should be about 3 gal in 1 min. Shake off excess water. Drain sample on the screen for 4 min. During the draining period, blot the bottom of the screen with paper towels. Weigh the residue. (This is most easily done by having the screen tared out on the scales before the test begins. Then you can weigh out the residue with the tared screen). To calculate integrity index, divide the weight of residue by the weight of hydrated sample. Example: If 50 g of sample were spray rinsed, and 55 g of sample was left on the screen after rinsing, 55/50 = 1.1 integrity index.

Discussion

Many tests have been performed with a wide variety of raw materials. Soybean-based raw materials such as defatted soy flour, soy concentrates, soy isolates and soy grits with various particle sizes, fat levels and fiber levels can all be used to produce texturized proteins. The new developments in processing techniques will allow work to continue with raw materials such as glandless cottonseed flour, rapeseed proteins or canola concentrates, defatted peanut flour and defatted sesame flour in order to allow a higher utilization of these protein sources. Initial work with these raw materials on single screw extruders resulted in texturized products, however, the marketability of these products was questionable. The use of the twin screw extrusion cookers should revitalize the interest in these raw materials.

Texturized proteins have been successfully marketed and produced throughout the world as a result of food scientists' ability to apply the extrusion process to produce products that fit into the individual society's normal eating habits. Interactions of equipment manufacturers, end users and food scientists will result in further improvements and new developments in the art of producing and utilizing texturized proteins.

Acknowledgment

J. McNab of the Agricultural Research Council, Roslin, Scotland, performed metabolizable energy and fat digestion studies used to prepare Table 4.

References

1. Simonsky, R.W., and D.W. Stanley, *Can. Inst. Food Sci. Technol. J. 15*:294 (1982).
2. DeGroot, A.P., P. Slump, V.J. Feron and L. VanBeek, *J. Nutr. 106*:1527 (1976).
3. Boison, G., M.V. Taranto and M. Cheryan, *J. Food Technol. 18*:719 (1983).
4. Smith, O.B., "Textured Vegetable Proteins" presented at the World Soybean Research Conference, University of Illinois, Aug. 3-8, 1975.
5. Harper, Judson, *Food Technol. 40*:70 (1986).
6. Altschul, A.M., *Basic Books*, New York (1965).
7. Smith, O.B., "Products of Textured Soy Proteins," 1st Latin America Soy Protein Conference, Mexico City (Nov. 1975).
8. Holay, S.H., and J.M. Harper, *J. Food Sci. 47*:1869 (1982).
9. Hagaiv, R.C., S.R. Dahl and R. Villota, *J. Food Sci. 51*:367 (1986).

Production and Utilization of Tempeh in Indonesian Foods

F.G. Winarno

Food Technology Development Center, Bogor Agricultural Institute, Kampus Darmaga, PO Box 61, Bogor, Indonesia

Abstract

Tempeh is one of the most popular fermented foods of Indonesia, serving as a major source of protein, calories and vitamins in diet. The total annual consumption of tempeh is about a half million tons. Tempeh also has been used and developed into foods for nutritional intervention purposes, including infant and mother foods. A well-made tempeh by definition is a compact cake completely covered and penetrated by the white mold mycelliusm of *Rhizopus sp*. A broad range of tempeh preparation methods have been reported for different places and countries. However, the major steps for preparation of tempeh are the same, i.e., cleaning the beans, dehulling, soaking, boiling, inoculation of starter, wrapping and incubating.

Tempeh is one of the most popular fermented foods of Indonesia. It is consumed by all socio-economic groups. Further, it serves as a major source of protein, calories, and vitamins in the diets of the people. Tempeh has been produced and consumed in Indonesia for centuries but there are no written records of its origin. Because Central and East Java of Indonesia are the major tempeh producers, one could assume that it may have originated first in these areas. From these areas, it was further spread into other parts of Indonesia.

Tempeh is relatively unknown in the surrounding countries such as Thailand, China, and Japan, where soybeans form an important part of the diet. Tempeh is produced in small quantities and consumed by immigrants from Indonesia in Malaysia, Surinam, Canada and The Netherlands. It is becoming a popular food for a number of vegetarians in the United States.

The technology of traditional tempeh making is simple with extremely low cost of production. A broad range of tempeh preparation methods have been reported from different places and countries. However, the major steps for preparation of tempeh are the same regardless of the places and countries of production, that is, similar to that of the Indonesia traditional method.

Tempeh formerly was considered an inferior food in part because of its cost, compared to other protein foods such as meats, fish, and eggs. Over the last 15 years the attitude toward tempeh has changed. Today, more attention has been given to tempeh because it is an inexpensive source of proteins, vitamins, and calories. The total annual production of tempeh is about 500,000 tons. In Indonesia, tempeh production is still a household art.

Most of the 41,000 small cottage industries that make fresh tempeh daily are family run and employ about 128,000 workers. Each small cottage industry employs about three workers and uses approximately 11 lb (5 kg) of dry soybeans per day to produce 21 lb (10 kg) of fresh tempeh. The large cottage industries employ 10 to 20 workers and use 600 to 1,100 lb (500 kg) of dry soybeans per day to produce tempeh. The average retail price of tempeh is about US $0.25 per kg (1).

A well-made tempeh by definition is a compact cake completely covered and penetrated by the white mold mycellium of *Rhizopus sp*. In describing tempeh, an additional word is added to tempeh to indicate the raw materials from which it was prepared. For example, *tempeh kedelai* is made from *kedelai* (soybeans) and *tempeh lamtoro* is made from *lamtoro* seeds. However, in this paper the word tempeh (without suffix) will be used instead of *tempeh kedelai* (1,2).

Tempeh

Depending on the raw material used tempeh may be grouped into several kinds i.e., *Tempeh kedelai*: using soybean as raw material; solid and white color; *Tempeh gembus*: using tofu residue as substrate; solid and white color; *Tempeh "bongkrek"*: using coconut presscake as raw material; solid and white color; or *Tempeh "benguk"*: using legumes *boro benguk* (Mucuna puriens); solid and white color.

In general, the word tempeh usually means *tempeh "kedelai"* or soybean tempeh. This particular product is the most popular among the Indonesia population. Regardless of the source of the raw material used as substrate, all tempeh products utilize *Rhizopus oligosporus* as the predominant microorganisms.

As shown in Table 1, *tempeh kedelai*, soybean tempeh, has the highest protein content and is more nutritious than tempeh of other kinds.

In Indonesia, first quality tempeh is made from whole dehulled soybeans and is not adulterated with any other substrate. In the market several tempeh can be found. Lower qualities of tempeh are adulterated with varying quantities of waste products such as waste from tapioca flour processing ("onggok"), soybean curd waste (*ampas tahu*), coconut press cake or grated coconut waste, and sometimes young papaya. The reason for adding the other ingredient is to reduce the cost.

In general, fermented foods are more attractive than the raw ingredients and one of the advantages of tempeh and other fermented foods is that they leave no waste. Although it is a firm cake, the specific gravity is only approximately 0.9, which makes the tempeh cake float in water. Tempeh has lost the beany flavor of cooked raw soybeans and, when fried in oil, it has a pleasant flavor, aroma and texture (1).

In Indonesia, particularly for people who live in the island of Java, tempeh is a key protein source for millions of people, who enjoy it daily in quantities of 1 to 4 ounces (30–120 gr) and it is served generally as a meat substitute together with their grain-centered diet. The conversion

TABLE 1

Typical Proximate Composition of Important Traditional Fermented Foods in Indonesia (per 100g)

Product	Moisture	Energy (kcal)	Protein (g)	Fat (g)	Carbohydrates (g)	Ash references
Tempeh						
Soybean tempeh	64	149	18.3	4.0	12.7	1.0
Tempeh gembus	81	—	4.9	2.3	—	0.8
Tempeh bongkrek	73	119	4.4	3.5	18.3	1.4
Tempeh benguk	64	141	10.2	1.3	23.2	—
Oncom						
Orange oncom						
(Peanut Presscake)	77	—	8.6	3.6	—	1.4
Black oncom						
(Peanut presscake)	57	187	13.0	6.0	22.6	1.4
Orange oncom						
(Tofu residue)	84	—	4.0	2.1	8.4	—
Tape						
Tape singkong						
(cassava root)	56	173	0.5	0.1	42.5	0.8
Tape ketan						
(glutinous rice)						
Tauco	65	165	10.5	4.9	24.1	7.4
	63	46	5.7	1.3	9.0	15.4

factor from dry soybean into tempeh of 61.3% moisture is 100:174 (1,2). Its protein content is over 40% on a dry basis. Due to its high nutritive value, tempeh makes a good substitute for meat. Tempeh fermentation is characterized by its simplicity and rapidity. After being fermented, soybeans require only 3.4 minutes deep fat frying (190°C) or 10 minutes boiling, as compared to raw soybeans which generally require about 6 hours boiling to prepare them for consumption. Tempeh is actually one of the "quick cooking" foods developed in the world (4).

Tempeh Making

A broad range of tempeh preparation methods have been reported from different places and countries. However, the major steps for preparation of tempeh are the same irrespective of the places and countries where it is produced. All methods yield an organoleptically satisfactory tempeh.

Even though in the old days most of the tempeh production still used banana leaves as a wrapping material, today several tempeh cottage industries have adopted the tray and plastic bag methods. The tray method is practiced by spreading the inoculated beans on a tray, covering them with layers of banana leaves or waxed paper and incubating them at room temperature. Martinelli and Hesseltine developed a new method of incubating tempeh in plastic bags or plastic tubes with perforations at 0.25 to 1.3 cm intervals to allow access of oxygen (1,3).

Microorganisms involved: Different mold species have been reported in the literature as the microorganisms responsible for the fermentation of soybeans to tempeh. In addition to the mold, numerous bacteria of both spore and nonspore forming types exist in tempeh. Yeast and other microorganisms were also found. Sano *et al.* (5) collected samples of tempeh from throughout West Java and isolated 69 mold species, 78 bacteria species, and 150 yeast species. The bacteria were reported as definitely undesirable, contributing off-odors to the tempeh if they were allowed to grow and develop (5).

In an extensive survey to determine which mold species generally are used by traditional Indonesia manufacturers to make good tempeh, 118 cultures were isolated from 81 tempeh samples collected from markets in various parts of Indonesia. Tempeh collections were made on the islands of Java and Sumatra and ranged from locations at sea level to places in the mountains up to 1000 m above sea level, with mean temperatures ranging from 15°C to above 30°C. Collection sites were selected intentionally to answer the question of whether mold cultures isolated from cooler regions would be different from those from warm places. Most of the cultures which could produce tempeh in pure culture, however, turned out to be *R. oligosporus* and it was proven from that finding that it does not depend on the place where tempeh is produced (5).

Starter for tempeh production: The availability of an appropriate starter culture is essential for producing a good quality tempeh. Traditionally, Indonesian people prepared starter cultures by collecting small pieces of a previously fermented tempeh. The collected pieces of tempeh are air-dried or sun-dried, ground to a smooth powder and used as an inoculum. In some cases, the surface of a previously fermented tempeh cake, where most of the mycellium is found, may be sliced, sun-dried, ground and used as an inoculum.

Usar (also called *waru* or *laru*) is the most popular starter culture used for tempeh preparation in Indonesia.

Usar is prepared by using leaves of either *Erythrina* sp, *Hibiscus dimilis* B or *Hibiscus tiliaceus linn* whose local name is *waru putih*. The lower sides of these leaves are covered with downy hairs (trichomos) to which the mold mycellium and spores can adhere.

Usar is prepared as follows: First place a Hibiscus leaf on a tray with lower hairy side facing up, sprinkle 30 to 40 inoculated soybeans over the surface of the leaf. Then place a second leaf about the same size with lower side down on top of the first leaf to form a sandwich with inoculated soybeans in between them. Likewise, several sandwiches are prepared and wrapped in a perforated plastic wrap and placed in a wooden tray. The wooden tray containing the sandwiches is covered with burlap and allowed to ferment for 5 to 6 hr. During fermentation, the mold will grow on the soybeans and hairy lower sides of the leaves. The leaves are removed, sun-dried and stored until they are used for inoculation (Fig. 1) (6).

Fig. 1. *Usar*, tempeh starter made of leaves.

A small leaf of *usar* inoculates about 3 kg of dry soybeans that have been previously soaked, dehulled, and cooked. This method is more popular because it produces the pure traditional starter culture. Futher, growing the mold between leaves eliminates contamination of microorganisms from outside. In some parts of Indonesia, other leaves with hairs, especially teak (*Tectona grandis*) also are used for producing mold starter culture (6).

Although "*usar*," a traditional inoculum or starter has been used for centuries in preparation of tempeh, very little is known about its characteristics and biochemistry. Therefore, it is important to study in detail the microbial and biochemical properties of *usar*.

Tempeh Preparation

Two methods (traditional and pilot plant production) for production of tempeh are presented here. For a broad range of tempeh preparation methods, the reader should consult Steinkraus (3) and Shurtleff and Aoyagi (7).

Traditional: The essential steps for the traditional preparation of tempeh are presented in Figure 2. First, the soybeans are soaked and cooked for 30 min in boiling water to loosen the soybean seed coats. The seed coats of cooked soybean are hand removed or rubbed with feet to loosen the seed coats and washed with water to separate dehulled beans from seed coats. The dehulled beans are again soaked overnight to hydrate and allow bacterial acid fermentation. The soaked dehulled beans are cooked again, drained, cooled and inoculated with a starter culture or an inoculum from a previous batch, wrapped in banana leaves or perforated plastic bags and incubated for up to 18 hrs at room temperature. A photograph of a traditional tempeh from Indonesia is shown in Figure 3.

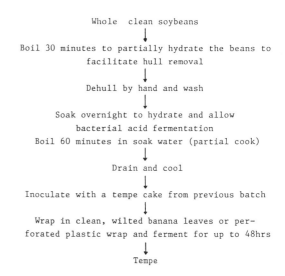

Fig. 2. Flow sheet for Indonesian household tempeh processing.

Fig. 3. Fresh tempeh (banana leaf wrapping, left; plastic bag wrapping, right).

Under natural conditions, the soybeans undergo a change (bacterial acid fermentation) during hydration prior to cooking and bacterial fermentation protects the tempeh from the growth of undesirable microorganisms (8,9).

When large amounts of tempeh need to be prepared, the soaked whole soybeans are placed in bamboo baskets at the edge of a stream and rubbed with the feet so as to float seed coats away from the dehulled beans in the running water. The dehulled beans are then boiled without pressure in excess water for 30 min (3). The cooked, dehulled soybeans are drained, cooled and inoculated with a starter culture or an inoculum from a previous batch. The mass of inoculated soybeans are wrapped in banana leaves or perforated plastic bags and tied with a string and incubated for 48 hrs. The freshly prepared tempeh is sold daily in Indonesia.

Several researchers have studied and reported the use of soybeans grits for preparation of tempeh with minor modifications.

Pilot plant production: Steinkraus et al. (10) developed a pilot plant method for the production of tempeh. This pilot plant method involved size grading, dry dehulling, hydration, acid fermentation, cooking, draining, cooling, and inoculation with mold and fermentation for 18 hr at 35–38°C and 75–85% relative humidity.

Hydration and acid fermentation will result in a drop of pH to 4.3–5.3; a further drop in pH prevents the growth of undesirable microorganisms that might contaminate tempeh. Soaking and discarding of soaked or cooked water helps in partial elimination or reduction of antinutritional or toxic factors (trypsin inhibitors, phytic acid, tannins, flatulence factors and others) from beans which subsequently improves the nutritional value of tempeh. Sometimes, the presence of water soluble antinutritional compounds in the soybeans inhibits mold growth.

About 3 g of dried, pulverized tempeh mold (*R. oligosporus*) culture containing both mycellium and spores were used to inoculate 1 kg of cooked, drained beans. The inoculum should be thoroughly mixed with the beans.

Losses and yields: The steps (dehulling, soaking, draining, and fermentation) used in the preparation of tempeh may contribute losses of soybean solids, protein, and nutrients (8,11). Average solids losses that occur during preparation of tempeh are presented below. The total losses of solids ranged from 24.5% to 48.3% depending on the type of soybeans (whole beans or grits) and variety used. It appears that the use of soybean grits in the preparation of tempeh results in greater losses of solids compared to whole soybeans. The solids losses of soybeans may be reduced partially by adding an equal amount of water to soak the beans (beans to water ratio of 1:1 w/v) and by not discarding the cook water. However, this procedure may result in poor quality of tempeh cake due to possible bacterial growth during fermentation. The solids losses and yield occur during tempeh production are shown in Figure 4. On a dry weight basis, the solids recovery was 78.1%, the total solids loss during tempeh production was 21.9% and attributed to various production steps: 8% due to hulls removal, 12.2% during soaking and cooking and 1.7% during fermentation (8).

Utilization and Storage Stability

Harvesting and Storage: Tempeh should be harvested as soon as the bean cotyledons have been overgrown completely and knitted into a compact cake (Fig. 4). Freshly made tempeh can be stored for a day or two at room temperature without changing many of its qualities and flavor characteristics. Traditionally in Indonesia, tempeh is consumed on the day it is made. If the fresh tempeh is

Fig. 4. Yields and losses in tempeh production.

stored for longer periods of time (more than two days) at room temperature it becomes unfit for consumption because of off-flavors and odors produced during storage by contaminating bacteria and others.

American vegetarians like "tempeh Burgers." These are prepared by cutting a fresh tempeh cake into either square or circle shaped pieces of about 1.5 cm thick. The cut pieces are placed in a covered skillet and steamed for 15 min with sufficient water to cover the bottom of the skillet. The steamed tempeh pieces are then fried in a cooking oil until both sides are golden brown. The fried tempeh is served in a bun with tomato, onion slices, lettuce and sauce (12).

Storage stability of tempeh can be extended by either drying, drying and frying, dehydration and freezing.

Tempeh in Indonesian Foods

In Indonesia the tempeh cake is cut into thin strips and sun-dried. The sun-dried strips are dipped in a spiced thin rice dough, deep fried in a cooking oil, and packed in sealed plastic bags for later consumption. These deep-fried strips remain tasty and acceptable for several weeks or months when stored at room temperature. Tempeh also is used in the preparation of an Indonesian dish, *sambal goreng kering tempeh*, where the half sun-dried thin tempeh strips are mixed with hot pepper and brown sugar and deep fried in a cooking oil. This product also remains tasty and acceptable for several days to weeks (1).

Freshly made tempeh is a white compact cake, has a yeasty odor and does not have a beany flavor. Tempeh is used as a main dish in Indonesia. Tempeh often is served in various ways, such as fried in oil, baked, or as a soup. Usually, the cake is sliced thin, dipped in a salt solution, and deep-fat fried in coconut oil until its surface is crisp and has a golden brown color. Following deep-fat frying, the flavor becomes nut-like and peppery due in part to the presence of free fatty acids. Deep-fat fried tempeh is highly acceptable even to people of the western world. Overripened tempeh is often used in small amounts as a seasoning in soups and dishes such as *Sambal tumpang* and *sayur loden*. Ching reported on the preparation of a shrimp flavored tempeh using a thermo-tolerant *Rhizopus* sp. T-3 isolated from a native tempeh collected from Bali Island (1).

Nutritive Value

Hackler et al. (13) studied the nutritional value of tempeh fermented for various periods of time (0, 12, 24, 36, 48, 60, and 72 hr). They found that the nutritional value of tempeh is reduced with an increase in fermentation time. Further, they reported that PER value remained fairly constant throughout the course of fermentation. Recently Zamora and Veum reported that fermentation of cooked soybeans with *R. oligosporus* greatly improved the apparent digestibility and net protein utilization in rats fed diets containing tempeh. The digestibility coefficient of tempeh is about 86.2%, which is lower than that of tofu (95%). Most of the foods prepared from dry beans are difficult to digest because they contain a variety of antinutritional and/or toxic factors that interfere in the availability of minerals and amino acids. Tempeh has a biological value of 58.7 (14).

Deep-fat frying may affect the digestibility of proteins and the nutritional value of tempeh. Stilling and Hackler reported that deep-fat frying of tempeh decreases the PER, weight gain in rats, and digestibility coefficients. Deep-fat frying of tempeh reduces the PER by 2.0% (15).

Storage Stability

Steinkraus et al. (10) developed a method for production of dehydrated tempeh. They cut freshly prepared tempeh into small squares (2.5 cm) and dehyrated them in a circulating hot air dryer at 69°C for 90–120 min. During drying, the moisture content of the tempeh squares was reduced to 2–4%. The dehydrated tempeh can be stored in a plastic storage bag for months at room temperature without changing the color and flavor. However, Steinkraus et al. found changes in reducing substances, soluble solids, and soluble nitrogen content of tempeh due to hot air dehydration. It is not known whether the reduction in these solubles represents a loss in nutrient value of tempeh.

Hesseltine et al. (16) suggested that the tempeh can be stored by a freezing method. This method involved blanching and freezing. The freshly prepared tempeh is cut into thin slices and blanched for 5 min in boiling water to inactivate the mold and proteolytic and lipolityc enzymes. The blanched slices are placed in cellophane packages and frozen in a deep freezer. They further reported that the blanched tempeh slices can be stored for up to 100 days by the freezing method without any changes in organoleptic characteristic (appearance, color, and taste).

Iljas et al. (17) studied the stored stability of tempeh in sealed cans. The cans containing blanched tempeh slices can be stored for up to 10 weeks without any significant changes in acceptability.

Another method to prolong the shelf life of tempeh is to defer the fermentation. In this method the plastic tubing (5x15) is filled with pre-inoculated tempeh and kept in a freezer. Whenever tempeh is needed, a small part of the plastic tubing containing pre-inoculated tempeh is cut, perforated for aeration and allowed to ferment in a warm place (30°C) for 20–22 hr. Similarly, pre-inoculated beans in small packages can be stored in the freezer in retail stores and sold as a ready-to-ferment tempeh.

Problems of Contamination

Tempeh contamination is always a serious problem in family and institutional consumption. Tempeh, which is a nice smooth, white-looking product, will change easily into an ugly looking product, get wet and its mycellium become weak and produce sweet acid smell from ammonia.

If tempeh becomes spoiled this is due to the predominant contaminant microorganisms most probably bacilli, mainly from *Bacillus subtilis* (18). This microorganism enters into product along with the soybeans, and grows and develops rapidly in an unsanitary environment. *B subtilis* can grow well in the warm and humid conditions which exist in many parts of Indonesia.

Even though *B. subtilis* is the main contaminant, there are other potential microorganisms which may cause undesired products in tempeh. Firstly, the yeast, rhodotorula, may produce red spots on the surface of tempeh after 2–3 days storage. The second microorganism is the *Psendomonas* sp; cool-loving microorganism which produces a yellow color like mustard or sometimes produces no color but produces strong undesired flavors. The third microorganism is mold, which produces slime and yellow color. None of these three microorganisms causes spoilage.

B. subtilis: This is not easily destroyed, as these microorganisms produce spores which are very heat resistant in nature. The vegetative form is easily destroyed within 1–4 minutes in boiling water, however, its spores sometime still survive after 20–30 min or 1 hr. in boiling water. During soaking, the spores germinate into vegetative form and can be easily destroyed by heat. But during more than 8 hrs of soaking at room temperature, the process of germination produces more spores. The chemical compounds that may be utilized in the food industry for sanitation purposes particularly to destroy spores are bleaching compound, i.e. 5% sodium hypochlorite or common bleach.

To avoid contamination, a strict sanitation program should be exercised religiously including strict contamination control on raw material (soybeans). Soybeans having a moisture content of 10% or more have a tendency toward high contamination. For that reason, storage at lower temperatures and humidities is recommended.

The tempeh starter should be checked routinely to avoid any contamination. If *B. subtilis* contamination is the cause of tempeh spoilage, larger amounts of starter (160%) than ordinary are used. In doing so the total amount of mold will be much higher than the contaminants.

The other way to reduce contamination is to reduce incubation temperature down to 28–29°C. By doing so, the growth of the *B. subtilis*, which has an optimum growth at 40°C, will be suppressed.

Sanitation program through disinfection should be conducted to all surfaces which have contact with food. Using bleach solution (bleach:water 7:3), spray the surface, and let it stay for four hours in order to have enough contact time to kill the spores and vegetative forms of *B. subtilis*. All the surface of the walls of incubator should be wiped with wet cloths which have been dipped in quaternary ammonium chloride solution (18).

References

1. Winarno, F.G., and N.R. Reddy, in *Legumes Based Fermented Foods*, edited by N.R. Reddy *et al.*, CRC Press p. 95 (1986).
2. Winarno, F.G., The Present Status of Soybeans in Indonesia. Fatemeta. Bogor Agricultural University (1976).
3. Saono, Susono *Berita Ilmu Pengetahuan dan Teknologi* 20(1) (1976).
4. Steinkraus, K.H., ed. *Handbook of Indigenous Fermented Food*, Marcel Dekker, New York (1983).
5. Saono, S., S. Brotonegoro, S. Abubakar, T.J. Basuki, and I.G.P. Badjra, *Microbiological Studies on Tempeh, Kecap and Tauco,* ASEAN Project on Soybean and Protein Rich Foods (1976).
6. Ko Swan Djien and C.W. Hesseltine, *Economic Microbiology* 4:115,197 (1979).
7. Shurtleff, W., and A. Aoyagi, *The Book of Tempeh: A superfood from Indonesia*, Harper & Row, New York (1979).
8. Steinkraus, K.H., B.H. Yap, J.P. van Buren, M.I. Provdenti and D.B. Hands, *Food Res.* 6:777 (1960).
9. Hesseltine, C.W., R. Decamargo and J.J. Rackis, *Nature.* 220:1226 (1963).
10. Steinkraus, K.H., J.P. van Buren, L.R. Hackler and D.B. Hand, *Food Technology* 19:63 (1965).
11. Smith, A.K., J.J. Rackis, C.W. Hesseltine, D.J. Robbins and A.N. Booth, *Cereal Chemistry.* 41:173 (1964).
12. Bates, C., A. Lyon, S. Sorenson, B. Keller and S. Jenkins, *Utilization of Tempeh in North America*, Symp. Indigenous Fermented Foods, Bangkok, Thailand (1977).
13. Heckler, L.R., K.H. Steinkraus, J.P. van Buren and D.B. Hand, *J. Nutrition* 82:452 (1964).
14. Winarno, F.G., *The Nutritional Potential of Fermented Foods in Indonesia in Traditional Food Fermentation as Industrial Resources in ASCA Countries*, The Indonesian Institute of Science (LIPI) Jakarta (1981).
15. Wang, H.L., D.I. Ruttle and S.W. Hesseltine, *J. Nutrition.* 96:109 (1968).
16. Hesseltine, C.W., M. Smith, B. Bradle and Ko Swan Djien, *Dev. Ind. Microbial,* 4:275 (1963).
17. Iljas, N., Master thesis, Ohio State University. Columbus, Ohio (1969).
18. Shurtleff, W., G. McBride, J. Robertson and T. Burgeson, *Soyfoods (Winter)*:29-31 (1982).

Miso Preparation and New Uses

Toshio Hanaoka

Hanamaruki Inc., 4-22-10 Himonya, Meguro-ku, Tokyo 152, Japan

Abstract

Miso is generally used as a soup ingredient, along with vegetable, seaweed, tofu and cooked meat and shellfish. Flaked dried bonito fish, sliced tangle (seaweed), monosodium glutamate (MSG) and 5'-ribonucleotides also are often used as seasonings in miso soup. A phosphatase in miso koji (soybeans initially cultured with *Aspergillus oryzae*) cleaves 5'-ribonucleotides but can be deactivated by heat treatment and cooling to develop new types of miso seasonings. Bacterial contamination is controlled by alcohol and miso yeasts. New miso consumer products emphasize ease of preparation and wholesomeness, and low-salt misos have been developed.

Miso is a semi-solid fermented food made from soybeans, rice or barley, and salt. Similar types also are produced in the other parts of east Asia. It is referred to as "chang" in China, "jang" in Korea, "taucho" in Indonesia and "tao-tsi" in the Philippines. The prototype of miso was introduced to Japan from Korea at least 1,300 years ago. Misho (the original type of miso) was first cited in Taihouryou in 701 A.D. Although the production method for misho was not described, it probably resembled soybean miso today. In the northeast district of China, it was called "misun," and in Korai (located in north Korea 935 to 1392 A.D.) "misso" and in south Korea "mijo." The process includes pounding and mashing of cooked soybeans, shaping into balls, wrapping in rice straw, hanging the wrapped materials under the eaves and mixing with salt and water and often was seen at farmhouses in prewar Japan. Its modified production process, employing a motor driven extruder for ball making, mold starter and koji fermenter, was widely used for making soybean miso in the prefectures of Aichi, Mie and Gifu located in the central part of Japan. This method has been changed by employing rice koji resulting in the development of the characteristic rice miso in Japan. Miso originally was made at Buddhist monasteries for internal and use in aristocratic circles.

Another route of introduction was from China to Japan by Buddhists and envoys who produced fermented soybean foods such as Kinzanji-miso. The difference of this process from the Korean process was to make koji from cooked whole soybeans with barley or wheat without pounding.

The industrial production of soybean miso was started in Aichi prefecture in 1625 and that of rice miso in 1645 at Sendai as the original of Sendai miso today, which is a typical variety of salty, red rice miso. Thereafter, the number of miso factories increased gradually in the Tokugawa period (1600–1867) and rapidly up to about 5,000 at the end of World War II. Recently, miso factories introduced many advanced facilities and equipment for economical and hygienic production. The number of miso factories was reduced to about 1,700, amalgamating to large scale factories of which capacity was over 3751 tons in 1984. They produced 353,000 tons equivalent to 59% of total industrial production. Employing 182,000 tons of soybeans, 104,000 tons of rice, 23,000 tons of barley and 72,000 tons of salt, *ca.* 568,000 tons of miso were produced by all factories in 1985. In addition, the amount of home-made miso is roughly estimated as 60,000 tons. The amount of miso supplied per capita was 4.9 kg in 1987.

Although there are many varieties, they can be classified into three major types on the basis of raw materials: rice miso made from rice, soybeans and salt; barley miso made from barley, soybeans and salt; and soybean miso made from soybeans and salt. Some of these three types are further divided by taste into sweet, medium salty, and salty groups and each group is further divided by color into white, light yellow and red (Table 1) (Fig. 1).

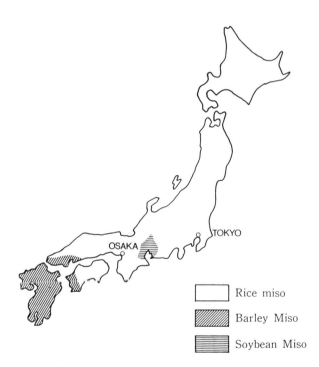

Fig. 1. Geographical distribution of miso in Japan.

Ninety per cent or more of miso is used as an ingredient of soup in Japan whereas most miso-like products are used as seasonings in most countries in south and east Asia. Miso soup is prepared simply by dissolving one part of chopped miso lump in ten parts of hot water in which suitable ingredients such as vegetables, tofu, fried tofu, mushroom, meat, fish, shellfish, etc. have been cooked previously. Daily per capita consumption was 28 to 30 g during the years from 1951 to 1959. From that time until 1985, it gradually decreased to the level of 16 g which is enough to prepare one bowl of miso soup.

TABLE 1

Varieties of Miso

Varieties	Taste	Color	Principal areas of production in Japan
Rice miso	Sweet	White	Kyoto, Kagawa, Hiroshima
		Red	Tokyo
	Medium salty	Light yellow	Shizuoka
	Salty	Light yellow	Nagano
		Red	Tohoku province, Hokkaido, Niigata, Kanto province, Hokuriku province, Sanin province
Barley miso	Sweet	Light yellow	Kyushu province, Chugoku province, Ehime
	Salty	Red	Kyushu province, Kanto province
Soybean miso	Salty	Red	Aichi, Mie, Gifu

The cost for one kg of packaged miso ranges from 300 yen to 1,300 yen. The wide range in the cost of miso is due principally to the cost of the raw materials, labor and duration of fermenting period which ranges from two weeks for sweet miso to three years for some varieties of soybean miso and salty rice miso.

Preparation

Raw Materials

Since the quality of miso depends markedly on the quality of raw materials, the selection of suitable raw materials is a key point to prepare good quality miso. The chemical composition is presented in Table 2.

Soybeans: Yellow soybeans rich in protein generally are used. Japanese domestic soybeans are more suitable for making miso than imported soybeans. Suitable soybeans are soft, smooth and sweet when cooked. Soybeans of light yellow color of seed coat and hilum are suitable particularly for making white or light yellow miso. Soybean varieties of higher water-absorbing and water-holding capacity, higher carbohydrate and lower oil and calcium content also are suitable (1).

Dehulled soybeans or soybean grits are often used for making white miso or light colored miso (2,3). Defatted soybeans are not suitable for miso preparation because they give poor color and texture to the final products (4).

Rice: Nonglutinous polished rice is used for making rice miso. Rice containing unripened kernels, different shapes, damaged kernels, or foreign matter is not suitable. Soft types or *Oryzae sativa var japonica* is suitable. On the contrary, hard types or *Oryzae sativa var indica* generally is not preferred.

Barley: Barley, including naked barley produced in western Japan, is used after milling to 70% yield. Barley with thin coats and brilliant yellow color is preferable for making barley miso.

Salt and water: Food grade salts with more than 95% purity generally are used. Salt containing less than 1 ppm iron is preferable. Water with reduced content of calcium and iron is preferable, since calcium often causes hard cooked soybeans and ferric ion accelerates the browning reaction of miso during fermentation and storage.

TABLE 2

Chemical Composition of Raw Materials

Raw material	(%) Water	(%) Protein	(%) Lipid	(%) Carbohydrate	(%) Ash	Ca[d]	P	K
Milled rice[a]	15.5	6.8	1.3	75.8	0.6	6	140	110
Milled barley[b]	14.0	8.8	2.1	74.2	0.9	24	140	200
Whole soybeans[c]	12.5	35.3	19.0	28.3	5.0	240	580	1900

[a] Milling yield: 90–92%
[b] Milling yield: 60–65%
[c] Dried Japanese whole soybeans
[d] Ca, P, and K: mg per 100g

Tane-koji (koji-starter): Tane-koji is olive green in color and consists of the spores of *Aspergillus oryzae* cultivated on cooked brown rice with small amount of ash from hard wood. One kg of tane-koji is enough to inoculate 1000kg rice or barley. There are many varieties of commercial tane-koji, each having different capability in breaking down protein, carbohydrate and lipid in raw materials. It is very important to select suitable variety for the variety of miso. For salty rice miso rich in protein, tane-koji of high proteolytic activity is suitable whereas for sweet rice miso rich in starch, that of high amylolytic activity is preferable.

Halophilic yeast and lactic acid bacteria: Halophilic yeast, *Zygosaccharomyces rouxii* and *Candida versatilis,* play a very important role in the fermentation of miso. Traditionally, soundly fermenting miso from a previous batch was used to inoculate the mixture of salted koji and cooked soybeans. Nowadays, the pure culture of the halophilic yeast generally is used. Culture media containing 10–12% raw soy sauce, 10% glucose and 10% NaCl are used for propagation of halophilic yeast. Incubation with aeration is carried out at 30°C until the count of viable yeast reaches the level of 10/ml. One liter of the culture is enough for the inoculation of one ton of miso. *Pediococcus halophilus* is used as a lactic starter. Culture media containing 20% raw soy sauce, 2% glucose and 5% NaCl are used. The pH value is maintained at 7 during cultivation. The count reaches 10/ml in about 60 hr. One liter of the culture is used for one ton of miso. These pure cultures are supplied by the laboratories under the miso industrial cooperative or are commercially available.

Commercial Preparation of Miso

There are many varieties of miso that differ from each other more or less in the manufacturing process. The manufacturing method for salty rice miso, the most typical one is described (Fig. 2).

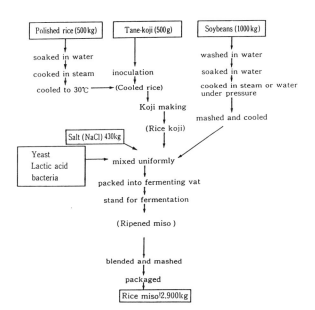

Fig. 2. Flow sheet for manufacture of rice miso.

Koji preparation: Polished and cleaned rice is washed and soaked in water at *ca.* 15°C over night. After draining the soaking water, the rice is cooked in steam for around 40 min in an open cooker. When cooled to 35°C, the cooked rice is inoculated with tane-koji and mixed well. The inoculated rice is incubated in a box designed to keep the temperature and humidity for the growth of the mold. Suitable temperature and relative humidity are 30°C and 96% respectively. In the koji-tray method, the incubated rice in the box for about 15 hr is transferred to the koji-trays and spread to a 4 cm depth and allowed to ferment in a koji-chamber. When the temperature goes up to 35°C or more, the young koji is turned over and stirred and allowed to grow until sponge-like koji is obtained after approximately 40 hr from inoculation. Recently the koji tray method has been replaced by a mechanical method with a koji fermenter equipped with an air conditioner and mechanical stirrer to break the sponge-like lump during cultivation. The complete koji is taken out from the koji-fermenter mechanically to be mixed with salt to halt any further mold development.

Cooking soybeans: Cleaned soybeans are washed and soaked at *ca.* 15°C overnight. It is possible to reduce the soaking time by elevating the temperature of the soaking water. There are two ways for cooking, one is in steam and another is in boiling water. To cook with steam, the soaking water is drained off from the soaking tank and the soybeans are transferred to a cooker to be cooked for 20–30 min under a pressure of $0.7 kg/cm^2$ to $1.0 kg/cm^2$. Generally, batch type cookers are used widely. Recently continuous cookers with rotary bulbs and belt conveyers have been used in the large scale factories. Soybeans are generally cooked in four parts of water for white miso or light colored miso to prevent browning during cooking. Cooking conditions are almost the same as cooking with steam. Cooking in water has an advantage in that cooked soybeans are soft and bright in color. On the other hand, it has disadvantages in increasing the amount of solids lost from soybeans to 10% to 20% of the dry matter. The desired hardness of cooked soybeans is 0.5 kg (weight necessary to break a cooked soybean) for salty miso. Cooked soybeans are cooled with a belt conveyer type cooler to room temperature. Sometimes, they are mashed with a chopper while hot and then cooled.

Mixing and mashing: Cooked soybeans are mixed with salted rice koji and water in which an inoculum of pure culture of yeast and/or lactic acid bacteria is inoculated. Then, the mixed materials are roughly mashed by passing them through a motor driven chopper with 5 mm perforations. Homogenous mixing is important to maintain the normal fermentation of miso. The mixture is packed tightly into a fermentation vat. The wooden vats used traditionally are now being replaced by steel vats coated with epoxy resin, stainless steel vats, or glass-lined resin vats. The surface of the miso is covered with a resin sheet. Weights equivalent to 5% to 10% of the miso are placed on the sheet. The young miso is allowed to ferment at ambient temperature for 6 months or more. Presently, it is often fermented in a fermenting chamber where the temperature is controlled automatically. The temperature is usually controlled to 30°C. It takes 10 to 14 days for sweet miso and 3 to 4 months for salty rice miso to ferment and ripen. During fermentation, miso is transferred from the original vat to another to retain the homogeneity of miso. The ripened miso is then blended if necessary, and mashed through a chopper with a plate cutter with perforations of 1 to 2 mm.

During mashing, miso is often pasteurized with a steam jacket. Finally, the miso is packaged in a resin bag or cubic container for market. When raw miso contains viable yeast, swelling of the bag is often caused during distribution. To prevent the swelling, raw miso is pasteurized or mixed with 2% ethyl alcohol or with less than 0.1% sorbic acid.

New Types and Uses of Miso
Miso with Seasonings

Miso generally is used as an ingredient of soup. Vegetables, seaweed, tofu, fried tofu, mushrooms, meat, fish, and shellfish are often used for making miso soup after it is cooked. Flaked dried bonito, dried small sardines, sliced tangle (kelp), monosodium glutamate (MSG) and 5'-ribonucleotide are often used as seasonings for making miso soup. However, it is difficult and takes much time to make the stock so complex seasonings that are a mixture of MSG and nucleic acid seasonings are now used in place of the stock. "Miso with seasonings," which is miso mixed with complex seasonings, was developed five years ago and now occupies 10% of total miso consumed. The main ingredient of the stock is a mixture of MSG, nucleic acid seasonings and processed goods that may be an extract or powder of natural materials such as tangle, bonito or mushroom. These natural products contain a lot of 5'-inosinic acid and 5'-guanilic acid as nucleic acids. These two substances and MSG, which is very tasteful, are utilized as seasonings. In miso, phosphatase which is produced mainly by Aspergillus oryzae hydrolyzes these nucleic acid seasonings to make them lose their taste effect. Phosphatase is inactivated by heating as shown in Fig. 3. From these data, we can prevent miso flavor from degradation by heating it at 90°C for five min followed by rapid cooling. The heat treatment of miso causes inactivation of enzymes and sterilization of harmful microorganisms. Thus, the taste of the nucleic acid seasonings can be maintained. Fig. 4 shows the result of the survival tests on nucleic acid seasonings in miso with seasonings.

Production of enzyme by microorganisms can be prevented after sterilization. Table 3 shows the composition of amino acids and nucleic acid seasonings of nonheated miso and miso with seasonings. The values of amino acids of nonheated miso are derived from material produced during fermentation.

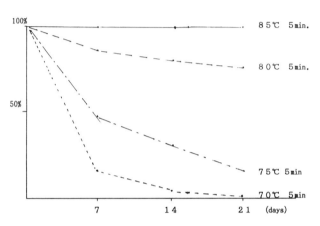

Fig. 4. Heat treatment temperatures and survival of nucleic acid.

TABLE 3

Amino Acids and Nucleic Acids in Miso and Miso with Seasoning

Amino acid	Miso mg/100g	Miso with seasoning mg/100g
ASP	375	349
THR	211	197
SER	204	179
GLU	448	980
PRO		
GLY	100	102
ALA	214	203
CYS		
VAL	182	162
MET	60	58
ILEU	155	138
LEU	206	194
TYR	144	130
PHE	129	127
HIS	110	118
LYS	148	126
ARG	150	143
Total	2836	3206
5'—IMP	3	57
5'—GMP	2	11
Total	6	68

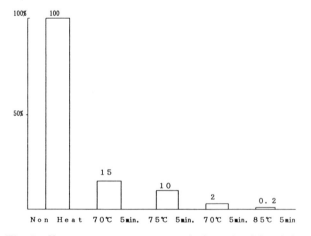

Fig. 3. Heat treatment temperature of miso and activity of phosphatase.

Low-salt Miso

It is a well-known fact that salt is essential for the human body, but it is said that the chronic excessive ingestion of salt causes hypertension and functional disorders of the heart and kidneys. Therefore, those who especially take good care of themselves and those who are restricting the ingestion of salt because of hypertension or kidney disease use low-salt miso. The concentration of salt in low-salt miso is 10% or less, as compared with that of general miso which is from about 11% to about 13%.

Main processes for low-salt miso: (a) Reduction of the moisture of miso: The moisture of miso is reduced mainly by reducing the moisture of steamed or boiled soybean. As a result the osmotic pressure of miso becomes higher so the growth of harmful microorganisms can be repressed.

(b) Addition of a large quantity of yeast: At the beginning of fermentation, *Zygosaccharomyces rouxii* is added to the miso at more than 10^6 cells per gram of miso. In addition, the growth of yeast is promoted at a temperature producing a large amount of alcohol in order to prevent abnormal fermentation.

(c) Addition of alcohol: At the beginning of fermentation, from 3 to 5 w% of alcohol is added to miso in order to repress the growth of harmful microorganisms.

(d) High temperature maturing: This is the process for white miso. The temperature of the mixture of rice koji, salt and boiled soybeans is held at 45° to 60°C for 12 to 24 hrs in order to promote enzymatic hydrolysis. Further, miso is matured at below 20°C, then refrigerated at below 5°C in order to promote enzymatic hydrolysis and repress the growth of microorganisms.

Instant Miso Soup

Dehydrated miso is prepared by freeze-drying or spray drying ordinary miso. The major use of dehydrated miso is for making instant miso soup as an ingredient with freeze dried vegetable food or animal foods and seasonings.

Composition

Chemical Composition

The fatty acid composition of several types of miso is given in Table 4. The moisture level ranges from 42.6% to 45.7%. Protein content ranges from 9.7% to 17.2%. Protein in the substrates is digested rapidly at early stage of fermentation by proteases to peptides and amino acids. Protein solubility degree (the ratio of water soluble nitrogen to total nitrogen) is approximately 60% and protein degrading degree (the ratio of formol-nitrogen to total nitrogen) is 25%. The amino acid content of miso is presented in the Table 3. Glutamic acid is the major amino acid in miso. It gives a meat-like flavor. Some amino acids including leucine, isoleucine, valine and methionine are decomposed to alcohols such as amylalcohol, butylalcohol, and methionol during fermentation by the yeasts. These alcohols or their ethyl esters give favorable aromas to miso. The level of fat in miso ranges from 3.0% to 10.5% depending on the amount of soybeans in the ingredients. According to Kiuchi (6), the fatty acid composition of most miso was similar to that of soybeans. During fermentation, lipids are hydrolyzed to free fatty acids and glycerol. Some of the fatty acids are esterified with ethanol to produce such esters as palmitate, oleinate and stearate (7). Linoleic acid is often decomposed by the yeasts. Carbohydrate content of miso ranges from 11.3% to 36.7%. The predominant sugar in ripened miso is glucose (10% to 13%) followed by isomaltose, fructose, galactose, galacturonic acid, stachyose, mannose, melibiose, arabinose and xylose (8).

Lactic, acetic, citric, pyroglutamic, succinic and pyruvic acids are the major organic acids in miso. The content of lactic acid in white miso, salty light yellow miso, salty red miso and soybean miso in 100g is 50, 40 to 150, 100 to 200, and 100 to 400 mg respectively. During fermentation, citric and malic acid decrease while others increase.

Ethanol and its esters are the most important aroma component in rice miso as well as barley miso. Isobutyl, butyl and isoamylalcohol are also important aromas of well-fermented miso. Vanillic acid and ferulic acid are the major components of the characteristic aroma of barley miso (9). Ferulic acid is often changed to 4-ethylguaiacol in fully fermented miso by *Torulopsis* (10).

Other Characteristics

Miso has a high buffer activity which maintains a constant pH value when miso is mixed with other foods (11). As the insoluble proteins in miso absorb unfavorable odors, miso is used as an ingredient to eliminate or mask the fishy or meaty odors in soups or dressings. Miso has a strong antioxidant activity and is used for storage of oily foods (12,13). According to Obata et al. (14), the biological value of nonsalty and low-salt miso is 73.2 and 70.7 respectively compared to that of casein, 80.1. The

TABLE 4

Fatty Acid Content of Miso[a]

Miso type	Palmitic acid	Stearic acid	Oleic acid	Linoleic acid	Linolenic acid
Salty light yellow rice miso	13.4	3.2	25.2	52.2	6.9
Salty red rice miso	11.5	2.8	22.6	54.6	8.3
Barley miso	14.1	2.8	20.1	55.2	7.1
Soybean miso	9.7	4.1	20.0	58.1	9.3

[a] Expressed as per cent of the total fatty acids

amount of unsaturated fatty acids including linoleic and linolenic acids is over 60% in miso (Table 4). The peroxide value of the fat in fully ripened miso is very low (0.04 to 0.06 meq/kg) indicating the presence of strong antioxidants (15). The content of sodium, potassium, and calcium in miso is comparatively high. Because the content of sodium chloride in miso soup ranges from 1.1 to 1.2%, one bowl of miso soup (150 ml) gives 1.65 to 1.8 g of sodium chloride. According to Yamori et al. (16), however, the Na/K ratio plays a very important role in the control of blood pressure. The ratio in miso can be lowered to 3 by preparing the miso soup using potassium rich foods such as potatoes, leafy vegetables and seaweed. In preparing miso soup, seaweed (wakame) is often used and the arginic acid in wakame has an interesting property of forming an undigestible complex with sodium in the intestines where the pH is alkaline (17).

Because of high calcium content, miso soup supplies about 15% of calcium intake for the inhabitants of Tohoku province (18). Among the vitamins in miso, the content of riboflavin and vitamin B_{12} is 0.1 mg and 0.17 g per 100 g of salty rice miso respectively. These are produced by microorganisms in miso (19,20,21).

Phytic acid is an antinutrient in soybeans and cereals. It reacts to form insoluble complexes with zinc, calcium, iron and copper resulting in their reduced bioavailability (22). Phytic acid, however, is easily hydrolyzed by the mold's enzymes during fermentation. According to Hirayama (23), a daily intake of miso soup reduces the mortality rate for gastric cancer in humans. The study was carried out on a large scale including 122,621 males and 142,857 females aged 40 or more in 29 health center districts of Japan between 1966 and 1978. Okazaki et al. (24) found a significant antimutagenic effect in the lipid fraction of miso. Ethyl linolenate formed in miso showed the highest antimutagenic effect.

Toxicology

There were no strains capable of producing aflatoxin among the 238 strains collected from factories which prepare koji. Furthermore, no aflatoxin was detected in 28 samples of rice koji from miso factories, 108 industrial miso samples, 30 home-made miso samples and 20 shoyu samples collected throughout Japan (25).

References

1. Ebine, H., Z. Matsushita, and H. Sasaki, *Rept. Natl. Food. Res. Insti., 26:*126 (1971).
2. Shibasaki, K., and C.W. Hesseltine, *J. Biochem. Microbiol. Technol. Eng. 3:*161 (1961).
3. Smith, A.K., C.W. Hesseltine and K. Shibasaki, U.S. Patent 2,967,108.
4. Ebine, H., K. Ito and M. Nakano, *Rept. Natl. Food. Res. Insti., 10:*133 (1955).
5. Ebine, H., "Integrated Research on Waste Reclamation in Bioconversion of Organic Waste for Rural Communities," at conference on State of Art of Bioconversion of Organic Residues for Communities, Guatemala City, 1978.
6. Kiuchi, K., T. Ohota and H. Ebine, *Nippon Shokuhin Kogy Gakkaishi 23:*455 (1976).
7. Yasuhira, H., I. Yonetani and T. Mochizuki, *Rept. Shinshuu Miso Inst. 11:*10 (1970).
8. Hondo, T., and T. Mochizuki, *Nippon Shokuhin Kogy Gakkaishi 26:*461 (1979).
9. Kuribayashi, Y., *Ibid. 14:*49 (1967).
10. Asao, Y., and T. Yokotsuka, *Nippon Nogei Kagaku Kaishi 32:*622 (1961).
11. Yamaguchi, N., Y. Yokoo and M. Fujimaki, *Nippon Shokuhin Kogyo Gakkaishi 26:*71 (1979).
12. Munesawa, T., N. Kosugi and T. Sagara, *Miso no Kagaku to Gijutsu:*8 (1976).
13. Watanabe, T., *Ibid.:*1 (1983).
14. Obata, Y., N. Matsuno and T. Tamura, in Rept. Natl. Nutrition Inst., No. 20, 1932, Tokyo, p. 20.
15. Ebine, H., unpublished data, 1975.
16. Yamori, M., R. Horie, Y. Nara and M. Kihara, in "Cerebrovascular Disease, New Trends in Surgical and Medical Aspects," edited by H. Banett, P. Paloletti, E. Flamm and G. Bramabiolla, Elsevier/North Holland Biomedical Press, Amsterdam, 1981.
17. Kawamura, M., Y. Nakagawa, K. Tuji and T. Ichikawa, "Na-Combining Capacity of Seaweed *in vitro*," presented at 38th Annual Meeting of Japanese Soc. of Nutrition and Food Sci., Kyoto, 1984.
18. Shimada, A., *Miso no Kagaku to Gijutsu,:*2 (1981).
19. Ebine, H., M. Nakajima and M. Nakano, *Rept. Natl. Food. Res. Inst., 10:*155 (1955).
20. Mogi, M., K. Murata, S. Iguchi and Y. Yoshida, *J. Ferment. Technol. (Japan) 29:*302 (1955).
21. Takahashi, J., *J. Food Nutrition (Japan), 8:*25 (1955).
22. Miyamoto, T., K. Murata and M. Kawamura, *Vitamin (Japan), 47:*233 (1973).
23. Sakamoto, M., and S. Iida, *J. Ferment. Technol. (Japan), 37:*11 (1959).
24. Wang, H.L., E.W. Swain and C.W. Hesseltine, *J. Food. Sci. 45:*1262 (1980).
25. Hirayama, T., "Does Daily Intake of Soybean Paste Reduce Gastric Cancer Risk?", presented to Japanese Cancer Association, Sapporo, 1981.
26. Okazaki, H., U. Kano and S. Kimura, presented at 37th convention of Japanese Soc. of Nutrition and Food Sci., Osaka, 1983.
27. Manabe, M., S. Matsuura and M. Nakano, *Nippon Shokuhin Kogyo Gakkaishi, 15:*341 (1968).

New Food Proteins, Extrusion Processes and Products in Japan

Akinori Noguchi and Seiichiro Isobe

Food Engineering Laboratory, National Food Research Institute, 2-1-2, Kannondai, Tsukuba, Ibaraki 305 Japan

Abstract

Drastic changes have been observed in textured vegetable protein (TVP) production with the extruder. Twin screw extruders (TSE), co-rotating and fully intermeshing type, are catching keen interest from the food manufacturers because of their wide acceptability of raw materials regardless of their water or oil content. Recent studies reveal that TSE equipped with a cooling die makes a meat analog with better texture even from wet, whole soybeans or highly denatured soy flour. DSC shows the melting point of soy protein near 130°C and high liquid content in raw materials increases the fluidity of melt materials above this temperature and allows them to have a well-aligned protein string matrix in the products with the aid of shearing effects at the die. The fibrous structure of extrudate is emphasized with egg white addition. Injection molding of the plastic industry can be applied also to soy protein and the special molding system attached to a TSE produces a larger block of meat analog than that of the usual extrusion die. SDS electrophoresis and the Kjeldahl method suggest that the major reaction among soy protein at dry extrusion is a kind of peptide bonding and that the reaction at wet extrusion is a typical -S-S-bonding. Extrusion cooking is thought to be the promising technology to produce a vegetable meat superior to usual "meat analogs" and can be used as a high potential, heat exchanger.

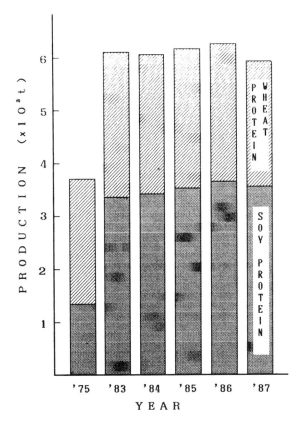

Fig. 1. Change of annual Japanese production of vegetable protein ingredients.

In Japan, soybeans are imported in amounts of 5 million metric tons annually and consumed for various purposes as shown in Figure 1. However, 4 million MT is consumed for edible oil production resulting in more than 3 million metric tons of defatted soybean. About 95% of these defatted soybeans is used for livestock feed because of heat denaturation of soy protein during the heat treatment in oil extraction. The remaining 1 million MT is consumed mainly in the production of traditional food like tofu, yuba, abura-age, natto and so on. Because Japan is short of protein resources, defatted soybeans can be important protein materials. However, this material has lost its protein functional properties and needs new processing methods. Extrusion cooking will be useful not only for this purpose but also for other unutilized or poorly-used proteinaceous materials.

Vegetable protein products (ingredients) were produced at 37,000 MT in 1975 and increased to about 60,000 MT in 1987, 59% of which was from soybeans and the remaining was from wheat protein (Fig. 1). Annual production for the past four or five years has not changed and seems to be saturated. Vegetable protein products currently distributed in the market are in the form of powders, paste (only from wheat protein), texturized granules and fibers. Figure 2 shows the production of these types except paste in 1987. It should be noticed that the texturized fiber and granule type of wheat protein are almost all frozen and that the reverse is true for soy protein. Both fiber and granule types of wheat protein have very similar texture to natural meat more than do those of soy protein and they maintain their texture while they are wet. Therefore, they are distributed frozen to have longer shelf life. There is still room for improvement in the fiber and granule types of soy protein.

Annual change of consumption of vegetable protein products clearly indicates that they have been used not by the consumer as dairy foods but by the food processors to improve the final food quality or sometimes to extend the food amount without changing the price (Figure 3). The limitation of current food processing technology leads to dry extrusion cooking and makes the cooking of vegetable protein products at home troublesome. The products should be rehydrated for a long time and squeezed

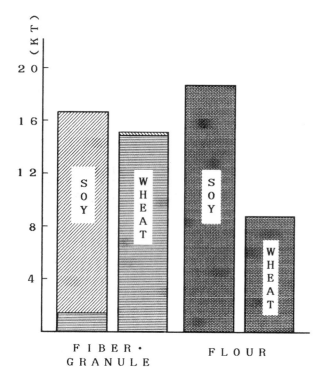

Fig. 2. Japanese 1987 production of various types of vegetable protein ingredients. Frozen types indicated by horizontal lines, others are dry types.

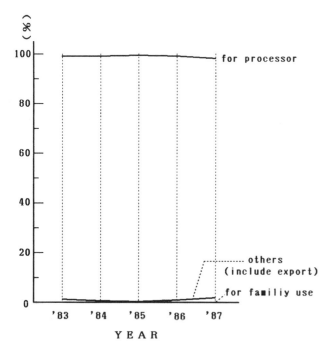

Fig. 3. Change in annual consumption.

to remove the excess water before use, which does not allow the manufacturers of these products to season them in advance. Further improved vegetable protein products with better texture have been desired.

Since Atkinson (1970) succeeded in the texturization of defatted soy flour (DSF) using a single screw extruder, many food manufacturers have used this technology for TVP production. The material transportation in the single screw extruder depends mainly on the friction between the material and metal. If the raw materials contain high amounts of water or fat, the materials will be co-rotated with the screw and cannot be processed. This transport phenomenon inevitably makes dry products with an expanded sponge-like structure and texture unsuitable for meat analogs. As Janssen (1978) has summarized, the principal differences between twin and single screw extruders is that the twin type is more suitable for the wet extrusion than the single screw type. When the material contains high amounts of water and comes out through the die after high temperature treatment, the excess heat accumulated in the materials changes the water to super-heated high pressure steam and leads to the explosive expansion or destruction of final products. Sair and Quass (1976) and Crocco (1976) suggested a cooling die to reduce expansion and to make more dense products. The combined equipment, the twin screw extruder with cooled die, seems to be one of promising ways to make more elastic and wet type of vegetable protein products with better texture.

The Japanese Research and Development Association on Twin Screw Extrusion Cooking was organized in 1984 by 26 companies including five machine manufacturers and sponsored for three years by the Japanese government's Ministry of Agriculture, Forestry and Fisheries. This paper will cover the selected results of the Association's studies on wet protein extrusion and discuss their application to the better production of meat analogs.

Wet Extrusion Cooking of Defatted Soy Flour

As there have been a few reports (Isobe and Noguchi, 1987, 1988; Noguchi, 1987) on wet extrusion of DSF in spite of many papers on the vegetable protein extrusion at moderate moisture conditions, preliminary experiments were required to obtain a steady wet extrusion.

A special cooled die with breaker plate (Fig. 4) was designed by developing the idea of Crocco and was attached to two twin screw extruders; Clextral BC-45 (L/D=12), France, and Mitsubishi FT-60N (L/D=20), Japan. Both extruders were fully intermeshing and corotating types. The breaker plate had many small straight holes and aimed to establish that sufficient materials filled

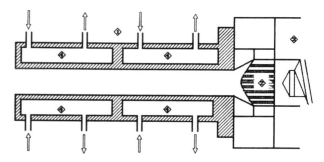

Fig. 4. Cooled die with breaker plate; twin screw extruder at right, cooling zone at left with cooling dies; breaker plate between extruder and cooling zone.

the heating barrel zone and also was employed to readjust the velocity distribution of melt fluid materials for the next cooling in the die. Heat was supplied to the barrels with electrical cartridge heaters (Mitsubishi) or induction heating unit (Clextral) because the dissipation of mechanical energy into heat could not be expected to do so when the materials contained large amounts of liquid. The necessary water to make the final water content more than 60% was supplied through the side wall of first barrel without conditioning because the adequate mixing function in twin screw extruder was well known. DSF used in this study was a type of highly denatured one and showed 50% of crude protein and 30% of Nitrogen Solubility Index (NSI).

Following is a description of materials change and protein texturization. The well mixed dough of DSF with water was moved into the heating section of barrel by the strong transportation ability of twin screw extruder and was heated homogeneously and quickly. The materials melted completely reacting each protein and were forced to pass through the numerous narrow holes in the breaker plate gaining homogeneous material flow. The presence of a high amount of water made the melt materials more fluid and prevented the components from fusing with each other, resulting in low viscoelasticity. Then the melt materials entered into the cooled die and obtained increasing shear force as its viscosity increased. The shear force deformed each elastic melt component parallel to the extrusion direction and, concurrently, the melt materials were fixed in shape, having an aligned fine structure.

Figure 5 shows the resulting flexible extrudate of wet DSF and any puffing and flashing have been not observed. The extrudate can be torn easily in the direction of extrusion. Saio et al. (1987) recently reported an effective method for detecting the protein, carbohydrate and fat distributions in the extrudate. Using the Saio method, Figure 6 clearly indicates that the protein components are changed to an aligned fibrous matrix and that the other components seem to be localized without mixing with protein components. These observations support the result of extrudate strength as shown in Figure 7. The lengthwise strength F_L, parallel to the extrusion direction,

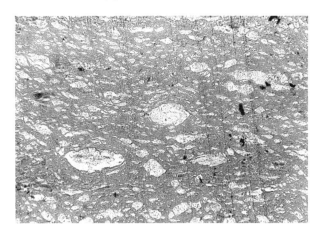

Fig. 6. Fine structure of DSF extrudate. Protein matric was stained with Coomassie Brilliant blue, which appears as black string in this photograph; extrusion direction is left to right; water content: 60%; barrel temperature, 150°C; machine: Mitsubishi FT-60N.

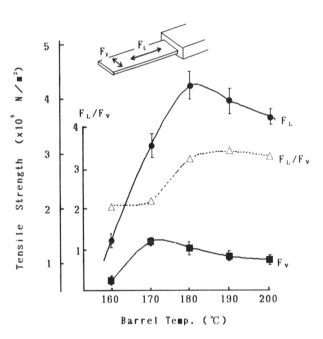

Fig. 7. Influence of barrel temperature on the Extrudate Strength of F_L and F_V. Material: DSF with 60% water; machine: Clextral BC-45; pH: adjusted to 7 with NaOH or NCl at the stage of extrudate.

Fig. 5. Twin screw extruder (Mitsubishi FT-60N) with cooled die and extrudate from wet DSF.

was increased to the maximum value around at 180°C as the barrel temperature increased, while the crosswise strength F_V was not so changed as F_L.

Influence of Barrel Temperature and pH on the Extrudate Strength

Differential Scanning Calorimetry (DSC) revealed an endothermic peak at 130°C for the used DSF containing 60% water. It is likely that the moist DSF would start to melt around 130°C and react the protein components. This temperature was low in comparison with 180°C of maximum value of F_L. Probably, the barrel length of the Clextral used in this study, would be too short (L/D = 12) for the DSF feed rate to raise up the temperature of DSF at a given barrel temperature and needed a higher barrel temperature than the result of DSC. In fact, the extruder with long barrel, Mitsubishi FT-60N (L/D = 20), showed the lower barrel temperature, 140°C, in order to obtain the firm extrudate. However, the extrudates prepared at 130°C with FT-60N were very fragile in spite of the fact that the Kjeldhal method and SDS electrophoresis showed the reduction of soluble protein fractions in phosphate buffer (pH 7) containing 1% SDS and 1% 2-mercaptoethanol. These results suggest that the better texturization of highly denatured DSF requires sufficient heating time at temperatures more than 130°C.

Texturization has been evaluated at times by the strength of obtained extrudate, but these results suggest to us that the "texturization" should be divided into the "reaction" and the following alignment or "reorganization" of reacted components and should then be examined separately. It is likely that the biopolymers crosslinked with each other will have a higher melting point than that of the original ones. If this is the case, the extrudate will be destroyed when it passes through the breaker plate or cooled die when the temperature of the final barrel section and breaker plate are maintained at too low a temperature for protein reaction.

The influence of pH on strength also was examined while keeping the barrel temperature at 180°C (Clextral BC-45). The strength, F_L, was more affected by pH than F_V and showed the maximum value at pH 7.

The protein in DSF generally exists in the form of very small granules called "protein bodies" with an average diameter of 10 mm. The protein matrix in the extrudate was observed to be composed of many fine fibrous protein strings. Their average diameter and length were 10 and 1 mm, respectively. Therefore, the average volume of each protein string can be easily calculated and found to correspond to 150 protein bodies. This calculation suggests that a kind of phase separation or component separation happens in the extruder during operation, as Tolstogusov et al. (1985) proposed for the formation of protein fiber near room temperature.

Thermoelasticity of Soy Protein and Application of Injection Molding

DSF used in this study has very low NSI and cannot be expected to be dissolved in the added water before it enters the heating zone of barrel. Therefore, the fine structure of extrudate observed in Figure 7 teaches us that the soy protein bodies are not dissolved but are melted and fused with each other during extrusion cooking. This thermoelasticity of soy protein also can be confirmed by multiple extrusion cooking.

The first extrudate was cut into small pieces and extruded for an additional two times. When the fine structure of extrudate was examined after the multiple extrusion cooking, the protein fibrous structure was observed to reform each time. Based on this observation, the injection molding method was investigated for its effectiveness in making a large lump of extrudate. The injection molding was designed as shown in Figure 8 and attached to the extruder in the place of the cooled die. The mold is heated to the same temperature as the final barrel section and the plunger in molding retreats with keeping constant back pressure as the materials are extruded. After the materials fill the mold, the valve is switched to change the material flow to another mold. The material in the mold is then pushed out by the plunger after the mold is cooled to room temperature. Figure 9 shows one of the products made by this injection molding method. Because of the small velocity of material flow in the mold, this method can not make a product with a fibrous structure differing from the cooled die. However, injection molding can be thought of as a promising method for making a large lump of TVP, which is impossible with the current extrusion cooking.

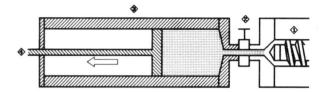

Fig. 8. Injection molding system; from right to left: twin screw extruder, valve, solid housing and plunger.

Fig. 9. The product of injection molding; material: DSF with 60% water; machine: Mitsubishi FT-60N; barrel temperature: 150°C. Cut product shows cross view.

Possible Model of Texturization Process of DSF in Extruder

It requires further investigation to determine the mechanism of protein texturization in the extruder. However, the results above mentioned allow us to imagine the texturization of DSF in the extruder as follows; after DSF is mixed with added water or an additive in the extruder, it is transported to the heating zone. There it will encounter the critical low temperature for its melting and/or reaction and also the critical high temperature for the decomposition of its components. After the melting of DSF, the proteins repeatedly fuse with each other and disperse in DSF with the screw movement and the phenomena of component separation. Then DSF is heated to a higher temperature in order to maintain its melting condition after the cross-linking of proteins. The satisfactory melt and reacted DSF is moved to the cooled die. As DSF is cooled, it increases the viscosity and shear force, which deform each component parallel to the material flow while the components keep their elasticity. Finally, DSF is cooled enough to have a fixed fine structure and is pushed out through the die mouth.

Some Examples Related to TVP Research by the Association of Research and Development on Extrusion Cooking

Japanese food companies and manufacturers of food processing equipment have keen interests in the possible multifunctions of twin screw extruders and the recent active research on extrusion cooking. As some oil mill companies already have been producing TVP with a single screw extruder, they are particularly interested in the twin screw extruder. This situation led them to organize the association for research on twin screw extruder and to establish it in 1984. The Ministry of Agriculture, Forestry and Fisheries has sponsored this association for three years and requested us to support its activity. In November 1987, the association had an open forum to show their selected results and published them as a book titled "Extrusion Cooking: Development of Twin Screw Extruder and Its Application." Table 1 shows the members of the association.

TABLE 1
Member List of Association of Research and Development on Extrusion Cooking

1. Food Manufacturer
   Cereals
       Kanro Co., Ltd.
       Nisshin Flour Milling Co., Ltd.
       Yamazaki Baking Co., Ltd.
       Asahi Denka Kogyo Co., Ltd.
       Risbon Confectionery Co., Ltd.
       Fuji Shokuhinn Kogyo Co., Ltd.
       Nippon Flour Mills Co., Ltd.
   Plant Proteins
       Nisshin Oil Mills Ltd.
       Gunma Beans Co., Ltd.
       Ajinomoto Co., Inc.
       Showa Sangyo Co., Ltd.
       Meiji Seika Kaisha, Ltd.
   Animal Products
       Nippon Meat Packers, Inc.
       Prima Meat Packers, Ltd.
       Meiji Milk Products Co., Ltd.
       Ito Ham Foods Inc.
       Q.P. Corporation
   Marine Products
       Taiyo Fishery Co., Ltd.
       Oriental Yeast Co., Ltd.
       Nippon Suisan Kaisha Ltd.
       Ajikan Co., Ltd.
2. Extruder Manufacturer
       Kobe Steel Ltd.
       Kowa Kogyo Co., Ltd.
       Kurimoto Ltd.
       Suehiro Iron Works Co., Ltd.
       Toshiba Machine Co., Ltd.

Following are a few related results on TVP in the book that are discussed regardless of wet or dry extrusion.

Preparation of Very Porous TVP from Soy Protein Concentrate

It is well known that the addition of protein regardless of its origin prevents the starchy extrudates from expanding. The proteins seem to have higher viscosity than that of starches while they are in melt condition and will act as a rigid network in the extrudate. Therefore, it is supposedly difficult to make a porous extrudate with low density from materials mainly composed of protein. The reason for making porous TVP is to improve the rate and amount of water absorption in order to enhance its consumption in the production of TVP foods. Nisshin Oil Mills Ltd. carried out this research using the twin screw extruder at the following conditions: (a) Machine: 2EXT-60 with 3 barrels (Kobe Steel Ltd.), fully intermeshing corotating type, L/D = 20; (b) Temperature profile of barrel: first, no heating; second, 110–120°C; third, 130–140°C; (c) Feed rate: 26–32 kg/hr; (d) Added water: 17–51 l/hr; (e) Screw speed: 100–120 rpm; (f) Die: 6 mm × 2–4; and (g) Materials: soy protein concentrate (SPC) or soy protein isolate (SPI) Additive: egg white powder, ethanol, sodium bicarbonate.

The experimental planning method was used for both SPC and SPI to find the best operational condition. Table 2 shows the relationships in the case of SPC among the operational parameters, apparent specific gravity, amount of water absorption and degree of expansion. The rate of water absorption was in the range of 160–600 sec. It is very interesting to have found very small coefficient of correlation ($r = -0.2724$) between apparent specific gravity and degree of expansion. Basing on these results, the best conditions for the target product were found as shown in Table 3 and the resulting extrudate was recognized to have very fine porous structure, showing 0.259 of apparent specific gravity and 463% of water absorption. When the material was SPI, the best result was 0.06 of apparent specific gravity and 769% of water absorption.

TABLE 2
Influence Degree of Operational Parameters on the Apparent Specific Gravity, Amount of Water Absorption and Degree of Expansion

Parameters	Apparent sp. gravity	Water absorption	Expansion degree
Feed rate	41.1	27.9	11.7
Added water	15.2	31.3	4.9
Screw speed	10.8	4.9	
Temp. of barrel	13.9	17.8	
Spacer (E)		0.9	
Number of die (F)	1.9		
Diameter of die (G)			75.5
Surfactant	3.4	7.5	1.3
E × F		1.8	
E × G			2.0
F × G	4.3	3.2	1.0

TABLE 3

Best Conditions for Making a Protein Extrudate with Fine Porous Structure

Parameters	Best conditions
Feed rate	40.0 (kg/hr)
Added water	20.0 (l/hr)
Screw speed	200 rpm
Barrel temp.	150 °C
Spacer	4 mm
Number of die	2
Diameter of die	5 mm
Surfactant	0.5 %

Sterilization and Granulation of Wet Okara for Making "Tenpei" and Culture Media for Mushroom

Soybean is a very important raw foodstuff in Japan and other Oriental countries as it can be processed into three major traditional foods including tofu, i.e., soy protein curd, miso and soy sauce. Tofu leaves the soybean draff, called "okara" in Japanese, during its processing. Draff contains large amounts of useful and nutritious but no extractable or less soluble components. The amount of okara is almost equal by weight to that of starting raw soybeans since okara contains water in large percentage. The estimated annual yield of okara including that from soy milk production in Japan was as large as about 70,000 MT in 1983. Okara typically contains 22% protein, 19% fat, 38% carbohydrate, 15% fibrous matter and 6% of ash on a dry basis. The protein utilization (NPU) of okara is 62%. A problem in utilizing okara is the difficulty in handling it due to the very high water content of 80% by weight or more. Dehydration of okara is technologically difficult and economically almost impossible. Okara as a food has the disadvantage of extremely high susceptibility to putrefaction as a consequence of its high nutrient and water content. In recent years, the trend has been that this nutritive food has less and less chance of use in the human foods. It is utilized almost exclusively as feed for domestic animals and poultry. A very large part of okara production occurs in locations near big cities. The amount of okara utilizable as a feed is limited because stock raising is usually not a principal industry in the vicinity of big cities. Therefore, a large part of okara can find no other way of disposal than to be discarded as a waste material involving a serious problem of possible environmental pollution in the near future.

Based on the above mentioned background, Gunma Beans Co., Ltd. tried to sterilize and granulate the okara for its stabilization and for making an Indonesian traditional food "tempeh," similar to Japanese "natto," or culture media for mushroom using a twin screw extruder. The machine this company used was a fully intermeshing corotating twin screw extruder (Suehiro, $L/D=20$) and the results obtained are shown in Tables 4 and 5.

More than 60% water was found necessary for the complete sterilization of okara at 140°C probably due to the better heat transfer from the barrels to okara. Eighty sec. was a remarkably short time for the complete sterilization which suggested that the extruder had a high potential as a heat exchanger. Okara was changed from a paste to a firm porous structure suitable for inoculation and fermentation.

This company summarized the superior points of extrusion cooking of okara to its current processing as follows; (a) complete and continuous sterilization within very short time, (b) very homogeneous heat treatment, and (c) porous granulation together with sterilization, and emphasized the high potential application to the processing of viscous foods like jam, vegetable paste, marmalade and so on.

In spite of recent active research on extrusion cooking, there are few research reports on the machines (barrels, screws, their shape and arrangement, die design and so on). In general, the processed protein materials are changed to insoluble products which make chemical analysis more difficult or sometimes impossible. In the place of current chemical analysis, physical analyses like NMR, DSC and NIR are used to obtain information on material changes, influence of machine design, etc., for the better understanding of extrusion cooking. Also, it should be pointed out that there still remain many problems in applying this technology to actual production; for example, automation, optimization and scale-up of extrusion cooking. However, these remaining problems should not decrease the high potential of this technology.

TABLE 4

Sterilization of Okara

Conditions	Test				
Feed rate (kg/hr)	30	30	30	65	65
Water content (%)	30	50	60	50	60
Residence time (sec)	120	120	120	85	85
Temperature (°C)	140	165	138	165	140
Remaining general microorganism	1×10^3	1×10^3	0	0	0
Remaining heat stable microorganism	1×10^3	1×10^3	0	1×10^2	0
Sterilization	X	X	G	X	G
Product rigidity	G	G	G	G	G
"Tempeh" quality	X	X	G	X	G

X : poor G : good
Authors modified the results for their simplification.

TABLE 5

Cultivation of Mushroom on the Extruded Mixture of Okara and Sawdust

Conditions	Test				
Feed rate (kg/hr)	30	30	30	65	65
Water content (%)	30	50	60	60	85
Residence time (sec)	120	120	120	85	85
Temperature (°C)	160	168	140	139	150
Remaining general microorganism	5×10^2	0	0	0	0
Remaining heat stable microorganism	4×10^2	1×10^2	0	0	0
Sterilization	X	X	G	G	G
Product rigidity	G	G	G	G	G
Growth of mushroom	X	X	G	G	G

X : poor   G : good
Authors modified the results for their simplification.

References

Atkinson, W.T., U.S. Patent 3,488,770 (1970).

Crocco, S.C., *Food Eng. Int.* 1:16 (1976).

Gwiazda, S., A. Noguchi and K. Saio, *Food Microstructure* 6:57 (1987).

Isobe, S., and A. Noguchi, *Nippon Shokuhin Kogyo Gakkaishi* 34:456 (1987).

Isobe, S., and A. Noguchi, *Nippon Shokuhin Kogyo Gakkaishi* 35:471 (1988).

Janssen, L.P.B.M., "Twin screw extrusion," Elsevier, Amsterdam, The Netherlands, 1978.

Noguchi, A., Extrusion cooking: Development of twin screw extruder and its application, Korin, Tokyo, 1987, pp. 45–58.

Sair, L., and D.W. Quass, U.S. Patent 3,968,268 (1976).

Tolstoguzov, V.B., V.Y. Grinberg, and A.N. Gurov, *J. Agric. Food Chem.* 33:151 (1989).

Traditional Chinese Soyfood

Susani K. Karta

American Soybean Association, 541 Orchard Road, #11-03 Liat Towers, Singapore 0923, Republic of Singapore

Abstract

The traditional Chinese soyfoods can be broadly divided into either nonfermented or fermented soyfood. The most popular are tofu and tofu related products, soymilk, yuba, soy sprouts, soy sauce and soy paste. Of these traditional Chinese soyfoods, tofu is the most widely consumed soyfood in the Far East countries. It plays an important role in protein nutrition. This paper will focus on the processing variables for making tofu and identify the critical factors affecting the product quality and quantity. Some of the important factors affecting tofu quality and quantity are soybean variety, grade, percentage of solids in soymilk, equipment used, water:bean ratio, heat processing, type and concentration of coagulant used, process of coagulation-temperature and pH, mode of mixing the coagulants, pressure applied to remove the whey and the hardness of water. Tofu is made with various firmness depending on type and regional preference. It can be categorized according to its protein content. Unfortunately, the shelf-life of tofu is very short even with refrigeration. Greater research is needed to develop methods for increasing its shelf-life.

Soybeans (*Glycine max*) have a history of cultivation of more than 5,000 years in China. The Chinese developed a wide variety of processed soybean foods that continue to play a part in their diet today. These traditional Chinese soyfoods can be divided broadly into either nonfermented or fermented soyfood. Soy sprouts, soymilk, protein-lipid film (*yuba*) and tofu are the principal nonfermented ones (1) (Table 1). Of these, tofu and soymilk are the most widely consumed (2).

Traditional Chinese fermented soyfoods are used primarily as flavoring agents (1) (Table 2). They are soy sauce, *miso* (soy paste), fermented whole soybeans (*tou-shih*), and *sufu* (fermented tofu). Of these, soy sauce is the most popular, as most of the Chinese dishes usually are seasoned and flavored with soy sauce.

Most of these traditional soyfoods are known to have originated in China and gradually been introduced or spread to Korea, Japan and other Asian countries. Table 3 is the estimated annual per capita consumption of soybeans in several Oriental countries. Taiwan has the highest per capita consumption (13.3 kg) for soyfood in the region and perhaps in the world. The per capita consumption of soybeans for food in China is about 6.9 kg which is lower than the other Far East countries, i.e. Taiwan (13.3 kg), Japan (9.3 kg) and Korea (7.8 kg). Thus the per capita consumption of soyfood in China is behind that of other countries; its industry is plagued by lack of raw food materials, inadequate distribution, and primitive processing equipment. In China, soyfood products have been rationed with the markets running out of supplies early in the morning.

The Chinese traditional soyfoods have become a vital part in the Oriental diet (2). The proximate nutritional

TABLE 1

Chinese Nonfermented Soy Food Products (1)

Food items	Chinese names	Description	Uses
Fresh green soybeans	*Mao-tou*	Picked green, large soft beans.	Cooked in pod, served as fresh vegetable or snacks, pod removed before eating.
Toasted soy powder	*Tou-fen*	Ground toasted dry beans, nutty flavor.	Used as filling or coating for pastries.
Soy sprouts	*Huang-tou-ya*	Bright yellow beans with 3–5 cm sprouts.	Cooked as vegetable or in soup.
Soy milk	*Tou-chiang*	Water extract of soy; heated and filtered.	Served hot for breakfast; beverage.
Soy milk film/*Yuba*	*Tou-fu-pi*	Creamy yellow film from surface of boiling soy milk; sheets, sticks or flakes.	Cooked with meat, soups, vegetables, edible. Packaging material; not eaten daily.
Soybean curd	*Tou-fu*	White protein curd form. Soy milk; bland taste; can be dried, frozen or fried.	Served with seasoning or after cooking as part of main meal or in soups.

TABLE 2
Chinese Fermented Soy Food Products (1)

Food items	Chinese names	Organisms used	Description	Uses
Fermented whole soybeans	*Tou-shih*	*Aspergillus, Streptococcus* and *Pediococcus*	Black and salty beans. Fermented with wheat flour.	A condiment, cooked with meat and vegetables.
Soy sauce	*Chiang-yu*	*Aspergillus, Pediococcus, Torulopsis, Saccharomyces*	Soy fermented with wheat flour.	Seasoning condiment.
Soy paste	*Chiang*	*Aspergillus, Pediococcus, Saccharomyces*	Light to dark brown paste, salty and soy sauce flavor.	Soup base and seasoning agent.
Fermented tofu	*So-fu*	*Actinomucor, Mucor*	Creamy cheese, mild flavor, salty.	Relish, also cooked with meat or vegetables.

TABLE 3
1987 Consumption of Soybeans as Foods in Asian Countries

Country and population (Millions)	Total soybean consumption[a] (1,000 M.T.)	Per Capita soybean consumption (kg)	Main soy foods
China (1,062)	7,325	6.9	Tofu, soy milk, yuba, soy sprouts, soy paste, soy sauce.
Indonesia (175)	1,575	9.0	Tofu, tempeh, soy sauce.
Japan (122.2)	1,141	9.3	Tofu, soy milk, natto, miso, soy sauce.
South Korea (42.1)	330	7.8	Tofu, soy milk, soy sprouts, soy sauce, yuba.
Malaysia (16.1)	55	3.4	Tofu, soy milk, soy sauce, yuba.
Philippines (61.5)	18	0.3	Tofu, soy sauce, soy milk.
Singapore (2.6)	20	7.7	Tofu, soy milk, soy sauce.
Taiwan (19.6)	260	13.3	Tofu, soy milk, yuba, soy sprouts, soy sauce, soy paste.
Thailand (53.6)	118	2.2	Tofu, soy sauce, soy milk.

[a] Author's estimate—for food.

TABLE 4
Nutritional Compositions of Traditional Fermented Foods (in 100 g)[a]

Nonfermented	Kcal	Prot g	Fat g	CHO g	Ca mg	P mg	Fe mg
Fresh green soybeans	355	31.3	16.7	26.1	197	456	1.2
Toasted soy powder	429	29.8	19.5	39.9	189	540	7.5
Soy sprouts, raw	46	6.2	1.4	5.3	43	48	0.9
Soy milk	33	3.4	1.5	2.2	18	36	1.2
Soy milk film/yuba	461	47.0	28.4	14.9	245	494	9.5
Tofu	72	7.8	4.2	2.4	150	104	2.2

[a] Food composition table for use in East Asia, 1978.

TABLE 5
Nutritional Composition of Traditional Fermented Foods (in 100 g)[a]

Fermented	Kcal	Prot g	Fat g	CHO g	Ca mg	P mg	Fe mg
Fermented soybeans	158	14.7	8.3	9.3	142	135	7.9
Soy sauce	39	5.3	1.3	2.5	59	100	4.9
Soy paste	148	10.9	4.8	17.1	86	97	4.8
Fermented tofu	175	13.5	8.4	14.8	165	182	5.7

[a] Food composition table for use in East Asia, 1978.

composition of these traditional soyfood products are given in Tables 4 and 5 (3). Among all the Chinese soyfood, *yuba* (*tou-pi*) contains the highest value for protein. A 100 g of dried *yuba* contains 47 gm protein. It is used extensively by Buddhist vegetarian cooks, to make stimulating meat-shaped delicacies such as chicken, roast duck, red-cooked pork, etc.

Of these traditional Chinese soyfoods, tofu is the most widely consumed soyfood in the region. Tofu is considered to be nutritionally equivalent to protein derived from animal protein (4). This paper will focus on the processing variables for making tofu and identify the critical factors affecting the product quality and quantity. The soymilk preparation method and processing equipment will not be covered in this presentation.

Variables in Manufacturing Tofu

Soybean Variables: Most soyfood processors recognize that in order to make a high quality tofu you must start with good quality soybeans. For example, the soybean specifications for tofu are as follows:

Seed size:	medium to large
Cleanliness:	U.S. Grade 1
Hilum:	white/yellow preferred
Hull:	thin, firm hull
Protein content:	high; high nitrogen solubility index (NSI)
Oil content:	low

The different soybean varieties were thought to affect the quality of tofu. This is considered to be due to differences in protein content of the soybeans and the ration of 7S and 11S proteins. Saio *et al.* (5) showed that 11S and 7S protein fractions could contribute to the textural characteristics of tofu. Tofu made from crude 11S protein was much harder than tofu made from 7S protein. The crude 11S fraction contributed greatly to the springiness, chewiness and gumminess of tofu. Murphy and Resurreccion (6) found a high correlation between 11S content of soybeans with the hardness, brittleness, elasticity and gumminess of tofu.

Wilson *et al.* (7) reported that tofu made from different varieties of soybeans grown under identical environmental conditions were found to consistently give statistically significant textural correlations with soybean 11S protein content. Hardness, brittleness and gumminess of the tofu were significantly correlated with 11S content. The 7S protein did not correlate well with textural characteristics under identical environmental conditions. These results are in concurrence with previous findings that 11S contributed to the hardness, gumminess and chewiness of tofu.

Wang *et al.* (8) studied the effect of soybean variety on the yield and quality of tofu using U.S. varieties and five Japanese varieties of soybeans grown under the same environmental conditions. They found that the composition and color of tofu were affected by soybean variety but that yield and texture were not significantly affected. Tofu made from a variety with high protein content had a higher protein/oil ratio than tofu made from a variety with less protein. Wilson et al. (9) found that the soybean variety and method of processing influenced the texture, color, flavor and yield of tofu. They felt that the method of processing greatly influenced tofu texture and quality.

Processing Variables: Tofu, silken tofu, firm tofu, deep-fried tofu and *yuba* are derived from soymilk. They are low-cost, high protein products which have been used widely in the Orient. The flow sheet for the manufacture of these foods is shown in Figure 1 (10).

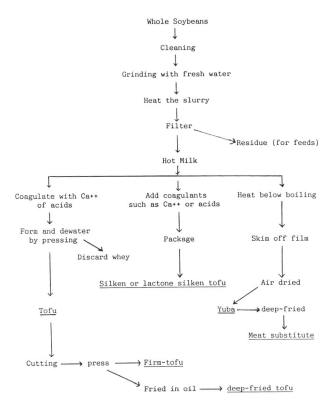

Fig. 1. Flowsheet for manufacture of tofu and products derived from soybean milk (10).

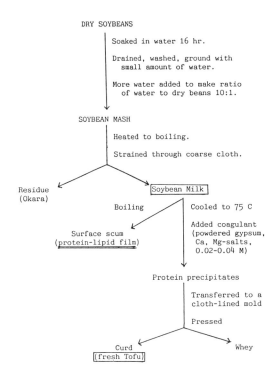

Fig. 2. Flow diagram for the preparation of tofu (1).

Traditionally, three main steps are involved in making tofu (1): the preparation of soybean milk, coagulation of protein and formation of tofu in a mold (Figure 2). Dry soybeans are washed and soaked in water overnight or until the beans are fully hydrated. The soaked beans are drained, rinsed and ground with water. The slurry is brought to a boil and kept at boiling point for 15 minutes. The boiled slurry is then filtered to yield a milk-like product known as soybean milk.

The making of a high-quality and reproducible tofu product is dependent on many factors. The variables in the preparation of tofu can be categorized into three phases (11–13):

Maceration and extraction (soaking and grinding) stages: The water:bean ratio was shown to be critical for optimum tofu yield and quality. The temperature and hardness of water, and the degree of maceration affected the amount of protein extracted. The traditional method using a 10:1, water:bean ratio gives the best protein yield. The maximum protein recovery also will depend on the maceration and extraction time and the grinding equipment. Thus, each processing system needs to be investigated for the optimal operating condition of the machine.

Filtration and heating stages: Better yields and quality of the product were observed when the bean slurry was filtered prior to heating. The temperature of soymilk, adding coagulant and the mode of mixing greatly affect the yield and texture of the resulting tofu. When the temperature at which coagulation takes place is too high, the process advances too fast, the curds become too fine, and the resulting product is not only low in yield but tough in texture. However, when the temperature is too low, coagulation is incomplete and the tofu becomes frail and lacks body. In this case too, yield is low. The optimum coagulation temperature recommended is *ca.* 75–80°C. As the temperature increases the gross weight and moisture content of the curd decrease; whereas its hardness increases. Increased mixing and stirring speed decrease tofu volume and increase the hardness.

Coagulation stages: It is the most significant in terms of yield and texture of tofu. When the soymilk is highly concentrated or dense, the milk will coagulate evenly, producing a smooth, soft tofu. When the density is low, coagulation is less thorough; the whey separates and the tofu becomes slightly tough and coarse. The right amount of coagulant and the proper rate of its addition are important factors necessary to attain a satisfactory soybean curd (1, 2). The ideal pH of the soymilk for coagulation occurred at or near 6.0 (14). Protein starts to coagulate when pH drops to about 6.0, therefore, the additions of coagulants should be stopped when the pH approaches 6.0.

Types and concentration of coagulants used in tofu manufacturing: The types of coagulants used in tofu manufacturing will influence the hardness and brittleness of the curds (14,15). Traditionally, the concentrated liquid obtained during the process of salt making from sea water, Nigari, which contains mainly magnesium chloride, is used as the coagulating agent. Tofu produced using nigari is said to taste good and the texture is fine and smooth. However, its yield is lower and it tends to be tough.

Today, calcium salts, such as sulfate, chloride, glucono-delta-lactone (GDL) and acids such as citric and acetic are used (14,15).

The characteristic of the five most used coagulants are: *magnesium chloride*: it has very rapid coagulation; the curd is less uniform and not as fine and continuous as those obtained with GDL, $CaSO_4$; *magnesium sulfate:* the curd is grainy and less cohesive; *calcium chloride:* very rapid coagulation at early stage; structure is not as fine and continuous as those obtained with GDL and $CaSO_4$; *calcium sulfate:* coagulates the protein slowly; more uniform and smoother than $CaCL_2$, $MgCL_2$ and $MgSO_4$; and *glucono-d-lactone* (GDL): commonly used for making silken tofu gives the best in texture on the basis of smoothness.

Calcium chloride usually is used in combination with other coagulating agents either as calcium sulfate or nigari. Generally, calcium chloride and magnesium chloride result in harder and bitter curd than calcium sulfate and magnesium sulfate. This suggests that anions have a stronger effect on water-holding capacity than do cations (2). De Man *et al.* (15) observed that calcium chloride and magnesium chloride coagulated soymilk instantly while calcium sulfate, magnesium sulfate, glucono-delta-lactone (GDL) acted comparatively slowly. Curd obtained with calcium chloride and magnesium chloride was coarse, granular and hard, whereas calcium sulfate and GDL (fresh solution) appeared to be most suitable for making tofu of soft, smooth and jelly-like texture. The amount of coagulant used to precipitate soy protein varied depending on the type of compound used (2).

Figure 3 shows that both concentration and type of coagulant affect the gross weight, moisture content, total solids and nitrogen recoveries (16). With an increase in coagulant concentration, there was a decrease in moisture content of the tofu. The total solid recovery steadily increased at a higher coagulant concentration. The percentage of nitrogen recovery increases as the concentration of salt increases, remains the same at 0.02–0.04 M, and then decreases at higher concentration. No curd is noted when the concentration of coagulant is higher than 0.1 M. Thus at the extent of higher concentration, the precipitation can decrease and the protein becomes soluble again which explains that nitrogen recovery decreases at higher coagulant concentrations (17). This figure shows that salt concentrations between 0.02–0.04 M have the least effect on the four types of coagulants investigated and is more likely to yield a reproducible product with high nitrogen recovery. Calcium sulfate is preferred when compared to the other three salts because of its limited solubility.

Making tofu is a relatively simple process but due to its bland nature, its textural properties play a big role in influencing quality and consumer acceptability. The textural characteristics of the curds are influenced also by the concentration and type of coagulant as shown in Figure 4 (16). When the concentration of coagulant is increased from 0.01 to 0.02 M, significant increases in hardness, brittleness, cohesiveness and elasticity are noted. No significant effect is observed at 0.2–0.04 M range concentration, but above this range, the measures for the curds decreased steadily. These data indicate that the use of coagulants at the level 0.2–0.04 M is more likely to produce firm products. Tsai et al. (18) found that based on coagulability and texture, coagulant concentrations between 0.025 to 0.03 M were the most suitable for making Chinese tofu.

With the increase in coagulant concentration, the structure of tofu became more porous separating more whey and leaving less moisture in the tofu. Hence the hardness of tofu increases as its water content decreases (8). Thus, if the coagulating agent is present in excess quantity, the tofu is tough and yield is low. If it is insufficient, the whey is inadequately clear; and the tofu lacks some of its characteristic features.

Thus, some of the important factors affecting tofu quality and quantity are soybean variety, grade, percentage of solids in soymilk, equipment used, water:bean ratio, heat processing, type and concentration of coagulant used, process of coagulation—temperature and pH, mode of mixing the coagulants, pressure applied to remove the whey, and hardness of water.

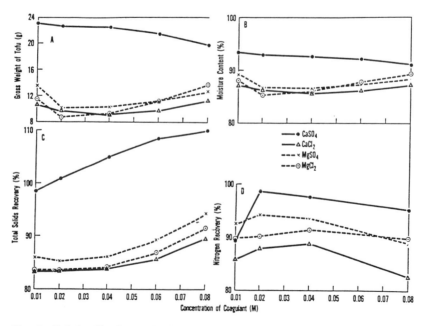

Fig. 3. Relationship of concentration and type of coagulant to the yield of tofu (16).

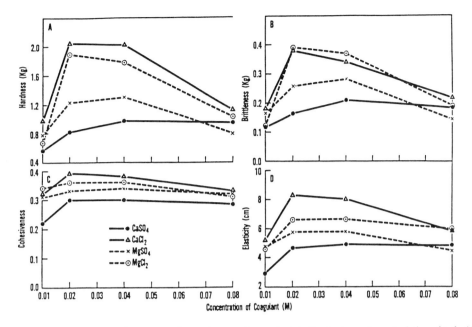

Fig. 4. Relationship of concentration and type of coagulant to the texture characteristics of tofu (16).

Tofu Products

Tofu is made with various firmness depending on type and regional preference.

In general, the Chinese soybean curd can be categorized according to the protein content: soft tofu contains 5–7.9%; regular tofu contains 8–10%; firm tofu contains 11–19.9%; and pressed tofu contains 20% or more.

Different types of Chinese soybean curd are: Soft/silken soybean cake, *Tou-Fu-Nao*; Soy pudding/jelly, *Tou-hwa*; Regular soybean cake, *Tou-Fu*; Firm soybean cake, *Tou-Kahn*, and Fried soy cake, *Tou-Pok*.

Soft/silken tofu is manufactured from soymilk of higher protein concentration coagulated in the container without pressing or removal of whey. For this soft/silken tofu, glucono-δ-lactose (GDL) is used as the precipitating agent. It is much more fragile than regular tofu, but somewhat firmer than Chinese soymilk pudding/jelly (*tou-hwa*) which normally is eaten with plain or almond-flavored syrup. This soy pudding is made from soymilk using a coagulant such as calcium sulfate and starch or agar-agar.

In preparing regular, firm or pressed tofu, soymilk is commonly mixed with a coagulating agent such as calcium sulfate, or a combination of coagulant calcium sulfate and magnesium chloride (Nigari) to give a better protein recovery during precipitation. After precipitation, it is placed in molding boxes to press out the water; the curd is then cut into sizes appropriate for cooking. The harder the tofu, more water is pressed from the curd. To make yellow firm tofu, tumeric is used to color the curd surface. To make black/dark brown tofu, pressed tofu is simmered in soy sauce and caramel solution. The yield average for soft/silken tofu is: 5–6 times weight of raw soybeans; regular tofu: 3–4 times weight of raw soybeans; firm tofu: 2–3 times weight of raw soybeans; and fried tofu: 1–1.5 times weight of raw soybeans.

References

1. Wang, H.L., in Handbook of Processing Utilization of Agriculture, Vol. II, Part 2, Plant Products, edited by I.A. Wolff, C.R.C. Press, Inc., Boca Raton, Florida, 1983, p. 91.
2. Wang, H.L., *J. Am. Oil Chem. Soc. 61*:528 (1984).
3. Food Composition Table for Use in East Asia, U.S. Dept. of Health Education and Welfare, 1978.
4. Muto, S., E. Takahari, M. Hara and J. Koruma, *J. Am. Diet. Assoc. 43*:451 (1963).
5. Saio, K., M. Kamiya and T. Watanabe, *Agric. Biol. Chem. 33*:1301 (1969).
6. Murphy, P.A. and A.P. Resurreccion, *J. Agric. Food Chem. 92*:911 (1984).
7. Wilson, L.A., A.P. Resurreccion, C.C. Hauck and P.M. Murphy in Proceedings of Soybean Utilization Alternatives, edited by L. McCann, The Center for Alternative Crops and Products, Minnesota, 1988, pp. 425–427.
8. Wang, H.L., E.W. Swan, W.F. Kwolek, and W.R. Fehr, *Cereal Chem. 60*:245 (1983).
9. Wilson, L.A., L.D. Johnson, and J.A. Love in Proceedings of The Sixth World Congress of Food Science and Technology, Dublin, Ireland, 1983, pp. 102–103.
10. Wei, L.S., in Proceedings of The Asian Seminar on Soybean Production for Local Utilization, Thailand, 1986, pp. 69–87.
11. Beddows, C.G., and J. Wong, *Int. J. of Food Sci. & Tech. 22*:15 (1987).
12. Beddows, C.G., and J. Wong, *Ibid. 22*:23 (1987).
13. Beddows, C.G., and J. Wong, *Ibid. 22*:29 (1987).
14. Lu, J.Y., E. Carter and R.A. Chung, *J. Food Sci. 45*:32 (1980).
15. de Man, J.M., L. de Man and S. Gupta, *Food Microstructure 5*:83 (1986).
16. Wang, H.L., and C.W. Hesseltine, *Process Biochem. 17*:7 (1982).
17. Appurao, A.G., and M.S. Narasingo Rao, *Cereal Chem. 52*:21 (1975).
18. Tsai, S.J., C.Y. Lan, C.S. Kao and S.C. Chen, *J. Food Sci. 46*:173 (1981).

Application for Vegetable Proteins in Processed Pet Foods

Douglas A. Thompson

Uncle Ben's of Australia, P.O. Box 153, Wodonga, Victoria 3690, Australia

Abstract

Vegetable proteins have been utilized for decades by the pet food industry for nutritional, cost or technological applications. During this time the total volume of pet food production has increased substantially while utilization of vegetable protein has remained almost static. The reason for this could be attributed to any of several factors, including palatability, functionality, cosmetic or cost/performance. However, the writer's view is that this is primarily a technical and marketing failure on the part of the grain/oilseed processing industry. Despite the range of vegetable protein available, technological applications for functional protein remain served by the food items of wheat gluten and soy protein concentrate. Soy meal and corn gluten meal find applications as a protein "filler" when competitively priced in the same way as in the stock feed industry. The functional proteins will continue to lose ground to competing technologies utilizing different raw materials. In recent years, advances in pet food technology have been supported by the developments in hydrocolloids, modified starches and even the traditional animal by-product proteins. Suppliers must revise the concepts of product design to identify and develop a tailor-made pet food-specific range of vegetable proteins to meet the diverse applications of a rapidly growing industry.

The worldwide pet food industry produces about 10 million metric tons of prepared pet food valued at over $12 billion U.S. In the major developed countries there are nearly 200 million dogs and cats in almost equal numbers, and their population has a compound growth rate of over 2%. However, the volume growth in processed pet food is actually much higher, about 6%, as the share of the total market (as measured in calories) increases at the expense of fresh meat and table scraps. The growth rate is lower in the United States where the market could be described as mature with 85% of the food calories required already provided by prepared pet food. Other markets have much potential for continued growth, for instance, the processed pet food share of total calories consumed by pets in the United Kingdom and Australia is 50–60%, Germany and France 40–50% and Japan only 20–30%.

The pet food industry provides a huge choice of products which can be classified into 3 basic types:

Moist or Canned Pet Foods

These products are designed to have a texture and moisture level similar to the fresh meat which they replace in the diet. The characteristic of the product is its high palatability. The preservation process is generally heat sterilization by retorting the product in an impervious container. The tinned steel can is still a dynamic package and with the steel industry offering tougher and lighter tinplate, as well as advances in seam welding, two-piece can manufacture and easy opening features, the can will be around for a long time.

At the super premium end of the moist pet food range are the high palatability, single-serve products, characterized by the Sheba brand, which feature easy opening and light weight packaging in aluminium or barrier plastic trays. The European style pet food has visible meaty chunks in a background of softer and smaller meat particles in jelly and is the leader in the moist product category. Although well known and of long standing, this product type is technologically advanced and offers the enthusiastic food scientist much scope for innovation.

Semi-Moist Pet Foods

Intermediate moisture pet foods (25–35% moisture) incorporate a level of water soluble solids such as sugar, sodium chloride, sorbates and propylene glycol sufficient to stabilize the product through control of water availability in order to prevent spoilage without the need for refrigeration. This product style has declined in popularity with tonnage in the United States falling by 40% between 1980–1985. However, there has been increasing growth in a similar technology which is often referred to as semi-dry where the moisture content is between 15–25%. This provides a softer version of the conventional dry, kibble products.

Dry Pet Foods

Usually of less than 12% moisture, this product type contains a starch source (typically wheat or corn), tallow and/or vegetable oil, and a protein source that could be soy or legumes, but is more typically meat and bone meal, poultry by-product meal or fish meal. Palatability is obtained by spraying a liquid digest or fat on to an extruded product. Fat also may be incorporated into the source material. Dry pet foods generally are produced by single screw cooker-extruders. In this process an extrusion screw, rotating in a tightly fitting barrel, conveys, heats and works food ingredients into a continuous plasticized mass, which is forced through a die and then puffing occurs due to the expansion of moisture as steam. This process is similar to that used for the production of expanded breakfast cereals and is not considered to be a texturizing process (which requires a higher protein content).

Nutrition

The most important ingredient in all types of processed pet food is the nutritional requirement for a source of protein. Domestic pets, like all animals, require specific nutrients rather than specific feedstuffs and therefore there are a lot of choices available in recipe selection. Dogs and cats require energy, amino acids, fatty acids, glucose precursors (provided by protein, carbohydrate or fat), minerals and vitamins. In order to supply an excess of all essential amino acids, the protein content should be around 25% of the dietary dry solids for growth, and 15% of the diet for maintenance of health and condition in adult animals. The quality of the protein should be similar to that provided by animal muscle and supply the amino acids arginine, histidine, isoleucine, leucine, lysine, methionine, phenylalanine, threonine, trytophan and valine.

There are some significant differences in the nutritional requirements of dogs and cats. Cats have evolved from a strictly carnivorous diet and require some nutrients that cannot be obtained from plants. These include arachidonic acid, taurine and vitamin A (preformed) as cats cannot utilize carotene. Dietary fiber is not considered to be essential for dogs and cats although the inclusion of some fiber in the diet is not unusual in commercial practice.

Recipe Flexibility

Although the wild cousins of domestic pets may be basically carnivorous, satisfactory diets containing significant amounts of vegetable matter can be developed by appropriate processing and supplementation. However, the presence of oligosaccharides in the diet does cause flatulence in pets.

Many types of vegetable protein derived from grains, legumes and oilseeds are used by the pet food industry and they can be used for a variety of purposes. The most commonly used items are wheat gluten, corn gluten, soybean meal and soy protein concentrate. These items may compete for use between themselves, directly with meat or meat by-products (including fish and poultry) as well as combinations of grain or grain by-product and meat protein meal. Where vegetable protein is utilized for a specific functional property rather than its protein content, it may compete for use with completely different ingredients utilizing different technologies.

Palatability

Cats tend to be individualistic in their feeding behavior and frequently exhibit food preferences that have been conditioned by previous dietary experience. Although diet preference should not be confused with nutrient requirement, to the pet food manufacturer it is of equal importance for as quoted by the industry, "the nutritional contribution of a food not eaten is nil."

Apart from several unique nutrient requirements for dogs and cats, it is their disdain of feedstuffs that other species consume readily that dictates the special attention that must be given to provide ingredients of exceptional palatability. It is the concept of palatability as much as the process or the package which differentiates the pet food industry from that of other feedstuffs. For recipe selection, stock feed manufacturers can utilize a large range of ingredients, driven by least cost formulation, and will supplement with the usual limiting amino acids, lysine and methionine.

In pet foods, the limiting factor of palatability is reached well before the need for amino acid fortification, and the inclusion of meat by-products provides palatability as well as a natural nutrient balance. To some extent, palatability can be recovered by special additives; however, commercial flavorings designed for human palates appear to have little attraction.

Dogs, unlike cats, are responsive to sweetness and both species respond to the taste of certain amino acids and peptides (from meat tissue digests). Animal fats also have positive acceptance properties for dogs and cats.

Within the balance of nutrition and palatability described above, it is possible to manipulate the protein source somewhat to provide limited, least-cost formulations. In this way it is possible to interchange soymeal, corn gluten meal and legumes within limits imposed by physical properties.

Physical Properties

In addition to the appeal any food must offer its consumer through palatability, pet foods are unusual in that they also must appeal to a purchaser who is not the actual consumer. Therefore, pet food manufacturers are also concerned with the physical appearance and odor of their products. Although it is fair to say that some odors which pet animals would find appealing would not be similarly enticing to their owners, there are some odors which are not appealing to either party. In the context of vegetable products, these would be burnt, rancid or strongly-cooked, cereal odors.

Pet foods also are designed to be attractive visually and the product color should be representative of the major constituent of the recipe. Therefore, chicken, fish and rabbit variety products are expected to be light to off-white in color and this reason alone would prevent the inclusion of dark meals such as rapeseed.

In addition to cosmetic appeal, the physical appearance and texture are of particular importance in all types of pet food. It is in this field that vegetable proteins have much to offer, but must compete with other ingredients and technologies.

Pet owners have demonstrated a strong purchasing preference for the chunky meat style of canned food over finely ground or meatloaf-type products. This preference has encouraged manufacturers to invest in the technologies of meat recombination and meat simulation. In traditional recombined meats, the technique relies on the utilization of natural binding properties. In this process mechanical or chemical treatment of muscle meat causes the soluble protein, myosin, to exude onto the meat surface. A subsequent heat process coagulates the protein, binding the meat chunks together into an appropriately shaped product.

For pet foods or food products using a nonbinding meat source such as soft tissue, the myosin can be replaced by other proteins which are capable of forming gels on heat treatment. These proteins are typically blood (plasma), wheat gluten or soy protein. The reformed meat chunks then are ingredients prepared by a sub-process and are not the final product. The sub-process must provide the chunks in a suitably robust form so that they do not lose their integrity during the mixing/filling (into cans) process or during the heat sterilization of the canned product.

The sub-process may involve a further technology to provide an interim binding step, such as the formation of a transportable chunk by a setting mechanism utilizing a calcium alginate gel. In this case, the chunk integrity in the final product is supplemented by the protein coagulation which occurs during the retorting process.

Other reformed meat processes may rely directly on heat with the formation of chunks and simultaneous heat denaturation of protein maintaining chunk integrity. These chunks may comprise nonbinding or organ meats, a cereal thickener or filler, a functional protein and in some applications, a texturizer. The heat process is applied by baking, blanching, frying or extruding. Where these processes are used in the food industry, particularly in the manufacture of pre-cooked small goods, an emulsifier to bind fat is advantageous. Although it is well known that soy protein concentrates perform well in reducing fat cook-out in sausages, this property has limited application in the pet food industry where the meat ingredients and offals tend to be of low fat content.

Texturization

As a variant to reformed meat, the pet food industry also utilizes products which may be considered as simulated meats. These products can be differentiated from other chunk forming processes by the inclusion of texturization. Vegetable proteins, usually wheat gluten or soya meal and concentrate, are commonly used for this purpose. The unique properties of wheat gluten in dough formation and extensibility enable it to provide a fibrous characteristic to comminuted meats. The gluten dough matrix within a meat emulsion is coagulated by heat as in the baking process and the product may be randomly cut or chopped to resemble meat pieces. Wheat gluten also can be used as an alternative to soy protein in the extruded type of simulated meat products.

Extrusion texturization of plant protein is created by a process in which the globular proteins are hydrated, unfolded, stretched, aligned and cross linked under the effect of heat, moisture and shear. The mixed ingredients are fed into the hollow barrel of the extruder, where a tapered screw having whorled ridges forces the material towards the exit orifice of the barrel. The temperature of the barrel may be controlled. The clearance between the flights on the screw and the inner barrel surface decreases as the material is moved along causing high shearing. The shearing action unravels and stretches the proteins and the rising temperature and internal pressures convert the ingredients to a viscous plastic state. At this point the starches gelatinize and proteins denature, the latter becoming realigned by the shearing action of the rotating screw.

In passing through a die, the proteins are compressed and laminated longitudinally into sheaths. On emergence from the die there is an instantaneous release of the high pressure superheated water as steam into the atmosphere, creating an expanded porous structure. The evaporation of moisture also causes a rapid cooling and consequent thermosetting of the product.

The physical chemistry of interactions between proteins and other components such as polysaccharides, lipids and electrolytes during texturization is also of importance. Because insoluble carbohydrates strengthen the structure to some extent, soluble carbohydrates are of no benefit in the development of texture or stabilization of the structure.

The fibrous structure of textured single vegetable protein still lacks some of the "feel" of meat, however this can be overcome partially by co-extrusion of different protein materials.

To obtain texturization, thermoplastically extruded products should be composed of ingredients totalling less than 1% fat in order to prevent slipping due to the increased lubrication, although fat itself does not interfere with the formation of the fibrillar matrix.

In extrusion, twin screws can tolerate a wider variety of ingredients and are less prone to surging. This enables more uniform processing as a result of the consistency of shear rate, narrower distribution of residence time, and better mixing and heat transfer, which allows the utilization of more viscous pastes and slurries.

Although a similar extrusion process is used for the manufacture of dry pet foods, that application rarely requires texturization. Rather, the process relies on the expansion of starch and thermosetting with any lamination being coincidental. The dry pet food formulation can therefore include a significant component of protein material which previously has been heat-coagulated, such as meat and bone meal, fish meal or expeller-produced vegetable protein meal. The limitation on expeller meal for this application tends to be the high level of associated fiber (which causes roughness), color, palatability or usually a combination of all three.

The untextured dry pet food formulations are not suitable for use in canned products where the additional strength and integrity provided by the protein texturization is needed in order to survive a wet retorting process.

The formulation parameters for textured pet food chunks are typically those required for the familiar textured soy protein products, i.e., minimum protein level around 40%, nitrogen solubility index above 15% and molecular weights above 10,000 Daltons. These requirements restrict ingredient selection, limiting the quantity of heat denatured, hydrolyzed or leguminous material that can be utilized. However, air classification of legume flour into proteinaceous and farinaceous components can provide a higher protein content fraction which is suitable for texturization.

Price

The textured chunk formulation in pet food is more complicated than its food industry cousin, the textured soy protein. Feeding tests have demonstrated clearly that cats, in particular, find soy flour to be unpalatable at levels much lower than can be detected by the human palate. Soy concentrate, soy isolate and wheat gluten being blander may be used to supplement the protein content of the meat/cereal/legume base, however, these items are purchased in direct competition with the human food industry, which perhaps does not suffer from some of the on-shelf price constraints of canned pet foods.

From the growth statistics mentioned earlier, it should be appreciated that the dynamic state of the pet food industry encourages the participation of several of the world's largest and most innovative food companies to compete directly in a market with smaller, often low-overhead, regional and entrepreneurial manufacturers, which is to the great benefit of the consumer.

In the battle for market share, it must also be appreciated that cross competition in the market is also a major consideration. In this area, the relatively low tech/low overhead dry products are striving to gain share from the traditional canned product. If this is not enough, many governments, in their wisdom, have decided to penalize the processed pet food industry by taxation. In Britain and some other European countries Value Added Tax on pet foods is 15% and in Australia pet owners are contributing 20% of their pet food dollar to government revenue via sales tax. The tax is truly a handicap to the industry, as its major competitors, i.e., unprocessed pet food, table scraps or butcher shop trimmings are, of course, tax free.

Therefore, in addition to the nutritional, palatability, cosmetic and functional properties, vegetable protein for use by the pet food industry must also be a relatively low cost item. This suggests that suppliers' price expectations should be lower than for sales to the food industry, without necessarily implying either a lower margin or a reduction in functionality, but perhaps a modification of the specification in a non-functional area such as particle size or the degree of refinement.

Cost/Performance

Suppliers of functional vegetable proteins to the pet food industry must recognize that their true competitors include suppliers of ingredients which generate a similar result from a different technology.

The use of sodium alginate in the sub-process for formation of chunks was mentioned previously. This, in fact, is only one of several polysaccharides which may be used to prepare recombined meat products. The list includes modified starches, carrageenan, pectin, xanthan gum and a host of other man made and natural galactomannans.

Also, changes in eating habits in the human food industry in the areas of convenience, portion size and restaurant growth have had a profound effect on the red meat, poultry and fish industries. These industries have recognized the benefits of by-product recovery and the adding of value through quality, consistency or functionality for supply to the pet food industry.

Although the amount of vegetable protein used by the worldwide pet food industry cannot be readily quantified, it is certainly less than half a million tons which represents under 5% of the processed pet food tonnage produced. Despite the large volume growth in pet food production, private estimates indicate that utilization of vegetable protein has actually declined in the last five years.

The vegetable protein industry has certainly missed a major opportunity to participate in the growth of pet foods. The beneficiaries have been the animal protein processors and the hydrocolloid producers. They have accomplished this by focusing on the specific needs of the pet food industry and developing products, processes and specifications most appropriate for the purpose intended. At present there appears to be no reason why this trend should not continue.

The vegetable protein industry offers meals for sale derived from traditional oil extraction techniques, such as coconut, linseed, rapeseed, cottonseed and sunflowerseed which are dark in color, have no functionality and are high in residual fiber which makes them fit only for stock feeds. The vegetable protein industry also offers some high quality, high functionality and high priced products for the food industry. The pet food industry does not really want these products and will continue to replace them as alternatives become available.

Vegetable proteins are failing then due to a number of reasons which may include quality, palatability, nutrition, appearance, functionality or cost/performance. Although specific reasons can be given to explain each lost opportunity, it is suggested that this is, in the main, a technical and marketing failure on behalf of the vegetable protein processing industry.

Product Design

The marketing process identifies and attempts to satisfy consumer wants through consistent products in a competitive situation. This requires that a specific consumer want is matched by an appropriately designed product. The product design defines all the characteristics needed and is the standard against which the product is measured.

To implement the product design, it is necessary to define product specifications and quality standards. These technical, working documents control the purchasing of materials and the manufacturing of the product. They should specify the best technical capability for delivering the product design, and are the criteria against which the finished product, its components, and its raw materials are measured. Most importantly, they help to ensure that the customers' usage experience of the product conforms consistently and optimally with the product design.

The effectiveness of product performance against the requirements of the product design must be examined at various levels: (a) How closely the specifications for materials to be purchased, the process used for their manufacture, and the product formulation conform to the product design requirements; (b) How consistently materials used, manufacturing performance and the manufactured product itself conform to product specifications; and (c) How closely the product at the point of usage by the consumer conforms to the product design.

Quality standards set the target values for specific attributes of materials, processes and products. Product specifications are technical definitions of what the product is and how it will be manufactured to meet the standards. Specifications also establish the operating tolerances caused by normal variation. Statistical control studies must be used to determine tolerances that are realistic.

The concept of product design is neither complicated nor new, and adoption of that process to identify and develop a "tailor-made" pet food specific range of products is necessary if the vegetable protein industry is to participate in the continued growth of processed pet foods.

Efficiency in Feed Resource Utilization and Animal Production

C. Devendra

Agriculture, Food and Nutrition Sciences Division, International Development Research Center, Tanglin P. O. Box 101, Singapore 9124

Abstract

Maximizing animal production necessitates efficiency in the utilization of the available feed resources: concentrates, forages, crop residues, agro-industrial by-products and non-conventional feeds. This is justified in the Asian region because of variable feed supplies, poor feeding systems, low per animal performance, production systems based on animals and mixed cropping systems that are unlikely to change, and sizeable animal populations. In terms of total world animal population, Asia has approximately 64% of ducks, 47% pigs, 34% chickens, 96% buffaloes, 47% goats, 30% cattle and 29% sheep. The productivity of non-ruminants (pigs and poultry) has been outstanding, but productivity with ruminants (buffaloes, cattle, goats and sheep), by comparison, has been less successful, resulting in an inability to meet requirements of national targets for meat and milk production. Dietary protein represents the principal limitation to high performance. Potential improvements are discussed with reference to the attributes of individual animal species, characteristics in the feeds, inherent limitations, innovative feeding systems and overcoming prevailing constraints. These strategies are consistent with increasing intensification, higher economic productivity from animals and the search for efficiency in the utilization of the available resources.

Efficiency in the utilization of available feed resources is an important prerequisite for maximizing animal production. The approach is justified in the Asian region by variable feed supplies, inefficient feeding systems, low per animal performance and sizable animal populations. The latter includes, in terms of total world population, approximately 97% buffaloes, 81% ducks, 48% goats, 47% pigs, 33% cattle, 32% chickens and 18% sheep. The productivity of nonruminants (pigs and poultry) has been particularly outstanding, but by comparison, ruminants (buffaloes, cattle, goats and sheep) have been less successful. The latter has resulted in problems concerned with access to food, national targets for animal proteins not being met, and doubts about the efficiency of existing animal production systems. As dietary protein represents the principal limiting factor to high performance, conservation and economic use of indigenous protein sources are essential. Potential possibilities for increasing current productivity from animals are discussed in the context of the attributes of individual animal species, characteristics of available feeds, inherent limitations, innovative feeding systems to support all year round feeding, and measures to overcome prevailing constraints. These strategies together provide for improved efficiency in the utilization of available feed resources, and increased future productivity from animals.

Maximizing food production in the developing countries assumes that all domestic animals of value to man will be fully exploited. This concept is consistent with the search for efficiency in the utilization of available resources, notably feeds and animals, and the need for self reliance. In comparison to crops, the position regarding animals, and particularly ruminants, is of concern. This is because the latter have failed to keep pace with the requirements of about 1.7% per annum growth rate in human population in the developing countries. The situation is such that national targets for animal proteins are far from being realized relative to the contribution by crops, which in turn has raised doubts about the efficiency of existing animal production systems, and the utilization of the available resources for food production. This implies that in situations where economic use of land is the main thrust in agriculture and animal proteins constitute a very important objective of food production, the role of animals and the priorities for their economic use need to be reexamined in the context of the opportunities for increasing their overall contribution.

The structure of the animal industries in the developing countries, definition of production objectives, the prevailing production systems and, in particular, the efficiency of utilization of the available feed resources are therefore important considerations. Both large scale, intensive, and small farm systems need to be considered relative to the choice and appropriateness of species, consumer preferences and demand, and realistic production targets. The reference to small farm systems is especially important because they constitute the pivot of traditional agriculture, involving many million hectares of land under crops, millions of farm animals and several millions of peasants, landless laborers and tenant farmers (1). These systems which combine animals and mixed cropping are a dominant feature of the agriculture of the developing countries except in Latin America and the Near East. Between regions, Asia has the largest land area under arable and permanent crops, within which there is a preponderance of small farms. It is also noteworthy that in Asia, more than 90% of the total population of buffaloes, cattle, goats and sheep and smaller populations of chickens, ducks and pigs are owned by small farmers, emphasizing that the animals represent important resources within the small farm operations.

An important aspect of maximizing productivity from animals to include both ruminants (buffaloes, cattle, goats and sheep) and nonruminants (chickens, ducks and pigs) concerns efficiency in the utilization of the available feed resources (concentrates, forages, crop residues, agro-industrial by-products and nonconventional feeds) in appropriate and economic feeding systems. The utilization of these feeds presently is generally inefficient and is reflected in low per animal performance, relatively lower contribution especially from the ruminants, and inefficient feeding systems.

This paper is concerned with current trends, scope for improved efficiency in feed resource utilization, the role of animals and, more particularly, strategies that are potentially valuable for increasing productivity from animals with specific reference to Asia.

Animal Populations in Asia

The Asian region has a large variety of animal populations of economic importance. The magnitude of these is reflected in Table 1. In addition to these, there also exist sizable populations of camels, horses, donkeys and mules, which make very valuable contributions in developing countries (3).

It is pertinent to draw attention to the fact that the ruminants (buffaloes, cattle, goats and sheep) are numerically more important than nonruminants and are generally also widely reared. Both species are, however, widely owned by small farmers, landless peasants and agricultural laborers. They are renewable resources and have varied functions from food production (meat, eggs and milk) to various miscellaneous benefits such as security, draught power, fertilizer (dung and urine), fuel, utilization of coarse crop residues, social values and recreation (1).

Over the period 1977–1987, the annual growth rate of individual animal populations suggests that among ruminants, the goat and buffalo populations grew the fastest, followed by cattle and sheep. Among nonruminants, chickens grew at a very rapid rate, followed by the duck population.

Table 2 presents data on food production from animals. The contribution by each species is expressed in percentage of total world output, including the annual rate of growth of the products between 1977–1987. The table indicates that the rate of growth of the nonruminant sector was distinctly higher than that of ruminants, and probably reflects the higher market demand for these products. Meat and milk production from buffaloes constituted a relatively high proportion of the total world output.

Between ruminants and nonruminants, pigs and poultry constitute advanced animal industries in many countries in Asia. The main reasons for this are associated with the availability and successful transfer of proven technology from industrialized countries mainly in temperate regions, support by large private feed mills, the ease of importing feedstuffs, a large and ready market for the products, credit facilities and the rapid turnover of capital investment. In most countries, the two industries have already assumed industrial proportions and usually are found in urban-fringe areas which can absorb the growing domestic market outlets for the products.

Value to man. Table 3 summarizes the value of each species to man. All species have primary, secondary and miscellaneous identifiable functions which need to be kept in perspective. Among ruminants, water buffaloes, cattle, goats and sheep are valued for meat and milk production. Goats are also useful for fiber [mohair and pashmina (cashmere)] and sheep for wool. Chickens and ducks are important for either meat or egg production or both, whereas pigs are only useful for meat. In all cases, several miscellaneous functions exist, and include value in recreation and culture.

Feed Resources

For purposes of this paper, with its specific focus on vegetable proteins, four categories of feed resources are identifiable: concentrates, crop residues, agro-industrial by-products and nonconventional feed resources (NCFR). These can, for convenience, be grouped into three categories: (a) energy rich feeds from bananas, citrus fruits, pineapple, sugarcane and root crops (e.g., banana waste and molasses); (b) protein supplements such as oilseed cakes and meals, animal by-products, by-products from the food industries, and fishmeals (e.g., coconut cake and feather meal); and (c) by-products from cereal milling and palm oil refining (e.g., rice bran and POME).

Priorities for feed resource use. Table 4 summarizes the priorities for using AIBP and NCFR in Asia according to their potential value and importance especially to individ-

TABLE 1

Animal Resources in Asia[a] (2)

Species	Population (10^6)	As % of total world population (%)	Annual growth rate (1977–87) (%)
Ruminants			
Buffalo	133.5	96.5	1.8
Cattle	426.1	33.3	1.4
Goats	242.5	48.3	1.9
Sheep	202.5	17.5	1.1
Nonruminants			
Chickens	2979.0	31.5	8.3
Ducks	403.0	80.6	2.9
Pigs	397.7	47.4	1.4

[a] Includes Iran and all countries east of it except the Pacific Islands, Japan, Australia and New Zealand.

TABLE 2

Food Productivity from Animals in Asia[a]

Category	Production (10^3 MT)	As % of total world output (%)	Annual growth rate (%) (1977–1987)
Ruminants			
Meat			
Buffalo beef	1031	83.4	4.7
Cattle beef	2499	19.3	4.0
Goat meat	1168	52.9	4.8
Mutton and lamb	1075	16.7	4.0
Milk			
From buffaloes	32.2	94.7	3.9
From cattle	29.2	15.8	5.7
Nonruminants			
Chicken meat	3720	12.1	7.1
Pig meat	21153	34.4	8.9
Hen eggs	9097	26.9	8.7

[a] Includes Iran and all countries east of it except the Pacific Islands, Japan, Australia and New Zealand.

ual species of animals. It categorizes the broad types of feeds, their essential characteristics and the main species which currently utilize them.

Priorities for the utilization of the available feeds are essential to ensure efficiency, expanded use of the available feeds, reduced reliance on imported feeds, spiralling feed costs, and excess capacity and inadequate use, especially of the more important NCFR. Such priorities are consistent, as well as ensure the well-known fact that the dairy cow has the highest efficiency of conversion of feed protein to food protein, followed by poultry- and egg-producing birds, pigs and ruminants producing meat. Concerning feed energy, pigs come first, followed by dairy cattle, poultry and meat-producing ruminants. Table 5 demonstrates typical data concerning various species.

Feed balance sheets. Feed balance sheets provide an important means to assess adequacy or the extent of inadequacy concerning the nutrition of the animal resources. More particularly, they enable the development of two

TABLE 3

The Value of Animals to Man

Species	Primary	Secondary	Miscellaneous
Ruminants			
Milch buffalo (river)	Milk	Draught	Dung, skin, recreational and cultural
Water buffalo (swamp)	Draught	Meat	Dung, skin, recreational and cultural
Cattle	Meat/milk	Draught	Dung, skin, recreational and cultural
Goats	Meat	Milk/fiber[a]	Skin, hair, dung, recreational and cultural
Sheep	Mutton/wool	Milk	Skin, dung, hair, recreational and cultural
Nonruminants			
Chickens	Meat/eggs	Dung	By-products, recreational
Ducks	Meat/eggs	Dung	By-products
Pigs	Meat	Dung	By-products

[a] Mohair, pashmina (cashmere) and coarse wool.

TABLE 4

Priorities for the Utilization by Animals of Agro-industrial By-products Priorities (AIBP) and Nonconventional Feed Resources (NCFR) in Asia (4)

Feed source	Characteristic	Species
Energy and protein concentrates (e.g., rice bran, coconut cake, soybean meal, poultry litter)	High energy High protein	Pigs, poultry, ducks, lactating ruminants
Good quality crop residues (e.g., cassava leaves)	High protein High energy	Pigs, ducks, lactating ruminants and use as supplements in meat animals
Medium quality crop residues (e.g., sweet potato vines)	Medium protein	Pigs, ruminants (meat and milk), camels and donkeys
Low quality crop residues (e.g., cereal straws and bagasse)	Low protein Very fibrous	Ruminants (meat and draught), camels and donkeys

[a] Ruminants refer to buffalo, cattle, goats and sheep.

TABLE 5

Efficiencies of Protein and Energy Conversion in Animals [Adapted from (5)]

Category	ME[a] (%)	Protein[b] (%)	Protein (g/Mcal ME)[c]
Beef	7.0	6.0	2.6
Lamb	3.0	3.0	1.3
Pork	23.0	12.0	6.0
Poultry	13.0	20.0	11.0
Eggs	15.0	18.0	11.0
Milk	21.0	23.0	10.0

[a] (Edible energy × 100) divided by (total metabolizable energy consumed).

[b] (Edible protein × 100) divided by (total feed proteins consumed).

[c] [Edible protein (g)] divided by (total metabolizable energy consumed, Mcal).

alternative strategies. One aims to increase feed production, its availability and the development of systems for their more intensive and efficient use. The alternative strategy is to expand animal production commensurate with excess, under-utilized feeds, and issues of conservation and feed security. These contrasting situations are exemplified by India and Pakistan in the first category, and Malaysia in the alternative situation. It is therefore appropriate to discuss these country comparisons briefly.

Table 6 summarizes the situation in India in 1984. The feed deficits in terms of metabolizable energy (ME) and digestible crude protein (DCP) for the animal resources were about 32% and 54%, respectively.

Table 7 provides a trend in the feed balance situation in India between 1870 and 1984. Two major conclusions are apparent. First, feed deficits and the malady of undernutrition were a continuing problem. Second, there has been a trend toward a reduced feed deficit despite increased animal populations over the last 14 years. The trend toward reduced deficits probably is reflective of improved feeding systems, more efficient use of the available feeds and increasingly intensive systems of production. Whether, in terms of scale and magnitude, these approaches are adequate and can be further improved is a matter of debate.

Table 8 illustrates a parallel situation in Pakistan also for the year 1984. The deficits in terms of total digestible nutrients (TDN) and DCP are about 25% and 41%, respectively.

By comparison, Table 9 presents an alternative situation in Malaysia where it has been estimated that land under native and cultivated grasses contributed a total annual dry matter (DM) production of about $3,838 \times 10^3$ tons, and from roughage by-products $2,935 \times 10^3$ tons, giving a total of 6,773 tons. This was in excess of the estimated total requirements by ruminants of 1,580 tons (11).

Imported feeds. Associated with feed balance sheets is that of dependence on import of feeds, notably animal proteins and mineral-vitamin supplements, especially for intensive poultry and pig production. The principal protein supplement is soybean meal. Over the last two decades, production has been increasing and, associated with this, exports as well. The meal also accounts for approximately 75% of the total world trade of oilcakes and meals. Cottonseed cake and fish meal are the next two most important protein feeds, but the rate of production of these does not compare with that of soybean production. Recently, the effects of drought and dependence on soybean meal utilization have resulted in a significant rise in the cost of the meal, which has necessitated more judicious use of the protein source in feeding systems, especially for nonruminants.

The magnitude of these imports varies among countries, and is influenced significantly by government policy and financial considerations. These countries can be grouped into two categories. The first group includes

TABLE 6

Feed Availability and Requirements in India in 1984 [Adapted from (6)]

Principal feed source	Availability[a]		Total requirements[b]	
	Energy (10^7 Mcal ME)	DCP (MT)	Energy (10^7 Mcal ME)	DCP (MT)
Crop residues and agro-industries by-products	5022.3	7437	—	—
Fodder crops	1228.0	3411	—	—
Grasses[c]	1149.0	2660	—	—
Total	7399.3	13508	10933.5	33396
% Deficit	—	—	32.3	54.0

[a] ME, Metabolizable energy; DCP, digestible crude protein.
[b] Of herbivores (buffalo, cattle, goats, sheep, asses, mules, yaks and chauri) and nonruminants (poultry and pigs).

TABLE 7

Trends in Feed Balances in India [Adapted from (6)]

	1970			1984		
Nutrient	Availability[a]	Requirement[b]	% Deficit	Availability	Requirement	% Deficit
Energy (10^7 Mcal ME)	6162.8	9877.9	37.6	7399.4	10933.5	32.3
DCP (10^4 MT)	113.2	297.8	61.9	135.1	344.0	54.0

[a] ME, Metabolizable energy; DCP, digestible crude protein.
[b] Of herbivores (buffalo, cattle, goats, sheep, asses, yaks and chauri) and nonruminants (poultry and pigs).

TABLE 8

Feed Availability and Requirements in Pakistan in 1984 (7, 8)

	Availability (10^3 MT)[a]		Total requirements[b] (10^3 MT)	
Principal feed source	TDN	DCP	TDN	DCP
Crop residues and agro-industries by-products	8359.9	947.5	—	—
Fodder crops	18059.5	692.8	—	—
Grasses[c]	11200.0	700.0	—	—
Total	37619.4	2340.3	50096	3951
% Deficit	—	—	24.9	40.7

[a] TDN, Total digestible nutrients; DCP, digestible crude protein.
[b] Of ruminants: buffalo, cattle, goats and sheep.
[c] From canals, banks, roadsides, orchards, flood plains and rangelands.

those countries which have controlled imports of feeds to sustain components of the animal industries. Examples in this group are Pakistan, India, Thailand, the Philippines and Indonesia. The other group represents those who have the capacity to export liberally to meet animal feed requirements, notwithstanding the availability of considerable supplies of indigenous feeds. These include countries such as Singapore and Malaysia.

There is no doubt that soybean meal and fish meal will continue to dominate efficient feeding systems asso-

TABLE 9

Total Availability of Dry Matter (DM) and Requirements by Ruminants in Malaysia (9)

Component	Availability (10^3 MT)
Grazing[a]	3839
Agro-industrial by-products[b]	3942
Requirements[c]	1580

[a] From herbage under plantation crops, grazing lands, roadsides and padi bunds.
[b] Includes nonconventional feeds.
[c] By ruminants (buffalo, cattle, goats and sheep).

ciated with use of superior genetic stock, sophisticated management systems and improvements in the environment, especially for nonruminants and ruminants. These factors have resulted in significant improvements to feed efficiency over the past four decades. Table 10 reports these improvements which are reflective of typical performance in several countries in the Asian region. Further improvements in feed efficiency are feasible, but they are likely to be small.

TABLE 10

Improvements in the Efficiency of Feed Conversion (EFC) in Pigs and Poultry

	EFC (units of feed required/unit live weight gain)	
Year	Pigs	Poultry
1945	4.1	3.5
1955	3.8	3.0
1965	3.4	2.8
1975	3.1	2.5
1985	2.5	2.2
1988	2.3	2.0

Preformed proteins especially are important to both nonruminants and ruminants because they are often the main limiting factor in the diet, in efficiency of feed utilization and in level of performance. With ruminants, a small amount of protein has a catalytic effect on rumen metabolism, manifesting a significant effect on intake and efficiency of feed utilization (10,11). The implication of this result is that protein resources within individual countries need to be conserved and used especially carefully in the context of exports of indigenous protein meals and imports of milk products. The significance of this is reflected in their use in lactating cattle, for example, fed on low quality, cereal straw diets supplemented with 100–500 g/day/cow of proteins. It has been calculated that this strategy and the potential availability of say 20,000 MT of proteins fed catalytically in any country would stimulate the additional production of 80 million liters of milk annually (13).

Impinging on the controlled utilization of imported feeds, notably maize, soybean meal, mineral-vitamin supplements, and the various problems associated with these, are the rising costs related to their use. Future approaches are thus likely to address and investigate more thoroughly those factors that can effectively reduce the cost of feeding with no loss in performance. Judicious and controlled use of imported feeds are thus likely to be coupled with more intensive use of such other protein feeds as cereal brans, coconut meal and cowpea seed meal.

Nonconventional feeds. Notable in this connection is greater attention to the use of indigenous feedstuffs, including nonconventional feeds produced in Asia. Table 11 indicates the magnitude of the contribution from the latter category, much of which is under-utilized. Table 12 gives examples of the utilization of five types of nonconventional feeds, with an indication of optimum dietary levels for feeding nonruminants and ruminants. Table 13 gives a specific example concerning the utilization of rice bran by poultry, to partially replace soybean meal. The results indicated that laying hens can be fed with up to 33.0% rice bran with no loss in performance on low energy diets (10.1 MJ/kg) with a minimum dietary protein level of 15%.

TABLE 11

The Availability of Nonconventional Feed Resources in Asia and the Pacific (13)

Category	Availability (10^6 MT)
Field crops	230.3
Tree crops	7.4
Total	237.7[a]

[a] Represents 46.3% of the total availability of feeds from field and plantation crops.

The utilization of various by-products and NCFR is not without problems of collection, transportation, storage, processing considerations, and deleterious components which affect animal performance. Table 14 provides a summary of various types of toxic substances in individual feeds. The list is not exhaustive but provides information on the type of toxic components and their approximate contents. With some of these, such as HCN in cassava, methods are now available to detoxify the substance to render the feed more useful. More information is required on these toxic components and, in particular, methods to reduce their deleterious effects on animals.

Intensifying Feed Resource Utilization

Two important prerequisites for intensifying the efficiency of feed resource utilization in the future concern iden-

TABLE 12

Optimum Level of Utilization of Some Nonconventional Dietary Vegetable Proteins for Farm Animals in Asia

Nonconventional feedstuff	Species	Location	Optimum level of dietary inclusion (%)	Reference
Castor				
Castor bean meal	Buffalo	India	30	(14)
	Sheep	India	10	(15)
Mango				
Mango seed kernel	Calves	India	20	(16)
	Bullocks	India	40	(17)
	Cows	India	10	(18)
Oil palm				
Palm oil mill effluent	Sheep	Malaysia	40	(19)
Palm oil mill effluent	Poultry	Malaysia	10–15	(20)
Palm oil solids	Poultry	Malaysia	10–15	(21)
Rubber				
Rubber seed meal	Pigs	Malaysia	20	(22)
	Poultry	Sri Lanka	20	(23)
	Poultry	Sri Lanka	20	(24)
	Calves and cows	India	20	(25)
	Calves	India	30	(18)
	Cows	India	25	(18)
	Pigs	India	40	(26)
Sal				
Sal seed meal (untreated)	Poultry	India	5	(27)
Sal seed meal (untreated)	Poultry	India	20	(28)
	Cows	India	30	(29)
	Bulls	India	40	(30)

TABLE 13

Effects of Feeding Rice Bran in Diets with Two Levels of Energy and Protein on Performance of Laying Hens[a] [Adapted from (31)]

Treatments	I	II	III	IV
Rice bran (%)	38.5	33.0	18.5	12.5
Feed intake (g/b/day)	117.6 ± 0^b	$120.1\pm.9^a$	$122.5\pm.6^a$	$119.1\pm1.3^{a,b}$
ME intake (kJ/b/day)	1182.0 ± 9^a	1232.0 ± 10^a	1384.0 ± 7^b	1345.0 ± 9^b
Protein intake (g/b/day)	$14.1\pm.1^a$	$18.0\pm.1^b$	$14.7\pm.1^a$	17.9 ± 2^b
H.D. Egg Prod. (%)	65.7 ± 2^a	$70.7\pm2^{a,b}$	73.4 ± 1^b	74.6 ± 2^b
FCR (kg feed/kg egg)	$3.11\pm.09^b$	$2.79\pm.09^a$	$2.76\pm.04^a$	$2.74\pm.04^a$
Egg weight (g)	58.4 ± 1.1^b	$61.6\pm.8^a$	60.1 ± 1^a	59.3 ± 1.3^a
Egg mass (g)	38.3 ± 1.2^a	43.5 ± 1.4^b	$44.1\pm.7^b$	44.2 ± 1.5^b

[a,b,c] Values in a row with different superscripts are significantly different (P < 0.01).
[a] Means for 12 weeks.

tification of their value in terms of priorities (Table 14) and, more particularly, wider efforts to include them in intensive feeding systems (4,13,32–34). These considerations also necessitate their definition into two categories. These are primary feedstuffs; ingredients that form the main base in a feeding system (these constitute about 70–80% in the diet); and secondary feedstuffs; minor ingredients that are supplements in the diet (these constitute up to 20–30% in the diet).

Table 15 sets out examples of the more important AIBP and NCFR which merit particular attention, and whose efficient utilization is likely to make a significant impact on the low level of animal performance prevailing in most countries. Associated with the utilization of these

TABLE 14

Examples of Toxic Principals in Some Common Nonconventional Feeds

Type of feed	Toxic principal
Banana waste, stems and leaves	Tannins
Cassava leaves, peeling and pomace	NCN (17.5 mg/100 g in leaves)
Castor seed meal	Ricinoleic acid
Cocoa seed husks	Theobromine (trace)
Coffee seed hulls, pulp	Caffeine and tannins (2.8% DM)
Cottonseed cake	Gossypol (0.05–0.20%)
Cowpea seed meal	Trypsin inhibitor
Guar meal	Trypsin inhibitor and gum
Kapok	Cyclopropenoid acid
Mango seed kernel	Tannin (5–10%)
Neem seed cake	Tannin
Palm oil mill effluent	High ash (12–16% DM)
Rubber seed meal	HCN (9 mg/100 g)
Sal seed meal	Tannin (6.2–13.7%)
Spent tea leaf	Tannin (12% DM)

TABLE 15

Some Examples of Primary Feeds for Intensive Utilization by Location (4)

Type of primary feed	Location	Species
Bananas	Philippines	Beef cattle, ducks
Cassava		
Leaves	Thailand, Indonesia, Philippines	Beef cattle, goats, swamp buffalo
Pomace	Thailand, Indonesia, Philippines	Pigs, ducks, lactating cattle and goats
Maize stover	Philippines, Indonesia	Beef cattle, swamp buffalo, goats and sheep
Oil palm		
POME, palm press fiber, palm kernel cake	Malaysia	Beef cattle and swamp buffalo
Rice		
Bran	Thailand, Indonesia, Philippines	Pigs, poultry and lactating ruminants
Straw	Thailand, Sri Lanka, Philippines	Beef cattle and swamp buffalo
Sugar cane		
Tops, bagasse	India, Pakistan, Thailand	Beef cattle and swamp buffalo
Wheat		
Bran	India, Pakistan	Pigs, poultry, lactating ruminants
Straw	India, Pakistan	Beef cattle and swamp buffalo

feed ingredients is the wider utilization of a variety of proteinaceous material whose potential value has been emphasized (35,36). Both ruminants and nonruminants are involved, and innovative feeding systems that can include these feeds in suitable proportions for all year round feeding systems can go a long way toward increasing the current contribution and future productivity of the animal resources.

References

1. Devendra, C., *Proc. 5th Wrld. Conf. Anim. Prod., Tokyo, Japan 1*:173 (1983).
2. F.A.O., *Selected indicators of food and agricultural development in Asia - Pacific region, 1977–87;* F.A.O. Regional Office, Bangkok, Thailand, pp. 199 (1987).
3. Devendra, C., in *Nutrition of Herbivores,* edited by J.B. Hacker and J.H. Ternouth, Academic Press, Sydney, Australia, 1987, p. 23.
4. Devendra, C., *Proc. APO Symp. on Animal Feed Resources,* Tokyo, Japan, 1987, in press.
5. Holmes, W., *Proc. Nutr. Soc. 29:*237 (1970).
6. Reddy, M.R., *Proc. APO Symp. on Animal Feed Resources,* Tokyo, Japan, 1987, in press.
7. Akram, M., *Proc. APO Symp. on Animal Feed Resources,* Tokyo, Japan, 1987, in press.
8. Mimeograph, Ministry of Food and Agriculture, *Government of Pakistan,* Islamabad, Pakistan, 1984.
9. Devendra, C., *Proc. Conf. on Exotic and Crossbred Livestock, Malays. Soc. Anim. Prod.,* Serdang, Malaysia, 1981, p. 235.
10. Kempton, T.J., J.V. Nolan and R.A. Leng, *Wrld. Anim. Rev., 22:*2 (1977).
11. Saadullah, M., M. Hague and F. Dolberg, in *Maximum Livestock Production from Minimum Land,* edited by C.H. Davis, T.R. Preston, M. Hague and M. Saadullah, James Cook University of North Queensland, Townsville, Queensland, Australia, 1983, p. 148.
12. Leng, R.A., and T.R. Preston, *Proc. V Wrld. Conf. Anim. Prod., Vol. 1:*301 (1983).
13. Devendra, C., *Nonconventional Feed Resources in Asia and the Pacific, APHCA/FAO Monograph,* 3rd edn, FAO Regional Office, Bangkok, Thailand, 1988, in press.
14. Reddy, M.V., M.R. Reddy and G.V.N. Reddy, *Indian J. Anim. Nutr. 3:*86 (1986).
15. Sungunakar Rao, M., N.P. Purushotham, G.V. Raghavan, M.R. Reddy and M. Mahendar, *Indian Vet. J. 63:*944 (1986).
16. Patel, B.M., and C.A. Patel, *Indian J. Nutr. and Dietetics 9:*157 (1971).
17. Patel, B.M., C.A. Patel and P.M. Talapada, *Ibid. 9:*347 (1972).
18. Annual Report, I.C.A.R., *College of Vet. Sci. and Anim. Husb.,* Jabalpur, India, 54 pp. (1983).
19. Devendra, C., and R.N. Muthurajah, *Reprint No. 8, Malays. Int. Symp. on Palm Oil Processing and Marketing,* Kuala Lumpur, Malaysia, 21 pp. (1976).
20. Yeong, S.W., and A. Azizah, *Proc. Advances in Animal Feeds and Feeding in the Tropics,* edited by R.I. Hutagalung, C.P. Chen, W.E. Wan Mohamed, A.T. Law and S. Sivarajasingam, Malays. Soc. Anim. Prod., Serdang, Malaysia, 1987, p. 302.
21. Devendra, C., S.W. Yeong and H.K. Ong, *Proc. PORIM Workshop on Oil Palm By-products Utilization,* Kuala Lumpur, Malaysia, 1982, p. 63.
22. Ong, H.K., and S.W. Yeong, *Proc. Symp. Feedingstuffs for Livestock in South-East Asia,* edited by C. Devendra and R.I. Hutagalung, Kuala Lumpur, 1977, p. 337.
23. Buvanendran, V., and J.A. de S. Siriwardene, *Ceylon Vet. J. 18:*33 (1970).
24. Rajaguru, A.S.B., *R.R.I.S.L. Bull. (Sri Lanka) 8:*39 (1973).
25. Annual Progress Reports, *Kerala Vet. College,* Trichur, India (1975-76).
26. Pathak, N.N., and S.K. Ranjhan, *Indian J. Anim. Sci. 43:*424 (1973).
27. Verma, S.V.S., M.V.Sc. Thesis, Agra University, India, 1970.
28. Sharma, K., C.S. Wah and M.G. Jackson, *Indian J. Anim. Sci. 47:*473 (1977).
29. Sonwane, S.N., and V.D. Mudgal, *Indian J. Dairy Sci. 27:*183 (1974).
30. Shukla, P.C., and P.M. Talapada, *Indian Vet. J. 50:*669 (1973).
31. Ramlah, H., and S. Jalaludin, *Proc. Advances in Animal Feeds and Feeding in the Tropics,* edited by R.I. Hutagalung, C.P. Chen, W.E. Wan Mohamed, A.T. Law and S. Sivarajasingam, Malays. Soc. Anim. Prod., Serdang, Malaysia, 1987, p. 307.
32. Devendra, C., in *Rice Straw and Related Feeds,* edited by M.N.M. Ibrahim and J.B. Schiere, *Straw Utilization Project Publ. No. 2,* Kandy, Sri Lanka, 1986, p. 11.
33. Doyle, P.T., C. Devendra and G.R. Pearce, *Rice Straw as a Feed for Ruminants, Int. Dev. Program of Australian Universities and Colleges,* Canberra, Australia, x + 117 pp. (1986).
34. *Nonconventional Feed Resources and Fibrous Agricultural Residues, Strategies for Expanded Utilization,* edited by C. Devendra, I.D.R.C./I.C.A.R., Singapore, in press, 1988.
35. Devendra, C., *Buffalo J. 2:*113 (1987).
36. Devendra, C., *Proc. 3rd. Crop-Animal Asian Rice Farming Systems Res. Workshop,* Int. Rice Res. Institute, Los Banos, Philippines, in press, 1988.

Mycotoxins in Oilseeds and Risks in Animal Production

H. Müschen and Karl Frank (photo)

Technical Services, Animal Nutrition, MEA/ETM - D 205, BASF Aktiengesellschaft, D-6700 Ludwigshafen, Federal Republic of Germany

Abstract

The main factors that will determine mycotoxin formation in feed components are: suitable environmental conditions for growth, the presence of spores of toxigenic molds and a suitable substrate to support mold growth. Oilseeds are an ideal substrate for mold growth, and thus mycotoxin production, since a moisture content of only 7% is necessary whereas in other raw materials growth starts at about 14% moisture. The best known mycotoxin in oilseeds is aflatoxin B_1, which was detected first in groundnut meal. Small quantities of mycotoxins in feedstuffs may cause severe problems in animal production after consumption of the feed. Reasons are (a) the direct effects of mycotoxins on metabolism of fat, carbohydrate and protein, and (b) a reduced immunity toward virulent organisms. In swine, mycotoxicosis induces liver damage, gross hemorrhages in many parts of the body, immune suppression and nephrosis. In poultry, aflatoxin is the most toxic mycotoxin and causes severe alterations in protein and lipid synthesis culminating in poor growth, impaired feed efficiency, altered immunity and increased susceptibility to bruising with impaired blood clotting. In ruminants, the major harmful effect is the contamination of milk. The relation of aflatoxin B_1 in feedstuffs to aflatoxin M_1 in milk is only 300:1. In order to protect the animal and subsequently man against mycotoxins, their formation in oilseeds should be inhibited by proper drying after harvest or by preservation with propionic acid. Once the mycotoxins are formed, only separation of the contaminated kernel or detoxification can reduce mycotoxin concentration. Detoxification with ammonia changes aflatoxin B_1 to aflatoxin D_1, while solvents such as methanol or ethanol make it possible to extract mycotoxins. The large scale detoxification, however, is expensive and may destroy nutrients such as amino acids.

TABLE 1

Moisture Levels Where No Aflatoxin Formation Should Occur

Commodity	"Safe" moisture levels, %
Groundnut in shell	8/9
Groundnut kernel	7
Groundnut cake	12
Cottonseed	10/11
Copra	≈ 7

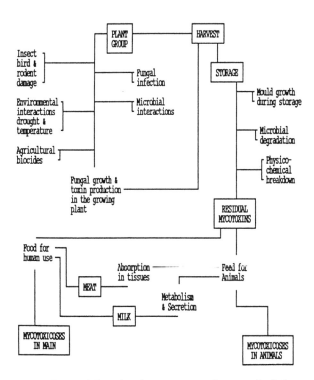

Fig. 1. Factors influencing the occurrence of mycotoxins in human food or animal feed.

Oilseeds are among the most thoroughly studied commodities associated with mycotoxins, especially aflatoxins. As we all know, mycotoxins are produced by fungi under certain environmental conditions, which support mold growth but which must be also favorable for mycotoxin production at the same time. By far the most critical factors determining whether a substrate will support mold growth are moisture content, temperature and time. Only few toxigenic fungi can grow at relative humidities below 70%, which correspond to moisture contents of about 7% for oil-containing commodities and 14% for starch-containing materials. Thus, oilseeds are an ideal substrate for mold growth and mycotoxin production. In Table 1 you will find those moisture contents which are generally regarded as safe levels, i.e. where almost no mycotoxin production during storage occurs. But the whole problem of mycotoxins is much more complex than looking only at proper storage conditions as you can see from Figure 1.

As do other mycotoxins, aflatoxin can be formed during the growing stages of certain crops. Many environmental conditions have been identified, which can promote mycotoxin formation and include insect infestation, drought conditions, variety susceptibility or resistance, mechanical damage, nutritional deficiencies and unseasonal temperatures and rainfall.

Attempts are now being made to develop commercially acceptable varieties of crops, e.g. peanuts, maize and cottonseed that will be more resistant to toxigenic molds or will inhibit toxin production. For example the so-called "hard seed coat" strain in cottonseed has been shown to be more resistant to *Aspergillus flavus* penetration. But, undoubtedly the most common period of mycotoxin production on plants or organic material is during storage.

The figures in Table 2 make it quite clear that man directly and indirectly *via* meat and milk from animals may suffer from mycotoxicoses. We know now, that aflatoxin is strongly carcinogenic and will be transferred from the feed to meat, milk and eggs.

TABLE 2

Most Likely Ratios of Aflatoxin B_1 Levels in Feed to Aflatoxin B_1 or M_1 in Food

Animal	Kind of food	Aflatoxin	Feed to food ratio
Beef cattle	liver	B_1	14,000 : 1
Dairy cattle	milk	M_1	300 : 1
Swine	liver	B_1	800 : 1
Layer	eggs	B_1	2,200 : 1
Broiler	liver	M_1	1,200 : 1

Adapted from Rodricks and Stoloff (1977).

This is one of the reasons why legislation in the EEC allows only a limited content of aflatoxin in feedstuffs for farm animals. For example in compound feed for milking cows a level of 10 ppb is accepted. But many dairies go much further and demand a level of only 5 ppb for those farmers who deliver their milk to the dairy. This is the reason why copra, for instance, though a very suitable component is no longer used in dairy rations. Now good quality copra imported into the Federal Republic of Germany contains 20 ppb aflatoxin and inferior quality contains about 100 ppb. Recently, a ship carrying groundnut cake (origin Gambia) was rejected from a Belgium harbor because aflatoxin levels of 350 to 600 ppb were much too high.

The effects of these aflatoxins on animal performance are quite negative and some data exist and document their negative impact. The LD_{50}-level of aflatoxins differs widely between different kinds of animals. See Table 3.

The most sensitive animal is the duckling and it is often used in trials to evaluate different decontamination methods. But, pigs also are quite sensitive to aflatoxin.

Mycotoxicosis in animals is divided into three different classes:

Acute primary mycotoxicosis: These effects are produced if high to moderate concentrations of mycotoxins are consumed causing a specific, observable, acute disease syndrome such as hepatitis, hemorrhage, nephritis, necrosis of oral and enteric epithelia or death.

Natural contamination levels of mycotoxins usually are not high enough to cause acute mycotoxicosis. Reports of naturally occurring outbreaks of acute mycotoxicosis are limited in both number and description. It is now becoming increasingly recognized that most field or feed contaminations with mycotoxins will cause chronic mycotoxin diseases.

TABLE 3

LD_{50} Levels of Aflatoxins for Selected Animals

Animal	LD_{50} (μg/kg)
Duckling	360
Cat	550
Pig	620
Dog	1,000
Turkey	1,860
Sheep	2,000
Chick	6,500
Rabbit	300

Chronic primary mycotoxicosis: This type occurs after a longer period with an intake of smaller to medium amounts of aflatoxin. There is a lack of microscopically visible changes in the infected animals and this prevents an easy diagnosis based on symptoms.

In chronic mycotoxicosis in pigs, poultry and ruminants, the effects appear in reduced productivity in the form of slower growth, reduced reproductive efficiency and inferior market quality.

In Table 4 the effects of chronic mycotoxicosis on pigs are listed.

TABLE 4

Responses of Pigs to Various Amounts of Aflatoxin

Animal	Quantity of toxin	Effect
20 kg pig	0.26 ppm	Decreased growth rate
20 kg pig	0.86 ppm	Impaired immuno genesis
22 kg pig	2–4 ppm	Acute fatal toxicosis

Adapted from: Pier et al. (1980).

Some data, however, indicate that the level of protein in the diet may counterbalance the aflatoxicosis. Sisk and co-workers (1968) reported that 10-week-old miniature swine on a 17% protein diet showed no effects of aflatoxins (99μg per kg bodyweight), but pigs on an 11.4% protein diet were affected. However, when 4-week-old pigs were provided feed containing 95μg aflatoxin per kg bodyweight, both those on 20.6% and those on 14.1% protein diets were adversely affected (Sisk et al., 1968).

The most thoroughly studied livestock with regard to aflatoxin is poultry. The effects of chronic aflatoxicosis are listed in Table 5.

Aflatoxin can cause a generalized loss of tissue strength and integrity which may cause an increased susceptibility to bruising and, with chicken, a decreased market acceptability. Furthermore, this problem is only seen after the birds have been killed and prepared for market. The levels of aflatoxin required in feed to cause this effect are less than half that required to reduce growth.

Aflatoxin also can cause a situation like the well-known fatty liver syndrome in laying hens. Egg production is de-

TABLE 5

Effects of Aflatoxins on Poultry

Animal	Quantity of toxin, ppm	Effect
Poult	1.00	Acute death, hepatic necrosis
Broiler	5.00–10.0	and haemorrage
Broiler	2.5	Coagulopathy
Broiler	0.6	Bruising
Poult	0.25	Impaired immunogenesis
Broiler	0.6–1.0	Reduced resistance
Broiler	1.5–2.5	Decreased gain
Poult	0.25	″
Laying hen	2.0–8.0	Decreased egg production

Pier et al. (1980).

creased, and the percentage of liver-lipids is doubled. Egg size may be decreased as well and these effects together mean heavy financial losses to the industry.

The malabsorption syndrome has caused a lot of trouble in poultry industry in recent years. It is characterized by bad feed conversion and, though not all reasons are quite clear yet, aflatoxin seems to play an important part. The depression in bodyweight gain can be reduced if the feed contains more protein, i.e. protein requirement is enhanced. But often it is just protein that is the growth limiting nutrient and that is quite expensive in many parts of the world.

In calves, hepatic changes observed with aflatoxin poisoning are bile duct proliferation, loss of hepatic glycogen, fatty infiltration, fibroplastic proliferation and perivascular edema.

The major ill effects of aflatoxins in dairy cows are the insidious abnormalities produced in animals when low levels of mycotoxins are ingested. In rumen metabolism, aflatoxin changes the proportions of volatile fatty acids. A single injection of 200 mg aflatoxin B_1 and 80 mg aflatoxin B_2 into ruminally fistulated cows resulted in decreased portions of acetic acid, increased butyric acid and variable portions of propionic acid. This points to an effect on the microbial population by aflatoxin contamination.

Secondary mycotoxin diseases: Consumption of low levels of certain mycotoxins can lead to impairment of the native and acquired resistance to infectious diseases causing health related economic losses and vaccine failure.

Aflatoxin consumption has been associated with increased susceptibility to candidiasis, Marek's disease, coccidiosis and salmonellosis in chicken, pasteurellosis and salmonellosis in turkeys, erysipelas and salmonellosis in pigs, fascioliasis and clostridial infections in calves and intraurinary infections in cows. The levels of intake of mycotoxins causing immunosuppressant effects are much lower than those which cause chronic effects such as retarded growth.

Which measures can be taken in order to protect animals and consequently man as well? Once the mycotoxin is built up only decontamination enables us to reduce the level of mycotoxin in a certain feedstuff. Decontamination methods are listed in Table 6.

Most of the decontamination methods listed are not suitable for large scale under practical conditions. Extrac-

TABLE 6

Decontamination Methods

Method	By means of
a) Separation	– Mechanical separation
	– Color sorters
b) Extraction of mycotoxins	– Organic solvents
	– Bleaching earth
c) Degradation or detoxification	– Irradiation
	– Heat
	– Chemical processes
	– Biological processes

tion with organic solvents is possible, but the need for special solvent-removing equipment and the loss of nutrients from the residual meals have made these processes of questionable economic value.

Ammonia used as an anhydrous gas at elevated temperature and pressure can cause a 95–98% reduction of total aflatoxin concentration in peanut meal. It is believed that the ammonia opens the lactone ring of aflatoxin B_1, forming an ammonium salt of the resulting hydroxyacid. Because the reaction is carried out at elevated temperature and pressure, it causes the decarboxylation of the beta-keto acid to the so-called aflatoxin D_1. This method is used on a commercial scale in the United States and although the protein efficiency rate values of the product are lowered, the chemical consumption of the meals altered, plus some degree of off-flavors and off-colors occurs, the final product may still be used for animal feed.

All decontamination measures are very difficult to carry out and cost some money. The better method therefore is to prevent mycotoxin production from the very beginning.

To influence mycotoxin production on living plants can be achieved only through measures like suitable crops and varieties and breeding of more resistant strains. Adequate agricultural management may also help to reduce the risk of strong fungal growth before harvest. The production of mycotoxins during storage can only be prevented through correct storage management i.e. rapid drying, avoiding rewetting, etc.

TABLE 7

Aflatoxin Inhibition During Storage with Propionic Acid Products (control 2.0 ppb)

Treatment	Application	ppb Aflatoxin during storage	
		after 14 days	after 40 days
Control		6.2	18.0
0.6 % Luprosil NC	watering can	3.4	23.0
0.8 % Luprosil NC	hand mixing	1.6	2.0
0.2 % Luprosil NC		1.1	4.0
0.3 % Luprosil NC	atomizer	1.0	4.0
0.4 % Luprosil NC	hand mixing	0.4	1.0
0.6 % Luprosil NC		0.9	1.0
0.3 % Luprosil	atomizer, mixer	1.9	2.0

A useful tool in order to create proper storage conditions and avoid mycotoxin production is the addition of propionic acid. We have carried out a trial with copra. Results look very promising as can be seen from Table 7 but more experience is necessary to find the optimum application level and a suitable application method especially for oil seeds. In small cereals the system has proven to fulfill the expectations of feedmills and farmers over many years. In comparison to the untreated control, all rates applied at all methods tested could reduce further aflatoxin formation remarkably if the material was distributed evenly.

Utilization of Vegetable Oilseed Cakes and Meals for Livestock Production in China

Shen Zai-Chun

Beijing, Agricultural Engineering University, Beijing, People's Republic of China

Abstract

China is a major world producer of vegetable oilseed cakes, which are important protein sources in feeding livestock, poultry and aquaculture species. Since 1978, with the increasing domestic demand for milk, meat and eggs, many collective and privately owned factory farms have been built near large cities for production of broilers, egg, swine, cattle and fish. Many compounded and mixed feeds are required and thus the feed industry in China has grown rapidly. Annual output of such feed has grown from about 1 million tons in 1980 to 24 million tons in 1987, and is expected to continue to grow. However, toxic and antinutritional compounds, found in soybean, cottonseed and rapeseed, need to be heated or chemically treated to make the meals suitable for feed use. This paper will summarize processing and use of oilseed cakes and meals for livestock, as well as the status of the Chinese feed industry.

China is one of the main countries in the world producing vegetable oilseed cakes and meals. Vegetable oilseed cakes and meals are an important protein source for feeding livestock, poultry and for aquaculture. Since 1978, with the increasing demand for milk, meat and eggs, many large farms of broilers, layers, swine, cattle and fish have been built in the great cities. A lot of collective and individual raisers have appeared rapidly in the countryside. Therefore, many formula and mixed feeds have been in demand recently in China and the growth of the feed industry has accelerated. The output of formula and mixed feed has increased rapidly annually. For example, in 1980, the annual total output of the formula and mixed feed was about one million tons. In 1987, it increased to 24 million tons. The average rate of increase in the annual output is 57.46%. Besides this, premix and concentrated feeds have increased quickly. The feed industry is expected to develop further in the future.

The development of the feed industry has helped to advance production of the livestock, poultry and aquaculture industries, which annually provide China with abundant products. For example, in 1986, the output of meat was 21.12 million tons; milk, 3.33 million tons; eggs, 5.56 million tons, and aquatic products, 8.24 million tons. The total output in the year is among the highest in the world, but the annual output per person is only, meat, 20 kg; milk, 3.16 kg; eggs, 5.27 kg, and aquatic products, 7.8 kg.

At present, there exists a striking problem about a serious lack of feed proteins in our feed industry. During 1986, the total output of formula and mixed feeds was about 20.10 million tons with the following ingredients, grains, 52.1%; rice polishings and wheat bran, 25.6%; vegetable oilseed cakes and meals, 13.6%; fish meals, meat and bone meals, 2.6%, and others. According to this proportion of ingredients, it is easy to see that feed proteins are lower. So there is a difference between feed conversion of our country and that of advanced countries. For example, the feed conversion is 2.8–3.2:1 for swine, 2.5–3:1 for layers, and 1.9–2:1 for broilers in advanced countries of the world. Recently the feed conversion in advanced livestock and poultry farms in China was 3.8–4:1 for swine, 2.5–3:1 for layers, and 2–2.6:1 for broilers. The level of feed conversion for layers and broilers has reached the world advanced level, but there exists a great difference for swine. During 1986, the annual output of meal was 21.12 million tons, of which swine's meal was 85.2%. Therefore, a great deal of feedstuffs can be saved yearly and production cost may be decreased, if the feed conversion for swine feedstuffs is raised to the world's advanced level. There are various reasons for the low feed conversion; the lack of feeds protein and a lack of efficient processing method are two of the most important reasons.

During 1986, the total output of vegetable oilseed cakes and meals was about 11.10 million tons, in which the soybean cake and meals were about 2.9 million tons, cottonseed cake about 2.9 million tons, rapeseed cake and meal about 2.8 million tons, peanut cake about 1.5 million tons and sunflower cake about 0.6 million tons. The sum of cottonseed and rapeseed cake was 5.7 million tons and it occupied 51.8% of the annual total output of vegetable oilseed cakes and meals. In fact, little of these were used in formula and mixed feed in 1986. Most of those were used directly on the soil as a fertilizer.

In China, soybean, cottonseed, rapeseed, peanut and sunflower are five major sources for processing oilseed cakes and meals. Each oilseed has its own requirement as to specific conditions and order of operations. The oilseed cakes and meals are a by-product during oil extraction.

Three alternative processing methods for oil extraction are available, i.e., hard press, prepress extraction, and direct solvent extraction.

Hard press has three kinds of methods for oil extractions. There are:

Wood block press, which is an old procedure for mechanical stressing of oilseed to rupture seed walls. The heated coarse ground oilseed is packed in burlap bags and pressed by wood blocks. The oil flows out of the enclosure of the burlap and yields a circular oilseed cake with a diameter between 300–600 mm and thickness between 30–60 cm. Residual oil content of the cakes pressed with a wood block usually is about 6–8%, with a high content of crude protein. It is very hard, laborious work, and is used less now than previously.

Hydraulic press is a processing method where overall dimensions of cakes are like the wood block press, but oil extraction is done with high-hydraulic presses. Residual oil content of hydraulic pressed cakes usually is about 5–7%, although some operations produce cakes with 4–5% residual oil. The cakes are processed with high pressure and middle temperature. The protein naturation has not changed with high quality, delicious taste and safety for use. Urease has lost its activity when processing soybean, so it is welcomed by feed millers and farmers. These cakes are broken to crumbles with a special crusher. The particle size of crumbles is now over 3 mm, and are mixed with other ingredients to become formula and mixed feeds.

Screw press, where the oilseed cake is like a crooked flake with a thickness of of 6–10 mm, uses seed that is usually well-cooked, or dried without cooking to a lower moisture content. Residual oil content of screw pressed cake usually is about 6–8%, although in some operations cakes have 4–5% residual oil. Screw-press operations usually inflict considerable heat damage to the protein. The protein has been denatured and urease has lost activity during soybean processing because this operation is carried out under high temperature and press. The flakes are tasty to eat. Before feed preparation, the flakes are ground with a common hammermill. Recently, extruder-cookers have been used to process the vegetable oilseed. Hard press operations are often used in processing cottonseed, rapeseed, soybean, peanut, oiltea-camillia seed, flaxseed and copra.

Prepress solvent oil extraction is used for processing high oil content seeds, especially rapeseed, peanut, sunflower, flaxseed and occasionally cottonseed. Conditioned oilseed is first pressed to an oil content of 15–18% and the intermediate meal is then reflaked and extracted with solvent.

Direct solvent extraction is used for processing low to medium oil content oilseeds. First, the seed is flaked by rolls to about 0.25 mm thickness, and then is extracted in a continuous extractor, usually with hexane. Residual oil content of 0.5–1.0% is sought in the meals. The meals have high crude protein.

In order to develop and utilize valuable oilseed cakes and meals, some comparatively efficient methods are being used in our country, as follows:

Soybean cakes: China is the soybean's home town. The annual output of soybeans reached about 10.064 million tons in 1936, 10.23 million tons in 1956, and 11.614 million tons in 1986. The average amount per person was only 11.02 kg of soybeans in 1986. This slow rate of production growth for soybeans is one of the important reasons seriously affecting the production of the Chinese food and feed industry.

Soybeans actually are a feed and food vegetable protein crop as well as an oil crop in our country. Soybean cake and meal contain abundant protein, which has been the main source of vegetable feed protein. The presence of antinutritional factors, such as trypsin inhibitor and urease, requires centralized processing or installation of on-farm or feed milling heat treatments, and extrusive equipment before soybean can be fed to livestock. Usually, the urease content in soybean cake and meal indicates quantities of antinutritional substances. The standard urease content in soybean cake and meal is defined as 0.05–0.5%, which can be measured with a fast method. Soybean meals with a high urease content may be used to make beancurd and fermented food.

Cottonseed cakes: Cottonseed cake contains a considerable amount of crude protein, amino acids and mineral substances. China produces about 2.9 million tons of cottonseed cake every year, which is 25% of the annual total output of vegetable oilseed cakes and meals. However, utilization of cottonseed cake in the feed industry is, in fact, quite limited, because a hindering factor in cottonseed has been the presence of the pigment gossypol in glanded varieties, which is toxic to nonruminant animals. Results of experiments have verified that swine grow normally if the gossypol content in the swine's diet is less than 0.01%. Swine appetites weaken and growth is abnormal if the gossypol content in the diet is from 0.01 to 0.02%. If the gossypol content in the diet is over 0.03%, swine develop illness symptoms, oedema and hyperaemia, and will become seriously ill and die.

At present, the following methods are being used to remove gossypol: a mechanical process uses single screw extruders comparatively efficiently to extrude cottonseed cakes with a high content of gossypol. Extruded cottonseed cake with a content of gossypol below 0.08 to 0.35% may be provided, without additional removal of gossypol. If cottonseed cake is added to the diet at about 20%, the gossypol content in this diet will reach the standard safe usage of 0.01%. Expanded product may be directly used in feeding chicken and fish. If the cottonseed cake has gossypol content above 0.08%, ferrous sulfate, calcium lime or urea must be added before extrusion to the cottonseed cakes. These products may reach the standard for safe usage.

The chemical processes involve use of ferrous sulfate, i.e., ferrous sulfate powder at five times the gossypol content put into the cottonseed cake to obtain comparatively satisfactory effects; soaking the cottonseed cakes in ferrous sulfate and calcium lime liquid; separating gossypol from cottonseed cake and meal with alcohol solvent soak (but this is very expensive), or separating gossypol from cottonseed cakes and meals with acetone-light petrol solvent, so as to acquire medicinal gossypol, pure cottonseed oil and pure protein without gossypol. This last method is used primarily in the medical industry and food processing industry.

China already has bred new glandless cottonseed varieties, such as Xiang cotton No. 11, in which the oil content of cottonseed is 35.2% (with a linoleic acid content of 58.5%), crude protein of 432.5, and gossypol content of only 0.0046%. The amino acid composition is complete and the production of cotton bolls is normal.

Rapeseed cake: Rapeseed cake is the third largest oilseed cake in China. Its average annual output is about 2.8 million tons. Rapeseed cake contains 35–45% crude protein, 1.8–5% crude fats, 11% crude fiber, 18–40% nitrogenfree extract (NFE) and some kind of mineral element. It will become a high protein feed, if properly used.

Though the nutrient component of rapeseed cakes is high, it has had limited use in animal feed, because rapeseed oil was high in erucic acid. Erucic acid limited digestibility characteristics and was considered to cause heart damage and other health disease. Also, rapeseed cake and meals contained an appreciable amount of glucosinolates, which were shown to be goitrogenic, and limited its use in animal feed to 10% or less. In addition, rapeseed cakes have a pungent scent, palatability is not good and animals don't like to eat it, so there must be attempts to remove the hindering factors before feeding it to animals.

Recently, we have used the following methods to remove hindering factors in rapeseed cake:

Chemical process: Ferrous sulfate is added to ground rapeseed cakes, generally, 0.003% was mixed fully, then 100 kg of fresh water was added, soaking was continued for 3–4 hours and the system was filtered. This ground cake may be fed to animals. The feed effect is best with addition of 0.2–0.3% lysin and methionine.

Method of soaking and filtering: Rapeseed cake is soaked in cold water, boiling water or in lime water and then filtered. In particular, soaking cake in boiling water is efficient. The residual content of glucosinolates remains 0.05%. The amount of removed toxic materials may reach 98.68%. This has been available as nontoxic rapeseed cake and this cake often controlled below 20% in a swine diet.

Other methods: Inoculating lactic acid bacteria into rapeseed cake to remove hindering factors is comparatively efficient and has achieved some good results; developing programs to research double-low varieties of rapeseed (with low content of erucic acid and glucosinolate per gram of dry cake and meal), and paying attention to feed efficiencies of mixtures, with reasonable proportions of soybean, cottonseed and rapeseed cakes.

Peanut cakes: Approximately 60% of peanut harvested is used as nuts, butter and other food, about 40% of peanut harvested is processed for oil. The amount of peanut cake for feed, in fact, is less and often is mixed with other oilseed cakes and meals to feed animals. Peanut often is contaminated with aflatoxins, which are formed by growth of the mold *Aspergillus flavus*. These compounds are carcinogenic to animals at low levels and toxic at high levels. So before feeding the meal to animals, the amount of aflatoxins must be assayed. Technical procedures to prevent growth of *Aspergillus flavus* in the peanut cake are needed.

Sunflower cake: Sunflower cake is a major vegetable protein feed in the northeast section of China. Annual production of sunflower cake is about 0.6 million tons. Dehulled sunflower cake contains about 30% crude protein and 20% crude fiber. The methionine content is 150% times and lysine content is about 66% that of soybean cake. A lot of Chinese sunflower cake contains appreciable amounts of hulls. However, the high-fiber content of cake limits its use to feeding animals and poultry. Recently, China has developed processing technology and equipment for manufacturing low-fiber (14% or less) content sunflower cake, with 40% or more crude protein and below 2% crude fat. Its nutritional value is equal to soybean cake. Sunflower cake is a good palatable feed, without toxins and various kinds of animals like to eat it. It is an excellent vegetable protein feedstuff and worthy of its popularity in the northeast and northwest of China.

The Use and Outlook of Full-Fat Soybean and Soybean Meal in Japanese Commercial Feeds

Karl H. G. Sera

American Soybean Association, Akasaka Tokyu Bldg., 2-14-3 Nagata-cho, Chiyoda-ku, Tokyo 100, Japan

Abstract

The use of soybean meal for commercial feeds in Japan has quadrupled between 1966 and 1986 to 2.6 million metric tons (MMT). The commercial feed production on the other hand showed less than a three-fold grown during the same period, to 26.1 MMT with approximately 46%, 37%, 12% and 5% for poultry, swine, dairy and beef, respectively. The average inclusion level of soybean meal for all commercial feeds during 1986 was 10%, with 8.3%, 13.4%, 13.5%, 12% and 4.4% for pullet/layer, broiler, swine, dairy and beef, respectively. The growth of commercial feeds is expected to be 1% to 2% annually. The growth of soybean meal quantity depends on its competitiveness per unit of essential nutrients. Though still small in quantity, the use of heat processed full-fat soybean is growing particularly for ruminants; growth also is expected for swine. Full-fat soybean processed with single-axle/dry extruder has grown seventeen-fold in four years to 5,000 MMT in 1987.

Commercial feed production in Japan was 9.9 million metric tons (MMT) in 1966. It increased to 25.8 MMT in 1986 and to 26.4 MMT in 1987 showing an increase in production of 2.7 fold between 1966 and 1987. Breakdown of this production was 4.4 MMT for layer feed in 1966 and 6.2 MMT in 1987 showing an increase of 1.4 fold during this period. The layer feed accounted for 49.9% of the total commercial feed production in 1966, and it dropped to 23.6% in 1987. Broiler feed production on the other hand was only 0.71 MMT in 1966, but it increased by 6.2 fold to 4.3 MMT in 1987. Broiler feed accounted for 7.1% of the total feed production of 1966 and it increased to 16.4% in 1987. All poultry feed production including approximately 1 MMT of pullet feed increased to 11.6 MMT in 1987 from 6.1 MMT in 1966 showing an increase of 1.9 fold. Share of the poultry feed against all other feeds dropped to 43.9% in 1987 from 61.6% in 1966.

Swine feed production increased its share from 25.8% to 26.7% during the 1966 to 1987 period, and production increased by 2.8 fold from 2.6 MMT in 1966 to 7.1 MMT in 1987. Dairy feed during the same period increased by 2.9 fold to 2.6 MMT from 0.9 MMT. Dairy feed production as a percentage of total feed production increased from 9.3% to 10.0%. Beef feed production increased dramatically by 24.3 fold from 0.13 MMT to 3.0 MMT during the 1986-1987 period. As the beef feed production was very small in 1966, it had only a 1.2% share of the total Japanese feed productions; it increased to 11.5% in 1987.

Mixed feed is basically a grain mixture, for example, 98% ground corn and 2% fish meal. This is purchased by livestock producers as a source of energy and used as an ingredient for on-the-farm mixing of feeds. It is ground and sold by licensed and bonded feed millers. As there is still tax for imported grains, such as corn to be used by food industry, and as tax is not placed on corn for feed use, silo, warehouse and feed milling operations are bonded to assure that nontaxed corn does not find its way into the food industry. Mixed feed production in 1966 was almost nil. It has increased during the past 10 years and the production in 1987 was 2.03 MMT accounting for 7.6% of the entire feed production.

Soybean Meal for Feed Use: Soybean meal for feed use was only 0.81 MMT in 1966. It increased by 3.3 fold to 2.65 MMT in 1987. The increase was greater for soybean meal than the increase of 2.7 fold for commercial feed during 1966–1987. The average percent of soybean meal use in all diets was 8.1% in 1966 and 10.0% in 1987.

Soybean meal use by tonnage for major classes of livestock in 1987 was 0.62 MMT for pullets and layers; 0.61 MMT for broilers; 0.94 MMT for swine; 0.31 MMT for dairy; and 0.13 MMT for beef. Soybean meal use on a percentage basis for the major classes of livestock was 46.4% for poultry, 23.5% for pullets and layers and 22.9% for broilers; 35.5% for swine; 11.7% for dairy, and 4.8% for beef. Average percent of soybean meal used in diets for each of the major classes of livestock was 8.6% for pullet/layer feed; 14.0% for broiler; 13.3% for swine; 11.7% for dairy; and 4.2% for beef in 1987.

Use of soybean meal for fish feed has grown by 45% since 1984 to 14.6 thousand metric tons in 1987. However, total fish feed production still remains very small and it is less than 1% of the entire feed production in Japan.

Soybean Meal, A Major Protein Ingredient in Feed: Soybean meal ranked second or third in 1987 among five energy and protein ingredients used for most of the major classes of livestock feeds. It was the major protein ingredient for all feeds except beef feed. In descending order of ingredient use for major classes of feeds, soybean meal ranked second for dairy feed; third for poultry and swine feeds; and fifth for beef feed. The first and second ingredient in most of the feeds were corn and sorghum grain. Fish meal, meat bone meal, rapeseed meal, corn gluten feed and wheat bran were the protein ingredients following soybean meal in most of the feeds. The five major ingredients generally made up 79 to 90% of the tonnage of poultry and swine feeds and 64 to 73% of the tonnage of ruminant feeds.

Full Fat Soybeans for Feed: Use of heat processed, full fat soybean for animal feeds was almost negligible until the mid-1970's. The interest in using full fat soybean started to grow gradually with the making of high density, high nondegradable protein-containing dairy feed for early lactation and for high energy swine feeds for late gestation and lactation.

Positive impact of the use of full fat soybean for dairy feeds has particularly become prominent after 1983 with the introduction of single cylinder dry extruders from the United States. Estimated use of total full fat soybean for feed in 1987 was between 60,000 and 80,000 metric tons. Dry extruded soybean for feed use in 1987 was 5,000 metric tons, and this was a 17 fold growth in four years.

Future of Feed and Livestock Industry

Several factors will have a definite impact on the future of the Japanese feed and livestock industry: (a) Imports of meat and meat products will increase. In 1987, imports of beef, pork, and chicken accounted for 36, 20, and 13% of supply, respectively. (b) The number of farms will continue to decrease with size of farm operations becoming larger. The number of layer farms decreased to 109,900 farms in 1987. This figure is one twenty-fifth of the number in 1966. The number of layers per farm increased to 1,240 birds in 1987, an increase by 30.2 fold from the number in 1966. Broiler farms decreased to one-third or 6,323 farms and the number of broilers per farm at any one time showed 24,500 birds, an increase by 21.4 fold during the same period. Swine farm numbers went down to 65,100 or one-eleventh of the number in 1966, and the number of hogs went up to 174.4 head, an increase by 24.2 fold. There were 74,500 dairy farms or a decrease by one fifth, and there were 27.5 milk cows per farm, an increase by 7.6 fold during the period of 1966 to 1987. The number of beef farms decreased to one fourth or 272,400 farms and there were 9.7 beef cattle per farm, an increase by 6.9 fold. No one is able to estimate the degree of the decrease in the number of farms in the future. (c) Placing too great an emphasis on production and market value added, too high priced brand products will result in increased imports; (d) The majority of meat, milk, and egg will be produced by larger, integrated operations in the future, and the majority of the feed industry will be absorbed by the food processing and marketing industry; and (e) The degree of free competition involving liberalization in all sectors of livestock, feed, and food industries will be the most vital factor in determining the future of the livestock and feed industry, both positively and negatively.

Considering the aforementioned factors, feed production may be expected to grow at 1 to 2% per annum until 1991 to 1992.

Future of Soybean Meal and Soybean Use for Feed

Several technical factors will have a positive impact on the growth of soybean meal and soybean utilization for feed: (a) further sophistication in feed ingredient analysis, feed formulation and technical sales services; (b) incorporating the concepts of available amino acids and their balances in feeds; (c) incorporating inorganic selenium in feeds; and (d) understanding the nutrition and practical methods of feeding full-fat soybean.

Sound incorporations of these four major technical factors may result in a total use of soybean meal for feed from 2.7 MMT in 1987 to 3.0 to 3.2 MMT in the future. Full fat soybean for feed may also result in a total estimated use of 60,000 to 80,000 metric tons in 1987 to 200,000 to 250,000 metric tons in the future.

Selected References

Feed Monthly Report, Japan Feed Association (1987, 6, 1974, 6).

Feeding, Chikusan Shuppan-sha (1988, 9, 1987, 6).

Feed Statistics, Agricultural Statistics Association (1972, 1974).

Thirty Years of Progress, Japan Feed Manufacturers' Association (1987).

Concentrate Feed Annual Report, Japan Feed Association (1963).

Lupins as an Energy-Rich Protein Source for Feed and Food

Rory Coffey

Grain Pool of Western Australia, Grain Pool Bldg, 172 St. George's Terrace, Perth 6000, Western Australia, Australia

Abstract

During 1988 in excess of 750,000 tons of lupinseed from the cultivar *Lupinus angustifolius* will be incorporated into livestock feed formulations and human food preparations in more than 20 countries throughout the world. In most instances the quantity used will be determined by supply restrictions rather than by consumer limitations, even though production has increased tenfold in the last decade. The appeal of lupins basically can be traced to the versatility of the seed in terms of nutritive qualities and the lack of antinutritional factors in its makeup. The composition of the seed makes lupins, with limited processing, an ideal ingredient for ruminant stockfeed. Once the fibrous testa is removed, the groat or kernel becomes more suited to monogastric feed use and to certain food preparations. Although the consumption of lupins historically has been dominated by the ruminant stockfeed industries, this situation is now rapidly changing due to the continued development of efficient processing technology.

Lupins have been grown and used in Western Australia for over 30 years; however, the type of lupin being produced today bears little resemblance to its predecessors. After extensive genetic engineering the composition of the present day varieties is far superior to those grown originally. Over 90% of the lupins grown in Western Australia and all that are available for export are varieties from the species *Lupinus angustifolius*, a sweet, white narrow leaf lupin (Table 1). Conversely, most of the lupins produced in Europe are varieties from the species *Lupinus albus* and *Lupinus luteus*, both of which have a number of characteristics dissimilar to Angustofolius varieties.

Although the area sown to the various species throughout the world is considerable (Table 2), a large proportion of the production outside of Australia is for purposes

TABLE 1
Lupin Types

Major species
European origin
Angustifolius
Albus
Luteus
South American origin
Mutabilis
Angustifolius cultivars produced in Western Australia
First released 1967, now obsolete
Unicrop
Uniharvest
Uniwhite
Currently used
Illyarri
Yandee
Chittick
New releases
Danju
Gungurru

other than the harvest of grain. Forage production and use of crops as green manure are the two major alternative purposes for growing lupin, particularly in the USSR and other Eastern Bloc countries.

Additionally, with the exception of Australia, only a very small percentage of the grain produced is consumed outside of the country of origin.

In Australia the bulk of lupin production is concentrated in the State of Western Australia. Over the past 20 years this production has increased fortyfold and in the process a

TABLE 2
Regional Distribution of Lupin Production, 1948–1986 ($\times$ 1000 ha)[a]

	1948/1952	1961/1965	1966/1970	1971/1975	1981/1984	1985
USSR	252	607	610	522	310	280
Europe	315	239	185	127	124	148
South Africa	87	185	200	104	20	20
Other African	11	9	5	5	6	7
Latin America	2	4	5	4	10	12
Australia	—	2	2	56	242	606
World total	667	1046	1007	818	712	1073

[a] FAO Production Yearbooks, Vols. 20–29 (Williams, 1986).

significant export industry has been developed (Table 3). The export marketing of lupins from Western Australia is controlled by a farmers cooperative, the Grain Pool, which operates under a State Act of Parliament.

TABLE 3
Western Australian Lupin Production

Season	Domestic use	Export sales
1971/72	21,000	1,000
72/73	11,000	4,000
73/74	32,000	17,000
74/75	33,000	44,000
75/76	37,000	51,000
76/77 Drought	23,000	nil
77/78 Drought	24,000	nil
78/79 Drought	18,000	6,000
79/80 Drought	25,000	nil
80/81	35,000	13,000
81/82	31,000	51,000
82/83	108,000	82,000
83/84	110,000	212,000
84/85	140,000	370,000
85/86	166,000	244,000
86/87	250,000	432,000
87/88	350,000	405,000
88/89 (est)	300,000	500,000

Lupin Composition and Usage

A valuable quality regarding the makeup of lupins is the very small incidence of fluctuation in chemical composition. This occurs despite the variations experienced in weather conditions from season to season and the different weather and soil characteristics experienced across a very broad cropping area. In turn, this aspect allows for the inclusion of lupins in ration formulations using a minimum of quality control.

In Western Australia the chemical composition of lupins is monitored annually. In order to gain an accurate and independent analysis this task is performed by the Western Australian State Government Chemical Laboratories with the sample analyzed being representative of the state average for each particular season.

The 1987/88 season's analysis is not available yet, but the average of previous seasons is listed to give an indication of the typical average composition of lupins (Table 4). Although the composition figures shown are relatively constant, the nutritive value that these levels sustain will be variable for different livestock types.

Ruminants. Because of their ability to overcome amino acid deficiencies by synthesis and their requirement of high levels of fiber, ruminants are the most efficient users of lupins. In addition, at least 30% of the protein in lupinseed by-passes fermentation by micro-organisms in the rumen and therefore is directly available to the animal for digestion. This compares with only about 1% of protein in cereal grain that has this by-pass capacity and means that lupins have a far greater efficiency of conversion of

TABLE 4
Western Australian Lupinseed Average Chemical Composition

	(As received) %
Crude protein	30
Crude fat	5
Crude fiber	16
Nitrogen free extractives	36
Ash	3
Calcium Ca	0.21
Phosphorus P	0.29
Gross energy value (MJ/kg)	19.5
Moisture	10

feedstuff to liveweight. Due to the low level of starch present (4.0%) in the lupin carbohydrate, the risk of lactic acidosis occurring is far lower than with cereal grains.

Because of the high level of readily digestible fiber and the relatively high fat and nitrogen free extractives (NFE) components of lupins they are also an excellent source of energy for ruminants. In relation to soybean meal and most cereals the metabolizable energy value of lupins for these livestock types is significantly superior.

Lupin usage has been dominated over the years by the ruminant sector, with sheep leading the way domestically and beef and dairy cattle being the major traditional users in the international markets. Feeding methods have been varied across these industries ranging from the use of whole, unprocessed lupins in the local sheep industry, inclusion in compound feed pellets and meals in Europe, and inclusion in flake form in Japan.

Swine. As with soybean meal, some of the essential amino acids are marginally deficient in lupins, necessitating the inclusion of some synthetic additives. Lysine and methionine are the two amino acids concerned, and the cost incurred in bringing these up to the required level is minimal (Table 5).

As with all other vegetable protein, yet unlike most animal protein, lupins do not contain sufficient levels of calcium and phosphorus to cover pig ration requirements, therefore necessitating a supplement of these two ingredients.

Although the fiber content of lupins is quite high, this material is quite digestible for pigs, with the energy level achieved being comparable to that of soybean meal. Studies conducted by M. R. Taverner in Australia disclosed that only 50% of the lupin is digested in the small intestine, but a further 30% is broken down in the large intestine. This is quite different from cereal grains such as wheat, in which 75% and approximately 15% disappear in the small and large intestines, respectively. It appears that the carbohydrates in lupins are poorly digested by enzymes within the stomach and small intestine, but they are susceptible to bacterial attack in the large intestine.

Poultry. The amino acid deficiency in lupins also holds for poultry; however, this is a problem that is easily rectified. By far the major problem confronting poultry with regard to the consumption of lupins is its indigestibility. Although

TABLE 5

Comparison of Amino Acid Content in Selected Materials

Amino acid	(% of protein)		
	Lupins	Soybean meal (solvent extracted)	Rapeseed meal (solvent extracted)
Lysine	5.08	6.55	5.90
Methionine	1.00	1.28	2.07
Cystine	1.72	1.51	0.85
Threonine	3.92	4.24	4.61
Glycine	4.33	5.23	5.24
Valine	4.02	5.27	5.56
Isoleucine	3.72	5.29	4.02
Leucine	6.81	8.21	7.30
Tyrosine	4.02	3.30	2.32
Phenylalanine	4.02	5.34	4.17
Histidine	2.17	2.75	3.04
Arginine	10.53	7.27	6.12
Tryptophan	1.20	1.26	1.51

the fibrous lupin hull can be digested by pigs, a large portion of its nutritive value is lost to poultry. A simple solution to this problem involves dehulling the lupin and discarding the fibrous outer coat. In most instances, the only economical form in which lupins can be used competitively in poultry rations is as milled kernel.

Dehulled Lupin

In 1985, in order to broaden the market application of lupins, the Grain Pool initiated a program to compare the qualities of lupins to stockfeed market requirements throughout the world. After an extended period investigating both the nutritional aspects and the market potential of lupins it became increasingly apparent that, although the monogastric sectors make up a very significant proportion of the world stockfeed industry, whole lupins do not have a strong application in this market. The price level at which whole lupins would go into monogastric rations was significantly lower than the price payable for ruminants and at a level that did not warrant the production of lupins.

Lupin kernel. An analysis of the component parts of lupins revealed the following information: Lupin kernel has a significantly reduced fiber content and as such is a very suitable ingredient for poultry rations. Not only does it have a crude protein level of around 40% as opposed to 30% for whole lupins, it also has a far superior metabolizable energy (ME) value for poultry. The ME of whole lupins for poultry is around 2,100 kcal/kg, whereas lupin kernel has a value of around 2,750 kcal/kg. This also compares favorably with its major competitor, soybean meal, which has an ME value of approximately 2,400 kcal/kg.

Table 6 illustrates the comparative advantage of lupin kernel over whole lupins for poultry.

Lupin hull. Lupin hull has a limited application for use as stockfeed material due to its low nutritional content; it contains a protein level of only 3 to 4%, a fiber level of

TABLE 6

Comparative Advantage of Lupin Kernel Over Whole Lupins for Poultry

	Crude protein (%)	ME (kcal/kg)	Crude fiber
Lupin kernel	40	2,750	3
Whole lupin	30	2,100	15
Soybean meal	44	2,400	2

around 50% and a very low energy value. This nutritional profile indicates that its only stockfeed application of any potential would be in ruminant rations.

Having identified a demand potential for dehulled lupins, the next task was to find a use for the by-products, the lupin hull material. The lupin hull represents about one quarter of the whole seed and, if this material is simply discarded, the premium required for the kernels would make it uneconomical to consumers.

In February 1985 a concept was developed involving the inclusion of a small amount of kernel with separated hull material. The resultant mixture was steam-pelleted. It was perceived that these pellets could be offered as an alternative to conventional pellets used in the sheep export industry. By a calculation of the variables involved it was determined that by altering the levels of inclusion of the kernel, nutritional specifications could then be made to order. The calculation was made as shown in Table 7.

Pellet mixture. Using Table 7 values, all of the hull (23% of whole lupin) and a proportion of kernel (7% of whole lupin) were combined theoretically. This mixture represented 30% of whole lupin leaving the remaining 70%, being all kernel, available for the poultry market.

The pellet mixture theoretically contained the following nutritional content: crude protein, 12.5%; ME (sheep), 2250 kcal/kg; and crude fiber, 36%. Having proved the

TABLE 7
Chemical Composition of Lupin Components

	Hull	Kernel
Percentage of whole lupin	23%	77%
Crude protein	4%	40%
Crude fiber	46%	4%
ME (sheep)	1900 kcal/kg	3400 kcal/kg

processing concept theoretically and practically at a laboratory level, the next stage was to evaluate the two products for nutritional, handling and storage qualities.

Assistance was sought from appropriate sources to conduct a range of trials which took place over a period of two years. The trials and the organizations involved are listed as follows: Pellet digestibility test, School of Agriculture, U.W.A.; shipper trial I (pellets), School of Agriculture, U.W.A.; shipper trial II (pellets), School of Agriculture, U.W.A. and W.A. Department of Agriculture Fares Rural; grain handling trial (kernel and pellets), Cooperative Bulk Handling; kernel storage and stability test, W.A. Department of Agriculture; pig and poultry feeding evaluation I (kernel), W.A. Department of Agriculture; pig and poultry feeding evaluation II (kernel), Japan Scientific Feed Association; and beef cattle feeding trial, W.A. Department of Agriculture.

Basically the trials were conducted in the order listed, although there was some overlap and the commencement of each test was dependent on the results achieved in the previous trial. In all instances, the results proved very positive and in some cases, particularly the beef and pig trials, the figures were much better than anticipated. In order to provide product material for the trials, large quantities of bulk kernel and pellets were made; the results achieved by the initial laboratory trial were reinforced. The pellets used in the sheep trials contained lower protein levels than the beef trial pellets, which proved the concept that pellets could be "tailor made" to suit various ruminant livestock requirements.

As a result of this work a number of lupin dehulling plants have been constructed in Western Australia, a large capacity plant was commissioned in Korea in 1987 and earlier this year a dehulling plant commenced operations in Indonesia. Effectively, the investigations initiated in 1985 have achieved the objective of broadening the scope of efficient applications of lupins in the stockfeed industry.

Human Consumption

Although there is evidence that different species of lupins have been used for human consumption for thousands of years very small quantities are being used for this purpose at the present time. Consumption basically is restricted to the Mediterranean region as a snack food and to certain parts of South America.

The Grain Pool in conjunction with others has pursued the prospects for a broader food use of lupins for a number of years with mixed success. Although many different prospective uses have been investigated, to date, the four major areas pursued have been the production of lupin concentrates and isolates; lupin miso soup; lupin tempeh and lupin flour.

Lupin concentrate. This work commenced under the supervision of C.H. Lee of Korea University in 1981 and by 1985 a 70% lupin protein concentrate (LPC) had been developed. The intended uses of the LPC were parallel to the applications for soy protein concentrate. Unfortunately, the 70% LPC development program was discontinued when numerous obstacles were encountered because of a residual bitter taste in the product caused by proteinase activity during the extraction process. Alterations have now centered on a 50% LPC that does not require the same enzymatic process to extract the carbohydrate fraction. Of concern, however, is whether the comparatively low protein level will render the product commercially nonviable. Personnel from the Western Australian Department of Agriculture are now endeavoring to establish the commercial effectiveness of the product by evaluating production costs and investigating potential return for the LPC and the by-products.

Recent discussions with a researcher at the Western Australian Department of Agriculture have revealed that he has been successful in developing a lupin protein isolate in the range of 90% protein from which a satisfactory milk product has been made.

Miso soup. A soybean-based food, miso soup is a staple dish in Japan. Although about 200,000 tons of soybeans are used annually in Japan to make miso soup, there is a problem with oxidation, which causes discoloring. In trying to solve this problem, the Shinshu Research Institute has investigated the use of lupins as a substitute. Oxidation does not occur with lupins and to date results have been encouraging.

The most recent work carried out by the Shinshu Research Institute was a sensory comparison of the two ingredients. The positive results are shown in Table 8.

TABLE 8
Sensory Evaluation of Miso (high score more favorable)

Product	Color	Taste	Aroma	Texture
Lupin miso	12	10	7	6
Soybean miso	0	2	5	6

Before this work can be taken further, however, approval must be gained from the Japanese Ministry of Health and Welfare to use lupins in foodstuffs. The Grain Pool, both directly and through the Australian Embassy, has taken initial steps to secure this approval.

Lupin tempeh. With the assistance of the Grain Pool and R. Yu of the Victorian Food Research Institute, investigations were commenced in 1986 by an Indonesian company into the feasibility of using lupins as a substitute for soybeans in the manufacture of tempeh.

According to Yu, tempeh is "an indigenous Asian method of texturizing or restructuring plant protein source such as soybeans is by cultivating the mold *Rhi-*

zopus oligosporus in the steam-cooked beans. The mold grows on the protein, lipid, and carbohydrate reserves of the seed endosperm and forms a dense mycelial mat binding together the beans, resulting in a compact cake with the concomitant flavor development and enhanced digestibility of the seed reserves. This compact cake product is known as tempeh and it serves as a major source of readily assimilatable protein in the Indonesian diet."

Despite a number of hiccups, the efforts by all three parties have proved successful and the Indonesian company is now using a significant tonnage of lupins in its tempeh production operations.

Lupin flour. In June 1987, the Grain Pool initiated an evaluation of lupin flour in human food products. Emphasis was placed in the first phase on establishing the product's nutritional aspects in order to determine its qualities compared with other flours in commercial use. Apart from the compositional evaluation, investigations have begun to determine the baking qualities of lupin flour. Investigations to date include a cooking competition conducted in conjunction with the Country Women's Association of WA. The competition proved highly successful in participation level and in the quality of products made. Information gained from the competition is being collated and analyzed.

An integral part of all the human consumption investigations has been the need to establish Australian Government Health Authority approval for the use of lupins for this purpose, and this has proved to be a drawn out process. Some years ago the Australian National Health and Medical Research Council (NH & MRC) placed a 10% limit on lupin derivatives in food products. This figure was set arbitrarily, without scientific foundation. With the growth of the lupin export trade, the presence of this Australian regulation has deterred some overseas users, who believed it was due to the presence of high alkaloid levels and other toxic elements.

In order for the 10% restriction to be lifted, the NH & MRC directed that a battery of toxicity tests be carried out by BIBRA, regarded as the foremost organization in the world for this type of research. The Grain Pool acted on this directive. Assisted by funds from the Grain Research Committee, the project commenced in 1985. Results of the lupin toxicity evaluation by the British Industrial & Biological Research Association were submitted to the NH & MRC in late 1986. On March 20, 1987, lupins were approved for unlimited inclusion in human food. This development is considered an important breakthrough for the lupin industry and will hopefully open the door for lupins use in food preparations throughout the world.

Utilization of Sunflower Meal for Swine and Poultry Feeds

Utahi Kanto

National Swine Research and Training Center, Kasetsart University, Kampaeng Saen, Nakorn Pathom, 73140, Thailand

Abstract

Sunflower meal has been included in pig and poultry diets in Thailand for many years. The material used was mainly imported from abroad and users were limited only to commercial feedmills. Commercial sunflower production in Thailand has been promoted recently and more domestic sunflower meal is available for feed industries. The meal produced is undecorticated and has crude protein, ether extractables, ash, calcium and phosphorus contents of 34.65, 3.50, 5.36, 0.56 and 0.88 per cent, respectively. According to the study at the National Swine Research and Training Center, Kasetsart University, sunflower meal can be included up to 15% in broken rice–soybean meal–fish meal diets for growing and finishing pigs without any significant adverse effects on production performance and carcass quality. The prices for conventional protein supplements have increased substantially during the past few years. Sunflower meal is an alternative which is widely accepted by pig and poultry farmers. However, the lack of uniformity in nutrient content and the problem of inadequate supply has limited the use of sunflower meal in Thailand.

Sunflower (*Helianhus annua*) is an oilseed crop that has been grown in many areas of the world. A major advantage of sunflowers is their ability to adapt to a variety of climates and soil conditions, from the heat of the southern United States to the frost of Central Asia. More than 13.1 million metric tons of seed are produced each year (1). More than 53% of the total world production was in the U.S.S.R. and the other centrally planned countries, whereas 20% and 13% were in Europe and South America, respectively. In the United States, hectarage of sunflowerseed production has been increased from 2,865 hectares in 1968 to more than 12.8 million hectares in 1980, most in the northern Great Plains region. Production during this time has increased from 110,000 tons to 1.73 million metric tons (2).

Nutritive Values of Sunflower Meal

Sunflowerseed is rich in oil content ranging from 25% to 32% which seems to be influenced by soil and climatic conditions. The protein content of whole seed is about 16% with a crude fiber content of 28%. However, new varieties of sunflowerseed with higher percentages of oil and lower percentages of hulls are continuously being developed, altering the chemical composition of the seed.

Sunflower meal is produced from sunflowerseed following oil extraction. The composition of the meal will vary according to the composition of the seed and, as with other oilseed proteins, the nutritive value of sunflower meal varies with the method of processing. Expeller-produced meals contain more fat and fiber and lower quantities of crude protein than do meals produced by solvent extraction (Table 1). The temperature involved in the process of oil extraction also influences quality of protein in the sunflower meal. Solvent extraction and low-temperature processing improve the quality of the protein by reducing the destruction and/or loss of lysine, methionine, threonine and tryptophan (3). High-temperature processing and especially dry heating causes a marked reduction in lysine content and availability. Other amino acids such as arginine and tryptophan also are reduced by high temperature process. Amos et al. (3) heated sunflowerseed at 0°, 75°, 100°, 115°, or 127°C for 1 hr before oil extraction and found that levels of essential amino acids tended to decrease as the heating temperature increased (Table 2). Decreases were greatest for lysine, arginine, and threonine. From rat growth studies they have found that the appropriate heating of the seed was 100°C for 1 hr (Table 3). Improvement of nutritive value of sunflower meal by mild heating may be due to destruction of chlorogenic acid, which is an effective trypsin inhibitor, present in the sunflowerseed. Milic et al. (4) have demonstrated that heating sunflowerseed to 100°C for 5 hr decreased chlorogenic acid by 42.3% and increased quinic acid, which had no effect of trypsin inhibition, by 30.4%.

TABLE 1

Average Composition of Sunflower Meal (2)

Components	Type of processing	
	Expeller	Solvent
Moisture (%)	7.0	7.0
Crude protein (%)	41.0	46.8
Crude fiber (%)	13.0	11.0
Ether extract (%)	7.6	2.9
Ash (%)	6.8	7.7
Calcium (%)	0.43	0.43
Phosphorus (%)	1.08	1.08

TABLE 2

Changes in Some Essential Amino Acid Content of Heat Treated Sunflower Meal (3)

Amino acid	Heat treatment (°C for 1 hr)				
	0	75	100	115	127
	(grams AA/16 g N)				
Lysine	4.03	3.77	3.94	3.88	3.88
Arginine	10.38	10.35	10.19	10.14	9.92
Threonine	5.33	5.10	5.02	4.88	4.97
Methionine	2.61	2.11	2.14	2.31	2.20
Isoleucine	4.28	4.16	4.23	4.17	4.32
Histidine	2.67	2.63	2.72	2.63	2.66

TABLE 3

Cumulative Growth and Feed Efficiency of Rats Receiving Heat Treated Sunflower Meal Supplemented Diets (for 28 days) (3)

Treatments	Growth (g)	Feed consumption (g)	Feed conversion ratio
Unheated SFM	42.86	291	6.8
75°C heated SFM	41.26	314	7.6
100°C heated SFM	54.5	308	5.7
115°C heated SFM	41.2	271	6.6
127°C heated SFM	38.8	286	7.4
Soybean control	157.6	440	2.8

The advances in oil extraction and meal processing technology has markedly lowered the crude fiber and ether extract content of the finished products. The meals with high-protein content can be produced by removing the largest quantity of hull possible before final processing, or by screening the meal after extraction. The technique improves nutritive value of the sunflower meal (Table 4).

Although high-protein sunflower meal can be produced, the quality of protein in the meal is inferior to those in soybean meal due to both the lower content of many essential amino acids, especially lysine (Table 5), and poorer amino acid availabilities than in soybean meal as determined at the end of ileum of growing pigs (Table 6). However, amino acid availabilities of sunflower meal are higher than that of meat and bone meal (5). Substitution of sunflower meal for soybean meal in animal diets, therefore, requires either lysine supplementation or the incor-

TABLE 4

Effect of Decortication on Nutrient Composition of Sunflower Meal (14)

Treatments	Protein (%)	Fiber (%)	Fat (%)	Ash (%)	NFE (%)
Undecorticated	36.30	18.22	0.60	7.06	13.62
Decorticated	40.00	11.70	0.86	7.50	32.54

TABLE 5

Nutrient Composition of Sunflower Meal, Soybean Meal, Fish Meal and Meat and Bone Meal (5)

Composition	Sunflower meal	Soybean meal	Fish meal	Meat and bone meal
Proximate analysis		%		
Crude protein, N × 6.25	46.0	52.0	60.2	56.0
Ether extract	1.9	3.9	14.3	9.0
Crude fiber	13.2	4.9	3.5	8.7
Ash	7.7	6.6	23.1	25.8
Amino acids		g/16 g N		
Arginine	7.6	7.2	6.3	6.8
Histidine	2.3	2.9	2.3	2.1
Isoleucine	3.8	4.6	3.8	2.9
Leucine	6.1	8.2	6.8	6.4
Lysine	3.6	6.7	6.8	5.6
Methionine	1.8	1.5	2.8	2.0
Phenylalanine	3.8	5.2	3.6	3.0
Threonine	3.5	4.1	4.1	3.5
Valine	4.0	4.1	3.7	4.6

TABLE 6

Apparent Ileal Availabilities of Amino Acids in Sunflower Meal, Soybean Meal, Fish Meal and Meat and Bone Meal for Growing Pigs (5)

Item	Sunflower meal	Soybean meal	Fish meal	Meat and bone meal
		%		
Dry matter	78.3	83.6	86.8	83.1
Organic matter	81.0	85.5	91.3	87.1
Crude protein	73.1	76.3	72.7	65.4
Amino acids				
Arginine	86.6	88.0	83.2	78.7
Histidine	81.0	84.4	82.0	75.8
Isoleucine	75.7	81.2	78.1	70.5
Leucine	75.6	80.8	79.2	72.3
Lysine	69.0	80.1	78.9	64.9
Methionine	82.7	79.9	82.4	81.6
Phenylalanine	71.8	78.9	68.2	62.8
Threonine	66.8	70.7	74.2	61.8
Valine	70.3	73.3	74.2	77.0

poration of a high-lysine feed ingredient such as fish meal into the diet. Amos et al. (3) have shown that performance of rats fed diets based on sunflower meal could be significantly ($p < 0.05$) improved by supplementation of crystalline lysine in the diets.

Utilization of Sunflower Meal for Swine Diets

There are very limited data available in English on the utilization of sunflower meal for swine diets. Seerley et al. (6) have studied the effects of substitution of sunflower meal for soybean meal either with or without crystalline lysine supplementation on the performances of growth-finishing pigs. In one experiment, 0, 25, 50 or 100% of crude protein from soybean meal was replaced by sunflower meal in the diet. They found that average daily gain, average daily feed intake as well as feed conversion ratio were depressed severely by replacing 50% or 100% of crude protein from soybean meal with sunflower meal without the deficient amino acid supplementation (Table 7). Crystalline lysine supplementation at a level of 0.3% in both of the diets improved average daily gain, feed/gain ratio and loineye area of the animals (Table 8). However, the improved feed/gain ratio and loineye area was not equal to the control. This study is in agreement with Baird (7) who used a sunflower meal containing 29.2% crude protein and 21.9% crude fiber to replace 25% or 50% of the soybean meal for growing-finishing pigs (19–84 kg) and reported no differences in either gain or feed efficiency. Lysine supplementation improved gain but no efficiency of feed utilization (Table 9). Very recently, Chinrasri (8) has studied nutritive value of sunflower meal produced in Thailand and its ability to replace crude protein from soybean meal in growing-finishing pig diets based on broken rice, soybean meal and fish meal. The sunflower meal used in the study was produced by

TABLE 7

Growth, Feed Conversion Ratio and Carcass Traits of Pigs Fed Levels of Sunflower Meal (SFM) (6)

	Levels of crude protein replaced by SFM (%)			
	0	25	50	100
Avg. daily gain (kg)	0.66[a]	0.66[a]	0.44[b]	0.38[c]
Avg. daily feed cons. (kg)	2.20	2.37	1.83	1.68
Feed/gain ratio	3.30	3.61	4.21	4.54
Dressing carcass (%)	71.4	72.6	71.8	71.8
Avg. backfat thickness (cm)	3.15	3.40	3.15	3.02
Loineye area (cm^2)	32.5[a]	30.7[a]	24.6[b]	19.9[c]

[a,b,c] Means with same superscript letters are not significantly different ($p < .05$).

TABLE 8

Effect of 0.3% Supplemental Lysine-HCl in Diets with Sunflower Meal on Growing Pigs (6)

	50% SFM		100% SFM	
	No lysine	0.3% lysine	No lysine	0.3% lysine
Avg. daily gain (kg)	0.37^a	0.77^b	0.31^a	0.66^b
Feed/gain ratio	4.94	3.46	4.85	3.74
Wt. off test (kg)	70.0	86.4	73.2	86.4
Dressing carcass (%)	70.9	69.5	70.2	70.0
Backfat thickness (cm)	2.97	2.97	3.12	3.48
Loineye area (cm^2)	23.7	28.2	18.5	23.2

a,b Mean with same superscript letters are not significantly different ($p < .05$).

TABLE 9

Effects of Substitution of Soybean Meal with Sunflower Meal for Growing-Finishing Pigs (7)

Treatments	Avg. daily gain (kg)	Avg. daily feed intake (kg)	Feed/gain ratio
1. Soybean meal 100%	0.67	1.94	2.86
2. Sunflower meal 25%	0.68	2.02	2.96
3. Sunflower meal 25% + 0.3% lysine	0.74	2.20	2.97
4. Sunflower meal 50%	0.68	2.02	2.99
5. Sunflower meal 50% + 0.3% lysine	0.76	2.28	2.99

solvent extraction and undecorticated. The contents of crude protein, ether extract, crude fiber, calcium and total phosphorus were 34.6, 3.50, 19.47, 0.56 and 0.88%, respectively. He found that increasing sunflower meal at levels of 0, 15, 25 and 35% in semi-purified diets significantly ($p < 0.05$) reduced digestibilities of dry matter of the diets in 30 or 60 kg pigs (Table 10). But digestibility of protein and net protein utilization were not influenced by the increasing levels of the meal in the diet. Pigs at 60 kg body weight tended to have better utilization

TABLE 10

Digestibilities of Dry Matter and Crude Protein, Net Protein Utilization, Digestible and Metabolizable Energy of Experimental Diets Containing Levels of Sunflower Meal (8)

	Levels of SFM in the diets (%)			
	0	15	25	35
Digestibility of dry matter				
–30 kg	97.64^a	90.14^b	86.99^c	83.15^d
–65 kg	97.05^a	91.34^b	88.56^c	84.51^d
Digestibility of crude protein				
–30 kg	—	82.66^a	85.40^a	75.68^b
–65 kg	—	86.13	85.07	84.87
Net protein utilization (NPU)				
–30 kg	—	55.99^a	59.96^a	48.83^b
–65 kg	—	67.60	60.43	62.40
Metabolizable energy (ME)				
–30 kg	—	3403^a	3284^a	2603^b
–65 kg	—	3025	3054	3233

a,b,c Means with same superscript letters are not significantly different ($p < .05$).

of sunflower meal protein in terms of both protein digestibility and net protein utilization either at the same or at higher level of sunflower meal inclusion in the diet than pigs at 30 kg body weight. Replacing 0, 25, and 100% of crude protein from soybean meal with sunflower meal did not produce any adverse effects on performance of growing-finishing (20–90 kg) pigs except a slightly ($p < 0.05$) growth depression was observed in pigs fed the diet in which 100% of protein from soybean meal was replaced with sunflower meal (Table 11). The results of this study are in agreement with those reported by Seeley et al. (6) and Baird (7). But feed conversion ratio of the pigs in this study did not change very much according to the increasing levels of sunflower meal in the diets. A possible explanation may be due to inclusion of fish meal (6% in growing and 4% in finishing pigs diets) which not only provides additional lysine but also supplies well-balanced amino acids to the diets (2). On the other hand, the results have demonstrated the quality of sunflower meal produced in Thailand. The data on utilization of sunflower meal in smaller pigs as well as in breeder pig diets are very limited. However, the meal seems to have potential in breeder pig diets.

Utilization of Sunflower Meal for Poultry Diets

Sunflower meal has been studied for utilization as poultry feed for many years. Pettit et al. (9) reported that sunflower meal may replace meat meal up to 14% in chicken starter ration. McGinnis et al. (10) demonstrated that sunflower protein is deficient in lysine, and the nutritive value of chick starter ration containing 31% sunflower meal was improved by addition of 0.6% DL-lysine monochloride. The ability of sunflower meal to replace other protein sources in the animal diets depends on the quality of the meal itself. Morrison et al. (11) have reported that low-temperature sunflower meal (processed in the cooker for 30 min at 200°F and in the conditioner for 3 min at 220°F) always gave better results in replacing crude protein from meat meal, soybean meal, fish meal or their combinations when compared to regular-temperature sunflower meal (processed for similar periods of time in the cooker at 240°F and in the conditioner at 260°F) at any levels of protein substitution in chick starter diets. Inclusion of a high quality protein ingredient, such as fish meal, even at a low level (2.5% in the diet) not only improved nutritive value of the sunflower meal-based diets, but also allowed sunflower meal to be utilized better in those diets. However, this study used only average weight gain of the chicks from 0 to 28 days of age as a main criterion of justification (Table 12). Cuca et al. (12) fed chicks aged 7 days with a 21% protein diet based on sunflower meal with various amino acids supplementation for 14 days. They found that methionine supplementation gave no response, leucine supplementation significantly ($p < 0.05$) reduced weight gain, whereas threonine supplementation significantly ($p < 0.05$) increased weight gain of the chicks. Threonine was suggested to be the second limiting amino acid in sunflower meal. Recently, Robertson (13) reported that replacing crude protein from soybean meal with sunflower meal decidedly reduced weight gain and feed conversion ratio of the broiler chicks. The addition of lysine to diets containing sunflower meal greatly improved performances, but lysine addition to soybean meal diets did not influence performances of the animals. They have concluded that use of half or all of the protein supplement in broiler diets as sunflower meal requires additional lysine to bring the dietary level up to the requirement of the chick. The results are in agreement with Rad and Keshavarz (14) who observed no significant changes in performances of broilers fed diets containing sunflower meal as a total protein substitute for soybean meal with adequate crystalline lysine supplementation in the diets (Table 13).

Utilization of sunflower meal for layer diets also has been studied. Walter et al. (15) replaced 50 and 100% of crude protein from meat meal with sunflower meal in layer diets containing fish meal at either 2.5 or 2.0% in the diet. They found that complete substitution of sunflower meal for meat meal, in a ration containing 2.5% fish meal, had no influence on egg production or body weight maintenance (Table 14). In another experiment, sunflower meal was used as a total protein supplement compared to soybean meal or meat meal in layer diets containing 2% fish meal. They also found that the use of sunflower meal as a complete substitute for meat meal or soybean meal had no effect on production and body weight maintenance of the hens (Table 14). Rose et al. (16) studied the effects of replacing 50 and 100% of crude protein from soybean meal with high-protein (containing 44.6% crude protein, 11.0% crude fiber and 3.0% lignin) or low-protein (containing 36.5%

TABLE 11

Performances of Growing-Finishing Pigs (20–90 kg) Fed Diets Containing Variable Levels of Protein Substitution of SFM for Soybean Meal (8)

	Levels of protein substitution by SFM (%)			
	0	25	50	100
Average daily gain (g/day)	683[a]	680[a]	691[a]	622[b]
Average daily feed intake (kg/day)	1.94	1.88	1.94	1.85
Feed conversion ratio	2.86	2.79	2.80	2.92
Backfat thickness (mm)	21.5	20.0	22.0	20.8
Loineye area (cm^2)	33.63	32.18	32.68	32.38

[a,b] Means with same superscript letters are not significantly different ($p < .05$).

TABLE 12

Replacement of Crude Protein from Fish Meal, Meat Meal, Soybean Meal or the Combination by Low and Regular-Temperature Sunflower Meal in the Chicks Starter (0–28 days) Diets (11)

	Diet no.						
	1	2	3	4	5	6	7
Experiment 1							
Basal + additives	82	80	78	76	80	78	76
Fish meal (76% protein)	6	4	2	—	4	2	—
Meat meal (55% protein)	12	8	4	—	8	4	—
SFM (low-temp)	—	8	16	24	—	—	—
SFM (high-temp)	—	—	—	—	8	16	24
Avg. wt. of chicks (g)	298.5	324.5	291.0	255.8	320.3	240.7	188.9
Experiment 2							
Basal + additives	79.5	78.5	77.5	76.5	78.5	77.5	76.5
Fish meal (76% protein)	2.5	2.5	2.5	2.5	2.5	2.5	2.5
Meat meal (55% protein)	18	12	6	—	12	6	—
SFM (low-temp)	—	7	14	21	—	—	—
SFM (high-temp)	—	—	—	—	7	14	21
Avg. wt. of chicks (g)	306.6	323.7	316.6	300.6	306.1	279.8	278.0
Experiment 3							
Basal + additives	76.5	76.5	76.5	76.5	76.5	76.5	76.5
Fish meal (76% protein)	2.5	2.58	2.5	2.5	2.5	2.5	2.5
Soybean meal (solvent)	21	14	7	—	14	7	—
SFM (low-temp)	—	7	14	21	—	—	—
SFM (high-temp)	—	—	—	—	7	14	21
Avg. wt. of chicks (g)	325.0	320.0	332.1	305.9	339.8	316.9	276.0
Experiment 4							
Basal + additives	79	77.5	76	74.5	77.54	76	74.5
Meat meal (55% protein)	21	14	7	—	14	7	—
SFM (low-temp)	—	8.5	17	25.5	—	—	—
SFM (high-temp)	—	—	—	—	8.5	17	25.5
Avg. wt. of chicks (g)	169.4	203.6	226.5	228.0	198.9	195.6	171.1
Experiment 5							
Basal + additives	73	73	73	73	73	73	73
Soybean meal (solvent)	27	18	9	—	18	9	—
SFM (low-temp)	—	9	18	27	—	—	—
SFM (high-temp)	—	—	—	—	9	18	27
Avg. wt. of chicks (g)	301.2	302.5	297.2	269.9	270.1	296.0	165.6

protein, 20.3% fiber and 6.4% lignin) sunflower meal in corn-soybean ration for laying hens. They reported that, generally speaking, sunflower meal could replace 50% of the soybean meal protein without adversely affecting hen performance. However, 100% replacement of soybean protein with sunflower meal resulted in decreased egg production and feed efficiency (Table 15). They also noticed a characteristic of egg shell stains that developed when sunflower meal was used in mash ration. However, the problem was reduced markedly when similar rations were fed in crumble form. A phenolic compound (chlorogenic acid) present in the sunflowerseed apparently is responsible for these stains. Sunflower meal, both high-protein (34%) and low-protein (28%), were used successfully as a total protein substitute for soybean meal in pullet developer diets for chickens reared in floor pens (Table 16). But, replacing 100% of crude protein from soybean meal with low-protein sunflower meal in the diet significantly ($p < 0.05$) depressed performance of pullets reared in cages (17).

TABLE 13

Performances of Broiler (0–4 weeks) Fed Diets Containing Levels of Substitution of Crude Protein from Soybean Meal with Sunflower Meal Supplemented With or Without Crystalline Lysine* (14)

	SFM in the diet (%)	Levels of substitution (%)	Gain (g)	FCR
T1 (control)	—	—	696^a	1.79^a
T2	8.74	25	641^{a,b}	1.83^a
T3	17.48	50	662^{a,b}	1.82^a
T4 (as T3 + lysine)	17.48	50	675^{a,b}	1.81^a
T5	26.22	70	614^b	1.85^a
T6 (as T5 + lysine)	26.28	70	646^{a,b}	1.84^a
T7	36.82	100	536^c	1.99^b
T8 (as T7 + lysine)	36.92	100	653^{a,b}	1.87^{a,b}

* All diets were isonitrogenous and isocaloric, and contained 6.55% of fish meal.
^{a,b} Means with same superscript letters are not significantly different ($p < .05$).

TABLE 14

Effects of Substitution of Sunflower Meal for Meat Meal and Soybean Meal in Layer Diets (15)

	Diet no.		
	1	2	3
Experiment I (301 days test)			
Protein supplements			
Fish meal (%)	2.5	2.5	2.5
Meat meal (%)	8.0	4.0	—
Sunflower meal (%)	—	4.75	9.5
Performances of layers			
Eggs per bird^a	175	178	181
Feed/dozen eggs (lb.)	5.9	6.1	6.2
Weight gained during test (lb.)	0.7	0.7	0.7
Experiment II (280 days test)			
Protein supplements			
Fish meal (%)	2.0	2.0	2.0
Meat meal (%)	8.5	—	—
Soybean meal (%)	—	11.5	—
Sunflower meal (%)	—	—	13.0
Performances of layers			
Eggs per bird^a	162	151	160
Feed/dozen eggs (lb.)	5.6	5.6	5.9
Weight gained during test (lb.)	0.8	0.8	0.9

^{a} Corrected for mortality.

Acceptability of Sunflower Meal for Pig and Poultry Feeds in Thailand

Rapid development of animal industries in Thailand in terms of both quantity and efficiency of production have created a great demand and price elevation of the commonly used feed ingredients: broken rice, rice bran, soybean meal and fish meal. Alternative basal and high protein feedstuffs were employed in animal feed formulas in order to decrease the cost of animal production. During the past five years, more and more pig and poultry, especially hen and duck layer farmers, therefore tended to

TABLE 15

Effects of High-Protein (44.6% Protein; A) and Low-Protein (36.5% Protein; B) Sunflower Meal Substitution for Soybean Meal Protein in Rations for Laying Hens (16)

		Levels of SFM Substitution			
		50%		100%	
Performances	SBM	A	B	A	B
Level of SFM in the diets (%)	—	12	15	24	32
Egg production (%)	83.4^a	$81.9^{a,b}$	$81.3^{b,c}$	$80.3^{b,c}$	80.0^c
Avg. daily feed intake (g)	$117^{a,b}$	$117^{a,b}$	118^b	116^a	$118^{a,b}$
Feed/dozen eggs (kg)	1.69^a	1.73^b	$1.75^{b,c}$	$1.74^{b,c}$	1.78^c
Mortality (%)	4.2	2.5	3.8	2.9	4.6
Body weight (kg)	2.21	2.20	2.25	2.16	2.19
Egg weight (g)	64	65	65	64	64
Egg shell staining scores	0	2.0	1.5	3.0	2.0

a,b,c Means with same superscript letters are not significantly different (p < .05).

TABLE 16

Effects of Complete Substitution of High-Protein (34% Protein) or Low-Protein (28% Protein) Sunflower Meal for Soybean Meal Protein in Pullet Developer Rations (17)

Treatments	SFM (%)	Wt. at 20 wk (kg)	Wt. gain 12–20 wk (kg)	Feed/gain 12–20 wk
Floor pens				
Control	—	1.35	0.42	9.5
High-protein SFM	14	1.35	0.44	9.8
High-protein SFM + lysine	14	1.34	0.43	8.8
Low-protein SFM	18	1.38	0.44	9.0
Low-protein SFM + lysine	18	1.33	0.41	9.8
Cages				
Control	—	1.24	0.36	12.2
High-protein SFM	14	1.29	0.41	10.6
High-protein SFM + lysine	14	1.29	0.41	13.6
Low-protein SFM	18	1.14	0.26	15.5
Low-protein SFM + lysine	18	1.16	0.26	17.0

use more feed ingredients locally available and mixed their own feeds at the farms. The activity was strongly supported by the National Swine Research and Training Center, Kasetsart University, where not only the studies on utilization of various feed ingredients both locally available and imported have been conducted continuously, but also an intensive farmer-training course on animal (especially swine) nutrition and feed production has been offered since 1982. Very recently, aquaculture production, especially in marine shrimp culture, in Thailand has increased tremendously. The great demand of high-protein aquaculture diets has created a shortage of good quality protein supplements especially fish meal and soybean meal. The situation seriously elevated not only the prices of those ingredients, but also the cost of pig and poultry production. Many alternative protein supplements were considered to be protein substitutes in animal rations.

Sunflower meal, which has been included in commercial diets even at a low level of approximately 5%, is of particular interest and is widely accepted by pig and poultry farmers. Although sunflower production in Thailand is still at very early stages, sunflower meal is available in the country but is mainly imported from the People's Republic of China. The meal was recommended as a moderate substitute for protein from soybean meal and fish meal in animal rations. The recommended levels of sunflower meal were higher than those normally employed by the commercial feed mills because a lesser number of lower quality protein supplements were included in the home-mixed rations. However, the acceptabilities and the results of the formulas by the farmers were satisfactory. A great variance in nutrient composition as well as quality of the meal, a limited availability and inadequate supply for crystalline lysine were factors limiting utilization of sunflower meal in pigs and poultry feeds in Thailand.

References

1. *FAO Production Yearbook*, FAO, Rome, Italy, 1980.
2. Pond, W.D., and J.H. Maner, "Swine Production and Nutrition," Avi Publishing Co., Westport, CT., 1984.
3. Amos, H.E., D. Birdick, and R.W. Seerley, *J. Anim. Sci. 40:*90 (1975).
4. Milic, B., S. Stojacnovic, N. Vucurevic and T. Turcic, *J. Sci. Fd. Agric. 19:*108 (1968).
5. Jorgensen, H., W.C. Sauer and P.A. Thacker, *J. Anim. Sci. 58:*926 (1984).
6. Seerley, R.W., D. Burdick, W.C. Rossom, R.S. Rowrey, H.C. McCampbell and H.E. Amos, *J. Anim. Sci. 38:*947 (1974).
7. Baird, D.M., *J. Anim. Sci. 51:*(Suppl.)185 (1980).
8. Chinrasri, T., M.Sc. thesis, Kasetsart University, 1988.
9. Pettit, J.H., S.J. Slinger, E.V. Evans and F.N. Marcellus, *Sci. Agr. 24:*201 (1944).
10. McGinnis, J., P.T. Hsi and J.S. Carver, *Poult. Sci. 27:*389 (1948).
11. Morrison, A.B., D.R. Clandinin and A.R. Robblee, *Poult. Sci. 32:*542 (1953).
12. Cuca, M., A. Avila and E. Sosa, *Poult. Sci. 52:*2016 (1973).
13. Robertson, R.H., *Poult. Sci. 65:*112 (1986).
14. Rad, F.H., and K. Keshavarz, *Poult. Sci. 55:*1757 (1976).
15. Walter, E.G., G.S. Lindbald and J.R. Aitken, *Can. J. Anim. Sci. 39:*45 (1959).
16. Rose, R.J., R.N. Coit and J.L. Sell, *Poult. Sci. 51:*960 (1972).
17. Michell, J.N., and M.L. Sunde, *Poult. Sci. 64:*669 (1985).

Uses of Soy Proteins in Bakery and Cereal Products

Richard W. Fulmer

Cargill, Inc., Research Department, P.O. Box 9300, Minneapolis, Minnesota 55440 USA

Abstract

Soy products having unique functional properties have found wide application in bakery foods. These soy products include enzyme-activity soy flour, full-fat soy flour, high-fat (refatted) soy flour, lecithinated soy flour, defatted soy flour, soy grits, and soy bran. Products used to a lesser extent, probably due to cost, are soy protein concentrates and isolates. The characteristics and functionalities of the various soy flours, concentrates and isolates make them desirable components in bakery formulations. Nutrition, cohesion, water/fat emulsification, enzymatic activity, and processability are only a few of the attributes contributed by soy to breads, cakes, rolls, crackers, pancakes and cereal products. The technology for the utilization of soy products in bakery foods is well established and reasonably simple.

Utilization of Soy Products in Bakery and Cereal Products

Many publications (1,2,3,4,8) have described the processing of soybean materials to achieve products with useful functional properties and nutritional value for bakery and cereal products. Improved processing techniques have been developed, resulting in products having a wide range of controlled properties useful in formulated food systems. In addition to functional properties, the excellent profile of soy protein can contribute both quantity and quality to nutrition.

Theoretical estimates of functional contribution must be proven in baking tests. Practical and reproducible uses must then be verified in repeated tests and finally in commercial acceptance by the bread makers and bakers around the world. Soy based products are more costly than wheat flour but may well economically replace milk solids in bakery formulations. Functionality may justify the extra cost. Soy protein products and particularly soy flours for bakery food applications are described on the basis of protein content, fat content, protein solubility, urease activity, lipoxidase activity and granulation. A typical analysis is given in Table 1.

A discussion of protein quality must focus on the amino acid content and in particular the essential amino acid content. Table 2 shows the essential amino acid content of defatted soy flour and grits.

In order to evaluate soy protein one must establish a desirable pattern. The Food and Agriculture Organization (FAO) of the United Nations has recommended the essential amino acid pattern of whole hen egg protein as the basis of comparison. Table 3 shows the ratio of the amino acid to the amount (total) of all essential amino acids present.

The degree of heat treatment to which the soy mate-

TABLE 1
Typical Analysis of Soy Flour

		Test method
Protein (N × 6.25)	52.5%	AOCS Bc 4-49
Moisture	6.0%	AOCS Bc 2-49
Fat (ether extract)	0.9%	AOCS Bc 3-49
Ash	6.0%	AOCS Bc 5-49
Fiber	2.5%	AOCS Bc 6-49
Carbohydrates	31.0%	By difference
Total bacterial count	< 10,000/gm	FDA aerobic test method
Salmonella	Negative	FDA standard test

90, 70 OR 20 PDI
AOCS BA 10-65
100 OR 200 mesh

rial has been subjected is related to functional properties for all food applications, including bakery products. Methods to define protein solubility include Nitrogen Solubility Index (NSI), Protein Dispersibility Index (PDI), and Protein Solubility Index (PSI). Each of these different tests indicates the percentage of total nitrogen soluble in water and is inversely related to the heat denaturing of the protein material. PDI values will range from 90+ for white flakes, to a lightly toasted soy flour of 60–80 PDI and a heavily toasted soy flour of 10–20 PDI. Table 4 relates heat treatment to various properties.

The enzyme activity of soy flour is related to protein solubility, in that the heat treatment also destroys enzyme activity. The primary enzymes in soy flour are lipoxidase, urease, amylase, lipase, and protease. Thus, if the bleaching effect of a lipoxidase active soy flour is desired, a relatively high PDI soy flour should be used (Table 5).

Tables 6, 7, and 8 list the uses of soy products in bakery goods, the contribution to the product, and the functional mode of action in food products.

Overall nutritional characteristics are also closely related to heat treatment. Soybeans contain some antinutritional factors which must be inactivated for maximum nutritional value. For example, an antitrypsin factor which retards the action of the trypsin enzyme in the human digestive system needs to be destroyed by heat. Thus, harsh protein denaturation (very low PDI, PSI, or NSI) indicates inactivation of these antinutritional factors, but

TABLE 2

Suggested Patterns for Amino Acid Requirements and Composition of Soy Protein Products [a,b] (6)

Essential amino acid	FAI/WHO [a] Child, age			FNB pattern [b]	Flours/grits (defatted) [c] mg/g protein
	2–5	10–12	Adult		
Histidine	19	19	16	17	26
Isoleucine	28	28	13	42	46
Leucine	66	44	19	70	78
Lysine	58	44	16	51	64
Methionine + Cystine	25	22	17	26	26
Phenylalanine + Tyrosine	63	22	19	73	88
Threonine	34	28	9	35	39
Tryptophan	11	9	5	11	14
Valine	35	25	13	48	46

[a] Amino acid requirements from FAO/WHO (1984).
[b] Food and Nutrition Board, National Academy of Sciences, 1980.
[c] Cavins et al. (1972).

TABLE 3

Mg/g Total Essential Amino Acids

	Defatted soy flour	Wheat flour	1:1 Soy:wheat	Egg protein
Isoleucine	119	116	118	129
Leucine	181	195	188	172
Lysine	161	82	121	125
Phenylalanine	117	140	128	114
Tyrosine	91	97	94	81
(Total aromatic amino acids)	(208)	(237)	(222)	(195)
Cystine	37	64	51	46
Methionine	37	49	43	61
(Total sulphur amino acids)	(74)	(113)	(94)	(107)
Threonine	101	88	95	99
Tryptophan	30	40	35	31
Valine	126	131	129	141

also implies a correlated destructive loss of the essential amino acid lysine, and reduction of protein quality, unless extreme care is taken.

As previously noted, defatted soy flours having a wide range of PDI are available, and the bakery food application for any particular soy flour depends on the characteristics imparted by the degree of heat treatment.

In bread and buns, soy flour at the rate of 1% to 3% (flour basis) increases absorption of about one pound additional water for each pound of soy flour, with improvement of crumb body and resilience, crust color and toasting characteristics. Defatted soy flour is an excellent economical, functional, and nutritional replacement for nonfat dry milk in these products.

In cakes, defatted soy flour is used where water absorption and film forming characteristics are desired. Usage level is 3% to 6% soy flour, flour weight basis, and benefits include a smoother batter having a more even dis-

TABLE 4

Processing and Nutritional Parameters of Heat-treated Soy Flours

Heat,[a] min	NSI[b]	TI, TIU/mg[c]	PER[d]	Pancreas wt, g/ 100 g body wt
0	97.2	96.9	1.13	0.68
1	78.2	74.9	1.35	0.58
3	69.6	45.0	1.75	0.51
6	56.5	28.0	2.07	0.52
9	51.3	20.5	2.19	0.48
20	37.9	10.1	2.08	0.49
30	28.2	8.0	—	—

[a] Live steam at 100 C.
[b] NSI = nitrogen solubility index.
[c] TI = trypsin inhibitor and TIU = trypsin inhibitor units.
[d] Protein efficiency ratio, corrected on a basis of PER = 2.50 for casein.

tribution of air cells, and a cake with a more even texture, a softer and more tender crumb. In sweet goods, 2% to 4% defatted soy flour improves water holding capacity and improves sheeting characteristics and finished-product quality. The same 2–4% levels should be used for yeast-raised doughnuts.

Cake doughnut quality is greatly improved by the addition of 2% to 6% defatted soy flour, based on wheat flour weight. The soy protein functions as a structure builder, producing a doughnut having an excellent star formation. Fat absorption during frying is significantly reduced, producing improved texture and eating properties. The increased absorption and moisture retention increases yield and extends shelf life.

In hard (snap) cookies, the use of 2% to 5% defatted soy flour improves machining and produces a cookie having a "crisp" bite.

Toasted defatted soy flours, having a PDI of about 20, add color to the crumb, and a nutty toasted flavor to whole grain, multi-grain, or natural grain bread. Studies substituting defatted soy flour for wheat flour in a chemically-leavened quick bread indicate that successful substitution at levels up to 15% can be made.

Enzyme active soy flour is defatted soy flour which has been processed to retain its lipoxidase enzyme activity, which results in the bleaching of carotenoid pigments and which produce peroxides that strengthen gluten proteins (5). A second type of enzyme-active soy flour is processed without removal of the fat, resulting in a full-fat, enzyme-active soy flour.

Enzyme-active soy flour is used primarily in white

TABLE 5

Food Uses of Soybean Products

Product	Function	Uses
90 PDI F	Bleaching agent	Bread
70 PDI F	Controlled fat and water absorption	Doughnut mix
20 PDI F	Nutrition, fat and water absorption, emulsification	Baby cereals, breakfast cereals, baked goods, comminuted meat products (sausage, bologna, luncheon loaves), soups, sauces and gravies
Soy grits	Nutrition, meat extender	Patties, meatballs and loaves, chili, sloppy joes, soups, sauces and gravies

TABLE 6

Bakery Food Applications

	White bread & rolls	Specialty bread & rolls	Cakes	Cake doughnuts	Yeast-raised doughnuts	Sweet goods	Cookies
Defatted soy flour	X	X	X	X	X	X	X
Enzyme-active soy flour	X						
Low-fat soy flour			X	X		X	
High-fat soy flour			X	X		X	
Full-fat soy flour			X	X		X	
Lecithinated soy flour			X	X			X
Soy grits		X					
Soy concentrates		X					
Soy isolates		X	X	X	X		
Soy fiber		X					

TABLE 7

Functional and Organoleptic Contribution by Soy Proteins to Bakery Products (6)

Improved texture
Moisture holding to create cake richness
Bread whitening
Extended shelf life
Reduced breakage and crumbling
Imparting better, more authentic color
Adding nutrition, especially as a lysine fortifier
Improvements in manufacturing, handling, and machineability
Improving desired bite, whether hard or soft
Improving overall quality, as perceived by the consumer

---

bread and bun production. The FDA Standards of Identity permit a maximum of 0.5% enzyme-active soy flour, flour weight basis, in standardized bakery foods. There are no limits for nonstandardized bakery foods.

One of the lipid components of soybeans, which is removed in producing defatted soy flour, is lecithin. Lecithin imparts functional properties useful in bakery foods. It is an emulsifier, a blending aid for dry and liquid ingredients and a pan release agent. Lecithin also acts as an anti-oxidant and enhances the stability of vitamins in bakery foods (3).

Re-lecithinated or refatted products (usually at 1–15% addition) are used where a lecithin-type emulsifier is beneficial, and where shortening sparing action is desired.

Full-fat soy flour is processed to retain all of the fat present in the soybean and is usually subjected to a mild heat treatment to minimize enzyme activity.

Dubois and Hoover (3) describe the use of full-fat, refatted and lecithinated soy flours in cakes for increased richness and moisture, in doughnuts with minimal egg yolk levels, and in short pastry items where lecithinated soy flour contributes machinability and retained freshness.

Soy grits will have the same chemical analysis as the flour, the only difference being in particle size. Grits provide less functionality than the same PDI flour. Heavily toasted grits, having a PDI of 20–30, are used in whole grain, multi-grain, and natural grain breads to add color and a nutty, toasted flavor. In this application, normal usage level is 2% to 4% toasted grits, flour weight basis. Where very high protein breads are desired, bakers should use toasted grits to minimize flavor and change to the texture/size of the loaf. The same nutty, toasted flavor could be imparted to specialty cookies through the use of these toasted grits.

Soy bran/fiber is produced by toasting and grinding the seed coat portion of the soybean, and has a crude fiber content of approximately 38%. It can be used in multi-grain breads, in blended fiber systems or can be used as the sole added fiber source in such breads. Usage in this application ranges from 5% to 20%, flour weight basis.

Soy protein concentrates and isolates, though higher in cost, are sometimes used in bakery products. The concentrates contribute water and fat absorption, emulsification and textural control. The residual polysaccharides in the concentrates also contribute to water absorption.

Isolates have specific functional properties that enable them to modify the physical properties of food products.

TABLE 8

Functional Properties of Soy Protein Products in Foods (9)

Functional property	Mode of action	Food system used	Product
Solubility	Protein solvation, pH dependent	Beverages	F,C,I,H
Water absorption and binding	Hydrogen-bonding of water, entrapment water (no drip)	Meats, sausages, breads, cakes	F,C
Viscosity	Thickening, water binding	Soups, gravies	F,C,I
Gelation	Protein matrix formation and setting	Meats, curds, cheeses	C,I
Cohesion-adhesion	Protein acts as an adhesive material	Meats, sausages, baked goods, pasta products	F,C,I
Elasticity	Disulfide links in deformable gels	Meats, bakery items	I
Emulsification	Formation and stabilization of fat emulsions	Sausages, bologna, soups, cakes	F,C,I
Fat absorption	Binding of free fat	Meats, sausages, doughnuts	F,C,I
Flavor-binding	Absorption, entrapment, release	Simulated meats, bakery items	C,I,H
Foaming	Forms film to entrap gas	Whipped toppings, chiffon desserts, angel cakes	I,W,H
Color control	Bleaching (lipoxygenase)	Breads	F

F,C,I,H,W denote soy flour, concentrate, isolate, hydrolyzed protein and soy whey, respectively.

Soy isolates are characterized by certain functional properties: solubility, gelation, emulsification, dispersibility, viscosity and retort stability. Solubility ranges from 5 NSI (Nitrogen Solubility Index) to 95 NSI. The emulsion capacity of soy protein isolates can vary from 10 to about 35 milliliters of oil per 100 milligrams of protein. Isolates have water absorption values of up to 400%.

Neutralized isolates are usually highly soluble; certain types will gel under appropriate aqueous conditions. They possess both emulsifying and emulsion-stabilizing properties, are excellent binders of fat and water and are good adhesive agents. They vary mainly in their dispersibility, gelling and viscosity characteristics. Onayemi and Lorenz (7) report that isolates can be used up to 5% flour weight basis without adversely affecting white bread quality.

The greatest usage for soy proteins in the bakery foods industry is in combination with other ingredients, such as sweet dairy whey, to replace nonfat dry milk. The particular blend is dictated by the functional and/or nutritional requirements of the particular product. Defatted soy flour is the primary soy product used in these blends, but concentrates and isolates also are used in combination with whey and sodium or calcium caseinate for special applications, including cake mixes. Bakers use these blends for economy, because dairy products are generally more expensive than soy flour.

Milk replacers are made either by dry blending the ingredients or wet blending them followed by spray drying. Dry blended products will perform as well as wet blended products in most bread and other yeast-raised bakery foods. However, in cakes and cake doughnuts, the wet blended products are more effective. Soy/whey blends appear to be more adaptable to yeast-raised bakery foods as milk replacers as compared to their use in cake products. In cake-type doughnuts, the structure forming characteristic of nonfat dry milk is more critical than in yeast-raised doughnuts.

Soy flours with minimum heat treatments (Protein Dispersibility Index [PDI] of 80) show high lipoxygenase activity, and are used at 0.5% to bleach flour and to impart flavor to bread. Soy flours with a PDI of approximately 60 possess a milder flavor and are most commonly used at 1% to 2% in standard applications.

Soy flour provides improved water absorption and dough handling properties, a tenderizing effect, body, and resiliency as does nonfat dry milk. Bread freshness is maintained because the soy protein retains free moisture during the baking cycle. And, soy protein products improve crust color and toasting characteristics in bread. The principles applied to white bread production also apply to buns and rolls. Nutritional studies indicate that the protein quality of commercial white bread containing 3% soy flour is equal to, or slightly superior to, bread containing 3% nonfat dry milk.

Ordinary white bread contains 8% to 9% protein. Specialty breads can be made with 13% to 14% protein by incorporating soy proteins into a formula along with vital wheat gluten and, if necessary, a lipid emulsifier. Without an emulsifier, incorporating high levels of soy protein depresses loaf volume and gives poor crumb characteristics. Hoover (4) reported that the addition of higher levels of soy flour brought about dramatic changes in the protein nutritive value of bread. The Protein Efficiency Ratio (PER) compares the nutritive value with casein at 2.5. By comparison, the PER of defatted soy flour, lightly toasted, is 2.03 to 2.3. The PER for white bread is about 0.7 and for bread with 3% soy flour added, about 0.83. When the soy flour is increased to the 6% level, the PER increases to 1.3, and at the 12% level, the PER is 1.95. In addition to the improvement in protein quality at the 12% level, the protein content is increased by 50%.

Several uses of soy protein products, including soy isolate-whey blends, have been reported in commercially acceptable pound cakes, devil's food cakes, yellow layer cakes, and sponge cakes, in which 50%, 75% and 100% of nonfat dry milk has been replaced without impairing quality. At a 50% replacement level, aside from an increased water absorption, no formula changes are necessary. With replacement levels at or above 75%, dextrose must be included in the total sugar used to improve color. Leavening must also be increased to obtain the desired volume, but the added cost of the leavening is offset by the increased yield of the batter.

Mixes for bread, pancakes, waffles, buns, and many other baked items are available containing defatted and full-fat soy flours and grits at a level of 2% to 15%.

Soy protein products help with the emulsification of fats and other ingredients. The resulting doughs are more uniform, smoother, more pliable and less sticky. The finished baked products have improved crust color, grain texture, and symmetry and will stay fresher longer due to effective moisture retention. The same functional properties of soy proteins also are utilized in cookies, crackers, biscuits, pancakes, sweet pastry and snacks. Incorporating a white soy flour (one that is lightly heated) or a mildly lecithinated soy flour in a pancake formulation at the 3% level will result in a product with improved texture.

Short-pastry items, such as pie crusts, fried pie crusts, and puff pastry, can be machined more easily and will retain freshness longer when lecithinated soy flour (lecithin content 0.5% to 15%) is used in the formula at levels of 2% to 4%, on a flour weight basis. Doughnuts containing soy protein absorb less fat during frying because the fat is prevented from penetrating into the interior. The result is a higher quality doughnut that is more economical due to lower frying oil use. Used in the range of 3% to 3.5% of the formula, soy flour also gives doughnuts a good crust color, improved shape, higher moisture absorption with resulting improvement in shelf-life, and a texture with shortness or tenderness. Lecithinated soy flour may be used to produce doughnut formulas containing minimal egg yolk levels because lecithin is the natural emulsifier in egg yolk.

High protein pasta products, such as spaghetti, can be prepared from durum semolina or hard wheat farina fortified with soy protein. All soy protein products increase the water absorption of spaghetti dough and affect its processing conditions. Pasta products, such as macaroni, spaghetti and vermicelli, also can be fortified with soy flour to increase nutritional value. Defatted soy flour and full-fat soy flour are most commonly used. These pasta products contain soy flour at 15% levels on a dry basis. Foods of these types have been accepted by the U.S. military, in government feeding programs, and in the National School Lunch Program. Expanded emphasis on nutrition in breakfast cereals has led to an increased use of soy protein to boost protein value and quantity. Soy proteins are used extensively as ingredients in hot cereal mixes and as components of compound breakfast bars.

Bibliography

1. Dubois, D.K., "Soy Products in Bakery Foods," American Institute of Baking Technical Bulletin, II, 9, September, 1980.
2. French, F., *Bakers Digest, 51*(5)pp. 98–103 (1977).
3. Dubois, D.K., and W.J. Hoover, *J. Am. Oil Chem. Soc., 58:*343–346 (1981).
4. Hoover, W.J., *Ibid., 56:*301–303 (1979).
5. Frazier, P.J., *Bakers Digest 53*(6):8 (1979).
6. Soy Protein Products, published by the Soy Protein Council, 1255 Twenty-Third St. N.W., Washington, D.C. 20037 (1987).
7. Onayemi, O., and K. Lorenz, *Bakers Digest, 52*(1):18 (1978).
8. Sipos, E.F., "Edible Uses of Soybean Protein," Soybean Utilization Alternatives Conference, Center for Alternate Crops and Products, University of Minnesota, Minneapolis. Proceedings pp. 57–95, Feb. 16, 1986.
9. Kinsella, J.E., *J. Am. Oil Chem. Soc. 56:*242 (1979).

The Utilization of Soy Proteins from Hot Dogs to Haramaki

Alexander T. Bonkowski

Archer Daniels Midland Company, PO Box 1470, Decatur, IL 62525, USA

Abstract

Utilization of vegetable protein in emulsion-type sausage products, and injection- and absorption-type meat systems are reviewed. This paper includes identification of market needs; equipment used; selection and analysis of raw materials; formulation to fixed constraints; grinding, mixing, emulsifying stuffing-linking, heat processing/smoking/cooking, showering/chilling, and peeling slicing where applicable; plus packaging, inventory control and shipping.

The Chinese understood the miracle of the soybean 4,000 years ago. The rest of the world has taken a little longer to fully appreciate this incredible resource. In the years to come, the soybean may become the single most important factor in good nutrition; perhaps survival itself for the earth's population. Malthus said, "It would take a miracle to feed the world" and here it is.

Soy protein made its greatest impact in the meat industry during the 1960's. The ability to texturize soy flour into meat-type particles and pieces with a mouthfeel and texture that greatly resembled meat found a marketplace in the ground meat systems of the meat industry. The evolution of soy concentrates and isolates opened the doors to the entire industry.

The meat industry manufactures products that can be separated into three basic categories: Whole muscle products, or primal cuts; a category that would include steaks, roasts, hams, chops, drumsticks, breasts, etc., that are easily recognized in the marketplace or on the dinner table. Fine and coarse ground emulsions and blends; these would include hot dogs, salamis, mortadella, goin chong, hamburgers, loaves, tacos, and the meat portion of haramaki. Restructured or engineered products that resemble their whole muscle counterparts, from cappacola to crabsticks.

Today's technology can make poultry taste like pork or pork like poultry, and fish like shrimp, crab or even calamari. Soy proteins contribute greatly to the quality, texture, nutrition and taste of them all. In fact, without soy proteins some of the products would be impossible to manufacture.

Meat Systems

To utilize and understand the miracle of the soybean in a functional, economical manner we must analyze the meat system. The manufacturing of meat products begins with the selection of proper materials for the product we intend to manufacture. Common sense tells us we certainly wouldn't choose mechanically deboned chicken to manufacture a bone-in smoked ham. We certainly could choose this type of material to produce the smooth texture of a hot dog.

In the meat industry, we must strive to maintain all of our selected meats at their highest economical level. From the time the animal is slaughtered, there is an ongoing struggle between the physical, microbiological and human elements to determine the final use of the meat. We certainly do not want the meats we selected to be utilized as organic fertilizer (at least until it has passed through the human digestive system). If we are to win the battle, we should understand some basic facts about the preservation of meat.

Microorganisms have been of concern to those processing, utilizing and eating meat since the beginning of history. Man has constantly battled the bug (microbe) to keep it from utilizing the meat before he can consume it. The ancient "kosher" or sanitary method of animal slaughter and dressing was developed by the Hebrews centuries ago to help preserve meats a little longer. The practice of drying meat to make jerky, or pemmican, was developed by the American Indians to preserve the meat. Removing moisture prevented the growth of microorganisms. The modern day practice of curing hams, fermenting salami and refrigeration have been developed through the years to preserve meat so man can use it before the microbes render it unacceptable.

Meat Preservation: Meat preservation treatments used in the meat industry include: the prevention of contamination, fermentation, heat treatment, refrigeration, drying and curing. The most important item on the list is of course the prevention of contamination. The equipment, from the sticking knife to the final box for storage, should be kept clean and sanitary. In the meat industry, sanitation must be a way of life.

The Emulsion System

All meat is composed primarily of three groups, i.e., fat, protein, and moisture. The most important part of this composition is the protein.

Protein: Proteins are complex molecular structures containing various amino acids with different molecular weights and different reactions to water, as well as salt solutions. In a broad grouping, we have water soluble proteins, salt soluble proteins, and of course the collagen group.

In sausage emulsions we are mainly interested in the salt soluble group because of their reactions with water and fats to produce desired texture, shape and stability. The water soluble proteins, or sarcoplasmic proteins, are the meat juices we often see in the bottom of the containers that are used to store meat. These proteins make up about 25–30% of the total protein. They are water soluble and coagulate when heated to form a soft gel. In this protein group are found the enzymes of the muscle tissue, as well as the important pigment myoglobin. This is the color-producing compound that reacts with the curing salts. How these sarcoplasmic proteins perform in the formation of a stable sausage emulsion is questionable. They may even interfere with the contractile or salt soluble proteins when forming a stable emulsion. In the final analysis, they probably are most important as contributors of protein as such.

The myofibrillar proteins, or contractile proteins, are the ones we are most concerned with. These proteins are soluble in salt. They will coagulate upon heating and form a very firm gel. The main components of the salt soluble group are actin and myosin. While the animal is alive the actin and myosin filaments slide past one another and give the muscle the ability to expand and contract. After slaughter, rigor sets in and the actin and the myosin react to form a complex called actomyosin. Myosin alone accounts for 35% of the muscle protein, with actin accounting for about 15%. The old sausage maker was simply trying to solubilize the myosin when he presalted the freshly slaughtered bull before the actomyosin complex was formed.

The third type of protein found in meat is collagen. Collagen is the connective tissue that forms the sheets that surround the muscle fibers and muscle bundles. They also make up the skin, intestine, sinew, etc. Collagen, when heated in the presence of moisture, becomes gelatin. It forms a stable gel only when it is cold. Collagen becomes less soluble as the animal becomes older. Collagen can be made fairly stable in the presence of salt.

The ability of proteins to bind fat and water into a stable matrix is the key to successful sausage production. Let us first consider the water-holding capacity. Because water is the least costly ingredient found in sausage, the object is to not only hold the water that is part of the meat itself, but also to hold the proper amount of added water to ensure the texture and characteristics of the sausage type. Soy proteins greatly enhance the ability of an emulsion to retain water by absorption and adsorption. The muscle proteins are able to bind some water chemically. Water also is trapped in the rope-like structure that forms the muscle fibers. When this structure shrinks, water is expelled. When this structure expands or swells, moisture is taken up.

A factor that effects the swelling of the protein fibers in the muscle tissue is the pH, or acidity. The meat muscle has a minimum water-holding capacity at its isoelectric point of about 5.4. Lower or higher pH increases this capacity. The muscle tissue at death has a pH of about 7 neutral. Water-holding ability at this time is good. After death, the muscle stores of glycogen are metabolized, resulting in the formation of lactic acid. This causes the pH to drop and the water-holding ability of the meat to decrease. When the meat reaches its isoelectric point, the positive and negative charges on the proteins are in balance, and there are no free electric charges to bind the water. When sodium chloride is added to the meat, the isoelectric point of meat shifts to a pH nearer to 4. The amount of salt added is critical. In normal lean meat, the addition of 4–4.5% salt gives the best water-holding capacity. If water is added with the salt, the effect is decreased. It is safe to say that if anything is added with the salt the efficiency of extracting the salt soluble proteins will be affected by simple dilution if nothing else. In an emulsion that requires the full benefit of all the functional proteins, I recommend the prehydration of the soy proteins separately, then adding back this material after the myosin extraction has been completed.

The procedure that I recommend in a highly stressed formula is as follows: (a) Chop the lean meat with the phosphates alone for 30–45 seconds. This will raise the pH, and also cause the muscle tissue to swell. The swollen tissue will allow quicker, easier penetration of the salt, causing a quicker, more efficient myosin extraction. (b) Add all the salt necessary to the cutter bowl, including the curing salts. (c) Add the prehydrated soy proteins. (d) Add the balance of the ice water. (e) Add the fat, spices and all other ingredients.

Temperature is also critical to water-holding capacity. Optimum water binding temperatures are about 2.8°C. This presents a problem, since to get fat into the correct plastic state for proper emulsion we must work with higher temperatures. As functional soy proteins are heat tolerant, a premade fat emulsion that has been made at the higher temperatures and then refrigerated until use can be added back to the lean portion that has been made at the lower temperatures that are needed. There is an added benefit to this type of process. Because functional soy proteins require no salt for extraction, all of the salt can be applied to the lean meats, making the myosin extraction more efficient.

Example: If the sausage emulsion contains 25% lean meat, and the finished product contains 2% salt, we will be adding all the salt to the lean portion. We would be adding what would be 8% salt to the myosin containing meat; more than enough to make the extraction quick and easy. The lean meat could be worked at the colder temperatures required, as the fat has been already emulsified at the higher temperature to achieve its plastic state by the heat tolerant soy protein.

All emulsion-type products have a common denominator: the particle size must be reduced from its normal state to one that is typical of the sausage type we are making. A sausage manufacturer will have in his sausage kitchen all or a combination of the following equipment: grinder, chopper (silent cutter), emulsion mill, mixer blender, stuffer, and thermal processing equipment for cooking or drying. The most important step in the preparation of any meat product is the selection of raw material and added ingredients. The factors to consider in the selection of the raw products are as follows: purpose, composition, soundness, flavor, i.e., oxidation or sex odor, microbiological, color, history (temperature and age) and cost.

After raw materials are selected, whether they be all of one animal type or a combination of all the species available, good manufacturing principles require an analysis for fat, protein and moisture. Establishing these values allows us to formulate wholesome, nutritious, tasty products that are priced economically and reasonably trouble-free, for further manufacturing (Fig. 1).

An accurate analysis performed on any of the three components of meat will allow a manufacturer to establish the other two with reasonable accuracy. The most common analysis used in the meat system is a Modified Babcock Fat Test.

Based upon this analysis, the other components are factored. In a beef system the protein is established by subtracting the fat percentage from 100, then multiplying the remainder by 0.22. The answer is a reasonably accurate per cent for the protein. By adding the fat percentage and the protein that was just established, and subtracting that figure from 100, the moisture percentage is established. The ash content of the meat is not considered. Working with the three components established, the manufacturer produces a consistent product.

The utilization of soy proteins allows the manufacturer

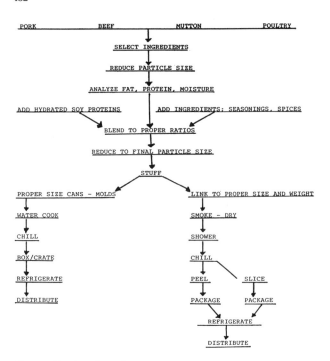

Fig. 1. Product flow chart.

to use the full carcass weight of the animal to greater economic and nutritional advantages. The sections of the animal that are analytically lacking in salt soluble proteins can now be fortified and used in combination with the soy protein group. The functional isolates and concentrates act as stabilizers and adhesives for the system, while providing excellent nutrition.

Pork skin emulsion can be used advantageously in products normally consumed cold, but can also be used in hot dog emulsions (Tables 1 and 2).

TABLE 1
Cooked Skin Emulsions

Formula	Parts	Percentage
Cooked pork skins	50.00	47.6100
Clean hot water	30.00	28.5660
Ice	20.00	19.0440
Isolated soy protein	3.00	2.8566
Salt	2.00	1.9044
Sodium nitrite	0.02	0.0190
Totals:	105.02	100.0000

Skin emulsion can be utilized in poultry products to increase whiteness and also to improve flavor. A table formula for an emulsion made with chicken, turkey or duck skin is shown in Table 3.

Additional systems are shown in Tables 4 and 5.

The soy protein groups allow a manufacturer greater flexibility in the field of nutrition. Functional proteins allow the meat manufacturers to play the caloric scales up or down with reasonable ease. An example is fatadella, described in Table 6.

TABLE 2
Raw Skin Emulsion

Formula	Parts	Percentage
Raw pork skin	50.00	47.466
Ice	25.00	23.740
Water	25.00	23.740
Isolated soy protein	3.00	2.850
Salt	2.00	1.900
Sodium tripolyphosphate	0.30	0.285
Sodium nitrite	0.02	0.019
Totals:	105.32	100.000

TABLE 3
Poultry Skin Emulsions[a]

	Percentage
Skin	50
Water	20
Ice	20
Isolated soy protein	10

[a] Preparation: (a) chop ice and isolated soy protein to disperse protein; (b) slowly add water and hydrate the protein until it is glossy in appearance; (c) add skin and continue chopping until a smooth white emulsion is formed; the temperature should not exceed 10°C (50°F). As an alternative the emulsion can be finished with an emulsion mill or emulsifier.

TABLE 4
Cold Pork Fat Emulsion[a]

Formula	Parts	Percentage
Pork fat	50.0	44.6
Water	25.0	22.3
Ice	25.0	22.3
Isolated soy protein	10.0	9.0
Salt	2.0	1.8
Totals:	112.0	100.0

[a] Place ice in bowl cutter and add isolated soy protein. Gradually add water and chop until protein is thoroughly hydrated. Add the pork fat and chop until a smooth emulsion is formed. Chop in 2% salt and place in pans no more than 3–4 inches (8–10 cm) deep and chill.

Restructured Meats

Webster defines "restructure" as a transitive verb meaning to reorganize or plan a new structure for, so in essence we are disassembling the present form and reassembling it into a more practical unit.

TABLE 5

Hot Beef Fat Emulsion[a]

Formula	Parts	Percentage
Cooked beef fat	50.0	44.6
Hot water	30.0	26.8
Ice	20.0	17.8
Isolated soy protein	10.0	9.0
Salt	2.0	1.8
Totals:	112.0	100.0

[a] Beef fat is first cooked in approximately 200°F (93°C) water for 30 min. The hot fat is placed in the bowl cutter and chopped to a small "pea" size. Add the isolated soy protein, disperse and slowly add the hot water and chop ISP until a smooth emulsion is formed. Add the ice to cool the emulsion and chop in the salt. Place mixture in pans 3–4 inches (8–10 cm) deep and chill.

TABLE 6

Fatadella (High Economy, High Calorie Special Sausage)[a]

	Kg
Pork 80% fat	100.00
Dispersion	
Non-gelling isolated soy protein	3.00
Gelling soy protein concentrate	4.50
Plasma	3.00
Nonfat dry milk	3.00
Starch	2.00
Sugar	1.00
Phosphate	0.70
Salt	1.00
Nitrite	0.04
Ascorbate	0.30
MSG	0.15
Flavor	0.10
Garlic powder	0.01
Water	31.20
Total dispersion:	50.00

[a] Procedure: Cut the pork into 50–80 gram pieces. In a regular meat mixer, make the dispersion listed above. Add fatty meats and mix until all dispersion is absorbed. Chill at 0°C overnight. Remix 10–12 minutes. Stuff into casings. Allow to equilibrate at 10°C 3–4 hours. Cook in staged cycle to 72°C. A light smoke should be applied.

In normal food service operations, restaurants, deli's, supermarkets, etc., sliced meats such as roast beef, corned beef, hams, etc., intended for sandwiches and for the lower priced meals, were prepared by cooking large bone-in or boneless cuts of meat on the premises, then slicing thin pieces from these primal cuts. This type of operation has a number of disadvantages. The meat must be trimmed and then roasted or cooked for long periods of time. Labor and facilities for cooking are therefore required, and substantial cooking losses are sustained. The slicing must be done by skilled carvers, and even then the slices are of nonuniform size and shape. They are certainly not well adapted to the portion control requirements of today's marketplace or food service operations. It has become necessary for the food service operator to provide to the consumer integral meat products in ham, roast beef or corned beef form, with the texture, appearance and general acceptance of intact meats. They also must be inexpensive and sliceable into portion control units with maximum yields.

Small chunks of meat of relatively lower value can be joined together into larger pieces of integral meats by a natural cement-like exudate. This exudate can be produced on the surfaces of meat by the mechanical working of the muscle tissue in mixers, massagers, tumblers and blenders. The properly worked meat can be compressed into a form and further processed by roasting or cooking. When this process is properly carried out, the exudate is not visually noticeable in the finished products. The joined surfaces do not separate or come apart during subsequent handling, such as slicing or dicing.

In order to hold down the cost of these products, the producer tried to use a high percentage of the more economical cuts. However, an upper limit must be placed upon the amount of the low cost cuts that can be used, simply because these cuts are high in undesirable material.

In the early 1960's, experimentation began with the idea of creating a perfectly portionable, easily sliced cured ham product to fill the needs of the marketplace. Fresh pork hams were trimmed reasonably free of skin, heavy fat, glandular material and connective tissue. A curing composition comprised of a solution containing 15 parts sodium chloride, 2 parts sugar, 5 parts sodium tripolyphosphate, 0.1 part sodium nitrate and 0.1 part sodium nitrite was prepared by dissolving these ingredients into 77.8 parts of water. The resulting solution was injected into the vascular system of the hams, using a hollow needle inserted into the femoral artery. Sufficient curing solution was introduced to increase the ham weight by 10%.

The injected hams were held in a cooler for five days at approximately 3.3°C, during which time they were kept covered with additional quantities of the above curing brine. At the end of five days they were deemed to have proper cure, flavor and appearance. They were removed from the curing cellar, washed to remove excess curing ingredients and then cut into angular chunks weighing from .22 kg to 1.4 kg. The chunks generally were cut with or parallel to the grain insofar as this was practical, instead of in a cross grain fashion. During the course of the chunking, the internal fats and undesirable material were further removed.

The chunks of meat were placed into a meat mixer and mixed for 2 minutes. After the 2 minute mix, a small amount of exudate was forming on the surface of the ham pieces. The ham pieces were mixed for an additional 8 minutes. The temperature of the ham had risen to 18°C and they were thoroughly coated with a creamy, sticky exudate. The ham chunks were also soft and pliable. The ham could be stuffed or pressed into any size or shape container. The meat was removed from the mixer and stuffed into a conventional 3 kg square pullman style can, vacuum closed, and cooked in hot water to an internal temperature of 68°C. After cooking and cooling the cans were opened for inspection. The ham chunks were firmly bonded together, and were easy to slice thinly, without breaking apart at the seams. The ham was visibly lean, uniform in size, smooth in texture, and greatly resembled whole muscle meat. The meat packer had solved two major problems, i.e., the appearance of the finished product was lean and showed none of the whole muscle's imperfections; he had altered the shape of the product so that it could be sliced thinly, with little or no waste for practical portion control packaging. The consumer benefited because of the improved quality, but paid the penalty of an increased cost.

In the mid-1970's, the modern stitch pump was fully utilized in the curing industry. By adjusting the amount of strokes per minute, regulating the belt speed and adjusting the pressure, whole muscle tissue could be injected to predictable increases accurately. The concept of injecting an isolated soy protein into sectioned and formed hams was in its infancy.

The problem of an economic product that would look, taste and slice similarly to the sectioned and formed ham described above was attacked by the soy industry. Because of the high viscosity of a soy fortified brine, problems with pocketing and visible striation occurred, leaving the end result more economical but less attractive to the consumer. With the onset of lower viscosity brines that were now lecithinated to improve dispersibility, the product became a saleable item. The first to realize the impact of reduced cost were probably the Japanese manufacturers. The cost of meat in Japan necessitated the utilization of these concepts.

Any commercial stitch pump can be used to pump the raw ham. It may require multiple passes through the machine, depending upon the machine. It is still almost impossible to use the artery pump system for this product, as the colloidal particles of the protein in suspension are larger than the inside diameters of the capillaries. If the pressure is increased, the capillaries generally burst and brine distribution throughout the muscle is not readily accomplished. The results are low yields and uncured or undercured areas in the finished product.

Procedures

Preparation of the Brine: (a) Hydrate the isolated soy protein in 80% of the water. Reserve 20% of the water as ice to add after brine has been made, to control temperature. (b) Add phosphates. Mix into solution. (c) Add sodium chloride. Mix into solution. (d) Add curing agents. (e) Add seasonings, flavorings. (f) Add ice to control temperature.

Preparation of Hams for Pumping: (a) Hams should be skinned and defatted. (b) Remove bones. (c) Remove heel, shank and butt kernel. In addition, remove star fat and seam fat.

After boning, trimming and sectioning, the hams are ready for pumping.

Products can be injected to 60% and higher. Cooking yields average 88–90% of injected weight. An extension average of 142% of raw weight can be expected. A brine formula is shown in Table 7.

Smaller sections of meat can be united into a solid, integral body of a larger size by virtue of the same natural exudate. The soy protein fortified brine (Table 7) can

TABLE 7

Brine Formula—160% Pump, 88–90% Cook and Chill Yield

Ingredients	Per cent
Water	81.8202
Isolated soy protein	10.0000
Salt	6.9333
Sodium phosphate	1.0666
Sodium erythorbate	0.1333
Sodium nitrite	0.0416
California ham spice	0.0050
	100.0000

be absorbed through mechanical action. Meat pieces from any section of the carcass of 113 to 400 grams will readily absorb a protein fortified brine in a mechanical mixer. The same process that releases the myosin protein exudate can be utilized. The mechanical action will cause penetration into the muscle tissue without the need for injection.

The advantages of utilizing this process may be summarized as follows: (a) Chunks of meat from less desirable sections may be combined into uniform bodies of meat of any desired shape. (b) Whole primal cuts can be shaped so that they conform to the various shapes and sizes of containers and forms. (c) Uniform weights may be obtained with primal cuts simply by adding one or more mechanically worked chunks to the mechanically worked whole muscle tissue. (d) The boning and trimming departments become more efficient through the utilization of all available whole muscle tissue. (e) The fat and undesirable portions of the meat muscle is economically replaced by the addition of an inexpensive soy protein fortified brine. (f) Noticeable tenderization occurs during the mechanical action and not only improves product by present standards, but also makes it practical to upgrade lesser quality meats that are tough, dry and unacceptable at this time.

Injection of Soy Protein Brines into Whole Muscle Tissue: Some of my colleagues refer to this application as recent technology. The sectioned and formed hams of the mid-70's employed the same concepts. The meat manufacturer selected a section of carcass that did not separate easily when mechanically worked in a tumbler. He injected the same types of brine into the flat portion of the beef bottom round, the trimmed pork loin, and the beef top rounds, and tumbled them under vacuum for longer periods of time with a gentler mechanical action. The muscles lost little of their original shape. The texture of the muscle improved and the economics of higher processing yields were attained. See Tables 8 and 9 for some brine examples.

Coarse-ground Blends: The coarse-ground blends contain a myriad of products: breakfast sausages, patties, salami, taco filling, and of course, the meat portion of haramaki. The textured soy proteins play an important role in products of this type. The texture and appearance of today's products easily adapt themselves to these manufactured meat types.

TABLE 8
Restructured Pastrami[a]

Brine ingredients	Percentage
Water	82.04
Salt	6.84
Isolated soy protein	7.81
Garlic powder	0.58
Coarse ground pepper, 16 mesh	0.25
Coriander	0.25
Brown sugar	0.25
Prague powder	0.50
Sodium erythorbate	0.12
H&R beef flavor	0.18
Sodium tripolyphosphate	1.18
Total:	100.00

[a] Beef must be trimmed free of excess fat, skin and gristle.

TABLE 9
Restructured Corn Beef

Brine ingredients	Percentage
Water	81.79
Salt	7.00
Isolated soy protein	8.00
Cane sugar	1.25
Garlic juice	0.60
Sodium tripolyphosphate	1.20
Sodium erythorbate	0.12
Sodium nitrite	0.04
Total:	100.00

Beef, Pork or Poultry Patties: A model patty operation can be used as a general guideline for most coarse-ground products. Temperatures are critical for proper machining, whether it be stuffing or extruding a breakfast sausage, forming a beef patty or any of the exotic shapes available, from chicken legs to pork chops.

A coarse-ground system must be monitored for temperatures closely. Temperature is critical to color and microbiological spoilage, as well as weights, shapes and sizing. The process is as follows: (a) Select meats for wholesomeness and temperature. (b) Analyze for fat content. (c) Hydrate textured soy products in ice water. (d) Pre-break raw meats to 80–90 mm. (e) Blend all ingredients, adjusting the temperature if necessary, using dry ice (CO_2) if necessary. (f) Regrind through proper size grinding plate. (g) Form shapes, stuff or extrude, batter and bread. (h) Freeze. (i) Package/store frozen.

The most common question asked by patty makers is, "How much water should be used to hydrate various nonmeat ingredients?" To balance beef patty formulas add water at the following ratios: Available lean meat × .05, isolated soy protein × 3.50, soy protein concentrate × 2.10, textured soy flour × 1.80, grain flours × 1.00, sweet dairy whey × .50 and dehydrated vegetables × 1.00.

Performance

Additional Information on Performance: Fried or broiled ground beef extended with textured vegetable protein appears to be juicier and have less shrinkage than all-beef patties prepared in a similar manner. This would indicate a difference in retention of water and fat in patties upon cooking. Patties composed of a mixture of 25% by weight of hydrated textured soy protein with beef retain a greater percentage of moisture and a lesser percentage of fat in cooking than all-beef patties of comparable fat levels. Studies on meat loaves which have been conducted employing a hydrated soy protein concentrate product at a 30% level of meat replacement show a 50% reduction in cooking losses in precooked, reheated beef soy loaves, compared with all-beef loaves.

The organoleptic quality of beef patties containing soy additives was compared to the quality of all-beef by a number of taste panels. The general conclusion was that the all-meat patties usually were rated better than beef patties containing concentrates at the 20 and 30% level when evaluated for flavor, appearance, odor, juiciness and overall acceptability. Patties containing 20% replacement were scored equal to all-beef patties when the usual condiments were added to the patties.

Other Products

Dry Sausage: The dry sausage industry generally uses nonfat dry milk (NFMD) as a binder or extender/filler in its product lines. Isolated soy proteins can be used to replace nonfat dry milk in dry sausage manufacture. Isolated soy proteins have similar properties to the salt soluble meat protein myosin. At the maximum permitted usage level, isolated soy protein will give a lower initial ingredient cost than does NFMD. (Based upon U.S.D.A. allowances 3.5% for NFMD, 2% for isolated soy protein.)

The U.S. Department of Agriculture requires that a moisture/protein ratio be based on meat protein. The amount of protein contributed by nonmeat sources must be deducted from the total finished protein. In an Italian style salami such as Genoa, the moisture/protein level is 1.9:1. In a pepperoni type product, the ratio is 1.6:1.

Semi-Dry Sausage Formulations

Basic Semi-dry Processing Schedule: All manufacturers of lactic acid starter cultures have instructions as to the proper use of their individual type products. The humidities and temperatures will vary depending upon the diameter of your product. Please follow the instructions carefully.

This schedule is useful for most products. On occasions it is necessary to make modifications to achieve proper color and to prevent case hardening.

1. Ferment product at 95°F (35°C) and 90% relative humidity to a pH of 4.8. Surface color should be red at the end of fermentation cycle. If not, reduce humidity to 85% when pH reaches 5.0.

2. Raise temperature of smokehouse to 145°F (63°C) and 65% relative humidity and apply smoke. Continue processing until internal temperature of 137°F (59°C) is attained.

3. Remove from smokehouse and allow to cool to 90°F (32°C) before moving to chill cooler.

4. For a reasonably shelf-stable product at nonrefrigerated temperatures, it will be necessary to modify this schedule to dry the product so that it has lost 25% of its weight. The water activity of a nonrefrigerated product should be A_w 0.85 or less. This type of product will be shelf-stable at a pH of less than 5 and a moisture protein ratio of 3.1 to 1.

Important Factors to be Watched: (a) Grinding temperatures $-2°C$ to $-1°C$; (b) Trimmings should be clean cuts, not smashed; (c) Grinding plates, grinding blades or silent cutter knives should be sharp and adjusted properly to insure clean cutting action; (d) Pork should be certified Trichina-free, if not, product must reach 59°C–60°C internal temperature; (e) Fat content should not exceed 30%; (f) Beef can be substituted for pork up to 60%; and (g) Lactic acid cultures feed and develop on dextrose. Substitute dextrose for sugar when using culture.

An example employing TVP is shown in Table 10.

pened, smoking, like curing, has been practiced throughout recorded history. Curing and smoking of meat and fish are closely interlaced and just seem to go together—i.e., cured meat is commonly smoked and smoked meat is commonly cured. The smoking of meat is also difficult to separate from cooking, because heat was traditionally applied with the smoke. However, the applications of smoke and heat are not necessarily closely allied, as smoke and heat can be applied separately.

Smoking, like curing, has a preservative effect on meat. But, it is only along the surface areas, and its preservative powers are negligible.

The primary purposes of smoking meat are: (a) The development of flavor and (b) The development of color.

There are over 200 different compounds found in wood smoke. The most important components of wood are generally considered to be phenols, acids, alcohols and hydrocarbons. A more complete understanding of the chemistry of wood smokes can be found in the scientific journals.

TABLE 10

Semi-dry Sausage Formulations—Cervelat[a]

Ingredient	Standard formula percentage	Soy protein variation percentage
Beef 30% fat	66.95	52.95
Pork trim 55% fat	19.13	19.13
Pork hearts 5% fat	9.56	9.56
Salt	2.86	2.86
Sugar	0.96	0.96
16 mesh black pepper	0.25	0.25
Sodium nitrate	0.12	0.12
Whole black pepper	0.12	0.12
Onion powder	0.05	0.05
Washed TVP		5.00
Isolated soy protein		1.50
Water		7.50
Totals:	100.00	100.00

[a] Processing instructions: Grind beef and pork hearts through a 0.25 inch grinding plate. Grind pork trimming through a 0.375 inch grinding plate. Mix all ingredients except whole black pepper until homogenous. Regrind through 0.125 inch grinding plate. Mix in whole peppercorns. Transfer to pans (ca. 8 inches deep) and hold at 38–42°F for 48 to 72 hours. Remix and stuff into No. 2.5 fibrous casings. Hang in drying room for 36 hours at 55°F. Place in smokehouse for 24 hours at 80°F. Slowly increase temperature to 115°F and hold for 6 hours to insure good color development. Cool at room temperature for 2 to 3 hours, then transfer to holding cooler. (If trichina-free trimmings are not used, 137°F must be reached during smoking.)

Smoking and Cooking

It is impossible to determine when man began to smoke meat or fish products. Probably, it was done accidentally and not by design. The nomadic hunter or fisherman probably left his day's catch or capture near the fire he used to warm himself and noticed that the burning wood imparted a flavor that was pleasant and tasty. He probably didn't even notice that the crudely smoked products did not spoil as readily as they did before. He liked the taste, so he continued the process. Regardless of how it hap-

In today's modern world we no longer refer to meats and fish that are heat treated as smoked or cooked. We now say they are "Thermally Processed." Thermal processing can be accomplished in a number of ways: water immersion (boiling), direct contact heating (frying), water molecule activation by electronic impulse wave (microwave), radiation heating (broiling), and air processing, which is a combination of convection and conduction. For this time and place we will narrow the subject matter by dealing primarily with "air" as the transfer medium in thermal processing.

Although air is not the most efficient medium for thermal transfer, it affords us the only medium in which the other processes can be incorporated easily. A quick examination shows us that: (a) Air temperature can be closely controlled between the required temperatures. (b) The water content of the air, within these temperatures, can be varied. This is called "Relative Humidity." By controlling the relative humidity at different temperatures, we can control specific effects that will develop the character of our products. (c) When required, additives can be diffused in the air, such as wood smoke, liquid smoke, and even some of the curing agents.

A good "Thermal Processing Chamber" must have a certain velocity and volume to achieve quick and even product processing. The velocity determines how far and at what speed the air will travel prior to losing its direction and power. The volume of air will determine the amount of heat and moisture transfer medium available to perform the required function. The "Thermal Processing Chamber" must be properly balanced for volume and velocity to avoid hot or cold spots, and process the product evenly and efficiently.

Humidification: Of all the values used in meat processing, the most misunderstood and most misused is humidity. Relative humidity is the percentage measurement of moisture present in the air as compared to the maximum moisture that can be contained in the air at a specific temperature. The higher the temperature, the greater amount of moisture that can be absorbed by the air. Any given relative humidity percentage must be mentioned with the air temperature to have any meaning at all.

The wet bulb temperature is the measurement in °F or °C of the air temperature when a wet sock is placed over the bulb of the thermometer. The evaporation of the water from the wet sock into the air cools the thermometer, resulting in a reading lower than the air temperature. The drier the air, the faster the evaporation, the greater the cooling effect, and the lower the temperature reading.

A comparison between the air temperature (dry bulb) and the evaporation temperature (wet bulb) on a prepared scale gives the relative humidity in percentage at that specific temperature. Air blowing at a specific velocity (2 m per sec) must be blowing at or around the wet bulb to give a correct reading.

Proper use of humidity controls texture, yields, and color, as well as shelf-life.

The Basic Steps for Finishing Meat Products

Reddening, preheating and equalization: This process is used to prepare product and the oven environment for processing. During this step, a gentle heating of the product is required to drive off the surface moisture, and equalize the temperature and humidity of the processing air and the product surface. The oven or house must have slow air movement, closed dampers, temperature required for the product (75–130°F) and, usually, no humidity added.

Hot air processing, baking, roasting: This process is to heat product with fast circulating air at any appropriate temperature and humidity with maximum BTU's transferred. The amount of humidity will determine the amount of moisture removed from the meat. It will also control the rate of skin formation and color. Some solid meats are processed this way. The oven or house must have fast air velocity, closed dampers, temperature as required and humidity as required.

Drying: Drying utilizes heat, air velocity and humidity control to remove moisture from the product. Any temperature can be used, as well as humidity of the air used to accomplish the desired result. Low humidities with elevated temperatures will remove surface water quickly and set the protein. Higher humidities with elevated temperatures will result in slower surface set and faster product dry-out. Low humidity with lower temperatures will remove surface water from product without excessive surface protein set. This process is used to pre-dry the surface of casing products (remove excess moisture and water droplets), to surface dry solid meat items which are too wet to apply smoke to, and to dry products which are to have a reduced moisture content. The oven or house must have (usually) fans at high speed, dampers open and temperatures as needed.

Drying (Gentle Mode): This process utilizes heat and a gentle air movement for drying of product and is also used to exhaust air saturated with steam or smoke outside. For dry or semi-dry sausage, this process gives the best results. The slow air movement will not remove surface moisture so rapidly as to cause case hardening that would prevent moisture being removed from the center. The oven or house must have slow air movement, dampers open or automatic, temperature as required for fermentation and humidity as required for fermentation.

Cooking, steaming and scalding: This process utilizes low pressure steam (4–7 psi) injected directly into the chamber for heat processing at any desired temperature. The low pressure steam is mixed with the chamber air and circulated throughout the oven. The amount of steam is controlled by the desired temperature setting. The air is saturated with humidity to the highest point.

This process is used to cook such items as corned beef, hams, and the larger types of sausages.

Steps for problem-free heat processing: (a) Fill the smoke wagons, trolleys or cages full, but avoid overcrowding, or touching, and allow space for adequate air movement. (b) Make sure the products are well stuffed. Loose or unevenly filled casings, or overfilled casings, will result in a poor product. Underfilled product causes excessive shrinks, while overfilled product will burst. (c) Avoid overstuffing cold meats. Meats will expand slightly before normal processing shrink occurs. Cold meats stuffed as tightly as warmer products will burst. (d) Unevenly cured products will cause color variations, as will products that have large temperature ranges or uneven surface moistures. A low temperature staging step will help prevent these problems.

References

1. Doerr, R.C., A.E. Wassermen and W. Fiddler, "The Composition of Hickory Smoke."
2. Kadane, V.V., *J. Am. Oil Chem. Soc.* 56:330.
3. Kotaula, A.W., G.G. Twiggs and E.P. Young, Project Oerder Amxred 72-92 Food Laboratory, U.S. Army Natick Laboratories, Natick, Massachusetts.
4. Kramlich, W.E., A.M. Pearson and F.W. Tauber, "Processed Meats," Avi Publishing Co. Inc., Westport, CT, 1982.

5. Kwasny, S., "Canned, Sectioned and Formed Ham Pieces," Griffith Labs, 1964.
6. Mass, R.H., "Fresh Meat Technology," Oscar Meyer Inc., 1963.
7. Rust, R.E., "Basic Chemistry of Meat Emulsions," in proceedings of Second Annual Sausage and Processed Meats Short Course, Iowa State University, Ames, Iowa, 1980.
8. Siegal, D.G., W.B. Tuley, H.W. Norton and G.R. Schmidt, *J. Food Sci. 44* (1979).
9. Young, L.S., "Soy Protein Products in Processed Meat and Dairy Foods," presented at World Soybean Research Conference, Ames, Iowa, 1984.
10. Young, L.S., G.A. Taylor and A.T. Bonkowski, "Use of Soy Protein Products in Injected and Absorbed Whole Muscle Meats," ACS Symposium Series, Plant Proteins Applications, Biological Effects and Chemistry, American Chemical Society, Washington, D.C., 1986.

Vegetable Protein Foods in Korea

Seung Ho Kim and Tai-Wan Kwon

Korea Food Research Institute, PO Box 131, Chongryang, Seoul, Korea

Abstract

Total consumption of soybeans for traditional foods in Korea is about 380,000 metric tons per year, showing only a slight increase over the last few years. However, soy curd (*doo bu*) and soy milk (*doo yoo*) showed rapid growth. The level of consumption of vegetable proteins in Korea is very low; however, they show an overall annual increase of about 15% recently. Three distinctively Korean vegetable protein foods, hot soy paste (*ko chu jang*), a kind of soy paste (*chung kuk jang*) and soybean sprout (*kong na mool*), along with more general vegetable protein foods, soy paste (*doen jang*) and soy sauce (*kan jang*) are described in some detail.

Korea, like other Oriental countries, has a long history of utilizing soybeans as essential protein foods in large quantities. Most of the protein foods such as soy sauce (*kan jang*), soy paste (*doen jang*) and soy curd (*doo bu*) are much like or identical to those from other Oriental countries, perhaps with the exception of hot soy paste (*ko chu jang*), a kind of soy paste (*chung kuk jang*) and soybean sprouts (*kong na mool*). A relatively new development is the use of vegetable proteins as ingredients for various food formulations, e.g., hamburgers, sausages, hams, fish pastes, dairy products, pastries and cookies in Korea, but only to a very limited extent.

Situation of Traditional Protein Foods

Table 1 shows the domestic production and import of soybeans in Korea and Table 2 shows how soybeans are consumed in the form of traditional foods. Soybeans produced

TABLE 1

Domestic Production and Import of Soybeans in Korea

	1,000 metric tons			
	Domestic production	Import		
		Food uses	Feed uses	Total
1982	233	104	479	816
1983	226	152	506	884
1984	253	119	603	975
1985	234	123	745	1102
1986	198	163	823	1184
1987	250	144	936	1330

TABLE 2

Consumption of Soybean as Traditional Korean Foods[a]

	1,000 metric tons					
	1982	1983	1984	1985	1986	1987
Soy sauce	69	67	68	69	71	70
(*Kan jang*)	(50)	(49)	(50)	(51)	(52)	(51)
Soy paste	51	49	47	49	50	50
(*Doen jang*)	(41)	(36)	(37)	(41)	(41)	(40)
Hot soy paste	5	5	5	5	5	5
(*Ko chu jang*)	(4)	(4)	(4)	(4)	(4)	(4)
Soy curd	60	60	83	96	93	114
(*Doo bu*)						
Soy milk	7	10	12	14	13	14
(*Doo yoo*)						
Others[b]	145	187	157	123	129	141
Total	337	378	372	356	361	394

[a] Figures represent soybeans consumed, not foods consumed. Figures in parentheses are for home-production.
[b] Others include home-produced soy curd (*doo bu*) and soy milk (*doo yoo*), *chung kuk jang* (a kind of soy paste), *choon jang* (a kind of soy paste described in the text), soybean sprouts (*kong na mool*), and direct consumption of soybeans.

domestically are used primarily for home-production of soy sauce, soy paste, hot soy paste and others as listed in Table 2. During recent years, the domestic production and import of soybeans for food uses decreased slightly and increased by a slightly larger extent, respectively. The resultant total consumption of soybeans for traditional foods has increased slightly for the past several years, and it now stands at about 380,000 metric tons per year.

A relative newcomer, soy milk (*doo yoo*), which shows a rapid increase in production in the last few years is also included in the table, although Koreans have enjoyed home-made soy milk for years. Soy paste, soy sauce, soy curd (*doo bu*) and other traditional soybean foods such as soybean sprouts (*kong na mool*), *choon jang* (soy paste used as sauce for Chinese-style noodles, *zha zang myon*) *kong ja baan* (cooked, seasoned whole soybeans) and *baap mit kong* (whole soybeans cooked with rice) have played important roles in the diets of Koreans as protein sources. Surprisingly, commercial production of soy paste, soy sauce and hot soy paste remained more or less the same for the past several years, contrary to our expectation that commercial production would replace home production as family income and product qualities increase.

The increase in total consumption is due to the growth of the soy milk market and, in particular, of the soy curd market (from 60,000 metric tons in 1982 to 114,000 metric tons in 1987 expressed as soybeans) and to consumers' preference for high-protein foods as income level increases. In the same time span, the meat and poultry markets enjoyed a spectacular growth of 48%, which is reflected in the rapid increase of soybean imports for feed uses (Table 1). Total consumption of meat and poultry was 654,000 metric tons in 1987.

This raises one important question to consumers and producers of vegetable protein foods: shall we let animal proteins take a major share of protein requirement for humans? This is especially important to Koreans, as rapidly increasing affluence began to make animal proteins more affordable than before. In the authors' opinions, we must stop this prevailing trend of animal protein preference of consumers before it becomes a habit like it has done to most Westerners, considering the ill implications of animal fats in coronary diseases. Korea has an advantage in that Koreans are accustomed to vegetable protein foods. Advocating soybean foods as a good source of protein and oil, enhancement of qualities of existing vegetable protein foods and development of new vegetable protein foods will certainly help revert the trend.

Situation of Vegetable Proteins

The current situation of vegetable protein production and import is shown in Table 3. This industry is at its infancy; only a few types are produced domestically with the rest imported. Their level of consumption still is very low in comparison with traditional foods.

Textured vegetable proteins from soybean are mainly used for hamburger, *maan doo* (similar to the Chinese snack food, *wan tan*) and soup base of ramyeon (instant fried noodles). This market is growing rapidly (about 20% yearly). Textured vegetable proteins from wheat gluten are used mainly as vegetarian foods and their market is not expected to grow much. Most of the wheat gluten is vital gluten and its main usage is in hamburger patties and as dough strengtheners of breads. Domestic production of

TABLE 3

Domestic Production and Import of Vegetable Proteins in Korea

		Estimated for 1987 (MT)
Soybean protein	Textured vegetable protein[a]	4,500
	Soy protein isolate[b]	1,800
	Soy protein concentrate[b]	150
Wheat protein	Textured vegetable protein[a]	900
	Vital gluten[a]	150
	Gluten (wet & dry)[b]	409
Total		7,909

[a] Domestic production.
[b] Import.

vital gluten is trying to replace the imported ones. Main uses of soy isolate and concentrate are in sausages, hams, ice creams and breads and this market is growing 10–15% annually.

Overall, the use of vegetable proteins as ingredients in processed foods has been low, but is increasing at a rapid pace. Current low usage reflects the low percentage (about 37%) of manufactured foods out of total food consumption. Prompted by the rapid growth of Korean food industry (about 8% annually), the need for vegetable proteins is expected to grow continuously and consequently their domestic production will be encouraged.

Since usage of vegetable protein as ingredients in processed foods is quite new to Koreans of conservative eating habits, manufacturers of vegetable proteins should initiate efforts in terms of labelling, standardization, education (of consumers and food processors) and government contact before significant and continuous growth in demand for vegetable protein can be realized. The government should enforce the yet-to-be established Korean Standard on vegetable proteins which will force the manufacturers to further improve the quality of the products. Present quality of vegetable proteins must be improved for the ultimate expansion of their demand, which requires ingenuity of and cooperation among scientists all over the world.

Traditional Protein Foods in Korea

Soy Sauce *(Kan Jang)* and Soy Paste *(Doen Jang):* The traditional fermentation processes for soy sauce, soy paste and hot soy paste in Korea are characterized by the use of *meju* as a starter for the fermentation. *Meju* is a solid matter in a form of rectangular block with its weight of around 1 kg each. *Meju* is made in autumn by allowing fermentation of steamed soybean mash on which molds are grown on the surface and bacteria are inoculated inside during the drying process in the air. The typical microorganisms found in *meju* are *Aspergillus oryzae*, *Aspergillus sojae* and *Bacillus subtilis*. Traditionally, homemade soy sauce and soy paste are obtainable simultaneously from the *meju*-brine mixtures after allowing further fermentation and ripening processes for several months in earthen jars. The supernatant dark brown liquid of the mixtures is soy sauce and the brownish solid residue is soy paste. Today, however, they are manufactured separately by industry, which uses a starter culture mass (*koji* in Japan) instead of *meju*. There has also been separate production of soy paste at home. Soy sauce manufactured by industry is suitable for most purposes but not for soup preparation, although industry imitates soy sauce for soup preparation by blending fermented soy sauce and chemical soy sauce.

The Korean government recognizes three types of soy sauce: fermented soy sauce, chemical soy sauce and blended soy sauce (blend of fermented soy sauce and chemical soy sauce). Blended soy sauce is the most popular in Korea. Manufacture of fermented soy sauce in industry generally follows that of Japan, the main type being *koikuchi* type. Korean soy paste is mainly soybean soy paste, which is different from the Japanese preference for rice soy paste.

Hot Soy Paste (Ko Chu Jang): Hot soy paste is a unique fermented product popular only in Korea, reflecting the hot spice preference of Koreans. It has a little of soy paste flavor and a combination of sweet, brothy, salty and hot taste. It is used as a soup base, usually used together with soy paste, and also as seasonings for various types of foods.

Hot soy paste can be classified into three major types on the basis of the raw materials used; that is, rice hot soy paste, barley hot soy paste and wheat hot soy paste. Among these, rice hot soy paste is the most popular, followed by barley hot soy paste.

The manufacturing method is shown in Figure 1. Rice is soaked for about 5 hr, drained and ground, whereas barley is ground as is and wheat flour is used as is. To the respective ground carbohydrate source, water is added while stirring (about 2–3 times water for rice and about 3–4 times water for barley and wheat). The suspension is cooked until gelatinization is complete, cooled to about 75°C and ground starter culture mass from soy sauce manufacture (30–40% of the carbohydrate source) is added while mixing to the final temperature of 60–65°C. The mixture is fermented for 3–5 hr at 60°C. To the fermented mixture, salt and powdered red pepper (both about 20–30% of the carbohydrate source) are added and the mixtures are aged at 30°C for about one month. The duration of aging is dependent upon the salt concentration.

Chung Kuk Jang (A Kind of Soy Paste): Chung kuk jang is another unique product popular only in Korea. Its uniqueness might suffer because there is a similar product, *natto*, in Japan. In fact, *natto* is further processed to *chung kuk jang*, i.e., *natto* is mixed with seasonings, ground and aged to make *chung kuk jang*.

The manufacturing method of *chung kuk jang* is shown in Figure 2. The precooked soybeans are cooked until tender, drained, cooled to 40°C, inoculated with a water suspension of *Bacillus natto*, put in a wide bowl, and incubated at 40–43°C for 12–20 hr. The *naap doo* (*natto*) produced has a characteristic odor and musty flavor and a slimy appearance since it is covered with a viscous and sticky polymer of glutamic acid produced by the microorganism. In Japan, *natto* is eaten with soy sauce and mustard, and often is used for breakfast and dinner along with cooked rice. In Korea, *naap doo* is further processed; it is mixed with salt and seasonings, ground, put in a big jar, and aged at a temperature below 30°C for about a month. The *chung kuk jang* produced is used as a soup base in Korea.

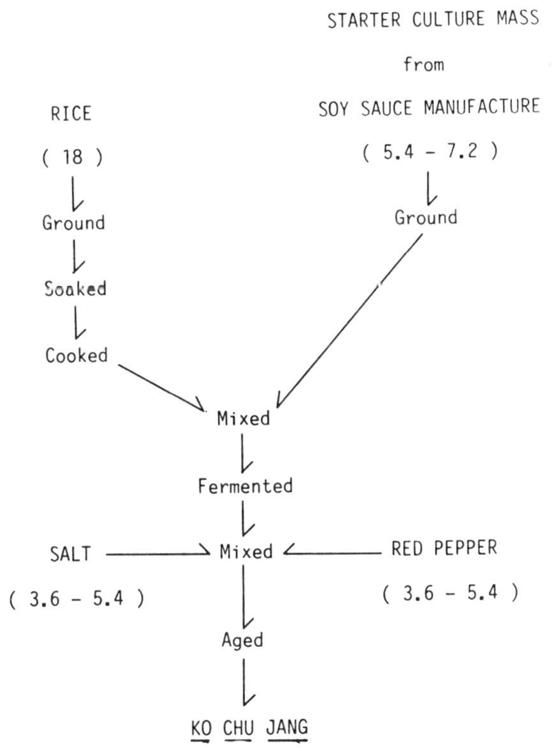

Fig. 1. Manufacturing method for rice hot soy paste (ko chu jang). Numbers represent typical ratio of ingredients in volume.

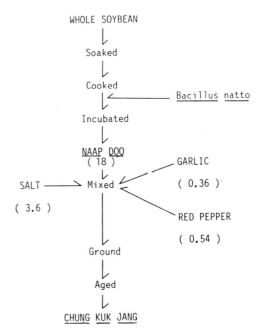

Fig. 2. Manufacturing method for *chung kuk jang* (a type of soy paste). Numbers represent typical ratio of ingredients in volume.

Kong Na Mool (Soybean Sprouts): Kong na mool is another soybean product consumed only in Korea in a large quantity. It is used for preparation of soup and *na mool* (cooked and seasoned soybean sprouts). Soybean sprouts resemble mungbean sprouts, but the cotyledons of soy sprouts are more distinct because of their large size, yellowish color and beany odor.

Soybean sprouts are a good source of vitamins as well as proteins (Table 4). And it has no so-called "flatulence factors" (stachyose and raffinose) since they are converted to vitamin C during sprouting. At the same time, trypsin inhibitor activity is decreased and protein digestibility is increased. The usual cooking practice of heating soybean sprouts in salt solutions reduces the thermal loss of heat-labile vitamins such as riboflavin and vitamin C. During cooking, it is desirable to minimize heating to maintain the inherent crisp texture and distinct taste and to minimize the destruction of the vitamins.

Selection of suitable soybean varieties for sprout production is important since different varieties vary in germination rate, growth rate and productivity. Koreans prefer sprouts made from small-size soybean varieties.

Impurities such as soil, straw and sand are removed from soybeans by washing them with water. The washed soybeans are soaked for germination. Generally, soaking is done for 4–6 hrs at 20–25°C. Soaking is accompanied by frequent stirring to incorporate air for promotion of germination. Longer than optimum soaking time reduces subsequent growth rate because of excessive metabolism. It takes about 24 hr for germination to occur after soaking is completed.

The germinated soybeans are covered in a dark room to avoid light and to prevent evaporation. They are grown for about 6 days with the temperature kept generally below 20°C. Frequent spraying of water, 100–150 times as much water as soybeans, is necessary not only to provide water for growth but also as a coolant. More frequent water spraying is needed during the early stage of growth when the respiration rate is highest. Less than optimum amount of sprayed water and spraying times raises the temperature of growing soybean sprouts so that their putrification due to microbial contamination occurs and the growth is retarded.

Control of microbial problems is critical in soybean sprout manufacture and hygienic conditions are maintained by such practices as treating germination vessels with formalin, holding containers and equipment in boiling water, and use of hypochlorites for water and equipment treatment. Currently, traditional and primitive equipment are employed for most of the small-scale Korean soybean sprout manufacture with excessive labor requirement and unstable production. Only a small number of larger scale manufacturers operate fully automated equipment. There is also small-scale, fully-automated equipment for household use.

TABLE 4

Nutritional Composition of Soybeans and Soybean Sprouts (per 100 g)

	Soybeans	Soybean sprouts
Water (g)	6.2	90.2
Protein (g)	41.3 (44.0)	4.2 (42.9)
Lipid (g)	17.6 (18.8)	1.0 (10.2)
Carbohydrate (g)		
Non-fiber	21.6 (23.0)	2.9 (29.6)
Fiber	3.5 (3.7)	0.5 (5.1)
Ash (g)	5.8 (6.2)	0.8 (8.2)
Calcium (mg)	127 (135)	32 (327)
Phosphorus (mg)	490 (522)	49 (500)
Iron (g)	7.6 (8.1)	0.8 (8.2)
Vitamin A (IU)	10 (11)	175 (1786)
beta-Carotene (μg)	6 (6.1)	105 (1071)
Thiamine (mg)	1.03 (1.10)	0.15 (1.53)
Riboflavin (mg)	0.30 (0.32)	0.13 (1.33)
Niacin (mg)	3.2 (3.4)	0.8 (8.2)
Vitamin C (mg)	—	16 (163)

Figures in parentheses give data on a dry weight basis.

Dealing with the Shortage of Protein Resources in China

Zhu Xiang-Yuan

Research Institute on Nutrition Resources, People's Republic of China

Abstract

This paper reviews Chinese plans for development of its food industry with special emphasis on the role that vegetable protein, which has been a part of the traditional diet, will play in future development of the food industry.

The original topic of my paper was "Assessment and development trends of Chinese food industry." but now I would like to change it to "Dealing with the shortage of protein resources in China." Everyone knows that China is a developing country and the nutritional standards should be improved. According to a national survey of nutrition conducted a few years ago, the average amount of caloric energy each person takes in per day had reached 2,485 kcal, but protein was only 67 g, with 12% animal protein. Therefore, the most important task in China is increasing the protein supplies, especially animal protein. By the end of the century we hope that the protein/person/day can be increased to 70 g with 22% animal protein.

In China, 95% of the animal protein mainly comes from the east agricultural region by changing from grain and protein feeds. By now, the feed industry in China has developed greatly, but the shortage of protein feed is still a serious problem. It is estimated that a 13 million ton shortage of protein feed will appear in China by the end of the 20th century. Both food and feed industries shall be faced with a lack of protein. The shortage of protein has become the general difficulty of China's food industry and animal husbandry.

The unreasonable crop planting structure results in the shortage of protein resources in China. In order to show this problem, I advanced the concept of C-N ratio in the grain industry in 1985. China's grain output consists of the following five categories: rice, wheat, coarse grain, various potatoes and soybean. The first four categories which are rich in carbohydrates are defined as nutritional carbon sources, and the last one, containing a lot of protein, as a nutritional nitrogen source. The ratio between the output of the first four categories and the last one is defined as the C-N ratio in the grain industry. According to this concept, the C-N ratio is about 100:18 in international grain trade, 100:18 in USA, 100:14 in Japan and 100:6 in the world except for China. To date, China's C-N ratio is only 100:3; in 1936 it was 100:8. Those figures indicate that the percentage of soybean is too low in China's crop planting structure to gain a balanced C-N ratio, resulting in a shortage of soybeans. In 1936, the total grain output was 150 million tons including 12 million tons of soybean.

At that time, China's soybean output was in first place in the world. But now, it has become third with a soybean output of 12 million tons although the total grain output has reached 400 million tons. By the way, China's soybean output was below 10 million tons for a rather long period. China must adjust the crop planting structure and increase soybean output quickly. It is estimated that China's total grain output will reach 480 million tons by the end of this century, and soybean output should be 36 million tons keeping the C-N ratio at 100:8 as in 1936.

We have proposed that our government enlarge the soybean planting area, increase its output and limit the exports of soybean and soybean meal. In the next few years, we should import several million tons of soybean in order to reduce the shortage of protein.

In 1984, I had advanced a "4F" sequence, which could help us to make a food use of the present protein resources. The "4F's" are the following: 1st, foodstuff; 2nd, feedstuff; 3rd, fuel; and 4th, fertilizer.

Various agricultural resources which contain protein should be utilized in this order as far as possible. Do not use a material as feedstuff if it is suitable for foodstuff. For example, fish should be used as foodstuff, because it is much better than using fish meal as a feedstuff. The material which can be used as feedstuff shouldn't be used as fuel or fertilizer and so on. In China, 700 million tons of dry grass and straw are used as fuel and fertilizer each year. Their heat energy utilization ratio is only about 17% when we make methane from grass and straw. If we do some pretreatment of the dry grass and straw then use the material as feedstuff so that part of nutrition of the feedstuff can be absorbed by animals, the drained waste of the animals can be used to make methane, thus the heat energy is not decreased.

China processes about seven million tons of cottonseed meal and rapeseed meal every year. Most of these meals are used as fertilizer, even though 6 kg of these only equal 1 kg of chemical fertilizer. If they were used to feed animals after processing, the 20% protein in them could be turned into animal protein and the animals' drained waste placed in generating pits still contains a lot of nitrogen, phosphorus and potassium which are very good fertilizer. Utilizing the agricultural resources in the "4F" order is a good way to save vegetable protein resources and to increase animal protein resources. This ecological agriculture is spreading now in China. Besides the way mentioned above, China has been engaged in single cell protein (SCP) research. Our institute is studying methane producers for SCP. But up to now, they have not been used yet in industrial production, because of higher cost which cannot compete economically with soybean meal.

Novel Traditional and Manufactured Soy Foods in Japan

Hitoshi Taniguchi

Fuji Oil Co. Ltd., 1-sumiyoshi-cho, Izumissano-shi, Osaka FU 590, Japan

Abstract

In Japan, not only many kinds of processed soybean foods such as tofu, aburage, ganmodoki and uba, but also various kinds of fermented soybean foods, such as miso, natto and shoyu, have been made since ancient times. For the past 30 years, we have produced isolated soy protein (ISP) with many functional properties. We have sold them as food improvers. At the same time, we have tried to develop novel traditional and manufactured soybean foods successfully by utilizing ISP and newly developed textured soy protein. This paper presents the outline of the recipes, processes and advantages and philosophy of these new soy protein foods.

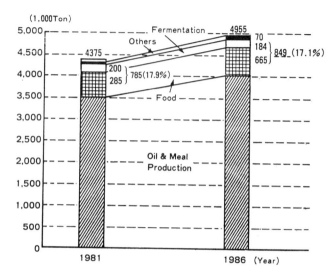

Fig. 1. Trends in annual consumption of soybeans by use.

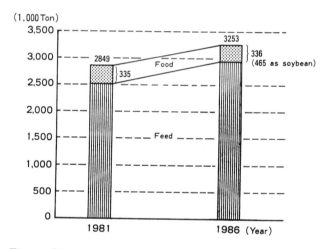

Fig. 2. Trends in annual consumption of soybean meal for feed.

I would like to discuss the recent situation of novel traditional and manufactured soy foods in Japan, which are produced by utilizing new soy protein isolate (SPI), soy protein concentrate (SPC) and textured soy protein (TSP) and its modified materials. Thus, while reviewing the functional properties of these soy processed raw materials, I would like to explain firstly, what kind of functional properties of the raw materials have been utilized to make these novel traditional and manufactured soy foods; secondly, what kind of properties come out consequently as the result of utilizing the raw materials; thirdly, how these new soy foods have been developed and accepted in the market; and finally, what kind of philosophy there has been, and so on.

New Soy Processed Materials in Soy Bean Consumption

The annual consumption of soybeans in Japan in 1987 was 4,955,000 metric tons, that is, about five million tons (Fig. 1). Of that quantity, 184,000 tons were used directly for fermented foods such as miso and natto, and 665,000 tons for processed foods such as tofu with a total quantity of 849,000 tons. This figure is greater by 64,000 tons than that of five years ago, but the ratio to the overall soybean consumption showed little change, moving from 17.9% to 17.1%.

In Figure 2 the amount of defatted soy bean meal which was used for soy sauce and new soy protein products was 336,000 tons. This figure is equal to 465,000 tons when calculated as whole soybeans, which corresponds to 9.4% of the overall consumption of soy beans for the year.

In Figure 3 all added figures which I have mentioned above total 1,314,000 tons. In other words, 26.5% of the overall consumption of soybeans was for food use; there is no other country in the world in which such a large amount of soy beans is used for food purposes.

Contemporary Soy Food Products

I would like to discuss various aspects of the popularization of new soy protein products and soy protein foods, so I am not going to say anything about traditional foods such as miso, soy sauce and natto.

The production of new soy protein products was initiated some 20 years ago, and standards for these products were established on the basis of the Japanese Agricultural Standards in 1976, when their consumption had become widespread. Figure 4 shows the trend of annual production of soy protein products for the decade after the establishment of the standards. For the past decade, the annual production increased steadily, and it reached 36,000 tons last year. Almost all of this amount is in the form of dry products, and this is equal to 140,000 tons when calculated in terms of hydrous products such as meat. This figure can be compared with one third of the annual production of ham and sausage products in Japan.

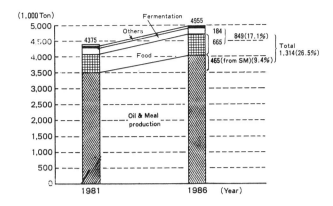

Fig. 3. Trends in annual consumption of soybean meal by use.

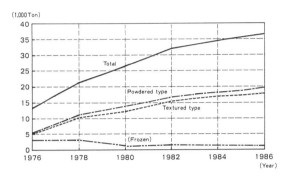

Fig. 4. Trends of annual production of soy protein products.

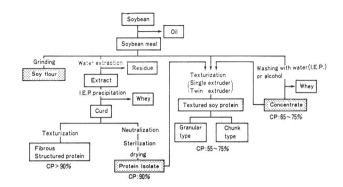

Fig. 5. Manufacturing process of soy protein products.

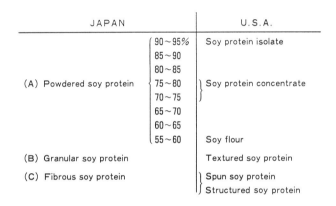

Fig. 6. Comparison of classification of soy protein products by Japanese Agricultural Standards and in United States.

The ratio of powdered protein products to textured ones is nearly 1 to 1, and as for the classification of the soy protein products, I will explain that in detail later on. The main product of powdered protein is soy protein isolate (SPI), and textured proteins are mainly of the granular type. However, the production of products of the nugget type has recently become noticeable.

About 100,000 tons of whole soy beans are now used for the production of these new soy protein products; this amount forms about 2% of the annual overall consumption of whole soy beans.

Classification of New Soy Protein Products in Japan

As shown in Figure 5, in Japan new soy protein products include soy flour (SF), soy protein concentrate, soy protein isolate, textured soy protein, fibrous soy protein, and others, that is, the new soy protein products which are produced in Japan are similar to those which are produced in other countries including the United States. However, the classification system and uses of these products in Japan are rather different from those in other countries.

In Japan, the classification of new soy protein products is based on the form of the products; the new soy protein products are classified into three groups, namely, powdered soy protein, granular soy protein and fibrous protein, and it is required that labelling of each product shall bear a protein content statement in 5% steps. Consequently, the terms SPI, SPC, SF, etc. do not appear in the JAS statistics, although these terms are used commercially. Figure 6 summarizes the relationship between the Japanese classification system which is based on the JAS and the American system. It is thought that the Japanese classification system is more convenient from the point of view of nutrition, because it presents clear information about protein contents.

Utilization and Popularization of New Soy Protein Products

It is necessary to separate this point into two categories: first, powdered soy protein products, specifically SPI, and second, textured soy protein products.

Utilization and popularization of soy protein isolate: Table 1 summarizes the relationship between the functional properties of various soy protein preparations such as SF, SPC, SPI and soy protein hydrolyzate, and their uses.

As is shown in this table, these new soy protein products have a wide variety of functions, and they are used mainly to improve food qualities. For example, SPI and SPC are used for sausage, soup, etc. on the basis of their emulsifying property; for soup, gravies, etc. on the basis of their viscosity: for "kamaboko," "chikuwa," etc., on the basis of their gelating property: for "gyoza," "shumai," bread, cake, etc. on the basis of their water absorbing property; and for TVP, etc. on the basis of their texturizing property. However, the amount of these new soy protein products shall not exceed the maximum levels prescribed for each soy protein product by JAS.

TABLE 1

Functional Properties Performed by Soy Protein Preparations in Actual Food Systems[a]

Functional property	Preparation used	Food system	Mode of action
Solubility	I,H	Beverages	Protein solvation, pH dependent
Emulsification	I,C,F	Sausages, age, gammo, soup, cakes	Formation and stabilization of fat emulsions
Fat adsorption	I,C,F	Meats, sausages, donuts	Binding of free fat
Viscosity	I,C,F	Soups, gravies	Thickening, H_2O (binding)
Gelation	I,C	Sausage, kamaboko, chikuwa, age, gammo, cheese	Protein matrix formation and setting
Water absorption and binding	I,C,F	Gyoza, shumai, sausages, breads, cakes	Hydrogen bonding of H_2O, entrapment of H_2O, no drip
Elasticity	I	Kamaboko, chikuwa, bakery	Disulfide links in gels deformable
Film formation	I	Yuba, edible film	
Cohesion-adhesion	I,C,F	Gyoza, shumai, hams, sausages, baked goods, pasta products	Protein acts as adhesive material
Texturization	I,C,F	T.V.P., karaage	
Flavor-binding	I,H,C	Simulated meats, bakery	Adsorption, entrapment, release
Foaming	I,H	Whipped toppings, chiffon desserts, angel cakes	Forms stable films to entrap gas
Color control	F	Breads	Bleaching of lipoxigenase

[a] F, C, I, H denote soy flour, concentrate, isolate and hydrolyzate respectively.

The utilization of soy protein isolate in Japan is also unique in that it has been used as a raw material for soy protein foods to produce entirely new foods as one of novel traditional and manufactured soy foods groups. As examples of such new foods, we may make mention of new age (abura-age), new gammo, new soy protein foods, frozen desserts, nutritionally balanced cakes, sashimi tofu, infant formulas, protein drinks, and mayonnaise, and now I would like to describe new age and new gammo as representative examples.

Figure 7 shows the differences in preparation procedures for gammo and age between the traditional process and the new process which heat rather than coagulate the SPI emulsion. The traditional process requires 10 to 15 hours to finish the products, while the new process requires at most only one hour. In addition, products obtained by the use of the new process do not change in texture even when they are frozen. As a result, it has become possible to market such products as frozen foods *via* the modern distribution system on a large scale. New age has also long been used as a garnish for instant Chinese noodles, as it is easy to use for instant foods because its freeze dried texture is excellent in shape-recovering quality in hot water.

Figure 8 shows a procedure of preparing new soy protein foods which employs this new process. New soy protein foods which range from those which are composed mainly of soy protein, such as gammodoki and abura-age,

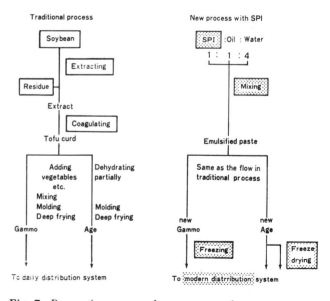

Fig. 7. Preparation process of new gammo and age.

to those which use soy protein and meat, fish paste or vegetables in combination. Products such as vegetable age and white loaf, which are marked with an asterisk in Fig. 8, are now produced on a large scale, in a short time,

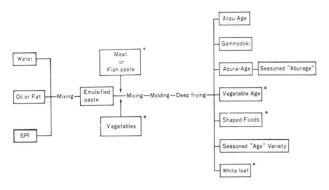

Fig. 8. Preparation process of new soy protein foods. Asterisk indicates products in large-scale production.

and under sanitary conditions by the modern process. As a result, the traditional process, which is classified as a household industry in terms of production scale, has been improved greatly.

Many other new foods have been produced due to the advent of SPI, and now I would like to choose five representative examples and show their formulas, i.e., in Table 2, formula of frozen dessert; in Table 3, formula of 'sashimi tofu'; in Table 4, formula of an infant formula; in Table 5, formula of a protein drink; and in Table 6, formula of a nutritionally balanced cake.

TABLE 2
Frozen Dessert (Sour type, Yogurt type)

Fermented SPI liquid	40.0%
Saccharide (sugar, dextrin)	24.0
Vegetable oil	3.0
Fruit juice	2.0
Stabilizer emulsifier	0.7
Flavor	0.2
Water	30.1

TABLE 3
Sashimi Tofu

Soy isolate emulsion (1:1:4)	50 parts
Surimi paste	50
Water	30
Seasoning	a proper quantity

TABLE 4
Infant Formula

Soy protein isolate	2.2%
Water	86.4
Saccharides	5.4
Vegetable oil	3.4
Minerals	a proper quantity
Vitamins	a proper quantity

TABLE 5
Protein Drink

Soy protein isolate	4%
Vegetable oil	3
Saccharides	8
Water	85
Minerals	a proper quantity
Vitamins	a proper quantity

TABLE 6
Nutritionally Balanced Cake

Seasoned SPI	4.6%
Vegetable margarine	23.0
Vegetable cheese	10.0
Wheat flour	40.0
Sugar	12.0
Egg	6.0
Non-fat dry milk	2.0
Seasoning	2.4

Utilization and popularization of textured soy protein: As another group of novel traditional and manufactured soy foods has been born by the improvement of textured vegetable protein and unique product development technology, I would like to introduce to you the examples as follows:

The production of textured soy protein was initiated some 20 years ago, utilizing machines, of which the extruder of Wenger International, Inc. is a representative example. At first, the textured soy protein products ranged from hydration that was nearly equivalent to that of meat, so it appeared that textured soy protein would be developed in the form of imitation meat. However, almost all but granular type products faded out because of incomplete texture, soybean odor and a lack of taste. Until 1986, mainly granular type products alone had been produced and marketed.

Uses of granular textured protein are shown in Figure 9. Its application is based on the excellent water absorption and oil retention of TSP. Granular protein holds gravies and oil and fat which ooze from ground meat in cooking so that the product is softened and recovery

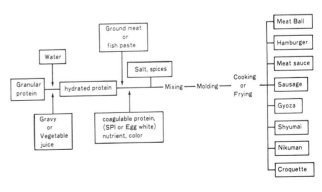

Fig. 9. Applications of textured granular protein.

is improved. That is, the addition of granular protein is conducive to efficient utilization of natural resources, but, in many cases, textured granular protein is used to reduce production cost of ground meat. It is commonly used in meatballs, hamburgers, gyaza, shumai, etc.

TSP is hydrated with water, gravy, or vegetable juice and the hydrated TSP is mixed with ground meat or fish paste. Salt and spices are then added and this is followed by uniform mixing. As a result, meat protein is solubilized with salt and it penetrates into TSP texture where meat protein or fish protein and TSP bind well with each other. Mixtures of ground meat or fish paste and TSP which are thus obtained are used for processed food in the same manner as in the case of ground meat.

On the other hand, investigations have been made of the application of the twin extruder to TSP production since around 1982 or some six years ago. About four years ago, the Japanese Research and Development Association for Extrusion Cooking was established to promote research activities for wider application of the twin extruder in the food industry field, by the united efforts of government, private enterprise and academic circles with the support of the Japanese Ministry of Agriculture, Forestry and Fisheries. This project has been completed this year with some generally satisfactory results, and a meeting to present the results was held on Nov. 10, 1987.

In Japan, Fuji Oil Co., Ltd. quickly succeeded in commercializing a new TSP product by using the twin extruder, and started to market the product late last year (Fig. 10). As the twin extruder is capable of texturizing the base material more uniformly and continuously than the single extruder does, it provides products having a more compact texture. In addition, products produced by using the twin extruder with choice raw materials and advanced seasoning technology have excellent taste and flavor. For example, very delicious soy nugget karaage has already been produced by deep frying the base ingredient which has been coated with a seasoned batter.

This soy nugget product is now marketed under the name of "soy KARAAGE" as an entirely new soybean

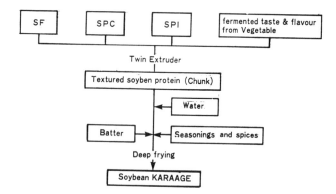

Fig. 10. Manufacturing process of soy nugget "KARAAGE."

food. Because this product is not a meat imitation food, it is not found on shelves for meat products, but it is available from tofu bean curd sellers along with tofu, abura-age, etc., and it sells very well. As a result of introducing this sales system, "Soy KARAAGE" has found general acceptance by Japanese consumers as a familiar soybean-based processed food. In producing this new "KARAAGE" product, a soybean-derived vegetable base material is fermented to provide the product with such taste and flavor that it is characterized as a vegetable food which can easily be distinguished from meat products.

I think that the future of new soy protein foods resides in that they will be developed as unique new foods which cannot be produced without soy protein, though they may also be developed as modifications of traditional foods. If soy protein is intended only for making imitation foods, it will not find any great development because it will not exceed the genuine products. It is important to develop unique soy protein foods which cannot be simulated by any other protein. I am fully convinced that, even if such unique foods are used in combination with other materials such as meat and fish, they will exist not as imitation foods but as foods which have their own significance.

Canola Meal for Livestock and Poultry

P.J. McKinnon[a] and D.A. Christensen (photo)[b]

[a] Canola Crushers of Western Canada. No. 170, 14315–118 Avenue, Edmonton, Alberta, Canada T5L 4S6 [b] Department of Animal and Poultry Science, University of Saskatchewan, Saskatoon, Saskatchewan, Canada S7N OWO

Abstract

Canola meal could be considered an early progeny of biotechnology. Canola was genetically developed from rapeseed to result in an oilseed that yields a protein meal low in sulphur containing compounds referred to as glucosinolates. Canola meal is defined as the product resulting from the crush and oil extraction of canola seed containing less than 30 micromoles of glucosinolates per gram of meal. The reduction in glucosinolates has resulted in the creation of an oilseed meal highly suited to its inclusion as a protein supplement in livestock and poultry rations yet possessing the trait of being a rich source of sulphur containing amino acids like methionine and cystine. Canola meal is traded on the basis of a minimum 35% protein and maximum 12% fiber and 11% moisture content. This presentation will discuss research undertaken in Canada and the USA as well as results from feeding trials undertaken in Canada, the USA, the Philippines and Mexico.

Canola meal (CM) is defined as the product resulting from the crushing and oil extraction of canola seed that has less than 30 micromoles of glucosinolates per gram of meal. CM is used as a protein source for livestock and has protein (36–38%), fiber (12%) and moisture (11%). Canola seed was developed in the 1970's by selective breeding of certain lines of Polish and Argentine rapeseed having both low glucosinolate level and low erucic acid level (less than 2%). At present, virtually all *Brassica napus* and *Brassica campestris* seed grown in Canada is of canola standard. Erucic acid was a component in rapeseed oil identified as causing certain cardiac lesions in rodents which raised concern about the possible effects of erucic acid in human nutrition. However, as the amount of residual oil in the meal is small (2–4%), erucic acid in the meal was never a concern in animal nutrition. The major area of concern in animal nutrition has been glucosinolates and their derivative products which may interfere to some extent with thyroid function. Glucosinolates are the compounds which give mustard and radishes, which are of the same plant family as rapeseed, their distinctive taste and smell. However, through selective plant breeding, levels of these compounds have been reduced in the last 10–15 years by 8–10 fold in canola as compared to rapeseed. Since glucosinolates and erucic acid were the major items of concern in rapeseed meal and oil, the dramatic reduction of levels of these substances did in fact create a new product, canola, a markedly superior product. CM is now used extensively in Canada in livestock feeds for pigs and poultry and is the main protein source of choice for dairy cattle diets in that market.

Nutrient Content of Canola Meal

Canola meal usually has been compared with soybean meal (SBM) but it is important to keep in mind that canola has certain attributes and characteristics in its own right and is competing with many other products including meat meal, sunflower meal, milk by-products, flour milling by-products, etc., in a highly competitive market for a place in livestock feeds. CM contains less protein than SBM but higher levels of most vitamins and minerals, particularly phosphorus and dramatically higher levels of selenium (Table 1), which may be an important consideration in selenium deficient areas. Bioavailability of minerals as shown in Table 2, indicate that these minerals are well utilized. Specifically, the high level of available phosphorus in CM should be considered when formulating diets.

Amino acid levels of CM (Table 3) and SBM are quite similar although lysine level in CM as a percentage of protein is approximately 5% less but methionine plus cystine level is higher than SBM. Thus, these two protein sources complement each other in livestock rations. Digestibility and amino acid availability will be discussed later in relation to the different livestock species but can be considered as about 5% lower in CM than SBM.

Canola meal has a fiber level of about 11–12% (Table 1) due largely to the hull on the small canola seed. A further implication of the small seed size is that canola cannot conveniently be dehulled, although technology is available and economically justified for high value products such as protein isolates or concentrates but not for livestock feedstuffs.

The ether extract of CM (Table 1) is higher than that usually found in SBM. This is a result of adding back the gums stripped out of the oil during processing to the meal. These gums represent 1–3% of the ether extract in Canadian CM. Canola gums have been added at levels up to 6% to canola meal in experimental diets with no detrimental effect on animal performance (1,2,3,4). Further, McCuaig and Bell (3) found canola gums did not affect digestibility of energy.

The energy values of CM are shown in Table 4. As a consequence of higher fiber level, CM has lower digestible energy than SBM; a fact that needs to be considered in formulating diets with CM. Numerous experiments have been conducted in different parts of the world on the value of CM in livestock diets (4,5). However, care must be exercised in evaluating the results of different experiments to ascertain that energy, protein, amino acid levels, etc. are uniform across experiments before conclusions are drawn. With the higher fiber level in CM, and, thus, lower digestible energy, nutrient levels and particularly energy levels frequently are not equivalent to rations containing SBM. High energy ingredients such as wheat, tallow or canola oil should be added to diets

TABLE 1
Canola Meal Analysis (41)

Proximate analysis	
Dry matter	89 %
Crude protein	36.5%
Ether extract	3.5%
Crude fiber	11.7%
Ash	6.8%

Major minerals		Trace minerals (mg/kg)	
Calcium	0.68%	Copper	10.4
Phosphorus	1.17%	Iron	159.2
Sodium	0.03%	Iodine	0.8
Chloride	0.02%	Manganese	53.9
Potassium	1.29%	Selenium	1.0
Magnesium	0.64%	Zinc	71.4

Vitamins (mg/kg)		Other	
Vitamin E (alpha-tocopherol)	14.5	Xanthophyll	2.7 mg/kg
Pantothenic acid	9.5	Linoleic acid	0.70%
Niacin	160	Bulk density per m^3	565 kg
Choline equivalents	6700		
Riboflavin	5.8		
Thiamine	5.2		
Biotin	1.1		
Folic acid	2.3		
Pyridoxine	7.2		

TABLE 2
Bioavailability of Minerals in Canola Meal (41)

Mineral	Average bioavailability (%)
Phosphorus	75
Calcium	68
Magnesium	62
Zinc	44
Copper	74
Manganese	54

TABLE 3
Amino Acid Composition (42)

	as fed %	in protein %
Arginine	2.26	6.08
Histidine	1.04	2.79
Isoleucine	1.28	3.44
Leucine	2.57	6.92
Lysine	2.21	5.95
Methionine	0.76	2.10
Cystine	0.90	2.48
Phenylalanine	1.45	3.90
Threonine	1.68	4.52
Tryptophan	0.44	1.19
Tyrosine	0.97	2.62
Valine	1.64	4.42

supplemented with high levels of CM to compensate for the energy lower than SBM.

Canola Meal for Pigs

Starting Pigs (weaning—20 kg): Results of experiments with starting pigs (5–20 kg body weight) have been variable. McKinnon and Bowland (6) found that feed intake, growth rate and feed conversion of weaner pigs was equivalent when either SBM was fed or one-half of the supplemental protein was provided from CM in isocaloric and isonitrogenous diets (Table 5). Complete replacement of SBM by CM in similar diets resulted in a significant reduction of feed intake. Other workers (7,8,9) have found a reduction of performance from starter pig diets with high levels of CM replacing SBM although MacIntosh and Aherne (8) found no significant effect on performance by inclusion of 9% (25% of protein) CM (Table 6). These authors in another study (43), found the palatability of CM to be less acceptable to young pigs than SBM which Bell and Aherne (10) concluded to be the cause of the reduction in performance when high levels of CM were fed to weaner pigs. The current recommended level of inclusion of CM in pig starter diets is 8%.

TABLE 4

Energy Content (41)

	kcal/kg		kcal/kg
ME: growing poultry	1900	DE: ruminants	2830
ME: adult poultry	2000	ME: ruminants	2400
TME: growing poultry	2200	NE_m: ruminants	1500
TME: adult poultry	2200	NE_g: ruminants	940
DE: swine	2900	NE_l: dairy	1510
ME: swine	2700		
TDN: swine	66%	Neutral detergent fiber:	ruminants 25.4%
TDN: ruminants	64%	Acid detergent fiber:	ruminants 20.2%

TABLE 5

Feed Intake, Average Daily Gain and Feed:Gain of Pigs 4–10 Wks. of Age (6)

Protein source:	SBM	CM	CM + SBM
Number of pigs:	15	16	16
Initial wt (kg)	5.3	5.5	5.3
10 wk wt (kg)	20.7	18.4	20.5
Avg daily feed (kg)	0.80^a	0.70^b	0.78^a
Avg daily gain (kg)	0.37	0.31	0.36
Feed/gain ratio	2.20	2.32	2.21

a,b Means with different letters differ significantly (P < 0.05).

Growing and Finishing Pigs (20–100 kg): The results of feeding trials with CM for growing pigs show much closer agreement than for starting pigs. Recent work shows that up to 75% of the supplemental protein from CM for growing pigs and all supplementary protein for finishers has not significantly affected pig performance (10). In an experiment with isocaloric and isonitrogenous diets, Narendran et al. (11) found no effect on gain, feed intake, feed efficiency or backfat thickness of 0, 5, 10, 15, 20 or 25% CM levels for growing and finishing pigs on ad libitum feeding (Table 7). Bell (12) and Bell and Aherne (10) report similar findings although the latter authors report a decrease of performance at levels of CM greater than 75% of supplementary protein. However, it is not clear from these reports if diets were isocaloric. Rundgren (5) in a review of literature reports that Scandinavian research in which net or metabolizable energy levels were equalized between treatments showed no differences in performance or carcass traits with inclusion of up to 24% CM for pigs that were limit fed. Bell (12) reported an interesting finding that suggested performance of pigs fed CM with a lower protein cereal, such as barley, and consequently requiring more supplemental protein to reach a standard protein level was in fact better than if a higher protein cereal such as wheat was used. In the latter case, lysine supplementation was beneficial (Table 8).

Experiments with corn (maize) as the basal cereal generally have shown similar results to experiments with barley and wheat. Bell (12) reported that substitution of CM for SBM at 15% of the ration resulted in a minor reduction of gain but no effect on feed conversion. An improvement of carcass backfat measurements also was found (Table 9). As mentioned above, Narendran et al. (11) found no effect on performance of up to 25% CM in corn-based diets.

The performance of pigs fed high levels of Canadian

TABLE 6

Performance of Piglets (130) Fed *ad libitum* on Diets Supplemented with Soybean Meal (SBM) and Canola Meal (CM) (43)

	#1 SBM control	#2 75% SBM- 25% CM	#3 50% SBM- 50% CM	#4 25% SBM- 75% CM	#5 100% CM
Level of CM in diet, %	0.0	9.0	18.0	27.0	36.0
No. pigs per treatment	26	26	26	26	26
No. days on test	28	28	28	28	28
Avg. initial wt., kg	6.9	6.9	6.9	6.9	6.9
Avg. final wt., kg	14.0	15.33	14.43	13.55	13.14
Avg. daily gain, g	295^a	301^a	$269^{a,b}$	$238^{b,c}$	223^c
Daily feed intake, g	570^a	$537^{a,b}$	$492^{b,c}$	447^c	433^c
Feed/gain ratio	1.94	1.79	1.85	1.89	1.98

1 SE = standard error of the mean.
a,b,c means with different letters differ significantly (P < 0.05).

TABLE 7

Effects of Diets Containing Different Levels of Tower Canola Meal (CM) and Sex on the Performance of Pigs Reared from 18 to 105 kg (Corn Based Diets) (11)

	Avg. daily gain (kg)	Avg. daily feed intake (kg)	Feed/ gain	Avg. maximum backfat thickness (cm)	Dressing (%)
Diet					
Diet 1 (0% CM)	.77	2.56	3.09	3.50	81.4
Diet 2 (5% CM)	.74	2.43	3.37	3.49	81.0
Diet 3 (10% CM)	.79	2.47	3.09	3.41	80.8
Diet 4 (15% CM)	.73	2.34	3.18	3.51	80.3
Diet 5 (20% CM)	.75	2.36	3.21	3.44	81.1
Diet 6 (25% CM)	.75	2.40	3.25	3.36	80.3
Pigs/treatment	16	16	16	16	16
SEM	.031	.100	.052	.110	.50
Significance	NS	NS	NS	NS	NS
Sex					
Barrows	.78	2.59	3.29	3.61	81.15
Gilts	.73	2.27	3.11	3.28	80.51
Pigs/treatment	8	8	8	8	8
SEM	.045	.073	.141	.155	.852
Significance	NS	**	*	**	NS

* = $P < 0.05$.
** = $P < 0.01$.
NS = Nonsignificant at 0.05 level of probability.
Crude protein = 17%; digestible energy = 3290 kcal/kg.

CM can be demonstrated by the findings of a large study conducted recently on pig producers' farms in the Province of Alberta, Canada (13). This study involved a total of 758 pigs on six farms in different parts of Alberta and was conducted by Swine Specialists of Alberta Department of Agriculture. Pigs were fed either SBM from 20–100 kg or one-half of the protein supplement from CM and the other half as SBM from 20–55 kg and then all CM from 55 kg to market at approximately 100 kg. An important point was that diets were isonitrogenous but not isocaloric. Results shown in Tables 10, 11, and 12

TABLE 8

Growth and Feed Utilization Responses of Pigs Fed Barley-Wheat Rations from 23 to 34 kg Liveweight

Treatment	Daily gain g	Feed/ gain	Feed/day kg
Protein supplement			
Soybean meal	624	2.51	1.57
Canola meal	572	2.69	1.53
Canola meal + lysine	602	2.44	1.46
Grain protein level			
High	591	2.62	1.53
Low	619	2.38	1.46

TABLE 9

Performance of Growing-Finishing Pigs Fed Corn Diets with Soybean (SBM) and Canola Meal (CM)

Treatment	Daily gain kg	Feed/ gain	Backfat[a] cm
Pigs 23 to 91 kg			
SBM	0.82	2.89	8.33
CM 15% (Tower)	0.79	2.90	7.59
CM 15% (Candle)	0.72	3.00	7.98
Pigs 18 to 105 kg			
SBM	0.79	2.94	
CM 15% (Tower)	0.76	3.03	

[a] Total of 3 measurements.

indicate that no significant differences were found in daily gain or feed conversion efficiency during the growing or finishing periods. However, during the overall period from 20–100 kg, slight, but significant ($P < .05$) differences of 20 gm/day in daily gain and six days longer on test were observed. No significant difference occurred in feed efficiency.

Shaw and Aherne (14) evaluated Tobin canola seed (B.

TABLE 10

Performance Traits for Pigs Fed a Canola Meal (CM) or a Soybean Meal (SMB)-based Ration During the Grower Period over All Farms[a] (17)

Ration	Total pigs	No. of pens	Starting age (days)	Initial weight (kg)	Average daily gain (kg)	Feed efficiency (kg feed/ kg gain)
CM	379	17	60	19.8	0.53	2.78
SBM	379	17	60	20.0	0.56	2.74
			NS	NS	NS	NS

[a] All values are adjusted for initial weight using analysis of covariance.

TABLE 11

Performance Traits for Pigs Fed a Canola Meal (CM)-based or a Soybean Meal (SBM)-based Ration During the Finisher Period over All Farms[a,b] (17)

Ration	Interim weight (kg)	Average daily gain (kg)	Feed efficiency (kg feed/kg gain)
CM	53.2	0.68	3.50
SBM	55.1	0.70	3.54
	(P < .1)	NS	NS

[a] All values are adjusted for initial weight using analysis of covariance.
[b] Figures in parentheses are probability levels for the ration effect.

TABLE 12

Performance Traits for Pigs Fed a Canola Meal (CM) or a Soybean Meal (SBM)-based Ration During the Grower and Finisher Periods over All Farms[a] (17)

Ration	Market weight (kg)	Average daily gain (kg)	Feed efficiency (kg feed/ kg gain)	Days on test
CM	100	0.61	3.17	137
SBM	100	0.63	3.14	131
	NS	*	NS	*

[a] All values are adjusted for initial weight using analysis of covariance.
*(P < .05).

campestris) as an ingredient in weanling pig diets. The seed was raw, ground and pelleted at levels of 7.5 and 15.0% or "jet-sploded" (200°C, 60–90 sec, or 107°C) at 15 and 30% of the diet. "Jet-sploding" did not improve pig performance. Feeding 7.5% or 15% canola seed had no significant effect on daily gain, feed intake or efficiency of feed utilization in isonitrogenous diets.

Breeding and Nursing Sows: Summaries of experiments with sows over several pregnancies have shown that CM can entirely replace SBM as a protein source in breeding pig and nursing sow diets (12). Recent field trial work (Table 13) in Alberta, Canada, by Alberta government staff has shown no meaningful differences in numbers of piglets born alive, birth weight or weaning weight of piglets from sows of up to four pregnancies (15).

TABLE 13

Overall Means of Various Traits from Birth to Weaning for Piglets from Sows Fed Canola or Soybean Meal Diets for up to Four Consecutive Reproductive Cycles (15)

	Canola	Soybean	
Number of piglets born	10.1	10.2	NS
Number of piglets born alive	9.5	9.4	NS
Pigs weaned per litter	8.2	8.2	NS
Birth weight of litter, kg			
Alive	13.6	14.1	NS
Dead	1.9	2.3	NS
Average birthweight of piglet, kg			
Alive	1.47	1.55	*
Dead	1.26	1.28	NS
Weaning weight of litter			
Total, kg	55.8	58.2	NS
Average, kg/pig	6.95	7.26	*

NS = not significant.
*(P < 0.05).

Digestibility of CM for Pigs: Recent work by Sauer et al. (16) with pigs fitted with ileocecal reentrant canulas to differentially determine digestibility at the end of the small intestine vs. total tract collection has greatly furthered the understanding of digestibility of CM. This is of particular importance in relation to the fiber level of CM and the known effects of bacterial fermentation of organic matter in the large intestine to organic acids which can be used by the pig as an energy source. There were significant dif-

TABLE 14

Apparent Ileal Availabilities of Amino Acids in Soybean Meal (SBM), Canola Meal (CM) and Rapeseed Meal (RSM) for Pigs (16)

Items Diets:	SBM	CM Regent %	CM Candle %	RSM Turret %	SE[b] %
Dry matter	78.3[c]	69.8[d]	67.2[d]	69.0[d]	1.4
Crude protein	80.6[c]	70.3[d]	68.2[d]	69.5[d]	1.4
Amino acid N recovery[a]	74.6[e]	68.1[f]	70.3[e,f]	66.1[f]	1.5
Amino acids:					
Indispensable					
Arginine	90.3[e]	81.2[f]	81.7[f]	85.6[f]	2.2
Histidine	82.1	79.8	82.9	85.1	1.8
Isoleucine	85.5[e]	76.1[f]	76.0[f]	78.0[f]	1.5
Leucine	84.2	79.8	79.3	81.3	1.3
Lysine	85.6[c]	75.4[d]	73.5[d]	73.5[d]	1.5
Methionine	86.3	82.2	81.4	84.3	2.7
Phenylalanine	86.3	77.5	79.8	81.8	1.9
Threonine	75.8	67.2	65.6	67.3	2.1
Valine	74.3	66.3	67.3	69.8	2.5
Dispensable					
Alanine	80.1	76.6	75.4	76.2	1.3
Aspartic acid	83.8[c]	71.4[d]	72.9[d]	73.6[d]	1.4
Cysteine	82.1	88.0	90.0	90.2	2.9
Glutamic acid	85.2	84.3	82.3	82.8	1.9
Glycine	72.3[e]	66.5[e,f]	63.0[f]	66.4[e,f]	1.5
Proline	75.1	62.4	75.2	74.6	3.1
Serine	81.9[e]	69.1[f]	70.8[f]	71.0[f]	1.9
Tyrosine	83.8[e]	71.4[f]	72.2[f]	73.6[f]	1.3

[a] Measured as a percentage of total N in ileal digesta.
[b] Standard error of the mean.
[c,d] Means in the same row with the same superscripts do not differ at $P < .01$.
[e,f] Means in the same row with the same superscripts do not differ at $P < .05$.

ferences in true ileal availabilities of amino acids from CM for lysine ($P < .01$) and isoleucine ($P < .05$) compared to SBM as well as for crude protein (Table 14). Apparent ileal digestibility of dry matter of CM was also lower ($P < .01$) than in SBM. These findings generally confirm the earlier work of McKinnon and Bowland (6) who found that digestible energy, protein and amino acids were reduced when CM completely replaced SBM in diets for young pigs but found no major effect from 50% replacement of CM for SBM.

The above findings of reduced digestibility of CM compared to SBM are not particularly surprising considering the higher fiber level of CM, although some authors have attributed a specific effect of lower available lysine in CM. Sauer et al. (16) using the protein efficiency ratio method with rats, showed that the quality of protein in CM was superior to that of SBM, despite a lower protein digestibility. Thus, the data suggest that the differences in digestibility between CM and SBM can be accounted for largely by the differences in crude fiber.

As mentioned previously, caution is required in evaluation of experimental results where CM and SBM are compared, to ensure that digestible or metabolizable energy levels are equivalent between diets. This concern is particularly relevant for growing and finishing pigs where simplified diets usually containing only a cereal grain, protein source and mineral-vitamin premix are used. It must be remembered that the cost of CM plus a source of high energy usually results in a lower cost of gain than using SBM alone as the protein supplement. The current recommended levels of CM usage for pigs are: starting diets, 8%; grower diets, 12%; finishing rations, all of the protein supplement; breeding pigs, all of the protein supplement to a maximum of 12% (10).

Canola Meal for Poultry

Canola meal has been used widely for poultry rations, particularly in layer rations and there are indications that because of the higher methionine content, CM complements SBM in poultry rations (17). However, the higher fiber level of CM and consequently lower metabolizable energy for poultry (Table 4) will result in a slight lowering of the metabolizable energy of the diet. In certain situations in Canada, this factor of lower energy is used in a strategic manner where a specifically lower energy level is required such as in pullet developer or beyond 30 weeks of age and after peak egg production is reached. In other cases, a high energy ingredient such as canola oil or tallow is added to attain the metabolizable energy level required.

The apparent metabolizable energy (AME) values for poultry published for CM in a recent review (18) are approximately 2000 kcal/kg for laying hens and 1900 kcal/kg for broilers. With a true metabolizable energy (TME) of about 110% of AME values, the corresponding values for TME would be about 2200 kcal/kg for layers and 2100 kcal/kg for broilers (19).

An important factor to consider in formulating diets with CM for poultry is the high level of phosphorus. Growing chickens appear to use the phosphorus in CM efficiently with an availability of about 30% (20), and as a result, the level of available phosphorus is higher than in SBM.

Broiler Chicken Diets: A great deal of research on CM for broilers has been conducted at the University of Alberta. Typical results are shown in Table 15 and indicate that up to 30% CM in isocaloric and isonitrogenous diets produced very good performance on wheat-based mash, broiler rations. Very recently, Nasser and Arscott (21) found similar excellent performance in broilers using high levels of CM in corn-based typical broiler rations with 7 week weights in the neighborhood of 2 kg and feed conversions of about 2 kg feed/kg gain (Table 16). Based on results such as the above, Robblee et al. (19) have concluded that up to 20% CM can be used in broiler rations.

Layer and Breeding Chicken Diets: Robblee et al. (19) recently summarized a number of experiments using CM in wheat-based diets for laying chickens and found no difference in performance of egg production, egg size, feed per dozen eggs or mortality with 10% CM (Tables 17 and 18). Thyroid weight was increased but this has not resulted in any deterioration of performance. Similarly Nasser et al. (22) found production was not affected by inclusion of high levels of CM (Table 19). Egg quality was also unaffected although thyroid enlargement was found. The work of Nasser et al. (22) also shows typical results of CM for hatching hens, with no effect on fertility or hatchability (Table 20).

One problem with use of CM for laying chickens that must be mentioned is the production of fishy-flavored eggs by brown-egg laying chickens. In the review of Robblee et al. (19), a number of references have been made to the nature of the problem which is related to an impairment of the enzyme system which metabolizes trimethylamine (TMA) in brown-egg layers. CM contains a rather high level of a choline derivative, sinapine, which is a precursor of TMA. In Canada, where almost all eggs sold in commercial trade are from white-egg layers, this is not a problem but can be where brown-egg layers predominate. As a result, 3% CM is suggested for brown-egg laying

TABLE 15

Performance of Broiler Chickens Fed Rations Containing Canola Meal (10)

	4 Week		8 Week		
Rations	Body wt g	Feed/ gain	Body wt g	Feed/ gain	Thyroid wt mg/ 100 g body wt
Control, wheat-soybean meal	617	1.68	1930	2.24	9.9
+10% canola meal (cv. Tower)	631	1.71	2104	2.20	12.9
+20% canola meal (cv. Tower)	641	1.69	1992	2.17	16.9
+30% canola meal (cv. Tower)	595	1.68	1974	2.16	17.6
+10% canola meal (cv. Candle)	650	1.65	2001	2.15	12.6
+20% canola meal (cv. Candle)	653	1.63	2015	2.16	14.3
+30% canola meal (cv. Candle)	621	1.69	1940	2.14	15.1

TABLE 16

Seven Week Broiler Performance* (21)

CM replacement level	BW (kg)	Feed consumption (kg)	Feed conversion	Leg abnorm. (%)
0	2.286^a	4.522^a	2.07^a	17.22^a
1/4	$2.150^{a,b}$	4.472^a	2.08^a	18.89^a
1/2	2.186^a	4.571^a	2.09^a	6.94^a
3/4	2.077^b	4.222^b	2.03^a	7.13^a
All	1.891^c	3.819^c	2.02^a	10.36^a

* Means within a column followed by different superscripts are significantly different ($P \leq .05$).

TABLE 17

Performance of Laying Chickens Fed Rations Containing Canola Meal[a] (19)

Experiment	Rations	Egg prod %[b]	Feed/doz kg	Egg wt g	Mortality %	Thyroid wt mg/100 g body wt
1	Control, wheat-soybean meal	79.6	1.86	59.7	3.4	7.8
	+5% canola meal[c]	78.5	1.83	59.8	5.8	13.5
	+10% canola meal[c]	81.3	1.78	59.0	4.0	19.4
	+5% canola meal[d]	81.5	1.82	59.4	1.8	10.2
	+10% canola meal[d]	80.1	1.80	59.1	5.4	11.8
2	Control, wheat-soybean meal	72.0	1.93	62.4	4.1	8.5
	+5% canola meal[d]	71.1	1.99	62.3	4.1	11.0
	+10% canola meal[d]	71.9	1.93	62.5	3.1	15.9
	+15% canola meal[d]	72.2	1.94	61.7	5.8	19.4
3	Control, wheat-soybean meal	81.6	1.74	59.2	1.1	7.9
	+10% canola meal[d]	81.6	1.72	59.6	1.4	14.6
	+12.5% canola meal[d]	79.1	1.76	59.4	2.1	17.9
	+15% canola meal[d]	78.6	1.70	59.0	4.3	16.7

[a] University of Alberta data.
[b] Hen-day production.
[c] Produced from canola (cv. Tower).
[d] Produced from canola (cv. Candle).

TABLE 18

Performance of Laying Chickens Fed Rations Containing Canola Meal[a] (19)

Rations	Egg prod %[b]	Feed intake g bird/day	Egg wt g	Egg shell deformation um
Control, corn-soybean meal	79.9	112.2	57.2	23.5
+5% canola meal (cv. Candle)	78.9	108.9	57.3	22.7
+10% canola meal (cv. Candle)	79.5	110.7	57.3	23.0
+15% canola meal (cv. Candle)	79.1	109.8	56.7	23.2

[a] University of Guelph data.
[b] Hen-day production.

TABLE 19

Performance of SCWL* Layers Fed Four Levels of Canola Meal (32)

Canola replacement level	Egg prod. (%)	Daily fd. cons. (g)	Av. fd/dz. eggs (kg)	Final B.W. (kg)	Final egg wt. (g)	Av. E.W. (g)
0	77.26[a]	108[a,b]	1.71[a]	1.86[a]	59.3[a]	56.9[a]
1/3	80.25[a]	110[a]	1.65[a]	2.00[a]	59.9[a]	56.9[a]
2/3	77.85[a]	106[b,c]	1.66[a]	1.91[a]	59.7[a]	56.5[a]
All	77.96[a]	106[c]	1.64[a]	1.91[a]	58.9[a]	55.6[a]

Means within a column with different superscripts are significantly different at ($P \leq .05$).
* SCWL = single comb white leghorn.

TABLE 20

Effect of Four Levels of Canola Meal on Fertility and Hatchability of SCWL* Layers (32)

Canola replacement level	Fert. per. 5 (eggs) (%)	Fert. per. 10 (eggs) (%)	Av. Fert. (%)	HFE per. 5 (%)	HFE per. 10 (%)	Av. HFE (%)
0	95.41^a	91.15^a	93.28^a	82.94^a	89.24^a	86.09^a
1/3	95.48^a	87.39^a	91.43^a	88.33^{a,b}	90.61^a	89.47^a
2/3	94.68^a	90.05^a	92.36^a	91.41^b	89.38^a	90.39^a
All	94.34^a	85.21^a	89.78^a	92.58^b	91.59^a	92.08^a

Means within a column with different superscripts are significantly different at (P ≤ .01).
* SCWL = single comb white leghorn.

chickens. For white-egg layers, a level of 10% CM can be used successfully.

Canola Meal for Dairy Cattle

Canola meal has been used extensively as a protein source for dairy cattle and is now the main protein source in dairy rations in Canada (23). Because the crude fiber level in CM is higher than in other protein sources such as SBM, much of the developmental research on CM has been focused on usage in cattle rations where the fiber level is not a detriment. In fact the fiber level in CM may be advantageous in certain dairy cattle diets, especially where high levels of cereal grain silage or concentrates are fed (24).

Calves: Canola meal has been used in calf starter-grower diets with very good results. Sharma et al. (25) and Fisher (26) found no differences in feed intake, growth rate or feed conversion in calves of one to six months of age fed high levels of CM or SBM. Sharma et al. (25) found digestibility of CM added to a basal ration (Table 21) was approximately 5% lower than SBM. However, a recent report by Claypool et al. (27) comparing CM, SBM and cottonseed meal, showed no differences in performance in calf rations. Although body weight gains were equivalent,

TABLE 21

Apparent Digestibilities and Nitrogen Retention of a Basal Ration to Which Either 50% Soybean Meal (SBM) or 50% Canola Meal (CM) was Added and Apparent Digestibility of Soybean Meal and Canola Meal as Determined by the Difference Method in Calves (24)

	Basal ration	50% Basal 50% SBM	50% Basal 50% CM	
			Tower	Candle
Dry matter intake, kg/day	4.3	3.8	3.9	3.9
Apparent digestibilities, %				
Dry matter	81.5	84.5	79.2	79.8
Crude protein	71.3	87.5	80.9	79.0
Acid detergent fiber	64.7	59.7	48.1	56.8
Energy	79.4	83.1	78.1	78.9
Nitrogen retention, % of total N intake	39.6	40.7	42.2	38.6

		CM	
	SBM	Tower	Candle
Apparent digestibilities by difference, %			
Dry matter	87.5	76.0	78.1
Crude protein	103.8	90.5	86.7
Acid detergent fiber	54.8	33.2	48.8
Energy	86.8	76.8	78.3

feed intake of calves fed CM was less, resulting in better feed conversion.

Dairy Cows: Many experiments have been conducted on the value of CM for dairy cows showing excellent milk production using CM as the main protein source (see reviews by Thomke (4) and Ingalls (28). Research reported by Laarveld and Christensen (29); Laarveld et al. (30); Sharma et al. (31); and DePeters and Bath (32) and others have shown that CM will either maintain or slightly increase milk yield. Recently McClean and Laarveld (33) have shown a substantial response (Table 22) in milk yield when 24% CM in the concentrate replaced soybean meal ($P < 0.05$). This trial was conducted to evaluate the milk response to injected somatotropin by cows fed soybean or canola based rations. The response to somatotropin was greater for the soybean or canola based rations. The response to somatotropin was greater for the soybean based rations but only because the CM control gave a substantially higher milk yield than the soybean ration. Somatotropin injection gave similar milk yields with both canola and soybean meal. In this research forage and concentrate were fed in equal amounts. Milk thiocyanate and iodine were reduced by feeding canola but were not influenced by somatotropin.

Most of this work has compared CM and SBM but recent reports comparing CM and cottonseed meal (CSM) are now available. Sanchez and Claypool (34) at Oregon State University reported no significant difference in milk production although cows fed CM tended to produce more milk than cows fed SBM or CSM in isonitrogenous and isocaloric diets (Table 23). Milk components (protein, fat, total solids and solids not fat) were not affected. Source of protein did not affect milk flavor and feed intake was similar for all diets. Also, DePeters and Bath (32) at the University of California, Davis, very recently reported no differences in milk production of cows fed CM or CSM (Table 24). As well, source of protein supplement did not affect fat, lactose, ash, total solids, total nitrogen, whey protein nitrogen or nonprotein nitrogen content. Volatile fatty acid composition and ammonia content of ruminal fluid were not affected and disappearance of nitrogen and dry matter incubated in the rumen was similar for both protein sources (Table 25).

Beef Cattle: Recent research at the University of Saskatchewan (35) has demonstrated the effectiveness of canola meal in beef growing-finishing rations at up to 22.5% of the ration dry matter. In this research, medium-framed steers (Hereford type) showed very little

TABLE 22

Effect of Somatotropin on Canola Meal and SBM Fed Cows (33)

	SBM	SBM + somatotropin	CM	CM + somatotropin
Milk yield, kg/day	28.9[c]	32.3[a]	30.7[b]	32.7[a]
Milk fat, %	2.42[a]	2.82[b]	2.84[b]	2.87[b]
Milk protein, %	2.83	2.89	2.90	2.84
Milk SCN, mg/l		1.12[a]		1.40[b]
Milk iodide, mg/l		.13[a]		.18[b]

Means in the same row with different superscripts differ ($P < .05$).

TABLE 23

Least Square Means for Lactation Performance and Body Weight Change for Cows Fed Rations Containing Soybean, Cottonseed, or Canola Meals as Supplemental Protein (34)

Variable	Number of animals	Soybean meal	Cottonseed meal	Canola meal	SE	P
Milk kg/day	10	34.45	36.50	37.67	.979	.120
Fat corrected milk (FCM), kg/day	10	28.07	29.71	32.13	.730	.103
Milk protein, kg/day	10	1.01	1.07	1.15	.0291	.180
Milk protein, %	10	2.95	3.02	2.96	.0374	.702
Milk fat, kg/day	10	.95	1.02	1.08	.0262	.129
Milk fat, %	10	2.70	2.76	2.63	.0429	.505
Milk total solids, kg/day	10	4.06	4.35	4.71	.122	.122
Milk total solids, %	10	12.02	12.06	12.01	.181	.993
Milk solids not fat, kg/day	10	3.13[a]	3.42[b]	3.70[c]	.0749	.024
Milk solids not fat, %	10	9.21	9.24	9.30	.165	.974
Milk flavor quality	10	7.96	8.09	7.74	.0861	.309
Body weight change, kg/day	10	.49	.36	.38	.0539	.582

Means in the same row with different superscripts differ ($P < .05$).

TABLE 24

Yield of Milk and Milk Components, Milk Composition, Feed Intake, and Body Weight Change of Cows During Early Lactation (Experiment I) (32)

	First lactation			Mature cows		
Item	Cottonseed	Canola	SE	Cottonseed	Canola	SE
Milk, kg/d	26.6	25.8	.99	39.8	41.4	1.18
4% Fat-corrected milk, kg/d	26.4	25.2	1.03	36.7	38.5	.99
Fat, kg/d	1.05	1.00	.05	1.38	1.46	.04
Fat, %	3.99	3.85	.11	3.52	3.54	.06
Nitrogen, kg/d	.12	.12	.005	.18	.19	.01
Nitrogen, %	.46	.48	.01	.46	.46	.01
Total solids, kg/d	3.40	3.29	.12	4.85	5.05	.12
Total solids, %	12.81	12.78	.12	12.28	12.24	.10
Milk energy, Mcal/d	19.8	19.0	.74	27.6	28.8	.69
Dry matter intake, kg/d	15.2	14.7	.50	21.0	20.8	.46
Organic matter intake, kg/d	14.1	13.6	.46	19.4	19.1	.42
Protein intake, kg/d	2.7	2.7	.08	3.7	3.7	.07
ADF intake,[a] kg/d	3.5	3.6	.13	5.0	4.9	.11
NDF intake,[b] kg/d	6.0	5.8	.22	8.2	8.1	.18
Estimated NE_1,[c] intake, Mcal/d	24.7	23.6	.82	33.5	33.2	.73
Body weight change,[d] kg	−10.6	−11.2	7.33	−7.6	−21.9	6.53

[a] Acid detergent fiber.
[b] Neutral detergent fiber.
[c] Estimated net energy for lactation (13).
[d] Final minus initial body weight.

TABLE 25

Parameter Estimates of Nitrogen (N) and Dry Matter (DM) Disappearance of Cottonseed Meal (CSM) and Canola Meal (CM) from Nylon Bags (Experiment 4) (32)

		Protein supplement	
Item	Constants[a]	CSM	CM
N	A	22.1	21.5
	B	66.5	70.2
	C	8.0	7.6
DM	A	22.6	23.2
	B	59.5	62.1
	C	6.2	9.0

[a] A, B, and C are constants for which A represents the rapidly soluble fraction, B the amount which in time will degrade, and C, the fractional rate constant (degradation per hour) of fraction B (14).

TABLE 26

Canola Meal in High and Low Energy Beef Rations[a]

	Ration Crude Protein (DM)		
	11.5	13.0	14.5
Canola level, % of DM[b]			
Low TDN (73.5%)	5.0	11.0	22.5
High TDN (78.5%)	3.0	6.0	14.3
Average daily gain, kg/day			
Medium frame steers			
Low TDN	1.28	1.30	1.28
High TDN	1.27	1.25	1.29
Large frame steers			
Low TDN	1.18a	1.24b	1.34c
High TDN	1.28a	1.35b	1.44c

[a] 270 to 320 kg initial wt; 420 kg to 500 kg final wt.
[b] Barley grain and silage based rations.
Source: J. J. McKinnon, 1988, unpublished.

response to protein supplementation from canola above 11.5% of the ration dry matter (Table 26). In contrast, large-framed steers (Charolais type) responded well to supplementary protein from canola meal at both high and low TDN levels in the ration.

Acidulated fatty acids arise from refining of canola oil. Fatty acids are removed from the oil by treatment with sodium hydroxide. The resulting soaps are treated with sulfuric acid resulting in an acidulated fatty acid product. These have been evaluated at the Agriculture Canada Melfort Research Station in beef cattle rations. Average daily gain was maintained or slightly increased when 2% or 4% acidulated fatty acids were included in the ration (Table 27). Feed conversion was improved by the inclusion of acidulated fatty acids in the barley/straw ration.

Protein Degradability-Digestibility: A number of products can be considered when examining ways of improving effi-

TABLE 27

Acidulated Fatty Acids for Beef Cattle (42)

	Level of AFA (% of DM)		
	0	2	4
Average daily gain, kg			
Barley silage/grain ration (65:35)	1.11	1.16	1.11
Barley/straw ration (90:10)	1.40	1.68	1.61
Feed gain ratio			
Barley silage/grain ration	9.5	9.2	9.8
Barley/straw ration	8.3	7.3	7.4

Average initial wt., 355 kg; average final wt., 535 kg.

ciency of use of canola protein. It is possible, for example, to use the presscake resulting from cooking and mechanical extraction of oil in the first part of the solvent extraction process. The various foots and gums and acidulated fatty acids arising from canola oil extraction and refining may be used in ruminant feeding. There are also various ways of processing whole seed, either by extrusion at relatively low temperature, or by exploding the seed (jet-sploded) by exposing the seed to temperatures over 300°C. The seed or meal may be treated in various ways to reduce the rate of digestion of the protein. Among these methods are treatment of seed or meal with formaldehyde or coating them with slowly degraded proteins such as whole blood.

Formaldehyde treated whole canola seed has been compared to a soy-tallow product (36) (Table 28). Formaldehyde treatment reduces the rumen degradation of canola protein and if this protein is used to coat the lipid component it also has a protective effect on the fat, allowing it to pass through the rumen without inhibiting rumen fermentation. When the Tower rapeseed and soy-tallow product were included in the dairy rations at a level of 8%, dry matter intake was slightly increased and milk yield was significantly increased by the Tower rapeseed product. Plasma calcium was slightly reduced by the soy tallow product but not by the Tower rapeseed product. The rations used in this trial were relatively low in forage content (35%). The high forage control (50%) tended to give the highest milk percentage compared to the low forage control and soy tallow product. The milk fat percentage on the Tower rapeseed product was similar to that in the high forage ration. In this trial protein intake was higher in the soy-tallow and Tower rapeseed rations. The favorable effects may have been due to the higher protein intake rather than to the higher energy and improved rumen bypass protein intakes.

Extensive research has been conducted at the University of Alberta in Edmonton, Alberta, on the rumen degradability of CM. Ha and Kennelly (37) reported rumen degradability of nitrogen was greater for CM than for SBM but dry matter disappearance from nylon bags suspended in the rumen was not significantly different for 0–24 hours incubation except at 12 hours in dairy cows in mid to late lactation. Recently, Ha et al. (38) have reported on earlier work with steers and found ruminal degradabilities of 66.3%, 57.9%, 63.3% and 41% for urea, SBM, CM and a dehydrated alfalfa + urea blend. As well, Kennelly and de Boer (39) have reported on using jet-sploded full-fat canola seed as a means of protecting protein from rumen degradation and increasing the bypass value of canola protein. Jet-sploding is a processing technique whereby the seed is fed by gravity into a heat exchanger with air heated to 316°C for a short period of time. The seed is then passed through a roller mill on exit from the chamber which ruptures the seed coat. The internal moisture in the seed is heated, as in popcorn, and the heat thus denatures and protects the protein from ruminal degradation. The protein is then digested in the small intestine along with the oil in the seed, which significantly adds to the energy supply of the animal. Both fat and protein in jet-sploded canola seed are well digested in the small intestine (39; Table 29).

Deacon et al. (40) demonstrated the effectiveness of heat in improving rumen bypass value in canola protein (Table 30). The processing involved in production of canola meal more than doubles the rumen bypass of canola seed protein. The use of the extrusion procedure does not improve bypass value over that of canola meal. The use of short term high temperature or jet-sploding of

TABLE 28

Protected Tallow and Rapeseed for Dairy Cattle (26)

	Low forage control	High forage control	Low forage	
			Soy-tallow	Tower rapeseed
Forage, % of ration DM	35	50	35	35
Protected lipid (40% lipid)				
% of ration	0	0	8	8
Dry matter intake, kg/day	19.7	19.7	20.3	20.6
CP intake, kg/day	2.82	2.87	3.14	3.33
Milk, kg/day	28.6	27.3	30.4	31.7
Milk fat, %	3.26	3.73	3.43	3.77
4% FCM, kg/day	25.5	26.0	27.8	30.4
Plasma calcium mg/dl	9.38	9.44	7.90	9.08
Plasma cholesterol mg/dl	157	153	230	289

TABLE 29

Influence of Jet-Sploding Temperature on Effective Dry Matter Degradability (EDDM), Effective Crude Protein Degradability (EDP) and Rumen Undegradable Protein in the Intestine (D-UDP) (39)

	Ground WCS[a]	\multicolumn{6}{c}{Jet-Sploding temperature °C}					
	Ground WCS[a]	116	127	138	140	160	177
EDDM	66.1	40.9	40.3	50.8	38.8	32.6	24.6
EDP	75.0	61.0	56.7	57.0	37.4	38.5	34.7
D-UDP	95.7	86.4	85.8	98.1	86.1	77.2	87.2

[a] WCS = whole canola seed.

TABLE 30

Bypass Protein and Rumen Escape Dry Matter (REDM) 5% Rumen Dilution Rate Per Hour (40)

	Bypass Protein	REDM
Canola meal	36	34
Canola seed	16	13
Extruded canola meal	35	35
Jet-sploded canola seed[a]	57	39
Soybean meal	23	32
Protec	46	57

[a] At 315°C air temp (seed temp 154°C for maximum bypass value).

canola seed significantly increases rumen bypass protein value without materially reducing the rumen escape dry matter content.

Research work is now in the planning stages using CM plus whole cottonseed as an alternate source of bypass protein. The possibility exists of these two protein sources complementing each other and may provide enhanced market opportunities for both CM and whole cottonseed.

In Canada it is anticipated that the jet-sploding technique will be applied to off-grade canola seed and provide substantial new market outlet for canola seed. This same process can be applied to canola seed for pig and poultry diets. Thus, the combination of CM and full-fat, jet-sploded canola seed can be used to meet the protein needs of all classes of livestock and make a substantial contribution to the overall energy supply of the animal as well.

Canola meal is a high quality protein supplement suitable for all classes of livestock. Particular attention should be given to digestible or metabolizable energy levels when formulating diets using CM, to ensure that the required energy levels are met. The costs and benefits of using CM can best be assessed by using a computer least-cost feed formulating system, with full confidence that the high quality CM on the market today is the result of intensive research and development.

As a result of the intensive research effort, the following levels of CM are recommended: Pigs: starter diets, 8%; grower diets, 12%; and finisher diets, all supplementary protein breeding and nursing diets and all supplementary protein to 12% maximum; poultry: starting and growing diets (broilers and layer replacement), 20%; white-egg layers, 10%; and brown-egg layers, 3%; and cattle: calves, 20% of diet; and dairy cows, 25% of grain concentrate portion.

References

1. Clandinin, D.R., A.R. Robblee, J.M. Bell and S.J. Slinger, "Composition of Canola Meal," *Canola Meal for Livestock and Poultry*, edited by D.R. Clandinin, Canola Council of Canada, Winnipeg, Manitoba, Canada (1986).
2. Mathison, G.W., *Can. J. Anim. Sci.* 61:131 (1978).
3. McCuaig, L.W., and J.M. Bell, *Can. J. Anim. Sci.* 61:463 (1981).
4. Thomke, S., *J. Am. Oil Chem. Soc.* 58:805 (1981).
5. Rundgren, M., *Anim. Feed Sci. Tech.* 9:239 (1983).
6. McKinnon, P.J., and J.P. Bowland, *Can. J. Anim. Sci.* 57:663 (1977).
7. Castell, A.J., *Can. J. Anim. Sci.* 57:313 (1977).
8. MacIntosh, M.K., and F.X. Aherne, "Canola Meal for Starter Pigs (4–8 weeks of age)," in *61st Annual Feeders Day Report*, Animal Science Department, University of Alberta, Edmonton, Alberta, Canada (1982), pp. 74–75.
9. Baidoo, S.K., and F.X. Aherne, "Canola Meal for Starter (10–20 kg) Pigs," in *62nd Annual Feeders Day Report*, Animal Science Department, the University of Alberta, Edmonton, Alberta, Canada, (1983) pp. 120.
10. Bell, J.M., and F.X. Aherne, "Canola Meal for Pigs" in *Canola Meal for Livestock and Poultry*, edited by D.R. Clandinin, Canola Council of Canada, Winnipeg, Manitoba, Canada (1986).
11. Narendran, R., G.H. Bowman, S. Lesson and W. Pfeiffer, *Can. J. Anim. Sci.* 61:213 (1981).
12. Bell, J.M., "Use of Canola Meal in Swine Rations," 19th Annual Convention of the Canola Council of Canada, Winnipeg, Manitoba, Canada, March 23–26, 1986.
13. Salomons, M., "Canola Meal as an Alternate Protein Supplement for Growing-Finishing Pigs," in *Proceedings of 12th Annual Alberta Pork Congress*, Red Deer, Alberta, Canada (1986).
14. Shaw, J., and F.X. Aherne, "An Evaluation of the Feeding Value of Full-Fat Canola Seed for Weanling Pigs," in *66th*

Annual Feeders Day Report, Animal Science Department, University of Alberta, Edmonton, Alberta, Canada (1987), pp. 7–9.
15. Jaikaran, S., "Reproductive Performance of Sows Fed Canola Meal," in *Proceedings of 12th Annual Alberta Pork Congress* (1986).
16. Sauer, W.C., R. Cichon and R. Misir, *J. Anim. Sci. 54*:292 (1980).
17. Baidoo, S.K. and F.X. Aherne, "Canola Meal for Livestock and Poultry," Agriculture Forestry Bulletin, the University of Alberta, Edmonton, Alberta, Canada *8*:(3)21 (1985).
18. Clandinin, D.R., and A.R. Robblee, *Feedstuffs 49*(5):20 (1983).
19. Robblee, A.R., D.R. Clandinin, J.D. Summers and S.J. Slinger, "Canola Meal for Poultry," in *Canola Meal for Livestock and Poultry*, edited by D.R. Clandinin, Canola Council of Canada, Winnipeg, Manitoba, Canada (1986).
20. Summers, J.D., B.D. Lee and S. Leeson, *Nutritional Reports International 28*:955 (1983).
21. Nasser, A.R., and G.H. Arscott, *Ibid. 34*:791 (1986).
22. Nasser, A.R., M.P. Goeger and G.H. Arscott, *Ibid. 31*:1349 (1985).
23. Patterson, R.J. "Where and How is Canola Meal Used" 19th Annual Convention, San Francisco, March 23–26, Canola Council of Canada, Winnipeg, Manitoba (1986).
24. Fisher, L.J.K., and J.R. Ingalls, "Canola Meal for Beef and Dairy Cattle" in *Canola Meal for Livestock and Poultry*, edited by D.R. Clandinin, Canola Council of Canada, Winnipeg, Manitoba, Canada (1986).
25. Sharma, H.R., J.R. Ingalls and T.J. Devlin, *Can. J. Anim. Sci. 60*:915 (1980).
26. Fisher, L.J., *Can. J. Anim. Sci. 60*:359 (1980).
27. Claypool, D.W., C.H. Hoffman, J.E. Oldfield and H.C. Adams, *J. Dairy Sci. 68*:67 (1985).
28. Ingalls, J.R., "Canola Meal for Dairy Cows," presented at 19th Annual Convention of Canola Council of Canada (1986).
29. Laarveld, B., and D.A. Christensen, *J. Dairy Sci. 59*:1929 (1976).
30. Laarveld, B., R.P. Brockman and D.A. Christensen, *Can. J. Anim. Sci. 61*:131 (1981).
31. Sharma, H.R., J.R. Ingalls and J.A. McKirdy, *Can. J. Anim. Sci. 57*:653 (1977).
32. DePeters, E.J., and D.L. Bath, *J. Dairy Sci. 69*:148 (1986).
33. McClean, C., and B. Laarveld, *J. Dairy Sci. 71 (Supp 1)*:122 (1988).
34. Sanchez, J.M., and D.W. Claypool, *J. Dairy Sci. 66*:80 (1983).
35. McKinnon, J.J., R.D.H. Cohen, D.A. Christensen and S.D.M. Jones, "Dietary Energy and Protein Requirements of Feedlot Cattle Differing in Mature Body Size," Canadian Society of Animal Science Annual meeting, Calgary, Alberta, Canada (1988).
36. Christensen, D.A., M. Cochran and M. Stacey, "Utilization of Protected and Unprotected Rapeseed by Lactating Dairy Cows" in *Proceedings of the 5th International Rapeseed Conference*, held June 12–16, 1978, pp. 217–219.
37. Ha, J.K., and J.J. Kennelly, *Can. J. Anim. Sci. 64*:443 (1986).
38. Ha, J.K., and J.J. Kennelly, *Anim. Feed Sci. Tech. 14*:117 (1986).
39. Kennelly, J.J., and G. De Boer, "Ruminal and Intestinal Disappearance of Whole Canola Seed as Influenced by Jet-sploding Temperature," in *65th Annual Feeders Day Reports*, Animal Science Department, University of Alberta, Edmonton, Alberta, Canada, 1986, pp. 82–83.
40. Deacon, M.A., G. De Boer and J.J. Kennelly, *J. Dairy Sci. 71*:745 (1988).
41. *Canadian Canola Meal*, Canola Council of Canada, 1984.
42. Agriculture Canada Melfort Research Station Annual Report, 1984.
43. MacIntosh, M.K., and Aherne, F.X., "Taste Preference of Piglets Fed Soybean Meal and Canola Meal Supplemented Diets," in *61st Annual Feeders Day Report*, Animal Science Department, University of Alberta, Edmonton, Alberta, Canada (1982), pp. 76–77.

Palm Kernel Cake as Ruminant Feed

D. Mohd. Jaafar (photo) and A. Yusof Hamali

Livestock Research Division, Malaysian Agricultural Research
And Development Institute, P.O. Box 12301, 50774 Kuala Lumpur, Malaysia

Abstract

The Malaysian oil palm industry produces three major by-products of which palm kernel cake (PKC) is the most popular by-product used as a feedstuff. Research conducted utilizing PKC has shown encouraging results. Many rations containing different levels of PKC were tested on major ruminant species giving average daily gains ranging from 0.37 to 0.76 kg/day and dry matter intakes of 3.57 to 5.81 kg/day. Supplementation of 3.5 kg of PKC/day to Sahiwal-Friesian cattle was reported to produce 7.5 liters of milk per day. Sheep fed with high level of PKC based diet recorded gains of 91 g/day but were reported to be prone to copper toxicity. Performance of cattle under feedlot management can be improved if mineral imbalance, contamination with shells, high levels of oil and the Vitamin A deficiency in PKC can be overcome.

An estimated 1,685,581 hectares of land was planted with oil palm in Malaysia in 1987. The crop produced 4.5 million metric tons of crude vegetable oil which accounted for 60% of the world output. The oil palm industry produces three major by-products of which palm kernel cake (PKC) is the most popular by-product used as a feedstuff. The success in utilizing PKC as ruminant feed is definitely boosting beef production in Malaysia. Never before has the local beef industry, which supplied roughly 55% of the demand in 1984 (1), looked so promising in reducing the country's reliance on imports. PKC is being used successfully in both large and small feedlot operations. In fact, raising cattle under a feedlot system is now favored by entrepreneurs as it requires less space and animals can be marketed within a short feeding period of about six months.

Although the price of PKC fluctuates a great deal, in most instances it is still attractive for use as a source of cattle feed. It is nutritious, easy to handle and store, and available in abundance throughout the year. With proper feeding of PKC to crossbred cattle, an impressive average daily gain (ADG) of 1.0 kg per day is very common compared to an ADG of 0.4 kg/day for animals raised on pasture. Rectification of the mineral imbalances and vitamin deficiencies possibly can induce even better performance. As such, PKC undoubtedly plays a major role in solving the feed constraints in beef and other ruminant production in Malaysia.

Government efforts to promote the fattening of cattle using PKC are numerous and cover various aspects such as supplying of feeder cattle, distribution of feed, providing health services and in the marketing of the finished animals. It is anticipated that with the interest in feedlots, the local utilization of PKC will increase tremendously.

Availability of PKC

Production of PKC, a by-product from palm kernel oil extraction has been on the increase commensurate with the increase in the production of palm oil resulting from the increase in planting. A tremendous increase in PKC production from 339,107 tons in 1981 to 725,993 tons in 1987 was reported by PORLA (2). During the same period the export value increased from $66 million to $158.4 million (Table 1).

Most of the PKC produced in Malaysia is exported to Europe and is used as an ingredient in compound

TABLE 1

Production of Palm Oil and PKC in Relation to Increase in Planting Hectarage and Export of PKC

Year	Planting hectarage	Production Oil (tons)	Production PKC (tons)	Export of PKC Quantity (tons)	Export of PKC Value $(mil)
1981	1,140,538	2,822,144	339,107	266,327	66.0
1982	1,215,812	3,510,920	538,256	385,911	101.5
1983	1,253,040	3,016,481	477,028	454,285	133.2
1984	1,349,192	3,714,795	530,041	476,056	115.5
1985	1,468,214	4,134,463	601,025	683,683	135.9
1986	1,599,311	4,540,277	708,518	792,393	182.7
1987	1,685,581	4,531,960	725,993	705,326	158.4

Source: PORLA (2).

feed. Reports from PORLA report that in 1987 the Netherlands and Federal Republic of Germany were the major importers of Malaysian PKC importing 480,243 and 225,071 tons, respectively.

Palm kernel crushers are distributed all over the major palm oil producing states with a high concentration in Selangor, Johore and Penang. Port facilities also determine the establishment of the crushers in these states. The capacities vary widely. In 1987, a total of 87 crushers were listed by PORLA (2) with an average of 48 crushers operating in each month.

Nutritive Value of PKC

Palm kernel cake can be grouped into two types depending on the method of oil extraction from the kernel: expeller pressed and solvent extracted. Expeller pressed PKC (PKC-E) usually contains 8 to 13% oil and solvent extracted contains less than 2%. Rations containing a high level of PKC-E were inferior to those rations containing a high level of solvent extracted PKC (PKC-S) (3). The poor performance probably was due to the higher oil content that suppressed the dry matter intake (4) and also attributed to the poor digestibility of fiber by the microbes in the rumen (5).

Chemical analysis of both PKC-E and PKC-S indicated in Table 2 revealed that the protein and gross energy and nitrogen free extract were higher in the PKC-S. TDN value of 71.9 for PKC-S was reported by Miyashige, et al. (7) in their digestibility study of oil palm by-products. However, the digestibility of PKC-E was reported to be slightly higher than PKC-S (8). Calcium content for both types of PKC was almost similar vis, 0.20% for PKC-E and 0.25% for PKC-S. Compared to the nutrient requirement in Table 3, the calcium level in PKC almost satisfies the beef cattle requirement of 0.3% for a steer of about 150 kg weight gaining more than 0.5 kg per day. The phosphorus levels 0.32% and 0.52% for PKC-E and PKC-S are far higher than the phosphorus requirement of cattle. The high level of phosphorus is not only in excess

TABLE 2

Chemical Analysis of PKC

Parameters	PKC-E %	PKC-S %
Dry matter	93.4	95.3
Crude protein	12.8	14.2
Ether extract	9.5	.7
Crude Fiber	20.7	20.2
Ash	3.9	3.2
Nitrogen Free Extract	53.5	61.7
Ca	0.20	0.21
P	0.41	0.46
Mg	.27	.22
Mn (ppm)	17.1	17.4
Cu (ppm)	.6	.2
Zn (ppm)	3.7	1.5
Gross Energy (MJ/kg)	18.5	22.2

Source: Devendra (6).

of the animal's requirement of around 0.2% but also is not in the right ratio to calcium level which is supposedly 2:1 calcium to phosphorus. A practical way to correct this imbalance is to supplement calcium in the diet. Ground limestone is used normally as it is easily available at a reasonable price.

Feeding high levels of PKC is beset by the problem of copper toxicity. Abd. Rahman, Mohd. Jaafar, Shariff and Faizah (9) stressed the possible occurrence of copper toxicity, if goats and sheep were fed a PKC diet. Sheep seemed to be more sensitive to excess copper than goats. As such, feeding of more than 30% PKC in the diet of goats and sheep is not advisable.

Some cattle fed only a PKC diet have been observed to experience blindness. Supplementation with a vitamin

TABLE 3

Nutrient Requirement of Beef Cattle Under Growing Stage

Body weight (kg)	Daily gain (kg)	Dry matter intake (kg)	Energy		Protein		Ca %	P %
			TDN %	ME (Kcal/kg)	Total %	Digestible %		
150	0.25	3.1	63	2.28	11.1	7.1	0.26	0.23
	0.50	3.2	72	2.61	12.2	8.1	0.38	0.31
	0.75	3.2	78	2.82	13.3	9.0	0.54	0.42
200	0.25	4.5	57	2.06	10.0	6.1	0.18	0.18
	0.50	4.9	63	2.28	11.1	7.1	0.26	0.20
	0.75	5.0	69	2.50	11.1	7.1	0.33	0.26
300	0.25	6.1	57	2.06	8.9	5.2	0.18	0.18
	0.50	7.7	57	2.06	10.0	6.1	0.18	0.18
	0.75	8.0	63	2.28	11.1	7.1	0.21	0.18
400	0.25	7.7	57	2.06	8.3	4.6	0.18	0.18
	0.50	9.7	57	2.06	8.9	5.2	0.18	0.18
	0.75	9.9	63	2.28	8.9	5.2	0.21	0.18

Source: NRC (8).

A source is a common practice to overcome this problem. Synthetic vitamin A is available in the market. It can be introduced in the drinking water, in feed or by intramuscular injection. Requirement of vitamin A for cattle is about 2.2×1000 IU/kg of feed. Supplementation of green forage is also a good alternative to overcome the vitamin A deficiency as it is easily available on the farm.

Another shortcoming in the utilization of PKC as animal feed is the wide variation in its chemical composition. Users have to monitor the quality regularly and adjust

TABLE 4

Composition of Rations Tested and Breed of Cattle Used

Ration	Composition % PKC	Others	Breed	Reference
1	100 S	—	Sahiwal × Friesian and LID × Jersey	(11)
2	100 S	Mineral Vitamin	Sahiwal × Friesian	(12)
3	50 S	50 POME	Sahiwal × Friesian and LID × Jersey	(11)
4	50 S	50 POME 1 kg grass/day	Sahiwal × Friesian LID × Jersey	(11)
5	50 S	50 POME 1 Limestone	Kedah Kelantan	(3)
6	50 E	50 POME 1 Limestone	Kedah Kelantan	(3)
7	78 S	10 PPF 10 Molasses 1 Limestone 1 Salt	Kedah Kelantan	(3)
8	42	55 Cocoa pod 1.5 Urea 1 Salt 0.5 Vitamin	Sahiwal × Friesian female	(13)
9	42	55 Coffee pulp 1.5 Urea 1 Salt 0.5 Vitamin	Sahiwal × Friesian	(13)
10	42	55 Pineapple bran 1.5 Urea 1 Salt 0.5 Vitamin	Sahiwal × Friesian female	(13)
11	25	40 Sago pith 30 Coffee pulp 2 Salt 1 Urea 1 Vit. premix	Kedah Kelantan (KK) Brahman × KK Hereford × KK Friesian × KK	(14)
12	50.5	15 PPF 30 POME 1 Urea 3.5 Mineral vit-premix	Buffalo	(15)

PPF = Palm Pressed Fiber
POME = Palm Oil Mill Effluent
S = Solvent extracted PKC
E = Expeller pressed PKC

the feed formulas accordingly. Quite often, PKC-E especially is contaminated with broken nut shells which can cause injury to the animal's digestive tract.

Utilization of PKC

Rarely does a single feedstuff satisfy all nutrient requirements of an animal. PKC is no exception in this aspect. For a practical ration, besides nutrients, other aspects such as cost, ease in handling and processing, availability and acceptability of the feed by the animal must also be considered. PKC seems to fit these criteria very well and it has been used as a single ingredient feed with some degree of success (10,11). However, many rations containing PKC have been tested and refined to suit specific locations and situations. PKC has been blended with other cheaper by-products that would otherwise be wasted to further reduce the feed cost. Composition of rations tested and breed of cattle used are listed in Table 4 and their performance in Table 5.

Average daily gain ranges between 0.35 to 0.76 kg/day, feed intake from 3.57 to 5.81 kg/day and feed conversion ratio from 5.83 to 9.71. Lately, preliminary results for a more refined formula of PKC-based diet have shown better body weight gain. Mohd. Yusoff (16), has reported average daily gains of almost 1.0 kg/day with a refined PKC diet. Generally, cross bred cattle performed much better than the local cattle, Kedah Kelantan. But, with improved PKC diets an average daily gain of 0.66 kg/day has been reported by Shamsuddin (17). This result is comparable to the performance of most crossbred cattle fed with other PKC-based diets. Feeding with 100% PKC diet might result in blindness and abnormal body condition. It has been observed that without calcium supplementation of 100% PKC diets, the body weight gain is lower as compared to those animals raised on improved pasture of good quality.

Economics of PKC

The economics of feeding PKC to livestock is highly influenced by its price. The price of PKC fluctuates tremendously from M$120/ton to M$350/ton depending on the export demand. Normally, demand is high during the winter period in Europe, i.e., from November to February.

TABLE 5

Performance of Cattle and Buffalo Fed with PKC Based Rations

Ration*	Breed	Average daily gain (kg)	Feed intake (kg)	Feed conversion ratio
1	SF/LJ	0.59	5.34	9.05
2	SF	0.75	4.77	6.36
3	SF/LJ	0.56	5.43	9.70
4	SF/LJ	0.60	5.63	9.38
5	KK	0.60	3.72	6.20
6	KK	0.38	2.69	7.08
7	KK	0.66	5.71	8.65
8	SF	0.69	5.81	8.42
9	SF	0.69	3.82	5.54
10	SF	0.76	4.76	6.26
11	KK	0.35	3.39	9.69
	BK	0.61	3.56	5.83
	HK	0.53	3.61	6.81
	FK	0.55	3.57	6.49
12	Buffalo	0.58	5.30	9.14

*Ration as in Table 4
KK = Kedah Kelantan   BK = Brahman × KK
SF = Sahiwal Friesian   FK = Friesian × KK
HK = Hereford × KK   LJ = LID × Jersey

In feedlot operation, feed cost is the second major cost (25%) after the cost of feeder cattle (70%) while other expenses only account for the remaining 5% of the total cost. With the present animal performance of average daily gains ranging from 0.6 to 0.8 kg, when the price of PKC is above 25 cents per kg then PKC is no longer an attractive choice as a feed ingredient for fattening cattle. Sensitivity analysis of PKC prices affecting the break-even selling price (BESP) per kg of liveweight of cattle under feedlot system was tabulated in Table 6 using the formula below:

BESP/kg = Cost of Feeder (A) + Cost of Feed (B) + Other Cost (C)/(Survivability × Selling Weight)

TABLE 6

Sensitivity Analysis of PKC Prices Affecting the Break-Even Selling Price and Profit

PKC price per Kg M$	Selling price/Kg liveweight M$	Feed cost M$	BESP per Kg M$	Profit/Loss per Kg M$	Profit/Loss per animal M$
0.10	3.10	66	2.45	0.65	169.00
0.15	3.10	99	2.60	0.50	130.00
0.20	3.10	132	2.73	0.37	96.20
0.25	3.10	165	2.87	0.23	59.80
0.30	3.10	198	3.01	0.09	23.40
0.35	3.10	231	3.14	−0.14	−36.40

= (Weight × Price) + (Gain × FCF × Feed Price) + 0.06 (A + B)/(0.98 × Selling Weight)

Assumptions in computing this are: weight of feeder, 150 kg; price of feeder/kg, M$3.50; animal gain, 110 kg; feed conversion factor (FCF), 6; other cost, 6% cost of feeder and feed; survivability, 98%; selling weight 260 kg; and selling price per kg, M$3.00.

Break-even selling price increases and profit decreases as the feed price increases. A reasonable profit of M$169 per animal can be achieved at PKC price of M$0.10 per kg. As the price of PKC increases to M$0.30 an operator can still make a marginal profit of M$23.40 per animal, but a loss of M$36.40 per animal can be expected at a PKC price of M$0.35 per kg.

Potential of PKC in Ruminant Industry

Proper feeding of PKC to livestock can result in excellent animal performance. In view of its simplicity in handling, processing, storage, and relatively cheap price, PKC use should be encouraged for animal production especially in the fattening program. In fact, at the present moment interest shown by the farmers in utilizing PKC for feedlots and dairying are very encouraging. Many achievements on the fattening of beef cattle using PKC have been highlighted in mass media lately. It is quite common for a farmer raising 10 to 20 head of cattle to achieve a profit of M$150.00 per animal after a fattening period of six months. Babjee (10) reported that a pilot project consisting of 10 farmers each with ten animals, received net returns ranging from M$2,000.00 to M$4,000.00 depending on the breed of cattle used and feed combination and quality.

An encouraging finding by MARDI recently in developing PKC based rations indicates that even Kedah Kelantan cattle can attain an average daily gain of 0.66 kg/day. With this ration, local cattle can be fattened at a much lower price as compared to crossbred cattle bought from local commercial farms or overseas. With the high demand for beef, the high production of PKC and the availability of cheap local feeder animals, the potential for livestock production in this country through feedlots, is very great indeed.

References

1. Nik Mahmood, M., "Potential of Ruminant Production in Malaysia," presented at Malaysian-French Symposium on Recent Advances in Livestock Production, 1986.
2. Anonymous, Oil Palm Update, Palm Oil Research and Licensing Authority, Kuala Lumpur, Malaysia (1988).
3. Shamsuddin, A.B., D. Mohd, Jaafar and Y.A. Wahab, "Performance of Kedah-Kelantan Cattle Fed with Different Combinations of Expeller Pressed and Solvent Extracted Palm Kernel Cake," in proceedings of Conferences on Advances in Animals Feeds and Feeding in the Tropics, pp. 287–291, (1987).
4. Palmquist, D.L., and H.R. Conrad, *J. Dairy Sci. 61*:890 (1980).
5. Harfoot, G.C., *Prog. Lipid Res. 17*:21 (1978).
6. Devendra, C., "Utilization of Feedingstuffs from the Oil Palm," in proceedings of the Conference on Feedingstuffs for Livestock in South East Asia, Kuala Lumpur, Malaysia, pp. 116–131 (1978).
7. Miyashige, T., O. Abu Hassan, D. Mohd. Jaafar and H.K. Wong, "Digestibility and Nutritive Value of Palm Kernel Cake, Palm Oil Mill Effluent, Palm Press Fiber and Rice Straw by Kedah Kelantan Bulls," in proceedings of the Conference on Advances in Animal Feeds and Feeding in the Tropics, pp. 226–229 (1987).
8. Anonymous, "Nutrient Requirement of Beef Cattle" in *Basic Animal Nutrition and Feeding*, edited by D.C. Church and W.G. Pond, pp. 282–283 (1976).
9. Suparjo Nordin, M., and M.Y. Abdul Rahman, "Digestibilities of Palm Kernel Cake, Dried Palm Oil Mill Effluent and Guinea Grass by Sheep," in proceedings of conference on Advances in Animal Feeds and Feeding in the Tropics, pp. 230–234 (1987).
10. Abd. Rahman, M.Y., D. Mohd. Jaafar, H. Shariff and M. Faizah, "Feedlot Performance of Goats and Sheep Fed with Oil Palm and Rice By-products," in proceedings of the Conference on Advances in Animal Feeds and Feeding in the Tropics, pp. 240–244 (1987).
11. Babjee, A.M., *Asian Livestock 11*:5 (1986).
12. Hamali, A.Y., "Potential in Utilizing Agricultural By-products for Growing and Fattening Beef Cattle," paper submitted for annual report, Livestock Research Division, Mardi, Serdang, Selangor, Malaysia (1984).
13. Babjee, A.M., M. Hussein and R.M. Lam, "Palmbeef: A Value Added Product of Palm Kernel Cake," presented at 8th MSAP Annual Conference, April 1984.
14. Dahlah, I., "Utilization of Agricultural By-products as a Feed for Feedlot Cattle," in annual report of Livestock Research Division, MARDI, Serdang, Selangor, Malaysia (1986).
15. Dahlan, I., "Performance of Kedah Kelantan Cattle and Its Crosses on Ration Containing Sago Pith, Coffee Pulp and Palm Kernel Cake, Reared under Smallholders Condition," in proceedings of the 5th International Conference of the Institute of Tropical Veterinary Medicine, Kuala Lumpur, Malaysia (1986).
16. Mohd. Sukri, H.J., M.N. Jariah and H. Rozali, "Fattening of Swamp Buffalo in Feedlot," in proceedings of Advances in Animal Feeds and Feeding in the Tropics, pp. 272–275 (1987).
17. Mohd. Yusoff, A., "Utilization of Oil Palm and Paddy By-Products for Fattening Cattle in Feedlot." Paper submitted for 1st Quarter Technical Report, Livestock Research Division, (1987).

Soybean Meal as the Only Supplementary Protein Source for Poultry Feeds in the People's Republic of China (PRC)

Don H. Bushman[a] and C.M. Collado[b]

[a]American Soybean Association, Beijing, People's Republic of China 100004, and [b]American Soybean Association, Singapore 0923

Abstract

Five broiler and three layer trials were conducted to determine the feasibility of eliminating fish meal from poultry diets in China, the effect of soybean meal quality, and the economic efficiency of using high energy diets for broilers. High quality soybean meal significantly increased weight gain of broilers and egg production in layers compared to similar diets containing fish meal. Variation in soybean meal quality is a major problem in China and performance was reduced when either overheated or under-processed soybean meal was utilized. In China the national standard for energy and protein for broiler feeds are 2,900 kcal/ME/kg and 21%, and 3,000 kcal/ME/kg and 19%, respectively, for starter and finisher diets. Increasing energy levels above the national standard improved the rate of gain but depressed economic efficiency due to the high cost of fat/vegetable oils in China.

It is a common practice in China that all feeds must contain fish meal. This may be based at least in part upon results obtained from feeding incomplete diets or diets containing high levels of antinutritional factors when plant proteins are utilized. For example, there is a wide variation in the quality of soybean meal and other plant proteins, as well as premixes, utilized in China. In addition, grains produced in Northeastern China are very deficient in selenium. However, the price of imported fish meal has essentially doubled in the past several months; currently it sells for 4.0–4.2/kg in local currency (RMB yen–People's Currency), or US $1,070–1,125/MT vs RMB yen 0.85–0.95/kg (US $230–255/MT) for soybean cake and meal. This has generated tremendous interest in the utilization of soybean meal to replace fish meal in diets for poultry, livestock and aquaculture.

The phenol red rapid test for urease was used to check processing adequacy of the soybean meals/cakes utilized in the five broiler and three layer trials. This method was selected because the trials were conducted on commercial farms. Consequently, there was little or no chance to control the quality of the meal or cake. It was also used to bring awareness to the farmers and feed millers of the wide variation existing in soybean meal quality.

Broiler Trials

The first broiler trial was conducted to compare simple corn-soybean meal and corn-soybean cake diets against a similar diet that contained 5% and 3% imported fish meal in the starter and finisher diets, respectively. In addition, two commercial feeds, one with fish meal and the other a mixture of plant proteins also were tested.

A total of 2,900 day-old Arbor Acre chicks were used in a completely randomized design. The chicks were brooded on the floor and fed a diet containing 2,900 kcal ME/kg and 21% protein from one day old through 28 days of age. On the 29th day they were placed in battery cages and fed a diet containing 3,000 kcal ME/kg and 19% protein through 56 days of age. These are China's national standards for energy and protein in broiler diets. All the diets were in mash form and fed *ad libitum*.

From the results presented in Table 1, the diet containing soybean meal as the only supplementary protein appeared to reduce weight gain and increase the amount of feed required to produce a kilogram of gain. When subjected to the phenol-red test, the soybean meal did not show any response even after 20 min. This indicated it was probably over-toasted. Nevertheless, it was used due to the unavailability of properly processed soybean meal. The same soybean meal was utilized in the commercial plant-protein feed; however, this diet contained additional lysine and methionine. When this diet was used, there was no significant difference in body weight or feed efficiency.

The phenol-red test on the soybean cake indicated acceptable processing. Feeding the diet containing soybean cake resulted in the heaviest body weight, and the feed efficiency was essentially equal to the fish meal diets and the commercial plant-protein diet. It also resulted in the most economical feed cost/kilogram of gain. The commercial plant-protein diet gave the second most economical feed cost/kilogram of weight gain. Although the soybean meal diet appeared to depress weight gain and feed efficiency, it still resulted in a feed cost/kilogram of weight gain equal to that of a similar diet containing fish meal. This of course was directly related to the large difference in price of the two ingredients.

Regarding economic efficiency, it is important to understand that China has a relatively complex pricing system including both a State and free market price for most ingredients. State prices are much lower than those found in the free market. However, the quantity of individual ingredients available at State prices varies and frequently is tied to the marketing of the finished product, i.e., eggs, pork, etc., to the State at the State price. To simplify calculations, representative free market prices at the time the trials were conducted are used throughout this paper.

The results from this trial indicated that soybean meal/cake can be used to replace fish meal in a balanced diet. However, they also indicated the importance of properly processed soybean meal.

Based upon the results of this trial, four additional trials were conducted to compare the economic and production efficiency when quality soybean meal was used to replace imported fish meal. In addition, varying energy levels were applied to determine if high energy diets would be economical in China.

In the second trial, 2,985 day-old Arbor Acre chicks were used in a 2 × 3 factorial design. Three energy lev-

TABLE 1

Effect of Replacing Fish Meal with Soybean Meal or Soybean Cake on 56-Day Growth and Feed Cost for Broilers

	Test diets			Commercial feeds	
	Fish meal	Soybean meal	Soybean cake	Plant[a] protein	Fish[b] meal
Number of birds	580	580	580	580	580
Performance data 0–56 days					
Body wt (kg)	1.98	1.88	2.00	1.93	1.98
Feed intake (kg)	4.37	4.43	4.50	4.33	4.44
Feed: gain ratio	2.25	2.41	2.30	2.29	2.29
Feed cost per kg of gain					
RMB yen	2.02	2.01	1.80	1.83	2.00
US$	0.54	0.54	0.48	0.49	0.54

[a] Commercial feed containing no fish meal and a mixture of domestic oilseed cakes and meals.

[b] Commercial feed containing imported fish meal and a mixture of domestic oilseed cakes and meals.

els, 2,890/2,990, 3,040/3,140 and 3,155/3,250 kcal ME/kg were applied, respectively, in the starter/finisher diets with and without fish meal. The fish meal diets were typical commercial diets containing 4.5% and 4.3% fish meal in the starter and finisher diets, respectively. Imported, U.S. high protein soybean meal (high pro, 48.5% protein) served as the only supplementary protein in the nonfish meal diets. All the feeds were in mash form and were fed *ad libitum*. The chicks were brooded on the floor from one day old through 28 days of age and received the starter diets. At 29 days of age they were placed in battery cages and fed the finisher diet through 56 days of age.

Results from the second trial are presented in Table 2. Both protein sources, fish meal and soybean meal, and energy level resulted in a highly significant difference ($P < .01$) in 56-day body weight. However, neither protein source nor energy level had a significant effect on feed efficiency.

It is apparent that the corn-soy diet resulted in greater weight gains at the two lower energy levels, but not at the highest energy level. In fact, the highest energy level appeared to decrease growth rate with both diets. The reason for this lower body weight for birds receiving the high energy diets is not known and did not occur in the other trials. In part, it may be attributed to a heat wave which resulted in the temperature inside the house exceeding 36°C during the last five days of the trial. According to the farm workers, the heavier birds, especially those fed the higher energy diets, were most

TABLE 2

Effect of Energy Level and Replacing Fish Meal with U.S. High Pro Soybean Meal on Broiler Performance (Trial 2)

Dietary treatments ME (kcal/kg)[a] Protein (%)[a]	2,890/2,990 20.9/19.0		3,040/3,140 22.2/20.0		3,150/3,250 22.7/20.6	
Protein source	Fish meal	US high pro SBM	Fish meal	US high pro SBM	Fish meal	US high pro SBM
Number of birds	497	497	498	498	498	497
Performance data 0–56 days						
Body wt (kg)	2.115[a]	2.174[b]	2.110[a]	2.260[c]	2.106[a]	2.145[a,b]
Feed intake (kg)	4.735	4.868	4.735	4.889	4.688	4.796
Feed:gain ratio	2.28	2.28	2.29	2.20	2.27	2.28
Feed cost per kg gain						
RMB yen	1.93	1.77	2.25	2.00	2.40	2.20
US$	0.52	0.48	0.61	0.54	0.65	0.59

[a] Starter/finisher diets, respectively.

[a-c] Means in a row with no common superscripts differ significantly ($P < .05$).

affected by the heat. Feed consumption dropped to almost nil for these five days. Because the study was conducted on a commercial farm this was only an observation; none of the birds were actually weighed for fear of high mortality through increased stress if they were handled.

With regard to feed cost/kilogram of gain, the corn-soy diets reduced cost by 8% or more at every energy level. This resulted from improved performance and lower feed cost.

In Trial 3, 2,400 Arbor Acre day-old chicks were used in a 2 × 3 factorial design similar to that employed in Trial 2. However, a much higher level of fish meal was used. In the starter diet, the fish meal level was 7%. This was increased to 9% in the finisher diet. This step-up style of fish meal usage was in agreement with the commercial practice being used at the farm. The soybean meal utilized in the nonfish meal diets was U.S. high pro. The birds were brooded and finished on concrete floors with litter. The energy levels used were: 2,950/3,130 (low), 3,075/3,240 (medium), and 3,175/3,340 Kcal ME/kg (high), respectively, in the starter/finisher diets. They were fed in mash form, *ad libitum*.

In this trial, protein source and energy level significantly affected ($P < .01$) 56-day body weight and feed efficiency. From the data presented in Table 3, it is apparent that the corn-soybean meal (corn-soy) diet resulted in heavier body weights than the fish meal control at each energy level. Also within each protein source increasing the energy level tended to increase weight gain. This difference was significant ($P < .05$) between the low and medium energy levels for the fish meal diets, and between the low and highest energy levels for the soybean meal diets. At the low and medium energy levels birds fed the corn-soy diets exhibited significantly better feed efficiency than those fed the fish meal control diets. In the case of soybean meal, the birds fed the medium energy level diet also had the best feed efficiency (significant $P < .05$) compared to other diets.

As in the second trial, the corn-soy diet resulted in much lower feed cost/kilogram of gain. Due to the very high level of imported fish meal used in this trial, the saving in feed cost was over 20% at each energy level. The results of Trials 2 and 3 definitely demonstrated that high quality soybean meal could replace imported fish meal. However, because these two trials utilized imported soybean meal, a question remained as to the performance when domestic soybean meal is used.

A fourth trial was conducted utilizing 3,072 day-old Shaver male chicks from a female breeding line in 4 × 4 factorial design. The chicks were brooded and finished in battery cages with 15 birds/cage. Each treatment was replicated 12 times. The protein sources tested were imported fish meal; full-fat soybean meal from domestic soybeans; domestic high protein soybean meal (48.6% protein); and imported U.S. high protein (48.6% protein) soybean meal. The energy levels applied were: 2,900/3,000, 3,000/3,100, 3,100/3,200 and 3,200/3,300 kcal ME/kg, respectively, in the starter/finisher mash diets. In this trial, the starter diets were fed for 21 days and the finisher diets from day 22 through 49 days of age.

Results from this trial are presented in Table 4. There was a significant difference ($P < .01$) in both 49-day body weight and feed efficiency among protein sources and energy level. In the case of protein source the U.S. high protein soybean meal diets significantly ($P < .05$) increased body weight over all other protein sources at each energy level. At the two lower energy levels there was no significant difference in weight gain between the other protein sources. However, at the two higher energy levels the fish meal resulted in significantly greater ($P < .05$) gain than full-fat soybean meal or the Chinese soybean meal.

Between treatments, there was no significant difference in feed efficiency within energy levels except at the lowest energy where the diet containing U.S. soybean meal improved efficiency ($P < .05$) compared to the diet containing full-fat soybean meal. Increasing the energy level appeared to improve feed efficiency in all treatments. However, the difference was not significant, except for the higher energy diets containing U.S. soybean meal

TABLE 3

Effect of Energy Level and Replacing Fish Meal with U.S. High Pro Soybean Meal on Broiler Performance (Trial 3)

Dietary treatments ME (kcal/kg)[a] Protein (%)[a]	2,950/3,130 21.1/19.9		3,075/3,240 22.3/20.4		3,175/3,345 23.1/21.0	
Protein source	Fish meal	US high pro SBM	Fish meal	US high pro SBM	Fish meal	US high pro SBM
Number of birds	400	400	400	400	400	400
Performance data 0–56 days						
Body wt (kg)	1.764[a]	1.925[b]	1.834[c]	1.964[b,d]	1.856[c]	2.007[d]
Feed intake (kg)	4.078	4.186	4.356	4.100	4.161	4.394
Feed:gain ratio	2.31[a]	2.17[b]	2.37[a]	2.09[c]	2.24[a,b]	2.19[b]
Feed cost per kg gain						
RMB yen	2.38	1.89	2.62	2.11	2.90	2.38
US$	0.64	0.51	0.71	0.57	0.78	0.64

[a] Starter/finisher diets, respectively.
[a-d] Means in a row with no common superscripts differ significantly ($P < .05$).

TABLE 4

Effect of Energy Level and Replacing Fish Meal with Full-fat, Chinese High Pro and U.S. High Pro Soybean Meal on Broiler Performance and Feed Cost

Dietary treatments	Performance data 0–49 days			Feed cost per kg gain	
	Body wt (kg)	Feed intake (kg)	Feed:gain ratio	RMB yen	US$
ME (kcal/kg) 2,900/3,000—Protein (%) 21.0/19.0[a]					
Fish meal	1.40[a]	3.10	2.20[a,b]	1.790	0.482
Full-fat	1.41[a]	3.26	2.31[a]	1.885	0.508
Chinese high pro	1.37[a]	3.01	2.20[a,b]	1.633	0.440
US high pro	1.59[g]	3.29	2.07[b,c]	1.634	0.440
ME (kcal/kg) 3,000/3,100—Protein (%) 21.0/18.0[a]					
Fish meal	1.49[c,d,e]	3.01	2.02[b,c]	1.797	0.484
Full-fat	1.46[c]	3.15	2.16[a,b]	1.851	0.499
Chinese high pro	1.45[b,c]	3.13	2.16[a,b]	1.782	0.480
US high pro	1.67[h]	3.36	2.01[b,c]	1.761	0.474
ME (kcal/kg) 3,100/3,200—Protein (%) 22.5/20.3					
Fish meal	1.54[e,f,g]	3.20	2.08[b,c]	2.030	0.547
Full-fat	1.47[c,d]	3.06	2.08[b,c]	1.966	0.530
Chinese high pro	1.44[b,c]	2.96	2.06[b,c]	1.878	0.506
US high pro	1.73[i]	3.98	1.93[c]	1.861	0.501
ME (kcal/kg) 3,200/3,300—Protein (%) 23.0/20.9					
Fish meal	1.58[g]	3.21	2.03[b,c]	2.149	0.579
Full-fat	1.55[f,g]	3.19	2.06[b,c]	2.115	0.570
Chinese high pro	1.52[d,e,f]	3.05	2.01[b,c]	1.997	0.538
US high pro	1.78[j]	3.41	1.91[c]	2.009	0.541

[a] Starter/finisher diets, respectively.
[a-j] Means in a column with no common superscripts differ significantly (P < .05).

compared to the other protein sources at the lowest energy level.

As in the previous trials, the corn-soy diets resulted in the most economical weight gain. There was very little difference in feed cost between the diets containing the U.S. and the domestic soybean meal. Both diets reduced feed cost/kilogram of gain by 7–9% compared with the fish meal diet except at the second energy level. This difference in cost of production agrees very well with the results from the second trial.

A fifth and concluding broiler trial was conducted using 2,000 day-old Arbor Acre chicks in a 2 × 2 factorial design to further test domestic soybean meal. The energy and protein levels were 2,900/3,000 kcal/ME/kg and 21/19% protein, and 3,000/2,100 kcal/ME/kg and 22/20% protein in the started/finisher diets, respectively. The fish meal diets contained 3% imported fish meal in both the starter and finisher. The nonfish meal diets utilized domestic soybean cake containing 43% protein. However, in contrast to the previous trials the feeds were pelleted in this trial.

The birds were brooded and finished on solid floors with litter. The starter diets were fed through 28 days of age and the finisher diets from 29 to 49 days of age. Each treatment was replicated four times with 125 birds/replication.

Results from this trial are presented in Table 5. There was no significant effect of protein source or energy level on either rate of gain or feed efficiency. This further indicates that good quality domestic soybean meal can give results equivalent to imported fish meal. In this trial, growth rate and feed efficiency were better than those obtained in the previous trials. This might be due in part to the fact that pelleted feeds were used in this trial and mash feeds in the other trials. However, in none of the trials were pellets vs mash compared. It could also be due in part to the milder temperatures when this trial was conducted.

Regarding cost efficiency, at the lower energy level the soybean meal diet reduced feed cost/kilogram of gain by 9%, or equal to results obtained in Trials 2 and 4. At the higher energy level, the savings in feed cost was 4%.

Layer Trials

Three layer trials were conducted on two commercial farms. The first trial conducted on farm A used 11,704 white Leghorn hybrids (Shaver × local strain) to compare production between birds fed a simple corn-soy diet vs the normal farm diet (control) containing 10% fish meal. Each treatment was replicated six times.

The second trial had 10,297 Dekalb brown egg layers, and the third trial 10,260 White Leghorn hybrids. Both trials were conducted on farm B. As in the first trial the comparison was between a simple corn-soy diet and the normal farm diet (control), which also contained 10% fish meal. In the second and third trials each treatment was replicated three times.

TABLE 5

Effect of Energy Level and Replacing Fish Meal with Domestic Soybean Meal on Broiler Performance (Trial 5)

Dietary treatments ME (kcal/kg)[a] Protein (%)[a]	2,900/3,000 21.0/19.0		3,000/3,100 22.0/20.0	
Protein source	Fish meal	Soybean meal	Fish meal	Soybean meal
Number of birds	500	500	500	500
Performance data 0–49 days				
Body wt (kg)	2.05	2.04	2.09	2.03
Feed intake (kg)	3.95	3.91	3.93	3.91
Feed:gain ratio	1.92	1.92	1.88	1.93
Feed cost per kg gain				
RMB yen	1.594	1.461	1.815	1.746
US$	0.430	0.394	0.489	0.471

[a] Starter/finisher diets, respectively.

In all trials the birds were maintained in cages, four birds/cage. Feeding was done with automatic feeders and the eggs were hand collected. The trials were conducted from December 1987 through September 1988. Trial 1 was conducted over nine 28-day periods and trials 2 and 3 over 10 28-day periods.

Results of the layer trials are presented in Table 6. Due to differences in management systems and strain of birds used, large variations in performance were obtained among the three trials. In trials 1 and 3 White leghorn hybrids were used, and purity of the lines is questionable. However, many layer farms in China use similar strains and it was deemed necessary to compare results on these birds as well as on high performing layers. The brown egg layers used in the second trial were from a pure line.

Feeding a simple corn-soy diet resulted in a significant increase in egg production and egg size in Trials 1 and 2. Consequently there was a large difference, 5%, in egg mass (significant P < .01). There was no significant difference in rate of lay or egg size between treatments in Trial 3. However, egg size tended to be larger from birds fed the corn-soy diet, and the difference in egg mass, 2.4%, was significant (P < .01). There was no significant difference in feed consumption between treatments within trials. However, feed consumption and soybean meal quality were associated with egg production (Fig. 1).

As noted in Figure 1, daily feed consumption exceeded 120 g for birds in both treatments during the first 30 weeks of the trial. However, at the start of the 30th week a heat wave resulted in a severe drop in feed intake and egg production. Temperatures returned to normal after about five weeks and daily feed intake also increased for both groups. As is also apparent from Figure 1 egg production in corn-soy fed birds consistently exceeded that of the control group during the first 31 weeks. However, after 32 weeks the difference in rate of lay was reduced; it fell below that of the controls in the 39th week. This decrease in rate of lay was accompanied by a reduced feed intake. However, feed consumption still exceeded 110 g. Based upon calculated amino acids this level of feed consumption would provide 913 mg of lysine, 670 mg of total sulfur amino acids daily which is well above the NRC (1984) requirements indicating that feed consumption *per se* was not the reason for the drop in egg production. In Trial 3 the same corn-soy diet supported a rate of lay equal to that of the control diet during peak lay even though

TABLE 6

Comparison of a Corn-Soy Diet with Commercial Layer Diets

	Trial 1		Trial 2		Trial 3	
	Corn-soy	Fish meal	Corn-soy	Fish-meal	Corn-soy	Fish meal
Number of birds	5,852	5,852	5,139	5,158	5,150	5,100
Egg prod, %	54.88[a]	53.21[b]	76.03[c]	73.97[d]	68.08[a]	68.14[a]
Egg size, g	54.43[c]	53.64[d]	62.70[c]	61.91[d]	59.63[a]	59.21[a]
Egg mass, kg	42,038[c]	39,913[d]	66,161[c]	63,119[d]	55,556[c]	54,230[d]
Feed consump, g/day	116.5[a]	120.4[a]	119.8[a]	120.0[a]	103.2[a]	103.1[a]
Kg feed/kg egg	3.90[a]	4.22[a]	2.51[c]	2.62[d]	2.54[a]	2.56[a]
Feed cost/kg egg, US$	0.683	1.042	0.439	0.673	0.413	0.658

[a,b] Means in a row within each trial with no common superscripts differ significantly (P < .05).
[c,d] Means in a row within each trial with no common superscripts differ significantly (P < .01).

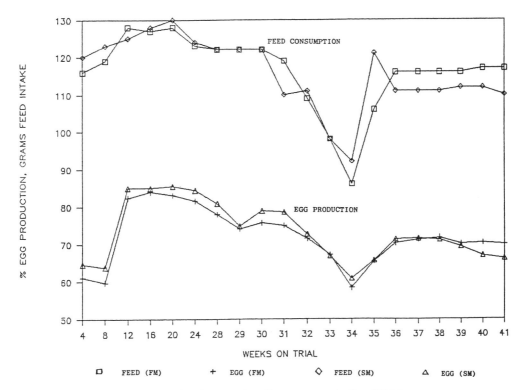

Fig. 1. Effect of temperature and soybean meal quality on egg production, Trial 2.

feed consumption was only 105 g compared to 110 g for the controls. Likewise when feed intake dropped to 93 g between the 29th and 30th week it still supported a rate of lay equal to the controls consuming 112 g. However, egg production was also consistently slightly lower for the corn-soy fed birds in Trial 3 after the 36th week of lay, while feed intake was essentially the same for both groups. Examination of the soybean meal utilized during the latter 8 to 10 weeks of these trials resulted in a completely negative test with phenol red. This indicated that a reduction in amino acid availability probably resulted in the reduced rate of lay. When feed consumption was over 120 g the daily intake of lysine and sulfur amino acids would be approximately 42% and 22%, respectively, above the NRC requirements. This could help compensate for reduced amino acid availability. A very similar response occurred in Trial 1. However, in this trial feed intake remained essentially equal for both groups. In this case the soybean meal gave a very strong positive test to phenol red, indicating that the meal was under-toasted. This was also supported by enlarged pancreases in posted birds. These results also illustrate the importance of utilizing properly processed soybean meal.

There was a tendency for improved feed efficiency when the corn-soy diets were fed. The difference was significant ($P < .01$) in Trial 2 and approached significance in Trial 1. The most notable difference between treatments was cost of production in all trials. As a result of reduced feed cost and increased egg mass, feeding the corn-soy diet reduced feed cost/kilogram of eggs by over 34% in each trial.

In summary, good quality soybean meal and/or cake can be efficiently and economically utilized as the only supplementary protein in poultry diets in China. In fact, high quality soybean meal appears to be superior to imported fish meal. However, as evidenced from these trials, improved processing techniques are needed to improve and stabilize domestic soybean meal quality. Until soybean meal/cake quality is stabilized, very close attention will have to be given to testing the level of toasting. In the case of over-toasted meal/cake it may be able to compensate for loss in available amino acids by either increasing the level of protein and/or additional use of synthetic amino acids. The choice will depend upon economic efficiency. In the case of under-toasted meal/cake it should be relegated to other uses.

Acknowledgements

This work was supported in part by the Animal Husbandry Division of Shanghai Municipal Agricultural Bureau, Shanghai, PRC; Cargill, Inc. Minneapolis, Minnesota; Conti Chia Tai, Hongkong; Elanco Products Co., Indianapolis, Indiana; Harbin Feedstuff Scientific Research Institute, Harbin, PRC; Heilongjiang Province Feedstuffs Company, Harbin, PRC; Hong Qiao Poultry Farm, Shanghai, PRC; Shanghai Poultry and Egg Co. Jin Qiao Layer Farm, Shanghai, PRC; Shanghai Scientific Research Institute, Shanghai, PRC; Shanghai Xinfeng Feedmill, Shanghai, PRC; Wuxi Feedmill, Wuxi, PRC; Xin Yang Farm, Shanghai, PRC.

References

1. Anonymous, *National Research Council Nutrient Requirements of Poultry*, 8th edn., National Academy Press, Washington, D.C., 1984.

Amino Acid Composition of Feedgrade Rice By-products From Several Countries

David Creswell

Monsanto Singapore Company PtE. LTD., Singapore

Abstract

Rice by-products from the rice milling industry are important livestock feed ingredients throughout Asia and Australia. The three major products are rice bran, extracted rice bran and broken rice. They contribute energy, amino acids and linoleic acid to diets. Past research as well as industry usage have defined the problems and limitations of these ingredients. Broken rice is well defined, is an excellent energy source, and has no nutritional factors limiting its use. Rice bran has problems of hydrolytic (and occasionally oxidative) rancidity, high phytate phosphorus and other undefined performance depressing components. These generally limit its inclusion to 10% and 20% in broiler and layer diets, respectively. Extracted rice bran has all the problems of rice bran except for rancidity. Ducks are able to utilize rice bran at high levels, as can finishing and breeding pigs. Availability of amino acids does not appear to be a problem in rice by-products. Rice by-products are variable in composition and little published work has appeared in recent years as to their amino acid composition. The present research was conducted to characterize the amino acid content of the three products from several nations (Australia, Indonesia, Malaysia, The Philippines and Thailand) and to define the variation. The data presented provide information that should be useful to the feed industry when formulating with these products. Regression equations enabling prediction of amino acid content from proximate analysis will be presented.

Large quantities of rice by-products are produced from the milling of rice in many countries. The primary use of these by-products is as animal feedstuffs. The three major by-products are rice bran (also called rice pollard or rice polish), extracted rice bran and broken rice.

Past research and industry usage have defined the problems and limitations of these materials. Rice bran has problems of hydrolytic (and occasionally oxidative) rancidity, high phytate phosphorus and other unidentified performance-depressing components. These factors generally limit its inclusion to 10% and 20% in broiler and layer diets, respectively. There is some evidence that ducks can utilize rice bran at high levels without showing depressed performance. The same may be true for finishing and breeding pigs.

From limited studies in the literature, availability of amino acids in rice bran does not appear to be a problem. Retention of calcium and other minerals is depressed by high levels of rice bran. Young chickens do not utilize the oil in rice bran well, but their ability to do so improves with age. Extracted rice bran has all the problems of full fat rice bran except for rancidity. Its energy value is reduced as a result of oil extraction. Broken rice is well defined, is an excellent energy source and has no nutritional factors limiting its use.

Rice by-products are notoriously variable in composition, and little published work has appeared in recent years on their amino acid composition. The present research was conducted to define the proximate and amino acid composition of these three products from several countries: Australia, Thailand, Indonesia, Malaysia and the Philippines. These data should be useful to the feed industry when formulating with these materials. Some examination of the data suggests that effects of proximate composition on energy value and amino acid content can be predicted accurately.

Methods

A total of 82 samples was collected from Indonesia, Malaysia, Thailand, The Philippines, India and Australia. Of these, 55 samples were rice bran, 11 were extracted rice bran and 16 were broken rice (Table 1). These samples generally were material that was made available to the commercial feed industry. Samples were sent to Monsanto's U.S. laboratories for analysis. Assays for proximate principles and amino acids were by standard methods. Tryptophan was analyzed by a colorimetric assay, and peroxidation prior to acid hydrolysis was used for sulfur amino acids (Table 2).

TABLE 1

Sample Numbers and Origin

Rice bran	55	
Extracted rice bran	11	
Broken rice	16	
Total	82	samples
Indonesia	29	(29 rice bran)
Malaysia	17	(7 rice bran, 3 extracted rice bran, 7 broken rice)
Thailand	19	(3 rice bran, 7 extracted rice bran, 9 broken rice)
The Philippines	6	(6 rice bran)
India	1	(extracted rice bran)
Australia	10	(10 rice bran)
Total	82	samples

Rice Bran

Analytical results are presented in Tables 3–7. Mean composition of Asian rice bran is 12.9% protein, 16% fat, 8.2% fiber and 9.4% ash. Note the fairly high standard deviation values which indicate the high degree of variability in this product. Differences in rice bran between countries did not appear consistent and so are not shown (with the exception of Australia).

Table 4 shows a breakout of the Asian rice brans into good and poor samples on the basis of protein and fiber levels. This indicates the potential for selecting and testing better quality material for purchase. Physical exami-

TABLE 2
Methods of Analysis

Proximate analysis	
Crude protein	Macro Kjeldahl
Moisture	Loss on drying
Fiber	Neutral detergent fiber
Fat	Goldfisch ether extraction
Ash	Residue at 500°C
Amino acid analysis	
Majority of amino acids	Method of Moore and Stein
Sulfur amino acids	Peroxidation before acid hydrolysis
Tryptophan	Colorimetric assay

TABLE 3
Composition of Asian Rice Bran[a] —All Samples[b]

	$\overline{X}$	SD
Moisture	10.7	1.32
Protein	12.9	1.10
Fat	16.0	2.45
Fiber	8.2	2.80
Ash	9.4	2.28
Methionine	.26	.02
Cystine	.27	.02
(M + C)	(.53)	.04
Lysine	.58	.07
Arginine	1.02	.10
Tryptophan	.10	.02
Tyrosine	.43	.05
Threonine	.49	.04
Serine	.59	.05
Phenylalanine	.59	.05
Aspartic acid	1.15	.09
Glutamic acid	1.79	.16
Proline	.55	.04
Glycine	.71	.07
Alanine	.80	.07
Valine	.67	.06
Isoleucine	.45	.04
Leucine	.89	.07
Histidine	.35	.03
Sum of amino acids	11.77	.95

[a] Values are in percentages on as is basis.
[b] 45 Samples of rice bran from Malaysia (7), Thailand (3), Indonesia (29), and The Philippines (6).

TABLE 4
Composition of Asian Rice Brans[a] —Selected Samples[b]

	Good samples	Poor samples
Moisture	10.3	10.8
Protein	13.7	10.9
Fat	17.0	13.9
Fiber	6.7	12.6
Ash	8.1	11.4
Methionine	.27	.22
Cystine	.29	.23
(M + C)	(.56)	(.45)
Lysine	.63	.53
Arginine	1.06	.84
Tryptophan	.10	.085
Tyrosine	.44	.35
Threonine	.50	.41
Serine	.61	.49
Phenylalanine	.62	.50
Aspartic acid	1.19	.97
Glutamic acid	1.87	1.51
Proline	.57	.48
Glycine	.73	.59
Alanine	.82	.67
Valine	.69	.57
Isoleucine	.46	.38
Leucine	.93	.76
Histidine	.37	.30
Sum of amino acids	12.17	9.96

[a] Values are in percentages on as is basis.
[b] 15 samples good quality and 15 samples poor quality rice bran.

operation. The effect of rice hull inclusion is seen in higher levels of fiber and ash (silica) and lower levels of fat, protein and amino acids. Rice hulls are easily detected by use of low power microscopy. Addition of sand or similar material is sometimes seen. High ash values will result.

Table 5 presents six samples of adulterated rice brans, detected from proximate composition. Sample 2 shows evidence of addition of sand or similar material while other samples have high levels of rice hulls.

TABLE 5
Composition of Asian Rice Bran—Adulterated Samples[a]

Sample no.	1	2	3	4	5	6
Moisture	8.7	8.1	9.7	6.9	11.1	11.0
Protein	10.4	11.0	11.0	2.1	8.5	9.5
Fat	12.3	14.2	12.1	0.3	8.2	11.9
Fiber	13.4	9.8	13.0	41.8	19.2	14.8
Ash	16.4	18.2	13.1	21.1	12.4	9.8
Origin[b]	M	T	P	P	I	I

[a] Values are in percentages on as is basis.
[b] Origin of samples: M, Malaysia; T, Thailand; P, The Philippines; I, Indonesia.

nation, physical and chemical testing allow this to be done. The magnitude of the difference in protein, fat, fiber, ash and amino acids between the two groups is quite remarkable.

Adulteration of rice by-products is common in Asia. Most often this is due to inclusion of rice hulls, which may be deliberate or due to the normal action of the dehulling

Five samples showing unusually low fat levels are presented in Table 6. High levels of hulls will reduce fat by a simple dilution effect but this does not explain all these samples. Note Sample 2, for example. These low fat levels clearly will reduce the energy value.

Australian rice brans (rice pollard) are shown separately in Table 7. They show less variability than Asian rice brans. Protein, fiber and ash values suggest there is little or no inclusion of hulls. The very high level of fat (19.7%) is noteworthy. Their composition is similar to the best of the Asian rice brans, although even in this comparison they are higher in fat and in level of lysine. In Australia, rice pollard is given a very high value by nutritionists and is widely used in broiler and layer diets. Its contribution of linoleic acid is considered particularly valuable.

Extracted Rice Bran

As a result of oil extraction, protein, fiber, ash and amino acids are all increased. Rice hull adulteration is common in some countries and this is clearly seen in the sample of Indian origin shown in Table 8. None of the samples collected from Thailand showed evidence of adulteration and so are presented separately in Table 9. These values can be assumed to represent "pure" extracted rice bran.

TABLE 6

Composition of Asian Rice Bran[a]—Low Fat Samples[b]

	Sample no.					
	1	2	3	4	5	Mean
Moisture	13.6	13.9	9.6	10.6	11.1	11.8
Protein	12.7	12.0	14.9	16.6	8.5	12.9
Fat	7.5	6.8	6.2	10.3	8.2	7.8
Fiber	12.2	8.0	15.2	11.8	19.2	13.3
Ash	9.9	3.6	11.2	14.6	12.4	10.3

[a] Values are in percentages on as is basis.
[b] All samples are of Indonesian origin.

TABLE 7

Composition of Australian Rice Bran[a,b]

	$\bar{X}$	SD
Moisture	9.3	.96
Protein	13.8	1.02
Fat	19.7	1.17
Fiber	7.6	.99
Ash	8.7	.89
Methionine	.26	.01
Cystine	.29	.02
(M + C)	(.55)	.03
Lysine	.74	.06
Arginine	1.05	.08
Tryptophan	.10	.01
Tyrosine	.42	.02
Threonine	.53	.04
Serine	.61	.04
Phenylalanine	.59	.03
Aspartic acid	1.31	.11
Glutamic acid	1.80	.09
Proline	.59	.03
Glycine	.75	.05
Alanine	.87	.06
Valine	.70	.05
Isoleucine	.47	.03
Leucine	.93	.04
Histidine	.38	.03
Sum of amino acids	12.44	.71

[a] Values are in percentages on as is basis.
[b] 10 Samples collected in 1987 and 1988.

TABLE 8

Composition of Extracted Rice Bran[a,b]

	$\bar{X}$	SD	Adulterated sample[c]
Moisture	10.4	.94	10.5
Protein	15.8	1.5	12.5
Fat	1.6	.98	.6
Fiber	12.4	4.1	19.6
Ash	12.0	2.3	17.1
Methionine	.31	.04	.22
Cystine	.32	.05	.22
(M + C)	(.63)	.09	(.44)
Lysine	.72	.14	.47
Arginine	1.18	.16	.88
Tryptophan	.15	.02	.10
Tyrosine	.48	.05	.35
Threonine	.59	.07	.45
Serine	.70	.08	.53
Phenylalanine	.74	.08	.57
Aspartic acid	1.37	.17	1.01
Glutamic acid	2.07	.24	1.57
Proline	.69	.08	.56
Glycine	.85	.10	.68
Alanine	.97	.12	.75
Valine	.83	.10	.67
Isoleucine	.54	.06	.44
Leucine	1.10	.13	.86
Histidine	.41	.06	.29
Sum of amino acids	14.07	1.64	10.70

[a] Values are in percentages on as is basis.
[b] 11 Samples collected in Malaysia (Indian origin) and Thailand.
[c] Adulterated sample is of Indian origin.

Broken Rice

Broken rice is a high starch material, low in fiber, fat and ash. It is rarely adulterated. Average values are shown in Table 10, and these are broken out into country of origin (Malaysia or Thailand) in Table 11. Interestingly, samples from Malaysia were higher in protein and most amino acids than samples from Thailand.

TABLE 9

Composition of Extracted Rice Bran[a]—Thailand[b]

	$\overline{X}$	SD
Moisture	10.3	.55
Protein	16.5	.51
Fat	2.0	.94
Fiber	10.5	.59
Ash	11.2	.46
Methionine	.32	.02
Cystine	.34	.02
(M + C)	(.66)	.05
Lysine	.78	.05
Arginine	1.25	.08
Tryptophan	.16	.01
Tyrosine	.50	.03
Threonine	.62	.04
Serine	.73	.04
Phenylalanine	.77	.04
Aspartic acid	1.45	.08
Glutamic acid	2.16	.12
Proline	.72	.05
Glycine	.89	.06
Alanine	1.01	.07
Valine	.85	.06
Isoleucine	.56	.03
Leucine	1.14	.07
Histidine	.44	.03
Sum of amino acids	14.71	.84

[a] Values are in percentages on as is basis.
[b] 7 Samples collected from Thailand.

TABLE 10

Composition of Broken Rice[a]—All Samples[b]

	$\overline{X}$	SD
Moisture	12.7	.79
Protein	7.7	.90
Fat	.9	.25
Fiber	.4	.33
Ash	.5	.15
Methionine	.20	.03
Cystine	.17	.02
(M + C)	(.37)	.06
Lysine	.28	.03
Arginine	.59	.07
Tryptophan	.08	.01
Tyrosine	.28	.05
Threonine	.25	.03
Serine	.34	.03
Phenylalanine	.41	.05
Aspartic acid	.66	.09
Glutamic acid	1.32	.18
Proline	.33	.04
Glycine	.33	.04
Alanine	.43	.05
Valine	.43	.06
Isoleucine	.32	.04
Leucine	.61	.07
Histidine	.17	.02
Sum of amino acids	7.21	.89

[a] Values are in percentages on as is basis.
[b] 16 Samples collected from Thailand and Malaysia.

TABLE 11

Composition of Broken Rice[a]—Malaysia and Thailand[b]

	Malaysia	Thailand		Malaysia	Thailand
Moisture	12.8	12.5	Serine	.35	.33
Protein	8.3	7.3	Phenylalanine	.44	.38
Fat	.9	.8	Aspartic acid	.73	.61
Fiber	.6	.2	Glutamic acid	1.45	1.21
Ash	.6	.5	Proline	.37	.30
Methionine	.23	.18	Glycine	.36	.30
Cystine	.20	.15	Alanine	.46	.40
(M + C)	(.43)	(.33)	Valine	.47	.40
Lysine	.31	.26	Isoleucine	.35	.30
Arginine	.64	.55	Leucine	.66	.57
Tryptophan	.08	.08	Histidine	.19	.16
Tyrosine	.32	.25	Sum of amino acids	7.89	6.67
Threonine	.28	.24			

[a] Values are in percentages on as is basis.
[b] 7 Samples from Malaysia and 9 from Thailand.

TABLE 12

Regression Equation for Calculating Metabolizable Energy (ME) of Rice Products

ME (kcal/kg product) = $(4759 - 8.86 \text{ g CP} - 12.77 \text{ g CF} + 5.21 \text{ g FAT}) \times \dfrac{\text{g DM}}{1000}$

where CP is crude protein,
CF is crude fiber,
FAT is crude fat and
DM is dry matter
(All expressed as g/kg product)

From *Feeding Values for Poultry*, W.M.M.A. Janssen, K. Terpstra, F.F.E. Beeking and A.J.N. Bisalsky, Spederholt Mededeling 303, 1979.

TABLE 13

Calculated ME[a] Values for Rice By-products

	kcal/kg Product
Asian rice bran—all samples (45)	3030
Rice bran—good samples (15)	3207
Rice bran—poor samples (15)	2594
Rice bran—adulterated sample no. 5	1761
Rice bran—low fat samples (5)	2050
Australian rice bran (10)	3258
Extracted rice bran—Thailand (7)	1848
Broken rice—all samples (16)	3555

[a] ME, metabolizable energy.

TABLE 14

Regressions for Rice Bran

Amino acid	BO	Protein	Moisture	Fat	Fiber	Ash	R^2
Methionine	.152	.006	.006	.001	−.002	−.002	.68
Cystine	.098	.006	.008	.003	−.003	0	.60
M + C	.276	.011	.012	.004	−.005	−.004	.69
Lysine	−.073	.012	.024	.017	0	0	.62
Arginine	−.068	.042	.044	.009	−.006	0	.73
Tryptophan	.052	.003	.004	−.001	−.002	0	.30
Threonine	.152	.025	0	0	0	0	.51

Example: % Methionine = .152 + .006% protein + .006% moisture + .001% fat − .002% fiber − .002% ash (R^2 = .68).

TABLE 15

Regressions for Extracted Rice Bran

Amino acid	BO	Protein	Moisture	Fat	Fiber	Ash	R^2
Methionine	−.304	.033	0	0	0	.007	.79
Cystine	.099	.028	−.021	0	0	0	.70
M + C	−.543	.064	0	0	0	.012	.72
Lysine	−.639	.112	−.070	0	0	.026	.89
Arginine	−1.416	.135	0	0	0	.038	.82
Tryptophan	−.035	.012	0	0	0	0	.51
Threonine	−.450	.057	0	0	0	.011	.89

TABLE 16

Regressions for Broken Rice

Amino acid	BO	Protein	Moisture	Fat	Fiber	Ash	R^2
Methionine	−.166	.028	.012	0	0	0	.85
Cystine	−.113	.019	.010	0	0	0	.81
M + C	−.279	.047	.022	0	0	0	.84
Lysine	−.105	.025	.013	.026	.018	0	.89
Arginine	−.078	.072	.008	0	.031	0	.91
Tryptophan	.039	.005	0	0	0	0	.33
Threonine	−.073	.023	.011	0	0	.027	.87

The high level of arginine in broken rice is worth noting. This is about 50% higher than in corn (as percent of protein) and should be considered when formulating.

Energy Values

An attempt was made to calculate metabolizable energy (ME) values of these rice by-products by use of a regression equation. The equation selected (see Table 12) was developed at the Spederhalt Institute in The Netherlands specifically for use with rice by-products.

Table 13 shows the ME values (poultry) for the different products calculated by use of this equation. In the opinion of the author, this equation overestimates the ME values of rice bran by about 5%. On the other hand, values for extracted rice bran and broken rice appear quite accurate. The use of such an equation illustrates quite dramatically the effect of proximate composition on energy value. It should be a useful tool to evaluate specific types of rice by-products. The high energy value of broken rice (slightly higher than corn) is shown in this calculation.

Amino Acid Predictions

Because proximate analysis is routinely conducted in most feedmill laboratories and amino acid analysis is not, prediction of amino acids from the proximate analysis may be a useful way of obtaining better amino acid values. Multiple linear regression equations were developed for this purpose.

Regressions for seven major amino acids in rice bran, extracted rice bran and broken rice are presented in Tables 14–16. The high R square values in most cases suggest these equations should predict the amino acid levels with a good level of accuracy.

Acknowledgments

Sincere thanks to F.J. Ivey and R.R. Ontiveros for statistical and analytical support.

DNA Sequences Controlling Nutritional and Functional Properties of Cereal Storage Proteins

M. Giband, B. Potier, S. Dukiandjiev[1], V. Burrows[2] and I. Altosaar

Biochemistry Department, University of Ottawa, Ottawa K1N 6N5, Canada;
[1]Biochemistry Department, Plovdiv University, Plovdiv, Bulgaria; [2]Plant Research Centre, Agriculture Canada, Ottawa K1A 0C6, Canada.

Abstract

Genes in cereals which code for industrially important proteins include those specifying nutritional and functional components like globulins and gluten respectively. Such genes have been identified, isolated, manipulated, and transferred to other plants in order to study their expression and to "engineer" the feed and food value of grain and grain products. Examples reviewed include lysine-rich DNA sequences that have been inserted into a zein gene of corn to improve the essential amino acid content of its protein, without disturbing the ability of the "engineered" protein to be synthesized and that are packaged properly into sub-cellular protein bodies. A hordein protein gene from barley and a high molecular weight glutenin gene from wheat have been expressed in transgenic tobacco plants, identifying the DNA regions responsible for tissue-specific and developmentally controlled expression in the endosperm of the grain. Bread-baking quality of wheat flour is now better understood through the studies of the molecular structures of the gliadins and glutenins. Specific protein folding domains and amino acid residues have been targeted for site-specific mutagenesis to increase the viscoelasticity of the dough.

The eight major cereal grains provide 60% of the energy and 50% of the proteins consumed in the world (1). Rice provides up to 80% of the total calories in Asia. The importance of any particular cereal is related to both its nutritional and functional properties. The nutritional value of food proteins is measured by their content of essential amino acids. Cereals are somewhat deficient in lysine, threonine (barley), tryptophane (corn) and to a lesser extent histidine and methionine. Among the cereals, the nutritional quality of proteins can be roughly classed as poor (sorghum, millet, corn) to fairly good (rice, oats) and this is determined by the relative amounts of the different classes of storage proteins in each grain. Unfortunately, those protein classes that commonly produce the desired functional quality, such as loaf volume in bread, are the proteins most deficient in essential amino acids. Therein lies the dilemma for cereal breeders: how to breed grains that are superior in both nutritional and functional quality.

The different storage proteins of cereals have been studied in groups or classes based on their extractibility and solubility properties in different solvents (2): the prolamin fraction is soluble in aqueous alcohol solutions, the globulins are solubilized in saline solutions, whereas the albumin fraction is soluble in water. The remainder of the grain protein (soluble in dilute acid or alkalies) has been referred to as the glutelin fraction. Plant breeding programs have aimed to improve both the nutritional and functional value of cereal storage proteins. However, it has been observed that increases in protein content are often associated with lower grain yields and an increase in the nutritionally poor prolamin class (3).

These and other limitations encountered by plant breeders have encouraged researchers to seek new methods for manipulating the composition of cereal proteins. Since the beginning of the development of plant molecular biology, cereal storage proteins have been the focus of recombinant DNA research, not only because of their importance for the food and feed industries, but also because they provide an interesting model for the study of gene regulation systems in plants.

Improving Nutritional and Functional Properties of Cereal Storage Proteins

For many years, breeders have been improving the nutritional and functional qualities of cereal grains through conventional breeding programs. Many successes have been achieved, like the development of high lysine-containing barley and sorghum. Similarly, maize varieties containing an improved lysine, tryptophan or methionine content have been discovered (3). Nevertheless, the conventional breeding approach depends on the isolation of mutants with improved traits. (For a review on cereal quality improvement by breeding, see Ref. 3.) In contrast, plant molecular biologists are gaining access to the genetic material responsible for these traits, and hope to modify it, although at present this is limited to traits governed by a single gene. The improvement of feeds at the molecular level, and barley in particular, has been reviewed by Shewry and Kreis (4).

Two approaches for improving seed protein quality can be considered. The quantitative approach consists of selectively overexpressing genes coding for nutritionally rich proteins. This may be achieved either by increasing the number of such existing genes and/or enhancing the level of expression of these genes. On the other hand, one can also envisage improvement of the quality of the grain protein either by modifying the existing genes to code for more essential amino acids in their respective proteins, or by introducing genes coding for nutritionally rich storage proteins from heterologous systems.

Target Sequences for the Genetic Engineering of Storage Proteins

Studies at the molecular level have allowed molecular biologists to define DNA sequences which play an important role in defining the nutritional or functional quality of cereal storage proteins. The determination of the primary sequence of cereal storage protein genes and that of their regulatory elements has revealed definite DNA sequences that could be targeted for site-specific mutagenesis.

Similar studies at the RNA level have highlighted sequences and structures that are important in the regulation of gene expression. Lastly, studies of the proteins themselves have given an insight into the structure and function of the different storage proteins and their various domains (Fig. 1).

Some potential alterations in DNA might include (a) deletions or insertions, with the latter shown as solid blocks in Fig. 1: (i), modification of the leader sequence to increase efficiency of translations (e.g. ribosome loading, scanning, initiation); (ii), to improve nutritional value, addition of DNA stretches coding for increased content of essential amino acids; (iii), to augment protein functionality, inclusion of new sequence coding for additional protein domains affecting solubility, visco-elasticity, or foaming, etc. Such insertions are indicated in Figure 1 by interrupted lines (._____._____.) in the mRNA transcript; the mature, processed mRNA; and the polypeptide, which are represented in succession below the gene; or (b) site-directed mutagenesis (*) of the following features: (i) transacting factors to enhance the DNA-affinity of binding proteins and affect gene activation, (ii) enhancers to alter time or tissue-specificity of gene expression, (iii) promoter sequences to modulate level of activity, (iv) 5′ leader to affect translational efficiency, (v) coding sequences to change nutritional impact and/or protein conformation, and (vi) the 3′-untranslated region to (de)stabilize the half-life, translatability or secondary structure of the mRNA.

DNA Primary Structure

Protein Coding Sequences: The DNA sequences coding for a number of storage proteins from various cereals (wheat, barley, rye, oat, rice, maize) have been determined (for a list, see Refs. 5,6). The sequencing of these genes or cDNAs shows that storage protein genes belong to multigene families rather than being single genes. The structure of storage protein genes has been reviewed elsewhere (7–11). Within these gene families, certain members exhibit characteristics that make them of particular interest for strategies aimed at improving the protein value of grains. Kirihara et al. (12) have isolated a cDNA clone corresponding to a maize zein (prolamin) which is unusually rich in methionine codons (up to 22.5% methionine in the predicted polypeptide). Similarly, high-lysine barley mutants have been discovered. The increase in lysine content is due to an increase in the lysine content of four salt-soluble storage proteins. These proteins have been identified as two chymotrypsin inhibitors (CI.1 and CI.2), a β-amylase, and a protein of unknown function (4). Recently, cDNA clones have been isolated for CI.1 (13), CI.2 and for the β-amylase (4). The cloning of these cDNAs should allow a better knowledge of the genes encoding these proteins.

Molecular cloning also has allowed a better understanding of the functional properties of storage proteins. For example, Halford et al. (14) and Flavell et al. (15) have sequenced high molecular weight (HMW) glutenin subunits from wheat, allowing the determination of amino acid residues and protein domains in part responsible for the elasticity of the dough, and the subsequent bread-baking quality of wheat flour.

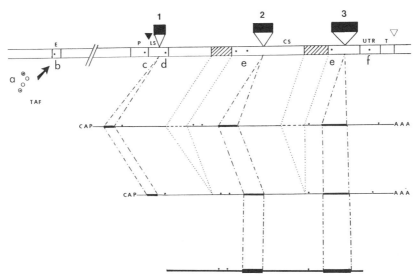

Fig. 1. Strategies and tactics for quantitative and qualitative manipulations of cereal storage proteins, showing regions of DNA sequence from a model gene which might be altered: E, enhancer-like element, involved in tissue specific expression; P, promoter, involved in start of messenger RNA transcription (solid inverted triangle); LS, leader sequence or 5′-untranslated region of mRNA; CS, coding sequence for the particular protein of interest, in this case, shown by two introns (cross-hatched regions) which are spliced out (_____ regions) during subsequence RNA processing (. . . .); UTR, 3″-untranslated or trailer region of the mRNA; T, terminator region, causing termination of mRNA transcription (open inverted triangle). TAF, transaction factors or DNA-binding proteins which help regulate enhancer elements during gene expression.

DNA Regulatory Sequences: The expression of seed storage protein genes is highly regulated in a tissue specific manner during the development of the cereal embryo. Before attempting to engineer the expression of the storage protein genes, one must first identify regulatory sequences and understand the mechanisms governing such regulation in detail. The isolation of storage protein genomic clones (5) has proven useful for the localization and characterization of such regulatory sequences. The storage protein genes contain the typical promoter TATA and CAAT boxes (16) which play a role in gene regulation. In some cases, when the CAAT box is absent, another consensus sequence, the AGGA box is found (8) and this might play the role of a CAAT box. While TATA and CAAT boxes are found in most animal and plant genes, storage protein genes contain additional specific sequences further upstream. Forde et al. (17) have identified a 28 bp consensus sequence, the "−300 element" which seems to be restricted to prolamin genes (B1-hordein, α-gliadin, and zein). These authors suggest a possible role for this sequence in the tissue-specific expression of prolamin genes in developing endosperm. Further transformation experiments are in agreement with this hypothesis.

In these experiments, constructions were made where various prolamin (barley B1-hordein, maize, zein) gene promoters were placed upstream of a reporter gene (chloramphenicol acetyl transferase (CAT) or β-glucuronidase (GUS)) whose activity can be assayed easily. These fusion constructs were used to transform tobacco plants, and the activity of the promoter under study was monitored by measuring the CAT or GUS activity in various tissues of the transformed tobacco. Such experiments showed that the reporter gene was expressed in a tissue specific and developmentally regulated manner (18,19). Only in one case, where a zein encoding gene was used to transform petunia plants, was the expression of the gene not limited to seeds (20). The "−300 element" might not be restricted to prolamin genes. Other storage protein gene promoters have been used in the same type of experiment [HMW and LMW (low-molecular weight) glutenin (21), HMW glutenin (22)]. These results show that 326 bp (for the LMW glutenin) and 433 bp (for the HMW glutenin) of upstream sequences are sufficient to confer endosperm-specific expression.

Maier et al. (23) have shown that a factor present in crude nuclear extracts of maize endosperm binds to a sequence located at −330, which includes a sequence conserved in all zein genes. Interestingly, this sequence overlaps the "−300 element" defined by Forde et al. (17), and also is present in the LMW glutenin gene (21). Furthermore, this protein binding site contains a sequence that is very homologous to the viral/animal enhancer core sequence (23). It is possible that other specific expression and/or enhancer sequences may be present further upstream or downstream of this protein binding site. Products from *trans*-acting genes (Fig. 1, TAF) and *cis*-acting elements (Fig. 1, E) are likely to be involved in the high level and specific expression of cereal seed storage proteins. Considerably more research is required before effective engineering tactics can be conceived to manipulate these features of cereal protein production.

Translational Control

If information on the DNA sequences regulating the expression of seed storage proteins is becoming available, contrastingly little is known about the regulation at the mRNA level. Heidecker and Messing (8) have reviewed the data available on the plant sequences surrounding the AUG initiation codon, as well as the polyadenylation signal. In all known cases, except one (pea lectin), it is the first AUG that is recognized as the initiation codon. These authors have been able to define a consensus sequence for the start of protein synthesis from plant genes: 5′ N N N N A N A/U N U/A A N N N N A N N AUG G C U.

The absence of a strict consensus sequence can be explained by the fact that although the initiation signal must be recognized by the ribosomes, different mRNAs must be translated with different efficiencies. A general feature of this consensus sequence is that it is rich in A/T, and the potential secondary structure could easily be melted by the ribosomal 40S subunit.

As it is the case for the initiation consensus sequence, the polyadenylation signal of plant mRNAs is less conserved than in animal mRNAs. The consensus sequence is A/U A A U A A Pu.

Most plant genes have more than one polyadenylation signal. The mechanism governing the choice of the signal is not known, although the involvement of a hairpin structure is suggested. Some evidence for the regulation of gene expression at the mRNA level is available (for a general view see Refs. 24,25). The primary sequence of the mRNA, and the secondary structure that the mRNA can adopt influence the level of transcription. Spena et al. (26) have shown that certain zein mRNAs have inverted repeats in their 5′- (leader) and 3′-untranslated regions. Further work (27) shows that the formation of an intra- or intermolecular hybrid between these two repeats is responsible for a translational block in a cell-free system. Similarly, oat globulin and prolamin cDNA clones have been characterized (6,28). It has been shown (unpublished work) that the mRNAs encoded by these cDNA clones are capable of forming an extensive secondary structure (Fig. 2); the role of these structures in translational control has not yet been elucidated.

As is the case in animal systems, phosphorylation of ribosomal proteins also is involved in translational control of seed storage protein expression. In studies aimed at understanding the high levels of globulin in oats, a protein of the small subunit of the ribosomes (protein S6) was preferentially phosphorylated in endosperm cells, when compared to leaf, stem or root cells (29). The phosphorylation of this particular ribosomal protein is linked to the tissue-specific expression of oat storage proteins, whereas the quantitative control seems to be under the control of polysomal proteins. By adding oat polysomal extracts to a wheat germ system, Fabijanski and Altosaar (30) were able to reproduce the selectivity of the translation machinery found in oat endosperm cells. These results are in agreement with other experiments showing the differential requirement for wheat germ initiation factors of different 5′-untranslated mRNA leaders (31).

Antisense RNA has been used as a tool to regulate the expression of mRNAs in various systems. Recently Rogers (32) was able to detect the presence of an antisense RNA, complementary to both the type A and type B α-amylase mRNAs in barley endosperm and aleurone tissue. Whether this antisense RNA plays a role in the regulation of α-amylase expression or not has yet to be confirmed.

It is probable that other regulatory mechanisms exist at the mRNA level. Such mechanisms could have effects on the stability of the mRNA, the level of expression,

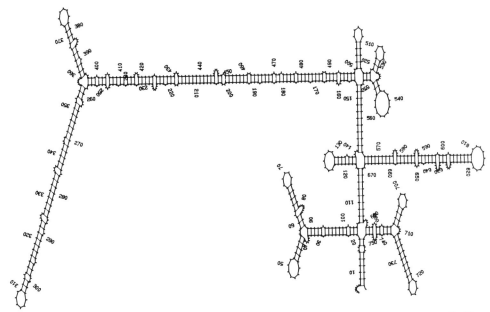

Fig. 2. Secondary structure of oat prolamin mRNA, predicted from 759 bps of cDNA sequence (6). The extensive tight stems might play a role in translational control by inhibiting ribosome-binding, initiation and/or scanning. The levels of prolamins in oat are low despite the abundance of its corresponding mRNA (30). Best-fit RNA-RNA hybridization was calculated by using the FOLD program from the Genetics Computer Group, University of Wisconsin, Madison.

and on the tissue and temporal specificity of expression of storage protein mRNAs. More knowledge must be generated to understand this area.

Protein Engineering in Cereals

The composition, structure and biochemistry of cereal storage proteins has been reviewed extensively (33–36). In general, the primary sequence of cereal storage proteins exhibits a series of repeated peptides. These repeats allow the formation of extensive secondary structures (α-helices or β-turns) and super secondary structures (α-helix barrels) giving a very compact feature to the molecule (37–39). This is consistent with the localization and function of these proteins.

The improvement of the nutritional and functional value of cereal storage proteins can be achieved either by regulating the expression of existing genes, or by modifying these genes to improve the value of their corresponding gene-products. The transfer of specific genes from heterologous systems, which code for proteins showing improved properties is also feasible. For an example of the use of molecular genetics for the improvement of wheat-grain quality, see (40).

Modifications of the ratios of the different components of the seed proteins can be a way to improve the nutritional value of cereal storage proteins. This is the case for two naturally occurring nutritionally improved cereals: the high lysine and tryptophane maize Opaque-2 and the high-lysine barley Hiproly varieties. It has been shown in both cases that the improvement is due to a reduced prolamin content (3). Similarly, rice and oats which show a better nutritional value have a higher globulin:prolamin ratio than the other cereals. A better understanding of the mechanisms regulating the expression of the different storage protein gene families would allow for some fine tuning and perhaps a shift toward the expression of nutritionally richer (globulin or albumin) protein fractions. It has been shown (41,42) that several cereals (wheat, rye, oats, corn) possess polypeptides homologous to the 12S pea legumin in their globulin fraction. The dicot, legumin-like peptides are in part responsible for the better nutritional value of legume proteins, compared to cereal protein. It would therefore be of interest to further characterize the cereal, legumin-like polypeptides in order to improve the nutritional value of cereal storage proteins.

If changing the balance between the different storage protein fractions is a way to increase the quality of grain proteins, the introduction of genes from heterologous systems may represent another way to improve cereals. Some monocot crops have proteins of interesting nutritional value (oats, rice), whereas others have proteins of interesting functional value (wheat). Because the DNA regulatory sequences and the overall protein structures are conserved in cereals, it is conceivable to produce "molecular hybrids" between different cereals. For example, the bread-baking quality of some nutritionally rich cereal flours could be improved by the introduction of specific wheat HMW glutenin and gliadin genes in these deficient species, whereas the good bread-baking quality of wheat flour could have an improved nutritional value if genes, effecting higher levels of sulfur-containing amino acids or essential amino acids, were transferred to wheat.

Several genes encoding dicot storage proteins showing improved nutritional value have been characterized. A Brazil nut albumin cDNA (43) and those for homologous proteins from pea (44) and sunflower (45) have been cloned. These proteins show elevated levels of cysteine and methionine (8% and 18% respectively). Although these proteins are naturally expressed in dicots, they are putative candidates to be transferred to monocots. Just as monocot storage protein genes have been expressed in dicots in a tissue specific and developmentally regulated manner (see above) showing that the monocot regulatory sequences can be recognized in dicots, it may be possible to transfer these superior albumin genes from dicots into monocots without encountering any problems regarding their expression.

A more direct and perhaps more feasible (with respect to the actual state of knowledge) approach for the qualitative improvement of cereal storage protein is the modification of the DNA sequences which code for storage proteins. Information on the structural and functional properties of storage proteins is available, and amino acids that are not essential in defining the structural properties of the proteins have been identified. Nutritionally important amino acids could be introduced by insertion of appropriate codons or site-directed mutagenesis to replace these nonessential amino acids. Recently, Bulleid and Freedman (46) have described an *in vitro* transcription and translation system which allowed them to demonstrate that a γ-gliadin synthesized *in vitro* is correctly processed and translocated into dog pancreatic microsomal vesicles. This system should allow the study of the effects of mutations or insertions on the processing of engineered proteins.

Such work has been done with maize prolamin (zein) by Wallace et al. (47). These authors have introduced lysine codons by site-directed mutagenesis, and have inserted lysine- and tryptophan-encoding oligonucleotides into the zein coding sequence. After injection of mRNAs for the modified zeins into *Xenopus* oocytes, the engineered proteins are correctly synthesized and self-aggregate into protein bodies. Although these experiments were performed in a heterologous system, they show that alterations in the coding sequence do not interfere with the processing of the protein, despite the addition of charged amino acid residues. This type of experiment represents a first step toward the production of nutritionally balanced maize, and could be conducted with selected proteins of other cereals as well.

Limitations and Biochemical Considerations

The engineering of cereal storage proteins is limited not only by the ability to modify existing genes or transfer genes from other sources. Storage proteins are encoded by multigene families and the information newly transferred may be "diluted," resulting in disappointing results. This situation can be avoided if strong promoters are used to regulate the expression of the newly introduced/modified gene. Furthermore, one must make sure that the modified proteins are expressed and processed correctly, especially when the genes come from a heterologous system. The expression of the foreign or modified gene should not interfere with the development of the seed, or reduce the yield of the crop.

In cereals, mutants have been described that accumulate certain free amino acids (for example, see Ref. 48). In some cases, this accumulation can account for an increase in the amino acid content of the grain. The enzymes involved in the biochemical pathway leading to the synthesis of the essential amino acids are also potential targets for the improvement of cereals by biotechnological methods. The cloning of wild-type and mutant forms of the gene coding for these enzymes would be of interest in the long-term scope of cereal amelioration. This type of experiment, the transfer of a foreign gene, or the modification of an existing gene in order to change the balance in the overall amino acid content of the grain is likely to have implications for the respective biochemical and metabolic pathways. The presence of pools of free amino acids during protein synthesis, the specific demand for each amino acid, the biosynthesis and the translocation of these amino acids are aspects that should be considered. These aspects, reviewed by Rao and Singh (49), will be interesting to test once gene-transfer in cereals becomes routine.

Prospects

Knowledge of the genes encoding storage proteins and of the mechanisms regulating their expression is being generated by molecular techniques. The following steps, the introduction of the new or altered genes into the plant genome and the regeneration of the transformed plant are procedures that are being currently studied. A number of transformation methods: electroporation (50); PEG mediated up-take (51); virus mediated transformation (52); pollen-tube pathway transformation (53); particle bombardment (54); injection of DNA into young tillers (55); agroinfection (56); and agrobacterium spheroplast mediated transformation (57) have been applied with success to cereals. These methods must be further developed to achieve reliability and high frequency of transformation. Regeneration of several monocots (rice (58–61), maize (62), millet (63), and sugarcane (64)) has been achieved, although in several species varieties of economic importance are still recalcitrant toward regeneration. Work is in progress and several commercial rice and maize varieties have been regenerated. Again, further developments are needed to achieve regeneration of important cereal crops.

Transformation, and regeneration of dicots has been achieved. Field tests of genetically engineered plants have been carried out. Transformation and regeneration protocols are available, especially for two important crops, rice and maize. Recently, transgenic rice plants have been obtained. The transformed plants, carrying a gene encoding aminoglycoside phosphotransferase II (APH (3')II) are tolerant to the antibiotic G418 (65). Genetically transformed maize, carrying a neomycin phosphotransferase (NPTII) gene have been regenerated and were shown to be resistant to kanamycin (66). Although these engineered maize plants are sterile, these experiments show that transformation and regeneration of cereals is possible. More recently Luo and Wu (67) have described a method for transformation of rice through the pollen-tube pathway. Because this method avoids the cell culture and regeneration steps, it is transferable to other monocots that are recalcitrant toward regeneration. With these recent advances in cereal transformation, it seems that genetic engineering of cereal storage proteins can become a reality. Nevertheless, more knowledge of the genes encoding the storage proteins, and the mechanisms regulating their expression is needed to improve the nutritional and functional properties of cereal storage proteins.

Acknowledgments

This research was funded in part by Operating and Strategic grants from NSERC. We thank NSERC for an International Scientific Exchange Award, the Ministry of External Affairs for a Government of Canada Postgraduate Scholarship and A. Sallafranque for preparing the manuscript.

References

1. Burrows, V.S., *J. Can. Dietetic Assoc. 44:*220 (1983).
2. Osborne, T.B., *The Vegetable Proteins*, 2nd ed., Longmans, Green and Co., London (1924).
3. Rhodes, A.P., and G. Jenkins, in *Plant Proteins*, edited by G. Norton, Butterworths, London, 1976, pp. 207–226.
4. Shewry, P.R., and M. Kreis, *Proc. Nutr. Soc. 46:*379 (1987).
5. Elliston, K.O., S. Imran and J. Messing, *Plant Mol. Biol. 6:*22 (1988).
6. Fabijanski, S., S.C. Chang, S. Dukiandjiev, M.B. Bahramian, P. Ferrara and I. Altosaar, *Biochem. Physiol. Pflanzen 183:*143 (1988).
7. Higgins, T.J.V., *Ann. Rev. Plant Physiol. 35:*191 (1984).
8. Heidecker, G., and J. Messing, *Ibid. 37:*439 (1986).
9. Casey, R., and C. Domoney, *Plant Mol. Biol. Rep. 5:*261 (1987).
10. Messing, J., in *Genetic Engineering*, Vol. 6, edited by P. Rigby, Academic Press, London, 1987, pp. 1–46.
11. Walbot, V., and J. Messing, in *Corn and Corn Improvement*, Amer. Soc. Agron., Madison, Wisconsin, 1988.
12. Kirihara, J.A., J.P. Hunsperger, W.C. Mahoney and J.W. Messing, *Mol. Gen. Genet. 211:*477 (1988).
13. Williamson, M.S., J. Forde and M. Kreis, *Plant Mol. Biol. 10:*521 (1988).
14. Halford, N.G., J. Forde, O.D. Anderson, F.C. Greene, and P.R. Shewry, *Theor. Appl. Genet. 75:*117 (1987).
15. Flavell, F.G., A. Goldsbrough, L.S. Robert, D. Schnick and R.D. Thompson, *Bio/Technology*, submitted (1988).
16. Messing, J., D. Geraghty, G. Heidecker, N.T. Hu, J. Kridl, and I. Rubenstein, in *Genetic Engineering of Plants*, edited by T. Kosuge et al., Plenum Press, New York, 1983, pp. 211–227.
17. Forde, B.G., A. Heyworth, J. Pywell and M. Kreis, *Nucl. Acids Res. 13:*7327 (1985).
18. Marris, C., P. Gallois, J. Copley and M. Kreis, *Plant Mol. Biol. 10:*359 (1988).
19. Schernthaner, J.P., M.A. Matzke, and A.J.M. Matzke, *EMBO J. 7:*1249 (1988).
20. Ueng, P., G. Galili, V. Sapanara, P.B. Goldsbrough, P. Dube, R.N. Beachy and B.A. Larkins, *Plant Physiol. 86:*1281 (1988).
21. Colot, V., L.S. Robert, T.A. Kavanagh, M.W. Bevan and R.D. Thompson, *EMBO J. 6:*3559 (1987).
22. Roberts, L.S., R.D. Thompson, and R.B. Flavell, *Genes Dev.*, submitted (1988).
23. Maier, U-G., J.W.S. Brown, C. Toloczyki and G. Feix, *EMBO J. 6:*17 (1987).
24. *Translational Control*, edited by M.B. Mathews, published by Cold Spring Harbor Laboratory, Cold Spring Harbor, MA, 1986.
25. *Translation Regulation of Gene Expression*, edited by J. Ilan, Plenum Publishing Corp., New York, 1987.
26. Spena, A., A. Viotti and V. Pirrotta, *EMBO J. 1:*1589 (1982).
27. Spena, A., E. Krausse and B. Dobberstein, *Ibid 4:*2153 (1985).
28. Shotwell, M.A., C. Alfonso, E. Davies, R.S. Chesnut and B.A. Larkins, *Plant Physiol. 87:*698 (1988).
29. Fabijanski, S., and I. Altosaar, *FEBS Lett.*, submitted (1988).
30. Fabijanski, S., and I. Altosaar, *Plant Mol. Biol. 4:*211 (1985).
31. Browning, K.S., S. Lax, J. Humphreys, J.M. Ravel, S.A. Jobling and L. Gehrke, *J. Biol. Chem. 263:*9630 (1988).
32. Rogers, J.C., *Plant Mol. Biol. 11:*125 (1988).
33. Laszitity, R., *The Chemistry of Cereal Proteins*, CRC Press, Boca Raton, Florida, 1984.
34. Kreis, M., P.R. Shewry, B.G. Forde, J. Forde and B.J. Miflin, *Oxford Surveys of Plant Mol. Cell. Biol. 2:*253 (1985).
35. Shewry, P.R., M. Kreis, M.M. Burrell and B.J. Miflin, in *Food Biotechnology*, edited by R.D. King and P.S.J. Cheetham, Elsevier, London, 1987, pp. 49–85.
36. Shotwell, M.A., and B.A. Larkins, in *The Biochemistry of Plants: A Comprehensive Treatise*, Vol. 15, edited by A. Marcus, 1988, in press.
37. Argos, P., K. Pedersen, M.D. Marks and B.A. Larkins, *J. Biol. Chem. 257:*9984 (1982).
38. Field, J.M., A.S. Tatham, A.M. Baker and P.R. Shewry, *FEBS Lett. 200:*76 (1986).
39. Plietz, P., B. Drescher and G. Damaschun, *Int. J. Biol. Macromol. 9:*161 (1987).
40. Flavell, R.B., P.I. Payne, R.D. Thompson and C.N. Law, *Biotech. and Genet. Eng. Rev. 2:*157 (1984).
41. Robert, L.S., C. Nozzolillo and I. Altosaar, *Biochem. J. 226:*847 (1985).
42. Fabijanski, S., I. Altosaar, M. Laurière, J-C. Pernollet and J. Mossé, *FEBS Lett. 182:*465 (1985).
43. Altenbach, S.B., K.W. Pearson, F.W. Leung, and S.S.M. Sun, *Plant Mol. Biol. 8:*239 (1987).
44. Higgins, T.J.V., L.R. Beach, D. Spencer, P.M. Chandler, P.J. Randall, R.J. Blagrove, A.A. Kortt and R.E. Gurthrie, *Ibid. 8:*37 (1987).
45. Lilley, G.G., J.B. Caldwell, T.J. Higgins, D. Spencer and A.A. Kortt, in *Proceedings of the World Congress on Vegetable Protein Utilization in Human Foods and Animal Feedstuffs*, edited by T.H. Applewhite, American Oil Chemists' Society, Champaign, IL, 1989.
46. Bulleid, N.J., and R.B. Freedman, *Biochem J. 254:*805 (1988).
47. Wallace, J.C., G. Galili, E.E. Kawata, R.E. Cuellar, M.A. Shotwell and B.A. Larkins, *Science 240:*662 (1988).
48. Hibberd, K.A., P.C. Anderson and M. Barker, U.S. Patent 4,581,847 (1986).
49. Rao, A.S., and R. Singh, *Trends Biotech. 4:*108 (1986).
50. Fromm, M., L.P. Taylor and V. Walbot, *Nature 319:*791 (1986).
51. Lörz, H., B. Baker and J. Schell, *Molec. Gen. Genet. 199:*178 (1985).
52. French, R., M. Janda and P. Ahlquist, *Science 231:*1294 (1986).
53. Duan, X., and S. Chen, *China Agri. Sci. 3:*6 (1985).
54. Klein, T.M., T. Gradziel, M.E. Fromm and J.C. Stanford, *Bio/Technology 6:*559 (1988).
55. de la Pena, A., H. Lörz and J. Schell, *Nature 325:*274 (1987).
56. Grimsley, N., B. Hohn, T. Hohn and R. Walden, *Proc. Natl. Acad. Sci. USA 83:*3282 (1986).

57. Baba, A., S. Hasézawa and K. Syona, *Plant Cell Physiol. 27:*463 (1986).
58. Fujimura, T., M. Sakurai, H. Akagi, T. Negishi and A. Hirose, *Plant Tissue Culture Lett. 2:*74 (1985).
59. Abdullah, R., E.C. Cocking and J.A. Thompson, *Bio/Technology 4:*1087 (1986).
60. Yamada, Y., Z.Q. Yang and D.T. Tang, *Plant Cell. Rep. 4:*85 (1986).
61. Kyozuka, J., Y. Hayashi and K. Shimamoto, *Molec. Gen. Genet. 206:*408 (1987).
62. Rhodes, C., K. Lowe and K. Ruby, *Bio/Technology 6:*56 (1988).
63. Vasil, V., and I.K. Vasil, *Ann. Bot. 47:*669 (1981).
64. Spinivasan, C., and I.K. Vasil, *J. Pl. Physiol. 126:*41 (1986).
65. Toriyama, K., Y. Arimoto, H. Uchimiya and H. Kokichi, *Bio/Technology 6:*1072 (1988).
66. Rhodes, C.A., D.A. Pierce, I.J. Mettler, D. Mascarenhas and J.J. Detmer, *Science 240:*204 (1988).
67. Luo, Z.X., and R. Wu, *Plant Mol. Biol. Rep. 6:*165 (1988).

In Vitro Modification and Assembly of Soybean Glycinin[1]

N.C. Nielsen

USDA/ARS and the Department of Agronomy, Purdue University, Lilly Hall of Life Sciences, West Lafayette, Indiana 47907-7899, USA

Abstract

Seeds from crop plants provide dietary protein, carbohydrate and oil for many populations. It is rare, however, that the physical and nutritional properties of these seed products conform precisely with those required or demanded by consumers. Biotechnology provides a mechanism whereby the properties of seeds might be tailored specifically for value-added end uses. The rationale we have been following, which is to develop soybeans with improved protein quality, will be described as an example of such an approach. These strategies have resulted in elucidation of assembly mechanisms for subunits of soybean glycinin into hexamers. They have also permitted an identification of regions within the subunits where changes can be made that have minimal effect on assembly of the glycinin complex. Subunits have been produced that have a considerably more balanced amino acid composition than those from which they are derived. These modifications appear suitable for use in many crop species because of the wide distribution of homologous proteins throughout the plant kingdom.

Soybean seed storage proteins are excellent candidates for genetic engineering to improve seed quality. The genes that encode these proteins are expressed only in seeds, and the proteins produced from them are present in very large amounts. On a dry weight basis, mature soybean seeds contain between 35 and 45% protein, the majority of which (60–70%) is either glycinin or β-conglycinin. A relatively small number of genes encode the subunits for each of these proteins. Because of the high concentration of glycinin and β-conglycinin in the seed, the properties of the two proteins make the major contribution to seed quality. Hence, manipulation of just a few genes has the potential to drastically affect the nutritional properties of soybean meal and the functional properties of foods made from soybeans.

In most monogastric animals, legume seeds are nutritionally limiting in the two sulfur amino acids, cysteine and methionine. As little genetic variability for sulfur amino acid content has been observed among soybeans (1), conventional plant breeding cannot be used to solve this problem. Although the limitation in sulfur amino acids can be overcome by including cereal or animal protein in the diet, a soybean which provides a complete protein source is potentially marketable, particularly in situations where expense is an important consideration. One approach to resolve the problem is to insert codons for additional sulfur amino acids into genes for the storage proteins. The insertion of amino acids into structurally neutral regions of the proteins is likely to be more straightforward than modifications directed toward improvement of more complex functional characteristics like digestibility or rheological properties. That is to say, improvement of the amino acid composition might be considered a paradigm for the manipulation of other, more complicated factors that affect protein quality.

The insertion of sulfur amino acids into storage protein subunits will still require that changes in their primary structures do not drastically alter vital biological functions. A number of such characteristics immediately come to mind: (a) Glycinin and β-conglycinin are targeted to specific subcellular organelles, the vacuolar protein bodies. Hence, mechanisms involved in the transport of precursors between the sites of synthesis and deposition in the cell will doubtless be dependent upon subunit structure. (b) The storage proteins are deposited in an insoluble, or osmotically neutral form within the seed, and this property should be conserved in order to maintain properties of seed desiccation and stability. (c) Because of space considerations within the seed, it can be anticipated that glycinin and β-conglycinin will be packaged with maximum efficiency. (d) Seed storage proteins must be protected from premature degradation, yet must be mobilized in a controlled fashion during germination so that the carbon and nitrogen reserves provided by the proteins are not used up too quickly. Alterations in subunit structure that perturb these properties can be expected to have a negative impact on protein accumulation and mobilization in the seed.

When considering the factors described above, it becomes apparent that the mechanisms of assembly that lead to the deposition of storage proteins in an insoluble form is of central importance to the overall process. Therefore, systematic efforts directed toward modification of legume storage proteins demand that the regions of the subunits involved in assembly be identified and preserved. The aim of this communication is two-fold, to review briefly what is known about the assembly properties of glycinin subunits, and to describe the effect certain changes in primary structure have on assembly. These two goals are very much related. Results from studies about the assembly of glycinin subunits point to new constraints that will need to be considered in attempts to engineer these proteins in a beneficial manner.

Glycinin Genes and Proteins

A considerable literature exists about the glycinin subunits and the nuclear genes that encode them. As this information has been reviewed (2–4), only features about the subunits that are pertinent to the ensuing discussion will be summarized.

The genomes of cultivated soybean varieties contain five major glycinin genes that contribute the majority of the storage protein subunits that can be isolated from mature seeds (5). The five major genes are divided into two groups based on differences in size, sulfur content and degree of sequence identities among the subunits they encode (Table 1). Subunits produced from group-1 genes (Gy1 thru Gy3) are smaller and contain more sulfur than those encoded by the group-2 genes (Gy4 and Gy5). Correspondence in primary structure between members of the same subunit group exceed 90%, but this is reduced to less than 50% when members of different groups are compared (5). Nevertheless, the five glycinin genes are clearly related to a common ancestral gene.

TABLE 1

Glycinin Gene Family

Gene	Group	Subunit[a]	M_r^b	Mol. wt.[c]
Gy1	1	$G_1(A_{1a}B_2)$	58,000	55,700
Gy2	1	$G_2(A_2B_{1a})$	58,000	54,400
Gy3	1	$G_3(A_{1b}B_{1b})$	58,000	54,300
Gy4	2	$G_4(A_5A_4B_3)$	69,000	63,700
Gy5	2	$G_5(A_3B_4)$	62,000	58,000

[a] Old nomenclature derived on basis of protein purification and sequencing of glycinin subunit chains purified from soybean line CX635-1-1-1 is shown in parentheses.

[b] The values for the apparent molecular weights are the sums of those determined by SDS-PAGE for the acidic and basic chains.

[c] Molecular weight calculated from primary structures as derived from nucleic acid sequences and rounded to nearest hundred.

Synthesis and Assembly of Glycinin *In Vivo*

The general pathway followed during synthesis and assembly of glycinin is outlined in Figure 1. Polyadenylated message derived from the five glycinin genes is translated in rough endoplasmic reticulum (7). A short signal sequence is removed from the initial translation products, presumably by a co-translational mechanism as the nascent polypeptides emerge into the lumen of the ER (8). The proglycinin precursors that result are assembled into 9S oligomers in the endoplasmic reticulum shortly after their synthesis (9). The 9S oligomers are considered to consist of trimers of proglycinin subunits. The trimers apparently move from the ER to vacuolar protein bodies, apparently via the Golgi apparatus (10). Upon their arrival in protein bodies the precursor subunits in the trimers are cleaved post-translationally to yield mature subunits, each of which consists of an acidic and basic chain. The two chains are linked to one another by a single disulfide bond (11). Post-translational cleavage occurs at a conserved ASN-GLY bond located between the NH_2-terminal acidic and COOH-terminal basic components in the precursor (8). The mature subunits can be isolated from seeds as stable 12S hexamers. The six subunits in the hexamers appear to be arranged at the vertices of a trigonal antiprism having 3,2 point symmetry (12,13). That is, the six subunits form two rings, each composed of three subunits. The two rings are stacked upon one another and twisted by 60°.

The organization of glycinin within protein bodies is unknown, although studies carried out using cucurbitin, an 11S seed globulin homologous to glycinin, bear on this issue. Crystals produced from cucurbitin have been studied using low angle x-ray diffraction methodology (14). The scatter patterns derived from these crystals bear a strong resemblance to those obtained using purified protein bodies. On the basis of these results it has been speculated that the arrangement of cucurbitin within protein bodies is highly organized, as it is within the crystals. These considerations lead one to conclude that the structures within protein bodies are arranged in an organized rather than a random fashion. Thus, domains in the protein that are required for this assembly will need to be maintained during manipulations to improve quality.

Synthesis and Assembly of Glycinin *In Vitro*

During the course of the past two years, methodology has been developed in my laboratory that permits the assembly of glycinin to be carried out *in vitro*. This has permitted processes that occur during assembly to be dissected in more detail. It has also resulted in a functional identification of regions in the subunits that are important for assembly. The methodologies are based on transcription-translation technology due to Melton et al. (15), as described by Dickinson et al. (15). Coding regions derived from full length cDNA clones, whose signal sequences have been removed, are transcribed from SP6 promoters. Translations are subsequently performed using rabbit reticulocyte lysates, and yield proglycinins that are presumed to be structurally equivalent to native proglycinins formed during translation across rough endoplasmic reticulum (16). Proglycinins produced in this manner can be used to test for assembly into oligomers.

One class of constructions that has been tested produces proglycinins whose primary structures are essentially the same as authentic subunits. For example, a G4 proglycinin has been produced *in vitro* that lacks the signal sequence, but whose remaining amino acid sequence is identical except for a MET for LEU substitution at

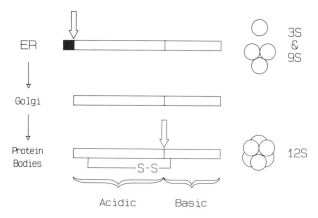

Fig. 1. Diagram summarizing several steps involved in the assembly of glycinin. Vertical arrows indicate points of proteolytic cleavage. The filled in region identifies the signal sequence considered to be removed co-translationally. Subunits are represented by spheres, which first assemble into trimers and then into hexamers. Post translational cleavage of proglycinin results in acidic and basic chains that are linked by a single disulfide bond.

the NH$_2$-terminal. These proglycinins are capable of self-assembly into 9S oligomers, a complex equivalent in size to those formed *in vivo* in endoplasmic reticulum. Similar results can be obtained using constructions that encode the other group-2 subunit, G5 (16).

Surprisingly, proglycinins produced from similarly constructed group-1 genes behave differently in the *in vitro* self-assembly assay. Whereas group-2 proglycinins are capable of self-assembly, those from group-1 do not carry out this reaction (17). The group-1 proglycinins are capable, however, of forming mixed 9S oligomers when self-assembly occurs in the presence of the group-2 precursors. The fact that the group-1 proglycinins efficiently co-assemble in the presence of group-2 precursors makes it unlikely that the deficiency in self-assembly of the former subunits is due to improper folding during their synthesis *in vitro*. It is more likely their inability to self-assemble reflects structural incompatibilities unique to the group-1 proglycinins. This point has been explored further using chimeric subunits constructed from reciprocal parts of group-1 and group-2 subunits (Scott and Nielsen, unpublished observations). Proglycinins with an acidic component from Gy2 (group-1) and a basic one from Gy4 (group-2) self-assemble into 9S oligomers. Those generated from a construction with the reciprocal orientation do not assemble. Thus, the basic components of group-1 proglycinins contain structures that block the self-assembly reaction.

Two important conclusions emerge as a consequence of the above results. The first has mechanistic and evolutionary implications. The fact that group-1 and group-2 proglycinins are capable of forming mixed trimers means that each type of precursor is apparently interchangeable in the complex. However, members of the two groups of subunits are not functionally equivalent because the group-1 precursors apparently have lost the ability to drive the formation of 9S oligomers. This deficiency is tolerated because all of the genes in the glycinin family are expressed coordinately, and the assembly of group-1 subunits into oligomers is potentiated by those from group-2.

The second conclusion has practical implications in terms of engineering the subunits to improve nutritional quality. Because the group-2 subunits are deficient in sulfur amino acids compared to those in group-1 (Table 1), we initially considered the best approach would be the introduction of additional methionine into group-1 subunits, accompanied by elimination of group-2 genes from the seed. Because products from the group-1 and group-2 genes apparently are not functionally equivalent, we have had to re-evaluate this strategy.

We considered it intriguing that assembly of the proglycinins stopped at the 9S trimer stage *in vitro*, and never proceeded to 12S hexamers like those isolated from seeds. The basis of this inability has turned out to be a consequence of structural differences between the precursor and mature forms of glycinin subunits. Two types of data justify this conclusion; both make use of a system that permits a complete and reversible dissociation of native glycinin hexamers into monomers and trimers (17). One approach relies on a mild treatment of proglycinin trimers with papain to artificially mimic the post-translational step that produces disulfide-linked acidic and basic chains. Radioactive proglycinin trimers re-assemble into hexamers along with unlabeled native trimers after they have been subjected to protease treatment. However, trimers are not incorporated into hexamers without this treatment. The result suggests that some type of a barrier to hexamer formation is removed by the treatment with protease.

The second approach relies on individual ^{3}H-acidic or ^{3}H-basic chains from the G4 subunit that are produced using the *in vitro* transcription-translation technology described by Dickinson et al. (16). When individual ^{3}H-acidic chains are introduced into a mixture of unlabeled monomers and trimers derived from native glycinin, a small proportion is recovered in the hexamer fraction upon re-assembly (17). None of these are observed as trimers, and the majority are recovered as free acidic polypeptide chains. Significantly, those ^{3}H-acidic chains recovered in the hexamer fraction are all disulfide bonded to a native basic chain. This result has important structural implications. It demonstrates that the basic chain is critical for assembly, because incorporation of the acidic chain does not occur unless it is associated with and has become cystine-linked to a basic one. The results obtained when ^{3}H-basic chains are re-associated in the presence of unlabeled glycinin monomers and trimers are consistent with this interpretation. In these experiments the majority of ^{3}H-basic chains are recovered as hexamers, a significant number are recovered as trimers, and very few are recovered as free basic chains. Only some of the basic chains recovered in the hexamers are disulfide bonded with acidic ones.

These data support two important concepts. First, the differences between the extent of reassembly of free acidic and basic polypeptides into hexamers indicate that the basic chains contain regions required for formation of oligomers. This conclusion is also consistent with the finding that the structure of the basic chains of group-1 subunits interferes with self-assembly of 9S oligomers *in vitro*. The second notion concerns the importance of the conserved ASN-GLY bonds that are cleaved post-translationally in protein bodies. The cleavage apparently regulates the ability of 9S oligomers to aggregate into an insoluble, ordered structure. Thus, in the absence of cleavage, assembly would not occur prior to arrival of the precursors in protein bodies. If this speculation is true, the expression of genes that encode a protease specific for the ASN-GLY bond would control a step of central importance in seed development and protein body formation.

Protein Modifications

The self-assembly and re-assembly techniques described above not only provide a means by which to dissect the assembly process, but they also furnish an opportunity to probe for regions in proglycinin that can tolerate structural changes. One recently completed set of experiments has focused on a domain in the primary structure of the subunits identified earlier as being extremely variable (18). This region, now referred to as the hypervariable region (HVR) (5), is located just NH$_2$-terminal to the highly conserved ASN-GLY cleavage site. It can vary in size by more than 100 amino acids, and accounts for the size variation encountered among the five glycinin subunits (cf. Table 1). It contains a high proportion of negatively charged ASP and GLU residues predicted to be arranged in either α-helical and turn secondary structures. By analogy to proteins whose three dimensional structures are known, this part of the protein is likely to be located on the surface of the subunits, perhaps as an exposed loop. The fact that the ASN-GLY cleavage site is located immediately after this region is consistent with this suggestion, because it must be accessible to proteases that carry out post-translational modification.

We have asked whether the HVR is required for self-assembly of 9S oligomers. To answer this question, six plasmids derived from the Gy4 coding region were prepared in which between 18 and nearly 100% of the HRV was removed (17). The modified precursors produced from these plasmids all self-assembled into oligomers, which indicates that assembly will occur in the absence of the HVR. To explore the effect of HVR removal further, we asked if removal of the HVR altered the rate of trimer assembly. For this purpose, an assay was developed in which the rate of mixed re-assembly of modified monomers into hexamers could be compared with those of G4-proglycinin (17). In each case, the rates of mixed re-assembly were equal to or greater than 80% of the rate observed with G4 proglycinin monomers.

We also considered whether the HVR in G4 would tolerate large insertions. Two approaches have been used to address this question. In one, tandem (ARG-MET)$_n$ units of increasing size were inserted into the HVR of Gy4 (17). Proglycinin subunits with five such units self-assembled to the same extent and at the same rate as G4 controls. To test the effect of even larger insertions, a number of synthetic oligonucleotides have been inserted into the HVR of Gy4. The largest of these insertions increased the molecular weight of the precursors by more than 2000. In several cases, precursors bearing these modifications self-assembled normally when the oligonucleotides were inserted in one orientation, but were incapable of assembly when inserted in the opposite orientation (Scott and Nielsen, unpublished observations). Thus, despite the fact that the G4 subunits contain the largest HVR of the five glycinin subunits, large insertions are tolerated in this region. However, the fact that some insertions are tolerated and others are not indicates that assembly properties of the subunits are sensitive to changes in the HVR.

Discussion

Arguments have been presented that an understanding of the relationship between subunit structure and the ability of glycinin subunits to assemble into a highly organized complex in protein bodies will be central to success in manipulating the properties of the major soybean storage proteins to improve quality. Assays were referred to that permit some reactions involved in assembly to be monitored *in vitro*. The significance of these assays is that they permit the effect of modifications in subunit structure to be tested rapidly and under controlled conditions. This avoids the use of cell transformation and plant regeneration techniques to test each construction, techniques which are extremely time consuming. While the validity of conclusions reached on the basis of *in vitro* tests will certainly need to be tested *in vivo*, one hopes that the less fruitful modifications can be eliminated from further consideration immediately using the *in vitro* assembly techniques. Because transformed soybeans have recently been reported (19,20), there is hope that lines that have seeds with improved nutritional quality are a realistic goal in the near term.

Acknowledgment

Cooperative research of the United States Department of Agriculture, Agricultural Research Service and the Indiana Agricultural Experiment Station.

References

1. Beversdorf, W.D., and D.J. Hume, in *Applications of Genetic Engineering to Crop Improvement*, edited by G.B. Collins and J.G. Petolino, Nijhoff/Junk, Boston, 1984, pp. 189–209.
2. Derbyshire, E., D.J. Wright and D. Boulter, *Phytochemistry 15*:3 (1976).
3. Nielsen, N.C., *J. Am. Oil Chem. Soc. 62*:1680 (1985).
4. Borroto, K., and L. Dure, *Plant Mol. Biol. 13*:113 (1987).
5. Nielsen, N.C., C.D. Dickinson, T-J. Cho, V.H. Thanh, B.J. Scallon, T.L. Sims, R.L. Fischer and R.B. Goldberg, publication pending.
6. Nielsen, N.C., in *Molecular Biology of Seed Storage Proteins and Lectins. Proceedings of the Ninth Annual Symposium in Plant Physiology*, edited by L.M. Shannon and M.J. Chrispeels, American Society of Physiologists, Rockville, Maryland, 1986, pp. 17–28.
7. Beachy, R.N., J.F. Thompson and J.T. Madison, *Plant Physiol. 61*:139 (1978).
8. Tumer, N.E., J.D. Richter and N.C. Nielsen, *J. Biol. Chem. 257*:4016 (1982).
9. Barton, K.A., J.F. Thompson, J.T. Madison, R. Rosenthal, N.P. Jarvis and R.N. Beachy, *Ibid. 257*:6089 (1982).
10. Chrispeels, M.J., *Oxford Survey of Plant Molecular and Cell Biology 2*:43 (1985).
11. Staswick, P.E., M.A. Hermodson and N.C. Nielsen, *J. Biol. Chem. 259*:13431 (1984).
12. Plietz, P., H. Damaschun, D. Zirwer, K. Gast, K.D. Schwenke, W. Paehtz and G. Damaschun, *FEBS Lett. 91*:227 (1978).
13. Plietz, P., D. Zirwer, B. Schleisier, K. Gast and G. Damaschun, *Biochim. Biophys. Acta 784*:140 (1984).
14. Coleman, P., E. Suzuki and A. van Donkelaar, *Eur. J. Biochem. 103*:585 (1980).
15. Melton, D.A., P.A. Krieg, M.R. Rebagliati, T. Maniatis, K. Zinn and M.R. Green, *Nucleic Acids Res. 12*:7035 (1984).
16. Dickinson, C.D., L.A. Floener, G.G. Lilley and N.C. Nielsen, *Proc. Natl. Acad. Science US 84*:5525 (1987).
17. Dickinson, C.D., *Development of a Novel in vitro Assembly System*, PhD Thesis, Purdue University, West Lafayette, IN, 1988.
18. Argos, P., S.V.L. Narayana and N.C. Nielsen, *EMBO J. 4*:1111 (1985).
19. Hinchee, M.A., D.V. Conner-Ward, C.A. Newell, R.E. McDonnell, S.J. Sato, C.S. Gasser, D.A. Fischhoff, D.B. Re RT Fraley and R.B. Horsch, *Bio/technology 6*:915 (1988).
20. McCabe, D.E., W.F. Swain, B.J. Martinell and P. Christou, *Bio/technology 6*:923 (1988).

The Biochemical Genetics of Lipoxygenases

Rod Casey, Claire Domoney, Paul Ealing and Helen North

John Innes Institute and AFRC Institute of Plant Science Research, Colney Lane, Norwich, NR4 7UH, UK

Abstract

Soybean seeds contain three major lipoxygenase isoenzymes. Two major polypeptides, possibly corresponding to soybean lipoxygenases 2 and 3, can be identified in pea seeds. The genes for these two major polypeptides are tightly linked and map to linkage group 4, near *Np* and *le*. The two major polypeptides appear at the same stage of seed development and have very similar patterns of synthesis. Fractionation of pea seed extracts and the use of specific antibodies, however, indicates the presence of other, minor, lipoxygenase polypeptides. Comparison of the amino acid sequences of soybean and pea lipoxygenases with those from humans highlights two areas of sequence conservation that may be important to the enzymes' function and specificity. A full-length pea seed lipoxygenase cDNA is being used, in collaboration with the Carlsberg laboratories, to transform bakers' yeast to produce strains with better bread-making properties.

Lipoxygenase(s) (LOX) (linoleate: oxygen oxidoreductase, E.C. 1.13.11.12) are a group of enzymes that catalyze the oxygenation of fatty acids containing a *cis,cis*-1,4-pentadiene system to form conjugated hydroperoxydienes (Fig. 1). LOX have the ability to catalyze cooxidation reactions and have been used commercially, in the form of LOX-rich soybean flour, as an additive in baking that promotes the bleaching of wheat flour pigments in white bread production. This addition of LOX also seems to enhance dough rheology and stability, possibly through a cooxidation of protein-SH groups in wheat gluten. Lipoxygenases from peas and *Phaseolus* beans and LOX-3 (see later) from soybean have high cooxidation potentials, whereas soybean LOX-1 and wheat LOX have poor cooxidation activity. It is for this reason that addition of soybean LOX to wheat flour achieves an increase in pigment bleaching. The high levels of LOX in soybean seeds have led to the implication of LOX in spoilage and off-flavor in soybean products, possibly through the production of short-chain-length volatile carbonyl compounds that bind to soya protein. LOX also are believed to be responsible for the development of rancid off-flavors in unblanched vegetables, including peas. LOX are therefore both desirable and undesirable components of legume seeds and their meals.

LOX have been isolated from a range of plant sources (1); the highest activities are found in legume seeds and LOX have been purified from peas, soybeans, faba beans, french beans, peanuts, lentils, winged beans, chickpeas and lupins. The amount of activity is a function of genotype; in soybeans an approximate two-fold range of activity exists across 19 different varieties (2). LOX also have been isolated from human and porcine peripheral blood leukocytes, rat basophilic leukemia cells and mouse mastocytoma cells; the 5-LOX activity found in mammalian cells catalyzes the first two reactions in the biosynthetic pathway leading to the formation of the leukotrienes from arachidonic acid. The amounts of LOX in such cells, however, are far lower than in legume seeds; in soybean they can approach 1% of the dry seed mass.

Plant LOX are not confined to seeds and the physiological role of plant LOX is poorly understood. They have been implicated in fruit ripening and abscission, senescence, resistance to plant pathogens and the production of possible plant growth regulators.

Isozymes in Soybean and Pea

The most comprehensively studied legume seed LOX are those from soybean and pea. The existence of multiple LOX isozymes in soybean has been established by gel electrophoresis, enzyme assays, enzyme fractionations, serological studies and genetics. A LOX from soybean which was purified and crystallized as early as 1947 (3) was later named LOX-1 (see Reference 1 for a review of LOX nomenclature). Four LOX isozymes (LOX-1, LOX-2, LOX-3a and LOX-3b) have been identified in soybean seeds (4), LOX-3a and -3b being so similar in behavior and composition that they may be considered as a single type (LOX-3). All are monomeric proteins of M_r approximately 10^5, containing one atom of tightly bound non-heme iron per molecule. Although the activities of LOX-1 and LOX-2/3 can be distinguished readily as purified components, the individual activities cannot easily be determined in mixtures such as are normally found in crude soybean extracts (5). Antibodies specific to each of the

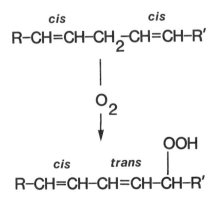

Fig. 1. Oxygenation of fatty acids containing *cis,cis*-1,4-pentadiene systems catalyzed by lipoxygenase.

three soybean isozymes (5–7) have proved useful in the estimation of the amounts of individual LOX species in soybean extracts. LOX-1 differs from the others in being heat stable, having a pH optimum of approximately 9 and a preference for anionic substrates (fatty acids); this latter occasionally has led to it misleadingly being described as an "acid LOX." LOX-2 and -3 are less heat stable, prefer esterified substrates and have pH optima close to neutrality (Table 1). In a standard assay with linoleic acid at 1.8 mM, LOX-2 shows maximum activity at pH 6.6 and negligible activity at pH 9.0. Addition of Ca^{2+}, or lowering of the linoleate concentration, however, considerably increases the LOX-2-catalyzed oxygenation of linoleic acid at pH 9.0 and shows that interpretations of isozyme status through activity measurements have to be treated with caution (see also Ref. 1 for discussion of LOX assays). The isozymes also differ in their product regiospecificity with linoleate as substrate, LOX-1 showing a preference for the 13 position as the site for hydroperoxidation, whereas LOX-2 and -3 use either position 9 or 13 (8).

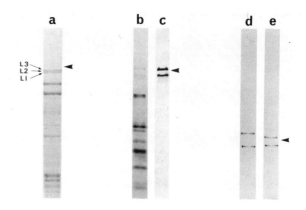

Fig. 2. Separation of lipoxygenase polypeptides from soybean and pea seeds by SDS-gel electrophoresis. a, b, Coomassie Blue stain; c-e, "Western" blots probed with antilipoxygenase: a, separation of soybean seed extract on a 7% gel (13) run for 5h at 160V; L1, L2 and L3 refer to the polypeptides of LOX-1, -2 and -3, respectively; b,c, separation of pea seed extracts from the cultivar "Birte" on a 10% gel (ref. 11) run for 16h at 50V; d,e, separation of pea seed extracts from genotypes JI 64 and JI 92, respectively, on a 10% gel run for 18h at 100V. The arrowheads indicate the positions adopted by markers of M_r 94,000; migration is from top to bottom in all tracks.

TABLE 1

Some Properties of Soybean and Pea Seed Lipoxygenase Isozymes

Isozyme[a]	pI	pH for optimal activity	Polypeptide Mr from sequence (amino acids)
Soybean			
LOX-1	5.5–5.7	9–9.5	94038 (838)
LOX-2	5.8–6.2	6–7	—
LOX-3	5.8–6.2	6–7	96663 (857)
Pea			
LOX-2	5.82	6–7	—
LOX-3	6.25	6-7	97628 (861)

[a] A widely used alternative nomenclature is to refer to LOX-1 as a "type-1" enzyme and LOX-2 and -3 as "type-2" enzymes. The PLI and PLII described by Yoon and Klein (9) correspond to LOX-3 and LOX-2, respectively (10), from peas.

Peas apparently have relatively low amounts of LOX-1 and the majority of pea seed LOX activity is attributable to LOX-2 and LOX-3 (9). Measurements of the apparent M_r of pea seed LOX polypeptides suggest a value of 90000-95000 (10,11); genetic variation for the apparent M_r of pea LOX polypeptides on SDS gels has been observed (see below). These values are in good agreement with those reported for soybean LOX-1, -2 and -3, which all have similar but distinct apparent M_r values on SDS/polyacrylamide gel electrophoresis (12,13). The soybean LOX polypeptides are relatively difficult to separate on SDS gels (13) (Fig. 2a), whereas the pea LOX polypeptides can easily be resolved (11) into two major polypeptides (Fig. 2c) that probably therefore correspond to LOX-2 and LOX-3 from soybean (9,10).

In addition to the three major LOX species in soybean and the two major polypeptides in peas, there exist a number of minor species. Several minor species from pea seeds have been reported (9,14-16), but little information exists on their biochemical properties. Fractionation of pea seed LOX by ammonium sulphate precipitation and gel filtration has demonstrated the existence of a number of minor species (Fig. 3) that can be distinguished from the major polypeptides shown in Fig. 2c on the basis of apparent molecular weight and reactions with specific antibodies. Their precise relationship to the major LOX polypeptides is unknown. Yoon and Klein (9) reported the isolation of two minor LOX isoenzymes from pea seeds that were present at very low concentrations and were active at pH 9, but not at pH 7.2. Such small quantities of these isozymes were obtained that no further characterization was possible.

A small peak of enzyme activity eluting as a shoulder on the LOX-1 peak from soybean seeds has been noted (13). It seemed to differ from LOX-1 in its enzymatic behavior, having higher activity at pH 7 than LOX-1, but was absent from the seeds of a LOX-1-null variant genotype.

The genetics of LOX: In genetic analyses of soybean LOX, isozyme status usually has been assigned through polypeptide patterns on SDS gels (17). Such assignments generally will prove correct, because they were initially based on activity and chromatographic analyses of null variants (13), but could be confounded by variants in which polypeptides have altered mobilities.

Soybean mutants, lacking either LOX-1 (18), LOX-2 (19) or LOX-3 (12) have been identified. Gel analysis of these null variants showed the absence from each type (13) of one of the three LOX polypeptides. Genetic studies (12,17,20) indicated that the absence of a given

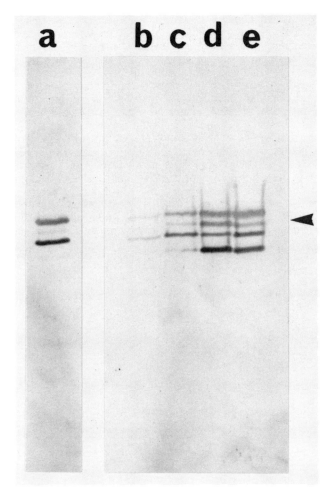

Fig. 3. Demonstration of minor pea seed lipoxygenase species. "Western" blots, of SDS-gel separations of pea seed extracts, probed with antilipoxygenase; a, 25–60% saturated ammonium sulphate cut of pea seed extract from the cultivar "Birte"; b-e, Successive cuts through the lipoxygenase-active peak obtained by gel filtration (Sephadex G200) of the material shown in a. Note the increase in prominence of the minor species as the analysis progresses through the peak. The arrowhead indicates the position adopted by a marker of $M_r \approx 94,000$.

If the two major pea LOX polypeptides correspond to LOX-2 and LOX-3, then this linkage of the LOX-2 and LOX-3 genes in the pea contrasts with soybean, where the genes for LOX-2 and LOX-3 are not tightly linked. Thus the relationship between enzyme activity, protein structure and genetics in the two legume species may not be straightforward. The larger of the two pea LOX polypeptides is more similar to soybean LOX-3 than to LOX-1 or -2 (21). Until a complete sequence for pea LOX-2 has been determined, however, it is premature to speculate on the significance of comparative genetic behavior. It appears that the soybean LOX genes that are closely linked (LOX-1 and LOX-2) encode sequences that are structurally more similar to each other than they are to homologous sequences (LOX-3) at an unlinked locus (22). A similar situation exists with the genes for the pea storage protein legumin (23), where genes at two loci on linkage group 1 are more similar to each other than to homologous genes at a locus on linkage group 7.

Soybean germplasm stocks, lacking both LOX-1 and LOX-3, or LOX-2 and LOX-3, have been derived from null variants; it was noted that the undesirable "grassy-beany" flavor of soybean seeds was very slight in the seeds of double-recessives lacking LOX-2 and LOX-3 (24). Studies with a LOX-2-null genotype suggest that LOX-2 makes a major contribution to the unacceptability of soymilk and soymeal in taste panel evaluations (N.C. Nielsen, pers. comm.).

The existence of null variants suggests that either there are very few genes in soybean for any given seed LOX isozyme, or that the genes for a given isozyme are either extremely tightly clustered or under common control such that a single genetic lesion can prevent the formation of all the polypeptides corresponding to a given isozyme.

The numbers of genes corresponding to the various soybean LOX polypeptides have not been reported, but it appears that there may be as many as five genes corresponding to one LOX cDNA per haploid pea genome (11).

Deposition and development regulation of the synthesis of seed lipoxygenases: There is little information on the activity or amount of lipoxygenases during seed development in soybeans or peas. The activities of all soybean LOX isozymes are maximal from about 5 to 20 days before seed maturity (i.e., fairly late in seed development) (25). Although the activities of LOX-2 and LOX-3 increased again between 4 and 7 days of germination, that of LOX-1 decreased. LOX-1 activity therefore seems to be regulated in a different fashion to the other two isozymes during germination.

In pea, the two major LOX polypeptides were first detected by "Western" blotting at precisely the same time during seed development (Fig. 4); i.e., 17 days after flowering where development of the seed from flowering to maximum fresh weight occurred in 30 days. This is a relatively late developmental stage compared to that at which the major storage proteins, legumin, vicilin, and convicilin, appear.

Studies of the intracellular localization of LOX-1 and LOX-2 using immunological techniques (26,27) have indicated that both isozymes are localized in the cytoplasm, not the protein bodies, of cotyledonary storage parenchyma cells from germinating soybean seeds.

isozyme was due to a single recessive allele (lx_1, lx_2, or lx_3). The genes for LOX-1 and LOX-3 segregate independently (12) whereas those for LOX-1 and LOX-2 are linked tightly (17). Linkage studies (20) showed that the Lx_1 locus is not linked to loci encoding seed lectin (Le), seed β-amylase (SP_1^b) or flower color (W_1); lx_3 is not closely linked to marker genes representing seven of the nine known soybean linkage groups (19).

Pea seed LOX variants displaying altered mobility on SDS gels have been observed (Fig. 2d,e). Inheritance studies suggest that the genes encoding the major pea seed LOX polypeptides are linked tightly and behave as single, codominant Mendelian genes. Linkage analyses show that this pea seed LOX locus maps to linkage group 4, between Np (which determines the presence of neoplastic growth on pods) and Le (responsible in peas for internode length) (North et al., submitted for publication).

over this sequence, whereas the soybean "Lox-1-like" (LOX-2?) sequence differs by one further amino acid (Fig. 5A). It has been suggested (29) that such a high degree of sequence conservation between enzymes from plants and animals indicates a role for this conserved region in enzymatic functions. The highly conserved area includes a histidine and may be part of the active site of the enzyme. There is little evidence at present to suggest a possible ligand for Fe within LOX, but histidine is a likely candidate. A further region of homology, although not as striking, exists (Fig. 5B) between all LOX sequences in the region corresponding to human 5-LOX residues 363–402 (29). This region, which includes five conserved histidyl residues, contains a peptide segment (human residues 368–382) that is related to the interface-binding domain of mammalian lipases and may possibly be a site for enzyme-substrate interaction.

Fig. 4. SDS-gel/"Western" blot analysis of the developmental appearance of pea seed lipoxygenase polypeptides. Extracts from cotyledons at various stages after flowering were separated by SDS-gel electrophoresis and the major lipoxygenase polypeptides located by "Western" blotting and detection with antilipoxygenase. Electrophoresis is from top to bottom; the arrowhead denotes the position adopted by a marker of $M_r \approx 94,000$.

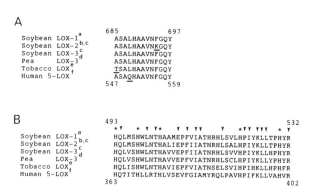

Fig. 5. Conserved amino acid sequences in lipoxygenases. A, The region corresponding to the soybean LOX-1 residues 685–697 and human 5-LOX residues 547–559 noted by Dixon et al. (29). Variant residues within this 13 amino acid sequence are underlined. B, The hydrophobic region corresponding to soybean LOX-1 residues 493–532 and human 5-LOX residues 363–402. Invariant histidines are indicated by an asterisk and other invariant residues by an arrowhead. (a, ref. 28; b, ref. 32 (shown here as corrected in c, ref. 22); d, ref. 21; e, D. Hildebrand, personal communication; f, refs. 29–31.)

Structural features of lipoxygenases: The isolation and analysis of cDNA and genomic clones, together with protein sequence data, have provided complete sequences for soybean LOX-1 (28) and LOX-3 (22), pea (probably) LOX-3 (21), and human 5-lipoxygenases from leukocytes, a leukemia cell line (29) and placenta (30,31). Partial sequences have also been determined for a LOX from tobacco (D. Hildebrand, pers. comm.) and for what is probably LOX-2 from soybean (32). The last sequence has been described as "LOX-1-like" (32) and needs a few minor changes to produce good homology with other soybean sequences (22,28). The complete sequences have permitted the determination of precise molecular weights for LOX polypeptides (Table 1). Soybean LOX-1 is appreciably shorter than LOX-3 from soybean and pea, this difference in size being almost entirely due to two small deletions of sequence near the N-terminus (21,22). The human sequences are considerably shorter than those from soybean and pea, but show such low homology to them that it is difficult to ascribe an exact basis to the difference. Despite this generally poor homology between the legume and human sequences, there are two areas of great similarity which may reflect conservation of amino acids that are important to the function of the enzymes. Dixon et al. (29) have noted a region in which 12 out of 13 amino acids are identical between soybean LOX-1 (residues 685–697) and human leukemia cell 5-LOX (residues 547–559). The sequence from pea and LOX-3 from soybean are both identical with soybean LOX-1

Experiments using a preparation of soybean seed LOX and a series of isomeric octadecanoic acids with the pentadiene system in different positions (33,34) suggested that there is only one orientation for the fatty acid moiety on the enzyme and that the important interaction of fatty acids with LOX is through the pentadiene system rather than the carboxyl group. Details of the specificity and reaction mechanisms of LOX have been reviewed by Galliard and Chan (1) and Veldink et al. (35).

The soybean and pea LOX sequences give no immediate clue as to why LOX-1 should differ in its properties from LOX-2 and LOX-3, but the existence of full-length clones that are capable of directing enzyme expression *in vitro* affords the opportunity to dissect the relationship between structure and function through a program of *in vitro* mutagenesis.

The pea LOX-3 clone can direct the *in vitro* transcription of an mRNA (Fig. 6a) which can be translated *in vitro* to a polypeptide of the correct size that is precipitable with antibody against soybean LOX (Fig. 6b) (21). We are using this clone in collaboration with the Carlsberg Laboratories, Copenhagen, as part of a project designed to genetically engineer strains of bakers' yeast to express pea LOX in addition to its intrinsic mitochondrial LOX activity (36); we hope that such an engineered strain might have improved bread-making qualities through the effect of LOX on dough rheology.

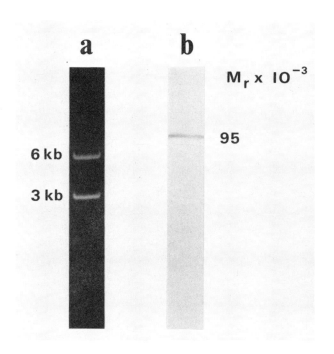

Fig. 6. Synthesis of pea seed lipoxygenase by the use of *in vitro* transcription and translation systems. (a) The glyoxalated products of transcription from a full-length pea seed LOX cDNA analyzed by agarose gel electrophoresis. The linearized vector + cDNA is at 6kb and the transcript RNA at 3kb. (b) Translation products *in vitro* of the RNA shown in (a) after immunoprecipitation with antiserum to lipoxygenase. For details see ref. 21.

Acknowledgments

This work was supported by the Agricultural and Food Research Council via a grant-in-aid to the John Innes Institute. Financial support was also received via the AFRC New Initiatives Scheme and the CEC Biotechnology Action Program (contract number BAP-0063-UK).

References

1. Galliard, T., and H.W.-S. Chan, in *"The Biochemistry of Plants,"* Vol. 4, edited by P.K. Stumpf, Academic Press, New York, pp. 131–161, 1980.
2. Hafez, Y.S., *Nutrition Reports International* 28:1197 (1983).
3. Theorell, H., S. Bergström and A. Åkeson, *Acta Chem. Scand. 1:*571 (1947).
4. Axelrod, B., T.M. Cheesbrough and S. Laakso, *Meths. Enzymol.* 71:441 (1981).
5. Yabuuchi, S., R.M. Lister, B. Axelrod, J.R. Wilcox and N.C. Nielsen, *Crop Science* 22:333 (1982).
6. Vernooy-Gerritsen, M., G.A. Veldink and J.F.G. Vliegenthart, *Biochim. Biophys. Acta 708:*330 (1982).
7. Vernooy-Gerritsen, M., A.L.M. Bos, G.A. Veldink and J.F.G. Vliegenthart, *Biochim. Biophys. Acta 748:*148 (1983a).
8. Christopher, J.P., and B. Axelrod, *Biochem. Biophys. Res. Commun.* 44:731 (1971).
9. Yoon, S., and B.P. Klein, *J. Agric. Food Chem.* 27:955 (1979).
10. Reynolds, P.A., and B.P. Klein, *J. Food Sci.* 47:1999 (1982).
11. Casey, R., C. Domoney and N.C. Nielsen, *Biochem. J.* 232:79 (1985).
12. Kitamura, K., C.S. Davies, N. Kaizuma and N.C. Nielsen, *Crop Science* 23:924 (1983).
13. Kitamura, K., *Agric. Biol. Chem.* 48:2339 (1984).
14. Eriksson, C.E., and S.G. Svensson, *Biochim. Biophys. Acta* 198:449 (1980).
15. Arens, D., W. Seilmeier, F. Weber, G. Kloos and W. Grosch, *Ibid.* 327:295 (1973).
16. Anstis, P.J.P., and J. Friend, *Planta* 115:329 (1974).
17. Davies, C.S., and N.C. Nielsen, *Crop Science* 26:460 (1986).
18. Hildebrand, D.F., and T. Hymowitz, *J. Am. Oil Chem. Soc.* 58:583 (1981).
19. Davies, C.S., K. Kitamura and N.C. Nielsen, *Agronomy Abstracts*, p.61 (1983).
20. Hildebrand, D.F., and T. Hymowitz, *Crop Science* 22:851 (1982).
21. Ealing, P.M., and R. Casey, *Biochem. J.* 253:915 (1988).
22. Yenofsky, R.L., M. Fine and C. Liu, *Mol. Gen. Genet.* 211:215 (1988).
23. Domoney, C., T.H.N. Ellis and D.R. Davies, *Ibid.* 202:280 (1986).
24. Kitamura, K., T. Kumagai and A. Kikuchi, *Jap. J. Breeding* 35:413 (1985).
25. Hildebrand, D., and T. Hymowitz, *Bot. Gaz.* 144:212 (1983).
26. Vernooy-Gerritsen, M., A.L.M. Bos, G.A. Veldink and J.F.G. Vliegenthart, *Plant Physiol.* 73:262 (1983b).
27. Vernooy-Gerritsen, M., J.L.M. Leunissen, G.A. Veldink and J.F.G. Vliegenthart, *Ibid.* 76:1070 (1984).
28. Shibata, D., J. Steczko, J.E. Dixon, M. Hermodson, R. Yazdanparast and B. Axelrod, *J. Biol. Chem.* 262:10080 (1987).
29. Dixon, R.A.F., R.E. Jones, R.E. Diehl, C.D. Bennett, S. Kargman and C.A. Rouzer, *Proc. Natl. Acad. Sci. U.S.A.* 85:416 (1988).
30. Matsumoto, T., C.D. Funk, O. Rådmark, J.-O. Höög, H. Jörnvall and B. Samuelsson, *Ibid.* 85:26 (1988a).
31. Matsumoto, T., C.D. Funk, O. Rådmark, J.-O. Höög, H. Jörnvall and B. Samuelsson, *Ibid.* 85:3406 (1988b).

32. Start, W.G., Y. Ma, J.C. Polacco, D.F. Hildebrand, G.A. Freyer and M. Altschuler, *Plant Mol. Biol. 7:*11 (1986).
33. Holman, R.T., P.O. Egwin and W.W. Christie, *J. Biol. Chem. 244:*1149 (1969).
34. Egmond, M.R., G.A. Veldink, J.F.G. Vliegenthart and J. Boldingh, *Biochim. Biophys. Acta 409:*399 (1975).
35. Veldink, G.A., J.F.G. Vliegenthart and J. Boldingh, *Prog. Chem. Fats other Lipids 15:*131 (1977).
36. Schecter, G., and S. Grossman, *Int. J. Biochem. 15:*1295 (1983).

Isolation and Primary Structure for a Novel, Methionine-rich Protein from Sunflowerseeds (*Helianthus annus.* L)

G.G. Lilley, J.B. Caldwell, A.A. Kortt

CSIRO, Division of Biotechnology, Parkville, 3052, Victoria, Australia

T.J. Higgins, D. Spencer

CSIRO, Division of Plant Industry, Canberra, 2601, ACT, Australia

Abstract

The seed proteins of dicotyledenous plants are divided by solubility properties into two major classes, the globulins and albumins. Many of the diverse 2S albumin class are relatively rich in essential sulfur-amino acids (methionine and cysteine). This class can be subdivided further into those which fulfill a storage role and those which have a defined biological activity (e.g., the protease inhibitors which are regarded as antinutritional factors). On the basis of sequence homology, several 2S seed proteins from diverse sources such as oilseeds and cereals have been grouped into a superfamily of proteins. Advances in gene technology and plant transgenesis now offer the possibility of improving the nutritive value of fodder and seed crops for both animal and human consumption by transfer of genes with desirable nutritional value (e.g., selected 2S sulfur-rich albumins).

That fraction of the 2S protein class in sunflowerseeds which is soluble in 60% methanol-water can be resolved by HPLC to give eight distinct components. Two of these components (M_r 10000) are closely related and contain 8 and 16 residues per molecule of 50% cystine and 50% methionine, respectively. These proteins are resistant to breakdown in the rumen fluid of sheep but are readily digested in the stomach by pepsin. The primary structure of the major sulfur-rich protein was determined by direct amino acid sequencing and confirmed by comparison with a cDNA clone isolated from a lambda-gt11 library. The protein shows extensive sequence homology with brazil nut albumin and other 2S seed albumins. The positions of the cysteine residues, and presumably the disulfide bridges which are thought to have provided high stability in this class of proteins, are conserved, supporting the idea that sunflower albumin 8 and other members of the 2S superfamily evolved from a common ancestral gene.

During the early 1960's and later, studies by Reis et al. and others (1–3) clearly demonstrated that sheep fed by abomasal infusion of a supplement of sulfur amino acids grew wool at rates in the order of 30–100% greater than normal. This affect was noted also when sheep were fed high quality proteins such as casein under conditions in which rumen fermentation was avoided or minimized. Dove and Robards (4) confirmed similar increases in wool growth rates and showed that the affect was enhanced if at the same time a high energy/quality fodder such as lucerne clover was fed by the normal route. Other experiments have shown that protein degradation in the rumen can be avoided by treatment of meals with chemical crosslinking agents (5).

It is known that dietary proteins ingested by sheep are rapidly degraded by bacteria in the rumen and converted to microbial proteins which form the main source of protein nutrient for the animal. Rumen fluid microbial proteins are typically low in sulfur amino acids (6), and it has been concluded that this lack of sulfur amino acids limits wool growth rates. The goal of supplementing fodder with proteins which are high in sulfur amino acids and resistant to rumen degradation can now be addressed by gene engineering. Genes of desirable proteins may be introduced into pasture plants such as lucerne and subterranean clovers as a more direct, cost-effective method of improving the ruminant diet.

Using techniques similar to those described by Nugent et al. (7), Spencer and his colleagues have identified pea and sunflower proteins which are resistant to rumen fluid degradation *in vitro* (8). PA1, a sulfur-rich, low molecular weight pea seed albumin, has been characterized and the respective gene isolated (9,10). As a model, the gene for another pea protein, vicilin (a 7S, low sulfur protein) has been transferred via the *Agrobacterium tumefaciens* system to both tobacco and lucerne and low level expression obtained in leaves and stems of regenerated plants (11). In a parallel study, the PA1 gene has been transferred to lucerne and messenger RNA detected in the leaves (12). Together these experiments give credence to the idea of improving pasture plants by genetic engineering techniques. The current study describes the isolation and characterization of a sulfur-rich protein from sunflowerseed that was resistant to degradation in the rumen. The primary structure of the protein has been determined, and a cDNA clone which codes for the protein has been identified and sequenced.

Preparation and Fractionation of Rumen Resistant Seed Proteins

SDS-PAGE of sunflowerseed protein showed a major group of high molecular weight polypeptides (M_r 40000-65000) and a less abundant low molecular weight group (M_r 10000-16000). This latter group of proteins was resistant to rumen degradation (8). Seeds of *Helianthus annus* cv. Hysun were ground to a meal which was defatted by extraction with petroleum ether and air-dried. Soluble proteins were extracted with buffered 0.5M salt (containing the protease inhibitor PMSF) at pH 8 for two hr at room temperature.

The extract was clarified and cooled to 0°C, and the high molecular weight polypeptides (seed globulins) were precipitated by the addition of methanol (60%). The low molecular weight components (seed albumins) were precipitated from the supernatant by the addition of acetone. The precipitate was dissolved in water and dialyzed extensively against water. The pigmented residue obtained on dialysis was removed by centrifugation, and the soluble proteins were recovered by lyophilization. An extract of 10 g of seed yielded approximately 200 mg of albumin material. Fractionation of the albumin preparation by reverse phase HPLC resolved the mixture into eight different components (Fig. 1). In the absence and presence of reducing agent, SDS-PAGE showed single subunits for each component (Fig. 2) which fell into three distinct size classes (albumins: 1–3, $M_r \approx 18000$; 4–6, $M_r \approx 14000$; 7–8, $M_r \approx 10000$). SDS-PAGE under nondenaturing conditions showed that all the albumins were basic, with albumin 6 having the most basic character.

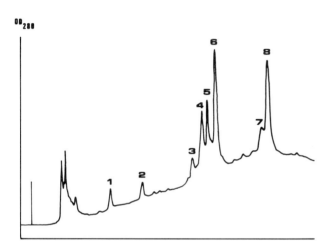

Fig. 1. Reverse phase HPLC fractionation of sunflowerseed albumins on a micro-Bondapak column (Waters P/L) with a gradient of acetonitrile in 0.1% (v/v) trifluoro-acetic acid.

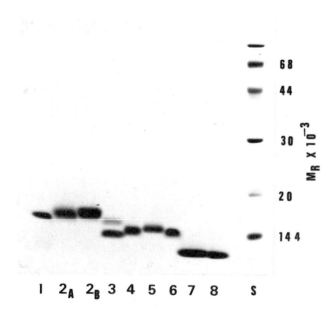

Fig. 2. SDS-PAGE of HPLC fractionated sunflowerseed Albumins (1,2a,2b,3-8) in a 12% acrylamide gel using a Laemmli buffer system.

Amino Acid Composition

Amino acid analysis was performed on a Waters HPLC amino acid analyzer after hydrolysis of protein by conventional methods (13). The amino acid compositions of the albumins 1–8 from sunflowerseed were similar to those of 2S albumins from other species (14–18) with characteristically high 1/2 cystine (up to 8%), and generally high lysine, arginine, glutamic and aspartic acids. Furthermore, albumins 7 and 8 (essentially identical in composition) were very rich in methionine (16%) in addition to the already elevated level of 1/2 cystine (7%) (Table 1). This favorable amino acid composition and their rumen stability suggested that they may be ideally suited as candidate proteins in studies which aim to improve the nutritive value of fodders.

TABLE 1

Amino Acid Composition of Sunflower Albumin 8 (SFA8)

Lys	4.0 (4)	Glu	19.3 (18)	Met	15.0 (16)
His	2.9 (3)	Pro	5.6 (6)	Ile	2.9 (3)
Arg	6.7 (7)	Gly	4.9 (5)	Leu	9.0 (9)
Asp	8.6 (9)	Ala	3.0 (3)	Tyr	3.0 (3)
Thr	1.0 (1)	1/2 Cys	6.7 (8)	Phe	0.0 (0)
Ser	4.5 (5)	Val	1.9 (2)	Trp	n.d (1)

Data is residues per 100 residues (sequence analysis in parentheses).

Amino Acid Sequence Determination of SFA8

Albumin 8 (SFA8) was alkylated with iodoacetic acid after reduction in SDS (13), and the N-terminal sequence of the intact protein (to residue 26) was determined in a gas-phase sequencer (Applied Biosystems). The peptides derived by digestion of the alkylated SFA8 preparation with trypsin (Worthington), α-chymotrypsin (Worthington) or *Staphylococcus aureus* V8 protease (Pierce) were fractionated by reverse phase HPLC and were sequenced either manually by a modified Edman procedure or on a gas-phase sequencer. The entire sequence of SFA8 (Fig. 3) was determined by sequencing the peptides derived from three enzymic digests of the protein. The C-terminal residue of SFA8 was identified as Met from the sequences of the C-terminal overlapping peptides from the three enzyme digests. This assignment was confirmed from the position of the translation termination codon identified in the corresponding cDNA sequence.

Preparation of Sunflowerseed cDNA Library

Total seed RNA was isolated (19) from mid-maturation, 19-day sunflower embryos and fractionated by passage through an oligo-d(T) cellulose (Collaborative Research Inc., USA) column to prepare messenger RNA (mRNA). Regions of the primary amino acid sequence of SFA8 (Fig. 3) were found with a minimum redundancy in the equivalent nucleotide sequence, and with these data two synthetic oligonucleotide probes were made to identify the cDNA coding for SFA8. The hybridization conditions for the oligonucleotide probes were as described (20), and

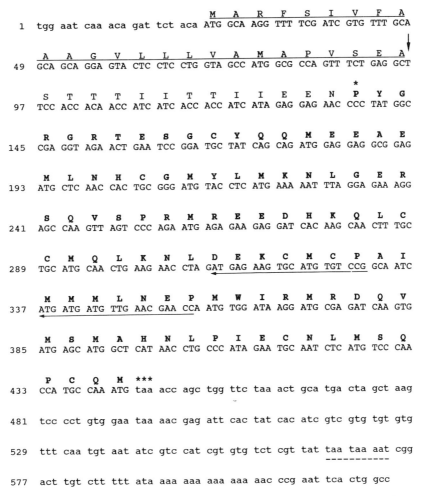

Fig. 3. (a) Nucleotide sequence and deduced pre-pro-protein sequence of clone pSF8: continuous underline, signal sequence; arrow, predicted signal cleavage site; *, mature protein N- terminus; ***, translation termination codon; left pointing arrows, regions for oligo probes; dashed underline, polyadenylation sequence. The deduced protein sequence was identical to that determined by protein sequencing (emphasized print).

hybridization temperatures were optimized by trial and error.

Northern analysis on a conventional formaldehyde gel system (21) (not shown) with a mixture of the end-labelled oligo probes showed a single dominant 0.8Kb component in the seed mRNA which hybridized strongly to the probes at 40°C. The mRNA preparations were judged to be sufficiently free of ribosomal material and RNAse contamination; ribosomal RNAs were not observed under any stringency conditions, and there was no evidence for degradation of the 0.8Kb message.

An Amersham kit (No. RPN 1285Y) was used both to prepare double stranded cDNA from total mRNA and to clone that cDNA (size selected on a Superose 12B column, Pharmacia) into lambda phage gt11. The library thus obtained contained less than 20% parental clones and a minimum of 2×10^5 unique recombinants.

Identification of cDNA coding for SFA8. Approximately 10^5 recombinant phage were screened with a mixture of the labelled oligos (20,22), one of several positive clones was sub-cloned into pUC119 (pSF16) and M13mp19 and sequenced by the Sanger dideoxy method. The insert from 0.56Kb clone (pSF16) was found to code for the entire sequence of mature SFA8. Using the purified insert from pSF16, labelled by incorporation of P^{32}-dATP by a standard random primer method, further clones were identified which encompassed the entire translated region of the SFA8 cDNA, and one of these (clone M13-SF8) was sequenced (Fig. 3); however, no full length clones were found. The length of the mRNA for SFA8, established by a primer extension method (23), showed that the SFA8 mRNA was 0.76Kb, in close agreement with the Northern analysis.

Nucleotide sequence of cDNA for SFA8. The amino acid sequence deduced from the nucleotide sequence of M13-SF8 agreed without exception with that determined by classical protein sequencing techniques for SFA8.

The deduced pre-pro-SFA8 sequence showed a typical (24) helical, hydrophobic signal sequence of 25 amino acids terminated by the sequence Pro-Val-Ser-Glu-Ala after which co-translational cleavage can be predicted (25) to be highly probable (although no direct evidence is yet available). Following the signal sequence was another quite hydrophobic repeating sequence which presumably is removed by posttranslational modification. The cleavage site between the latter sequence and the N-terminus of the mature protein was homologous to processing sites in other 2S albumins and similar to that in ricin from *Ricinus communis* (26) (Fig. 4) being characteristically charge rich. The hydrophobic region N-terminal to the processing site in pre-pro-SFA8 is also a feature of both brazil nut albumin and ricin. There was no evidence that SFA8 undergoes further post-translational processing like that observed in 2S proteins from lupinseed, castor bean, pea, brazil nut and rapeseed, although there is some homology between brazil nut albumin and SFA8 at the processing site in the brazil nut protein.

Secondary structure. The circular dichroic spectrum of SFA8 was measured in a Dichrographe III (Jobin Yvon)

	--	+ + -
Sunflower albumin 8	IITTIIEEN	PYGRGRTESGC
	---	-- +- +
Brazilnut albumin	TTTVVEEEN	QEECREQMQRQ
	- --- -	+
Rapeseed Napin	FDFEDDMEN	PQGPQQRPPLL
	--	+
Castor bean Ricin	WSFTLEDNN	IFPKQYPIINF

Fig. 4. A comparison of amino acid sequences surrounding putative post-translational processing sites. Sites in SFA8 and ricin are adjacent to the N-terminus of the mature protein. The site in Brazil nut albumin is adjacent to the N-terminus of the small subunit, while that in 2S napin is adjacent to the N-terminus of the large subunit.

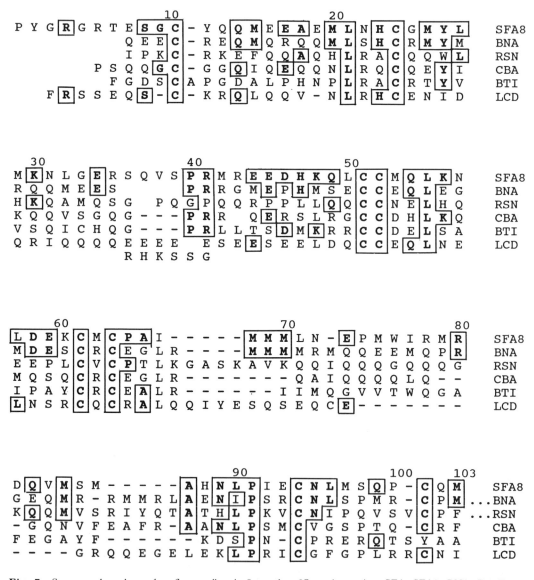

Fig. 5. Sequence homology of sunflower albumin 8 to other 2S seed proteins: SFA, SFA8; BNA, Brazil nut albumin; RSN, rapeseed napin; CBA, castor bean 2S protein; BTI, barley trypsin inhibitor; LCD, lupin conglutin delta.

using a 2-mm pathlength cell containing a 0.009% solution of the albumin in water. The ratio of α-helix (12%), β-structure (45%) and turn/random coil (43%), estimated by the curve fitting procedure of Provencher and Glockner (27) and predicted with appropriate constraints from the primary sequence data (28), were in agreement. A predictive method (29) also highlighted the striking hydrophobicity of the pre-pro sequence region.

Homologies between SFA8 and other 2S albumins. The accumulation of both protein and nucleotide sequence data for the 7-12S seed storage globulins and the 2S seed albumins from diverse plant species has allowed some interesting conclusions to be drawn with regard to their ancestral origins and functions. The availability of primary structural data for the 2S albumin-like proteins from legumes, grasses and other species (15–18,32) has led to a proposal by Kreis et al. (30) that these proteins fall into a superfamily of homologous proteins which have a common ancestral gene(s). The amino acid sequence of SFA8 shows key elements of homology with other 2S albumins (Fig. 5) and presumably therefore it can be included as a member of the proposed superfamily; the most significant feature of this homology is the conservation of cysteine residues. Unlike other 2S seed proteins studied, SFA8 is not processed to a disulfide bonded large and small subunit but presumably is still interlinked by disulfide bonds. The conservation of key elements of structure such as the disposition of cysteine residues is thought to maintain and to add stability (31,32) to a three-dimensional structure important for biological activity observed in some members of the superfamily (16,18,33,34) but not in others (14,15,17,31,32).

SFA8 shows no tryptic, chymotryptic or α-amylase inhibitor activity although, for example, the site implicated in the inhibitory activity of the barley trypsin inhibitor (Pro-Arg-Leu[34]) (34) is conserved in certain members of the superfamily of 2S proteins and is present in the SFA8 sequence except for the substitution of Leu[34] for a Met in SFA8. Presumably that and other structural changes have led to loss of activity in SFA8 if it ever existed. It is interesting that even with the similarities of sequence and the proposal that this is due to the conservation of key structure that these proteins as a family show distinct differences in secondary structure. Some are highly helical, such as the lupin 2S protein (31) and wheat 0.28 α-amylase inhibitor (33), while SFA8 was found to have 45% β-sheet structure with only about 12% α-helix.

Certain members of the 2S superfamily have biological activities and are thought to fulfill a protective role against various seed predators. It is also generally accepted that this group of proteins functions as a pool of sulfur and other amino acids for utilization by the seed upon germination. The interesting question is whether these proteins evolved to perform a role as protective agents or whether the structure that we now see conserved in various species evolve to suit the role of a storage protein. Nevertheless, having a near full length cDNA clone and the respective sequence data enables us to contemplate the isolation of the gene for this interesting sunflowerseed albumin and to undertake the transfer of the gene to other fodders such as lucerne and clover with the aim of improving the nutritive value of these pasture plants.

Acknowledgments

The authors thank L.C. Gruen for circular dichroism measurements, P.M. Strike for gas-phase sequence analysis and N. Bartone for amino acid analyses.

References

1. Reis, P.J., and P.G. Schinckel, *Aust. J. Biol. Sci. 16:*218 (1963).
2. Reis, P.J., *Ibid. 20:*809 (1967).
3. Langlands, J.P., *Aust. J. Exp. Agric. Anim. Husb. 10:*665 (1970).
4. Dove, H., and G.E. Robards, *Aust. J. Agric. Sci. 25:*945 (1974).
5. Ferguson, K.A., J.A. Hemsley and P.J. Reis, *Aust. J. Sci. 30:*215 (1967).
6. Thomson, D.J., in *Forage Protein in Ruminant Animal Production,* edited by D.J. Thomson, D.E. Beever and R.G. Gunn, Occasional Publ. No. 6, British Society for Animal Production, 1982, pp. 53–56.
7. Nugent, J.H.A., W.T. Jones, D.J. Jordan and J.L. Mangan, *Brit. J. Nutr. 50:*357 (1983).
8. Spencer, D., T.J.V. Higgins, M. Freer, H. Dove and J.B. Coombe, *Ibid.* (1988) in press.
9. Schroeder, H.E., *J. Sci. Food Agric. 35:*191 (1984).
10. Higgins, T.J.V., P.M. Chandler, D. Spencer, L.R. Beach, R.B. Blagrove, A.A. Kortt and A.S. Inglis, *J. Biol. Chem. 261:*11124 (1986).
11. Higgins, T.J.V., E.J. Newbiggin, D. Spencer, D.J. Llewellyn and S. Craig, *Plant Mol. Biol.* (1988) in press.
12. Higgins, T.J.V., P.A. O'Brien, D. Spencer, H.E. Schroeder, H. Dove and M. Freer, in *Symp. Biol. Wool Hair,* edited by K.H. Ward and P.J. Reis, Croom Helm, UK, in press, 1988.
13. Kortt, A.A., A.S. Inglis, A.I. Fleming and C.A. Appleby, *FEBS Lett. 231:*341 (1988).
14. Sun, S.S.M., F.W. Leung and J.C. Tomic, *J. Agric. Food Chem. 35:*232 (1987).
15. Ericson, M.L., J. Rodin, M. Lenman, K. Glimelius, L.G. Josefsson and L. Rask, *J. Biol. Chem. 261:*14576 (1986).
16. Kashlan, N., and M. Richardson, *Phytochemistry 20:*1781 (1981).
17. Lilley, G.G., and A.S. Inglis, *FEBS Lett. 195:*235 (1986).
18. Sharief, F., and S. S-L. Li, *J. Biol. Chem. 257:*14753 (1982).
19. Chandler, P.M., T.J.V. Higgins, P.J. Randall and D. Spencer, *Plant Physiol. 71:*47 (1984).
20. Woods, D., *Focus—BRL Research Laboratories 6*, No. 3, 1–3 (1984).
21. Gonda, T.J., D.K. Sheiness and J.M. Bishop, *Mol. Cell Biol. 2:*617 (1982).
22. Maniatis, T., E.F. Fritsch and J. Sambrook, in *Molecular Cloning—A Laboratory Manual,* 12th ed., Cold Spring Harbor Laboratory Publication, New York, 1982.
23. Hudson, P., J. Haley, M. Cronk, J. Shine and H. Niall, *Nature 291:*127 (1981).
24. Von Heijne, G., *Eur. J. Biochem. 133:*17 (1983).
25. Folz, R.J., and J.I. Gordon, *Biochem. Biophys. Res. Commun. 146:*870 (1987).

26. Halling, K.C., A.C. Halling, E.E. Murray, B.F. Ladin, L.L. Houston and R.F. Weaver, *Nucl. Acids Res. 13:*8019 (1985).
27. Provencher, S.W., and J. Glockner, *Biochemistry 20:*33 (1981).
28. Garnier, J., D.J. Osguthorpe and B.J. Robson, *J. Mol. Biol. 120:*97 (1978).
29. Hopp, T.P., and K.R. Woods, *Proc. Natl. Acad. Sci. 78:*3824 (1981).
30. Kreis, M., B.G. Forde, S. Rahman, B.J. Miflin and P.R. Shewry, *J. Mol. Biol. 183:*499 (1985).
31. Lilley, G.G., *J. Sci. Food Agric. 37:*895 (1986).
32. Ampe, C., J. Van Damme, L.A.B. deCastro, MJ. A.M. Sampaio, M. Van Montagu and J. Vandekerchove, *Eur. J. Biochem. 159:*597 (1986).
33. Petrucci, T., G. Sannia, R. Parlamenti and V. Silano, *Biochem. J. 173:*229 (1978).
34. Odani, S., T. Koide, T. Ono and K. Ohnishi, *Ibid. 213:*543 (1983).

Enzyme Use in the Food Industry with Potential Applications to Vegetable Protein Utilization in Human Foods

Svend Eriksen

Food Ingredients Team, Novo Industri A/S, DK 2880 Bagsværd, Denmark

Abstract

The use of enzymes in the food industry is a well established and accepted procedure for both the manufacturing of food ingredients and as a processing aid. In the vegetable protein industry the use of proteolytic enzymes for protein modification serves mainly two purposes: (a) in the production of functional ingredients, e.g. whipping agents, and (b) as nutritional ingredients for application in certain medical foods. In the area of enteral nutrition, where elemental diets are required for medical reasons, the use of enzymatic protein hydrolyzates from soy is a real possibility that is economically attractive to the current practice of using amino acid mixtures. The increasing concern of food allergy creates a demand for hypo- or nonallergenic foods leaving opportunities for ingredients with reduced allergenicity. Enzyme-modified ingredients have great possibilities due to the low molecular weight average of the resultant peptide mixture. As processing aids, enzymes have potential in the vegetable protein industry, for example, in the soy milk industry as a yield improver or in the manufacture of acidified or cultured milk-like drinks from soy. Finally an area of increasing importance is the market for athletes and health-conscious people demanding high quality liquid nutrition with improved performance.

The products of enzymatic reactions have been utilized for ages in the manufacture of traditional food products, such as cheese, beer, and other alcoholic beverages. The application of microbial enzymes, however, produced on an industrial scale has only taken place since the early 1960's when amyloglucosidase was first used to saccharify liquid starch. This enzymatic conversion has been further developed to the point where a whole array of enzymes changes starch into high-fructose corn syrup. This process is the No. 1 application of individual enzymes in the food industry.

In the protein industry to date, hydrolyzates are the only enzyme derived products of commercial interest. Protein hydrolyzates have been used as food ingredients for over 70 years (1) and since the 1940's the production of soy-based whipping agents has been an established technology (2).

Protein hydrolyzates are produced mainly for two different purposes: The production of functional ingredients (e.g., whipping agents) and nutritional ingredients for application in clinical nutrition.

Enzyme Technology

Enzymes are ideal catalysts in food processing. They require little energy and simple equipment. Enzymes work at mild conditions with respect to both temperature and pH, and because they are specific and efficient some of the undesirable results of regular processing are avoided, e.g. racemization of amino acids in the production of protein hydrolyzates for dietetic feeding.

The production of protein hydrolyzates for nutritional fortification presents a difficult problem, because the protein hydrolyzate becomes a dominant ingredient and the taste of the ingredient thus becomes crucial for general acceptance.

Taste of protein hydrolyzates: The taste of enzymatic hydrolyzates from vegetable sources is closely related to the formation of bitterness, which we must accept as a necessary consequence of proteolysis; this is associated with hydrophobic peptides. It should be stressed, though, that it is the organoleptic quality of the food product, rather than of the hydrolyzate itself, which ultimately determines the acceptability of soy protein hydrolyzates as food ingredients, and an evaluation of the taste of enzymatic soy protein hydrolyzates must therefore be seen in the light of their end use (3). Various measures have been suggested in order to improve the flavor of protein hydrolyzates. They can be divided into three different groups: Masking, removal, or prevention (4). Masking has been achieved using polyphosphates, organic acids, addition of glycine or simultaneous hydrolysis with gelatin. Removal of bitter peptides involves chromatographic procedures, solvent extraction with butanol, or other methods, which are too expensive or involve compounds, which have not been approved for food use. Therefore, due to economic constraints or toxicological considerations, commercial methods for removal of bitter peptides are not available. Carbon treatment is commonly used in protein hydrolyzate manufacture to decolorize, but the right choice of carbon will also improve taste (3,5).

Prevention of or minimizing the formation of bitter peptides can be obtained by a combination of the choice of raw material, strict control of the proteolytic reaction (6) and by choosing the appropriate unit operations. No single parameter mentioned above is responsible for obtaining a nonbitter hydrolyzate. Enzyme specificity is also important because it is the combined effect of enzyme and substrate that generally determines the resultant bitterness (7).

Nutritional Ingredients

Protein hydrolyzates for general nutrition: The production of a soy hydrolyzate for fortification of acidic soft drinks has been made possible by controlled enzymatic hydrolysis. With this technique the hydrolysis process can be continuously monitored and terminated at a well defined stage or degree of hydrolysis before bitterness develops (8).

In the early 1980's this process was commercialized for the production of protein enriched juice drinks (9) by a Danish fruit juice manufacturing company. Today, several enzymatic soy protein hydrolyzates for protein enrichment of various juice type products are commercially available. The essential amino acid composition of a soluble protein ingredient developed by Novo is listed in Table 1 together with the most recent recommendations (10). The comparison shows an excellent amino acid score. The new recommendations have resulted in the conclusion that soy is of much better nutritional quality for humans than has been commonly appreciated. This is due primarily to a substantial lowering of the previous FAO recommendation for sulfur amino acids.

Some concern has recently been expressed that the relatively low requirements for essential amino acids that have been proposed for adults are probably significantly lower than the true need for maintenance (11) and that a more appropriate recommendation for adults to be similar to those proposed for the 2-year-old age group. In consequence of that, the soy ingredient has to be supplemented with tryptophan and methionine to fulfill the requirements for high quality proteins (12). This can either be done by simple addition of the two amino acids or by addition of a protein hydrolyzate from another vegetable source with an amino acid composition that complements that of soy (e.g. sesame).

We have prepared hydrolyzates from soy and sesame protein concentrates, and the results of the amino acid analyses and biological values of mixing the two in various ratio are presented in Table 2. The predictive biological value is in each case calculated according to Mørup and Olesen (13) based upon the amino acid analyses. This published mathematical model is based on the exhaustive studies by Kofranyi et al. (14) in the late 1960's, when they determined the minimum nitrogen requirement for equilibrium in young men for a number of food proteins including various two-component mixtures. Varying the ratios in these mixtures showed a minimum amount of

TABLE 1

Comparison of Suggested Pattern of Amino Acid Requirements with Soy Protein Hydrolyzate

	FAO/WHO/UNU 1985			Experimental
	Children			
Amino acid	2 years	10–12 years	Adults	SPI-40L
Isoleucine	2.8	2.8	1.3	4.1
Leucine	6.6	4.4	1.9	6.7
Lysine	5.8	4.4	1.6	6.8
Methionine and cystine	2.5	2.2	1.7	2.2
Phenylalanine and tyrosine	6.3	2.2	1.9	7.8
Threonine	3.4	2.8	0.9	3.9
Tryptophan	1.1	0.9	0.5	0.7
Valine	3.5	2.5	1.3	4.4
Total	32.0	22.2	11.1	36.6

Values are in g/100g protein.

TABLE 2

Amino Acid Analyses and Predictive Biological Values (PV) of Mixtures of Sesame and Soy Protein Hydrolyzates

Sesame (%)	0	25	40	45	50	75	100
Soy (%)	100	75	60	55	50	25	0
Amino acid (g/100g protein)							
ile	3.9	3.8	3.7	3.6	3.6	3.5	3.3
leu	7.1	6.9	6.8	6.7	6.7	6.5	6.3
lys	6.3	5.4	4.8	4.6	4.4	3.4	2.4
phe + tyr	7.6	7.6	7.6	7.5	7.5	7.5	7.5
met + cys	2.3	2.7	3.0	3.1	3.1	3.6	4.0
thr	3.9	3.8	3.7	3.7	3.7	3.6	3.4
try	0.7	0.8	0.8	0.8	0.9	0.9	1.0
val	4.0	4.1	4.2	4.2	4.2	4.3	4.3
Total	35.8	35.1	34.6	34.2	34.1	33.3	32.2
PV	74	93	104	106	102	84	67

protein to maintain N-balance, meaning that there exists a given ratio between two different proteins where the biological value will be equal to or greater than that of either protein. That is also the case with mixtures of soy and sesame proteins.

Srihara and Alexander (15) in a recent study of protein quality of five blends of plant protein sources reported that one blend was superior in protein quality to the other blends when evaluated biologically in rats even though the blends were formulated to provide similar quality of protein based on their amino acid scores, which ranged from 63 to 69, and to give comparable energy values. Not surprisingly, this superior blend consisted of equal amounts of soybean meal and sesame meal and some peanut meal.

Protein hydrolyzates for clinical use: Protein hydrolyzates are generally used in enteral nutrition for patients with impaired digestion or malabsorption. Because milk proteins are usually the basis for such products they have to be tube fed due to the bitterness of milk protein hydrolyzates. The use of a nonbitter, acid-soluble hydrolyzate from soy would create a platform for alternative products that could be administered orally.

The concept of protein enriched juices for that purpose already has been mentioned. It has been shown in normal human subjects that amino-N absorption from enzymatic hydrolyzates is greater than from the respective free amino acid (16). It thus seems obvious to use products containing enzymatic hydrolyzates where prescription of elemental diets is warranted, e.g. patients suffering from cystic fibrosis, short gut syndrome, Chrohn's disease or nonspecific malabsorption.

In addition to medical benefits and better organoleptic properties, enzymatic hydrolyzates containing peptides and low levels of free amino acids contribute less to the osmotic load than free amino acid, elemental diets resulting in less gastrointestinal discomfort. This is of particular importance when such products are being used as the sole source of nutrition.

Applying our enzyme technology, we have just introduced an isotonic, predigested, nutritionally complete, liquid diet for enteral nutrition. Due to the fact that the protein source is soy and that the enzymatic process is carefully controlled, the organoleptic qualities of that product allows oral use. The product, called Top Up, now is only available in Australia.

Protein allergenicity: Food proteins may induce allergy and this is a known phenomenon. The practice of using low heat treatment in the production of whey-based infant formulas to maintain high protein solubility upon rehydration has established these products as commercial sources of whey protein allergens (17).

During enzymatic hydrolysis, cleavage of polypeptide chains at specific sites leads to a collapse of the antigenic structure of the protein molecules, so selective enzyme hydrolysis as a means of reducing the allergenicity of whey protein has been developed (18). Hypoallergenic peptide formulas based on either casein or whey protein hydrolyzates are marketed in the form of semi-elemental diets and have been shown to be generally well tolerated by infants with allergy. Their use is however complicated by low palatability and high cost (18). The use of enzymatic hydrolyzates from soy in these products is an obvious possibility because the consumer acceptance undoubtedly would be greatly improved.

Functional Ingredients

One of the ways of expanding the range of functional properties offered by protein flours, concentrates and isolates is enzymatic modification. For years enzymatic modification has been used only commercially to a limited extent, e.g. in the production of whipping agents (2). Limited hydrolysis has been demonstrated to improve several functional properties such as emulsification capacity, whipping expansion and foam stability (19). The quality of the starting material greatly influences the extent of modification.

At a low degree of hydrolysis the emulsification property of soy isolate is greatly improved, in contrast to the emulsification capacity of soy concentrate that is only slightly improved (Fig. 1). The improvement in whipping expansion is similar for both isolate and concentrate but the foam of the modified concentrate is remarkably stable in contrast to that of the isolate (Fig. 2). The observed effects most probably are due to the presence of polysaccharides in the concentrate. In fact, this product can substitute for egg white in the making of meringue. Modification of soy protein with microbial rennet produces a whipping protein that is as stable as egg white but expands three times greater. This product can completely replace egg white in frozen and whipped products and substitute half of the egg white in certain baked goods, such as chiffon cake (20).

In our laboratory, we have demonstrated that an enzymatic soy protein hydrolyzate in low concentrations exerted a significant bulking effect in dietetic soft drink formulations. This opens the possibility for a fully digestible and nutritious mouth-feel agent.

Processing aid: Finally, an example of the use of enzymes as processing aids in the vegetable protein industry is the application of a neutral protease as a yield improver in soy milk production (22). The traditional soy milk process includes soaking, grinding, filtration, and baking. In order

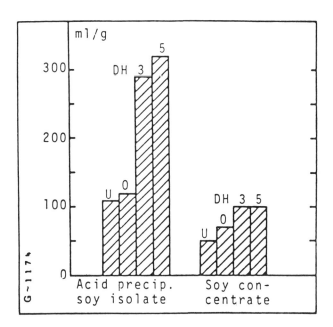

Fig. 1. Effect of proteolysis on emulsifying properties.

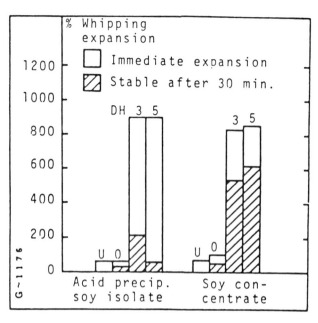

Fig. 2. Effect of proteolysis on whipping properties.

to avoid the development of beany flavor, the beans can be heat treated prior to grinding, but this will decrease the yield of protein in the milk. A short hydrolysis step will solubilize the protein and restore the yield. Due to the limited hydrolysis, it is possible to stabilize this milk at acid pH creating possibilities for a new type of product such as a cultured milk-like drink or yogurt-milk.

Discussion

The general application of enzymes within the food industry certainly will continue and expand in the future. The present review has shown that there exist great potentials within the vegetable protein industry for enzyme-derived ingredients, particularly in the area of nutritional ingredients, where products containing these ingredients already have been developed to satisfy the protein requirement of specific consumer groups. The current awareness of healthy living and keeping fit creates demands for new type of products for general consumption. The application of enzymatic hydrolyzates of vegetable origin from either one or more sources allows formulation of liquid sport-type drinks of high nutritional value. The long time practice of enzymatic modification for improving the functional properties of food proteins will continue but the future will, without doubt, bring new special ingredients to be formulated into both traditional and new food types.

References

1. Kilara, A., *Proc. Biochem. 20*:149 (1985).
2. Gunther, R.C., *J. Am. Oil Chem. Soc. 56:345 (1979)*.
3. Adler-Nissen, J., and H.S. Olsen in *"Chemistry of Foods and Beverages—Recent Developments"* edited by G. Charalambous and G. Inglett, Academic Press, N.Y., (1982), p. 149.
4. Cowan, D., in *"Industrial Enzymology. The application of enzymes in industry"* edited by T. Godfrey and J. Reichelt, MacMillan, London (1983), p. 352.
5. Adler-Nissen, J., *Enzymic Hydrolysis of Food Proteins*, Elsevier Applied Science Publisher, (1986), p. 243.
6. Adler-Nissen, J., *J. Chem. Tech. Biotechnol. 34B*:215 (1984).
7. Adler-Nissen, J., in *"Protein Tailoring For Food and Medical Uses"* edited by R.E. Feeney and J.R. Whitaker, Marcel Dekker, N.Y. (1986), p. 97.
8. Eriksen, S., *Biochem. Soc. Trans. 10*:585 (1982).
9. Møller, O., *Confructa 32:*16 (1988).
10. FAO/WHO/UNU, *Energy and Protein Requirements*, Tech. Rep. Ser. No. 724, WHO, Geneva (1985).
11. Young, V.R., and P.L. Pellett, *Am. J. Clin. Nutr. 45*:1323 (1987).
12. Food and Nutrition Board, *Recommended Dietary Allowances (9th ed.)*, Nat. Academy of Science, Washington, D.C. (1980).
13. Mørup, I.-L., and E.S. Olesen, *Nutr. Rep. Int. 13*:355 (1976).
14. Kofranyi, E., F. Jekat and H. Müller-Weckel, *Z. Physiol. Chem. 387:*1485 (1970).
15. Srihara, P., and J.C. Alexander, *J. Can. Inst. Food Sci. Technol. 17*:237 (1984).
16. Silk, D.B.A., P.D. Fairclough, M.L. Clark, J.E. Hegarty, T.C. Marrs, J.M. Addison, D. Burston, K.M. Clegg and D.M. Matthews, *JPEN 4:*548 (1980).
17. Eastham, E.J., T. Lichauco, M.I. Grady and W.A. Walker, *J. Pediatrics 93:*561 (1978).
18. Jost, R., J.C. Monti, and J.J. Pahud, *Food Technol. 44:*118 (1987).
19. Adler-Nissen, J., S. Eriksen, and H.S. Olsen, *Qual. Plant: Plant Foods Hum. Nutr. 32:*411 (1983).
20. LaBell, F., *Food Processing 47*(12):56 (1987).
21. Adler-Nissen, J., and S. Eriksen in *"The Shelf-Life of Foods and Beverages,"* edited by G. Charalambous, Elsevier, N.Y. (1986), p. 551.
22. Eriksen, S., *J. Food Sci. 48:*445 (1983).

Soy Protein Fractionation and Applications

Paul W. Gibson and Walter C. Yackel

A. E. Staley Mfg. Co., 2200 E. Eldorado, Decatur, Illinois 62525, USA

Abstract

The major soybean globulins were separated into two portions on commercial scale equipment and spray dried. Enzymatic modification of the 7S globulin resulted in an egg white-like protein which, upon frying, became brittle and turned white, like egg white.

Soy protein isolate may be fractionated into its various globulins by a variety of techniques. Davidson et al. (1) describe extraction followed by cooling in which 11S globulins precipitate; Shemer (2) describes an extraction at pH 5.1–5.9 followed by pH 4.5 adjustment with phosphoric acid to obtain a protein which is more than 70% 7S globulin; Thanh et al. (3), Eldridge et al. (4) and Wolf et al. (5) describe techniques for preparing pure 7S and 11S globulins.

Research in our laboratories on protein fractionation was directed toward separation techniques that were suitable for commercial processing equipment and operating conditions, and that would provide product of food grade quality. Furthermore, we were interested in enzymatic modification of the protein globulins and we were interested in the applications of the globulins and enzyme-modified globulins.

Fractionation of the two major globulins, 7S and 11S, was accomplished by taking advantage of their differential solubility between pH 5.0 and pH 6.3. Additionally, adjustment of the ionic strength, for example, with sodium chloride, enhanced separation. Furthermore, addition of sulfite provided even more help in the separation of these two major globulins. Howard et al. (6) and Lehnhardt et al. (7) describe the two basic approaches taken to achieve commercial separation of globulins.

The first approach, according to Howard et al. (6), involved extraction of protein from defatted soy flakes at pH 8 with an aqueous solution of 0.03–0.06 M sodium chloride and 0.5–0.8 mM sodium bisulfite. The clarified extract was adjusted to pH 5.5 with HCl and the precipitated protein separated by centrifugation. The supernatant was adjusted to pH 4.5 and the additional precipitated protein separated from whey protein and sugars by centrifugation. The proteins which precipitated above pH 5.5 consisted of about 90% 11S. The remaining proteins which precipitated at pH 4.5, and were recovered, consisted of about 70% 7S and 30% 11S. Higher purity 11S was obtained by first precipitating and separating the precipitate closer to pH 6. Increased 7S purity was obtained by first precipitating and separating the precipitate closer to pH 5.3. Proper selection of ionic strength and sulfite concentration aids in obtaining desirable purity of either 11S or 7S.

The second approach to fractionation, described by Lehnhardt et al. (7), involves extraction of defatted soy flakes at pH 8, clarification of the extract, precipitation of the proteins at pH 4.3 with HCl, and separation of the protein curd from whey protein and sugars. The curd was resuspended in an aqueous medium of 0.1 M sodium chloride and 7.5 mM sodium bisulfite. The suspension was brought to pH 5.3 with NaOH. At this pH, the 7S globulins dissolved and the remaining insoluble proteins were mostly 11S. Separation of the two fractions was accomplished by centrifugation. Under these conditions, a protein consisting of 90% 7S and 10% 11S was obtained, and the other fraction consisted of about 30% 7S and 70% 11S.

Figure 1 shows the effect of sodium chloride concentration on the amount of precipitated protein at pH 6 and its effect on the amount of 7S and 11S globulin in the curd. The data indicate a broad optimum salt concentration from 0.02 M to 0.08 M to achieve maximum 11S, minimum 7S and maximum curd.

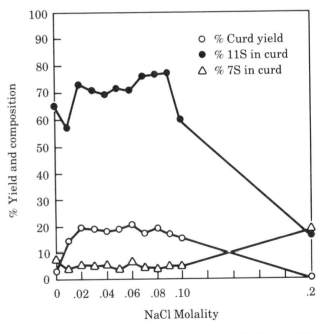

Fig. 1. Effect of NaCl on yield and composition of isolate at pH 6 with 0.48 mM sodium bisulfite.

Figure 2 shows the effect of bisulfite concentration on protein solubility with 0.02 M sodium chloride. As the bisulfite concentration is increased to about 0.5 mM, protein solubility increases. Excessive solubility would decrease pH 6 yield; however, some bisulfite is important to aid globulin separation.

Figure 3 shows the effect of bisulfite concentration on 7S and 11S globulin solubility at different pH's. The small amount of bisulfite caused the 7S to remain substan-

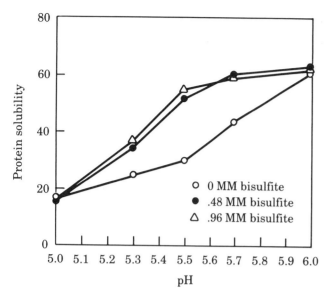

Fig. 2. Effect of sodium bisulfite on protein solubility with pH and 0.02 M NaCl.

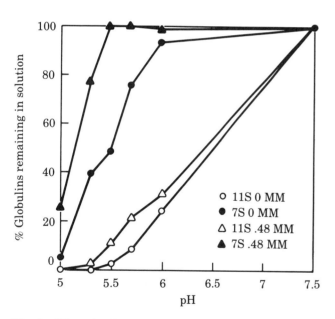

Fig. 3. Effect of sodium bisulfite on solubility of 7S and 11S globulins with pH at 0.02 M NaCl.

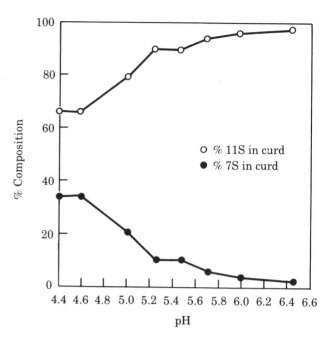

Fig. 4 Composition of precipitated curd as pH is reduced from pH 7.5 at .06 M NaCl and 0.48 mM bisulfite.

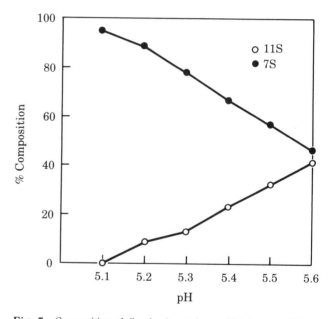

Fig. 5. Composition of dissolved protein as pH is increased from pH 4.3 at 0.1 M NaCl and 7.5 mM sodium bisulfite.

tially dissolved at pH 5.5 and only 10% of 11S remained dissolved. Therefore, separation near pH 5.5 under these conditions resulted in good fractionation.

Figure 4 gives the curd composition in terms of its 7S and 11S percentage, ignoring the small amount of other globulins. Very enriched 11S curd was obtained at pH 6, about 90% 11S curd was found at pH 5.5, and 80% 11S curd was obtained at pH 5.0. Using this scheme, a wide variety of fractionation compositions can be obtained. This scheme is from Howard et al. (7).

Figure 5 gives the dissolved protein composition as curd was dissolved after its formation at pH 4.3. At pH 5.1, nearly pure 7S globulin extract was obtained. Less 7S-enriched protein was dissolved at higher pH. The dissolved protein is separated from the insoluble protein and recovered as an enriched product. This scheme follows Lehnhardt et al. (8).

Functional Properties

The 7S Globulins: Highly enriched 7S protein exhibits sticky, viscous properties. It is difficult to wash from equipment, utensils and hands. Interestingly, if it is cooled to about 2–4°C, its properties change and it then dissolves and washes off easily. If the 7S material contains about 10% 11S globulin, the 7S-enriched substance does not present solubility difficulties. As a practical matter, the 7S fraction was designed to contain about 10% 11S so that it behaved well during processing on commercial equipment and during spray drying, and so that it reconstituted easily in water.

7S proteins heat set to a strong gel at about 150°F. The 7S material was tested as a heat settable binder in a number of applications and it easily replaced egg white.

The 11S enriched protein fraction was not very viscous. It heat sets at about 180–190°F, and it had a cleaner flavor than normal soy isolate. The 11S material behaved more like milk protein than normal soy isolate.

The Protein Efficiency Ratio (PER) of 7S enriched protein was determined as 1.3 and the PER of the 11S enriched material was found to be 1.9.

Enzymatic Modifications of 7S and 11S

Lehnhardt and Orthoefer (8) described a pepsin treatment of 7S enriched protein in which there is a partial hydrolysis of the subunits and the formation of a new subunit of about 46,700 daltons. This enzyme modified product showed unique heat gel properties very much like egg white; and, it imparted stability to foams made with highly modified soy protein.

The enzyme-modified 7S gelled rapidly at 85°C and resulted in white, brittle and strong gels with a mouthfeel texture very much like egg white.

The 11S globulins were enzymatically digested with pepsin to prepare a whipping protein. The modified 11S protein was compared with whipping agents described by Gunther (9). Functionally, the modified 11S were much like the Gunther whipping agents. Furthermore, the modified 11S proteins had better color and flavor than the traditional products. Additionally, the modification process is more easily accomplished with 11S and yields are better. Waste disposal is substantially reduced.

References

1. Davidson, R.M., and R.E. Sands, U.S. Patent 4,172,828 (1979).
2. Shemer, M., U.S. Patent 4,188,399 (1980).
3. Thanh, V.H., and K. Shibasaki, *J. Agr. Food Chem.* 24:(6):1117–1121 (1976).
4. Eldridge, A.C., *Cereal Chem.* 44:645–654 (1976).
5. Wolf, W.J., J.J. Rackis, A.K. Smith, H.A. Sasame, and G.E. Babcock, *Archives of Biochem. Biophys.* 85:186–189 (1959).
6. Howard, P.A., W.F. Lehnhardt and F.T. Orthoefer, U.S. Patent 4,368,151 (1983).
7. Lehnhardt, W.F., P.W. Gibson and F.T. Orthoefer, U.S. Patent 4,370,267 (1983).
8. Lehnhardt, W.F., and F.T. Orthoefer, U.S. Patent 4,409,248 (1983).
9. Gunther, R.C., U.S. Patent 3,814,816 (1974).

Biotechnology in Rice Breeding

Timothy P. Croughan

Louisiana Agricultural Experiment Station, Louisiana State University Agricultural Center, PO Box 1429, Crowley, Louisiana 70527-1429, USA

Abstract

Rice generally is considered the most important food crop from a nutritional standpoint. While wheat exceeds rice in terms of total world production, more than a third of the world's population depends on rice as a dietary staple. Rice flourishes in tropical climates where population pressures frequently are greatest. Excellent progress has been made in breeding better rice varieties, but there remains a pressing need for continued genetic improvements. Biotechnology offers a number of new approaches to facilitate rice breeding. Two important areas are the development of new genetic diversity (novel germplasm) and the replacement of time-consuming field techniques with relatively rapid laboratory procedures. Rice already is the cereal crop in which the greatest progress has been made in applying biotechnology to breeding. China has produced several new varieties through a technique termed anther culture, in which plants developed in a few months in the laboratory substitute for plants requiring several years to develop through conventional breeding techniques. Many rice breeding programs worldwide are now adopting this adjunct technique. Excellent progress also has been made in recent years toward the goal of genetically engineering rice, including the development of plants from single rice cells lacking cell walls (protoplasts). These membrane-enclosed cells are prime targets for the introduction of foreign genes to engineer cells from which genetically altered plants potentially can be developed. Even more recently, successes are being reported in developing genetically engineered (transformed) rice plants through other techniques, such as the injection of DNA solutions into stems containing developing flowers. The prospects are bright for further rapid progress in utilizing biotechnology in rice breeding. It appears that the world's most important food crop also may prove to be a model system for applying biotechnology to the improvement of cereal crops.

Rice is the most important food crop from a nutritional standpoint. More people depend on rice as a dietary mainstay than any other cereal, and most of the world's rice is grown in the tropics, where population densities are frequently high and food shortages a problem.

The majority of cultivated rice is *Oryza sativa*. Rice is most productive when grown in shallow water, but also can be grown as an unflooded, upland crop. One type of rice, termed floating rice, responds to deepening water by undergoing extensive stem elongation, allowing it to be cultivated in floodlands where water will reach several meters in depth. Breeding programs around the world are continuing to develop improved varieties specifically bred for cultivation in rainfed upland fields, paddies flooded to shallow and intermediate depths, and deeply flooded basin situations.

As an adjunct to the successful conventional plant breeding procedures used to improve rice, biotechnology offers a set of laboratory techniques with both short and long term promise for contributing to rice improvement. Several new varieties of rice have already been developed from laboratory-cultured plant cells in China, and genetic engineering techniques will allow the transfer of individual genes to tailor future varieties with specific traits.

To expedite the development of new varieties in the near term, two techniques termed anther culture and somaculture are being used. Anther culture entails the development of plants from pollen grains. A conventional plant breeding program requires as many as 10 years to develop a new variety. Several years of this time are spent replanting successive generations in order to obtain homozygous (true breeding) lines following the initial hybridization step. Anther culture produces homozygous lines in a few months rather than the few years normally required.

Anthers are the botanical structure in flowers which produce pollen. Each rice flower has six individual anthers containing as many as several hundred pollen grains apiece. Anther culture is used to expedite breeding through the use of anthers from hybrid plants. Each of the individual pollen grains from these hybrid plants is genetically different, and possesses varying combinations of the original parental characters. As the male haploid reproductive units, pollen contains only half of the normal number of chromosomes. The key to producing true breeding plants through anther culture is to duplicate this single set of pollen chromosomes, resulting in the production of two identical sets. When both of each pair of chromosomes are identical, homozygosity, or true-breeding status, is achieved.

The sequence of steps actually utilized are as follows. A hybridization is made through cross-pollination of two parental plants having complementary desirable characteristics. The hybrid seed which forms gives rise to a hybrid plant. As this plant approaches flowering but before emergence of the seed head, the stem is harvested and brought into the laboratory. The seed head is aseptically removed from the stem, and the anthers excised from the individual flowers. The anthers are placed in dishes containing a combination of nutrients and plant hormones which induces the pollen grains within the anther to initiate cell proliferation. This pollen-derived cellular growth is then transferred to a second dish in which the nutrient and hormone composition triggers plants to sprout from the proliferating clumps of cells.

Because the plants which develop trace back to pollen grains, approximately half still contain only a single set of chromosomes, and in this condition cannot produce seeds. These plants, however, can be treated with the chemical colchicine to induce chromosome doubling. The other half of the plants produced have two sets of chromosomes, having undergone a spontaneous chromosome doubling to the fertile diploid state. Either way that homozygosity is achieved, the plants obtained from anther culture in a few months provide the true breeding material normally requiring several years to develop through conventional methods.

A second approach to contributing to the development of new varieties in the near term is the use of somaculture. In this case plants are derived from the culture of somatic plant tissue, such as roots, stems, and leaves. As with anther culture, the plant material to be cultured is placed on a nutrient medium which induces cell proliferation, and the cell clumps are then transferred to another nutrient medium which triggers the cells to form plants. Researchers utilizing these procedures with somatic tissues originally expected that the resulting plants would be cloned copies of the plant from which the somatic tissue was originally obtained. It is now well understood, however, that the plants produced frequently differ genetically from the original donor plant. Mutations frequently induced in rice through this procedure include changes in plant height, days to maturity, tillering, leaf shape and display, grain size, panicle size, and degree of sterility. Somaculture therefore serves as a rich source of mutants potentially useful for plant breeding. The procedure followed is to utilize established varieties or superior breeding lines, and from this material attempt to develop a mutant with one or more improved characteristics. For example, a leading rice variety may be excellent in most respects, but susceptible to a disease that occurs under certain climatic conditions. Somaculture is used to produce laboratory-derived lines from this material, which are then tested for their disease susceptibility. A mutant which is equally productive but more disease resistant may offer an improved version of the original variety. An advantage of somaculture is that it may rapidly provide further improvements in already superior varieties.

Over the longer term, genetic engineering techniques offer the potential for transferring individual traits to design plants with particular characteristics. Genes coding for improved traits could potentially be utilized from almost any organism, and the entire biosphere could essentially function as a repository of genetic traits. Additionally, DNA pieces can be synthesized in the laboratory, so that new genes not already occurring in nature can be created. For example, DNA coding for a protein containing elevated levels of the nutritionally important amino acid lysine can be synthesized on laboratory equipment for transfer into plants.

Overall, genetic engineering offers the potential to revolutionize plant breeding. In the near term, however, serious limitations remain to be overcome. Identification of the genes responsible for important characteristics is taking place, but only a small fraction of this effort is accomplished to date. Transferring DNA between organisms has been accomplished and is progressing, but again, this technology is far from complete. Environmental concerns seriously limit the release of genetically engineered organisms into the environment at present, and acquiring the required approvals in many countries is currently a major effort in itself. But advances on all of these fronts are being made, and biotechnology as a useful plant breeding tool offers exciting potentials for contributing to the genetic improvement of rice.

Applications of Biotechnology to Soybean Improvement

Suzan S. Croughan

Rice Research Station, Louisiana Agricultural Experiment Station, Louisiana State University Agricultural Center

Plant tissue culture and genetic engineering are two approaches of biotechnology which have been used in crop improvement programs. Tissue culture involves stimulating plant cells to divide and grow into a mass of undifferentiated tissue called callus. Once formed, the callus can be triggered to differentiate into organized plants. Plants regenerated from callus frequently exhibit structural, physiological, or biochemical characteristics which distinguish them from the plant from which they were derived. This phenomenon, termed somaclonal variation, has been observed in several species and variation has been found in many traits (1–4). For example, regenerated sugarcane plants have varied in level of disease resistance and sugar content (5–7), and variation in plant height and heading date are typically seen in plants regenerated from rice cultures (8). Mutants arising through somaclonal variation can be used subsequently in germplasm improvement programs.

Genetic engineering is gaining importance as a method of germplasm enhancement as techniques for DNA isolation and transfer are refined and important genetic material is identified. In this approach, DNA that codes for a particular trait is introduced into either protoplasts, cells, tissue or even whole plants. This approach to crop improvement offers the ability to design plants with specific traits. Isolated DNA has been moved successfully into several agronomically important species including rice (9) and corn (10). As plant scientists gain a better understanding of gene expression and action, the transfer of important traits should become relatively routine.

The use of somaclonal variation and genetic engineering techniques for soybean improvement has been hindered by the inability of plant scientists to routinely regenerate plants from soybean cells. Until routine plant regeneration is possible, neither technique has value in soybean improvement programs. Methods for soybean plant regeneration have been developed recently which should provide the mechanism for developing novel germplasm through tissue culture and genetic engineering. This paper will review the progress that has been made in using these techniques for the improvement of soybean germplasm.

Regeneration in Soybeans

Soybean callus can be initiated from a number of explant sources, but repeatable and reliable methods for plant regeneration have only recently been developed. In early reports, different levels of differentiation were observed. For example, Kimball and Bingham (11) reported adventitious bud development from hypocotyl sections. Embryo-like structures (12), buds (13), and heart-shaped embryos (14) have been observed in soybean cultures. Cheng et al. (15) and Saka et al. (16) stimulated soybean node segments to produce multiple shoots. Christianson et al. (17) de-

ments to rived a morphogenetically competent soybean culture from embryonic axes and Lippmann and Lippmann (18) obtained somatic embryos from immature embryos of soybean.

Since these early reports, there have been a number of papers which describe routine regeneration of soybean plants from tissue culture (20–27). Most methods initiate cultures from immature embryos, while others use primary leaf tissue, cotyledonary nodes, or epicotyl-derived tissue.

Somaclonal Variation

Because plant scientists have succeeded only recently in regenerating soybeans from callus, few observations have been reported on the amount of somaclonal variation in regenerated plants. Barwale and Widholm (28) studied the variation in 263 regenerated plants and some of their progeny produced from immature embryos of nine soybean genotypes. A number of phenotypic differences were observed, including chlorophyll deficiency, sterility, abnormal leaf morphology, and dwarf stature. Graybosch et al. (29) obtained regenerated plants from cotyledonary node tissue of three soybean varieties and then evaluated the progeny from several of these regenerated plants for yield, height, lodging, and maturity dates in field tests. Significant variation was observed only in plant height. These workers suggested that one reason for the low amount of variation was the small number of plants tested. They recommended that larger numbers of regenerated plants and progeny be screened. Hildebrand et al. (personal comm.) observed variation in the lipid composition of regenerated soybean plants but found that the changes in lipid content did not appear to be stable.

Genetic Engineering

Some techniques of genetic engineering involve the transfer of foreign DNA into plant protoplasts. Soybean protoplasts have been derived from a number of plant parts including suspension cultures (30), pods (31), leaf tissue (32) and cotyledons and immature seeds (33). Two methods, electroporation and chemical-mediated transfer with polyethylene glycol, allow DNA to move into plant protoplasts through holes which develop in the membrane after electrical or osmotic treatment.

Cutler and Saleem (34) showed that soybean protoplasts will take up macromolecules following electroporation. Lin et al. (35) reported the uptake of DNA in soybean protoplasts and Saxena et al. (36) were able to cause *Datura innoxia* protoplasts to take up nuclei isolated from soybeans using polyethylene glycol. Christou et al. reported the transformation of soybean callus through electroporation of protoplasts.

Regeneration of plants from protoplasts is an important step in utilizing the techniques mentioned above for germplasm enhancement. Newell and Luu (37) were able to regenerate plants from protoplasts derived from *Glycine canescens*, and Hammatt et al. (38) regenerated plants from *G. canescens* and *G. clandestina* protoplasts, however there are no reports of regeneration from protoplasts derived from *G. max*. This will need to be accomplished before electroporation and polyethylene glycol can be used to genetically engineer soybeans.

Particle acceleration and biological vectors are two methods of genetic engineering which do not use protoplasts. Particle acceleration uses an electric discharge to "shoot" DNA coated particles into cells, while biological vectors use bacterium to transfer DNA. The introduction of DNA into plants through particle acceleration has been used successfully with soybeans. McCabe et al. (39) used this method on immature embryos. Some plants regenerated from these immature embryos and a few progeny from the regenerated plants were found to express the introduced genes. *Agrobacterium*-mediated DNA transfer also has been used successfully with soybeans. Hinchee et al. (40) produced soybean plants which have incorporated a gene confering tolerance to glyphosate, a widely used herbicide.

Soybean tissue culture techniques are becoming fairly well established. Differences in varietal response to tissue culture manipulation are known to exist in other crops, and it remains to be seen which soybean varieties will easily regenerate plants and produce useful variants for soybean improvement programs. Genetic engineering offers the possibility for designing soybeans with specific traits. Techniques need to be refined for identifying and transferring useful genes into plants, but biotechnology holds much promise for becoming a useful tool in soybean varietal improvement programs.

References

1. Larkin, P.J., and W.R. Scowcroft, *Theor. Appl. Genet. 60:*197 (1981).
2. Reisch, B., "Genetic Variability in Regenerated Plants," in *Handbook of Plant Cell Culture*, Vol. 1, Techniques for Propagation and Breeding, edited by D.A. Evans, W.R. Sharp, P.V. Ammirato and Y. Yamada, Macmillan, New York, 1983, pp. 748-769.
3. Evans, D.A., and W.R. Sharp, "Somaclonal and Gametoclonal Variation," in *Handbook of Plant Cell Culture*, Vol. 4, Techniques and Applications, edited by D. A. Evans, W.R. Sharp and P.V. Ammirato, Macmillan, New York, 1986, pp. 97-132.
4. Larkin, P.J., *Iowa State J. Res. 61:*393 (1987).
5. Heinz, D.J., M. Krishnamurthi, L.G. Nickell and A. Maretzki, "Cell, Tissue and Organ Culture in Sugar Cane Improvement," in *Applied and Fundamental Aspects of Plant Cell Tissue Organ Culture*, edited by J. Reinert and Y.P.S. Bajaj, Springer-Verlag, Berlin, 1977, pp. 3-17.
6. Liu, M.C., and W.H. Chen, *Euphytica 27:*273 (1978).
7. Larkin, P.J., and W.R. Scowcroft, *Plant Cell Tissue Organ Culture 2:*111 (1983).
8. Oono, K., *Test Tube Breeding Rice by Tissue Culture. Proceedings Symposium Methods in Crop Breeding.* Tropical Agriculture Research Series No. 11, Ministry of Agriculture and Forestry, Ibaraki, Japan (1978).
9. Toriyama, K., Y. Arimoto, H. Uchimiya and K. Hinata, *Bio/Technology 6:*1072 (1988).
10. Rhodes, C.A., D.A. Pierce, I.J. Mettler, D. Mascarenhas and J.J. Detmer, *Science 240:*204 (1988).
11. Kimball, S.L., and E.T. Bingham, *Crop Sci. 13:*758 (1973).
12. Beversdorf, W.D., and E.T. Bingham, *Crop Sci. 17:*307 (1977).
13. Oswald, T.H., A.E. Smith and D.V. Phillips, *Physiol. Plant 39:*129 (1977).
14. Phillips, G.C., and G.B. Collins, *Plant Cell Tissue Organ Culture 1:*123 (1981).
15. Cheng, T., H. Saka and T.H. Voqui-Dinh, *Plant Sci. Letters 19:*91 (1980).
16. Saka, H., T.H. Voqui-Dinh and T. Cheng, *Plant Sci. Letters 19:*193 (1980).
17. Christianson, M.L., D.A. Warnick and P.S. Carlson, *Science 222:*632 (1983).
18. Lippmann, B., and G. Lippmann, *Plant Cell Reports 3:*215 (1984).
19. Ranch, J.P., L. Oglesby and A.C. Zielinski, *In Vitro Cellular and Development Biology 21:*653 (1985).
20. Ranch, J.P., L. Oglesby and A.C. Zielinski, "Plant Regeneration from Tissue Cultures of Soybeans by Somatic Embryogenesis," in *Cell Culture and Somatic Cell Genetics of Plants*, Vol. 3, edited by I.K. Vasil, Academic Press Inc., New York, 1986.
21. Barwale, U.B., H.J.R. Kerns and J.M. Widholm, *Planta 167:*473 (1986).
22. Wright, M.S., M.G. Carnes, M.A. Hinchee, G.C. Davis, S.M. Koehler, M.H. Williams, S.M. Colburn and P.E. Pierson, "Plant Regeneration from Tissue Culture of Soybean by Organogenesis," in *Cell Culture and Somatic Cell Genetics of Plants*, Vol. 3, edited by I.K. Vasil, Academic Press Inc., New York, 1986.
23. Wright, M.S., D.V. Ward, M.A. Hinchee, M.G. Carnes and R.J. Kaufman, *Plant Cell Reports 6:*83 (1987).
24. Wright, M.S., M.H. Williams, P.E. Pierson and M.G. Carnes, *Plant Cell Tissue Organ Culture 8:*83 (1987).
25. Lazzeri, P.A., D.F. Hildebrand and G.B. Collins, *Plant Cell, Tissue Organ Culture 10:*197 (1987).
26. Lazzeri, P.A., D.F. Hildebrand and G.B. Collins, *Ibid. 10:*209 (1987).
27. Hammatt, N., and M.R. Davey, *J. Plant Physiol. 128:*219 (1987).
28. Barwale, U.B., and J.M. Widholm, *Plant Cell Reports 6:*365 (1987).
29. Graybosch, R.A., M.E. Edge and X. Delannay, *Crop Sci. 27:*803 (1987).
30. Kao, K.N., W.A. Keller and R.A. Miller, *Experimental Cell Res. 62:*338 (1970).
31. Zieg, R.G., and D.E. Outka, *Plant Sci. Letters 18:*105 (1980).
32. Schwenk, F.W., C.A. Pearson and M.R. Roth, *Plant Sci. Letters 23:*153 (1981).
33. Lu, D.Y., S. Cooper-Bland, D. Pental, E.C. Cocking and M.R. Davey, *Z. Pflanzenphysiol. Bd. 111:*389 (1983).
34. Cutler, A.J., and M. Salem, *Plant Physiol. 83:*24 (1987).
35. Lin, W., J.T. Odell and R.M. Schreiner, *Plant Physiol. 84:*856 (1987).

36. Saxena, P.K., Y. Liu and J. King, *J. Plant Physiol.* *128:*153 (1981).
37. Newell, C.A., and H.T. Luu, *Plant Cell Tissue Organ Culture 4:*145 (1985).
38. Hammatt, N., H.I. Kim, M.R. Davey, R.S. Nelson and E. C. Cocking, *Plant Sci. 48:*129 (1987).
39. McCabe, D.E., W.F. Swain, B.J. Martinell and P. Christou, *Bio/Technology 6:*923 (1988).
40. Hinchee, M.A.W., D.V. Connor-Ward, C.A. Newell, R. E. McDonnell, S.J. Sato, C.S. Gasser, D.A. Fischhoff, D.B. Re, R.T. Fraley and R.B. Horsch, *Bio/Technology 6:*915 (1988).

Biotechnology and Livestock Systems with an Emphasis on Porcine Somatotropin

David Meisinger

Pitman-Moore Inc., PO Box 207, Terre Haute, Indiana 47808-9990, USA

Abstract

Porcine somatotropin represents one of several new products that will dramatically affect food production in the future. Through the use of genetic engineering, new biochemical processing techniques, innovative formulation and delivery technology, and creative marketing, industry will provide livestock producers in the near term the opportunity to select from an array of those products that best fit their production and marketing needs to produce leaner meat at less cost. This talk provides an historical perspective on where we are and where we are going.

We live in a world of change. Dramatic changes are occurring in the way we live our lives. We drank more, smoked more, and ate more. Then we realized that the price for that unhealthy lifestyle—heart disease, cancer and obesity—was too high; so today we drink, smoke and eat less; (There are 50 million obese people) and we exercise more, or, at least, we talk about exercising more.

It is in this change of lifestyle that those of us in the business of agriculture find our challenge. Along with this change in lifestyle, consumers worldwide are demanding changes in the foods they eat. We see some of those demands being met through dramatic advances in science and technology, e.g. biotechnology; through the introduction and adoption of bioengineered products for livestock production.

One such product is porcine somatotropin. But before we talk about that product and its potential to revolutionize the pork industry, let's look at some of the trends that have already taken place and will continue, probably at an increasing rate.

Economic pressures are forcing consolidation in the industry throughout the world. Packing industry consolidation has placed the majority of slaughter operations in fewer hands, and there has been a continuing increase in the number of hogs sold directly to packers. There is, likewise, consolidation in the feed industry. Fewer feed manufacturers will serve fewer and larger pork operations interested in buying increased tonnages of bulk ingredients and premixes. Economists also see the feed industry becoming more involved in the ownership of swine operations.

Consolidation also is occurring on the pork production side. Here it means that individual units continue to grow in size while shrinking in number in most countries. Vertical integration is increasing as well, but likely not to the same extent or with the same speed as has occurred in the poultry industry. As a result of increased size, pork producers will benefit from economies of scale. As a result of specialization, there will be less emphasis placed upon the on-farm production of feedstuffs.

Another trend for producers worldwide is the increased need for business analysis in response to market pressures. Growing in sophistication, producers are moving away from "selling corn through pigs" and to "marketing pork." We are seeing fewer pig farmers and swine producers and more pork producers. These producers are doing more than keeping income tax records. They are keeping production records on a monthly and, often, weekly basis. They also are adopting new technologies to gain further production efficiencies. These include techniques to manage the risk of adverse price movements and adverse interest rate movements. Watch for continued growth of computer technology with applications in production, financial, and marketing decision support systems.

A third trend is the mounting consumer demand for leaner pork; one that will force pork producers to change their styles of management. Pork production, therefore, is shifting from a supply-driven to a demand-driven industry. In just the last few years we have seen consumer preferences being communicated through wholesalers, processors and packing houses back to producers. Thus, through nutrition management and genetic advances, the industry has responded with leaner, more uniform hogs. The industry is also doing a better job of marketing its product, The U.S. National Pork Producers Council's current "other white mean" campaign being a prime example.

This has come in response to consumer demand. But with static per capita consumption figures for the world's most popular meat, the message is clear: consumers want even leaner pork.

And now there is the fourth trend. The application of new technologies will cause dramatic changes in the production efficiencies of producing lean pork. In fact, we think that porcine somatotropin is one example of a new technology that will virtually revolutionize the pork industry.

These revolutionary changes will come on the heels of centuries of evolutionary changes that got us to where we are today. The world's breeds of pigs were all thought to originate from some combination of the European wild boar *Sus screta* and the Far Eastern wild pig *Sus indicus*. But we did not evolve directly to our modern breeds.

Through selection for high-energy content, we went through a period of extremely lardy pigs known in their day as cob rollers because of the way their bellies rolled the corn cobs from their last feeding. Finally we moved to the modern pig, the result of great progress over the decades and centuries. In fact, we have made some substantial strides in recent years. Almost every advanced country can tout similar figures of progress, but let me discuss some from the U.S.A.

A recently completed study shows that average pork cuts averaged 4.3% fat or less, far below the USDA statistics from 20 years ago. In fact, that is well below the American Heart Association's guidelines for food that should be considered for inclusion in a prudent diet. Hogs today are 50% leaner in the U.S. than they were 20 years ago. But, there are still challenges. Pork consumption has still declined. In the most recent study conducted for Pitman-Moore by an independent research company, 57% of the respondents said they have decreased consumption of pork in the last five years. That decrease has leveled out, but the pork industry needs to grow and to increase use of its product.

Are there other challenges? Yes! Although pork is leaner than ever before, consumers still do not regard it as lean enough. In fact, in the study the only meat that consumers spontaneously referred to as "fatty" was pork. Obviously, this is partly a problem of perception calling for the pork industry to continue consumer reeducation. But it is also a continuing challenge to the industry to redouble efforts to produce leaner animals and give those increasingly health-conscious consumers the meat they want.

Pitman-Moore also looked more deeply into exactly what the desired meat is. The most recent study determined that 69% of meat buyers would buy more pork if it were 10% leaner. They told us that the pork products they want to buy would have little or no visible fat. But it would still be as tender, juicy and tasty as pork has always been. If they could get this kind of a pork product, consumers said they would pay up to 24 cents more per pound for it.

They want leaner pork. And they're willing to pay for it. What other challenges are facing the pork industry? Updating the marketing of hogs.

Most pork in the United States is still marketed on a live weight basis. That means that in the majority of cases, there is little incentive to produce the lean, high-quality pork that consumers want. It's only realistic to expect this situation to change, and it is changing. Carcass merit incentive programs are being used by the more progressive packers, and leading producers are working harder to produce pork that is competitive with the leanest of meats. Yet these trends will have to become "business-as-usual" in order to meet the challenges of the future.

Today the pork industry is readying itself for the opportunities ahead... opportunities that will accompany, among other things, the commercialization of porcine somatotropin, or PST.

What is somatotropin? Somatotropin is the growth-regulating protein or hormone produced continually in the pituitary gland of all animals. This protein is relatively species-specific, so that the avian somatotropin, for instance, would have no effect if given to pigs. The pig's metabolism would simply not respond to the avian hormone, but rather break it down into constituent amino acids and assimilate them as with any other protein.

Somatotropin in all species regulates the utilization of nutrients—by speeding fat breakdown, and protein deposition, while inhibiting fat deposition and protein breakdown. Somatotropin, working through specific organ tissues, mediates cell growth in bone, muscle, and connective tissue. The specific active substances at a tissue/cell level are called somatomedins, which are released by somatotropin's action on specific tissue receptors.

Finally, the growth process in animals is controlled by the presence of another protein hormone, somatostatin, also produced in the pituitary, which balances the actions of somatotropin.

Somatotropins aren't a recent discovery. Back in the 1920's, animal scientists knew that milk production could be increased by giving dairy cows extra amounts of their own protein hormone. The problem in that approach was in getting enough somatotropin economically, since the only method for obtaining bovine somatotropin was to extract small amounts from the pituitaries of cows sent to slaughter.

With genetic engineering techniques, we have the means to isolate the porcine somatotropin gene, introduce the gene into plasmid, insert the plasmid into a strain of bacteria, and then produce PST in large quantities. These bacteria in large-scale fermentation vessels produce PST as they grow under controlled conditions.

Pitman-Moore and other companies are working to bring porcine somatotropin products to market. A measure of Pitman-Moore's confidence in PST is the construction, now underway, of a facility to produce PST in commercial quantities.

Will the benefits of PST justify the pharmaceutical industry's investment in research and facilities? In a word, yes. And actually, we predict that PST will permit continued growth for the pork industry, providing real answers to real needs. Furthermore, besides producing the pork that fits consumer demand for leanness, PST answers the producer's need for lower cost of production and higher profit margins.

First, PST supplementation enhances the growth rate of pigs and improves their feed conversion efficiency— more lean gain per pound of feed. Breeding selection and improved management have already produced leaner, more efficient pigs. PST builds on those advances gained in recent decades with animals that are leaner still, that grow faster and that make better use of nutrients. It makes these improvements faster. Genetic selection usually takes years to make big changes in hogs. And sometimes undesirable traits can be carried forward along with desirable traits. PST will provide benefits to the producers immediately. PST won't, however, replace genetic selection or balanced feeding programs. The best situation may be to breed for more lean, feed for more lean, and use PST.

When high-performing finishing pigs have been treated with PST over the weight range from 100 pounds to market weight, feed efficiency improved by up to 24%. There was a 16 to 18% improvement in gain. Fewer days were required by the hogs to reach market weight, and 12% less feed was required. Including the estimated cost of PST and of its administration, the per-pig production costs dropped about 7% with PST.

But, what about safety? That's an obvious question when discussing any new technology, no matter how economically promising it is. Indeed, the Food Marketing Institute's trends report for 1987 indicates that safety is a very real concern among our ultimate customers at the supermarket meat case. The good news here is that PST is safe to the animal and safe to the consumer. PST adds nothing to the challenges of maintaining a safe, wholesome meat supply.

PST, first of all, is safe to the target animal. In our work at Pitman-Moore, there have been no abscesses at the injection sites on the 2,000+ pigs already treated.

There has been no toxic level identified for PST. And, no hypersensitivity or diabetes has been induced in pigs by repeated administration of PST. Additionally, there is no degree of danger to anyone eating pork from PST-supplemented hogs. Residues are not an issue, since PST is an easily degradable protein. The half-life of PST in pig serum is only about 20 minutes. The pig's metabolism has evolved to clear itself rapidly of these protein regulators.

Pork from pigs supplemented with PST is the same as pork from pigs not given additional levels of PST. The only difference is the amount of fat present. I believe that PST will result in revolutionary changes in pork products and pork demand over the next decade. This compound has the potential to make pork the number one meat in most countries as well as a premium source of protein.

Hot Dehulling System: "Popping"

W. Fetzer

Buhler Brothers Ltd., Uzwil, Switzerland

Abstract

The dehulling of soybeans by means of fluid-bed technology in oil mills and other industries has become more and more popular. Depending on the seed properties the soybeans are heat treated for three to seven minutes in the fluid-bed in order to reduce the binding forces between hulls and seed kernels. In the meantime, a new system has been developed which allows shorter treatment time in the fluid-bed. This new system features a conditioning column installed above the fluid-bed. By means of contact heat transfers, the soybeans are conditioned and fed into the fluid-bed where the temperature of the soybeans is further increased by means of hot air. For the dehulling process, the beans need to be treated in the fluid-bed for only one minute. This has the advantage that the fluid-bed is smaller for a given capacity. The short-heat treatment in the fluid-bed does hardly reduce the water dissolubility index of the protein which is of great importance for the production of white flakes.

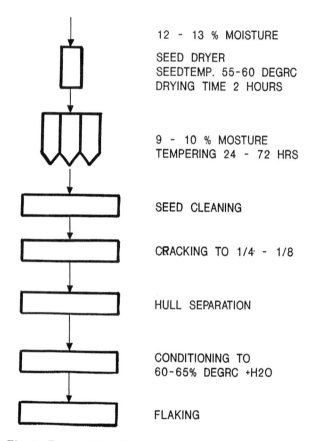

Fig. 1. Front-end dehulling.

To start my presentation on the subject of Hot Dehulling System: "Popping," please allow me to make some general comments on the dehulling of soybeans. The reasons for the dehulling are: in the oil milling industry, to yield HP solvent extraction meal, white flakes, concentrates and isolates; and, in the food industry, to yield full-fat soy meal, soya drinks, soya milk, and imitation dairy products, e.g., ice cream, yogurt and cheese, plus tofu.

In order to better understand and especially to fully comprehend the advantages of the "popping" system it is necessary to elaborate in short on the other known systems. Today the following systems for the dehulling of soybeans are commonly applied: front-end dehulling with a seed dryer or hot dehulling with short-time drying. These schemes are shown in Figures 1 through 5 and the result from the various methods are given in Table 1.

Fig. 2. Soybean drying and tempering.

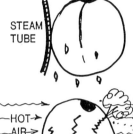

Fig. 4. Soybean fluid-bed drying.

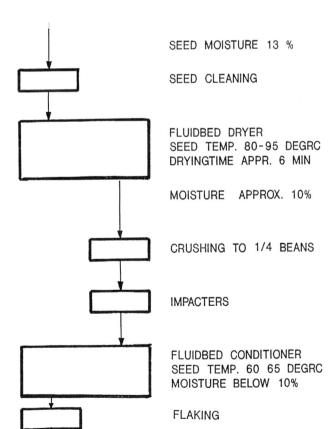

Fig. 3. Hot dehulling.

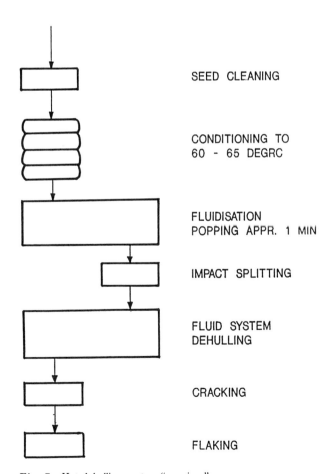

Fig. 5. Hot dehulling system "popping."

TABLE 1

Comparison Between Common Systems for Dehulling Soybeans

	Conventional system	Hot dehulling "popping"	Hot dehulling short-time drying
Drying	2 hours, 50° to 55°C		5 to 6 minutes, 80° to 95°C
NSI Red.	Appr. 5–8%	Appr. 5–8%	30–50% (heat damage)
Husk in dehulled kernels	Max. 0.9% or less loose husks		Depending on seed properties
Preparation quality	Gentle conditioning resulting in even distribution in the kernels results in good flakes		Uneven moisture and temperature distribution in the kernels
Flexibility	Fixed to HP meal with Kon. heat treatment	Very flexible, all product qualities possible	Fixed to profat—possibly HP meal
Power: based on 1000 TPD project steam cons	105% KW inst. 100–170%	100% KW inst. 100%	120% KW inst. 120%

A New Generation of Flaking Mills

W. Fetzer

Buhler Brothers Ltd., Uzwil, Switzerland

Abstract

Flaking is still the most important process in the oilseed preparation for prepressing or solvent extraction. During the past decades, flakers have developed to a high mechanical standard allowing high throughput rates. Besides all the achieved improvement, flaking rolls tend to develop damage on the roll ends and also hot spots on the roll surface. These problems are caused mainly by the uneven wear of the rolls as well as by deviation in the feeding rate and the seed conditioning. Roll end damages can be avoided by roll end grinding, which means service costs and downtime for flakers. To avoid these problems, or to at least minimize them to a great extent, a new flaker has been developed featuring a special feeder arrangement with incorporated mixing device resulting in a constant feeding of a homogeneous particle size distribution over the whole roll length. A new roll end sealing system incorporating a trouble-free roll end shape and this in combination with the above-mentioned feeder-mixer reduces the above stated problems to a great extent.

Flaking is still the best method to open the cells of oilseeds for ease of the oil removal by expeller pressing or solvent extraction. Because of the importance of this process, flaking machines have developed to a high technical standard. In spite of all achievements made to date, flaker operators even today are still afraid of: cracks and pitting on the roll surface, spallings at the roll ends, broken out roll ends, and hot spots. Best results in solvent extraction plants are achieved when feeding flakes of regular size and thickness and this in addition at a constant capacity. This means flakers must produce flakes of equal thickness over the whole length of the flaking rolls at a constant throughput rate. To achieve this the rolls of the flaker should be cylindrical over the whole working range and the material fed into the roller nip should be homogeneous and of constant bulk density.

Experience shows, however, that the material to be flaked will never be uniform in respect of: moisture content of the individual particles, sizes of seed particles (ranging from fine powder to half beans), more or less hull content, and temperature and contamination. Because of these conditions the bulk density changes within small product quantities, thus making it impossible to guarantee an equal feed over the whole roll length within such small limits as to avoid the development of hot spots.

Screw conveyors and inlet chutes feeding the flakers contribute a great deal to this. Heavy particles flow faster in the inlet chute than light ones. This results in segregation with the consequence of higher wear in the roll center. This again results in higher roll ends. In order to compensate for the uneven wear it is common practice to frequently grind the roll ends and to keep the latter tapered. If this is not followed, the roll ends will run dry and develop spallings followed by breaking off of roll ends (Figs. 1 and 2).

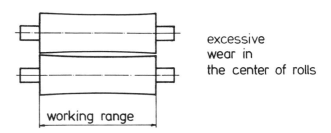

Fig. 1. Flaker rolls with excessive wear.

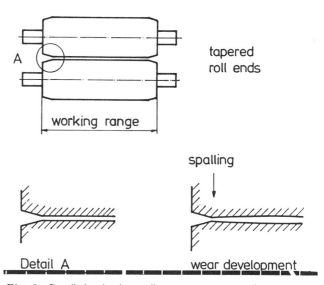

Fig. 2. Detail showing how rollers may wear unevenly.

Hot spots are developed by uneven feeding. Once this occurs the only way to overcome this problem is to reduce the throughput rate or to open the roll gap. Both actions have adverse consequences for the performance of the plant.

Summing up the above facts we can conclude that the causes of the problems lie to a certain degree in the properties of the materials to be flaked and on the other hand in the construction and installation of the flaker itself.

As a consequence future flaker models should be designed and built in order to be: insensitive to the product properties of the materials to be flaked, trouble free with regard to spalling and breaking of roll ends, low in maintenance, i.e. no roll end grinding, and easy to install.

Approaches to these problems are described below:

Mixer-feeder: In order to overcome the problems caused by the heterogeneous product and other properties of the material to be flaked. New generation flakers feature an integral mixer in the flaker inlet. Here the incoming material is homogeneously mixed and distributed over the whole width of the machine. The mixer is designed in such a way that material lumps are disintegrated before the material is discharged into the integrated roll feeder. The corrugated feeder roll in combination with the segment regulating slide dose the material into the roller nip. As the product to be fed is a homogeneous one, the bulk density is equal over the whole length of the feeder roll. This condition makes it possible to guarantee an equal and constant feeding of a homogeneous material into the roller nip. The result of this is even wear and tear over the whole working range of the flaking rolls, avoiding development of hot spots (Fig. 3).

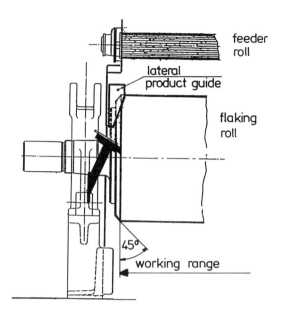

Fig. 4. Lateral product guide.

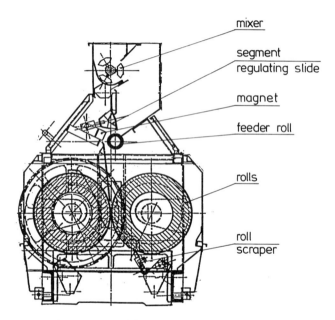

Fig. 3. Cross section of flaker.

Cylindrical flaking rolls (Fig. 4): Because the wear and tear over the whole working range of the flaking rolls is equal, roll ends need no longer be tapered. Because there are no tapers, spalling cannot develop. In order to eliminate the roll end problems, the working range of the rolls does not extend over the whole roll length. The distance between end of working range and roll shoulder is bevelled at an angle of 45 degrees.

Low maintenance: Special lateral product guides, adapted to the above described shape of the roll ends, leading from the feeder to the roll nip are designed in such a way that the roll ends always are covered with product and no unflaked material can pass. Because of this arrangement the end of the working range of the rolls wear out to exactly the same degree as the remaining part of the roll cylinder. Because of this no roll end grinding is required. The prototype of this flaker has been in operation for one year processing soybeans. During this period no roll or roll end grinding was required.

Easy installation: The inlet opening of the flaker extends about to half of the machine width only. The material distribution over the whole width is affected by the above described integral mixer. This design allows the installation of small feeding chutes leading from the conveyor above the flaker. This saves height even if the flakers are installed crosswise to the conveyor above the flaker.

The flaker and the main motor are mounted on a common base frame which rests on vibroabsorbers. There are no bolts to fix the base frame on the floor. This arrangement prevents the transmission of vibrations to the building and at the same time also contributes to a smooth running of the flaker itself.

The installation of this type of flaker requires a flat floor with the floor opening for the flaker outlet only. No separate foundations for the machine and drive motor are necessary.

Studies on the Expansion of Vegetable Proteins in China

Yang Mingduo

Department of Food Engineering, Heilongjiang Commercial College, Harbin, China

Abstract

Expansion as a food processing technique has a long history of use in China. Recent research has demonstrated its applicability to a wide variety of foodstuffs. Examples are described together with the new methods now being commercialized.

Brief Review: Expansion Techniques Developed in China

Expanding is one of the Chinese traditional food processing techniques. Expanded rice popcorn made with a handspun, hot air extruder is loved especially by the Chinese children. Unfortunately, this technology developed very slowly, had never surpassed handicraft with low output volume, and few brands for various reasons. Real changes didn't occur until the end of 1970's. Wide-spread research and application of expansion technology in food processing speeded up throughout the nation with China's economic reform development. In 1978, Heilongjiang Commercial College conducted a project from the Science and Technology Bureau of the Ministry of Commerce, and was the first to start "Research on Expanded Food Processing"(1). The project passed expert technical evaluation in 1979. Soon after this, Beijing (2), Jilin (3), Inner Mongolia (4), (5), Jiangsu (6), and Liaoning (7) reported their studies in the field. These researchers made expanding experiments with wheat, rice, waxed rice, broomcorn millet, sorghum, corn, soybean and other agricultural products. After smashing, they produced bread, steamed bread, baked cake, seasoned millet mush, compressed biscuits, cakes, sticky cakes, iced cakes, and ice stick additives. In the meanwhile, coconut powder additive and dried granular kelp also were made into concentrated expanded food (8). At the time when coarse grain accounted for most of the grain supply to both the rural and urban residents, these efforts greatly improved food palatability and the degree of digestion and absorption, and as a result, contributed to the people's acceptance of coarse food grains. However, with steady economic development and rising living standards, we have seen a decline in human consumption of coarse food grains but an increase in other utilization; expanded foods made from coarse food grains have become less and less. By 1979, children foods were the only expanded foods left in the market.

In 1981, extruded texture protein was developed (9-12). As more than 95% of the Chinese oil processing mills were using heating technology for oil extraction, the success of highly variable, textured vegetable proteins opened a new field in the application of plant proteins to human food processing (13). More than 20 provinces, cities, and autonomous regions and about 400 factories are using the results to produce "Plant Protein Meat."

Because the expanding process has some very unique advantages over other food processing techniques and operations, application of expanding technology in the food processing field has been developing rapidly. Some of the features are as following: multiple function; higher productivity; lower cost; lower energy consumption; pollution free; different product patterns; quality products; new food product creation; easy to retain; and higher digestion rate. It is because of these features that expanding technology is widely applied not only in China, but around the world.

Expanding technology and its applications in the food processing field has stimulated research in the area of expanding principles of cereals (14–16). Research emphasis has expanded into such areas as handling of raw materials in food processing and reform of the traditional brewing techniques. The theoretical basis of application of the expanding technology in food processing and brewing industry has been established (14,17). In 1984, the applied expanding technique in Yellow Wine research was quite a success (14,18). Expert evaluation gave it positive appraisal, and the product, Yellow Wine, won the silver cup prize in the Fourth National Wine Competition. The same product subsequently received the Science and Progress Prize from Heilongjiang Province, the Ministry of Light Industry, and the Ministry of Commerce respectively. Later, research on expanded vinegar (19), expanded rum (20), expanded starch sugar (21), and expanded alcohol (22) also succeeded. Research on expanded bean meal and expanded corn in soy sauce making began in 1987 and positive results have been made; some research results have become Chinese patents. The application of expanding technology in the fermentation industry is a significant reform to traditional fermentation techniques. It has the following advantages: it raises raw material utilization rate; it simplifies working process and is energy-saving; it shortens the production cycle; and it reduces the volume of leaven (and wine yeast) required.

Expansion technology and its application fields and scale is closely related with expanding equipment. In 1974, Heilongjiang Commercial College designed a new batch explosion-puffing extruder, which has a capacity of 10–15 times of the original. Some Chinese firms also were involved in research and development on extrusion machines. Shanghai had talks with Japanese experts and imported a small food extrusion extruder (23); Beijing, Shanghai, Heilongjiang, Jilin, Shandong and a few other places manufactured various types of cereal extrusion and textured protein extruders. Dalian Industrial Oil Plant used a plastic extruder to replace the bean textured protein extruder (24). In less than a year, extruder manufacturers exceeded 100 (25). In 1985, Qiqihar Light Industry Institute designed a 100 Kg/hour continuous air pumping extruder (26). In 1987, the same institute announced a 200 Kg/hour central heating, surface air compressing, and continuous feeding line extruder. In May 1988, Heilongjiang Commercial College, in cooperation with Qinghua Machinery Factory in Beian, designed a 100 Kg/hour extrusion extruder. At the same time of designing and manufacturing this equipment, the various organizations also monitored the foreign development of a large, complete, piece of equipment. Hubei province imported an X-series extrusion extruder from the American Wenger Company and used it for animal feed production. Triple-River Food Corporation in Heilongjiang bought comprehensive bean processing equipment from the West Germany Bagoka Groups and the American 3 I Corp (27).

During research some Chinese scholars translated and compiled many foreign patents and publications; the book "Extrusion of Foods" by J.M. Harper was one of them. Some scholars made analysis of the heating process of the textured protein, extruding steam machine, and discovered that the heat generated from the process of extrusion (28), expanding, and cooling on the plant protein texturing belonged, in essence, to a heating process similar to that of explosion-puffing. Yang Mingduo, from Heilongjiang Commercial College, was the first among all Chinese colleges and universities who opened a "theory and technique of cereal expanding" course to undergraduate and graduate students. In China today the expanding principles and techniques are used widely in food processing, fermentation, animal feed, medical research, construction and many other areas.

Research on Textured Protein Extrusion (9,12,13,29,30)

Production Process and Technical Requirements

Production Process as below: Bean meal → strip and clean → smash → sieve (60-100 MU) → blend (with water, salt, baking soda, pH 7. 6) → extrude → dry → pack → press

Raw Materials: Odorless, contains no grit, water content less than 13%, fat below 9%, total protein above 40%, in which water soluble nitrogen (NSI) can be more than 10%. Bean granules passed 100 MU if used for high temperature, or passed 60 MU for low temperature. pH value to be set at protein medium level.

Water Added: Bean meal at high temperatures and low temperatures have different water absorption rates. Bean meal at high temperature would have protein variety change, decreased water absorption rate, water content in blended material can be around 40–60% while the water content in low temperature bean meal could be around 20-30%. The relations of water content of bean meal material in high temperature to bean meal water soluble protein (NSI) is given in Table 1; the relations of water content of bean meal material in low temperature to bean meal absorption rate and the product taste is given in Table 2.

TABLE 1

The Relations of Water Content in Bean Meal Material to NSI

Water content (%)	25–30	30–35	35–50
NSI value	> 50	50–30	30–10

Temperature in Expanding Extruder: After the other conditions are set, such as rotation rate, feed-in volume and water content, too low temperatures result in "uncooked" texture protein and may also leave a "hard core" in it. Too high temperature results in dark color and burnt product. Spiral temperature should be set as follows: heating range 1: 70–90°C; heating range 2: 110–150°C; heating range 3: 160–190°C; and heating range 4 (machine head part): 190–220°C.

TABLE 2

The Relations of Water Content of Bean Meal Material in Low Temperature to Bean Meal Absorption Rate and the Product's Taste

Material water content	Water absorption rate	Review of taste
23.5	2.80	Color brown, burnt, pops during extrusion time loose spread
24.0	2.90	Same as above
28.5	4.00	Easy to feed in material, light color, even granular, join together after extrusion, surface even, spongy structure
33.5	3.20	Material sticky, hard to feed in, strong explosion force, broken up after extrusion
35.0	3.05	Material sticky, feed nearly impossible, huge explosion force, broken up after extrusion

Expanding Extruder Spiral Rate: Spiral rate has direct effect on production output; their relations are listed in Table 3.

Other conditions being equal and in its normal range, higher spiral rate results in higher output volume.

TABLE 3

Relations Between Spiral Rate and Output Volume

Spiral rate (RPM)	Output volume (Kg/hour)
65	50
54	42
42	34
28	30

Nutrition Value of Extruded Texture Protein: Soybean has a relatively high protein content (37.9%). Its protein could reach as high as 50% after being de-fatted; this is higher than milk, eggs, and meat. Table 4 shows the amino acid results of extruded texture protein from high temperature bean meal and low temperature bean meal.

Table 4 demonstrates clearly that extruded texture protein retains most of the amino acid components of soybean protein except that all the amino acid amounts are less than bean meal. However, nutritious contents in bean meal can be preserved as much as possible with improved processing methods.

tyrosine protein. The experiment ended in 4 weeks. The nutritious effect of protein used in the experiment (i.e., PER) was generated from the total protein mass (gram) consumed by each animal in 4 weeks divided by its weight increase (gram). The experiment results are given in Table 5.

From Table 5, we see that nutritious effect of extruded textured protein is equivalent to more than 65% of tyrosine protein. Since extruded textured protein has fat removed from soybeans, its protein concentration has increased by one third. Therefore, in consideration of its nutritious effect, it can be said to be an excellent protein-adding food. As adding animal serum cannot increase nutritious value and restricts scope of usage, it becomes unnecessary from the nutrition point. Better nutritional effects can be achieved with improved processing methods.

Application of Extruded Textured Protein in Food Engineering (31–34)

Canned and filled products: Textured proteins of soy beans may be used as additives to the filling materials for luncheon meat, pork braised in soy sauce, sausage, ham, hotdog, meat balls, dumplings, pies and many other products. The general steps are: (A) put the textured protein of soy bean in warm water and soak it for about 10 minutes so that it becomes expanded; (B) mix it with pork, beef, or fish filling at 10-30% ratio to the meat; and (C) follow the usual recipe to add needed seasonings. Due to its high water absorption rate, textured protein of soy bean can absorb the fresh taste from meat juice and keep the original flavor and taste unchanged.

TABLE 4

Amino Acid Analysis of Extruded Textured Protein (Mg/100mg sample)

Amino acid	High-temperature bean meal material		Low-temperature bean meal material		Soybeans	Bean meal in low temperature
	Normal	210°C	USA ADM	Dalian Oil Factory Product		
Lys.	3.37	3.12	3.2	3.01	6.05	5.94
Met.	0.71	0.70	0.6	0.57	1.08	1.19
Thr.	2.21	1.96	2.2	1.91	4.34	3.95
Val.	2.61	2.55	2.6	0.11	4.75	4.38
Leu.	4.57	4.11	4.0	4.26	—	—
Ile.	2.64	2.33	2.4	2.52	—	—
Phe.	0.51	0.39	0.6	—	4.75	5.08
Cys.	1.83	2.50	0.7	5.76	1.10	1.55
Arg.	4.09	3.83	3.8	4.09	8.30	7.58
His.	1.32	1.26	1.4	1.26	2.68	2.50
Tyr.	1.92	1.83	1.6	1.52	—	2.39

A comparative experiment was made on animal growth after they were fed extruded textured protein. Some weanling male mice, weighing about 70 g, from the same parents were randomly divided into groups. They were fed with compound feed of 10% protein content, but from different sources, and the contrast group was fed with

Fried dish and salad: Textured protein of soy beans may be used as part or all of the meat substitute material after it has been processed. The general method is to use soaked textured protein of soy beans, squeeze the water out, prepare some meat soup containing the seasonings required (such as Chinese prickly ash, aniseed, soy

TABLE 5

Protein Nutrition Evaluation (PER)[a]

Protein category	Coarse protein content (%)	# of animals	Weight (g) increase in 4 wks	Protein (g) consumed in 4 wks	PER	PER alt.	PER in %
Tyrosine protein group	92.45	7	123.5 ± 48.3	41.3 ± 14.4	29.6 ± 0.16	2.50	100
Extruded textured protein group	48.92	7	83.9 ± 23.6	43.6 ± 4.4	1.91 ± 0.39	1.63	65.2
Semiserum sugar group	76.93	6	87.5 ± 22.2	38.9 ± 3.8	2.23 ± 0.42	2.01	79
Semitotal serum sugar group	96.93	4	−19.3 ± 8.1	77.0 ± 5.2	—	—	—
Extruded textured protein + serum 16.13% group	52.22	7	64.4 ± 16.8	29.8 ± 4.9	2.13 ± 0.26	1.56	62.4
Extruded textured protein + serum 30.3% group	53.30	8	80.4 ± 25.1	34.0 ± 6.3	2.35 ± 0.40	1.70	68.0
Bean meal powder group	50.09	6	110.6 ± 21.8	47.4 ± 5.8	2.32 ± 0.26	1.96	78.4
Corn powder group	9.79	6	9.7 ± 8.4	25.8 ± 4.3	0.38 ± 0.35	0.32	12.8
Deficiency group	0	7	−24.7	—	0	—	—

[a] Basic prescription of the compound feed: protein 10%, corn starch powder 47%, sucrose 32%, vegetable oil 4%, mixed inorganic salt 4%, cod-liver oil 2%, and vitamin sugar 1%.

sauce, gourmet powder, salt, sugar, cooking wine, vinegar and some others), soak the substitute material in the soup or boil it in the soup, then bring it out and dry it a little in the air, and it is ready to serve for a fried dish or salad.

Convenience food: Use soybean protein as the major material, plus various kinds of seasonings to make a variety of flavored, convenience foods; or add pork skin and other seasonings to get nutritionally-balanced convenience dishes. Examples are: (a) dry-fried protein. Squeeze water out of the soaked textured protein of soybeans, add in such seasonings as gourmet powder, Chinese prickly ash and sugar, etc., fry with oil to get dry-fried protein. It may either be used to make convenient snack foods, or as meat substitute materials to make fried dishes. By changing the seasoning recipe, we may also make "Strange taste protein," "Protein tenderloin," "Spicy protein," "5-flavor protein," etc. (b) Meat and vegetable dishes: Use textured protein of soybeans as the main material, add pork skin, potato catch-up, starch, bone soup, plus other needed seasonings, after the process of recipe making, blending, pattern/shape forming, cooling, packing, to realize industrial volume production. Analysis on this shows that the product contains protein, fat, amino acid, and vitamin; its protein and essential amino acid contents are much higher than that in pork, beef, eggs, and milk; its vitamin content is only lower than eggs, but higher than pork, beef, and milk.

Of all the proteins consumed by Chinese people, plant protein accounts for the major part, as high as 89%, according to statistics. Judging by China's current population and production status, the existing food consumption structure is unlikely to have any dramatic changes in a relatively short period ahead. With this situation, therefore, it is very important to make full use of China's plant protein resources. Soybean is one of the most important food protein source; its history began as far back as 4,000 years ago. There have been a variety of valuable nutritious foods made from soybeans. Soybean protein has very rich nutrients and does not contain cholesterol.

Soybean meal cake is inexpensive after its oil has been extracted out. Using expanding technique to produce textured protein of soy beans is a very efficient way to alleviate the protein deficiency. At present, we should adopt large-scale, continuous extruding equipment to achieve healthy development of extruded protein of soy beans, and other extrusion techniques, in China's food industry.

References

1. Expanded Food Study Group, Heilongjiang Commercial College, *Commercial Study*, 2nd ed. p. 6 (1979).
2. Intelligence Office of Food Science Research Institute, Beijing, *Food and Oil Study of Beijing*, 1st ed. p. 32 (1981).
3. Food Science Research Institute, Changcun, *Oil and Food Science and Technology of Jilin*, 6th ed. p. 117 (1980).
4. Food Science Research Institute, Inner Mongolia, *Oil and Food Science and Technology of Inner Mongolia*, 1st ed. p. 26 (1980).
5. Food Science Research Institute, Baotou, *Science and Technology Bulletin* (May, 1980).
6. Grain Bureau of Suzhou, "A Report on Expanded Food Study" (1980).
7. Zhang Chunjie, Wang Junling, *Commercial Science and Technology*, 4th ed. p. 10 (1980).
8. Shi Yuchuan, Liu Jun, *Food Science*, 6th ed. p. 51 (1982).
9. Wang Zhenyuan, Du Housan, *Food Science*, 9th ed. p. 25 (1982).
10. Plant protein factory of Qichun Commune, *Science and Technology of Jujube Village*, 2nd ed. p. 14 (1983).
11. Ma Jie, *Science and Technology of Food Industry of Sichuan*, 4th ed. p. 1 (1984).
12. Xu Bangjie, *Food Science*, 4th ed. p. 35 (1984).
13. Li Chao, Li Ronghe, *Journal of Northeast Normal University (Natural Science Edition)*, 4th ed. p. 41 (1982).
14. Yang Mingduo, "A Probe into the Principles of Cereal

Expansion and its Application in Yellow Wine Brewing" (Master Degree Thesis), (October 1984).
15. Yang Mingduo, *Food Fermenting Industry*, 4th ed. (82) (1988).
16. Yang Mingduo, *Chinese Grain and Oil Journal*, 1st ed. (3) p. 53 (1988).
17. Yang Mingduo, *Chinese Brewing*, 2nd ed. p. 21 (1988).
18. Heilongjiang Commercial College, "A Research Report on Application of Extrusion Technology to Yellow Wine Brewing" (September 1984).
19. Heilongjiang Commercial College, "A Research Report on Application of Extrusion Technology to Food Vinegar Brewing" (May 1986).
20. Heilongjiang Commercial College, "A Research Report on Application of Extrusion Technology to Spirit Production" (September 1987).
21. Heilongjiang Commercial College, "A Research Report on Application of Extrusion Technology to Starch Syrup Production" (September 1987).
22. He Zhicun, *Industry Microorganism*, 1st ed. p. 9 (1984).
23. Zhang Ruilin, *Foreign Food Technology*, 3rd ed. p. 13 (1980).
24. Grain Bureau of Yantai District, "A Report on HDP-95 Model Extruder Design Experiment" (1981).
25. Guan Daguang, *Food Science*, 9th ed. p. 44 (1984).
26. Qiqihar Light Industry Institute, "A Technical Report on Continuous Extruding Method Via Air Circulation Transmission System and Design of Mechanical Equipment" (September 1985).
27. Ma Zhensheng, *Science and Technology of Food Industry*, 4th ed. p. 5 (1987).
28. Chen Tingqiang, *Food Fermenting Industry*, 2d ed. (74) p. 32 (1987).
29. Gu Jingfan (as the main author), *Food Science*, 4th ed. p. 1 (1983).
30. Yang Shengkui, *Oil and Food Science and Technology of Tianjing*, 4th ed. (1982).
31. Zhao Qingzheng, *Science and Technology of Food Industry*, 4th ed. p. 13 (1987).
32. Office of Commercial Science and Technology Information, Lu Da, *Science and Technology of Food Industry*, 3rd ed. p. 1 (1981).
33. Yuan Miaole, *Food Science*, 11th ed. p. 43 (1984).
34. Fermentation Food Research Office, Ministry of Light Industry, *Food Science and Technology* of Wuhan, 2.3d ed. p. 4 (1982).

Aqueous, Membrane and Adsorptive Separations of Vegetable Proteins

S.S. Köseoğlu and E.W. Lusas

Food Protein Research and Development Center, Texas Engineering Experiment Station, Texas A&M University System, College Station, Texas 77843-2476, U.S.A.

Abstract

During the past ten years, separations processes have been improved greatly by the introduction of better ultrafiltration/reverse osmosis membranes, adsorbents, alternative solvents and front- and down-stream processing equipment. This paper reviews experimental processes in membrane technology as applied to isolation of vegetable proteins from oilseeds, cereals and nuts. Alternative aqueous extraction methods for simultaneous separation of oil and proteins are also discussed with emphasis on product quality and economics. Advantages and disadvantages of aqueous and nonaqueous extraction processing techniques, process flows, and product quality are characterized, and opportunities for using adsorptive protein isolation/purification methods alone or in conjunction with membrane processing methods, are discussed.

World needs for food, fiber and energy are escalating rapidly. Today's world population of 5 billion is expected to increase to about 6 billion by the year 2000. Most nations have accepted the principle that population control measures are necessary; but world population is likely to reach at least 12 billion (with most of the increase in third world countries) before hopes of zero population-growth might be achieved. More consumers will place additional demands on the world's food protein production capabilities. In addition to fully utilizing conventional protein sources, technologies may have to be developed for processing proteins from novel sources such as microorganisms and leaf sources to provide sufficient quantities of protein products for the growing population of the world. The year 2000 is not far away, and the majority of persons, 55 years of age or under today, are likely to witness it.

Improved protein isolation techniques are needed for several reasons: Currently, substantial amounts of protein are under-utilized because of technical and economic limitations of removing fiber and undesirable components (1); most of the protein isolation and/or fractionation techniques are based on alkaline, salt or acidic extractions and subsequent precipitation by pH adjustments. During and after their formation, protein precipitates are sensitive to shear which can alter their physical structures; efficient and economic extraction and purification processes may enable processors to produce improved protein fractions. Because of the diversity of animal and vegetable raw materials, a variety of processes is needed to remove toxins, coloring pigments and off-flavor compounds.

Proteins can be separated from crude materials by various methods, including: precipitation; membrane processing; adsorptive/chromatography including continuous and affinity/specific adsorption; supercritical gas extraction; liquid/liquid extraction; and electrical processes (1). Most of these techniques work well in laboratory scale purifications, but only a few have been utilized to produce commercial quantities of proteins economically.

Principles of Membrane Processing

During the last decade a new technology, membrane processing, has evolved, enabling scientists to isolate, purify, and/or fractionate complex mixtures. Separations by this process are based on the molecular/particle sizes and shapes of individual components, and is somewhat dependent on their interactions with the membrane surface and other components of the mixture. This energy efficient separations technology has found industrial scale uses in waste water treatment, fruit and vegetable juice processing, purifications of products, and recovery and purification of proteins, oils and other materials.

Membrane Processing

Principles

The basis of membrane separations is illustrated in simplified form in Figure 1 (3). When a feed stream containing a mixture of large and small molecules passes across a membrane under pressure, only those molecules of a smaller size than the membrane pores permeate through the membrane. The concentrated fraction containing mainly large molecules is called the "retentate." In practice, separation mechanisms are more complex; and are influenced by variables such as composition of membranes, methods of manufacturing, shapes and configurations of permeate molecules, their interaction with each other and the membrane surface, fluid dynamics of the membrane unit, and pressure, temperature and velocity of the mixture.

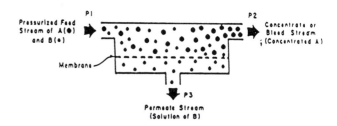

Fig. 1. Schematic representation of ultrafiltration process (3).

Membrane separations are divided into three different categories: microfiltration (MF) is concerned with separation of micron or sub-micron size particles in the 0.02–2.0 μ range. Perpendicular flow mode is common in this separation method, but, recently, interest has grown in cross flow configurations; ultrafiltration (UF) covers the 0.002 to 0.2 μ range, which corresponds to 500 to 300,000 molecular-weight. Both structure and performance of UF membranes are different from micro filtration membranes, and these separations always are run in cross flow mode; and reverse osmosis (RO) is the most complex separation technique in membrane technology. It is affected by salt concentration of the mixture, composition of membrane materials, and interactions between the molecules, the membrane surface and associated charges. Pore sizes of reverse osmosis membranes range from 5 to 15 Å. The lower limit corresponds to a molecular weight cut off (MWCO) of less then 250.

Applications

A simplified diagram of a batch UF system is shown in Figure 2. Membrane cartridges are the heart of the UF system. Other components include pump(s), tanks, instrumentation, manifolding in various arrangements, valves and cleaning subunits. UF systems can be designed in several modes, including: single pass; batch; feed and bleed; and feed and bleed multistage.

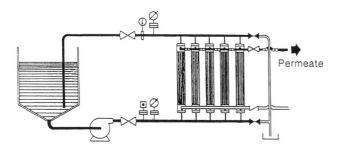

Fig. 2. Batch ultrafiltration process (3).

Earlier UF/RO membranes were made primarily of cellulose acetate and various polyamide materials that are sensitive to chlorine and organic solvents and were limited to aqueous separations (4). In recent years, new membranes with increased chemical stability, and pressure and temperature ranges, have become available. These are made of polysulfones, zirconium coated carbon, extruded alumina "ceramics" and alumina coating on inert support materials (5). However, they are not yet available in RO pore sizes.

Commercial membranes are available in various forms such as tubular, spiral wound, hollow fiber, flat, and flat leaf. The proper membrane selection is made by considering concentration and type of feed, pressure and temperature range of the separation, chemical stability of the membrane and the required hygienity of the products.

Membrane technology has received considerable attention from researchers in isolation of proteins from oilseeds, cereals and nuts for several reasons including simplicity, selectivity and energy efficiency of separation. Whey solutions containing proteins are not produced during membrane isolation processes, and nearly complete recovery of proteins is possible. Spray-dried products from UF have better functional properties and nitrogen solubility indexes (NSIs), and less color and undesirable flavors than those made by typical alkali solubilization and precipitation commercial procedures (6). The inclusion of

RO techniques in protein recovery processes improves the economic feasibility of the process by reducing waste treatment costs, re-utilizing the extraction water, and recovering by-products that can be used as food ingredients.

Aqueous Extraction

Advantages of aqueous extractions are multi-fold. Problems of safety, as a result with flammable organic solvents, are avoided. AEP processes have the flexibility of easy start-up and shut-down. Capital investment costs for an AEP plant is lower than for conventional solvent extraction plants since there is no need for explosion proof pumps, electrical motors and connections. Another advantage is the ability to deactivate undesirable components, such as toxins and allergens, during the process using water soluble chemicals.

On the other hand, AEPs are less efficient than hexane for recovering oil, which is an important factor for seeds that are grown mainly for their oil content. Strict sanitation is required because of rapid growth of microorganisms in aqueous mixtures. Also, due to extraction of emulsifying agents, special techniques are needed to recover good yields of high quality oils.

Aqueous extraction (AEP) is a process for removing oil from seeds by hot water with the assistance of mechanical centrifuges and/or decanters, and can be adapted to most oilseeds, legumes, and nuts (7). AEP consists of three major steps: extraction; phase separation (demulsification); and drying to produce final product. The oil and protein contents of the spray-dried product are dependent on both extraction and separation procedures (Fig. 3).

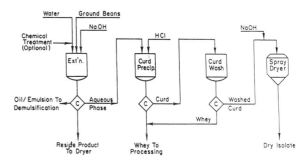

Fig. 3. Simplified flow diagram of aqueous extraction process applied to soybeans (7).

Factors that influence the efficiency of extraction, including solid-to-water ratio, pH of the extraction, temperature and residence time are shown in Table 1 for major oil seeds. Because the protein solubility curves of individual oilseeds differ, pH optimization in both the extraction and precipitation steps is critical to recover maximum protein (Fig. 4) (8). For example, soybean is more efficiently extracted at pH-9.0 but cottonseed requires ph 9.5–10.0. Temperature is more important for maximum extraction of oil than protein. Temperatures of 40–60°C and residence times of 45–90 minutes have been found effective for most of the oilseeds. Solids-to-water dispersion ratios are controlled by the type of separator used, solubility of the proteins, particle size of the flour or

TABLE 1

Summary of Important Extraction Parameters for the Aqueous Extraction of Proteins (Pilot Plant Scale)

Material	References	Temperature (°C)	pH	Time (min)	Solid-to-water ratio	Pressure
Soybean	(9)	60	9.0	30	1/12	Atm.
Peanut	(12,13)	60	8.0	30	1/6	Atm.
Rapeseed[a]	(23)	70	6.6	60	1/3	Atm.
Sunflower	(19)	53	9.0	30	1/20	Vacuum
Lupin seed	(31)	65	8.0	40	1/15	Atm.
Jojoba seed	(35)	60	9.0	45	1/10	Atm.
Sesame seed	(40)	45	9.0	40	1/15	Atm.
Coconut	(38)	80	7.0	—	—	Atm.
Macadamia nut	(35)	60	9.25	45	1/12	Atm.

[a] Laboratory scale.

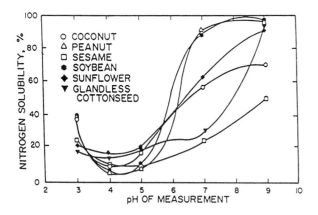

Fig. 4. Protein solubility of six defatted oilseed flours at various pH's (8).

the flake, and economics of the process. At the end of the extraction, the added processing water must be removed to obtain a dry product. A solid-to-water ratio of about 1:15 is optimal in most cases.

Figure 5 illustrates how the oil content of the product is manipulated by use of different separators. If a low-fat product is desired (Figure 5a), separation can be best accomplished by first separating the full-fat extract into two fractions, solid and liquid phases, followed by three-phase separation of liquid phase. Full-fat extraction can be achieved by using single stage solid-liquid separation (Figure 5b); but the most efficient method is single step three-phase centrifugation. Optionally, the solid residue can be re-extracted to increase the yields of protein and oil (9). Other types of separators such as basket centrifuges, decanters and fine vibratory screens also can be employed in solid-liquid separations.

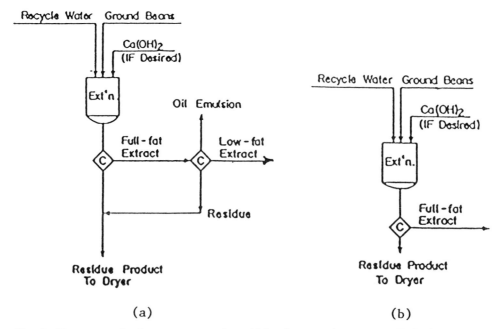

Fig. 5. Phase separation in aqueous extractions. (a) low-fat extraction process; (b) full-fat extraction process (7).

The three streams from three-phase separation are further processed to recover both oil and protein. Demulsification is a critical unit operation in AEP. Dissolved proteins and other natural emulsifiers from oilseeds cause emulsification of the oil during extraction and separation. The extent of emulsification varies from one oilseed to another. In some cases, it is not possible to break the emulsion with conventional methods. Lawhon et al. (7) found that the moisture content of the "cream" is the most critical parameter for demulsification in AEP processing of soybeans; and should be reduced to 20–23% by the addition of soybean oil. The optimal conditions for other variables: mode, speed and duration of agitation, speed and duration of centrifugation and pH are summarized in Table 2.

TABLE 2
Optimum Conditions for Demulsification (7)

Allowable moisture level	20% with a maximum of 23%
pH	4.5 with a range of 4–6
Mode of agitation	Shearing action
Speed of agitation	Slow
Length of agitation	1–3 minutes
Speed of centrifugation	1,000 rpm or higher
Length of centrifugation	1–3 minutes
Temperature	40°C or higher

Combined Aqueous Extraction and Membrane Processing

Principles

Investigators in our center have combined AEP and membrane separation techniques to obtain advantages of both processes (Fig. 6) (9,10). In the resulting process, dehulled, ground full-fat flour is extracted with hot water. After the solid-liquid phase separation, the aqueous extract containing dissolved proteins, is first processed by UF to separate the smaller protein molecules which are tightly bound with the flavor and color components (6). The most commonly used molecular weight cut off membrane range for removing the smaller molecular weight proteins is 10,000 to 100,000 daltons. The retentate may be concentrated further by RO before spray drying. The permeate also is treated by reverse osmosis to recover the small molecular weight proteins and soluble carbohydrates as byproducts and water is reused in the process.

By applying membrane processing, proteins can be directly recovered from oilseed flour extracts and the generation of wheys, which result from traditional acid precipitation procedures is avoided (10). With the combination of ultrafiltration and reverse-osmosis technology, almost all of the proteins extracted from oilseed are recovered. First ultrafiltration is used to remove smaller molecular weight compounds such as salt, sugars, coloring pigments and flavors which are not bound to proteins, and other undesirable dissolved components of the oilseeds. The permeate from ultrafiltration is further membrane processed using an RO unit to recover the remaining total solids. In addition, AEPs offer safety and discontinuous operations and provide opportunities for removal or deactivation of undesirable raw material components using appropriate water-soluble chemicals (10). Also, increased

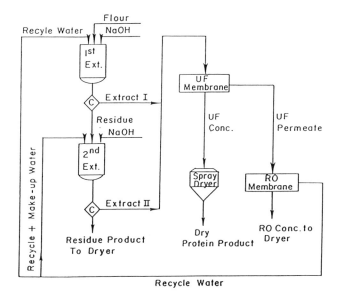

Fig. 6. Simplified flow diagram for membrane isolation of oilseed proteins (6).

isolate yields, combined with desirable functional properties, makes membrane technology an attractive process.

Applications

Soybean, Peanut and Cottonseed Protein: AEP has been applied to soybeans (11), peanut (12,13) and cottonseed (14). Analysis of the products obtained from aqueous extraction processing are given in Table 3. Peanut isolates and concentrates had the highest protein contents whereas soybeans had the lowest. Soybean concentrates were slightly higher in oil and ash contents. However, glandless cottonseed concentrate contained up to 19.1% total sugars.

Processes that incorporate UF in producing soy protein isolates with low phytic acid content have been reported

TABLE 3
Proximate Analysis of AEP Protein Concentrates and Isolates (7)

Product type	Protein (N × 6.25) %	Oil %	Total sugars %	Crude fiber %	Ash %
Protein isolate					
Peanut	95.0	0.84	2.1	0.63	1.5
Soybean	89.6	3.20	3.6	0.40	1.6
Glandless cottonseed	90.8	5.10	4.2	1.30	3.7
Protein concentrate					
Peanut	74.3	6.41	15.1	2.20	2.0
Soybean	70.2	8.40	13.8	0.31	7.3
Glandless cottonseed	67.0	7.20	19.1	1.80	4.9

(15,16), and UF has been used for reduction of carbohydrates and mineral constituents of soybean extracts (17). Peanut proteins also have been extracted by combinations of AEP and membrane isolation processes (MIP) (10). Glandless cottonseed proteins can be separated into two fractions, storage and nonstorage proteins by applying precipitation at selected pH conditions (Fig. 7), and can be further processed by UF and RO techniques.

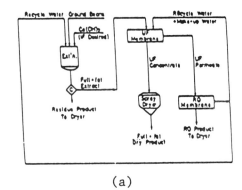

(a)

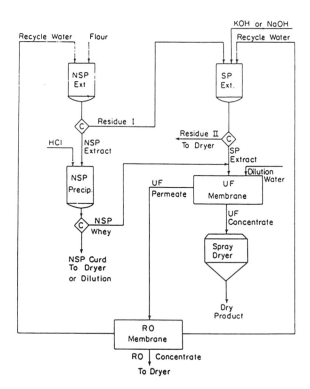

Fig. 7. Isolation of glandless cottonseed proteins by AEP-MIP process (14).

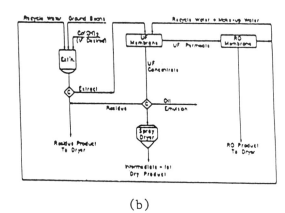

(b)

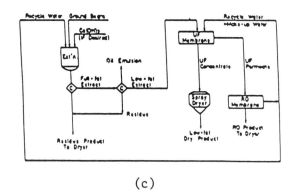

(c)

Fig. 8. Simplified flow diagram for production of protein products from undefatted soybean. (a) low-fat; (b) intermediate-fat; and (c) full-fat (9,10).

A simplified flow diagram for production of low-fat, intermediate-fat, and full-fat products from unfatted soybeans or peanuts using UF and RO membranes is shown in Figures 8a, b and c. Only low- and full-fat products were produced from blanched peanuts by processes shown in Figures 8a and c. Approximately 86% of the original nitrogen was recovered when processing full-fat soybeans, and the lowest recovery was obtained in preparing the low-fat product. Total solids contents also showed similar trends (Fig. 9). During soybean processing, higher water ratios extracted more nitrogen, and water ratios influenced nitrogen extractability more significantly than did pH of the extraction.

Peanut processing produced a low-fat UF isolate and 79.5% of the original peanut nitrogen was retained in the product. Full fat product contained 88% of the original protein. Low-fat isolate contained 29.6% of the peanut solid as compared to 74.6% in the full-fat product (10). Data on UF membrane performance indicate that mean flux was higher when processing soy and peanut extracts prepared with higher water ratios (Table 4). Per cent mean retentions of nitrogen and total solids were in the range of 95.7–96.7% to 83.1–90.1% respectively, and 38.5–51.1% of the sugars was retained in the permeate fraction. Mean fluxes of peanut extracts were much higher than soybean extracts.

Composition of products obtained from processes 8a, b and c are summarized in Table 5. Residue product had a protein content of 19.17%, oil content of 12.3% and crude fiber content of 18.4%. Protein contents of other products from full-fat, intermediate-fat and low-fat process ranged from 66.0% to 78.9%. Protein content of low-fat peanut isolate was exceptionally high (90.9%). RO effluents were lower in total solids than area tap water, indicating its suitability for reuse without further treatment. The spray dried RO product essentially consisted of sugar and ash,

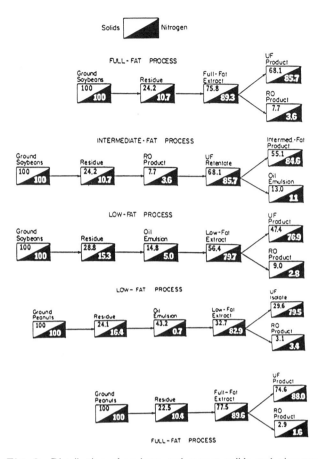

Fig. 9. Distribution of soybean and peanut solids and nitrogen during AEP-MIP process (9,10).

Sunflower Seeds: During the past decade, researchers at the Food Protein Research and Development Center (FPRDC) have investigated the possibility of utilizing alternative solvents which can simultaneously extract both proteins and oils at the same time from sunflower seed. In the initial work of Hagenmaier (18), whole seed was used rather than dehulled seed. Seed was cleaned by aspiration and screening. During the next step, seed was comminuted by flaking, then was passed through a disc attrition mill set to grind as fine as possible (70% material through 20 mesh) without overheating the product. Extraction was conducted at a water/flour ratio of 10:1, temperature at 25°C and various pH values, with pH adjusted by using NaOH and/or HCl solutions. The initial separation of solid phase from liquid extract was made by basket centrifuge. The liquid recovered from the basket centrifuge was passed through a clarifying centrifuge to remove remaining solids. Combination of the solids from both centrifugations resulted in a residue which contained 15% total protein and 14% of the total oil.

In an alternative separation process, the mixture was separated into three different phases: solid residue; oil emulsion; and aqueous phase. The oil-water emulsion was then broken by an inversion technique of adding sufficient oil to reduce moisture and agitating the mixture. The dissolved protein in the aqueous phase was precipitated by reducing the pH of the solution to 5.0.

A crucial step in AEP of full-fat sunflower seed was the solid-liquid separation which follows aqueous extraction. The best results were obtained at high pH values where less darkening of the product occurred. When whole sunflower seed with hull particles is used, it produces an unattractive colored product. Optimum extraction was achieved at pH 10 with 85% of the protein dissolved. The protein at this condition is darkened by exposure to basic pH. Despite the use of sodium sulfite, the protein isolate still had a brownish color. The color change of the product and efficiency of oil extraction at pH values of 3–10, were dependent on the type of separator used. They ranged from 79–85% and 27–56% for basket centrifuge and batch centrifuge respectively. For the oilseeds, like sunflower seed, that are basically grown for their oil content, the efficiency levels of 85% were not acceptable. However, the brown colored isolates can be utilized as food ingredients in food applications in which colored protein would not be objectionable.

and only contained 11.5% protein. No lipoxygenase activity was observed in any of the products. Ash and sugar contents of the products were acceptably low, and it is suggested they could be further reduced by increasing the diafiltration time. The products had satisfactorily high L-scale color readings. Lower values of nitrogen solubility of high-fat products appeared to be related to high oil content.

TABLE 4

Summary of UF Membrane Performance on Full-fat and Low-fat Extract from Ground Soybeans and Peanuts (9,10)

Extract type	Mean water-to ratio	Flux (gfd)	% Mean retentions				
			Solids	Nitrogen	NPN	Ash	Sugars
Full-fat (soybean)	30 to 1	30.6	90.1	95.7	39.3	59.6	50.6
	30 to 1	42.6	89.0	96.7	43.2	64.2	47.8
Low-fat (soybean)	25 to 1	29.4	85.7	96.6	48.6	50.6	51.1
	12 to 1	20.0	83.1	96.3	—	—	38.5
Full-fat (peanut)	30 to 1	76.9	96.3	98.2	41.1	74.0	67.1
Low-fat (peanut)	20 to 1	68.6	91.0	97.3	36.8	62.1	72.7

TABLE 5

Analytical Data on Full-fat, Intermediate and Low-fat AEP-MIP Products from Soybeans and Peanuts (9,10)

Product type	Nitrogen total %	NPN %	Protein (N × 6.25) %	Ash %	Total sugars %	Crude fiber %	Oil %	NSI %
Soybean								
Low-fat	11.8	0.3	78.8	4.2	7.5	0.1	1.9	100.0
Intermediate-fat	12.6	0.2	78.9	3.5	4.8	—	9.8	95.9
Full-fat	10.6	0.4	66.0	3.9	5.5	0.1	32.3	100.0
Residue product	3.1	0.3	19.2	3.9	2.3	18.4	12.3	56.3
Peanut								
Low-fat	14.5	0.1	90.9	2.3	6.7	0.1	2.4	93.3
Residue product	2.8	0.9	17.2	5.4	4.1	9.0	36.6	—
Full-fat	5.7	0.1	35.5	1.1	1.6	0.1	60.9	66.9
Residue product	5.6	0.3	35.3	2.1	6.7	0.1	57.4	—

Lawhon et al. (19) prepared a white colored protein isolate by aqueous extraction of full-fat, ground, sunflower kernels. Procedures to inactive polyphenoloxidase enzymes and exclude and expel oxygen during extraction and precipitation of sunflower protein are shown in a simplified flow diagram in Figure 10. Protein isolate was prepared from ground kernels under a nitrogen atmosphere in a stirred glass lined reactor. Oxygen was purged from the extraction water at 53°C and minimum of 23-in Hg vacuum. A water-to-flour ratio of 20 to 1, pH of 9.0 and 30 minutes of reaction time were found to be the optimum conditions for the extraction. After the extraction, nitrogen gas was bled into the reactor with the vacuum line closed to lower the vacuum to around 5-in Hg. The slurry was pumped directly to a three-phase Westfalia separator to separate oil from the aqueous protein mixture which then was acidified to a pH of 4.5 to precipitate the curd. This was washed twice to produce a slurry that was spray-dried to produce white protein isolate.

The essential amino acids were noticeably higher in the OEEP isolate than that made by the conventional process. A possible explanation for the higher recoveries is that the polyphenolic compounds may not have reacted with the protein in the isolate. Nitrogen solubility of the isolate ranged from 2.1 to 37.7 over pH range of 2.0 to 9.0. The aqueous extract from the three phase separation contained 77.9% of the original proteins but only 53.1%

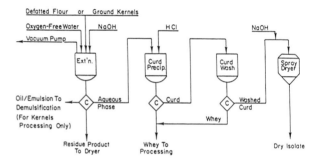

Fig. 10. Simplified flow diagram of the OEEP process (19).

of the original proteins were recovered in the isolate after precipitation. Table 6 shows the composition of the protein product obtained by OEEP-AEP. The protein and oil contents were 71.88% and 18.8% respectively.

Rapeseed: During recent years, rapeseed, especially a cultivar called "canola" has become a commercially important oil seed because of high oil content of the seed, composition of the oil, and the well-balanced amino acid composition of the protein. However, as other oilseeds, its commercial use has been limited by undesirable impurities, in this case by residual sulfur and glucosinolates in the oil and meal respectively. The high sulfur content of the

TABLE 6

Analytical Data on Sunflower Kernels and Isolate Prepared Using a Combination of OEEP and AEP (19)

Total product	Nitrogen %	Nonprotein nitrogen %	Protein (N × 6.25) %	Ash %	Total sugars %	Crude fibers %	Oil %
Kernels	4.0	0.2	24.8	3.9	5.9	2.5	52.5
Isolate	11.5	0.2	71.9	3.6	1.6	0.5	18.8

oil causes a significant increase in usage of catalysts during hydrogenation. Antinutrients, such as glucosinolates, erucic acid and phytate limit use of the meal and isolated proteins in food and feed formulations. The development of new processes for reduction of these undesirable components is as important as plant breeding efforts.

Many methods have been proposed for inactivation or removal of glucosinolates, including detoxification by heat treatment, enzyme deactivation, and aqueous extraction. Because glucosinolates are water soluble, they are easily removed by water leaching processes (20). However, because of the accompanying solubility of rapeseed proteins, water washing procedures are not economically feasible unless the proteins from the wash streams are recovered by membrane technology (21) and can be further purified for use as food ingredients.

The aqueous extraction of rapeseed oil was developed by Embong and Jelen (22,23). The procedure utilized in their laboratory-scale trials is shown in Figure 11. Dry rapeseed was ground in a rotating disc type Quaker City Mill and was then mixed with boiling water. In contrast to other oilseeds, boiling for 5 minutes was necessary for deactivation of the myrosinase enzyme. After cooling it, the pH of the slurry was adjusted to 7.3 with sodium hydroxide. The slurry then was reground using a blender to produce a very fine powder. After blending the slurry, it was stirred in a flask fitted with a condenser and held in hot water. The slurry then was centrifuged and the oil emulsion was separated from the solid residue. This method does not appear to be applicable to pilot plant size trials since the repeated grindings are labor intensive and also increase losses during processing. However the intention was to demonstrate the principles of aqueous extraction and evaluate the oil for improvement in oil quality. The optimum conditions for their experiments were: temperature 70°C; solid/water ratio, 1:3, pH, 6.6 and stirring time, 1 h. Approximately 90% of the oil was extracted that contained reduced amounts of sulfur (4–6 ppm), phospholipids (0.03–0.05%) and free fatty acids (0.2%). The residual oil content of the meal was 4–7%. Rapeseed proteins were more soluble in a combination of alkaline and salt solution, and less soluble in water and salt solution. During the aqueous extraction process, 58 to 79% of the glucosinolates were removed from the meal. The aqueous extract contained 0.5% protein of which only 65% could be precipitated by acid.

Another study by Yehya and Jelen (24) investigated the possibility of using one- and two-step aqueous-alkaline extraction of rapeseed meal for recovery of oil and protein from rapeseed. The first step of the process was enzyme deactivation, and this was followed by blending for 10 minutes. The slurry was mixed for another 10 minutes at the original pH of 5.8, 80°C and a solid-to-water ratio of 1:3.5. The liquid and solid fractions were separated from the mixture by centrifugation. The liquid fraction contained about 52% of the total oil, mostly in free state, and most of the recoverable glucosinolates. The second step consisted of re-extraction and drying of the solid residue. The conditions for this step were a pH of 11, an extraction time of 30 min at 25°C. The solids residue, after drying and solvent extraction, had a protein content of 69%, was light-colored and bland, and contained 15% of the original protein. The yield of extracted oil was very low compared to AEP of other oilseeds.

The first application of membrane processing was by Bockelman et al. (25) who used a single stage ultrafiltration system with a 10,000 molecular weight cut off (MWCO) membrane to remove 93% of the glucosinolates from the extracted rapeseed protein. However, the protein content of the dry product was low due to the large amount of carbohydrates in the retentate fraction.

Siy and Talbot (20) prepared low-phytate content rapeseed protein with ultrafiltration by extracting phytate- and nitrogen- and phosphorous-containing compounds from Tower and Candle variety rapeseed meals and flours using aqueous sodium chloride solution. In a single stage extraction, using a solvent-to-flour ratio of 1/50, 0–10% salt concentration and pH of 3–10, they were able to remove all of the phytate- and phosphorous-containing-compounds.

Diosady et al. (21) used a different approach in applying membrane technology to produce rapeseed proteins (Fig. 12). In this method, protein isolates were prepared from the extract by a two-stage ultrafiltration process. In the first stage, the water extract was ultrafiltered, in the second stage, the retentate was diluted by the second extract and ultrafiltered again. A membrane with a MWCO range of 50,000 gave the best results and removed about 87% of the original glucosinolates, nonprotein nitrogen and nitrogen free material (Table 7). The overall process of water leaching and ultrafiltration produced an isolate and meal containing 0.42 mg/g and < 0.2 mg/g glucosinolates, respectively. The protein contents were 80.4% for the isolate and 41.6% for the meal. The permeate stream contained only 1.3% protein, and 15% of the original solids content was in the waste stream.

Corn Proteins: Corn endosperm protein generally is of limited interest to human and animal nutritionists as a protein source due to its relatively low lysine content and PER values. However, recent improvements in lysine and protein content, and, more importantly, the growth of the corn sweeteners and grain ethanol industries has resulted in availability of appreciable amounts of good-quality corn germ, gluten feed, meal, and dried distillers grains and solubles.

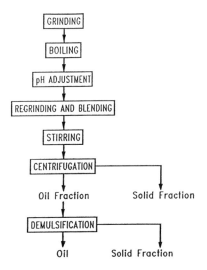

Fig. 11. General flow diagram of the aqueous rapeseed oil extraction (23).

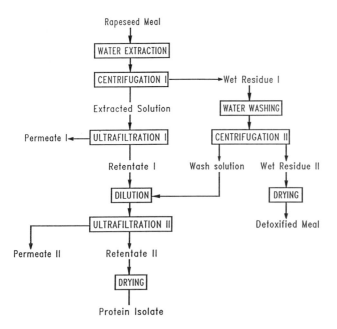

Fig. 12. Flowsheet for the laboratory production of low glucosinolate rapeseed protein concentrate and isolate (21).

fractions. Figure 13 shows the processing scheme during their evaluations. Reverse osmosis of the solubles fraction of stillage from fermentation, using recycled stillage solubles containing high salt, yielded 56–62% of the original volume as permeate. This fraction contained 10–18% of the total solids and 3.2–10% of the nitrogen in the solubles fraction. In later research, Wu and Sexson (28,29) recovered corn proteins from stillage solubles of dry milled fractions of corn, grits, flour, degerminated meal and hominy feeds. The solubles fraction was membrane processed using ultrafiltration and reverse osmosis membranes with MWCOs and 1000 and 200, respectively. After production of alcohol, the stillage was separated into solubles and insoluble fractions by filtration and centrifugation. The solubles fraction contained 0.036 to 0.08% nitrogen and 1.4 to 7.2% total solids. The recoveries of permeate, nitrogen and solids are summarized in Table 8. Permeates from stillage solubles contained 90 to 97% of the original volume, 38 to 65% of the total solids and 23 to 60% of the total nitrogen.

Several pathways exist for recovering protein products from corn (26). The first method is mechanical grinding of the corn by dry milling to produce grits and flours. The germ is removed by sieving and aspiration and/or gravity table, and defatted usually by prepress-solvent extraction. Protein concentrates and isolates are obtained from the defatted meal by milling and air classification, or alkali extraction-acid precipitation process. Dry-milled fractions of corn such as grits, flour, degerminated meal and hominy feeds are different in composition. The second process, wet milling technique, includes presoaking, degermination, defatting of the germ by solvent extraction and separating the starch from the corn gluten. Alcohol is produced from the purified starch by fermentation, and attempts have been made to recover protein from the wet-milling for food uses.

The third method includes grinding and fermentation of the whole corn, followed by recovery of protein from the stillage. Wu et al. (27) reported that ultrafiltration of the "thin" corn stillage (solubles fraction) from fermentation of ground whole corn, contained 1.1% solids yielded 76% of the initial volume as permeate which contained 4.6% of the total solids and 3.2% in the nitrogen in the solubles

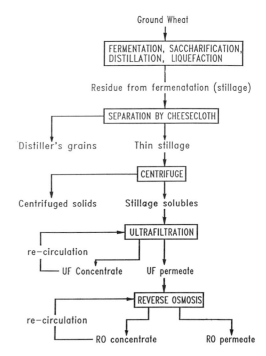

Fig. 13. Schematic diagram of the whole process including the fermentation, fractionation of the stillage, and ultrafiltration and reverse osmosis recovery process (27).

TABLE 7

Permeability of 50,000 MWCO UF Membrane for the Major Components of Defatted Tower Canola Meal (21)

Stage	Total nitrogen %	Nonprotein nitrogen %	Protein nitrogen %	Glucosinolate %	Nitrogen-free matter %
First	13.9	62.4	2.8	66.0	63.1
Second	13.5	76.8	2.7	74.0	75.0
Overall	25.9	88.0	5.7	87.0	88.2

TABLE 8

Ultrafiltration and Reverse Osmosis Permeate Volumes and Composition of Corn and Corn Fractions Stillage Solubles (29)

Stillage solubles membrane[a] (pressure)	Volume %	Nitrogen %	Solids %	Ash %
Corn				
UF	90	37.0	38.0	70.0
RO (5,440 k Pa)	80	0.3	0.2	0.2
RO (1,360 k Pa)[b]	71	18.0	30.0	37.0
Corn grits				
UF	94	60.0	55.0	92.0
RO (5,440 k Pa)	86	0.6	0.3	0.1
RO (1,360 k Pa)[c]	85	12.0	9.2	5.8
Corn Flour				
UF	97	28.0	65.0	68.0
RO (5,440 k Pa)	77	1.3	0.1	0.0
RO (1,360 k Pa)[c]	85	12.0	9.2	5.8
Degerminator meal				
UF	96	26.0	39.0	68.0
RO (5,440 k Pa)	83	0.3	0.4	0.3
RO (1,360 k Pa)[c]	85	12.0	9.2	5.8
Hominy feed				
UF	95	23.0	38.0	62.0
RO (5,440 k Pa)	80	0.5	0.3	0.2
RO (1,360 k Pa)[c]	70	4.6	4.5	3.4

[a] UF = Ultrafiltration; RO = reverse osmosis.
[b] Wu et al. (27).
[c] Wu and Sexson (28).

The reverse osmosis permeate of the ultrafiltration permeate contained 70 to 86% of the original twin stillage volume, 0.1 to 30% of the total solids, and 0.3 to 18% of the total nitrogen. The protein recovery process would be simpler if only reverse osmosis was used. However, the combination of UF and RO processes has potential advantages for fractionating proteins with distinct functional properties.

Food use trials of the protein products obtained from stillage solubles have been reported. Their large scale utilization in the food industry may be limited by flavor, food regulatory agency and esthetic concerns. Thorough cleaning of grains and sanitary handling has been implemented by corn millers but not by grain alcohol distillers, and isolation of the proteins from corn and/or its fractions before fermentation is desirable.

The wet extraction of protein from whole corn or ground degermed endosperm, by ultrafiltration process is described as the "FPRDC" process by Lawhon (30). Germ, separated by dry milling also can be solvent defatted and the meal converted into protein concentrate and isolate by the same process. Figures 14 and 15 illustrate corn protein extraction and recovery processes with degermed corn meal and defatted corn germ.

In the FPRDC process, ground corn is dispersed in a selected solvent, e.g., a dilute alkali or alkali-alcohol solution using about 6 to 15 parts (w) of solvent for each weight of corn. The resulting dispersion is stirred for 75 min at pH of 11.7, and 120°F. Sonication increases the yield by enhancing extraction and reduces extraction time at 130°F. In the succeeding step, the mixture is separated into two fractions by centrifugation. Optionally, the solids phase can be resuspended to extract the remaining proteins and followed by solid-liquid separation. The pH of the solution prior to prefiltration and ultrafiltration is adjusted to pH levels of 7 to 10. Protein is recovered from the combined extract either by conventional acid precipitation or membrane processing methods. After acid precipitation, the pH of the product may be adjusted before spray drying. In membrane processing, the mixture is prefiltered and pumped through a suitable UF membrane with a molecular weight cut off range of 10,000 to 30,000 daltons. The protein in the ultrafiltration retentate is precipitated and spray dried to produce an isolate. The permeate from the ultrafiltration process, or the whey if

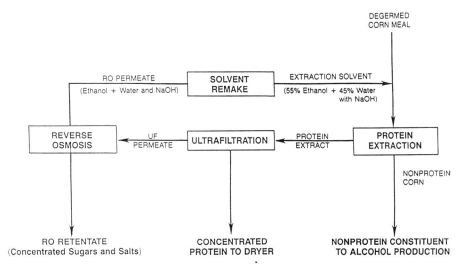

Fig. 14. Corn protein extractions and recovery process (30).

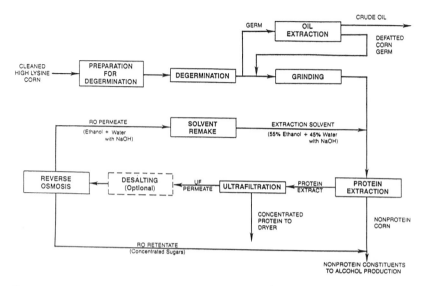

Fig. 15. Corn protein extraction and recovery process with defatted corn germ included (30).

the acid precipitation process is used, is then processed by a RO system to concentrate the sugars, salts and other constituents in the solution. This fraction can be mixed with nonprotein residue from the extraction and utilized in production of alcohol or mixed with other food ingredients. The purified water from RO (permeate) can be reused in subsequent extractions. This reduces water usage and treatment costs considerably.

Recoveries of solids and protein from undegermed corn meal extracts by UF and/or acid precipitation is shown in Table 9. The UF technique recovered 77 to 90% of the extracted protein, but only 40 to 78% of protein was recovered by acid precipitation. Proximate analyses of undegermed and degermed corn meal and its membrane-processed products indicate that a higher percentage of solids and nitrogen was extracted from the undegermed corn meal (Table 10). However, the solids extracted from degermed corn meal were higher in protein content than those from undegermed corn. The high oil content of solids extracted from undegermed corn lowered the protein content of the product resulting in a concentrate (77.6% dry weight basis) compared to 94.10% protein in the isolate.

Lupin: The use of "blue lupin" (*Lupinus mutabiclis*) has been limited by the high content of bitter, poisonous alkaloids which can range from 1–2%. Protein isolates have been made from white lupin (*L. albunus*) and contain only 0.01–0.05% alkaloids. The oil and protein content were 8–13% and 40%, respectively. Solvent extraction of oil from white lupin is not economical. Pilot-plant scale aqueous extractions of sweet and bitter seeds have been reported by Aquilera et al. (31) (Fig. 16).

Sweet lupin was dehulled with a Carver dehuller and

TABLE 9

Recovery of Solids and Protein from Undegermed Corn Meal and Extracts by Ultrafiltration and/or Acid Precipitation (28)

	Ultrafiltration		Acid precipitation	
	Solids in products	Protein in products	Solids in products	Protein in products
Processing	% of original		% of original	
Acid precipitations				
Example # 1	—	—	11.2	57.1
Example # 2	—	—	4.8	33.4
Ultrafiltration				
MWCO–10,000	13.6	66.3	—	—
MWCO–30,000	8.3	63.6	—	—
MWCO–10,000[a]	5.2	53.7	7.2	35.3

[a] Extracted mixture divided into two fractions and one is acid precipitated and the other ultrafiltered.

TABLE 10

Proximate Analyses of Undegermed and Degermed Corn Meal and Their Protein Isolates (30)

Corn products	Moist. %	Oil %	Nitrogen total %	Protein (N × 6.25) %	Ash %	Total sugars %	Color dry
Undegermed							
Meal	11.3	3.3	1.6	9.8	1.3	2.8	—
Isolate	2.6	8.5	12.4	77.6	6.0	7.9	65.0
Degermed							
Meal	12.9	0.9	1.3	8.3	0.5	1.4	—
Isolate	2.8	1.1	15.0	94.1	4.1	3.9	78.9

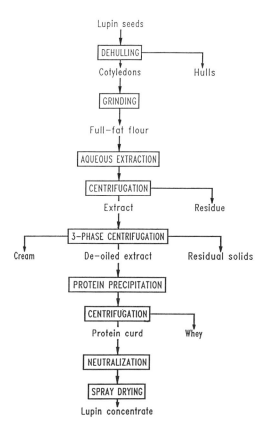

Fig. 16. Flow diagram for aqueous processing of lupins (31).

separator and ground in an Alpine Contraplex pin mill to obtain clean cotyledon flour with 90% passing through a 100 mesh screen. Aqueous extraction was performed at 1:15 (w/w) flour:solvent ratio, a pH range of 8.0–9.0, and a temperature of 65°C. Either a Westfalia separator or a Sharpless centrifugal decanter were used for mechanical separations. The Westfalia, three-phase separator separated the slurry into an oil emulsion, and aqueous and solid phases. Lupin proteins were precipitated from the aqueous phase at pH 4.5, and protein curd was separated by the Westfalia separator. The curd was neutralized to pH 6.8 with NaOH and spray dried to produce protein concentrate.

Bitter lupin seeds were cracked using corrugated rolls, and the hulls were removed by air aspiration. The cotyledons were divided into two lots and one portion was ground as in processing sweet seed. The other portion was flaked to 0.010–0.020 inch thickness. The ground or flaked bitter seed cotyledons were extracted the same way as sweet seed. The only difference was the re-extraction of solid residue and the washing of curd with acidified water.

Recovery of protein from flaked or ground lupin seed ranged from 69 to 75% of the original content of the seed. Only about 71% of the solubilized protein was recovered in the precipitate. The variation in oil extraction was much wider and ranged from a low of 36% to a high of 70% of the initial oil content. The lower pH achieved higher oil recovery but reduced protein yields. Protein concentrates of lupin contain 3.6–9.9% oil, 71.1–77.5% protein and < 0.01–0.002% alkaloids (Table 11). Re-extraction of the solid residue recovered 20% more protein and 33.3% more oil in the final products. The precipitation pH had important effects on recovery of protein curds. A pH change from 4.5 to 4.4 increased the yield by 14%. Washing with water is essential for removal of the alkaloids. Aqueous extraction reduced the alkaloids levels from about 2% to 0.002% in a single step in blue (bitter) lupin isolate, thereby enabling the use of the protein isolate as food ingredients.

Wheat: Wheat is the basic cereal food of the Western world. Wheat flour typically contains about 12% to 13% protein by weight. Gluten accounts for about 90% of the protein, which consists of gliadin 39.09%, glutenin 35.07% and globulin 6.75%. Because of its rubber-like cohesive properties, wheat gluten has been used in bread making. However, there are many opportunities for utilizing wheat proteins in food processing. For example, wheat gluten has been used as a substitute for casein which is derived from milk and relatively expensive. Wheat gluten can be blended with soy and other oilseed flours and protein concentrates and isolates to produce more nutritious products than if used alone. However, wheat gluten produced by some manufacturing processes can have undesirable cereal flavors, components causing color, and excessive salt that limit their applications.

TABLE 11
Analyses of Lupin Seed and Products (31)

Sample	Moist. %	Oil %	Protein %	Crude fiber %	Ash %	Alkaloids %
Sweet lupin						
Whole seed	8.4	11.2	38.0	4.7	4.1	
Hulls	7.6	2.6	2.4	56.0	2.1	
Full-fat flour	7.1	11.0	39.4	2.1	4.5	0.020
Concentrate A	5.3	5.0	79.4	0.1	4.3	<0.010
Concentrate B	7.3	6.1	71.1	0.1	5.5	<0.010
Concentrate C	5.1	3.7	72.0	0.1	4.1	<0.010
Bitter lupin						
Full-fat flour	7.8	19.1	41.4	1.7	3.6	2.000
Concentrate D	6.0	3.6	77.5	0.1	3.8	0.002
Concentrate C	4.1	9.9	73.5	0.1	5.0	0.003

A method developed by Lawhon (32) produces a bland wheat gluten product that is substantially free of cereal and salty tastes. In this process, a clear wheat flour was suspended in a mixture of 35:65% isopropyl alcohol and dilute sodium hydroxide solution at a solvent to flour ratio of 20:1, pH of 12.00 and temperature range of 100–110°F. After 40 minutes of extraction 93% of the flour nitrogen was recovered. Only 75% of the original flour nitrogen was recovered by the UF membrane. The stirred suspension had uniform distribution of solids, and was ultrafiltered through a 30,000 MWCO membrane. A hollow fiber type membrane was preferred because of capabilities to backflush and recover the original feed rates. Diafiltration was used until most of the undesirable cereal flavors, colors and salt were removed. The solution was either freeze dried or spray dried to produce the product. The 65% isopropyl alcohol extracted much less protein from the flour than at the optimum alcohol ratio of 1:1.85. Table 12 shows ultrafiltration membrane performance on clear wheat flour extracts. Extract containing more alcohol had higher flow rates and ash retention in contrast to lower nitrogen, sugars and total solids retention (Table 13).

The composition of wheat flour protein extracts is shown in Table 12. Products made with different solvent mixtures had different compositions. Solvents with higher alcohol contents extracted more oil and ash and less sugar. The product from the solvent with the higher alcohol content had more soluble nitrogen.

Cereal grains such as corn and rice sometimes are used to produce ethyl alcohol by fermentation. A thin

TABLE 12
Composition Protein Products Recovered from Wheat Flour Extractions[a] (32)

Procedure no.	Moist.	Oil	Nitrogen Total	Nitrogen NPN	Protein (N × 6.25)	Ash	Total sugars	Color Wet	Color Dry
65% IPA + NaOH	1.6	24.4	10.1	0.2	63.1	4.4	4.2	46.6	73.2
35% IPA + NaOH	1.2	6.9	11.0	0.3	68.4	3.6	9.2	61.9	72.5

[a] Expressed as %.

TABLE 13
Ultrafiltration Membrane Performance on Clear Wheat Flour Extracts (32)

Parameters	Extraction solvent 65% IPA and NaOH	Extraction solvent 35% IPA and NaOH
Mean flux (gal/ft^2/day)	43.9	29.0
Mean solids retention (%)	47.8	74.2
Mean ash retention (%)	21.6	15.1
Mean total sugars retention (%)	35.3	56.6
Mean nitrogen retention (%)	40.0	75.0

stillage is obtained from the fermentation as a liquid waste stream after distillation of alcohol and contains 5–10% total solids consisting of solid particles and compounds, including proteins. Thin stillage is evaporated to make syrup type products and dried distiller's solubles. Evaporation of the low solids stillage is very costly and the protein is denatured at the evaporation temperatures. An alternative process developed by Wu (33) recovers distillage proteins with a minimal denaturation (Fig. 13). A combination of UF and RO is used to separate the thin stillage into a syrup like concentrate and an aqueous permeate that can be reused. In this process, after fermentation, the entire stillage is centrifuged to remove solid particles. The liquid, which contains soluble proteins, carbohydrates and other small molecular weight compounds is ultrafiltered using a 1,000 MWCO membrane, made either of polysulfone or cellulose acetate. Depending on the type of membrane used, ultrafiltration of stillage solubles from wheat flours recovered 91–98% of the total volume, 10–25% of the total nitrogen, 20–37% of the total solids and 70–80% of the total ash in the permeate fractions (Table 14). The structure and chemical composition of the membrane had an important effect on permeation of the solubles through the membrane. The wheat flour thin stillages passed through a cellulose acetate membrane at 1,360 k Pa permeated considerable amounts of solids (5.8–8.1%), nitrogen (4.0–4.8%) and ash (7.2–9.6%). In contrast, the polyamide membrane retained 99 + % of all solids, nitrogen and ash that was originally present in the feed.

Calculations based on Gregor and Jeffries (34) work show that UF and RO processing of stillage solubles is 67% less expensive than evaporation processes. The method described by Wu et al. (33) is claimed to improve the economics of wheat fermentation for ethanol production by recovering protein that otherwise would be discarded. This process produces food-grade byproducts and also reduces the amount of waste streams.

Jojoba Seed: Commercial interest in jojoba oil and its derivatives has gathered momentum since the ban imposed in 1971 on imported sperm whale products which are critically important in certain lubrication applications. Jojoba oil is a pure straight chain wax of fatty acids and alcohols and contains no glycerol. Jojoba oil and its hydrogenated products are used in cosmetic, pharmaceutical and other applications. Economic and effective separations of jojoba oil are very important to producers because of their high value.

A flow chart for aqueous extraction of jojoba seed is shown in Figure 17. Dehulled seed was broken into coarse particles with peanut butter mill and ground with a Bauer

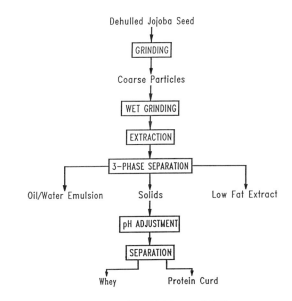

Fig. 17. Aqueous processing of jojoba seed (35).

TABLE 14

Permeate from Ultrafiltration and Reverse Osmosis as Percentage of Original Material (33)[a]

Stillage solubles	Membrane[b] (pressure)	Volume	Nitrogen	Solids	Ash
Hard wheat flour	UF-PS	94	20.00	31.00	80.00
	UF-CA	91	10.00	20.00	71.00
	RO-CA (1,360 k Pa)	90	4.40	8.10	7.20
	RO-PA (5,440 k Pa)	82	0.08	0.23	0.30
Hard wheat	UF-PS	92	25.00	37.00	74.00
	UF-CA	98	20.00	35.00	71.00
	RO-CA (1,360 k Pa)	89	4.80	5.80	9.60
	RO-PA (5,440 k Pa)	85	0.11	0.12	0.15
Soft wheat flour	UF-PS	91	24.00	36.00	70.00
	UF-CA	97	13.00	26.00	76.00
	RO-CA (1,360 k Pa)	77	4.00	6.60	8.60
	RO-PA (5,440 k Pa)	80	0.13	0.15	0.03

[a] Per cent of stillage solubles for ultrafiltration; per cent of ultrafiltration permeate for reverse osmosis.
[b] UF = Ultrafiltration; RO = reverse osmosis; PS = polysulfone membrane; CA = cellulose acetate; and PA = polyamide membrane.

Bros. laboratory grinding mill. Water at 140°F (60°C) was pumped into the downspout of the mill simultaneously with the jojoba seed, at a water-to-seed ratio of 5:1. The resulting slurry was placed in a processing tank, and additional water was added to bring the water:seed ratio to 10:1. The pH of the slurry was adjusted to 9.0, and the extraction was continued at 60°C for 45 min. (35).

After extraction, the slurry was separated into oil emulsion, low fat-extract and residual solids fractions using a Westfalia separator set configured for three-phase separation. Protein was precipitated from the low fat-extract at pH 4.45 with HCl, and the whey and curd were separated by recentrifugation. The protein curd and residual solids fractions from this trial were discarded, and the oil emulsion was saved for demulsification studies.

A combination of aqueous extraction and membrane isolation process was employed in the second method (Fig. 18). Jojoba seed was ground as for the first method. After extraction, the slurry was separated into oil emulsion, low-fat extract, and residual solids by three-phase centrifugation. The residual solids fraction was resuspended in water (3:1 water-to-original seed weight) at pH 9.0 and 60°C for 15 min., then recentrifuged. The low-fat extracts were combined, and ultrafiltered using a Romicon PM30 hollow fiber noncellulosic UF membrane system having 26.5 sq. ft. of 30,000 molecular weight cutoff. The UF retentate and residual solids fractions were adjusted to near neutral pH with HCl and spray dried. The oil emulsion fractions were sampled and combined for demulsification studies.

Analyses of dry products from processing jojoba seed by aqueous extraction and isoelectric precipitation are summarized in Table 15. The spray dried product contained 37.2% oil and 32.4% protein.

Jojoba oil emulsions did not respond to the emulsion-breaking techniques generally used with emulsions from aqueous processing of oilseeds. Among the methods employed, freezing the emulsion followed by thawing proved to be the most effective method.

The flux rate of the permeate from the ultrafiltration of the aqueous extract was 76.5 gfd which is very acceptable. The membrane retained 50.9% of the solids, 59.3% of the nitrogen, 81.1% of the oil and 14.9% of the sugars (Table 16).

Coconuts: Coconut protein, readily available in many tropical countries, currently is not usable as food because of unsanitary processing of the dried copra, and initial processes to recover food-grade protein from fresh coconuts

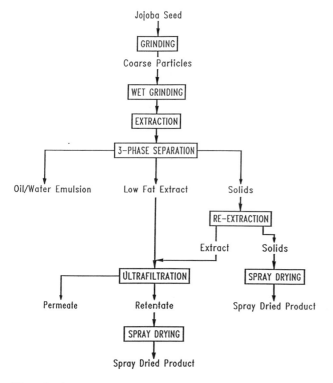

Fig. 18. Aqueous and membrane processing of jojoba seed (35).

TABLE 16

Membrane Performance While Processing Low-fat Extract from Ground Jojoba Seed (35)

Performance measurements	
Mean flux, gfd	76.5
% Mean retention	
Solids	50.9
Nitrogen	59.3
Oil	81.1
Sugars	14.9

TABLE 15

Analytical Data for Dry Products from Processing Macadamia Nut and Jojoba Seed by Aqueous Processing, Membrane Processing and Isoelectric Precipitation (35)

Product	Moist.	Oil	Nitrogen	Protein (N × 6.25)	Total sugars	Ash
			% by weight			
Jojoba seed						
SD residue[a]	3.8	32.1	1.5	9.3	2.1	2.9
SD UF retentate	4.1	37.2	5.2	32.4	8.4	3.7

[a] SD = Spray dried.

were inefficient (36). A pilot plant based on the processing scheme developed at Texas A&M University was built at San Carlos University, Philippines, to demonstrate that coconuts can be efficiently processed by aqueous extraction process. Details of the process and the various products obtained from fresh coconut meat are summarized by Hagenmaier (37,38).

A simplified flow diagram of the process is shown in Figure 19. Husked coconuts are shelled and pared manually. The washed meats are ground by hammer mill and then disk attrition mill. The mixture containing the ground coconuts is mixed with heated (80°C) coconut water and/or tap water, and is pressed at 50 psi in a pulp press with 0.01 dia. screen. Twin-screw presses also were utilized in these studies. The extraction and pressing were repeated several times to increase the total recovery.

TABLE 17
Typical Composition of Dried Products (38)

Products	Moist. %	Fat %	Protein %	Ash %
Coconut water	96.0	0.0	0.1	0.1
Coconut milk	52.0	38.0	3.5	0.9
Coconut cream	18.0	79.0	1.2	—
Coconut flour from prepressed cakes	5.0	14.0	7.6	9.0
Coconut skim milk	85.5	2.0	8.3	13.6
Ultrafiltration Permeate	90.7	0.0	8.3	13.6
Retentate	4.0	3.1	61.0	5.4

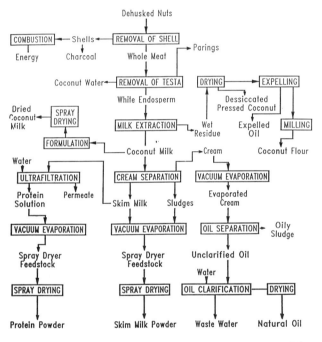

Fig. 19. Flow diagram for aqueous processing of coconuts (38).

The resulting coconut milk is filtered using a vibrating screen to remove the remaining solid residue, which then is dried by drum driers to yield dried prepressed coconut which can be fed into an expeller to recover the residual oil. A dried coconut beverage can be formulated as a mixture of low-dextrose syrup and the dried coconut milk. An alternative method, after adjustment of the pH with sodium hydroxide, the coconut milk can be pasteurized and passed through a three phase Westfalia separator. The lightest phase called "coconut cream," has an oil content of approximately 65%. Coconut oil is recovered from this phase by breaking the emulsion and can be further clarified by centrifugation. The aqueous phase, coconut skim milk, is a solution of soluble coconut components with suspended particles and is combined with the aqueous extract from inversion of the coconut cream. This mixture is spray-dried to produce coconut skim milk powder. Table 17 shows the moisture, protein and oil contents of the principal food products. Their composition and yield can vary depending on the processing conditions used. Food-quality products can be obtained by mixing these ingredients with casein, corn syrup solids and/or maltodextrins.

Coconut skim milk has two protein fractions (38) with molecular weights of 24,000 and 150,000 respectively. Attempts were made to separate the minor constituents of coconut by employing membranes with appropriate pore sizes. Compositions of permeates from these trials are summarized in Table 18. Permeates from six different membranes contained 8 to 10% protein, dry weight basis. There were no differences in protein content of permeates from 10,000 or 30,000 MWCO membranes. The differences in protein contents of permeates from membranes of different manufacturers were believed related to protein concentration layers on the surface of the membrane. Permeate flux was increased by increasing the temperatures from 50°C to 70°C but was reduced by increasing the acidity. The flux was lower when the skim milk was not re-centrifuged. As no prefilters were used before ultrafiltration, these variations also may be a result of the variations in the product from the separator. The permeate was a clear, amber-colored liquid containing about 9.3% total solids. As can be seen from the high ash content (Table 17) the permeate contains more minerals, salts and sugar than the retentate. The permeate had a

TABLE 18
Composition of Coconut Skim-Milk Permeate from Different Membranes (38)

Type membrane	MWCO[a]	Permeate protein (% dry basis)
Romicon PM10	10,000	7.9
Romicon PM30	30,000	8.0
Abcor HFM-100	10,000	7.9
Abcor HFM-100	18,000	9.2
Dorr-Oliver XP-18	18,000	8.4
Dorr-Oliver XP-24	24,000	10.0

[a] Molecular weight cut off ranges, daltons.

pronounced coconut flavor, and the protein product had high protein content and high solubility in water. The spray-dried, ultrafiltered product had less flavor than the skim milk, but retained some of the fresh coconut aroma. Protein digestibility and the protein efficiency ratio were 94% and 2.3 respectively.

Macadamia Nuts: The macadamia nut is a pleasant, delicately flavored confectionery nut. Significant commercial quantities of macadamia nuts are produced in Hawaii, California, the western seaboard, and Australia in the Pacific area. Oil produced from low grade and high oil nuts is used mainly as a replacement for applications that use coconut frying oils.

In determining the feasibility of recovering oil by aqueous extraction, two forms of macadamia nuts were processed: "fines" fraction, which results from conventional macadamia nut processing; and a mixture of macadamia nut culls and fragments.

The first method (Fig. 20) utilized macadamia nut fines which were ground using a peanut butter mill. The resulting material then was dispersed in water (12:1 water-to-nut ratio by weight), heated to 60°C in a steam-jacketed 300 gal. stainless steel processing tank. The ground fines were extracted at pH 9.25 for 45 minutes with slow agitation.

trifuged to separate a second extract from the residual solids. The first and second extracts were combined and passed through the Westfalia separator, this time set for three-phase operation, to separate the oil emulsion from the protein-water fraction (low-fat extract). The pH of the low-fat extract was adjusted to 4.5 with HCl to precipitate protein curd, which then was separated from the whey by centrifugation.

The residual solids fraction and the protein curd were adjusted to approximately pH 6.7 and spray dried separately. The oil emulsion fraction was passed through the separator again to reduce its water content, then saved for demulsification studies.

In the second method (Fig. 21), macadamia nut culls and broken pieces were ground by the same procedure used to process nut fines. The ground material then was extracted once, using a water-to-nuts ratio of 10:1 by weight. The dispersion was adjusted to pH 9.0 with 15% (w/w) sodium hydroxide solution. Extraction was continued for 45 minutes at 60°C and separated into oil emulsion, low-fat extract and residual solids fractions using a Westfalia separator set for three-phase operation. Protein curd was precipitated and separated from the low fat extract, and the curd and residual solids fractions were spray dried separately at neutral pH. The oil emulsion fraction was passed through the separator twice more to increase its oil content by removing water, then saved for demulsification.

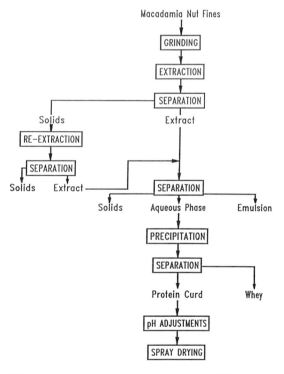

Fig. 20. Aqueous processing of macadamia nuts (35).

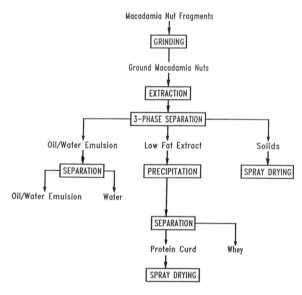

Fig. 21. Aqueous processing of macadamia nuts by three-phase separations (35).

After extraction, the slurry was separated into full-fat extract and solids using a Westfalia continuous separator set for two-phase operation. The solids fraction was resuspended in water (6:1 water-to-original nut weight), and adjusted to pH 9.15 with sodium hydroxide. After holding the slurry for 15 min., the suspension was cen-

The spray-dried residue and protein curd products obtained from AEP of macadamia nuts contained considerably more oil than had been expected, 53.66% and 30.32% respectively (Table 19). The process yielded almost three times more residue than protein "isolate." However, the "isolate" was much higher in protein (43.16%) than the residue product (6.67%).

Using the double-extraction process 80.6% of the original oil was recovered from the ground fines, with single

TABLE 19
Analytical Data for Dry Products from Processing Macadamia Nut Aqueous Processing and Isoelectric Precipitation (35)

Product	Moist.	Oil	Nitrogen	Protein (N × 6.25)	Total sugars	Ash
			% by weight			
Macadamia nuts						
SD[a] residue	2.9	53.7	1.1	6.7	0.5	2.6
SD protein curd	4.1	30.3	6.9	43.2	7.5	6.2

[a] SD = Spray dried.

extraction of the ground macadamia nuts recovered 88% of the original oil. Macadamia nut oil emulsions were very stable and could not be broken by the inversion techniques tried.

Sesame Seed: Sesame seed contains approximately 50% oil which has a desirable mild taste and excellent stability due to natural antioxidants such as sesamol, sesamolin and sesamin (39). Because of its high methionine content, sesame protein is considered to be a good complement to soy proteins.

Sesame seed is an important crop in India, Mexico, China, Venezuela and Ethiopia. Simultaneous recovery of protein and oil from sesame seeds by aqueous extraction process might be an economical process for countries with limited foreign exchange resources which import most of their petroleum.

Chen (39) examined the applicability of the AEP to sesame seed. The best combination of extraction conditions for maximum recovery of oil and protein were a seed:water ratio of 1:3, 80°C extraction temperature, 30 minutes residence time and dispersion at pH 10.5. Protein "isolate" produced by her method contained only 78% protein, and had an oil content of 2.1%. The method developed showed a good prospect for commercial application and requires separation of the oil phase before precipitation. This reduces the oil content of the product, and increases protein yields. The dry products had a light beige color, and high methionine and low lysine and isoleucine contents.

Suleiman (40) examined the use of salts such as NaCl, Na_2SO_3, and $CaCl_2$ for extracting protein from sesame seed. His data show that 4% NaCl can increase the protein yield from 38% to 77%; the phytate and oxalate contents were reduced from 5% and 4% in the flour to 0.1% in the final product, respectively. The process used in this study is shown in Figure 22, and includes a desalting option by UF processing.

The product from pilot plant scale trials, using extraction conditions of pH 9, 45°C, and solubilization $Ca(OH)_2$ in the presence of 4% NaCl, recovered only 55–58% of the original protein, an amount considerably higher than the 37% obtained by conventional isolation techniques. The product from these trials had the following composition: 0.1% phytate, 0.1% oxalate, 0.4% crude free fat, 2.9% total fat, 20.2% residual NaCl and 78.8% protein. The solubility of the protein isolate obtained from this process was lower than commercially available soy protein isolates. Although UF technique was tried, the salt content was not significantly reduced in the protein concentrate.

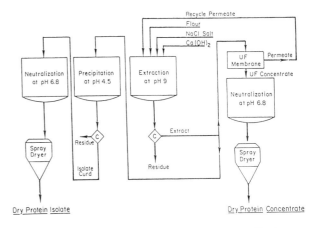

Fig. 22. Sesame seed extraction and membrane purification process (40).

Adsorptive Separation of Proteins

Gel filtration, ion exchange, chromatofocusing, hydrophobic interaction, and affinity chromatography methods have been used in research laboratories to purify proteins. However, applicability of these techniques to large-scale separations is dependent on the desired degree of product purity, yield and commercial value of the product. Most of chromatographic methods are limited by the flow rates, the amount of the solvents required, and high costs of the adsorbents.

Adsorption/Chromatography: Highly purified proteins can be produced at the laboratory scale with ion exchange celluloses and cross-linked polysaccharide. The high costs of molecular sieves or gels have prevented large scale utilization of adsorption/chromatography and gel permeation separations in food processing. The ultimate purification in food processing is not as necessary in foods as in the pharmaceutical industry. However, adsorptive processes in conjunction with membrane processing might offer acceptable throughput rates and products with improved functional properties.

Ion exchange resins have been used in the pharmaceutical industry for years to separate and purify a variety of products. They can be generally applied to demineralization of solutions, concentration of specific components and separation of products with different affinities (41). Proteins can be separated by ion exchanges, but the number of applications is restricted because of the sizes of protein molecules. In recent years, efforts have been concentrated on development of rigid, noncompressible inorganic spheroids with pores large enough to accommodate proteins. These can be manufactured from a wide range of inorganic materials. Silica-based ion exchanges are the best candidates for recovery or purification of proteins, and are incompressible, nonbiodegradable and have substantial capacities.

Figure 23 shows a flow sheet for the "Vistec Process" (42) for isolation of protein by use of ion exchange cellulose. This technique has been perfected for recovery of high purity, undenatured, functional protein from cheese whey. The process consists of a holding tank and a filter bottom stirred tank reactor in which the ion exchange takes place. Processing includes the following steps: The feed solution is fed into the reactor to enable interaction of the protein with ion exchange resins. The reactor is drained, and the resins and adsorbed proteins are washed to remove any remaining impurities. Then proteins are desorbed from the resins and concentrated to produce protein products. The reactor containing the adsorbent is refreshed by washing and reused in the following runs. This process is claimed to be economical and offers good prospects for purifications of vegetable proteins.

Continuous Chromatography: One of the disadvantages of chromatographic methods is its speed and low loads. Numerous attempts have been made to devise and continuously operate chromatographic systems. One such development is described here as its applicability has been demonstrated in various food processes.

In conventional chromatographic operations, feed is introduced into the system and one of the components is retained by the adsorbent, while the other components pass through the column. In the next step, a desorbant solution is added and retained component is desorbed from the column.

The "Sorbex system" developed by Universal Oil Products (43) utilizes a simulated moving bed concept (Fig. 24) and employs a single column with several zones. Automated operation of valves and flow control results in continuous movement of the solvent, feed, the adsorbent zones, and desorbents in one direction. The Sorbex process operates continuously, reduces adsorbent and desorbent requirements, and is seemingly attractive although commercial applications to separating proteins have not been published.

Affinity/Specific Adsorption: This process is based on the specific interaction of a ligand with a protein. This is an attractive method for purifying medium to high value proteins such as enzymes, substrates and antibodies. Its

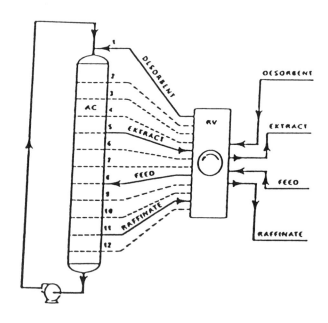

Fig. 24. Flow diagram for sorbex simulated moving bed for adsorptive separation (43).

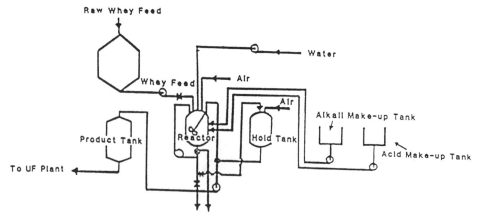

Fig. 23. Vistec protein isolation process using ion exchange cellulose as adsorbents (42).

application to vegetable proteins also is dependent on development of less compressible inorganic supports (1).

References

1. Thomson, A.R., *J. Chem. Tech. Biotechnol. 34B:*190 (1984).
2. Hoare, M., and P. Dunnill, *Ibid. 34B:*199 (1984).
3. Romicon, "Ultrafiltration Handbook," Woburn, MA (1983).
4. Applegate, L.E., *Chem. Eng. 64:* (1984).
5. Haggin, J., *Chem. Eng. News* 7: (1988).
6. Lawhon, J.T., and E.W. Lusas, *Food Technol. 38:*97 (1987).
7. Lusas, E.W., J.T. Lawhon and K.C. Rhee, *Oil Mill Gazetteer 28* (1982).
8. Lusas, E.W., and K.C. Rhee, "Applications of Vegetable Food Proteins in Traditional Foods," in Plant Proteins: Applications, Biological Effects and Chemistry, p. 32, ACS, Washington (1986).
9. Lawhon, J.T., L.J. Manak, K.C. Rhee, K.S. Rhee and E.W. Lusas, *J. Food Sci. 46:*912 (1981).
10. Lawhon, J.T., L.J. Manak, K.C. Rhee and E.W. Lusas, *Ibid. 46:*391 (1981).
11. Lawhon, J.T., U.S. Patent, 4,420,425 (1984).
12. Rhee, K.C., C.M. Cater and K.F. Mattil, *J. Food Sci. 37:*90 (1972).
13. Rhee, K.C., C.M. Cater and K.F. Mattil, *Ibid. 38:*126 (1973).
14. Lawhon, J.T., L.J. Manak and E.W. Lusas, *Ibid. 45:*197 (1980).
15. Goodnight, K.C., G.H. Hartman and R.F. Marquardt, U.S. Patent 3,995,071 (1976).
16. Iacabucci, G.A., D.B. Mayers and K. Okubo, U.S. Patent 3,736,147 (1973).
17. Goodnight, K.C., G.H. Hartman and R.F. Marquardt, U.S. Patent 4,091,120 (1978).
18. Hagenmaier, R.D., *J. Am. Oil Chem. Soc. 51:*470 (1974).
19. Lawhon, J.T., R.W. Glass, L.J. Manak and E.W. Lusas, *Food Technol.* 76 (1982).
20. Siy, R.D., and F.D.F. Talbot, *J. Am. Oil. Chem. Soc. 59:*191 (1982).
21. Diosady, L.L., Y-M. Tzeng and L.J. Rubin, *J. Food Sci. 49:*768 (1984).
22. Embong, M.S.B., and P. Jelen, *J. Am. Oil Chem. Soc. 57:*75 (1980).
23. Embong, M.S.B., Ph.D. Thesis, University of Alberta, Edmonton (1977).
24. Yehya, M.N., M.Sc. Thesis, University of Alberta, Edmonton (1981).
25. Von Bockelmann, I.V., P. Dejmek, G. Eriksson and B. Hallstrom, "Potential Applications in Food Processing" in Reverse Osmosis and Synthetic Membranes. Editor, Sourirajan, S., National Research Council of Canada (1977).
26. Lusas, E.W., K.C. Rhee and J.T. Lawhon, "Recovery and Utilization Potential of Proteins from Corn and Other Cereals in Alcohol Production," Proceedings of International Conference on Ethanol from Biomass, Phillipsburg, Saint Maarten, (1985).
27. Wu, Y.V., K.R. Sexson and J.S. Wall, *Cereal Chem. 60:*248 (1983).
28. Wu, Y.V., K.R. Sexson, *J. Am. Oil Chem. Soc. 62:*92 (1985).
29. Wu, Y.V., *Cereal Chem. 65:*345 (1988).
30. Lawhon, J.T., U.S. Patent 4,624,805 (1986).
31. Aquilera, J.M., M.F. Gerngross and E.W. Lusas, *J. Food Technol. 17:*1345 (1983).
32. Lawhon, J.T., U.S. Patent 4,645,831 (1987).
33. Wu, Y.V., *Cereal Chem. 64:*260 (1987).
34. Gregor, H.P. and T.W. Jeffries, *Ann. N.Y. Acad. Sci. 326:*273 (1979).
35. Lawhon, J.T., K.C. Rhee and E.W. Lusas, Annual Progress Report, Food Protein R&D Center, Texas A&M University, College Station (1985).
36. Rajasekharan, N. and A. Sreenivasan, *J. Food Sci. Tech. 4:*59 (1967).
37. Hagenmaier, R., C.M. Cater and K.F. Mattil, *J. Food Sci. 38:*516 (1973).
38. Hagenmaier, R., "Coconut Aqueous Processing" San Carlos Publications, Cebu City, Philippines (1980).
39. Chen, S-L., M. Sc. Thesis, Texas A&M University, College Station (1976).
40. Suleiman, T.M., Ph.D. Thesis, Texas A&M University, College Station, (1982).
41. Herve, D., Process Biochemistry Jan. 24, 1975.
42. Palmer, D.E., *Ibid.* June 24, 1977.
43. Gembicki, S.A., D.W. Penner and K.U. Johnson, "Continuous Countercurrent Adsorptive Separation of Mono and Diglycerides" presented at 77th American Oil Chemists' Society Meeting, 1986.

Drying of Vegetable Food Proteins

Chan Kwee Chew

Niro Atomizer Food and Dairy Inc., No. 11 Joo Kon Circle, Singapore

Traditional Drying Methods

The traditional approach to the spray drying of vegetable food proteins in the United States has been developed around the soy protein isolate business. The largest concentration of spray drying has been found in this industry and presently the capacity requirement for industrial use has been met by eight to ten existing plants. Most of these plants date back to the mid- and late-1970's and there has been very little development of improvements to this traditional method of drying.

The developments in protein dryers principally have come from the dairy sector where there are some 3,000+ industrial installations around the world. The demand for improved energy efficiency stemming from the oil crisis days brought about significant changes in design, also with the added benefit of improved functionality of powders. The various types of spray dryers presently being used are: tall form drying; conventional wide body; conventional wide body–two-stage with vibrating fluid bed; conventional wide body–with integrated static fluid bed; and multi-stage dryer with vibro-fluidizer. With an ever growing demand for more and more difficult product to be spray dried, the limits of what can be spray dried have been expanded. As a consequence of all of this, we at Niro Atomizer have developed a multi-stage dryer.

Tall Form Drying: As mentioned previously, the traditional approach to the spray drying of soy protein isolate and hydrolyzed vegetable protein in the United States has been by the so-called tall form drying method. This principle can be seen in Figure 1.

The flowsheet in Figure 1 is of the latest plants we have built which require that soy protein isolate be flash-cooled, so that some control over the viscosity can be gained. From the point where the product is flash cooled, it is then passed to a high pressure pump operating in the range of 5,000–8,000 psi. The liquid is then pumped to a spray nozzle device, either a single- or multi-nozzle cluster, located in the dryer neck. The transition from the sprayed liquid to powder occurs at the point where the hot air enters the cylindrical tall form chamber; the inlet air is approximately 450°F. At the point of atomization, powder is formed and the powder and drying air flow concurrently downward in the tall form chamber.

The tall form dryer is characterized by having a ratio of cylindrical height to diameter of 3:1. The residence time within the chamber is relatively short and in the context of spray drying vegetable proteins, this is vital.

The general principle of fast drying around the atomization zone allows the moisture to be driven from these protein products rather rapidly, therefore diminishing chances of undried material securing themselves to the chamber walls, in turn forming deposits. Removing powder and air from the drying chamber is either by way of a U-tube, as indicated at the bottom of the flowsheet, or by a bustle which gives some primary separation of powder and air within the drying chamber itself. Once the volume of air and powder exits the drying chamber it is at that point dried down to its required final moisture, which traditionally in soy protein isolate is approximately 5% residual moisture and in the case of hydrolyzed vegetable proteins, fractionally lower.

The separation of the powder from the air takes place in high efficiency cyclones that separate approximately 90–95% of the powder from the air. The resulting 5%

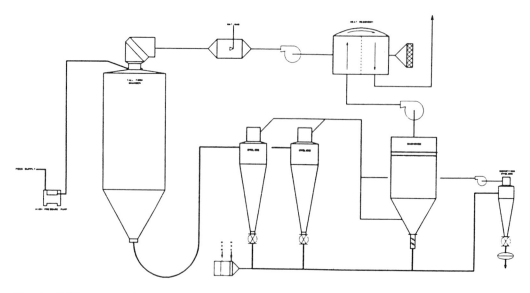

Fig. 1. Tall form spray dryer.

fine material is captured in a baghouse. The drying air is exhausted from the baghouse through an air-to-air heat recovery system.

The powder is collected from the bottom of the cyclones and baghouses, and is conveyed pneumatically to a powder cyclone. The transport air is cooled to remove the remaining heat from the powder.

In the case of hydrolyzed vegetable protein (HVP) products that are extremely hygroscopic and thermoplastic, it is necessary to dehumidify the pneumatic transport air so as not to reintroduce moisture in the product.

In some instances where the vegetable protein products are extremely thermoplastic, it is necessary to introduce conditioned air into the drying chamber in order to form an air curtain to inhibit or to limit the contact between powder and the metallic surface of the dryer.

Furthermore, it should be noted that with soy protein isolate, a baghouse is required because of the low bulk density of the product. However, in the case of HVP this product normally is separated from the final air by a wet scrubber. The numbers indicated in Table 1 give some typical drying temperatures for soya isolate as well as HVP and also indicate energy consumptions relative to pounds of water evaporated.

The length of time and type of hydrolysis has radical influence on the drying characteristics of each type of HVP. After extensive testwork, a graphic representation of the moisture equilibrium is established and a determination of the thermoplasticity and hygroscopicity is established. Upon gaining this information, the design criteria for the plant is then set out. This essentially dimensions the plants in terms of residence time, drying temperature, inlet and outlet temperature, as well as the air flow through the system. In general it could be established that most of the soya based products are thermoplastic and sticky under drying conditions, even when dried to the final moisture content. In such products, two-stage drying, which will be further described below, cannot be applied successfully. These products have to be dried to the final moisture content in the spray drying chamber and sometimes because of their tendency to deposit, the duration of continuous operation is limited.

In conclusion, there has been very little development in the drying of soy protein isolate as the tall form dryer system seems to be optimal. Numerous tests in both pilot scale and commercial scale on the tangential inlet wide body designed dryer (Fig. 2) have not given as good result principally because of the nature of the air flow and longer residence time within the two different drying systems.

Wide Bodies Single Stage Drying: This is another principle of drying that also can be applied to some food protein products. The traditionally wide bodies design is characterized by a cylindrical height to diameter ratio of 1:1. At the bottom of the cylindrical part can be varying angles between 40° and 60°; this again is predetermined by the type of product and the characteristics of that particular to be dried. This system can be seen in Figure 2.

This type of dryer has a tangential air flow which is introduced through spiral shaped distributor housing at the top of the dryer. The product to be dried is pumped to an atomizing unit that could be either a rotating high

TABLE 1

Typical Running Conditions for Soy Protein Isolate and HVP

	Soy protein isolate	HVP
Inlet temperature °F	430–460	460–500
Outlet temperature °F	190–210	210–230
T.S., %	13–17	37–40
Final moisture, %	4–5	2–3
Approx. specific consumption:		
BTU/lb evaporated water	1,400	1,500
BTU/lb final product	8,050	1,335

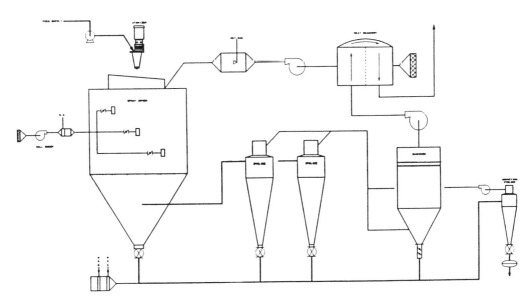

Fig. 2. Spray dryer with pneumatic conveying system.

speed atomizing disc or, as in the case of the tall form dryer, spray nozzles with the pressure being supplied by a high pressure pump. The air flow through the dryer is cyclonic, which invariably sends the semi-dried and dried material to the chamber walls in very much the way a cyclone operates. The exhaust air is removed from the middle of the chamber with the powder predominantly being discharged at the bottom.

The thermoplastic nature of food proteins in general again demands introduction of some curtain of air as can be seen on this flowsheet, to remove the possibilities of powder coming in contact with the metallic surfaces of the drying chamber.

The residence type in this type of dryer is approximately three times as long as the tall form dryer. The separation of the powder and the air is the same as described for the tall form dryer. In this case, the powder is collected at the bottom of the drying chamber and cyclones and is transported pneumatically to a powder collection system. Again, the principle is to dry the powder to its final moisture within the drying system.

It should be noted that this type of dryer is unsuitable for the drying of high viscous materials such as soy protein isolate, because of the difficult atomization and relatively long drying period at the point of atomization, which tends to give a build up of deposits within the chamber.

A traditional method of cooling products can be seen in Figure 3.

However, in the drying of vegetable proteins with low bulk density and the benefit of the long contact time between cooling air and product is negated because of the tendency of the powder to blow itself off the fluid bed screen. However, on products of heavier bulk density, this method of cooling gives a very good efficiency.

The thermoplasticity of the product to be dried will determine if secondary drying utilizing a fluid bed is possible. As mentioned, if the product is thermoplastic at higher moisture, then the product should be dried to its final residual moisture in a one-stage drying operation, and then removed from the drying system.

In both Figures 2 and 3 it has been noted that the final exhausting from the plant of the drying air can either be passed through a wet scrubber as indicated or a baghouse which can be seen on the soy protein isolate (Fig. 1). The commercial value of the products today as well as environmental requirements demand certain emission standards. These standards vary not only state-to-state, but country-to-country. As a function of the efficiency of the operation, it also is necessary to collect the final powder in one form or another as the low bulk density tends to give, even with high efficiency cyclones, a greater than normal carryover in the exhaust air.

Compact Dryer: A further development of the wide body to single stage design mentioned above has been the compact dryer. This principle can be seen in Figure 4.

This spraying dryer chamber has a downward rotary air flow the same as a conventional single stage. The chamber shape is similar to the conventional rotary air dryer, however, the outlet duct draws the air out through the base of the chamber which is surrounded by a conical integrated fluid bed. The compact dryer operates preferably with rotary atomization, but pressure nozzles can be used if the varying powder properties are required. The stationary fluid bed is a concentric ring shape placed at the bottom of the chamber around the outlet duct. There are no parts in the chamber obstructing the rotary air flow. The benefit of this system is that we have integrated a

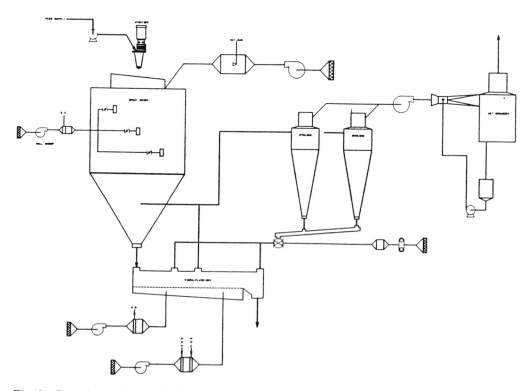

Fig. 3. Spray dryer with vibro-fluidizer.

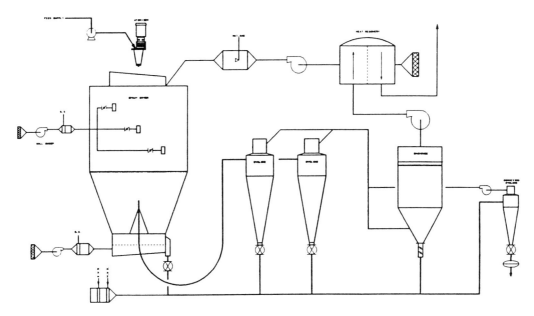

Fig. 4. Compact spray dryer with pneumatic conveying system.

further drying step in the drying chamber, at the same time reducing the overall height of such a drying tower.

New Developments—Spray Dryers with Integrated Fluid Bed

The Multi-Stage Dryer: Economy of production and product quality can be achieved with this type of dryer, because of its ability to handle higher moisture powders in the first stages, at the same time avoiding the problems of sticking of thermoplastic and hygroscopic materials. This, of course, allows an increase in the drying inlet temperature without damaging the product quality. The general principle of this type of dryer was to avoid the contact of wet powder with any metal surfaces in the primary drying stage by transferring it directly into a fluidized powder layer, which is the second stage. To achieve this requirement, a fluid bed is installed directly into the base of the spray drying chamber. This principle can be seen on Figure 5.

The primary drying air enters the chamber via an annular parallel flow air disperser in the center of the chamber roof. The stationary back-mix fluid bed is built into the bottom section of the dryer and fed with secondary drying air. Both primary and secondary air leaves the system via the outlet ducts in the chamber roof. Atomization is achieved by pressure nozzles that are located centrally in the air disperser. The product leaves the system via the rotary valve at the base of the bed. The fine powder that is generated in the system is separated via a scalping cyclone and return to the bed for further agglomeration.

The degree of agglomeration can be controlled to produce a dust-free granular material with somewhat instant properties. The secondary cyclone in the system separates the remaining fines. Final separation can be by baghouse or wet scrubber.

Final drying and cooling is performed in a conventional vibro-fluidizer attached to the spray drying chamber. This may or may not be required depending upon the type of products to be run. The typical running parameters for a

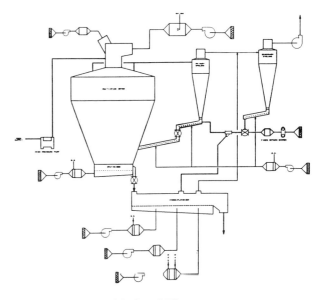

Fig. 5. Multi-stage with vibro-fluidizer.

vegetable product can be seen in Table 2. The advantages are: very good heat economy; low space requirements; no parts disturbing the air flow; no moving parts; no nonmetallic materials in the chamber; produces good agglomerates and an extraordinarily low amount of fines (of a size smaller than 100 micron); good instant properties, that is, good dispersibility and wettability; it is possible to lecithinate directly into the fluid bed at the bottom; it has the ability to handle thermoplastic products that cannot be produced successfully in a conventional dryer; has extraordinarily good flowability even with products with very high fat content; and no recycling of fines to the atomization zone.

Because of its exceptional ability to handle thermoplastic products, this dryer is most suitable for the drying of

TABLE 2

MSD Typical Running Conditions on a Vegetable Protein

Inlet temperature	
First stage, °F	570–580
Second stage, °F	210–230
Third stage, °F	140–200
Feed concentration, %	20
Final moisture, %	5
Final powder temperature, °F	85–90

TABLE 3

Typical Particle Size Distribution for a Protein Product

	MSD	Wide bodies single-stage
% 500 microns	1	0
% 315 microns	14	0
% 250 microns	44	0
% 200 microns	64	0
% 149 microns	89	3
% 90 microns	99	14
% 75 microns	100	32
% 53 microns	100	56
% 20 microns	100	95

fat filled powders and has performed exceptionally well in the production of instant whole milk powders, for example.

Listed in Table 3 is a comparison of particle size distribution gained on a hydrolyzed protein product between a multi-stage dryer and a wide bodied dryer/tall form dryer.

A further development has taken place in the production of soya milk and soya milk powder. This has been achieved also using the multi-stage dryer to give a powder with the necessary functional properties. This soya milk composition principally emulates that of cow's milk. The market potential for this type of product lies in the health food/nutritional market as a product that would be accepted widely by people that have for many generations used soybeans as a source of edible vegetable protein.

Preparation and Uses of Vegetable Food Proteins Made by Dry Processes

Frank W. Sosulski

Department of Crop Science and Plant Ecology, University of Saskatchewan, Saskatoon, Sk., Canada, S7N OWO

Abstract

As an alternative to wet processes, fine grinding and air classification were effective in separation and concentration of protein and starch. The efficiencies of protein and starch recoveries in the fine and coarse fractions from the air classifier were 75–85% and 93–98%, respectively, for field peas. The functional properties of the products generally were proportional to the degrees of protein and starch concentration. For break flour from hard wheats, abrasive milling was more effective than pin milling in reduction of particle size with less starch damage. The fine fraction from air classification of fifth-break flour contained 28–31% protein, 14.0% moisture basis, and functioned satisfactorily as a gluten replacement in bakery formulations. The coarse fraction contained nearly 14% protein and could be blended into patent flours because of its low ash level.

Rapid growth in the consumption of convenience foods has increased the usage of functional ingredients to provide the desired texture, structure and organoleptic properties in formulated foods. Cereal and oilseed flours are economical and widely used sources of functional protein and starch. When problems arise with texture, color, flavor or antinutrients, then concentrates or isolates from these sources come into use. The processes employed in the concentration and extraction of protein and starch normally require large volumes of water, ethanol or hexane. Regulations on pollution and high energy requirements for drying have made the wet processes for seed extraction relatively expensive.

During the past decade, there has been renewed interest in utilizing dry methods for disintegration of seed materials and separation of the components on the basis of their differential densities and mass. This principle has been employed to a limited extent in soft wheat milling where fine grinding to the subcellular level releases sufficient starch and protein to permit air classification into a fine, proteinaceous fraction and a coarser, starchy fraction. Because of the low protein content in soft wheats, the degrees of protein or starch concentration are small, and commercial production of the air-classified fractions is limited.

The full potential of fine grinding and air classification were not realized until the discovery of Youngs (1) that finely ground field peas could be air classified into a coarse fraction which contained essentially all of the starch granules while the majority of the other seed constituents, including protein, were concentrated in the fines. In this case, the concentrated starch and protein fractions exhibited greatly enhanced functional properties and small commercial plants for processing field peas, fababean and common beans have been established in Canada, France and Denmark.

The purpose of this paper is to contrast the relative efficiencies of wet and dry processing of field peas, and to compare the functional properties of the air-classified protein fraction with that of soybean. In addition, data are presented on the air-classified protein concentrates from hard wheat flours which may extend this technology more widely in flour milling.

Air Classification of Peas

The protein contents of grain legumes are comparable to those of most full-fat oilseeds (i.e., 20–30%) and their lysine contents are usually in the range of soybean and animal proteins. Legume flours have not been utilized widely in foods because of the poor functionality of the starch (2). Therefore, fine grinding of dehulled field pea and separation of the lighter protein particles from the denser starch granules (3) serve several purposes. The pea hulls have found a ready market in high-fiber breads (4,5). The protein fraction has enhanced functional properties in potential food applications (6) and as a protein supplement in blends with cereals (7). The starch fraction is utilized almost exclusively for industrial applications such as in paper coatings, adhesives and as a floatation agent in ore refining.

After dehulling the peas between corrugated rolls and aspiration, the fine grinding of pea cotyledons is accomplished by a single pass through a pin mill (Fig. 1). The first cut on the air classifier yields 20–30% of fines (8) depending on the cut point selected on the classifier (9). The remaining protein in the starch fraction can be recov-

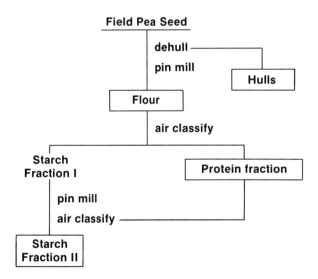

Fig. 1. Flow diagram for pin milling and air classification of field pea into protein and starch fractions.

ered via a second grinding and air classification to obtain another 10% of protein fraction and a low-protein starch fraction. Usually the two protein fractions are combined and represent about one-third of the original pin-milled flour.

In a comparison of dry and wet processing of the same starting material, a pea flour containing 22.2% protein and 55.0% starch, Sosulski and Sosulski (10) obtained 31.8% yield of protein fraction and 61.7% of starch fraction (Table 1). The combined protein fractions contained 52.7% protein (Table 2). Due to the 6.4% residual protein in the large starch fraction the protein recovery in the fines was only 75.5%, somewhat lower than the 82–85% recoveries reported by Tyler et al. (9,11). The coarse and fine fractions contained 83.2 and 8.3% starch, respectively, for a starch recovery of 93.3%, also lower than the 98% separation efficiency of earlier studies (9,11). Much of the crude fat, crude fiber, ash and simple sugars concentrated into the fine fraction with the protein. Therefore, the starch fraction was a much purer product than the protein fraction (Fig. 2).

of these fine particles ($< 15\mu$m diameter) would greatly improve their dispersibility in water. Another problem associated with the air-classified fines was their bitter, beany flavor. The approximately 3.0% lipid in this fraction (Table 2) would create additional flavor problems during storage. Heat inactivation of lipoxygenase in the field pea before processing might control the development of these adverse flavors. The concentration of simple sugars into the fines fraction should be taken into account in formulating foods containing the protein fraction. Dark colored crusts due to Maillard reactions were observed in pea protein-supplemented breads (12). Antinutritional factors and oligosaccharides also concentrated into the protein fraction but the levels were no greater than those of soybean flour (13,14).

Wet Processing of Peas

Vose (15) proposed a general scheme for the wet milling of field peas and production of protein isolate, purified starch and refined fiber. This basic process, as illustrated

TABLE 1

Yield and Processing Efficiency of Dry and Wet Milling of Pea Flour, Dry Basis (10)

Process	Protein product	Starch product	Fiber product	D.M.[a] losses	Recovery of	
					Protein	Starch
	g product/100 g flour				%	%
Dry	31.8	61.7	—	6.5	75.5	93.3
Wet	18.2	43.4	9.7	28.7	72.7	79.2

[a] Dry matter.

TABLE 2

Composition of Proximate Constituents and Carbohydrates in Flour and Dry and Wet Processed Products from Dehulled Field Pea, % Dry Basis (10)

Process and product	Crude protein[a]	Crude fat	Crude fiber	Ash	Starch	Other	Total sugars
Flour	22.2	1.3	1.3	2.8	55.0	17.4	7.0
Dry process							
Protein fraction	52.7	2.9	2.9	5.7	8.3	27.5	12.7
Starch fraction	6.4	0.6	0.5	1.3	83.2	8.0	3.5
Wet-process							
Proteinate	87.7	3.0	0.2	5.8	0	3.3	—
Refined starch	0.4	0.1	0.8	0.2	94.0	4.5	—
Refined fiber	0.8	1.2	47.3[b]	1.7	0	49.0	—

[a] N × 6.25.
[b] Insoluble dietary fiber.

A portion of the differences in protein and starch recoveries shown in Table 1 and those of previous studies was due to the dry matter losses of 6.5% during the dry milling operation. The fines were sticky, low in bulk density and lacked flow properties. Agglomeration

in Fig. 3, was employed to obtain the yield data in Table 1. The recovery of flour proteins in the isolate was 72.7% based on the proteinate yield of 18.2 g/100 g flour and the protein content of 87.7% (Table 2). The purified starch, 43.4 g/100 g flour, was 94.0% starch and 0.4%

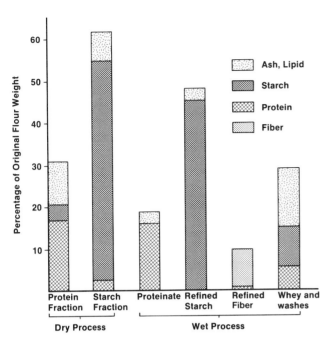

Fig. 2. Yields of products from dry and wet processing of field pea flour and their compositions of crude protein, starch, fiber and other constituents (ash, lipid) (10).

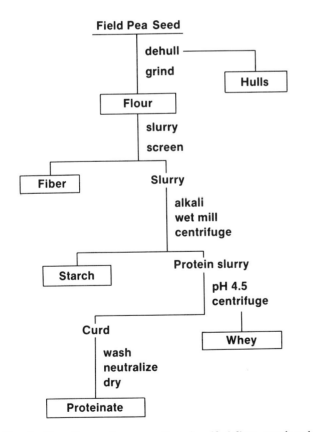

Fig. 3. Flow diagram for preparation of purified fiber, starch and proteinate from field pea by wet processing techniques (10).

protein. Almost 10% yield of a fiber product was recovered but solids losses of 18.6 g/100 g in the whey and 10.1 g/100 g in the starch and fiber washes were relatively high (Fig. 2).

Based on the high solubility of pea proteins at pH 2–3 (Fig. 4), Nickel (16) patented an acid extraction process in which the protein solution was separated from starch on hydrocyclones. The low pH resulted in a lower lipid content in the protein isolate, 1.7% (6), as compared to the alkaline isolation process (Table 2), and reduced the oxidative enzyme activity associated with the development of beany flavors. The high solubility of pea proteins in pea flour at the isoelectric ph of 4.5, nearly 18% (Fig. 4), would account for the high whey losses during protein isolation (Table 1).

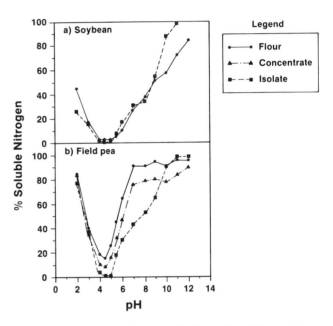

Fig. 4. Nitrogen solubility curves for flours (●—●) protein fraction (▲···▲) and isolates (■---■) of (a) soybean and (b) field pea (6).

Functionality of Pea Protein Products

The nitrogen solubilities of the commercial soybean flour and protein isolate were quite low, between pH 3–7, reflecting the degree of denaturation during processing (Fig. 4). The nitrogen solubility curve for the air-classified protein fraction of field peas was slightly lower than that of the flour. The acid-extracted protein isolate prepared by the Nickel (16) procedure showed low solubility in alkaline pH. Still, at the natural pH of the products in distilled water (pH 6.6), the nitrogen solubility indices for field pea products were generally higher than those of soybean products (Table 3).

Water holding and oil absorption capacities increased in proportion to protein contents of the flours, air-classified protein fraction or protein isolate of field pea and soybean, but oil emulsification properties were uniformly high among all products (Table 3). The protein fraction exhibited excellent whippability and foam stability compared to

TABLE 3

Functional Properties of Flours, Protein Fractions and Protein Isolates from Soybean and Field Pea, Dry Basis[a] (6).

Legume and product	Crude protein[a] (%)	Nitrogen solubility index (%)	Water-holding capacity		Oil Absorption capacity		Oil emulsification capacity (mL oil/0.1 g sample)
			21°C (g H$_2$O/g sample)	70°C	21°C (g oil/g sample)	70°C	
Soybean							
flour	48.2	20.6	1.75	1.85	0.56	0.69	37.2
isolate	82.3	30.6	2.65	3.71	1.03	1.09	45.1
Field pea							
flour	25.0	80.3	0.78	1.02	0.41	0.34	34.6
protein fraction	47.2	65.1	1.09	1.65	0.59	0.72	37.2
isolate	80.3	38.1	2.52	2.97	0.98	0.98	36.6

[a] N × 5.7. Average of duplicate determinations.

the soybean controls and the product was lighter in color than soybean flour and isolate (6). Despite the strong beany flavor of the protein fraction, pea breads were nearly as acceptable as soy breads in sensory evaluations because the adverse flavors were volatilized by moist heat treatment (17).

Air Classification of Wheat Flours

Bran layers of cereals are fused to endosperm tissues and air classification studies have been limited to roller-milled flours. Flour particles are agglomerates of endosperm cells packed with starch granules, the amorphous protein layers between the granules being thin and only concentrated in interstitial regions. In soft wheats, pin millings of roller-milled flours was effective in releasing wedge protein (1–5 micron diameter) and free starch granules (17–30 microns) from the larger cell agglomerates (18). The objective of air classification of these reground soft wheat flours has been to obtain a high yield of the intermediate-sized fraction which is rich in starch and depleted in protein content (19). This starch fraction is of superior quality for cake mixes and other pastry applications but has not served as a substitute for wheat starch in industrial applications because residual protein levels were at least 5%.

Hard wheat flours require more intense grinding to separate the gluten proteins from starch granules. The extensive starch damage created starchy fines which shifted with the numerous small starch granules into the fine fraction and interfered with the functionality of gluten-forming proteins in bread formulations (20,21). Among the hard wheat cultivars investigated by Dick et al. (22), a hard red spring (HRS) variety with known strong gluten characteristics provided a functional protein fraction in bread-making systems. Tipples and Kilborn (23) and, more recently, Dexter et al. (24) demonstrated that the higher water absorption associated with excessive starch damage in HRS wheat resulted in greater bread yields in pan breads. This effect was especially prominent in the short-time bread-making systems being adopted widely in small bakeries.

TABLE 4

Production Rates and Composition of the Break Flours in a Commercial HRS Flour Mill in Percent, 14.0% Moisture Basis (25)

Flour stream	Proportion of total flour	Moisture	Protein[a]	Ash
Pre-Break	0.7	15.1	18.3	0.70
1st-Break	2.6	15.0	16.2	0.49
2nd-Break	7.5	14.7	16.6	0.41
3rd-Break	4.1	14.2	19.2	0.51
4th-Break	1.9	14.2	20.4	0.74
5th-Break	2.1	13.7	20.8	0.84
1st & 2nd-Break cuts	2.5	15.0	16.4	0.42
3rd-Break cuts	0.5	14.7	18.0	0.69
4th-Break cuts	0.6	12.7	17.3	1.00
Break duster	1.1	13.8	19.9	1.16

[a] N × 5.7.

Generally, the patent flours obtained during the roller-milling of HRS wheats are too expensive for use as base flour for protein displacement milling. But clears flours from the break system of the flour mill are least suitable for bread-making and constitute a disposal problem for flour millers unless a gluten-starch extraction plant is associated with the mill. Clears constituted 20–25% of the total flour streams in a typical HRS mill (25) and contained 16.2–20.8% protein and 0.41–1.16% ash, 14.0% moisture basis (Table 4).

The fifth-break flour arises from the outer endosperm of the wheat kernel and is characterized by thick cell walls and uniform, medium-sized starch granules in a thicker protein matrix (19). This clear fraction was selected for experiments on protein displacement milling by Nowakowski et al. (26) because it represented the most difficult fraction to process, yet offered the greatest potential for protein concentration. The flour contained 19.8% protein, 58.6% starch, 1.13% ash, 2.2% lipid and 3.1% dietary fiber on a 14.0% moisture basis.

Particle size distributions of the roller-milled, fifth-break flour (Fig. 5) was determined by image analysis which converted the areas of the irregular particles visualized on a video monitor into spheres of equivalent mass (27). The number of particles was multiplied by their volumes to obtain the percentages of flour volume represented by particles within the size ranges listed in Table 5. Note that 30% of the roller-milled flour particles were in the 0–15 μm diameter size range but, due to their small mass, they represented only 0.3% of the flour volume. Thus, only 10.3% of the roller-milled flour was < 45 μm diameter, and potentially classifiable into starch and protein fractions. Two passes through a pin mill, an Alpine 250 Contraplex CW (Alpine AG, Augsburg, W. Germany), were required to reduce 94.4% of the fifth-break flour to < 45 μm particle size diameter (Fig. 5). During the two passes, starch damage in the flour increased from 10.9 to 20.2% (Table 5). The fifth-break flour was also reground on a Hurricane Pulverizer-Classifier Vertical Attrition Mill (C-E Raymond—A Division of Combustion Engineering Inc., Chicago, IL) and all of the flour was reduced to < 45 μm diameter with less starch damage, 18.3%.

The pin- and attrition-milled flours were passed through an Alpine Microplex 132 MP spiral air classifier at progressively wider vane settings (Fig. 5) to select the optimal cuts for separating protein and starch (25,27). For pin-milled, fifth-break flour, cut points at vane setting 20 and 34 gave two high protein fractions (HPF) containing 28.2 and 21.0% protein in yield of 31.3 and 14.7%, respectively (Table 6). For the attrition-milled flour, it was possible to separate HPF fractions containing 30.7 and 24.8% protein at lower yields of 27.8 and 9.7%, respectively. In both separations, the major fraction was medium in protein content (MPF), 13.8–13.9%, and high in starch 66.0%–66.5%. The attrition-milled flour appeared to provide a greater separation of protein into HPF and more starch into MPF at a lower level of starch damage than was obtained with pin-milled flour.

The HPF fraction with 1.9% ash could be a raw material for gluten extraction or used more widely as an economical replacement for in gluten in bakery goods that require strong dough properties (27). Nowakowski et al. (25) demonstrated that the MPF fraction, due to its low ash level, could be blended successfully into Patent or Bakers' Patent flours. The coarse fraction from this second cut on the classifier (Table 6) was a high protein,

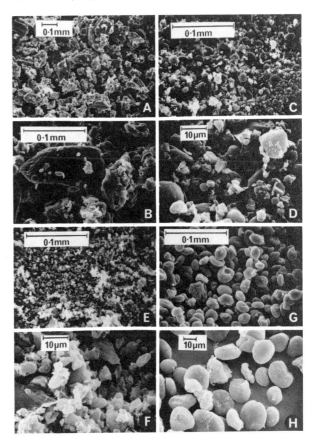

Fig. 5. Scanning electron micrographs of fifth break flour after roller milling (A,B), two passes through the pin mill (C,D), and air classification into fine (E,F) and coarse (G,H) fractions.

TABLE 5

Particle Size Distributions of Roller-, Pin- and Attrition-Milled Fifth Break Flours Based on Volume (27).

Particle size range based on diameter, μm	Roller-milled flour %	Pin-milled flour (2×) %	Attrition-milled flour %
0–15	0.3	21.2	28.1
16–30	3.2	51.6	62.1
31–45	6.8	21.6	9.8
46–60	12.2	2.5	0.0
61–75	16.6	3.1	0.0
76–90	17.2	0.0	0.0
90–105	14.0	0.0	0.0
>105	29.7	0.0	0.0
Starch damage	10.9	20.2	18.3

low ash flour suitable for upgrading low protein patent flours intended for household use because of its uniform, intermediate particle size. Thus, one of the lowest quality flours from the break system could be fully upgraded

TABLE 6

Yield and Composition of Fine and Coarse Fractions from Two-Stage Air Classification of Pin- and Attrition-Milled, Fifth-Break Flour, 14.0% Moisture Basis (25,27)

Vane setting	Fraction yield %	Fraction composition				Starch damage %	Designation of product
		Protein %	Starch %	Ash %	Lipid %		
		Fifth-break flour					
—	100	19.8	58.6	1.13	2.2	10.9	—
		Air-classified fractions from pin-milled flour					
20 fines	31.3	28.2	45.9	1.92	3.8	20.1	HPF
34 fines	54.2	13.8	66.5	0.40	1.4	20.9	MPF
34 coarse	14.7	21.0	59.2	0.40	1.2	17.6	HPF-2
		Air-classified fractions from attrition-milled flour					
12 fines	27.8	30.7	41.3	1.97	4.0	17.8	HPF
34 fines	62.5	13.9	66.0	0.65	1.6	17.2	MPF
34 coarse	9.7	24.8	58.2	0.47	1.4	15.2	HPF-2

into high-value blending flours by dry remilling and air classification. Although the fifth-break flour constituted only 2.1% of the total flour, there were several other high protein streams (Table 4), representing another 20% of the total flour, that could be processed into value-added flour products by this technique.

References

1. Youngs, C.G., "Oilseeds and Pulse Crops in Western Canada—A Symposium"; Harapiak, J.T., Ed., Western Cooperative Fertilizers Ltd., Calgary, AB. 1975, Chap. 27.
2. Hoover, R., and F. Sosulski, *Starch/Stärke 38:*149 (1986).
3. Vose, J.R., M.J., Basterrechea, P.A.J. Gorin, A.J. Finlayson and C.G. Youngs, *Cereal Chem. 53:*928 (1976).
4. Satin, M., B. McKeown and C. Finlay, *Cereal Foods World 23:*676 (1978).
5. Sosulski, F.W., and K.K. Wu, *Cereal Chem. 65:*186 (1988).
6. Sosulski, F.W., and A.R. McCurdy, *J. Food Sci. 52:*1010 (1987).
7. Fleming, S.E., and F.W. Sosulski, *Cereal Chem. 54:*1238 (1977).
8. Sosulski, F.W., and C.G. Youngs, *J. Am. Oil Chem. Soc. 56:*292 (1979).
9. Tyler, R.T., C.G. Youngs and F.W. Sosulski, *Can. Inst. Food Sci. Technol. J. 17:*71 (1984).
10. Sosulski, F.W., and K. Sosulski, "Plant Proteins: Applications, Biological Effects, and Chemistry", Ory, R.L. Ed. The American Chemical Society, Wash. D.C., 1986, p. 176.
11. Tyler, R.T., C.G. Youngs and F.W. Sosulski, *Cereal Chem. 58:*144 (1981).
12. Fleming, S.E., and F.W. Sosulski, *Cereal Chem. 54:*1124 (1977).
13. Elkowicz, K., and F.W. Sosulski, *J. Food Sci. 47:*1301 (1982).
14. Sosulski, F.W., K. Elkowicz and R.D. Reichert, *J. Food Sci. 47:*498 (1982).
15. Vose, J.R., *Cereal Chem. 57:*406 (1980).
16. Nickel, G.B., Canadian Patent, 1, 104, 871 (1981).
17. Sosulski, F.W., and S.E. Fleming, *Bakers Digest, 53*(3) 20 (1979).
18. Jones, C.R., P. Halton and D.J. Stevens, *J. Biochem. Microbiol. Technol. Eng. 1:*77 (1959).
19. Kent, N.L., *Cereal Chem. 42:*125 (1966).
20. Hayashi, M., B.L. D'Appolonia and W.C. Shuey, *Cereal Chem. 53:*525 (1976).
21. Dick, J.W., W.C. Shuey and O.J. Banasik, *Cereal Chem. 54:*246 (1977).
22. Dick, J.W., W.C. Shuey and O.J. Banasik, *Cereal Chem. 56:*480 (1979).
23. Tipples, K.H., and R.H. Kilborn, *Cereal Sci. Today 13:*331 (1968).
24. Dexter, J.E., K.H. Preston, A.R. Tweed, R.H. Kilborn and K.H. Tipples, *Cereal Foods World 30:*514 (1985).
25. Nowakowski, D.M., F.W. Sosulski and R.D. Reichert, *Cereal Chem. 64:*363 (1987).
26. Nowakowski, D.M., F.W. Sosulski and R. Hoover, *Starch/Stärke 38:*253 (1986).
27. Sosulski, F.W., D.M. Nowakowski and R.D. Reichert, *Starch/Stärke 40:*100 (1988).

Novel Soybean Food in China

Shen Zai-Chun

Department of Agricultural Products Processing Engineering, Beijing Agricultural Engineering University, Qinghua Donglu, Beijing, China

Xu Jing-Kuan

ChangChun Power Component Factory, ChangChun City, Jilin Province, China

Jiang Xue-Li

Tong Xian Machinery Factory, Beijing, China

Abstract

Soybean foods are traditional in the Chinese diet. A new process is described for producing a sheet-like food using full-fat and defatted soybean flours. The product is tasty and inexpensive; its tenderness, softness and toughness can be controlled in cooking.

Recently a novel soybean food has been manufactured in China. This vegetable protein food preserves the whole protein and natural ingredients of soybeans. It is called dried soybean milk cream, or "FuPi" for short; another vegetable protein food is called dried soybean milk cream in tight rolls, or "Fushu" for short. These two foods, made from half defatted soybean flour, contain more than 45% protein, 6–8% vegetable fat, 2% nutritious cellulose, vitamin E and other trace elements. This food is soft, tender and tough while eaten, as well as very inexpensive. Hence it has become one of the main modern foods and is deeply welcomed by the consumer at home.

The flow sheet for producing FuPi and Fushu are shown in Figure 1.

After cleaning, dehulling, crushing and oil extraction with oil press, the thickness of the resulting soybean meal (cake) should be not less than 1.5 mm and ground into fine soybean flour with particle size of not less than 80 mesh. Oil extraction rate is about 10–12%. The best way is to adopt one-time extraction. The thickness of the soybean flake is about 3 mm and the oil extraction rate is about 6–8% in the first oil extraction if two times oil extraction is adopted. The thickness of soybean flake of the second oil extraction should be no less than 1.5 mm. Total oil extraction rate is about 8–12%. Soybean flour must be mixed with water and other trace elements in a short time with a batch mixer. A thick, wide, cooked and texturized slice of vegetable protein food is formed. These products are dried and packaged.

The cooking extruder is important equipment. It has been a supplier of short time and high temperature (STHT) extrusion systems for soybean milk cream, in particular, FuPi. The extruder may incorporate the following: feeder with bin discharger, extruder section with single screw, electrical heating element, metering device for pressure, gear reducer, belt conveyors and motor. The extrusion screw is the central portion of an extruder. It accepts the ingredients at the feeding port, conveys, works and forces them through the discharge. The extrusion screw is of increasing root diameter and decreasing pitch screw in a consistent diameter barrel. The screw is divided into three sections: feed section, compression section and metering section. The compression ratio is 46:1. The length (L) to diameter (D) ratio of screw (L/D) is 18:1. The barrel is like a metal tube; the electrical resistance band heaters are used, placed directly around the barrel.

The theoretical and experimental research on the behavior of this extruder has been done on the basis of analysis of movement and dynamics of half defatted soybean flour in this extruder. The principle of the extruder was studied and an extrusion flow model has been established. The results of the experiments are in good agreement with theoretical predictions.

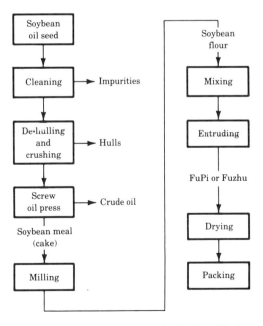

Fig. 1. Flow diagram for producing FuPi and Fushu.

Production and Markets for Soy Protein Industrial Ingredients in Brazil

Wilson L. Canto and Jane M. Turatti

Instituto de Tecnologia de Alimentos, Av. Brasil, 2880, 13073, Campinas-SP, Brasil

Abstract

The purpose of this study was to collect and to make available information regarding the identification of soy protein ingredients in the food industry, their manufacturers, their industrial users, the existent reasons for their use and barriers, and the current markets in Brazil for these ingredients and for the resulting final products. The following soy industrial ingredients are being produced by four companies and used by the Brazilian food industry: pre-cooked full fat-flour, defatted flour, textured flours, powdered soy milk and its residue, concentrates and isolates. Their total production was around 80,000 tons in 1985, destined almost entirely for the domestic market. Their main users are: the industries involved in the school feeding program and the meat processors. The secondary markets are constituted by the bakery, chocolate, candy, ice cream and pharmaceutical industries. The functional properties and low cost were shown to be the promising factors for the successful introduction of soy protein in human feeding in Brazil.

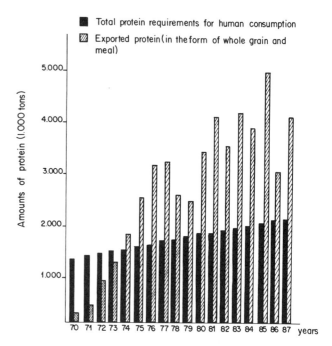

Fig. 1. Total annual protein requirements for human food in Brazil compared to the amount of protein contained in the Brazilian soybean and soy meal exports (1970/87).

Although malnutrition can be found in many places in Brazil, the country is one of the major exporters of high quality protein in the form of soy grain and meal. The protein content in these two goods has represented, for some years, more than twice the quantity required to feed the entire population of the country in terms of protein (Fig. 1) (1,2).

Several national and regional campaigns, as well as a significant amount of official research, have been undertaken in Brazil during the last four decades aimed at increasing the use of soybean products as a source of protein in human feeding. Most of the efforts were focused on the nutritional value of soy, with relatively modest overall results.

Nevertheless, one national industrial segment, which produces some soy protein derivatives, also called soy intermediate products here, has increased very quickly by itself, i.e., by the intrinsic forces operating in the industrial area and in the market, during the last few years. These products have been destined to the food industries which use them as just one of the ingredients in the final products. This growing process resulted not just from nutritional aspects, but from cost considerations and from the functional property contributions of these ingredients to many final products. In this way, the people are constantly consuming more soy protein without realizing the fact, because these final products do not mention their soybean content. Today, the most important of these ingredients produced in Brazil are: defatted soy flour, textured flours, powdered soy milk and its residue, concentrates, isolates and pre-cooked full fat soy flour.

The evidence is growing that the easiest path to success in integrating soy protein into the Brazilian diet may be through the contribution of soy intermediate products. This implies some redefinition in research priorities. Such a decision also requires the existence of information about what is happening in this sector, which previously was unavailable.

The purpose of this study was to collect and make available information regarding the identification of soy protein intermediate producers and their products, industrial users, final products, existent reasons for their use and barriers, and their present domestic markets in Brazil. Information was obtained through a survey in which experts from private and government agencies were interviewed.

Results

Producers

There are four soy protein derivatives (SPD) producers in Brazil: Samrig S.A., factory located in Esteio-RS; Olvebra, factory in Guaíba-RS; Noval, factory in Guaíba-RS; and Nutrimental, through its filial Energe, located in Guaramirim-SC.

Samrig produces defatted soy flour, 30 types of textured flours of different colours and granulation, and five types of isolates and concentrates with distinct functional properties. Samrig is the only industrial producer of isolates in Brazil and its products are marketed under the brand Maxten, Escol, Proteimax, Alipro and Samprosoy

Olvebra produces defatted soy flour, powdered soy milk and its residue, textured flour and others. They are sold under the brand names Provesol and Novomilk.

The products of Noval are: pre-cooked full fat soy flour, powdered soy milk and its residue, whole soy flakes, and others.

Nutrimental produces defatted soy flour, the base of its textured products, sold under the brand name Energe.

An estimation of Brazilian production of SPD was 80,000 tons in 1985, 75% of which was just textured flour, followed by powdered soy milk. The largest producer was Samrig, followed by Olvebra. In 1986 some producers expanded their factory capacities significantly to fulfill the growing demand for their products. Their production has been almost totally directed to domestic markets. A small portion has been exported in the form of powdered soy milk, textured flours, isolates and concentrates. The international markets for these products are very competitive, mainly in the most developed countries, where the multinational company, Purina, is the leader. The main difficulty found by Brazilian exporters is that of meeting the requirement of maintaining an intensive technological assistance structure abroad, close to the clients, which is very costly, and thus only viable for large companies.

Some goods, which are not the main purpose of our study, but which have some importance, are the protein hydrolyzed products used to intensify the flavour potential of final products. They are produced in Brazil from a mixture of defatted, protein-rich meals obtained from soy, groundnut, cotton, wheat germ, corn gluten, etc. Today, the production of these hydrolyzed products is divided between three firms: Nestlé (50%); Laboratórios Griffith do Brasil (40%) and Indústrias Reunidas Jaraguá do Sul (branch Duas Rodas) (10%).

Industrial Users

The most important industrial users of SPD in Brazil are the producers of enriched foods destined for institutional markets, and the meat industry. Of less importance are the bakery, pharmaceutical, chocolate, candy and ice cream industries.

The main institutional market is represented by the school feeding program, coordinated by the Fundação de Assistência ao Estudante—FAE (Student Assistance Foundation) operating under the Brazilian Ministry of Education and Culture. The FAE program is large and has increased during the last decade, including 32.5 million children in 1987 (Fig. 2). If one considers the number of working days in schools as being 180 days per year, and that one meal is served per student per working day, one can estimate that around 5.8 billion meals were consumed under this program in 1987. The final products bought consist mostly of dried, instant food.

Another important institutional market is constituted by the Programa de Complementação Alimentar-PCA (Food Complementary Program) working under the Legião Brasileira de Assistência—LBA. The LBA program is destined for children of up to 3 years of age, wet-nurses and pregnant women of low income.

Among the important institutional food suppliers of these programs are: Nutrimental, the largest, with a share of 25% of the market, followed by Nutricia (from Rio de Janeiro-RJ), Bel Prato (Barra do Piraí-RJ), Prátika

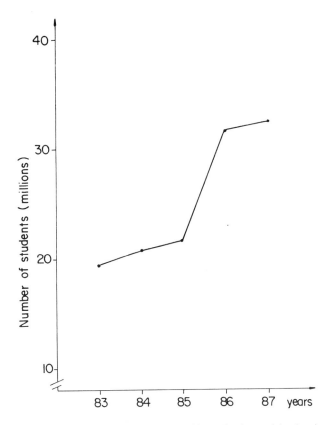

Fig. 2. Number of students at Brazilian schools participating in FAE school feeding program (1983/87).

(Taquara-RS), Liotécnica (São Paulo-SP), Olvebra, Protisa (Curitiba-PR), Bhering (Rio de Janeiro), and others. These firms sell to other less important institutional markets such as the restaurants located in large industrial and commercial establishments. The Swift company sells some meat products to the school lunch market, such as meat balls and sausages which contain soy protein derivatives.

The FAE institutional program is centralized in Brasilia and covers the whole country except the States of São Paulo and Rio de Janeiro, where the buying, storage and distribution of food are performed, respectively, at the municipal and schools level. FAE has promoted, in the last few years, a decentralization of its operations to the municipal level in around 120 cities. The basic idea was to serve locally produced foods in the schools of these cities, more fitted to the consumption habits of each region.

It was initially hoped that with this change the menu would eventually consist mainly of basic foods and even some in the fresh form. However, this objective has been shown to be difficult to attain since the schools are not, in general, sufficiently well equipped to deal with basic food, being obliged to work with the easy-to-prepare instant foods. This is an important point to be considered, since the available products for use in the school feeding program containing soy protein in their formulation, currently belong to the group of dried instant foods. These products show the additional advantages of being less perishable, light, of high protein content, easy to store and require lower transportation costs.

The meat industry is another considerably large market for SPD. Almost all processing firms operating in this sector in the country are using these ingredients in several sausage type products as well as in hamburgers, meat balls, etc. Among the largest industrial producers of meat products in which soy intermediate products could enter are: Frigobrás (Sadia), Perdigão, Swift, Bordon, Comabra (Wilson), Chapecó, Eder, Seara, Aurora, Ideal and many others. The Brazilian legislation permits, today, the use of soy textured flour in meat products up to 7.5% (dry basis), or up to 22.5% in the hydrated form.

Data on the total Brazilian industrial meat production are still not available. What could be found were figures on the production occurring in establishments under federal government inspection, located in São Paul State. Even in this state it is thought that a significant amount of meat industrial production is not under federal control. The total observed production of important meat products in which SPD could enter in those industries inspected in São Paulo State was 106.3 thousand tons in 1987.

During the economic crisis called the "cruzado crisis" which occurred in Brazil in 1986, the industry faced serious fresh meat supply difficulties, having to learn and master the use of soy intermediate products as a partial substitute for meat. Today, practically all the firms in this industry dominate the technology related to the use of SPD and use them in their final products.

In the bakery industry some large companies are using SPD in industrial breads such as the sandwich loaf; in hamburger rolls; in crackers; in biscuits; in Swiss-rolls to facilitate rolling operation; in cookies to improve crispness and color; in cakes as milk substitute (powdered soy milk); and in all kind of crackers, biscuits, macaroni, breads, etc., as well as in large proportions in the final products destined for child feeding programs. A significant barrier confronted by the bakery industry is that the most popular bread is the locally produced French bread, in whose composition SPD does not enter.

Up to the beginning of 1988 the price of wheat flour was much lower than that of soy derivatives, due to the government subsidies to wheat. Today, with no subsidies the price of wheat flour has risen significantly, and is now close to that of defatted soy flour. If the price of SPD declines a bit, this should improve the possibilities for its use in bakery, specifically in those cases in which its use would really improve the quality of the final products.

Some SPD products are being used in the chocolate industry, such as powdered soy milk, and including the full fat soy flour used as a source of fat. The firm Bhering is testing a chocolate whose crispness characteristic is provided by a soy product. Soy protein enriched instant food beverages, chocolate based, are widely used in the national school feeding program.

A survey carried out among the industrial users showed that powdered soy milk is being used as a milk substitute in candies and ice creams, although still on a small scale.

The use of SPD openly in the dairy industry will occur only in expressive amounts when and if the current very restrictive legislation changes significantly. If this occurs some day, it may well open a large market for SPD in Brazil.

The industry of pharmaceutical and dietetic products based on SPD is rather small. The prices of the final products are high and the market is still little developed. The available products are: Nutrogast and Isolac produced by Nutrícia and used in hospitals; Sobee from Laborterápica Bristol obtained from full fat soy flour by re-processing it to obtain a soluble fraction of a pharmaceutical quality level; Novo Milk from Olvebra, based on powdered soy milk; and Vealit from Veafarm. The firm Abott, which leads the production of pharmaceutical and dietetic products in some countries, is studying the feasibility of producing a soy protein product in Brazil.

Discussion

The necessary technology to produce the most important SPD for human consumption such as full-fat soy flour, defatted soy flour, textured, concentrates, isolates, soy milk, etc., is available in Brazil.

The growth of the institutional market constituted by the schools, and by other similar programs depends, in the last instance, on government political and social tendencies with respect to these programs at federal, state and city levels, which means dependence on the amount of official resources allocated to this purpose. The trend of progress presented by this market in the last decade probably will continue in the future, although some short term variations could occur. The effects of the recent administrative decentralization that has taken place in the execution of these programs reduced to a certain extent the participation of final products containing SPD in the school menus. It is still too early to visualize the extent of the effects of such process in the future.

The use of SPD in the meat industry is already well established, and its industrial consumption levels will depend directly on the fluctuations which may occur in the consumer market of meat products.

The meat industry, as well as the suppliers of the school feeding program have a reasonably clear understanding of the functional properties of SPD. The perception level of these properties in the bakery industry is still rather low and the same is true in the chocolate, ice cream and candy industries, although some of these firms are already observing benefits in using SPD in their final products in terms of quality improvement as well as costs. The expansion of these markets requires an intensive extension activity in the chocolate, ice cream and candy industries.

The development of the market for pharmaceutical and dietetic products depends on the introduction of a diversified line of products to be used in particular clinical situations; the quality level of those products would have to be high. The investment required for this would, consequently, also be high. Currently, this industry is reluctant to promote new investments. One apparently interesting alternative would be to use the services and capabilities of the current SPD producers existent in the country by the pharmaceutical and dietetic industries, with the former supplying SPD products to the latter under contract according to required technical specifications.

It is known that the adult absorption capacity of nutrients reduces with advance of age. The availability of food with a high protein content and presenting easy absorption by the intestinal tract could provide a meaningful contribution to the improvement of the nutritional condition of the aged. This group of people comprises a large but still unaided and almost forgotten market in Brazil. It is foreseen that the population of people over 60 years old will reach 10.6 million in 1990 (3). The purchasing power of this group will soon rise, as a consequence of the new Brazilian Constitution promulgated in October 1988, which raised the pensions of the retired to the levels of the corresponding labor force. SPD may have an important role to be performed in this case.

The possible applications of SPD in the food industry have still been studied little in Brazil, although SPD are currently contributing in a significant and increasing way toward the use of soy protein in human feeding, a target extensively sought for many years by many soybean researchers in the country. The findings of this paper indicate that it would be a rewardable endeavour to start a redefinition in the direction of soybean research in Brazil, emphasizing the uses of SPD.

Acknowledgments

The authors are indebted to FAPESP (S. Paulo, Brazil) for their financial support (Grant Eng. Al. 86/1073–2) given during this research. We acknowledge CNPq for the fellowship (nr. 302411-82-8) given to one of the authors of this work.

References

1. Canto, W.L., and L.C. Santos, *Informe Agropecuário, 8 (94)*:62 (1982).
2. Canto, W.L., V.A. Moretti, J. Gasparino Filho, L.A.S.B. Almeida, and L.C. Santos, "Leite" de Soja Líquido: Uma Opção Alimentar," Estudos Econômicos Séries 13, edited by ITAL, Campinas, S.P., 1982, p. 27.
3. I.B.G.E. Anuário Estatístico do Brasil, 47:53 (1986).

Sunflower Proteins in Dairy Products

E. Baldi, L. Lencioni, A. M. Pisanelli, R. Fiorentini and C. Galoppini

Istituto Industrie Agrarie, Università degli Studi, Via S. Michele, 4-56100 Pisa, Italy.

Abstract

Defatted sunflower meal is a protein source suitable for human consumption because of its nutritional/functional properties and its addition to traditional and widely used foods, such as dairy products, should be of great interest. With this thought in mind, three defatted meals (whole from hexane and dehulled from hexane and ethanol extraction) were tested in large scale cheese making, using sheep milk added with 1% of meal. The material and nitrogen balances, the evolution of dry and nitrogen matter and firmness during cheese ripening were considered, together with a panel test to evaluate the acceptability of the final products.

Three defatted sunflower meals (whole from hexane and dehulled from hexane and ethanol extraction) were tested in large scale cheese-making, using sheep's milk with 1% added meal. The addition of sunflower meal to cheese produced an increase in protein content of 1.1–2.6% and an average firmness in the ripened product extended to double that of the control. The influence of sunflower meal addition on cheese proteolysis seems to depend on the degree of protein denaturation. In particular, the cheese extended with dehulled hexane-extracted meal showed the highest values of total soluble nitrogen and unprecipitable soluble nitrogen. The acceptability of the extended products was generally good and in the case of the cheese enriched with dehulled and ethanol-extracted meal, it was close to the control.

Defatted sunflower meal is a potential protein source for human consumption, because of its nutritional and functional properties. As a part of the research on vegetable protein technology carried out at this Institute (1–4), we have tested the possibility of adding sunflower defatted meal to traditional and widely used foods, such as dairy products. Three sunflower protein meals: whole, hexane-extracted of commercial origin (WHM), dehulled hexane-extracted (DHM) and dehulled ethanol-extracted (DEM), were tested in large-scale cheese making using sheep milk with 1% added meal. This paper reports the evolution of dry matter (DM), firmness, pH, total titratable acidity (TTA), total soluble nitrogen (TSN) and unprecipitable soluble nitrogen (USN) during cheese ripening, together with the acceptability of the final products.

Materials and Methods

Control and extended cheeses were produced in a commercial dairy according to the protocol reported in Table 1.

The bacterial starter was prepared from pasteurized whey inoculated with 1% of a commercial culture containing *Streptococcus lactis, S. termophilus* and *Lactobacillus bulgaricus*. After inoculation, the whey was incubated at 40°C for 12 hours. The pasteurized milk was rapidly cooled to 36°C and added with the bacterial starter and the protein meal.

Milk was sampled before the lactic starter addition and control and extended cheeses (two cheeses at a time)

TABLE 1

Experimental Protocol for the Manufacture and Ripening of Control and Extended Cheeses

Parameter	Value
Milk centrifugation ($\times g$)	2500
Centrifugation temperature (°C)	40
Pasteurization temperature (°C)	70–72
Pasteurization time (min)	1
Bacterial starter added (% V/V)	1.2
Curdling volume (l)	100
Sunflower meal added in the extended trials (kg)	1
Rennet titre	1/100000
Rennet added (g)	4
Curdling pH	6.5
Curdling temperature (°C)	36
Curdling time (min)	30
Curdling cutting time (min)	5
Curd pH	6.5
Molding and pressing	by hand
Fresh cheese average diameter (cm)	16
Fresh cheese height (cm)	10
Salting (hours after curdling)	36
Ripening time (days)	30
Ripening temperature (°C)	8–10
Ripening relative humidity (%)	88–90

were sampled at 6 hours and 3, 5, 8, 12, 16, 21, 26 and 31 days of ripening.

The chemical composition (protein, lipid, ash, fiber and N-free extracts) of milk, sunflower meals and cheeses and TTA of cheeses were determined according to AOAC standard methods (5). TSN and USN were determined as described by Calzolari et al. (6).

The pH determination was done directly on the cheese body using a needle electrode (Ingold, mod. 406 M3) equipped with a 3-mm-diameter plunger.

The development of firmness during ripening was followed by a hand-held pressure tester (Effegi, mod. FT 011, full scale 5 kg) equipped with an 8-mm-diameter plunger.

The sensory acceptability of the products was evaluated by 20 trained panelists according to the scoring testing system reported by Amerine et al. (7).

Results and Discussion

The chemical composition of raw materials and ripened cheeses is reported in Table 2. The addition of sunflower meal to cheese produces an increase in protein content of 1.1–2.6%. Concerning the changes in dry matter, no relevant differences were found, even if at full ripening

TABLE 2

Dry Matter (DM) and Chemical Composition (% DM) of Raw Materials and Ripened Cheeses

	DM %	Protein (N × 6.38) %	Lipids %	Ash %	Fiber %	N-free extracts %
Raw materials						
Milk	15.4	29.7	40.5	5.2	—	24.6
DHM	100.0	59.6	3.6	9.3	5.6	21.9
WHM	100.0	34.1	4.6	6.7	17.8	36.8
DEM	100.0	63.8	7.1	9.1	5.6	14.4
Cheeses						
Control	53.6	38.1	52.8	6.6	—	2.5
DHM-enriched	56.7	39.2	49.8	6.1	trace	4.9
WHM-enriched	55.2	39.2	47.3	6.5	1.7	5.3
DEM-enriched	55.6	40.7	47.5	6.7	trace	5.1

the dry matter of the control was 1.6–3.1% lower than that of the extended cheeses.

The development of firmness during ripening is reported in Figure 1. During the first days the extended cheeses showed a rapid increase of inside firmness, noticeably higher than the control. At full ripening the average inside firmness of the extended cheeses was double with respect to the control and the WHM-enriched cheese was the most firm.

Figures 2 and 3 report respectively pH and TTA development of cheeses during ripening. Results show a particularly low pH of DHM-enriched samples as confirmed by the high level of TTA.

Concerning TSN and USN, the curves reported in Figure 4 evidence the particularly high values of the two parameters for the DHM-enriched cheese. Considering that this cheese showed a high proteolysis degree just after curdling (6 hours) and that DHM is the less denatured meal, it can be hypothesized that hydrolysis is enhanced by sunflower proteolytic enzymes. TSN and USN of control cheese show a time-evolution very close to that reported by Lencioni et al. (2).

The panel test, carried out on the basis of organoleptic properties (color, flavor, taste and texture), showed different acceptabilities of the final products. In particular the DEM-enriched sample showed the highest acceptability, close to that of the control. DHM-enriched cheese showed, on the contrary, the worse acceptability and a marked bitter taste, indirectly confirming the high proteolysis occurred. The WHM-extended cheese, finally, was rated just acceptable, mainly for its unusual grey color.

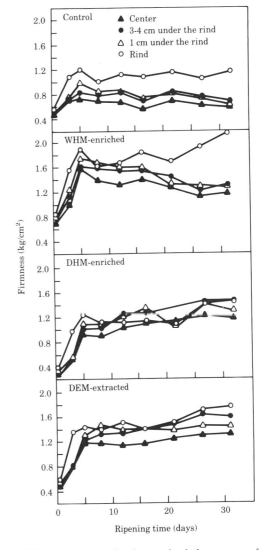

Fig. 1. Firmness in control and extendend cheeses as a function of ripening time.

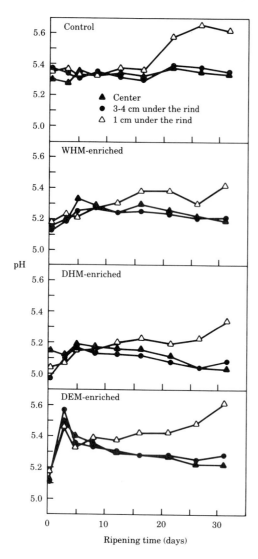

Fig. 2. Development of pH during ripening.

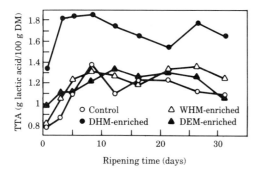

Fig. 3. Development of TTA during cheese ripening.

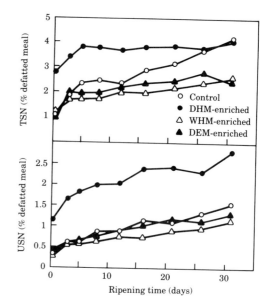

Fig. 4. TSN and USN evolution during cheese ripening.

References

1. Lencioni L., A.M. Pisanelli, E. Baldi, R. Fiorentini and C. Galoppini, *Lebensm. -Wiss. u. -Technol. 20*:227 (1987).
2. Lencioni L., A.M. Pisanelli, E. Baldi, R. Fiorentini and C. Galoppini, *Ibid.* in press (1988).
3. Baldi E., L. Lencioni, R. Fiorentini and C. Galoppini, *Congresso su Stato Attuale e Prospettive delle Colture Oleaginose in Italia*, Feb. 24–26, Pisa, Italy (1988).
4. Baldi E., L. Lencioni, A.M. Pisanelli, R. Fiorentini, *12th Int. Sunflower Conference.*, July 25–29, Novi Sad, Yugoslavia (1988).
5. AOAC Official Methods of Analysis, 14th Ed. Arlington, VA (1984).
6. Calzolari G., G. Bruni and B. Stancher, *Rassegna Chimica,2*:71 (1980).
7. Amerine M.A., R.M. Pangborn and E.B. Roessler, *Food Sci. and Techn.*, p. 354, Academic Press, N.Y. (1965).

Leaf Protein in Human Diet: Potential and Perspectives

C. Galoppini, R. Fiorentini and F. Favati

Istituto di Industrie Agrarie dell' Università, Via S. Michele, 4-56100 Pisa, Italy.

Abstract

The nature of leaf proteins and the principles of the wet fractionation of green crops are reviewed, together with the criteria for the preparation of edible leaf protein concentrates (LPC). The essential amino acid pattern and good digestibility of leaf proteins make LPC of high biological value. The commercial value of LPC is strictly related to the functional properties. Good hydrophilic and surface properties are revealed when the LPC is suitably prepared. This makes leaf proteins particularly useful for incorporation in bakery products, vegetable soups and meat-like products.

Contrary to the traditional idea that leaf proteins are only useful for animal feeding, they are also of high nutritional value for man. The major part of proteins found in leaves are enzymes involved in photosynthesis and other biological activities. This means that leaf proteins have a uniform amino acid composition and that all leaves can be considered as a potential source for the preparation of edible leaf protein concentrates (LPC).

Much research has been carried out in the last 30 years on the production techniques and nutritional properties of LPC, with more than 1,000 bibliographical references (1–7). This scientific effort is justified when one considers that leaves represent the world's largest source of proteins, and that the production of protein per hectare for some of the most widespread forage crops, e.g., alfalfa, is much greater than that for the seeds of the most important cereals and legumes.

Digestibility and safety of leaf proteins can be severely affected by the high fiber level of the green vegetation, and by its content of other antinutritional factors. Therefore, to be used as food in the human field, leaf proteins must first be extracted and purified.

Technological Aspects

The extraction of leaf protiens to obtain food-grade LPC is possible by wet, green crop fractionation. This technology basically consists in a mechanical operation of grinding and pressing which separates the green material into two fractions: a protein juice and an extracted residue (Fig. 1). When produced by partial or exhaustive extraction, the juice shows a high dry matter content, usually more than 10%.

The proteins found in the juice (30–40% of dry matter) are made up of two quite different fractions. The first, called chloroplastic, consists of insoluble lipoproteins which can be easily destabilized and, once isolated, have a dark green color and strong grassy flavor. The second fraction partially originates in the cytoplasm and for this reason is called cytoplasmatic. It consists of relatively stable soluble proteins that when recovered are tasteless, odorless and white or cream in color.

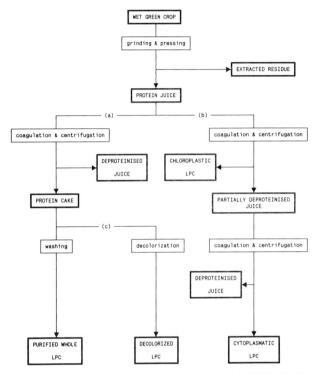

Fig. 1. General outline for the production of edible LPC. (a), (b) and (c) represent three different technological solutions.

Although there are several different methods to recover the juice proteins (heat coagulation, acidification, addition of polyelectrolytes or polar solvents, ultracentrifugation, anaerobic fermentation, etc.) (8–11), the processes developed industrially for the production of feed-grade LPC use heat coagulation by means of steam injection.

This technology is fairly standardized and involves the heat coagulation at 85–90°C of the proteins contained in the press juice of alfalfa, by far the most commonly used plant. With this process the cytoplasmatic and the chloroplastic proteins are recovered together (whole LPC) and can be consumed directly by man after being washed (purified whole LPC). The purified, whole LPC is the only product that has been tested widely in humans (12), but has a number of disadvantages, above all the organoleptic characteristics which severely limit its acceptability. The drawbacks can be partially overcome by removing the pigments and other undesirable substances (decolorized LPC).

The color of the LPC is due primarily to plant pigments and, secondarily, to tannic substances and to the products from the Maillard reaction. All these substances can be removed to a large extent using suitable polar solvents that, moreover, may significantly lessen the grassy taste (13). Further, as a result of the lipid removal, solvent extraction has the beneficial effect of increasing the protein level and the shelf life of the product. Although decolorized, whole LPC shows good acceptability, it has not been produced industrially, principally because of the high operational costs.

A fractionated recovery of chloroplastic and cytoplasmatic proteins can be achieved by exploiting their different physico-chemical properties. The chloroplastic fraction in suspension can be directly removed by ultracentrifugation, but at present there are no industrial devices capable of generating high centrifugal forces at reasonable prices. Consequently the chloroplastic proteins need to be destabilized to enable their removal and this can be achieved by heating them at ca. 60°C or using other methods (9,11,14). After the chloroplastic material (chloroplastic LPC) has been removed, the soluble proteins (cytoplasmatic LPC) can be recovered in several ways, e.g., acid or heat coagulation, ultrafiltration, etc.

Chloroplastic LPC is suitable for feeding monogastrics with an efficient protein conversion (pigs, rabbits, poultry), and cytoplasmatic LPC can be used directly in human diets.

Table 1 reports the material balance for a pilot plant production of a chloroplastic feed-grade LPC and a cytoplasmatic food-grade LPC. The process has been developed at the Istituto di Industrie Agrarie of the University of Pisa, and is characterized by the use of polyelectrolytes to coagulate the green lipoproteins of the juice and by the acid coagulation (pH 4) of the soluble white proteins (11). Data show that starting from fresh alfalfa containing 23.9% solids and 20.9% proteins on a dry basis and using a moderate degree of extraction, the yields of chloroplastic and cytoplasmatic LPC are respectively 8.4 and 1.7% of the alfalfa dry matter. The total yield of LPC is around 10% of the original material (dry basis), and the total amount of proteins recovered in the two fractions is 23% of that found in fresh alfalfa. The ratio of cytoplasmatic LPC/chloroplastic LPC is 1:5.

TABLE 1

Material Balance for 1,000 kg of Fresh Alfalfa (11)

	Weight (kg)		
	As is basis	Dry basis	Crude protein
Inflow			
fresh alfalfa	1,000	239.0	50.0
wash water	2,750	—	—
bisulfite solution (5% w/v)	10	0.5	—
NaOH 2N	40	3.0	—
Polyelectrolyte solution	40	tr.	—
HCl 2N	110	7.6	—
Total	3,950	250.1	50.0
Outflow			
alfalfa extracted residue	550	169.5	26.5
chloroplastic protein cake	40	20.0	8.5
cytoplasmatic protein cake	8	4.1	3.0
deproteinised juice	1,190	55.4	11.7
wash solution	2,162	1.1	0.3
Total	3,950	250.1	50.0

Nutritional Value, Functional Properties and Uses of LPC

The proximate composition of alfalfa food-grade LPC is reported in Table 2. The high protein level of the different LPC's, especially of the cytoplasmatic one, makes them useful as protein integrators in food formulations.

The amino acid composition is one of the most common criteria to evaluate the nutritional value of a protein. Sulfur-containing amino acids, together with lysine, threonine and tryptophan, which are the essential amino acids most commonly lacking in the diet, are all present in high levels, especially in the cytoplasmatic LPC. Data regarding the chemical score and E/T ratio reveal that LPC have a nutritional value comparable with that of meat and eggs.

As far as the digestibility of the LPC is concerned, studies carried out *in vivo* and *in vitro* indicate that for whole LPC the value is 80–90%, while for cytoplasmatic LPC it is usually higher than 95%.

All these data, together with the research undertaken during the last years to study the influence of agronomical and technological factors on the nutritional value of LPC, indicate that LPC's, when properly prepared, have a high biological value and that good proteins, suitable for human consumption, can be obtained from several green crops.

The commercial value of the proteins is strictly related to their functional properties, which largely affect the physical behavior in foods (16). The proteins extracted without being severely denatured generally show good functional properties, which are affected adversely by strong chemical and physical treatments.

The nitrogen solubility profile, determined as a function of pH and/or ionic strength, is one of the best protein functionality indices, as it reflects how severely the protein has been denatured. Nitrogen solubility is higher in the case of LPC obtained using acid rather than heat coagulation (Fig. 2). It increases more when the proteins are recovered by membrane filtration and further improvements are obtained by enzymatic controlled hydrolysis or succinylation of the lysine ε-amino group. On the other hand, the extraction of pigments with solvents has deleterious effects on this property.

As far as the surface properties are concerned, LPC emulsifying activity is largely affected by the drying process, whereas the coagulation method has little influence. LPC's have good emulsifying properties, especially at pH values close to their isoelectric point. Freeze-dried cytoplasmatic LPC's have a very high fat absorption capacity, far higher than that of soya protein concentrates and isolates. Foaming capacity and foam stability are properties directly related to the protein solubility, and the least-denatured or succinylated LPC provide the best results. The lipid extraction considerably improves both foam capacity and stability.

In summary, LPC, when correctly recovered and dried, show good functional properties, especially regarding hydrophilic and surface properties. This makes the LPC useful as protein integrators in meat or meat like-products.

The incorporation of leaf proteins into food formulation has been studied by several authors for a wide range of products (4,7,15–18). LPC of different vegetable origin obtained by various processes have been employed. Vegetable soups have proved to be the best way to introduce purified, whole LPC into the human diet, because their green color and grassy taste do not provide a strong

TABLE 2

Chemical Composition and Nutritional Value of Alfalfa LPC for Human Consumption (15)

	Purified whole LPC	Decolorized whole LPC	Cytoplasmatic LPC
Proximate composition (% dry basis)			
Protein (N × 6.25)	55 ÷ 65	65 ÷ 75	80 ÷ 95
Lipid	15 ÷ 20	2 ÷ 5	0.5 ÷ 1.0
Ash	2 ÷ 5	2 ÷ 5	0.5 ÷ 1.0
Fiber	0.5 ÷ 1.5	0.5 ÷ 1.5	—
N-free extract	10 ÷ 25	15 ÷ 30	5 ÷ 20
Essential amino acids (g/16g N)			
Isoleucine	5.24	—	5.28
Leucine	8.92	—	9.42
Lysine	6.31	—	6.43
Methionine + Cystine	3.27	—	4.33
Phenylalanine + Tyrosine	10.36	—	11.99
Threonine	5.01	—	5.70
Tryptophan	1.65	—	2.45
Valine	6.41	—	6.91
Amino acid score	99.0	—	100
E/T ratio	3.02	—	3.28

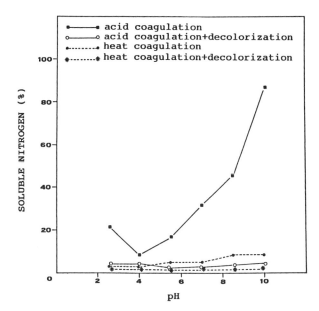

Fig. 2. Soluble nitrogen profile of edible LPC obtained with different techniques.

contrast with the characteristics of the soup. Results from various studies indicate that the optimum level of whole LPC to be used in these products is 5–10%. The almost neutral organoleptic qualities of cytoplasmatic LPC allow its useful incorporation in a much wider range of foods, especially bakery, meat and dairy products. In the case of bakery products their acceptability is still good at levels of around 3%, and even higher (5–10%) for meat and dairy products.

Although leaf proteins have high nutritional value and good acceptability, at present the production of food-grade LPC is not of industrial interest, remaining restricted to laboratory and pilot-plant scale. This failure on commercial scale is related to economical factors and mainly to the availability of inexpensive seed proteins (e.g., soya) having good functional properties.

References

1. Koegel, R.G., H.D. Bruhn, N. Kai Kong and H.W. Ream, *Juice Protein and Moisture Expression from Organic Materials. A Bibliography.*, Res. Div. College of Agric. and Life Sciences, University of Wisconsin, Madison, U.S.A., 1976.
2. Bray, W.J. and C. Humphries, in *Human Food*, Dept. Food Science, University of Reading, Reading, U.K., 1978.
3. Costes, C., A. Chominot, J. Culioli, R. Douillard, I. Gastineau, P. Guy, O. de Mathan, B. Monties, A. Mouranche, L. Petit and D. Sauvant. *Protéines Foliaires et Alimentation*, Gauthier-Villars, Paris, France, 1981.
4. Telek, L. and H.D. Graham, *Leaf Protein Concentrates*, Avi Publ. Co., Westport, Connecticut, U.S.A., 1983.
5. Sing, N., ed., *Progress in Leaf Protein Research*, Today and Tomorrow's Printers and Publishers, New Delhi, India, 1984.

6. Tasaki, I., ed., *Recent Advances in Leaf Protein Research,* Nagoya University Farm, Nagoya, Japan, 1985.
7. Pirie, N.W., *Leaf Protein and Its By-products in Human and Animal Nutrition,* Cambridge University Press, Cambridge, U.K., 1987.
8. Huang, K.H., M.C. Tao, M. Boulet, R.R. Riel, J.P. Julien and G.J. Brisson, *J. Inst. Can. Technol. Aliment.,4*:85, (1971).
9. Edwards, R.H., R.E. Miller, D. De Fremery, B.E. Knukles, E.M. Bickoff and G.O. Kohler, *J. Agric. Food Chem.,4*:621, (1975).
10. Knukles, B.E., R.H. Edwards, G.O. Kohler and L.F. Whitney, *J. Agric. Food Chem., 28*:32, (1980).
11. Fiorentini, R. and C. Galoppini, *J. Food Sci., 46*:1514, (1981).
12. Devadas, R.P., P. Vijayalakshmi and S. Vijaya, in *Progress in Leaf Protein Research,* edited by N. Singh, Today and Tomorrow's Printers and Publishers, New Delhi, India, (1984).
13. Favati, F., R. Fiorentini and C. Galoppini, *Acta Alimentaria, 3,* (1988), (in press).
14. Gwiazda, S., and K. Saio, *Agric. Biol. Chem., 45*:2659, (1981).
15. Galoppini, C., and R. Fiorentini, in *Proceedings of the 2nd International Conference on Leaf Protein Research,* edited by I. Tasaki, Nagoya University Farm, Nagoya, Japan, (1985).
16. Meimbam, E.J., J.G. Bautista and M.R. Soriano, *Philippine J. Nutr., 35*:82, (1982).
17. Lencioni, L., R. Fiorentini and C. Galoppini, *Industrie Alimentari, 23*:106, (1984).
18. Lencioni, L., A.M. Pisanelli, E. Baldi, R. Fiorentini and C. Galoppini, *Lebensm. Wissen. Technol., 20*:277, (1987).

Enzymatic Incorporation of Arginine into Soybean Meal by the Plastein Reaction

S. Divakaran

Oceanic Institute, Makapuu Point, PO Box 25280 Honolulu, Hawaii 96825. U.S.A.

Abstract

The incorporation of arginine residues into peptides is difficult due to its highly basic character. Amino acids like methionine can be covalently bonded to proteins by the plastein reaction as methionine ethyl ester without any further modifications. The bonding of arginine was attempted by the plastein reaction using N^2-Carbobenzyloxy-L-arginine ethyl ester hydrochloride. Triethanolamine was used as the alkalizing agent for the plastein reaction with the enzyme papain. The water insoluble plastein enriched with arginine was prepared using soybean meal as the protein substrate. The amino acid enriched soybean meal will be used for feeding studies in shrimp to understand their amino acid requirements.

The plastein reaction is the formation of a gel-like proteinaceous substance from a concentrated protein hydrolysate which has been incubated with a protease under appropriate conditions (1). During the plastein reaction an enrichment of the hydrophobic amino acids occur for some of the peptides through transpeptidation reactions. These hydrophobic peptides are sparingly soluble in water and condense into small particles forming the insoluble plastein (5). The plastein reaction using papain as the enzyme catalyst to covalently bond methionine to soy protein has been reported earlier (2,3,7,9). Our interest in plastein reaction was to incorporate arginine into whole grain and seed proteins, in which they are limiting for aquaculture diets especially for shrimp feeds. The plastein reaction is useful for the production of aquaculture feeds, as proteins modified by this reaction are insoluble in water.

Materials and Methods

The one-step process for enzymatic incorporation of amino acids into proteins (9) was used with some modification as described. Soybean meal was ground and sieved through 20 mesh (0.8 mm opening) nylon screen. The sieved flour (150 grams) was moistened with water (80–95 ml) and alkalized with 8–10 ml of triethanolamine (Aldrich) to make a dough, pH 8.5–9.0. Approximately 900 mg of papain (Papain crude, Type-1, 2.5 units/mg solid, Sigma) and L-cysteine 45 mg were dissolved in 4–5 ml of water and mixed with the alkalized dough by kneading.

The mixed dough was then divided into three portions. To one portion was added 5 grams of N^2-Carbobenzyloxy-L-arginine (Aldrich) that had been converted into its ethyl ester prepared by the procedure described by Boissonnas et al. (4). To 20 ml of ethyl alcohol (200 proof) at $-10°C$; 2.5 ml of thionyl chloride ($SOCl_2$) plus 5 grams of N^2-Carbobenzyloxy-L-arginine were added. The mixture was refluxed for two hours in a 100 ml round bottom ground joint flask fitted vertically with a Liebing condenser, heated by a water bath and circulating ice cold water as condenser coolant. At the end of two hours reflux time, the unreacted solvent was evaporated to dryness using a rotary vacuum evaporator. The dry N^2-Carbenzyloxy-L-arginine ethyl ester hydrochloride thus formed was reconstituted in about 5 ml of water and then added to the alkalized dough and kneaded thoroughly to achieve uniform distribution.

To the second portion of the dough, methionine ethyl ester hydrochloride prepared from 5 grams of commercial feed grade D-L-Methionine by the same procedure as described earlier (4) was added and kneaded well into the dough. No amino acid was added to the third portion which served as the control.

All samples were incubated at 37°C in covered 10 cm Petri dishes in a vacuum dessicator under vacuum (about 90 kPa) for 16 hours. At the end of the incubation period, the three dough samples were washed separately by repeated suspension in 100 ml of water and centifugation at 3000 RPM. (about 100 to 120 G). The centrifuged supernatant wash water was tested for the presence of free amino acids (not bonded to the protein by plastein reaction) by the Ninhydrin color reaction (8). The washing was repeated (usually 5 to 6 washes) until no free amino acid was detected in the wash water.

Each dough sample was freeze dried in a Virtis freeze drier, Model #12 overnight (12–16 hours) at $-70°C$ with a vacuum of 100 millitorr. The dried samples (10 mg each) were hydrolyzed with 10 ml of 6N.HCl in sealed flasks under vacuum at 105°C for 16 hours. The hydrolysates were then evaporated with 10 ml of a standard buffer pH 2.2 (Sodium diluent, Pickering Labs. CA) and analyzed in Perkin Elmer Series 4 HPLC. The column (25cm × 4.6 mm ID) was packed with a sulfonated polystyrene cation exchange resin. The eluates were measured for amino acid content by post column derivatization with orthophenanthroline (OPA).

Results and Discussion

The amino acid composition of the plastein control, arginine and methionine plastein incorporated soybean meal are given in Table 1. The arginine content was increased by almost 2.0 times (range 1.7 to 2.0 times) that of the control sample when N2-carbenzyloxy-L-arginine-ethyl ester hydrochloride was used. The established procedure (9) for plastein reaction on soybean meal with methionine ethyl ester hydrocholoride, with the same modifications as applied to arginine served as the control to confirm the validity of the plastein reaction. Methionine levels were increased (Table 1) by 6.8 times (range 5 to 8 times).

Earlier attempts to covalently bond nitro arginine and nitro arginine ethyl ester hydrochloride using the plastein reaction were not successful. It was expected that nitro arginine or its ester would bond due to the introduction

TABLE 1

Amino Acid Composition of Soybean Meal Enriched with Arginine and Methionine by Plastein Reaction Using Papain as the Enzyme Catalyst

Amino acid	Control	Methionine enriched	Arginine enriched
Threonine	4.5 ± 0.2	4.7 ± 0.5	4.2 ± 0.2
Valine	6.0 ± 0.4	4.9 ± 1.2	5.1 ± 0.1
Methionine	1.2 ± 0.3	8.2 ± 1.5	0.7 ± 0.2
Isoleucine	4.9 ± 1.0	4.8 ± 0.3	4.2 ± 0.2
Leucine	8.0 ± 0.6	6.7 ± 1.5	6.8 ± 0.5
Phenylalanine	3.6 ± 0.6	3.6 ± 0.2	3.3 ± 0.5
Lysine	5.8 ± 0.1	6.2 ± 1.3	9.1 ± 2.2
Histidine	4.2 ± 0.6	4.3 ± 1.5	10.3 ± 0.9
Arginine	5.6 ± 0.3	4.9 ± 0.6	11.2 ± 2.4

Only the essential amino acids are listed. Tryptophan was not estimated. The amino acids are listed in the order of elution. (Control n = 5, methionine n = 4, and arginine n = 3.)

of the strongly electronegative nitro group to depress the basic character of the guanidino group in arginine (6) although such preparations have been successfully used in the preparation of polypeptides containing arginine through carbobenzoxy reactions.

Besides soybean meal other proteins like gelatine and casein could also be used as substrates for the bonding of these amino acids by the plastein reaction. Attempts to bond more than one acid by using a physical mixture of amino acid ethyl esters to protein in a single plastein reaction resulted in no specific increase in content of any amino acid.

Soybean meal bonded with amino acids by plastein reaction serves as an ideal water insoluble enriched protein for nutritional studies in aquaculture. The methionine and arginine plastein bonded soybean meal will be used for feeding studies in shrimp to understand their amino acid requirements. Although these plastein enriched seed proteins and other proteins may find use for such experimental studies, the commercial viability of these procedures to produce amino acid enriched grain proteins remains to be demonstrated.

Acknowledgments

This project was funded through grant No. 59-32u4-4-25 from the U.S. Department of Agriculture, Agricultural Research Service.

References

1. Alder-Nissen, J., *Enzymic hydrolysis of food proteins*. Elsevier Applied Science Publishers. NY. p. 427, (1986).
2. Arai, S., M. Yamashita and M. Fujikami., *Agric. Biol. Chem. 43:*1069–1074 (1979).
3. Arai, S., A. Hiroshi., and H. Kimura., *Ibid. 47:*2115–2117 (1983).
4. Boissonnas, R.A., St. Guttmann., P.A. Jaquenoud. and J.P. Waller., *Helv. Chem. Acta. 39:*1421–1427 (1956).
5. Gololobov, M.Y., V.M. Belikov., S.V. Vitt., E.A. Paskonova. , E.F. Titova., *Nahrung. 25:*961–967 (1981).
6. Hofmann, K. and H. Yajima., "Synthesis of complex polypeptides possessing a defined arrangement of their amino acid residues." In, *Polyamino acids, Polypeptides and Proteins*. M.A. Stahmann Ed. The University of Wisconsin Press, Madison. USA. p. 394 (1962).
7. Monti, J.C., and R. Jost., *J. Agric. Food Chem. 27:*1281–1285 (1979).
8. Rosen, H., *Arch. Biochem. Biophys. 67:*10-15 (1957).
9. Yamashita, M., S. Arai., and Y. Amano., *Agric. Biol. Chem. 43:*1065–1068 (1979).

A New Process for Protein Protection

J.Å.H. Dahlén, L.A. Lindh and C.G.S. Münter

Svensk Oljeextraktion AB, PO Box 3, S-374 21 Karlshamn, Sweden; Svensk Exergiteknik AB, Stampgatan 38, S-411 01 Goteborg, Sweden

Abstract

A method to reduce the rumen degradability of vegetable protein has been developed by Svensk Oljeextraktion AB-Exab-in cooperation with Svensk Exergiteknik AB. The method, which could be used for any kind of protein containing vegetable feedstuff, has been successfully tested in a continuously working pilot plant. A full size plant with a rated capacity of 25 tons of rapeseed meal per hour has been erected by Exab and will be started up in October 1988.

In 1980 a group of specialists representing the Nordic countries started an investigation with the aim of developing a new protein evaluation system for ruminants. The result of the collaboration, the AAT-PBV system was presented in 1985 (1). The new evaluation system considers the microbial demand for degradable protein as well as protein escaping rumen degradation. It has been decided to practice the new system in Sweden starting in 1989 when it will replace the present method of expressing the protein requirement as digestible crude protein.

Because rapeseed protein has low resistance to ruminal degradation the Nordic system is unfavorable to Swedish rapeseed meal. This fact together with positive results from a number of reported feeding experiments with protected rapeseed meal caused Exab to start a research project in 1981 concerning heat treatment of rapeseed meal.

An extensive survey on the effect of different process conditions for treatment of rapeseed meal has been performed. The most interesting observation is that the degradability *in sacco* could be reduced to about 50% of the original value even following extremely short periods of treatment.

Based on this research a process technology had to be found to implement the effects on an industrial scale. The aim was to achieve the reduced degradability through the mildest treatment possible and with uniform treatment of each individual protein containing particle.

The Process

In 1987 the idea of modifying the technology for back pressure drying in steam atmosphere was shown to be successful in achieving the desired protein degradability. A collaboration with Svensk Exergiteknik AB, the company behind the back pressure steam dryer, was established. Based on experience from test runs a full scale plant for treating 25 tons of rapeseed meal per hour has been projected. The principle of the process is shown in Figure 1.

The specific, protein containing feedstuff is treated in suspension by pressurized, superheated steam. Steam is recirculated and necessary heat is introduced indirectly by convection. Two different types of tubular heat exchangers are used to control the heat treatment temperature

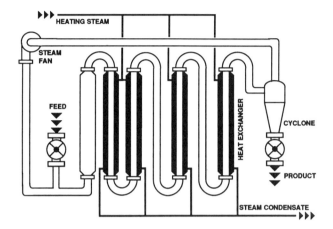

Fig. 1. The principle of the process.

and to cover heat losses. Specially designed feeding and discharge equipment makes it possible to operate under high pressure without significant leakage of steam. The dilute suspension of particles in a plug-like flow enables the treatment to be identical for all particles. After rapid passage through the thermal reactor the feedstuff is separated in a cyclone and steam is recirculated by a centrifugal fan. The treated material is discharged before cooling and further handling. Through the unique approach of using convective heat transfer the moisture content of the product can be accurately controlled and if desired reduced by drying in the system. Rapid start up, shutdown and response are some benefits of using an indirectly heated steam system. Only steam pressure control valves are necessary for controlling the operation and there are only a few moving parts in the installation.

Typical data for consumption of utilities, when treating 25 tons of rapeseed meal per hour, are: steam, 45 kg/ton of product; power, 4 kWh/ton of product; and water, 6m^3/hr.

The Product

Table 1 shows laboratory analysis data from a test run when four tons of ordinary Swedish rapeseed meal was treated in a pilot plant. The fat content of the meal was 2.5% on dry matter. The duration of treatment was less than 30 seonds. Five different samples, each representing 20% of the total amount, were analyzed separately. The treated meal is henceforth called ExPro-meal.

The values in Table 1 show that the treatment had hardly any influence on the contents of pepsine soluble crude protein or ADF-nitrogen. The buffer solubility test clearly demonstrates that the protein degradability has been reduced in an optimal way.

TABLE 1

Analysis Data from a Test Run with Rapeseed Meal

Sample	Protein (% of dm)	Pepsine soluble crude protein (% of dm)	Buffer solubility (2) (% of crude prot.)	ADF-N (% of tot. N)
Reference	40.3	35.8	33.2	6.0
Sample 1	41.1	36.1	7.5	—
Sample 2	39.5	34.4	7.3	—
Sample 3	41.7	36.6	7.4	—
Sample 4	40.4	35.2	7.3	—
Sample 5	39.6	34.6	7.2	—
Average 1–5	40.4	35.4	7.3	6.1

The more precise *in situ* nylon bag (35 μm) technique (3) was used on a representative sample from the five batches. Results are shown in Figure 2.

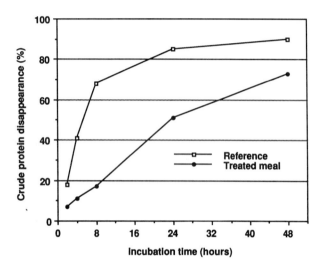

Fig. 2. Crude protein disappearance versus time *in sacco*.

The crude protein disappearance is in close agreement with results from tests with Protec and, to some degree, Jet-Sploded material (4).

The contents of some amino acids of special interest are not significantly influenced by the treatment as shown in Table 2.

TABLE 2

Amino Acid Content in Untreated Meal and ExPro-meal

Amino acid	Untreated meal (g/kg of dm)	ExPro-meal (g/kg of dm)
Lysine	23.2	22.8
Methionine	8.6	8.4
Arginine	24.9	24.6
Histidine	9.7	9.8
Cystine	9.2	9.2

The effect of the pilot plant treatment on whole rapeseeds, rapeseed expeller and soybean meal has also been tested. The values in Table 3 show that the treatment had a remarkable effect.

TABLE 3

The Buffer Solubility of Treated Rapeseed Expeller Cake, Whole Rapeseeds and Soymeal

Sample	Protein (% of dm)	Buffer solubility (% of crude prot.)
Expeller cake reference	32.3	53.3
Expeller cake treated	32.4	10.3
Rapeseed reference	22.4	53.3
Rapeseed treated	22.8	10.2
Soymeal reference	52.7	15.6
Soymeal treated	53.8	6.6

Feeding Tests

Feeding tests, using meal treated in the pilot plant, were performed at the Swedish University of Agricultural Sciences. The results will be published at the EAAP congress in Dublin in August 1989 (J. Bertilsson, Personal Communication). By courtesy of the author it can be presently disclosed that the production of milk increased by nearly 1 kg FCM per cow per day. In addition the contents of milk fat and milk protein increased and the feed efficiency was influenced in a positive way.

Conclusions

There are many patented methods for reducing the degradability of proteins in the rumen by using chemicals. Some of those are being used industrially. Present and future environmental concerns as well as legislation prevent the use of some of those chemicals in Sweden and many other countries.

As far as the authors know there is only one commercial plant in operation producing protected protein without using chemicals.

A major advantage of using the process technology described here is that several large installations already exist that successfully handle several materials at various origins and structures. These installations are using the same basic technology and mechanical components as in the protein protection process.

Acknowledgement

R.Håkansson and L. Karlsson assisted in the experimental work.

References

1. *Acta Agric. Scand.*, Protein evaluation for ruminants, Supplementum 25:9–20 (1985).
2. J.E. Lindberg, C. Clason, P. Ciszuk and E. Den Braver, *Swedish J. Agric. Res. 12*:77–82 (1982).
3. *Acta Agric. Scand.*, Protein evaluation for ruminants, Supplementum 25:64–97 (1985).
4. M.A. Deacon, G.De Boer, and J.J. Kennelly, *J. Dairy Sci. 71:*745–753 (1988).